Blackburn
College

Library
01254 292120

Please return this book on or before the last date below

3 0 OCT 2006	0 5 FEB 2007
	3 - MAY 2007
2 7 NOV 2006	- 7 JUN 2007
0 8 JAN 2007	- 2 MAY 2008
	1 7 NOV 2008
1 9 APR 2007	
1 5 FEB 2008	

DIGITAL COMMUNICATIONS

Fundamentals and Applications

Second Edition

BERNARD SKLAR

Communications Engineering Services, Tarzana, California
and
University of California, Los Angeles

Prentice Hall P T R
Upper Saddle River, New Jersey 07458
www.phptr.com

ISBN 0-13-084788-7

9 780130 847881

90000

Library of Congress Cataloging-in-Publication Data
SKLAR, BERNARD
  Digital communications, 2/e/Bernard Sklar—2nd ed.

  Includes index and bibliographical references.
  ISBN: 0-13-084788-7
  1. Digital communications.   I. Title.
TK5103.7.855   2000   00-7476
621.38′0413

Editorial/Production Supervision: *Rose Kernan*
Publisher: *Bernard Goodwin*
Marketing Manager: *Bryan Gambrel*
Manufacturing Manager: *Alexis Heydt*
Cover Design Director: *Jerry Votta*
Cover Designer: *Nina Scuderi*
Interior Formatting: *Pine Tree Composition*

To my wife Gwen, and our children,
Debra, Sharon, and Dean, and to
the memory of my mother and father,
Ruth and Julius Sklar

Contents

2   FORMATTING AND BASEBAND MODULATION   55

9   MODULATION AND CODING TRADE-OFFS   520

10   SYNCHRONIZATION   598

11   MULTIPLEXING AND MULTIPLE ACCESS   656

12   SPREAD-SPECTRUM TECHNIQUES   718

13 SOURCE CODING   803

14 ENCRYPTION AND DECRYPTION   890

Preface

This second edition of *Digital Communications: Fundamentals and Applications* represents an update of the original publication. The key features that have been updated are:

- The error-correction coding chapters have been expanded, particularly in the areas of Reed–Solomon codes, turbo codes, and trellis-coded modulation.
- A new chapter on fading channels and how to mitigate the degrading effects of fading has been introduced.
- Explanations and descriptions of essential digital communication concepts have been amplified.
- End-of-chapter problem sets have been expanded. Also, end-of-chapter question sets (and where to find the answers), as well as end-of-chapter CD exercises have been added.
- A compact disc (CD) containing an educational version of the design software SystemView by ELANIX® accompanies the textbook. The CD contains a workbook with over 200 exercises, as well as a concise tutorial on digital signal processing (DSP). CD exercises in the workbook reinforce material in the textbook; concepts can be explored by viewing waveforms with a windows-based PC and by changing parameters to see the effects on the overall system. Some of the exercises provide basic training in using SystemView; others provide additional training in DSP techniques.

The teaching of a one-semester university course proceeds in a very different manner compared with that of a short-course in the same subject. At the university, one has the luxury of time—time to develop the needed skills and mathematical tools, time to practice the ideas with homework exercises. In a short-course, the treatment is almost backwards compared with the university. Because of the time factor, a short-course teacher must "jump in" early with essential concepts and applications. One of the vehicles that I found useful in structuring a short course was to start by handing out a check list. This was not merely an outline of the curriculum. It represented a collection of concepts and nomenclature that are not clearly documented, and are often misunderstood. The short-course students were thus initiated into the course by being challenged. I promised them that once they felt comfortable describing each issue, or answering each question on the list, they would be well on their way toward becoming knowledgeable in the field of digital communications. I have learned that this list of essential concepts is just as valuable for teaching full-semester courses as it is for short courses. Here then is my "check list" for digital communications.

1. What mathematical dilemma is the cause for there being several definitions of bandwidth? (See Section 1.7.2.)

2. Why is the ratio of bit energy-to-noise power spectral density, E_b/N_0, a natural figure-of-merit for digital communication systems? (See Section 3.1.5.)

3. When representing timed events, what dilemma can easily result in confusing the most-significant bit (MSB) and the least-significant bit (LSB)? (See Section 3.2.3.2.)

4. The error performance of digital signaling suffers primarily from two degradation types. a) loss in signal-to-noise ratio, b) distortion resulting in an irreducible bit-error probability. How do they differ? (See Section 3.3.2.)

5. Often times, providing more E_b/N_0 will not mitigate the degradation due to intersymbol interference (ISI). Explain why. (See Section 3.3.2.)

6. At what location in the system is E_b/N_0 defined? (See Section 4.3.2.)

7. Digital modulation schemes fall into one of two classes with opposite behavior characteristics. a) orthogonal signaling, b) phase/amplitude signaling. Describe the behavior of each class. (See Sections 4.8.2 and 9.7.)

8. Why do binary phase shift keying (BPSK) and quaternary phase shift keying (QPSK) manifest the same bit-error-probability relationship? Does the same hold true for M-ary pulse amplitude modulation (M-PAM) and M^2-ary quadrature amplitude modulation (M^2-QAM) bit-error probability? (See Sections 4.8.4 and 9.8.3.1.)

9. In orthogonal signaling, why does error-performance improve with higher dimensional signaling? (See Section 4.8.5.)

10. Why is *free-space loss* a function of wavelength? (See Section 5.3.3.)

11. What is the relationship between received signal to noise (S/N) ratio and carrier to noise (C/N) ratio? (See Section 5.4.)

12. Describe four types of trade-offs that can be accomplished by using an error-correcting code. (See Section 6.3.4.)

13. Why do traditional error-correcting codes yield error-performance degradation at low values of E_b/N_0? (See Section 6.3.4.6)

14. Of what use is the *standard array* in understanding a block code, and in evaluating its capability? (See Section 6.6.5.)

15. Why is the Shannon limit of −1.6 dB not a useful goal in the design of real systems? (See Section 8.4.5.2.)

16. What are the consequences of the fact that the Viterbi decoding algorithm does not yield *a posteriori* probabilities? What is a more descriptive name for the Viterbi algorithm? (See Section 8.4.6.)

17. Why do binary and 4-ary orthogonal frequency shift keying (FSK) manifest the same bandwidth-efficiency relationship? (See Section 9.5.1.)

18. Describe the subtle energy and rate transformations of received signals: from data-bits to channel-bits to symbols to chips. (See Section 9.7.7.)

19. Define the following terms: Baud, State, Communications Resource, Chip, Robust Signal. (See Sections 1.1.3 and 7.2.2, Chapter 11, and Sections 12.3.2 and 12.4.2.)

20. For the case of a mobile system that experiences fading, why is signal dispersion independent of fading rapidity? (See Section 15.4.1.1.)

I hope you find it useful to be challenged in this way. Now, let us describe the purpose of the book in a more methodical way. This second edition is intended to provide a comprehensive coverage of digital communication systems for senior level undergraduates, first year graduate students, and practicing engineers. Though the emphasis is on digital communications, necessary analog fundamentals are included since analog waveforms are used for the radio transmission of digital signals. The key feature of a digital communication system is that it deals with a finite set of discrete messages, in contrast to an analog communication system in which messages are defined on a continuum. The objective at the receiver of the digital system is not to reproduce a waveform with precision; it is instead to determine from a noise-perturbed signal, which of the finite set of waveforms had been sent by the transmitter. In fulfillment of this objective, there has arisen an impressive assortment of signal processing techniques.

The book develops these techniques in the context of a unified structure. The structure, in block diagram form, appears at the beginning of each chapter; blocks in the diagram are emphasized, when appropriate, to correspond to the subject of that chapter. Major purposes of the book are to add organization and structure to a field that has grown and continues to grow rapidly, and to insure awareness of the "big picture" even while delving into the details. Signals and key processing steps are traced from the information source through the transmitter, channel, receiver, and ultimately to the information sink. Signal transformations are organized according to nine functional classes: Formatting and source coding, Baseband signaling, Bandpass signaling, Equalization, Channel coding, Muliplexing and multiple access, Spreading, Encryption, and Synchronization. Throughout the book, emphasis is placed on system goals and the need to trade off basic system parameters such as signal-to-noise ratio, probability of error, and bandwidth expenditure.

ORGANIZATION OF THE BOOK

Chapter 1 introduces the overall digital communication system and the basic signal transformations that are highlighted in subsequent chapters. Some basic ideas of random variables and the *additive white Gaussian noise* (AWGN) model are reviewed. Also, the relationship between power spectral density and autocorrelation, and the basics of signal transmission through linear systems are established. Chapter 2 covers the signal processing step, known as *formatting,* in order to render an information signal compatible with a digital system. Chapter 3 emphasizes *baseband signaling,* the detection of signals in Gaussian noise, and receiver optimization. Chapter 4 deals with *bandpass signaling* and its associated modulation and demodulation/detection techniques. Chapter 5 deals with *link analysis,* an important subject for providing overall system insight; it considers some subtleties that are often missed. Chapters 6, 7, and 8 deal with *channel coding*—a cost-effective way of providing a variety of system performance trade-offs. Chapter 6 emphasizes *linear block codes,* Chapter 7 deals with *convolutional codes,* and Chapter 8 deals with *Reed–Solomon codes* and *concatenated codes* such as *turbo codes.*

Chapter 9 considers various modulation/coding system *trade-offs* dealing with probability of bit-error performance, bandwidth efficiency, and signal-to-noise ratio. It also treats the important area of coded modulation, particularly *trellis-coded modulation.* Chapter 10 deals with *synchronization* for digital systems. It covers phase-locked loop implementation for achieving carrier synchronization. It covers bit synchronization, frame synchronization, and network synchronization, and it introduces some ways of performing synchronization using digital methods.

Chapter 11 treats *multiplexing* and *multiple access.* It explores techniques that are available for utilizing the communication resource efficiently. Chapter 12 introduces *spread spectrum* techniques and their application in such areas as multiple access, ranging, and interference rejection. This technology is important for both military and commercial applications. Chapter 13 deals with *source coding* which is a special class of data formatting. Both formatting and source coding involve digitization of data; the main difference between them is that source coding additionally involves data redundancy reduction. Rather than considering source coding immediately after formatting, it is purposely treated in a later chapter so as not to interrupt the presentation flow of the basic processing steps. Chapter 14 covers basic *encryption/decryption* ideas. It includes some classical concepts, as well as a class of systems called public key cryptosystems, and the widely used E-mail encryption software known as *Pretty Good Privacy* (PGP). Chapter 15 deals with *fading channels.* Here, we deal with applications, such as mobile radios, where characterization of the channel is much more involved than that of a nonfading one. The design of a communication system that will withstand the degradation effects of fading can be much more challenging than the design of its nonfading counterpart. In this chapter, we describe a variety of techniques that can mitigate the effects of fading, and we show some successful designs that have been implemented.

It is assumed that the reader is familiar with Fourier methods and convolution. Appendix A reviews these techniques, emphasizing those properties that are particularly useful in the study of communication theory. It also assumed that the reader has a knowledge of basic probability and has some familiarity with random variables. Appendix B builds on these disciplines for a short treatment on statistical decision theory with emphasis on hypothesis testing—so important in the understanding of detection theory. A new section, Appendix E, has been added to serve as a short tutorial on s-domain, z-domain, and digital filtering. A concise DSP tutorial also appears on the CD that accompanies the book.

If the book is used for a two-term course, a simple partitioning is suggested; the first seven chapters can be taught in the first term, and the last eight chapters in the second term. If the book is used for a one-term introductory course, it is suggested that the course material be selected from the following chapters: 1, 2, 3, 4, 5, 6, 7, 9, 10, 12.

ACKNOWLEDGMENTS

It is difficult to write a technical book without contributions from others. I have received an abundance of such assistance, for which I am deeply grateful. For their generous help, I want to thank Dr. Andrew Viterbi, Dr. Chuck Wheatley, Dr. Ed Tiedeman, Dr. Joe Odenwalder, and Serge Willinegger of Qualcomm. I also want to thank Dr. Dariush Divsalar of Jet Propulsion Laboratory (JPL), Dr. Bob Bogusch of Mission Research, Dr. Tom Stanley of the Federal Communications Commission, Professor Larry Milstein of the University of California, San Diego, Professor Ray Pickholtz of George Washington University, Professor Daniel Costello of Notre Dame University, Professor Ted Rappaport of Virginia Polytechnic Institute, Phil Kossin of Lincom, Les Brown of Motorola, as well as Dr. Bob Price and Frank Amoroso.

I also want to acknowledge those people who played a big part in helping me with the first edition of the book. They are: Dr. Maurice King, Don Martin and Ned Feldman of The Aerospace Corporation, Dr. Marv Simon of JPL, Dr. Bill Lindsey of Lincom, Professor Wayne Stark of the University of Michigan, as well as Dr. Jim Omura, Dr. Adam Lender, and Dr. Todd Citron.

I want to thank Dr. Maurice King for contributing Chapter 10 on Synchronization, and Professor Fred Harris of San Diego State University for contributing Chapter 13 on Source Coding. Also, thanks to Michelle Landry for writing the sections on Pretty Good Privacy in Chapter 14, and to Andrew Guidi for contributing end-of-chapter problems in Chapter 15.

I am particularly indebted to my friends and colleagues Fred Harris, Professor Dan Bukofzer of California State University at Fresno, and Dr. Maury Schiff of Elanix, who put up with my incessant argumentative discussions anytime that I called on them. I also want to thank my very best teachers—they are my students at the University of California, Los Angeles, as well as those students all over the world who attended my short courses. Their questions motivated me and provoked me to write this second edition. I hope that I have answered all their questions with clarity.

I offer special thanks for technical clarifications that my son, Dean Sklar, suggested; he took on the difficult role of being his father's chief critic and "devil's advocate." I am particularly indebted to Professor Bob Stewart of the University of Strathclyde, Glasgow, who contributed countless hours of work in writing and preparing the CD and in authoring Appendix E. I thank Rose Kernan, my editor, for watching over me and this project, and I thank Bernard Goodwin, Publisher at Prentice Hall, for indulging me and believing in me. His recommendations were invaluable. Finally, I am extremely grateful to my wife, Gwen, for her encouragement, devotion, and valuable advice. She protected me from the "slings and arrows" of everyday life, making it possible for me to complete this second edition.

BERNARD SKLAR

Tarzana, California

CHAPTER 1

Signals and Spectra

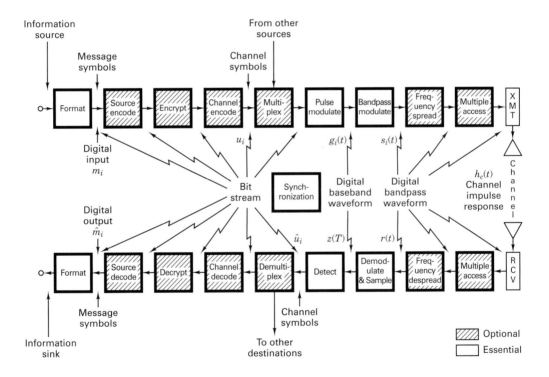

This book presents the ideas and techniques fundamental to digital communication systems. Emphasis is placed on system design goals and on the need for trade-offs among basic system parameters such as signal-to-noise ratio (SNR), probability of error, and bandwidth expenditure. We shall deal with the transmission of information (voice, video, or data) over a path (channel) that may consist of wires, waveguides, or space.

Digital communication systems are becoming increasingly attractive because of the ever-growing demand for data communication and because digital transmission offers data processing options and flexibilities not available with analog transmission. In this book, a digital system is often treated in the context of a satellite communications link. Sometimes the treatment is in the context of a mobile radio system, in which case signal transmission typically suffers from a phenomenon called *fading*. In general, the task of characterizing and mitigating the degradation effects of a fading channel is more challenging than performing similar tasks for a nonfading channel.

The principal feature of a digital communication system (DCS) is that during a finite interval of time, it sends a waveform from a finite set of possible waveforms, in contrast to an analog communication system, which sends a waveform from an infinite variety of waveform shapes with theoretically infinite resolution. In a DCS, the objective at the receiver is *not* to reproduce a transmitted waveform with precision; instead, the objective is to determine from a noise-perturbed signal which waveform from the finite set of waveforms was sent by the transmitter. An important measure of system performance in a DCS is the probability of error (P_E).

1.1 DIGITAL COMMUNICATION SIGNAL PROCESSING

1.1.1 Why Digital?

Why are communication systems, military and commercial alike, "going digital"? There are many reasons. The primary advantage is the ease with which digital signals, compared with analog signals, are regenerated. Figure 1.1 illustrates an ideal binary digital pulse propagating along a transmission line. The shape of the waveform is affected by two basic mechanisms: (1) as all transmission lines and circuits have some nonideal frequency transfer function, there is a distorting effect on the ideal pulse; and (2) unwanted electrical noise or other interference further distorts the pulse waveform. Both of these mechanisms cause the pulse shape to degrade as a function of line length, as shown in Figure 1.1. During the time that the transmitted pulse can still be reliably identified (before it is degraded to an ambiguous state), the pulse is amplified by a digital amplifier that recovers its original ideal shape. The pulse is thus "reborn" or regenerated. Circuits that perform this function at regular intervals along a transmission system are called *regenerative repeaters.*

Digital circuits are less subject to distortion and interference than are analog circuits. Because binary digital circuits operate in one of two states—fully on or fully off—to be meaningful, a disturbance must be large enough to change the circuit operating point from one state to the other. Such two-state operation facilitates signal regeneration and thus prevents noise and other disturbances from accumulating in transmission. Analog signals, however, are *not* two-state signals; they can take an *infinite variety* of shapes. With analog circuits, even a small disturbance can render the reproduced waveform unacceptably distorted. Once the analog signal is distorted, the distortion cannot be removed by amplification. Because accumulated noise is irrevocably bound to analog signals, they cannot be perfectly regenerated. With digital techniques, extremely low error rates producing

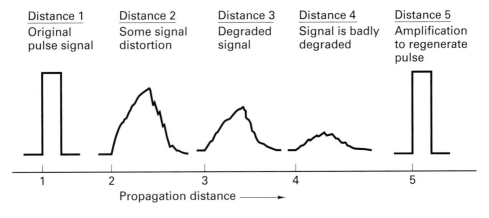

Figure 1.1   Pulse degradation and regeneration.

high signal fidelity are possible through error detection and correction but similar procedures are not available with analog.

There are other important advantages to digital communications. Digital circuits are *more reliable* and can be produced at a lower cost than analog circuits. Also, digital hardware lends itself to *more flexible* implementation than analog hardware [e.g., microprocessors, digital switching, and large-scale integrated (LSI) circuits]. The combining of digital signals using time-division multiplexing (TDM) is *simpler* than the combining of analog signals using frequency-division multiplexing (FDM). Different types of digital signals (data, telegraph, telephone, television) can be treated as identical signals in transmission and switching—*a bit is a bit.* Also, for convenient switching, digital messages can be handled in autonomous groups called *packets.* Digital techniques lend themselves naturally to signal processing functions that protect against interference and jamming, or that provide encryption and privacy. (Such techniques are discussed in Chapters 12 and 14, respectively.) Also, much data communication is from computer to computer, or from digital instruments or terminal to computer. Such digital terminations are naturally best served by digital communication links.

What are the costs associated with the beneficial attributes of digital communication systems? Digital systems tend to be very signal-processing intensive compared with analog. Also, digital systems need to allocate a significant share of their resources to the task of synchronization at various levels. (See Chapter 10.) With analog systems, on the other hand, synchronization often is accomplished more easily. One disadvantage of a digital communication system is *nongraceful degradation.* When the signal-to-noise ratio drops below a certain threshold, the quality of service can change suddenly from very good to very poor. In contrast, most analog communication systems degrade more gracefully.

1.1.2 Typical Block Diagram and Transformations

The functional block diagram shown in Figure 1.2 illustrates the signal flow and the signal-processing steps through a typical digital communication system (DCS). This figure can serve as a kind of road map, guiding the reader through the chapters of this book. The upper blocks—format, source encode, encrypt, channel encode, multiplex, pulse modulate, bandpass modulate, frequency spread, and multiple access— denote signal transformations from the source to the transmitter (XMT). The lower blocks denote signal transformations from the receiver (RCV) to the sink, essentially reversing the signal processing steps performed by the upper blocks. The *modulate* and *demodulate/detect* blocks together are called a *modem.* The term "modem" often encompasses several of the signal processing steps shown in Figure 1.2; when this is the case, the modem can be thought of as the "brains" of the system. The transmitter and receiver can be thought of as the "muscles" of the system. For wireless applications, the transmitter consists of a frequency up-conversion stage to a radio frequency (RF), a high-power amplifier, and an antenna. The receiver portion consists of an antenna and a low-noise amplifier (LNA). Frequency down-conversion is performed in the front end of the receiver and/or the demodulator.

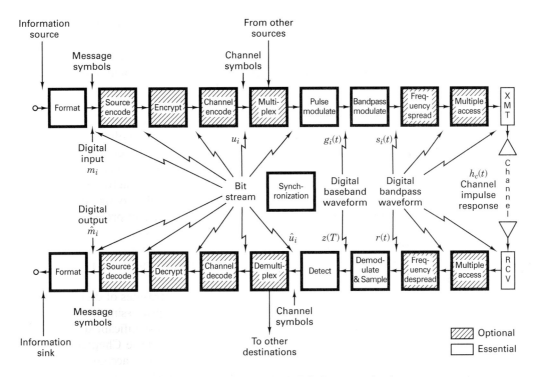

Figure 1.2 Block diagram of a typical digital communication system.

Figure 1.2 illustrates a kind of reciprocity between the blocks in the upper transmitter part of the figure and those in the lower receiver part. The signal processing steps that take place in the transmitter are, for the most part, reversed in the receiver. In Figure 1.2, the input information source is converted to binary digits (*bits*); the bits are then grouped to form *digital messages* or *message symbols*. Each such symbol (m_i, where $i = 1, \ldots, M$) can be regarded as a member of a *finite alphabet* set containing M members. Thus, for $M = 2$, the message symbol m_i is binary (meaning that it constitutes just a single bit). Even though binary symbols fall within the general definition of *M*-ary, nevertheless the name *M*-ary is usually applied to those cases where $M > 2$; hence, such symbols are each made up of a sequence of two or more bits. (Compare such a finite alphabet in a DCS with an analog system, where the message waveform is typically a member of an infinite set of possible waveforms.) For systems that use *channel coding* (error correction coding), a sequence of message symbols becomes transformed to a sequence of *channel symbols* (code symbols), where each channel symbol is denoted u_i. Because a message symbol or a channel symbol can consist of a single bit or a grouping of bits, a sequence of such symbols is also described as a *bit stream,* as shown in Figure 1.2.

Consider the key signal processing blocks shown in Figure 1.2; only formatting, modulation, demodulation/detection, and synchronization are essential for a DCS. *Formatting* transforms the source information into bits, thus assuring compatibility between the information and the signal processing within the DCS. From this point in the figure up to the pulse-modulation block, the information remains in the form of a *bit stream*. Modulation is the process by which message symbols or channel symbols (when channel coding is used) are converted to *waveforms* that are compatible with the requirements imposed by the transmission channel. *Pulse modulation* is an essential step because each symbol to be transmitted must first be transformed from a binary representation (voltage levels representing binary ones and zeros) to a *baseband* waveform. The term baseband refers to a signal whose spectrum extends from (or near) dc up to some finite value, usually less than a few megahertz. The pulse-modulation block usually includes filtering for minimizing the transmission bandwidth. When pulse modulation is applied to binary symbols, the resulting binary waveform is called a pulse-code-modulation (PCM) waveform. There are several types of PCM waveforms (described in Chapter 2); in telephone applications, these waveforms are often called *line codes*. When pulse modulation is applied to nonbinary symbols, the resulting waveform is called an *M*-ary pulse-modulation waveform. There are several types of such waveforms, and they too are described in Chapter 2, where the one called *pulse-amplitude modulation* (PAM) is emphasized. After pulse modulation, each message symbol or channel symbol takes the form of a baseband waveform $g_i(t)$, where $i = 1, \ldots, M$. In any electronic implementation, the bit stream, prior to pulse-modulation, is represented with voltage levels. One might wonder why there is a separate block for pulse modulation when in fact different voltage levels for binary ones and zeros can be viewed as impulses or as ideal rectangular pulses, each pulse occupying one bit time. There are two important differences between such voltage levels and the baseband waveforms used for modulation. First, the pulse-modulation block allows for a variety of binary and *M*-ary pulse-waveform types. Section 2.8.2 describes the different useful attributes of these types of waveforms. Second, the filtering within the pulse-modulation block yields pulses that occupy more than just one-bit time. Filtering yields pulses that are spread in time, thus the pulses are "smeared" into neighboring bit-times. This filtering is sometimes referred to as pulse shaping; it is used to contain the transmission bandwidth within some desired spectral region.

For an application involving RF transmission, the next important step is *bandpass modulation;* it is required whenever the transmission medium will not support the propagation of pulse-like waveforms. For such cases, the medium requires a bandpass waveform $s_i(t)$, where $i = 1, \ldots, M$. The term *bandpass* is used to indicate that the baseband waveform $g_i(t)$ is frequency translated by a carrier wave to a frequency that is much larger than the spectral content of $g_i(t)$. As $s_i(t)$ propagates over the channel, it is impacted by the channel characteristics, which can be described in terms of the channel's *impulse response* $h_c(t)$ (see Section 1.6.1). Also, at various points along the signal route, additive random noise distorts the received signal $r(t)$, so that its reception must be termed a corrupted version of the signal $s_i(t)$ that was launched at the transmitter. The received signal $r(t)$ can be expressed as

$$r(t) = s_i(t) * h_c(t) + n(t) \qquad i = 1, \ldots, M \qquad (1.1)$$

where $*$ represents a convolution operation (see Appendix A), and $n(t)$ represents a noise process (see Section 1.5.5).

In the reverse direction, the receiver front end and/or the demodulator provides frequency down-conversion for each bandpass waveform $r(t)$. The demodulator restores $r(t)$ to an optimally shaped baseband pulse $z(t)$ in preparation for detection. Typically, there can be several filters associated with the receiver and demodulator—filtering to remove unwanted high frequency terms (in the frequency down-conversion of bandpass waveforms), and filtering for pulse shaping. Equalization can be described as a filtering option that is used in or after the demodulator to reverse any degrading effects on the signal that were caused by the channel. Equalization becomes essential whenever the impulse response of the channel, $h_c(t)$, is so poor that the received signal is badly distorted. An equalizer is implemented to compensate for (i.e., remove or diminish) any signal distortion caused by a nonideal $h_c(t)$. Finally, the sampling step transforms the shaped pulse $z(t)$ to a sample $z(T)$, and the detection step transforms $z(T)$ to an estimate of the channel symbol $\hat{u}_i$ or an estimate of the message symbol $\hat{m}_i$ (if there is no channel coding). Some authors use the terms "demodulation" and "detection" interchangeably. However, in this book, *demodulation* is defined as recovery of a waveform (baseband pulse), and *detection* is defined as decision-making regarding the digital meaning of that waveform.

The other signal processing steps within the modem are design options for specific system needs. *Source coding* produces analog-to-digital (A/D) conversion (for analog sources) *and* removes redundant (unneeded) information. Note that a typical DCS would either use the *source coding* option (for both digitizing and compressing the source information), or it would use the simpler *formatting* transformation (for digitizing alone). A system would not use both source coding and formatting, because the former already includes the essential step of digitizing the information. Encryption, which is used to provide communication privacy, prevents unauthorized users from understanding messages and from injecting false messages into the system. *Channel coding,* for a given data rate, can reduce the probability of error, P_E, or reduce the required signal-to-noise ratio to achieve a desired P_E at the expense of transmission bandwidth or decoder complexity. *Multiplexing* and *multiple-access procedures* combine signals that might have different characteristics or might originate from different sources, so that they can share a portion of the communications resource (e.g., spectrum, time). Frequency spreading can produce a signal that is relatively invulnerable to interference (both natural and intentional) and can be used to enhance the privacy of the communicators. It is also a valuable technique used for multiple access.

The signal processing blocks shown in Figure 1.2 represent a typical arrangement; however, these blocks are sometimes implemented in a different order. For example, multiplexing can take place prior to channel encoding, *or* prior to modulation, *or*—with a two-step modulation process (subcarrier and carrier)—it can be performed between the two modulation steps. Similarly, frequency spreading can take place at various locations along the upper portion of Figure 1.2; its precise location depends on the particular technique used. Synchronization and its key element, a clock signal, is involved in the control of all signal processing within the

DCS. For simplicity, the synchronization block in Figure 1.2 is drawn without any connecting lines, when in fact it actually plays a role in regulating the operation of almost every block shown in the figure.

Figure 1.3 shows the basic signal processing functions, which may be viewed as transformations, classified into the following nine groups:

1. Formatting and source coding
2. Baseband signaling
3. Bandpass signaling
4. Equalization
5. Channel coding
6. Multiplexing and multiple access
7. Spreading
8. Encryption
9. Synchronization

Although this organization has some inherent overlap, it provides a useful structure for the book. Beginning with Chapter 2, the nine basic transformations are considered individually. In Chapter 2, the basic formatting techniques for transforming the source information into message symbols are discussed, as well as the selection of baseband pulse waveforms and pulse filtering for making the message symbols compatible with baseband transmission. The reverse steps of demodulation, equalization, sampling, and detection are described in Chapter 3. Formatting and source coding are similar processes, in that they both involve data digitization. However, the term "source coding" has taken on the connotation of data compression in addition to digitization; it is treated later (in Chapter 13), as a special case of formatting.

In Figure 1.3, the *Baseband Signaling* block contains a list of binary choices under the heading of PCM waveforms or line codes. In this block, a nonbinary category of waveforms called *M*-ary pulse modulation is also listed. Another transformation in Figure 1.3, labeled *Bandpass Signaling* is partitioned into two basic blocks, coherent and noncoherent. Demodulation is typically accomplished with the aid of *reference* waveforms. When the references used are a measure of all the signal attributes (particularly phase), the process is termed *coherent;* when phase information is not used, the process is termed *noncoherent.* Both techniques are detailed in Chapter 4.

Chapter 5 is devoted to *link analysis.* Of the many specifications, analyses, and tabulations that support a developing communication system, link analysis stands out in its ability to provide overall system insight. In Chapter 5 we bring together all the link fundamentals that are essential for the analysis of most communication systems.

Channel coding deals with the techniques used to enhance digital signals so that they are less vulnerable to such channel impairments as noise, fading, and jamming. In Figure 1.3 channel coding is partitioned into two blocks, waveform coding

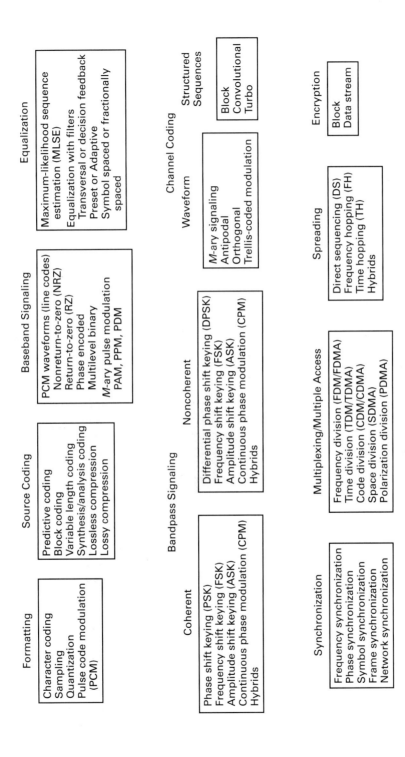

Figure 1.3 Basic digital communication transformations.

and structured sequences. *Waveform coding* involves the use of new waveforms, yielding improved detection performance over that of the original waveforms. *Structured sequences* involve the use of redundant bits to determine whether or not an error has occurred due to noise on the channel. One of these techniques, known as *automatic repeat request* (ARQ), simply recognizes the occurrence of an error and requests that the sender retransmit the message; other techniques, known as *forward error correction* (FEC), are capable of automatically correcting the errors (within specified limitations). Under the heading of structured sequences, we shall discuss three prevalent techniques—block, convolutional, and turbo coding. In Chapter 6, we primarily consider *linear block coding.* In Chapter 7 we consider *convolutional coding,* Viterbi decoding (and other decoding algorithms), and hard versus soft decoding procedures. Chapter 8 treats concatenated coding, which has led to the class of codes known as *turbo* codes, and it also examines the details of *Reed-Solomon* codes.

In Chapter 9 we summarize the design goals for a communication system and present various modulation and coding trade-offs that need to be considered in the design of a system. Theoretical limitations, such as the Nyquist criterion and the Shannon limit, are discussed. Also, *bandwidth-efficient* modulation schemes, such as trellis-coded modulation, are examined.

Chapter 10 deals with *synchronization.* In digital communications, synchronization involves the estimation of both time and frequency. The subject is divided into five subcategories as shown in Figure 1.3. Coherent systems need to synchronize their frequency reference with the carrier (and possibly subcarrier) in both frequency and phase. For noncoherent systems, phase synchronization is not needed. The fundamental time-synchronization process is symbol synchronization (or bit synchronization for binary symbols). The demodulator and detector need to know when to start and end the process of symbol detection and bit detection; a timing error will degrade detection performance. The next time-synchronization level, frame synchronization, allows the reconstruction of the message. Finally, network synchronization allows coordination with other users so resources may be used efficiently. In Chapter 10, we are concerned with the alignment of the timing of spatially separated periodic processes.

Chapter 11 deals with *multiplexing* and *multiple access.* The two terms mean very similar things. Both involve the idea of resource sharing. The main difference between the two is that multiplexing takes place locally (e.g., on a printed circuit board, within an assembly, or even within a facility), and multiple access takes place remotely (e.g., multiple users need to share the use of a satellite transponder). Multiplexing involves an algorithm that is known a priori; usually, it is hardwired into the system. Multiple access, on the other hand, is generally adaptive, and may require some overhead to enable the algorithm to operate. In Chapter 11, we discuss the classical ways of sharing a communications resource: frequency division, time division, and code division. Also, some of the multiple-access techniques that have emerged as a result of satellite communications are considered.

Chapter 12 introduces a transformation originally developed for military communications called *spreading.* The chapter deals with the spread spectrum techniques that are important for achieving interference protection and privacy.

Signals can be spread in frequency, in time, or in both frequency and time. This chapter primarily deals with frequency spreading. The chapter also illustrates how frequency-spreading techniques are used to share the bandwidth-limited resource in commercial cellular telephony.

Chapter 13 treats *source coding,* which involves the efficient description of source information. It deals with the process of compactly describing a signal to within a specified fidelity criterion. Source coding can be applied to digital or analog signals; by reducing data redundancy, source codes can reduce a system's data rate. Thus, the main advantage of source coding is to decrease the amount of required system resources (e.g., bandwidth).

Chapter 14 deals with *encryption* and *decryption,* the basic goals of which are communication privacy and authentication. Maintaining privacy means preventing unauthorized persons from extracting information (eavesdropping) from the channel. Establishing authentication means preventing unauthorized persons from injecting spurious signals (spoofing) into the channel. In this chapter we highlight the data encryption standard (DES) and the basic ideas regarding a class of encryption systems called *public key cryptosystems.* We also examine the novel scheme of Pretty Good Privacy (PGP) which is an important file-encryption method for sending data via electronic mail.

The final chapter of the book, Chapter 15, deals with fading channels. In it, we address fading that affects mobile systems such as cellular and personal communication systems (PCS). The chapter itemizes the fundamental fading manifestations, types of degradation, and methods to mitigate the degradation. Two particular mitigation techniques are examined: the Viterbi equalizer implemented in the Global System for Mobile Communication (GSM), and the Rake receiver used in CDMA systems.

1.1.3 Basic Digital Communication Nomenclature

The following are some of the basic digital signal nomenclature that frequently appears in digital communication literature:

> *Information source.* This is the device producing information to be communicated by means of the DCS. Information sources can be *analog* or *discrete.* The output of an analog source can have any value in a continuous range of amplitudes, whereas the output of a discrete information source takes its value from a finite set. Analog information sources can be transformed into digital sources through the use of *sampling* and *quantization.* Sampling and quantization techniques called formatting and source coding (see Figure 1.3) are described in Chapters 2 and 13.
>
> *Textual message.* This is a sequence of characters. (See Figure 1.4a.) For digital transmission, the message will be a sequence of digits or symbols from a finite symbol set or alphabet.
>
> *Character.* A character is a member of an alphabet or set of symbols. (See Figure 1.4b.) Characters may be mapped into a sequence of binary digits.

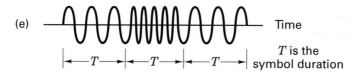

(a)
```
HOW ARE YOU?
OK
$9, 567, 216.73
```

(b)
```
A
9
&
```

(c) H   O   W

0 0 0 1 0 0 1 1 1 1 1 0 0 1 1 1 1 0 1 0 1

(d)
```
  1   Binary symbol (k = 1, M = 2)
 10   Quaternary symbol (k = 2, M = 4)
011   8-ary symbol (k = 3, M = 8)
```

(e)   Time

$\longmapsto T \longrightarrow\!\!\longmapsto T \longrightarrow\!\!\longmapsto T \longrightarrow\!$ T is the symbol duration

Figure 1.4 Nomenclature examples. (a) Textual messages. (b) Characters. (c) Bit stream (7-bit ASCII). (d) Symbols m_i, $i = 1, \ldots, M$, $M = 2^k$. (e) Bandpass digital waveform $s_i(t)$, $i = 1, \ldots, M$.

There are several standardized codes used for character encoding, including the American Standard Code for Information Interchange (ASCII), Extended Binary Coded Decimal Interchange Code (EBCDIC), Hollerith, Baudot, Murray, and Morse.

Binary digit (bit). This is the fundamental information unit for all digital systems. The term *bit* also is used as a unit of information content, as described in Chapter 9.

Bit stream. This is a sequence of binary digits (ones and zeros). A bit stream is often termed a *baseband* signal, which implies that its spectral content extends from (or near) dc up to some finite value, usually less than a few megahertz. In Figure 1.4c, the message, "HOW," is represented with the 7-bit ASCII character code, where the bit stream is shown by using a convenient picture of 2-level pulses. The sequence of pulses is drawn using very stylized (ideal-rectangular) shapes with spaces between successive pulses. In a real system, the pulses would never appear as they are depicted here, because such spaces would serve no useful purpose. For a given bit rate, the spaces would increase the bandwidth needed for transmission; or, for a given bandwidth, they would increase the time delay needed to receive the message.

Symbol (digital message). A symbol is a group of k bits considered as a unit. We refer to this unit as a *message symbol m_i ($i = 1, \ldots, M$)* from a finite symbol set or alphabet. (See Figure 1.4d.) The size of the alphabet, M, is $M = 2^k$, where k is the number of bits in the symbol. For *baseband* transmission, each m_i symbol will be represented by one of a set of baseband pulse waveforms $g_1(t), g_2(t), \ldots, g_M(t)$. When transmitting a sequence of such pulses, the unit *Baud* is sometimes used to express pulse rate (symbol rate). For typical *bandpass* transmission, each $g_i(t)$ pulse will then be represented by one of a set of bandpass waveforms $s_1(t), s_2(t), \ldots, s_M(t)$. Thus, for wireless systems, the symbol m_i is sent by transmitting the digital waveform $s_i(t)$ for T seconds, the symbol-time duration. The next symbol is sent during the next time interval, T. The fact that the symbol set transmitted by the DCS is finite is a primary difference between a DCS and an analog system. The DCS receiver need only decide which of the M waveforms was transmitted; however, an analog receiver must be capable of accurately estimating a continuous range of waveforms.

Digital waveform. This is a voltage or current waveform (a pulse for baseband transmission, or a sinusoid for bandpass transmission) that represents a digital symbol. The waveform characteristics (amplitude, width, and position for pulses or amplitude, frequency, and phase for sinusoids) allow its identification as one of the symbols in the finite symbol alphabet. Figure 1.4e shows an example of a bandpass digital waveform. Even though the waveform is sinusoidal and consequently has an analog appearance, it is called a *digital waveform* because it is encoded with digital information. In the figure, during each time interval, T, a preassigned frequency indicates the value of a digit.

Data rate. This quantity in bits per second (bits/s) is given by $R = k/T = (1/T)$ $\log_2 M$ bits/s, where k bits identify a symbol from an $M = 2^k$-symbol alphabet, and T is the k-bit symbol duration.

1.1.4 Digital versus Analog Performance Criteria

A principal difference between analog and digital communication systems has to do with the way in which we evaluate their performance. Analog systems draw their waveforms from a continuum, which therefore forms an infinite set—that is, a receiver must deal with an infinite number of possible waveshapes. The figure of merit for the performance of analog communication systems is a fidelity criterion, such as signal-to-noise ratio, percent distortion, or expected mean-square error between the transmitted and received waveforms.

By contrast, a digital communication system transmits signals that represent digits. These digits form a finite set or alphabet, and the set is known a priori to the receiver. A figure of merit for digital communication systems is the probability of incorrectly detecting a digit, or the probability of error (P_E).

1.2 CLASSIFICATION OF SIGNALS

1.2.1 Deterministic and Random Signals

A signal can be classified as *deterministic,* meaning that there is no uncertainty with respect to its value at any time, or as *random,* meaning that there is some degree of uncertainty before the signal actually occurs. Deterministic signals or waveforms are modeled by explicit mathematical expressions, such as $x(t) = 5 \cos 10t$. For a random waveform it is *not* possible to write such an explicit expression. However, when examined over a long period, a random waveform, also referred to as a *random process,* may exhibit certain regularities that can be described in terms of probabilities and statistical averages. Such a model, in the form of a probabilistic description of the random process, is particularly useful for characterizing signals and noise in communication systems.

1.2.2 Periodic and Nonperiodic Signals

A signal $x(t)$ is called *periodic in time* if there exists a constant $T_0 > 0$ such that

$$x(t) = x(t + T_0) \qquad \text{for } -\infty < t < \infty \qquad (1.2)$$

where t denotes time. The smallest value of T_0 that satisfies this condition is called the *period* of $x(t)$. The period T_0 defines the duration of one complete cycle of $x(t)$. A signal for which there is no value of T_0 that satisfies Equation (1.2) is called a *nonperiodic signal.*

1.2.3 Analog and Discrete Signals

An *analog signal* $x(t)$ is a continuous function of time; that is, $x(t)$ is uniquely defined for all t. An electrical analog signal arises when a physical waveform (e.g., speech) is converted into an electrical signal by means of a transducer. By comparison, a *discrete signal* $x(kT)$ is one that exists only at discrete times; it is characterized by a sequence of numbers defined for each time, kT, where k is an integer and T is a fixed time interval.

1.2.4 Energy and Power Signals

An electrical signal can be represented as a voltage $v(t)$ or a current $i(t)$ with instantaneous power $p(t)$ across a resistor $\mathcal{R}$ defined by

$$p(t) = \frac{v^2(t)}{\mathcal{R}} \qquad (1.3a)$$

or

$$p(t) = i^2(t)\mathcal{R} \qquad (1.3b)$$

In communication systems, power is often normalized by assuming $\mathcal{R}$ to be 1 Ω, although $\mathcal{R}$ may be another value in the actual circuit. If the actual value of the power is needed, it is obtained by "denormalization" of the normalized value. For the normalized case, Equations 1.3a and 1.3b have the same form. Therefore, regardless of whether the signal is a voltage or current waveform, the normalization convention allows us to express the instantaneous power as

$$p(t) = x^2(t) \tag{1.4}$$

where $x(t)$ is either a voltage or a current signal. The energy dissipated during the time interval $(-T/2, T/2)$ by a real signal with instantaneous power expressed by Equation (1.4) can then be written as

$$E_x^T = \int_{-T/2}^{T/2} x^2(t) \, dt \tag{1.5}$$

and the average power dissipated by the signal during the interval is

$$P_x^T = \frac{1}{T} E_x^T = \frac{1}{T} \int_{-T/2}^{T/2} x^2(t) \, dt \tag{1.6}$$

The performance of a communication system depends on the received signal *energy;* higher energy signals are detected more reliably (with fewer errors) than are lower energy signals—the received *energy does the work.* On the other hand, *power* is the *rate* at which energy is delivered. It is important for different reasons. The power determines the voltages that must be applied to a transmitter and the intensities of the electromagnetic fields that one must contend with in radio systems (i.e., fields in waveguides that connect the transmitter to the antenna, and fields around the radiating elements of the antenna).

In analyzing communication signals, it is often desirable to deal with the *waveform energy.* We classify $x(t)$ as an *energy signal* if, and only if, it has nonzero but finite energy $(0 < E_x < \infty)$ for all time, where

$$
\begin{aligned}
E_x &= \lim_{T \to \infty} \int_{-T/2}^{T/2} x^2(t) \, dt \\
&= \int_{-\infty}^{\infty} x^2(t) \, dt
\end{aligned}
\tag{1.7}
$$

In the real world, we always transmit signals having finite energy $(0 < E_x < \infty)$. However, in order to describe *periodic signals,* which by definition [Equation (1.2)] exist for all time and thus have infinite energy, and in order to deal with random signals that have infinite energy, it is convenient to define a class of signals called *power signals.* A signal is defined as a power signal if, and only if, it has finite but nonzero power $(0 < P_x < \infty)$ for all time, where

$$P_x = \lim_{T \to \infty} \frac{1}{T} \int_{-T/2}^{T/2} x^2(t) \, dt \tag{1.8}$$

The energy and power classifications are mutually exclusive. An energy signal has finite energy but *zero average power,* whereas a power signal has finite average power but *infinite energy.* A waveform in a system may be constrained in either its power or energy values. As a general rule, periodic signals and random signals are classified as power signals, while signals that are both deterministic and nonperiodic are classified as energy signals [1, 2].

Signal energy and power are both important parameters in specifying a communication system. The classification of a signal as either an energy signal or a power signal is a convenient model to facilitate the mathematical treatment of various signals and noise. In Section 3.1.5, these ideas are developed further, in the context of a digital communication system.

1.2.5 The Unit Impulse Function

A useful function in communication theory is the unit impulse or *Dirac delta function* $\delta(t)$. The impulse function is an abstraction—an infinitely large amplitude pulse, with zero pulse width, and unity weight (area under the pulse), concentrated at the point where its argument is zero. The unit impulse is characterized by the following relationships:

$$\int_{-\infty}^{\infty} \delta(t) \, dt \; = \; 1 \tag{1.9}$$

$$\delta(t) \; = \; 0 \qquad \text{for } t \neq 0 \tag{1.10}$$

$$\delta(t) \text{ is unbounded at } t \; = \; 0 \tag{1.11}$$

$$\int_{-\infty}^{\infty} x(t)\delta(t \, - \, t_0) \, dt \; = \; x(t_0) \tag{1.12}$$

The unit impulse function $\delta(t)$ is not a function in the usual sense. When operations involve $\delta(t)$, the convention is to interpret $\delta(t)$ as a unit-area pulse of finite amplitude and nonzero duration, after which the limit is considered as the pulse duration approaches zero. $\delta(t - t_0)$ can be depicted graphically as a spike located at $t = t_0$ with height equal to its integral or area. Thus $A\delta(t - t_0)$ with A constant represents an impulse function whose area or weight is equal to A, that is zero everywhere except at $t = t_0$.

Equation (1.12) is known as the *sifting* or *sampling property* of the unit impulse function; the unit impulse multiplier selects a sample of the function $x(t)$ evaluated at $t = t_0$.

1.3 SPECTRAL DENSITY

The *spectral density* of a signal characterizes the distribution of the signal's energy or power in the frequency domain. This concept is particularly important when considering filtering in communication systems. We need to be able to evaluate the signal and noise at the filter output. The energy spectral density (ESD) or the power spectral density (PSD) is used in the evaluation.

1.3.1 Energy Spectral Density

The total energy of a real-valued energy signal $x(t)$, defined over the interval, $(-\infty, \infty)$, is described by Equation (1.7). Using Parseval's theorem [1], we can relate the energy of such a signal expressed in the time domain to the energy expressed in the frequency domain, as

$$E_x = \int_{-\infty}^{\infty} x^2(t)\, dt = \int_{-\infty}^{\infty} |X(f)|^2\, df \qquad (1.13)$$

where $X(f)$ is the Fourier transform of the nonperiodic signal $x(t)$. (For a review of Fourier techniques, see Appendix A.) Let $\psi_x(f)$ denote the squared magnitude spectrum, defined as

$$\psi_x(f) = |X(f)|^2 \qquad (1.14)$$

The quantify $\psi_x(f)$ is the waveform *energy spectral density* (ESD) of the signal $x(t)$. Therefore, from Equation (1.13), we can express the total energy of $x(t)$ by integrating the spectral density with respect to frequency:

$$E_x = \int_{-\infty}^{\infty} \psi_x(f)\, df \qquad (1.15)$$

This equation states that the energy of a signal is equal to the area under the $\psi_x(f)$ versus frequency curve. Energy spectral density describes the signal energy per unit bandwidth measured in joules/hertz. There are equal energy contributions from both positive and negative frequency components, since for a real signal, $x(t)$, $|X(f)|$ is an even function of frequency. Therefore, the energy spectral density is symmetrical in frequency about the origin, and thus the total energy of the signal $x(t)$ can be expressed as

$$E_x = 2 \int_{0}^{\infty} \psi_x(f)\, df \qquad (1.16)$$

1.3.2 Power Spectral Density

The average power P_x of a real-valued power signal $x(t)$ is defined in Equation (1.8). If $x(t)$ is a *periodic signal* with period T_0, it is classified as a power signal. The expression for the average power of a periodic signal takes the form of Equation (1.6), where the time average is taken over the signal period T_0, as follows:

$$P_x = \frac{1}{T_0} \int_{-T_0/2}^{T_0/2} x^2(t)\, dt \qquad (1.17a)$$

Parseval's theorem for a real-valued periodic signal [1] takes the form

$$P_x = \frac{1}{T_0} \int_{-T_0/2}^{T_0/2} x^2(t) \, dt = \sum_{n=-\infty}^{\infty} |c_n|^2 \qquad (1.17b)$$

where the c_n terms are the complex Fourier series coefficients of the periodic signal. (See Appendix A.)

To apply Equation (1.17b), we need only know the magnitude of the coefficients, $|c_n|$. The *power spectral density* (PSD) function $G_x(f)$ of the periodic signal $x(t)$ is a real, even, and nonnegative function of frequency that gives the distribution of the power of $x(t)$ in the frequency domain, defined as

$$G_x(f) = \sum_{n=-\infty}^{\infty} |c_n|^2 \, \delta(f - nf_0) \qquad (1.18)$$

Equation (1.18) defines the power spectral density of a periodic signal $x(t)$ as a succession of the weighted delta functions. Therefore, the PSD of a periodic signal is a discrete function of frequency. Using the PSD defined in Equation (1.18), we can now write the average normalized power of a real-valued signal as

$$P_x = \int_{-\infty}^{\infty} G_x(f) \, df = 2 \int_{0}^{\infty} G_x(f) \, df \qquad (1.19)$$

Equation (1.18) describes the PSD of periodic (power) signals only. If $x(t)$ is a nonperiodic signal it *cannot* be expressed by a Fourier series, and if it is a nonperiodic power signal (having infinite energy) it *may not* have a Fourier transform. However, we may still express the power spectral density of such signals in the *limiting sense*. If we form a *truncated version* $x_T(t)$ of the nonperiodic power signal $x(t)$ by observing it only in the interval $(-T/2, T/2)$, then $x_T(t)$ has finite energy and has a proper Fourier transform $X_T(f)$. It can be shown [2] that the power spectral density of the nonperiodic $x(t)$ can then be defined in the limit as

$$G_x(f) = \lim_{T \to \infty} \frac{1}{T} |X_T(f)|^2 \qquad (1.20)$$

Example 1.1   Average Normalized Power

(a) Find the average normalized power in the waveform, $x(t) = A \cos 2\pi f_0 t$, using time averaging.

(b) Repeat part (a) using the summation of spectral coefficients.

Solution

(a) Using Equation (1.17a), we have

$$P_x = \frac{1}{T_0} \int_{-T_0/2}^{T_0/2} A^2 \cos^2 2\pi f_0 t \, dt$$

$$= \frac{A^2}{2T_0} \int_{-T_0/2}^{T_0/2} (1 + \cos 4\pi f_0 t) \, dt$$

$$= \frac{A^2}{2T_0} (T_0) = \frac{A^2}{2}$$

(b) Using Equations (1.18) and (1.19) gives us

$$G_x(f) = \sum_{n=-\infty}^{\infty} |c_n|^2 \delta(f - nf_0)$$

$$\left.\begin{array}{l} c_1 = c_{-1} = \dfrac{A}{2} \\[2mm] c_n = 0 \qquad \text{for } n = 0, \pm2, \pm3, \dots \end{array}\right\} \quad \text{(see Appendix A)}$$

$$G_x(f) = \left(\frac{A}{2}\right)^2 \delta(f - f_0) + \left(\frac{A}{2}\right)^2 \delta(f + f_0)$$

$$P_x = \int_{-\infty}^{\infty} G_x(f)\, df = \frac{A^2}{2}$$

1.4 AUTOCORRELATION

1.4.1 Autocorrelation of an Energy Signal

Correlation is a matching process; *autocorrelation* refers to the matching of a signal with a delayed version of itself. The autocorrelation function of a real-valued energy signal $x(t)$ is defined as

$$R_x(\tau) = \int_{-\infty}^{\infty} x(t)x(t + \tau)\, dt \qquad \text{for } -\infty < \tau < \infty \qquad (1.21)$$

The autocorrelation function $R_x(\tau)$ provides a measure of how closely the signal matches a copy of itself as the copy is shifted τ units in time. The variable τ plays the role of a scanning or searching parameter. $R_x(\tau)$ is not a function of time; it is only a function of the time difference τ between the waveform and its shifted copy.

The autocorrelative function of a real-valued *energy* signal has the following properties:

1. $R_x(\tau) = R_x(-\tau)$ symmetrical in τ about zero
2. $|R_x(\tau)| \le R_x(0)$ for all τ maximum value occurs at the origin
3. $R_x(\tau) \leftrightarrow \psi_x(f)$ autocorrelation and ESD form a Fourier transform pair, as designated by the double-headed arrows
4. $R_x(0) = \int_{-\infty}^{\infty} x^2(t)\, dt$ value at the origin is equal to the energy of the signal

If items 1 through 3 are satisfied, $R_x(\tau)$ satisfies the properties of an autocorrelation function. Property 4 can be derived from property 3 and thus need not be included as a basic test.

1.4.2 Autocorrelation of a Periodic (Power) Signal

The autocorrelation function of a real-valued power signal $x(t)$ is defined as

$$R_x(\tau) = \lim_{T \to \infty} \frac{1}{T} \int_{-T/2}^{T/2} x(t)x(t + \tau)\, dt \qquad \text{for} -\infty < \tau < \infty \qquad (1.22)$$

When the power signal $x(t)$ is periodic with period T_0, the time average in Equation (1.22) may be taken over a *single period* T_0, and the autocorrelation function can be expressed as

$$R_x(\tau) = \frac{1}{T_0} \int_{-T_0/2}^{T_0/2} x(t)x(t + \tau)\, dt \qquad \text{for} -\infty < \tau < \infty \qquad (1.23)$$

The autocorrelation function of a real-valued *periodic* signal has properties similar to those of an energy signal:

1. $R_x(\tau) = R_x(-\tau)$ symmetrical in τ about zero
2. $|R_x(\tau)| \le R_x(0)$ for all τ maximum value occurs at the origin
3. $R_x(\tau) \leftrightarrow G_x(f)$ autocorrelation and PSD form a Fourier transform pair
4. $R_x(0) = \dfrac{1}{T_0} \displaystyle\int_{-T_0/2}^{T_0/2} x^2(t)\, dt$ value at the origin is equal to the average power of the signal

1.5 RANDOM SIGNALS

The main objective of a communication system is the transfer of information over a channel. All useful message signals appear random; that is, the receiver does not know, a priori, which of the possible message waveforms will be transmitted. Also, the noise that accompanies the message signals is due to random electrical signals. Therefore, we need to be able to form efficient descriptions of random signals.

1.5.1 Random Variables

Let a *random variable* $X(A)$ represent the functional relationship between a random event A and a real number. For notational convenience, we shall designate the random variable by X, and let the functional dependence upon A be implicit. The random variable may be discrete or continuous. The *distribution function* $F_X(x)$ of the random variable X is given by

$$F_X(x) = P(X \le x) \qquad (1.24)$$

where $P(X \le x)$ is the probability that the value taken by the random variable X is less than or equal to a real number x. The distribution function $F_X(x)$ has the following properties:

1. $0 \le F_X(x) \le 1$

2. $F_X(x_1) \le F_X(x_2)$ if $x_1 \le x_2$

3. $F_X(-\infty) = 0$

4. $F_X(+\infty) = 1$

Another useful function relating to the random variable X is the *probability density function* (pdf), denoted

$$p_X(x) = \frac{dF_X(x)}{dx} \tag{1.25a}$$

As in the case of the distribution function, the pdf is a function of a real number x. The name "density function" arises from the fact that the probability of the event $x_1 \le X \le x_2$ equals

$$
\begin{aligned}
P(x_1 \le X \le x_2) &= P(X \le x_2) - P(X \le x_1) \\
&= F_X(x_2) - F_X(x_1) \\
&= \int_{x_1}^{x_2} p_X(x)\, dx
\end{aligned}
\tag{1.25b}
$$

From Equation (1.25b), the probability that a random variable X has a value in some very narrow range between x and $x + \Delta x$ can be approximated as

$$P(x \le X \le x + \Delta x) \approx p_X(x)\Delta x \tag{1.25c}$$

Thus, in the limit as Δx approaches zero, we can write

$$P(X = x) = p_X(x)dx \tag{1.25d}$$

The probability density function has the following properties:

1. $p_X(x) \ge 0$.

2. $\int_{-\infty}^{\infty} p_X(x)\, dx = F_X(+\infty) - F_X(-\infty) = 1$.

Thus, a probability density function is always a nonnegative function with a total area of one. Throughout the book we use the designation $p_X(x)$ for the probability density function of a *continuous* random variable. For ease of notation, we will often omit the subscript X and write simply $p(x)$. We will use the designation $p(X = x_i)$ for the probability of a random variable X, where X can take on *discrete* values only.

1.5.1.1 Ensemble Averages

The *mean value m_X*, or *expected value* of a random variable X, is defined by

$$m_X = \mathbf{E}\{X\} = \int_{-\infty}^{\infty} x p_X(x)\, dx \tag{1.26}$$

where $\mathbf{E}\{\cdot\}$ is called the *expected value operator*. The *nth moment* of a probability distribution of a random variable X is defined by

$$\mathbf{E}\{X^n\} = \int_{-\infty}^{\infty} x^n p_X(x) \, dx \qquad (1.27)$$

For the purposes of communication system analysis, the most important moments of X are the first two moments. Thus, $n = 1$ in Equation (1.27) gives m_X as discussed above, whereas $n = 2$ gives the mean-square value of X, as follows:

$$\mathbf{E}\{X^2\} = \int_{-\infty}^{\infty} x^2 p_X(x) \, dx \qquad (1.28)$$

We can also define *central moments,* which are the moments of the difference between X and m_X. The second central moment, called the *variance* of X, is defined as

$$\text{var}\,(X) = \mathbf{E}\{(X - m_X)^2\} = \int_{-\infty}^{\infty} (x - m_X)^2 p_X(x) \, dx \qquad (1.29)$$

The variance of X is also denoted as σ_X^2, and its square root, σ_X, is called the *standard deviation* of X. Variance is a measure of the "randomness" of the random variable X. By specifying the variance of a random variable, we are constraining the width of its probability density function. The variance and the mean-square value are related by

$$\begin{aligned}
\sigma_X^2 &= \mathbf{E}\{X^2 - 2m_X X + m_X^2\} \\
&= \mathbf{E}\{X^2\} - 2m_X \mathbf{E}\{X\} + m_X^2 \\
&= \mathbf{E}\{X^2\} - m_X^2
\end{aligned}$$

Thus, the variance is equal to the difference between the mean-square value and the square of the mean.

1.5.2 Random Processes

A random process $X(A, t)$ can be viewed as a function of two variables: *an event A* and *time.* Figure 1.5 illustrates a random process. In the figure there are *N sample functions* of time, $\{X_j(t)\}$. Each of the sample functions can be regarded as the output of a different noise generator. For a specific event A_j, we have a single time function $X(A_j, t) = X_j(t)$ (i.e., a sample function). The totality of all sample functions is called an *ensemble.* For a specific time t_k, $X(A, t_k)$ is a *random variable* $X(t_k)$ whose value depends on the event. Finally, for a specific event, $A = A_j$ and a specific time $t = t_k$, $X(A_j, t_k)$ is simply a *number.* For notational convenience we shall designate the random process by $X(t)$, and let the functional dependence upon A be implicit.

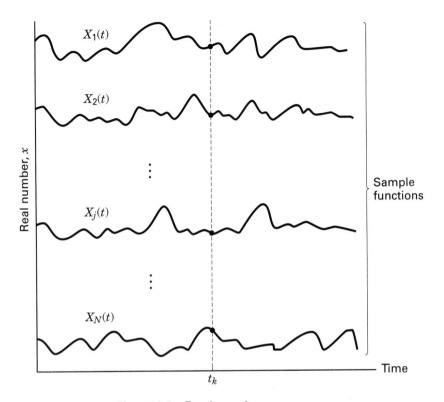

Figure 1.5   Random noise process.

1.5.2.1 Statistical Averages of a Random Process

Because the value of a random process at any future time is unknown (since the identity of the event A is unknown), a random process whose distribution functions are continuous can be described statistically with a probability density function (pdf). In general, the form of the pdf of a random process will be different for different times. In most situations it is not practical to determine empirically the probability distribution of a random process. However, a partial description consisting of the mean and autocorrelation function are often adequate for the needs of communication systems. We define the mean of the random process $X(t)$ as

$$\mathbf{E}\{X(t_k)\} = \int_{-\infty}^{\infty} x p_{X_k}(x)\,dx = m_X(t_k) \tag{1.30}$$

where $X(t_k)$ is the random variable obtained by observing the random process at time t_k and the pdf of $X(t_k)$, the density over the ensemble of events at time t_k, is designated $p_{X_k}(x)$.

We define the autocorrelation function of the random process $X(t)$ to be a function of two variables, t_1 and t_2, given by

$$R_X(t_1, t_2) = \mathbf{E}\{X(t_1)X(t_2)\} \tag{1.31}$$

where $X(t_1)$ and $X(t_2)$ are random variables obtained by observing $X(t)$ at times t_1 and t_2, respectively. The autocorrelation function is a measure of the degree to which two time samples of the same random process are related.

1.5.2.2 Stationarity

A random process $X(t)$ is said to be *stationary* in the *strict sense* if none of its statistics are affected by a shift in the time origin. A random process is said to be *wide-sense stationary* (WSS) if two of its statistics, its mean and autocorrelation function, do not vary with a shift in the time origin. Thus, a process is WSS if

$$\mathbf{E}\{X(t)\} = m_X = \text{a constant} \tag{1.32}$$

and

$$R_X(t_1, t_2) = R_X(t_1 - t_2) \tag{1.33}$$

Strict-sense stationary implies wide-sense stationary, but not vice versa. Most of the useful results in communication theory are predicated on random information signals and noise being wide-sense stationary. From a practical point of view, it is not necessary for a random process to be stationary for all time but only for some observation interval of interest.

For stationary processes, the autocorrelation function in Equation (1.33) does not depend on time but only on the difference between t_1 and t_2. That is, all pairs of values of $X(t)$ at points in time separated by $\tau = t_1 - t_2$ have the same correlation value. Thus, for stationary systems, we can denote $R_X(t_1, t_2)$ simply as $R_X(\tau)$.

1.5.2.3 Autocorrelation of a Wide-Sense Stationary Random Process

Just as the variance provides a measure of randomness for random variables, the autocorrelation function provides a similar measure for random processes. For a wide-sense stationary process, the autocorrelation function is only a function of the *time difference* $\tau = t_1 - t_2$; that is,

$$R_X(\tau) = \mathbf{E}\{X(t)X(t + \tau)\} \qquad \text{for } -\infty < \tau < \infty \tag{1.34}$$

For a zero mean WSS process, $R_X(\tau)$ indicates the extent to which the random values of the process separated by τ seconds in time are statistically correlated. In other words, $R_X(\tau)$ gives us an idea of the frequency response that is associated with a random process. If $R_X(\tau)$ changes slowly as τ increases from zero to some value, it indicates that, on average, sample values of $X(t)$ taken at $t = t_1$ and $t = t_1 + \tau$ are nearly the same. Thus, we would expect a frequency domain representation of $X(t)$ to contain a preponderance of low frequencies. On the other hand, if $R_X(\tau)$ decreases rapidly as τ is increased, we would expect $X(t)$ to change rapidly with time and thereby contain mostly high frequencies.

Properties of the autocorrelation function of a real-valued wide-sense stationary process are as follows:

1. $R_X(\tau) = R_X(-\tau)$ symmetrical in τ about zero
2. $|R_X(\tau)| \le R_X(0)$ for all τ maximum value occurs at the origin
3. $R_X(\tau) \leftrightarrow G_X(f)$ autocorrelation and power spectral density form a Fourier transform pair
4. $R_X(0) = E\{X^2(t)\}$ value at the origin is equal to the average power of the signal

1.5.3 Time Averaging and Ergodicity

To compute m_X and $R_X(\tau)$ by ensemble averaging, we would have to average across all the sample functions of the process and would need to have complete knowledge of the first- and second-order joint probability density functions. Such knowledge is generally not available.

When a random process belongs to a special class, known as an *ergodic process*, its time averages equal its ensemble averages, and the statistical properties of the process can be determined by *time averaging over a single sample function* of the process. For a random process to be ergodic, it must be stationary in the strict sense. (The converse is not necessary.) However, for communication systems, where we are satisfied to meet the conditions of wide-sense stationarity, we are interested only in the mean and autocorrelation functions.

We can say that a random process is *ergodic in the mean* if

$$m_X = \lim_{T \to \infty} 1/T \int_{-T/2}^{T/2} X(t)\,dt \tag{1.35}$$

and it is *ergodic in the autocorrelation function* if

$$R_X(\tau) = \lim_{T \to \infty} 1/T \int_{-T/2}^{T/2} X(t)X(t + \tau)\,dt \tag{1.36}$$

Testing for the ergodicity of a random process is usually very difficult. In practice one makes an intuitive judgment as to whether it is reasonable to interchange the time and ensemble averages. A reasonable assumption in the analysis of most communication signals (in the absence of transient effects) is that the random waveforms are ergodic in the mean and the autocorrelation function. Since time averages equal ensemble averages for ergodic processes, fundamental electrical engineering parameters, such as dc value, rms value, and average power can be related to the moments of an ergodic random process. Following is a summary of these relationships:

1. The quantity $m_X = E\{X(t)\}$ is equal to the dc level of the signal.
2. The quantity m_X^2 is equal to the normalized power in the dc component.
3. The second moment of $X(t)$, $E\{X^2(t)\}$, is equal to the total average normalized power.

4. The quantity $\sqrt{\mathbf{E}\{X^2(t)\}}$ is equal to the root-mean-square (rms) value of the voltage or current signal.

5. The variance σ_X^2 is equal to the average normalized power in the time-varying or ac component of the signal.

6. If the process has zero mean (i.e., $m_X = m_X^2 = 0$), then $\sigma_X^2 = \mathbf{E}\{X^2\}$ and the variance is the same as the mean-square value, or the variance represents the total power in the normalized load.

7. The standard deviation σ_X is the rms value of the ac component of the signal.

8. If $m_X = 0$, then σ_X is the rms value of the signal.

1.5.4 Power Spectral Density and Autocorrelation of a Random Process

A random process $X(t)$ can generally be classified as a power signal having a power spectral density (PSD) $G_X(f)$ of the form shown in Equation (1.20). $G_X(f)$ is particularly useful in communication systems, because it describes the distribution of a signal's power in the frequency domain. The PSD enables us to evaluate the signal power that will pass through a network having known frequency characteristics. We summarize the principal features of PSD functions as follows:

1. $G_X(f) \ge 0$ and is always real valued

2. $G_X(f) = G_X(-f)$ for $X(t)$ real-valued

3. $G_X(f) \leftrightarrow R_X(\tau)$ PSD and autocorrelation form a Fourier transform pair

4. $P_X = \displaystyle\int_{-\infty}^{\infty} G_X(f)\, df$ relationship between average normalized power and PSD

In Figure 1.6, we present a visualization of autocorrelation and power spectral density functions. What does the term *correlation* mean? When we inquire about the correlation between two phenomena, we are asking how closely do they correspond in behavior or appearance, how well do they match one another. In mathematics, an autocorrelation function of a signal (in the time domain) describes the correspondence of the signal to itself in the following way. An exact copy of the signal is made and located in time at minus infinity. Then we move the copy an increment in the direction of positive time and ask the question, "How well do these two (the original versus the copy) match"? We move the copy another step in the positive direction and ask, "How well do they match now?" And so forth. The correlation between the two is plotted as a function of time, denoted τ, which can be thought of as a scanning parameter.

Figure 1.6a–d highlights some of these steps. Figure 1.6a illustrates a single sample waveform from a WSS random process, $X(t)$. The waveform is a binary random sequence with unit-amplitude positive and negative (bipolar) pulses. The positive and negative pulses occur with equal probability. The duration of each binary digit is T seconds, and the average or dc value of the random sequence is zero. Figure 1.6b shows the same sequence displaced τ_1 seconds in time; this sequence is therefore denoted $X(t - \tau_1)$. Let us assume that $X(t)$ is ergodic in the autocorrela-

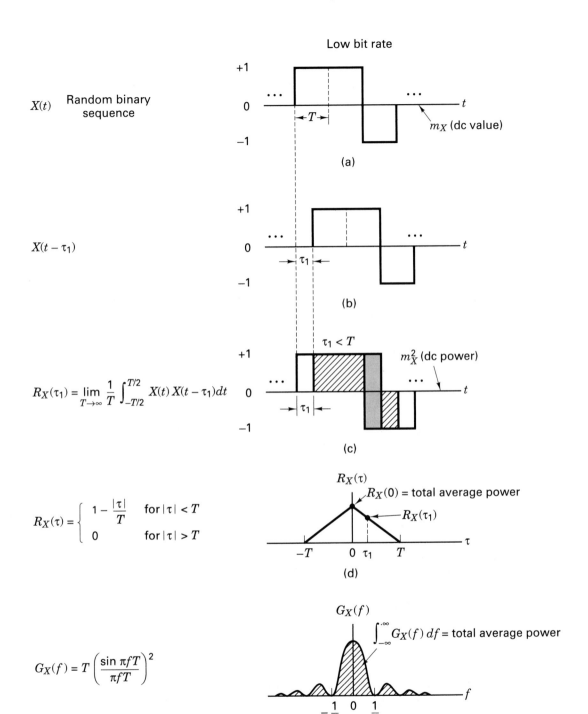

Low bit rate

$X(t)$ Random binary sequence

$+1$
0 ⋯ ⋯ t
-1

$\leftarrow T \rightarrow$

m_X (dc value)

(a)

$X(t - \tau_1)$

$+1$
0 ⋯ ⋯ t
-1

$\rightarrow \tau_1 \leftarrow$

(b)

$R_X(\tau_1) = \lim_{T \to \infty} \frac{1}{T} \int_{-T/2}^{T/2} X(t)\, X(t - \tau_1)dt$

$\tau_1 < T$

$+1$
0 ⋯ ⋯ t
-1

$\rightarrow |\tau_1| \leftarrow$

m_X^2 (dc power)

(c)

$R_X(\tau) = \begin{cases} 1 - \dfrac{|\tau|}{T} & \text{for } |\tau| < T \\ 0 & \text{for } |\tau| > T \end{cases}$

$R_X(\tau)$

$R_X(0)$ = total average power

$R_X(\tau_1)$

$-T$ $\quad$ 0 τ_1 $\quad$ T $\quad$ τ

(d)

$G_X(f) = T \left(\dfrac{\sin \pi fT}{\pi fT} \right)^2$

$G_X(f)$

$\int_{-\infty}^{\infty} G_X(f)\, df$ = total average power

$-\dfrac{1}{T}$ $\quad$ 0 $\quad$ $\dfrac{1}{T}$ $\quad$ f

Figure 1.6 Autocorrelation and power spectral density.

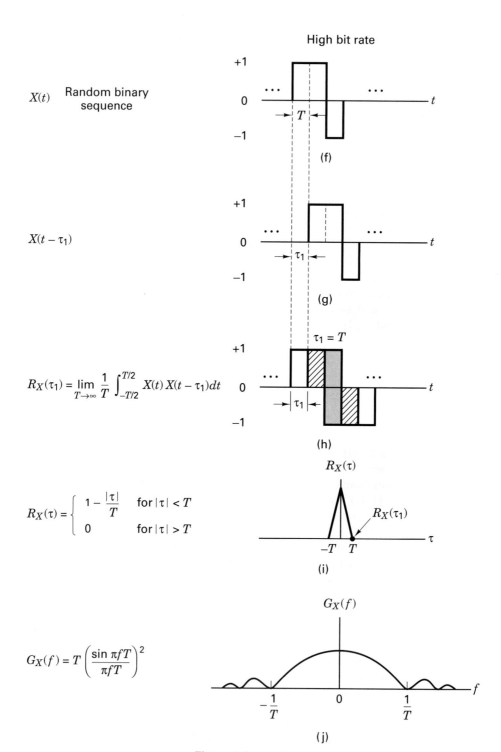

High bit rate

$X(t)$ Random binary sequence

(f)

$X(t - \tau_1)$

(g)

$R_X(\tau_1) = \lim\limits_{T \to \infty} \dfrac{1}{T} \displaystyle\int_{-T/2}^{T/2} X(t)\,X(t - \tau_1)\,dt$

$\tau_1 = T$

(h)

$R_X(\tau)$

$R_X(\tau) = \begin{cases} 1 - \dfrac{|\tau|}{T} & \text{for } |\tau| < T \\ 0 & \text{for } |\tau| > T \end{cases}$

$R_X(\tau_1)$

(i)

$G_X(f)$

$G_X(f) = T \left(\dfrac{\sin \pi f T}{\pi f T} \right)^2$

(j)

Figure 1.6   continued

tion function so that we can use time averaging instead of ensemble averaging to find $R_X(\tau)$. The value of $R_X(\tau_1)$ is obtained by taking the product of the two sequences $X(t)$ and $X(t - \tau_1)$ and finding the average value using Equation (1.36). Equation (1.36) is accurate for ergodic processes *only in the limit*. However, integration over an integer number of periods can provide us with an estimate of $R_X(\tau)$. Notice that $R_X(\tau_1)$ can be obtained by a positive or negative shift of $X(t)$. Figure 1.6c illustrates such a calculation, using the single sample sequence (Figure 1.6a) and its shifted replica (Figure 1.6b). The cross-hatched areas under the product curve $X(t)X(t - \tau_1)$ contribute to positive values of the product, and the grey areas contribute to negative values. The integration of $X(t)\ X(t - \tau_1)$ over several pulse times yields a net value of area which is one point, the $R_X(\tau_1)$ point of the $R_X(\tau)$ curve. The sequences can be further shifted by $\tau_2, \tau_3, \ldots,$ each shift yielding a point on the overall autocorrelation function $R_X(\tau)$ shown in Figure 1.6d. Every random sequence of bipolar pulses has an autocorrelation plot of the general shape shown in Figure 1.6d. The plot peaks at $R_X(0)$ [the best match occurs when τ equals zero, since $R(\tau) \leq R(0)$ for all τ], and it declines as τ increases. Figure 1.6d shows points corresponding to $R_X(0)$ and $R_X(\tau_1)$.

The analytical expression for the autocorrelation function $R_X(\tau)$ shown in Figure 1.6d, is [1]

$$
R_X(\tau) = \begin{cases} 1 - \dfrac{|\tau|}{T} & \text{for } |\tau| \leq T \\ 0 & \text{for } |\tau| > T \end{cases} \tag{1.37}
$$

Notice that the autocorrelation function gives us frequency information; it tells us something about the bandwidth of the signal. Autocorrelation is a time-domain function; there are no frequency-related terms in the relationship shown in Equation (1.37). How does it give us bandwidth information about the signal? Consider that the signal is a very slowly moving (low bandwidth) signal. As we step the copy along the τ axis, at each step asking the question, "How good is the match between the original and the copy?" the match will be quite good for a while. In other words, the triangular-shaped autocorrelation function in Figure 1.6d and Equation (1.37) will ramp down gradually with τ. But if we have a very rapidly moving (high bandwidth) signal, perhaps a very small shift in τ will result in there being zero correlation. In this case, the autocorrelation function will have a very steep appearance. Therefore, the relative shape of the autocorrelation function tells us something about the bandwidth of the underlying signal. Does it ramp down gently? If so, then we are dealing with a low bandwidth signal. Is the function steep? If so, then we are dealing with a high bandwidth signal.

The autocorrelation function allows us to express a random signal's power spectral density directly. Since the PSD and the autocorrelation function are Fourier transforms of each other, the PSD, $G_X(f)$, of the random bipolar-pulse sequence can be found using Table A.1 as the transform of $R_X(\tau)$ in Equation (1.37). Observe that

$$G_X(f) = T\left(\frac{\sin \pi f T}{\pi f T}\right)^2 = T \operatorname{sinc}^2 fT \qquad (1.38)$$

where

$$\operatorname{sinc} y = \frac{\sin \pi y}{\pi y} \qquad (1.39)$$

The general shape of $G_X(f)$ is illustrated in Figure 1.6e.

Notice that the area under the PSD curve represents the average power in the signal. One convenient measure of *bandwidth* is the width of the main spectral lobe. (See Section 1.7.2.) Figure 1.6e illustrates that the bandwidth of a signal is inversely related to the symbol duration or pulse width, Figures 1.6f–j repeat the steps shown in Figures 1.6a–e, except that the pulse duration is shorter. Notice that the shape of the shorter pulse duration $R_X(\tau)$ is narrower, shown in Figure 1.6i, than it is for the longer pulse duration $R_X(\tau)$, shown in Figure 1.6d. In Figure 1.6i, $R_X(\tau_1) = 0$; in other words, a shift of τ_1 in the case of the shorter pulse duration example is enough to produce a zero match, or a complete decorrelation between the shifted sequences. Since the pulse duration T is shorter (pulse rate is higher) in Figure 1.6f, than in Figure 1.6a, the bandwidth occupancy in Figure 1.6j is greater than the bandwidth occupancy shown in Figure 1.6e for the lower pulse rate.

1.5.5 Noise in Communication Systems

The term *noise* refers to *unwanted* electrical signals that are always present in electrical systems. The presence of noise superimposed on a signal tends to obscure or mask the signal; it limits the receiver's ability to make correct symbol decisions, and thereby limits the rate of information transmission. Noise arises from a variety of sources, both man made and natural. The *man-made noise* includes such sources as spark-plug ignition noise, switching transients, and other radiating electromagnetic signals. *Natural noise* includes such elements as the atmosphere, the sun, and other galactic sources.

Good engineering design can eliminate much of the noise or its undesirable effect through filtering, shielding, the choice of modulation, and the selection of an optimum receiver site. For example, sensitive radio astronomy measurements are typically located at remote desert locations, far from man-made noise sources. However, there is one natural source of noise, called *thermal* or *Johnson noise,* that cannot be eliminated. Thermal noise [4, 5] is caused by the thermal motion of electrons in all dissipative components—resistors, wires, and so on. The same electrons that are responsible for electrical conduction are also responsible for thermal noise.

We can describe thermal noise as a zero-mean *Gaussian* random process. A Gaussian process $n(t)$ is a random function whose value n at any arbitrary time t is statistically characterized by the Gaussian probability density function

$$p(n) = \frac{1}{\sigma\sqrt{2\pi}} \exp\left[-\frac{1}{2}\left(\frac{n}{\sigma}\right)^2\right] \qquad (1.40)$$

where σ^2 is the variance of n. The *normalized* or *standardized Gaussian density function* of a zero-mean process is obtained by assuming that $\sigma = 1$. This normalized pdf is shown sketched in Figure 1.7.

We will often represent a random signal as the sum of a Gaussian noise random variable and a dc signal. That is,

$$z = a + n$$

where z is the random signal, a is the dc component, and n is the Gaussian noise random variable. The pdf $p(z)$ is then expressed as

$$p(z) = \frac{1}{\sigma\sqrt{2\pi}} \exp\left[-\frac{1}{2}\left(\frac{z-a}{\sigma}\right)^2 \right] \tag{1.41}$$

where, as before, σ^2 is the variance of n. The Gaussian distribution is often used as the system noise model because of a theorem, called the *central limit theorem* [3], which states that under very general conditions the probability distribution of the sum of j statistically independent random variables approaches the Gaussian distribution as $j \to \infty$, no matter what the individual distribution functions may be. Therefore, even though individual noise mechanisms might have other than Gaussian distributions, the aggregate of many such mechanisms will tend toward the Gaussian distribution.

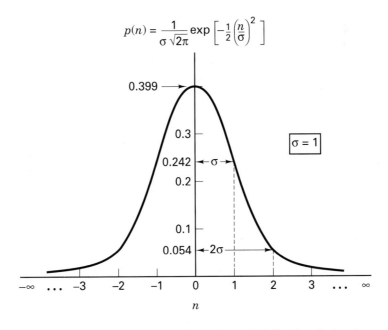

Figure 1.7 Normalized ($\sigma = 1$) Gaussian probability density function.

1.5.5.1 White Noise

The primary spectral characteristic of thermal noise is that its power spectral density is *the same* for all frequencies of interest in most communication systems; in other words, a thermal noise source emanates an equal amount of noise power per unit bandwidth at all frequencies—from dc to about 10^{12} Hz. Therefore, a simple model for thermal noise assumes that its power spectral density $G_n(f)$ is flat for all frequencies, as shown in Figure 1.8a, and is denoted as

$$G_n(f) = \frac{N_0}{2} \qquad \text{watts/hertz} \tag{1.42}$$

where the factor of 2 is included to indicate that $G_n(f)$ is a *two-sided* power spectral density. When the noise power has such a uniform spectral density we refer to it as *white noise*. The adjective "white" is used in the same sense as it is with white light, which contains equal amounts of all frequencies within the visible band of electromagnetic radiation.

The autocorrelation function of white noise is given by the inverse Fourier transform of the noise power spectral density (see Table A.1), denoted as follows:

$$R_n(\tau) = \mathscr{F}^{-1}\{G_n(f)\} = \frac{N_0}{2}\delta(\tau) \tag{1.43}$$

Thus the autocorrelation of white noise is a delta function weighted by the factor $N_0/2$ and occurring at $\tau = 0$, as seen in Figure 1.8b. Note that $R_n(\tau)$ is zero for $\tau \neq 0$; that is, any two different samples of white noise, no matter how close together in time they are taken, are uncorrelated.

The average power P_n of white noise is *infinite* because its bandwidth is infinite. This can be seen by combining Equations (1.19) and (1.42) to yield

$$P_n = \int_{-\infty}^{\infty} \frac{N_0}{2}\,df = \infty \tag{1.44}$$

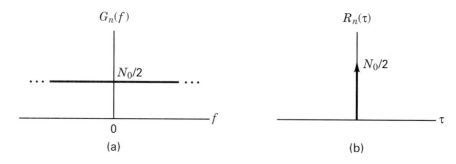

Figure 1.8 (a) Power spectral density of white noise. (b) Autocorrelation function of white noise.

Although white noise is a useful abstraction, no noise process can truly be white; however, the noise encountered in many real systems can be assumed to be approximately white. We can only observe such noise after it has passed through a real system which will have a finite bandwidth. Thus, as long as the bandwidth of the noise is appreciably larger than that of the system, the noise can be considered to have an infinite bandwidth.

The delta function in Equation (1.43) means that the noise signal $n(t)$ is totally decorrelated from its time-shifted version, for any $\tau > 0$. Equation (1.43) indicates that *any* two different samples of a white noise process are uncorrelated. Since thermal noise is a Gaussian process and the samples are uncorrelated, the noise samples are also independent [3]. Therefore, the effect on the detection process of a channel with *additive white Gaussian noise* (AWGN) is that the noise affects each transmitted symbol *independently*. Such a channel is called a *memoryless channel.* The term "additive" means that the noise is simply superimposed or added to the signal—that there are no multiplicative mechanisms at work.

Since thermal noise is present in all communication systems and is the prominent noise source for most systems, the thermal noise characteristics—additive, white, and Gaussian—are most often used to model the noise in communication systems. Since zero-mean Gaussian noise is completely characterized by its *variance,* this model is particularly simple to use in the detection of signals and in the design of optimum receivers. In this book we shall assume, unless otherwise stated, that the system is corrupted by *additive zero-mean white Gaussian noise,* even though this is sometimes an oversimplification.

1.6 SIGNAL TRANSMISSION THROUGH LINEAR SYSTEMS

Having developed a set of models for signals and noise, we now consider the characterization of systems and their effects on such signals and noise. Since a system can be characterized equally well in the time domain or the frequency domain, techniques will be developed in both domains to analyze the response of a linear system to an arbitrary input signal. The signal, applied to the input of the system, as shown in Figure 1.9, can be described either as a time-domain signal, $x(t)$, or by its Fourier transform, $X(f)$. The use of time-domain analysis yields the time-domain output $y(t)$, and in the process, $h(t)$, the characteristic or *impulse response* of the network will be defined. When the input is considered in the frequency domain, we shall define a *frequency transfer function $H(f)$* for the system, which will determine the frequency-domain output $Y(f)$. The system is assumed to be linear and time invariant. It is also assumed that there is no stored energy in the system at the time the input is applied.

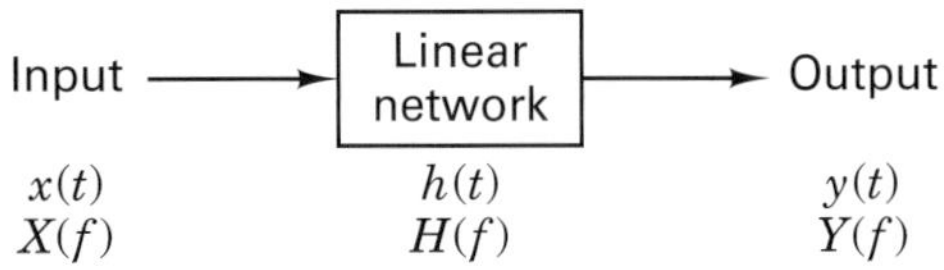

Input $\longrightarrow$ [Linear network] $\longrightarrow$ Output

Figure 1.9 Linear system and its key parameters.

$x(t)$ $h(t)$ $y(t)$
$X(f)$ $H(f)$ $Y(f)$

1.6.1 Impulse Response

The linear time invariant system or network illustrated in Figure 1.9 is characterized in the time domain by an impulse response $h(t)$, which is the response when the input is equal to a unit impulse $\delta(t)$; that is,

$$h(t) = y(t) \qquad \text{when } x(t) = \delta(t) \tag{1.45}$$

Consider the name *impulse response*. That is a very appropriate name for this event. Characterizing a linear system in terms of its impulse response has a straightforward physical interpretation. At the system input, we apply a unit impulse (a nonrealizable signal, having infinite amplitude, zero width, and unit area), as illustrated in Figure 1.10a. Applying such an impulse to the system can be thought of as giving the system "a whack." How does the system respond to such a force (impulse) at the input? The output response $h(t)$ is the system's impulse response. (A possible shape is depicted in Figure 1.10b.)

The response of the network to an arbitrary input signal $x(t)$ is found by the convolution of $x(t)$ with $h(t)$, expressed as

$$y(t) = x(t) * h(t) = \int_{-\infty}^{\infty} x(\tau)\, h(t - \tau)\, d\tau \tag{1.46}$$

where $*$ denotes the convolution operation. (See Section A.5.) The system is assumed to be *causal*, which means that there can be *no* output prior to the time, $t = 0$, when the input is applied. Therefore, the lower limit of integration can be changed to zero, and we can express the output $y(t)$ in either the form

$$y(t) = \int_{0}^{\infty} x(\tau)h(t - \tau)\, d\tau \tag{1.47a}$$

or the form

$$y(t) = \int_{0}^{\infty} x(t - \tau)\, h(\tau)\, d\tau \tag{1.47b}$$

Each of the expressions in Equations (1.46) and (1.47) is called the *convolution integral.* Convolution is a basic mathematical tool that plays an important role in understanding all communication systems. Thus, the reader is urged to review

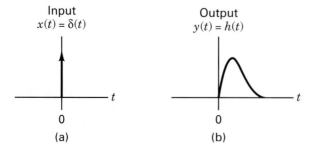

Input
$x(t) = \delta(t)$

Output
$y(t) = h(t)$

0

(a)

0

(b)

Figure 1.10 (a) Input signal $x(t)$ is a unit impulse function. (b) Output signal $y(t)$ is the system's impulse response $h(t)$.

section A.5, where one can see that Equations (1.46) and (1.47) are the results of a straightforward process.

1.6.2 Frequency Transfer Function

The frequency-domain output signal $Y(f)$ is obtained by taking the Fourier transform of both sides of Equation (1.46). Since convolution in the time domain transforms to multiplication in the frequency domain (and vice versa), Equation (1.46) yields

$$Y(f) = X(f)H(f) \qquad (1.48)$$

or

$$H(f) = \frac{Y(f)}{X(f)} \qquad (1.49)$$

provided, of course, that $X(f) \neq 0$ for all f. Here $H(f) = \mathcal{F}\{h(t)\}$, the Fourier transform of the impulse response function, is called the *frequency transfer function* or the *frequency response* of the network. In general, $H(f)$ is complex and can be written as

$$H(f) = |H(f)| \, e^{j\theta(f)} \qquad (1.50)$$

where $|H(f)|$ is the magnitude response. The phase response is defined as

$$\theta(f) = \tan^{-1} \frac{\text{Im}\,\{H(f)\}}{\text{Re}\,\{H(f)\}} \qquad (1.51)$$

where the terms "Re" and "Im" denote "the real part of" and "the imaginary part of," respectively.

The frequency transfer function of a linear time-invariant network can easily be measured in the laboratory with a sinusoidal generator at the input of the network and an oscilloscope at the output. When the input waveform $x(t)$ is expressed as

$$x(t) = A \cos 2\pi f_0 t$$

the output of the network will be

$$y(t) = A \, |H(f_0)| \, \cos \left[2\pi f_0 t + \theta(f_0) \right] \qquad (1.52)$$

The input frequency f_0 is stepped through the values of interest; at each step, the amplitude and phase at the output are measured.

1.6.2.1 Random Processes and Linear Systems

If a random process forms the input to a time-invariant linear system, the output will also be a random process. That is, each sample function of the input process yields a sample function of the output process. The input power spectral density $G_X(f)$ and the output power spectral density $G_Y(f)$ are related as follows:

$$G_Y(f) = G_X(f) \, |H(f)|^2 \tag{1.53}$$

Equation (1.53) provides a simple way of finding the power spectral density out of a time-invariant linear system when the input is a random process.

In Chapters 3 and 4 we consider the detection of signals in Gaussian noise. We will utilize a fundamental property of a Gaussian process applied to a linear system, as follows. It can be shown that if a Gaussian process $X(t)$ is applied to a time-invariant linear filter, the random process $Y(t)$ developed at the output of the filter is also Gaussian [6].

1.6.3 Distortionless Transmission

What is required of a network for it to behave like an *ideal* transmission line? The output signal from an ideal transmission line may have some time delay compared with the input, and it may have a different amplitude than the input (just a scale change), but otherwise it must have no distortion—it must have the same shape as the input. Therefore, for ideal distortionless transmission, we can describe the output signal as

$$y(t) = Kx(t - t_0) \tag{1.54}$$

where K and t_0 are constants. Taking the Fourier transform of both sides (see Section A.3.1), we write

$$Y(f) = KX(f)e^{-j2\pi ft_0} \tag{1.55}$$

Substituting the expression (1.55) for $Y(f)$ into Equation (1.49), we see that the required system transfer function for distortionless transmission is

$$H(f) = Ke^{-j2\pi ft_0} \tag{1.56}$$

Therefore, to achieve *ideal distortionless transmission,* the overall system response must have a constant magnitude response and its phase shift must be linear with frequency. It is not enough that the system amplify or attenuate all frequency components equally. All of the signal's frequency components must also arrive with identical time delay in order to add up correctly. Since the time delay t_0 is related to the phase shift θ and the radian frequency $\omega = 2\pi f$ by

$$t_0 \text{ (seconds)} = \frac{\theta \text{ (radians)}}{2\pi f \text{ (radians/second)}} \tag{1.57a}$$

it is clear that phase shift must be proportional to frequency in order for the time delay of all components to be identical. A characteristic often used to measure delay distortion of a signal is called *envelope delay* or *group delay,* which is defined as

$$\tau(f) = -\frac{1}{2\pi} \frac{d\theta(f)}{df} \tag{1.57b}$$

Therefore, for distortionless transmission, an equivalent way of characterizing phase to be a linear function of frequency is to characterize the envelope delay $\tau(f)$ as a constant. In practice, a signal will be distorted in passing through some parts of a system. Phase or amplitude correction (*equalization*) networks may be introduced elsewhere in the system to correct for this distortion. It is the overall input–output characteristic of the system that determines its performance.

1.6.3.1 Ideal Filter

One cannot build the ideal network described in Equation (1.56). The problem is that Equation (1.56) implies an infinite bandwidth capability, where the bandwidth of a system is defined as the interval of positive frequencies over which the magnitude $|H(f)|$ remains within a specified value. In Section 1.7 various measures of bandwidth are enumerated. As an approximation to the ideal infinite-bandwidth network, let us choose a truncated network that passes, without distortion, all frequency components between f_ℓ and f_u, where f_ℓ is the lower cutoff frequency and f_u is the upper cutoff frequency, as shown in Figure 1.11. Each of these networks is called an *ideal filter*. Outside the range $f_\ell < f < f_u$, which is called the *passband*, the ideal filter is assumed to have a response of zero magnitude. The effective width of the passband is specified by the filter bandwidth $W_f = (f_u - f_\ell)$ hertz.

When $f_\ell \neq 0$ and $f_u \neq \infty$, the filter is called a *bandpass filter* (BPF), shown in Figure 1.11a. When $f_\ell = 0$ and f_u has a finite value, the filter is called a *low-pass filter* (LPF), shown in Figure 1.11b. When f_ℓ has a nonzero value and when $f_u \to \infty$, the filter is called a *high-pass filter* (HPF), shown in Figure 1.11c.

Following Equation (1.56) and letting $K = 1$, for the ideal low-pass filter transfer function with bandwidth $W_f = f_u$ hertz, shown in Figure 1.11b, we can write the transfer function as

$$H(f) = |H(f)| \, e^{-j\theta(f)} \tag{1.58}$$

where

$$|H(f)| = \begin{cases} 1 & \text{for } |f| < f_u \\ 0 & \text{for } |f| \geq f_u \end{cases} \tag{1.59}$$

and

$$e^{-j\theta(f)} = e^{-j\,2\pi f t_0} \tag{1.60}$$

The impulse response of the ideal low-pass filter, illustrated in Figure 1.12, is

$$h(t) = \mathcal{F}^{-1}\{H(f)\} = \int_{-\infty}^{\infty} H(f) e^{j\,2\pi f t} \, df \tag{1.61}$$

$$= \int_{-f_u}^{f_u} e^{-j2\pi f t_0} e^{j2\pi f t} \, df$$

or

$$= \int_{-f_u}^{f_u} e^{j\,2\pi f(t\,-\,t_0)}\,df$$

$$= 2f_u \frac{\sin 2\pi f_u(t\,-\,t_0)}{2\pi f_u(t\,-\,t_0)}$$

$$= 2f_u \operatorname{sinc} 2f_u(t\,-\,t_0) \tag{1.62}$$

where sinc x is as defined in Equation (1.39). The impulse response shown in Figure 1.12 is noncausal, which means that it has a nonzero output prior to the

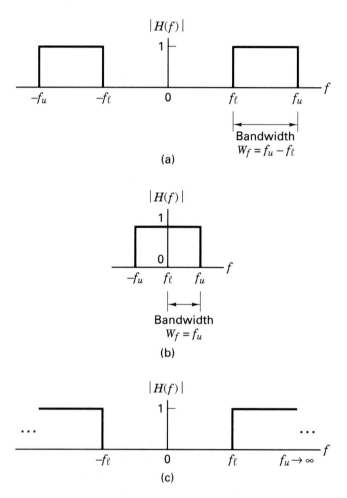

Figure 1.11 Ideal filter transfer function. (a) Ideal bandpass filter. (b) Ideal low-pass filter. (c) Ideal high-pass filter.

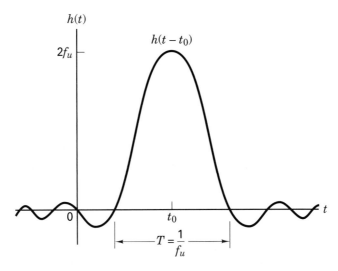

Figure 1.12   Impulse response of the ideal low-pass filter.

application of an input at time $t = 0$. Therefore, it should be clear that the ideal filter described in Equation (1.58) is not realizable.

Example 1.2   Effect of an Ideal Filter on White Noise

White noise with power spectral density $G_n(f) = N_0/2$, shown in Figure 1.8a, forms the input to the ideal low-pass filter shown in Figure 1.11b. Find the power spectral density $G_Y(f)$ and the autocorrelation function $R_Y(\tau)$ of the output signal.

Solution

$$G_Y(f) = G_n(f)\,|H(f)|^2$$

$$= \begin{cases} \dfrac{N_0}{2} & \text{for } |f| < f_u \\ 0 & \text{otherwise} \end{cases}$$

The autocorrelation is the inverse Fourier transform of the power spectral density and is given by (see Table A.1)

$$R_Y(\tau) = N_0 f_u \frac{\sin 2\pi f_u \tau}{2\pi f_u \tau}$$
$$= N_0 f_u \operatorname{sinc} 2f_u \tau$$

Comparing this result with Equation (1.62), we see that $R_Y(\tau)$ has the same shape as the impulse response of the ideal low-pass filter shown in Figure 1.12. In this example the ideal low-pass filter transforms the autocorrelation function of white noise (defined by the delta function) into a sinc function. After filtering, we no longer have white noise. The output noise signal will have zero correlation with shifted copies of itself, only at shifts of $\tau = n/2f_u$, where n is any integer other than zero.

1.6.3.2 Realizable Filters

The very simplest example of a realizable low-pass filter is made up of resistance ($\mathcal{R}$) and capacitance (C), as shown in Figure 1.13a; it is called an $\mathcal{R}C$ *filter*, and its transfer function can be expressed as [7]

$$H(f) = \frac{1}{1 + j2\pi f \mathcal{R}C} = \frac{1}{\sqrt{1 + (2\pi f \mathcal{R}C)^2}} e^{-j\theta(f)} \qquad (1.63)$$

where $\theta(f) = \tan^{-1} 2\pi f \mathcal{R}C$. The magnitude characteristic $|H(f)|$ and the phase characteristic $\theta(f)$ are plotted in Figures 1.13b and c, respectively. The low-pass filter bandwidth is defined to be its half-power point; this point is the frequency at which the output signal power has fallen to one-half of its peak value, or the frequency at which the magnitude of the output voltage has fallen to $1/\sqrt{2}$ of its peak value.

The half-power point is generally expressed in decibel (dB) units as the −3-dB point, or the point that is 3 dB down from the peak, where the decibel is defined as the ratio of two amounts of power, P_1 and P_2, existing at two points. By definition,

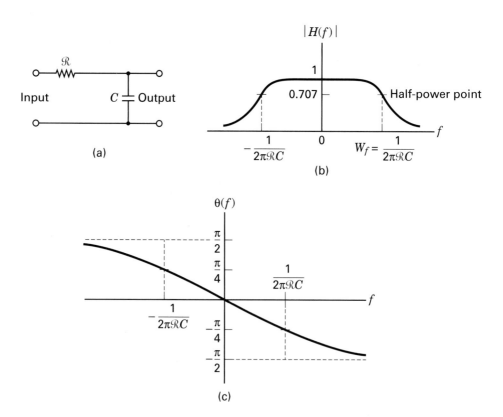

(a)

(b)

(c)

Figure 1.13 $\mathcal{R}C$ filter and its transfer function. (a) $\mathcal{R}C$ filter. (b) Magnitude characteristic of the $\mathcal{R}C$ filter. (c) Phase characteristic of the $\mathcal{R}C$ filter.

$$\text{number of dB} = 10 \log_{10} \frac{P_2}{P_1} = 10 \log_{10} \frac{V_2^2/\mathscr{R}_2}{V_1^2/\mathscr{R}_1} \tag{1.64a}$$

where V_1 and V_2 are voltages and $\mathscr{R}_1$ and $\mathscr{R}_2$ are resistances. For communication systems, *normalized power* is generally used for analysis; in this case, $\mathscr{R}_1$ and $\mathscr{R}_2$ are set equal to 1 Ω, so that

$$\text{number of dB} = 10 \log_{10} \frac{P_2}{P_1} = 10 \log_{10} \frac{V_2^2}{V_1^2} \tag{1.64b}$$

The amplitude response can be expressed in decibels by

$$|H(f)|_{\text{dB}} = 20 \log_{10} \frac{V_2}{V_1} = 20 \log_{10} |H(f)| \tag{1.64c}$$

where V_1 and V_2 are the input and output voltages, respectively, and where the input and output resistances have been assumed equal.

From Equation (1.63) it is easy to verify that the half-power point of the low-pass $\mathscr{R}C$ filter corresponds to $\omega = 1/\mathscr{R}C$ radians per second or $f = 1/(2\pi\mathscr{R}C)$ hertz. Thus the bandwidth W_f in hertz is $1/(2\pi\mathscr{R}C)$. The filter *shape factor* is a measure of how well a realizable filter approximates the ideal filter. It is typically defined as the ratio of the filter bandwidths at the -60-dB and -6-dB amplitude response points. A sharp-cutoff bandpass filter can be made with a shape factor as low as about 2. By comparison, the shape factor of the simple $\mathscr{R}C$ low-pass filter is almost 600.

There are several useful approximations to the ideal low-pass filter characteristic. One of these, the *Butterworth filter,* approximates the ideal low-pass filter with the function

$$|H_n(f)| = \frac{1}{\sqrt{1 + (f/f_u)^{2n}}} \qquad n \geq 1 \tag{1.65}$$

where f_u is the upper -3-db cutoff frequency and n is referred to as the *order* of the filter. The higher the order, the greater will be the complexity and the cost to implement the filter. The magnitude function, $|H(f)|$, is sketched (single sided) for several values of n in Figure 1.14. Note that as n gets larger, the magnitude characteristics approach that of the ideal filter. Butterworth filters are popular because they are the best approximation to the ideal, in the sense of *maximal flatness* in the filter passband.

Example 1.3   Effect of an $\mathscr{R}C$ Filter on White Noise

White noise with spectral density $G_n(f) = N_0/2$, shown in Figure 1.8a, forms the input to the $\mathscr{R}C$ filter shown in Figure 1.13a. Find the power spectral density $G_Y(f)$ and the autocorrelation function $R_Y(\tau)$ of the output signal.

Solution

$$G_Y(f) = G_n(f) \, |H(f)|^2$$
$$= \frac{N_0}{2} \frac{1}{1 + (2\pi f \mathscr{R}C)^2}$$
$$R_Y(\tau) = \mathscr{F}^{-1}\{G_Y(f)\}$$

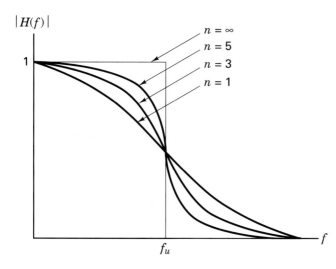

$|H(f)|$

$n = \infty$

$n = 5$

$n = 3$

$n = 1$

1

f_u

f

Figure 1.14 Butterworth filter magnitude response.

Using Table A.1, we find that the inverse Fourier transform of $G_Y(f)$ is

$$R_Y(\tau) = \frac{N_0}{4\mathcal{R}C} \exp\left(-\frac{|\tau|}{\mathcal{R}C}\right)$$

As might have been predicted, we no longer have white noise after filtering. The $\mathcal{R}C$ filter transforms the input autocorrelation function of white noise (defined by the delta function) into an exponential function. For a narrowband filter (a large $\mathcal{R}C$ product), the output noise will exhibit higher correlation between noise samples of a fixed time shift than will the output noise from a wideband filter.

1.6.4 Signals, Circuits, and Spectra

Signals have been described in terms of their spectra. Similarly, networks or circuits have been described in terms of their spectral characteristics or frequency transfer functions. How is a signal's bandwidth affected as a result of the signal passing through a filter circuit? Figure 1.15 illustrates two cases of interest. In Figure 1.15a (case 1), the input signal has a narrowband spectrum, and the filter transfer function is a wideband function. From Equation (1.48), we see that the output signal spectrum is simply the product of these two spectra. In Figure 1.15a we can verify that multiplication of the two spectral functions will result in a spectrum with a bandwidth approximately equal to the smaller of the two bandwidths (when one of the two spectral functions goes to zero, the multiplication yields zero). Therefore, for case 1, the output signal spectrum is constrained by the input signal spectrum alone. Similarly, we see that for case 2, in Figure 1.15b, where the input signal is a wideband signal but the filter has a narrowband transfer function, the bandwidth of the output signal is constrained by the filter bandwidth; the output signal will be a filtered (distorted) rendition of the input signal.

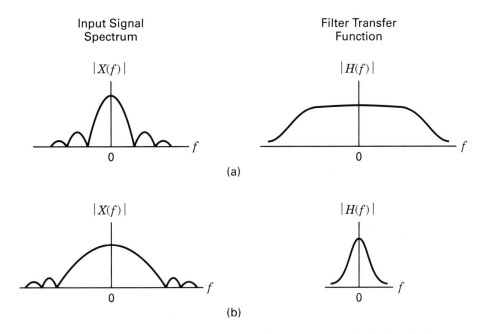

Input Signal
Spectrum

$|X(f)|$

0

Filter Transfer
Function

$|H(f)|$

0

(a)

$|X(f)|$

0

$|H(f)|$

0

(b)

Figure 1.15 Spectral characteristics of the input signal and the circuit contribute to the spectral characteristics of the output signal. (a) Case 1: Output bandwidth is constrained by input signal bandwidth. (b) Case 2: Output bandwidth is constrained by filter bandwidth.

The effect of a filter on a waveform can also be viewed in the time domain. The output $y(t)$ resulting from convolving an ideal input pulse $x(t)$ (having amplitude V_m and pulse width T) with the impulse response of a low-pass $\mathcal{R}C$ filter can be written as [8]

$$y(t) = \begin{cases} V_m(1 - e^{-t/\mathcal{R}C}) & \text{for } 0 \le t \le T \\ V'_m e^{-(t-T)/\mathcal{R}C} & \text{for } t > T \end{cases} \qquad (1.66)$$

where

$$V'_m = V_m(1 - e^{-T/\mathcal{R}C}) \qquad (1.67)$$

Let us define the pulse bandwidth as

$$W_p = \frac{1}{T} \qquad (1.68)$$

and the $\mathcal{R}C$ filter bandwidth, as

$$W_f = \frac{1}{2\pi\mathcal{R}C} \qquad (1.69)$$

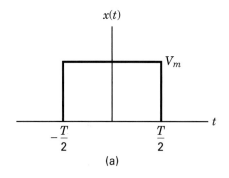

(a)

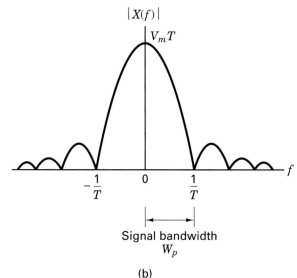

Signal bandwidth
W_p

(b)

Figure 1.16 (a) Ideal pulse. (b) Magnitude spectrum of the ideal pulse.

The ideal input pulse $x(t)$ and its magnitude spectrum $|X(f)|$ are shown in Figure 1.16. The $\mathcal{R}C$ filter and its magnitude characteristic $|H(f)|$ are shown in Figures 1.13a and b, respectively. Following Equations (1.66) to (1.69), three cases are illustrated in Figure 1.17. Example 1 illustrates the case where $W_p \ll W_f$. Notice that the output response $y(t)$ is a reasonably good approximation of the input pulse $x(t)$, shown in dashed lines. This represents an example of *good fidelity*. In example 2, where $W_p \approx W_f$, we can still recognize that a pulse had been transmitted from the output $y(t)$. Finally, example 3 illustrates the case in which $W_p \gg W_f$. Here the presence of the pulse is barely perceptible from $y(t)$. Can you think of an application where the large filter bandwidth or good fidelity of example 1 is called for? A *precise ranging application,* perhaps, where the pulse time of arrival translates into distance, necessitates a pulse with a steep rise time. Which example characterizes the binary digital communications application? *It is example 2.* As we pointed out earlier regarding Figure 1.1, one of the principal features of binary digital communications is that each received pulse need only be accurately *perceived* as being in one of its two states; a high-fidelity signal need not be maintained. Example 3 has

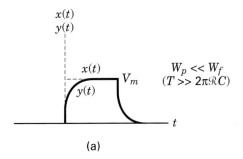

(a)

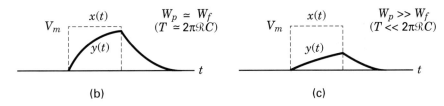

(b)   (c)

Figure 1.17   Three examples of filtering an ideal pulse. (a) Example 1: Good-fidelity output. (b) Example 2: Good-recognition output. (c) Example 3: Poor-recognition output.

been included for completeness; it would not be used as a design criterion for a practical system.

1.7 BANDWIDTH OF DIGITAL DATA

1.7.1 Baseband versus Bandpass

An easy way to translate the spectrum of a low-pass or baseband signal $x(t)$ to a higher frequency is to multiply or *heterodyne* the baseband signal with a carrier wave $\cos 2\pi f_c t$, as shown in Figure 1.18. The resulting waveform, $x_c(t)$, is called a *double-sideband* (DSB) *modulated signal* and is expressed as

$$x_c(t) = x(t) \cos 2\pi f_c t \qquad (1.70)$$

From the frequency shifting theorem (see Section A.3.2), the spectrum of the DSB signal $x_c(t)$ is given by

$$X_c(f) = \tfrac{1}{2}[X(f - f_c) + X(f + f_c)] \qquad (1.71)$$

The magnitude spectrum $|X(f)|$ of the baseband signal $x(t)$ having a bandwidth f_m and the magnitude spectrum $|X_c(f)|$ of the DSB signal $x_c(t)$ having a bandwidth W_{DSB} are shown in Figure 1.18b and c, respectively. In the plot of $|X_c(f)|$, spectral

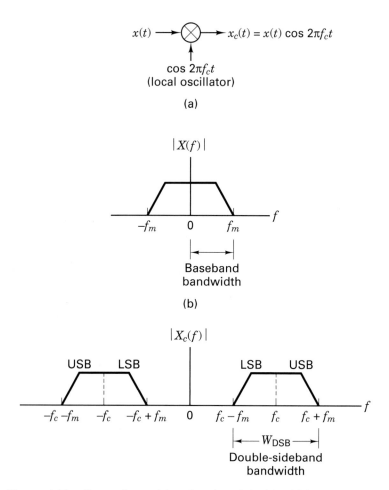

Figure 1.18 Comparison of baseband and double-sideband spectra. (a) Heterodyning. (b) Baseband spectrum. (c) Double-sideband spectrum.

components corresponding to positive baseband frequencies appear in the range f_c to $(f_c + f_m)$. This part of the DSB spectrum is called the *upper sideband* (USB). Spectral components corresponding to negative baseband frequencies appear in the range $(f_c - f_m)$ to f_c. This part of the DSB spectrum is called the *lower sideband* (LSB). Mirror images of the USB and LSB spectra appear in the negative-frequency half of the plot. The *carrier wave* is sometimes referred to as a *local oscillator* (LO) *signal*, a *mixing signal*, or a *heterodyne signal*. Generally, the carrier wave frequency is much higher than the bandwidth of the baseband signal; that is,

$$f_c \gg f_m$$

From Figure 1.18, we can readily compare the bandwidth f_m required to transmit the baseband signal with the bandwidth W_{DSB} required to transmit the DSB signal; we see that

$$W_{\text{DSB}} = 2f_m \qquad\qquad (1.72)$$

That is, we need twice as much transmission bandwidth to transmit a DSB version of the signal than we do to transmit its baseband counterpart.

1.7.2 The Bandwidth Dilemma

Many important theorems of communication and information theory are based on the assumption of *strictly bandlimited* channels, which means that no signal power whatever is allowed outside the defined band. We are faced with the dilemma that strictly bandlimited signals, as depicted by the spectrum $|X_1(f)|$ in Figure 1.19b, are not realizable, because they imply signals with infinite duration, as seen by $x_1(t)$ in Figure 1.19a (the inverse Fourier transform of $X_1(f)$). Duration-limited signals, as seen by $x_2(t)$ in Figure 1.19c, can clearly be realized. However, such signals are just as unreasonable, since their Fourier transforms contain energy at arbitrarily high frequencies as depicted by the spectrum $|X_2(f)|$ in Figure 1.19d. In summary, for all bandlimited spectra, the waveforms are not realizable, and for all realizable waveforms, the absolute bandwidth is infinite. The mathematical description of a real signal does not permit the signal to be strictly duration limited and strictly bandlimited. Hence, the mathematical models are abstractions; it is no wonder that there is no single universal definition of bandwidth.

All bandwidth criteria have in common the attempt to specify a measure of the width, W, of a nonnegative real-valued spectral density defined for all frequen-

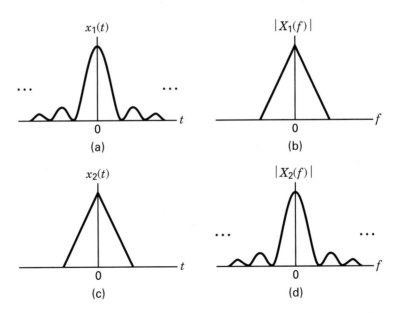

Figure 1.19 (a) Strictly bandlimited signal in the time domain. (b) In the frequency domain. (c) Strictly time limited signal in the time domain. (d) In the frequency domain.

cies $|f| < \infty$. Figure 1.20 illustrates some of the most common definitions of bandwidth; in general, the various criteria are not interchangeable. The single-sided power spectral density for a single heterodyned pulse $x_c(t)$ takes the analytical form

$$G_x(f) = T \left[\frac{\sin \pi(f - f_c)T}{\pi(f - f_c)T} \right]^2 \tag{1.73}$$

where f_c is the carrier wave frequency and T is the pulse duration. This power spectral density, whose general appearance is sketched in Figure 1.20, also characterizes a *random pulse sequence*, assuming that the averaging time is long relative to the pulse duration. The plot consists of a main lobe and smaller symmetrical sidelobes. The general shape of the plot is valid for most digital modulation formats; some formats, however, do not have well-defined lobes. The bandwidth criteria depicted in Figure 1.20 are as follows:

(a) *Half-power bandwidth.* This is the interval between frequencies at which $G_x(f)$ has dropped to half-power, or 3 dB below the peak value.

(b) *Equivalent rectangular or noise equivalent bandwidth, W_N.* This bandwidth is defined by the relationship $W_N = P_x/G_x(f_c)$, where P_x is the total signal power over all frequencies and $G_x(f_c)$ is the maximum value (assumed at the band center) of $G_x(f)$. W_N is the bandwidth of a fictitious (ideal rectangular) filter,

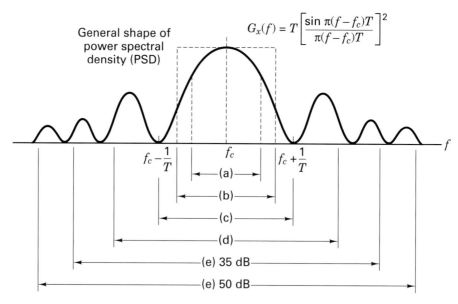

Figure 1.20 Bandwidth of digital data. (a) Half-power. (b) Noise equivalent. (c) Null to null. (d) 99% of power. (e) Bounded PSD (defines attentuation outside bandwidth) at 35 and 50 dB.

Signals and Spectra   Chap. 1

with the same band-center gain as an actual system, that would pass as much white-noise power as the actual system. The concept of W_N facilitates describing or comparing practical linear systems by using idealized equivalents.

(c) *Null-to-null bandwidth.* The most popular measure of bandwidth for digital communications is the width of the main spectral lobe, where most of the signal power is contained. This criterion lacks complete generality since some modulation formats lack well-defined lobes.

(d) *Fractional power containment bandwidth.* This bandwidth criterion has been adopted by the Federal Communications Commission (FCC Rules and Regulations Section 2.202) and states that the occupied bandwidth is the band that leaves exactly 0.5% of the signal power above the upper band limit and exactly 0.5% of the signal power below the lower band limit. Thus 99% of the signal power is inside the occupied band.

(e) *Bounded power spectral density.* A popular method of specifying bandwidth is to state that everywhere outside the specified band, $G_x(f)$ must have fallen at least to a certain stated level below that found at the band center. Typical attenuation levels might be 35 or 50 dB.

(f) *Absolute bandwidth.* This is the interval between frequencies, outside of which the spectrum is zero. This is a useful abstraction. However, for all realizable waveforms, the absolute bandwidth is infinite.

Example 1.4   Strictly Bandlimited Signals

The concept of a signal that is strictly limited to a band of frequencies is not realizable. Prove this by showing that a *strictly bandlimited* signal must also be a signal of *infinite time duration*.

Solution

Let $x(t)$ be a signal with Fourier transform $X(f)$ that is strictly limited to the band of frequencies centered at $\pm f_c$ and of width $2W$. We may express $X(f)$ in terms of an ideal filter transfer function $H(f)$, illustrated in Figure 1.21a, as

$$X(f) = X'(f)H(f) \tag{1.74}$$

where $X'(f)$ is the Fourier transform of a signal $x'(t)$, not necessarily bandlimited, and

$$H(f) = \text{rect}\left(\frac{f - f_c}{2W}\right) + \text{rect}\left(\frac{f + f_c}{2W}\right) \tag{1.75}$$

in which

$$\text{rect}\left(\frac{f}{2W}\right) = \begin{cases} 1 & \text{for } -W < f < W \\ 0 & \text{for } |f| > W \end{cases}$$

We can express $X(f)$ in terms of $X'(f)$ as

$$X(f) = \begin{cases} X'(f) & \text{for } (f_c - W) \le |f_c| \le (f_c + W) \\ 0 & \text{otherwise} \end{cases}$$

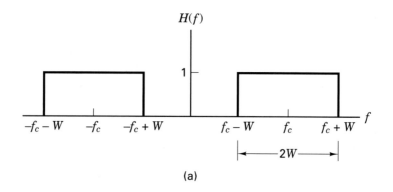

(a)

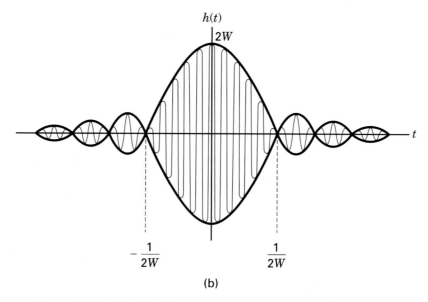

(b)

Figure 1.21 Transfer function and impulse response for a strictly bandlimited signal. (a) Ideal bandpass filter. (b) Ideal bandpass impulse response.

Multiplication in the frequency domain, as seen in Equation (1.74), transforms to convolution in the time domain as

$$x(t) = x'(t) * h(t) \tag{1.76}$$

where $h(t)$, the inverse Fourier transform of $H(f)$, can be written as (see Tables A.1 and A.2)

$$h(t) = 2W \, (\text{sinc } 2Wt) \cos 2\pi f_c t$$

and is illustrated in Figure 1.21b. We note that $h(t)$ is of *infinite duration*. It follows, therefore, that $x(t)$ obtained in Equation (1.76) by convolving $x'(t)$ with $h(t)$ is also of infinite duration and therefore is *not realizable*.

1.8 CONCLUSION

In this chapter, the goals of the book have been outlined and the basic nomenclature has been defined. The fundamental concepts of time-varying signals, such as classification, spectral density, and autocorrelation, have been reviewed. Also, random signals have been considered, and white Gaussian noise, the primary noise model in most communication systems, has been characterized, statistically and spectrally. Finally, we have treated the important area of signal transmission through linear systems and have examined some of the realizable approximations to the ideal case. We have also established that the concept of an absolute bandwidth is an abstraction, and that in the real world we are faced with the need to choose a definition of bandwidth that is useful for our particular application. In the remainder of the book, each of the signal processing steps introduced in this chapter will be explored in the context of the typical system block diagram appearing at the beginning of each chapter.

REFERENCES

1. Haykin, S., *Communication Systems,* John Wiley & Sons, Inc., New York, 1983.
2. Shanmugam, K. S., *Digital and Analog Communication Systems,* John Wiley & Sons, Inc., New York, 1979.
3. Papoulis, A., *Probability, Random Variables, and Stochastic Processes,* McGraw-Hill Book Company, New York, 1965.
4. Johnson, J. B., "Thermal Agitation of Electricity in Conductors," *Phys. Rev.,* vol. 32, July 1928, pp. 97–109.
5. Nyguist, H., "Thermal Agitation of Electric Charge in Conductors," *Phys. Rev.,* vol. 32, July 1928, pp. 110–113.
6. Van Trees, H. L., *Detection, Estimation, and Modulation Theory,* Part 1, John Wiley & Sons, New York, 1968.
7. Schwartz, M., *Information Transmission, Modulation, and Noise,* McGraw-Hill Book Company, New York, 1970.
8. Millman, J., and Taub, H., *Pulse, Digital, and Switching Waveforms,* McGraw-Hill Book Company, New York, 1965.

PROBLEMS

1.1. Classify the following signals as energy signals or power signals. Find the normalized energy or normalized power of each.

(a) $x(t) = A \cos 2\pi f_0 t$ for $-\infty < t < \infty$

(b) $x(t) = \begin{cases} A \cos 2\pi f_0 t & \text{for } -T_0/2 \leq t \leq T_0/2, \text{ where } T_0 = 1/f_0 \\ 0 & \text{elsewhere} \end{cases}$

(c) $x(t) = \begin{cases} A \exp(-at) & \text{for } t > 0, a > 0 \\ 0 & \text{elsewhere} \end{cases}$

(d) $x(t) = \cos t + 5 \cos 2t$ for $-\infty < t < \infty$

1.2. Determine the energy spectral density of a square pulse $x(t) = $ rect (t/T), where rect (t/T) equals 1, for $-T/2 \leq t \leq T/2$, and equals 0, elsewhere. Calculate the normalized energy E_x in the pulse.

1.3. Find an expression for the average normalized power in a periodic signal in terms of its complex Fourier series coefficients.

1.4. Using time averaging, find the average normalized power in the waveform $x(t) = 10 \cos 10t + 20 \cos 20t$.

1.5. Repeat Problem 1.4 using the summation of spectral coefficients.

1.6. Determine which, if any, of the following functions have the properties of autocorrelation functions. Justify your determination. [*Note:* $\mathcal{F}\{R(\tau)\}$ must be a nonnegative function. Why?]

(a) $x(\tau) = \begin{cases} 1 & \text{for } -1 \leq \tau \leq 1 \\ 0 & \text{otherwise} \end{cases}$

(b) $x(\tau) = \delta(\tau) + \sin 2\pi f_0 \tau$

(c) $x(\tau) = \exp(|\tau|)$

(d) $x(\tau) = 1 - |\tau| \quad$ for $-1 \leq \tau \leq 1$, 0 elsewhere

1.7. Determine which, if any, of the following functions have the properties of power spectral density functions. Justify your determination.

(a) $X(f) = \delta(f) + \cos^2 2\pi f$

(b) $X(f) = 10 + \delta(f - 10)$

(c) $X(f) = \exp(-2\pi |f - 10|)$

(d) $X(f) = \exp[-2\pi(f^2 - 10)]$

1.8. Find the autocorrelation function of $x(t) = A \cos (2\pi f_0 t + \phi)$ in terms of its period, $T_0 = 1/f_0$. Find the average normalized power of $x(t)$, using $P_x = R(0)$.

1.9. (a) Use the results of Problem 1.8 to find the autocorrelation function $R(\tau)$ of waveform $x(t) = 10 \cos 10t + 20 \cos 20t$.
   (b) Use the relationship P_x $R(0)$ to find the average normalized power in $x(t)$. Compare the answer with the answers to Problems 1.4 and 1.5.

1.10. For the function $x(t) = 1 + \cos 2\pi f_0 t$, calculate **(a)** the average value of $x(t)$; **(b)** the ac power of $x(t)$; **(c)** the rms value of $x(t)$.

1.11. Consider a random process given by $X(t) = A \cos (2\pi f_0 t + \phi)$, where A and f_0 are constants and ϕ is a random variable that is uniformly distributed over $(0, 2\pi)$. If $X(t)$ is an ergodic process, the time averages of $X(t)$ in the limit as $t \to \infty$ are equal to the corresponding ensemble averages of $X(t)$.
   (a) Use time averaging over an integer number of periods to calculate the approximations to the first and second moments of $X(t)$.
   (b) Use Equations (1.26) and (1.28) to calculate the ensemble-average approximations to the first and second moments of $X(t)$. Compare the results with your answers in part (a).

1.12. The Fourier transform of a signal $x(t)$ is defined by $X(f) = $ sinc f, where the sinc function is as defined in Equation (1.39). Find the autocorrelation function, $R_x(\tau)$, of the signal $x(t)$.

1.13. Use the sampling property of the unit impulse function to evaluate the following integrals.

(a) $\displaystyle\int_{-\infty}^{\infty} \cos 6t\delta(t - 3)\, dt$

(b) $\displaystyle\int_{-\infty}^{\infty} 10\delta(t)(1 + t)^{-1}\, dt$

(c) $\displaystyle\int_{-\infty}^{\infty} \delta(t + 4)(t^2 + 6t + 1)\, dt$

(d) $\displaystyle\int_{-\infty}^{\infty} \exp(-t^2)\delta(t - 2)\, dt$

1.14. Find $X_1(f) * X_2(f)$ for the spectra shown in Figure P1.1.

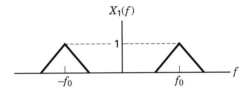

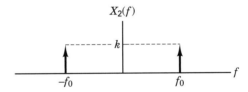

Figure P1.1

1.15. The two-sided power spectral density, $G_x(f) = 10^{-6} f^2$, of a waveform $x(t)$ is shown in Figure P1.2.

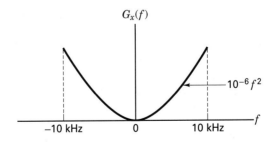

Figure P1.2

(a) Find the normalized average power in $x(t)$ over the frequency band from 0 to 10 kHz.

(b) Find the normalized average power contained in the frequency band from 5 to 6 kHz.

1.16. Decibels are logarithmic measures of *power ratios,* as described in Equation (1.64a). Sometimes, a similar formulation is used to express nonpower measurements in decibels (referenced to some designated unit). As an example, calculate how many decibels of hamburger meat you would buy to feed 2 hamburgers each to a group of 100 people. Assume that you and the butcher have agreed on the unit of "½ pound of meat" (the amount in one hamburger) as a reference unit.

1.17. Consider the Butterworth low-pass amplitude response given in Equation (1.65).
   (a) Find the value of n so that $|H(f)|^2$ is constant to within ± 1 dB over the range $|f| \le 0.9 f_u$.
   (b) Show that as n approaches infinity, the amplitude response approaches that of an ideal low-pass filter.

1.18. Consider the network in Figure 1.9, whose frequency transfer function is $H(f)$. An impulse $\delta(t)$ is applied at the input. Show that the response $y(t)$ at the output is the inverse Fourier transform of $H(f)$.

1.19. An example of a *holding circuit,* commonly used in pulse systems, is shown in Figure P1.3. Determine the impulse response of this circuit.

1.20. Given the spectrum

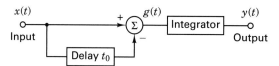

Figure P1.3

$$G_x(f) = 10^{-4} \left\{ \frac{\sin\left[\pi(f - 10^6)10^{-4}\right]}{\pi(f - 10^6)10^{-4}} \right\}^2$$

find the value of the signal bandwidth using the following bandwidth definitions:
   (a) Half-power bandwidth.
   (b) Noise equivalent bandwidth.
   (c) Null-to-null bandwidth.
   (d) 99% of power bandwidth. (Hint: Use numerical methods.)
   (e) Bandwidth beyond which the attenuation is 35 dB.
   (f) Absolute bandwidth.

QUESTIONS

1.1. How does the plot of a signal's autocorrelation function reveal its bandwidth occupancy? (See Section 1.5.4.)

1.2. What two requirements must be fulfilled in order to insure distortionless transmission through a linear system? (See Section 1.6.3.)

1.3. Define the parameter *envelope delay* or *group delay.* (See Section 1.6.3.)

1.4. What mathematical dilemma is the cause for there being several different definitions of bandwidth? (See Section 1.7.2.)

EXERCISES

Using the Companion CD, run the exercises associated with Chapter 1.

CHAPTER 2

Formatting
and
Baseband Modulation

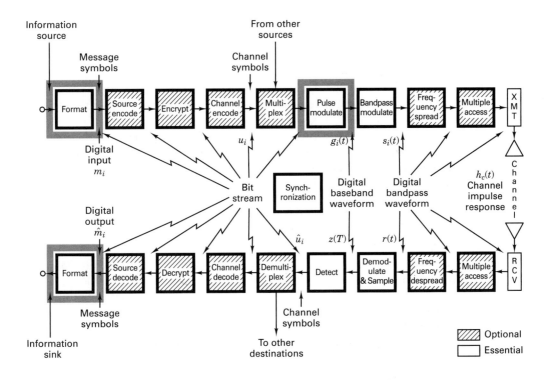

The goal of the first essential signal processing step, *formatting,* is to insure that the message (or source signal) is compatible with digital processing. *Transmit formatting* is a transformation from source information to digital symbols. (It is the reverse transformation in the receive chain.) When data compression in addition to formatting is employed, the process is termed *source coding.* Some authors consider formatting a special case of source coding. We treat formatting (and baseband modulation) in this chapter, and treat source coding as a special case of the *efficient description* of source information in Chapter 13.

In Figure 2.1, the highlighted block labeled "formatting" contains a list of topics that deal with transforming information to digital messages. The digital messages are considered to be in the logical format of binary ones and zeros until they are transformed by the next essential step, called pulse modulation, into *baseband* (pulse) waveforms. Such waveforms can then be transmitted over a cable. In Figure 2.1, the highlighted block labeled "baseband signaling" contains a list of pulse modulating waveforms that are described in this chapter. The term baseband refers to a signal whose spectrum extends from (or near) dc up to some finite value, usually less than a few megahertz. In Chapter 3, the subject of baseband signaling is continued with emphasis on demodulation and detection.

2.1 BASEBAND SYSTEMS

In Figure 1.2 we presented a block diagram of a typical digital communication system. A version of this functional diagram, focusing primarily on the formatting and transmission of *baseband* signals, is shown in Figure 2.2. Data already in a digital

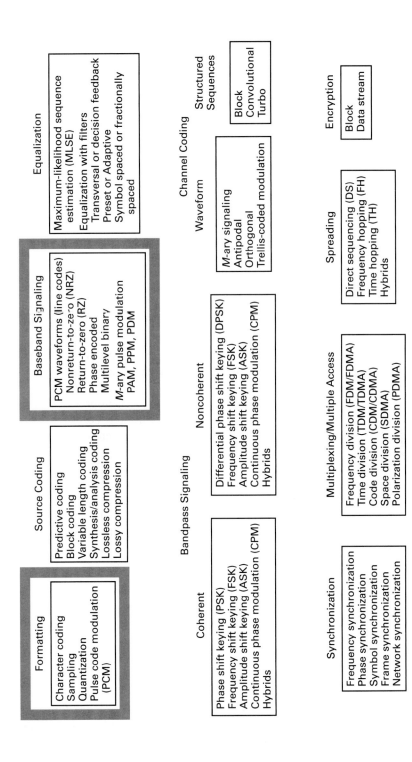

Figure 2.1 Basic digital communication transformations

57

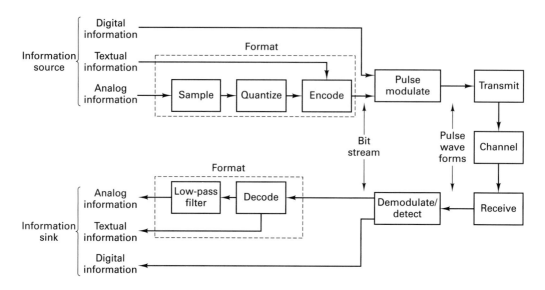

Figure 2.2 Formatting and transmission of baseband signals.

format would bypass the formatting function. Textual information is transformed into binary digits by use of a coder. Analog information is formatted using three separate processes: sampling, quantization, and coding. In all cases, the formatting step results in a sequence of binary digits.

These digits are to be transmitted through a *baseband channel,* such as a pair of wires or a coaxial cable. However, no channel can be used for the transmission of binary digits without first transforming the digits to *waveforms* that are compatible with the channel. For baseband channels, compatible waveforms are pulses.

In Figure 2.2, the conversion from a bit stream to a sequence of pulse waveforms takes place in the block labeled pulse modulate. The output of the modulator is typically a sequence of pulses with characteristics that correspond to the digits being sent. After transmission through the channel, the pulse waveforms are recovered (demodulated) and detected to produce an estimate of the transmitted digits; the final step, (reverse) formatting, recovers an estimate of the source information.

2.2 FORMATTING TEXTUAL DATA (CHARACTER CODING)

The original form of most communicated data (except for computer-to-computer transmissions) is either textual or analog. If the data consist of alphanumeric text, they will be character encoded with one of several standard formats; examples include the American Standard Code for Information Interchange (ASCII), the Extended Binary Coded Decimal Interchange Code (EBCDIC), Baudot, and Hollerith. The textual material is thereby transformed into a digital format. The ASCII format is shown in Figure 2.3; the EBCDIC format is shown in Figure 2.4.

Figure 2.3 Seven-bit American standard code for information interchange (ASCII).

Bits				5 →	0	1	0	1	0	1	0	1
				6 →	0	0	1	1	0	0	1	1
1	2	3	4	7 →	0	0	0	0	1	1	1	1
0	0	0	0		NUL	DLE	SP	0	@	P	`	p
1	0	0	0		SOH	DC1	!	1	A	Q	a	q
0	1	0	0		STX	DC2	"	2	B	R	b	r
1	1	0	0		ETX	DC3	#	3	C	S	c	s
0	0	1	0		EOT	DC4	$	4	D	T	d	t
1	0	1	0		ENQ	NAK	%	5	E	U	e	u
0	1	1	0		ACK	SYN	&	6	F	V	f	v
1	1	1	0		BEL	ETB	'	7	G	W	g	w
0	0	0	1		BS	CAN	(	8	H	X	h	x
1	0	0	1		HT	EM	)	9	I	Y	i	y
0	1	0	1		LF	SUB	*	:	J	Z	j	z
1	1	0	1		VT	ESC	+	;	K	[	k	{
0	0	1	1		FF	FS	,	<	L	\	l	|
1	0	1	1		CR	GS	-	=	M	]	m	}
0	1	1	1		SO	RS	.	>	N	^	n	~
1	1	1	1		SI	US	/	?	O	_	o	DEL

NUL Null, or all zeros
SOH Start of heading
STX Start of text
ETX End of text
EOT End of transmission
ENQ Enquiry
ACK Acknowledge
BEL Bell, or alarm
BS   Backspace
HT   Horizontal tabulation
LF   Line feed
VT   Vertical tabulation
FF   Form feed
CR   Carriage return
SO   Shift out
SI   Shift in
DLE Data link escape

DC1 Device control 1
DC2 Device control 2
DC3 Device control 3
DC4 Device control 4
NAK Negative acknowledge
SYN Synchronous idle
ETB End of transmission
CAN Cancel
EM   End of medium
SUB Substitute
ESC Escape
FS   File separator
GS   Group separator
RS   Record separator
US   Unit separator
SP   Space
DEL Delete

Figure 2.4 EBCDIC character code set.

Legend:

PF	Punch off
HT	Horizontal tab
LC	Lower case
DEL	Delete
SP	Space
UC	Upper case
RES	Restore
NL	New line
BS	Backspace
IL	Idle
PN	Punch on
EOT	End of transmission
BYP	Bypass
LF	Line feed
EOB	End of block
PRE	Prefix (ESC)
RS	Reader stop
SM	Start message
DS	Digit select
SOS	Start of significance
IFS	Interchange file separator
IGS	Interchange group separator
IRS	Interchange record separator
IUS	Interchange unit separator
Others	Same as ASCII

Character code table (rows = bits 1 2 3 4; columns = bits 5 6 7 8):

Bits 1234 \ 5678	0000	0001	0010	0011	0100	0101	0110	0111	1000	1001	1010	1011	1100	1101	1110	1111
0000	NUL	DLE	DS		SP	&	-									0
0001	SOH	DC1	SOS				/		a	j			A	J		1
0010	STX	DC2	FS	SYN					b	k	s		B	K	S	2
0011	ETX	DC3							c	l	t		C	L	T	3
0100	PF	RES	BYP	PN					d	m	u		D	M	U	4
0101	HT	NL	LF	RS					e	n	v		E	N	V	5
0110	LC	BS	EOB	UC					f	o	w		F	O	W	6
0111	DEL	IL	PRE	EOT					g	p	x		G	P	X	7
1000		CAN							h	q	y		H	Q	Y	8
1001		EM							i	r	z		I	R	Z	9
1010	SMM	CC	SM		¢	!	|	:								
1011	VT				.	$	,	#								
1100	FF	IFS		DC4	<	*	%	@								
1101	CR	IGS	ENQ	NAK	(	)	_	'								
1110	SO	IRS	ACK		+	;	>	=								
1111	SI	IUS	BEL	SUB	|	¬	?	"								

The bit numbers signify the order of serial transmission, where bit number 1 is the first signaling element. Character coding, then, is the step that transforms text into binary digits (bits). Sometimes existing character codes are modified to meet specialized needs. For example, the 7-bit ASCII code (Figure 2.3) can be modified to include an added bit for error detection purposes. (See Chapter 6.) On the other hand, sometimes the code is truncated to a 6-bit ASCII version, which provides capability for only 64 characters instead of the 128 characters allowed by 7-bit ASCII.

2.3 MESSAGES, CHARACTERS, AND SYMBOLS

Textual messages comprise a sequence of alphanumeric characters. When digitally transmitted, the characters are first encoded into a sequence of bits, called a *bit stream* or *baseband signal*. Groups of k bits can then be combined to form new digits, or *symbols*, from a finite symbol set or alphabet of $M = 2^k$ such symbols. A system using a symbol set size of M is referred to as an *M-ary system*. The value of k or M represents an important initial choice in the design of any digital communication system. For $k = 1$, the system is termed *binary*, the size of the symbol set is $M = 2$, and the modulator uses one of the two different waveforms to represent the binary "one" and the other to represent the binary "zero." For this special case, the symbol and the bit are the same. For $k = 2$, the system is termed *quaternary* or 4-*ary* ($M = 4$). At each symbol time, the modulator uses one of the four different waveforms that represents the symbol. The partitioning of the sequence of message bits is determined by the specification of the symbol set size, M. The following example should help clarify the relationship between the following terms: "message," "character," "symbol," "bit," and "digital waveform."

2.3.1 Example of Messages, Characters, and Symbols

Figure 2.5 shows examples of bit stream partitioning, based on the system specification for the values of k and M. The textual message in the figure is the word "THINK." Using 6-bit ASCII character coding (bit numbers 1 to 6 from Figure 2.3) yields a bit stream comprising 30 bits. In Figure 2.5a, the symbol set size, M, has been chosen to be 8 (each symbol represents an 8-ary digit). The bits are therefore partitioned into groups of three ($k = \log_2 8$); the resulting 10 numbers represent the 10 octal symbols to be transmitted. The transmitter must have a repertoire of eight waveforms $s_i(t)$, where $i = 1, \ldots, 8$, to represent the possible symbols, any one of which may be transmitted during a symbol time. The final row of Figure 2.5a lists the 10 waveforms that an 8-ary modulating system transmits to represent the textual message "THINK."

In Figure 2.5b, the symbol set size, M, has been chosen to be 32 (each symbol represents a 32-ary digit). The bits are therefore taken five at a time, and the resulting group of six numbers represent the six 32-ary symbols to be transmitted. Notice that there is no need for the symbol boundaries and the character boundaries to coincide. The first symbol represents $\frac{5}{6}$ of the first character, "T." The second symbol

Message (text):   "THINK"

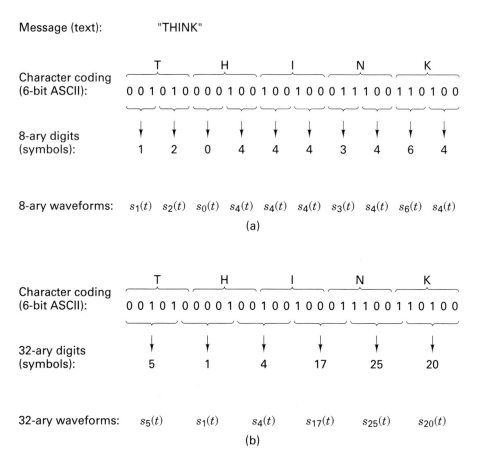

Figure 2.5   Messages, characters, and symbols. (a) 8-ary example. (b) 32-ary example.

represents the remaining $\frac{1}{6}$ of the character "T" and $\frac{4}{6}$ of the next character, "H," and so on. It is not necessary that the characters be partitioned more aesthetically. The system sees the characters as a string of digits to be transmitted; only the end user (or the user's teleprinter machine) ascribes textual meaning to the final delivered sequence of bits. In this 32-ary case, a transmitter needs a repertoire of 32 waveforms $s_i(t)$, where $i = 1, \ldots , 32$, one for each possible symbol that may be transmitted. The final row of the figure lists the six waveforms that a 32-ary modulating system transmits to represent the textual message "THINK."

2.4 FORMATTING ANALOG INFORMATION

If the information is analog, it cannot be character encoded as in the case of textual data; the information must first be transformed into a digital format. The process of transforming an analog waveform into a form that is compatible with a digital communication system starts with sampling the waveform to produce a discrete pulse-amplitude-modulated waveform, as described below.

2.4.1 The Sampling Theorem

The link between an analog waveform and its sampled version is provided by what is known as the *sampling process.* This process can be implemented in several ways, the most popular being the *sample-and-hold* operation. In this operation, a switch and storage mechanism (such as a transistor and a capacitor, or a shutter and a filmstrip) form a sequence of samples of the continuous input waveform. The output of the sampling process is called *pulse amplitude modulation* (PAM) because the successive output intervals can be described as a sequence of pulses with amplitudes derived from the input waveform samples. The analog waveform can be approximately retrieved from a PAM waveform by simple low-pass filtering. An important question: how closely can a filtered PAM waveform approximate the original input waveform? This question can be answered by reviewing the *sampling theorem,* which states the following [1]: A bandlimited signal having no spectral components above f_m hertz can be determined uniquely by values sampled at uniform intervals of

$$T_s \leq \frac{1}{2f_m} \text{ sec} \qquad (2.1)$$

This particular statement is also known as the *uniform sampling theorem.* Stated another way, the upper limit on T_s can be expressed in terms of the sampling rate, denoted $f_s = 1/T_s$. The restriction, stated in terms of the sampling rate, is known as the *Nyquist criterion.* The statement is

$$f_s \geq 2f_m \qquad (2.2)$$

The sampling rate $f_s = 2f_m$ is also called the *Nyquist rate.* The Nyquist criterion is a theoretically sufficient condition to allow an analog signal to be *reconstructed completely* from a set of uniformly spaced discrete-time samples. In the sections that follow, the validity of the sampling theorem is demonstrated using different sampling approaches.

2.4.1.1 Impulse Sampling

Here we demonstrate the validity of the sampling theorem using the frequency convolution property of the Fourier transform. Let us first examine the case of *ideal sampling* with a sequence of unit impulse functions. Assume an analog waveform, $x(t)$, as shown in Figure 2.6a, with a Fourier transform, $X(f)$, which is zero outside the interval $(-f_m < f < f_m)$, as shown in Figure 2.6b. The sampling of $x(t)$ can be viewed as the product of $x(t)$ with a periodic train of unit impulse functions $x_\delta(t)$, shown in Figure 2.6c and defined as

$$x_\delta(t) = \sum_{n=-\infty}^{\infty} \delta(t - nT_s) \qquad (2.3)$$

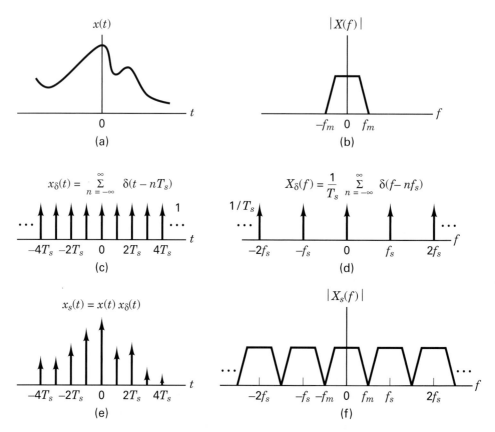

Figure 2.6 Sampling theorem using the frequency convolution property of the Fourier transform.

where T_s is the sampling period and $\delta(t)$ is the unit impulse or Dirac delta function defined in Section 1.2.5. Let us choose $T_s = 1/2f_m$, so that the Nyquist criterion is just satisfied.

The *sifting property* of the impulse function (see Section A.4.1) states that

$$x(t)\delta(t - t_0) = x(t_0)\delta(t - t_0) \qquad (2.4)$$

Using this property, we can see that $x_s(t)$, the sampled version of $x(t)$ shown in Figure 2.6e, is given by

$$\begin{aligned} x_s(t) = x(t)x_\delta(t) &= \sum_{n=-\infty}^{\infty} x(t)\delta(t - nT_s) \\ &= \sum_{n=-\infty}^{\infty} x(nT_s)\delta(t - nT_s) \end{aligned} \qquad (2.5)$$

Using the *frequency convolution property* of the Fourier transform (see Section A.5.3), we can transform the time-domain product $x(t)x_\delta(t)$ of Equation (2.5) to the frequency-domain convolution $X(f) * X_\delta(f)$, where

$$X_\delta(f) = \frac{1}{T_s} \sum_{n=-\infty}^{\infty} \delta(f - nf_s) \qquad (2.6)$$

is the Fourier transform of the impulse train $x_\delta(t)$ and where $f_s = 1/T_s$ is the sampling frequency. Notice that the Fourier transform of an impulse train is another impulse train; the values of the periods of the two trains are reciprocally related to one another. Figures 2.6c and d illustrate the impulse train $x_\delta(t)$ and its Fourier transform $X_\delta(f)$, respectively.

Convolution with an impulse function simply shifts the original function as follows:

$$X(f) * \delta(f - nf_s) = X(f - nf_s) \qquad (2.7)$$

We can now solve for the transform $X_s(f)$ of the sampled waveform:

$$
\begin{aligned}
X_s(f) = X(f) * X_\delta(f) &= X(f) * \left[\frac{1}{T_s} \sum_{n=-\infty}^{\infty} \delta(f - nf_s) \right] \\
&= \frac{1}{T_s} \sum_{n=-\infty}^{\infty} X(f - nf_s)
\end{aligned}
\qquad (2.8)
$$

We therefore conclude that within the original bandwidth, the spectrum $X_s(f)$ of the sampled signal $x_s(t)$ is, to within a constant factor $(1/T_s)$, exactly the same as that of $x(t)$. In addition, the spectrum repeats itself periodically in frequency every f_s hertz. The sifting property of an impulse function makes the convolving of an impulse train with another function easy to visualize. The impulses act as sampling functions. Hence, convolution can be performed graphically by sweeping the impulse train $X_\delta(f)$ in Figure 2.6d past the transform $|X(f)|$ in Figure 2.6b. This sampling of $|X(f)|$ at each step in the sweep replicates $|X(f)|$ at each of the frequency positions of the impulse train, resulting in $|X_s(f)|$, shown in Figure 2.6f.

When the sampling rate is chosen, as it has been here, such that $f_s = 2f_m$, each spectral replicate is separated from each of its neighbors by a frequency band exactly equal to f_s hertz, and the analog waveform can theoretically be completely recovered from the samples, by the use of filtering. However, a filter with infinitely steep sides would be required. It should be clear that if $f_s > 2f_m$, the replications will move farther apart in frequency, as shown in Figure 2.7a, making it easier to perform the filtering operation. A typical low-pass filter characteristic that might be used to separate the baseband spectrum from those at higher frequencies is shown in the figure. When the sampling rate is reduced, such that $f_s < 2f_m$, the replications will overlap, as shown in Figure 2.7b, and some information will be lost. The phenomenon, the result of undersampling (sampling at too low a rate), is called *aliasing*. The Nyquist rate, $f_s = 2f_m$, is the sampling rate below which aliasing occurs; to avoid aliasing, the Nyquist criterion, $f_s \geq 2f_m$, must be satisfied.

As a matter of practical consideration, neither waveforms of engineering interest nor realizable bandlimiting filters are strictly bandlimited. Perfectly bandlimited signals do not occur in nature (see Section 1.7.2); thus, realizable signals, even though we may think of them as bandlimited, always contain some aliasing. These signals and filters can, however, be considered to be "essentially" bandlimited. By

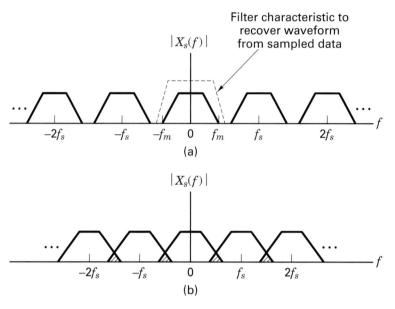

Figure 2.7 Spectra for various sampling rates. (a) Sampled spectrum ($f_s > 2f_m$). (b) Sampled spectrum ($f_s < 2f_m$).

this we mean that a bandwidth can be determined beyond which the spectral components are attenuated to a level that is considered negligible.

2.4.1.2 Natural Sampling

Here we demonstrate the validity of the sampling theorem using the frequency shifting property of the Fourier transform. Although instantaneous sampling is a convenient model, a more practical way of accomplishing the sampling of a bandlimited analog signal $x(t)$ is to multiply $x(t)$, shown in Figure 2.8a, by the pulse train or switching waveform $x_p(t)$, shown in Figure 2.8c. Each pulse in $x_p(t)$ has width T and amplitude $1/T$. Multiplication by $x_p(t)$ can be viewed as the opening and closing of a switch. As before, the sampling frequency is designated f_s, and its reciprocal, the time period between samples, is designated T_s. The resulting sampled-data sequence, $x_s(t)$, is illustrated in Figure 2.8e and is expressed as

$$x_s(t) = x(t)x_p(t) \tag{2.9}$$

The sampling here is termed *natural sampling,* since the top of each pulse in the $x_s(t)$ sequence retains the shape of its corresponding analog segment during the pulse interval. Using Equation (A.13), we can express the periodic pulse train as a Fourier series in the form

$$x_p(t) = \sum_{n=-\infty}^{\infty} c_n e^{j\,2\pi n f_s t} \tag{2.10}$$

Formatting and Baseband Modulation   Chap. 2

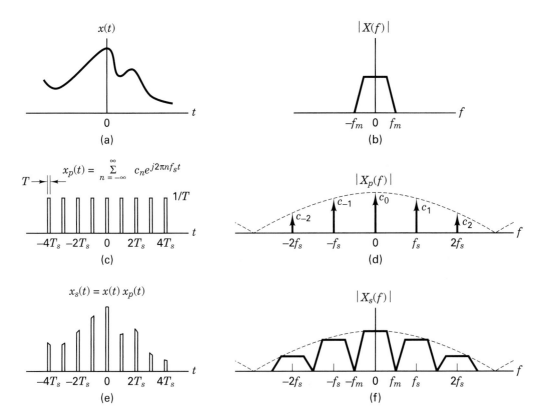

Figure 2.8 Sampling theorem using the frequency shifting property of the Fourier transform.

where the sampling rate, $f_s = 1/T_s$, is chosen equal to $2f_m$, so that the Nyquist criterion is just satisfied. From Equation (A.24), $c_n = (1/T_s)$ sinc (nT/T_s), where T is the pulse width, $1/T$ is the pulse amplitude, and

$$\text{sinc } y = \frac{\sin \pi y}{\pi y}$$

The envelope of the magnitude spectrum of the pulse train, seen as a dashed line in Figure 2.8d, has the characteristic sinc shape. Combining Equations (2.9) and (2.10) yields

$$x_s(t) = x(t) \sum_{n=-\infty}^{\infty} c_n e^{j\,2\pi n f_s t} \tag{2.11}$$

The transform $X_s(f)$ of the sampled waveform is found as follows:

$$X_s(f) = \mathcal{F}\left\{ x(t) \sum_{n=-\infty}^{\infty} c_n e^{j\,2\pi n f_s t} \right\} \tag{2.12}$$

For linear systems, we can interchange the operations of summation and Fourier transformation. Therefore, we can write

$$X_s(f) = \sum_{n=-\infty}^{\infty} c_n \mathcal{F}\{x(t)e^{j\,2\pi n f_s t}\} \tag{2.13}$$

Using the *frequency translation* property of the Fourier transform (see Section A.3.2), we solve for $X_s(f)$ as follows:

$$X_s(f) = \sum_{n=-\infty}^{\infty} c_n X(f - n f_s) \tag{2.14}$$

Similar to the unit impulse sampling case, Equation (2.14) and Figure 2.8f illustrate that $X_s(f)$ is a replication of $X(f)$, periodically repeated in frequency every f_s hertz. In this natural-sampled case, however, we see that $X_s(f)$ is weighted by the Fourier series coefficients of the pulse train, compared with a constant value in the impulse-sampled case. It is satisfying to note that *in the limit,* as the pulse width, T, approaches zero, c_n approaches $1/T_s$ for all n (see the example that follows), and Equation (2.14) converges to Equation (2.8).

Example 2.1   Comparison of Impulse Sampling and Natural Sampling

Consider a given waveform $x(t)$ with Fourier transform $X(f)$. Let $X_{s1}(f)$ be the spectrum of $x_{s1}(t)$, which is the result of sampling $x(t)$ with a unit impulse train $x_\delta(t)$. Let $X_{s2}(f)$ be the spectrum of $x_{s2}(t)$, the result of sampling $x(t)$ with a pulse train $x_p(t)$ with pulse width T, amplitude $1/T$, and period T_s. Show that in the limit, as T approaches zero, $X_{s1}(f) = X_{s2}(f)$.

Solution

From Equation (2.8),

$$X_{s1}(f) = \frac{1}{T_s} \sum_{n=-\infty}^{\infty} X(f - n f_s)$$

and from Equation (2.14),

$$X_{s2}(f) = \sum_{n=-\infty}^{\infty} c_n X(f - n f_s)$$

As the pulse with $T \to 0$, and the pulse amplitude approaches infinity (the area of the pulse remains unity), $x_p(t) \to x_\delta(t)$. Using Equation (A.14), we can solve for c_n in the limit as follows:

$$
\begin{aligned}
c_n &= \lim_{T \to 0} \frac{1}{T_s} \int_{-T_s/2}^{T_s/2} x_p(t) e^{-j\,2\pi n f_s t}\, dt \\
&= \frac{1}{T_s} \int_{-T_s/2}^{T_s/2} x_\delta(t) e^{-j\,2\pi n f_s t}\, dt
\end{aligned}
$$

Since, within the range of integration, $-T_s/2$ to $T_s/2$, the only contribution of $x_\delta(t)$ is that due to the impulse at the origin, we can write

$$c_n = \frac{1}{T_s} \int_{-T_s/2}^{T_s/2} \delta(t)e^{-j\,2\pi nf_s t}\,dt = \frac{1}{T_s}$$

Therefore, in the limit, $X_{s1}(f) = X_{s2}(f)$ for all n.

2.4.1.3 Sample-and-Hold Operation

The simplest and thus most popular sampling method, *sample and hold*, can be described by the convolution of the sampled pulse train, $[x(t)x_\delta(t)]$, shown in Figure 2.6e, with a unity amplitude rectangular pulse $p(t)$ of pulse width T_s. This time, convolution results in the *flattop* sampled sequence

$$
\begin{aligned}
x_s(t) &= p(t) * [x(t)x_\delta(t)] \\
&= p(t) * \left[x(t) \sum_{n=-\infty}^{\infty} \delta(t - nT_s) \right]
\end{aligned}
$$
(2.15)

The Fourier transform, $X_s(f)$, of the time convolution in Equation (2.15) is the frequency-domain product of the transform $P(f)$ of the rectangular pulse and the periodic spectrum, shown in Figure 2.6f, of the impulse-sampled data:

$$
\begin{aligned}
X_s(f) &= P(f)\mathscr{F}\left\{ x(t) \sum_{n=-\infty}^{\infty} \delta(t - nT_s) \right\} \\
&= P(f)\left\{ X(f) * \left[\frac{1}{T_s} \sum_{n=-\infty}^{\infty} \delta(f - nf_s) \right] \right\} \\
&= P(f)\frac{1}{T_s} \sum_{n=-\infty}^{\infty} X(f - nf_s)
\end{aligned}
$$
(2.16)

Here, $P(f)$ is of the form $T_s \operatorname{sinc} fT_s$. The effect of this product operation results in a spectrum similar in appearance to the natural-sampled example presented in Figure 2.8f. The most obvious effect of the hold operation is the significant attenuation of the higher-frequency spectral replicates (compare Figure 2.8f to Figure 2.6f), which is a desired effect. Additional analog postfiltering is usually required to finish the filtering process by further attenuating the residual spectral components located at the multiples of the sample rate. A secondary effect of the hold operation is the nonuniform spectral gain $P(f)$ applied to the desired baseband spectrum shown in Equation (2.16). The postfiltering operation can compensate for this attenuation by incorporating the inverse of $P(f)$ over the signal passband.

2.4.2 Aliasing

Figure 2.9 is a detailed view of the positive half of the baseband spectrum and one of the replicates from Figure 2.7b. It illustrates aliasing in the frequency domain. The overlapped region, shown in Figure 2.9b, contains that part of the spectrum which is aliased due to *undersampling*. The aliased spectral components represent ambiguous data that appear in the frequency band between $(f_s - f_m)$ and f_m. Figure 2.10 illustrates that a higher sampling rate f'_s, can eliminate the aliasing by separat-

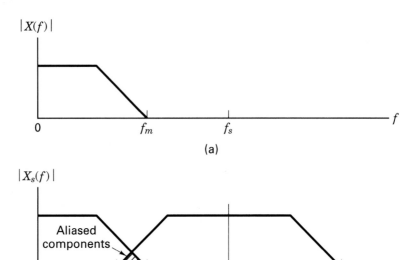

Figure 2.9 Aliasing in the frequency domain. (a) Continuous signal spectrum. (b) Sampled signal spectrum.

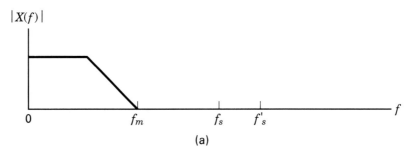

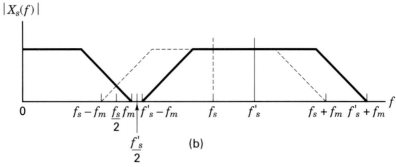

Figure 2.10 Higher sampling rate eliminates aliasing. (a) Continuous signal spectrum. (b) Sampled signal spectrum.

ing the spectral replicates; the resulting spectrum in Figure 2.10b corresponds to the case in Figure 2.7a. Figures 2.11 and 2.12 illustrate two ways of eliminating aliasing using *antialiasing filters*. In Figure 2.11 the analog signal is *prefiltered* so that the new maximum frequency, f'_m, is reduced to $f_s/2$ or less. Thus there are no aliased components seen in Figure 2.11b, since $f_s > 2f'_m$. Eliminating the aliasing terms prior to sampling is good engineering practice. When the signal structure is well known, the aliased terms can be eliminated after sampling, with a low-pass filter operating on the sampled data [2]. In Figure 2.12 the aliased components are removed by *postfiltering* after sampling; the filter cutoff frequency, f''_m, removes the aliased components; f''_m needs to be less than ($f_s - f_m$). Notice that the filtering techniques for eliminating the aliased portion of the spectrum in Figures 2.11 and 2.12 *will result in a loss* of some of the signal information. For this reason, the sample rate, cutoff bandwidth, and filter type selected for a particular signal bandwidth are all interrelated.

Realizable filters require a nonzero bandwidth for the transition between the passband and the required out-of-band attenuation. This is called the *transition bandwidth*. To minimize the system sample rate, we desire that the antialiasing filter have a small transition bandwidth. Filter complexity and cost rise sharply with narrower transition bandwidth, so a trade-off is required between the cost of a small transition bandwidth and the costs of the higher sampling rate, which are those of more storage and higher transmission rates. In many systems the answer has been to make the transition bandwidth between 10 and 20% of the signal band-

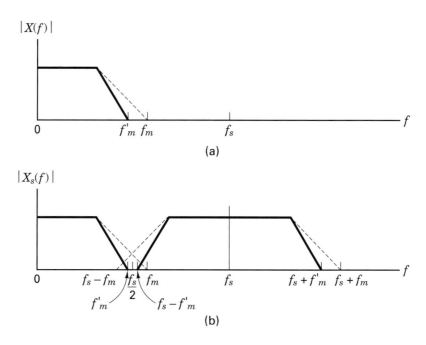

Figure 2.11 Sharper-cutoff filters eliminate aliasing. (a) Continuous signal spectrum. (b) Sampled signal spectrum.

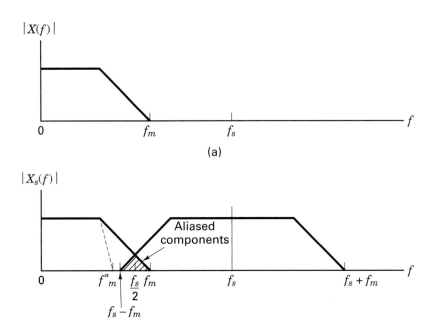

Figure 2.12 Postfilter eliminates aliased portion of spectrum. (a) Continuous signal spectrum. (b) Sampled signal spectrum.

width. If we account for the 20% transition bandwidth of the antialiasing filter, we have an *engineer's version* of the Nyquist sampling rate:

$$f_s \geq 2.2 f_m \qquad (2.17)$$

Figure 2.13 provides some insight into aliasing as seen in the time domain. The sampling instants of the solid-line sinusoid have been chosen so that the sinusoidal signal is undersampled. Notice that the resulting ambiguity allows one to draw a totally different (dashed-line) sinusoid, following the undersampled points.

Example 2.2   Sampling Rate for a High-Quality Music System

We wish to produce a high-quality digitization of a 20-kHz bandwidth music source. We are to determine a reasonable sample rate for this source. By the engineer's version of the Nyquist rate, in Equation (2.17), the sampling rate should be greater than 44.0 ksamples/s. As a matter of comparison, the standard sampling rate for the compact disc digital audio player is 44.1 ksamples/s, and the standard sampling rate for studio-quality audio is 48.0 ksamples/s.

2.4.3 Why Oversample?

Oversampling is the most economic solution for the task of transforming an analog signal to a digital signal, or the reverse, transforming a digital signal to an analog signal. This is so because signal processing performed with high performance ana-

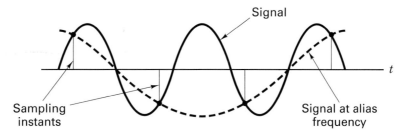

Figure 2.13   Alias frequency generated by sub-Nyquist sampling rate.

log equipment is typically much more costly than using digital signal processing equipment to perform the same task. Consider the task of transforming analog signals to digital signals. When this task is performed without the benefit of over-sampling, the process is characterized by three simple steps, performed in the order that follows.

Without Oversampling

1. The signal passes through a high performance analog lowpass filter to limit its bandwidth.
2. The filtered signal is sampled at the Nyquist rate for the (approximated) bandlimited signal. As described in Section 1.7.2, a strictly bandlimited signal is not realizable.
3. The samples are processed by an analog-to-digital converter that maps the continuous-valued samples to a finite list of discrete output levels.

When this task is performed with the benefit of over-sampling, the process is best described as five simple steps, performed in the order that follows.

With Oversampling

1. The signal is passed through a low performance (less costly) analog low-pass filter (prefilter) to limit its bandwidth.
2. The pre-filtered signal is sampled at the (now higher) Nyquist rate for the (approximated) bandlimited signal.
3. The samples are processed by an analog-to-digital converter that maps the continuous-valued samples to a finite list of discrete output levels.
4. The digital samples are then processed by a high performance digital filter to reduce the bandwidth of the digital samples.
5. The sample rate at the output of the digital filter is reduced in proportion to the bandwidth reduction obtained by this digital filter.

The next two sections examine the benefits of over-sampling.

2.4.3.1 Analog Filtering, Sampling, and Analog to Digital Conversion

The analog filter that limits the bandwidth of an input signal has a passband frequency equal to the signal bandwidth, followed by a transition to a stop band. The bandwidth of the transition region results in an increase in bandwidth of the output signal by some amount f_t. The Nyquist rate f_s for the filtered output, nominally equal to $2f_m$ (twice the highest frequency in the sampled signal) must now be increased to $2f_m + f_t$. The transition bandwidth of the filter represents an overhead in the sampling process. This additional spectral interval does not represent useful signal bandwidth but rather protects the signal bandwidth by reserving a spectral region for the aliased spectrum due to the sampling process. The aliasing stems from the fact that real signals cannot be strictly bandlimited. Typical transition bandwidths represent a 10- to 20-percent increase of the sample rate relative to that dictated by the Nyquist criterion. Examples of this overhead are seen in the compact disc (CD) digital audio system, for which the two-sided bandwidth is 40 kHz and the sample rate is 44.1 kHz, and also in the digital audio tape (DAT) system, which also has a two-sided bandwidth of 40 kHz with a sample rate of 48.0 kHz.

Our intuition and initial impulse is to keep the sample rate as low as possible by building analog filters with narrow transition bandwidths. However, analog filters can exhibit two undesirable characteristics. First, they can exhibit distortion (nonlinear phase versus frequency) due to narrow transition bandwidths. Second, the cost can be high because narrow transition bandwidths dictate high-order filters (see Section 1.6.3.2) requiring a large number of high-quality components. Our quandary is that we wish to operate the sampler at the lowest possible rate to reduce the data-storage cost. To meet this goal we might build a sophisticated analog filter with a narrow transition bandwidth. But such a filter is not only expensive, it also distorts the very signal it has been designed to protect (from undesired aliasing).

The solution (oversampling) is elegant—having been given a problem that we can't solve, we convert it to one that we can solve. We elect to use a low-cost, less sophisticated analog prefilter to limit the bandwidth of the input signal. This analog filter has been simplified by choosing a wider transition bandwidth. With a wider transition bandwidth, the required sample rate must now be increased to accommodate this larger spectrum. We typically start by selecting the higher sample rate to be 4 times the original sample rate, and then we design the analog filter to have a transition bandwidth that matches the increased sample rate. As an example, rather than sampling a CD signal at 44.1 kHz with a transition bandwidth of 4.1 kHz implemented with a sophisticated 10th order elliptic filter (implying that the filter includes 10 energy storage elements, such as capacitors and inductors), we might choose the option to employ oversampling. In that case, we could operate the sampler at 176.4 kHz with a transition bandwidth of 136.4 kHz implemented with a simpler 4th-order elliptic filter (having only 4 energy storage elements).

2.4.3.2 Digital Filtering and Resampling

Now that we have the sampled data, with its higher-than-desired sample rate, we pass the sampled data through a high-performance, low-cost, digital filter to perform the desired anti-alias filtering. The digital filter can realize the narrow transition bandwidth without the distortion associated with analog filters, and it can operate at low cost. We next reduce the sample rate of the signal (resample) after the digital filtering operation that had reduced the transition bandwidth. Good digital signal processing techniques combine the filtering and the resampling in a single structure.

Now we address a system consideration to further improve the quality of the data collection process. The analog prefilter induces some amplitude and phase distortion. We know precisely what this distortion is, and we design the digital filter so that it not only completes the anti-aliasing task of the analog prefilter, but also compensates for its gain and phase distortion. The composite response can be made as good as we want it to be. Thus we obtain a collected signal of higher quality (less distortion) at reduced cost. Digital signal processing hardware, an extension of the computer industry, is characterized by significantly lower prices each year, which has not been the case with analog processing.

In a similar fashion, oversampling is employed in the process of converting the digital signal to an analog signal (DAC). The analog filter following the DAC suffers from distortion if it has a sharp transition bandwidth. But the transition bandwidth will not be narrow if the output data presented to the DAC has been digitally oversampled.

2.4.4 Signal Interface for a Digital System

Let us examine four ways in which analog source information can be described. Figure 2.14 illustrates the choices. Let us refer to the waveform in Figure 2.14a as the *original analog waveform.* Figure 2.14b represents a sampled version of the original waveform, typically referred to as *natural-sampled data* or PAM (pulse amplitude modulation). Do you suppose that the sampled data in Figure 2.14b are compatible with a digital system? No, they are not, because the amplitude of each natural sample still has an infinite number of possible values; a digital system deals with a finite number of values. Even if the sampling is flat-top sampling, the possible pulse values form an infinite set, since they reflect all the possible values of the continuous analog waveform. Figure 2.14c illustrates the original waveform represented by discrete pulses. Here the pulses have flat tops *and* the pulse amplitude values are limited to a finite set. Each pulse is expressed as a level from a finite number of predetermined levels; each such level can be represented by a symbol from a finite alphabet. The pulses in Figure 2.14c are referred to as *quantized samples;* such a format is the obvious choice for interfacing with a digital system. The format in Figure 2.14d may be construed as the output of a sample-and-hold circuit. When the sample values are quantized to a finite set, this format can also interface with a digital system. After quantization, the analog waveform can still be recovered, but not precisely; improved reconstruction fidelity of the analog waveform can be achieved by increasing the number of quantization levels (requiring increased system bandwidth). Signal distortion due to quantization is treated in the following sections (and later in Chapter 13).

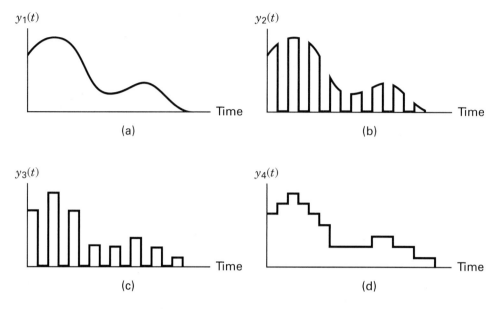

Figure 2.14 Amplitude and time coordinates of source data. (a) Original analog waveform. (b) Natural-sampled data. (c) Quantized samples. (d) Sample and hold.

2.5 SOURCES OF CORRUPTION

The analog signal recovered from the sampled, quantized, and transmitted pulses will contain corruption from several sources. The sources of corruption are related to (1) sampling and quantizing effects, and (2) channel effects. These effects are considered in the sections that follow.

2.5.1 Sampling and Quantizing Effects

2.5.1.1 Quantization Noise

The distortion inherent in quantization is a round-off or truncation error. The process of encoding the PAM signal into a quantized PAM signal involves discarding some of the original analog information. This distortion, introduced by the need to approximate the analog waveform with quantized samples, is referred to as *quantization noise;* the amount of such noise is inversely proportional to the number of levels employed in the quantization process. (The signal-to-noise ratio of quantized pulses is treated in Sections 2.5.3 and 13.2.)

2.5.1.2 Quantizer Saturation

The quantizer (or analog-to-digital converter) allocates L levels to the task of approximating the continuous range of inputs with a finite set of outputs. The range of inputs for which the difference between the input and output is small is called the *operating range* of the converter. If the input exceeds this range, the difference between the input and the output becomes large, and we say that the converter is operating in *saturation.* Saturation errors, being large, are more objectionable than quantizing noise. Generally, saturation is avoided by the use of automatic gain control (AGC), which effectively extends the operating range of the converter. (Chapter 13 covers quantizer saturation in greater detail.)

2.5.1.3 Timing Jitter

Our analysis of the sampling theorem predicted precise reconstruction of the signal based on uniformly spaced samples of the signal. If there is a slight jitter in the position of the sample, the sampling is no longer uniform. Although exact reconstruction is still possible if the sample positions are accurately known, the jitter is usually a random process and thus the sample positions are not accurately known. The effect of the jitter is equivalent to frequency modulation (FM) of the baseband signal. If the jitter is random, a low-level wideband spectral contribution is induced whose properties are very close to those of the quantizing noise. If the jitter exhibits periodic components, as might be found in data extracted from a tape recorder, the periodic FM will induce low-level spectral lines in the data. Timing jitter can be controlled with very good power supply isolation and stable clock references.

2.5.2 Channel Effects

2.5.2.1 Channel Noise

Thermal noise, interference from other users, and interference from circuit switching transients can cause errors in detecting the pulses carrying the digitized samples. Channel-induced errors can degrade the reconstructed signal quality quite quickly. This rapid degradation of output signal quality with channel-induced errors is called a *threshold effect.* If the channel noise is small, there will be no problem detecting the presence of the waveforms. Thus, small noise does not corrupt the reconstruct signals. In this case, the only noise present in the reconstruction is the quantization noise. On the other hand, if the channel noise is large enough to affect our ability to detect the waveforms, the resulting detection error causes reconstruction errors. A large difference in behavior can occur for very small changes in channel noise level.

2.5.2.2 Intersymbol Interference

The channel is always bandlimited. A bandlimited channel disperses or spreads a pulse waveform passing through it (see Section 1.6.4). When the channel bandwidth is much greater than the pulse bandwidth, the spreading of the pulse will be slight. When the channel bandwidth is close to the signal bandwidth, the spreading will exceed a symbol duration and cause signal pulses to overlap. This overlapping is called *intersymbol interference* (ISI). Like any other source of interference, ISI causes system degradation (higher error rates); it is a particularly insidious form of interference because raising the signal power to overcome the interference will not always improve the error performance. (Details of how ISI is handled are presented in the next chapter, in Sections 3.3 and 3.4.)

2.5.3 Signal-to-Noise Ratio for Quantized Pulses

Figure 2.15 illustrates an L-level linear quantizer for an analog signal with a peak-to-peak voltage range of $V_{pp} = V_p - (-V_p) = 2V_p$ volts. The quantized pulses assume positive and negative values, as shown in the figure. The step size between quantization levels, called the *quantile interval,* is denoted q volts. When the quantization levels are uniformly distributed over the full range, the quantizer is called a *uniform or linear quantizer.* Each sample value of the analog waveform is approximated with a quantized pulse; the approximation will result in an error no larger than $q/2$ in the positive direction or $-q/2$ in the negative direction. The degradation of the signal due to quantization is therefore limited to half a quantile interval, $\pm q/2$ volts.

A useful figure of merit for the uniform quantizer is the quantizer variance (mean-square error assuming zero mean). If we assume that the quantization error, $e,$ is uniformly distributed over a single quantile interval q-wide (i.e., the analog input takes on all values with equal probability), the quantizer error variance is found to be

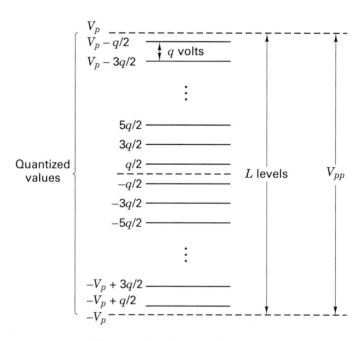

Figure 2.15   Quantization levels.

$$\sigma^2 = \int_{-q/2}^{+q/2} e^2 p(e) \, de \tag{2.18a}$$

$$= \int_{-q/2}^{+q/2} e^2 \frac{1}{q} \, de = \frac{q^2}{12} \tag{2.18b}$$

where $p(e) = 1/q$ is the (uniform) probability density function of the quantization error. The variance, σ^2, corresponds to the *average quantization noise power.* The peak power of the analog signal (normalized to 1 Ω) can be expressed as

$$V_p^2 = \left(\frac{V_{pp}}{2}\right)^2 = \left[\frac{q(L-1)}{2}\right]^2 \approx \left(\frac{Lq}{2}\right)^2 = \frac{L^2 q^2}{4} \tag{2.19}$$

where L is the number of quantization levels. Equations (2.18) and (2.19) combined yield the ratio of *peak* signal power to *average* quantization noise power

$$\left(\frac{S}{N}\right)_q = \frac{L^2 q^2 / 4}{q^2 / 12} = 3L^2 \tag{2.20}$$

where N is the average quantization noise power, σ^2. We see that $(S/N)_q$ increases as a function of the number of quantization levels squared. In the limit (as $L \to \infty$), the signal becomes analog (infinite quantization and zero quantization noise). Note that for random signals, the important $(S/N)_q$ parameter deals with *average* rather than *peak* signal power. In that case, to obtain average signal power, one needs to know the signal's pdf.

2.6 PULSE CODE MODULATION

Pulse code modulation (PCM) is the name given to the class of baseband signals obtained from the quantized PAM signals by encoding each quantized sample into a *digital word* [3]. The source information is sampled and quantized to one of L levels; then each quantized sample is digitally encoded into an ℓ-bit ($\ell = \log_2 L$) codeword. For baseband transmission, the codeword bits will then be transformed to pulse waveforms. The essential features of binary PCM are shown in Figure 2.16. Assume that an analog signal $x(t)$ is limited in its excursions to the range -4 to $+4$ V. The step size between quantization levels has been set at 1 V. Thus, eight quantization levels are employed; these are located at $-3.5, -2.5, \ldots, +3.5$ V. We assign the code number 0 to the level at -3.5 V, the code number 1 to the level at -2.5 V, and so on, until the level at 3.5 V, which is assigned the code number 7. Each code number has its representation in binary arithmetic, ranging from 000 for code number 0 to 111 for code number 7. Why have the voltage levels been chosen in this manner, compared with using a sequence of consecutive integers, 1, 2, 3, ... ? The choice of voltage levels is guided by two constraints. First, the quantile intervals between the levels should be equal; and second, it is convenient for the levels to be symmetrical about zero.

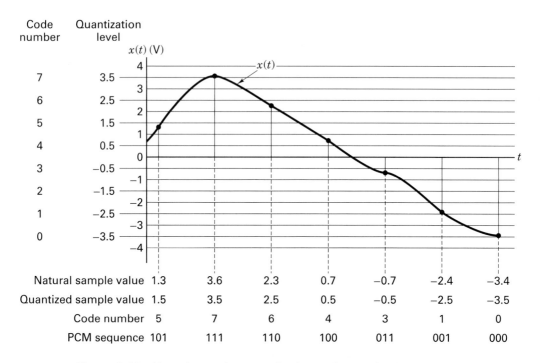

Natural sample value	1.3	3.6	2.3	0.7	−0.7	−2.4	−3.4
Quantized sample value	1.5	3.5	2.5	0.5	−0.5	−2.5	−3.5
Code number	5	7	6	4	3	1	0
PCM sequence	101	111	110	100	011	001	000

Figure 2.16 Natural samples, quantized samples, and pulse code modulation. (Reprinted with permission from Taub and Schilling, *Principles of Communications Systems,* McGraw-Hill Book Company, New York, 1971, Fig. 6.5-1, p. 205.)

The ordinate in Figure 2.16 is labeled with quantization levels and their code numbers. Each sample of the analog signal is assigned to the quantization level closest to the value of the sample. Beneath the analog waveform $x(t)$ are seen four representations of $x(t)$, as follows: the natural sample values, the quantized sample values, the code numbers, and the PCM sequence.

Note, that in the example of Figure 2.16, each sample is assigned to one of eight levels or a three-bit PCM sequence. Suppose that the analog signal is a musical passage, which is sampled at the Nyquist rate. And, suppose that when we listen to the music in digital form, it sounds terrible. What could we do to improve the fidelity? Recall that the process of quantization replaces the true signal with an approximation (i.e., adds quantization noise). Thus, increasing the number of levels will reduce the quantization noise. If we double the number of levels to 16, what are the consequences? In that case, each analog sample will be represented as a four-bit PCM sequence. Will that cost anything? In a real-time communication system, the messages must not be delayed. Hence, the transmission time for each sample must be the same, regardless of how many bits represent the sample. Hence, when there are more bits per sample, the bits must move faster; in other words, they must be replaced by "skinnier" bits. The data rate is thus increased, and the cost is a greater transmission bandwidth. This explains how one can generally obtain better fidelity at the cost of more transmission bandwidth. Be aware, however, that there are some communication applications where delay is permissible. For example, consider the transmission of planetary images from a spacecraft. The Galileo project, launched in 1989, was on such a mission to photograph and transmit images of the planet Jupiter. The Galileo spacecraft arrived at its Jupiter destination in 1995. The journey took several years; therefore, any excess signal delay of several minutes (or hours or days) would certainly not be a problem. In such cases, the cost of more quantization levels and greater fidelity need not be bandwidth; it can be time delay.

In Figure 2.1, the term "PCM" appears in two places. First, it is a formatting topic, since the process of analog-to-digital (A/D) conversion involves sampling, quantization, and ultimately yields binary digits via the conversion of quantized PAM to PCM. Here, PCM digits are just binary numbers—a baseband carrier wave has not yet been discussed. The second appearance of PCM in Figure 2.1 is under the heading *Baseband Signaling.* Here, we list various PCM waveforms (line codes) that can be used to "carry" the PCM digits. Therefore, note that the difference between PCM and a PCM waveform is that the former represents a bit sequence, and the latter represents a particular waveform conveyance of that sequence.

2.7 UNIFORM AND NONUNIFORM QUANTIZATION

2.7.1 Statistics of Speech Amplitudes

Speech communication is a very important and specialized area of digital communications. Human speech is characterized by unique statistical properties; one such property is illustrated in Figure 2.17. The abscissa represents speech signal magnitudes, normalized to the root-mean-square (rms) value of such magnitudes through a typical communication channel, and the ordinate is probability. For most voice

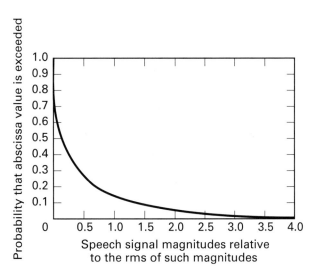

Figure 2.17 Statistical distribution of single-talker speech signal magnitudes.

communication channels, very low speech volumes predominate; 50% of the time, the voltage characterizing detected speech energy is less than one-fourth of the rms value. Large amplitude values are relatively rare; only 15% of the time does the voltage exceed the rms value. We see from Equation (2.18b) that the quantization noise depends on the step size (size of the quantile interval). When the steps are uniform in size the quantization is known as *uniform quantization.* Such a system would be wasteful for speech signals; many of the quantizing steps would rarely be used. In a system that uses equally spaced quantization levels, the quantization noise is the same for all signal magnitudes. Therefore, with uniform quantization, the signal-to-noise (SNR) is worse for low-level signals than for high-level signals. *Nonuniform quantization* can provide fine quantization of the weak signals and coarse quantization of the strong signals. Thus in the case of nonuniform quantization, quantization noise can be made proportional to signal size. The effect is to improve the overall SNR by reducing the noise for the predominant weak signals, at the expense of an increase in noise for the rarely occurring strong signals. Figure 2.18 compares the quantization of a strong versus a weak signal for uniform and nonuniform quantization. The staircase-like waveforms represent the approximations to the analog waveforms (after quantization distortion has been introduced). The SNR improvement that nonuniform quantization provides for the weak signal should be apparent. Nonuniform quantization can be used to make the SNR a constant for all signals within the input range. For voice signals, the typical input signal

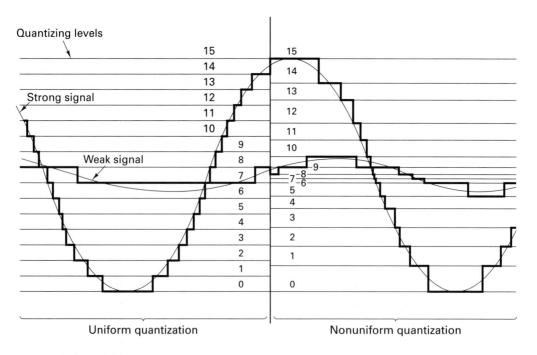

Figure 2.18   Uniform and nonuniform quantization of signals.

dynamic range is 40 decibels (dB), where a decibel is defined in terms of the ratio of power P_2 to power P_1:

$$\text{number of dB} = 10\log_{10}\frac{P_2}{P_1} \qquad (2.21)$$

With a uniform quantizer, weak signals would experience a 40-dB-poorer SNR than that of strong signals. The standard telephone technique of handling the large range of possible input signal levels is to use a *logarithmic-compressed* quantizer instead of a uniform one. With such a nonuniform compressor the output SNR is independent of the distribution of input signal levels.

2.7.2 Nonuniform Quantization

One way of achieving nonuniform quantization is to use a nonuniform quantizer characteristic, shown in Figure 2.19a. More often, nonuniform quantization is achieved by first distorting the original signal with a logarithmic compression characteristic, as shown in Figure 2.19b, and then using a uniform quantizer. For small magnitude signals the compression characteristic has a much steeper slope than for large magnitude signals. Thus, a given signal change at small magnitudes will carry the uniform quantizer through more steps than the same change at large magnitudes. The compression characteristic effectively changes the distribution of the

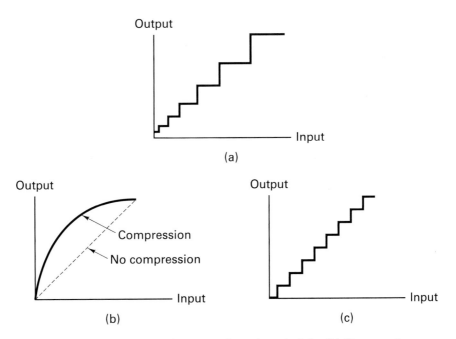

Figure 2.19 (a) Nonuniform quantizer characteristic. (b) Compression characteristic. (c) Uniform quantizer characteristic.

input signal magnitudes so that there is not a preponderance of *low* magnitude signals at the output of the compressor. After compression, the distorted signal is used as the input to a uniform (linear) quantizer characteristic, shown in Figure 2.19c. At the receiver, an inverse compression characteristic, called *expansion,* is applied so that the overall transmission is not distorted. The processing pair (compression and expansion) is usually referred to as *companding.*

2.7.3 Companding Characteristics

The early PCM systems implemented a smooth logarithmic compression function. Today, most PCM systems use a piecewise linear approximation to the logarithmic compression characteristic. In North America, a μ-law compression characteristic

$$y = y_{\max} \frac{\log_e[1 + \mu(|x|/x_{\max})]}{\log_e(1 + \mu)} \text{ sgn } x \qquad (2.22)$$

is used, where

$$\text{sgn } x = \begin{cases} +1 & \text{for } x \geq 0 \\ -1 & \text{for } x < 0 \end{cases}$$

and where μ is a positive constant, x and y represent input and output voltages, and $x_{\max}$ and $y_{\max}$ are the maximum positive excursions of the input and output voltages, respectively. The compression characteristic is shown in Figure 2.20a for several values of μ. In North America, the standard value for μ is 255. Notice that μ = 0 corresponds to linear amplification (uniform quantization).

Another compression characteristic, used mainly in Europe, is the *A*-law characteristic, defined as

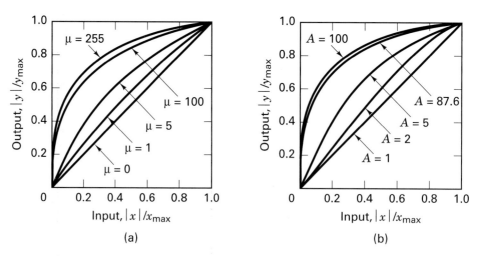

Figure 2.20 Compression characteristics. (a) μ-law characteristic. (b) *A*-law characteristic.

$$y = \begin{cases} y_{max} \dfrac{A(|x|/x_{max})}{1 + \log_e A} \operatorname{sgn} x & 0 < \dfrac{|x|}{x_{max}} \leq \dfrac{1}{A} \\[4mm] y_{max} \dfrac{1 + \log_e[A(|x|/x_{max})]}{1 + \log_e A} \operatorname{sgn} x & \dfrac{1}{A} < \dfrac{|x|}{x_{max}} < 1 \end{cases} \tag{2.23}$$

where A is a positive constant and x and y are defined the same way as they were for Equation (2.22). The A-law compression characteristic is shown in Figure 2.20b for several values of A. A standard value for A is 87.6. (The subjects of uniform and nonuniform quantization are treated further in Chapter 13, Section 13.2.)

2.8 BASEBAND TRANSMISSION

2.8.1 Waveform Representation of Binary Digits

In Section 2.6, it was shown how analog waveforms are transformed into binary digits via the use of PCM. There is nothing "physical" about the digits resulting from this process. Digits are just abstractions—a way to describe the message information. Thus, we need something physical that will represent or "carry" the digits.

We will represent the binary digits with electrical pulses in order to transmit them through a baseband channel. Such a representation is shown in Figure 2.21. Codeword time slots are shown in Figure 2.21a, where the codeword is a 4-bit representation of each quantized sample. In Figure 2.21b, each binary one is represented by a pulse and each binary zero is represented by the absence of a pulse. Thus a sequence of electrical pulses having the pattern shown in Figure 2.21b can be used to transmit the information in the PCM bit stream, and hence the information in the quantized samples of a message.

At the receiver, a determination must be made as to the presence or absence of a pulse in each bit time slot. It will be shown in Section 2.9 that the likelihood of correctly detecting the presence of a pulse is a function of the received pulse energy (or area under the pulse). Thus there is an advantage in making the pulse width T' in Figure 2.21b as wide as possible. If we increase the pulse width to the maximum possible (equal to the bit time T), we have the waveform shown in Figure 2.21c. Rather than describe this waveform as a sequence of present or absent pulses (unipolar), we can describe it as a sequence of transitions between two levels (bipolar). When the waveform occupies the upper voltage level it represents a binary one; when it occupies the lower voltage level it represents a binary zero.

2.8.2 PCM Waveform Types

When pulse modulation is applied to a *binary* symbol, the resulting binary waveform is called a pulse-code modulation (PCM) waveform. There are several types of PCM waveforms that are described below and illustrated in Figure 2.22; in telephony applications, these waveforms are often called *line codes*. When pulse modulation is applied to a *nonbinary* symbol, the resulting waveform is called an *M*-ary

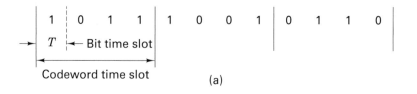

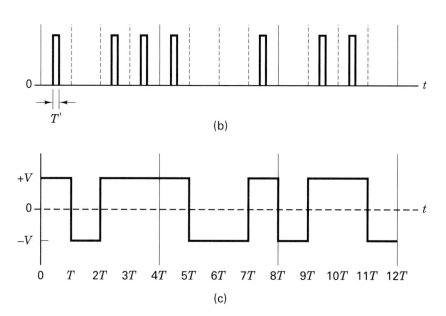

Figure 2.21 Example of waveform representation of binary digits. (a) PCM sequence. (b) Pulse representation of PCM. (c) Pulse waveform (transition between two levels).

pulse-modulation waveform, of which there are several types. They are described in Section 2.8.5, where one of them, called pulse-amplitude modulation (PAM), is emphasized. In Figure 2.1, the highlighted block, labeled *Baseband Signaling,* shows the basic classification of the PCM waveforms and the *M*-ary pulse waveforms. The PCM waveforms fall into the following four groups.

1. Nonreturn-to-zero (NRZ)
2. Return-to-zero (RZ)
3. Phase encoded
4. Multilevel binary

The NRZ group is probably the most commonly used PCM waveform. It can be partitioned into the following subgroups: NRZ-L (L for level), NRZ-M (M for mark), and NRZ-S (S for space). NRZ-L is used extensively in digital logic circuits.

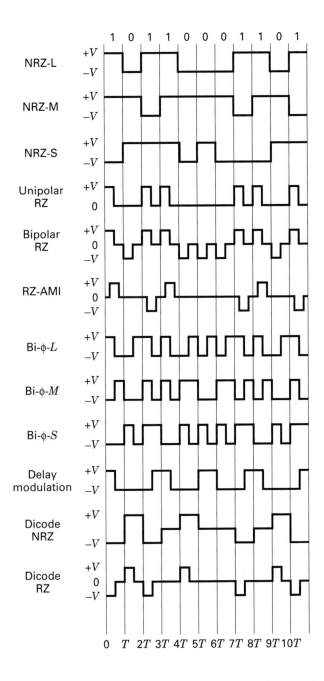

Figure 2.22 Various PCM waveforms.

A binary one is represented by one voltage level and a binary zero is represented by another voltage level. There is a change in level whenever the data change from a one to a zero or from a zero to a one. With NRZ-M, the one, or *mark,* is represented by a change in level, and the zero, or *space,* is represented by no change in level. This is often referred to as *differential encoding.* NRZ-M is used primarily in

2.8   Baseband Transmission

magnetic tape recording. NRZ-S is the complement of NRZ-M: A one is represented by no change in level, and a zero is represented by a change in level.

The RZ waveforms consist of unipolar-RZ, bipolar-RZ, and RZ-AMI. These codes find application in baseband data transmission and in magnetic recording. With unipolar-RZ, a one is represented by a half-bit-wide pulse, and a zero is represented by the absence of a pulse. With bipolar-RZ, the ones and zeros are represented by opposite-level pulses that are one-half bit wide. There is a pulse present in each bit interval. RZ-AMI (AMI for "alternate mark inversion") is a signaling scheme used in telephone systems. The ones are represented by equal-amplitude alternating pulses. The zeros are represented by the absence of pulses.

The phase-encoded group consists of bi-ϕ-L (bi-phase-level), better known as *Manchester coding;* bi-ϕ-M (bi-phase-mark); bi-ϕ-S (bi-phase-space); and *delay modulation* (DM), or *Miller coding.* The phase-encoding schemes are used in magnetic recording systems and optical communications and in some satellite telemetry links. With bi-ϕ-L, a one is represented by a half-bit-wide pulse positioned during the first half of the bit interval; a zero is represented by a half-bit-wide pulse positioned during the second half of the bit interval. With bi-ϕ-M, a transition occurs at the beginning of every bit interval. A one is represented by a second transition one-half bit interval later; a zero is represented by no second transition. With bi-ϕ-S, a transition also occurs at the beginning of every bit interval. A one is represented by no second transition; a zero is represented by a second transition one-half bit interval later. With delay modulation [4], a one is represented by a transition at the midpoint of the bit interval. A zero is represented by no transition, unless it is followed by another zero. In this case, a transition is placed at the end of the bit interval of the first zero. Reference to the illustration in Figure 2.22 should help to make these descriptions clear.

Many binary waveforms use three levels, instead of two, to encode the binary data. Bipolar RZ and RZ-AMI belong to this group. The group also contains formats called *dicode* and *duobinary.* With dicode-NRZ, the one-to-zero or zero-to-one data transition changes the pulse polarity; without a data transition, the zero level is sent. With dicode-RZ, the one-to-zero or zero-to-one transition produces a half-duration polarity change; otherwise, a zero level is sent. The three-level duobinary signaling scheme is treated in Section 2.9.

One might ask why there are so many PCM waveforms. Are there really so many unique applications necessitating such a variety of waveforms to represent digits? The reason for the large selection relates to the differences in performance that characterize each waveform [5]. In choosing a PCM waveform for a particular application, some of the parameters worth examining are the following:

1. *Dc component.* Eliminating the dc energy from the signal's power spectrum enables the system to be ac coupled. Magnetic recording systems, or systems using transformer coupling, have little sensitivity to very low frequency signal components. Thus low-frequency information could be lost.
2. *Self-Clocking.* Symbol or bit synchronization is required for any digital communication system. Some PCM coding schemes have inherent synchronizing

or clocking features that aid in the recovery of the clock signal. For example, the Manchester code has a transition in the middle of every bit interval whether a one or a zero is being sent. This guaranteed transition provides a clocking signal.

3. *Error detection.* Some schemes, such as duobinary, provide the means of detecting data errors without introducing additional error-detection bits into the data sequence.

4. *Bandwidth compression.* Some schemes, such as multilevel codes, increase the efficiency of bandwidth utilization by allowing a reduction in required bandwidth for a given data rate; thus there is more information transmitted per unit bandwidth.

5. *Differential encoding.* This technique is useful because it allows the polarity of differentially encoded waveforms to be inverted without affecting the data detection. In communication systems where waveforms sometimes experience inversion, this is a great advantage. (Differential encoding is treated in greater detail in Chapter 4, Section 4.5.2.)

6. *Noise immunity.* The various PCM waveform types can be further characterized by probability of bit error versus signal-to-noise ratio. Some of the schemes are more immune than others to noise. For example, the NRZ waveforms have better error performance than does the unipolar RZ waveform.

2.8.3 Spectral Attributes of PCM Waveforms

The most common criteria used for comparing PCM waveforms and for selecting one waveform type from the many available are spectral characteristics, bit synchronization capabilities, error-detecting capabilities, interference and noise immunity, and cost and complexity of implementation. Figure 2.23 shows the spectral characteristics of some of the most popular PCM waveforms. The figure plots power spectral density in watts/hertz versus normalized bandwidth, WT, where W is bandwidth, and T is the duration of the pulse. WT is often referred to as the *time-bandwidth product,* of the signal. Since the pulse or symbol rate R_s is the reciprocal of T, normalized bandwidth can also be expressed as W/R_s. From this latter expression, we see that the units of normalized bandwidth are hertz/(pulse/s) or hertz/(symbol/s). This is a relative measure of bandwidth; it is valuable because it describes how efficiently the transmission bandwidth is being utilized for each waveform of interest. Any waveform type that requires less than 1.0 Hz for sending 1 symbol/s is relatively bandwidth efficient. Examples would be delay modulation and duobinary (see Section 2.9). By comparison, any waveform type that requires more than 1.0 Hz for sending 1 symbol/s is relatively bandwidth inefficient. An example of this would be bi-phase (Manchester) signaling. From Figure 2.23, we can also see the spectral concentration of signaling energy for each waveform type. For example, NRZ and duobinary schemes have large spectral components at dc and low frequency, while bi-phase has no energy at dc.

An important parameter for measuring *bandwidth efficiency* is R/W having units of bits/s/hz. This measure involves data rate rather than symbol rate. For a

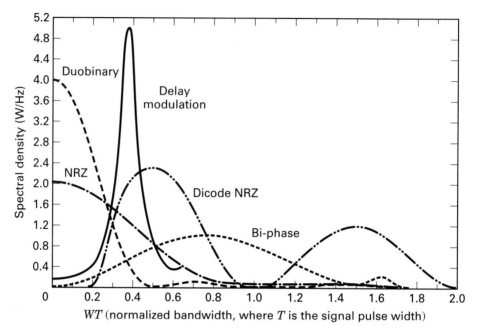

Figure 2.23 Spectral densities of various PCM waveforms.

given signaling scheme, R/W describes how much data throughput can be transmitted for each Hertz of available bandwidth. (Bandwidth efficiency is treated in greater detail in Chapter 9.)

2.8.4 Bits per PCM Word and Bits per Symbol

Throughout Chapters 1 and 2, the idea of binary partitioning $(M = 2^k)$ is used to relate the grouping of bits to form symbols for the purpose of signal processing and transmission. We now examine an analogous application where the $M = 2^k$ concept is also applicable. Consider the process of formatting analog information into a bit steam via sampling, quantization, and coding. Each analog sample is transformed into a PCM word made up of groups of bits. The PCM word size can be described by the number of quantization levels allowed for each sample; this is identical to the number of values that the PCM word can assume. Or, the quantization can be described by the number of bits required to identify that set of levels. The relationship between the number of levels per sample and the number of bits needed to represent those levels is the same as the $M = 2^k$ relationship between the size of a set of message symbols and the number of bits needed to represent the symbol. To distinguish between the two applications, the notation is changed for the PCM case. Instead of $M = 2^k$, we use $L = 2^\ell$, where L is the number of quantization levels in the PCM word, and ℓ is the number of bits needed to represent those levels.

2.8.4.1 PCM Word Size

How many bits shall we assign to each analog sample? For digital telephone channels, each speech sample is PCM encoded using 8 bits, yielding 2^8 or 256 levels per sample. The choice of the number of levels, or bits per sample, depends on how much quantization distortion we are willing to tolerate with the PCM format. It is useful to develop a general relationship between the required number of bits per analog sample (the PCM word size), and the allowable quantization distortion. Let the magnitude of the quantization distortion error, $|e|$, be specified as a fraction p of the peak-to-peak analog voltage V_{pp} as follows:

$$|e| \leq p V_{pp} \qquad (2.24)$$

Since the quantization error can be no larger than $q/2$, where q is the quantile interval, we can write

$$|e|_{max} = \frac{q}{2} = \frac{V_{pp}}{2(L-1)} \approx \frac{V_{pp}}{2L} \qquad (2.25)$$

where L is the number of quantization levels. For most applications the number of levels is large enough so that $L - 1$ can be replaced by L, as was done above. Then, from Equations (2.24) and (2.25), we can write

$$\frac{V_{pp}}{2L} \leq p V_{pp} \qquad (2.26)$$

$$2^{\ell} = L \geq \frac{1}{2p} \quad \text{levels} \qquad (2.27)$$

and

$$\ell \geq \log_2 \frac{1}{2p} \quad \text{bits} \qquad (2.28)$$

It is important that we do not confuse the idea of bits per PCM word, denoted by ℓ in Equation (2.28), with the M-level transmission concept of k data bits per symbol. (Example 2.3, presented shortly, should clarify the distinction.)

2.8.5 *M-ary Pulse-Modulation Waveforms*

There are three basic ways to modulate information on to a sequence of pulses: we can vary the pulse's amplitude, position, or duration, which leads to the names *pulse-amplitude modulation* (PAM), *pulse-position modulation* (PPM), and *pulse-duration modulation* (PDM), respectively. PDM is sometimes called pulse-width modulation (PWM). When information samples without any quantization are modulated on to pulses, the resulting pulse modulation can be called *analog pulse modulation*. When the information samples are first quantized, yielding symbols from an M-ary alphabet set, and then modulated on to pulses, the resulting pulse modulation is digital and we refer to it as *M-ary pulse modulation*. In the case of *M*-ary PAM, one of M allowable amplitude levels are assigned to each of the M possible symbol values. Earlier we described PCM waveforms as binary waveforms having two amplitude values (e.g., NRZ, RZ). Note that such PCM waveforms requiring only two levels represent the special case ($M = 2$) of the general M-ary PAM that requires M levels. In this book, the PCM waveforms are grouped separately (see Figure 2.1 and Section 2.8.2) and are emphasized because they are the most popular of the pulse-modulation schemes.

In the case of M-ary PPM waveforms, modulation is effected by delaying (or advancing) a pulse occurrence, by an amount that corresponds to the value of the information symbols. In the case of M-ary PDM waveforms, modulation is effected by varying the pulse width by an amount that corresponds to the value of the symbols. For both PPM and PDM, the pulse amplitude is held constant. Baseband modulation with pulses have analogous counterparts in the area of bandpass modulation. PAM is similar to amplitude modulation, while PPM and PDM are similar to phase and frequency modulation respectively. In this section, we only address M-ary PAM waveforms as they compare to PCM waveforms.

The transmission bandwidth required for binary digital waveforms such as PCM may be very large. What might we do to reduce the required bandwidth? One possibility is to use *multilevel signaling*. Consider a bit stream with data rate, R bits per second. Instead of transmitting a pulse waveform for each bit, we might first partition the data into k-bit groups, and then use ($M = 2^k$)-level pulses for transmission. With such multilevel signaling or M-ary PAM, each pulse waveform can now represent a k-bit symbol in a symbol stream moving at the rate of R/k symbols per second (a factor k slower than the bit stream). Thus for a given data rate, multilevel signaling, where $M > 2$, can be used to reduce the number of symbols transmitted per second; or, in other words, M-ary PAM as opposed to binary PCM can be used to reduce the transmission bandwidth requirements of the channel. Is there a price to be paid for such bandwidth reduction? Of course, and that is discussed below.

Consider the task that the pulse receiver must perform: It must distinguish between the possible levels of each pulse. Can the receiver distinguish among the eight possible levels of each octal pulse in Figure 2.24a as easily as it can distinguish between the two possible levels of each binary pulse in Figure 2.24b? The transmission of an 8-level (compared with a 2-level) pulse requires a greater amount of energy for equivalent detection performance. (It is the amount of received E_b/N_0 that determines how reliably a signal will be detected). For equal average power in the binary and the octal pulses, it is easier to detect the binary pulses because the detector has more signal energy per level for making a binary decision that an 8-level decision. What price does a system designer pay if he or she chooses the transmission waveform to be the easier-to-detect binary PCM rather than the 8-level PAM? The engineer pays the price of needing three times as much transmission bandwidth for a given data rate, compared with the octal pulses, since each octal pulse must be replaced with three binary pulses (each one-third as wide as the octal pulses). One might ask, Why not use binary pulses with the same pulse duration as the original octal pulses and suffer the information delay? For some cases, this might be appropriate, but for real-time communication systems, such an increase in delay cannot be tolerated—the 6 o'clock news *must* be received at 6 o-clock. (In Chapter 9, we examine in detail the trade-off between signal power and transmission bandwidth.)

Amplitude

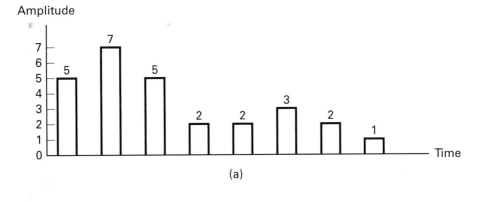

(a)

Amplitude

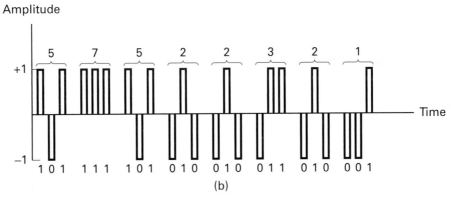

(b)

Figure 2.24 Pulse code modulation signaling. (a) Eight-level signaling. (b) Two-level signaling.

Example 2.3   Quantization Levels and Multilevel Signaling

The information in an analog waveform, with maximum frequency $f_m = 3$ kHz, is to be transmitted over an M-ary PAM system, where the number of pulse levels is $M = 16$. The quantization distortion is specified not to exceed $\pm 1\%$ of the peak-to-peak analog signal.

(a) What is the minimum number of bits/sample, or bits/PCM word that should be used in digitizing the analog waveform?
(b) What is the minimum required sampling rate, and what is the resulting bit transmission rate?
(c) What is the PAM pulse or symbol transmission rate?
(d) If the transmission bandwidth (including filtering) equals 12 kHz, determine the bandwidth efficiency for this system.

In this example we are concerned with two types of *levels:* the number of quantization levels for fulfilling the distortion requirement and the 16 levels of the multilevel PAM pulses.

Solution

(a) Using Equation (2.28), we calculate

$$\ell \geq \log_2 \frac{1}{0.02} = \log_2 50 \approx 5.6.$$

Therefore, use $\ell = 6$ bits/sample to meet the distortion requirement.

(b) Using the Nyquist sampling criterion, the minimum sampling rate $f_s = 2f_m = 6000$ samples/second. From part (a), each sample will give rise to a PCM word composed of 6 bits. Therefore the bit transmission rare $R = \ell f_s = 36,000$ bits/sec.

(c) Since multilevel pulses are to be used with $M = 2^k = 16$ levels, then $k = \log_2 16 = 4$ bits/symbol. Therefore, the bit stream will be partitioned into groups of 4 bits to form the new 16-level PAM digits, and the resulting symbol transmission rate R_s is $R/k = 36,000/4 = 9000$ symbols/s.

(d) Bandwidth efficiency is described by data throughput per hertz, R/W. Since $R = 36,000$ bits/s, and $W = 12$ kHz, then $R/W = 3$ bits/s/hz.

2.9 CORRELATIVE CODING

In 1963, Adam Lender [6, 7] showed that it is possible to transmit $2W$ symbols/s with zero ISI, using the theoretical minimum bandwidth of W hertz, without infinitely sharp filters. Lender used a technique called *duobinary signaling,* also referred to as *correlative coding* and *partial response signaling.* The basic idea behind the duobinary technique is to introduce some controlled amount of ISI into the data stream rather than trying to eliminate it completely. By introducing correlated interference between the pulses, and by changing the detection procedure, Lender, in effect, "canceled out" the interference at the detector and thereby achieved the ideal symbol-rate packing of 2 symbols/s/Hz, an amount that had been considered unrealizable.

2.9.1 Duobinary Signaling

To understand how duobinary signaling introduces controlled ISI, let us look at a model of the process. We can think of the duobinary coding operation as if it were implemented as shown in Figure 2.25. Assume that a sequence of binary symbols $\{x_k\}$ is to be transmitted at the rate of R symbols/s over a system having an ideal rectangular spectrum of bandwidth $W = R/2 = 1/2T$ hertz. You might ask: How is this rectangular spectrum, in Figure 2.25, different from the unrealizable Nyquist characteristic? It has the same ideal characteristic; but we are not trying to implement the ideal rectangular filter. It is only the part of our equivalent model that is used for developing a filter that is easier to approximate. Before being shaped by the ideal filter, the pulses pass through a simple digital filter, as shown in the figure. The digital filter incorporates a one-digit delay; to each incoming pulse, the filter adds the value of the previous pulse. In other words, for every pulse into the digital filter, we get the summation of two pulses out. Each pulse of the sequence $\{y_k\}$ out of the digital filter can be expressed as

$$y_k = x_k + x_{k-1} \tag{2.29}$$

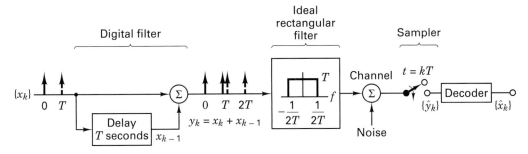

Figure 2.25 Duobinary signaling.

Hence, the $\{y_k\}$ amplitudes are not independent; each y_k digit carries with it the *memory* of the prior digit. The ISI introduced to each y_k digit comes only from the preceding x_{k-1} digit. This correlation between the pulse amplitudes of $\{y_k\}$ can be thought of as the controlled ISI introduced by the duobinary coding. Controlled interference is the essence of this novel technique because at the detector, such controlled interference can be removed as easily as it was added. The $\{y_k\}$ sequence is followed by the ideal Nyquist filter that does not introduce any ISI. In Figure 2.25, at the receiver sampler, we would expect to recover the sequence $\{y_k\}$ exactly in the absence of noise. Since all systems experience noise contamination, we shall refer to the *received* $\{y_k\}$ as the estimate of $\{y_k\}$ and denote it $\{\hat{y}_k\}$. Removing the controlled interference with the duobinary decoder yields an estimate of $\{x_k\}$ which we shall denote as $\{\hat{x}_k\}$.

2.9.2 Duobinary Decoding

If the binary digit x_k is equal to ± 1, then using Equation (2.29), y_k has one of three possible values: $+2$, 0, or -2. The duobinary code results in a three-level output: in general, for M-ary transmission, partial response signaling results in $2M - 1$ output levels. The decoding procedure involves the inverse of the coding procedure, namely, subtracting the x_{k-1} decision from the y_k digit. Consider the following coding/decoding example.

Example 2.4   Duobinary Coding and Decoding

Use Equation (2.29) to demonstrate duobinary coding and decoding for the following sequence: $\{x_k\} = 0\,0\,1\,0\,1\,1\,0$. Consider the first bit of the sequence to be a startup digit, not part of the data.

Solution

Binary digit sequence $\{x_k\}$:	0	0	1	0	1	1	0
Bipolar amplitudes $\{x_k\}$:	-1	-1	$+1$	-1	$+1$	$+1$	-1
Coding rule: $y_k = x_k + x_{k-1}$:		-2	0	0	0	2	0

Decoding decision rule:	If $\hat{y}_k = 2$, decide that $\hat{x}_k = +1$ (or binary one).
	If $\hat{y}_k = -2$, decide that $\hat{x}_k = -1$ (or binary zero).
	If $\hat{y}_k = 0$, decide opposite of the previous decision.

Decoded bipolar sequence $\{\hat{x}_k\}$:	−1	+1	−1	+1	+1	−1
Decoded binary sequence $\{\hat{x}_k\}$:	0	1	0	1	1	0

The decision rule simply implements the subtraction of each $\hat{x}_{k-1}$ decision from each $\hat{y}_k$. One drawback of this detection technique is that once an error is made, it tends to propagate, causing further errors, since present decisions depend on prior decisions. A means of avoiding this error propagation is known as *precoding.*

2.9.3 Precoding

Precoding is accomplished by first differentially encoding the $\{x_k\}$ binary sequence into a new $\{w_k\}$ binary sequence by means of the equation:

$$w_k = x_k \oplus w_{k-1} \qquad (2.30)$$

where the symbol $\oplus$ represents modulo-2 addition (equivalent to the logical *exclusive-or* operation) of the binary digits. The rules of modulo-2 addition are as follows:

$$0 \oplus 0 = 0$$
$$0 \oplus 1 = 1$$
$$1 \oplus 0 = 1$$
$$1 \oplus 1 = 0$$

The $\{w_k\}$ binary sequence is then converted to a bipolar pulse sequence, and the coding operation proceeds in the same way as it did in Example 2.4. However, with precoding, the detection process is quite different from the detection of ordinary duobinary, as shown below in Example 2.5: The precoding model is shown in Figure 2.26; in this figure it is implicit that the modulo-2 addition producing the precoded $\{w_k\}$ sequence is performed on the *binary* digits, while the digital filtering producing the $\{y_k\}$ sequence is performed on the *bipolar* pulses.

Example 2.5   Duobinary Precoding

Illustrate the duobinary coding and decoding rules when using the differential precoding of Equation (2.30). Assume the same $\{x_k\}$ sequence as that given in Example 2.4.

Solution

Binary digit sequence $\{x_k\}$:	0	0	1	0	1	1	0
Precoded sequence $w_k = x_k \oplus w_{k-1}$:	0	0	1	1	0	1	1
Bipolar sequence $\{w_k\}$:	−1	−1	+1	+1	−1	+1	+1
Coding rule: $y_k = w_k + w_{k-1}$:		−2	0	+2	0	0	+2

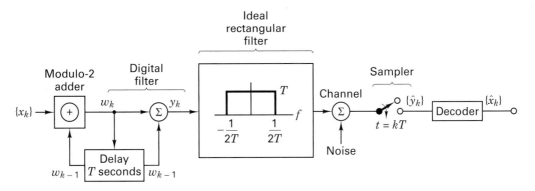

Figure 2.26 Precoded duobinary signaling.

Decoding decision rule:

If $\hat{y}_k = \pm 2$, decide that $\hat{x}_k$ = binary zero.
If $\hat{y}_k = 0$, decide that $\hat{x}_k$ = binary one.

Decoded binary sequence $\{x_k\}$:

0   1   0   1   1   0

The differential precoding enables us to decode the $\{\hat{y}_k\}$ sequence by making a decision on each received sample singly, without resorting to prior decisions that could be in error. The major advantage is that in the event of a digit error due to noise, such an error does not propagate to other digits. Notice that the first bit in the differentially precoded binary sequence $\{w_k\}$ is an arbitrary choice. If the startup bit in $\{w_k\}$ had been chosen to be a binary one instead of a binary zero, the decoded result would have been the same.

2.9.4 Duobinary Equivalent Transfer Function

In Section 2.9.1, we described the duobinary transfer function as a digital filter incorporating a one-digit delay followed by an ideal rectangular transfer function. Let us now examine an equivalent model. The Fourier transform of a delay can be described as $e^{-j2\pi fT}$ (see Section A.3.1); therefore, the input digital filter of Figure 2.25 can be characterized as the frequency transfer function

$$H_1(f) = 1 + e^{-j2\pi fT} \tag{2.31}$$

The transfer function of the ideal rectangular filter, is

$$H_2(f) = \begin{cases} T & \text{for } |f| < \dfrac{1}{2T} \\ 0 & \text{elsewhere} \end{cases} \tag{2.32}$$

The overall equivalent transfer function of the digital filter cascaded with the ideal rectangular filter is then given by

$$H_e(f) = H_1(f)H_2(f) \qquad \text{for } |f| < \frac{1}{2T}$$

$$= (1 + e^{-j2\pi fT})T \tag{2.33}$$

$$= T(e^{j\pi fT} + e^{-j\pi fT})e^{-j\pi fT}$$

so that

$$|H_e(f)| = \begin{cases} 2T\cos \pi fT & \text{for } |f| < \dfrac{1}{2T} \\ 0 & \text{elsewhere} \end{cases} \tag{2.34}$$

Thus $H_e(f)$, the composite transfer function for the cascaded digital and rectangular filters, has a gradual roll-off to the band edge, as can be seen in Figure 2.27a. The transfer function can be approximated by using realizable analog filtering; a separate digital filter is not needed. The duobinary equivalent $H_e(f)$ is called a *cosine filter* [8]. The cosine filter should not be confused with the *raised* cosine filter (described in Chapter 3, Section 3.3.1). The corresponding impulse response $h_e(t)$, found by taking the inverse Fourier transform of $H_e(f)$ in Equation (2.33) is

$$h_e(t) = \text{sinc}\left(\frac{t}{T}\right) + \text{sinc}\left(\frac{t-T}{T}\right) \tag{2.35}$$

and is plotted in Figure 2.27b. For every impulse $\delta(t)$ at the input of Figure 2.25, the output is $h_e(t)$ with an appropriate polarity. Notice that there are only two nonzero samples at T-second intervals, giving rise to controlled ISI from the adjacent bit. The introduced ISI is eliminated by use of the decoding procedure discussed in Section 2.9.2. Although the cosine filter is noncausal and therefore nonrealizable, it can be easily approximated. The implementation of the precoded duobinary technique described in Section 2.9.3 can be accomplished by first differentially encoding the binary sequence $\{x_k\}$ into the sequence $\{w_k\}$ (see Example 2.5). The pulse sequence $\{w_k\}$ is then filtered by the equivalent cosine characteristic described in Equation (2.34).

2.9.5 Comparison of Binary with Duobinary Signaling

The duobinary technique introduces correlation between pulse amplitudes, whereas the more restrictive Nyquist criterion assumes that the transmitted pulse amplitudes are independent of one another. We have shown that duobinary signaling can exploit this introduced correlation to achieve zero ISI signal transmission, using a smaller system bandwidth than is otherwise possible. Do we get this performance improvement without paying a price? No, such is rarely the case with engineering design options—there is almost always a trade-off involved. We saw that duobinary coding requires three levels, compared with the usual two levels for binary coding. Recall our discussion in Section 2.8.5, where we compared the performance and the required signal power for making eight-level PAM decisions versus two-level (PCM) decisions. For a fixed amount of signal power, the ease of making reliable decisions is inversely related to the number of levels that must be

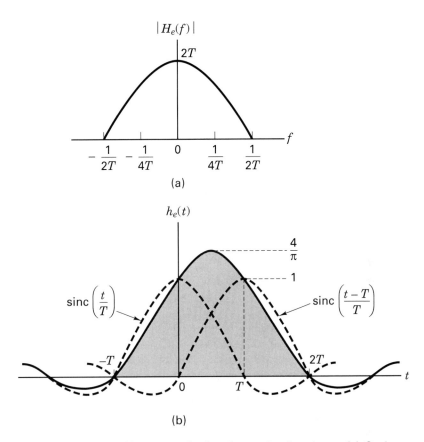

Figure 2.27 Duobinary transfer function and pulse shape. (a) Cosine filter. (b) Impulse response of the cosine filter.

distinguished in each waveform. Therefore, it should be no surprise that although duobinary signaling accomplishes the zero ISI requirement with minimum bandwidth, duobinary signaling also requires more power than binary signaling, for equivalent performance against noise. For a given probability of bit error (P_B), duobinary signaling requires approximately 2.5 dB greater SNR than binary signaling, while using only $1/(1 + r)$ the bandwidth that binary signaling requires [7], where r is the filter roll-off.

2.9.6 Polybinary Signaling

Duobinary signaling can be extended to more than three digits or levels, resulting in greater bandwidth efficiency; such systems are called *polybinary* [7, 9]. Consider that a binary message with two signaling levels is transformed into a signal with j signaling levels numbered consecutively from zero to $(j - 1)$. The transformation from binary to polybinary takes place in two steps. First, the original sequence $\{x_k\}$, consisting of binary ones and zeros, is converted into another binary sequence $\{y_k\}$, as follows. The present binary digit of sequence $\{y_k\}$ is formed from the modulo-2 addition of the $(j - 2)$ immediately preceding digits of sequence $\{y_k\}$ and the present digit x_k. For example, let

$$y_k = x_k \oplus y_{k-1} \oplus y_{k-2} \oplus y_{k-3} \tag{2.36}$$

Here x_k represents the input binary digit and y_k the kth encoded digit. Since the expression involves $(j - 2) = 3$ bits preceding y_k, there are $j = 5$ signaling levels. Next, the binary sequence $\{y_k\}$ is transformed into a polybinary pulse train $\{z_k\}$ by adding *algebraically* the present bit of sequence $\{y_k\}$ to the $(j - 2)$ preceding bits of $\{y_k\}$. Therefore, z_k modulo-2 $= x_k$, and the binary elements one and zero are mapped into even- and odd-valued pulses in the sequence $\{z_k\}$. Note that each digit in $\{z_k\}$ can be independently detected despite the strong correlation between bits. The primary advantage of such a signaling scheme is the redistribution of the spectral density of the original sequence $\{x_k\}$, so as to favor the low frequencies, thus improving system bandwidth efficiency.

2.10 CONCLUSION

In this chapter we have considered the first important step in any digital communication system, transforming the source information (both textual and analog) to a form that is compatible with a digital system. We treated various aspects of sampling, quantization (both uniform and nonuniform), and pulse code modulation (PCM). We considered the selection of pulse waveforms for the transmission of baseband signals through the channel. We also introduced the duobinary concept of adding a controlled amount of ISI to achieve an improvement in bandwidth efficiency at the expense of an increase in power.

REFERENCES

1. Black, H. S., *Modulation Theory,* D. Van Nostrand Company, Princeton, N.J., 1953.
2. Oppenheim, A. V., *Applications of Digital Signal Processing,* Prentice-Hall, Inc., Englewood Cliffs, N.J., 1978.
3. Stiltz, H., ed., *Aerospace Telemetry,* Vol. 1, Prentice-Hall, Inc., Englewood Cliffs, N.J., 1961, p. 179.
4. Hecht, M., and Guida, A., "Delay Modulation," *Proc. IEEE,* vol. 57, no. 7, July 1969, pp. 1314–1316.
5. Deffebach, H. L., and Frost, W. O., "A Survey of Digital Baseband Signaling Techniques," *NASA Technical Memorandum NASA TM X-64615,* June 30, 1971.
6. Lender, A., "The Duobinary Technique for High Speed Data Transmission," *IEEE Trans. Commun. Electron.,* vol. 82, May 1963, pp. 214–218.
7. Lender, A., "Correlative (Partial Response) Techniques and Applications to Digital Radio Systems," in K. Feher, *Digital Communications: Microwave Applications,* Prentice-Hall, Inc., Englewood Cliffs, N.J., 1981, Chap. 7.

8. Couch, L. W., II, *Digital and Analog Communication Systems,* Macmillan Publishing Company, New York, 1982.

9. Lender, A., "Correlative Digital Communication Techniques," *IEEE Trans. Commun. Technol.,* Dec. 1964, pp. 128–135.

PROBLEMS

2.1. You want to transmit the word "HOW" using an 8-ary system.

   (a) Encode the word "HOW" into a sequence of bits, using 7-bit ASCII coding, followed by an eighth bit for error detection, per character. The eighth bit is chosen so that the number of ones in the 8 bits is an even number. How many total bits are there in the message?

   (b) Partition the bit stream into $k = 3$ bit segments. Represent each of the 3-bit segments as an octal number (symbol). How many octal symbols are there in the message?

   (c) If the system were designed with 16-ary modulation, how many symbols would be used to represent the word "HOW"?

   (d) If the system were designed with 256-ary modulation, how many symbols would be used to represent the word "HOW"?

2.2. We want to transmit 800 characters/s, where each character is represented by its 7-bit ASCII codeword, followed by an eighth bit for error detection, per character, as in Problem 2.1. A multilevel PAM waveform with $M = 16$ levels is used.

   (a) What is the effective transmitted bit rate?

   (b) What is the symbol rate?

2.3. We wish to transmit a 100-character alphanumeric message in 2 s, using 7-bit ASCII coding, followed by an eighth bit for error detection, per character, as in Problem 2.1. A multilevel PAM waveform with $M = 32$ levels is used.

   (a) Calculate the effective transmitted bit rate and the symbol rate.

   (b) Repeat part (a) for 16-level PAM, eight-level PAM, four-level PAM, and PCM (binary) waveforms.

2.4. Given an analog waveform that has been sampled at its Nyquist rate, f_s, using natural sampling, prove that a waveform (proportional to the original waveform) can be recovered from the samples, using the recovery techniques shown in Figure P2.1. The parameter mf_s is the frequency of the local oscillator, where m is an integer.

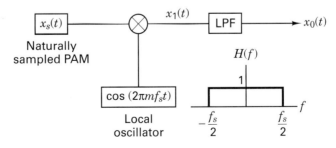

Figure P2.1

2.5. An analog signal is sampled at its Nyquist rate $1/T_s$, and quantized using L quantization levels. The derived digital signal is then transmitted on some channel.

(a) Show that the time duration, T, of one bit of the transmitted binary encoded signal must satisfy $T \le T_s/(\log_2 L)$.

(b) When is the equality sign valid?

2.6. Determine the number of quantization levels that are implied if the number of bits per sample in a given PCM code is (a) 5; (b) 8; (c) x.

2.7. Determine the minimum sampling rate necessary to sample and perfectly reconstruct the signal $x(t) = \sin{(6280t)}/(6280t)$.

2.8. Consider an audio signal with spectral components limited to the frequency band 300 to 3300 Hz. Assume that a sampling rate of 8000 samples/s will be used to generate a PCM signal. Assume that the ratio of peak signal power to average quantization noise power at the output needs to be 30 dB.

(a) What is the minimum number of uniform quantization levels needed, and what is the minimum number of bits per sample needed?

(b) Calculate the system bandwidth (as specified by the main spectral lobe of the signal) required for the detection of such a PCM signal.

2.9. A waveform, $x(t) = 10 \cos{(1000t + \pi/3)} + 20 \cos{(2000t + \pi/6)}$ is to be uniformly sampled for digital transmission.

(a) What is the maximum allowable time interval between sample values that will ensure perfect signal reproduction?

(b) If we want to reproduce 1 hour of this waveform, how many sample values need to be stored?

2.10. (a) A waveform that is bandlimited to 50 kHz is sampled every 10 μs. Show graphically that these samples uniquely characterize the waveform. (Use a sinusoidal example for simplicity. Avoid sampling at points where the waveform equals zero.)

(b) If samples are taken 30 μs apart instead of 10 μs, show graphically that waveforms other than the original can be characterized by the samples.

2.11. Use the method of convolution to illustrate the effect of undersampling the waveform $x(t) = \cos{2\pi f_0 t}$ for a sampling rate of $f_s = \frac{3}{2} f_0$.

2.12. Aliasing will not occur if the sampling rate is greater than twice the signal bandwidth. However, perfectly bandlimited signals do not occur in nature. Hence, there is always some aliasing present.

(a) Suppose that a filtered signal has a spectrum described by a Butterworth filter with order $n = 6$, and upper cutoff frequency $f_u = 1000$ Hz. What sampling rate is required so that aliasing is reduced to the −50 dB point in the power spectrum?

(b) Repeat for a Butterworth filter with order $n = 12$.

213. (a) Sketch the complete $\mu = 10$ compression characteristic that will handle input voltages in the range −5 to +5 V.

(b) Plot the corresponding expansion characteristic.

(c) Draw a 16-level nonuniform quantizer characteristic that corresponds to the $\mu = 10$ compression characteristic.

2.14. The information in an analog waveform, whose maximum frequency $f_m = 4000$ Hz, is to be transmitted using a 16-level PAM system. The quantization distortion must not exceed ±1% of the peak-to-peak analog signal.

(a) What is the minimum number of bits per sample or bits per PCM word that should be used in this PAM transmission system?

(b) What is the minimum required sampling rate, and what is the resulting bit rate?

(c) What is the 16-ary PAM symbol transmission rate?

2.15. A signal in the frequency range 300 to 3300 Hz is limited to a peak-to-peak swing of 10 V. It is sampled at 8000 samples/s and the samples are quantized to 64 evenly spaced levels. Calculate and compare the bandwidths and ratio of peak signal power to rms quantization noise if the quantized samples are transmitted either as binary pulses or as four-level pulses. Assume that the system bandwidth is defined by the main spectral lobe of the signal.

2.16. In the compact disc (CD) digital audio system, an analog signal is digitized so that the ratio of the peak-signal power to the peak-quantization noise power is at least 96 dB. The sampling rate is 44.1 kilosamples/s.
   (a) How many quantization levels of the analog signal are needed for $(S/N_q)_{peak} = $ 96 dB?
   (b) How many bits per sample are needed for the number of levels found in part (a)?
   (c) What is the data rate in bits/s?

2.17. Calculate the difference in required signal power between two PCM waveforms, unipolar RZ and bipolar RZ (see Figure 2.22), assuming that each signaling scheme has the same requirements for data-rate and bit-error probability. Also assume equally likely signaling, and that the difference between the high-voltage and low-voltage levels is the same for both schemes. If there is a power advantage in using one of the signaling schemes, what, if any, is the disadvantage in using it?

2.18. In the year 1962, AT&T first offered digital telephone transmission referred to as T1 service. With this service, each T1 frame is partitioned into 24 channels or time slots. Each time slot contains 8 bits (one speech sample), and there is one additional bit per frame for alignment. The frame is sampled at the Nyquist rate of 8000 samples/s, and the bandwidth used for transmitting the composite signal is 386 kHz. Find the bandwidth efficiency (bits/s/Hz) for this signaling scheme.

2.19. **(a)** Consider that you desire a digital transmission system, such that the quantization distortion of any audio source does not exceed ± 2% of the peak-to-peak analog signal voltage. If the audio signal bandwidth and the allowable transmission bandwidth are each 4000 Hz, and sampling takes place at the Nyquist rate, what value of bandwidth efficiency (bits/s/Hz) is required?
   (b) Repeat part (a) except that the audio signal bandwidth is 20 kHz (high fidelity), yet the available transmission bandwidth is still 4000 Hz.

QUESTIONS

2.1 What are the similarities and differences between the terms "formatting" and "source coding"? (See Chapter 2, introduction.)

2.2 In the process of *formatting* information, why is it often desirable to perform *over-sampling*? (See Section 2.4.3.)

2.3 In using pulse code modulation (PCM) for digitizing analog information, explain how the parameters *fidelity, bandwidth,* and *time delay* can be traded off. (See Section 2.6.)

2.4 Why is it often preferred to use units of normalized bandwidth, *WT* (or time-bandwidth product), compared with bandwidth alone? (See Section 2.8.3.)

EXERCISES

Using the Companion CD, run the exercises associated with Chapter 2.

CHAPTER 3

Baseband
Demodulation/Detection

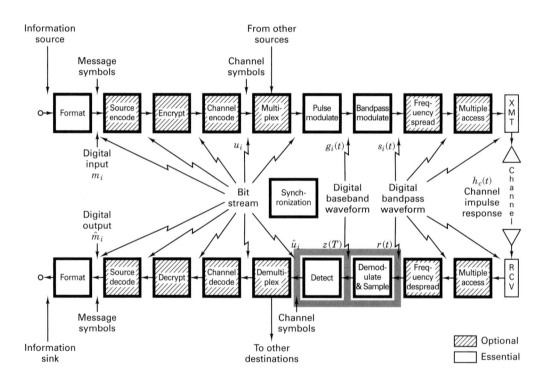

In the case of baseband signaling, the received waveforms are already in a pulse-like form. One might ask, why then, is a demodulator needed to recover the pulse waveforms? The answer is that the arriving baseband pulses are not in the form of ideal pulse shapes, each one occupying its own symbol interval. The filtering at the transmitter and the channel typically cause the received pulse sequence to suffer from intersymbol interference (ISI) and thus appear as an amorphous "smeared" signal, not quite ready for sampling and detection. The goal of the demodulator (receiving filter) is to recover a baseband pulse with the best possible signal-to-noise ration (SNR), free of any ISI. Equalization, covered in this chapter, is a technique used to help accomplish this goal. The equalization process is not required for every type of communication channel. However, since equalization embodies a sophisticated set of signal-processing techniques, making it possible to compensate for channel-induced interference, it is an important area for many systems.

The bandpass model of the detection process, covered in Chapter 4, is virtually identical to the baseband model considered in this chapter. That is because a received bandpass waveform is first transformed to a baseband waveform before the final detection step takes place. For linear systems, the mathematics of detection is unaffected by a shift in frequency. In fact, we can define an *equivalence theorem* as follows: Performing bandpass linear signal processing followed by heterodyning the signal to baseband, yields the same results as heterodyning the bandpass signal to baseband, followed by baseband linear signal processing. The term "heterodyning" refers to a frequency *conversion* or *mixing* process that yields a spectral shift in the signal. As a result of this equivalence theorem, all linear signal-processing simulations can take place at baseband (which is preferred for simplicity) with the same results as at bandpass. This means that the performance of most digital communication systems will often be described and analyzed as if the transmission channel is a baseband channel.

3.1 SIGNALS AND NOISE

3.1.1 Error-Performance Degradation in Communication Systems

The task of the detector is to retrieve the bit stream from the received waveform, as error free as possible, notwithstanding the impairments to which the signal may have been subjected. There are two primary causes for error-performance degradation. The first is the effect of filtering at the transmitter, channel, and receiver, discussed in Section 3.3, below. As described there, a nonideal system transfer function causes symbol "smearing" or *intersymbol interference* (ISI).

Another cause for error-performance degradation is electrical noise and interference produced by a variety of sources, such as galaxy and atmospheric noise, switching transients, intermodulation noise, as well as interfering signals from other sources. (These are discussed in Chapter 5.) With proper precautions, much of the noise and interference entering a receiver can be reduced in intensity or even eliminated. However, there is one noise source that cannot be eliminated, and that is the noise caused by the thermal motion of electrons in any conducting media. This motion produces *thermal noise* in amplifiers and circuits, and corrupts the signal in an additive fashion. The statistics of thermal noise have been developed using quantum mechanics, and are well known [1].

The primary statistical characteristic of thermal noise is that the noise amplitudes are distributed according to a normal or Gaussian distribution, discussed in Section 1.5.5, and shown in Figure 1.7. In this figure, it can be seen that the most probable noise amplitudes are those with small positive or negative values. In theory, the noise can be infinitely large, but very large noise amplitudes are rare. The primary spectral characteristic of thermal noise in communication systems, is that its two-sided power spectral density $G_n(f) = N_0/2$ is flat for all frequencies of interest. In other words, the thermal noise, on the average, has just as much power per hertz in low-frequency fluctuations as in high-frequency fluctuations—up to a frequency of about 10^{12} hertz. When the noise power is characterized by such a constant-power spectral density, we refer to it as *white noise*. Since thermal noise is present in all communication systems and is the predominant noise source for many systems, the thermal noise characteristics (additive, white, and Gaussian, giving rise to the name AWGN) are most often used to model the noise in the detection process and in the design of receivers. Whenever a channel is designated as an AWGN channel (with no other impairments specified), we are in effect being told that its impairments are limited to the degradation caused by this unavoidable thermal noise.

3.1.2 Demodulation and Detection

During a given signaling interval T, a binary baseband system will transmit one of two waveforms, denoted $g_1(t)$ and $g_2(t)$. Similarly, a binary bandpass system will transmit one of two waveforms, denoted $s_1(t)$ and $s_2(t)$. Since the general treatment of demodulation and detection are essentially the same for baseband and bandpass systems, we use $s_i(t)$ here as a generic designation for a transmitted waveform, whether the system is baseband or bandpass. This allows much of the baseband demodulation/detection treatment in this chapter to be consistent with similar bandpass descriptions in Chapter 4. Then, for any binary channel, the transmitted signal over a symbol interval $(0, T)$ is represented by

$$s_i(t) = \begin{cases} s_1(t) & 0 \leq t \leq T & \text{for a binary 1} \\ s_2(t) & 0 \leq t \leq T & \text{for a binary 0} \end{cases}$$

The received signal $r(t)$ degraded by noise $n(t)$ and possibly degraded by the impulse response of the channel $h_c(t)$ was described in Equation (1.1) and is re-written as

$$r(t) = s_i(t) * h_c(t) + n(t) \qquad i = 1, \dots, M \tag{3.1}$$

where $n(t)$ is here assumed to be a zero mean AWGN process, and $*$ represents a convolution operation. For binary transmission over an ideal distortionless channel where convolution with $h_c(t)$ produces no degradation (since for the ideal case $h_c(t)$ is an impulse function), the representation of $r(t)$ can be simplified to

$$r(t) = s_i(t) + n(t) \qquad i = 1, 2, \qquad 0 \leq t \leq T \tag{3.2}$$

Figure 3.1 shows the typical demodulation and detection functions of a digital receiver. Some authors use the terms "demodulation" and "detection" interchangeably. This book makes a distinction between the two. We define *demodulation* as recovery of a waveform (to an undistorted baseband pulse), and we designate *detection* to mean the decision-making process of selecting the digital meaning of that waveform. If error-correction coding is *not* present, the detector output consists of estimates of message symbols (or bits), $\hat{m}_i$ (also called *hard decisions*). If error-correction coding is used, the detector output consists of estimates of channel symbols (or coded bits) $\hat{u}_i$, which can take the form of *hard* or *soft decisions* (see Section 7.3.2). For brevity, the term "detection" is occasionally used loosely to encompass all the receiver signal-processing steps through the decision making step. The *frequency down-conversion* block, shown in the demodulator portion of Figure 3.1, performs frequency translation for bandpass signals operating at some radio frequency (RF). This function may be configured in a variety of ways. It may take place within the front end of the receiver, within the demodulator, shared between the two locations, or not at all.

Within the *demodulate* and *sample* block of Figure 3.1 is the *receiving filter* (essentially the demodulator), which performs waveform recovery in preparation for the next important step—detection. The filtering at the transmitter and the channel typically cause the received pulse sequence to suffer from ISI, and thus it is

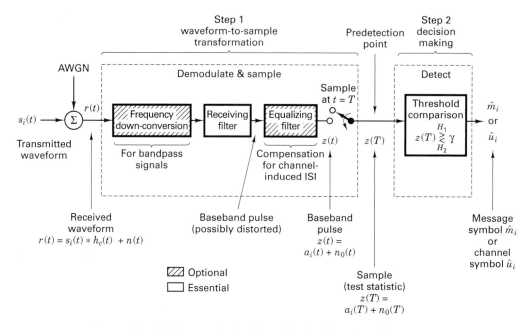

Figure 3.1 Two basic steps in the demodulation/detection of digital signals.

not quite ready for sampling and detection. The goal of the receiving filter is to re-cover a baseband pulse with the best possible signal-to-noise ratio (SNR), free of any ISI. The optimum receiving filter for accomplishing this is called a *matched filter or correlator,* described in Sections 3.2.2 and 3.2.3. An optional *equalizing filter* follows the receiving filter; it is only needed for those systems where channel-induced ISI can distort the signals. The receiving filter and equalizing filter are shown as two separate blocks in order to emphasize their separate functions. In most cases, however, when an equalizer is used, a single filter would be designed to incorporate both functions and thereby compensate for the distortion caused by both the transmitter and the channel. Such a composite filter is sometimes referred to simply as the *equalizing filter* or the *receiving and equalizing filter.*

Figure 3.1 highlights two steps in the demodulation/detection process. Step 1, the waveform-to-sample transformation, is made up of the demodulator followed by a sampler. At the end of each symbol duration T, the output of the sampler, the *predetection point,* yields a sample $z(T)$, sometimes called the test statistic. $z(T)$ has a voltage value directly proportional to the energy of the received symbol and that of the noise. In step 2, a decision (detection) is made regarding the digital meaning of that sample. We assume that the input noise is a Gaussian random process and that the receiving filter in the demodulator is linear. A linear operation performed on a Gaussian random process will produce a second Gaussian random process [2]. Thus, the filter output noise is Gaussian. The output of step 1 yields the test statistic

$$z(T) = a_i(T) + n_0(T) \qquad i = 1, 2 \tag{3.3}$$

where $a_i(T)$ is the desired signal component, and $n_0(T)$ is the noise component. To simplify the notation, we sometimes express Equation (3.3) in the form of $z = a_i + n_0$. The noise component n_0 is a zero mean Gaussian random variable, and thus $z(T)$ is a Gaussian random variable with a mean of either a_1 or a_2 depending on whether a binary one or binary zero was sent. As described in Section 1.5.5, the probability density function (pdf) of the Gaussian random noise n_0 can be expressed as

$$p(n_0) = \frac{1}{\sigma_0 \sqrt{2\pi}} \exp\left[-\frac{1}{2}\left(\frac{n_0}{\sigma_0} \right)^2 \right] \tag{3.4}$$

where σ_0^2 is the noise variance. Thus it follows from Equations (3.3) and (3.4) that the conditional pdfs $p(z|s_1)$ and $p(z|s_2)$ can be expressed as

$$p(z|s_1) = \frac{1}{\sigma_0 \sqrt{2\pi}} \exp\left[-\frac{1}{2}\left(\frac{z - a_1}{\sigma_0} \right)^2 \right] \tag{3.5}$$

and

$$p(z|s_2) = \frac{1}{\sigma_0 \sqrt{2\pi}} \exp\left[-\frac{1}{2}\left(\frac{z - a_2}{\sigma_0} \right)^2 \right] \tag{3.6}$$

These conditional pdfs are illustrated in Figure 3.2. The rightmost conditional pdf, $p(z|s_1)$, called the *likelihood* of s_1, illustrates the probability density function of the random variable $z(T)$, given that symbol s_1 was transmitted. Similarly, the leftmost conditional pdf, $p(z|s_2)$, called the *likelihood* of s_2, illustrates the pdf of $z(T)$, given that symbol s_2 was transmitted. The abscissa, $z(T)$, represents the full range of possible sample output values from step 1 of Figure 3.1.

After a received waveform has been transformed to a sample, the actual shape of the waveform is no longer important; all waveform types that are transformed to the same value of $z(T)$ are identical for detection purposes. Later it is shown that an optimum receiving filter (matched filter) in step 1 of Figure 3.1 maps all signals of equal energy into the same point $z(T)$. Therefore, the received *signal energy* (not its shape) is the important parameter in the detection process. This is why the detection analysis for baseband signals is the same as that for bandpass sig-

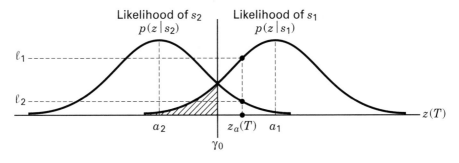

Figure 3.2   Conditional probability density functions: $p(z|s_1)$ and $p(z|s_2)$.

nals. Since $z(T)$ is a voltage signal that is proportional to the energy of the received symbol, the larger the magnitude of $z(T)$, the more error free will be the decision-making process. In step 2, detection is performed by choosing the hypothesis that results from the threshold measurement

$$z(T) \underset{H_2}{\overset{H_1}{\gtrless}} \gamma \qquad (3.7)$$

where H_1 and H_2 are the two possible (binary) hypotheses. The inequality relationship indicates that hypothesis H_1 is chosen if $z(T) > \gamma$, and hypothesis H_2 is chosen if $z(T) < \gamma$. If $z(T) = \gamma$, the decision can be an arbitrary one. Choosing H_1 is equivalent to deciding that signal $s_1(t)$ was sent and hence a binary 1 is detected. Similarly, choosing H_2 is equivalent to deciding that signal $s_2(t)$ was sent, and hence a binary 0 is detected.

3.1.3 A Vectorial View of Signals and Noise

We now present a geometric or vectorial view of signal waveforms that are useful for either baseband or bandpass signals. We define an N-dimensional *orthogonal space* as a space characterized by a set of N linearly independent functions $\{\psi_j(t)\}$, called *basis functions*. Any arbitrary function in the space can be generated by a linear combination of these basis functions. The basis functions must satisfy the conditions

$$\int_0^T \psi_j(t)\psi_k(t)\, dt = K_j \delta_{jk} \qquad 0 \leq t \leq T \quad j, k = 1, \ldots, N \qquad (3.8a)$$

where the operator

$$\delta_{jk} = \begin{cases} 1 & \text{for } j = k \\ 0 & \text{otherwise} \end{cases} \qquad (3.8b)$$

is called the *Kronecker delta function* and is defined by Equation (3.8b). When the K_j constants are nonzero, the signal space is called *orthogonal*. When the basis functions are normalized so that each $K_j = 1$, the space is called an *orthonormal* space. The principal requirement for orthogonality can be stated as follows. Each $\psi_j(t)$ function of the set of basis functions must be independent of the other members of the set. Each $\psi_j(t)$ must not interfere with any other members of the set in the detection process. From a geometric point of view, each $\psi_j(t)$ is mutually perpendicular to each of the other $\psi_k(t)$ for $j \neq k$. An example of such a space with $N = 3$ is shown in Figure 3.3, where the mutually perpendicular axes are designated $\psi_1(t)$, $\psi_2(t)$, and $\psi_3(t)$. If $\psi_j(t)$ corresponds to a real-valued voltage or current waveform component, associated with a 1-Ω resistive load, then using Equations (1.5) and (3.8), the normalized energy in joules dissipated in the load in T seconds, due to ψ_j, is

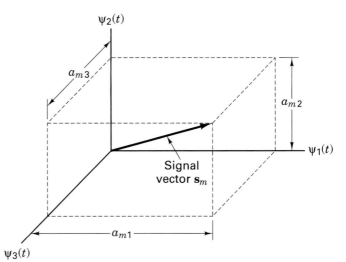

Figure 3.3 Vectorial representation of the signal waveform $s_m(t)$.

Figure 3.3 Vectorial representation of the signal waveform $s_m(t)$.

$$E_j = \int_0^T \psi_j^2(t)\, dt = K_j \tag{3.9}$$

One reason we focus on an *orthogonal signal space* is that Euclidean distance measurements, fundamental to the detection process, are easily formulated in such a space. However, even if the signaling waveforms do not make up such an orthogonal set, they can be transformed into linear combinations of orthogonal waveforms. It can be shown [3] that *any arbitrary* finite set of waveforms $\{s_i(t)\}$ $(i = 1, \ldots, M)$, where each member of the set is physically realizable and of duration T, can be expressed as a linear combination of N orthogonal waveforms $\psi_1(t)$, $\psi_2(t), \ldots, \psi_N(t)$, where $N \leq M$, such that

$$s_1(t) = a_{11}\psi_1(t) + a_{12}\psi_2(t) + \cdots + a_{1N}\psi_N(t)$$
$$s_2(t) = a_{21}\psi_1(t) + a_{22}\psi_2(t) + \cdots + a_{2N}\psi_N(t)$$
$$\vdots \qquad\qquad\qquad\qquad\qquad \vdots$$
$$s_M(t) = a_{M1}\psi_1(t) + a_{M2}\psi_2(t) + \cdots + a_{MN}\psi_N(t)$$

These relationships are expressed in more compact notation as

$$s_i(t) = \sum_{j=1}^{N} a_{ij}\, \psi_j(t) \quad i = 1, \ldots, M \tag{3.10}$$
$$N \leq M$$

where

$$a_{ij} = \frac{1}{K_j} \int_0^T s_i(t)\psi_j(t)\, dt \quad \begin{aligned} i &= 1, \ldots, M \quad 0 \leq t \leq T \\ j &= 1, \ldots, N \end{aligned} \tag{3.11}$$

The coefficient a_{ij} is the value of the $\psi_j(t)$ component of signal $s_i(t)$. The form of the $\{\psi_j(t)\}$ is not specified; it is chosen for convenience and will depend on the form of the signal waveforms. The set of signal waveforms, $\{s_i(t)\}$, can be viewed as a set of vectors, $\{s_i\} = \{a_{i1}, a_{i2}, \ldots, a_{iN}\}$. If, for example, $N = 3$, we may plot the vector s_m corresponding to the waveform

$$s_m(t) = a_{m1}\psi_1(t) + a_{m2}\psi_2(t) + a_{m3}\psi_3(t)$$

as a point in a three-dimensional Euclidean space with coordinates (a_{m1}, a_{m2}, a_{m3}), as shown in Figure 3.3. The orientation among the signal vectors describes the relation of the signals to one another (with respect to phase or frequency), and the amplitude of each vector in the set $\{s_i\}$ is a measure of the signal energy transmitted during a symbol duration. In general, once a set of N orthogonal functions has been adopted, each of the transmitted signal waveforms, $s_i(t)$, is completely determined by the vector of its coefficients,

$$s_i = (a_{i1}, a_{i2}, \ldots, a_{iN}) \qquad i = 1, \ldots, M \tag{3.12}$$

We shall employ the notation of signal vectors, $\{s\}$, or signal waveforms, $\{s(t)\}$, as best suits the discussion. A typical detection problem, conveniently viewed in terms of signal vectors, is illustrated in Figure 3.4. Vectors s_j and s_k represent *prototype* or *reference signals* belonging to the set of M waveforms, $\{s_i(t)\}$. The receiver knows, a priori, the location in the signal space of each prototype vector belonging to the M-ary set. During the transmission of any signal, the signal is perturbed by noise so that the resultant vector that is actually received is a perturbed version (e.g., $s_j + n$ or $s_k + n$) of the original one, where n represents a noise vector. The noise is additive and has a Gaussian distribution; therefore, the resulting distribution of possible received signals is a cluster or cloud of points around s_j and s_k. The cluster is dense in the center and becomes sparse with increasing distance from the prototype. The arrow marked "r" represents a signal vector that might arrive at the receiver during some symbol interval. The task of the receiver is to decide whether r has a close "resem-

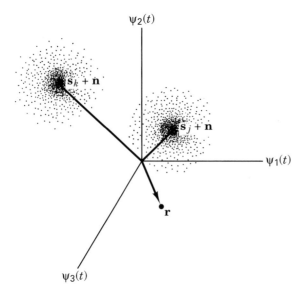

Figure 3.4 Signals and noise in a three-dimensional vector space.

Baseband Demodulation/Detection   Chap. 3

blance" to the prototype $\mathbf{s}_j$, whether it more closely resembles $\mathbf{s}_k$, or whether it is closer to some other prototype signal in the *M*-ary set. The measurement can be thought of as a *distance* measurement. The receiver or detector must decide which of the prototypes within the signal space is *closest* in distance to the received vector $\mathbf{r}$. The analysis of all demodulation or detection schemes involves this concept of *distance* between a received waveform and a set of possible transmitted waveforms. A simple rule for the detector to follow is to decide that $\mathbf{r}$ belongs to the same class as its nearest neighbor (nearest prototype vector).

3.1.3.1 Waveform Energy

Using Equations (1.5), (3.10), and (3.8), the normalized energy E_i, associated with the waveform $s_i(t)$ over a symbol interval T can be expressed in terms of the orthogonal components of $s_i(t)$ as follows:

$$E_i = \int_0^T s_i^2(t)\, dt = \int_0^T \left[\sum_j a_{ij}\, \psi_j(t) \right]^2 dt \tag{3.13}$$

$$= \int_0^T \sum_j a_{ij}\, \psi_j(t) \sum_k a_{ik} \psi_k(t)\, dt \tag{3.14}$$

$$= \sum_j \sum_k a_{ij}\, a_{ik} \int_0^T \psi_j(t)\, \psi_k(t)\, dt \tag{3.15}$$

$$= \sum_j \sum_k a_{ij}\, a_{ik} K_j \delta_{jk} \tag{3.16}$$

$$= \sum_{j=1}^N a_{ij}^2 K_j \qquad i = 1, \dots, M \tag{3.17}$$

Equation (3.17) is a special case of Parseval's theorem relating the integral of the square of the waveform $s_i(t)$ to the sum of the square of the orthogonal series coefficients. If orthonormal functions are used (i.e., $K_j = 1$), the normalized energy over a symbol duration T is given by

$$E_i = \sum_{j=1}^N a_{ij}^2 \tag{3.18}$$

If there is equal energy E in each of the $s_i(t)$ waveforms, we can write Equation (3.18) in the form

$$E = \sum_{j=1}^N a_{ij}^2 \qquad \text{for all } i \tag{3.19}$$

3.1.3.2 Generalized Fourier Transforms

The transformation described by Equations (3.8), (3.10), and (3.11) is referred to as the *generalized Fourier transformation*. In the case of ordinary Fourier transforms, the $\{\psi_j(t)\}$ set comprises sine and cosine harmonic functions. But in the case of generalized Fourier transforms, the $\{\psi_j(t)\}$ set is not constrained to any specific form; it must only satisfy the orthogonality statement of Equation (3.8). *Any arbitrary integrable waveform set, as well as noise, can be represented as a linear combination of orthogonal waveforms* through such a generalized Fourier transformation [3]. Therefore, in such an orthogonal space, we are justified in using distance (Euclidean distance) as a decision criterion for the detection of *any* signal set in the presence of AWGN. The most important application of this orthogonal transformation has to do with the way in which signals are actually transmitted and received. The transmission of a nonorthogonal signal set is generally accomplished by the appropriate weighting of the orthogonal carrier components.

Example 3.1   Orthogonal Representation of Waveforms

Figure 3.5 illustrates the statement that any arbitrary integrable waveform set can be represented as a linear combination of orthogonal waveforms. Figure 3.5a shows a set of three waveforms, $s_1(t)$, $s_2(t)$, and $s_3(t)$.

(a) Demonstrate that these waveforms *do not* form an orthogonal set.
(b) Figure 3.5b shows a set of two waveforms, $\psi_1(t)$ and $\psi_2(t)$. Verify that these waveforms form an orthogonal set.
(c) Show how the nonorthogonal waveform set in part (a) can be expressed as a linear combination of the orthogonal set in part (b).
(d) Figure 3.5c illustrates another orthogonal set of two waveforms, $\psi_1'(t)$ and $\psi_2'(t)$. Show how the nonorthogonal set in Figure 3.5a can be expressed as a linear combination of the set in Figure 3.5c.

Solution

(a) $s_1(t)$, $s_2(t)$, and $s_3(t)$ are clearly not orthogonal, since they do not meet the requirements of Equation (3.8); that is, the time integrated value (over a symbol duration) of the product of any two of the three waveforms is not zero. Let us verify this for $s_1(t)$ and $s_2(t)$:

$$\int_0^T s_1(t)s_2(t)\ dt = \int_0^{T/2} s_1(t)s_2(t)\ dt + \int_{T/2}^T s_1(t)s_2(t)\ dt$$

$$= \int_0^{T/2} (-1)(2)\ dt + \int_{T/2}^T (-3)(0)\ dt = -T$$

Similarly, the integral over the interval T of each of the cross-products $s_1(t)s_3(t)$ and $s_2(t)s_3(t)$ results in nonzero values. Hence, the waveform set $\{s_i(t)\}$ ($i = 1, 2, 3$) in Figure 3.5a is not an orthogonal set.

(b) Using Equation (3.8), we verify that $\psi_1(t)$ and $\psi_2(t)$ form an orthogonal set as follows:

$$\int_0^T \psi_1(t)\psi_2(t)\ dt = \int_0^{T/2} (1)(1)\ dt + \int_{T/2}^T (-1)(1)\ dt = 0$$

(c) Using Equation (3.11) with $K_j = T$, we can express the nonorthogonal set $\{s_i(t)\}$ ($i = 1, 2, 3$) as a linear combination of the orthogonal basis waveforms $\{\psi_j(t)\}$ ($j = 1, 2$):

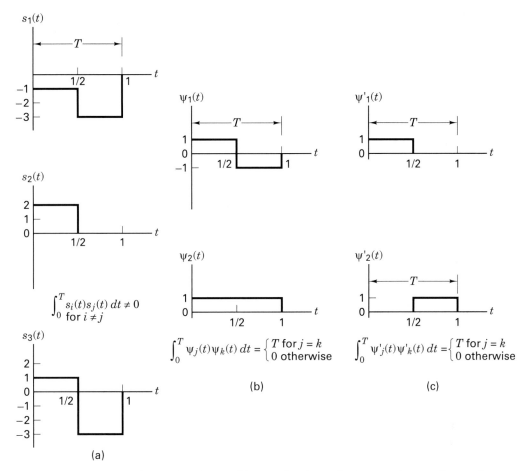

Figure 3.5 Example of an arbitrary signal set in terms of an orthogonal set. (a) Arbitrary signal set. (b) A set of orthogonal basis functions. (c) Another set of orthogonal basis functions.

$$s_1(t) = \psi_1(t) - 2\psi_2(t)$$
$$s_2(t) = \psi_1(t) + \psi_2(t)$$
$$s_3(t) = 2\psi_1(t) - \psi_2(t)$$

(d) Similar to part (c), the nonorthogonal set $\{s_i(t)\}$ ($i = 1, 2, 3$) can be expressed in terms of the simple orthogonal basis set $\{\psi'_j(t)\}$ ($j = 1, 2$) in Figure 3.5c, as follows:

$$s_1(t) = -\psi'_1(t) - 3\psi'_2(t)$$
$$s_2(t) = 2\psi'_1(t)$$
$$s_3(t) = \psi'_1(t) - 3\psi'_2(t)$$

These relationships illustrate how an arbitrary waveform set $\{s_i(t)\}$ can be expressed as a linear combination of an orthogonal set $\{\psi_j(t)\}$, as described in Equations (3.10) and (3.11). What are the practical applications of being able to describe $s_1(t)$, $s_2(t)$, and $s_3(t)$ in terms of $\psi_1(t)$, $\psi_2(t)$, and the appropriate coeffi-

3.1   Signals and Noise   **115**

cients? If we want a system for transmitting waveforms $s_1(t)$, $s_2(t)$, and $s_3(t)$, the transmitter and the receiver need only be implemented using the two basis functions $\psi_1(t)$ and $\psi_2(t)$ instead of the three original waveforms. The *Gram–Schmidt orthogonalization procedure* provides a convenient way in which an appropriate choice of a basis function set $\{\psi_j(t)\}$, can be obtained for any given signal set $\{s_i(t)\}$. (It is described in Appendix 4A of Reference [4].)

3.1.3.3 Representing White Noise with Orthogonal Waveforms

Additive white Gaussian noise (AWGN) can be expressed as a linear combination of orthogonal waveforms in the same way as signals. For the signal detection problem, the noise can be partitioned into two components,

$$n(t) = \hat{n}(t) + \tilde{n}(t) \tag{3.20}$$

where

$$\hat{n}(t) = \sum_{j=1}^{N} n_j \psi_j(t) \tag{3.21}$$

is taken to be the noise within the signal space, or the projection of the noise components on the signal coordinates $\psi_1(t), \ldots, \psi_N(t)$, and

$$\tilde{n}(t) = n(t) - \hat{n}(t) \tag{3.22}$$

is defined as the noise outside the signal space. In other words, $\tilde{n}(t)$ may be thought of as the noise that is effectively tuned out by the detector. The symbol $\hat{n}(t)$ represents the noise that will interfere with the detection process. We can express the noise waveform $n(t)$ as

$$n(t) = \sum_{j=1}^{N} n_j \psi_j(t) + \tilde{n}(t) \tag{3.23}$$

where

$$n_j = \frac{1}{K_j} \int_0^T n(t)\psi_j(t)\, dt \qquad \text{for all } j \tag{3.24}$$

and

$$\int_0^T \tilde{n}(t)\psi_j(t)\, dt = 0 \qquad \text{for all } j \tag{3.25}$$

The interfering portion of the noise, $\hat{n}(t)$, expressed in Equation (3.21) will henceforth be referred to simply as $n(t)$. We can express $n(t)$ by a vector of its coefficients similar to the way we did for signals in Equation (3.12). We have

$$\mathbf{n} = (n_1, n_2, \ldots, n_N) \tag{3.26}$$

where $\mathbf{n}$ is a random vector with zero mean and Gaussian distribution, and where the noise components n_i $(i = 1, \ldots, N)$ are independent.

3.1.3.4 Variance of White Noise

White noise is an *idealized process* with two-sided power spectral density equal to a constant $N_0/2$, for all frequencies from $-\infty$ to $+\infty$. Hence, the noise variance (average noise power, since the noise has zero mean) is

$$\sigma^2 = \text{var}\,[n(t)] = \int_{-\infty}^{\infty} \left(\frac{N_0}{2}\right) df = \infty \qquad (3.27)$$

Although the variance for AWGN is infinite, the variance for *filtered* AWGN is finite. For example, if AWGN is correlated with one of a set of orthonormal functions $\psi_j(t)$, the variance of the correlator output is given by

$$\sigma^2 = \text{var}\,(n_j) = \mathbf{E}\left\{\left[\int_0^T n(t)\psi_j(t)\, dt\right]^2\right\} = \frac{N_0}{2} \qquad (3.28)$$

The proof of Equation (3.28) is given in Appendix C. Henceforth we shall assume that the noise of interest in the detection process is the output noise of a correlator or matched filter with variance $\sigma^2 = N_0/2$ as expressed in Equation (3.28).

3.1.4 The Basic SNR Parameter for Digital Communication Systems

Anyone who has studied analog communications is familiar with the figure of merit, *average signal power to average noise power ratio (S/N or SNR)*. In digital communications, we more often use E_b/N_0, a normalized version of SNR, as a figure of merit. E_b is bit energy and can be described as signal power S times the bit time T_b. N_0 is noise power spectral density, and can be described as noise power N divided by bandwidth W. Since the bit time and bit rate R_b are reciprocal, we can replace T_b with $1/R_b$ and write

$$\frac{E_b}{N_0} = \frac{S\,T_b}{N/W} = \frac{S/R_b}{N/W} \qquad (3.29)$$

Data rate, in units of bits per second, is one of the most recurring parameters in digital communications. We therefore simplify the notation throughout the book, by using R instead of R_b to represent bits/s, and we rewrite Equation (3.29) to emphasize that E_b/N_0 is just a version of S/N normalized by bandwidth and bit rate, as follows:

$$\frac{E_b}{N_0} = \frac{S}{N}\left(\frac{W}{R}\right) \qquad (3.30)$$

One of the most important metrics of performance in digital communication systems is a plot of the bit-error probability P_B versus E_b/N_0. Figure 3.6 illustrates the "waterfall-like" shape of most such curves. For $E_b/N_0 \geq x_0$, $P_B \leq P_0$. The dimensionless ratio E_b/N_0 is a standard quality measure for digital communications system performance. Therefore, required E_b/N_0 can be considered a metric that characterizes the performance of one system versus another; the smaller the required E_b/N_0, the more efficient is the detection process for a given probability of error.

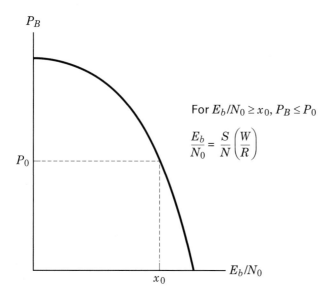

For $E_b/N_0 \geq x_0$, $P_B \leq P_0$

$$\frac{E_b}{N_0} = \frac{S}{N}\left(\frac{W}{R}\right)$$

Figure 3.6 General shape of the P_B versus E_b/N_0 curve.

3.1.5 Why E_b/N_0 Is a Natural Figure of Merit

A newcomer to digital communications may question the usefulness of the parameter E_b/N_0. After all, S/N is a useful figure of merit for analog communications—the numerator represents a power measurement of the signal we wish to preserve and deliver, and the denominator represents electrical noise degradation. Moreover, S/N is intuitively acceptable as a metric of goodness. Thus, why can't we continue to use S/N as a figure of merit for digital communications? Why has a different metric for digital systems—the ratio of bit energy to noise power spectral density—arisen? The explanation is given below.

In Section 1.2.4, a power signal was defined as a signal having finite average power and infinite energy. An energy signal was defined as a signal having zero average power and finite energy. These classifications are useful in comparing analog and digital waveforms. We classify an analog waveform as a power signal. Why does this make sense? We can think of an analog waveform as having infinite duration that need not be partitioned or windowed in time. An infinitely long electrical waveform has an infinite amount of energy; hence, energy is not a useful way to characterize this waveform. Power (or rate of delivering the energy) is a more useful parameter for analog waveforms.

However, in a digital communication system we transmit (and receive) a symbol by using a transmission waveform within a window of time, the symbol time T_s. Focusing on one symbol, we can see that the power (averaged over all time) goes to zero. Hence, power is not a useful way to characterize a digital waveform. What we need for such waveforms is a metric of the "good stuff" within the window. In other words, the symbol energy (power integrated over T_s) is a more useful parameter for characterizing digital waveforms.

The fact that a digital signal is best characterized by its received energy doesn't yet get to the crux of why E_b/N_0 is a natural metric for digital systems, so let us continue. The digital waveform is a vehicle that represents a digital message. The message may contain one bit (binary), two bits (4-ary), . . . , 10 bits (1024-ary). In analog systems, there is nothing akin to such a discretized message structure. An analog information source is an infinitely quantized continuous wave. For digital systems, a figure of merit should allow us to compare one system with another at the bit level. Therefore, a description of the digital waveform in terms of *S/N* is virtually useless, since the waveform may have a one-bit meaning, a two-bit meaning, or a 10-bit meaning. For example, suppose we are told that for a given error probability, the required *S/N* for a digital binary waveform is 20 units. Think of the waveform as being interchangeable with its meaning. Since the binary waveform has a one-bit meaning, then the *S/N* requirement per bit is equal to the same 20 units. However, suppose that the waveform is 1024-ary, with the same 20 units of required *S/N*. Now, since the waveform has a 10-bit meaning, the *S/N* requirement per bit is only 2 units. Why should we have to go through such computational manipulations to find a metric that represents a figure of merit? Why not immediately describe the metric in terms of what we need—an energy-related parameter at the bit level, E_b/N_0? Just as *S/N* is a dimensionless ratio, so too is E_b/N_0. To verify this, consider the following units of measure:

$$\frac{E_b}{N_0} = \frac{\text{Joule}}{\text{Watt per Hz}} = \frac{\text{Watt-s}}{\text{Watt-s}}$$

3.2 DETECTION OF BINARY SIGNALS IN GAUSSIAN NOISE

3.2.1 Maximum Likelihood Receiver Structure

The decision-making criterion shown in step 2 of Figure 3.1 was described by Equation (3.7) as

$$z(T) \underset{H_2}{\overset{H_1}{\gtrless}} \gamma$$

A popular criterion for choosing the threshold level γ for the binary decision in Equation (3.7) is based on minimizing the probability of error. The computation for this *minimum error* value of $\gamma = \gamma_0$ starts with forming an inequality expression between the ratio of conditional probability density functions and the signal a priori probabilities. Since the conditional density function $p(z|s_i)$ is also called the *likelihood* of s_i, the formulation

$$\frac{p(z|s_1)}{p(z|s_2)} \underset{H_2}{\overset{H_1}{\gtrless}} \frac{P(s_2)}{P(s_1)} \tag{3.31}$$

is called the *likelihood ratio test.* (See Appendix B.) In this inequality, $P(s_1)$ and $P(s_2)$ are the a priori probabilities that $s_1(t)$ and $s_2(t)$, respectively, are transmitted, and H_1 and H_2 are the two possible hypotheses. The rule for minimizing the error probability states that we should choose hypothesis H_1 if the ratio of likelihoods is greater than the ratio of a priori probabilities, as shown in Equation (3.31).

It is shown in Section B.3.1, that if $P(s_1) = P(s_2)$, and if the likelihoods, $p(z|s_i)$ ($i = 1, 2$), are symmetrical, the substitution of Equations (3.5) and (3.6) into (3.31) yields

$$z(T) \underset{H_2}{\overset{H_1}{\gtrless}} \frac{a_1 + a_2}{2} = \gamma_0 \tag{3.32}$$

where a_1 is the signal component of $z(T)$ when $s_1(t)$ is transmitted, and a_2 is the signal component of $z(T)$ when $s_2(t)$ is transmitted. The threshold level γ_0, represented by $(a_1 + a_2)/2$, is the *optimum threshold* for minimizing the probability of making an incorrect decision for this important special case. This strategy is known as the *minimum error criterion.*

For equally likely signals, the optimum threshold γ_0 passes through the intersection of the likelihood functions, as shown in Figure 3.2. Thus by following Equation (3.32), the decision stage effectively selects the hypothesis that corresponds to the signal with the *maximum likelihood.* For example, given an arbitrary detector output value $z_a(T)$, for which there is a nonzero likelihood that $z_a(T)$ belongs to either signal class $s_1(t)$ or $s_2(t)$, one can think of the likelihood test as a comparison of the likelihood values $p(z_a|s_1)$ and $p(z_a|s_2)$. The signal corresponding to the maximum pdf is chosen as the most likely to have been transmitted. In other words, the detector chooses $s_1(t)$ if

$$p(z_a|s_1) > p(z_a|s_2) \tag{3.33}$$

Otherwise, the detector chooses $s_2(t)$. A detector that minimizes the error probability (for the case where the signal classes are equally likely) is also known as a *maximum likelihood detector.*

Figure 3.2 illustrates that Equation (3.33) is just a "common sense" way to make a decision when there exists statistical knowledge of the classes. Given the detector output value $z_a(T)$, we see in Figure 3.2 that $z_a(T)$ intersects the likelihood of $s_1(t)$ at a value ℓ_1, and it intersects the likelihood of $s_2(t)$ at a value ℓ_2. What is the most reasonable decision for the detector to make? For this example, choosing class $s_1(t)$, which has the greater likelihood, is the most sensible choice. If this was an M-ary instead of a binary example, there would be a total of M likelihood functions representing the M signal classes to which a received signal might belong. The maximum likelihood decision would then be to choose the class that had the greatest likelihood of all M likelihoods. (Refer to Appendix B for a review of decision theory fundamentals.)

3.2.1.1 Error Probability

For the binary decision-making depicted in Figure 3.2, there are two ways errors can occur. An error e will occur when $s_1(t)$ is sent, and channel noise results in the receiver output signal $z(t)$ being less than γ_0. The probability of such an occurrence is

$$P(e \mid s_1) = P(H_2 \mid s_1) = \int_{-\infty}^{\gamma_0} p(z \mid s_1)\, dz \tag{3.34}$$

This is illustrated by the shaded area to the left of γ_0 in Figure 3.2. Similarly, an error occurs when $s_2(t)$ is sent, and the channel noise results in $z(T)$ being greater than γ_0. The probability of this occurrence is

$$P(e \mid s_2) = P(H_1 \mid s_2) = \int_{\gamma_0}^{\infty} p(z \mid s_2)\, dz \tag{3.35}$$

The probability of an error is the sum of the probabilities of all the ways that an error can occur. For the binary case, we can express the probability of bit error as

$$P_B = \sum_{i=1}^{2} P(e, s_i) = \sum_{i=1}^{2} P(e \mid s_i)\, P(s_i) \tag{3.36}$$

Combining Equations (3.34) to (3.36), we can write

$$P_B = P(e \mid s_1)P(s_1) + P(e \mid s_2)P(s_2) \tag{3.37a}$$

or equivalently,

$$P_B = P(H_2 \mid s_1)P(s_1) + P(H_1 \mid s_2)P(s_2) \tag{3.37b}$$

That is, given that signal $s_1(t)$ was transmitted, an error results if hypothesis H_2 is chosen; or given that signal $s_2(t)$ was transmitted, an error results if hypothesis H_1 is chosen. For the case where the a priori probabilities are equal [that is, $P(s_1) = P(s_2) = \frac{1}{2}$],

$$P_B = \tfrac{1}{2} P(H_2 \mid s_1) + \tfrac{1}{2} P(H_1 \mid s_2) \tag{3.38}$$

and because of the symmetry of the probability density functions,

$$P_B = P(H_2 \mid s_1) = P(H_1 \mid s_2) \tag{3.39}$$

The probability of a bit error, P_B, is numerically equal to the area under the "tail" of either likelihood function, $p(z \mid s_1)$ or $p(z \mid s_2)$, falling on the "incorrect" side of the threshold. We can therefore compute P_B by integrating $p(z \mid s_1)$ between the limits $-\infty$ and γ_0, or by integrating $p(z \mid s_2)$ between the limits γ_0 and ∞:

$$P_B = \int_{\gamma_0 = (a_1 + a_2)/2}^{\infty} p(z \mid s_2)\, dz \tag{3.40}$$

Here, $\gamma_0 = (a_1 + a_2)/2$ is the optimum threshold from Equation (3.32). Replacing the likelihood $p(z \mid s_2)$ with its Gaussian equivalent from Equation (3.6), we have

$$P_B = \int_{\gamma_0 = (a_1 + a_2)/2}^{\infty} \frac{1}{\sigma_0 \sqrt{2\pi}} \exp\left[-\frac{1}{2}\left(\frac{z - a_2}{\sigma_0}\right)^2\right] dz \qquad (3.41)$$

where σ_0^2 is the variance of the noise out of the correlator.

Let $u = (z - a_2)/\sigma_0$. Then $\sigma_0\, du = dz$ and

$$P_B = \int_{u=(a_1-a_2)/2\sigma_0}^{u=\infty} \frac{1}{\sqrt{2\pi}} \exp\left(-\frac{u^2}{2}\right) du = Q\left(\frac{a_1 - a_2}{2\sigma_0}\right) \qquad (3.42)$$

where $Q(x)$, called the *complementary error function* or *co-error function*, is a commonly used symbol for the probability under the tail of the Gaussian pdf. It is defined as

$$Q(x) \approx \frac{1}{\sqrt{2\pi}} \int_x^{\infty} \exp\left(-\frac{u^2}{2}\right) du \qquad (3.43)$$

Note that the co-error function is defined in several ways (see Appendix B); however, all definitions are equally useful for determining probability of error in Gaussian noise. $Q(x)$ cannot be evaluated in closed form. It is presented in tabular form in Table B.1. Good approximations to $Q(x)$ by simpler functions can be found in Reference [5]. One such approximation, valid for $x > 3$, is

$$Q(x) \approx \frac{1}{x\sqrt{2\pi}} \exp\left(-\frac{x^2}{2}\right) \qquad (3.44)$$

We have optimized (in the sense of minimizing P_B) the threshold level γ, but have not optimized the receiving filter in block 1 of Figure 3.1. We next consider optimizing this filter by maximizing the argument of $Q(x)$ in Equation (3.42).

3.2.2 The Matched Filter

A matched filter is a linear filter designed to provide the maximum signal-to-noise power ratio at its output for a given transmitted symbol waveform. Consider that a known signal $s(t)$ plus AWGN $n(t)$ is the input to a linear, time-invariant (receiving) filter followed by a sampler, as shown in Figure 3.1. At time $t = T$, the sampler output $z(T)$ consists of a signal component a_i and a noise component n_0. The variance of the output noise (average noise power) is denoted by σ_0^2, so that the ratio of the instantaneous signal power to average noise power, $(S/N)_T$, at time $t = T$, out of the sampler in step 1, is

$$\left(\frac{S}{N}\right)_T = \frac{a_i^2}{\sigma_0^2} \qquad (3.45)$$

We wish to find the filter transfer function $H_0(f)$ that *maximizes* Equation (3.45). We can express the signal $a_i(t)$ at the filter output in terms of the filter transfer function $H(f)$ (before optimization) and the Fourier transform of the input signal, as

$$a_i(t) = \int_{-\infty}^{\infty} H(f)S(f)e^{j2\pi ft}\, df \tag{3.46}$$

where $S(f)$ is the Fourier transform of the input signal, $s(t)$. If the two-sided power spectral density of the input noise is $N_0/2$ watts/hertz, then, using Equations (1.19) and (1.53), we can express the output noise power as

$$\sigma_0^2 = \frac{N_0}{2} \int_{-\infty}^{\infty} |H(f)|^2\, df \tag{3.47}$$

We then combine Equations (3.45) to (3.47) to express $(S/N)_T$, as follows:

$$\left(\frac{S}{N}\right)_T = \frac{\left| \int_{-\infty}^{\infty} H(f)S(f)e^{j2\pi fT}\, df \right|^2}{N_0/2 \int_{-\infty}^{\infty} |H(f)|^2\, df} \tag{3.48}$$

We next find that value of $H(f) = H_0(f)$ for which the maximum $(S/N)_T$ is achieved, by using *Schwarz's inequality*. One form of the inequality can be stated as

$$\left| \int_{-\infty}^{\infty} f_1(x)f_2(x)\, dx \right|^2 \leq \int_{-\infty}^{\infty} |f_1(x)|^2\, dx \int_{-\infty}^{\infty} |f_2(x)|^2\, dx \tag{3.49}$$

The equality holds if $f_1(x) = kf_2^*(x)$, where k is an arbitrary constant and $*$ indicates complex conjugate. If we identify $H(f)$ with $f_1(x)$ and $S(f)\, e^{j2\pi fT}$ with $f_2(x)$, we can write

$$\left| \int_{-\infty}^{\infty} H(f)S(f)e^{j2\pi fT} df \right|^2 \leq \int_{-\infty}^{\infty} |H(f)|^2\, df \int_{-\infty}^{\infty} |S(f)|^2\, df \tag{3.50}$$

Substituting into Equation (3.48) yields

$$\left(\frac{S}{N}\right)_T \leq \frac{2}{N_0} \int_{-\infty}^{\infty} |S(f)|^2\, df \tag{3.51}$$

or

$$\max \left(\frac{S}{N}\right)_T = \frac{2E}{N_0} \tag{3.52}$$

where the energy E of the input signal $s(t)$ is

$$E = \int_{-\infty}^{\infty} |S(f)|^2\, df \tag{3.53}$$

Thus, the maximum output $(S/N)_T$ depends on the input *signal energy* and the power spectral density of the noise, *not on the particular shape of the waveform* that is used.

The equality in Equation (3.52) holds only if the optimum filter transfer function $H_0(f)$ is employed, such that

$$H(f) = H_0(f) = kS*(f)e^{-j2\pi f T} \tag{3.54}$$

or

$$h(t) = \mathcal{F}^{-1}\{kS*(f)e^{-j2\pi f T}\} \tag{3.55}$$

Since $s(t)$ is a real-valued signal, we can write, from Equations (A.29) and (A.31),

$$h(t) = \begin{cases} ks(T-t) & 0 \le t \le T \\ 0 & \text{elsewhere} \end{cases} \tag{3.56}$$

Thus, the impulse response of a filter that produces the maximum output signal-to-noise ratio is the mirror image of the message signal $s(t)$, *delayed* by the symbol time duration T. Note that the delay of T seconds makes Equation (3.56) *causal;* that is, the delay of T seconds makes $h(t)$ a function of positive time in the interval $0 \le t \le T$. Without the delay of T seconds, the response $s\ (-t)$ is unrealizable because it describes a response as a function of negative time.

3.2.3 Correlation Realization of the Matched Filter

Equation (3.56) and Figure 3.7a illustrate the matched filter's basic property: The impulse response of the filter is a delayed version of the mirror image (rotated on the $t = 0$ axis) of the signal waveform. Therefore, if the signal waveform is $s(t)$, its mirror image is $s(-t)$, and the mirror image delayed by T seconds is $s(T - t)$. The output $z(t)$ of a causal filter can be described in the time domain as the convolution of a received input waveform $r(t)$ with the impulse response of the filter (see Section A.5):

$$z(t) = r(t) * h(t) = \int_0^t r(\tau)h(t-\tau)\,d\tau \tag{3.57}$$

Substituting $h(t)$ of Equation (3.56) into $h(t-\tau)$ of Equation (3.57) and arbitrarily setting the constant k equal to unity, we get

$$z(t) = \int_0^t r(\tau)s[T - (t - \tau)]\,d\tau$$
$$= \int_0^t r(\tau)s(T - t + \tau)\,d\tau \tag{3.58}$$

When $t = T$, we can write Equation (3.58) as

$$z(T) = \int_0^T r(\tau)s(\tau)\,d\tau \tag{3.59}$$

The operation of Equation (3.59), the product integration of the received signal $r(t)$ with a replica of the transmitted waveform $s(t)$ over one symbol interval is known as the *correlation* of $r(t)$ with $s(t)$. Consider that a received signal $r(t)$ is correlated with each prototype signal $s_i(t)$ $(i = 1, \ldots, M)$, using a bank of M correlators. The signal $s_i(t)$ whose product integration or correlation with $r(t)$ yields the maximum

Baseband Demodulation/Detection   Chap. 3

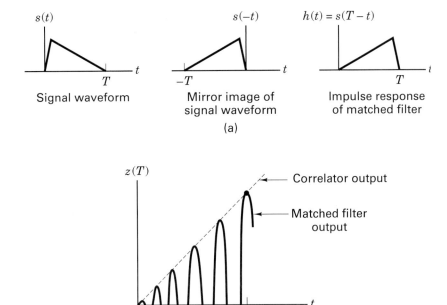

Figure 3.7 Correlator and matched filter. (a) Matched filter characteristic. (b) Comparison of correlator and matched filter outputs.

output $z_i(T)$ is the signal that matches $r(t)$ better than all the other $s_j(t)$, $j \neq i$. We will subsequently use this correlation characteristic for the optimum detection of signals.

3.2.3.1 Comparison of Convolution and Correlation

The mathematical operation of a matched filter (MF) is *convolution;* a signal is convolved with the impulse response of a filter. The mathematical operation of a correlator is *correlation;* a signal is correlated with a replica of itself. The term "matched filter" is often used synonymously with "correlator." How is that possible when their mathematical operations are different? Recall that the process of convolving two signals reverses one of them in time. Also, the impulse response of an MF is defined in terms of a signal that is reversed in time. Thus, convolution in the MF with a time-reversed function results in a second time-reversal, making the output (at the end of a symbol time) appear to be the result of a signal that has been correlated with its replica. Therefore, it is valid to implement the receiving filter in Figure 3.1 with either a matched filter or a correlator. It is important to note that the correlator output and the matched filter output are the same *only at time*

$t = T$. For a sine-wave input, the output of the correlator, $z(t)$, is approximately a linear ramp during the interval $0 \le t \le T$. However, the matched filter output is approximately a sine wave that is amplitude modulated by a linear ramp during the same time interval. The comparison is shown in Figure 3.7b. Since for comparable imputs, the MF output and the correlator output are identical at the sampling time $t = T$, the matched filter and correlator functions pictured in Figure 3.8 are often used interchangeably.

3.2.3.2 Dilemma in Representing Earliest versus Latest Event

A serious dilemma exists in representing timed events. This dilemma is undoubtedly the cause of a frequently made error in electrical engineering—confusing the most significant bit (MSB) with the least significant bit (LSB). Figure 3.9a illustrates how a function of time is typically plotted; the earliest event appears leftmost, and the latest event rightmost. In western societies, where we read from left to right, would there be any other way to plot timed events? Consider Figure 3.9b, where pulses are shown entering (and leaving) a network or circuit. Here, the earliest events are shown rightmost, and the latest are leftmost. From the figure, it should be clear that whenever we denote timed events, there is an inference that we are following one of the two formats described here. Often, we have to provide some descriptive words (e.g., the rightmost bit is the earliest bit) to avoid confusion.

Mathematical relationships often have built-in features guaranteeing the proper alignment of time events. For example, in Section 3.2.3, a matched filter is defined as having an impulse response $h(t)$ that is a delayed version of the time-reversed copy of the signal. That is, $h(t) = s(T - t)$. Delay of one symbol time T is needed for the filter to be causal (the output must occur in positive time). Time reversal can be thought of as a "precorrection" where the rightmost part of the time plot will now correspond to the earliest event. Since convolution dictates another time reversal, the arriving signal and the filter's impulse response will be "in step" (earliest with earliest, and latest with latest).

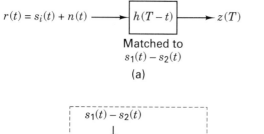

$r(t) = s_i(t) + n(t) \longrightarrow \boxed{h(T - t)} \longrightarrow z(T)$

Matched to
$s_1(t) - s_2(t)$

(a)

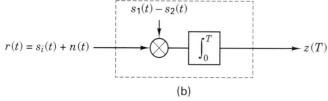

$r(t) = s_i(t) + n(t) \longrightarrow$

(b)

Figure 3.8 Equivalence of matched filter and correlator. (a) Matched filter. (b) Correlator.

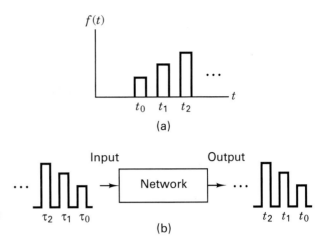

Figure 3.9 Dilemma in representing earliest versus latest events.

3.2.4 Optimizing Error Performance

To optimize (minimize) P_B in the context of an AWGN channel and the receiver shown in Figure 3.1, we need to select the optimum receiving filter in step 1 and the optimum decision threshold in step 2. For the binary case, the optimum decision threshold has already been chosen in Equation (3.32), and it was shown in Equation (3.42) that this threshold results in $P_B = Q\left[(a_1 - a_2)/2\sigma_0\right]$. Next, for minimizing P_B, it is necessary to choose the filter (matched filter) that maximizes the argument of $Q(x)$. Thus, we need to determine the linear filter that maximizes $(a_1 - a_2)/2\sigma_0$, or equivalently, that maximizes

$$\frac{(a_1 - a_2)^2}{\sigma_0^2} \qquad (3.60)$$

where $(a_1 - a_2)$ is the difference of the desired signal components at the filter output at time $t = T$, and the square of this difference signal is the instantaneous power of the difference signal. In Section 3.2.2, a matched filter was described as one that maximizes the output signal-to-noise ratio (SNR) for a given known signal. Here, we continue that development for binary signaling, where we view the optimum filter as one that maximizes the difference between two possible signal outputs. Starting with Equations (3.45) and (3.47), it was shown in Equation (3.52) that a matched filter achieves the maximum possible output SNR equal to $2E/N_0$. Consider that the filter is matched to the input difference signal $[s_1(t) - s_2(t)]$; thus, we can write an output SNR at time $t = T$ as

$$\left(\frac{S}{N}\right)_T = \frac{(a_1 - a_2)^2}{\sigma_0^2} = \frac{2E_d}{N_0} \qquad (3.61)$$

where $N_0/2$ is the two-sided power spectral density of the noise at the filter input and

$$E_d = \int_0^T [s_1(t) - s_2(t)]^2 \, dt \qquad (3.62)$$

is the energy of the difference signal at the filter input. Note that Equation (3.61) does not represent the SNR for any single transmission, $s_1(t)$ or $s_2(t)$. This SNR yields a metric of signal difference for the filter output. By maximizing the output SNR as shown in Equation (3.61), the matched filter provides the maximum distance (normalized by noise) between the two candidate outputs—signal a_1 and signal a_2.

Next, combining Equations (3.42) and (3.61) yields

$$P_B = Q\left(\sqrt{\frac{E_d}{2N_0}}\right) \qquad (3.63)$$

For the matched filter, Equation (3.63) is an important interim result in terms of the energy of the difference signal at the filter's input. From this equation, a more general relationship in terms of received bit energy can be developed. We start by defining a time cross-correlation coefficient ρ as a measure of similarity between two signals, $s_1(t)$ and $s_2(t)$. We have

$$\rho = \frac{1}{E_b} \int_0^T s_1(t) \, s_2(t) \, dt \qquad (3.64a)$$

and

$$\rho = \cos \theta \qquad (3.64b)$$

where $-1 \le \rho \le 1$. Equation (3.64a) is the classical mathematical way of expressing correlation. However, when $s_1(t)$ and $s_2(t)$ are viewed as signal vectors, $\mathbf{s}_1$ and $\mathbf{s}_2$, respectively, then ρ is conveniently expressed by Equation (3.64b). This vector view provides a useful image. The vectors $\mathbf{s}_1$ and $\mathbf{s}_2$ are separated by the angle θ; for small angular separation, the vectors are quite similar (highly correlated) to each other, and for large angular separation, they are quite dissimilar. The cosine of this angle gives us the same normalized metric of correlation as Equation (3.64a).

Expanding Equation (3.62), we get

$$E_d = \int_0^T s_1^2(t) \, dt + \int_0^T s_2^2(t) \, dt - 2 \int_0^T s_1(t) \, s_2(t) \, dt \qquad (3.65)$$

Recall that each of the first two terms in Equation (3.65) represents the energy associated with a bit, E_b; that is,

$$E_b = \int_0^T s_1^2(t) \, dt = \int_0^T s_2^2(t) \, dt \qquad (3.66)$$

Substituting Equations (3.64a) and (3.66) into Equation (3.65), we get

$$E_d = E_b + E_b - 2\rho E_b = 2E_b(1 - \rho) \qquad (3.67)$$

Substituting Equation (3.67) into (3.63), we obtain

Since the waveforms depicted in Figure 3.11 are antipodal and are detected with a matched filter, we use Equation (3.70) to find the bit-error probability, as

$$Q\left(\sqrt{\frac{12 \times 10^{-12}}{10^{-12}}}\right) = Q(\sqrt{12}) = Q(3.46)$$

From Table B.1, we find that $P_B = 3 \times 10^{-4}$. Or, since the argument of $Q(x)$ is greater than 3, we can also use the approximate relationship in Equation (3.44) which yields $P_B \approx 2.9 \times 10^{-4}$.

Because the received signals are antipodal and are received by an MF, these are sufficient prerequisites such that Equation (3.70) provides the proper relationship for finding bit-error probability. The waveforms $s_1(t)$ and $s_2(t)$ could have been pictured in a much more bizarre fashion, but as long as they are antipodal and detected by an MF, their shapes do not enter into the P_B computations. The shapes of the waveforms, of course, *do matter* when it comes to specifying the impulse response of the MF needed to detect these waveforms.

3.2.5 Error Probability Performance of Binary Signaling

3.2.5.1 Unipolar Signaling

Figure 3.12a illustrates an example of baseband orthogonal signaling— namely, unipolar signaling, where

$$\begin{aligned} s_1(t) &= A & 0 \le t \le T & \quad \text{for binary 1} \\ s_2(t) &= 0 & 0 \le t \le T & \quad \text{for binary 0} \end{aligned} \tag{3.72}$$

and where $A > 0$ is the amplitude of symbol $s_1(t)$. The definition of orthogonal signaling described by Equation (3.69) requires that $s_1(t)$ and $s_2(t)$ have zero correlation over each symbol time duration. Because in Equation (3.72), $s_2(t)$ is equal to zero during the symbol time, this set of unipolar pulses clearly fulfills the condition

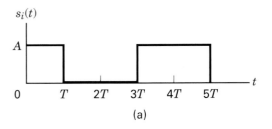

(a)

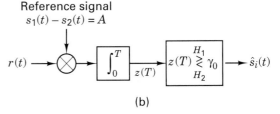

Figure 3.12 Detection of unipolar baseband signaling. (a) Unipolar signaling example. (b) Correlator detector.

(b)

shown in Equation (3.69), and hence, they form an orthogonal signal set. Consider such unipolar signaling, as illustrated in Figure 3.12a, as well as the correlator shown in Figure 3.12b, which can be used for detecting such pulses. The correlator multiplies and integrates the incoming signal $r(t)$ with the difference of the proto-type signals, $[s_1(t) - s_2(t)] = A$. After a symbol duration T, a sampler (inferred by the upper limit of integration) yields the test statistic $z(T)$, which is then compared with the threshold γ_0. For the case of $s_1(t)$ plus AWGN being received—that is, when $r(t) = s_1(t) + n(t)$—the signal component of $z(T)$ is found, using Equation (3.59), to be

$$a_1(T) = \mathrm{E}\{z(T)|s_1(t)\} = \mathrm{E}\left\{ \int_0^T A^2 + An(t)\, dt \right\} = A^2 T$$

where $\mathrm{E}\{z(T)|s_1(t)\}$ is the *expected value* of $z(T)$, given that $s_1(t)$ was sent. This follows since $\mathrm{E}\{n(t)\} = 0$. Similarly, when $r(t) = s_2(t) + n(t)$, then $a_2(T) = 0$. Thus, in this case, the optimum decision threshold, from Equation (3.32), is given by $\gamma_0 = (a_1 + a_2)/2 = 1/2\, A^2 T$. If the test statistic $z(T)$ is greater than γ_0, the signal is declared to be $s_1(t)$; otherwise, it is declared to be $s_2(t)$.

The energy difference signal, from Equation (3.62), is given by $E_d = A^2 T$. Then the bit-error performance at the output is obtained from Equation (3.63) as

$$P_B = Q\left(\sqrt{\frac{E_d}{2N_0}} \right) = Q\left(\sqrt{\frac{A^2 T}{2N_0}} \right) = Q\left(\sqrt{\frac{E_b}{N_0}} \right) \tag{3.73}$$

where, for the case of equally likely signaling, the *average* energy per bit is $E_b = A^2 T/2$. Equation (3.73) corroborates Equation (3.71) where this relationship was established for orthogonal signaling in a more general way.

For the multiplier circuit in Figure 3.12b, with a voltage signal at each of its two inputs, we expect an output having units of volt-squared. But, the measurable output of such a circuit is a voltage signal; thus its gain must be 1/volt. The multiplier output can be seen as a normalized voltage—with units of volt/volt-squared. For the integrator circuit in Figure 3.12b, with a voltage signal input, we expect an output having units of volt-seconds. Again, the measurable output of such a circuit is a voltage; thus its gain must be 1/second. Similarly, the overall gain of the product-integrator must be 1/volt-second, and its output $z(T)$ can be seen as a normalized voltage signal proportional to received signal energy—that is, volt/volt-squared-second or volt/joule.

3.2.5.2 Bipolar Signaling

Figure (3.13a) illustrates an example of baseband antipodal signaling—namely, bipolar signaling, where

$$s_1(t) = +A \qquad 0 \leq t \leq T \qquad \text{for binary 1}$$

and $\tag{3.74}$

$$s_2(t) = -A \qquad 0 \leq t \leq T \qquad \text{for binary 0}$$

Baseband Demodulation/Detection   Chap. 3

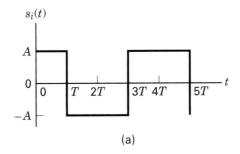

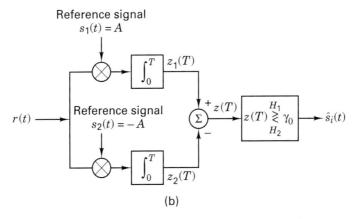

(b)

Figure 3.13 Detection of bipolar baseband signaling.
(a) Bipolar signaling example. (b) Correlator detector.

As defined earlier, the term "antipodal" refers to binary signals that are mirror images of one another; that is, $s_1(t) = -s_2(t)$. A correlator receiver for such antipodal waveforms can be configured as shown in Figure 3.13b. One correlator multiplies and integrates the incoming signal $r(t)$ with the prototype signal $s_1(t)$; the second correlator multiplies and integrates $r(t)$ with $s_2(t)$.

Figure 3.13b captures the essence of a digital receiver's main function. That is, during each symbol interval, a noisy input signal is sent down multiple "avenues" in an effort to correlate it with each of the possible candidates. The receiver then seeks the largest output voltage (the best match) to make a detection. For the binary example, there are just two possible candidates. For a 4-ary example, there would be four candidates, and so forth. In Figure 3.13b, the correlator outputs are designated $z_i(T)$ $(i = 1, 2)$. The test statistic, formed from the difference of the correlator outputs, is

$$z(T) = z_1(T) - z_2(T) \tag{3.75}$$

and the decision is made using the threshold shown in Equation (3.32). For antipodal signals, $a_1 = -a_2$; therefore, $\gamma_0 = 0$. Thus, if the test statistic $z(T)$ is positive, the signal is declared to be $s_1(T)$; and if it is negative, it is declared to be $s_2(T)$.

3.2   Detection of Binary Signals in Gaussian Noise   **133**

From Equation (3.62), the energy-difference signal is $E_d = (2A)^2 T$. Then, the bit-error performance at the output can be obtained from Equation (3.63) as

$$P_B = Q\left(\sqrt{\frac{E_d}{2N_0}}\right) = Q\left(\sqrt{\frac{2A^2 T}{N_0}}\right) = Q\left(\sqrt{\frac{2E_b}{N_0}}\right) \qquad (3.76)$$

where the average energy per bit is given by $E_b = A^2 T$. Equation (3.76) corroborates Equation (3.70) where this relationship was established for antipodal signaling in a more general way.

3.2.5.3 Signaling Described with Basis Functions

Instead of designating $s_i(t)$ as the reference signals in the correlator of Figure 3.13b, we can use the concept of *basis functions* described in Section 3.1.3. Binary signaling with unipolar or bipolar pulses provides particularly simple examples for doing this, because the entire signaling space can be described by just one basis function. If we normalize the space by choosing $K_j = 1$ in Equation (3.9), then it should be clear that the basis function $\psi_1(t)$ must be equal to $\sqrt{1/T}$.

For unipolar pulse signaling, we could then write

$$s_1(t) = a_{11}\psi_1(t) = A\sqrt{T} \times \left(\sqrt{\frac{1}{T}}\right) = A$$

and

$$s_2(t) = a_{21}\psi_1(t) = 0 \times \left(\sqrt{\frac{1}{T}}\right) = 0$$

where the coefficients a_{11} and a_{21} equal $A\sqrt{T}$ and 0, respectively.

For bipolar pulse signaling, we would write

$$s_1(t) = a_{11}\psi_1(t) = A\sqrt{T} \times \left(\sqrt{\frac{1}{T}}\right) = A$$

and

$$s_2(t) = a_{21}\psi_1(t) = -A\sqrt{T} \times \left(\sqrt{\frac{1}{T}}\right) = -A$$

where the coefficients a_{11} and a_{21} equal $A\sqrt{T}$ and $-A\sqrt{T}$, respectively. For the case of antipodal pulses, we can envision the correlator receiver taking the form of Figure 3.12b with the reference signal equal to $\sqrt{1/T}$. Then, for the case of $s_1(t) = A$ being sent, we can write

$$a_1(T) = \mathbf{E}\{z(T)|s_1(t)\} = \mathbf{E}\left\{\int_0^T \frac{A}{\sqrt{T}} + \frac{n(t)}{\sqrt{T}}\, dt\right\} = A\sqrt{T}$$

This follows because $\mathbf{E}\{n(t)\} = 0$, and since, for antipodal signaling, $E_b = A^2 T$, it follows that $a_1(T) = \sqrt{E_b}$. Similarly, for a received signal $r(t) = s_2(t) + n(t)$, it follows that $a_2(T) = -\sqrt{E_b}$. When reference signals are treated in this way, then the expected value of $z(T)$ has a magnitude of $\sqrt{E_b}$, which has units of normalized volts proportional to received energy. This basis-function treatment of the correlator yields a convenient value of $z(T)$ that is consistent with units of volts out of multipliers and integrators. We therefore repeat an important point: At the output of the sampler (the predetection point), the test statistic $z(T)$ is a voltage signal that is proportional to received signal energy.

Figure 3.14 illustrates curves of P_B versus E_b/N_0 for bipolar and unipolar signaling. There are only two fair ways to compare such curves. By drawing a vertical line at some given E_b/N_0, say, 10 dB, we see that the unipolar signaling yields P_B in the order of 10^{-3}, but that the bipolar signaling yields P_B in the order of 10^{-6}. The lower curve is the better performing one. Also, by drawing a horizontal line at some required P_B, say 10^{-5}, we see that with unipolar signaling each received bit would require an E_b/N_0 of about 12.5 dB, but with bipolar signaling, we could get away with requiring each received bit to have an E_b/N_0 of only about 9.5 dB. Of course, the smaller requirement of E_b/N_0 is better (using less power, smaller batteries). In general, the better performing curves are the ones closest to the axes, lower and leftmost. In examining the two curves in Figure 3.14, we can see a 3-dB error-performance improvement for bipolar compared with unipolar signaling. This

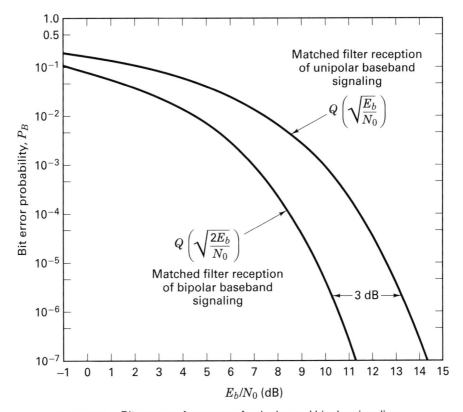

Figure 3.14 Bit error performance of unipolar and bipolar signaling.

difference could have been predicted by the factor-of-2 difference in the coefficient of E_b/N_0 in Equation (3.70) compared with (3.71). In Chapter 4, it is shown that, with MF detection, *bandpass* antipodal signaling (e.g., binary phase-shift keying) has the same P_B performance as *baseband* antipodal signaling (e.g., bipolar pulses). It is also shown that, with MF detection, *bandpass* orthogonal signaling (e.g., orthogonal frequency-shift keying) has the same P_B performance as *baseband* orthogonal signaling (e.g., unipolar pulses).

3.3 INTERSYMBOL INTERFERENCE

Figure 3.15a introduces the filtering aspects of a typical digital communication system. There are various filters (and reactive circuit elements such as inductors and capacitors) throughout the system—in the transmitter, in the receiver, and in the channel. At the transmitter, the information symbols, characterized as impulses or voltage levels, modulate pulses that are then filtered to comply with some bandwidth constraint. For baseband systems, the channel (a cable) has distributed reactances that distort the pulses. Some bandpass systems, such as wireless systems, are characterized by fading channels (see Chapter 15), that behave like undesirable filters manifesting signal distortion. When the receiving filter is configured to compensate for the distortion caused by *both* the transmitter and the channel, it is often referred to as an *equalizing filter* or a *receiving/equalizing filter*. Figure 3.15b illustrates a convenient model for the system, lumping all the filtering effects into one overall equivalent system transfer function

$$H(f) = H_t(f)\ H_c(f)\ H_r(f) \tag{3.77}$$

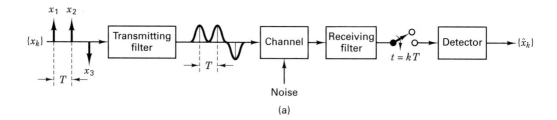

(a)

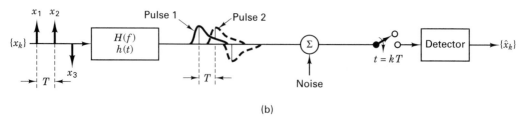

(b)

Figure 3.15 Intersymbol interference in the detection process. (a) Typical baseband digital system. (b) Equivalent model.

where $H_t(f)$ characterizes the transmitting filter, $H_c(f)$ the filtering within the channel, and $H_r(f)$ the receiving/equalizing filter. The characteristic $H(f)$, then, represents the composite system transfer function due to all the filtering at various locations throughout the transmitter/channel/receiver chain. In a binary system with a common PCM waveform, such as NRZ-L, the detector makes a symbol decision by comparing a sample of the received pulse to a threshold; for example, the detector in Figure 3.15 decides that a binary one was sent if the received pulse is positive, and that a binary zero was sent, if the received pulse is negative. Due to the effects of system filtering, the received pulses can overlap one another as shown in Figure 3.15b. The tail of a pulse can "smear" into adjacent symbol intervals, thereby interfering with the detection process and degrading the error performance; such interference is termed *intersymbol interference* (ISI). Even in the absence of noise, the effects of filtering and channel-induced distortion lead to ISI. Sometimes $H_c(f)$ is specified, and the problem remains to determine $H_t(f)$ and $H_r(f)$, such that the ISI is minimized at the output of $H_r(f)$.

Nyquist [6] investigated the problem of specifying a received pulse shape so that no ISI occurs at the detector. He showed that the theoretical minimum system bandwidth needed in order to detect R_s symbols/s, without ISI, is $R_s/2$ hertz. This occurs when the system transfer function $H(f)$ is made rectangular, as shown in Figure 3.16a. For baseband systems, when $H(f)$ is such a filter with single-sided bandwidth $1/2T$ (the *ideal Nyquist filter*), its impulse response, the inverse Fourier transform of $H(f)$ (from Table A.1) is of the form $h(t) = \text{sinc}(t/T)$, shown in Figure 3.16b. This sinc (t/T)-shaped pulse is called the *ideal Nyquist pulse;* its multiple lobes comprise a mainlobe and sidelobes called pre- and post-mainlobe *tails* that are infinitely long. Nyquist established that if each pulse of a received sequence is of the form sinc (t/T), the pulses can be detected without ISI. Figure 3.16b illustrates how ISI is avoided. There are two successive pulses, $h(t)$ and $h(t-T)$. Even though $h(t)$ has long tails, the figure shows a tail passing through zero amplitude at the instant $(t = T)$ when $h(t-T)$ is to be sampled, and likewise all tails pass through zero amplitude when any other pulse of the sequence $h(t-kT)$, $k = \pm1$, $\pm2, \ldots$ is to be sampled. Therefore, assuming that the sample timing is perfect, there will be no ISI degradation introduced. For baseband systems, the bandwidth

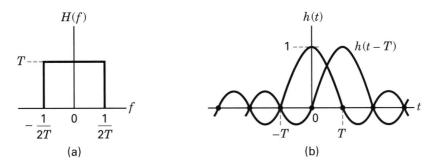

Figure 3.16 Nyquist channel for zero ISI. (a) Rectangular system transfer function $H(f)$. (b) Received pulse shape $h(t) = \text{sinc}(t/T)$.

required to detect $1/T$ such pulses (symbols) per second is equal to $1/2T$; in other words, a system with bandwidth $W = 1/2T = R_s/2$ hertz can support a maximum transmission rate of $2W = 1/T = R_s$ symbols/s (*Nyquist bandwidth constraint*) without ISI. Thus, for ideal Nyquist filtering (and zero ISI), the maximum possible symbol transmission rate per hertz, called the *symbol-rate packing,* is 2 symbols/s/Hz. It should be clear from the rectangular-shaped transfer function of the ideal Nyquist filter and the infinite length of its corresponding pulse, that such ideal filters are not realizable; they can only be approximately realized.

The names "Nyquist filter" and "Nyquist pulse" are often used to describe the general class of filtering and pulse-shaping that satisfy zero ISI at the sampling points. A Nyquist filter is one whose frequency transfer function can be represented by a rectangular function convolved with any real even-symmetric frequency function. A Nyquist pulse is one whose shape can be represented by a sinc (t/T) function multiplied by another time function. Hence, there are a countless number of Nyquist filters and corresponding pulse shapes. Amongst the class of Nyquist filters, the most popular ones are the raised cosine and the root-raised cosine, treated below.

A fundamental parameter for communication systems is *bandwidth efficiency, R/W,* whose units are bits/s/Hz. As the units imply, R/W represents a measure of data throughput per hertz of bandwidth and thus measures how efficiently any signaling technique utilizes the bandwidth resource. Since the Nyquist bandwidth constraint dictates that the theoretical maximum symbol-rate packing without ISI is 2 symbols/s/Hz, one might ask what it says about the maximum number of bits/s/Hz. It says nothing about bits, directly; the constraint deals only with pulses or symbols, and the ability to detect their amplitude values without distortion from other pulses. To find R/W for any signaling scheme, one must know how many bits each symbol represents, which is a separate issue. Consider an M-ary PAM signaling set. Each symbol (comprising k bits) is represented by one of M-pulse amplitudes. For $k = 6$ bits per symbol, the symbol set size is $M = 2^k = 64$ amplitudes. Thus with 64-ary PAM, the theoretical maximum bandwidth efficiency that is possible without ISI is 12 bits/s/Hz. (Bandwidth efficiency is treated in greater detail in Chapter 9.)

3.3.1 Pulse Shaping to Reduce ISI

3.3.1.1 Goals and Trade-offs

The more compact we make the signaling spectrum, the higher is the allowable data rate or the greater is the number of users that can simultaneously be served. This has important implications to communication service providers, since greater utilization of the available bandwidth translates into greater revenue. For most communication systems (with the exception of spread-spectrum systems, covered in Chapter 12), our goal is to reduce the required system bandwidth as much as possible. Nyquist has provided us with a basic limitation to such bandwidth reduction. What would happen if we tried to force a system to operate at smaller bandwidths than the constraint dictates? The pulses would become spread in time, which would degrade the system's error performance due to increased ISI. A prudent goal is to compress the bandwidth of the data impulses to some reasonably small bandwidth greater than the Nyquist minimum. This is accomplished by pulse-shaping with a Nyquist filter. If the band edge of the filter is steep, approaching the rectangle in Figure 3.16a, then the signaling spectrum can be made most compact. However, such a filter has an impulse response duration approaching infinity, as indicated in Figure 3.16b. Each pulse extends into every pulse in the entire sequence. Long time responses exhibit large-amplitude tails nearest the main lobe of each pulse. Such tails are undesirable because, as shown in Figure 3.16b, they contribute zero ISI *only* when the sampling is performed *at exactly* the correct sampling time; when the tails are large, small timing errors will result in ISI. Therefore, although a compact spectrum provides optimum bandwidth utilization, it is very susceptible to ISI degradation induced by timing errors.

3.3.1.2 The Raised-Cosine Filter

Earlier, it was stated that the receiving filter is often referred to as an *equalizing filter,* when it is configured to compensate for the distortion caused by both the transmitter and the channel. In other words, the configuration of this filter is chosen so as to optimize the composite system frequency transfer function $H(f)$, shown in Equation (3.77). One frequently used $H(f)$ transfer function belonging to the Nyquist class (zero ISI at the sampling times) is called the *raised-cosine filter.* It can be expressed as

$$
H(f) = \begin{cases} 1 & \text{for } |f| < 2W_0 - W \\ \cos^2\left(\frac{\pi}{4}\frac{|f| + W - 2W_0}{W - W_0}\right) & \text{for } 2W_0 - W < |f| < W \\ 0 & \text{for } |f| > W \end{cases} \quad (3.78)
$$

where W is the absolute bandwidth and $W_0 = 1/2T$ represents the minimum Nyquist bandwidth for the rectangular spectrum and the –6-dB bandwidth (or half-amplitude point) for the raised-cosine spectrum. The difference $W - W_0$ is termed the "excess bandwidth," which means additional bandwidth beyond the Nyquist minimum (i.e., for the rectangular spectrum, W is equal to W_0). The *roll-off factor* is defined to be $r = (W - W_0)/W_0$, where $0 \leq r \leq 1$. It represents the excess bandwidth divided by the filter –6-dB bandwidth (i.e., the fractional excess bandwidth). For a given W_0, the roll-off r specifies the required excess bandwidth as a fraction of W_0 and characterizes the steepness of the filter roll off. The raised-cosine characteristic is illustrated in Figure 3.17a for roll-off values of $r = 0$, $r = 0.5$, and $r = 1$. The $r = 0$ roll-off is the Nyquist minimum-bandwidth case. Note that when $r = 1$, the required excess bandwidth is 100%, and the tails are quite small. A system with such an overall spectral characteristic can provide a symbol rate of R_s symbols/s using a bandwidth of R_s hertz (twice the Nyquist minimum bandwidth), thus yielding a symbol-rate packing of 1 symbol/s/Hz. The corresponding impulse response for the $H(f)$ of Equation (3.78) is

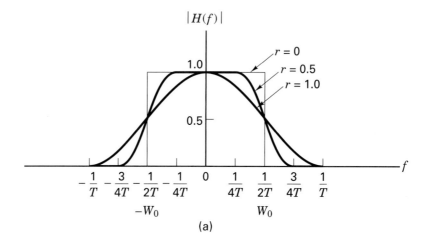

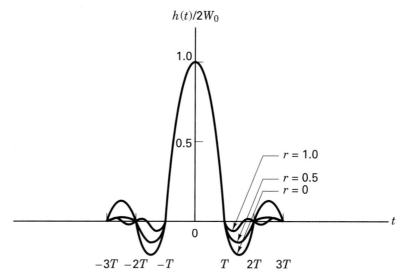

Figure 3.17 Raised-cosine filter characteristics. (a) System transfer function. (b) System impulse response.

$$h(t) = 2W_0(\text{sinc } 2W_0 t)\frac{\cos\left[2\pi(W - W_0)t\right]}{1 - \left[4(W - W_0)t\right]^2} \qquad (3.79)$$

and is plotted in Figure 3.17b for $r = 0$, $r = 0.5$, and $r = 1$. The tails have zero value at each pulse-sampling time, regardless of the roll-off value.

We can only approximately implement a filter described by Equation (3.78) and a pulse shape described by Equation (3.79), since, strictly speaking, the raised-cosine spectrum is not physically realizable (for the same reason that the ideal Nyquist filter is not realizable). A realizable filter must have an impulse response of finite duration and exhibit a zero output prior to the pulse turn-on time (see Section 1.7.2), which is not the case for the family of raised-cosine characteristics. These unrealizable filters are *noncausal* (the filter impulse response has infinite duration, and the filtered pulse begins at time $t = -\infty$). A pulse-shaping filter should satisfy two requirements. It should provide the desired roll-off, and it should be realizable (the impulse response needs to be truncated to a finite length).

Starting with the Nyquist bandwidth constraint that the minimum required system bandwidth W for a symbol rate of R_s symbols/s without ISI is $R_s/2$ hertz, a more general relationship between required bandwidth and symbol transmission rate involves the filter roll-off factor r and can be stated as

$$W = \frac{1}{2}(1 + r)R_s \qquad (3.80)$$

Thus, with $r = 0$, Equation (3.80) describes the minimum required bandwidth for ideal Nyquist filtering. For $r > 0$, there is a bandwidth expansion beyond the Nyquist minimum; thus, for this case, R_s is now less than twice the bandwidth. If the demodulator outputs one sample per symbol, then the Nyquist sampling theorem has been violated, since we are left with too few samples to reconstruct the analog waveform unambiguously (aliasing is present). However, for digital communication systems, we are not interested in reconstructing the analog waveform. Since the family of raised-cosine filters is characterized by zero ISI at the times that the symbols are sampled, we can still achieve unambiguous detection.

Bandpass-modulated signals (see Chapter 4), such as amplitude shift keying (ASK) and phase-shift keying (PSK), require twice the transmission bandwidth of the equivalent baseband signals. (See Section 1.7.1.) Such frequency-translated signals, occupying twice their baseband bandwidth, are often called double-sideband (DSB) signals. Therefore, for ASK- and PSK-modulated signals, the relationship between the required DSB bandwidth W_{DSB} and the symbol transmission rate R_s is

$$W_{\text{DSB}} = (1 + r)R_s \qquad (3.81)$$

Recall that the raised-cosine frequency transfer function describes the composite $H(f)$ that is the "full round trip" from the inception of the message (as an impulse) at the transmitter, through the channel, and through the receiving filter. The filtering at the receiver is the compensating portion of the overall transfer function to help bring about zero ISI with an overall transfer function, such as the raised cosine. Often this is accomplished by choosing (matching) the receiving filter and the transmitting filter so that each has a transfer function known as a root-raised cosine (square root of the raised cosine). Neglecting any channel-induced ISI, the product of these root-raised-cosine functions yields the composite raised-cosine system transfer function. Whenever a separate equalizing filter is introduced to mitigate the effects of channel-induced ISI, the receiving and equalizing filters together should be configured to compensate for the distortion caused by both the transmitter and the channel so as to yield an overall system transfer function characterized by zero ISI.

Let's review the trade-off that faces us in specifying pulse-shaping filters. The larger the filter roll-off, the shorter will be the pulse tails (which implies smaller tail amplitudes). Small tails exhibit less sensitivity to timing errors and thus make for small degradation due to ISI. Notice in Figure 3.17b, for $r = 1$, that timing errors can still result in some ISI degradation. However, the problem is not as serious as it is for the case in which $r = 0$, because the tails of the $h(t)$ waveform are of much smaller amplitude for $r = 1$ than they are for $r = 0$. The cost is more excess bandwidth. On the other hand, the smaller the filter roll-off, the smaller will be the excess bandwidth, thereby allowing us to increase the signaling rate or the number of users that can simultaneously use the system. The cost is longer pulse tails, larger pulse amplitudes, and thus, greater sensitivity to timing errors.

3.3.2 Two Types of Error-Performance Degradation

The effects of error-performance degradation in digital communications can be partitioned into two categories. The first is due to a decrease in received signal power or an increase in noise or interference power, giving rise to a loss in signal-to-noise ratio or E_b/N_0. The second is due to signal distortion, such as might be caused by intersymbol interference (ISI). Let us demonstrate how different are the effects of these two degradation types.

Suppose that we require a communication system with a bit-error probability P_B versus E_b/N_0 characteristic corresponding to the solid-line curve plotted in Figure 3.18a. Suppose that after the system is configured and measurements are taken, we find, to our disappointment, that the performance does not follow the theoretical curve, but in fact follows the dashed line plot. A loss in E_b/N_0 has come about because of some signal losses or an increased level of noise or interference. For a

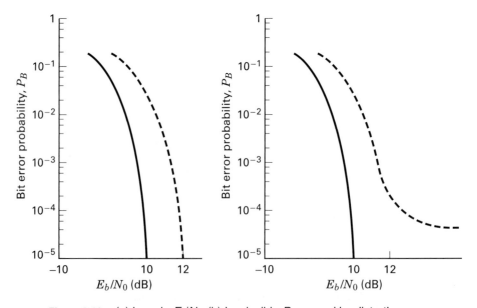

Figure 3.18 (a) Loss in E_b/N_0. (b) Irreducible P_B caused by distortion.

desired bit-error probability of 10^{-5}, the theoretical required E_b/N_0 is 10 dB. Since our system performance falls short of our goal, we can see from the dashed-line curve that, for the same bit-error probability of 10^{-5}, the required E_b/N_0 is now 12 dB. If there were no way to remedy this problem, how much more E_b/N_0 would have to be provided in order to meet the required bit-error probability? The answer is 2 dB, of course. It might be a serious problem—especially if the system is power-limited, and it is difficult to come up with the additional 2 dB. But that loss in E_b/N_0 is not *so* terrible when compared with the possible effects of degradation caused by a distortion mechanism.

In Figure 3.18b, again imagine that we do not meet the desired performance of the solid-line curve. But instead of suffering a simple loss in signal-to-noise ratio, there is a degradation effect brought about by ISI (plotted with the dashed line). If there were no way to remedy this problem, how much more E_b/N_0 would be required in order to meet the desired bit-error probability? It would require an infinite amount—or, in other words, there is no amount of E_b/N_0 that will ameliorate this problem. More E_b/N_0 cannot help when the curve manifests such an irreducible P_B (assuming that the bottoming-out point is located above the system's required P_B). Undoubtedly, every P_B-versus-E_b/N_0 curve bottoms out somewhere, but if the bottoming-out point is well below the region of interest, it will be of no consequence.

More E_b/N_0 may not help the ISI problem (it won't help at all if the P_B curve has reached an irreducible level). This can be inferred by looking at the overlapped pulses in Figure 3.15b; if we increase the E_b/N_0, the ratio of that overlap does not change. The pulses are subject to the same distortion. What, then, is the usual cure for the degradation effects of ISI? The cure is found in a technique called equalization. (See Section 3.4.) Since the distortion effects of ISI are caused by filtering in the transmitter and the channel, equalization can be thought of as the process that reverses such nonoptimum filtering effects.

Example 3.3 Bandwidth Requirements

(a) Find the minimum required bandwidth for the baseband transmission of a four-level PAM pulse sequence having a data rate of $R = 2400$ bits/s if the system transfer characteristic consists of a raised-cosine spectrum with 100% excess bandwidth ($r = 1$).

(b) The same 4-ary PAM sequence is modulated onto a carrier wave, so that the baseband spectrum is shifted and centered at frequency f_0. Find the minimum required DSB bandwidth for transmitting the modulated PAM sequence. Assume that the system transfer characteristic is the same as in part (a).

Solution

(a) $M = 2^k$; since $M = 4$ levels, $k = 2$.

$$\text{Symbol or pulse rate } R_s = \frac{R}{k} = \frac{2400}{2} = 1200 \text{ symbols/s;}$$

$$\text{Minimum bandwidth } W = \tfrac{1}{2}(1 + r)R_s = \tfrac{1}{2}(2)(1200) = 1200 \text{ Hz.}$$

Figure 3.19a illustrates the baseband PAM received pulse in the time domain—an approximation to the $h(t)$ in Equation (3.79). Figure 3.19b illustrates the Fourier

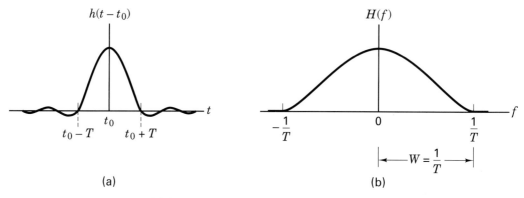

(a)   (b)

Figure 3.19   (a) Shaped pulse. (b) Baseband raised cosine spectrum.

transform of $h(t)$—the raised cosine spectrum. Notice that the required bandwidth, W, starts at zero frequency and extends to $f = 1/T$; it is twice the size of the Nyquist theoretical minimum bandwidth.

(b) As in part (a),

$$R_s = 1200 \text{ symbols/s};$$

$$W_{\text{DSB}} = (1 + r)R_s = 2(1200) = 2400 \text{ Hz}.$$

Figure 3.20a illustrates the modulated PAM received pulse. This waveform can be viewed as the product of a high-frequency sinusoidal carrier wave and a waveform with the pulse shape of Figure 3.19a. The single-sided spectral plot in Figure 3.20b illustrates that the modulated bandwidth is

$$W_{\text{DSB}} = \left(f_0 + \frac{1}{T}\right) - \left(f_0 - \frac{1}{T}\right) = \frac{2}{T}.$$

When the spectrum of Figure 3.19b is shifted up in frequency, the negative and positive halves of the baseband spectrum are shifted up in frequency, thereby dou-

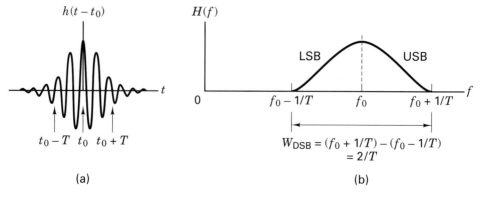

(a)   (b)

Figure 3.20   (a) Modulated shaped pulse. (b) DSB-modulated raised cosine spectrum.

bling the required transmission bandwidth. As the name implies, the DSB signal has two sidebands: the upper sideband (USB), derived from the baseband positive half, and the lower sideband (LSB), derived from the baseband negative half.

Example 3.4   Digital Telephone Circuits

Compare the system bandwidth requirements for a terrestrial 3-kHz analog telephone voice channel with that of a digital one. For the digital channel, the voice is formatted as a PCM bit stream, where the sampling rate for the analog-to-digital (A/D) conversion is 8000 samples/s and each voice sample is quantized to one of 256 levels. The bit stream is then transmitted using a PCM waveform and received with zero ISI.

Solution

The result of the sampling and quanitization process yields PCM words such that each word (representing one sample) has one of $L = 256$ different levels. If each sample were sent as a 256-ary PAM pulse (symbol), then from Equation (3.80) we can write that the required system bandwidth (without ISI) for sending R_s symbols/s would be

$$W \geq \frac{R_s}{2} \text{ hertz}$$

where the equality sign holds true only for ideal Nyquist filtering. Since the digital telephone system uses PCM (binary) waveforms, each PCM word is converted to $\ell = \log_2 L = \log_2 256 = 8$ bits. Therefore, the system bandwidth required to transmit voice using PCM is

$$W_{\text{PCM}} \geq (\log_2 L) \frac{R_s}{2} \text{ hertz}$$

$$\geq \frac{1}{2} (8 \text{ bits/symbol}) (8000 \text{ symbols/s}) = 32 \text{ kHz}$$

The 3-kHz analog voice channel will generally require approximately 4-kHz of bandwidth, including some bandwidth separation between channels, called *guard bands*. Therefore, the PCM format, using 8-bit quantization and binary signaling with a PCM waveform, requires at least eight times the bandwidth required for the analog channel.

3.3.3 Demodulation/Detection of Shaped Pulses

3.3.3.1 Matched Filters versus Conventional Filters

Conventional filters screen out unwanted spectral components of a received signal while maintaining some measure of fidelity for signals occupying a selected span of the spectrum, called the *pass-band*. These filters are generally designed to provide approximately uniform gain, a linear phase-versus-frequency characteristic over the pass-band, and a specified minimum attenuation over the remaining spectrum, called the *stop-band(s)*. A matched filter has a different "design priority," namely that of maximizing the SNR of a known signal in the presence of AWGN. Conventional filters are applied to random signals defined only by their bandwidth, while matched filters are applied to *known signals* with random parameters (such as amplitude and arrival time). The matched filter can be considered to be a *template* that is matched to the known shape of the signal being processed. A conventional filter tries to preserve the temporal or spectral structure of the signal of interest. On the other hand, a matched filter significantly modifies the temporal structure by gathering the signal energy matched to its template, and, at the end of each symbol time, presenting the result as a peak amplitude. In general, a digital communications receiver processes received signals with both kinds of filters. The task of the conventional filter is to isolate and extract a high-fidelity estimate of the signal for presentation to the matched filter. The matched filter gathers the received signal energy, and when its output is sampled (at $t = T$), a voltage proportional to that energy is produced for subsequent detection and post-detection processing.

3.3.3.2 Nyquist Pulse and Square-Root Nyquist Pulse

Consider a sequence of data impulses at a transmitter input compared with the corresponding output sequence from a raised-cosine transfer function (before sampling). In Figure 3.21, transmitted data is represented by impulse waveforms that occur at times $\tau_0, \tau_1, \ldots$ Filtering spreads the input waveforms, and thus delays them in time. We use the notation, $t_0, t_1, \ldots$, to denote received time. The impulse event that was transmitted at time τ_0 arrives at the receiver at time t_0 corresponding to the start of the output pulse event. The premainlobe tail of a demodulated pulse is referred to as its *precursor.* For a real system with a fixed system-time reference, causality dictates that $t_0 \geq \tau_0$, and the time difference between τ_0 and t_0 represents any propagation delay in the system. In this example, the time duration from the start of a demodulated pulse precursor until the appearance of its mainlobe or peak amplitude is $3T$ (three pulse-time durations). Each output pulse in the sequence is superimposed with other pulses; each pulse has an effect on the main lobes of three earlier and three later pulses. When a pulse is filtered (shaped) so that it occupies more than one symbol time, we define the pulse *support time* as the total number of symbol intervals over which the pulse persists. In Figure 3.21, the pulse support time consists of 6-symbol intervals (7 data points with 6 intervals between them).

The impulse response of a root-raised cosine filter, called the *square-root Nyquist pulse,* is shown in Figure 3.22a (normalized to a peak value of unity, with a filter rolloff of $r = 0.5$). The impulse response of the raised-cosine filter, called the

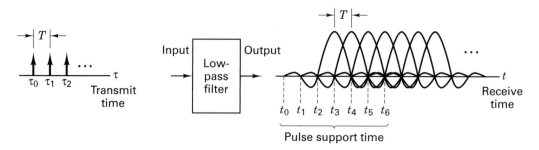

Figure 3.21 Filtered impulse sequence: output versus input.

Nyquist pulse, is shown in Figure 3.22b (with the same normalization and filter rolloff). Inspecting these two pulse shapes, we see that they have a similar appearance, but the square-root Nyquist pulse makes slightly faster transitions, thus its spectrum (root-raised cosine) does not decay as rapidly as the spectrum (raised cosine) of the Nyquist pulse. Another subtle but important difference is that the square-root Nyquist pulse *does not* exhibit zero ISI (you can verify that the pulse tails in Figure 3.22a do not go through zero amplitude at the symbol times). However, if a root-raised cosine filter is used at both the transmitter and the receiver, the product of these transfer functions being a raised cosine, will give rise to an output having zero ISI.

It is interesting to see how the square-root Nyquist pulses appear at the output of a transmitter and how they appear after demodulation with a root-raised cosine MF. Figure 3.23a illustrates an example of sending a sequence of message symbols {+1 +1 −1 +3 +1 +3} from a 4-ary set, where the members of the alphabet set are: {±1, ±3}. Consider that the pulse modulation is 4-ary PAM, and that the pulses have been shaped with a root-raised cosine filter, having a roll-off value of 0.5. The analog waveform in this figure represents the transmitter output. Since the output waveform from any filter is delayed in time, then in Figure 3.23a, the input message symbols (shown as approximate impulses) have been delayed the same amount as the output waveform in order to align the message sequence with its corresponding filtered waveform (the square-root Nyquist shaped-pulse sequence). This is just a visual convenience so that the reader can compare the filter input with its output. It is, of course, only the output analog waveform that is transmitted (or modulated) onto a carrier wave.

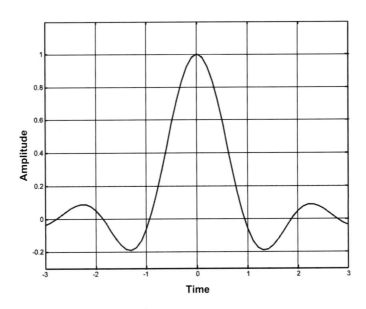

Figure 3.22a   Square-root Nyquist pulse.

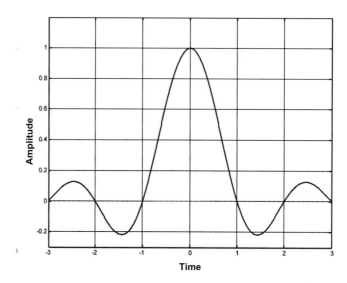

Figure 3.22b Nyquist pulse.

Figure 3.23b shows the same delayed message samples together with the output waveform from the root-raised cosine MF, yielding a raised-cosine transfer function for the overall system. Let us describe a simple test to determine if the filtered output (assuming no noise) contains ISI. It is only necessary to sample the filtered waveform at the times corresponding to the original input samples; if

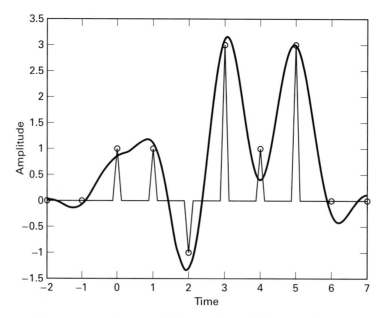

Figure 3.23a Square-root Nyquist-shaped *M*-ary waveform and delayed-input sample values.

Baseband Demodulation/Detection   Chap. 3

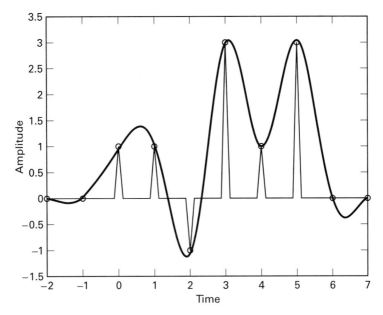

Figure 3.23b Output of raised-cosine matched filter and delayed-input sample values.

the resulting sample values are unchanged from those of the original message, then the filter output has zero ISI (at the sample times). When Figures 3.23a and 3.23b are compared with regard to ISI, it should be apparent that sampling the square-root Nyquist waveform of Figure 3.23a (transmitter output) *will not* yield the exact original samples; however, sampling the Nyquist waveform in Figure 3.23b (MF output) *will* yield the exact original samples. This supports the statement that a Nyquist filter yields zero ISI at the sample points, while any other filter does not do so.

3.4 EQUALIZATION

3.4.1 Channel Characterization

Many communication channels (e.g., telephone, wireless) can be characterized as band-limited linear filters with an impulse response $h_c(t)$ and a frequency response

$$H_c(f) = |H_c(f)|e^{j\theta_c(f)} \tag{3.82}$$

where $h_c(t)$ and $H_c(f)$ are Fourier transform pairs, $|H_c(f)|$ is the channel's amplitude response, and $\theta_c(f)$ is the channel's phase response. In order to achieve ideal (nondistorting) transmission characteristics over a channel, it was shown in Section 1.6.3, that within a signal's bandwidth W, $|H_c(f)|$ *must be constant.* Also, $\theta_c(f)$ *must be a linear function of frequency,* which is tantamount to saying that the delay must be constant for all spectral components of the signal. If $|H_c(f)|$ is not constant within W, then the effect is amplitude distortion. If $\theta_c(f)$ is not a linear function of frequency within W, then the effect is phase distortion. For many channels that exhibit distortion of this type, such as fading channels, amplitude and phase distortion typically occur together. For a transmitted sequence of pulses, such distortion manifests itself as a signal dispersion or "smearing" so that any one pulse in the received demodulated sequence is not well defined. The overlap or smearing, known as *intersymbol interference* (ISI), described in Section 3.3, arises in most modulation systems; it is one of the major obstacles to reliable high-speed data transmission over bandlimited channels. In the broad sense, the name "equalization" refers to any signal processing or filtering technique that is designed to eliminate or reduce ISI.

In Figure 2.1, equalization is partitioned into two broad categories. The first category, *maximum-likelihood sequence estimation* (MLSE), entails making measurements of $h_c(t)$ and then providing a means for adjusting the receiver to the transmission environment. The goal of such adjustments is to enable the detector to make good estimates from the demodulated distorted pulse sequence. With an MLSE receiver, the distorted samples are not reshaped or directly compensated in any way; instead, the mitigating technique for the MLSE receiver is to adjust itself in such a way that it can better deal with the distorted samples. (An example of this method, known as Viterbi equalization, is treated in Section 15.7.1.) The second category, *equalization with filters,* uses filters to compensate the distorted pulses. In this second category, the detector is presented with a sequence of demodulated samples that the equalizer has modified or "cleaned up" from the effects of ISI. Equalizing with filters, the more popular approach and the one described in this section, lends itself to further partitioning. The filters can be described as to whether they are linear devices that contain only feedforward elements *transversal equalizers*), or whether they are nonlinear devices that contain both feedforward and feedback elements (*decision feedback equalizers*). They can be grouped according to the automatic nature of their operation, which may be either *preset* or *adaptive*. They also can be grouped according to the filter's resolution or update rate. Are predetection samples provided only on symbol boundaries, that is, one sample per symbol? If so, the condition is known as *symbol spaced*. Are multiple samples provided for each symbol? If so, this condition is known as *fractionally spaced*.

We now modify Equation (3.77) by letting the receiving/equalizing filter be replaced by a separate receiving filter and equalizing filter, defined by frequency transfer functions $H_r(f)$ and $H_e(f)$, respectively. Also, let the overall system transfer function $H(f)$ be a raised-cosine filter, designated $H_{RC}(f)$. Thus, we write

$$H_{RC}(f) = H_t(f)\, H_c(f)\, H_r(f)\, H_e(f) \qquad (3.83)$$

In practical systems, the channel's frequency transfer function $H_c(f)$ and its impulse response $h_c(t)$ are not known with sufficient precision to allow for a receiver design to yield zero ISI for all time. Usually, the transmit and receive filters are chosen to be matched so that

$$H_{RC}(f) = H_t(f)\, H_r(f) \qquad (3.84)$$

In this way, $H_t(f)$ and $H_r(f)$ each have frequency transfer functions that are the square root of the raised cosine (root-raised cosine). Then, the equalizer transfer function needed to compensate for channel distortion is simply the inverse of the channel transfer function:

$$H_e(f) = \frac{1}{H_c(f)} = \frac{1}{|H_c(f)|} e^{-j\theta_c(f)} \tag{3.85}$$

Sometimes a system frequency transfer function manifesting ISI at the sampling points is purposely chosen (e.g., a Gaussian filter transfer function). The motivation for such a transfer function is to improve bandwidth efficiency, compared with using a raised-cosine filter. When such a design choice is made, the role of the equalizing filter is not only to compensate for the channel-induced ISI, but also to compensate for the ISI brought about by the transmitter and receiver filters [7].

3.4.2 Eye Pattern

An eye pattern is the display that results from measuring a system's response to baseband signals in a prescribed way. On the vertical plates of an oscilloscope we connect the receiver's response to a random pulse sequence. On the horizontal plates we connect a sawtooth wave at the signaling frequency. In other words, the horizontal time base of the oscilloscope is set equal to the symbol (pulse) duration. This setup superimposes the waveform in each signaling interval into a family of traces in a single interval $(0, T)$. Figure 3.24 illustrates the eye pattern that results for binary antipodal (bipolar pulse) signaling. Because the symbols stem from a random source, they are sometimes positive and sometimes negative, and the persistence of the cathode ray tube display allows us to see the resulting pattern shaped as an eye. The width of the opening indicates the time over which sampling for detection might be performed. Of course, the optimum sampling time corresponds to the maximum eye opening, yielding the greatest protection against noise.

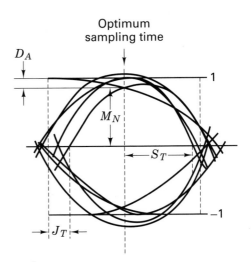

Figure 3.24   Eye Pattern.

If there were no filtering in the system—that is, if the bandwidth corresponding to the transmission of these data pulses were infinite—then the system response would yield ideal rectangular pulse shapes. In that case, the pattern would look like a box rather than an eye. In Figure 3.24, the range of amplitude differences labelled D_A is a measure of distortion caused by ISI, and the range of time differences of the zero crossings labelled J_T is a measure of the timing jitter. Measures of noise margin M_N and sensitivity-to-timing error S_T are also shown in the figure. In general, the most frequent use of the eye pattern is for qualitatively assessing the extent of the ISI. As the eye closes, ISI is increasing; as the eye opens, ISI is decreasing.

3.4.3 Equalizer Filter Types

3.4.3.1 Transversal Equalizer

A training sequence used for equalization is often chosen to be a noise-like sequence, "rich" in spectral content, which is needed to estimate the channel frequency response. In the simplest sense, training might consist of sending a single narrow pulse (approximating an ideal impulse) and thereby learning the impulse response of the channel. In practice, a pseudonoise (PN) signal is preferred over a single pulse for the training sequence because the PN signal has larger average power and hence larger SNR for the same peak transmitted power. For describing the transversal filter, consider that a single pulse was transmitted over a system designated to have a raised-cosine transfer function $H_{RC}(f) = H_t(f) H_r(f)$. Also consider that the channel induces ISI, so that the received demodulated pulse exhibits distortion, as shown in Figure 3.25, such that the pulse sidelobes do not go through zero at sample times adjacent to the mainlobe of the pulse. The distortion can be viewed as positive or negative echoes occurring both before and after the mainlobe. To achieve the desired raised-cosine transfer function, the equalizing filter should have a frequency response $H_e(f)$, as shown in Equation (3.85), such that the actual channel response when multiplied by $H_e(f)$ yields $H_{RC}(f)$. In other words, we would like the equalizing filter to generate a set of canceling echoes.

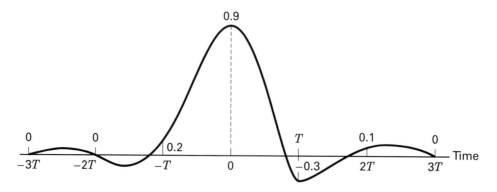

Figure 3.25   Received pulse exhibiting distortion.

Since we are interested in sampling the equalized waveform at only a few predetermined sampling times, then the design of an equalizing filter can be a straightforward task.

The transversal filter, depicted in Figure 3.26, is the most popular form of an easily adjustable equalizing filter consisting of a delay line with T-second taps (where T is the symbol duration). In such an equalizer, the current and past values of the received signal are linearly weighted with equalizer coefficients or tap weights $\{c_n\}$ and are then summed to produce the output. The main contribution is from a central tap, with the other taps contributing echoes of the main signal at symbol intervals on either side of the main signal. If it were possible for the filter to have an infinite number of taps, then the tap weights could be chosen to force the system impulse response to zero at all but one of the sampling times, thus making $H_e(f)$ correspond exactly to the inverse of the channel transfer function in Equation (3.85). Even though an infinite length filter is not realizable, one can still specify practical filters that approximate the ideal case.

In Figure 3.26, the outputs of the weighted taps are amplified, summed, and fed to a coefficient-adjustment device. The tap weights $\{c_n\}$ need to be chosen so as to subtract the effects of interference from symbols adjacent in time to the desired symbol. Consider that there are $(2N + 1)$ taps with weights $c_{-N}, c_{-N+1}, \ldots, c_N$. Output samples $\{z(k)\}$ of the equalizer are then found by convolving the input samples $\{x(k)\}$ and tap weights $\{c_n\}$ as follows:

$$z(k) = \sum_{n=-N}^{N} x(k - n)\, c_n \quad k = -2N, \ldots, 2N \quad n = -N, \ldots, N \qquad (3.86)$$

where $k = 0, \pm 1, \pm 2, \ldots$ is a time index that is shown in parentheses. (Time may take on any range of values.) The index n is used two ways—as a time offset, and as a filter coefficient identifier (which is an address in the filter). When used in the latter

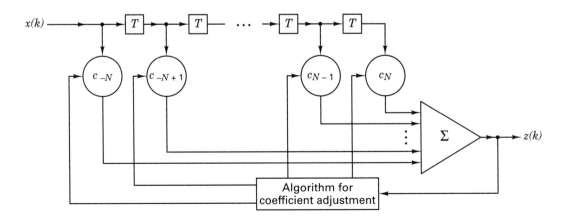

Figure 3.26   Transversal filter.

sense, it is shown as a subscript. By defining the vectors $\mathbf{z}$ and $\mathbf{c}$ and the matrix $\mathbf{x}$ as, respectively,

$$\mathbf{z} = \begin{bmatrix} z(-2N) \\ \vdots \\ z(0) \\ \vdots \\ z(2N) \end{bmatrix} \qquad \mathbf{c} = \begin{bmatrix} c_{-N} \\ \vdots \\ c_0 \\ \vdots \\ c_N \end{bmatrix} \tag{3.87}$$

and

$$\mathbf{x} = \begin{bmatrix} x(-N) & 0 & 0 & \cdots & 0 & 0 \\ x(-N+1) & x(-N) & 0 & \cdots & \cdots & \cdots \\ \vdots & & & & \vdots & \vdots \\ x(N) & x(N-1) & x(N-2) & \cdots & x(-N+1) & x(-N) \\ \vdots & & & & & \vdots \\ 0 & 0 & 0 & \cdots & x(N) & x(N-1) \\ 0 & 0 & 0 & \cdots & 0 & x(N) \end{bmatrix} \tag{3.88}$$

we can describe the relationship among $\{z(k)\}$, $\{x(k)\}$, and $\{c_n\}$ more compactly as

$$\mathbf{z} = \mathbf{x}\,\mathbf{c} \tag{3.89a}$$

And whenever the matrix $\mathbf{x}$ is square, with its rows and columns each having the same dimension as the number of elements in $\mathbf{c}$, we can find $\mathbf{c}$ by solving the following equation:

$$\mathbf{c} = \mathbf{x}^{-1}\mathbf{z} \tag{3.89b}$$

Notice that the size of the vector $\mathbf{z}$ and the number of rows in the matrix $\mathbf{x}$ may be chosen to be any value, because one might be interested in the ISI at sample points far removed from the mainlobe of the pulse in question. In Equations (3.86) through (3.88), the index k was arbitrarily chosen to allow for $4N + 1$ sample points. The vectors $\mathbf{z}$ and $\mathbf{c}$ have dimensions $4N + 1$ and $2N + 1$, respectively, and the matrix $\mathbf{x}$ is nonsquare with dimensions $4N + 1$ by $2N + 1$. Such equations are referred to as an overdetermined set (i.e., there are more equations than unknowns). One can solve such a problem in a deterministic way known as the *zero-forcing* solution, or, in a statistical way, known as the *minimum mean-square error* (MSE) solution.

Zero-Forcing Solution. This solution starts by disposing of the top N and bottom N rows of the matrix $\mathbf{x}$ in Equation (3.88), thereby transforming $\mathbf{x}$ into a square matrix of dimension $2N + 1$ by $2N + 1$, transforming $\mathbf{z}$ into a vector of dimension $2N + 1$, and yielding in Equation (3.89a), a deterministic set of $2N + 1$ simultaneous equations. This zero-forcing solution minimizes the peak ISI distortion by selecting the $\{c_n\}$ weights so that the equalizer output is forced to zero at N sample points on either side of the desired pulse. In other words, the weights are chosen so that

$$z(k) = \begin{cases} 1 & \text{for } k = 0 \\ 0 & \text{for } k = \pm 1, \pm 2, \dots, \pm N \end{cases} \qquad (3.90)$$

Equation (3.89) is used to solve the $2N + 1$ simultaneous equations for the set of $2N + 1$ weights $\{c_n\}$. The required length of the filter (number of tap weights) is a function of how much smearing the channel may introduce. For such an equalizer with finite length, the peak distortion is guaranteed to be minimized only if the eye pattern is initially open. However, for high-speed transmission and channels introducing much ISI, the eye is often closed before equalization [8]. Since the zero-forcing equalizer neglects the effect of noise, it is not always the best system solution.

Example 3.5   A Zero-Forcing Three-Tap Equalizer

Consider that the tap weights of an equalizing transversal filter are to be determined by transmitting a single impulse as a training signal. Let the equalizer circuit in Figure 3.26 be made up of just three taps. Given a received distorted set of pulse samples $\{x(k)\}$, with voltage values 0.0, 0.2, 0.9, −0.3, 0.1, as shown in Figure 3.25, use a zero-forcing solution to find the weights $\{c_{-1}, c_0, c_1\}$ that reduce the ISI so that the equalized pulse samples $\{z(k)\}$ have the values, $\{z(-1) = 0, z(0) = 1, z(1) = 0\}$. Using these weights, calculate the ISI values of the equalized pulse at the sample times $k = \pm 2, \pm 3$. What is the largest magnitude sample contributing to ISI, and what is the sum of all the ISI magnitudes?

Solution

For the channel impulse response specified, Equation (3.89) yields

$$\mathbf{z} = \mathbf{x}\,\mathbf{c}$$

or

$$\begin{bmatrix} 0 \\ 1 \\ 0 \end{bmatrix} = \begin{bmatrix} x(0) & x(-1) & x(-2) \\ x(1) & x(0) & x(-1) \\ x(2) & x(1) & x(0) \end{bmatrix} \begin{bmatrix} c_{-1} \\ c_0 \\ c_1 \end{bmatrix}$$

$$= \begin{bmatrix} 0.9 & 0.2 & 0 \\ -0.3 & 0.9 & 0.2 \\ 0.1 & -0.3 & 0.9 \end{bmatrix} \begin{bmatrix} c_{-1} \\ c_0 \\ c_1 \end{bmatrix}$$

Solving these three simultaneous equations results in the following weights:

$$\begin{bmatrix} c_{-1} \\ c_0 \\ c_1 \end{bmatrix} = \begin{bmatrix} -0.2140 \\ 0.9631 \\ 0.3448 \end{bmatrix}$$

The values of the equalized pulse samples $\{z(k)\}$ corresponding to sample times $k = -3, -2, -1, 0, 1, 2, 3$ are computed by using the preceding weights in Equation (3.89a), yielding

$$0.0000, \ -0.0428, \ 0.0000, \ 1.0000, \ 0.0000, \ -0.0071, \ 0.0345$$

The sample of greatest magnitude contributing to ISI equals 0.0428, and the sum of all the ISI magnitudes equals 0.0844. It should be clear that this three-tap equalizer has forced the sample points on either side of the equalized pulse to be zero. If the equalizer is made longer than three taps, more of the equalized sample points can be forced to a zero value.

Minimum MSE Solution. A more robust equalizer is obtained if the $\{c_n\}$ tap weights are chosen to minimize the mean-square error (MSE) of all the ISI terms plus the noise power at the output of the equalizer [9]. MSE is defined as the expected value of the squared difference between the desired data symbol and the estimated data symbol. One can use the set of overdetermined equations to obtain a minimum MSE solution by multiplying both sides of Equation (3.89a) by $\mathbf{x}^T$, which yields [10]

$$\mathbf{x}^T\mathbf{z} = \mathbf{x}^T\mathbf{x}\mathbf{c} \tag{3.91a}$$

and

$$\mathbf{R}_{xz} = \mathbf{R}_{xx}\mathbf{c} \tag{3.91b}$$

where $\mathbf{R}_{xz} = \mathbf{x}^T\mathbf{z}$ is called the *cross-correlation* vector and $\mathbf{R}_{xx} = \mathbf{x}^T\mathbf{x}$ is called the *autocorrelation* matrix of the input noisy signal. In practice, $\mathbf{R}_{xz}$ and $\mathbf{R}_{xx}$ are unknown *a priori,* but can be approximated by transmitting a test signal over the channel and using time average estimates to solve for the tap weights from Equation (3.91), as follows:

$$\mathbf{c} = \mathbf{R}_{xx}^{-1}\mathbf{R}_{xz} \tag{3.92}$$

In the case of the deterministic zero-forcing solution, the $\mathbf{x}$ matrix must be square. But to achieve the minimum MSE (statistical) solution, one starts with an overdetermined set of equations and hence a *nonsquare* $\mathbf{x}$ matrix, which then gets transformed to a *square* autocorrelation matrix $\mathbf{R}_{xx} = \mathbf{x}^T\mathbf{x}$, yielding a set of $2N + 1$ simultaneous equations, whose solution leads to tap weights that minimize the MSE. The size of the vector $\mathbf{c}$ and the number of columns of the matrix $\mathbf{x}$ correspond to the number of taps in the equalizing filter. Most high-speed telephone-line modems use an MSE weight criterion because it is superior to a zero-forcing criterion; it is more robust in the presence of noise and large ISI [8].

Example 3.6   A Minimum MSE 7-Tap Equalizer

Consider that the tap weights of an equalizing transversal filter are to be determined by transmitting a single impulse as a training signal. Let the equalizer circuit in Figure 3.26 be made up of seven taps. Given a received distorted set of pulse samples $\{x(k)\}$, with values 0.0108, −0.0558, 0.1617, 1.0000, −0.1749, 0.0227, 0.0110, use a minimum MSE solution to find the value of the weights $\{c_n\}$ that will minimize the ISI. With these weights, calculate the resulting values of the equalized pulse samples at the following times: $\{k = 0, \pm1, \pm2, \ldots, \pm6\}$. What is the largest magnitude sample contributing to ISI, and what is the sum of all the ISI magnitudes?

Solution

For a seven-tap filter ($N = 3$), one can form the **x** matrix in Equation (3.88) that has dimensions $4N + 1$ by $2N + 1 = 13 \times 7$:

$$
\mathbf{x} = \begin{bmatrix}
0.0110 & 0 & 0 & 0 & 0 & 0 & 0 \\
0.0227 & 0.0110 & 0 & 0 & 0 & 0 & 0 \\
-0.1749 & 0.0227 & 0.0110 & 0 & 0 & 0 & 0 \\
1.0000 & -0.1749 & 0.0227 & 0.0110 & 0 & 0 & 0 \\
0.1617 & 1.0000 & -0.1749 & 0.0227 & 0.0110 & 0 & 0 \\
-0.0558 & 0.1617 & 1.0000 & -0.1749 & 0.0227 & 0.0110 & 0 \\
0.0108 & -0.0558 & 0.1617 & 1.0000 & -0.1749 & 0.0227 & 0.0110 \\
0 & 0.0108 & -0.0558 & 0.1617 & 1.0000 & -0.1749 & 0.0227 \\
0 & 0 & 0.0108 & -0.0558 & 0.1617 & 1.0000 & -0.1749 \\
0 & 0 & 0 & 0.0108 & -0.0558 & 0.1617 & 1.0000 \\
0 & 0 & 0 & 0 & 0.0108 & -0.0558 & 0.1617 \\
0 & 0 & 0 & 0 & 0 & 0.0108 & -0.0558 \\
0 & 0 & 0 & 0 & 0 & 0 & 0.0108
\end{bmatrix}
$$

Using this **x** matrix, one can form the autocorrelation matrix $\mathbf{R}_{xx}$ and the cross-correlation vector $\mathbf{R}_{xz}$, defined in Equation (3.91). With the help of a computer to invert $\mathbf{R}_{xx}$ and perform matrix multiplication, the solution for the tap weights $\{c_{-3}, c_{-2}, c_{-1}, c_0, c_1, c_2, c_3\}$ shown in Equation (3.92) yields

$$-0.0116, \ 0.0108, \ 0.1659, \ 0.9495, \ -0.1318, \ 0.0670, \ -0.0269$$

Using these weights in Equation (3.89a), we solve for the 13 equalized samples $\{z(k)\}$ at times $k = -6, -5, \ldots, 5, 6$:

$$-0.0001, \ -0.0001, \ 0.0041, \ 0.0007, \ 0.0000, \ -0.0000, \ 1.0000, \ 0.0003,$$
$$-0.0007, \ 0.0015, \ -0.0095, \ 0.0022, \ -0.0003$$

The largest magnitude sample contributing to ISI equals 0.0095, and the sum of all the ISI magnitudes equals 0.0195.

3.4.3.2 Decision Feedback Equalizer

The basic limitation of a linear equalizer, such as the transversal filter, is that it performs poorly on channels having spectral nulls [11]. Such channels are often encountered in mobile radio applications. A decision feedback equalizer (DFE) is a nonlinear equalizer that uses previous detector decisions to eliminate the ISI on pulses that are currently being demodulated. The ISI being removed was caused by the tails of previous pulses; in effect, the distortion on a current pulse that was caused by previous pulses is subtracted.

Figure 3.27 shows a simplified block diagram of a DFE where the forward filter and the feedback filter can each be a linear filter, such as a transversal filter. The figure also illustrates how the filter tap weights are updated adaptively. (See the following section.) The nonlinearity of the DFE stems from the nonlinear

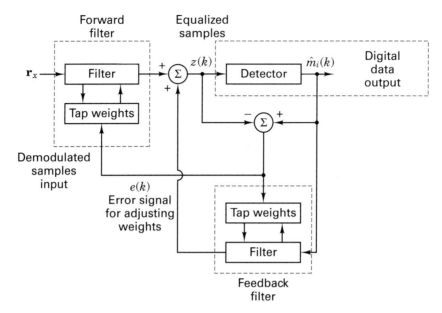

Figure 3.27 Decision Feedback Equalizer.

characteristic of the detector that provides an input to the feedback filter. The basic idea of a DFE is that if the values of the symbols previously detected are known (past decisions are assumed to be correct), then the ISI contributed by these symbols can be canceled out exactly at the output of the forward filter by subtracting past symbol values with appropriate weighting. The forward and feedback tap weights can be adjusted simultaneously to fulfill a criterion such as minimizing the MSE.

When only a forward filter is used, the output of the filter contains channel noise contributed from every sample in the filter. The advantage of a DFE implementation is that the feedback filter, which is additionally working to remove ISI, operates on noiseless quantized levels, and thus its output is free of channel noise.

3.4.4 Preset and Adaptive Equalization

On channels whose frequency responses are known and time invariant, the channel characteristics can be measured and the filter's tap weights adjusted accordingly. If the weights remain fixed during transmission of data, the equalization is called *preset* equalization; one very simple method of preset equalization consists of setting the tap weights $\{c_n\}$ according to some average knowledge of the channel. This was used for data transmission over voice-grade telephone lines at less than 2400 bit/s. Another preset method consists of transmitting a training sequence that is compared at the receiver with a locally generated sequence. The differences between the two sequences are used to set $\{c_n\}$. The significant aspect of any preset method is that it is done once at the start of transmission or seldom (when transmission is broken and needs to be reestablished).

Another type of equalization, capable of tracking a slowly time-varying channel response, is known as *adaptive* equalization. It can be implemented to perform tap-weight adjustments periodically or continually. Periodic adjustments are accomplished by periodically transmitting a preamble or short training sequence of digital data that is known in advance by the receiver. The receiver also uses the preamble to detect start of transmission, to set the automatic gain control (AGC) level, and to align internal clocks and local oscillator with the received signal. Continual adjustments are accomplished by replacing the known training sequence with a sequence of data symbols estimated from the equalizer output and treated as known data. When performed continually and automatically in this way, the adaptive procedure (the most popular) is referred to as *decision directed* [11]. The name "decision directed" is not to be confused with decision feedback (DFE). Decision directed only addresses how filter tap weights are adjusted—that is, with the help of a signal from the detector. DFE, however, refers to the fact that there exists an additional filter that operates on the detector output and recursively feeds back a signal to the detector input. Thus, with DFE there are two filters, a feed-forward filter and a feedback filter that process the data and help mitigate the ISI.

A disadvantage of preset equalization is that it requires an initial training period that must be invoked at the start of any new transmission. Also, a time-varying channel can degrade system performance due to ISI, since the tap weights are fixed. Adaptive equalization, particularly *decision-directed* adaptive equalization, successfully cancels ISI when the initial probability of error exceeds one percent, (rule of thumb). If the probability of error exceeds one percent, the decision directed equalizer might not converge. A common solution to this problem is to initialize the equalizer with an alternate process, such as a preamble to provide good channel-error performance, and then switch to the decision-directed mode. To avoid the overhead represented by a preamble, many systems designed to operate in a continuous broadcast mode use *blind equalization* algorithms to form initial channel estimates. These algorithms adjust filter coefficients in response to sample statistics rather than in response to sample decisions [11].

Automatic equalizers use iterative techniques to estimate the optimum coefficients. The simultaneous equations described in Equation (3.89) do not include the affects of channel noise. To obtain a stable solution to the filter weights, it is necessary that the data be averaged to obtain stable signal statistics, or the noisy solutions obtained from the noisy data must be averaged. Considerations of algorithm complexity and numerical stability most often lead to algorithms that average noisy solutions. The most robust of this class of algorithm is the least-mean-square (LMS) algorithm. Each iteration of this algorithm uses a noisy estimate of the error *gradient* to adjust the weights in the direction to reduce the average mean-square error. The noisy gradient is simply the product $e(k)\,\mathbf{r}_x$ of an error scalar $e(k)$ and the data vector $\mathbf{r}_x$. The vector $\mathbf{r}_x$ is the vector of noise-corrupted channel samples residing in the equalizer filter at time k. Earlier, an impulse was transmitted and the equalizing filter operated on a sequence of samples (a vector) that represented the impulse response of the channel. We displayed these received samples (in time-shifted fashion) as the matrix **x.** Now, rather than dealing with the response to an impulse, consider that data is sent and thus the vector of received samples $\mathbf{r}_x$ at the input to the filter (Figure 3.27) represents the data response of the channel. The error is formed as the difference between the desired output signal and the filter output signal and is given by

$$e(k) = z(k) - \hat{z}(k) \tag{3.93}$$

where $z(k)$ is the desired output signal (a sample free of ISI) and $\hat{z}(k)$ is an estimate of $z(k)$ at time k out of the filter (into the quantizer of Figure 3.27), which is obtained as follows:

$$\hat{z}(k) = \mathbf{c}^T\mathbf{r}_x = \sum_{n=-N}^{N} x(k-n)c_n \tag{3.94}$$

In Equation (3.94), the summation represents a convolution of the input data samples with the $\{c_n\}$ tap weights, where c_n refers to the nth tap weight at time k, and $\mathbf{c}^T$ is the transpose of the weight vector at time k. We next show the iterative process that updates the set of weights at each time k as

$$\mathbf{c}(k+1) = \mathbf{c}(k) + \Delta e(k)\mathbf{r}_x \tag{3.95}$$

where $\mathbf{c}(k)$ is the vector of filter weights at time k, and Δ is a small term that limits the coefficient step size and thus controls the rate of convergence of the algorithm as well as the variance of the steady state solution. This simple relationship is a consequence of the orthogonality principle that states that the error formed by an optimal solution is orthogonal to the processed data. Since this is a recursive algorithm (in the weights), care must be exercised to assure algorithm stability. Stability is assured if the parameter Δ is smaller than the reciprocal of the energy of the data in the filter. When stable, this algorithm converges in the mean to the optimal solution but exhibits a variance proportional to the parameter Δ. Thus, while we want the convergence parameter Δ to be large for fast convergence but not so large as to be unstable, we also want it to be small enough for low variance. The parameter Δ is usually set to a fixed small amount [12] to obtain a low-variance steady-state tap-weight solution. Schemes exist that permit Δ to change from large values during initial acquisition to small values for stable steady-state solutions [13].

Note that Equations (3.93) through (3.95) are shown in the context of real signals. When the receiver is implemented in quadrature fashion, such that the signals appear as real and imaginary (or inphase and quadrature) ordered pairs, then each line in Figure 3.27 actually consists of two lines, and Equations (3.93) through (3.95) need to be expressed with complex notation. (Such quadrature implementation is treated in greater detail in Sections 4.2.1 and 4.6.)

3.4.5 Filter Update Rate

Equalizer filters are classified by the rate at which the input signal is sampled. A transversal filter with taps spaced T seconds apart, where T is the symbol time, is called a *symbol-spaced* equalizer. The process of sampling the equalizer output at a rate $1/T$ causes aliasing if the signal is not strictly bandlimited to $1/T$ hertz—that is, the signal's spectral components spaced $1/T$ hertz apart are folded over and superimposed. The aliased version of the signal may exhibit spectral nulls [8]. A filter update rate that is greater than the symbol rate helps to mitigate this difficulty. Equalizers using this technique are called *fractionally-spaced* equalizers. With a fractionally spaced equalizer, the filter taps are spaced at

$$T' \leq \frac{T}{(1 + r)} \tag{3.96}$$

seconds apart, where r denotes the excess bandwidth. In other words, the received signal bandwidth is

$$W \leq \frac{(1 + r)}{T} \tag{3.97}$$

The goal is to choose T' so that the equalizer transfer function $H_e(f)$ becomes sufficiently broad to accommodate the whole signal spectrum. Note that the signal at the output of the equalizer is still sampled at a rate $1/T$, but since the tap weights are spaced T' seconds apart (the equalizer input signal is sampled at a rate $1/T'$), the equalization action operates on the received signal before its frequency components are aliased. Equalizer simulations over voice-grade telephone lines, with $T' = T/2$, confirm that such fractionally-spaced equalizers outperform symbol-spaced equalizers [14].

3.5 CONCLUSION

In this chapter, we described the detection of binary signals plus Gaussian noise in terms of two basic steps. In the first step the received waveform is reduced to a single number $z(T)$, and in the second step a decision is made as to which signal was transmitted, on the basis of comparing $z(T)$ to a threshold. We discussed how to best choose this threshold. We also showed that a linear filter known as a matched filter or correlator is the optimum choice for maximizing the output signal-to-noise ratio and thus minimizing the probability of error.

We defined intersymbol interference (ISI) and explained the importance of Nyquist's work in establishing a theoretical minimum bandwidth for symbol detection without ISI. We partitioned error-performance degradation into two main types. The first is a simple loss in signal-to-noise ratio. The second, resulting from distortion, is a bottoming-out of the error probability versus the E_b/N_0 curve.

Finally, we described equalization techniques that can be used to mitigate the effects of ISI.

REFERENCES

1. Nyquist, H., "Thermal Agitation of Electric Charge in Conductors," *Phys. Rev.,* vol. 32, July 1928, pp. 110–113.

2. Van Trees, H. L., *Detection, Estimation, and Modulation Theory,* Part 1, John Wiley & Sons, Inc., New York, 1968.

3. Arthurs, E., and Dym, H., "On the Optimum Detection of Digital Signals in the Presence of White Gaussian Noise—A Geometric Interpretation of Three Basic Data Transmission Systems," *IRE Trans. Commun. Syst.,* December 1962.

4. Wozencraft, J. M. and Jacobs, I. M., *Principles of Communication Engineering,* John Wiley & Sons, Inc., New York, 1965.

5. Borjesson, P. O., and Sundberg, C. E., "Simple Approximations of the Error Function $Q(x)$ for Communications Applications," *IEEE Trans. Commun.,* vol. COM27, Mar. 1979, pp. 639–642.

6. Nyquist, H., "Certain Topics of Telegraph Transmission Theory," *Trans. Am. Inst. Electr. Eng.,* vol. 47, Apr. 1928, pp. 617–644.

7. Hanzo, L. and Stefanov, J., "The Pan-European Digital Cellular Mobile Radio System—Known as GSM," *Mobile Radio Communications,* edited by R. Steele, Chapter 8, Pentech Press, London, 1992.

8. Qureshi, S. U. H., "Adaptive Equalization," *Proc. IEEE,* vol. 73, no. 9, September 1985, pp. 1340–1387.

9. Lucky, R. W., Salz, J., and Weldon, E. J., Jr., *Principles of Data Communications,* McGraw Hill Book Co., New York, 1968.

10. Harris, F., and Adams, B., "Digital Signal Processing to Equalize the Pulse Response of Non Synchronous Systems Such as Encountered in Sonar and Radar," *Proc. of the Twenty-Fourth Annual ASILOMAR Conference on Signals, Systems, and Computers,* Pacific Grove, California, November 5–7, 1990.

11. Proakis, J. G., *Digital Communications,* McGraw-Hill Book Company, New York, 1983.

12. Feuer, A., and Weinstein, E., "Convergence Analysis of LMS Filters with Uncorrelated Gaussian Data, "*IEEE Trans. on ASSP,* vol. V-33 pp. 220–230, 1985.

13. Macchi, O., *Adaptive Processing: Least Mean Square Approach With Applications in Transmission,* John Wiley & Sons, New York, 1995.

14. Benedetto, S., Biglieri, E., and Castellani, V., *Digital Transmission Theory,* Prentice Hall, 1987.

PROBLEMS

3.1. Determine whether or not $s_1(t)$ and $s_2(t)$ are orthogonal over the interval $(-1.5T_2 < t < 1.5T_2)$, where $s_1(t) = \cos(2\pi f_1 t + \phi_1)$, $s_2(t) = \cos(2\pi f_2 t + \phi_2)$, and $f_2 = 1/T_2$ for the following cases.

    (a) $f_1 = f_2$ and $\phi_1 = \phi_2$

    (b) $f_1 = \frac{1}{3}f_2$ and $\phi_1 = \phi_2$

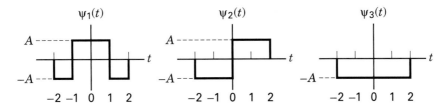

$\psi_1(t)$ $\psi_2(t)$ $\psi_3(t)$

Figure P3.1

(c) $f_1 = 2f_2$ and $\phi_1 = \phi_2$
(d) $f_1 = \pi f_2$ and $\phi_1 = \phi_2$
(e) $f_1 = f_2$ and $\phi_1 = \phi_2 + \pi/2$
(f) $f_1 = f_2$ and $\phi_1 = \phi_2 + \pi$

3.2. **(a)** Show that the three functions illustrated in Figure P3.1 are pairwise orthogonal over the interval $(-2, 2)$.

   (b) Determine the value of the constant, A, that makes the set of functions in part (a) an orthonormal set.

   (c) Express the following waveform, $x(t)$, in terms of the orthonormal set of part (b).

$$x(t) = \begin{cases} 1 & \text{for } 0 \le t \le 2 \\ 0 & \text{otherwise} \end{cases}$$

3.3. Consider the functions

$$\psi_1(t) = \exp(-|t|) \quad \text{and} \quad \psi_2 = 1 - A \exp(-2|t|)$$

Determine the constant, A, such that $\psi_1(t)$ and $\psi_2(t)$ are orthogonal over the interval $(-\infty, \infty)$.

3.4. Assume that in a binary digital communication system, the signal component out of the correlator receiver is $a_i(T) = +1$ or -1 V with equal probability. If the Gaussian noise at the correlator output has unit variance, find the probability of a bit error.

3.5. A bipolar binary signal, $s_i(t)$, is a +1- or -1-V pulse during the interval $(0, T)$. Additive white Gaussian noise having two-sided power spectral density of 10^{-3} W/Hz is added to the signal. If the received signal is detected with a matched filter, determine the maximum bit rate that can be sent with a bit error probability of $P_B \le 10^{-3}$.

3.6. Bipolar pulse signals, $s_i(t)$ $(i = 1, 2)$, of amplitude ± 1 V are received in the presence of AWGN that has a variance of 0.1 V^2. Determine the optimum (minimum probability of error) detection threshold, γ_0, for matched filter detection if the a priori probabilities are: **(a)** $P(s_1) = 0.5$; **(b)** $P(s_1) = 0.7$; **(c)** $P(s_1) = 0.2$. **(d)** Explain the effect of the a priori probabilities on the value of γ_0. {*Hint:* Refer to Equations (B.10) to (B.12).}

3.7. A binary communication system transmits signals $s_i(t)$ $(i = 1, 2)$. The receiver test statistic $z(T) = a_i + n_0$, where the signal component a_i is either $a_1 = +1$ or $a_2 = -1$ and the noise component n_0 is uniformly distributed, yielding the conditional density functions $p(z|s_i)$ given by

$$p(z|s_1) = \begin{cases} \frac{1}{2} & \text{for } -0.2 \le z \le 1.8 \\ 0 & \text{otherwise} \end{cases}$$

and

$$p(z|s_2) = \begin{cases} \frac{1}{2} & \text{for } -1.8 \leq z \leq 0.2 \\ 0 & \text{otherwise} \end{cases}$$

Find the probability of a bit error, P_B, for the case of equally likely signaling and the use of an optimum decision threshold.

3.8. (a) What is the theoretical minimum system bandwidth needed for a 10-Mbits/s signal using 16-level PAM without ISI?

(b) How large can the filter roll-off factor be if the allowable system bandwidth is 1.375 MHz?

3.9. A voice signal (300 to 3300 Hz) is digitized such that the quantization distortion $\leq \pm 0.1\%$ of the peak-to-peak signal voltage. Assume a sampling rate of 8000 samples/s and a multilevel PAM waveform with $M = 32$ levels. Find the theoretical minimum system bandwidth that avoids ISI.

3.10. Binary data at 9600 bits/s are transmitted using 8-ary PAM modulation with a system using a raised cosine roll-off filter characteristic. The system has a frequency response out to 2.4 kHz.

(a) What is the symbol rate?

(b) What is the roll-off factor of the filter characteristic?

3.11. A voice signal in the range 300 to 3300 Hz is sampled at 8000 samples/s. We may transmit these samples directly as PAM pulses or we may first convert each sample to a PCM format and use binary (PCM) waveforms for transmission.

(a) What is the minimum system bandwidth required for the detection of PAM with no ISI and with a filter roll-off characteristic of $r = 1$?

(b) Using the same filter roll-off characteristic, what is the minimum bandwidth required for the detection of binary (PCM) waveforms if the samples are quantized to eight levels?

(c) Repeat part (b) using 128 quantization levels.

3.12. An analog signal is PCM formatted and transmitted using binary waveforms over a channel that is bandlimited to 100 kHz. Assume that 32 quantization levels are used and that the overall equivalent transfer function is of the raised cosine type with roll-off $r = 0.6$.

(a) Find the maximum bit rate that can be used by this system without introducing ISI.

(b) Find the maximum bandwidth of the original analog signal that can be accommodated with these parameters.

(c) Repeat parts (a) and (b) for transmission with 8-ary PAM waveforms.

3.13. Assume that equally-likely RZ binary pulses are coherently detected over a Gaussian channel with $N_0 = 10^{-8}$ Watt/Hz. Assume that synchronization is perfect, and that the received pulses have an amplitude of 100 mV. If the bit-error probability specification is $P_B = 10^{-3}$, find the largest data rate that can be transmitted using this system.

3.14. Consider that NRZ binary pulses are transmitted along a cable that attenuates the signal power by 3 dB (from transmitter to receiver). The pulses are coherently detected at the receiver, and the data rate is 56 kbit/s. Assume Gaussian noise with $N_0 = 10^{-6}$ Watt/Hz. What is the minimum amount of power needed at the transmitter in order to maintain a bit-error probability of $P_B = 10^{-3}$?

3.15. Show that the Nyquist minimum bandwidth for a random binary sequence sent with ideal-shaped bipolar pulses is the same as the noise equivalent bandwidth. Hint: the power spectral density for a random bipolar sequence is given in Equation (1.38) and the noise equivalent bandwidth is defined in Section 1.7.2.

3.16. Consider the 4-ary PAM-modulated sequence of message symbols {+1 +1 −1 +3 +1 +3}, where the members of the alphabet set are: {±1, ±3}. The pulses have been shaped with a root-raised cosine filter such that the support time of each filtered pulse is 6-symbol times, and the transmitted sequence is the analog waveform shown in Figure 3.23a. Note that the waveform appears "smeared" due to the filter-induced ISI. Show how a bank of N correlators can be implemented to perform matched-filter demodulation of the received pulse sequence, $r(t)$, where N corresponds to the number of symbols in the pulse-support time. [*Hint:* For the bank of correlators, use reference signals of the form $s_1(t − kT)$, where $k = 0, \ldots, 5$ and T is the symbol time.]

3.17. A desired impulse response of a communication system is the ideal $h(t) = \delta(t)$, where $\delta(t)$ is the impulse function. Assume that the channel introduces ISI so that the overall impulse response becomes $h(t) = \delta(t) + \alpha\delta(t − T)$, where $\alpha < 1$, and T is the symbol time. Derive an expression for the impulse response of a zero-forcing filter that will equalize the effects of ISI. Demonstrate that this filter suppresses the ISI. If the resulting suppression is deemed inadequate, how can the filter design be modified to increase the ISI suppression further?

3.18. The result of a single pulse (impulse) transmission is a received sequence of samples (impulse response), with values 0.1, 0.3, −0.2, 1.0, 0.4, −0.1, 0.1, where the leftmost sample is the earliest. The value 1.0 corresponds to the mainlobe of the pulse, and the other entries correspond to adjacent samples. Design a 3-tap transversal equalizer that forces the ISI to be zero at one sampling point on each side of the mainlobe. Calculate the values of the equalized output samples at times $k = 0, \pm1, \ldots, \pm4$. After equalization, what is the largest magnitude sample contributing to ISI, and what is the sum of all the ISI magnitudes?

3.19. Repeat problem 3.18 for the case of a channel impulse response described by the following received samples: 0.01, 0.02, −0.03, 0.1, 1.0, 0.2, −0.1, 0.05, 0.02. Use a computer to find the weights of a nine-tap transversal equalizer to meet the minimum MSE criterion. Calculate the values of the equalized output pulses at times $k = 0, \pm1, \ldots, \pm8$. After equalization, what is the largest magnitude sample contributing to ISI, and what is the sum of all the ISI magnitudes?

3.20. In this chapter, it has been emphasized that signal-processing devices, such as multipliers and integrators, typically deal with signals having units of *volts*. Therefore, the gains of such processors must accommodate these units. Draw a block diagram of a product-integrator showing the signal units on each of the wires, and the device gain in each of the blocks (Hint: see Section 3.2.5.1).

QUESTIONS

3.1. In the case of *baseband* signaling, the received waveforms are already in a pulse-like form. Why then, is a demodulator needed to recover the pulse waveform? (See Chapter 3, introduction.)

3.2. Why is E_b/N_0 a natural figure-of-merit for digital communication systems? (See Section 3.1.5.)

3.3. When representing timed events, what dilemma can easily result in confusing the most-significant bit (MSB) and the least-significant bit (LSB)? (See Section 3.2.3.1.)

3.4. The term *matched-filter* is often used synonymously with *correlator*. How is that possible when their mathematical operations are different? (See Section 3.2.3.1.)

3.5. Describe the two fair ways of comparing different curves that depict bit-error probability versus E_b/N_0. (See Section 3.2.5.3.)

3.6. Are there other pulse-shaping filter functions, besides the *raised-cosine,* that exhibit zero ISI? (See Section 3.3.)

3.7. Describe a reasonable goal in endeavoring to *compress bandwidth* to the minimum possible, without incurring ISI. (See Section 3.3.1.1.)

3.8. The error performance of digital signaling suffers primarily from two degradation types: *loss* in signal-to-noise ratio, and *distortion* resulting in an irreducible bit-error probability. How do they differ? (See Section 3.3.2.)

3.9. Often times, providing more E_b/N_0 will not mitigate the degradation due to *intersymbol interference* (ISI). Explain why this is the case. (See Section 3.3.2.)

3.10. Describe the difference between equalizers that use a *zero-forcing* solution, and those that use a *minimum mean-square error* solution? (See Section 3.4.3.1.)

EXERCISES

Using the Companion CD, run the exercises associated with Chapter 3.

CHAPTER 4

Bandpass Modulation
and Demodulation

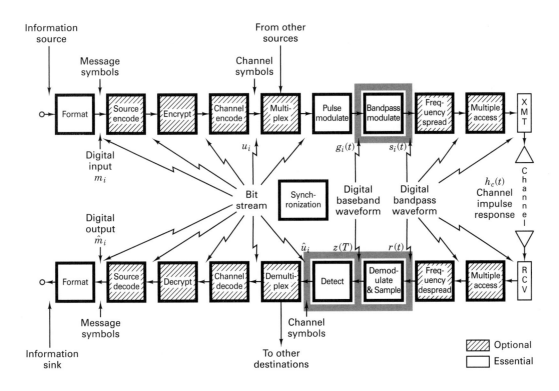

4.1 WHY MODULATE?

Digital modulation is the process by which digital symbols are transformed into waveforms that are compatible with the characteristics of the channel. In the case of baseband modulation, these waveforms usually take the form of shaped pulses. But in the case of *bandpass modulation* the shaped pulses modulate a sinusoid called a *carrier wave,* or simply a *carrier;* for radio transmission the carrier is converted to an electromagnetic (EM) field for propagation to the desired destination. One might ask why it is necessary to use a carrier for the radio transmission of baseband signals. The answer is as follows. The transmission of EM fields through space is accomplished with the use of antennas. The size of the antenna depends on the wavelength λ and the application. For cellular telephones, antennas are typically $\lambda/4$ in size, where wavelength is equal to c/f, and c, the speed of light, is 3×10^8 m/s. Consider sending a baseband signal (say, $f = 3000$ Hz) by coupling it to an antenna directly without a carrier wave. How large would the antenna have to be? Let us size it by using the telephone industry benchmark of $\lambda/4$ as the antenna dimension. For the 3,000 Hz baseband signal, $\lambda/4 = 2.5 \times 10^4$ $m \approx 15$ miles. To transmit a 3,000 Hz signal through space, *without carrier-wave modulation,* an antenna that spans 15 miles would be required. However, if the baseband information is first modulated on a higher frequency carrier, for example a 900 MHz carrier, the equivalent antenna diameter would be about 8 cm. For this reason, carrier-wave or bandpass modulation is an essential step for all systems involving radio transmission.

Bandpass modulation can provide other important benefits in signal transmission. If more than one signal utilizes a single channel, modulation may be used to separate the different signals. Such a technique, known as *frequency-division multiplexing,* is discussed in Chapter 11. Modulation can be used to minimize the effects of interference. A class of such modulation schemes, known as *spread-spectrum modulation,* requires a system bandwidth much larger than the minimum bandwidth that would be required by the message. The trade-off of bandwidth for interference rejection is considered in Chapter 12. Modulation can also be used to place a signal in a frequency band where design requirements, such as filtering and amplification, can be easily met. This is the case when radio-frequency (RF) signals are converted to an intermediate frequency (IF) in a receiver.

4.2 DIGITAL BANDPASS MODULATION TECHNIQUES

Bandpass modulation (either analog or digital) is the process by which an information signal is converted to a sinusoidal waveform; for digital modulation, such a sinusoid of duration T is referred to as a digital symbol. The sinusoid has just three features that can be used to distinguish it from other sinusoids: amplitude, frequency, and phase. Thus bandpass modulation can be defined as the process whereby the amplitude, frequency, or phase of an RF carrier, or a combination of them, is varied in accordance with the information to be transmitted. The general form of the carrier wave is

$$s(t) = A(t) \cos \theta(t) \qquad (4.1)$$

where $A(t)$ is the time-varying amplitude and $\theta(t)$ is the time-varying angle. It is convenient to write

$$\theta(t) = \omega_0 t + \phi(t) \qquad (4.2)$$

so that

$$s(t) = A(t) \cos [\omega_0 t + \phi(t)] \qquad (4.3)$$

where ω_0 is the *radian frequency* of the carrier and $\phi(t)$ is the *phase.* The terms f and ω will each be used to denote frequency. When f is used, frequency in hertz is intended; when ω is used, frequency in radians per second is intended. The two frequency parameters are related by $\omega = 2\pi f$.

The basic *bandpass modulation/demodulation* types are listed in Figure 4.1. When the receiver exploits knowledge of the carrier's phase to detect the signals, the process is called *coherent detection;* when the receiver does not utilize such phase reference information, the process is called *noncoherent detection.* In digital communications, the terms *demodulation* and *detection* are often used interchangeably, although demodulation emphasizes waveform recovery, and detection entails the process of symbol decision. In ideal coherent detection, there is available at the receiver a prototype of each possible arriving signal. These prototype waveforms attempt to duplicate the transmitted signal set in every respect, even RF phase. The

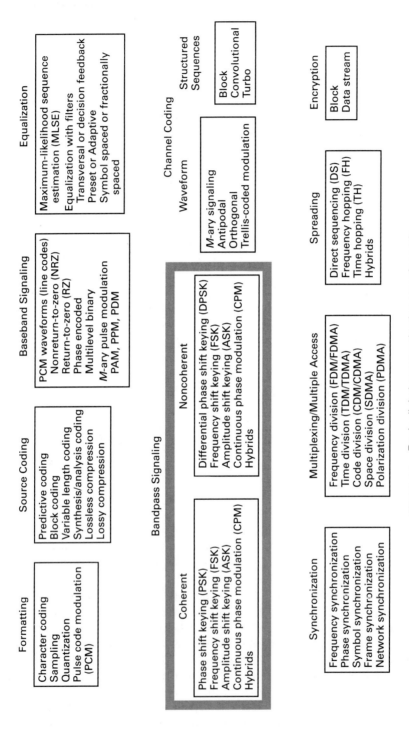

Figure 4.1 Basic digital communication transformations.

receiver is then said to be *phase locked* to the incoming signal. During demodulation, the receiver multiplies and integrates (correlates) the incoming signal with each of its prototype replicas. Under the heading of coherent modulation/demodulation in Figure 4.1 are listed phase shift keying (PSK), frequency shift keying (FSK), amplitude shift keying (ASK), continuous phase modulation (CPM), and hybrid combinations. The basic bandpass modulation formats are discussed in this chapter. Some specialized formats, such as offset quadrature PSK (OQPSK), minimum shift keying (MSK) belonging to the CPM class, and quadrature amplitude modulation (QAM), are treated in Chapter 9.

Noncoherent demodulation refers to systems employing demodulators that are designed to operate without knowledge of the absolute value of the incoming signal's phase; therefore, phase estimation is not required. Thus the advantage of noncoherent over coherent systems is reduced complexity, and the price paid is increased probability of error (P_E). In Figure 4.1 the modulation/demodulation types that are listed in the noncoherent column, DPSK, FSK, ASK, CPM, and hybrids, are similar to those listed in the coherent column. We had implied that phase information is not used for noncoherent reception; how do you account for the fact that there is a form of phase shift keying under the noncoherent heading? It turns out that an important form of PSK can be classified as noncoherent (or differentially coherent) since it does not require a reference in phase with the received carrier. This "pseudo-PSK," termed *differential PSK* (DPSK), utilizes phase information of the prior symbol as a phase reference for detecting the current symbol. This is described in Sections 4.5.1 and 4.5.2.

4.2.1 Phasor Representation of a Sinusoid

Using a well-known trigonometric identity called Euler's theorem, we introduce the complex notation of a sinusoidal carrier wave as follows:

$$e^{j\omega_0 t} = \cos \omega_0 t + j \sin \omega_0 t \qquad (4.4)$$

One might be more comfortable with the simpler, more straightforward notation $\cos \omega_0 t$ or $\sin \omega_0 t$. What possible benefit can there be with the complex notation? We will see (in Section 4.6) that this notation facilitates our description of how real-world modulators and demodulators are implemented. For now, let us point to the general benefits of viewing a carrier wave in the complex form of Equation (4.4).

First, within this compact from, $e^{j\omega_0 t}$, is contained the two important quadrature components of any sinusoidal carrier wave, namely the inphase (real) and the quadrature (imaginary) components that are orthogonal to each other. Second, the unmodulated carrier wave is conveniently represented in a polar coordinate system as a unit vector or phasor rotating counterclockwise at the constant rate of ω_0 radians/s, as depicted in Figure 4.2. As time is increasing (i.e., from t_0 to t_1) we can visualize the time-varying projections of the rotating phasor on the inphase (I) axis and the quadrature (Q) axis. These cartesian axes are usually referred to as the I channel and Q channel respectively, and the projections on them represent the

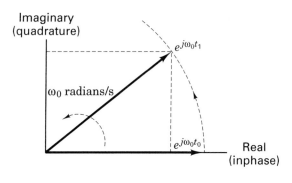

Figure 4.2 Phasor representation of a sinusoid.

signal components (orthogonal to each other) associated with those channels. Third, when it comes time to modulate the carrier wave with information, we can view this modulation as a methodical perturbation of the rotating phasor (and its projections).

For example, consider a carrier wave that is *amplitude modulated* (AM) with a sinusoid having an amplitude of unity and a frequency ω_m, where $\omega_m \ll \omega_0$. The analytical form of the transmitted waveform is

$$s(t) = \mathrm{Re}\left\{ e^{j\omega_0 t}\left(1 + \frac{e^{j\omega_m t}}{2} + \frac{e^{-j\omega_m t}}{2} \right) \right\} \qquad (4.5)$$

where Re{x} is the real part of the complex quantity {x}. Figure 4.3 illustrates that the rotating phasor $e^{j\omega_0 t}$ of Figure 4.2 is now perturbed by two sideband terms—$e^{j\omega_m t}/2$ rotating counterclockwise and $e^{-j\omega_m t}/2$ rotating clockwise. The sideband phasors are rotating at a much slower speed than the carrier-wave phasor. The net result of the composite signal is that the rotating carrier-wave phasor now appears to be growing longer and shorter pursuant to the dictates of the sidebands, but its frequency stays constant—hence, the term "amplitude modulation."

Another example to reinforce the usefulness of the phasor view is that of *frequency modulating* (FM) the carrier wave with a similar sinusoid having a frequency of ω_m radians/s. The analytical representation of *narrowband* FM (NFM) has an appearance similar to AM and is represented by

$$s(t) = \mathrm{Re}\left\{ e^{j\omega_0 t}\left(1 - \frac{\beta}{2} e^{-j\omega_m t} + \frac{\beta}{2} e^{j\omega_m t} \right) \right\} \qquad (4.6)$$

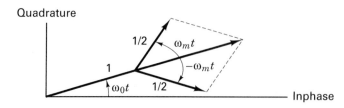

Figure 4.3 Amplitude modulation.

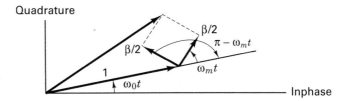

Figure 4.4 Narrowband frequency modulation.

where β is the modulation index [1]. Figure 4.4 illustrates that the rotating carrier-wave phasor is again perturbed by two sideband terms, but because one of the sideband terms carries a minus sign in Equation (4.6), the clockwise and counterclockwise rotating sideband phasors have a different symmetry than in the case of AM. In the case of AM the sideband symmetry results in the carrier-wave phasor growing longer and shorter with time. In NFM, the sideband symmetry (90° different than AM) results in the carrier-wave phasor speeding up and slowing down according to the dictates of the sidebands, but the amplitude says essentially constant—hence, the term "frequency modulation."

Figure 4.5 illustrates examples of the most common digital modulation formats: PSK, FSK, ASK, and a hybrid combination of ASK and PSK (ASK/PSK or APK). The first column lists the analytic expression, the second is a typical pictorial of the waveform versus time, and the third is a vector (or phasor) schematic, with the orthogonal axes labeled $\{\psi_j(t)\}$. In the general M-ary signaling case, the processor accepts k source bits (or channel bits if there is coding) at a time and instructs the modulator to produce one of an available set of $M = 2^k$ waveform types. Binary modulation, where $k = 1$, is just a special case of M-ary modulation.

In Figure 4.2, we represented a carrier wave as a phasor rotating in a plane at the speed of the carrier-wave frequency ω_0 radians/s. In Figure 4.5, the phasor schematic for each digital-modulation example represents a constellation of information signals (vectors or points in the signaling space), where time is not represented. In other words, the constantly rotating aspect of the unmodulated carrier wave has been removed, and only the information-bearing phasor positions, relative to one another, are presented. Each example in Figure 4.5 uses a particular value of M, the set size.

4.2.2 Phase Shift Keying

Phase shift keying (PSK) was developed during the early days of the deep-space program; PSK is now widely used in both military and commercial communications systems. The general analytic expression for PSK is

$$s_i(t) = \sqrt{\frac{2E}{T}} \cos\left[\omega_0 t + \phi_i(t)\right] \qquad \begin{matrix} 0 \le t \le T \\ i = 1, \dots, M \end{matrix} \qquad (4.7)$$

where the phase term, $\phi_i(t)$, will have M discrete values, typically given by

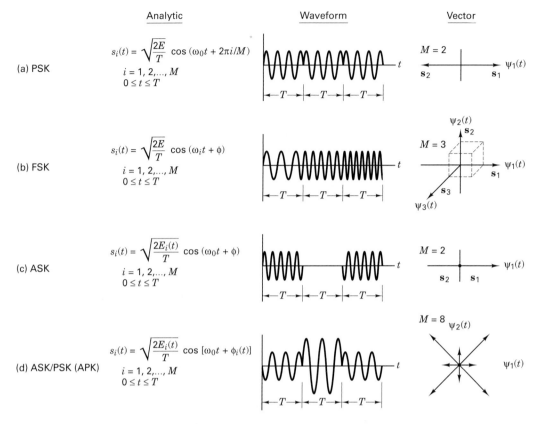

	Analytic	Waveform	Vector

(a) PSK

$$s_i(t) = \sqrt{\frac{2E}{T}} \cos{(\omega_0 t + 2\pi i/M)}$$

$i = 1, 2, ..., M$
$0 \le t \le T$

(b) FSK

$$s_i(t) = \sqrt{\frac{2E}{T}} \cos{(\omega_i t + \phi)}$$

$i = 1, 2, ..., M$
$0 \le t \le T$

(c) ASK

$$s_i(t) = \sqrt{\frac{2E_i(t)}{T}} \cos{(\omega_0 t + \phi)}$$

$i = 1, 2, ..., M$
$0 \le t \le T$

(d) ASK/PSK (APK)

$$s_i(t) = \sqrt{\frac{2E_i(t)}{T}} \cos{[\omega_0 t + \phi_i(t)]}$$

$i = 1, 2, ..., M$
$0 \le t \le T$

Figure 4.5 Digital modulations. (a) PSK. (b) FSK. (c) ASK. (d) ASK/PSK (APK).

$$\phi_i(t) = \frac{2\pi i}{M} \qquad i = 1, \dots, M$$

For the binary PSK (BPSK) example in Figure 4.5a, M is 2. The parameter E is symbol energy, T is symbol time duration, and $0 \le t \le T$. In BPSK modulation, the modulating data signal shifts the phase of the waveform $s_i(t)$ to one of two states, either zero or π (180°). The waveform sketch in Figure 4.5a shows a typical BPSK waveform with its abrupt phase changes at the symbol transitions; if the modulating data stream were to consist of alternating ones and zeros, there would be such an abrupt change at each transition. The signal waveforms can be represented as vectors or phasors on a polar plot; the vector length corresponds to the signal amplitude, and the vector direction for the general M-ary case corresponds to the signal phase relative to the other $M - 1$ signals in the set. For the BPSK example, the vector picture illustrates the two 180° opposing vectors. Signal sets that can be depicted with such opposing vectors are called *antipodal signal sets*.

4.2.3 Frequency Shift Keying

The general analytic expression for FSK modulation is

$$s_i(t) = \sqrt{\frac{2E}{T}} \cos{(\omega_i t + \phi)} \qquad \begin{matrix} 0 \le t \le T \\ i = 1, \dots, M \end{matrix} \qquad (4.8)$$

where the frequency term ω_i has M discrete values, and the phase term ϕ is an arbitrary constant. The FSK waveform sketch in Figure 4.5b illustrates the typical frequency changes at the symbol transitions. At the symbol transitions, the figure depicts a gentle shift from one frequency (tone) to another. This behavior is only true for a special class of FSK called continuous-phase FSK (CPFSK) which is described in Section 9.8. In the general MFSK case, the change to a different tone can be quite abrupt, because there is no requirement for the phase to be continuous. In this example, M has been chosen equal to 3, corresponding to the same number of waveform types (3-ary); note that this $M = 3$ choice for FSK has been selected to emphasize the mutually perpendicular axes. In practice, M is usually a nonzero power of 2 (2, 4, 8, 16, ...). The signal set is characterized by Cartesian coordinates, such that each of the mutually perpendicular axes represents a sinusoid with a different frequency. As described earlier, signal sets that can be characterized with such mutually perpendicular vectors are called *orthogonal* signals. Not all FSK signaling is orthogonal. For any signal set to be orthogonal, it must meet the criterion set forth in Equation (3.69). For an FSK signal set, in the process of meeting this criterion, a condition arises on the spacing between the tones in the set. The necessary frequency spacing between tones to fulfill the orthogonality requirement is discussed in Section 4.5.4.

4.2.4 Amplitude Shift Keying

For the ASK example in Figure 4.5c, the general analytic expression is

$$s_i(t) = \sqrt{\frac{2E_i(t)}{T}} \cos{(\omega_0 t + \phi)} \qquad \begin{matrix} 0 \le t \le T \\ i = 1, \dots, M \end{matrix} \qquad (4.9)$$

where the amplitude term $\sqrt{2E_i(t)/T}$ will have M discrete values, and the phase term ϕ is an arbitrary constant. In Figure 4.5c, M has been chosen equal to 2, corresponding to two waveform types. The ASK waveform sketch in the figure can describe a radar transmission example, where the two signal amplitude states would be $\sqrt{2E/T}$ and zero. The vector picture utilizes the same phase–amplitude polar coordinates as the PSK example. Here we see a vector corresponding to the maximum-amplitude state, and a point at the origin corresponding to the zero-amplitude state. Binary ASK signaling (also called on–off keying) was one of the earliest forms of digital modulation used in radio telegraphy in the early 1900s. Simple ASK is no longer widely used in digital communications systems, and thus it will not be treated in detail here.

4.2.5 Amplitude Phase Keying

For the combination of ASK and PSK (APK) example in Figure 4.5d, the general analytic expression

$$s_i(t) = \sqrt{\frac{2E_i(t)}{T}} \cos\left[\omega_0 t + \phi_i(t)\right] \qquad \begin{array}{l} 0 \le t \le T \\ i = 1, \ldots, M \end{array} \qquad (4.10)$$

illustrates the indexing of both the signal amplitude term and the phase term. The APK waveform picture in Figure 4.5d illustrates some typical simultaneous phase and amplitude changes at the symbol transition times. For this example, M has been chosen equal to 8, corresponding to eight waveforms (8-ary). The figure illustrates a hypothetical eight-vector signal set on the phase-amplitude plane. Four of the vectors are at one amplitude, and the other four vectors are at a different amplitude. Each of the vectors is separated by 45°. When the set of M symbols in the two-dimensional signal space are arranged in a rectangular constellation, the signaling is referred to as quadrature amplitude modulation (QAM); examples of QAM are considered in Chapter 9.

The vector picture for each of the modulation types described in Figure 4.5 (except the FSK case) is characterized on a plane whose *polar* coordinates represent signal *amplitude* and *phase*. The FSK case assumes orthogonal FSK (see Section 4.5.4) and is characterized in a *Cartesian* coordinate space, with each axis representing a *frequency tone* ($\cos \omega_i t$) from the M-ary set of orthogonal tones.

4.2.6 Waveform Amplitude Coefficient

The waveform amplitude coefficient appearing in Equations (4.7) to (4.10) has the same general form $\sqrt{2E/T}$ for all modulation formats. The derivation of this expression begins with

$$s(t) = A \cos \omega t \qquad (4.11)$$

where A is the peak value of the waveform. Since the peak value of a sinusoidal waveform equals $\sqrt{2}$ times the root-mean-square (rms) value, we can write

$$s(t) = \sqrt{2} A_{\text{rms}} \cos \omega t$$
$$= \sqrt{2 A_{\text{rms}}^2} \cos \omega t$$

Assuming the signal to be a voltage or a current waveform, A_{rms}^2 represents average power P (normalized to 1 Ω). Therefore, we can write

$$s(t) = \sqrt{2P} \cos \omega t \qquad (4.12)$$

Replacing P watts by E joules/T seconds, we get

$$s(t) = \sqrt{\frac{2E}{T}} \cos \omega t \qquad (4.13)$$

We shall use either the amplitude notation A in Equation (4.11) or the designation $\sqrt{2E/T}$ in Equation (4.13). Since the *energy* of a received signal is the key parameter in determining the error performance of the detection process, it is often more convenient to use the amplitude notation in Equation (4.13) because it facilitates solving directly for the probability of error P_E as a function of signal energy.

4.3 DETECTION OF SIGNALS IN GAUSSIAN NOISE

The bandpass model of the detection process is virtually identical to the baseband model considered in Chapter 3. That is because a received bandpass waveform is first transformed to a baseband waveform before the final detection step takes place. For linear systems, the mathematics of detection is unaffected by a shift in frequency. In fact, we can define an *equivalence theorem* as follows: Performing bandpass linear signal processing, followed by heterodyning the signal to baseband yields the same results as heterodyning the bandpass signal to baseband, followed by baseband linear signal processing. The term "heterodyning" refers to a frequency *conversion* or *mixing* process that yields a spectral shift in the signal. As a result of this equivalence theorem, all linear signal-processing simulations can take place at baseband (which is preferred for simplicity), with the same results as at bandpass. This means that the performance of most digital communication systems will often be described and analyzed as if the transmission channel is a baseband channel.

4.3.1 Decision Regions

Consider that the two-dimensional signal space in Figure 4.6 is the locus of the noise-perturbed prototype binary vectors $(\mathbf{s}_1 + \mathbf{n})$ and $(\mathbf{s}_2 + \mathbf{n})$. The noise vector, $\mathbf{n}$, is a zero-mean random vector; hence the received signal vector, $\mathbf{r}$, is a random vector with mean $\mathbf{s}_1$ or $\mathbf{s}_2$. The detector's task after receiving $\mathbf{r}$ is to decide which of the signals ($\mathbf{s}_1$ or $\mathbf{s}_2$,) was actually transmitted. The method is usually to decide on the signal classification that yields the minimum expected P_E, although other strategies are possible [2]. For the case where M equals 2, with $\mathbf{s}_1$ and $\mathbf{s}_2$ being equally likely and with the noise being an additive white Gaussian noise (AWGN) process, we will see that the minimum-error decision rule is equivalent to choosing the signal class such that the distance $d(\mathbf{r}, \mathbf{s}_i) = \| \mathbf{r} - \mathbf{s}_i \|$ is minimized, where $\| \mathbf{x} \|$ is called the *norm* or *magnitude* of vector $\mathbf{x}$. This rule is often stated in terms of decision regions. In Figure 4.6, let us construct decision regions in the following way. Draw a line connecting the tips of the prototype vectors $\mathbf{s}_1$ and $\mathbf{s}_2$. Next, construct the perpendicular bisector of the connecting line. Notice that this bisector passes through the origin of the space if $\mathbf{s}_1$ and $\mathbf{s}_2$ are equal in amplitude. For this $M = 2$ example, in Figure 4.6, the constructed perpendicular bisector represents the locus of points equidistant between $\mathbf{s}_1$ and $\mathbf{s}_2$; hence, the bisector describes the boundary between decision region 1 and decision region 2. The *decision rule* for the detector, stated in terms of *decision regions,* is as follows: Whenever the received signal $\mathbf{r}$ is located

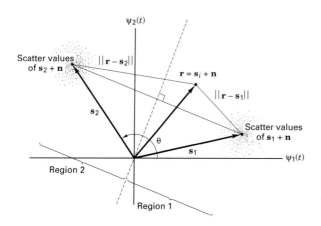

Region 2

Region 1

Figure 4.6 Two-dimensional signal space, with arbitrary equal-amplitude vectors s_1 and s_2.

in region 1, choose signal s_1; when it is located in region 2, choose signal s_2. In Figure 4.6, if the angle θ equals 180°, then the signal set s_1 and s_2 represents BPSK. However, in this figure, θ was purposely chosen to be less than 180° in order to emphasize the idea of decision regions in general.

4.3.2 Correlation Receiver

In Section 3.2 we treated the detection of *baseband* binary signals in Gaussian noise. Since the detection of *bandpass* signals employs the same concepts, we shall summarize the key findings of that section. We focus particularly on that realization of a matched filter known as a *correlator*. In addition to binary detection, we also consider the more general case of *M*-ary detection. We assume that the only performance degradation is due to AWGN. The received signal is the sum of the transmitted prototype signal plus the random noise:

$$r(t) = s_i(t) + n(t) \qquad \begin{aligned} 0 &\leq t \leq T \\ i &= 1, \ldots, M \end{aligned} \qquad (4.14)$$

Given such a received signal, the detection process consists of *two basic steps* as shown in Figure 3.1. In the first step, the received waveform, $r(t)$, is reduced to a *single random variable* $z(T)$, or to a *set of random variables* $z_i(T)$ ($i = 1, \ldots, M$), formed at the output of the demodulator and sampler at time $t = T$, where T is the symbol duration. In the second step, a symbol decision is made on the basis of comparing $z(T)$ to a threshold or on the basis of choosing the maximum $z_i(T)$. Step 1 can be thought of as transforming the waveform into a point in the decision space. We call this point the *predetection point,* and view it as the "location" of the received E_b/N_0 (or other related receiving-system SNR parameter); beyond here, signals are replaced by bits. Often, such SNR parameters are modeled with reference to the output of the receiving antenna. Note that although the signal power and noise power is each different at different points in the system, a receiving-system SNR can be modeled to yield the same value at various reference points (see Section 5.5.5). At the predetection point, there will be a new sample of a baseband pulse during each symbol time. Note that E_b/N_0 is defined at a place where there are no bits yet. Bits will only appear after completion of the detection process. Perhaps a better name for E_b/N_0 would be energy per *effective* bit versus N_0. Step 2 can be thought of as determining *in which decision region* the point is located. For the detector to be optimized (in the sense of minimizing the error probability), it is necessary to optimize the waveform-to-random-variable transformation, by using matched filters or correlators in step 1, and by also optimizing the decision criterion in step 2.

In Sections 3.2.2 and 3.2.3 we found that the matched filter provides the maximum signal-to-noise ratio at the filter output at time $t = T$. We described a correlator as one realization of a matched filter. We can define a *correlation receiver* comprised of M correlators, as shown in Figure 4.7a, that transforms a received waveform, $r(t)$, to a sequence of M numbers or correlator outputs, $z_i(T)$ $(i = 1, \ldots, M)$. Each correlator output is characterized by the following product integration or correlation with the received signal:

$$z_i(T) = \int_0^T r(t)s_i(t)\, dt \qquad i = 1, \ldots, M \qquad (4.15)$$

The verb "to correlate" means "to match." The correlators attempt to match the incoming received signal, $r(t)$, with each of the candidate prototype waveforms, $s_i(t)$, known a priori to the receiver. A reasonable decision rule is to choose the waveform, $s_i(t)$, that *matches best* or has the *largest correlation* with $r(t)$. In other words, the decision rule is

$$\text{Choose the } s_i(t) \text{ whose index} \\ \text{corresponds to the max } z_i(T) \qquad (4.16)$$

Following Equation (3.10), any signal set, $\{s_i(t)\}$ $(i = 1, \ldots, M)$, can be expressed in terms of some set of basis functions, $\{\psi_j(t)\}$ $(j = 1, \ldots, N)$, where $N \le M$. Then the bank of M correlators in Figure 4.7a may be replaced with a bank of N correlators, shown in Figure 4.7b, where the set of basis functions $\{\psi_j(t)\}$ form *reference signals*. The decision stage of this receiver consists of logic circuitry for choosing the signal, $s_i(t)$. The choice of $s_i(t)$ is made according to the best match of the coefficients, a_{ij}, seen in Equation (3.10), with the set of outputs $\{z_j(T)\}$. When the prototype waveform set, $\{s_i(t)\}$, is an orthogonal set, the receiver implementation in Figure 4.7a is identical to that in Figure 4.7b (differing perhaps by a scale factor). However, when $\{s_i(t)\}$ is *not* an orthogonal set, the receiver in Figure 4.7b, using N correlators instead of M, with reference signals $\{\psi_j(t)\}$, can represent a cost-effective implementation. We examine such an application for the detection of multiple phase shift keying (MPSK) in Section 4.4.3.

In the case of *binary detection,* the correlation receiver can be configured as a single matched filter or product integrator, as shown in Figure 4.8a, with the reference signal being the difference between the binary prototype signals, $s_1(t) - s_2(t)$. The output of the correlator, $z(T)$, is fed directly to the decision stage.

For binary detection, the correlation receiver can also be drawn as two matched filters or product integrators, one of which is matched to $s_1(t)$, and the

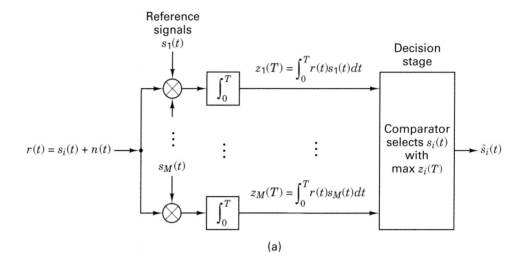

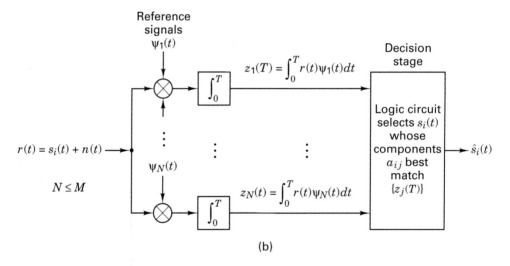

Figure 4.7 (a) Correlator receiver with reference signals $\{s_i(t)\}$.
(b) Correlator receiver with reference signals $\{\psi_j(t)\}$.

other is matched to $s_2(t)$. (See Figure 4.8b.) The decision stage can then be configured to follow the rule in Equation 4.16, or the correlator outputs $z_i(T)$ $(i = 1, 2)$ can be differenced to form

$$z(T) = z_1(T) - z_2(T) \qquad (4.17)$$

as shown in Figure 4.8b. Then, $z(T)$, called the *test statistic*, is fed to the decision stage, as in the case of the single correlator. In the *absence of noise*, an input waveform $s_i(t)$ yields the output $z(T) = a_i(T)$, a signal-only component. The input noise

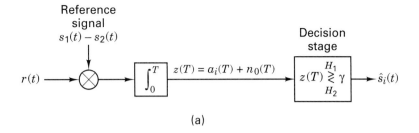

(a)

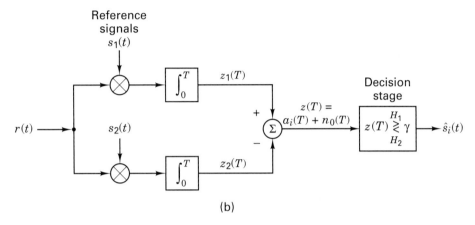

(b)

Figure 4.8 Binary correlator receiver. (a) Using a single correlator. (b) Using two correlators.

$n(T)$ is a Gaussian random process. Since the correlator is a *linear* device, the output noise is also a Gaussian random process [2]. Thus, the output of the correlator, sampled at $t = T$, yields

$$z(T) = a_i(T) + n_0(T) \qquad i = 1, 2$$

where $n_0(T)$ is the noise component. To shorten the notation we sometimes express $z(t)$ as $a_i + n_0$. The noise component n_0 is a zero-mean *Gaussian random variable,* and thus $z(T)$ is a *Gaussian random variable* with a mean of either a_1 or a_2, depending on whether a binary one or binary zero was sent.

4.3.2.1 Binary Decision Threshold

For the random variable $z(T)$, Figure 4.9 illustrates the two conditional probability density functions (pdfs), $p(z|s_1)$ and $p(z|s_2)$, with mean value of a_1 and a_2, respectively. These pdfs, also called the *likelihood* of s_1 and the *likelihood* of s_2, respectively, were presented in Section 3.1.2, and are rewritten as

$$p(z|s_1) = \frac{1}{\sigma_0 \sqrt{2\pi}} \exp\left[-\frac{1}{2}\left(\frac{z - a_1}{\sigma_0}\right)^2\right] \qquad (4.18a)$$

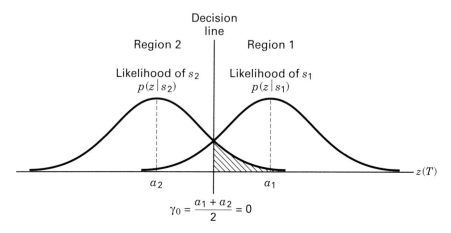

Region 2   Decision line   Region 1

Likelihood of s_2
$p(z|s_2)$

Likelihood of s_1
$p(z|s_1)$

a_2 a_1 $z(T)$

$$\gamma_0 = \frac{a_1 + a_2}{2} = 0$$

Figure 4.9   Conditional probability density functions: $p(z/s_1)$, $p(z/s_2)$.

and

$$p(z|s_2) = \frac{1}{\sigma_0 \sqrt{2\pi}} \exp\left[-\frac{1}{2}\left(\frac{z - a_2}{\sigma_0} \right)^2 \right] \tag{4.18b}$$

where σ_0^2 is the noise variance. In Figure 4.9 the rightmost likelihood $p(z|s_1)$ illustrates the probability density of the detector output $z(T)$, given that $s_1(t)$ was transmitted. Similarly, the leftmost likelihood $p(z|s_2)$ illustrates the probability density of $z(T)$, given that $s_2(t)$ was transmitted. The abscissa $z(T)$ represents the full range of possible sample output values from the correlation receiver shown in Figure 4.8.

    With regard to optimizing the binary decision threshold for deciding in which region a received signal is located, we found in Section 3.2.1 that the *minimum error* criterion for equally likely binary signals corrupted by Gaussian noise can be stated as

$$z(T) \underset{H_2}{\overset{H_1}{\gtrless}} \frac{a_1 + a_2}{2} = \gamma_0 \tag{4.19}$$

where a_1 is the signal component of $z(T)$ when $s_1(t)$ is transmitted, and a_2 is the signal component of $z(T)$ when $s_2(t)$ is transmitted. The threshold level γ_0 represented by $(a_1 + a_2)/2$ is the *optimum threshold* for minimizing the probability of making an incorrect decision given equally likely signals and symmetrical likelihoods. The decision rule in Equation (4.19) states that hypothesis H_1 should be selected [equivalent to deciding that signal $s_1(t)$ was sent] if $z(T) > \gamma_0$, and hypothesis H_2 should be selected [equivalent to deciding that $s_2(t)$ was sent] if $z(T) < \gamma_0$. If $z(T) = \gamma_0$, the decision can be an arbitrary one. For equal-energy, equally likely antipodal signals, where $s_1(t) = - s_2(t)$ and $a_1 = - a_2$, the optimum decision rule becomes

$$z(T) \overset{H_1}{\underset{H_2}{\gtrless}} = 0 \qquad (4.20a)$$

or

$$\begin{array}{ll} \text{decide } s_1(t) & \text{if } z_1(T) > z_2(T) \\ \text{decide } s_2(t) & \text{otherwise} \end{array} \qquad (4.20b)$$

4.4 COHERENT DETECTION

4.4.1 Coherent Detection of PSK

The detector shown in Figure 4.7 can be used for the coherent detection of any digital waveforms. Such a correlating detector is often referred to as a *maximum likelihood detector*. Consider the following binary PSK (BPSK) example: Let

$$s_1(t) = \sqrt{\frac{2E}{T}} \cos(\omega_0 t + \phi) \qquad 0 \le t \le T \qquad (4.21a)$$

$$s_2(t) = \sqrt{\frac{2E}{T}} \cos(\omega_0 t + \phi + \pi)$$

$$= -\sqrt{\frac{2E}{T}} \cos(\omega_0 t + \phi) \qquad 0 \le t \le T \qquad (4.21b)$$

and

$$n(t) = \text{zero-mean white Gaussian random process}$$

where the phase term ϕ is an arbitrary constant, so that the analysis is unaffected by setting $\phi = 0$. The parameter E is the signal energy per symbol, and T is the symbol duration. For this antipodal case, only a single basis function is needed. If an orthonormal signal space is assumed in Equations (3.10) and (3.11) (i.e., $K_j = 1$), we can express a basis function $\psi_1(t)$ as

$$\psi_1(t) = \sqrt{\frac{2}{T}} \cos \omega_0 t \qquad \text{for } 0 \le t \le T \qquad (4.22)$$

Thus, we may express the transmitted signals $s_i(t)$ in terms of $\psi_1(t)$ and the coefficients $a_{i1}(t)$ as follows:

$$s_i(t) = a_{i1}\psi_1(t) \qquad (4.23a)$$
$$s_1(t) = a_{11}\psi_1(t) = \sqrt{E}\,\psi_1(t) \qquad (4.23b)$$
$$s_2(t) = a_{21}\psi_1(t) = -\sqrt{E}\,\psi_1(t) \qquad (4.23c)$$

Assume that $s_1(t)$ was transmitted. Then the expected values of the product integrators in Figure 4.7b, with reference signal $\psi_1(t)$, are found as

$$\mathbf{E}\{z_1|s_1\} = \mathbf{E}\left\{ \int_0^T \sqrt{E}\,\psi_1^2(t) + n(t)\psi_1(t)\,dt \right\} \tag{4.24a}$$

$$\mathbf{E}\{z_2|s_1\} = \mathbf{E}\left\{ \int_0^T -\sqrt{E}\,\psi_1^2(t) + n(t)\psi_1(t)\,dt \right\} \tag{4.24b}$$

$$\mathbf{E}\{z_1|s_1\} = \mathbf{E}\left\{ \int_0^T \frac{2}{T}\sqrt{E}\cos^2\omega_0 t + n(t)\sqrt{\frac{2}{T}}\cos\omega_0 t\,dt \right\} = \sqrt{E} \tag{4.25a}$$

and

$$\mathbf{E}\{z_2|s_1\} = \mathbf{E}\left\{ \int_0^T -\frac{2}{T}\sqrt{E}\cos^2\omega_0 t + n(t)\sqrt{\frac{2}{T}}\cos\omega_0 t\,dt \right\} = -\sqrt{E} \tag{4.25b}$$

where $\mathbf{E}\{\cdot\}$ denotes the ensemble average, referred to as the *expected value*. Equation (4.25) follows because $\mathbf{E}\{n(t)\} = 0$. The decision stage must decide which signal was transmitted by determining its location within the signal space. For this example, the choice of $\psi_1(t) = \sqrt{2/T}\cos\omega_0 t$ normalizes $\mathbf{E}\{z_i(T)\}$ to be $\pm\sqrt{E}$. The prototype signals $\{s_i(t)\}$ are the same as the reference signals $\{\psi_j(t)\}$ except for the normalizing scale factor. The decision stage chooses the signal with the largest value of $z_i(T)$. Thus, the received signal in this example is judged to be $s_1(t)$. The error performance for such coherently detected BPSK systems is treated in Section 4.7.1.

4.4.2 Sampled Matched Filter

In Section 3.2.2, we discussed the basic characteristic of the matched filter—namely, that its impulse response is a delayed version of the mirror image (rotated on the $t = 0$ axis) of the input signal waveform. Therefore, if the signal waveform is $s(t)$, its mirror image is $s(-t)$, and the mirror image delayed by T seconds is $s(T-t)$. The impulse response $h(t)$ of a filter matched to $s(t)$ is then described by

$$h(t) = \begin{cases} s(T-t) & 0 \le t \le T \\ 0 & \text{elsewhere} \end{cases} \tag{4.26}$$

Figures 4.7 and 4.8 illustrate the basic function of a correlator to product-integrate the received noisy signal with each of the candidate reference signals and determine the best match. The schematics in these figures imply the use of analog hardware (multipliers and integrators) and continuous signals. They do not reflect the way that the correlator or matched filter (MF) can be implemented using digital techniques and sampled waveforms. Figure 4.10 shows how an MF can be implemented using digital hardware. The input signal $r(t)$ comprises a prototype signal $s_i(t)$, plus noise $n(t)$, and the bandwidth of the signal is $W = 1/2T$, where T is the symbol time. Thus, the minimum Nyquist sampling rate is $f_s = 2W = 1/T$, and the sampling time T_s needs to be equal to or less than the symbol time. In other words, there must be at least one sample per symbol. In real systems, such sampling is usually performed at a rate that exceeds the Nyquist minimum by a factor of 4 or more. The only cost is processor speed, not transmission bandwidth. At the clock

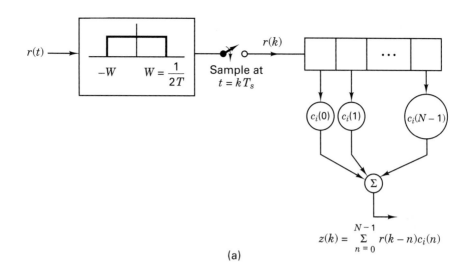

$$z(k) = \sum_{n=0}^{N-1} r(k-n)c_i(n)$$

(a)

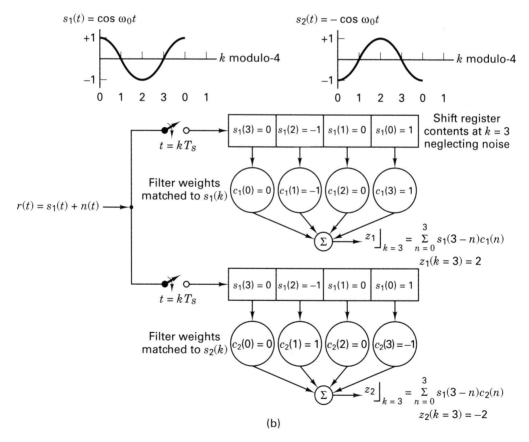

$s_1(t) = \cos \omega_0 t$

$s_2(t) = -\cos \omega_0 t$

k modulo-4

k modulo-4

Shift register contents at $k = 3$ neglecting noise

$t = kT_s$

$s_1(3) = 0$ | $s_1(2) = -1$ | $s_1(1) = 0$ | $s_1(0) = 1$

Filter weights matched to $s_1(k)$ $\quad c_1(0) = 0 \quad c_1(1) = -1 \quad c_1(2) = 0 \quad c_1(3) = 1$

$r(t) = s_1(t) + n(t)$

$$z_1 \Big|_{k=3} = \sum_{n=0}^{3} s_1(3-n)c_1(n)$$

$$z_1(k=3) = 2$$

$t = kT_s$

$s_1(3) = 0$ | $s_1(2) = -1$ | $s_1(1) = 0$ | $s_1(0) = 1$

Filter weights matched to $s_2(k)$ $\quad c_2(0) = 0 \quad c_2(1) = 1 \quad c_2(2) = 0 \quad c_2(3) = -1$

$$z_2 \Big|_{k=3} = \sum_{n=0}^{3} s_1(3-n)c_2(n)$$

$$z_2(k=3) = -2$$

(b)

Figure 4.10 (a) Sampled matched filter. (b) Sampled matched filter detection example, neglecting noise.

4.4   Coherent Detection   **185**

times of $t = kT_s$, the samples are shifted into the register (as shown in Figure 4.10a) so that earlier samples are located to the right of later samples. Once the received signal has been sampled, the continuous time notation t is changed to kT_s or simply to k to allow use of the simple discrete notation

$$r(k) = s_i(k) + n(k) \qquad i = 1, 2 \qquad k = 0, 1, \ldots$$

where the index i identifies a particular symbol out of the M-ary set (binary in this example), and k is the sampling-time index. In Figure 4.10, the shift register, with its coefficients or weights $c_i(n)$ approximate an MF, where $n = 0, \ldots, N - 1$ is the time index of the weights and register stages. In this example, $N = 4$ represents the number of stages in the register and the number of samples per symbol. Hence, the summation shown in the figure takes place over the time $n = 0$ to $n = 3$. Following the summer, a symbol decision will be made after 4 time samples have entered the registers. Note that for simplicity, the example in Figure 4.10b uses only 3-level amplitude values $(0, \pm 1)$ for the $s_i(k)$ samples. In real systems, each sample (and weight) would be represented by 6–10 bits, depending on the density of the signal constellation. The set of filter weights $\{c_i(n)\}$ constitutes the filter impulse response; the weights are matched to signal samples according to a discrete form of Equation (4.26), as follows:

$$c_i(n) = s_i[(N - 1) - n] = s_i(3 - n) \tag{4.27}$$

By using a discrete form of the *convolution integral* in Equation (A.44b), the output at a time corresponding to the kth sample can be expressed as

$$z_i(k) = \sum_{n=0}^{N-1} r(k - n) \, c_i(n) \qquad k = 0, 1, \ldots, \text{modulo-}N \tag{4.28}$$

where x modulo-y is defined as the remainder of dividing x by y, the index k represents time for both the received samples and the filter output, and n is a dummy time variable. In the expression $r(k - n)$ of Equation (4.28), it is helpful to think of n as the "age" of the sample (how long has it been sitting in the filter). In the expression $c_i(n)$, it is helpful to think of n as the address of the weight. We assume that because the system is synchronized, the symbol timing is known. Also, we assume the noise to have a zero mean, so that the expected value of a received sample is

$$\mathbf{E}\{r(k)\} = s_i(k) \qquad i = 1, 2$$

Thus, if $s_i(t)$ is transmitted, the expected matched filter output is

$$\mathbf{E}\{z_i(k)\} = \sum_{n=0}^{N-1} s_i(k - n) \, c_i(n) \qquad k = 0, 1, \ldots, \text{modulo-}N \tag{4.29}$$

In Figure 4.10b, where the prototype waveforms are plotted as a function of time, we see that the leftmost sample (amplitude equals $+1$) of the $s_1(t)$ plot represents the earliest sample at time $k = 0$. Assuming that $s_1(t)$ was sent, and the noise is neglected for notational simplicity, we then denote the received samples of $r(k)$ as $s_1(k)$. As these samples fill the stages of the MF, at the end of each symbol time the

$k = 0$ sample is located in the rightmost stage of each register. In Equations (4.28) and (4.29), notice that the time indexes n of the reference weights are in reverse order compared with the time index $k - n$ of the samples, which is a key aspect of the convolution integral. The fact that the earliest time sample now corresponds to the rightmost weight will ensure a meaningful correlation. Even though we describe the mathematical operation of an MF to be *convolution* of a signal with the impulse response of the filter, the end result appears to be the *correlation* of a signal with a replica of that same signal. That is why it is valid to describe a correlator as an implementation of a matched filter.

In Figure 4.10b, detection will follow the MF in the usual way. For the binary decision, the $z_i(k)$ outputs are examined at each value of $k = N - 1$ corresponding to the end of a symbol. Under the condition that $s_1(t)$ had been transmitted and noise is neglected, we combine Equations (4.27) through (4.29) to express the correlator outputs at time $k = N - 1 = 3$ as

$$z_1(k{=}3) = \sum_{n=0}^{3} s_1(3 - n)\, c_1(n) = 2 \qquad (4.30a)$$

and

$$z_2(k{=}3) = \sum_{n=0}^{3} s_1(3 - n)\, c_2(n) = -2 \qquad (4.30b)$$

Since $z_1(k = 3)$ is greater than $z_2(k = 3)$, the detector chooses $s_1(t)$ as the transmitted symbol.

One might ask, "What is the difference between the MF in Figure 4.10b and the correlator in Figure 4.8?" In the case of the MF, a new output value is available in response to each new input sample; thus the output will be a time series such as the MF output seen in Figure 3.7b (a succession of increasing positive and negative correlations to an input sine wave). Such an MF output sequence can be equated to several correlators operating at different starting points of the input time series. Note that a correlator only computes an output once per symbol time, such as the value of the peak signal at time T in Figure 3.7b. If the timing of the MF and correlator are aligned, then their outputs at the end of a symbol time are identical. An important distinction between the MF and correlator is that since the correlator yields a single output value per symbol, it must have side information, such as the start and stop times over which the product integration should take place. If there are timing errors in the correlator, then the sampled output fed to the detector may be badly degraded. On the other hand, since the MF yields a *time series* of output values (reflecting time-shifted input samples multiplied by fixed weights), then with the use of additional circuitry, the best time for sampling the MF output can be learned.

Example 4.1   Sampled Matched Filter

Consider the waveform set

$$s_1(t) = At \qquad 0 \le t \le kT$$

and

$$s_2(t) = -At \qquad 0 \le t \le kT$$

where $k = 0, 1, 2, 3$.

Illustrate how a *sampled* matched filter as shown in Figure 4.10 can be used to detect a received signal, say $s_1(t)$, from this sawtooth waveform set in the absence of noise.

Solution

First, the waveform is sampled so that $s_1(t)$ is transformed into the set of samples $\{s_1(k)\}$. The sampled matched filter receiver will be shown with two branches, following the implementation in Figure 4.10b. The top branch is made up of shift registers and coefficients matched to the $\{s_1(k)\}$ sample points. The bottom branch is similarly matched to the $\{s_2(k)\}$ sample points. The four equally spaced sample points ($k = 0, 1, 2, 3$) for each of the $\{s_i(k)\}$ are as follows:

$$s_1(k = 0) = 0 \qquad s_1(k = 1) = \frac{A}{4} \qquad s_1(k = 2) = \frac{A}{2} \qquad s_1(k = 3) = \frac{3A}{4}$$

$$s_2(k = 0) = 0 \qquad s_2(k = 1) = -\frac{A}{4} \qquad s_2(k = 2) = -\frac{A}{2} \qquad s_2(k = 3) = -\frac{3A}{4}$$

The $c_i(n)$ coefficients represent the delayed mirror-image rotation of the signal to which the filter is matched. Therefore, $c_i(n) = s_i(N - 1 - n)$, where $n = 0, \ldots, N - 1$, and we can write $c_i(0) = s_i(3)$, $c_i(1) = s_i(2)$, $c_i(2) = s_i(1)$, $c_i(3) = s_i(0)$.

Consider the top branch in Figure 4.10b. At the $k = 0$ clock time, the first sample $s_1(k = 0) = 0$ enters the leftmost stage of each register. At the next clock time, the second sample $s_1(k = 1) = A/4$ enters the leftmost stage of each register; at this same time the first sample has been shifted to the next right stage in each register, and so on. At the $k = 3$ clock time the sample $s_1(k = 3) = 3A/4$ enters the leftmost stage; by this time the first sample has been shifted into the rightmost stage. The four signal samples are now located in the registers in mirror-image arrangement compared with how they would be plotted in time. Hence, the convolution operation is an appropriate expression for describing the alignment of the incoming waveform samples with the reference coefficients to maximize the correlation in the proper branch.

4.4.3 Coherent Detection of Multiple Phase-Shift Keying

Figure 4.11 illustrates the signal space for a multiple phase-shift keying (MPSK) signal set; the figure describes a four-level (4-ary) PSK or quadriphase shift keying (QPSK) example ($M = 4$). At the transmitter, binary digits are collected two at a time, and for each symbol interval, the two sequential digits instruct the modulator as to which of the four waveforms to produce. For typical coherent M-ary PSK (MPSK) systems, $s_i(t)$ can be expressed as

$$s_i(t) = \sqrt{\frac{2E}{T}} \cos\left(\omega_0 t - \frac{2\pi i}{M}\right) \qquad \begin{array}{l} 0 \leq t \leq T \\ i = 1, \ldots, M \end{array} \qquad (4.31)$$

where E is the received energy of such a waveform over each symbol duration T, and ω_0 is the carrier frequency. If an orthonormal signal space is assumed in Equations (3.10) and (3.11), we can choose a convenient set of axes, such as

$$\psi_1(t) = \sqrt{\frac{2}{T}} \cos \omega_0 t \qquad (4.32a)$$

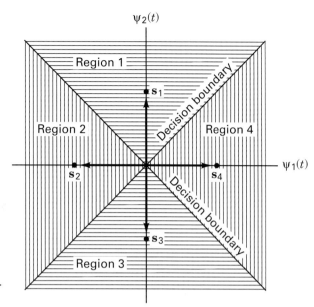

Figure 4.11 Signal space and decision regions for a QPSK system.

and

$$\psi_2(t) = \sqrt{\frac{2}{T}} \sin \omega_0 t \qquad (4.32b)$$

where the amplitude $\sqrt{2/T}$ has been chosen to normalize the expected output of the detector, as was done in Section 4.4.1. Now $s_i(t)$ can be written in terms of these orthonormal coordinates, giving

$$s_i(t) = a_{i1}\psi_1(t) + a_{i2}\psi_2(t) \qquad \begin{matrix} 0 \le t \le T \\ i = 1, \ldots, M \end{matrix} \qquad (4.33a)$$

$$= \sqrt{E} \cos\left(\frac{2\pi i}{M}\right) \psi_1(t) + \sqrt{E} \sin\left(\frac{2\pi i}{M}\right) \psi_2(t) \qquad (4.33b)$$

Notice that Equation (4.33) describes a set of M multiple phase waveforms (intrinsically nonorthogonal) in terms of only two orthogonal carrier-wave components. The $M = 4$ (QPSK) case is unique among MPSK signal sets in the sense that the QPSK waveform set is represented by a combination of antipodal and orthogonal members. The decision boundaries partition the signal space into $M = 4$ regions; the construction is similar to the procedure outlined in Section 4.3.1 and Figure 4.6 for $M = 2$. The decision rule for the detector (see Figure 4.11) is to decide that $s_1(t)$ was transmitted if the received signal vector falls in region 1, that $s_2(t)$ was transmitted if the received signal vector falls in region 2, and so on. In other words, the decision rule is to choose the ith waveform if $z_i(T)$ is the largest of the correlator outputs (seen in Figure 4.7).

The form of the correlator shown in Figure 4.7a implies that there are always M product correlators used for the demodulation of MPSK signals. The figure infers that for each of the M branches, a reference signal with the appropriate phase shift is configured. In practice, the implementation of an MPSK demodulator follows Figure 4.7b, requiring only $N = 2$ product integrators regardless of the size of the signal set M. The savings in implementation is possible because any arbitrary integrable waveform set can be expressed as a linear combination of orthogonal waveforms, as shown in Section 3.1.3. Figure 4.12 illustrates such a demodulator. The received signal $r(t)$ can be expressed by combining Equations (4.32) and (4.33) as

$$r(t) = \sqrt{\frac{2E}{T}} \left(\cos \phi_i \cos \omega_0 t + \sin \phi_i \sin \omega_0 t \right) + n(t) \qquad \begin{array}{l} 0 \leq t \leq T \\ i = 1, \ldots, M \end{array} \qquad (4.34)$$

where $\phi_i = 2\pi i/M$, and $n(t)$ is a zero-mean white Gaussian noise process. Notice that in Figure 4.12, there are only two reference waveforms or basis functions, $\psi_1(t) = \sqrt{2/T} \cos \omega_0 t$ for the upper correlator and $\psi_2(t) = \sqrt{2/T} \sin \omega_0 t$ for the lower correlator. The upper correlator computes

$$X = \int_0^T r(t)\psi_1(t) \, dt \qquad (4.35)$$

and the lower correlator computes

$$Y = \int_0^T r(t)\psi_2(t) \, dt \qquad (4.36)$$

Figure 4.13 illustrates that the computation of the received phase angle $\hat{\phi}$ can be accomplished by computing the arctan of Y/X, where X can be thought of as the in-phase component of the received signal, Y is the quadrature component, and $\hat{\phi}$ is a

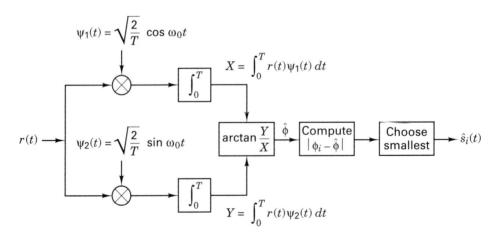

Figure 4.12 Demodulator for MPSK signals.

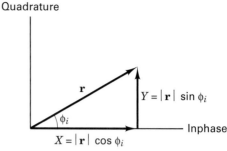

Quadrature

$Y = |\mathbf{r}| \sin \phi_i$

Inphase

$X = |\mathbf{r}| \cos \phi_i$

Figure 4.13 In-phase and quadrature components of the received signal vector **r**.

$\hat{\phi} = \arctan(Y/X)$ $\left\{ \begin{array}{l} \text{Noisy estimate} \\ \text{of transmitted } \phi_i \end{array} \right.$

noisy estimate of the transmitted ϕ_i. In other words, the upper correlator of Figure 4.12 produces an output X, the magnitude of the in-phase projection of the vector **r**, and the lower correlator produces an output Y, the magnitude of the quadrature projection of the vector **r**, where **r** is the vector representation of $r(t)$. The X and Y outputs of the correlators feed into the block marked arctan (Y/X). The resulting value of the angle $\hat{\phi}$ is compared with each of the prototype phase angles, ϕ_i. The demodulator selects the ϕ_i that is closest to the angle $\hat{\phi}$. In other words, the demodulator computes $|\phi_i - \hat{\phi}|$ for each of the ϕ_i prototypes and chooses the ϕ_i yielding the smallest output.

4.4.4 Coherent Detection of FSK

FSK modulation is characterized by the information being contained in the frequency of the carrier. A typical set of FSK signal waveforms was described in Equation (4.8) as

$$s_i(t) = \sqrt{\frac{2E}{T}} \cos(\omega_i t + \phi) \qquad \begin{array}{l} 0 \le t \le T \\ i = 1, \ldots, M \end{array}$$

where E is the energy content of $s_i(t)$ over each symbol duration T, and $(\omega_{i+1} - \omega_i)$ is typically assumed to be an integral multiple of π/T. The phase term ϕ is an arbitrary constant and can be set equal to zero. Assuming that the basis functions $\psi_1(t)$, $\psi_2(t), \ldots, \psi_N(t)$ form an orthonormal set, the most useful form for $\{\psi_j(t)\}$ is

$$\psi_j(t) = \sqrt{\frac{2}{T}} \cos \omega_j t \qquad j = 1, \ldots, N \qquad (4.37)$$

where, as before, the amplitude $\sqrt{2/T}$ normalizes the expected output of the matched filter. From Equation (3.11), we can write

$$a_{ij} = \int_0^T \sqrt{\frac{2E}{T}} \cos(\omega_i t) \sqrt{\frac{2}{T}} \cos \omega_j t \; dt \qquad (4.38)$$

Therefore,

$$a_{ij} = \begin{cases} \sqrt{E} & \text{for } i = j \\ 0 & \text{otherwise} \end{cases} \tag{4.39}$$

In other words, the ith prototype signal vector is located on the ith coordinate axis at a displacement $\sqrt{E}$ from the origin of the signal space. In this scheme, for the general M-ary case and a given E, the distance between any two prototype signal vectors $\mathbf{s}_i$ and $\mathbf{s}_j$ is constant:

$$d(\mathbf{s}_i, \mathbf{s}_j) = \|\mathbf{s}_i - \mathbf{s}_j\| = \sqrt{2E} \qquad \text{for } i \neq j \tag{4.40}$$

Figure 4.14 illustrates the prototype signal vectors and the decision regions for a 3-ary ($M = 3$) coherently detected orthogonal FSK signaling scheme. A natural choice for the size M of a signaling set is any power-of-two value. However in this case, the reasons for the unorthodox selection of $M = 3$, are: We desire to examine a signaling set that is greater than binary, and the signaling space for orthogonal signaling is best visualized as having mutually perpendicular axes. It is not possible, beyond 3-axes, to convey the notion of mutual perpendicularity in some visually accurate way. As in the PSK case, the signal space is partitioned into M distinct regions, each containing one prototype signal vector; in this example, because the decision region is three-dimensional, the decision boundaries are planes instead of lines. The optimum decision rule is to decide that the transmitted signal belongs to the class whose index corresponds to the region where the received signal is

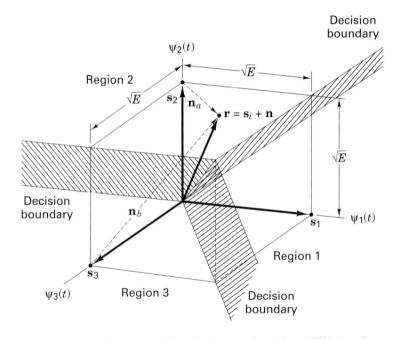

Figure 4.14 Partitioning the signal space for a 3-ary FSK signal.

found. In Figure 4.14, a received signal vector **r** is shown in region 2. Using the decision rule stated above, the detector classifies **r** as signal s_2. Since the noise is a Gaussian random vector, there is a probability greater than zero that **r** could have been produced by some signal other than s_2. For example, if the transmitter had sent s_2, then **r** would be the sum of signal plus noise, $s_2 + n_a$, and the decision to choose s_2 is correct; however, if the transmitter had actually sent s_3, then **r** would be the sum of signal plus noise, $s_3 + n_b$, and the decision to select s_2 is an error. The error performance of coherently detected FSK systems is treated in Section 4.7.3.

Example 4.2   Received Phase as a Function of Propagation Delay

(a) In Figure 4.8, the schematics fail to indicate where the correlator reference signals come from. One might think that they are known for all time and stored in memory until needed. Under some controlled circumstances, it is conceivable that the receiver might predict some expected value of the arriving signal's amplitude or its frequency within tolerable limits. But the one parameter that cannot be estimated without special help is the received signal phase. The most popular way to achieve phase estimation is through the use of circuitry called a *phase-locked loop* (PLL). This carrier-recovery method locks on to the arriving carrier wave (or recreates it) and estimates its phase. To show how futile it would be to predict the phase without a PLL, consider the mobile radio link shown in Figure 4.15. In the figure, a mobile user is positioned at point A, at a distance d from the base station, such that the propagation delay is T_d. Using complex notation, the transmitted waveform emanating from the transmitter can be described as $s(t) = \exp(j2\pi f_0 t)$. Let us take f_0 equal to 1 GHz. Neglecting noise, the waveform received at the base station can be denoted as $r(t) = \exp[j2\pi f_0(t + T_d)]$. If the mobile user moves in-line away from the base station to point B, or in-line toward the base station to point C, what is the minimum distance of movement that will cause a 2π phase rotation of the received waveform?

(b) Do we really care about a 2π phase rotation? Of course not, because the received phasor would then be located at exactly the point at which it was initially postulated with the user located at point A. But let us ask, What is the minimum distance that will cause a $\pi/2$ phase rotation (say a lag of $\pi/2$)? Had the receiver predicted that the received phasor would correspond to $r(t)$, as denoted in part (a),

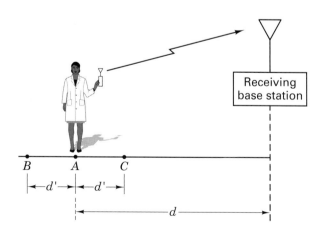

Figure 4.15   Mobile radio link.

4.4   Coherent Detection

193

but instead the lag caused the received phasor to appear as $\exp\left[j2\pi f_0(t + T_d) - \pi/2\right]$, then the detecting correlator would yield a zero output. This is so because

$$\int_0^T \cos\omega_0 t \cos\left(\omega_0 t - \frac{\pi}{2}\right) dt = \int_0^T \cos\omega_0 t \sin\omega_0 t \; dt = 0$$

Find the user's minimum distance movement that will cause a $\pi/2$ phase rotation.

Solution

(a) Initially, let $t = 0$, so that when the mobile user is located at point A, the received phasor at the base station can be expressed as $r(t) = \exp(j2\pi f_0 T_d)$. Then, after the user's movement to point B, the received (further delayed) phasor $r_d(t = T_d + T_d')$, can be written as $r_d(t) = \exp\left[j2\pi f_0(T_d + T_d')\right]$. The minimum delay time $T_{d'}$ corresponding to a 2π (one wavelength) phasor rotation is $T_d' = 1/f_0 = 10^{-9}$ second. Therefore, the minimum distance for such a rotation (assuming ideal electromagnetic propagation at the speed of light) is

$$d' = \frac{c}{f_0} = 3 \times 10^8 \text{ m/s} \times 10^{-9}\text{ s} = 0.3 \text{ m}$$

(b) Thus, for a $\pi/2$ phasor rotation, the minimum distance is

$$d'' = \frac{d'}{4} = \frac{0.3 \text{ m}}{4} = 7.5 \text{ cm}$$

It should be clear that even if a transmitter and receiver are located on fixed towers, a small amount of wind movement can bring about complete uncertainty regarding phase. If we scale our example from a frequency of 1 GHz to that of 10 GHz, the minimum distance scales from 7.5 cm to 0.75 cm. Very often we might want to avoid building receivers with PLLs for carrier recovery. The results of this example might then motivate us to ask, How will the error performance suffer if phase information is not used in the detection process? In other words, how will the system fare if the detection is performed noncoherently? We address this question in the sections that follow.

4.5 NONCOHERENT DETECTION

4.5.1 Detection of Differential PSK

The name *differential* PSK (DPSK) sometimes needs clarification because two separate aspects of the modulation/demodulation format are being referred to: the encoding procedure and the detection procedure. The term *differential encoding* refers to the procedure of encoding the data differentially; that is, the presence of a binary one or zero is manifested by the symbol's similarity or difference when compared with the preceding symbol. The term *differentially coherent detection* of differentially encoded PSK, the usual meaning of DPSK, refers to a detection scheme often classified as noncoherent because it does not require a reference in phase with the received carrier. Occasionally, differentially encoded PSK is *coherently* detected. This will be discussed in Section 4.7.2.

With noncoherent systems, no attempt is made to determine the actual value of the phase of the incoming signal. Therefore, if the transmitted waveform is

$$s_i(t) = \sqrt{\frac{2E}{T}} \cos\left[\omega_0 t + \theta_i(t)\right] \qquad \begin{array}{l} 0 \le t \le T \\ i = 1, \dots, M \end{array}$$

the received signal can be characterized by

$$r(t) = \sqrt{\frac{2E}{T}} \cos\left[\omega_0 t + \theta_i(t) + \alpha\right] + n(t) \qquad \begin{array}{l} 0 \le t \le T \\ i = 1, \dots, M \end{array} \qquad (4.41)$$

where α is an arbitrary constant and is typically assumed to be a random variable uniformly distributed between zero and 2π, and $n(t)$ is an AWGN process.

For coherent detection, matched filters (or their equivalents) are used; for noncoherent detection, this is not possible because the matched filter output is a function of the unknown angle α. However, if we assume that α varies slowly relative to two period times $(2T)$, the phase difference between two successive waveforms $\theta_j(T_1)$ and $\theta_k(T_2)$ is independent of α; that is,

$$\left[\theta_k(T_2) + \alpha\right] - \left[\theta_j(T_1) + \alpha\right] = \theta_k(T_2) - \theta_j(T_1) = \phi_i(T_2) \qquad (4.42)$$

The basis for *differentially coherent detection* of differentially encoded PSK (DPSK) is as follows. The carrier phase of the previous signaling interval can be used as a phase reference for demodulation. Its use requires *differential encoding* of the message sequence at the transmitter since the information is carried by the difference in phase between two successive waveforms. To send the ith message ($i = 1, 2, \dots, M$), the present signal waveform must have its phase advanced by $\phi_i = 2\pi i/M$ radians over the previous waveform. The detector, in general, calculates the coordinates of the incoming signal by correlating it with locally generated waveforms, such as $\sqrt{2/T} \cos \omega_0 t$ and $\sqrt{2/T} \sin \omega_0 t$. the detector then measures the angle between the currently received signal vector and the previously received signal vector, as illustrated in Figure 4.16.

In general, DPSK signaling performs less efficiently than PSK, because the errors in DPSK tend to propagate (to adjacent symbol times) due to the correlation between signaling waveforms. One way of viewing the difference between PSK and

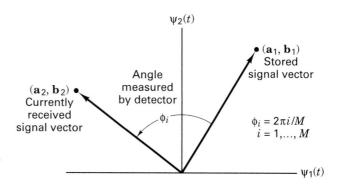

Figure 4.16 Signal space for DPSK.

DPSK is that the former compares the received signal with a clean reference; in the latter, however, two noisy signals are compared with each other. We might say that there is twice as much noise associated with DPSK signaling compared to PSK signaling. Consequently, as a first guess, we might estimate that the error probability for DPSK is approximately two times (3 dB) worse than PSK; this degradation decreases rapidly with increasing signal-to-noise ratio. The trade-off for this performance loss is reduced system complexity. The error performance for the detection of DPSK is treated in Section 4.7.5.

4.5.2 Binary Differential PSK Example

The essence of differentially coherent detection in DPSK is that the identity of the data is inferred from the changes in phase from symbol to symbol. Therefore, because the data are detected by differentially examining the waveform, the transmitted waveform must first be encoded in a differential fashion. Figure 4.17a illustrates a differential encoding of a binary message data stream $m(k)$, where k is the sample time index. The differential encoding starts (third row in the figure) with the first bit of the code-bit sequence $c(k = 0)$, chosen arbitrarily (here taken to be a one). Then the sequence of encoded bits $c(k)$ can, in general, be encoded in one of two ways:

$$c(k) = c(k - 1) \oplus m(k) \tag{4.43}$$

or

$$c(k) = \overline{c(k - 1) \oplus m(k)} \tag{4.44}$$

where the symbol $\oplus$ represents modulo-2 addition (defined in Section 2.9.3) and the overbar denotes complement. In Figure 4.17a the differentially encoded message was obtained by using Equation (4.44). In other words, the present code bit $c(k)$ is a one if the message bit $m(k)$ and the prior coded bit $c(k - 1)$ are the same, otherwise, $c(k)$ is a zero. The fourth row translates the coded bit sequence $c(k)$ into the phase shift sequence $\theta(k)$, where a one is characterized by a 180° phase shift, and a zero is characterized by a 0° phase shift.

Figure 4.17b illustrates the binary DPSK detection scheme in block diagram form. Notice that the basic product integrator of Figure 4.7 is the essence of the demodulator; as with coherent PSK, we are still attempting to correlate a received signal with a reference. The interesting difference here is that the reference signal is simply a delayed version of the received signal. In other words, during each symbol time, we are matching a received symbol with the prior symbol and looking for a correlation or an anticorrelation (180° out of phase).

Consider the received signal with phase shift sequence $\theta(k)$ entering the correlator of Figure 4.17b, in the absence of noise. The phase $\theta(k = 1)$ is matched with $\theta(k = 0)$; they have the same value, π; hence the first bit of the detected output is $\hat{m}(k = 1) = 1$. Then $\theta(k = 2)$ is matched with $\theta(k = 1)$; again they have the same value, and $\hat{m}(k = 2) = 1$. Then $\theta(k = 3)$ is matched with $\theta(k = 2)$; they are different, so that $\hat{m}(k = 3) = 0$, and so on.

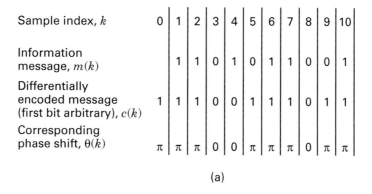

Sample index, k	0	1	2	3	4	5	6	7	8	9	10
Information message, $m(k)$		1	1	0	1	0	1	1	0	0	1
Differentially encoded message (first bit arbitrary), $c(k)$	1	1	1	0	0	1	1	1	0	1	1
Corresponding phase shift, $\theta(k)$	π	π	π	0	0	π	π	π	0	π	π

(a)

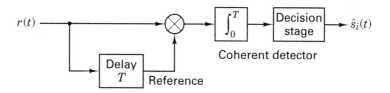

Coherent detector

Detected message, $\hat{m}(k)$ 1 1 0 1 0 1 1 0 0 1

(b)

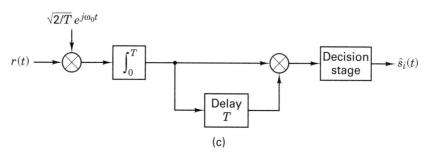

(c)

Figure 4.17 Differential PSK (DPSK). (a) Differential encoding. (b) Differentially coherent detection. (c) Optimum differentially coherent detection.

It must be pointed out that the detector in Figure 4.17b is suboptimum [3] in the sense of error performance. The optimum differential detector for DPSK requires a reference carrier in frequency but not necessarily in phase with the received carrier. Hence the optimum differential detector is shown in Figure 4.17c [4]. Its performance is treated in Section 4.7.5. In Figure 4.17c, the reference signal is shown in complex form ($\sqrt{2/T}\ e^{j\omega_0 t}$) to indicate that a quadrature implementation using both I and Q references are required (see Section 4.6.1).

44.5   Noncoherent Detection

4.5.3 Noncoherent Detection of FSK

A detector for the *noncoherent* detection of FSK waveforms described by Equation (4.8) can be implemented with correlators similar to those shown in Figure 4.7. However, the hardware must be configured as an *energy detector,* without exploiting phase measurements. For this reason, the noncoherent detector typically requires twice as many channel branches as the coherent detector. Figure 4.18 illustrates the in-phase (I) and quadrature (Q) channels used to detect a binary FSK (BFSK) signal set noncoherently. Notice that the upper two branches are configured to detect the signal with frequency ω_1; the reference signals are $\sqrt{2/T}$ cos $\omega_1 t$ for the I branch and $\sqrt{2/T}$ sin $\omega_1 t$ for the Q branch. Similarly, the lower two branches are configured to detect the signal with frequency ω_2; the reference signals are $\sqrt{2/T}$ cos $\omega_2 t$ for the I branch and $\sqrt{2/T}$ sin $\omega_2 t$ for the Q branch. Imagine that the received signal $r(t)$, by chance alone, is exactly of the form cos $\omega_1 t$ + $n(t)$; that is, the phase is exactly zero, and thus the signal component of the received signal exactly matches the top-branch reference signal with regard to frequency and phase. In that event, the product integrator of the top branch should yield the maximum output. The second branch should yield a near-zero output (integrated zero-mean noise), since its reference signal $\sqrt{2/T}$ sin $\omega_1 t$ is orthogonal to the sig-

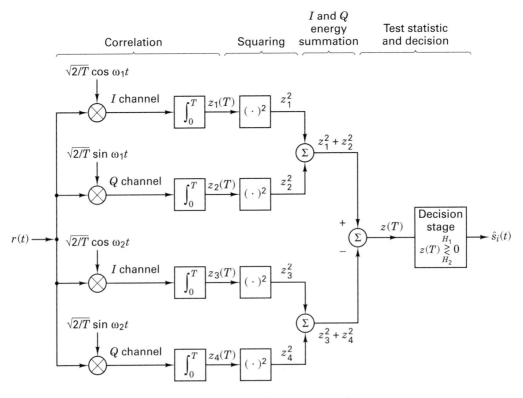

Figure 4.18 Quadrature receiver.

nal component of $r(t)$. For orthogonal signaling (see Section 4.5.4), the third and fourth branches should also yield near-zero outputs, since their ω_2 reference signals are also orthogonal to the signal component of $r(t)$.

Now imagine a different scenario: Suppose that by chance alone, the received signal $r(t)$ is of the form $\sin \omega_1 t + n(t)$. In that event, the second branch in Figure 4.18 should yield the maximum output, while the others should yield near-zero outputs. In actual practice, the most likely scenario is that $r(t)$ is of the form $\cos (\omega_1 t + \phi) + n(t)$; that is, the incoming signal will *partially* correlate with the $\cos \omega_1 t$ reference and *partially* correlate with the $\sin \omega_1 t$ reference. Hence a noncoherent quadrature receiver for orthogonal signals requires an I and Q branch for each candidate signal in the signaling set. In Figure 4.18, the blocks following the product integrators perform a squaring operation to prevent the appearance of any negative values. Then for each signal class in the set (two in this binary example), z_1^2 from the I channel and z_2^2 from the Q channel are added. The final stage forms the test statistic $z(T)$ and chooses the signal with frequency ω_1 or the signal with frequency ω_2, depending upon which pair of energy detectors yields the maximum output.

Another possible implementation for noncoherent FSK detection uses band-pass filters—centered at $f_i = \omega_i/2\pi$ with bandwidth $W_f = 1/T$—followed by *envelope detectors,* as shown in Figure 4.19. An envelope detector consists of a rectifier and a low-pass filter. The detectors are matched to the *signal envelopes* and not to the signals themselves. The phase of the carrier is of no importance in defining the envelope; hence no phase information is used. In the case of binary FSK, the decision as to whether a one or a zero was transmitted is made on the basis of which of two envelope detectors has the largest amplitude at the moment of measurement. Similarly, for a multiple frequency shift-keying (MFSK) system, the decision as to which of M signals was transmitted is made on the basis of which of the M envelope detectors has the maximum output.

Even though the envelope detector block diagram of Figure 4.19 looks functionally more simple than the quadrature receiver shown in Figure 4.18, the use of

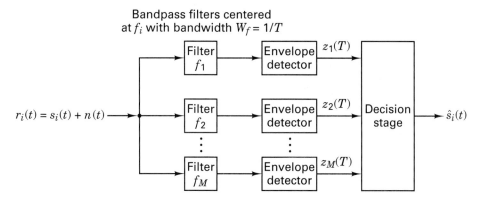

Figure 4.19   Noncoherent detection of FSK using envelope detectors.

4.5   Noncoherent Detection

(analog) filters usually results in the envelope detector design having greater weight and cost than the quadrature receiver. Quadrature receivers can be implemented digitally; thus, with the advent of large-scale integrated (LSI) circuits, they are often the preferred choice for noncoherent detectors. The detector shown in Figure 4.19 can also be implemented digitally by performing discrete Fourier transformations instead of using analog filters, but such a design is usually more complex than a digital implementation of the quadrature receiver.

4.5.4 Required Tone Spacing for Noncoherent Orthogonal FSK Signaling

Frequency shift keying (FSK) is usually implemented as orthogonal signaling, but not all FSK signaling is orthogonal. How can we tell if the tones in a signaling set form an orthogonal set? For example, suppose we have two tones $f_1 = 10,000$ Hz and $f_2 = 11,000$ Hz. Are they orthogonal to each other? In other words, do they fulfill the criterion (as set forth in Equation 3.69) that they are uncorrelated over a symbol time T? We do not have enough information to answer that question yet. Tones f_1 and f_2 manifest orthogonality if, for a transmitted tone at f_1, the sampled envelope of the receiver output filter tuned to f_2 is zero (i.e., no crosstalk). A property that insures such orthogonality between tones in an FSK signaling set states that any pair of tones in the set must have a frequency separation that is a multiple of $1/T$ hertz. (This will be proven below, in Example 4.3.) A tone with frequency f_i that is switched on for a symbol duration of T seconds and then switched off, such as the FSK tone described in Equation (4.8), can be analytically described by

$$s_i(t) = (\cos 2\pi f_i t) \text{ rect } (t/T)$$

$$\text{where rect } (t/T) = \begin{cases} 1 & \text{for } -T/2 \leq t \leq T/2 \\ 0 & \text{for } |t| > T/2 \end{cases}$$

From Table A.1, the Fourier transform of $s_i(t)$ is

$$\mathcal{F}\{s_i(t)\} = T \text{ sinc } (f - f_i)T$$

where the sinc function is as defined in Equation (1.39). The spectra of two such adjacent tones—tone 1 with frequency f_1 and tone 2 with frequency f_2—are plotted in Figure 4.20.

4.5.4.1 Minimum Tone Spacing and Bandwidth

In order for a noncoherently detected tone to manifest a maximum output signal at its associated receiver filter and a zero-output signal at any neighboring filter (as implemented in Figure 4.19), the peak of the spectrum of tone 1 must coincide with one of the zero crossings of the spectrum of tone 2, and similarly, the peak of the spectrum of tone 2 must coincide with one of the zero crossings of the spectrum of tone 1. The frequency difference between the center of the spectral main lobe and the first zero crossing represents the *minimum required spacing*. With noncoherent detection, this corresponds to a minimum tone separation of $1/T$ hertz, as can be seen in Figure 4.20. Even though FSK signaling entails the trans-

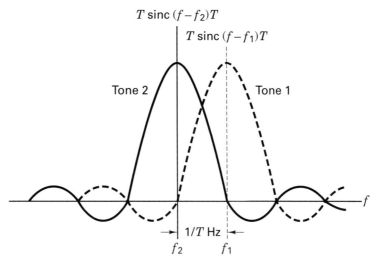

Figure 4.20 Minimum tone spacing for noncoherently detected orthogonal FSK signaling.

mission of just one single-sideband tone during each symbol time, when we speak of the signal bandwidth, we mean the amount of spectrum that needs to be available for the entire range of all the tones in the M-ary set. Thus for FSK, the bandwidth requirements are related to the spectral spacing between the tones. From a group of adjacent tones, the spectrum associated with just one tone can be considered to make up one-half of the tone spacing on each side of the tone. Thus, for the binary FSK case shown in Figure 4.20, the signaling bandwidth is equal to the spectrum that separates the tone centers plus one-half the tone spacing on each side of the set. This amounts to two times the tone spacing. Extrapolating this approach to the M-ary case, the bandwidth of noncoherently detected orthogonal MFSK is equal to M/T.

Thus far, we have only addressed noncoherently detected orthogonal FSK. Is the minimum tone-spacing criterion (and hence the bandwidth) different if the signaling is detected coherently? Indeed, it is. As we will see below, in Example 4.3, the minimum tone spacing is reduced to $1/2T$ when coherent detection is used.

4.5.4.2 Dual Relationships

The engineering concept of duality can be defined as follows: Two processes (functions, elements, or systems) are *dual* to each other if their mathematical relationships are identical, even though they are described with different variables (e.g., time and frequency). Consider the case of FSK signaling, where rectangular-shaped keying corresponds to sinc (fT) shaped tones, as seen in Figure 4.20. A given tone duration gives rise to the minimum frequency spacing between tones that would be needed to achieve orthogonality. This frequency-domain relationship has as its dual counterpart a time-domain relationship—the case of pusle signaling, where a rectangular-shaped bandwidth corresponds to sinc (t/T) shaped pulses, as seen in Figure 3.16b. A given bandwidth gives rise to the minimum time spacing between pulses that would be needed to achieve zero ISI.

Example 4.3   Minimum Tone Spacing for Orthogonal FSK

Consider two waveforms cos $(2\pi f_1 t + \phi)$ and cos $2\pi f_2 t$ to be used for *noncoherent* FSK-signaling, where $f_1 > f_2$. The symbol rate is equal to $1/T$ symbols/s, where T is the symbol duration and ϕ is a constant arbitrary angle from 0 to 2π.

(a) Prove that the minimum tone spacing for *noncoherently detected* orthogonal FSK-signaling is $1/T$.
(b) What is the minimum tone spacing for *coherently detected* orthogonal FSK signaling?

Solution

(a) For the two waveforms to be orthogonal, they must fulfill the orthogonality constraint of Equation (3.69):

$$\int_0^T \cos (2\pi f_1 t + \phi) \cos 2\pi f_2 t \; dt = 0 \tag{4.45}$$

Using the basic trigonometric identities shown in Equations (D.6) and (D.1) to (D.3), we can write Equation (4.45) as

$$\cos \phi \int_0^T \cos 2\pi f_1 t \; \cos 2\pi f_2 t \; dt \tag{4.46}$$

$$- \sin \phi \int_0^T \sin 2\pi f_1 t \; \cos 2\pi f_2 t \; dt = 0$$

so that

$$\cos \phi \int_0^T [\cos 2\pi(f_1 + f_2)t + \cos 2\pi(f_1 - f_2)t] \, dt$$

$$- \sin \phi \int_0^T [\sin 2\pi(f_1 + f_2)t + \sin 2\pi(f_1 - f_2)t] \, dt = 0 \tag{4.47}$$

which, in turn, yields

$$\cos \phi \left[\frac{\sin 2\pi(f_1 + f_2)t}{2\pi(f_1 + f_2)} + \frac{\sin 2\pi(f_1 - f_2)t}{2\pi(f_1 - f_2)} \right]_0^T$$

$$+ \sin \phi \left[\frac{\cos 2\pi(f_1 + f_2)t}{2\pi(f_1 + f_2)} + \frac{\cos 2\pi(f_1 - f_2)t}{2\pi(f_1 - f_2)} \right]_0^T = 0 \tag{4.48}$$

or

$$\cos \phi \left[\frac{\sin 2\pi(f_1 + f_2)T}{2\pi(f_1 + f_2)} + \frac{\sin 2\pi(f_1 - f_2)T}{2\pi(f_1 - f_2)} \right]$$

$$+ \sin \phi \left[\frac{\cos 2\pi(f_1 + f_2)T - 1}{2\pi(f_1 + f_2)} + \frac{\cos 2\pi(f_1 - f_2)T - 1}{2\pi(f_1 - f_2)} \right] = 0 \tag{4.49}$$

We can assume that $f_1 + f_2 \gg 1$ and can thus make the following approximation:

$$\frac{\sin 2\pi(f_1 + f_2)T}{2\pi(f_1 + f_2)} \approx \frac{\cos 2\pi(f_1 + f_2)T}{2\pi(f_1 + f_2)} \approx 0 \qquad (4.50)$$

Then, combining Equations (4.49) and (4.50), we can write

$$\cos \phi \sin 2\pi(f_1 - f_2)T + \sin \phi \left[\cos 2\pi(f_1 - f_2)T - 1\right] \approx 0 \qquad (4.51)$$

Note that for arbitrary ϕ, the terms in Equation (4.51) can sum to zero only when $\sin 2\pi(f_1 - f_2)T = 0$, and simultaneously $\cos 2\pi(f_1 - f_2)T = 1$.

Since

$$\sin x = 0 \qquad \text{for } x = n\pi$$

and

$$\cos x = 1 \qquad \text{for } x = 2k\pi$$

where n and k are integers, then both $\sin x = 0$ and $\cos x = 1$ occur simultaneously when $n = 2k$. From Equation (4.51), for arbitrary ϕ, we can therefore write

$$2\pi(f_1 - f_2)T = 2k\pi$$

or $\qquad\qquad\qquad\qquad\qquad\qquad\qquad\qquad\qquad\qquad\qquad\qquad\qquad$ (4.52)

$$f_1 - f_2 = \frac{k}{T}$$

Thus the minimum tone spacing for *noncoherent* FSK signaling occurs for $k = 1$, in which case we write

$$f_1 - f_2 = \frac{1}{T} \qquad (4.53)$$

Recall the question posed earlier. Having two tones $f_1 = 10,000$ hertz and $f_2 = 11,000$ hertz, we asked, Are they orthogonal? Now we have sufficient information to answer that question. The answer depends on the speed of the FSK signaling. If the tones are being keyed (switched) at the rate of 1,000 symbols/s and noncoherently detected, then they are orthogonal. If they are being keyed faster—say, at the rate of 10,000 symbols/s—they are not orthogonal.

(b) For the noncoherent detection in part (a) of this example, the tone spacings that rendered the signals orthogonal was found by satisfying Equation (4.45) for any arbitrary phase. In this case, however, for coherent detection, the tone spacings needed for orthogonality are found by setting $\phi = 0$. This is because we know the phase of the received signal (from our phase-locked loop estimate). This received signal will be correlated with each reference signal using the same phase estimate for the reference signals. We can now rewrite Equation (4.51) with $\phi = 0$, which gives

$$\sin 2\pi(f_1 - f_2)T = 0 \qquad (4.54)$$

or

$$f_1 - f_2 = \frac{n}{2T} \qquad (4.55)$$

Thus the minimum tone spacing for *coherent* FSK signaling occurs for $n = 1$ as follows:

$$f_1 - f_2 = \frac{1}{2T} \tag{4.56}$$

Therefore, for the same symbol rate, coherently detected FSK can occupy less bandwidth than noncoherently detected FSK and still retain orthogonal signaling. We can say that coherent FSK is more *bandwidth efficient*. (The subject of bandwidth efficiency is addressed in greater detail in Chapter 9.)

      The required tone spacings are now closer than in part (a) because when we align two periodic waveforms so that their starting phases are the same, we achieve orthogonality by virtue of an even-versus-odd symmetry in the respective waveforms over one symbol time. This is unlike the way orthogonality was achieved in part (a), where we paid no attention to phase. In the coherent case here, the phase alignment in the correlator stages means that we can bring the tones closer together in frequency and still maintain orthogonality among the set of FSK tones. Prove it to yourself by plotting two sine waves (or cosine waves or square waves). Start them off at the same phase (0 radians is convenient). Using quadrille paper, choose a simple time scale to represent one symbol time T. Then, plot a tone at one cycle per T. Below that, with the same starting phase, plot another tone at one-and-a-half cycles per T. Perform product-summation over period T, and verify that these waveforms (spaced $1/2T$ hertz apart) are indeed orthogonal.

4.6 COMPLEX ENVELOPE

The description of real-world modulators and demodulators is facilitated by the use of complex notation which began in Section 4.2.1 and continues here. Any real bandpass waveform $s(t)$ can be represented using complex notation as

$$s(t) = \text{Re}\{g(t)e^{j\omega_0 t}\} \tag{4.57}$$

where $g(t)$ is known as the *complex envelope,* expressed as

$$g(t) = x(t) + jy(t) = |g(t)|e^{j\theta(t)} = R(t)e^{j\theta(t)} \tag{4.58}$$

The magnitude of the complex envelope is then

$$R(t) = |g(t)| = \sqrt{x^2(t) + y^2(t)} \tag{4.59}$$

and its phase is

$$\theta(t) = \tan^{-1}\frac{y(t)}{x(t)} \tag{4.60}$$

With respect to Equation (4.57), we can call $g(t)$ the baseband message or data in complex form, and $e^{j\omega_0 t}$ the carrier wave in complex form. The product of these two represents modulation, and $s(t)$, the real part of this product, is the transmitted waveform. Therefore, using Equations (4.4), (4.57), and (4.58), we can express $s(t)$ as follows:

$$s(t) = \text{Re}\{[x(t) + jy(t)][\cos \omega_0 t + j \sin \omega_0 t]\}$$
$$= x(t) \cos \omega_0 t - y(t) \sin \omega_0 t \qquad (4.61)$$

Note that the modulation of signals expressed in the general form of $(a + jb)$ times $(c + jd)$ yields a waveform with a sign inversion (at the quadrature term of the carrier wave) of the form *ac-bd*.

4.6.1 Quadrature Implementation of a Modulator

Consider an example of a baseband waveform $g(t)$, described by a sequence of ideal pulses $x(t)$ and $y(t)$ appearing at discrete times $k = 1, 2, \ldots$. Thus, $g(t)$, $x(t)$, and $y(t)$ in Equation (4.58) can be written as g_k, x_k, and y_k, respectively. Let the pulse-amplitude values be $x_k = y_k = 0.707A$. The complex envelope can then be expressed in discrete form as

$$g_k = x_k + jy_k = 0.707A + j0.707A \qquad (4.62)$$

We know from complex algebra that $j = \sqrt{-1}$, but from a practical point of view, the j term can be regarded as a "flag," reminding us that we may not use ordinary addition in combing the terms in Equation (4.62). In preparation for the next step, inphase and quadrature modulation, x_k and y_k are treated as an ordered pair. A quadrature-type modulator is shown in Figure 4.21, where we see that the x_k pulse is multiplied by $\cos \omega_0 t$ (inphase component of the carrier wave), and the y_k pulse is multiplied by $\sin \omega_0 t$ (quadrature component of the carrier wave). The modulation process can be described succinctly as multiplying the complex envelope by $e^{j\omega_0 t}$, and then transmitting the real part of the product. Hence, we write

$$
\begin{aligned}
s(t) &= \text{Re}\{g_k e^{j\omega_0 t}\} \\
&= \text{Re}\{(x_k + jy_k)(\cos \omega_0 t + j \sin \omega_0 t)\} \\
&= x_k \cos \omega_0 t - y_k \sin \omega_0 t \qquad (4.63) \\
&= 0.707A \cos \omega_0 t - 0.707A \sin \omega_0 t \\
&= A \cos\left(\omega_0 t + \frac{\pi}{4}\right)
\end{aligned}
$$

Again, note that the quadrature term of the carrier wave has undergone a sign inversion in the modulation process. If we use $0.707A \cos \omega_0 t$ as a reference, then the transmitted waveform $s(t)$ in Equation (4.63) leads the reference waveform by $\pi/4$.

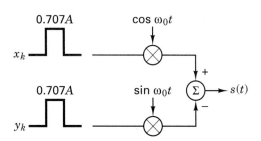

Figure 4.21 Quadrature type modulator.

Similarly, if we consider $-0.707A \sin \omega_0 t$ as the reference, then $s(t)$ in Equation (4.63) lags the reference waveform by $\pi/4$. These relationships are illustrated in Figure 4.22.

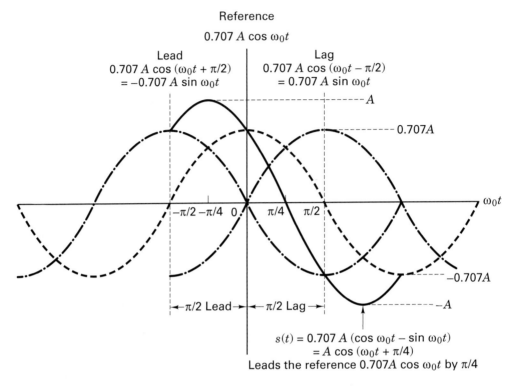

Figure 4.22 Lead/Lag relationships of sinusoids.

4.6.2 D8PSK Modulator Example

Figure 4.23 depicts a quadrature implementation of a differential 8-PSK (D8PSK) modulator. Because the modulation is 8-ary, we assign a 3-bit message (x_k, y_k, z_k) to each phase $\Delta\phi_k$. Because the modulation is differential, at each kth transmission time we send a data phasor ϕ_k, which can be expressed as

$$\phi_k = \Delta\phi_k + \phi_{k-1} \tag{4.64}$$

The process of adding the current message-to-phase assignment $\Delta\phi_k$ to the prior data phase ϕ_{k-1} provides for the differential encoding of the message. A sequence of phasors created by following Equation (4.64) yields a similar differential encoding as the procedures described in Section 4.5.2. You might notice in Figure 4.23 that the assignment of 3-bit message sequences to $\Delta\phi_k$ does not proceed along the natural binary progression from 000 to 111. There is a special code being used here called a *Gray code*. (The benefits that such a binary to M-ary assignment provides is explained in Section 4.9.4.)

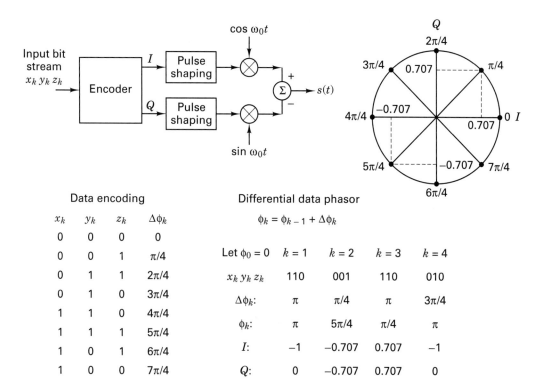

Data encoding				Differential data phasor
x_k	y_k	z_k	$\Delta\phi_k$	$\phi_k = \phi_{k-1} + \Delta\phi_k$

x_k	y_k	z_k	$\Delta\phi_k$
0	0	0	0
0	0	1	$\pi/4$
0	1	1	$2\pi/4$
0	1	0	$3\pi/4$
1	1	0	$4\pi/4$
1	1	1	$5\pi/4$
1	0	1	$6\pi/4$
1	0	0	$7\pi/4$

Let $\phi_0 = 0$	$k = 1$	$k = 2$	$k = 3$	$k = 4$
$x_k\, y_k\, z_k$	110	001	110	010
$\Delta\phi_k$:	π	$\pi/4$	π	$3\pi/4$
ϕ_k:	π	$5\pi/4$	$\pi/4$	π
I:	-1	-0.707	0.707	-1
Q:	0	-0.707	0.707	0

Figure 4.23 Quadrature implementation of a D8PSK modulator.

For the modulator in Figure 4.23, let the input data sequence at times $k = 1, 2,$ 3, 4, be equal to 110, 001, 110, 010, respectively. Next, we use the date-encoding table shown in Figure 4.23 and Equation (4.64), with the starting phase at time $k = 0$ to be $\phi_0 = 0$. At time $k = 1$, the differential data phase corresponding to $x_1\, y_1\, z_1$ = 110 is $\phi_1 = 4\pi/4 = \pi$. Taking the magnitude of the rotating phasor to be unity, the inphase (I) and the quadrature (Q) baseband pulses are -1 and 0, respectively. As indicated in Figure 4.23, these pulses are generally shaped with a filter (such as a root-raised-cosine).

For time $k = 2$, the table in Figure 4.23 shows that the message 001 is assigned $\Delta\phi_2 = \pi/4$. Therefore, following Equation (4.64), the second differential data phase is $\phi_2 = \pi + \pi/4 = 5\pi/4$, and for time $k = 2$, the I and Q baseband pulses are $x_k = -0.707$ and $y_k = -0.707$, respectively. The transmitted waveform follows the form of Equation (4.61), rewritten as

$$s(t) = \mathrm{Re}\{(x_k + jy_k)(\cos\omega_0 t + j\sin\omega_0 t)\}$$
$$= x_k \cos\omega_0 t - y_k \sin\omega_0 t \qquad (4.65)$$

For a signaling set that can be represented on a phase-amplitude plane, such as MPSK or MQAM, Equation (4.65) provides an interesting observation. That is, quadrature implementation of the transmitter transforms all such signaling types

4.6 Complex Envelope

to simple amplitude modulation. Any phasor on the plane is transmitted by amplitude-modulating its inphase and quadrature projections onto the cosine and sine wave components of the carrier, respectively. For ease of notation, we neglect the pulse shaping; that is, we assume that the data pulses have ideal rectangular shapes. Then, using Equation (4.65), for time $k = 2$, where $x_k = -0.707$ and $y_k = -0.707$, the transmitted $s(t)$ can be written as follows:

$$s(t) = -0.707 \cos \omega_0 t + 0.707 \sin \omega_0 t$$

$$= \sin \left(\omega_0 t - \frac{\pi}{4} \right)$$

(4.66)

4.6.3 D8PSK Demodulator Example

In the previous section, the quadrature implementation of a modulator began with multiplying the complex envelope (baseband message) by $e^{j\omega_0 t}$, and transmitting the real part of the product $s(t)$, as described in Equation (4.63). Using a similar quadrature implementation, demodulation consists of reversing the process—that is, multiplying the received bandpass waveform by $e^{-j\omega_0 t}$ in order to recover the baseband waveform. The left side of Figure 4.24 shows the modulator of Figure 4.23 in simplified form, and the waveform $s(t) = \sin (\omega_0 t - \pi/4)$ that was launched at time $k = 2$ for that example. We continue this same example in this section, and also show a quadrature implementation of the demodulator on the right side of Figure 4.24.

Notice the subtle difference between the $- \sin \omega_0 t$ term at the modulator and the $- \sin \omega_0 t$ term at the demodulator. At the modulator, the minus sign stems from taking the real part of the complex waveform (product of the complex envelope and complex carrier wave). At the demodulator, $- \sin \omega_0 t$ stems from multiplying the bandpass waveform by the conjugate $e^{-j\omega_0 t}$ of the modulator carrier wave; demodulation is coherent if phase is recovered. To simplify writing the basic relationships of the process, the noise is neglected. After the inphase multiplication by $\cos \omega_0 t$ in the demodulator, we get the signal at point A:

$$A = (-0.707 \cos \omega_0 t + 0.707 \sin \omega_0 t) \cos \omega_0 t$$

$$= -0.707 \cos^2 \omega_0 t + 0.707 \sin \omega_0 t \cos \omega_0 t$$

(4.67)

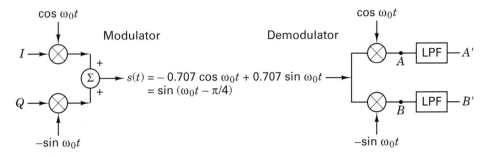

Figure 4.24 Modulator/demodulator example.

Using the trigonometric identities shown in Equation (D.7) and (D.9), we get

$$A = \frac{-0.707}{2}(1 + \cos 2\omega_0 t) + \frac{0.707}{2}\sin 2\omega_0 t \qquad (4.68)$$

After filtering with a low-pass filter (LPF) we recover at point A' an ideal negative pulse, as follows:

$$A' = -0.707 \text{ (times a scale factor)} \qquad (4.69)$$

Similarly, after the quadrature multiplication by $-\sin \omega_0 t$ in the demodulator, we get the signal at point B, as follows:

$$\begin{aligned}B &= (-0.707 \cos \omega_0 t + 0.707 \sin \omega_0 t)(-\sin \omega_0 t)\\ &= \frac{0.707}{2}\sin 2\omega_0 t - \frac{0.707}{2}(1 - \cos 2\omega_0 t)\end{aligned} \qquad (4.70)$$

After the LPF we recover at point B' an ideal negative pulse, as follows:

$$B' = -0.707 \text{ (times a scale factor)} \qquad (4.71)$$

Thus, we see from the demodulated and filtered points A' and B' that the differential (ideal) data pulses for the I and Q channels are each equal to -0.707. Since the modulator/demodulator is differential, then for this $k = 2$ example,

$$\Delta\phi_{k=2} = \phi_{k=2} - \phi_{k=1} \qquad (4.72)$$

We presume that the demodulator at the earlier time $k = 1$ had properly recovered the signal phase to be π. Then, from Equation (4.72), we can write

$$\Delta\phi_{k=2} = \frac{5\pi}{4} - \pi = \frac{\pi}{4} \qquad (4.73)$$

Referring back to the data-encoding table in Figure 4.23, we see that the detected data sequence is $x_2\, y_2\, z_2 = 001$, which corresponds to the data that was sent at time $k = 2$.

4.7 ERROR PERFORMANCE FOR BINARY SYSTEMS

4.7.1 Probability of Bit Error for Coherently Detected BPSK

An important measure of performance used for comparing digital modulation schemes is the probability of error, P_E. For the correlator or matched filter, the calculations for obtaining P_E can be viewed geometrically. (See Figure 4.6.) They involve finding the probability that given a particular transmitted signal vector, say s_1, the noise vector $\mathbf{n}$ will give rise to a received signal falling outside region 1. The probability of the detector making an incorrect decision is termed the *probability of symbol error* (P_E). It is often convenient to specify system performance by the probability of bit error (P_B), even when decisions are made on the basis of symbols for which $M > 2$. The relationship between P_B and P_E is treated in Section 4.9.3 for orthogonal signaling and in Section 4.9.4 for multiple phase signaling.

For convenience, this section is restricted to the coherent detection of BPSK modulation. For this case the symbol error probability is the bit error probability. Assume that the signals are equally likely. Also assume that when signal $s_i(t)$ $(i = 1, 2)$ is transmitted, the received signal $r(t)$ is equal to $s_i(t) + n(t)$, where $n(t)$ is an AWGN process, and any degradation effects due to channel-induced ISI or circuit-induced ISI have been neglected. The antipodal signals $s_1(t)$ and $s_2(t)$ can be characterized in a one-dimensional signal space as described in Section 4.4.1, where

$$\left. \begin{array}{l} s_1(t) = \sqrt{E}\, \psi_1(t) \\ s_2(t) = -\sqrt{E}\, \psi_1(t) \end{array} \right\} 0 \le t \le T \qquad (4.74)$$

and

The detector will choose the $s_i(t)$ with the largest correlator output $z_i(T)$; or, in this case of equal-energy antipodal signals, the detector, using the decision rule in Equation (4.20), decides on the basis of

$$\begin{array}{ll} s_1(t) & \text{if } z(T) > \gamma_0 = 0 \\ s_2(t) & \text{otherwise} \end{array} \qquad (4.75)$$

and

Two types of errors can be made, as shown in Figure 4.9: The first type of error takes place if signal $s_1(t)$ is transmitted but the noise is such that the detector measures a negative value for $z(T)$ and chooses hypothesis H_2, the hypothesis that signal $s_2(t)$ was sent. The second type of error takes place if signal $s_2(t)$ is transmitted but the detector measures a positive value for $z(T)$ and chooses hypothesis H_1, the hypothesis that signal $s_1(t)$ was sent.

In Section 3.2.1.1, an expression was developed in Equation (3.42) for the probability of a bit error P_B, for this binary *minimum error* detector. We rewrite this relationship as

$$P_B = \int_{(a_1 - a_2)/2\sigma_0}^{\infty} \frac{1}{\sqrt{2\pi}} \exp\left(-\frac{u^2}{2}\right) du = Q\left(\frac{a_1 - a_2}{2\sigma_0}\right) \qquad (4.76)$$

where σ_0 is the standard deviation of the noise out of the correlator. The function $Q(x)$, called the *complementary error function* or *co-error function.* is defined as

$$Q(X) = \frac{1}{\sqrt{2\pi}} \int_x^{\infty} \exp\left(-\frac{u^2}{2}\right) du \qquad (4.77)$$

and is described in greater detail in Sections 3.2 and B.3.2.

For equal-energy antipodal signaling, such as the BPSK format in Equation (4.74), the receiver output signal components are $a_1 = \sqrt{E_b}$ when $s_1(t)$ is sent and $a_2 = -\sqrt{E_b}$ when $s_2(t)$ is sent, where E_b is the signal energy per binary symbol. For AWGN we can replace the noise variance σ_0^2 out of the correlator with $N_0/2$ (see Appendix C), so that we can rewrite Equation (4.76) as follows:

$$P_B = \int_{\sqrt{2E_b/N_0}}^{\infty} \frac{1}{\sqrt{2\pi}} \exp\left(-\frac{u^2}{2}\right) du \qquad (4.78)$$

$$= Q\left(\sqrt{\frac{2E_b}{N_0}}\right) \qquad (4.79)$$

This result for bandpass antipodal BPSK signaling is the same as the results that were developed earlier in Equation (3.70) for the matched-filter detection of antipodal signaling in general, and in Equation (3.76) for the matched-filter detection of baseband antipodal signaling in particular. This is an example of an *equivalence theorem,* described earlier. For linear systems the equivalence theorem establishes that the mathematics of detection is unaffected by a shift in frequency. Hence in this chapter, the use of matched filters or correlators in the detection of bandpass signals yields the same relationships as those developed for comparable signals at baseband.

Example 4.4 Bit Error Probability for BPSK Signaling

Find the bit error probability for a BPSK system with a bit rate of 1 Mbit/s. The received waveforms $s_1(t) = A \cos \omega_0 t$ and $s_2(t) = -A \cos \omega_0 t$, are coherently detected with a matched filter. The value of A is 10 mV. Assume that the single-sided noise power spectral density is $N_0 = 10^{-11}$ W/Hz and that signal power and energy per bit are normalized relative to a 1-Ω load.

Solution

$$A = \sqrt{\frac{2E_b}{T}} = 10^{-2} \text{ V} \qquad T = \frac{1}{R} = 10^{-6} \text{ s}$$

Thus,

$$E_b = \frac{A^2}{2} T = 5 \times 10^{-11} \text{ J} \qquad \text{and} \qquad \sqrt{\frac{2E_b}{N_0}} = 3.16$$

Also,

$$P_B = Q\left(\sqrt{\frac{2E_b}{N_0}}\right) = Q(3.16)$$

Using Table B.1 or Equation (3.44), we obtain

$$P_B = 8 \times 10^{-4}$$

4.7.2 Probability of Bit Error for Coherently Detected, Differentially Encoded Binary PSK

Channel waveforms sometimes experience inversion; for example, when using a coherent reference generated by a phase-locked loop, one may have phase ambiguity. If the carrier phase were reversed in a DPSK modulation application, what would be the effect on the message? The only effect would be an error in the bit during

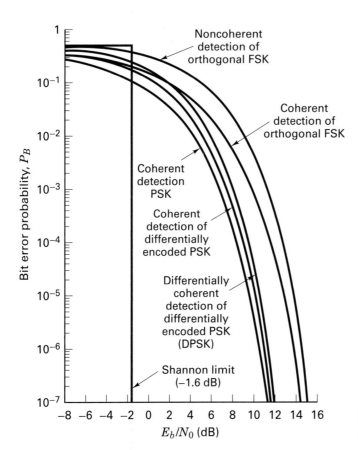

Bit error probability, P_B

Noncoherent detection of orthogonal FSK

Coherent detection of orthogonal FSK

Coherent detection PSK

Coherent detection of differentially encoded PSK

Differentially coherent detection of differentially encoded PSK (DPSK)

Shannon limit (−1.6 dB)

E_b/N_0 (dB)

Figure 4.25 Bit error probability for several types of binary systems.

which inversion occurred or the bit just after inversion, since the message information is encoded in the similarity or difference between adjacent symbols. The similarity or difference quality remains unchanged if the carrier is inverted. Sometimes messages (and their assigned waveforms) are *differentially encoded* and *coherently detected* simply to avoid these phase ambiguities.

The probability of bit error for coherently detected, differentially encoded PSK is given by [5]

$$P_B = 2Q\left(\sqrt{\frac{2E_b}{N_0}}\right)\left[1 - Q\left(\sqrt{\frac{2E_b}{N_0}}\right)\right] \qquad (4.80)$$

This relationship is plotted in Figure 4.25. Notice that there is a small degradation of error performance compared with the coherent detection of PSK. This is due to the differential encoding, since any single detection error will usually result in two decision errors. Error performance for the more popular differentially coherent detection (DPSK) is covered in Section 4.7.5.

4.7.3 Probability of Bit Error for Coherently Detected Binary Orthogonal FSK

Equations (4.78) and (4.79) describe the probability of bit error for coherent antipodal signals. A more general treatment for binary coherent signals (not limited to antipodal signals) yields the following equation for P_B [6]:

$$P_B = \frac{1}{\sqrt{2\pi}} \int_{\sqrt{(1-\rho)E_b/N_0}}^{\infty} \exp\left(-\frac{u^2}{2}\right) du \qquad (4.81)$$

From Equation (3.64b), $\rho = \cos\theta$ is the time cross-correlation coefficient between signal $s_1(t)$ and $s_2(t)$, where θ is the angle between signal vectors $\mathbf{s}_1$ and $\mathbf{s}_2$ (see Figure 4.6). For antipodal signals such as BPSK, $\theta = \pi$, thus $\rho = -1$.

For orthogonal signals such as binary FSK (BFSK), $\theta = \pi/2$, since the $\mathbf{s}_1$ and $\mathbf{s}_2$ vectors are perpendicular to each other; thus $\rho = 0$, as can be verified with Equation (3.64a), and Equation (4.81) can then be written as

$$P_B = \frac{1}{\sqrt{2\pi}} \int_{\sqrt{E_b/N_0}}^{\infty} \exp\left(-\frac{u^2}{2}\right) du = Q\left(\sqrt{\frac{E_b}{N_0}}\right) \qquad (4.82)$$

where the *co-error function* $Q(x)$ is described in greater detail in Sections 3.2 and B.3.2. The result in Equation (4.82) for the coherent detection of orthogonal BFSK plotted in Figure 4.25 is the same as the results that were developed earlier in Equation (3.71) for the matched-filter detection of orthogonal signaling, in general, and in Equation (3.73) for the matched-filter detection of baseband orthogonal signaling (unipolar pulses), in particular. The details of *on-off keying* (OOK) are not treated in this book. However, it is worth noting that on-off keying is an orthogonal signaling set (unipolar pulse signaling is the baseband equivalent of OOK). Thus the relationship in Equation (4.82) applies to the matched-filter detection of OOK, as it does to any coherent detection of orthogonal signaling.

The relationship in Equation (4.82) can also be confirmed by noting that the energy difference between the orthogonal signal vectors $\mathbf{s}_1$ and $\mathbf{s}_2$, with amplitudes of $\sqrt{E_b}$, as shown in Figure 3.10b, can be computed as the square of the distance between the heads of the orthogonal vectors, which will be $E_d = 2E_b$. Using this result in Equation (3.63) also yields Equation (4.82). If we compare Equation (4.82) with Equation (4.79), we can see that 3-dB more E_b/N_0 is required for BFSK to provide the same performance as BPSK. It should not be surprising that the performance of BFSK signaling is 3-dB worse than BPSK signaling, since for a given signal power, the distance-squared between orthogonal vectors is a factor of two less than the distance squared between antipodal vectors.

4.7.4 Probability of Bit Error for Noncoherently Detected Binary Orthogonal FSK

Consider the equally likely binary orthogonal FSK signal set $\{s_i(t)\}$, defined in Equation (4.8) as follows:

$$s_i(t) = \sqrt{\frac{2E}{T}} \cos{(\omega_i t + \phi)} \qquad 0 \le t \le T, \qquad i = 1, 2$$

The phase term ϕ is unknown and assumed constant. The detector is characterized by $M = 2$ channels of bandpass filters and envelope detectors, as shown in Figure 4.19. The input to the detector consists of the received signal $r(t) = s_i(t) + n(t)$, where $n(t)$ is a white Gaussian noise process with two-sided power spectral density $N_0/2$. Assume that $s_1(t)$ and $s_2(t)$ are separated in frequency sufficiently that they have negligible overlap. For $s_1(t)$ and $s_2(t)$ being equally likely, we start the bit-error probability P_B computation with Equation (3.38) as we did for baseband signaling:

$$P_B = \tfrac{1}{2}P(H_2|s_1) + \tfrac{1}{2}P(H_1|s_2) \tag{4.83}$$

$$= \frac{1}{2} \int_{-\infty}^{0} p(z|s_1)\, dz + \frac{1}{2} \int_{0}^{\infty} p(z|s_2)\, dz$$

For the binary case, the *test statistic* $z(T)$ is defined by $z_1(T) - z_2(T)$. Assume that the bandwidth of the filter W_f is $1/T$, so that the envelope of the FSK signal is (approximately) preserved at the filter output. If there was no noise at the receiver, the value of $z(T) = \sqrt{2E/T}$ when $s_1(t)$ is sent, and $z(T) = -\sqrt{2E/T}$ when $s_2(t)$ is sent. Because of this symmetry, the optimum threshold is $\gamma_0 = 0$. The pdf $p(z|s_1)$ is similar to $p(z|s_2)$; that is,

$$p(z|s_1) = p(-z|s_2) \tag{4.84}$$

Therefore, we can write

$$P_B = \int_{0}^{\infty} p(z|s_2)\, dz \tag{4.85}$$

or

$$P_B = P(z_1 > z_2|s_2) \tag{4.86}$$

where z_1 and z_2 denote the outputs $z_1(T)$ and $z_2(T)$ from the envelope detectors shown in Figure 4.19. For the case in which the tone $s_2(t) = \cos \omega_2 t$ is sent, such that $r(t) = s_2(t) + n(t)$, the output $z_1(T)$ is a *Gaussian noise random variable only*; it has no signal component. A Gaussian distribution into the *nonlinear envelope detector* yields a Rayleigh distribution at the output [6], so that

$$p(z_1|s_2) = \begin{cases} \dfrac{z_1}{\sigma_0^2} \exp\left(-\dfrac{z_1^2}{2\sigma_0^2}\right) & z_1 \ge 0 \\ 0 & z_1 < 0 \end{cases} \tag{4.87}$$

where σ_0^2 is the noise at the filter output. On the other hand, $z_2(T)$ has a Rician distribution, since the input to the lower envelope detector is a sinusoid plus noise [6]. The pdf $p(z_2|s_2)$ is written as

$$p(z_2|s_2) = \begin{cases} \dfrac{z_2}{\sigma_0^2} \exp\left[-\dfrac{(z_2^2 + A^2)}{2\sigma_0^2}\right] I_0\left(\dfrac{z_2 A}{\sigma_0^2}\right) & z_2 \geq 0 \\ 0 & z_2 < 0 \end{cases} \tag{4.88}$$

where $A = \sqrt{2E/T}$, and as before, σ_0^2 is the noise at the filter output. The function $I_0(x)$, known as the modified zero-order Bessel function of the first kind [7], is defined as

$$I_0(x) = \frac{1}{2\pi} \int_0^{2\pi} \exp\left(x \cos\theta\right) d\theta \tag{4.89}$$

When $s_2(t)$ is transmitted, the receiver makes an error whenever the envelope sample $z_1(T)$ obtained from the upper channel (due to noise alone) exceeds the envelope sample $z_2(T)$ obtained from the lower channel (due to signal plus noise). Thus the probability of this error can be obtained by integrating $p(z_1|s_2)$ with respect to z_1 from z_2 to infinity, and then averaging over all possible values of z_2. That is,

$$
\begin{aligned}
P_B &= P(z_1 > z_2|s_2) \\
&= \int_0^\infty p(z_2|s_2) \left[\int_{z_2}^\infty p(z_1|s_2)\, dz_1\right] dz_2 \tag{4.90} \\
&= \int_0^\infty \frac{z_2}{\sigma_0^2} \exp\left[-\frac{(z_2^2 + A^2)}{2\sigma_0^2}\right] I_0\left(\frac{z_2 A}{\sigma_0^2}\right) \left[\int_{z_2}^\infty \frac{z_1}{\sigma_0^2} \exp\left(-\frac{z_1^2}{2\sigma_0^2}\right) dz_1\right] dz_2 \tag{4.91}
\end{aligned}
$$

where $A = \sqrt{2E/T}$ and where the inner integral is the conditioned probability of an error for a fixed value of z_2, given that $s_2(t)$ was sent, and the outer integral averages this conditional probability over all possible values of z_2. This integral can be evaluated [8], yielding

$$P_B = \frac{1}{2} \exp\left(-\frac{A^2}{4\sigma_0^2}\right) \tag{4.92}$$

Using Equation (1.19), we can express the filter output noise as

$$\sigma_0^2 = 2\left(\frac{N_0}{2}\right) W_f \tag{4.93}$$

where $G_n(f) = N_0/2$ and W_f is the filter bandwidth. Thus Equation (4.92) becomes

$$P_B = \frac{1}{2} \exp\left(-\frac{A^2}{4N_0 W_f}\right) \tag{4.94}$$

Equation (4.94) indicates that the error performance depends on the bandpass filter bandwidth, and that P_B becomes smaller as W_f is decreased. The result is valid only when the intersymbol interference (ISI) is negligible. The minimum W_f allowed (i.e., for no ISI) is obtained from Equation (3.81) with the filter roll-off factor $r = 0$. Thus $W_f = R$ bits/s $= 1/T$, and we can write Equation (4.94) as

$$P_B = \frac{1}{2} \exp\left(-\frac{A^2 T}{4N_0}\right) \tag{4.95}$$

$$= \frac{1}{2} \exp\left(-\frac{E_b}{2N_0}\right) \tag{4.96}$$

where $E_b = (1/2)A^2 T$ is the energy per bit. When comparing the error performance of noncoherent FSK with coherent FSK (see Figure 4.25), it is seen that for the same P_B, noncoherent FSK requires approximately 1 dB more E_b/N_0 than that for coherent FSK (for $P_B \leq 10^{-4}$). The noncoherent receiver is easier to implement, because coherent reference signals need not be generated. Therefore, almost all FSK receivers use noncoherent detection. It can be seen in the following section that when comparing noncoherent orthogonal FSK to noncoherent DPSK, the same 3-dB difference occurs as for the comparison between coherent orthogonal FSK and coherent PSK.

As mentioned earlier, the details of on-off keying (OOK) are not treated in this book. However, it is worth noting that the bit error probability P_B described in Equation (4.96) is identical to the P_B for the noncoherent detection of OOK signaling.

4.7.5 Probability of Bit Error for Binary DPSK

Let us define a BPSK signal set as follows:

$$x_1(t) = \sqrt{\frac{2E}{T}} \cos(\omega_0 t + \phi) \qquad 0 \leq t \leq T$$

$$x_2(t) = \sqrt{\frac{2E}{T}} \cos(\omega_0 t + \phi \pm \pi) \qquad 0 \leq t \leq T \tag{4.97}$$

A characteristic of DPSK is that there are no fixed decision regions in the signal space. Instead, the decision is based on the phase difference between successively received signals. Then for DPSK signaling we are really transmitting each bit with the binary signal pair

and
$$\begin{aligned} s_1(t) &= (x_1, x_1) \quad \text{or} \quad (x_2, x_2) \qquad 0 \leq t \leq 2T \\ s_2(t) &= (x_1, x_2) \quad \text{or} \quad (x_2, x_1) \qquad 0 \leq t \leq 2T \end{aligned} \tag{4.98}$$

where (x_i, x_j) $(i, j = 1, 2)$ denotes $x_i(t)$ followed by $x_j(t)$ defined in Equation (4.97). The first T seconds of each waveform are actually the last T seconds of the previous waveform. Note that $s_1(t)$ and $s_2(t)$ can each have either of two possible forms and that $x_1(t)$ and $x_2(t)$ are antipodal signals. Thus the correlation between $s_1(t)$ and $s_2(t)$ for *any combination* of forms can be written as

$$z(2T) = \int_0^{2T} s_1(t)s_2(t)\,dt$$

$$\tag{4.99}$$

$$= \int_0^T [x_1(t)]^2\,dt - \int_0^T [x_1(t)]^2\,dt = 0$$

Therefore, pairs of DPSK signals can be represented as orthogonal signals $2T$ seconds long. Detection could correspond to noncoherent envelope detection with four channels matched to each of the possible envelope outputs, as shown in Figure 4.26a. Since the two envelope detectors representing each symbol are negatives of each other, the envelope sample of each will be the same. Hence we can implement the detector as a single channel for $s_1(t)$ matched to either (x_1, x_1) or (x_2, x_2), and a single channel for $s_2(t)$ matched to either (x_1, x_2) or (x_2, x_1), as shown in Figure 4.26b. The DPSK detector is therefore reduced to a standard two-channel noncoherent detector. In reality, the filter can be matched to the difference signal so that only one channel is necessary. In Figure 4.26, the filters are matched to the signal envelopes (over two symbol times). What does this mean in light of the fact that

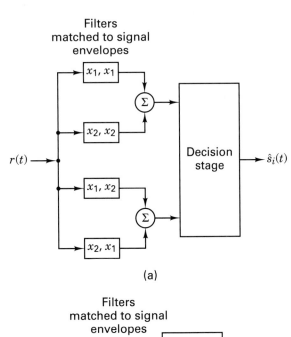

Figure 4.26 DPSK detection. (a) Four-channel differentially coherent detection of binary DPSK. (b) Equivalent two-channel detector for binary DPSK.

DPSK is a constant envelope signaling scheme? It means that we need to implement an energy detector, similar to the quadrature receiver in Figure 4.18, with each of the I and Q reference signals occuring in pairs during the time interval $(0 \leq t \leq 2T)$, as follows:

$$s_1(t)\ I \text{ reference:} \quad \sqrt{2/T} \cos \omega_0 t,\ \sqrt{2/T} \cos \omega_0 t$$
$$s_1(t)\ Q \text{ reference:} \quad \sqrt{2/T} \sin \omega_0 t,\ \sqrt{2/T} \sin \omega_0 t$$

$$s_2(t)\ I \text{ reference:} \quad \sqrt{2/T} \cos \omega_0 t,\ -\sqrt{2/T} \cos \omega_0 t$$
$$s_2(t)\ Q \text{ reference:} \quad \sqrt{2/T} \sin \omega_0 t,\ -\sqrt{2/T} \sin \omega_0 t$$

Since pairs of DPSK signals are orthogonal, such noncoherent detection operates with the bit-error probability given by Equation (4.96). However, since the DPSK signals have a bit interval of $2T$, the $s_i(t)$ signals defined in Equation (4.98) have twice the energy of a signal defined over a single-symbol duration. Thus, we may write P_B as

$$P_B = \frac{1}{2} \exp \left(-\frac{E_b}{N_0} \right) \tag{4.100}$$

Equation (4.100) is plotted in Figure 4.25, designated as differentially coherent detection of differentially encoded PSK, or simply DPSK. This expression is valid for the optimum DPSK detector shown in Figure 4.17c. For the detector shown in Figure (4.17b), the error probability will be slightly inferior to that given in Equation (4.100) [3]. When comparing the error performance of Equation (4.100) with that of coherent PSK (see Figure 4.25), it is seen that for the same P_B, DPSK requires approximately 1 dB more E_b/N_0 than does BPSK (for $P_B \leq 10^{-4}$). It is easier to implement a DPSK system than a PSK system, since the DPSK receiver does not need phase synchronization. For this reason, DPSK, although less efficient than PSK, is sometimes the preferred choice between the two.

4.7.6 Comparison of Bit Error Performance for Various Modulation Types

The P_B expressions for the best known of the binary modulation schemes discussed above are listed in Table 4.1 and are illustrated in Figure 4.25. For $P_B = 10^{-4}$, it can be seen that there is a difference of approximately 4-dB between the best (coherent PSK) and the worst (noncoherent orthogonal FSK) that were discussed here. In some cases, 4 dB is a small price to pay for the implementation simplicity gained in going from coherent PSK to noncoherent FSK; however, for other cases, even a 1-dB saving is worthwhile. There are other considerations besides P_B and system complexity; for example, in some cases (such as a randomly fading channel), a noncoherent system is more desirable because there may be difficulty in establishing and maintaining a coherent reference. Signals that can withstand significant degradation before their ability to be detected is affected are clearly desirable in military and space applications.

TABLE 4.1   Probability of Error for Selected Binary Modulation Schemes

Modulation	P_B
PSK (coherent)	$Q\left(\sqrt{\dfrac{2E_b}{N_0}}\right)$
DPSK (differentially coherent)	$\dfrac{1}{2}\exp\left(-\dfrac{E_b}{N_0}\right)$
Orthogonal FSK (coherent)	$Q\left(\sqrt{\dfrac{E_b}{N_0}}\right)$
Orthogonal FSK (noncoherent)	$\dfrac{1}{2}\exp\left(-\dfrac{1}{2}\dfrac{E_b}{N_0}\right)$

4.8 *M*-ARY SIGNALING AND PERFORMANCE

4.8.1 Ideal Probability of Bit Error Performance

The typical probability of error versus E_b/N_0 curve was shown to have a waterfall-like shape in Figure 3.6. The probability of bit error (P_B) characteristics of various binary modulation schemes in AWGN also display this shape, as shown in Figure 4.25. What should an *ideal* P_B versus E_b/N_0 curve look like? Figure 4.27 displays the ideal characteristic as the *Shannon limit*. The limit represents the threshold E_b/N_0 below which reliable communication cannot be maintained. Shannon's work is described in greater detail in Chapter 9.

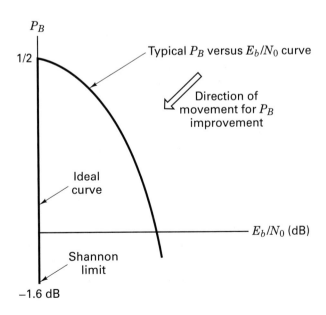

Figure   4.27 P_B versus E_b/N_0 curves: ideal and typical.

We can describe the ideal curve in Figure 4.27 as follows. For all values of E_b/N_0 above the Shannon limit of -1.6 dB, P_B is zero. Once E_b/N_0 is reduced below the Shannon limit, P_B degrades to the worst-case value of $\frac{1}{2}$. (Note that $P_B = 1$ is not the worst case for binary signaling, since that value is just as good as $P_B = 0$; if the probability of making a bit error is 100%, the bit stream could simply be inverted to retrieve the correct data.) It should be clear, by comparing the typical P_B curve with the ideal one in Figure 4.27 that the large arrow in the figure describes the desired direction of movement to achieve improved P_B performance.

4.8.2 *M*-ary Signaling

Let us review *M*-ary signaling. The processor considers k bits at a time. It instructs the modulator to produce one of $M = 2^k$ waveforms; binary signaling is the special case where $k = 1$. Does *M*-ary signaling improve or degrade error performance? (Be careful with your answer—the question is a loaded one.) Figure 4.28 illustrates the probability of bit error $P_B(M)$ versus E_b/N_0 for coherently detected *orthogonal* *M*-ary signaling over a Gaussian channel. Figure 4.29 similarly illustrates $P_B(M)$ versus E_b/N_0 for coherently detected *multiple phase* *M*-ary signaling over a Gaussian channel. In which direction do the curves move as the value of k (or M) increases? From Figure 4.27 we know the directions of curve movement for improved and degraded error performance. In Figure 4.28, as k increases, the curves move in the direction of improved error performance. In Figure 4.29, as k increases, the curve move in the direction of degraded error performance. Such movement tells us that *M*-ary signaling produces improved error performance with orthogonal signaling and degraded error performance with multiple phase signaling. Can that be true? Why would anyone ever use multiple phase PSK signaling if it provides degraded error performance compared to binary PSK signaling? It *is* true, and many systems do use multiple phase signaling. The question, as stated, is loaded because it implies that error probability versus E_b/N_0 is the *only* performance criterion; there are many others (e.g., bandwidth, throughput, complexity, cost), but in Figures 4.28 and 4.29 error performance is the characteristic that stands out explicitly.

A performance characteristic that is not explicitly seen in Figures 4.28 and 4.29 is the required system bandwidth. For the curves characterizing *M*-ary orthogonal signals in Figure 4.28, as k increases, the required bandwidth also increases. For the *M*-ary multiple phase curves in Figure 4.29, as k increases, a larger bit rate can be transmitted within the same bandwidth. In other words, for a fixed data rate, the required bandwidth is decreased. Therefore, *both* the orthogonal and multiple phase error performance curves tell us that *M*-ary signaling represents a vehicle for performing a system trade-off. In the case of orthogonal signaling, error performance improvement can be achieved at the expense of bandwidth. In the case of multiple phase signaling, bandwidth performance can be achieved at the expense of error performance. Error performance versus bandwidth performance, a fundamental communications trade-off, is treated in greater detail in Chapter 9.

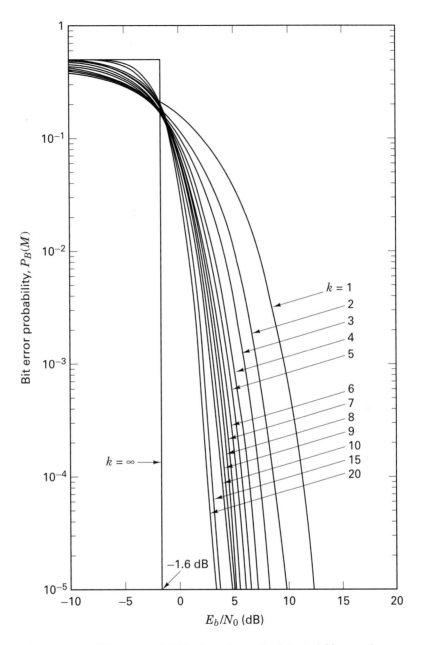

Figure 4.28 Bit error probability for coherently detected *M*-ary orthogonal signaling. (Reprinted from W. C. Lindsey and M. K. Simon, *Telecommunication Systems Engineering,* Prentice Hall, Inc., Englewood Cliffs, N.J., 1973, courtesy of W. C. Lindsey and Marvin K. Simon).

4.8   *M*-ary Signaling and Performance

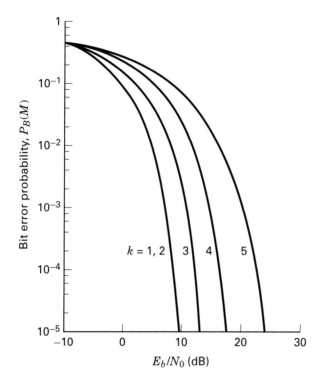

Figure 4.29 Bit error probability for coherently detected multiple phase signaling.

4.8.3 Vectorial View of MPSK Signaling

Figure 4.30 illustrates MPSK signal sets for $M = 2$, 4, 8, and 16. In Figure 4.30a we see the binary ($k = 1$, $M = 2$) antipodal vectors $\mathbf{s}_1$ and $\mathbf{s}_2$ positioned 180° apart. The decision boundary is drawn so as to partition the signal space into two regions. On the figure is also shown a noise vector $\mathbf{n}$ equal in magnitude to $\mathbf{s}_1$. The figure establishes the magnitude and orientation of the minimum energy noise vector that would cause the detector to make a symbol error.

In Figure 4.30b we see the 4-ary ($k = 2$, $M = 4$) vectors positioned 90° apart. The decision boundaries (only one line is drawn) divide the signal space into four regions. Again a noise vector $\mathbf{n}$ is drawn (from the head of a signal vector, normal to the closest decision boundary) to illustrate the minimum energy noise vector that would cause the detector to make a symbol error. Notice that the minimum energy noise vector of Figure 4.30b is smaller than that of Figure 4.30a, illustrating that the 4-ary system is more vulnerable to noise than the 2-ary system (signal energy being equal for each case). As we move on to Figure 4.30c for the 8-ary case and Figure 4.30d for the 16-ary case, it should be clear that for multiple phase signaling, as M increases, we are crowding more signal vectors into the signal plane. As the vectors are moved closer together, a smaller amount of noise energy is required to cause an error.

Figure 4.30 adds some insight as to why the curves of Figure 4.29 behave as they do as k is increased. Figure 4.30 also provides some insight into a basic trade-

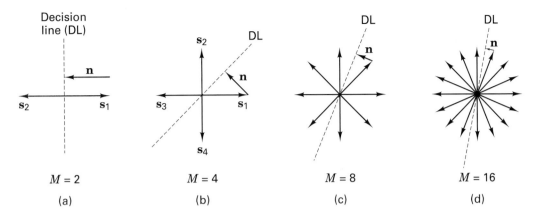

Figure 4.30 MPSK signal sets for $M = 2, 4, 8, 16$.

off in multiple phase signaling. Crowding more signal vectors into the signal space is tantamount to increasing the data rate without increasing the system bandwidth (the vectors are all confined to the same plane). In other words, we have increased the bandwidth utilization at the expense of error performance. Look at Figure 4.30d, where the error performance is worse than any of the other examples in Figure 4.30. How might we "buy back" the degraded error performance? In other words, what can we trade-off so that the distance between neighboring signal vectors in Figure 4.30d is increased to that in Figure 4.30a? We can increase the signal strength (make the signal vectors larger) until the minimum distance from the head of a signal vector to a decision line equals the length of the noise vector in Figure 4.30a. Therefore, in a multiple phase system, as M is increased, we can either achieve improved bandwidth performance at the expense of error performance, or if we increase the E_b/N_0 so that the error probability is not degraded, we can achieve improved bandwidth performance at the expense of increasing E_b/N_0.

Note that Figure 4.30 has been sketched so that all phasors have the same length for any of the M-ary cases. This is tantamount to saying that the comparisons are being considered for a fixed E_s/N_0, where E_s is symbol energy. The figure can also be drawn for a fixed E_b/N_0, in which case the phasor magnitudes would increase with increasing M. The phasors for $M = 4, 8$, and 16 would then have lengths greater than the $M = 2$ case by the factors $\sqrt{2}$, $\sqrt{3}$, and 2 respectively. We would still see crowding and increased vulnerability to noise, with increasing M, but the appearance would not be as pronounced as it is in Figure 4.30.

4.8.4 BPSK and QPSK Have the Same Bit Error Probability

In Equation (3.30) we stated the general relationship between E_b/N_0 and S/N which is rewritten

$$\frac{E_b}{N_0} = \frac{S}{N}\left(\frac{W}{R}\right) \tag{4.101}$$

where S is the average signal power and R is the bit rate. A BPSK signal with the available E_b/N_0 found from Equation (4.101) will perform with a P_B that can be read from the $k = 1$ curve in Figure 4.29. QPSK can be characterized as two orthogonal BPSK channels. The QPSK bit stream is usually partitioned into an even and odd (I and Q) stream; each new stream modulates an orthogonal component of the carrier at half the bit rate of the original stream. The I stream modulates the $\cos \omega_0 t$ term and the Q stream modulates the $\sin \omega_0 t$ term. If the magnitude of the original QPSK vector has the value A, the magnitude of the I and Q component vectors each has a value of $A/\sqrt{2}$, as shown in Figure 4.31. Thus, each of the quadrature BPSK signals has half of the average power of the original QPSK signal. Hence if the original QPSK waveform has a bit rate of R bits/s and an average power of S watts, the quadrature partitioning results in each of the BPSK waveforms having a bit rate of $R/2$ bits/s and an average power of $S/2$ watts.

Therefore, the E_b/N_0 characterizing each of the orthogonal BPSK channels, making up the QPSK signal, is equivalent to the E_b/N_0 in Equation (4.101), since it can be written as

$$\frac{E_b}{N_0} = \frac{S/2}{N_0}\left(\frac{W}{R/2}\right) = \frac{S}{N_0}\left(\frac{W}{R}\right) \tag{4.102}$$

Thus each of the orthogonal BPSK channels, and hence the composite QPSK signal, is characterized by the same E_b/N_0 and hence the same P_B performance as a BPSK signal. The natural orthogonality of the 90° phase shifts between adjacent QPSK symbols results in the *bit error probabilities* being equal for both BPSK and QPSK signaling. It is important to note that the symbol *error probabilities* are *not* equal for BPSK and QPSK signaling. The relationship between bit error probability and symbol error probability is treated in Sections 4.9.3 and 4.9.4. We see that, in effect, QPSK is the equivalent of two BPSK channels in quadrature. This same idea can be extended to any symmetrical M-ary amplitude/phase signaling, such as quadrature amplitude modulation (QAM) described in Section 9.8.3.

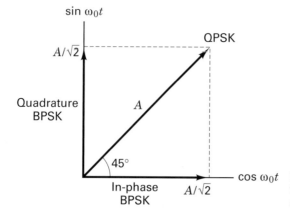

Figure 4.31 In-phase and quadrature BPSK components of QPSK signaling.

4.8.5 Vectorial View of MFSK Signaling

In Section 4.8.3, Figure 4.30 provides some insight as to why the error performance of MPSK signaling degrades as k (or M) increases. It would be useful to have a similar vectorial illustration for the error performance of orthogonal MFSK signaling as seen in the curves of Figure (4.28). Since the MFSK signal space is characterized by M mutually perpendicular axes, we can only conveniently illustrate the cases $M = 2$ and $M = 3$. In Figure 4.32a we see the binary orthogonal vectors s_1 and s_2 positioned $90°$ apart. The decision boundary is drawn so as to partition the signal space into two regions. On the figure is also shown a noise vector $\mathbf{n}$, which represents the minimum noise vector that would cause the detector to make an error.

In Figure 4.32b we see a 3-ary signal space with axes positioned $90°$ apart. Here decision planes partition the signal space into three regions. Noise vectors $\mathbf{n}$ are shown added to each of the prototype signal vectors s_1, s_2, and s_3; each noise vector illustrates an example of the minimum noise vector that would cause the detector to make a symbol error. The minimum noise vectors in Figure 4.32b are the same length as the noise vector in Figure 4.32a. In Section 4.4.4 we stated that for a given level of received energy, the distance between any two prototype signal vectors s_i and s_j in an M-ary orthogonal space is constant. It follows that the minimum distance between a prototype signal vector and any of the decision boundaries remains fixed as M increases. Unlike the case of MPSK signaling, where adding new signals to the signal set makes the signals vulnerable to smaller noise vectors, here, in the case of MFSK signaling, adding new signals to the signal set does *not* make the signals vulnerable to smaller noise vectors.

It would be convenient to illustrate the point by drawing higher dimensional orthogonal spaces, but of course this is not possible. We can only use our "mind's eye" to understand that increasing the signal set M—by adding additional axes, where each new axis is mutually perpendicular to all the others—does not crowd

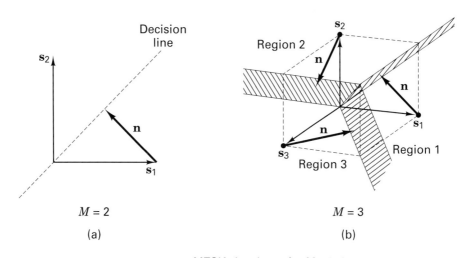

$M = 2$ $M = 3$

(a)   (b)

Figure 4.32 MFSK signal sets for $M = 2, 3$.

the signal set more closely together. Thus, a transmitted signal from an orthogonal set is *not* more vulnerable to a noise vector when the set is increased in size. In fact, we see from Figure 4.28 that as k increases, the bit error performance improves.

Understanding the error performance improvement of orthogonal signaling, as illustrated in Figure 4.28, is facilitated by comparing the probability of symbol error (P_E) versus unnormalized SNR, with P_E versus E_b/N_0. Figure 4.33 represents a set of P_E performance curves plotted against unnormalized SNR for coherent FSK signaling. Here we see that P_E degrades as M is increased. Didn't we say that a signal from an orthogonal set is *not* made more vulnerable to a given noise vector, as the orthogonal set is increased in size? It is correct that for orthogonal signaling, with a given SNR it takes a fixed size noise vector to perturb a transmitted signal into an error region; the signals do not become vulnerable to smaller noise vectors as M increases. However, as M increases, more neighboring decision regions are introduced; thus the number of ways in which a symbol error can be made increases. Figure 4.33 reflects the degradation in P_E versus unnormalized SNR as M is increased; there are $(M - 1)$ ways to make an error. Examining performance under the condition of a fixed SNR (as M increases) is not very useful for digital communications. A fixed SNR means a fixed amount of energy per symbol; thus as M increases, there is a fixed amount of energy to be apportioned over a larger number of bits, or there is less energy per bit. The most useful way of comparing one digital system with another is on the basis of *bit-normalized SNR* or E_b/N_0. The error performance improvement with increasing M (see Figure 4.28) manifests itself only when error probability is plotted against E_b/N_0. For this case, as M increases, the required E_b/N_0 (to meet a given error probability) is reduced for a fixed SNR; therefore, we need to map the plot shown in Figure 4.33 into a new plot, similar to that shown in Figure 4.28, where the abscissa represents E_b/N_0 instead of SNR. Figure 4.34 illustrates such a mapping; it demonstrates that curves manifesting degraded P_E with increasing M (such as Figure 4.33) are transformed into curves manifesting improved P_E with increasing M. The basic mapping relationship is expressed in Equation (4.101), repeated here as

$$\frac{E_b}{N_0} = \frac{S}{N} \left(\frac{W}{R} \right)$$

where W is the detection bandwidth. Since

$$R = \frac{\log_2 M}{T} = \frac{k}{T}$$

where T is the symbol duration, we can then write

$$\frac{E_b}{N_0} = \frac{S}{N} \left(\frac{WT}{\log_2 M} \right) = \frac{S}{N} \left(\frac{WT}{k} \right) \tag{4.103}$$

For FSK signaling, the detection bandwidth W (in hertz) is typically equal in value to the symbol rate $1/T$; in other words, $WT \approx 1$. Therefore,

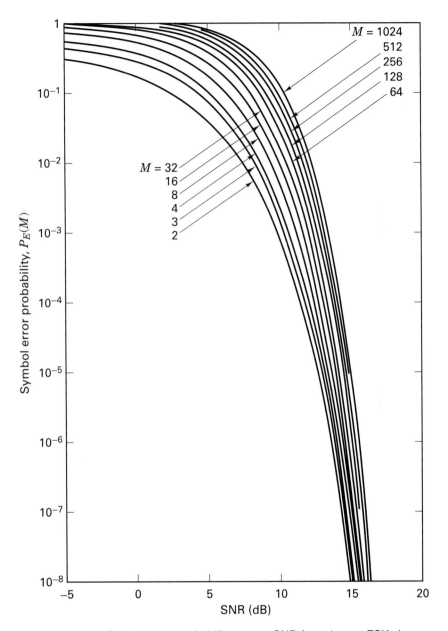

Figure 4.33 Symbol error probability versus SNR for coherent FSK signaling. (From Bureau of Standards, *Technical Note 167,* March 1963.) (Reprinted from *Central Radio Propagation Laboratory Technical Note 167,* March 25, 1963, Fig. 1, p. 5, courtesy of National Bureau of Standards.)

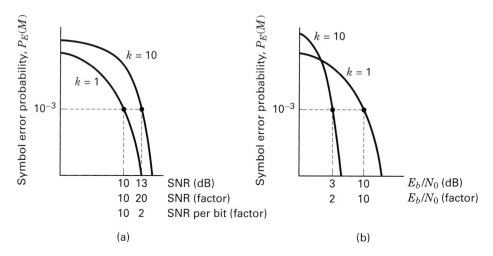

Figure 4.34 Mapping P_E versus SNR into P_E versus E_b/N_0 for orthogonal signaling. (a) Unnormalized. (b) Normalized.

$$\frac{E_b}{N_0} \approx \frac{S}{N}\left(\frac{1}{k}\right) \qquad (4.104)$$

Figure 4.34 illustrates the mapping from P_E versus SNR to P_E versus E_b/N_0 for coherently detected M-ary orthogonal signaling, with "ballpark" numbers on the axes. In Figure 4.34a, on the $k = 1$ curve is shown an operating point corresponding to $P_E = 10^{-3}$ and SNR = 10 dB. On the $k = 10$ curve is shown an operating point at the same $P_E = 10^{-3}$ but with SNR = 13 dB (approximate values taken from Figure 4.33). Here we clearly see the degradation in error performance as k increases. To appreciate where the performance improvement comes from, let us convert the abscissa from the nonlinear scale of SNR in decibels to a linear one—SNR expressed as a factor. This is shown in Figure 4.34a as the factors 10 and 20 for the $k = 1$ and $k = 10$ cases, respectively. Next, we further convert the abscissa scale to SNR per bit (expressed as a factor). This is shown in Figure 4.34a as the factors 10 and 2 for the $k = 1$ and $k = 10$ cases, respectively. It is convenient to think of the 1024-ary symbol or waveform ($k = 10$ case) as being interchangeable with its 10-bit meaning. Thus, if the symbol requires 20 units of SNR then the 10 bits belonging to that symbol require that same 20 units; or, in other words, each bit requires 2 units.

Rather than performing such computations, we can simply map these same $k = 1$ and $k = 10$ cases onto the Figure 4.34b plane, representing P_E versus E_b/N_0. The $k = 1$ case looks exactly the same as it does in Figure 4.34a. But for the $k = 10$ case, there is a dramatic change. We can immediately see that signaling with the $k = 10$-bit symbol requires only 2 units (3 dB) of E_b/N_0 compared with 10 units (10 dB) for the binary symbol. The mapping that gives rise to the required E_b/N_0 for the $k = 10$ case is obtained from Equation (4.104) as follows: $E_b/N_0 = 20\ (1/10) = 2$ (or 3-dB), which shows the error performance improvement as k is increased. In digital communication systems, error performance is almost always considered in terms of

E_b/N_0, since such a measurement makes for a meaningful comparison between one system's performance and another. Therefore, the curves shown in Figures 4.33 and 4.34a are hardly ever seen.

Although Figure 4.33 is not often seen, we can still use it for gaining insight into why orthogonal signaling provides improved error performance as M or k increases. Let us consider the analogy of purchasing a commodity—say, grade A cottage cheese. The choice of the grade corresponds to some point on the P_E axis of Figure 4.33—say, 10^{-3}. From this point, construct a horizontal line through all of the curves (from $M = 2$ through $M = 1024$). At the grocery store we buy the very smallest container of cottage cheese, containing 2 ounces and costing \$1. On Figure 4.33 we can say that this purchase corresponds to our horizontal construct intercepting the $M = 2$ curve. We look down at the corresponding SNR and call the intercept on this axis our cost of \$1. The next time we purchase cottage cheese, we remember that the first purchase seemed expensive at 50 cents an ounce. So, we decide to buy a larger carton, containing 8 ounces and costing \$2. On Figure 4.33, we can say that this purchase corresponds to the point at which our horizontal construct intercepts the $M = 8$ curve. We look down at the corresponding SNR, and call this intercept our cost of \$2. Notice that we bought a larger container so the price went up, but because we bought a greater quantity, the price per ounce went down (the unit cost is now only 25 cents per ounce). We can continue this analogy by purchasing larger and larger containers so that the price of the container (SNR) keeps going up, but the price per ounce keeps going down. This is the age-old story called the *economy of scale.* Buying larger quantities at a time is commensurate with purchasing at the wholesale level; it makes for a lower unit price. Similarly, when we use orthogonal signaling with symbols that contain more bits, we need more power (more SNR), but the requirement per bit (E_b/N_0) is reduced.

4.9 SYMBOL ERROR PERFORMANCE FOR *M*-ARY SYSTEMS (*M* > 2)

4.9.1 Probability of Symbol Error for MPSK

For large energy-to-noise ratios, the symbol error performance $P_E(M)$, for equally likely, coherently detected M-ary PSK signaling, can be expressed [7] as

$$P_E(M) \approx 2Q\left(\sqrt{\frac{2E_s}{N_0}}\sin\frac{\pi}{M}\right) \qquad (4.105)$$

where $P_E(M)$ is the probability of symbol error, $E_s = E_b(\log_2 M)$ is the energy per symbol, and $M = 2^k$ is the size of the symbol set. The $P_E(M)$ performance curves for coherently detected MPSK signaling are plotted versus E_b/N_0 in Figure 4.35.

The symbol error performance for differentially coherent detection of M-ary DPSK (for large E_s/N_0) is similarly expressed [7] as

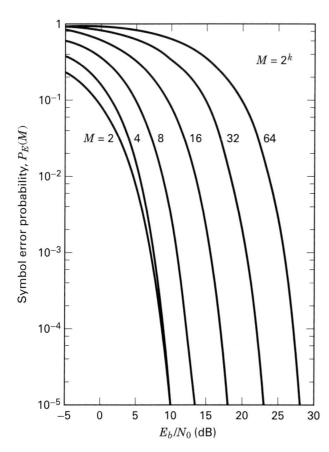

$$M = 2^k$$

$$M = 2 \quad 4 \quad 8 \quad 16 \quad 32 \quad 64$$

Symbol error probability, $P_E(M)$

E_b/N_0 (dB)

Figure 4.35 Symbol error probability for coherently detected multiple phase signaling. (Reprinted from W. C. Lindsey and M. K. Simon, *Telecommunication Systems Engineering,* Prentice-Hall, Inc. Englewood Cliffs, N.J., 1973, courtesy of W. C. Lindsey and Marvin K. Simon.)

$$P_E(M) \approx 2Q\left(\sqrt{\frac{2E_s}{N_0}} \sin \frac{\pi}{\sqrt{2}M}\right) \tag{4.106}$$

4.9.2 Probability of Symbol Error for MFSK

The symbol error performance $P_E(M)$, for equally likely, *coherently* detected M-ary orthogonal signaling can be upper bounded [5] as

$$P_E(M) \leq (M-1)Q\left(\sqrt{\frac{E_s}{N_0}}\right) \tag{4.107}$$

where $E_s = E_b(\log_2 M)$ is the energy per symbol and M is the size of the symbol set. The $P_E(M)$ performance curves for coherently detected M-ary orthogonal signaling are plotted versus E_b/N_0 in Figure 4.36.

The symbol error performance for equally likely, *noncoherently* detected M-ary orthogonal signaling is [9]

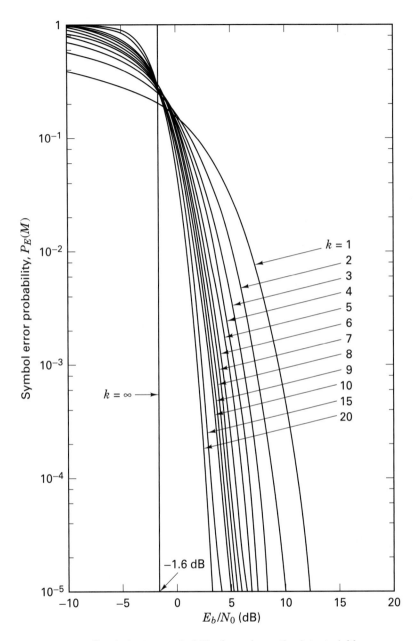

Figure 4.36   Symbol error probability for coherently detected *M*-ary orthogonal signaling. (Reprinted from W. C. Lindsey and M. K. Simon, *Telecommunication Systems Engineering,* Prentice-Hall, Inc., Englewod Cliffs, N.J., 1973, courtesy of W. C. Lindsey and Marvin K. Simon.)

4.9   Symbol Error Performance for *M*-ary Systems (*M* > 2)   **231**

$$P_E(M) = \frac{1}{M} \exp\left(-\frac{E_s}{N_0}\right) \sum_{j=2}^{M} (-1)^j \binom{M}{j} \exp\left(\frac{E_s}{jN_0}\right) \qquad (4.108)$$

where

$$\binom{M}{j} = \frac{M!}{j!\,(M-j)!} \qquad (4.109)$$

is the standard binomial coefficient yielding the number of ways in which j symbols out of M may be in error. Note that for binary case, Equation (4.108) reduces to

$$P_B = \frac{1}{2} \exp\left(-\frac{E_b}{2N_0}\right) \qquad (4.110)$$

which is the same result as that described by Equation (4.96). The $P_E(M)$ performance curves for noncoherently detected M-ary orthogonal signaling are plotted versus E_b/N_0 in Figure 4.37. If we compare this noncoherent orthogonal $P_E(M)$ performance with the corresponding $P_E(M)$ results for the coherent detection of orthogonal signals in Figure 4.36, it can be seen that for $k > 7$, there is a negligible difference. An upper bound for coherent as well as noncoherent reception of orthogonal signals is [9]

$$P_E(M) < \frac{M-1}{2} \exp\left(-\frac{E_s}{2N_0}\right) \qquad (4.111)$$

where E_s is the energy per symbol and M is the size of the symbol set.

4.9.3 Bit Error Probability versus Symbol Error Probability for Orthogonal Signals

It can be shown [9] that the relationship between probability of bit error (P_B) and probability of symbol error (P_E) for an M-ary orthogonal signal set is

$$\frac{P_B}{P_E} = \frac{2^{k-1}}{2^k - 1} = \frac{M/2}{M-1} \qquad (4.112)$$

In the limit as k increases, we get

$$\lim_{k \to \infty} \frac{P_B}{P_E} = \frac{1}{2}$$

A simple example will make Equation (4.112) intuitively acceptable. Figure 4.38 describes an octal message set. The message symbols (assumed equally likely) are to be transmitted on orthogonal waveforms such as FSK. With orthogonal signaling, a decision error will transform the correct signal into any one of the $(M - 1)$ incorrect signals with equal probability. The example in Figure 4.38 indicates that the symbol comprising bits 0 1 1 was transmitted. An error might occur in any one of the other $2^k - 1 = 7$ symbols, with equal probability. Notice that just because a symbol error is made does not mean that all the bits within the symbol will be in error. In Figure 4.38, if the receiver decides that the transmitted symbol is the bot-

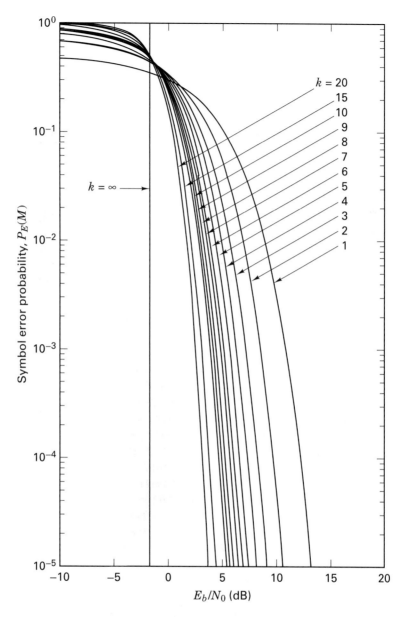

Figure 4.37 Symbol error probability for noncoherently detected *M*-ary orthogonal signaling. (Reprinted from W. C. Lindsey and M. K. Simon, *Telecommunication Systems Engineering,* Prentice-Hall, Inc., Englewood Cliffs, N.J., 1973, courtesy of W. C. Lindsey and Marvin K. Simon.)

                            Bit
                          position

                        0   0   0

                        0   0   1

                        0   1   0

Transmitted   ┌─────────┐
 symbol   │ 0   1   1 │
                      └─────────┘
                        1   0   0

                        1   0   1

                        1   1   0

                        1   1   1   **Figure 4.38** Example of P_B
                                        versus P_E.

tom one listed, comprising bits 1 1 1, two of the three transmitted symbol bits will be correct; only one bit will be in error. It should be apparent that for nonbinary signaling, P_B will always be less than P_E (keep in mind that P_B and P_E reflect the frequency of making errors on the *average*.)

Consider any of the bit-position columns in Figure 4.38. For each bit position, the digit occupancy consists of 50% ones and 50% zeros. In the context of the first bit position (rightmost column) and the transmitted symbol, how many ways are there to cause an error to the binary one? There are $2^{k-1} = 4$ ways (four places where zeros appear in the column) that a bit error can be made; it is the same for each of the columns. The final relationship P_B/P_E, for orthogonal signaling, in Equation (4.112), is obtained by forming the following ratio: the number of ways that a bit error can be made (2^{k-1}) divided by the number of ways that a symbol error can be made $(2^k - 1)$. For the Figure 4.38 example, $P_B/P_E = 4/7$.

4.9.4 Bit Error Probability Versus Symbol Error Probability for Multiple Phase Signaling

For the case of MPSK signaling, P_B is less than or equal to P_E, just as in the case of MFSK signaling. However, there is an important difference. For orthogonal signaling, selecting any one of the $(M - 1)$ erroneous symbols is equally likely. In the case of MPSK signaling, each signal vector is not equidistant from all of the others. Figure 4.39a illustrates an 8-ary decision space with the pie-shaped decision regions denoted by the 8-ary symbols in binary notation. If symbol (0 1 1) is transmitted, it is clear that should an error occur, the transmitted signal will most likely be mistaken for one of its closest neighbors, (0 1 0) or (1 0 0). The likelihood that (0 1 1) would get mistaken for (1 1 1) is relatively remote. If the assignment of bits to symbols follows the binary sequence shown in the symbol decision regions of Figure 4.39a, some symbol errors will usually result in two or more bit errors, even with a large signal-to-noise ratio.

For nonorthogonal schemes, such as MPSK signaling, one often uses a binary-to-M-ary code such that binary sequences corresponding to adjacent sym-

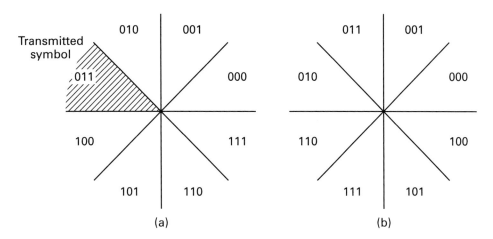

Figure 4.39 Binary-coded versus Gray-coded decision regions in an MPSK signal space. (a) Binary coded. (b) Gray coded.

bols (phase shifts) differ in only one bit position; thus when an *M*-ary symbol error occurs, it is more likely that only one of the *k* input bits will be in error. A code that provides this desirable feature is the Gray code [7]; Figure 4.39b illustrates the bit-to-symbol assignment using a Gray code for 8-ary PSK. Here it can be seen that neighboring symbols differ from one another in only one bit position. Therefore, the occurrence of a multibit error, for a given symbol error, is much reduced compared to the uncoded binary assignment seen in Figure 4.39a. Implementing such a Gray code, represents one of the few cases in digital communications where a benefit can be achieved without incurring any cost. The Gray code is simply an assignment that requires no special or additional circuitry. Utilizing the Gray code assignment, it can be shown [5] that

$$P_B \approx \frac{P_E}{\log_2 M} \qquad (\text{for } P_E \ll 1) \qquad (4.113)$$

Recall from Section 4.8.4 that BPSK and QPSK signaling have the same bit error probability. Here, in Equation (4.113), we verify that they do not have the same symbol error probability. For BPSK, $P_E = P_B$. However, for QPSK, $P_E \approx 2P_B$.

An exact closed-form expression for the bit-error probability P_B of 8-ary PSK, together with tight upper and lower bounds on P_B for *M*-ary PSK with larger *M*, may be found in Lee [10].

4.9.5 Effects of Intersymbol Interference

In the previous sections and in Chapter 3 we have treated the detection of signals in the presence of AWGN under the assumption that there is no intersymbol interference (ISI). Thus the analysis has been straightforward, since the zero-mean AWGN process is characterized by its variance alone. In practice we find that ISI is often a second source of interference which must be accounted for. As explained in Section 3.3, ISI can be generated by the use of bandlimiting filters at the transmitter output, in the channel, or at the receiver input. The result of this additional interference is to degrade the error probabilities for coherent as well as for noncoherent reception. Calculating error performance in the presence of ISI in addition to AWGN is much more complicated since it involves the impulse response of the channel. The subject will not be treated here; however, for those readers interested in the details of the analysis, References [11–16] should prove interesting.

4.10 CONCLUSION

We have catalogued some basic bandpass digital modulation formats, particularly phase shift keying (PSK) and frequency shift keying (FSK). We have considered a geometric view of signal vectors and noise vectors, particularly antipodal and orthogonal signal sets. This geometric view allows us to consider the detection problem in the light of an orthogonal signal space and signal regions. This view of the space, and the effect of noise vectors causing transmitted signals to be received in the incorrect region, facilitates the understanding of the detection problem and the performance of various modulation and demodulation techniques. In Chapter 9 we reconsider the subjects of modulation and demodulation, and we investigate some bandwidth-efficient modulation techniques.

REFERENCES

1. Schwartz, M., *Information, Transmission, Modulation, and Noise,* McGraw-Hill Book Company, New York, 1970.
2. Van Trees, H. L., *Detection, Estimation, and Modulation Theory,* Part 1, John Wiley & Sons, Inc., New York, 1968.
3. Park, J. H., Jr., "On Binary DPSK Detection," *IEEE Trans. Commun.,* vol. COM26, no. 4, Apr. 1978, pp. 484–486.
4. Ziemer, R. E., and Peterson, R. L., *Digital Communications and Spread Spectrum Systems,* Macmillan Publishing Company, Inc., New York, 1985.
5. Lindsey, W. C., and Simon, M. K., *Telecommunication Systems Engineering,* Prentice-Hall, Inc., Englewood Cliffs, N.J., 1973.
6. Whalen, A. D., *Detection of Signals in Noise,* Academic Press, Inc., New York, 1971.
7. Korn, I., *Digital Communications,* Van Nostrand Reinhold Company, Inc., New York, 1985.
8. Couch, L. W. II, *Digital and Analog Communication Systems,* Macmillan Publishing Company, New York, 1983.
9. Viterbi, A. J., *Principles of Coherent Communications,* McGraw-Hill Book Company, New York, 1966.
10. Lee, P. J., "Computation of the Bit Error Rate of Coherent M-*ary* PSK with Gray Code Bit Mapping," *IEEE Trans. Commun.,* vol. COM34, no. 5, May 1986, pp. 488–491.

11. Hoo, E. Y., and Yeh, Y. S., "A New Approach for Evaluating the Error Probability in the Presence of the Intersymbol Interference and Additive Gaussian Noise," *Bell Syst. Tech. J.,* vol. 49, Nov. 1970, pp. 2249–2266.

12. Shimbo, O., Fang, R. J., and Celebiler, M., "Performance of *M*-ary PSK Systems in Gaussian Noise and Intersymbol Interference," *IEEE Trans. Inf. Theory,* vol. IT19, Jan. 1973, pp. 44–58.

13. Prabhu, V. K., "Error Probability Performance of *M*-ary CPSK Systems with Intersymbol Interference," *IEEE Trans. Commun.,* vol. COM21, Feb. 1973, pp. 97–109.

14. Yao, K., and Tobin, R. M., "Moment Space Upper and Lower Error Bounds for Digital Systems with Intersymbol Interference," *IEEE Trans. Inf. Theory,* vol. IT22, Jan. 1976, pp. 65–74.

15. King, M. A., Jr., "Three Dimensional Geometric Moment Bounding Techniques," *J. Franklin Inst.,* vol. 309, no. 4, Apr. 1980, pp. 195–213.

16. Prabhu, V. K., and Salz, J., "On the Performance of Phase-Shift Keying Systems," *Bell Syst. Tech. J.,* vol. 60, Dec. 1981, pp. 2307–2343.

PROBLEMS

4.1. Find the expected number of bit errors made in one day by the following continuously operating coherent BPSK receiver. The data rate is 5000 bits/s. The input digital waveforms are $s_1(t) = A \cos \omega_0 t$ and $s_2(t) = -A \cos \omega_0 t$ where $A = 1$ mV and the single-sided noise power spectral density is $N_0 = 10^{-11}$ W/Hz. Assume that signal power and energy per bit are normalized relative to a 1-Ω resistive load.

4.2. A continuously operating coherent BPSK system makes errors at the average rate of 100 errors per day. The data rate is 1000 bits/s. The single-sided noise power spectral density is $N_0 = 10^{-10}$ W/Hz.
  (a) If the system is ergodic, what is the average bit error probability?
  (b) If the value of received average signal power is adjusted to be 10^{-6} W, will this received power be adequate to maintain the error probability found in part (a)?

4.3. If a system's main performance criterion is bit error probability, which of the following two modulation schemes would be selected for an AWGN channel? Show computations.

Binary noncoherent orthogonal FSK with $E_b/N_0 = 13$ dB

Binary coherent PSK with $E_b/N_0 = 8$ dB

4.4. The bit stream

1 0 1 0 1 0 1 1 1 1 0 1 0 1 0 1 0 0 0 0 1 1 1 1

is to be transmitted using DPSK modulation. Show four different differentially encoded sequences that can represent the data sequence above, and explain the algorithm that generated each.

4.5. (a) Calculate the minimum required bandwidth for a noncoherently detected orthogonal binary FSK system. The higher-frequency signaling tone is 1 MHz and the symbol duration is 1 ms.
  (b) What is the minimum required bandwidth for a noncoherent MFSK system having the same symbol duration?

4.6. Consider a BPSK system with equally likely waveforms $s_1(t) = \cos \omega_0 t$ and $s_2(t) = -\cos \omega_0 t$. Assume that the received $E_b/N_0 = 9.6$ dB, giving rise to a bit-error probability of 10^{-5}, when the synchronization is perfect. Consider that carrier recovery with the PLL suffers some fixed error ϕ associated with the phase estimate, so that the reference signals are expressed as $\cos (\omega_0 t + \phi)$ and $-\cos (\omega_0 t + \phi)$. Note that the error-degradation effect of a fixed known bias can be computed by using the closed-end relationships presented in this chapter. However, if the phase error were to consist of a random jitter, computing its effect would then require a stochastic treatment. (See Chapter 10.)

(a) How badly does the bit-error probability degrade when $\phi = 25°$?

(b) How large a phase error would cause the bit-error probability to degrade to 10^{-3}?

4.7. Find the probability of bit error, P_B, for the coherent matched filter detection of the equally likely binary FSK signals

$$s_1(t) = 0.5 \cos 2000\pi t$$

and

$$s_2(t) = 0.5 \cos 2020\pi t$$

where the two-sided AWGN power spectral density is $N_0/2 = 0.0001$. Assume that the symbol duration is $T = 0.01$ s.

4.8. Find the optimum (minimum probability of error) threshold γ_0, for detecting the equally likely signals $s_1(t) = \sqrt{2E/T} \cos \omega_0 t$ and $s_2(t) = \sqrt{\frac{1}{2} E/T} \cos (\omega_0 t + \pi)$ in AWGN, using a correlator receiver as shown in Figure 4.7b. Assume a reference signal of $\psi_1(t) = \sqrt{2/T} \cos \omega_0 t$.

4.9. A system using matched filter detection of equally likely BPSK signals, $s_1(t) = \sqrt{2E/T} \cos \omega_0 t$ and $s_2(t) = \sqrt{2E/T} \cos (\omega_0 t + \pi)$, operates in AWGN with a received E_b/N_0 of 6.8 dB. Assume that $\mathbf{E}\{z(T)\} = \pm \sqrt{E}$.

(a) Find the minimum probability of bit error, P_B, for this signal set and E_b/N_0.

(b) If the decision threshold is $\gamma = 0.1 \sqrt{E}$, find P_B.

(c) The threshold of $\gamma = 0.1 \sqrt{E}$ is optimum for a particular set of a priori probabilities, $P(s_1)$ and $P(s_2)$. Find the values of these probabilities (refer to Section B.2).

4.10. (a) Describe the impulse response of a matched filter for detecting the discrete signal shown in Figure P4.1. With this signal at the input to the filter, show the output as a function of time. Neglect the effects of noise. What is the maximum output value?

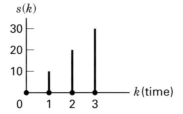

Figure P4.1

(b) In a matched filter (MF), a signal is convolved with a time-reversed function of the signal (impulse response of the MF). Convolution reverses the function again; thus, the MF yields the correlation of a signal and its "look alike" copy (even though the MF operation is designated as convolution). In implementing a

MF, suppose that you accidently connect the circuitry so as to yield the correlation of a signal and its time-reversed copy; the output would then yield a signal convolved with itself. Show the output as a function of time. What is the maximum output value? Note that the maximum output value for part (a) will occur at a different time index compared with that of part (b).

(c) By examining the filter's output values from the flawed circuit of part (b), compared with the correct values in part (a), can you find a clue that can help you predict when such an output sequence appears to be a valid matched filter output, and when it does not?

(d) If noise were added to the signal, compare the output SNR for the correlator versus the convolver. Also, if the input consists of noise only, compare the output of the correlator versus the convolver.

4.11. A binary source with equally likely symbols controls the switch position in a transmitter operating over an AWGN channel, as shown in Figure P4.2. The noise has two-sided spectral density $N_0/2$. Assume antipodal signals of time duration T seconds and energy E joules. The system clock produces a clock pulse every T seconds, and the binary source rate is $1/T$ bits/s. Under *normal* operation, the switch is up when the source produces a binary zero, and it is down when the source produces a binary one. However, the switch is *faulty*. With probability p, it will be thrown in the wrong direction during a given T-second interval. The presence of a switch error during any interval is independent of the presence of a switch error at any other time. Assume that $\mathbf{E}\{z(T)\} = \pm\sqrt{E}$.

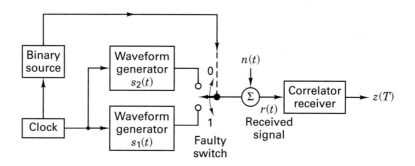

Figure P4.2

(a) Sketch the conditional probability functions, $p(z|s_1)$ and $p(z|s_2)$.

(b) The correlator receiver observes $r(t)$ in the interval $(0, T)$. Sketch the block diagram of an optimum receiver for minimizing the bit error probability when it is known that the switch is faulty with probability, p.

(c) Which one of the following two systems would you prefer to have?

$$p = 0.1 \text{ and } \frac{E_b}{N_0} = \infty$$

or

$$p = 0 \text{ and } \frac{E_b}{N_0} = 7 \text{ dB}$$

4.12. (a) Consider a 16-ary PSK system with symbol error probability $P_E = 10^{-5}$. A Gray code is used for the symbol to bit assignment. What is the approximate bit error probability?

(b) Repeat part (a) for a 16-ary orthogonal FSK system.

4.13. Consider a coherent orthogonal MFSK system with $M = 8$ having the equally likely waveforms $s_i(t) = A \cos 2\pi f_i t$, $i = 1, \ldots, M$, $0 \leq t \leq T$, where $T = 0.2$ ms. The received carrier amplitude, A, is 1 mV, and the two-sided AWGN spectral density, $N_0/2$, is 10^{-11} W/Hz. Calculate the probability of bit error, P_B.

4.14. A bit error probability of $P_B = 10^{-3}$ is required for a system with a data rate of 100 kbits/s to be transmitted over an AWGN channel using coherently detected MPSK modulation. The system bandwidth is 50 kHz. Assume that the system frequency transfer function is a raised cosine with a roll-off characteristic of $r = 1$ and that a Gray code is used for the symbol to bit assignment.

(a) What E_s/N_0 is required for the specified P_B?

(b) What E_b/N_0 is required?

4.15. A differentially coherent MPSK system operates over an AWGN channel with an E_b/N_0 of 10 dB. What is the symbol error probability for $M = 8$ and equally likely symbols?

4.16. If a system's main performance criterion is bit-error probability, which of the following two modulation schemes would be selected for transmission over an AWGN channel?

$$\text{coherent 8-ary orthogonal FSK with } \frac{E_b}{N_0} = 8 \text{ dB}$$

or

$$\text{coherent 8-ary PSK with } \frac{E_b}{N_0} = 13 \text{ dB}$$

(Assume that a Gray code is used for the MPSK symbol-to-bit assignment, and show computations.)

4.17. Consider that a BPSK demodulator/detector has a synchronization error consisting of a time bias pT, where p is a fraction ($0 \leq p \leq 1$) of the symbol time T. In other words, the detection of a symbol starts early (late) and concludes early (late) by an amount pT. Assume equally likely signaling and perfect frequency and phase synchronization. Note that the error-degradation effect of a fixed known bias can be computed by using the closed-end relationships presented in this chapter. However, if the timing error were to consist of a random jitter, computing its effect would then require a stochastic treatment. (See Chapter 10.)

(a) Find the general expression for bit-error probability P_b as a function of p.

(b) If the received $E_b/N_0 = 9.6$ dB and $p = 0.2$, compute the value of degraded P_b due to the timing bias.

(c) If one did not compensate for the timing bias in this example, how much additional E_b/N_0 in dB must be provided in order to restore the P_b that exists when $p = 0$?

4.18. Using all of the stated specifications, repeat problem 4.17 for the case of coherently detected, binary frequency-shift-keying (BFSK) modulation.

4.19. Consider that a BPSK demodulator/detector has a synchronization error consisting of a time bias pT, where p is a fraction ($0 \leq p \leq 1$) of the symbol time T. Consider that there is also a constant phase-estimation error ϕ. Assume equally-likely signaling and perfect frequency synchronization.

    (a) Find the general expression for bit-error probability P_b as a function of p and ϕ.

    (b) If received $E_b/N_0 = 9.6$ dB, $p = 0.2$, and $\phi = 25°$, compute the value of degraded P_b due to the combined effects of timing and phase bias.

    (c) If one did not compensate for the biases in this example, how much additional E_b/N_0 in dB must be provided in order to restore the P_b that exists when $p = 0$ and $\phi = 0°$?

4.20. Correlating to a known Barker sequence is an often used synchronization technique, since the Barker sequence yields a prominent correlation peak when properly synchronized, and a small correlation output when not synchronized. Using the short Barker sequence 1 0 1 1 1, where the leftmost bit is the earliest bit, devise a discrete matched filter similar to the one in Figure 4.10 that is matched to this sequence. Verify its usefulness for synchronization by plotting the output versus input as a function of time, when the input is the 1 0 1 1 1 sequence.

QUESTIONS

4.1. At what location in the system is E_b/N_0 defined? (See Section 4.3.2.)

4.2. Amplitude- or phase-shift keying is visualized as a constellation of points or phasors on a plane. Why can't we use a similarly simple visualization for orthogonal signaling such as FSK? (See Section 4.4.4.)

4.3. In the case of MFSK signaling, what is the minimum tone spacing that insures signal *orthogonality?* (See Section 4.5.4.)

4.4. What benefits are there in using *complex notation* for representing sinusoids? (See Sections 4.2.1 and 4.6.)

4.5. Digital modulation schemes fall into one of the two classes with opposite behavior characteristics: *orthogonal* signaling, and *phase/amplitude* signaling. Describe the behavior of each class. (See Sections 4.8.2.)

4.6. Why do binary phase shift keying (BPSK) and quaternary phase shift keying (QPSK) manifest the same bit-error-probability relationship? (See Section 4.8.4.)

4.7. In the case of multiple-phase shift keying (MPSK), why does *bandwidth efficiency* improve with higher dimensional signaling? (See Sections 4.8.2 and 4.8.3.)

4.8. In the case of orthogonal signaling such as MFSK, why does *error-performance* improve with higher dimensional signaling? (See Section 4.8.5.)

4.9. The use of a Gray code for assigning bits to symbols, represents one of the few cases in digital communications where a benefit can be achieved *free-of-charge*. Explain why there is no cost. (See Section 4.9.4.)

EXERCISES

Using the Companion CD, run the exercises associated with Chapter 4.

CHAPTER 5

Communications Link
Analysis

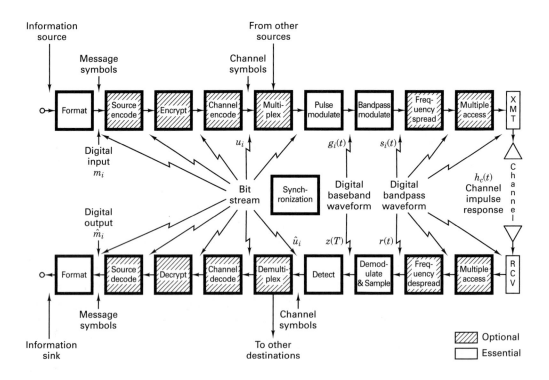

5.1 WHAT THE SYSTEM LINK BUDGET TELLS THE SYSTEM ENGINEER

When we talk about a communications *link*, to what part of the system are we referring? Is it simply the channel or region between the transmitter and receiver? No, it is far more than that. The link encompasses the entire communications path, from the information source, through all the encoding and modulation steps, through the transmitter and the channel, up to and including the receiver with all its signal processing steps, and terminating at the information sink.

What is a link analysis, and what purpose does it serve in the development of a communication system? The link analysis, and its output, the *link budget,* consist of the calculations and tabulation of the useful signal power and the interfering noise power available at the receiver. The link budget is a balance sheet of gains and losses; it outlines the detailed apportionment of transmission and reception resources, noise sources, signal attenuators, and effects of processes throughout the link. Some of the budget parameters are statistical (e.g., allowances for the fading of signals as described in Chapter 15). The budget is an *estimation* technique for evaluating communication system error performance. In Chapters 3 and 4 we examined probability of error versus E_b/N_0 curves having a "waterfall-like" shape, such as the one shown in Figure 3.6. We thereby related error probability to E_b/N_0 for various modulation types in Gaussian noise. Once a modulation scheme has been chosen, the requirement to meet a particular error probability dictates a particular operating point on the curve; in other words, the required error performance dictates the value of E_b/N_0 that must be made available at the receiver in order to meet that performance. The primary purpose of a link analysis is to determine the *actual* system operating point in Figure 3.6 and to establish that the error probability associated with that point is less than or equal to the system requirement. Of the many specifications, analyses, and tabulations that are used in the development of a communication system, the link budget stands out as a basic tool for providing the system engineer with overall system insight.

By examining the link budget, one can learn many things about overall system design and performance. For example, from the link margin, one learns whether the system will meet many of its requirements comfortably, marginally, or not at all. The link budget may reveal if there are any hardware constraints, and whether such constraints can be compensated for in other parts of the link. The link budget is often used as a "score sheet" in considering system trade-offs and configuration changes, and in understanding subsystem nuances and interdependencies. From a quick examination of the link budget and its supporting documentation, one can judge whether the analysis was done precisely or if it represents a rough estimate. Together with other modeling techniques, the link budget can help predict equipment weight, size, prime power requirements, technical risk, and cost. The link budget is one of the system manager's most useful documents; it represents the "bottom-line" tally in the search for optimimum system performance.

5.2 THE CHANNEL

The propagating medium or electromagnetic path connecting the transmitter and receiver is called the *channel*. In general, a communications channel might consist of wires, coaxial cables, fiber optic cables, and in the case of radio-frequency (RF) links, waveguides, the atmosphere, or empty space. For most terrestrial communication links, the channel space is occupied by the atmosphere and partially bounded by the earth's surface. For satellite links, the channel is occupied mostly by empty space. Although some atmospheric effects occur at altitudes up to 100 km, the *bulk* of the atmosphere extends to an altitude of 20 km. Therefore, only a small part (0.05%) of the total synchronous altitude (35,800 km) path is occupied by significant amounts of atmosphere. Most of this chapter presents link analysis in the context of such a satellite communications link. In Chapter 15, the link budget issues are extended to terrestrial wireless links.

5.2.1 The Concept of Free Space

The concept of *free space* assumes a channel free of all hindrances to RF propagation, such as absorption, reflection, refraction, or diffraction. If there is any atmosphere in the channel, it must be perfectly uniform and meet all these conditions. Also, we assume that the earth is infinitely far away or that its reflection coefficient is negligible. The RF energy arriving at the receiver is assumed to be a function only of distance from the transmitter (following the inverse-square law as used in optics). A free-space channel characterizes an ideal RF propagation path; in practice, propagation through the atmosphere and near the ground results in absorption, reflection, diffraction and scattering, which modify the free-space transmission. Atmospheric absorption is treated in later sections. Reflection, diffraction, and scattering, which play an important role in determining terrestrial communications performance, are treated in Chapter 15. Also, Panter [1] provides a comprehensive treatment of these mechanisms.

5.2.2 Error-Performance Degradation

In Chapter 3, it was established that there are two primary causes for degradation of error performance. The first is loss in signal-to-noise ratio. The second is signal distortion as might be caused by intersymbol interference (ISI). In Chapters 3 and 15, some of the equalization techniques to counter the degradation effect of ISI are treated. In this chapter, we are concerned with the "bookkeeping" of the gains and losses for signal power and interfering power. ISI will not be included in the link budget because the uniqueness of ISI is that an increase in signal power will not always mitigate the degradation that it causes. (See Section 3.3.2.)

For digital communications, error performance depends on the received E_b/N_0, which was defined in Equation (3.30) as

$$\frac{E_b}{N_0} = \frac{S}{N}\left(\frac{W}{R}\right)$$

In other words, E_b/N_0 is a measure of normalized signal-to-noise ratio (*S/N* or SNR). Unless otherwise stated, SNR refers to *average* signal power and *average* noise power. The signal can be an information signal, a baseband waveform, or a modulated carrier. The SNR can degrade in two ways: (1) through the decrease of the desired signal power, and (2) through the increase of noise power, or the increase of interfering signal power. Let us refer to these degradations as *loss* and *noise* (or *interference*), respectively. Losses occur when a portion of the signal is absorbed, diverted, scattered, or reflected along its route to the intended receiver; thus a portion of the transmitted energy does not arrive at the receiver. There are several sources of interfering electrical noise and interference produced by a variety of mechanisms and sources, such as thermal noise, galaxy noise, atmospheric noise, switching transients, intermodulation noise, and interfering signals from other sources. Industry usage of the terms *loss* and *noise* frequently confuses the underlying degradation mechanism; however, the net effect on the SNR is the same.

5.2.3 Sources of Signal Loss and Noise

Figure 5.1 is a block diagram of a satellite communications link, emphasizing the sources of signal loss and noise. In the figure a signal loss is distinguished from a noise source by a dot pattern or line pattern, respectively. The contributors of *both* signal loss *and* noise are identified by a crosshatched line pattern. The following list of 21

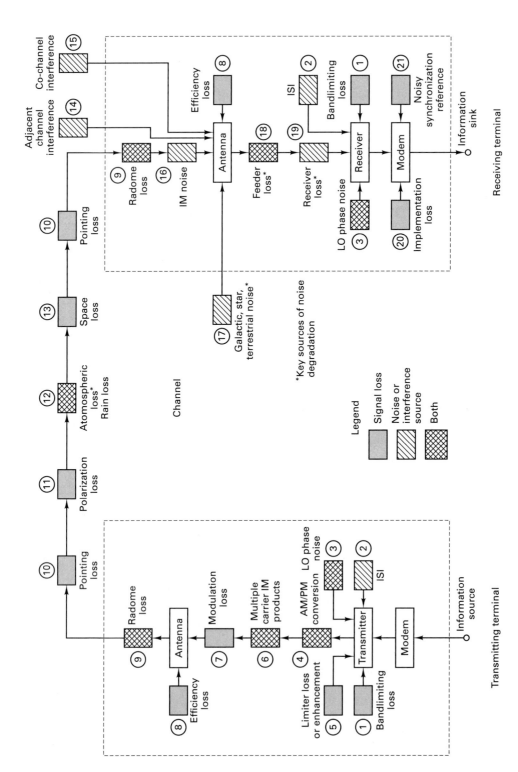

Figure 5.1 Satellite transmitter-to-receiver link with typical loss and noise sources.

sources of degradation represents a partial catalog of the major contributors to SNR degradation. The numbers correspond to the numbered circles in Figure 5.1.

1. *Bandlimiting loss.* All systems use filters in the transmitter to ensure that the transmitted energy is confined to the allocated or assigned bandwidth. This is to avoid interfering with other channels or users and to meet the requirements of regulatory agencies. Such filtering reduces the total amount of energy that would otherwise have been transmitted; the result is a *loss* in signal.

2. *Intersymbol interference (ISI).* As discussed in Chapter 3, filtering throughout the system—in the transmitter, in the receiver, and in the channel—can result in ISI. The received pulses overlap one another; the tail of one pulse "smears" into adjacent symbol intervals so as to *interfere* with the detection process. Even in the absence of thermal noise, imperfect filtering, system bandwidth constraints, and fading channels lead to ISI degradation.

3. *Local oscillator (LO) phase noise.* When an LO is used in signal mixing, phase fluctuations or jitter adds phase *noise* to the signal. When used as the reference signal in a receiver correlator, phase jitter can cause detector degradation and hence signal *loss*. At the transmitter, phase jitter can cause out-of-band signal spreading, which, in turn, will be filtered out and cause a *loss* in signal.

4. *AM/PM conversion.* AM-to-PM conversion is a phase *noise* phenomenon occurring in nonlinear devices such as traveling-wave tubes (TWT). Signal amplitude fluctuations (amplitude modulation) produce phase variations that contribute phase *noise* to signals that will be coherently detected. AM-to-PM conversion can also cause extraneous sidebands, resulting in signal *loss*.

5. *Limiter loss or enhancement.* A hard limiter can enhance the stronger of two signals, and suppress the weaker; this can result in either a signal *loss* or a signal *gain* [2].

6. *Multiple-carrier intermodulation (IM) products.* When several signals having different carrier frequencies are simultaneously present in a nonlinear device, such as a TWT, the result is a multiplicative interaction between the carrier frequencies which can produce signals at all combinations of sum and difference frequencies. The energy apportioned to these spurious signals (intermodulation or IM products) represents a *loss* in signal energy. In addition, if these IM products appear within the bandwidth region of these or other signals, the effect is that of added *noise* for those signals.

7. *Modulation loss.* The link budget is a calculation of received useful power (or energy). Only the power associated with information-bearing signals is useful. Error performance is a function of energy per transmitted symbol. Any power used for transmitting the carrier rather than the modulating signal (symbols) is a modulation *loss*. (However, energy in the carrier may be useful in aiding synchronization.)

8. *Antenna efficiency.* Antennas are transducers that convert electronic signals into electromagnetic fields, and vice versa. They are also used to focus the

electromagnetic energy in a desired direction. The larger the antenna aperture (area), the larger is the resulting signal power density in the desired direction. An antenna's efficiency is described by the ratio of its effective aperture to its physical aperture. Mechanisms contributing to a reduction in efficiency (*loss* in signal strength) are known as amplitude tapering, aperture blockage, scattering, re-radiation, spillover, edge diffraction, and dissipative loss [3]. Typical efficiencies due to the combined effects of these mechanisms range between 50 and 80%.

9. *Radome loss and noise.* A radome is a protective cover, used with some antennas, for shielding against weather effects. The radome, being in the path of the signal, will scatter and absorb some of the signal energy, thus resulting in a signal *loss*. A basic law of physics holds that a body capable of absorbing energy also radiates energy (at temperatures above 0 K). Some of this energy falls in the bandwidth of the receiver and constitutes injected *noise.*

10. *Pointing loss.* There is a *loss* of signal when either the transmitting antenna or the receiving antenna is imperfectly pointed.

11. *Polarization loss.* The polarization of an electromagnetic (EM) field is defined as the direction in space along which the field lines point, and the polarization of an antenna is described by the polarization of its radiated field. There is a *loss* of signal due to any polarization mismatch between the transmitting and receiving antennas.

12. *Atmospheric loss and noise.* The atmosphere is responsible for signal loss and is also a contributor of unwanted noise. The bulk of the atmosphere extends to an altitude of approximately 20 km; yet within that relatively short path, important loss and noise mechanisms are at work. Figure 5.2 is a plot of the theoretical one-way attenuation from a specified height to the top of the atmosphere. The calculations were made for several heights (0 km is sea level) and for a water vapor content of 7.5 g/m^3 at the earth's surface. The magnitude of signal *loss* due to oxygen (O_2) and water vapor absorption is plotted as a function of carrier frequency. Local maxima of attenuation occur in the vicinities of 22 GHz (water vapor), and 60 and 120 GHz (O_2). The atmosphere also contributes *noise* energy into the link. As in the case of the radome, molecules that absorb energy also radiate energy. The oxygen and water vapor molecules radiate noise throughout the RF spectrum. The portion of this noise that falls within the bandwidth of a given communication system will degrade its SNR. A primary atmospheric cause of signal *loss* and contributor of *noise* is rainfall. The more intense the rainfall, the more signal energy it will absorb. Also, on a day when rain passes through the antenna beam, there is a larger amount of atmospheric noise radiated into the system receiver than there is on a clear day. More will be said about atmospheric noise in later sections.

13. *Space loss.* There is a decrease in the electric field strength, and thus in signal strength (power density or flux density), as a function of distance. For a satellite communications link, the space loss is the largest single *loss* in the

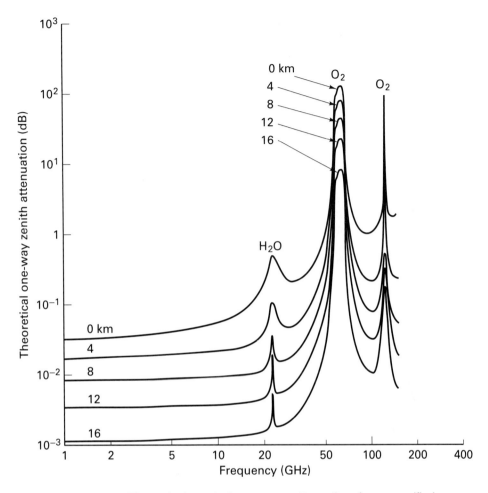

Figure 5.2 Theoretical vertical one-way attenuation from specified height to top of atmosphere for 7.5 g/m^3 of water vapor at the surface. (Does not include effect of rain or cloud attenuation.) (Reprinted from NASA Reference Publication 1082(03), "Propagation Effects Handbook for Satellite Systems Design," June 1983, Fig. 6.2-1, p. 218, courtesy of the National Aeronautics and Space Administration.)

system. It is a loss in the sense that all the radiated energy is not focused on the intended receiving antenna.

14. *Adjacent channel interference.* This *interference* is characterized by unwanted signals from other frequency channels "spilling over" or injecting energy into the channel of interest. The proximity with which channels can be located in frequency is determined by the modulation spectral roll-off and the width and shape of the main spectral lobe.

15. *Co-channel interference.* This *interference* refers to the degradation caused by an interfering waveform appearing within the signal bandwidth. It can be intro-

duced by a variety of ways, such as accidental transmissions, insufficient vertical and horizontal polarization discrimination, or by radiation spillover from an antenna sidelobe (low-energy beam surrounding the main antenna beam). It can be brought about by other authorized users of the same spectrum.

16. *Intermodulation (IM) noise.* The IM products described in item 6 result from multiple-carrier signals interacting in a nonlinear device. Such IM products are sometimes called *active intermods;* as described in item 6, they can either cause a loss in signal energy or be responsible for noise injected into a link. Here we consider *passive intermods;* these are caused by multiple-carrier transmission signals interacting with nonlinear components at the transmitter output. These nonlinearities generally occur at the junction of waveguide coupling joints, at corroded surfaces, and at surfaces having poor electrical contact. When large EM fields impinge on surfaces that have a diode-like transfer function (work potential), they cause multiplicative products, and hence *noise.* If such noise radiates into a closely located receiving antenna, it can seriously degrade the receiver performance.

17. *Galactic or cosmic, star, and terrestrial noise.* All the celestial bodies, such as the stars and the planets, radiate energy. Such *noise* energy in the field of view of the antenna will degrade the SNR.

18. *Feeder line loss.* The level of the received signal might be very small (e.g., 10^{-12} W), and thus will be particularly susceptible to noise degradation. The receiver front end, therefore, is a region where great care is taken to keep the noise as small as possible until the signal has been suitably amplified. The waveguide or cable (feeder line) between the receiving antenna and the receiver front end contributes both signal *attenuation* and thermal *noise;* the details are treated in Section 5.5.3.

19. *Receiver noise.* This is the thermal *noise* generated within the receiver; the details are treated in Sections 5.5.1 to 5.5.4.

20. *Implementation loss.* This *loss* in performance is the difference between theoretical detection performance and the actual performance due to imperfections such as timing errors, frequency offsets, finite rise and fall times of waveforms, and finite-value arithmetic.

21. *Imperfect synchronization reference.* When the carrier phase, the subcarrier phase, and the symbol timing references are all derived perfectly, the error probability is a well-defined function of E_b/N_0 discussed in Chapters 3 and 4. In general, they are not derived perfectly, resulting in a system *loss.*

5.3 RECEIVED SIGNAL POWER AND NOISE POWER

5.3.1 The Range Equation

The main purpose of the link budget is to verify that the communication system will operate according to plan—that is, the message quality (error performance) will meet the specifications. The link budget monitors the "ups" and "downs" (gains and losses) of the transmitted signal, from its inception at the transmitter until it is ultimately received at the receiver. The compilation yields how much E_b/N_0 is received and how much safety margin exists beyond what is required. The process starts with the *range equation,* which relates power received to the distance between transmitter and receiver. We develop the range equation below.

In radio communication systems, the carrier wave is propagated from the transmitter by the use of a transmitting antenna. The transmitting antenna is a transducer that converts electronic signals into electromagnetic (EM) fields. At the receiver, a receiving antenna performs the reverse function; it converts EM fields into electronic signals. The development of the fundamental power relationship between the receiver and transmitter usually begins with the assumption of an omni-directional RF source, transmitting uniformly over 4π steradians. Such an ideal source, called an *isotropic radiator,* is illustrated in Figure 5.3. The power density $p(d)$ on a hypothetical sphere at a distance d from the source is related to the transmitted power P_t by

$$p(d) = \frac{P_t}{4\pi d^2} \qquad \text{watts/m}^2 \tag{5.1}$$

since $4\pi d^2$ is the area of the sphere. For d much greater than the propagation wavelength (known as the *far field*), the power extracted at a receiving antenna is

$$P_r = p(d) A_{er} = \frac{P_t A_{er}}{4\pi d^2} \tag{5.2}$$

where the parameter A_{er} is the absorption cross section (effective area) of the receiving antenna, defined by

$$A_{er} = \frac{\text{total power extracted}}{\text{incident power flux density}} \tag{5.3}$$

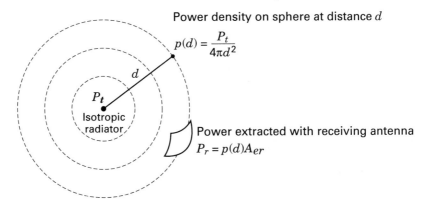

Figure 5.3 Range equation. Expresses received power in terms of distance.

5.3   Received Signal Power and Noise Power

If the antenna under consideration is a transmitting antenna, its effective area is designated by A_{et}. If the antenna in question is unspecified as to its receiving or transmitting function, its effective area is designated simply by A_e.

An antenna's effective area A_e and physical area A_p are related by an efficiency parameter η as

$$A_e = \eta A_p \qquad (5.4)$$

which accounts for the fact that the total incident power is not extracted; it is lost through various mechanisms [3]. Nominal values for η are 0.55 for a dish (parabolic-shaped reflector) and 0.75 for a horn-shaped antenna.

The antenna parameter that relates the power output (or input) to that of an isotropic radiator as a purely geometric ratio is the antenna directivity or *directive gain*

$$G = \frac{\text{maximum power intensity}}{\text{average power intensity over } 4\pi \text{ steradians}} \qquad (5.5)$$

In the absence of any dissipative loss or impedance mismatch loss, the antenna *gain* (in the direction of maximum intensity) is defined simply as the directive gain in Equation (5.5). However, in the event that there exists some dissipative or impedance mismatch loss, the antenna gain is then equal to the directive gain times a loss factor to account for these losses [4]. In this chapter we shall assume that the dissipative loss is zero and that the impedances are perfectly matched. Therefore, Equation (5.5) describes the *peak antenna gain;* it can be viewed as the result of concentrating the RF flux in some restricted region less than 4π steradians, as shown in Figure 5.4. Now we can define an *effective radiated power,* with respect to an isotropic radiator (EIRP), as the product of the transmitted power P_t and the gain of the transmitting antenna G_t, as follows:

$$\text{EIRP} = P_t G_t \qquad (5.6)$$

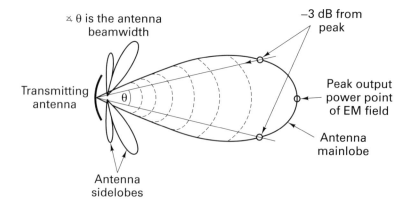

Figure 5.4 Antenna gain is the result of concentrating the isotropic RF flux.

Communications Link Analysis   Chap. 5

Example 5.1   Effective Isotropic Radiated Power

Show that the same value of EIRP can be produced equally well by using a transmitter with $P_t = 100$ W or with $P_t = 0.1$ W, by employing the appropriate antenna in each case.

Solution

Figure 5.5a depicts a 100-W transmitter coupled to an isotropic antenna; the EIRP = $P_t G_t = 100 \times 1 = 100$ W. Figure 5.5b depicts a 0.1-W transmitter coupled to an antenna with gain $G_t = 1000$; the EIRP = $P_t G_t = 0.1 \times 1000 = 100$ W. If field-strength meters were positioned, as shown, to measure the effective power, the measurements could not distinguish between the two cases.

5.3.1.1 Back to the Range Equation

For the more general case in which the transmitter has some antenna gain relative to an isotropic antenna, we replace P_t with EIRP in Equation (5.2) to yield

$$P_r = \text{EIRP}\, \frac{A_{er}}{4\pi d^2} \tag{5.7}$$

The relationship between antenna gain G and antenna effective area A_e is [4]

$$G = \frac{4\pi A_e}{\lambda^2} \quad (\text{for } A_e \gg \lambda^2) \tag{5.8}$$

where λ is the wavelength of the carrier. Wavelength λ and frequency f are reciprocally related by $\lambda = c/f$, where c is the speed of light ($\approx 3 \times 10^8$ m/s). Similar expressions apply for both the transmitting and receiving antennas. The *reciprocity*

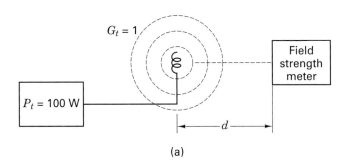

(a)

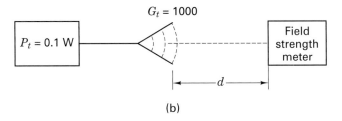

(b)

Figure 5.5 The same value of EIRP produced two different ways.

theorem states that for a given antenna and carrier wavelength, the transmitting and receiving gains are identical [4].

The antenna field of view is a measure of the solid angle into which most of the field power is concentrated. Field of view is a measure of the directional properties of the antenna; it is inversely related to antenna gain—high-gain antennas are commensurate with narrow fields of view. Instead of using the solid-angle field of view, we often deal with the planar angle *beamwidth* measured in radians or degrees. Figure 5.4 pictures a directive antenna pattern and illustrates the common definition of the antenna beamwidth. The beamwidth is the angle that subtends the points at which the peak field power is reduced by 3 dB. How does the antenna beamwidth vary with signal frequency? How does the beamwidth vary with antenna size? As can be seen from Equation (5.8), the antenna gain increases with decreased wavelength (increased frequency); antenna gain also increases with increased effective area. Increasing antenna gain is tantamount to focusing the flux density into a more restricted cone angle; hence, increasing either the signal frequency or the antenna size results in a *narrower beamwidth.*

We can calculate the effective area of an isotropic antenna by setting $G = 1$ in Equation (5.8) and solving for A_e as follows:

$$A_e = \frac{\lambda^2}{4\pi} \tag{5.9}$$

Then to find the power received, P_r, when the receiving antenna is isotropic, we substitute Equation (5.9) into Equation (5.7) to get

$$P_r = \frac{\text{EIRP}}{(4\pi d/\lambda)^2} = \frac{\text{EIRP}}{L_s} \tag{5.10}$$

where the collection of terms $(4\pi d/\lambda)^2$, called the *path loss* or *free-space loss,* is designated by L_s. Notice that Equation (5.10) states that the power received by an isotropic antenna is equal to the effective transmitted power, reduced only by the path loss. When the receiving antenna is not isotropic, replacing A_{er} in Equation (5.7) with $G_r\lambda^2/4\pi$ from Equation (5.8) yields the more general expression

$$P_r = \frac{\text{EIRP } G_r\lambda^2}{(4\pi d)^2} = \frac{\text{EIRP } G_r}{L_s} \tag{5.11}$$

where G_r is the receiving antenna gain. Equation (5.11) can be termed the *range equation.*

5.3.2 Received Signal Power as a Function of Frequency

Since the transmitting antenna and the receiving antenna can each be expressed as a gain or an area, P_r can be expressed four different ways:

$$P_r = \frac{P_t G_t A_{er}}{4\pi d^2} \tag{5.12}$$

$$P_r = \frac{P_t A_{et} A_{er}}{\lambda^2 d^2} \qquad (5.13)$$

$$P_r = \frac{P_t A_{et} G_r}{4\pi d^2} \qquad (5.14)$$

$$P_r = \frac{P_t G_t G_r \lambda^2}{(4\pi d)^2} \qquad (5.15)$$

In these equations, A_{er} and A_{et} are the effective areas of the receiving and transmitting antennas, respectively.

In Equations (5.12) to (5.15) the dependent variable is received signal power, P_r, and the independent variables involve parameters such as transmitted power, antenna gain, antenna area, wavelength, and range. Suppose that we ask the question: How does received power vary as wavelength is decreased (or as frequency is increased), all other independent variables remaining constant? From Equations (5.12) and (5.14) it appears that P_r and wavelength are not related at all. From Equation (5.13), P_r appears to be inversely proportional to wavelength squared, and from Equation (5.15), P_r appears to be directly proportional to wavelength squared. Is there a paradox here? Of course there is not; Equations (5.12) to (5.15) seem to conflict only because antenna gain and antenna area are wavelength related, as stated in Equation (5.8). When should one use each of the Equations (5.12) to (5.15) for determining P_r as a function of wavelength? Consider a system that is already configured; that is, the antennas have already been built or their dimensions are fixed (A_{et} and A_{er} are fixed). Then Equation (5.13) is the appropriate choice for calculating the P_r performance. Equation (5.13) states that for fixed-size antennas, the received power increases as the wavelength is decreased.

Consider the use of Equation (5.12), where G_t and A_{er} are independent variables. We want G_t and A_{er} held fixed over the range of P_r versus wavelength calculations. What happens to the gain of a fixed-dimension transmitting antenna as the independent variable λ is decreased? G_t increases [see Equation 5.8]. But we cannot have G_t increasing in Equation (5.12)—we want G_t held fixed. In other words, to ensure that G_t remains fixed, we would need to reduce the transmitting antenna size as wavelength decreases. It should be apparent that Equation (5.12) is the appropriate equation when starting with a *fixed transmitting antenna gain* (or beamwidth) requirement and the parameter A_{et} is not fixed. For similar reasons, Equation (5.14) is used when A_{et} and G_r are fixed, and Equation (5.15) is used when both the transmitting and receiving antenna gains (or beamwidths) are fixed.

Figure 5.6 illustrates a satellite application where the downlink antenna beam is required to provide earth coverage (a beamwidth of approximately $17°$ from synchronous altitude). Since the satellite antenna gain G_t must be fixed, the resulting P_r is independent of wavelength, as shown in Equation (5.12). If the transmission at some frequency f_1 ($= c/\lambda_1$) provides earth coverage, then a frequency change to f_2, where $f_2 > f_1$, will result in reduced coverage (since for a given antenna, G_t will increase); hence the antenna size must be reduced to maintain the required earth

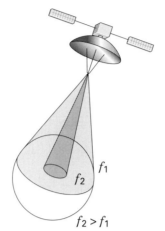

$f_2 > f_1$

Figure 5.6 Received power as a function of frequency.

coverage or beamwidth. Thus earth coverage antennas become smaller as the carrier frequency is increased.

5.3.3 Path Loss Is Frequency Dependent

From Equation (5.10) it can be seen that path loss, L_s, is wavelength (frequency) dependent. The question is often asked: Why should path loss, which is just a geometric inverse-square loss, be a function of frequency? The answer is that path loss, as characterized in Equation (5.10), is a *definition* predicated on the use of an isotropic receiving antenna ($G_r = 1$). Hence path loss is a convenient tool; it represents a hypothetical received-power loss that *would occur if the receiving antenna were isotropic.* Figure 5.3 and Equation (5.1) have established that power density, $p(d)$, is a function of distance—a purely geometric consideration; $p(d)$ is *not* a function of frequency. However, since path loss is predicated on $G_r = 1$, when we attempt to collect some P_r with an *isotropic antenna,* the result is characterized by Equation (5.10). Again let us emphasize that L_s can be viewed as a convenient collection of terms that have been assigned the unfortunate name *path loss.* The name conjures up an image of a purely geometric effect and fails to emphasize the requirement that $G_r = 1$. A better choice of a name would have been *unity-gain propagation loss.* In a radio communication system, path loss accounts for the largest loss in signal power. In satellite systems, the path loss for a C-band (6-GHz) link to a synchronous altitude satellite is typically 200 dB.

Example 5.2   Antenna Design for Measuring Path Loss

Design a hypothetical experiment to measure path loss L_s, at frequencies $f_1 = 30$ MHz and $f_2 = 60$ MHz, when the distance between the transmitter and receiver is 100 km. Find the effective area of the receiving antenna, and calculate the path loss in decibels for each case.

Solution

Figure 5.7 illustrates the two links for measuring L_s at frequencies f_1 and f_2, respectively. The power density, $p(d)$, at each receiver is identical and equal to

$$p(d) = \frac{\text{EIRP}}{4\pi d^2}$$

This reduction in power density is due *only* to the inverse-square law. The actual power received at each receiver is found by multiplying the power density $p(d)$ at the receiver by the effective area, A_{er}, of the collecting antenna, as shown in Equation (5.7). Since path loss is predicated on $G_r = 1$, we compute the effective area A_{er1} at frequency f_1, and A_{er2} at frequency f_2, using Equation (5.9):

$$A_{er} = \frac{\lambda^2}{4\pi} = \frac{(c/f)^2}{4\pi}$$

$$A_{er1} = \frac{(3 \times 10^8/30 \times 10^6)^2}{4\pi} \approx 8 \text{ m}^2$$

$$A_{er2} = \frac{(3 \times 10^8/60 \times 10^6)^2}{4\pi} \approx 2 \text{ m}^2$$

The path loss for each case in decibels is

$$L_{s1} = 10 \times \log_{10}\left(\frac{4\pi d}{\lambda_1}\right)^2 = 10 \times \log_{10}\left(\frac{4\pi \times 10^5}{3 \times 10^8/30 \times 10^6}\right)^2$$
$$= 102 \text{ dB}$$

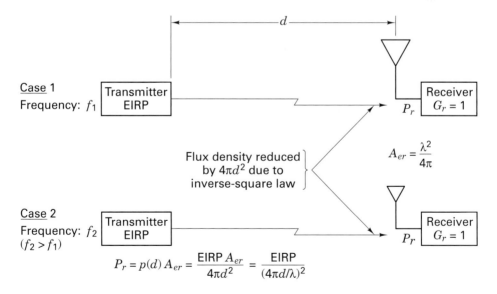

Figure 5.7 Path loss versus frequency. Hypothetical experiment to measure path loss at two different frequencies.

$$L_{s2} = 10 \times \log_{10} \left(\frac{4\pi d}{\lambda_2} \right)^2 = 10 \times \log_{10} \left(\frac{4\pi \times 10^5}{3 \times 10^8 / 60 \times 10^6} \right)^2$$

$$= 108 \text{ dB}$$

5.3.4 Thermal Noise Power

Thermal noise is caused by the thermal motion of electrons in all conductors. It is generated in the lossy coupling between an antenna and receiver and in the first stages of the receiver. The noise power spectral density is constant at all frequencies up to about 10^{12} Hz, giving rise to the name *white noise*. The thermal noise process in communication receivers is modeled as an additive white Gaussian noise (AWGN) process, as described in Section 1.5.5. The physical model [5, 6] for thermal or Johnson noise is a noise generator with an open-circuit mean-square voltage of $4\kappa T^\circ W \mathcal{R}$, where

$$\kappa = \text{Boltzmann's constant} = 1.38 \times 10^{-23} \text{ J/K or W/K-Hz}$$

$$= -228.6 \text{ dBW/K-Hz},$$

$T^\circ = \text{temperature, kelvin,}$

$W = \text{bandwidth, hertz,}$

and

$\mathcal{R} = \text{resistance, ohms.}$

The maximum thermal noise power N that could be coupled from the noise generator into the front end of an amplifier is

$$N = \kappa T^\circ W \qquad \text{watts} \tag{5.16}$$

Thus, the maximum single-sided noise power spectral density N_0 (noise power in a 1-Hz bandwidth), available at the amplifier input is

$$N_0 = \frac{N}{W} = \kappa T^\circ \qquad \text{watts/hertz} \tag{5.17}$$

It might seem that the noise power should depend on the magnitude of the resistance—but it does not. Consider an intuitive argument to verify this. Electrically connect a large resistance to a small one, such that they form a closed path and such that their physical temperatures are the same. If noise power were a function of resistance, there would be a net power flow from the large resistance to the small one; the large resistance would become cooler and the small one would become warmer. This violates our experience, not to mention the second law of thermodynamics. Therefore, the power delivered from the large resistance to the small one must be equal to the power it receives.

The available power from a thermal noise source is dependent on the ambient temperature of the source (the *noise temperature*), as is seen in Equation (5.16). This leads to the useful concept of an *effective noise temperature* for noise sources that are not necessarily thermal in origin (e.g., galactic, atmospheric, interfering signals) that can be introduced into the receiving antenna. The effective noise temperature of such a noise source is defined as the temperature of a hypothetical thermal noise source that would give rise to an equivalent amount of interfering power. The subject of noise temperature is treated in greater detail in Section 5.5.

Example 5.3 Maximum Available Noise Power

Using a noise generator with mean-square voltage equal to $4\kappa T^{\circ}W\mathcal{R}$, demonstrate that the maximum amount of noise power that can be coupled from this source into an amplifier is $N_i = \kappa T^{\circ}W$.

Solution

A theorem from network theory states that maximum power is delivered to a load when the value of the load impedance is made equal to the complex conjugate of the generator impedance [7]. In this case the generator impedance is a pure resistance, $\mathcal{R}$; therefore, the condition for maximum power transfer is fulfilled when the input resistance of the amplifier equals $\mathcal{R}$. Figure 5.8 illustrates such a network. The input thermal noise source is represented by an electrically equivalent model consisting of a noiseless source resistor in series with an ideal voltage generator whose rms noise voltage is $\sqrt{4\kappa T^{\circ}W\mathcal{R}}$. The input resistance of the amplifier is made equal to $\mathcal{R}$. The noise voltage delivered to the amplifier input is just one-half the generator voltage, following basic circuit principles. The noise power delivered to the amplifier input can accordingly be expressed as

$$N_i = \frac{(\sqrt{4\kappa T^{\circ}W\mathcal{R}}/2)^2}{\mathcal{R}} = \frac{4\kappa T^{\circ}W\mathcal{R}}{4\mathcal{R}}$$
$$= \kappa T^{\circ}W$$

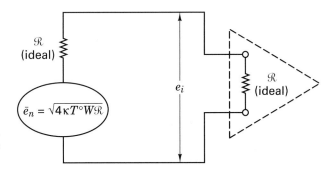

Figure 5.8 Electrical model of maximum available thermal noise power at amplifier input.

5.4 LINK BUDGET ANALYSIS

In link budget calculations, the quantity of greatest interest is the receiving-system signal-to-noise-ratio (SNR), often modeled at the output of the receiving antenna. This SNR is sometimes called *carrier-to-noise* ratio (*C/N*), where $N = kT_s^{\circ}W$, k is Boltzmann's constant, T_s° is system temperature in Kelvin (see Section 5.5.5), and W is bandwidth. *C/N* is often used in many satellite link-budget calculations. The custom probably arose because satellite signals are usually suppressed-carrier signals, wherein the carrier can be viewed as being totally modulated (transformed into inforamtion-carrying sidelobes). The information-bearing SNR at the predetection point, usually denoted P_r/N or S/N, is the parameter of interest for obtaining E_b/N_0. For suppressed-carrier signaling, P_r/N and C/N have the same value; thus the following expressions are sometimes used interchangeably.

$$\frac{P_r}{N} \equiv \frac{S}{N} \equiv \frac{C}{N} \equiv \frac{C}{\kappa T^\circ W}$$

Is P_r/N always the same as C/N? No, the signal power and the carrier power are only the same when the carrier is totally modulated (e.g., wideband angle modulation). For example, consider a frequency-modulated (FM) carrier wave, as

$$s(t) = A \cos(\omega_0 t + K \int m(t)\, dt)$$

where $m(t)$ is a modulating signal, and K is a system constant. The average power in $m(t)$ is $\overline{m^2(t)}$. Increasing this modulating power only serves to increase the frequency deviation of $s(t)$, which means that the carrier is spread over a wider spectrum, but its average power $\overline{s^2(t)}$ remains equal to $A^2/2$, regardless of the power in the modulating signal. With such wideband angle-modulated signals, the carrier is "carrying" the information over an increased bandwidth, explaining why the information-bearing signal power is sometimes referred to as the carrier power.

For linear modulation, such as amplitude modulation (AM), the power in the carrier is quite different than the power in the modulating signal. For example, consider an AM carrier wave in terms of the modulating signal $m(t)$:

$$s(t) = [1 + m(t)]\, A \cos \omega_0 t$$

$$\overline{s^2(t)} = [1 + m(t)]^2 \frac{A^2}{2}$$

$$= \frac{A^2}{2} [1 + \overline{m^2(t)} + \overline{2\, m(t)}]$$

If we assume that $m(t)$ has a zero mean, then the average carrier power can be written as

$$\overline{s^2(t)} = \frac{A^2}{2} + \frac{A^2}{2} \overline{m^2(t)}$$

In this expression, there is an unmodulated carrier component; hence, the carrier power is not the same as the signal power. In summary, the parameters C/N and P_r/N are the same for suppressed-carrier signals (e.g., PSK or FSK), but this is not the case for a signal having an unmodulated carrier component represented by a spectral line at the carrier frequency (e.g., amplitude modulation).

We obtain P_r/N by dividing Equation (5.11) by noise power N, as follows:

$$\frac{P_r}{N} = \frac{\text{EIRP } G_r/N}{L_s} \tag{5.18}$$

Equation (5.18) applies to any one-way RF link. With *analog receivers,* the noise bandwidth (generally referred to as the effective or equivalent noise bandwidth) seen by the demodulator is usually greater than the signal bandwidth, and P_r/N is the main parameter for measuring signal detectability and performance quality. With *digital receivers,* however, correlators or matched filters are usually implemented, and signal bandwidth is taken to be equal to noise bandwidth. Rather than consider input noise power, a common formulation for digital links is to replace noise power with *noise power spectral density.* We can use Equation (5.17) with $T°$ taken to be system effective temperature (developed later) to rewrite Equation (5.18) as

$$\frac{P_r}{N_0} = \frac{\text{EIRP } G_r/T°}{\kappa L_s L_o} \tag{5.19}$$

where the system effective temperature $T°$ is a function of the noise radiated into the antenna and the thermal noise generated within the receiver. Note that the receiving antenna gain G_r and system temperature $T°$ are grouped together. The grouping $G_r/T°$ is sometimes called the *receiver figure-of-merit.* The reason for treating these terms in this way is explained in Section 5.6.2.

It is important to emphasize that the system effective temperature $T°$ is a parameter that *models* all the noise in the receiving system; the subject is treated in greater detail in Section 5.5. In Equation (5.19) we have introduced a term L_o to represent all other losses and degradation factors not specifically addressed by the other terms of Equation (5.18). The factor L_o allows for the large assortment of different losses and noise sources cataloged earlier. Equation (5.19) summarizes the key parameters of any link analysis: the received signal power-to-noise power spectral density (P_r/N_0), the effective transmitted power (EIRP), the receiver figure-of-merit ($G_r/T°$), and the losses (L_s, L_0). We are developing a methodical way to keep track of the gains and losses in a communications link. By starting with some source of power we can compute, using Equation (5.19), the net SNR arriving at the "face" of the detector (predetection point). We are working toward a "bookkeeping" system, much like the type used in a business that earmarks assets and liabilities and tallies up the bottom-line of profit (or loss). Equation (5.19) takes on such an entrepreneurial appearance. All of the numerator parameters (effective radiated power, receiver figure-of-merit) are like the assets of a business, and all the denominator parameters (thermal noise, space loss, other losses) are like the liabilities of a business.

Assuming that all the received power P_r is in the modulating (information-bearing) signal, we now relate E_b/N_0 and SNR from Equation (3.30) and write

$$\frac{E_b}{N_0} = \frac{P_r}{N}\left(\frac{W}{R}\right) \tag{5.20a}$$

$$\frac{E_b}{N_0} = \frac{P_r}{N_0}\left(\frac{1}{R}\right) \tag{5.20b}$$

and

$$\frac{P_r}{N_0} = \frac{E_b}{N_0}R \tag{5.20c}$$

where R is the bit rate. If some of the received power is carrier power (a signal-power loss), we can still employ Equation (5.20), except that the carrier power contributes to the loss factor, L_o in Equation (5.19). This fundamental relationship between E_b/N_0 and P_r/N_0 in Equation (5.20) will be required frequently in designing and evaluating systems. (See Chapter 9.)

5.4.1 Two E_b/N_0 Values of Interest

We have referred to E_b/N_0 as that value of bit energy per noise power spectral density required to yield a specified error probability. To facilitate calculating a margin or safety factor M, we need to differentiate between the *required* E_b/N_0 and the actual or *received* E_b/N_0. From this point on we will refer to the former as $(E_b/N_0)_{\text{reqd}}$ and to the latter as $(E_b/N_0)_r$. Figure 5.9 depicts an example with two operating points. The first is associated with $P_B = 10^{-3}$; let us call this operating point the system required error performance. Let us assume that an $(E_b/N_0)_{\text{reqd}}$ value of 10 dB will yield this required performance. Do you suppose we would build this system so that the demodulator received this 10-dB value *exactly*? Of course not; we would specify and design the system to have a safety margin, so that the $(E_b/N_0)_r$ actually received would be somewhat larger than the $(E_b/N_0)_{\text{reqd}}$. Thus we might design the system to operate at the second operating point on Figure 5.9; here $(E_b/N_0)_r = 12$ dB and $P_B = 10^{-5}$. For this example we can describe the safety margin or *link margin*, as providing a two-order-of-magnitude improved P_B, or as is more usual, we can describe the link margin in terms of providing 2 dB more E_b/N_0

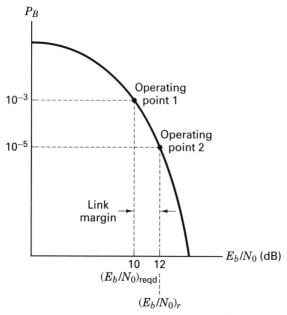

Figure 5.9 Two E_b/N_0 values of interest.

than is required. We can rewrite Equation (5.20c), introducing the link margin parameter M as

$$\frac{P_r}{N_0} = \left(\frac{E_b}{N_0}\right)_r R = M\left(\frac{E_b}{N_0}\right)_{\text{reqd}} R \qquad (5.21)$$

The difference in decibels between $(E_b/N_0)_r$ and $(E_b/N_0)_{\text{reqd}}$ yields the link margin:

$$M(\text{dB}) = \left(\frac{E_b}{N_0}\right)_r (\text{dB}) - \left(\frac{E_b}{N_0}\right)_{\text{reqd}} (\text{dB}) \qquad (5.22)$$

The parameter $(E_b/N_0)_{\text{reqd}}$ reflects the differences from one system design to another; these might be due to differences in modulation or coding schemes. A larger than expected $(E_b/N_0)_{\text{reqd}}$ may be due to a suboptimal RF system, which manifests large timing errors or which allows more noise into the detection process than does an ideal matched filter.

Combining Equations (5.19) and (5.21) and solving for the link margin M yields

$$M = \frac{\text{EIRP } G_r/T^\circ}{(E_b/N_0)_{\text{reqd}} R\kappa L_s L_o} \qquad (5.23)$$

Equation (5.23), the *link margin equation,* contains all of the parameters contributing to the link's error performance. Some of these parameters are defined with reference to particular locations. For example, E_b/N_0 is defined at the input to the receiver. More precisely, it is defined at the input to the detector (predetection point), where a demodulated waveform having a voltage amplitude proportional to received energy is the basis of a symbol decision. Similarly, any parameter describing received energy or power, whether useful or degrading, is also defined with reference to this predetection point. The receiver's figure-of-merit G_r/T° is defined at the input to the receiving antenna, where G_r is the gain of the receiving antenna, and T° is the effective system temperature. (See Section 5.5.5.) The effective radiated power EIRP is defined as the power associated with the electromagnetic wave at the output of the transmitting antenna. Each of these parameters, E_b/N_0, G_r/T°, and EIRP is defined with reference to a particular system location, and nowhere else.

5.4.2 Link Budgets Are Typically Calculated in Decibels

Since link budget analysis is typically calculated in decibels, we can express Equation (5.23) as

$$M(\text{dB}) = \text{EIRP}(\text{dBW}) + G_r(\text{dBi}) - \left(\frac{E_b}{N_0}\right)_{\text{reqd}} (\text{dB}) - R(\text{dB-bit/s})$$

$$- \kappa T^\circ (\text{dBW/Hz}) - L_s(\text{dB}) - L_o(\text{dB}) \qquad (5.24)$$

Transmitted signal power EIRP is expressed in decibel-watts (dBW); noise power spectral density N_0 is in decibel-watts per hertz (dBW/Hz); antenna gain G_r is in decibels referenced to isotropic gain (dBi); data rate R is in decibels referenced to

1 bit/s (dB-bit/s); and all other terms are in decibels (dB). The numerical values of the Equation (5.24) parameters constitute the link budget, a useful tool for allocating communications resources. In an effort to maintain a positive margin, we might trade off any parameter with any other parameter; we might choose to reduce transmitter power by giving up excess margin, or we might elect to increase the data rate by reducing $(E_b/N_0)_{\text{reqd}}$ (through the selection of improved modulation and coding). Any one of the Equation (5.24) decibels, regardless of the parameter from which it stems, is just as good as any other decibel—a dB is a dB is a dB. The transmission system "does not know and does not care" where the decibels come from. As long as the needed amount of E_b/N_0 arrives at the receiver, the desired system error performance can be met. Well, let us add two other conditions for achieving a required error performance—synchronization must be maintained, and ISI distortion must be minimized or equalized. One might ask, Since the system has no preference as to where the decibels of E_b/N_0 come from, how shall we prioritize the search for an adequate number of decibels? The answer is, we should look for the most cost-effective decibels. This goal will steer us toward reading the next several chapters on error-correction coding, since this discipline has historically provided a continual reduction in the cost of electronics to achieve error-performance improvements.

5.4.3 How Much Link Margin Is Enough?

The question of how much link margin should be designed into a system is asked frequently. The answer is that if all sources of gain, loss, and noise have been rigorously detailed (worst case), and if the link parameters with large variances (e.g., fades due to weather) match the statistical requirements for link availability, very little additional margin is needed. The margin needed depends on how much confidence one has in each of the link budget entries. For systems employing new technology or new operating frequencies, one needs more margin than for systems that have been repeatedly built and tested. Sometimes the link budget provides an allowance for fades due to weather directly, as a line item. Other times, however, the required value of margin reflects the link requirements for a given rain degradation. For satellite communications at C-band (uplink at 6 GHz, downlink at 4 GHz), where the parameters are well known and fairly well behaved, it should be possible to design a system with only 1 dB of link margin. Receive-only television stations operating with 16-ft-diameter dishes at C-band are frequently designed with only a fraction of a decibel of margin. However, telephone communications via satellite using standards of 99.9% availability require considerably more margin; some of the INTELSAT systems have 4 to 5 dB of margin. When nominal rather than worst-case computations are performed, allowances are usually made for unit-to-unit equipment variations over the operating temperature range, line voltage variations, and mission duration. Also, for space communications, there may be an allowance for errors in tracking a satellite's location.

Designs using higher frequencies (e.g., 14/12 GHz) generally call for larger (weather) margins because atmospheric losses increase with frequency and are highly variable. It should be noted that a by-product of the attenuation due to atmospheric loss is greater antenna noise. With low-noise amplifiers, small weather changes can result in increases of 40 to 50 K in antenna temperature. Table 5.1 represents a link analysis proposed to the Federal Communications Commission (FCC) by Satellite Television Corporation for the Direct Broadcast Satellite (DBS) service. Notice that the downlink budget is tabulated for two alternative weather conditions: clear, and 5-dB loss due to rain. The signal loss due to atmospheric attenuation is only a small fraction of a decibel for clear weather and is the full 5 dB during rain. The next item in the downlink tabulation, home receiver $G/T°$, illustrates the additional degradation caused by the rain; additional thermal noise irradiates the receiving antenna, making the effective system noise temperature, $T°$, increase, and the home receiver $G/T°$ decrease (from 9.4 dB/K to 8.1 dB/K). Therefore, when extra margin is allowed for weather loss, additional margin should simultaneously be added to compensate for the increase in system noise temperature.

With regard to satellite links, in industry one often hears such expressions as "the link *can* be closed," meaning that the margin, in decibels, has a positive value and the required error performance will be satisfied, or "the link *cannot* be closed," meaning that the margin has a negative value and the required error performance will *not* be satisfied. Even though the words "the link closes" or "the link does not close" give the impression of an "on-off" condition, it is worth emphasizing that lack of link closure, or a negative margin, means that the error performance falls short of the system requirement; it does not necessarily mean that communications cease. For example, consider a system whose $(E_b/N_0)_{\text{reqd}} = 10$ dB, as shown in Figure 5.9, but whose $(E_b/N_0)_r = 8$ dB. Assume that 8 dB corresponds to $P_B = 10^{-2}$.

TABLE 5.1 Proposed Direct Broadcast Satellite (DBS) from Satellite Television Corp.

Uplink	
Earth station EIRP	86.6 dBW
Free-space loss (17.6 GHz, 48° elevation)	208.9 dB
Assumed rain attenuation	12.0 dB
Satellite $G/T°$	7.7 dB/K
Uplink $C/\kappa T°$	102.0 dB-Hz

	Atmospheric Condition	
Downlink	Clear	5-dB Rain Attenuation
Satellite EIRP	57.0 dBW	57.0 dBW
Free-space loss (12.5 GHz, 30° elevation)	206.1 dB	206.1 dB
Atmospheric attenuation	0.14 dB	5.0 dB
Home receiver $G/T°$ (0.75 m dish)	9.4 dB/K	8.1 dB/K
Receiver pointing loss (0.5° error)	0.6 dB	0.6 dB
Polarization mismatch loss (average)	0.04 dB	0.04 dB
Downlink $C/\kappa T°$	88.1 dB-Hz	82.0 dB-Hz
Overall $C/\kappa T°$	87.9 dB-Hz	82.0 dB-Hz
Overall C/N (in 16 MHz)	15.9 dB	10.0 dB
Reference threshold C/N	10.0 dB	10.0 dB
Margin over threshold	5.9 dB	0.0 dB

Thus there is a margin of –2 dB, and a bit error probability of 10 times the specified error probability. The link may still be useful, though degraded.

5.4.4 Link Availability

Link availability is usually a measure of long-term link utility stated on an average annual basis; for a given geographical location, the link availability measures the percentage of time the link can be closed. For example, for a particular link between Washington, D.C., and a satellite repeater, the long-term weather pattern may be such that a 10-dB weather margin is adequate for link closure 98% of the time; for 2% of the time, heavy rains result in greater than 10 dB SNR degradation, so that the link does not close. Since the effect of rain on SNR degradation is a function of signal frequency, link availability and required margin must be examined in the context of a particular transmission frequency.

Figure 5.10 summarizes worldwide satellite link availability at a frequency of 44 GHz. The plot illustrates percentage of the earth visible (the link closes, and a prescribed probability of error is met) as a function of margin for the case of three equispaced geostationary satellites. A *geostationary satellite* is located in a circular orbit in the same plane as the earth's equatorial plane and at the synchronous altitude of 35,800 km. The satellite's orbital period is identical with that of the earth's rotational period, and therefore the satellite appears stationary when viewed from the earth. Figure 5.10 shows a family of visibility curves with different required link availabilities, ranging from benign (95% availability) to fairly stringent (99% availability). In general, for a fixed link margin, visibility is inversely proportional to required availability, and for a fixed availability, visibility increases monotonically with margin [8]. Figures 5.11 to 5.13 illustrate, by

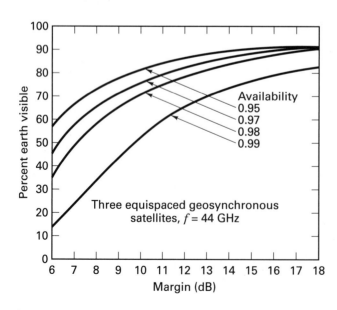

Figure 5.10 Earth coverage versus link margin for various values of link availability. (Reprinted from L. M. Schwab, "World-Wide Link Availability for Geostationary and Critically Inclined Orbits Including Rain Effects," *Lincoln Laboratory, Rep. DCA-9,* Jan. 27, 1981, Fig. 14, p. 38, courtesy of Lincoln Laboratory.)

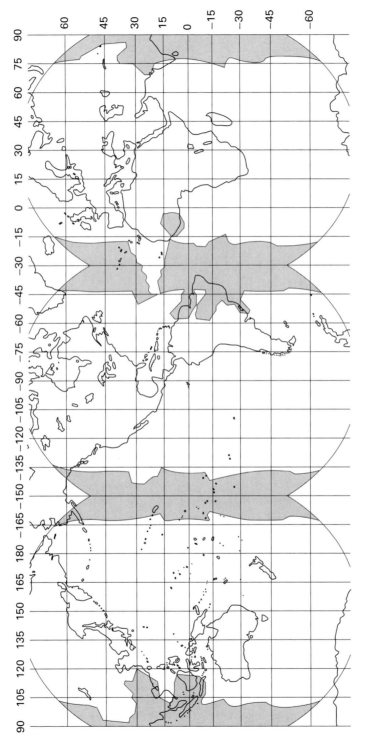

Figure 5.11 Earth coverage (unshaded) for 0.99 link availability for three equi-spaced geostationary satellites, $f = 44$ GHz, link margin = 14 dB. (Reprinted from L. M. Schwab, "World-Wide Link Availability for Geostationary and Critically Inclined Orbits Including Rain Effects," *Lincoln Laboratory, Rep. DCA-9,* Jan. 27, 1981, Fig. 17, p. 42, courtesy of Lincoln Laboratory.)

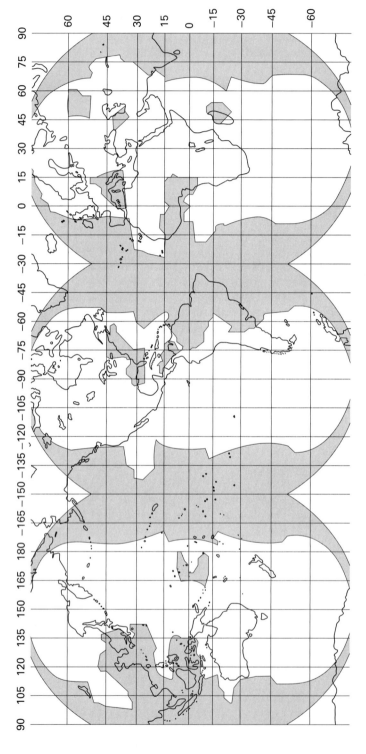

Figure 5.12 Earth coverage (unshaded) for 0.99 link availability for three equi-spaced geostationary satellites, $f = 44$ GHz, link margin = 10 dB. (Reprinted from L. M. Schwab, "World-Wide Link Availability for Geostationary and Critically In-clined Orbits Including Rain Effects," *Lincoln Laboratory, Rep. DCA-9*, Jan. 27, 1981, Fig. 18, p. 43, courtesy of Lincoln Laboratory.)

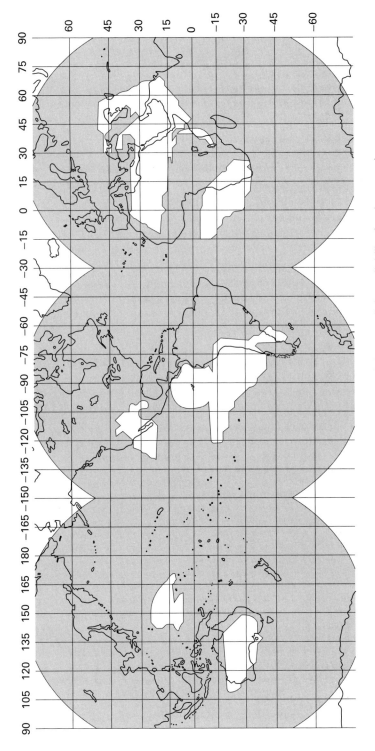

Figure 5.13 Earth coverage (unshaded) for 0.99 link availability for three equi-spaced geostationary satellites, $f = 44$ GHz, link margin = 6 dB. (Reprinted from L. M. Schwab, "World-Wide Link Availability for Geostationary and Critically Inclined Orbits Including Rain Effects," *Lincoln Laboratory, Rep. DCA-9,* Jan. 27, 1981, Fig. 19, p. 44, courtesy of Lincoln Laboratory.)

unshaded and shaded areas, the parts of the earth from which the 44-GHz link can and cannot be closed 99% of the time for three different values of link margin. Figure 5.11 illustrates the link coverage of such locations for a margin value of 14 dB. Notice that this figure can be used to pinpoint the regions of heaviest rainfall, such as Brazil and Indonesia. The figure represents the result of a link calculation performed in concert with a weather model of the earth.

In Figure 5.11 there are shaded strips on the east and west boundaries of each satellite's field of view. Why do you suppose the link availability is not met in these regions? At the edge of the earth the propagation path between the satellite and ground is longer than the path directly beneath the satellite. Degradation occurs in three ways: (1) the longer path results in reduced power density at the receiving antenna; (2) the edge of coverage sites will experience reduced satellite antenna gain, unless the satellite antenna pattern is designed to be uniform over its entire field of view (typically the pattern is −3 dB at the beam edge compared to the peak gain at the beam center); and (3) propagation to the edge of the earth traverses a thicker atmospheric layer because of the oblique path and the earth's curvature. The third item is of prime importance at those signal frequencies that are most attenuated by the atmosphere. Why do you suppose you do not see the same shaded areas near the north and south poles in Figure 5.11? Snowfall does not have the same deleterious effect on signal propagation as does rainfall; the phenomenon is known as the *freeze effect*.

Figure 5.12 illustrates the parts of the earth that can and cannot close the 44-GHz link 99% of the time with 10-dB link margin. Notice that the shaded areas have grown considerably compared to the 14-dB margin case; now, the east coast of the United States, the Mediterranean, and most of Japan cannot close the link 99% of the time. Figure 5.13 illustrates similar link performance for a margin of 6 dB. Whereas Figure 5.11 could be used to locate the regions of greatest rainfall, Figure 5.13 can be used to locate the driest weather regions on the earth; such areas are seen to be the southwestern part of the United States, most of Australia, the coast of Peru and Chile, and the Sahara desert in Africa.

5.5 NOISE FIGURE, NOISE TEMPERATURE, AND SYSTEM TEMPERATURE

5.5.1 Noise Figure

Noise figure, F, relates the SNR at the input of a network to the SNR at the output of the network. Thus noise figure measures the SNR degradation caused by the network. Figure 5.14 illustrates such an example. Figure 5.14a depicts the SNR at an *amplifier input* $(SNR)_{in}$ as a function of frequency. At its peak, the signal is 40 dB above the noise floor. Figure 5.14b depicts the SNR at the *amplifier output* $(SNR)_{out}$. The amplifier gain has increased the signal by 20 dB; however, the amplifier has added its own additional noise. The output signal, at its peak, is only 30 dB above the noise floor. Since the SNR degradation from input to output is 10 dB,

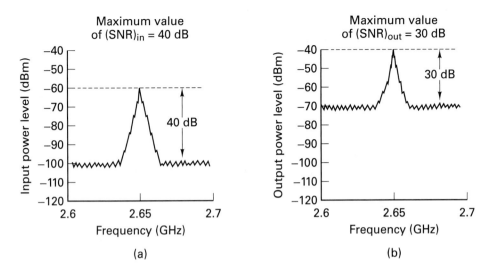

Figure 5.14 Amplifier signal and noise levels as a function of frequency. (a) Amplifier input. (b) Amplifier output.

this is tantamount to describing the amplifier as having a 10-dB noise figure. Noise figure is a parameter that expresses the noisiness of a two-port network or device, such as an amplifier, compared with a reference noise source at the input port. It can be written as

$$F = \frac{(\text{SNR})_{\text{in}}}{(\text{SNR})_{\text{out}}} = \frac{S_i/N_i}{GS_i/G(N_i + N_{ai})} \tag{5.25}$$

where

$$S_i = \text{signal power at the amplifier input port,}$$

$$N_i = \text{noise power at the amplifier input port,}$$

$$N_{ai} = \text{amplifier noise referred to the input port,}$$

and

$$G = \text{amplifier gain.}$$

Figure 5.15 is an example illustrating Equation (5.25). Figure 5.15a represents a *realizable amplifier* example with a gain $G = 100$, and internal noise power $N_a = 10 \,\mu\text{W}$. The source noise, external to the amplifier, is $N_i = 1 \,\mu\text{W}$. In Figure 5.15b we assume that the *amplifier is ideal,* and we ascribe the noisiness of the real amplifier, from part (a) of the figure, to an external source N_{ai} in series with the original source N_i. The value of N_{ai} is obtained by reducing N_a by the amplifier gain. As shown in Figure 5.15b, Equation (5.25) references all noise to the amplifier input, whether the noise is actually present at the input or is internal to the device. As can

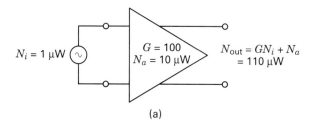

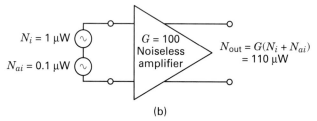

Figure 5.15 Example of noise treatment in amplifiers.

(a)

(b)

be seen in Figure 5.15, the noise power output from the real amplifier is identical to that of its electrically equivalent model.

Equation (5.25) reduces to

$$F = \frac{N_i + N_{ai}}{N_i} = 1 + \frac{N_{ai}}{N_i} \qquad (5.26)$$

Notice from Equation (5.26) that the noise figure expresses the noisiness of a network relative to an input source noise; noise figure is *not* an absolute measure of noise. An ideal amplifier or network, one that contributes no noise ($N_{ai} = 0$), has a noise figure equal to unity (0 dB).

For the concept of noise figure to have utility, we need to be able to make equitable comparisons among devices on the basis of Equation (5.26). We must, therefore, choose a value of N_i as a *reference*. The noise figure of any device will then represent a measure of how much noisier the device is than the reference. In 1944, Friis [9] suggested that noise figure be defined for a noise source at a reference temperature of $T_0^\circ = 290$ K. That suggestion was subsequently adopted by the IEEE as part of its standard definition for noise figure [10]. From Equation (5.17) we see that the maximum available noise power spectral density from any source resistance is established by specifying its temperature. The value of 290 K was selected as the reference because it is a reasonable approximation of the source temperature for many links. Also, with T_0° chosen to be 290 K, the value of noise spectral density N_0 at T_0° results in an aesthetically pleasing number:

$$N_0 = \kappa T_0^\circ = 1.38 \times 10^{-23} \times 290 = 4.00 \times 10^{-21} \text{ W/Hz}$$

Or, expressed in decibels,

$$N_0 = -204 \text{ dBW/Hz}$$

Now that noise figure F has been defined with reference to a 290 K noise source, it is important to emphasize that the noise figure relationships in Equations (5.25) and (5.26) are only accurate when N_i is a 290 K noise source. For those cases where N_i is other than a 290 K noise source, we must rename F in Equations (5.25) and (5.26) to be termed *operational noise figure* F_{op}. The relationship between F_{op} and F is shown later in Equation (5.48).

5.5.2 Noise Temperature

Rearranging Equation (5.26), we can write

$$N_{ai} = (F - 1)N_i \qquad (5.27)$$

From Equation (5.16) we can replace N_i with $\kappa T_0^{\circ}W$ and N_{ai} with $\kappa T_R^{\circ}W$, where T_0° is the reference temperature of the source and T_R° is called the *effective noise temperature* of the receiver (or network). We can then write

$$\kappa T_R^{\circ}W = (F - 1)\kappa T_0^{\circ}W$$

or

$$T_R^{\circ} = (F - 1)\ T_0^{\circ}$$

Or, since T_0° has been chosen to be 290 K,

$$T_R^{\circ} = (F - 1)290 \text{ K} \qquad (5.28)$$

Equation (5.26) uses the concept of noise figure to characterize the noisiness of an amplifier. Equation (5.28) represents an alternative but equivalent characterization known as *effective noise temperature*. Note that the noise figure is a measurement relative to a reference. However, noise temperature has no such constraint.

We can think of available noise power spectral density and effective noise temperature, in the context of Equation (5.17), as equivalent ways of characterizing noise sources. Equation (5.28) tells us that the noisiness of an amplifier can be modeled as if it were caused by an additional noise source, as seen in Figure 5.15b, operating at some effective temperature called T_R°. For a purely resistive termination, T_R° is never less than ambient temperature unless it is cooled. It is important to note that for reactive terminations, such as uncooled parametric amplifiers or other low-noise devices, T_R° can be much less than 290 K, even though the ambient temperature is higher [11]. For the output of an amplifier as a function of its effective temperature, we can use Equations (5.16), (5.25), and (5.28) to write

$$N_{out} = GN_i + GN_{ai} \qquad (5.29a)$$

$$= G\kappa T_g^{\circ}W + G\kappa T_R^{\circ}W = G\kappa(T_g^{\circ} + T_R^{\circ})\ W \qquad (5.29b)$$

$$= G\kappa T_g^{\circ}W + (F - 1)\ G\kappa T_0^{\circ}\ W \qquad (5.29c)$$

where T_g° is the temperature of the source, and T_0° is 290 K.

5.5.3 Line Loss

The difference between amplifier networks and lossy line networks can be viewed in the context of the degradation mechanisms *loss* and *noise,* described earlier. Noisy networks in Sections 5.5.1 and 5.5.2 were discussed with amplifiers in mind. We saw that SNR degradation resulted from injecting additional (amplifier) noise into the link, as shown in Figure 5.15. However, in the case of a lossy line, we shall show that the SNR degradation results from the signal being attenuated while the noise remains fixed (for the case where the line temperature is equal to or less than the source temperature). The degradation effect will nonetheless be measured as an increase in noise figure or effective noise temperature.

Consider the lossy line or network shown in Figure 5.16. Assume the line is matched with its characteristic impedance at the source and at the load. We shall define power loss as

$$L = \frac{\text{input power}}{\text{output power}}$$

Then, the network gain G equals $1/L$ (less than unity for a lossy line). Let all components be at temperature T_g°. The total output noise power flowing from the network into the load is

$$N_{\text{out}} = \kappa T_g^\circ W$$

since the network output appears as a pure resistance at the temperature T_g°. The total power flowing from the load back into the network must also equal N_{out} to ensure thermal equilibrium. Recall that available noise power $\kappa T^\circ W$ is dependent only on temperature, bandwidth, and impedance matching; it is not dependent on the resistance value. N_{out} can be considered to be made up of two components, N_{go} and GN_{Li}, such that

$$N_{\text{out}} = \kappa T_g^\circ W = N_{go} + GN_{Li} \tag{5.30}$$

where

$$N_{go} = G\kappa T_g^\circ W \tag{5.31}$$

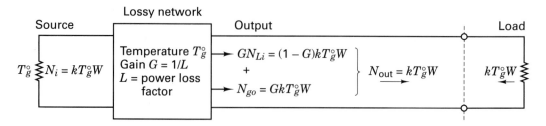

Figure 5.16 Lossy line: impedance matched and temperature matched at both ends.

is the component of output noise power due to the source and GN_{Li} is the component of output noise power due to the lossy network, where N_{Li} is the network noise relative to its input. Combining Equations (5.30) and (5.31), we can write

$$\kappa T_g^\circ W = G\kappa T_g^\circ W + GN_{Li} \tag{5.32}$$

then, solving for N_{Li} yields

$$N_{Li} = \frac{1-G}{G}\kappa T_g^\circ W = \kappa T_L^\circ W \tag{5.33}$$

Therefore, the effective noise temperature of the line is

$$T_L^\circ = \frac{1-G}{G}T_g^\circ \tag{5.34}$$

and since $G = 1/L$,

$$T_L^\circ = (L-1)T_g^\circ \tag{5.35}$$

Choosing $T_g^\circ = 290$ K as the reference temperature, we can write

$$T_L^\circ = (L-1)290 \text{ K} \tag{5.36}$$

Using Equations (5.28) and (5.36), the *noise figure for a lossy line* can be expressed as

$$F = 1 + \frac{T_L^\circ}{290} = L \tag{5.37}$$

When the network is a lossy line, such that $F = L$ and $G = 1/L$, then N_{out} in Equation (5.29c) takes the following form:

$$N_{out} = \frac{\kappa T_g^\circ W}{L} + \left(1 - \frac{1}{L}\right)\kappa T_0^\circ W \tag{5.38}$$

Note that some authors use the parameter L to mean the reciprocal of the loss factor defined here. In such cases, noise figure $F = 1/L$.

Example 5.4   Lossy Line

A line at temperature $T_0^\circ = 290$ K is fed from a source whose noise temperature is $T_g^\circ = 1450$ K. The input signal power S_i is 100 picowatts (pW) and the signal bandwidth W is 1 GHz. The line has a loss factor $L = 2$. Calculate the $(SNR)_{in}$, the effective line temperature T_L°, the output signal power S_{out}, and the $(SNR)_{out}$.

Solution

$$N_i = \kappa T_g^\circ W$$
$$= 1.38 \times 10^{-23} \text{ W/K-Hz} \times 1450 \text{ K} \times 10^9 \text{ Hz}$$
$$= 2 \times 10^{-11} \text{ W} = 20 \text{ pW}$$

$$(SNR)_{in} = \frac{100 \text{ pW}}{20 \text{ pW}} = 5 \text{ (7 dB)}$$

$$T_L^\circ = (L - 1)\,290\,K = 290\,K$$

$$S_{\text{out}} = \frac{S_i}{L} = \frac{100\text{ pW}}{2} = 50\text{ pW}$$

Using Equation (5.39), we obtain

$$
\begin{aligned}
N_{\text{out}} &= \frac{\kappa T_g^\circ W}{L} + \left(1 - \frac{1}{L}\right)\kappa T_0^\circ W \\
&= \frac{2 \times 10^{-11}}{2}\,W + \frac{1}{2}(4 \times 10^{-12})\,W = 12\text{ pW}
\end{aligned}
$$

and

$$(\text{SNR})_{\text{out}} = \frac{50\text{ pW}}{12\text{ pW}} = 4.17\ (6.2\text{ dB})$$

5.5.4 Composite Noise Figure and Composite Noise Temperature

When two networks are connected in series, as shown in Figure 5.17a, their composite noise figure can be written as

$$F_{\text{comp}} = F_1 + \frac{F_2 - 1}{G_1} \tag{5.39}$$

where G_1 is the gain associated with network 1. When n networks are connected in series the relationship between stages expressed in Equation (5.39) continues, so that the *composite noise figure* for a sequence of n stages is written as

$$F_{\text{comp}} = F_1 + \frac{F_2 - 1}{G_1} + \frac{F_3 - 1}{G_1 G_2} + \cdots + \frac{F_n - 1}{G_1 G_2 \cdots G_{n-1}} \tag{5.40}$$

Can you guess from Equation (5.40) what the design goals for the front end of the receiver (especially the first stage or the first couple of stages) should be? At the front end of the receiver, the signal is most susceptible to added noise; therefore, the first stage should have as low a noise figure F_1 as possible. Also, because the noise figure of each subsequent stage is reduced by the gains of the prior stages, it behooves us to strive for as high a gain G_1 as possible. Simultaneously achieving the lowest F_1 and the highest G_1 represents conflicting goals; therefore, compromises are always necessary.

Equations (5.40) and (5.28) can be combined to express the composite effective noise temperature of a sequence of n stages:

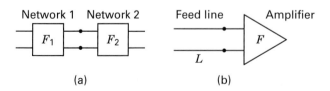

(a)   (b)

$$T^\circ_{\text{comp}} = T^\circ_1 + \frac{T^\circ_2}{G_1} + \frac{T^\circ_3}{G_1 G_2} + \cdots + \frac{T^\circ_n}{G_1 G_2 \cdots G_{n-1}} \tag{5.41}$$

Figure 5.17b illustrates a feed line in series with an amplifier; this is a typical arrangement following a receiving antenna. Using Equation (5.39) to find F_{comp} for such a lossy line and amplifier arrangement, we can write

$$F_{\text{comp}} = L + L(F - 1) = LF \tag{5.42}$$

since the noise figure of the lossy line is L and the gain of the line is $1/L$. By analogy with Equation (5.36), we can write the composite temperature as

$$T^\circ_{\text{comp}} = (LF - 1)290 \text{ K} \tag{5.43}$$

We can also write the composite temperature of line and amplifier as follows:

$$\begin{aligned}
T^\circ_{\text{comp}} &= (LF - 1 + L - L)290 \text{ K} \\
&= [(L - 1) + L(F - 1)]290 \text{ K} \\
&= T^\circ_L + LT^\circ_R
\end{aligned} \tag{5.44}$$

5.5.4.1 Comparison of Noise Figure and Noise Temperature

Since noise figure F and effective noise temperature T° characterize the noise performance of devices, some engineers feel compelled to select one of these measures as the more useful. However, they each have their place. For terrestrial applications, F is almost universally used; the concept of SNR degradation for a 290 K source temperature makes sense, because terrestrial source temperatures are typically close to 290 K. Terrestrial noise figure values typically fall in the convenient range 1 to 10 dB.

For space applications, T° is the more common figure of merit. The range of values for commercial systems is typically between 30 and 150 K, giving adequate resolution for comparing performance between systems. A disadvantage of using noise figures for such low-noise networks is that the values obtained are all close to unity (0.5 to 1.5 dB), which makes it difficult to compare devices. For low-noise applications, F (in decibels) would need to be expressed to two decimal places to provide the same resolution or precision as does T°. For space applications, a reference temperature of 290 K is not as appropriate as it is for terrestrial applications. When using effective temperature, no reference temperature (other than absolute zero K) is needed for judging degradation. The effective input noise temperature is simply compared to the effective source noise temperature. In general, applications involving very low noise devices seem to favor the effective temperature measure over the noise figure.

5.5.5 System Effective Temperature

Figure 5.18 represents a simplified schematic of a receiving system, identifying those areas—the antenna, the line, and the preamplifier—that play a primary role in SNR degradation. We have already discussed the degradation role of the pream-

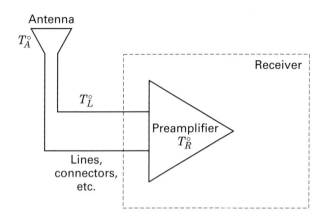

Antenna

T_A°

T_L°

Preamplifier
T_R°

Lines,
connectors,
etc.

Receiver

Figure 5.18 Major noise contributors of a receiving system.

plifier—additional noise is injected into the link. And we have discussed line loss—the signal is attenuated, while the noise is held fixed (for the case where the line temperature is less than or equal to the source temperature). The remaining source of degradation stems from both natural and man-made sources of noise and interference that enter the receiving antenna. Natural sources include lightning, celestial radio sources, atmospheric sources, and thermal radiation from the ground and other physical structures. Man-made noise includes radiation from automobile ignition systems and any other electrical machinery, and radio transmissions from other users that fall within the receiver bandwidth. The total noise contributed by these external sources can be characterized by $kT_{ant}W$, where T_{ant} is known as the *antenna temperature*. An antenna is like a lens. Its noise contributions are dictated by what the antenna is "looking at." If an antenna is pointed at a cool portion of the sky, very little thermal noise is introduced. The antenna temperature is a measure of the effective temperature integrated over the entire antenna pattern.

We now find *system temperature* T_S° by adding together all the system noise contributors (in terms of effective temperature). The summation is expressed as

$$T_S^\circ = T_A^\circ + T_{comp}^\circ \tag{5.45}$$

where T_A° is the antenna temperature and T_{comp}° is the composite temperature of the line and the preamplifier. Equation (5.45) illustrates the two primary sources of noise and interference degradation at a receiver. One source, characterized by T_A°, represents degradation from the "outside world" arriving via the antenna. The second source, characterized by T_{comp}°, is thermal noise caused by the motion of electrons in all conductors. Since the system temperature T_S° is a new composite, made up of T_A° and the composite effective temperature of the line and preamp, one might ask: Why doesn't Equation (5.45) appear to have the same sequential gain reduction factors as those in Equation (5.41)? We have assumed that the antenna has *no dissipative parts;* its gain, unlike an amplifier or attenuator, can be thought of as a processing gain. Whatever effective temperature is introduced at the antenna comes through, unaltered by the antenna; the antenna represents the source noise, or source temperature, at the input to the line.

Using Equation (5.44), we can modify Equation (5.45) as follows:

$$T_S^{\circ} = T_A^{\circ} + T_L^{\circ} + LT_R^{\circ} \qquad (5.46)$$

$$= T_A^{\circ} + (L-1)290 \text{ K} + L(F-1)290 \text{ K}$$

$$= T_A^{\circ} + (LF-1)290 \text{ K} \qquad (5.47)$$

If LF is provided in units of decibels, we must first convert LF to a ratio, so that T_S° takes the form

$$T_S^{\circ} = T_A^{\circ} + (10^{LF/10} - 1)290 \text{ K}$$

Equations (5.10), (5.11), and (5.45) through (5.47) describe power received P_r and system temperature T_S respectively. In each case, the parameters are referred to the output of the receiving antenna, a popular reference, preferred by system and antenna designers, and those working at the transmitter side of the link. Another convention, often used by receiver designers, describes received power P_r' and system temperature T_S' referenced to the input of the receiver. Assuming that the antenna and receiver are connected via a lossy line, then P_r and P_r' (also T_S and T_S') are related by the line-loss factor L. That is, $P_r = LP_r'$, and $T_S = LT_S'$. Note that the ratio of received power to system temperature, a receiving-system SNR parameter, is the same for both refereences. That is so because, $P_r/T_S = LP_r'/LT_S'$. One can similarly choose a reference anywhere in the receiver circuit-string (from the output of the receiving antenna to the predetection point) that results in the same receiving-system SNR model.

Example 5.5   Noise Figure and Noise Temperature

A receiver front end, shown in Figure 5.19a, has a noise figure of 10 dB, a gain of 80 dB, and a bandwidth of 6 MHz. The input signal power, S_i, is 10^{-11} W. Assume that the line is lossless and the antenna temperature is 150 K. Find T_R°, T_S°, N_{out}, (SNR)$_{\text{in}}$, and (SNR)$_{\text{out}}$.

Solution

First convert all decibel values to ratios:

$$T_R^{\circ} = (F-1)290 \text{ K} = 2610 \text{ K}$$

Using Equation (5.46) with $L = 1$ for a lossless line yields

$$T_S^{\circ} = T_A^{\circ} + T_R^{\circ} = 150 \text{ K} + 2610 \text{ K} = 2760 \text{ K}$$

$$N_{\text{out}} = G\kappa T_A^{\circ} W + G\kappa T_R^{\circ} W = G\kappa T_S^{\circ} W$$

$$= 10^8 \times 1.38 \times 10^{-23} \times 6 \times 10^6 (150 \text{ K} + 2610 \text{ K})$$

$$= \underbrace{1.2 \text{ } \mu\text{W}}_{\substack{\text{source} \\ \text{contribution}}} + \underbrace{21.6 \text{ } \mu\text{W}}_{\substack{\text{front-end} \\ \text{contribution}}} = 22.8 \text{ } \mu\text{W}$$

$$(\text{SNR})_{\text{in}} = \frac{S_i}{\kappa T_A^{\circ} W} = \frac{10^{-11}}{1.24 \times 10^{-14}} = 806.5 \text{ } (29.1 \text{ dB})$$

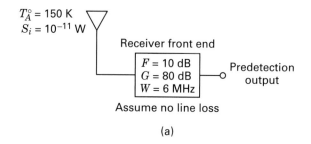

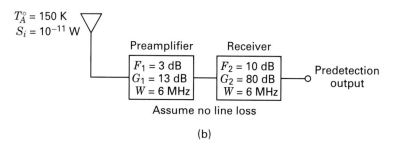

Figure 5.19 Improving a receiver front end with a low-noise preamplifier.

$$(\text{SNR})_{\text{out}} = \frac{S_{\text{out}}}{N_{\text{out}}} = \frac{10^8 \times 10^{-11}}{22.8 \times 10^{-6}} = 43.9 \ (16.4 \text{ dB})$$

Notice in this example that the amplifier noise is significantly larger than the source noise and represents the major cause of SNR degradation.

Example 5.6   Improving SNR with a Low-Noise Preamplifier

Use a preamplifier, as shown in Figure 5.19b, with a noise figure of 3 dB, a gain of 13 dB, and a bandwidth of 6 MHz to improve the SNR of the receiver in Example 5.5. Find T°_{comp} for the composite preamplifier and receiver. Find T°_{S}, F_{comp}, N_{out}, and $(\text{SNR})_{\text{out}}$. Assume zero line loss.

Solution

Again, convert all decibel values to ratios before proceeding:

$$T^{\circ}_{R1} = (F_1 - 1)290 \text{ K} = 290 \text{ K}$$

$$T^{\circ}_{R2} = (F_2 - 1)290 \text{ K} = 2610 \text{ K}$$

$$T^{\circ}_{\text{comp}} = T^{\circ}_{R1} + \frac{T^{\circ}_{R2}}{G_1} = 290 \text{ K} + \frac{2610 \text{ K}}{20} = 420.5 \text{ K}$$

$$T^{\circ}_{S} = T^{\circ}_{A} + T^{\circ}_{\text{comp}} = 150 \text{ K} + 420.5 \text{ K} = 570.5 \text{ K}$$

$$F_{\text{comp}} = F_1 + \frac{F_2 - 1}{G_1} = 2 + \frac{9}{20} = 2.5 \ (4 \text{ dB})$$

$$N_{\text{out}} = G\kappa T_A^{\circ} W + G\kappa T_{\text{comp}}^{\circ} W = G\kappa T_S^{\circ} W$$

$$= 20 \times 10^8 \times 1.38 \times 10^{-23} \times 6 \times 10^6 (150 \text{ K} + 420.5 \text{ K})$$

$$= \underbrace{24.8 \ \mu\text{W}}_{\substack{\text{source} \\ \text{contribution}}} + \underbrace{69.6 \ \mu\text{W}}_{\substack{\text{front-end} \\ \text{contribution}}} = 94.4 \ \mu\text{W}$$

$$(\text{SNR})_{\text{out}} = \frac{S_{\text{out}}}{N_{\text{out}}} = \frac{10^{-11} \times 20 \times 10^8}{94.4 \times 10^{-6}} = 212.0 \ (23.3 \text{ dB})$$

With the added preamplifier the (predetection) output noise has increased (from 22.8 μW in Example 5.5) to 94.4 μW. Even though the noise power has increased, the lower system temperature has resulted in a 6.9-dB improvement in SNR (from 16.4 dB in Example 5.5, to 23.3 dB here). The price we pay for this improvement is the need to provide an F_{comp} improvement of 6 dB (from 10 dB in Example 5.5, to 4 dB in this example).

The unwanted noise is, in part, *injected via the antenna* ($\kappa T_A^{\circ} W$), and in part, *generated internally* in the receiver front end ($\kappa T_{\text{comp}}^{\circ} W$). The amount of system improvement that can be rendered via front-end design depends on what portion of the total noise the front end contributes. We saw in Example 5.5 that the front end contributed the major portion of the noise. Therefore, in Example 5.6, providing a low-noise preamplifier improved the system SNR significantly. In the next example, we show the case where the major portion of the noise is injected via the antenna; we shall see that introducing a low-noise preamplifier in such a case will not help the SNR appreciably.

Example 5.7   Attempting SNR Improvement When the Value of T_A° Is Large

Repeat Examples 5.5 and 5.6 with one change: let $T_A^{\circ} = 8000$ K. In other words, the preponderant amount of noise is being injected via the antenna; the antenna might have a very hot body (the sun) fully occupying its field of view. Calculate the SNR improvement that would be provided by the preamplifier used in Example 5.6 and Figure 5.19b, and compare the result with that of Example 5.6.

Solution

Without preamplifier

$$N_{\text{out}} = G\kappa W(T_A^{\circ} + T_R^{\circ})$$

$$= 10^8 \times 1.38 \times 10^{-23} \times 6 \times 10^6 (8000 \text{ K} + 2610 \text{ K})$$

$$= \underbrace{66.2 \ \mu\text{W}}_{\substack{\text{source} \\ \text{contribution}}} + \underbrace{21.6 \ \mu\text{W}}_{\substack{\text{front-end} \\ \text{contribution}}} = 87.8 \ \mu\text{W}$$

$$(\text{SNR})_{\text{out}} = \frac{S_{\text{out}}}{N_{\text{out}}} = \frac{10^8 \times 10^{-11}}{87.8 \times 10^{-6}} = 11.4 \ (10.6 \text{ dB})$$

With preamplifier

$$N_{\text{out}} = 20 \times 10^8 \times 1.38 \times 10^{-23} \times 6 \times 10^6 (8000 \text{ K} = 420.5 \text{ K})$$

$$= \underbrace{1324.8 \ \mu W}_{\substack{\text{source} \\ \text{contribution}}} + \underbrace{69.6 \ \mu W}_{\substack{\text{front-end} \\ \text{contribution}}} = 1394.4 \ \mu W$$

$$(\text{SNR})_{\text{out}} = \frac{20 \times 10^8 \times 10^{-11}}{1.39 \times 10^{-3}} = 14.4 \ (11.6 \ \text{dB})$$

Therefore, for this case, the SNR improvement is only 1 dB, a far cry from the 6.9 dB accomplished earlier. When the noise is mostly due to devices within the receiver, it is possible to improve the SNR by introducing low-noise devices. However, when the noise is mostly due to external causes, improving the receiver front end will not help much.

Noise figure is a definition, predicated on a reference temperature of 290 K. When the source temperature is other than 290 K, as is the case in Examples 5.5, 5.6, and 5.7, it is necessary to define an *operational* or *effective noise figure* that describes the actual $(\text{SNR})_{\text{in}}$ versus $(\text{SNR})_{\text{out}}$ relationship. Starting with Equations (5.25) and (5.27), the operational noise figure can be found as follows:

$$
\begin{aligned}
F_{\text{op}} &= \frac{S_i/kT_A W}{GS_i/G(kT_A W + N_{ai})} \\
&= \frac{kT_A W + N_{ai}}{kT_A W} = 1 + \frac{(F-1)kT_0 W}{kT_A W} \\
&= 1 + \frac{T_0}{T_A}(F-1)
\end{aligned}
\qquad (5.48)
$$

5.5.6 Sky Noise Temperature

The receiving antenna collects random noise emissions from galactic, solar, and terrestrial sources, constituting the sky background noise. The sky background appears as a combination of galactic effects that decrease with frequency, and atmospheric effects that start becoming significant at 10 GHz and increase with frequency. Figure 5.20 illustrates the sky temperature, as measured from the earth, due to both these effects. Notice that there is a region, between 1 and 10 GHz, where the temperature is lowest; the galaxy noise has become quite small at 1 GHz, and for satellite communications the blackbody radiation noise due to the absorbing atmosphere is not significant below 10 GHz. (For other applications, e.g., passive radiometry, it is still a problem.) This region, known as the *microwave window* or *space window,* is particularly useful for satellite or deep-space communication. The low sky noise is the principal reason that such systems primarily use carrier frequencies in this part of the spectrum. The galaxy and atmospheric noise curves in Figure 5.20 are made up of a family of curves each at a different elevation angle θ. When θ is zero, the receiving antenna points at the horizon and the propagation path encompasses the longest possible atmospheric layer; when θ is 90°, the receiving antenna points to the zenith, and the resulting propagation path contains the shortest possible atmospheric layer. Thus the upper curve of the family repre-

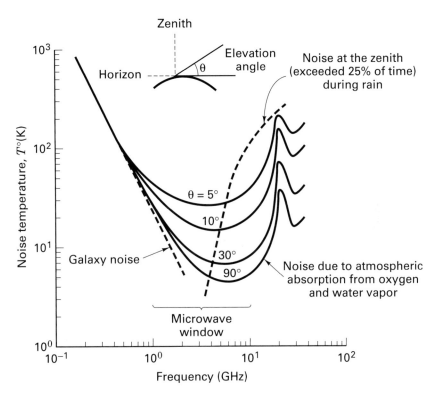

Figure 5.20 Sky noise temperature.

sents the near worst-case (clear-weather) sky noise temperature versus frequency, and the lower curve represents the most benign case. Also shown in Figure 5.20 is a plot of noise temperature versus frequency *due to rain.* Since the intensity of any rainstorm can only be expressed statistically, the noise temperatures shown are values that are exceeded 25% of the time (at the zenith). Which spectral region appears the most benign for space communications when rainfall is taken into account? It is the region at the low end of the space window. For this reason, systems such as the Space Ground Link Subsystem or SGLS (military) and the Unified S-Band Telemetry, Tracking, and Control System (NASA) are located in the 1.8 to 2.4-GHz band.

5.5.6.1 Radio Maps of the Sky

Various researchers have mapped the galactic noise radiation as a function of frequency. Figure 5.21 is such a radio temperature map, after Ko and Kraus [12], indicating the temperature contours of the sky in the region of 250 MHz when viewed from the earth. In general, the sky is composed of localized galactic sources (sun, moon, planets, etc.), each having its own temperature. The map is effectively

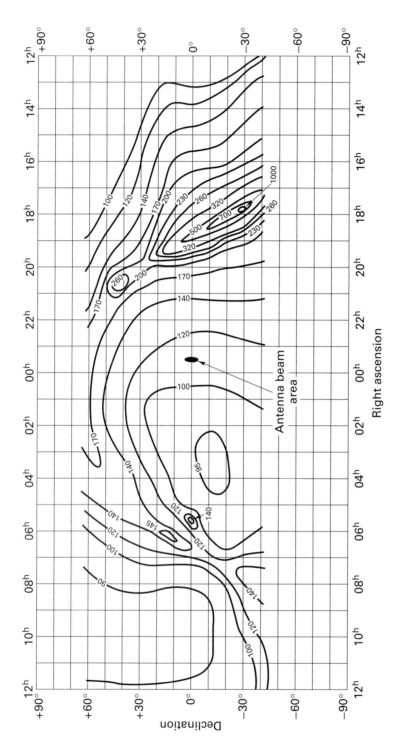

Figure 5.21 Radio map of the sky background at 250 MHz. (Reprinted from H. C. Ko and J. D. Kraus, "A Radio Map of the Sky at 1.2 Meters," *Sky Telesc.*, vol. 16, Feb. 1957, p. 160, with permission from *Sky and Telescope* astronomy magazine, Cambridge, Mass.)

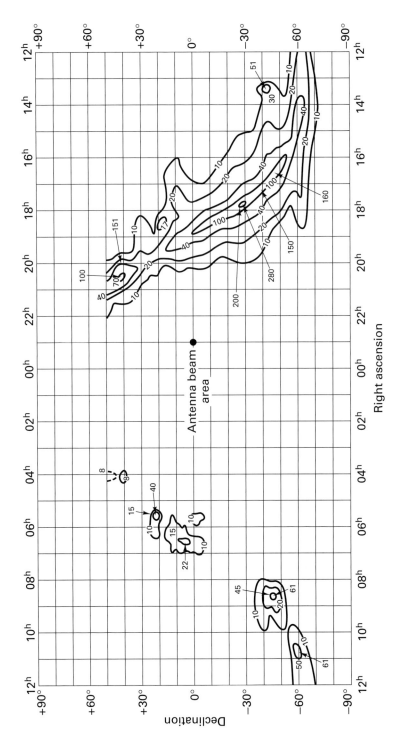

Figure 5.22 Radio map of the sky background at 600 MHz. (Reprinted with permission from J. H. Piddington and G. H. Trent, "A Survey of Cosmic Radio Emission at 600 Mc/s," *Aust. J. Phys.,* vol. 9, Dec. 1956, Fig. 1, pp. 483–486.)

a weighted sum of the individual galactic source temperatures plus a constant sky background. The coordinates of the map, *declination and right ascension,* can be thought of as celestial latitude and longitude with an earth reference (right ascension is calibrated in units of hour-angle, where 24 hours corresponds to a complete rotation of the earth). On Figure 5.21, the temperature contours range from a low of 90 K to a high of 1000K. The measurements were made so as to exclude the sun (night sky). The antenna beam in the center of the map indicates the size of the sky area over which measurements are made (each measurement is an average over that beam area). The narrower the beam, the finer the resolution of the temperature contours; the wider the beam, the coarser the resolution.

Figure 5.22 is another such radio map at 600 MHz, after Piddington and Trent [13]. At this frequency the galaxy noise is reduced compared to Figure 5.21, as predicted in Figure 5.20; the low is 8 K and the high is 280 K. If you examine Figures 5.21 and 5.22 for the region of greatest noise radiation, where on the map do you see the most activity, and what is its significance? It is seen as an elongated region in the right-hand midsection of each map; the longitudinal axis of the elongation designates the location of our *galactic plane,* where such cosmic noise radiation is most intense.

5.6 SAMPLE LINK ANALYSIS

In Section 5.4 we developed the basic link parameter relationships. In this section we use these relationships to calculate a sample link budget, as shown in Table 5.2. The table may appear to house a formidable listing of terms; one can get the false impression that the link budget represents a complex compilation. Just the opposite is true, and we introduce Figure 5.23 to underscore this assertion. In this figure we have reduced the set of line items from Table 5.2 to a few key parameters. The goal of a link analysis is to determine whether or not the required error performance is met, by examining the E_b/N_0 actually received and comparing it with the E_b/N_0 required to meet the system specification. The principal items needed for this determination are the EIRP (how much effective power is transmitted), the $G/T°$ figure of merit (how much capability the receiver has for collecting this power), L_s (the largest single loss, the space loss), and L_o (other contributing losses and degradations). That is *all* there is to it!

5.6.1 Link Budget Details

The link budget example in Table 5.2 consists of three columns of numbers. Only the middle column represents the link budget. The other columns consist of ancillary information, such as antenna beamwidth, or computations to support the main tabulation. Losses are bracketed in the usual bookkeeping way. If a value is not bracketed, it represents a gain. Subtotals are shown enclosed in a box. Starting from the top of the middle column, we algebraically sum all of the gains and losses. The final link margin result is shown in a double box in item 21. The computations

TABLE 5.2 Earth Terminal to Satellite Link Budget Example: Frequency = 8 GHz, Range = 21,915 Nautical Miles.

1. Transmitter power (dBW)	(100.00W)	20.0	P_t
2. Transmitter circuit loss (dB)		⟨2.0⟩	L_o
3. Tramsmitter antenna gain (peak dBi)		51.6	G_t
Dish diameter (ft)	20.00		
Half-power beamwidth (degrees)	0.45		
4. Terminal EIRP (dBW)		69.6	EIRP
5. Path loss (dB)	(10° elev.)	⟨202.7⟩	L_s
6. Fade allowance (dB)		⟨4.0⟩	L_o
7. Other losses (dB)		⟨6.0⟩	L_o
8. Received isotropic power (dBW)		−143.1	
9. Receiver antenna gain (peak dBi)		35.1	G_r
Dish diameter (ft)	3.00		
Half-power beamwidth (degrees)	2.99		
10. Edge-of-coverage loss (dB)		⟨2.0⟩	L_o
11. Received signal power (dBW)		−110.0	P_r
Receiver noise figure at antenna port (dB)			11.5
Receiver temperature (dB-K)			35.8 (3806 K)
Receiver antenna temperature (dB-K)			24.8 (300 K)
12. System temperature (dB-K)			36.1 (4106 K)
13. System $G/T°$ (dB/K)	−1.0		$G/T°$
14. Boltzmann's constant (dBW/K-Hz)			−228.60
15. Noise spectral density (dBW/Hz)		⟨−192.5⟩	$N_0 = kT°$
16. Received P_r/N_0 (dB-Hz)		82.5	$(P_r/N_0)_r$
17. Data rate (dB-bit/s)	(2 Mbits/s)	⟨63.0⟩	R
18. Received E_b/N_0 (dB)		19.5	$(E_b/N_0)_r$
19. Implementation loss (dB)		⟨1.5⟩	L_o
20. Required E_b/N_0 (dB)		⟨10.0⟩	$(E_b/N_0)_{reqd}$
21. Margin (dB)		8.0	M

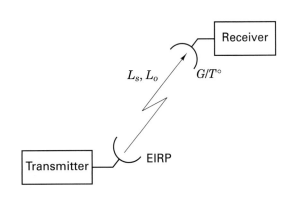

Figure 5.23 Key parameters of a link analysis.

are performed as in Equation (5.24), which is repeated here, with the exception that the terms G_r and $T°$ are grouped together at $G_r/T°$ instead of being listed separately:

$$M(\text{dB}) = \text{EIRP}(\text{dBW}) + \frac{G_r}{T°}(\text{dB/K}) - \left(\frac{E_b}{N_0}\right)_{\text{reqd}}(\text{dB}) - R(\text{dB-bits/s})$$

$$- \kappa(\text{dBW/K-Hz}) - L_s(\text{dB}) - L_o(\text{dB})$$

Let us examine the 21 line items listed in Table 5.2.

1. Transmitter power is 100 W (20 dBW).
2. Circuit loss between the transmitter and antenna is 2 dB.
3. Transmitting antenna gain is 51.6 dBi.
4. The net tally of items 1 to 3 yields the EIRP = 69.6 dBW.
5. The path loss has been calculated for the range shown in the table title, corresponding to a $10°$ elevation angle at the earth terminal.
6 **and 7.** Here are allowances made for weather fades and a variety of other, unspecified losses.
8. Received isotropic power refers to the power that would be received, −143.1 dBW, if the receiving antenna were isotropic.
9. The peak gain of the receiving antenna is 35.1 dBi.
10. Edge-of-coverage loss is due to the off-axis antenna gain (compared to peak gain) and to the increased range for users at the extreme edge of communication coverage (a nominal 2-dB loss is shown here.)
11. The input power to the receiver, tallied from items 8, 9, and 10, is −110 dBW.
12. System temperature is found using Equation (5.46). However, in this example we are assuming a lossless line from the receiver antenna to the front end, so that the line loss factor L is equal to 1, and system temperature is computed in column 3, as $T°_S = T_A + T_R$.
13. We form the receiver figure-of-merit ratio $G/T°$ by combining the gain of the receiver antenna G_r (see item 9) with system temperature T_S. This ratio is placed in the left column as a parameter of interest, rather than in the middle column. This is because G_r is accounted for in link budget item 9, and T_S is accounted for in item 15. If $G/T°$ were to be placed in the center column, it would represent a double tabulation.
14. Boltzmann's constant is −228.6 dBW/K-Hz.
15. Boltzmann's constant in decibels (item 14), plus system temperature in decibels (item 12), yields noise power spectral density.
16. Finally, we can form the received signal-to-noise spectral density, 82.5 dB-Hz, by subtracting noise spectral density in decibels (item 15), from received signal power in decibels (item 11).
17. The data rate is listed in dB-bit/s.

18. Since $E_b/N_0 = (1/R)(P_r/N_0)$, we need to subtract R in decibels (item 17), from P_r/N_0 in decibels (item 16), yielding $(E_b/N_0)_r = 19.5$ dB.

19. An implementation loss, here taken to be 1.5 dB, accounts for the difference between theoretically predicted detection performance, and the performance of the actual detector.

20. This is our required E_b/N_0, a result of the modulation and coding chosen, and the probability of error specified.

21. The difference between the received and the required E_b/N_0 in decibels (taking implementation loss into account), yields the final margin.

The gain or loss items shown in a link budget, generally follow the convention of first presenting an *ideal* or *simplistic* result, followed by a gain or loss term that modifies the simplistic yielding an *actual result.* In other words, the link budget typically follows a *modular* approach to partitioning the gains and losses in a way that easily adapts to the needs of any system. Consider the following examples of this format. In Table 5.2, item 1 gives the transmitter power that would be launched from a transmitter with an isotropic transmitting antenna (the simplistic). However, only after applying the modules of circuit loss and transmitter-antenna gain do we see, in item 4, the EIRP (actually) launched. Similarly, item 8 shows the power received by an isotropic antenna (the simplistic). However, only after applying the modules of receiver-antenna gain and edge-of-coverage loss do we see, in item 11, the (actually) received signal power.

5.6.2 Receiver Figure of Merit

An explanation of why receiving antenna gain and system temperature are often grouped together as $G/T°$, is as follows. In the early days of satellite communications development, the G_r and the $T_S°$ were specified separately. A contractor who agreed to meet these specifications would need to allow himself some safety margin for meeting each specification. Even though the user was generally only interested in the "bottom-line" performance, and not in the explicit value of G_r or $T_S°$, the contractor would not be able to exploit potential trade-offs. The net result was an overspecified (more costly) system than was necessary. Recognition of such overspecification resulted in specifying the antenna and receiver front end as a single figure-of-merit parameter $G/T°$ (sometimes called the *receiver sensitivity*), such that cost-effective trade-offs between the antenna design and the receiver design might be employed.

5.6.3 Received Isotropic Power

Another recognized area of overspecification in receiver design is in the separate specification of the required P_r/N_0 (or E_b/N_0) and receiver $G/T°$. If P_r/N_0 and $G/T°$ are specified separately, the system contractor is forced to meet each value. The contractor will plan for a margin in both places. As in the $G/T°$ case of the preceding section, there are advantages in specifying P_r/N_0 and $G/T°$ as one parameter;

this new parameter, called the *received isotropic power* (RIP), can be written as follows:

$$\text{RIP (dBW)} = \frac{P_r}{N_0} \text{(dB-Hz)} - \frac{G}{T^\circ} \text{(dB/K)} + \kappa \text{ (dBW/K-Hz)} \qquad (5.49)$$

Or, in terms of ratios, it also can be written as

$$\text{RIP} = \frac{P_r}{\kappa T^\circ} \left(\frac{\kappa T^\circ}{G_r} \right) = \frac{P_r}{G_r} \qquad (5.50)$$

It is important to note that P_r/N_0 refers to the predetection signal-to-noise spectral density ratio (SNR) *required* for a particular error probability when using a particular modulation scheme (it usually includes an allowance for *detector implementation losses*). Let us designate the theoretically required SNR to yield a particular P_B as $(P_r/N_0)_{\text{th-rq}}$. We can therefore write

$$\frac{P_r}{N_0} = L'_o \left(\frac{P_r}{N_0} \right)_{\text{th-rq}} \qquad (5.51)$$

where L'_o is called the implementation loss and accounts for the hardware and operational losses in the detection process. Combining Equations (5.50) and (5.51), we can write

$$\text{RIP} = L'_o \left(\frac{P_r}{\kappa T^\circ} \right)_{\text{th-rq}} \frac{\kappa T^\circ}{G_r} \qquad (5.52)$$

Specifying the RIP required to meet the system error performance allows the contractor to commit to meeting a single parameter value. The contractor is allowed to trade off P_r/N_0 versus G/T° and L'_o performance. As G/T° is improved, the detector performance can be degraded, and vice versa.

5.7 SATELLITE REPEATERS

Satellite repeaters retransmit the messages they receive (with a translation in carrier frequency). A *regenerative* (digital) repeater regenerates, that is, demodulates and reconstitutes the digital information embedded in the received waveforms before retransmission; however, a *nonregenerative* repeater only amplifies and retransmits. A nonregenerative repeater, therefore, can be used with many different modulation formats (simultaneously or sequentially without any switching), but a regenerative repeater is usually designed to operate with only one, or a very few, modulation formats. A link analysis for a regenerative satellite repeater treats the uplink and downlink as two separate point-to-point analyses. To calculate the overall bit error performance of a regenerative repeater link, it is necessary to determine separately the bit error probability on the uplink and downlink. Let P_u and P_d be the probability of a bit being in error on the uplink and downlink, respectively. A bit will be correct in the end-to-end link if either the bit is correct on both the up- and downlink, or if it is in error on both the up- and downlink. Therefore, the overall probability that a bit is correct is

$$P_c = (1 - P_u)(1 - P_d) + P_u P_d \qquad (5.53)$$

and the overall probability that a bit is in error is

$$P_B = 1 - P_c = P_u + P_d - 2P_u P_d \qquad (5.54)$$

For low values of P_u and P_d, the overall bit error performance is approximated simply by summing the individual uplink and downlink bit error probabilities:

$$P_B \approx P_u + P_d \qquad (5.55)$$

5.7.1 Nonregenerative Repeaters

Link analysis for a nonregenerative repeater treats the entire "round trip" (uplink transmission to the satellite and downlink retransmission to an earth terminal) as a single analysis. Features that are unique to nonregenerative repeaters, are the dependence of the overall SNR on the uplink SNR and the sharing of the repeater downlink power in proportion to the uplink power from each of the various uplink signals and noise. Henceforth, reference to a repeater or transponder will mean a *nonregenerative repeater,* and for simplicity, we will assume that the transponder is operating in its linear range.

A satellite transponder is limited in transmission capability by its downlink power, the earth terminal's uplink power, satellite and earth terminal noise, and channel bandwidth. One of these usually is a dominant performance constraint; most often the downlink power or the channel bandwidth proves to be the major system limitation. Figure 5.24 illustrates the important link parameters of a linear satellite repeater channel. The repeater transmits all uplink signals (or noise, in the absence of signal) without any processing beyond amplification and frequency translation. Let us assume that there are multiple simultaneous uplinks within the receiver's bandwidth W and that they are separated from one another through the use of a technique called *frequency-division multiple access* (FDMA). FDMA is a communications resource-sharing technique whereby different users occupy disjoint portions of the transponder bandwidth; FDMA is treated in Chapter 11. The satellite effective downlink power $EIRP_s$ is constant and since we are assuming a linear transponder, $EIRP_s$ is shared among the multiple uplink signals (and noise) in proportion to their respective input power levels.

The transmission starts from a ground station (bandwidth $\leq W$), say terminal *i,* with a terminal $EIRP_{ti} = P_{ti} G_{ti}$. Simultaneously, other signals are being transmitted to the satellite (from other terminals). The EIRP from the *k*th terminal will henceforth be referred to simply as P_k. At the satellite, a total signal power $P_T = \Sigma A_k P_k$ is received, where A_k reflects the uplink propagation loss and the satellite receive antenna gain for each terminal. $N_s W$ is the satellite uplink noise power, where N_s is the composite noise power spectral density due to noise radiated into the satellite antenna *and* generated in the satellite receiver. The total satellite downlink $EIRP_s = P_s G_{ts}$, where P_s is the satellite transponder output power and G_{ts}

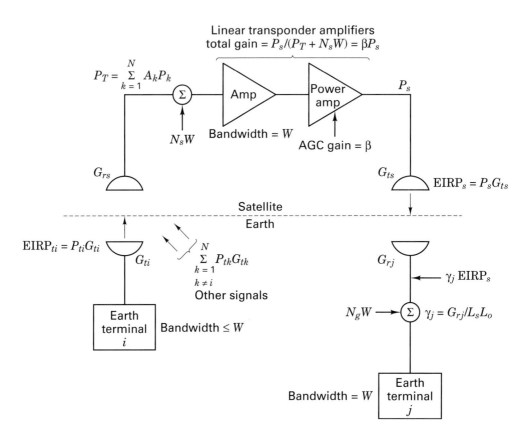

Figure 5.24 Nonregenerative satellite repeater.

is the satellite transmitting antenna gain, can be expressed by the following identity
[14]:

$$\text{EIRP}_s = \text{EIRP}_s \beta [A_i P_i + (P_T - A_i P_i) + N_s W] \tag{5.56}$$

Both the left and right sides of Equation (5.56) express the total satellite EIRP. On
the right side, the term $\beta[A_i P_i + (P_T - A_i P_i) + N_s W]$ constitutes the fractional apportionment of EIRP_s for the various users and uplink noise, such that the composite expression equals unity. The usefulness of this identity should become clear
shortly. The total power gain of the transponder can be expressed as βP_s. Since P_s
is fixed and the input signals can vary, $\beta = 1/(P_T + N_s W)$ represents an automatic
gain control (AGC) term. The total received uplink signal power, P_T, has purposely
been written as $A_i P_i + (P_T - A_i P_i)$ to separate signal i power from the remainder of
the simultaneous signals in the transponder. The total power received at the jth
earth terminal, with bandwidth W, can be written as

$$P_{rj} = \text{EIRP}_s \gamma_j \beta [A_i P_i + (P_T - A_i P_i) + N_s W] + N_g W \tag{5.57}$$

where $\gamma_j = G_{rj}/L_s L_o$ accounts for downlink losses and receiving antenna gain for the jth earth terminal. $\text{EIRP}_s\gamma_j$ represents the portion of EIRP_s that is received by the jth earth terminal, and N_g is the downlink noise power spectral density generated and introduced into that terminal receiver. Equation (5.57) describes the essence of downlink power apportionment among the various users and noise in a repeater. Let us rewrite Equation (5.57) by replacing β with its equivalent $1/(P_T + N_s W)$, as follows:

$$P_{rj} = \text{EIRP}_s\gamma_j \left(\frac{A_i P_i}{P_T + N_s W} + \frac{P_T - A_i P_i}{P_T + N_s W} + \frac{N_s W}{P_T + N_s W} \right) + N_g W \quad (5.58)$$

To facilitate our discussion, let us amplify Equation (5.58) with words, yielding

$$P_{rj} = \text{EIRP}_s\gamma_j \left(\frac{S_i \text{ U/L power}}{\text{total } (S + N) \text{ U/L power}} + \frac{\text{balance } S \text{ U/L power}}{\text{total } (S + N) \text{ U/L power}} \right.$$

$$\left. + \frac{\text{U/L noise power}}{\text{total } (S + N) \text{ U/L power}} \right) + N_g W$$

where S stands for signal power, N for noise power, and U/L for uplink.

From Equation (5.58), can you recognize an important relationship that must exist among multiple users sharing a nonregenerative transponder? The users must *cooperate with one another,* by not exceeding agreed-upon uplink transmission power levels. Equation (5.58) states that the portion of the downlink EIRP dedicated to any one user (or to uplink noise) is determined by the ratio of that user's uplink power to the total uplink signal plus noise power. Hence if one of the sharing users should choose to "cheat" by increasing his or her uplink power, the effect would be an enhancement of this user's downlink signal level, at the expense of the other users' downlink signal levels. Notice from Equation (5.58) that the uplink noise shares the downlink resource along with the other users. This coupling of uplink noise onto the downlink is a feature unique to nonregenerative repeaters.

From Equation (5.58) we can express the P_r/N for signal i received at the jth terminal as

$$\left(\frac{P_r}{N} \right)_{ij} \approx \frac{\text{EIRP}_s\gamma_j[A_i P_i/(P_T + N_s W)]}{\text{EIRP}_s\gamma_j[N_s W/(P_T + N_s W)] + N_g W} \quad (5.59)$$

and we can write the overall P_r/N_0 for signal i received at the jth terminal as [14]

$$\left(\frac{P_r}{N_0} \right)_{ij} = \frac{\text{EIRP}_s\gamma_j \beta A_i P_i}{\text{EIRP}_s\gamma_j \beta N_s + N_g} \quad (5.60)$$

Equations (5.58) to (5.60) illustrate that the uplink repeater noise degrades the overall SNR in two ways—it "steals" downlink EIRP, and it contributes to the total system noise. When the satellite uplink noise dominates—that is, when $P_T \ll N_s W$, the link is said to be *uplink limited,* and most of the downlink EIRP_s is wastefully allocated to uplink noise power. When this is the case and when $\text{EIRP}_s\gamma_j \gg N_g W$, we can rewrite Equation (5.60) as

$$\left(\frac{P_r}{N_0}\right)_{ij} \approx \frac{\text{EIRP}_s \gamma_j A_i P_i / N_s W}{(\text{EIRP}_s \gamma_j / W) + N_g} \approx \frac{A_i P_i}{N_s} \qquad (5.61)$$

Equation (5.61) illustrates that in the case of an uplink limited channel, the overall P_r/N_0 ratio essentially follows the uplink SNR. The more common situation is the *downlink limited* channel, in which case $P_T \gg N_s W$, and the satellite EIRP is limited. In this case, Equation (5.60) can be rewritten as

$$\left(\frac{P_r}{N_0}\right)_{ij} \approx \frac{\text{EIRP}_s \gamma_j A_i P_i / P_T}{N_g} \qquad (5.62)$$

The power of the transponder is then shared primarily among the various uplink transmitted signals; very little uplink noise is transmitted on the downlink. The performance of the repeater, in this case, is constrained only by its downlink parameters.

Table 5.3 illustrates a link analysis example (full round trip) for a nonregenerative repeater. The uplink portion by itself does not constitute a link budget since the transmission is not demodulated at the satellite. Without demodulation, *there are no bits* and therefore there is no way to measure the bit-error performance. After the full round trip, the signal is demodulated at the earth terminal; only then does the link analysis yield the margin. The example in Table 5.3 represents a case where the satellite transponder is servicing 10 simultaneous users. In the block marked "A" is shown the ratio $P_r/(P_T + N_s W)$, which dictates the apportionment of the downlink EIRP for the signal of interest. In this example, with all users transmitting the same power level, each of the signals is allocated 9.8% of the downlink EIRP. In the block marked "B" we see the apportionment of the downlink EIRP. The total is 1514.7 W; the user of interest gets 148.5 W; the other nine users get a total of 1336.1 W; and the uplink noise is apportioned 30.1 W.

An estimate of the performance described in Equation (5.60) can be obtained by using the uplink and downlink values of E_b/N_0 (or P_r/N_0), combined as follows, in the *absence of intermodulation noise* [15].

$$\left(\frac{E_b}{N_0}\right)_{ov}^{-1} = \left(\frac{E_b}{N_0}\right)_u^{-1} + \left(\frac{E_b}{N_0}\right)_d^{-1} \qquad (5.63)$$

where the subscripts ov, u, and d, indicate overall, uplink, and downlink values of E_b/N_0, respectively.

Most commercial satellite transponder designs are nonregenerative. However, it seems clear that future commercial systems will require on-board processing, switching, or selective message addressing, and will use regenerative repeaters to transform the received waveforms to message bits. Besides the potential for sophisticated data processing, one of the principal advantages of regenerative compared to nonregenerative repeaters is that the uplink is decoupled from the downlink so that the uplink noise is not retransmitted on the downlink. There are significant performance improvements possible with regenerative satellite repeaters in terms of the E_b/N_0 values needed on the uplinks and downlinks, relative to the values needed for the conventional nonregenerative designs in use today. Improvements of as much as 5 dB on the uplink and 6.8 dB on the downlink (using coherent QPSK modulation, with $P_B = 10^{-4}$) have been demonstrated [16].

TABLE 5.3 Link Budget Example For a Nonregenerative Satellite Repeater with 10 Users: Uplink Frequency = 375 MHz, Downlink Frequency = 275 MHz, Range = 22,000 Nautical Miles

		Uplink			Downlink	
Transmitter power (dBW)		27.0	(500.0 W)		13.0	(20.0 W)
Transmitter circuit losses (dB)		1.0			1.0	
Transmitter antenna gain (peak-dBi)		19.0			19.8	
Dish diameter (ft)	10.00			15.00		
Half-power beamwidth (degrees)	19.16			17.42		
EIRP (dBW)		45.0			31.8	(1514.7 W)
Path loss (dB)		176.1			173.4	
Transmitted signal power (dBW)					21.7	(148.5 W) **B**
Transmitted other signal power (dBW)					31.3	(1336.1 W)
Transmitted U/L noise power (dBW)					14.8	(30.1 W)
Other losses (dB)		2.0			2.0	
Received isotropic signal power (dBW)		−133.1			−153.7	
Received isotropic U/L noise power (dBW)					−160.6	
Receiver antenna gain (peak dBi)		22.5			16.3	
Dish diameter (ft)	15.00			10.00		
Half-power beamwidth (degrees)	12.77			26.13		
Received signal power (dBW)		−110.6			−137.4	
Received U/L noise power (dBW)					−144.3	
Receiver antenna temperature (dB-K)		24.6	(290 K)		20.0	(100 K)
Receiver noise figure at antenna port (dB)		10.8			2.0	
Receiver temperature (dB-K)		35.1	(3197 K)		22.3	(170 K)
System temperature (dB-K)		35.4	(3487 K)		24.3	(270 K)
System G/T° (dB/K)		−12.9			−8.0	
Boltzmann's constant (dBW/K-Hz)		−228.6			−228.6	
Noise spectral density (dBW/Hz)		−193.2			−204.3	
System bandwidth (dB-Hz)		75.6	(36.0 MHz)		75.6	(36.0 MHz)
Noise power (dBW)		−117.6			−128.7	
U/L noise + D/L noise power (dBW)					−128.6	
Simultaneous accesses	10					
Received other signal power (dBW)		−101.1				
Other signals + noise (dBW)		−101.0				
$P_r/(P_T + N_s W)$ (dB)		−10.1	(0.098) **A**			
P_r/N (dB)		7.0			−8.7	
Overall P_r/N (dB)					−8.8	
P_r/N_0 (dB-Hz)		82.6			66.9	
Overall P_r/N_0 (dB-Hz)					66.8	
Data rate (dB-bit/s)					50.0	(100,000 bits/s)
Available E_b/N_0 (dB)					16.8	
Required E_b/N_0 (dB)					10.0	
Margin (dB)					6.8	

5.7.2 Nonlinear Repeater Amplifiers

Power is severely limited in most satellite communication systems, and the ineffi-
ciencies associated with linear power amplification stages are expensive to bear.
For this reason, many satellite repeaters employ nonlinear power amplifiers. Efficient power amplification is obtained at the cost of signal distortion due to nonlinear operation. The major undesirable effects of the repeater nonlinearities are:

1. Intermodulation (IM) noise due to the interaction of different carriers. The
   harm is twofold; useful power can be lost from the channel as IM energy
   (typically 1 to 2 dB), and spurious IM products can be introduced into the
   channel as interference. The latter problem can be quite serious.
2. AM-to-AM conversion is a phenomenon common to nonlinear devices such
   as traveling wave tubes (TWT). At the device input, any signal-envelope fluctuations (amplitude modulation) undergo a nonlinear transformation and
   thus result in amplitude distortion at the device output. Hence, a TWT operating in its nonlinear region would not be the optimum power-amplifier
   choice for an amplitude-based modulation scheme (such as QAM).
3. AM-to-PM conversion is another phenomenon common to nonlinear devices.
   Fluctuations in the signal envelope produce phase variations that can affect
   the error performance for any phase-based modulation scheme (such as PSK
   or DPSK).
4. In hard limiters, weak signals can be suppressed, relative to stronger signals,
   by as much as 6 dB [2]. In saturated TWTs, the suppression of weak signals is
   due not only to limiting, but also to the fact that the signal coupling mechanism of the tube is optimized in favor of the stronger signals. The effect can
   cause weak signals to be suppressed by as much as 18 dB [17].

Conventional nonregenerative repeaters are generally operated *backed-off*
from their highly nonlinear saturated region; this is done to avoid appreciable IM
noise and thus to allow efficient utilization of the system's entire bandwidth. However, backing off to the linear region is a compromise; some level of IM noise must
be accepted to achieve a useful level of output power.

5.8 SYSTEM TRADE-OFFS

The link budget example in Table 5.3 is a resource allocation document. With such
a link tabulation, one can examine potential system trade-offs and attempt to optimize system performance. The link budget is a natural starting point for considering all sorts of potential trade-offs: margin versus noise figure, antenna size versus
transmitter power, and so on. Table 5.4 represents an example of a computer exercise for examining a possible trade-off between the earth station transmitting
power and the system noise margin at the receiving terminal. The first row in the
table is taken from the Table 5.3 link budget. Suppose a system engineer is concerned that a 500-W transmitter is not practical because of some physical con-

TABLE 5.4 Potential Trade-Off: P_t versus Margin

P_t (W)	$(P_r/N_0)_u$ (dB-Hz)	$(P_r/N_0)_d$ (dB-Hz)	$(P_r/N_0)_{ov}$ (dB-Hz)	Margin (dB)
500.0	82.6	66.9	66.8	6.8
250.0	79.6	66.8	66.6	6.6
125.0	76.6	66.6	66.2	6.2
62.5	73.6	66.3	65.5	5.5
31.3	70.5	65.7	64.5	4.5
15.6	67.5	64.8	62.9	2.9
7.8	64.5	63.3	60.8	0.8
3.9	61.5	61.4	58.4	−1.6
2.0	58.4	59.0	55.7	−4.3
1.0	55.4	56.4	52.9	−7.2
0.5	52.4	53.6	49.9	−10.1

straints within the transmitting earth terminal or that such a transmitter makes the system "uplink rich" (a poor design point). The engineer might then consider a trade-off of transmitter power versus thermal noise margin. The listing of candidate trade-offs is a trivial task for a computer. Table 5.4 was generated by repeating the link budget computation multiple times, and at each iteration, reducing P_t by one-half.

The result is a selection of transmitters (in steps of 3 dB) and uplink, downlink, and overall SNRs, and margin, associated with each transmitter value. The system engineer need only peruse the list to find a likely candidate. For example, if the engineer were satisfied with a margin of 3 to 4 dB, it appears he could reduce the transmitter from 500 W to 20 or 30 W. Or, he might be willing to provide a transmitter with, say, $P_t = 100$ W, since he may want to consider additional trade-offs (perhaps because of having misgivings about one of the other subsystems, say the antenna size). The engineer would then start a new tabulation with $P_t = 100$ W, and again perform a succession of link budget computations, to produce a similar enumeration of other possible trade-offs.

Notice from Table 5.4 that one can recognize the uplink-limited and downlink-limited regions, discussed earlier. In the first few rows, where the uplink SNR is high, a 3-dB degradation in uplink SNR results in only a few tenths of a decibel degradation to the overall SNR. Here the system is *downlink limited;* that is, the system is constrained primarily by its downlink parameters and is hardly affected by the uplink parameters. In the bottom few rows of the table, we see that a 3-dB degradation to the uplink affects the overall SNR by almost 3 dB. Here the system is *uplink limited;* that is, the system is constrained primarily by the uplink parameters.

5.9 CONCLUSION

Of the many analyses that support a developing communication system, the link budget stands out in its ability to provide overall system insight. By examining the link budget, one can learn many things about the overall system design and performance. For example, from the link margin, one learns whether the system will meet its requirements comfortably, marginally, or not at all. It will be evident if there are any hardware constraints, and whether such constraints can be compensated for in other parts of the link. The link budget is often used for considering system trade-offs and configuration changes, and in understanding subsystem nuances and inter-dependencies. Together with other modeling techniques, the link budget can help predict weight, size, and cost. We have considered how to formulate this budget and how it might be used for system trade-offs. The link budget is one of the system manager's most useful documents; it represents a "bottom-line" tally in the search for optimum error performance of the system.

REFERENCES

1. Panter, P. F., *Communication Systems Design: Line-of-Sight and Tropo-Scatter Systems,* R. E. Krieger Publishing Co, Inc., Melbourne, Fla., 1982.

2. Jones, J. J., "Hard Limiting of Two Signals in Random Noise," *IEEE Trans. Inf. Theory,* vol. IT9, January 1963.

3. Silver, S., *Microwave Antenna Theory and Design,* MIT Radiation Laboratory Series, Vol. 12, McGraw-Hill Book Company, New York, 1949.

4. Kraus, J. D., *Antennas,* McGraw-Hill Book Company, New York, 1950.

5. Johnson, J. B., "Thermal Agitation of Electricity in Conductors," *Phys. Rev.,* vol. 32, July 1928, pp. 97–109.

6. Nyquist, H., "Thermal Agitation of Electric Charge in Conductors," *Phys. Rev.,* vol. 32, July 1928, pp. 110–113.

7. Desoer, C. A., and Kuh, E. S., *Basic Circuit Theory,* McGraw-Hill Book Company, New York, 1969.

8. Schwab, L. M., "World-Wide Link Availability for Geostationary and Critically Inclined Orbits Including Rain Attenuation Effects," *Lincoln Laboratory, Rep. DCA-9,* Jan. 27, 1981.

9. Friis, H. T., "Noise Figure of Radio Receivers," *Proc. IRE,* July 1944, pp. 419–422.

10. IRE Subcommittee 7.9 on Noise, "Description of the Noise Performance of Amplifiers and Receiving Systems," *Proc. IEEE,* Mar. 1963, pp. 436–442.

11. Blackwell, L. A., and Kotzebue, K. L., *Semiconductor Diode Parametric Amplifiers,* Prentice-Hall, Inc., Englewood Cliffs, N.J., 1961.

12. Ko, H. C., and Kraus, J. D., "A Radio Map of the Sky at 1.2 Meters," *Sky Telesc.,* vol. 16, Feb. 1957, pp. 160–161.

13. Piddington, J. H., and Trent, G. H., "A Survey of Cosmic Radio Emission at 600 Mc/s," *Aust. J. Phys.,* vol. 9, Dec. 1956, pp. 481–493.

14. Spilker, J. J., *Digital Communications by Satellite,* Prentice-Hall, Inc., Englewood Cliffs, N.J., 1977.

15. Pritchard, W. L., and Sciulli, J. A., *Satellite Communication Systems Engineering,* Prentice-Hall, Inc., Englewood Cliffs, N.J., 1986.

16. Campanella, S. J., Assal, F., and Berman, A., "Onboard Regenerative Repeaters," *Int. Conf. Commun.,* Chicago, vol. 1, 1977, pp. 6.2-121–66.2-125.

17. Wolkstein, H. J., "Suppression and Limiting of Undesired Signals in Travelling-Wave-Tube Amplifiers," Publication ST-1583, *RCA Rev.,* vol. 22, no. 2, June 1961, pp. 280–291.

PROBLEMS

5.1. (a) What is the value in decibels of the free-space loss for a carrier frequency of 100 MHz and a range of 3 miles?

(b) The transmitter output power is 10 W. Assume that both the transmitting and receiving antennas are isotropic and that there are no other losses. Calculate the received power in dBW.

(c) If in part (b) the EIRP is equal to 20 W, calculate the received power in dBW.

(d) If the diameter of a dish antenna is doubled, calculate the antenna gain increase in decibels.

(e) For the system of part (a), what must the diameter of a dish antenna be in order for the antenna gain to be 10 dB? Assume an antenna efficiency of 0.55.

5.2. A transmitter has an output of 2 W at a carrier frequency of 2 GHz. Assume that the transmitting and receiving antennas are parabolic dishes each 3 ft in diameter. Assume that the efficiency of each antenna is 0.55.

(a) Evaluate the gain of each antenna.

(b) Calculate the EIRP of the transmitted signal in units of dBW.

(c) If the receiving antenna is located 25 miles from the transmitting antenna over a free-space path, find the available signal power out of the receiving antenna in units of dBW.

5.3. From Table 5.1 we see that the proposal from Satellite Television Corporation called for a direct broadcast satellite (DBS) EIRP of 57 dBW and a downlink transmission frequency of 12.5 GHz. Assume that the only loss is the downlink space loss shown. Suppose that the downlink information consists of a digital signal with a data rate of 5×10^7 bits/s. Assume that the required E_b/N_0 is 10 dB, the system temperature at your home receiver is 600 K, and that your rooftop dish has an efficiency of 0.55. What is the minimum dish diameter that you can use in order to close the link? Do you think the neighbors will object?

5.4. An amplifier has an input and output resistance of 50 Ω, a 60-dB gain, and a bandwidth of 10 kHz. When a 50-Ω resistor at 290 K is connected to the input, the output rms noise voltage is 100 μV. Determine the effective noise temperature of the amplifier.

5.5. An amplifier has a noise figure of 4 dB, a bandwidth of 500 kHz, and an input resistance of 50 Ω. Calculate the input signal voltage needed to yield an output SNR = 1 when the amplifier is connected to a signal source of 50 Ω at 290 K.

5.6. Consider a communication system with the following specifications: transmission frequency = 3 GHz, modulation format is BPSK, bit-error probability = 10^{-3}, data rate = 100 bits/s, link margin = 3 dB, EIRP = 100 W, receiver antenna gain = 10 dB, distance between transmitter and receiver = 40,000 km. Assume that the line loss between the receiving antenna and the receiver is negligible.

(a) Calculate the maximum permissible noise power spectral density in watts/hertz referenced to the receiver input.

(b) What is the maximum permissible effective noise temperature in kelvin for the receiver if the antenna temperature is 290 K?

(c) What is the maximum permissible noise figure in dB for the receiver?

5.7. A receiver preamplifier has a noise figure of 13 dB, a gain of 60 dB, and a bandwidth of 2 MHz. The antenna temperature is 490 K, and the input signal power is 10^{-12} W.

(a) Find the effective temperature, in kelvin, of the preamplifier.

(b) Find the system temperature in kelvin.

(c) Find the output SNR in decibels.

5.8. Assume that a receiver has the following parameters: gain = 50 dB, noise figure = 10 dB, bandwidth = 500 MHz, input signal power = 50×10^{-12} W, source temperature, $T_A^\circ = 10$ K, line loss = 0 dB. You are asked to insert a preamplifier between the antenna and the receiver. The preamplifier is to have a gain of 20 dB and a bandwidth of 500 MHz. Find the preamplifier noise figure that would be required to provide a 10-dB improvement in overall system SNR.

5.9. Find the maximum allowable effective system temperature, T_S°, required to *just close* a particular link with a bit error probability of 10^{-5} for a data rate of $R = 10$ kbits/s. The link parameters are as follows: transmission frequency = 12 GHz, EIRP = 10 dBW, receiver antenna gain = 0 dB, modulation type is noncoherently detected BFSK, other losses = 0 dB, and the distance between transmitter and receiver = 100 km.

5.10. Consider a receiver made up of the following three stages: The input stage is a preamplifier with a gain of 20 dB and a noise figure of 6 dB. The second stage is a 3-dB lossy network. The output stage is an amplifier with a gain of 60 dB and a noise figure of 16 dB.
  (a) Find the composite noise figure for the receiver.
  (b) Repeat part (a) with the preamplifier removed.

5.11. (a) Find the effective input noise temperature, T_R°, of a receiver comprised of three amplifier stages connected in series with power gains, from input to output, of 10, 16, and 20 dB, and effective noise temperatures, from input to output, of 1800, 2700, and 4800 K.
  (b) What would the gain of the first stage have to be to reduce the contribution to T_R° of all stages after the first to 10% of the first-stage contribution?

5.12. The effective temperature of a particular multiple-stage receiver is required to be 300 K. Assume that the effective temperatures and gains of stages 2 through 4 are as follows: $T_2^\circ = 600$ K, $T_3^\circ = T_4^\circ = 2000$ K, $G_2 = 13$ dB, and $G_3 = G_4 = 20$ dB.
  (a) Compute the required gain, G_1, for the first stage, under the conditions that $T_1^\circ = 200, 230, 265, 290, 295,$ and 300 K.
  (b) Plot the G_1 versus T_1° trade-off.
  (c) Regarding contributions to the effective temperature of the receiver, why is it reasonable in this case to ignore all stages beyond the fourth stage?
  (d) In a practical engineering trade-off between T_1° and G_1 what range of T_1° values do you think should be considered?

5.13. A receiver consists of a preamplifier followed by multiple amplifier stages. The composite effective temperature of all the amplifier stages is 1000 K, referenced to the preamplifier output.
  (a) Compute the receiver effective noise temperature, referenced to the preamplifier input, for a single-stage preamplifier with a noise temperature of 400 K and gains of 3, 6, 10, 16, and 20 dB.
  (b) Repeat part (a) for a two-stage preamplifier with 400-K noise per stage and gains of 3, 6, 10, and 13 dB per stage.
  (c) Plot the receiver effective temperature versus the gain of the first stage for parts (a) and (b).

5.14. (a) Equation (5.42) shows the composite noise figure for a network made up of a lossy line followed by an amplifier. Develop a general expression for the composite noise figure of three such networks connected in series.
  (b) Consider a network that is made up of an amplifier followed by a lossy line. Develop a general expression for the composite noise figure of three such networks connected in series.

(c) A receiver is made up of the following component parts in series: Receiving antenna with temperature $T_A = 1160$ K, lossy line 1 with $L_1 = 6$ dB, amplifier 1 with noise figure $F_1 = 3$ dB and gain $G_1 = 13$ dB, lossy line 2 with $L_2 = 10$ dB, and amplifier 2 with noise figure $F_2 = 6$ dB and gain $G_2 = 10$ dB. The input signal is 80 picowatts (pW) and the signal bandwidth is 0.25 GHz. Trace the values of signal power, noise power, and SNR throughout the system.

5.15. (a) An amplifier having a gain of 10 dB and a noise figure of 3 dB is connected to the output of a receiving antenna directly (no line loss between them). Following the amplifier is a lossy line with a loss factor of 10 dB. Consider that the input signal power is 10 pW, the antenna temperature is 290 K, and the signal bandwidth is 0.25 GHz. Find the SNR into and out of the amplifier, and out of the lossy line.

(b) Repeat part (a) with the antenna temperature equal to 1450 K.

5.16. A receiver with 80-dB gain and an effective noise temperature of 3000 K is connected to an antenna that has a noise temperature of 600 K.

(a) Find the noise power that is available from the source over a 40-MHz band.

(b) Find the receiver noise power referenced to the receiver input.

(c) Find the receiver output noise power over a 40-MHz band.

5.17. An antenna is pointed in a direction such that it has a noise temperature of 50 K. It is connected to a preamplifier that has a noise figure of 2 dB and an available gain of 30 db over an effective bandwidth of 20 MHz. The input signal to the preamplifier has a value of 10^{-12} W.

(a) Find the effective input noise temperature of the preamplifier.

(b) Find the SNR out of the preamplifier.

5.18. A receiver with a noise figure of 13 dB is connected to an antenna through 75 ft of 300-Ω transmission line that has a loss of 3 dB per 100 ft.

(a) Evaluate the composite noise figure of the line and the receiver.

(b) If a 20-dB preamplifier with a 3-dB noise figure is inserted between the line and the receiver, evaluate the composite noise figure of the line, the preamplifier, and the receiver.

(c) Evaluate the composite noise figure if the preamplifier is inserted between the antenna and the transmission line.

5.19. A satellite communication system uses a transmitter that produces 20 W of RF power at a carrier frequency of 8 GHz that is fed into a 2-ft parabolic antenna. The distance to the receiving earth station is 20,000 nautical miles. The receiving system uses an 8-ft parabolic antenna and has a 100-K system noise temperature. Assume that each antenna has an efficiency of 0.55. Also assume that the incidental losses amount to 2 dB.

(a) Calculate the maximum data rate that can be used if the modulation is differentially coherent PSK (DPSK) and the bit error probability is not to exceed 10^{-5}.

(b) Repeat part (a) assuming that the downlink transmission is at a carrier frequency of 2 GHz.

5.20. Consider that an unmanned spacecraft with a carrier frequency of 2 GHz and a 10-W transponder is in the vicinity of the planet Saturn (a distance of 7.9×10^8 miles from the earth). The receiving earth station has a 75-ft antenna and a system noise temperature of 20 K. Calculate the size of the spacecraft antenna that would be required to just close a 100-bits/s data link. Assume that the required E_b/N_0 is 10 dB and that there are incidental losses amounting to 3 dB. Also assume that each antenna has an efficiency of 0.55.

5.21. (a) Assume a receiver front end with the following parameters: gain = 60 dB, band-width = 500 MHz, noise figure = 6 dB, input signal power = 6.4×10^{-11} W, source temperature, $T_A^\circ = 290$ K, line loss = 0 dB. A preamplifier with the following characteristics is inserted between the antenna and the receiver: gain = 10 dB, noise figure = 1 dB. Find the composite receiver noise figure, in decibels. How much noise figure improvement, in decibels, has been realized?

(b) Find the output SNR improvement, in decibels, as a result of the improved noise figure.

(c) Repeat part (b) for $T_A^\circ = 6000$ K. What is the output SNR improvement in decibels?

(d) Repeat part (b) for $T_A^\circ = 15$ K. What is the output SNR improvement in decibels?

(e) What conclusions can you draw from your answers with regard to how the improvement in output SNR tracks the improvement in noise figure? Explain.

5.22. (a) Given the following link parameters, find the maximum allowable receiver noise figure. The modulation is coherent BPSK with a bit-error probability of 10^{-5} for a data rate of 10 Mbits/s. The transmission frequency is 12 GHz. The EIRP is 0 dBW. The receiving antenna diameter is 0.1 m (assume an efficiency of 0.55), and the antenna temperature is 800 K. The distance between the transmitter and receiver is 10 km. The margin is 0 dB and the incidental losses are assumed to be 0 dB.

(b) If the data rate is doubled, how will that affect the value of the noise figure in part (a)?

(c) If the antenna diameter is doubled, how will that affect the value of the noise figure in part (a)?

5.23. (a) Ten users simultaneously access a nonregenerative satellite repeater with a 50-MHz bandwidth using an FDMA access scheme. Assume that each user's EIRP is 10 dBW; also assume that each user's coefficient $A_i = G_{rs}/L_sL_o = -140$ dB. What is the total power P_T received by the satellite receiver?

(b) Assume that the satellite system noise temperature is 2000 K. What is the value of the satellite receiver noise power in watts, referenced to the receiver input?

(c) What is the uplink SNR at the satellite receiver for each user's signal?

(d) Assuming the received power at the satellite from each user is the same, what fraction of the satellite EIRP is allocated to each of the 10 users' signals? If the satellite downlink $\text{EIRP}_s = 1000$ W, how many watts per user is downlinked?

(e) How much of the satellite EIRP is allocated to the transmission of uplink thermal noise?

(f) Is the satellite uplink limited or downlink limited? Explain.

(g) At the earth station, the receiver noise temperature is 800 K. What is the resultant (overall) average signal-to-noise power spectral density (P_r/N_0) for a single user's transmission across a 50-MHz band? Assume that the coefficient $\gamma = G_r/L_sL_o = -140$ dB.

(h) Recalculate P_r/N_0 for a single user's transmission, using an approximation resulting from your answer to part (f).

(i) In the absence of intermodulation noise, the following repeater relationship is often used:

$$\text{overall} \left(\frac{P_r}{N_0} \right)^{-1} = \text{uplink} \left(\frac{P_r}{N_0} \right)^{-1} + \text{downlink} \left(\frac{P_r}{N_0} \right)^{-1}$$

Recalculate P_r/N_0 using this relationship, and compare the result with your answers to parts (g) and (h):

5.24. How many users can simultaneously access a nonregenerative satellite repeater with a 100-MHz bandwidth, such that each user is allocated 50 W of the satellite's EIRP of 5000 W? At the satellite, the effective system temperature $T_S^\circ = 3500$ K. Assume that each user's uplink EIRP is 10 dBW and that the $G_r/L_s L_o$ term that reduces this EIRP at the satellite receiver is equal to -140 dB for each user.

5.25. An AWGN channel has the following parameters and requirements: The data rate = 2.5 Mbits/s; the modulation is coherent BPSK with perfect frequency-, carrier-, and timing-synchronization, and the required bit-error probability is 10^{-5}; the carrier frequency is 300 MHz; the distance between transmitter and receiver is 100 km; the transmitter power is 10^{-3} Watt; the transmit and receive antennas each have a diameter of 2 m and an efficiency of 0.55; the receiver-antenna temperature is 290 K; the line, from the receiver-antenna output to the receiver input has a loss factor, of 1 dB; there are no other losses. Find the maximum receiver noise figure in dB that can be used to *just* close the link.

5.26. A wristwatch radio is to transmit and receive 1 Mbit/s data with a bit-error probability of 10^{-7}. It is to operate over a range of 10 km at a carrier frequency of 3 GHz. The modulation is DPSK, and the G/T° is -30 dB/K. Such a radio might be used in a moving vehicle and subject to a fading-signal loss. The radio designer wants to examine the trade-off between minimizing the required EIRP and maximizing the fading loss that can be sustained. Produce a table showing several EIRP versus fading-loss values to help in selecting the needed battery. Consider the EIRP values of interest to be in the range of 300 mW to 10 W. Is it possible to meet the system specifications with a fading loss of 20 dB and an EIRP under 10 W?

5.27. The designer decides that the wristwatch radio in Problem 5.26 does not have to meet the stated specifications while in a moving vehicle, and hence the fading loss can be established as 0 dB. Assume that the minimum allowable EIRP associated with this 0-dB loss is chosen for the transmitter (from the solution to Problem 5.26). What is the minimum value of transmitter power that can be used if the effective area of the transmitting antenna is 25 cm^2?

QUESTIONS

5.1. Why is *free-space loss* a function of wavelength? (See Section 5.3.3.)

5.2. What is the relationship between received signal-to-noise (*S/N*) ratio and carrier-to-noise (*C/N*) ratio? (See Section 5.4.)

5.3. How much *link margin* is enough? (See Section 5.4.3.)

5.4. There are two primary sources of noise and interference degradation at the input of a receiver. What are they? (See Section 5.5.5.)

5.5. In order to achieve equitable sharing of a *nonregenerative* satellite repeater, what important relationship must exist among multiple users? (See Section 5.7.1.)

EXERCISES

Using the Companion CD, run the exercises associated with Chapter 5.

CHAPTER 6

Channel Coding:
Part 1

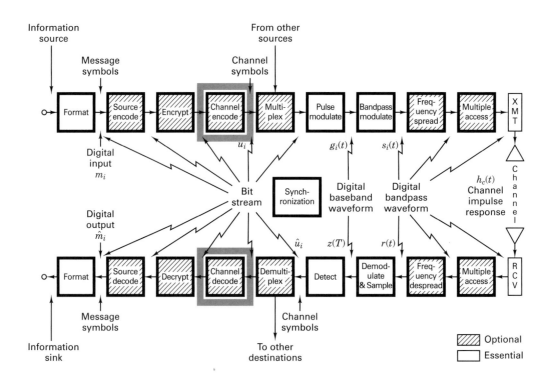

Channel coding refers to the class of signal transformations designed to improve communications performance by enabling the transmitted signals to better withstand the effects of various channel impairments, such as noise, interference, and fading. These signal-processing techniques can be thought of as vehicles for accomplishing desirable system trade-offs (e.g., error-performance versus bandwidth, power versus bandwidth). Why do you suppose channel coding has become such a popular way to bring about these beneficial effects? The use of large-scale integrated circuits (LSI) and high-speed digital signal processing (DSP) techniques have made it possible to provide as much as 10 dB performance improvement through these methods, at much less cost than through the use of most other methods such as higher power transmitters or larger antennas.

6.1 WAVEFORM CODING AND STRUCTURED SEQUENCES

Channel coding can be partitioned into two study areas, waveform (or signal design) coding and structured sequences (or structured redundancy), as shown in Figure 6.1. *Waveform coding* deals with transforming waveforms into "better waveforms," to make the detection process less subject to errors. *Structured sequences* deals with transforming data sequences into "better sequences," having structured redundancy (redundant bits). The redundant bits can then be used for the detection and correction of errors. The encoding procedure provides the coded signal (whether waveforms or structured sequences) with better distance properties

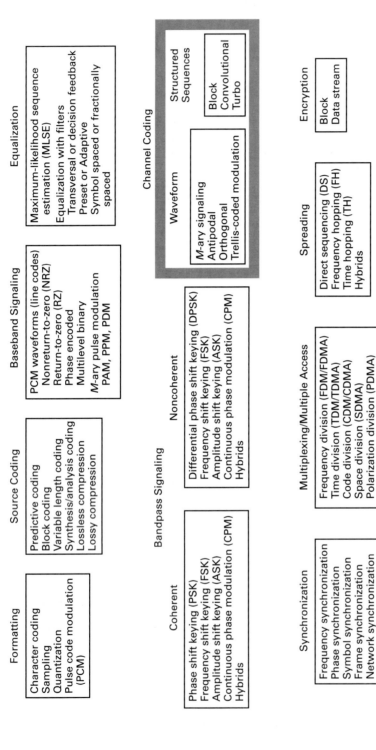

Figure 6.1 Basic digital communication transformations.

than those of their uncoded counterparts. First, we consider some waveform coding techniques. Then, starting with Section 6.3, we treat the subject of structured sequences.

6.1.1 Antipodal and Orthogonal Signals

Antipodal and orthogonal signals have been discussed earlier; we shall repeat the paramount features of these signal classes. The example shown in Figure 6.2 illustrates the analytical representation, $s_1(t) = -s_2(t) = \sin \omega_0 t$, $0 \le t \le T$, of a sinusoidal antipodal signal set, as well as its waveform representation and its vector representation. What are some synonyms or analogies that are used to describe *antipodal signals*? We can say that such signals are mirror images, or that one signal is the negative of the other, or that the signals are 180° apart.

The example shown in Figure 6.3 illustrates an orthogonal signal set made up of pulse waveforms, described by

$$s_1(t) = p(t) \qquad 0 \le t \le T$$

and

$$s_2(t) = p\left(t - \frac{T}{2}\right) \qquad 0 \le t \le T$$

where $p(t)$ is a pulse with duration $\tau = T/2$, and T is the symbol duration. Another orthogonal waveform set frequently used in communication systems is $\sin x$ and $\cos x$. In general, a set of equal-energy signals $s_i(t)$, where $i = 1, 2, \ldots, M$, constitutes an orthonormal (orthogonal, normalized to unity) set if and only if

$$z_{ij} = \frac{1}{E} \int_0^T s_i(t) s_j(t) \, dt = \begin{cases} 1 & \text{for } i = j \\ 0 & \text{otherwise} \end{cases} \qquad (6.1)$$

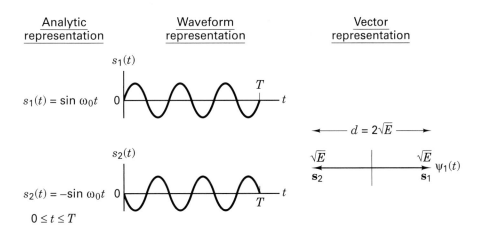

Figure 6.2 Example of an antipodal signal set.

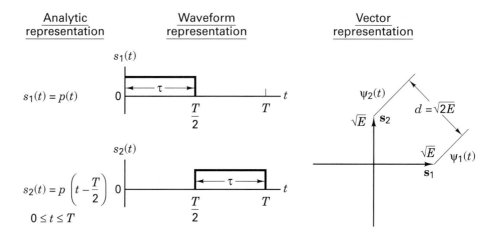

Analytic
representation

Waveform
representation

Vector
representation

$s_1(t) = p(t)$

$s_1(t)$

$s_2(t) = p\left(t - \dfrac{T}{2}\right)$

$s_2(t)$

$0 \le t \le T$

Figure 6.3 Example of a binary orthogonal signal set.

where z_{ij} is called the *cross-correlation coefficient*, and where E is the signal energy, expressed as

$$E = \int_0^T s_i^2(t)\, dt \qquad (6.2)$$

The waveform representation in Figure 6.3 illustrates that $s_1(t)$ and $s_2(t)$ cannot interfere with one another because they are disjoint in time. The vector representation illustrates the perpendicular relationship between orthogonal signals. Consider some alternative descriptions of orthogonal signals or vectors. We can say that the inner or dot product of two different vectors in the orthogonal set must equal zero. In a two- or three-dimensional Cartesian coordinate space, we can describe the signal vectors, geometrically, as being mutually perpendicular to one another. We can say that one vector has zero projection on the other, or that one signal cannot interfere with the other, since they do not share the same *signal space*.

6.1.2 *M*-ary Signaling

With *M*-ary signaling, the processor accepts k data bits at a time. It then instructs the modulator to produce one of $M = 2^k$ waveforms; binary signaling is the special case where $k = 1$. For $k > 1$, *M*-ary signaling alone can be regarded as a *waveform coding* procedure. For orthogonal signaling (e.g., MFSK), as k increases there is improved error performance or a reduction in required E_b/N_0, at the expense of bandwidth; nonorthogonal signaling (e.g., MPSK) manifests improved bandwidth efficiency, at the expense of degraded error performance or an increase in required E_b/N_0. By the appropriate choice of signal waveforms, one can trade off error probability, F_b/N_0, and bandwidth efficiency. Such trade-offs are treated in greater detail in Chapter 9.

6.1.3 Waveform Coding

Waveform coding procedures transform a waveform set (representing a message set) into an improved waveform set. The improved waveform set can then be used to provide improved P_B compared to the original set. The most popular of such *waveform codes* are referred to as *orthogonal* and *biorthogonal codes.* The encoding procedure endeavors to make each of the waveforms in the coded signal set as unalike as possible; the goal is to render the cross-correlation coefficient z_{ij}—among all pairs of signals, as described by the integral term in Equation 6.1—as small as possible. The smallest possible value of the cross-correlation coefficient occurs when the signals are anticorrelated ($z_{ij} = -1$); however, this can be achieved only when the number of symbols in the set is two ($M = 2$) and the symbols are *antipodal.* In general, it is possible to make all the cross-correlation coefficients equal to zero [1]. The set is then said to be *orthogonal.* Antipodal signal sets are optimum in the sense that each signal is most distant from the other signal in the set; this is seen in Figure 6.2 where the distance d between signal vectors is seen to be $d = 2\sqrt{E}$, where E represents the signal energy during a symbol duration T, as expressed in Equation (6.2). Compared with antipodal signals, the distance properties of orthogonal signal sets can be thought of as "pretty good" (for a given level of waveform energy). In Figure 6.3 the distance between the orthogonal signal vectors is seen to be $d = \sqrt{2E}$.

The *cross-correlation* between two signals is a measure of the *distance* between the signal vectors. The smaller the cross-correlation, the more distant are the vectors from each other. This can be verified in Figure 6.2, where the antipodal signals (whose $z_{ij} = -1$) are represented by vectors that are most distant from each other; and in Figure 6.3, where the orthogonal signals (whose $z_{ij} = 0$) are represented by vectors that are closer to one another than the antipodal vectors. It should be obvious that the distance between two identical waveforms (whose $z_{ij} = 1$) is zero.

The orthogonality condition in Equation (6.1) is presented in terms of waveforms $s_i(t)$ and $s_j(t)$, where $i, j = 1, \ldots, M,$ and M is the size of the waveform set. Each waveform in the set $\{s_i(t)\}$ may consist of a sequence of pulses, where each pulse is designated with a level +1 or −1, which in turn represents the binary digit 1 or 0, respectively. When the set is expressed in this way, Equation (6.1) can be simplified by stating that $\{s_i(t)\}$ constitutes an orthogonal set if and only if

$$z_{ij} = \frac{\text{number of digit agreements} - \text{number of digit disagreements}}{\text{total number of digits in the sequence}} \quad (6.3)$$

$$= \begin{cases} 1 & \text{for } i = j \\ 0 & \text{otherwise} \end{cases}$$

6.1.3.1 Orthogonal Codes

A one-bit data set can be transformed, using *orthogonal codewords* of two digits each, described by the rows of matrix $\mathbf{H}_1$ as follows:

<table>
<tr><td style="text-align:center">Data set</td><td style="text-align:center">Orthogonal codeword set</td><td></td></tr>
<tr><td style="text-align:center">0
1</td><td style="text-align:center">$\mathbf{H}_1 = \begin{bmatrix} 0 & 0 \\ 0 & 1 \end{bmatrix}$</td><td style="text-align:right">(6.4a)</td></tr>
</table>

For this, and the following examples, use Equation (6.3) to verify the orthogonality of the codeword set. To encode a 2-bit data set, we extend the foregoing set both horizontally and vertically, creating matrix $\mathbf{H}_2$.

<table>
<tr><td style="text-align:center">Data set</td><td style="text-align:center">Orthogonal codeword set</td><td></td></tr>
<tr><td style="text-align:center">0 0
0 1

1 0
1 1</td><td style="text-align:center">$\mathbf{H}_2 = \left[\begin{array}{cc:cc} 0 & 0 & 0 & 0 \\ 0 & 1 & 0 & 1 \\ \hdashline 0 & 0 & 1 & 1 \\ 0 & 1 & 1 & 0 \end{array}\right] = \begin{bmatrix} \mathbf{H}_1 & \mathbf{H}_1 \\ \mathbf{H}_1 & \overline{\mathbf{H}_1} \end{bmatrix}$</td><td style="text-align:right">(6.4b)</td></tr>
</table>

The lower right quadrant is the complement of the prior codeword set. We continue the same construction rule to obtain an orthogonal set $\mathbf{H}_3$ for a 3-bit data set.

<table>
<tr><td style="text-align:center">Data Set</td><td style="text-align:center">Orthogonal codeword set</td><td></td></tr>
<tr><td style="text-align:center">0 0 0
0 0 1
0 1 0
0 1 1

1 0 0
1 0 1
1 1 0
1 1 1</td><td style="text-align:center">$\mathbf{H}_3 = \left[\begin{array}{cccc:cccc} 0 & 0 & 0 & 0 & 0 & 0 & 0 & 0 \\ 0 & 1 & 0 & 1 & 0 & 1 & 0 & 1 \\ 0 & 0 & 1 & 1 & 0 & 0 & 1 & 1 \\ 0 & 1 & 1 & 0 & 0 & 1 & 1 & 0 \\ \hdashline 0 & 0 & 0 & 0 & 1 & 1 & 1 & 1 \\ 0 & 1 & 0 & 1 & 1 & 0 & 1 & 0 \\ 0 & 0 & 1 & 1 & 1 & 1 & 0 & 0 \\ 0 & 1 & 1 & 0 & 1 & 0 & 0 & 1 \end{array}\right] = \begin{bmatrix} \mathbf{H}_2 & \mathbf{H}_2 \\ \mathbf{H}_2 & \overline{\mathbf{H}_2} \end{bmatrix}$</td><td style="text-align:right">(6.4c)</td></tr>
</table>

In general, we can construct a codeword set $\mathbf{H}_k$, of dimension $2^k \times 2^k$, called a *Hadamard matrix,* for a k-bit data set from the $\mathbf{H}_{k-1}$ matrix, as follows:

$$\mathbf{H}_k = \begin{bmatrix} \mathbf{H}_{k-1} & \mathbf{H}_{k-1} \\ \mathbf{H}_{k-1} & \overline{\mathbf{H}_{k-1}} \end{bmatrix} \qquad (6.4d)$$

Each pair of words in each codeword set $\mathbf{H}_1, \mathbf{H}_2, \mathbf{H}_3, \ldots, \mathbf{H}_k, \ldots$ has as many digit agreements as disagreements [2]. Hence, in accordance with Equation (6.3), $z_{ij} = 0$ (for $i \neq j$), and each of the sets is orthogonal.

Just as *M*-ary signaling with an orthogonal modulation format (such as MFSK) improves the P_B performance, waveform coding with an orthogonally constructed signal set, in combination with coherent detection, produces *exactly the same* improvement. For equally likely, equal-energy orthogonal signals, the probability of codeword (symbol) error, P_E, can be upper bounded as [2]

$$P_E(M) \le (M-1) \, Q\left(\sqrt{\frac{E_s}{N_0}}\right) \tag{6.5}$$

where the codeword set M equals 2^k, and k is the number of data bits per codeword. The function $Q(x)$ is defined by Equation (3.43), and $E_s = kE_b$ is the energy per codeword. For a fixed *M*, as E_b/N_0 is increased, the bound becomes increasingly tight; for $P_E(M) \le 10^{-3}$, Equation (6.5) is a good approximation. For expressing the bit-error probability, we next use the relationship between P_B and P_E, given in Equation (4.112) and repeated here:

$$\frac{P_B(k)}{P_E(k)} = \frac{2^{k-1}}{2^k - 1} \quad \text{or} \quad \frac{P_B(M)}{P_E(M)} = \frac{M/2}{(M-1)} \tag{6.6}$$

Combining Equations (6.5) and (6.6), the probability of bit error can be bounded as follows:

$$P_B(k) \le (2^{k-1}) \, Q\left(\sqrt{\frac{kE_b}{N_0}}\right) \quad \text{or} \quad P_B(M) \le \frac{M}{2} \, Q\left(\sqrt{\frac{E_s}{N_0}}\right) \tag{6.7}$$

6.1.3.2 Biorthogonal Codes

A *biorthogonal* signal set of *M* total signals or codewords can be obtained from an orthogonal set of *M*/2 signals by augmenting it with the negative of each signal as follows:

$$\mathbf{B}_k = \left[\frac{\mathbf{H}_{k-1}}{\overline{\mathbf{H}}_{k-1}}\right]$$

For example, a 3-bit data set can be transformed into a biothogonal codeword set as follows:

Data set	Biorthogonal codeword set

$$
\begin{array}{ccc}
\text{Data set} & & \text{Biorthogonal codeword set} \\
\begin{array}{ccc}
0 & 0 & 0 \\
0 & 0 & 1 \\
0 & 1 & 0 \\
0 & 1 & 1 \\
\\
1 & 0 & 0 \\
1 & 0 & 1 \\
1 & 1 & 0 \\
1 & 1 & 1 \\
\end{array}
&
\mathbf{B}_3 =
&
\begin{bmatrix}
0 & 0 & 0 & 0 \\
0 & 1 & 0 & 1 \\
0 & 0 & 1 & 1 \\
0 & 1 & 1 & 0 \\
\hline
1 & 1 & 1 & 1 \\
1 & 0 & 1 & 0 \\
1 & 1 & 0 & 0 \\
1 & 0 & 0 & 1 \\
\end{bmatrix}
\end{array}
$$

The biorthogonal set is really two sets of orthogonal codes such that each codeword in one set has its antipodal codeword in the other set. The biorthogonal set consists of a *combination of orthogonal and antipodal* signals. With respect to z_{ij} of Equation (6.1), biorthogonal codes can be characterized as

$$z_{ij} = \begin{cases} 1 & \text{for } i = j \\ -1 & \text{for } i \neq j, \; |i - j| = \dfrac{M}{2} \\ 0 & \text{for } i \neq j, \; |i - j| \neq \dfrac{M}{2} \end{cases} \tag{6.8}$$

One advantage of a biorthogonal code over an orthogonal one for the same data set is that the biorthogonal code requires *one-half* as many code bits per codeword (compare the columns of the $\mathbf{B}_3$ matrix with those of the $\mathbf{H}_3$ matrix presented earlier). Thus the bandwidth requirements for biorthogonal codes are one-half the requirements for comparable orthogonal ones. Since antipodal signal vectors have better distance properties than orthogonal ones, it should come as no surprise that biorthogonal codes perform slightly better than orthogonal ones. For equally likely, equal-energy biorthogonal signals, the probability of codeword (symbol) error can be upper bounded, as follows [2]:

$$P_E(M) \leq (M - 2)Q\left(\sqrt{\frac{E_s}{N_0}}\right) + Q\left(\sqrt{\frac{2E_s}{N_0}}\right) \tag{6.9}$$

which becomes increasingly tight for fixed M as E_b/N_0 is increased. $P_B(M)$ is a complicated function of $P_E(M)$; we can approximate it with the relationship [2]

$$P_B(M) \approx \frac{P_E(M)}{2}$$

The approximation is quite good for $M > 8$. Therefore, we can write

$$P_B(M) \lesssim \frac{1}{2}\left[(M - 2)Q\left(\sqrt{\frac{E_s}{N_0}}\right) + Q\left(\sqrt{\frac{2E_s}{N_0}}\right)\right] \tag{6.10}$$

These biorthogonal codes offer improved P_B performance, compared with the performance of the orthogonal codes, and require only *half the bandwidth* of orthogonal codes.

6.1.3.3 Transorthogonal (Simplex) Codes

A code generated from an orthogonal set by deleting the first digit of each codeword is called a *transorthogonal* or *simplex code*. Such a code is characterized by

$$z_{ij} = \begin{cases} 1 & \text{for } i = j \\ \dfrac{-1}{M - 1} & \text{for } i \neq j \end{cases} \tag{6.11}$$

A simplex code represents the *minimum energy* equivalent (in the error-probability sense) of the equally likely orthogonal set. In comparing the error performance of orthogonal, biorthogonal, and simplex codes, we can state that simplex coding requires the minimum E_b/N_0 for a specified symbol error rate. However, for a *large value of M*, all three schemes are *essentially identical* in error performance. Biorthogonal coding requires half the bandwidth of the others. But for each of these codes, bandwidth requirements (and system complexity) grow exponentially with the value of M; therefore, such coding schemes are attractive only when large bandwidths are available.

6.1.4 Waveform-Coding System Example

Figure 6.4 illustrates an example of assigning a k-bit message from a message set of size $M = 2^k$, with a coded-pulse sequence from a code set of the same size. Each k-bit message chooses one of the generators yielding a coded-pulse sequence or codeword. The sequences in the coded set that replace the messages form a waveform set with good distance properties (e.g., orthogonal, biorthogonal). For the orthogonal code described in Section 6.1.3.1, each codeword consists of $M = 2^k$ pulses (representing code bits). Hence 2^k code bits replace k message bits. The chosen sequence then modulates a carrier wave using binary PSK, such that the phase ($\phi_j = 0$ or π) of the carrier during each code-bit duration, $0 \le t \le T_c$, corresponds to the amplitude ($j = -1$ or 1) of the jth bipolar pulse in the codeword. At the receiver in Figure 6.5, the signal is demodulated to baseband and fed to M correlators (or matched filters). For orthogonal codes, such as those characterized by the Hadamard matrix in Section 6.1.3.1, correlation is performed over a codeword duration that can be expressed as $T = 2^k T_c$. For a real-time communication system, messages may not be delayed; hence, the codeword duration must be the same as the message duration, and thus, T can also be expressed as $T = (\log_2 M) T_b = k T_b$,

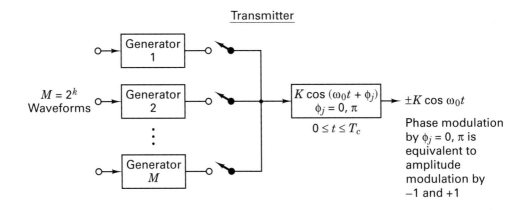

Figure 6.4 Waveform-encoded system (transmitter).

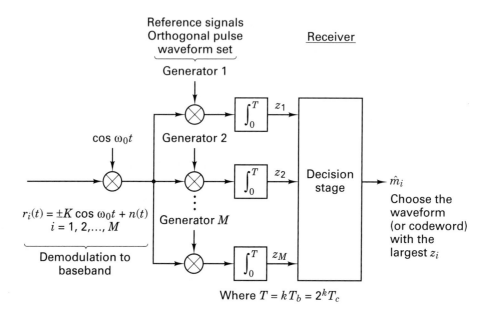

Reference signals
Orthogonal pulse
waveform set

Generator 1

$\cos \omega_0 t$

Generator 2

z_1

$\int_0^T$

z_2

Generator M

Decision
stage

$\hat{m}_i$

Choose the
waveform
(or codeword)
with the
largest z_i

$r_i(t) = \pm K \cos \omega_0 t + n(t)$
$i = 1, 2, ..., M$

Demodulation to
baseband

$\int_0^T$

z_M

$\int_0^T$

Where $T = kT_b = 2^k T_c$

Figure 6.5 Waveform-encoded system with coherent detection (receiver).

where T_b is the message-bit duration. Note that the time duration of a message bit is M/k times longer than that of a code bit. In other words, the code bits or coded pulses (which are PSK modulated) must move at a rate M/k faster than the message bits. For such orthogonally coded waveforms and an AWGN channel, the expected value at the output of each correlator, at time T, is zero, except for the correlator corresponding to the transmitted codeword.

What is the advantage of such orthogonal waveform coding compared with simply sending one bit or one pulse at a time? One can compare the bit-error performance with and without such coding by comparing Equation (4.79) for coherent detection of antipodal signals with Equation (6.7) for the coherent detection of orthogonal codewords. For a given size k-bit message (say, $k = 5$) and a desired bit-error probability (say, 10^{-5}), the detection of orthogonal codewords (each having a 5-bit meaning) can be accomplished with about 2.9 dB less E_b/N_0 than the bit-by-bit detection of antipodal signals. (The demonstration is left as an exercise for the reader in Problem 6.28.) One might have guessed this result by comparing the performance curves for orthogonal signaling in Figure 4.28 with the binary (antipodal) curve in Figure 4.29. What price do we pay for this error-performance improvement? The cost is more transmission bandwidth. In this example, transmission of an uncoded message consists of sending 5 bits. With coding, how many coded pulses must be transmitted for each message sequence? With the waveform coding of this example, each 5-bit message sequence is represented by $M = 2^k = 2^5 = 32$ code bits or coded pulses. The 32 coded pulses in a codeword must be sent in the same time duration as the corresponding 5 bits from which they stem. Thus, the required transmission bandwidth is 32/5 times that of the uncoded case. In general, the bandwidth needed for such orthogonally coded signals is M/k times greater than that needed for the uncoded case. Later, more efficient ways to trade off the benefits of coding versus bandwidth [3, 4] will be examined.

6.2 TYPES OF ERROR CONTROL

Before we discuss the details of structured redundancy, let us describe the two basic ways such redundancy is used for controlling errors. The first, *error detection and retransmission,* utilizes *parity bits* (redundant bits added to the data) to detect that an error has been made. The receiving terminal does not attempt to correct the error; it simply requests that the transmitter retransmit the data. Notice that a two-way link is required for such dialogue between the transmitter and receiver. The second type of error control, *forward error correction* (FEC), requires a one-way link only, since in this case the parity bits are designed for both the detection and correction of errors. We shall see that not all error patterns can be corrected; error-correcting codes are classified according to their error-correcting capabilities.

6.2.1 Terminal Connectivity

Communication terminals are often classified according to their connectivity with other terminals. The possible connections, shown in Figure 6.6, are termed *simplex* (not to be confused with the simplex or transorthogonal codes), *half-duplex,* and *full-duplex.* The simplex connection in Figure 6.6a is a one-way link. Transmissions

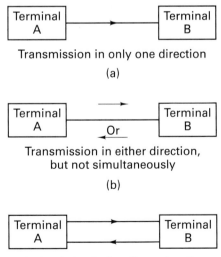

Figure 6.6 Terminal connectivity classifications. (a) Simplex. (b) Half-duplex. (c) Full-duplex.

are made from terminal A to terminal B only, never in the reverse direction. The half-duplex connection in Figure 6.6b is a link whereby transmissions may be made in either direction but not simultaneously. Finally, the full-duplex connection in Figure 6.6c is a two-way link, where transmissions may proceed in both directions simultaneously.

6.2.2 Automatic Repeat Request

When the error control consists of error detection only, the communication system generally needs to provide a means of alerting the transmitter that an error has been detected and that a retransmission is necessary. Such error control procedures are known as *automatic repeat request* or automatic retransmission query (ARQ) methods. Figure 6.7 illustrates three of the most popular ARQ procedures. In each of the diagrams, time is advancing from left to right. The first procedure, called *stop-and-wait ARQ*, is shown in Figure 6.7a. It requires a half-duplex connection only, since the transmitter waits for an acknowledgment (ACK) of each transmis-

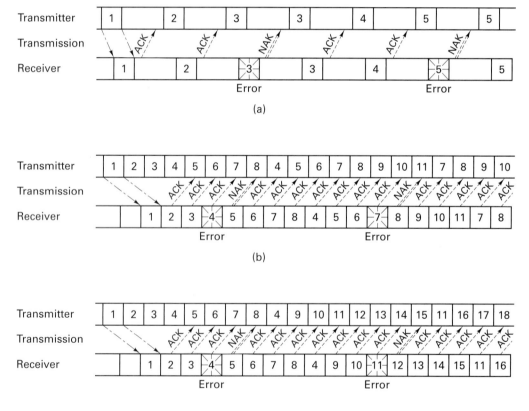

Figure 6.7 Automatic repeat request (ARQ). (a) Stop-and-wait ARQ (half-duplex). (b) Continuous ARQ with pullback (full-duplex). (c) Continuous ARQ with selective repeat (full-duplex).

sion before it proceeds with the next transmission. In the figure, the third transmission block is received in error; therefore, the receiver responds with a negative acknowledgment (NAK), and the transmitter retransmits this third message block before transmitting the next in the sequence. The second ARQ procedure, called *continuous ARQ with pullback,* is shown in Figure 6.7b. Here a full-duplex connection is necessary. Both terminals are transmitting simultaneously; the transmitter is sending message data and the receiver is sending acknowledgment data. Notice that a sequence number has to be assigned to each block of data. Also, the ACKs and NAKs need to reference such numbers, or else there needs to be *a priori* knowledge of the propagation delays, so that the transmitter knows which messages are associated with which acknowledgments. In the example of Figure 6.7b, there is a fixed separation of four blocks between the message being transmitted and the acknowledgment being simultaneously received. For example, when message 8 is being sent, a NAK corresponding to the corrupted message 4 is being received. In the ARQ procedure, the transmitter "pulls back" to the message in error and retransmits all message data, starting with the corrupted message. The final method, called *continuous ARQ with selective repeat,* is shown in Figure 6.7c. Here, as with the second ARQ procedure, a full-duplex connection is needed. In this procedure, however, only the corrupted message is repeated; then, the transmitter continues the transmission sequence where it had left off instead of repeating any subsequent correctly received messages.

The choice of which ARQ procedure to choose is a trade-off between the requirements for efficient utilization of the communications resource and the need to provide full-duplex connectivity. The half-duplex connectivity required in Figure 6.7a is less costly than full-duplex; the associated inefficiency can be measured by the blank time slots. The more efficient utilization illustrated in Figures 6.7b and c requires the more costly full-duplex connectivity.

The major advantage of ARQ over forward error correction (FEC) is that error detection requires much simpler decoding equipment and much less redundancy than does error correction. Also, ARQ is adaptive in the sense that information is retransmitted only when errors occur. On the other hand, FEC may be desirable in place of, or in addition to, error detection, for any of the following reasons:

1. A reverse channel is not available or the delay with ARQ would be excessive.
2. The retransmission strategy is not conveniently implemented.
3. The expected number of errors, without corrections, would require excessive retransmissions.

6.3 STRUCTURED SEQUENCES

In Section 4.8 we considered digital signaling by means of $M = 2^k$ signal waveforms (*M*-ary signaling), where each waveform contains k bits of information. We saw that in the case of orthogonal *M*-ary signaling, we can decrease P_B by increasing M

(expanding the bandwidth). Similarly, in Section 6.1 we showed that it is possible to decrease P_B by encoding k binary digits into one of M orthogonal codewords. The major disadvantage with such orthogonal coding techniques is the associated inefficient use of bandwidth. For an orthogonally coded set of $M = 2^k$ waveforms, the required transmission bandwidth is M/k times that needed for the uncoded case. In this and subsequent sections we abandon the need for antipodal or orthogonal properties and focus on a class of encoding procedures known as *parity-check codes*. Such channel coding procedures are classified as *structured sequences* because they represent methods of inserting structured redundancy into the source data so that the presence of errors can be detected or the errors corrected. Structured sequences are partitioned into three subcategories, as shown in Figure 6.1: *block, convolutional,* and *turbo*. Block coding (primarily) is treated in this chapter, and the others are treated in Chapters 7 and 8 respectively.

6.3.1 Channel Models

6.3.1.1 Discrete Memoryless Channel

A *discrete memoryless channel* (DMC) is characterized by a discrete input alphabet, a discrete output alphabet, and a set of conditional probabilities $P(j|i)$ ($1 \leq i \leq M$, $1 \leq j \leq Q$), where i represents a modulator M-ary input symbol, j represents a demodulator Q-ary output symbol, and $P(j|i)$ is the probability of receiving j given that i was transmitted. Each output symbol of the channel depends only on the corresponding input, so that for a given input sequence $\mathbf{U} = u_1, u_2, \ldots, u_m, \ldots, u_N$, the conditional probability of a corresponding output sequence $\mathbf{Z} = z_1, z_2, \ldots, z_m, \ldots, z_N$ may be expressed as

$$P(\mathbf{Z}|\mathbf{U}) = \prod_{m=1}^{N} P(z_m|u_m) \tag{6.12}$$

In the event that the channel *has memory* (i.e., noise or fading that occurs in bursts), the conditional probability of the sequence $\mathbf{Z}$ would need to be expressed as the *joint* probability of all the elements of the sequence. Equation (6.12) expresses the *memoryless* condition of the channel. Since the channel noise in a memoryless channel is defined to affect each symbol independently of all the other symbols, the conditional probability of $\mathbf{Z}$ is seen as the product of the independent element probabilities.

6.3.1.2 Binary Symmetric Channel

A *binary symmetric channel* (BSC) is a special case of a DMC; the input and output alphabet sets consist of the binary elements (0 and 1). The conditional probabilities are symmetric:

$$P(0|1) = P(1|0) = p$$

and

$$P(1|1) = P(0|0) = 1 - p \tag{6.13}$$

Equation (6.13) expresses the channel *transition probabilities*. That is, given that a channel symbol was transmitted, the probability that it is received in error is p (related to the symbol energy), and the probability that it is received correctly is $(1 - p)$. Since the demodulator output consists of the discrete elements 0 and 1, the demodulator is said to make a firm or *hard decision* on each symbol. A commonly used code system consists of BPSK modulated coded data and hard decision demodulation. Then the channel symbol error probability is found using the methods discussed in Section 4.7.1 and Equation (4.79) to be

$$p = Q \left(\sqrt{\frac{2E_c}{N_0}} \right)$$

where E_c/N_0 is the channel symbol energy per noise density, and $Q(x)$ is defined in Equation (3.43).

When such hard decisions are used in a binary coded system, the demodulator feeds the two-valued *code symbols* or *channel bits* to the decoder. Since the decoder then operates on the hard decisions made by the demodulator, decoding with a BSC channel is called *hard-decision decoding*.

6.3.1.3 Gaussian Channel

We can generalize our definition of the DMC to channels with alphabets that are not discrete. An example is the *Gaussian channel* with a discrete input alphabet and a continuous output alphabet over the range $(-\infty, \infty)$. The channel adds noise to the symbols. Since the noise is a Gaussian random variable with zero mean and variance σ^2, the resulting probability density function (pdf) of the received random variable z, conditioned on the symbol u_k (the likelihood of u_k), can be written as

$$p(z \,|\, u_k) = \frac{1}{\sigma \sqrt{2\pi}} \exp \left[\frac{-(z - u_k)^2}{2\sigma^2} \right] \tag{6.14}$$

for all z, where $k = 1, 2, \ldots, M$. For this case, *memoryless* has the same meaning as it does in Section 6.3.1.1, and Equation (6.12) can be used to obtain the conditional probability for the sequence $\mathbf{Z}$.

When the demodulator output consists of a continuous alphabet or its quantized approximation (with greater than two quantization levels), the demodulator is said to make *soft decisions*. In the case of a coded system, the demodulator feeds such quantized code symbols to the decoder. Since the decoder then operates on the soft decisions made by the demodulator, decoding with a Gaussian channel is called *soft-decision decoding*.

In the case of a hard-decision channel, we are able to characterize the detection process with a channel symbol error probability. However, in the case of a soft-decision channel, the detector makes the kind of decisions (soft decisions) that cannot be labeled as correct or incorrect. Thus, since there are no firm decisions, there cannot be a probability of making an error; the detector can only formulate a family of conditional probabilities or likelihoods of the different symbol types.

It is possible to design decoders using soft decisions, but block code soft-decision decoders are substantially more complex than hard-decision decoders; therefore, block codes are usually implemented with hard-decision decoders. For convolutional codes, both hard- and soft-decision implementations are equally popular. In this chapter we consider that the channel is a binary symmetric channel (BSC), and hence the decoder employs hard decisions. In Chapter 7 we further discuss channel models, as well as hard- versus soft-decision decoding for convolutional codes.

6.3.2 Code Rate and Redundancy

In the case of block codes, the source data are segmented into blocks of k data bits, also called information bits or message bits; each block can represent any one of 2^k distinct messages. The encoder transforms each k-bit data block into a larger block of n bits, called code bits or channel symbols. The $(n - k)$ bits, which the encoder adds to each data block, are called *redundant bits, parity bits,* or *check bits;* they carry no new information. The code is referred to as an (n, k) code. The ratio of redundant bits to data bits, denoted $(n - k)/k$, within a block is called the *redundancy* of the code; the ratio of data bits to total bits, k/n, is called the *code rate.* The code rate can be thought of as the portion of a code bit that constitutes information. For example, in a rate $\frac{1}{2}$ code, each code bit carries $\frac{1}{2}$ bit of information.

In this chapter and in Chapters 7 and 8 we consider those coding techniques that provide redundancy by increasing the required transmission bandwidth. For example, an error control technique that employs a rate 1/2 code (100% redundancy) will require double the bandwidth of an uncoded system. However, if a rate 3/4 code is used, the redundancy is 33% and the bandwidth expansion is only 4/3. In Chapter 9 we consider modulation/coding techniques for bandlimited channels where complexity instead of bandwidth is traded for error performance improvement.

6.3.2.1 Code-Element Nomenclature

Different authors describe an encoder's output elements in a variety of ways: code bits, channel bits, code symbols, channel symbols, parity bits, parity symbols. The terms are all very similar. In this text, for a binary code, the terms "code bit," "channel bit," "code symbol," and "channel symbol" have exactly the same meaning. The terms "code bit" and "channel bit" are most descriptive for binary codes only. The more generic names "code symbol" and "channel symbol" are often preferred because they can be used to describe binary or nonbinary codes equally well. Note that such code symbols or channel symbols are not to be confused with the grouping of bits to form transmission symbols that was done in previous chapters. The terms "parity bit" and "parity symbol" are used to identify only those code elements that represent the redundancy components added to the original data.

6.3.3 Parity-Check Codes

6.3.3.1 Single-Parity-Check Code

Parity-check codes use linear sums of the information bits, called *parity symbols* or *parity bits,* for error detection or correction. A single-parity-check code is constructed by adding a single-parity bit to a block of data bits. The parity bit takes on the value of one or zero as needed to ensure that the summation of all the bits in the codeword yields an even (or odd) result. The summation operation is performed using modulo-2 arithmetic (exclusive-or logic), as described in Section 2.9.3. If the added parity is designed to yield an even result, the method is termed *even parity;* if it is designed to yield an odd result, the method is termed *odd parity.* Figure 6.8a illustrates a serial data transmission (the rightmost bit is the earliest bit). A single-parity bit is added (the leftmost bit in each block) to yield even parity.

At the receiving terminal, the decoding procedure consists of testing that the modulo-2 sum of the codeword bits yields a zero result (even parity). If the result is found to be one instead of zero, the codeword is known to contain errors. The rate of the code can be expressed as $k/(k+1)$. Do you suppose the decoder can automatically *correct* a digit that is received in error? No, it cannot. It can only *detect* the presence of an odd number of bit errors. (If an even number of bits are inverted, the parity test will appear correct, which represents the case of an *undetected error.*) Assuming that

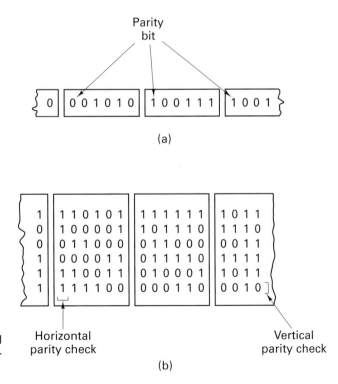

Figure 6.8 Parity checks for serial and parallel code structures. (a) Serial structure. (b) Parallel structure.

all bit errors are equally likely and occur independently, we can write the probability of j errors occurring in a block of n symbols as

$$P(j, n) = \binom{n}{j} p^j (1 - p)^{n-j} \tag{6.15}$$

where p is the probability that a *channel symbol* is received in error, and where

$$\binom{n}{j} = \frac{n!}{j!(n - j)!} \tag{6.16}$$

is the number of various ways in which j bits out of n may be in error. Thus, for a single-parity error-detection code, the probability of an undetected error P_{nd} with a block of n bits is computed as follows:

$$P_{nd} = \sum_{j=1}^{\substack{n/2 \text{ (for } n \text{ even)} \\ (n-1)/2 \text{ (for } n \text{ odd)}}} \binom{n}{2j} p^{2j} (1 - p)^{n-2j} \tag{6.17}$$

Example 6.1   Even-Parity Code

Configure a (4, 3) even-parity error-detection code such that the parity symbol appears as the leftmost symbol of the codeword. Which error patterns can the code detect? Compute the probability of an undetected message error, assuming that all symbol errors are independent events and that the probability of a channel symbol error is $p = 10^{-3}$.

Solution

Message	Parity	Codeword	
000	0	0	000
100	1	1	100
010	1	1	010
110	0	0	110
001	1	1	001
101	0	0	101
011	0	0	011
111	1	1	111

parity   message

The code is capable of detecting all single- and triple-error patterns. The probability of an undetected error is equal to the probability that two or four errors occur anywhere in a codeword.

$$P_{nd} = \binom{4}{2} p^2 (1 - p)^2 + \binom{4}{4} p^4$$

$$= 6p^2 (1 - p)^2 + p^4$$

$$= 6p^2 - 12p^3 + 7p^4$$

$$= 6(10^{-3})^2 - 12(10^{-3})^3 + 7(10^{-3})^4 \approx 6 \times 10^{-6}$$

6.3.3.2 Rectangular Code

A *rectangular code,* also called a *product code,* can be thought of as a parallel code structure, depicted in Figure 6.8b. First we form a rectangle of message bits comprising M rows and N columns; then, a horizontal parity check is appended to each row and a vertical parity check is appended to each column, resulting in an augmented array of dimensions $(M + 1) \times (N + 1)$. The rate of the rectangular code k/n can then be written as

$$\frac{k}{n} = \frac{MN}{(M + 1)(N + 1)}$$

How much more powerful is the rectangular code than the single-parity code, which is only capable of error detection? Notice that any single bit error will cause a parity check failure in one of the array columns *and* in one of the array rows. Therefore, the rectangular code can correct any single error pattern since such an error is uniquely located at the intersection of the error-detecting row and the error-detecting column. For the example shown in Figure 6.8b, the array dimensions are $M = N = 5$; therefore, the figure depicts a (36, 25) code that can correct a single error located anywhere in the 36-bit positions. For such an error-correcting block code, we compute the probability that the decoded block has an uncorrected error by accounting for all the ways in which a *message error* can be made. Starting with the probability of j errors in a block of n symbols, expressed in Equation (6.15), we can write the probability of a message error, also called a *block error* or *word error,* for a code that can correct all t and fewer error patterns as

$$P_M = \sum_{j=t+1}^{n} \binom{n}{j} p^j (1 - p)^{n-j} \tag{6.18}$$

where p is the probability that a *channel symbol* is received in error. For the example in Figure 6.8b, the code can correct all single error patterns ($t = 1$) within the rectangular block of $n = 36$ bits. Hence, the summation in Equation (6.18) starts with $j = 2$:

$$P_M = \sum_{j=2}^{36} \binom{36}{j} p^j (1 - p)^{36-j}$$

When p is reasonably small, the first term in the summation is the dominant one; Therefore, for this (36, 25) rectangular code example, we can write

$$P_M \approx \binom{36}{2} p^2 (1 - p)^{34}$$

The *bit error probability* P_B depends on the particular code and decoder. An approximation for P_B is given in Section 6.5.3.

6.3.4 Why Use Error-Correction Coding?

Error-correction coding can be regarded as a vehicle for effecting various system trade-offs. Figure 6.9 compares two curves depicting bit-error performance versus

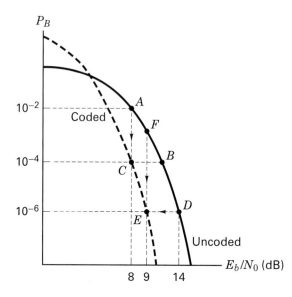

P_B

10^{-2}

Coded

A

F

10^{-4}

C

B

10^{-6}

E

D

Uncoded

E_b/N_0 (dB)

8 9 14

Figure 6.9 Comparison of typical coded versus uncoded error performance.

E_b/N_0. One curve represents a typical modulation scheme without coding. The second curve represents the same modulation with coding. Demonstrated below are four benefits or trade-offs that can be achieved with the use of channel coding.

6.3.4.1 Trade-Off 1: Error Performance versus Bandwidth

Imagine that a simple, inexpensive voice communication system has just been developed and delivered to a customer. The system does not use error-correction coding. Consider that the operating point of the system can be depicted by point A in Figure 6.9 ($E_b/N_0 = 8$ dB, and $P_B = 10^{-2}$). After a few trials, there are complaints about the voice quality; the customer suggests that the bit-error probability should be lowered to 10^{-4}. The usual way of obtaining better error performance in such a system would be by effecting an operating point movement from point A to, say, point B in Figure 6.9. However, suppose that the E_b/N_0 of 8 dB is the most that is available in this system. The figure suggests that one possible trade-off is to move the operating point from point A to point C. That is, "walking" down the vertical line to point C on the coded curve can provide the customer with improved error performance. What does it cost? Aside from the new components (encoder and decoder) needed, the price is more transmission bandwidth. Error-correction coding needs redundancy. If we assume that the system is a real-time communication system (such that the message may not be delayed), the addition of redundant bits dictates a faster rate of transmission, which of course means more bandwidth.

6.3.4.2 Trade-Off 2: Power versus Bandwidth

Consider that a system without coding, operating at point D in Figure 6.9 ($E_b/N_0 = 14$ dB, and $P_B = 10^{-6}$), has been delivered to a customer. The customer has no complaints about the quality of the data, but the equipment is having some reliability problems as a result of providing an E_b/N_0 of 14 dB. In other words, the equipment keeps breaking down. If the requirement on E_b/N_0 or power could be reduced, the reliability difficulties might also be reduced. Figure 6.9 suggests a trade-off by moving the operating point from point D to point E. That is, if error-correction coding is introduced, a reduction in the required E_b/N_0 can be achieved. Thus, the trade-off is one in which the same quality of data is achieved, but the coding allows for a reduction in power or E_b/N_0. What is the cost? The same as before—more bandwidth.

Notice that for *non-real-time* communication systems, error-correction coding can be used with a somewhat different trade-off. It is possible to obtain improved bit-error probability or reduced power (similar to trade-off 1 or 2 above) by paying the price of *delay* instead of bandwidth.

6.3.4.3 Coding Gain

The trade-off example described in the previous section has allowed a reduction in E_b/N_0 from 14 dB to 9 dB, while maintaining the same error performance. In the context of this example and Figure 6.9, we now define *coding gain*. For a *given bit-error probability,* coding gain is defined as the "relief" or reduction in E_b/N_0 that can be realized through the use of the code. Coding gain G is generally expressed in dB, such as

$$G(\text{dB}) = \left(\frac{E_b}{N_0}\right)_u (\text{dB}) - \left(\frac{E_b}{N_0}\right)_c (\text{dB}) \qquad (6.19)$$

where $(E_b/N_0)_u$ and $(E_b/N_0)_c$ represent the required E_b/N_0, uncoded and coded, respectively.

6.3.4.4 Trade-Off 3: Data Rate versus Bandwidth

Consider that a system without coding, operating at point D in Figure 6.9 ($E_b/N_0 = 14$ dB, and $P_B = 10^{-6}$) has been developed. Assume that there is no problem with the data quality and no particular need to reduce power. However, in this example, suppose that the customer's data rate requirement increases. Recall the relationship in Equation (5.20b):

$$\frac{E_b}{N_0} = \frac{P_r}{N_0} \left(\frac{1}{R}\right)$$

If we do nothing to the system except increase the data rate R, the above expression shows that the received E_b/N_0 would decrease, and in Figure 6.9, the operating point would move upwards from point D to, let us say, some point F. Now, envision "walking" down the vertical line to point E on the curve that represents coded modulation. Increasing the data rate has degraded the quality of the data. But, the use of error-correction coding brings back the same quality at the same power level (P_r/N_0). The E_b/N_0 is reduced, but the code facilitates getting the same error probability with a lower E_b/N_0. What price do we pay for getting this higher data rate or greater capacity? The same as before—increased bandwidth.

6.3.4.5 Trade-Off 4: Capacity versus Bandwidth

Trade-off 4 is similar to trade-off 3 because both achieve increased capacity. A spread-spectrum multiple access technique, called code-division multiple access (CDMA) and described in Chapter 12, is one of the schemes used in cellular telephony. In CDMA, where users simultaneously share the same spectrum, each user is an interferer to each of the other users in the same cell or nearby cells. Hence, the capacity (maximum number of users) per cell is inversely proportional to E_b/N_0. (See Section 12.8.) In this application, a lowered E_b/N_0 results in a raised capacity; the code achieves a reduction in each user's power, which in turn allows for an increase in the number of users. Again, the cost is more bandwidth. But, in this case, the signal-bandwidth expansion due to the error-correcting code is small compared with the more significant spread-spectrum bandwidth expansion, and thus, there is no impact on the transmission bandwidth.

In each of the above trade-off examples, a "traditional" code involving redundant bits and faster signaling (for a real-time communication system) has been assumed; hence, in each case, the cost was expanded bandwidth. However, there exists an error-correcting technique, called *trellis-coded modulation,* that does not require faster signaling or expanded bandwidth for real-time systems. (This technique is described in Section 9.10.)

Example 6.2   Coded versus Uncoded Performance

Compare the message error probability for a communications link with and without the use of error-correction coding. Assume that the uncoded transmission characteristics are: BPSK modulation, Gaussian noise, $P_r/N_0 = 43,776$, data rate $R = 4800$ bits/s. For the coded case, also assume the use of a (15, 11) error-correcting code that is capable of correcting any single-error pattern within a block of 15 bits. Consider that the demodulator makes hard decisions and thus feeds the demodulated code bits directly to the decoder, which in turn outputs an estimate of the original message.

Solution

Following Equation (4.79), let $p_u = Q\sqrt{2E_b/N_0}$ and $p_c = Q\sqrt{2E_c/N_0}$ be the uncoded and coded channel symbol error probabilities, respectively, where E_b/N_0 is the bit energy per noise spectral density and E_c/N_0 is the code-bit energy per noise spectral density.

Without coding

$$\frac{E_b}{N_0} = \frac{P_r}{N_0}\left(\frac{1}{R}\right) = 9.12 \ (9.6 \text{ dB})$$

and

$$p_u = Q\left(\sqrt{\frac{2E_b}{N_0}}\right) = Q(\sqrt{18.24}) = 1.02 \times 10^{-5} \tag{6.20}$$

where the following approximation of $Q(x)$ from Equation (3.44) was used:

$$Q(x) \approx \frac{1}{x\sqrt{2\pi}}\exp\left(\frac{-x^2}{2}\right) \qquad \text{for } x > 3$$

The probability that the uncoded message block P_M^u will be received in error is 1 minus the product of the probabilities that each bit will be detected correctly. Thus,

$$
\begin{aligned}
P_M^u &= 1 - (1 - p_u)^k \\
&= 1 - \underbrace{(1 - p_u)^{11}}_{\substack{\text{probability that all} \\ \text{11 bits in uncoded} \\ \text{block are correct}}} \qquad = \underbrace{1.12 \times 10^{-4}}_{\substack{\text{probability that at} \\ \text{least 1 bit out of} \\ \text{11 is in error}}}
\end{aligned} \tag{6.21}
$$

With coding

Assuming a *real-time* communication system such that delay is unacceptable, the channel-symbol rate or code-bit rate R_c is 15/11 times the data bit rate:

$$
R_c = 4800 \times \tfrac{15}{11} \approx 6545 \text{ bps}
$$

and

$$
\frac{E_c}{N_0} = \frac{P_r}{N_0}\left(\frac{1}{R_c}\right) = 6.69 \,(8.3 \text{ dB})
$$

The E_c/N_0 for each code bit is less than that for the data bit in the uncoded case because the channel-bit rate has increased, but the transmitter power is assumed to be fixed:

$$
p_c = Q\left(\sqrt{\frac{2E_c}{N_0}}\right) = Q(\sqrt{13.38} = 1.36 \times 10^{-4} \tag{6.22}
$$

It can be seen by comparing the results of Equation 6.20 with those of Equation 6.22 that because redundancy was added, the channel bit-error probability has degraded. More bits must be detected during the same time interval and with the same available power; the performance improvement due to the coding *is not yet apparent.* We now compute the coded message error rate P_M^c, using Equation 6.18:

$$
P_M^c = \sum_{j=2}^{n=15} \binom{15}{j} (p_c)^j (1 - p_c)^{15-j}
$$

The summation is started with $j = 2$, since the code corrects all single errors within a block of $n = 15$ bits. An approximation is obtained by using only the first term of the summation. For p_c, we use the value calculated in Equation 6.22:

$$
P_M^c \approx \binom{15}{2} (p_c)^2 (1 - p_c)^{13} = 1.94 \times 10^{-6} \tag{6.23}
$$

By comparing the results of Equation (6.21) with (6.23), we can see that the probability of message error has improved by a factor of 58 due to the error-correcting code used in this example. This example illustrates the typical behavior of all such real-time communication systems using error-correction coding. Added redundancy means faster signaling, less energy per channel symbol, and more errors out of the demodulator. The benefits arise because the behavior of the decoder will (at reasonable values of E_b/N_0) more than compensate for the poor performance of the demodulator.

6.3.4.6 Code Performance at Low Values of E_b/N_0

The reader is urged to solve Problem 6.5, which is similar to Example 6.2. In part (a) of Problem 6.5, where an E_b/N_0 of 14 dB is given, the result is a message-error performance improvement through the use of coding. However, in part (b) where the E_b/N_0 has been reduced to 10 dB, coding provides no improvement; in fact, there is a degradation. One might ask, Why does part (b) manifest a degradation? After all, the same procedure is used for applying the code in both parts of the problem. The answer can be seen in the coded-versus-uncoded pictorial shown in Figure 6.9. Even though Problem 6.5 deals with message-error probability, and Figure 6.9 displays bit-error probability, the following explanation still applies. In all such plots, there is a crossover between the curves (usually at some low value of E_b/N_0). The reason for such crossover (threshold) is that every code system has some fixed error-correcting capability. If there are more errors within a block than the code is capable of correcting, the system will perform poorly. Imagine that E_b/N_0 is continually reduced. What happens at the output of the demodulator? It makes more and more errors. Therefore, such a continual decrease in E_b/N_0 must eventually cause some threshold to be reached where the decoder becomes overwhelmed with errors. When that threshold is crossed, we can interpret the degraded performance as being caused by the redundant bits consuming energy but giving back nothing beneficial in return. Does it strike the reader as a paradox that operating in a region (low values of E_b/N_0), where one would best like to see an error-performance improvement, is where the code makes things worse? There is, however, a class of powerful codes called *turbo codes* that provide error-performance improvements at low values of E_b/N_0; the crossover point is lower for turbo codes compared with conventional codes. (These are treated in Section 8.4.)

6.4 LINEAR BLOCK CODES

Linear block codes (such as the one described in Example 6.2) are a class of parity-check codes that can be characterized by the (n, k) notation described earlier. The encoder transforms a block of k message digits (a message vector) into a longer block of n codeword digits (a code vector) constructed from a given alphabet of elements. When the alphabet consists of two elements (0 and 1), the code is a binary code comprising binary digits (bits). Our discussion of linear block codes is restricted to binary codes unless otherwise noted.

The k-bit messages form 2^k distinct message sequences, referred to as *k-tuples* (sequences of k digits). The n-bit blocks can form as many as 2^n distinct sequences, referred to as *n-tuples*. The encoding procedure assigns to each of the 2^k message k-tuples *one* of the 2^n n-tuples. A block code represents a one-to-one assignment, whereby the 2^k message k-tuples are *uniquely* mapped into a new set of 2^k code-word n-tuples; the mapping can be accomplished via a look-up table. For *linear codes,* the mapping transformation is, of course *linear.*

6.4.1 Vector Spaces

The set of all binary n-tuples, V_n, is called a *vector space* over the binary field of two elements (0 and 1). The binary field has two operations, addition and multiplication, such that the results of all operations are in the same set of two elements. The arithmetic operations of addition and multiplication are defined by the conventions of the algebraic field [4]. For example, in a binary field, the rules of addition and multiplication are as follows:

Addition	Multiplication
$0 \oplus 0 = 0$	$0 \cdot 0 = 0$
$0 \oplus 1 = 1$	$0 \cdot 1 = 0$
$1 \oplus 0 = 1$	$1 \cdot 0 = 0$
$1 \oplus 1 = 0$	$1 \cdot 1 = 1$

The addition operation, designated with the symbol $\oplus$, is the same modulo-2 operation described in Section 2.9.3. The summation of binary n-tuples always entails modulo-2 addition. However, for notational simplicity the ordinary + sign will often be used.

6.4.2 Vector Subspaces

A subset S of the vector space V_n is called a *subspace* if the following two conditions are met:

1. The all-zeros vector is in S.
2. The sum of any two vectors in S is also in S (known as the *closure property*).

These properties are fundamental for the algebraic characterization of *linear block codes*. Suppose that $\mathbf{V}_i$ and $\mathbf{V}_j$ are two codewords (or code vectors) in an (n, k) binary block code. The code is said to be *linear* if, and only if $(\mathbf{V}_i \oplus \mathbf{V}_j)$ is also a code vector. A linear block code, then, is one in which vectors outside the subspace cannot be created by the addition of legitimate codewords (members of the subspace).

For example, the vector space V_4 is totally populated by the following $2^4 =$ sixteen 4-tuples:

0000	0001	0010	0011	0100	0101	0110	0111
1000	1001	1010	1011	1100	1101	1110	1111

An example of a subset of V_4 that forms a subspace is

$$0000 \quad 0101 \quad 1010 \quad 1111$$

It is easy to verify that the addition of any two vectors in the subspace can only yield one of the other members of the subspace. A set of 2^k n-tuples is called a *linear block code* if, and only if, it is a subspace of the vector space V_n of all n-tuples. Figure 6.10 illustrates, with a simple geometric analogy, the structure behind linear

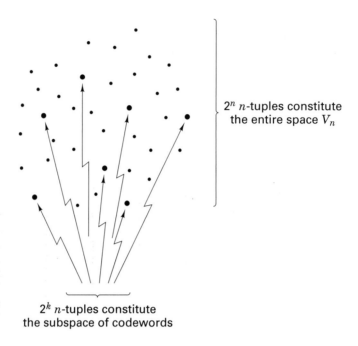

2^n n-tuples constitute
the entire space V_n

2^k n-tuples constitute
the subspace of codewords

Figure 6.10   Linear block-code structure.

block codes. We can imagine the vector space V_n comprising 2^n n-tuples. Within this vector space there exists a subset of 2^k n-tuples making up a subspace. These 2^k vectors or points, shown "sprinkled" among the more numerous 2^n points, represent the legitimate or allowable codeword assignments. A message is encoded into one of the 2^k allowable code vectors and then transmitted. Because of noise in the channel, a perturbed version of the codeword (one of the other 2^n vectors in the n-tuple space) may be received. If the perturbed vector is not too unlike (not too distant from) the valid codeword, the decoder can decode the message correctly. The basic goals in choosing a particular code, similar to the goals in selecting a set of modulation waveforms, can be stated in the context of Figure 6.10 as follows:

1. We strive for coding efficiency by packing the V_n space with as many codewords as possible. This is tantamount to saying that we only want to expend a *small amount of redundancy* (excess bandwidth).
2. We want the codewords to be as *far apart from one another* as possible, so that even if the vectors experience some corruption during transmission, they may still be correctly decoded, with a high probability.

6.4.3 A (6, 3) Linear Block Code Example

Examine the following coding assignment that describes a (6, 3) code. There are $2^k = 2^3 = 8$ message vectors, and therefore eight codewords. There are $2^k = 2^6 =$ sixty-four 6-tuples in the V_6 vector space:

It is easy to check that the eight codewords shown in Table 6.1 form a subspace of V_6 (the all-zeros vector is present, and the sum of any two codewords yields another codeword member of the subspace). Therefore, these codewords represent a *linear block code,* as defined in Section 6.4.2. A natural question to ask is How was the codeword-to-message assignment chosen for this (6, 3) code? A unique assignment for a particular (*n, k*) code does not exist; however, neither is there complete freedom of choice. In Section 6.6.3, we examine the requirements and constraints that drive a code design.

TABLE 6.1 Assignment of Codewords to Messages

Message vector	Codeword
0 0 0	0 0 0 0 0 0
1 0 0	1 1 0 1 0 0
0 1 0	0 1 1 0 1 0
1 1 0	1 0 1 1 1 0
0 0 1	1 0 1 0 0 1
1 0 1	0 1 1 1 0 1
0 1 1	1 1 0 0 1 1
1 1 1	0 0 0 1 1 1

6.4.4 Generator Matrix

If *k* is large, a *table look-up* implementation of the encoder becomes prohibitive. For a (127, 92) code there are 2^{92} or approximately 5×10^{27} code vectors. If the encoding procedure consists of a simple look-up table, imagine the size of the memory necessary to contain such a large number of codewords. Fortunately, it is possible to reduce complexity by generating the required codewords as needed, instead of storing them.

Since a set of codewords that forms a linear block code is a *k*-dimensional subspace of the *n*-dimensional binary vector space ($k < n$), it is always possible to find a set of *n*-tuples, fewer than 2^k, that can generate all the 2^k codewords of the subspace. The generating set of vectors is said to *span* the subspace. The smallest *linearly independent* set that spans the subspace is called a *basis* of the subspace, and the number of vectors in this basis set is the dimension of the subspace. Any basis set of *k* linearly independent *n*-tuples $\mathbf{V}_1, \mathbf{V}_2, \ldots, \mathbf{V}_k$ can be used to generate the required linear block code vectors, since each code vector is a linear combination of $\mathbf{V}_1, \mathbf{V}_2, \ldots, \mathbf{V}_k$. That is, each of the set of 2^k codewords {**U**} can be described by

$$\mathbf{U} = m_1 \mathbf{V}_1 + m_2 \mathbf{V}_2 + \cdots + m_k \mathbf{V}_k$$

where $m_i = (0 \text{ or } 1)$ are the message digits and $i = 1, \ldots, k$.

In general, we can define a *generator matrix* by the following $k \times n$ array:

$$\mathbf{G} = \begin{bmatrix} \mathbf{V}_1 \\ \mathbf{V}_2 \\ \vdots \\ \mathbf{V}_k \end{bmatrix} = \begin{bmatrix} v_{11} & v_{12} & \cdots & v_{1n} \\ v_{21} & v_{22} & \cdots & v_{2n} \\ \vdots & & & \\ v_{k1} & v_{k2} & \cdots & v_{kn} \end{bmatrix} \tag{6.24}$$

Code vectors, by convention, are usually designated as row vectors. Thus, the message $\mathbf{m}$, a sequence of k message bits, is shown below as a row vector ($1 \times k$ matrix having one row and k columns):

$$\mathbf{m} = m_1, m_2, \ldots, m_k$$

The generation of the codeword $\mathbf{U}$ is written in matrix notation as the product of $\mathbf{m}$ and $\mathbf{G}$, and we write

$$\mathbf{U} = \mathbf{mG} \tag{6.25}$$

where, in general, the matrix multiplication $\mathbf{C} = \mathbf{AB}$ is performed in the usual way by using the rule

$$c_{ij} = \sum_{k}^{n} a_{ik} b_{kj} \qquad i = 1, \ldots, l \quad j = 1, \ldots, m$$

where $\mathbf{A}$ is an $l \times n$ matrix, $\mathbf{B}$ is an $n \times m$ matrix, and the result $\mathbf{C}$ is an $l \times m$ matrix. For the example introduced in the preceding section, we can fashion a generator matrix as

$$\mathbf{G} = \begin{bmatrix} \mathbf{V}_1 \\ \mathbf{V}_2 \\ \mathbf{V}_3 \end{bmatrix} = \begin{bmatrix} 1 & 1 & 0 & 1 & 0 & 0 \\ 0 & 1 & 1 & 0 & 1 & 0 \\ 1 & 0 & 1 & 0 & 0 & 1 \end{bmatrix} \tag{6.26}$$

where $\mathbf{V}_1$, $\mathbf{V}_2$, and $\mathbf{V}_3$ are three *linearly independent vectors* (a subset of the eight code vectors) that can generate all the code vectors. Notice that the sum of any two generating vectors does not yield any of the other generating vectors (opposite of closure). Let us generate the codeword $\mathbf{U}_4$ for the fourth message vector 1 1 0 in Table 6.1, using the generator matrix of Equation (6.26):

$$\mathbf{U}_4 = \begin{bmatrix} 1 & 1 & 0 \end{bmatrix} \begin{bmatrix} \mathbf{V}_1 \\ \mathbf{V}_2 \\ \mathbf{V}_3 \end{bmatrix} = 1 \cdot \mathbf{V}_1 + 1 \cdot \mathbf{V}_2 + 0 \cdot \mathbf{V}_3$$

$$= 1\ 1\ 0\ 1\ 0\ 0 + 0\ 1\ 1\ 0\ 1\ 0 + 0\ 0\ 0\ 0\ 0\ 0$$

$$= 1\ 0\ 1\ 1\ 1\ 0 \quad \text{(codeword for the message vector 1 1 0)}$$

Thus, the code vector corresponding to a message vector is a linear combination of the rows of $\mathbf{G}$. Since the code is totally defined by $\mathbf{G}$, the encoder need only store the k rows of $\mathbf{G}$ instead of the total 2^k vectors of the code. For this example, notice that the generator array of dimension 3×6 in Equation (6.26) replaces the original codeword array of dimension 8×6 in Table 6.1, representing a reduction in system complexity.

6.4.5 Systematic Linear Block Codes

A systematic (n, k) linear block code is a mapping from a k-dimensional message vector to an n-dimensional codeword in such a way that part of the sequence generated coincides with the k message digits. The remaining $(n - k)$ digits are parity digits. A systematic linear block code will have a generator matrix of the form

$$\mathbf{G} = \left[\begin{array}{c|c} \mathbf{P} & \mathbf{I}_k \end{array}\right]$$

$$= \begin{bmatrix} p_{11} & p_{12} & \cdots & p_{1,(n-k)} & 1 & 0 & \cdots & 0 \\ p_{21} & p_{22} & \cdots & p_{2,(n-k)} & 0 & 1 & \cdots & 0 \\ \vdots & & & & & & \vdots \\ p_{k1} & p_{k2} & \cdots & p_{k,(n-k)} & 0 & 0 & \cdots & 1 \end{bmatrix} \qquad (6.27)$$

where $\mathbf{P}$ is the parity array portion of the generator matrix $p_{ij} = (0 \text{ or } 1)$, and $\mathbf{I}_k$ is the $k \times k$ identity matrix (ones on the main diagonal and zeros elsewhere). Notice that with this systematic generator, the encoding complexity is further reduced since it is not necessary to store the identity matrix portion of the array. By combining Equations (6.26) and (6.27), each codeword is expressed as

$$u_1, u_2, \ldots, u_n = [m_1, m_2, \ldots, m_k] \times \begin{bmatrix} p_{11} & p_{12} & \cdots & p_{1,(n-k)} & 1 & 0 & \cdots & 0 \\ p_{21} & p_{22} & \cdots & p_{2,(n-k)} & 0 & 1 & \cdots & 0 \\ \vdots & & & & & & \vdots \\ p_{k1} & p_{k2} & \cdots & p_{k,(n-k)} & 0 & 0 & \cdots & 1 \end{bmatrix}$$

where

$$u_i = m_1 p_{1i} + m_2 p_{2i} + \cdots + m_k p_{ki} \qquad \text{for } i = 1, \ldots, (n - k)$$
$$= m_{i-n+k} \qquad \qquad \qquad \qquad \text{for } i = (n - k + 1), \ldots, n$$

Given the message k-tuple

$$\mathbf{m} = m_1, m_2, \ldots, m_k$$

and the general code vector n-tuple

$$\mathbf{U} = u_1, u_2, \ldots, u_n$$

the systematic code vector can be expressed as

$$\mathbf{U} = \underbrace{p_1, p_2, \ldots, p_{n-k}}_{\text{parity bits}}, \underbrace{m_1, m_2, \ldots, m_k}_{\text{message bits}} \qquad (6.28)$$

where

$$p_1 = m_1 p_{11} + m_2 p_{21} + \cdots + m_k p_{k1}$$

$$p_2 = m_1 p_{12} + m_2 p_{22} + \cdots + m_k p_{k2} \qquad (6.29)$$

$$p_{n-k} = m_1 p_{1,(n-k)} + m_2 p_{2,(n-k)} + \cdots + m_k p_{k,(n-k)}$$

Systematic codewords are sometimes written so that the message bits occupy the left-hand portion of the codeword and the parity bits occupy the right-hand portion. This reordering has no effect on the error detection or error correction properties of the code, and will not be considered further.

For the (6, 3) code example in Section 6.4.3, the codewords are described as follows:

$$\mathbf{U} = [m_1, m_2, m_3] \underbrace{\begin{bmatrix} 1 & 1 & 0 \\ 0 & 1 & 1 \\ 1 & 0 & 1 \end{bmatrix}}_{\mathbf{P}} \underbrace{\begin{matrix} 1 & 0 & 0 \\ 0 & 1 & 0 \\ 0 & 0 & 1 \end{matrix}}_{\mathbf{I}_3} \qquad (6.30)$$

$$= \underbrace{m_1 + m_3}_{u_1}, \underbrace{m_1 + m_2}_{u_2}, \underbrace{m_2 + m_3}_{u_3}, \underbrace{m_1}_{u_4}, \underbrace{m_2}_{u_5}, \underbrace{m_3}_{u_6} \qquad (6.31)$$

Equation (6.31) provides some insight into the structure of linear block codes. We see that the redundant digits are produced in a variety of ways. The first parity bit is the sum of the first and third message bits; the second parity bit is the sum of the first and second message bits; and the third parity bit is the sum of the second and third message bits. Intuition tells us that such structure, compared with single-parity checks or simple digit-repeat procedures, may provide greater ability to detect and correct errors.

6.4.6 Parity-Check Matrix

Let us define a matrix $\mathbf{H}$, called the *parity-check matrix,* that will enable us to decode the received vectors. For each $(k \times n)$ generator matrix $\mathbf{G}$, there exists an $(n-k) \times n$ matrix $\mathbf{H}$, such that the rows of $\mathbf{G}$ are orthogonal to the rows of $\mathbf{H}$; that is, $\mathbf{GH}^T = \mathbf{0}$, where $\mathbf{H}^T$ is the *transpose* of $\mathbf{H}$, and $\mathbf{0}$ is a $k \times (n-k)$ all-zeros matrix. $\mathbf{H}^T$ is an $n \times (n-k)$ matrix whose rows are the columns of $\mathbf{H}$ and whose columns are the rows of $\mathbf{H}$. To fulfill the orthogonality requirements for a systematic code, the components of the $\mathbf{H}$ matrix are written as

$$\mathbf{H} = [\mathbf{I}_{n-k} \mid \mathbf{P}^T] \qquad (6.32)$$

Hence, the $\mathbf{H}^T$ matrix is written as

$$\mathbf{H}^T = \begin{bmatrix} \mathbf{I}_{n-k} \\ ----- \\ \mathbf{P} \end{bmatrix} \qquad (6.33\text{a})$$

$$= \begin{bmatrix} 1 & 0 & \cdots & 0 \\ 0 & 1 & \cdots & 0 \\ \vdots & & & \\ 0 & 0 & \cdots & 1 \\ p_{11} & p_{12} & \cdots & p_{1,(n-k)} \\ p_{21} & p_{22} & \cdots & p_{2,(n-k)} \\ \vdots & & & \\ p_{k1} & p_{k2} & \cdots & p_{k,(n-k)} \end{bmatrix} \qquad (6.33b)$$

It is easy to verify that the product $\mathbf{UH}^T$ of each codeword $\mathbf{U}$, generated by $\mathbf{G}$ and the $\mathbf{H}^T$ matrix, yields

$$\mathbf{UH}^T = p_1 + p_1, \; p_2 + p_2, \; \ldots, \; p_{n-k} + p_{n-k} = \mathbf{0}$$

where the parity bits $p_1, p_2, \ldots p_{n-k}$ are defined in Equation (6.29). Thus, once the *parity-check matrix* $\mathbf{H}$ is constructed to fulfill the foregoing orthogonality requirements, we can use it to test whether a received vector is a valid member of the codeword set. $\mathbf{U}$ is a codeword generated by matrix $\mathbf{G}$ if, and only if $\mathbf{UH}^T = \mathbf{0}$.

6.4.7 Syndrome Testing

Let $\mathbf{r} = r_1, r_2, \ldots, r_n$ be a received vector (one of 2^n n-tuples) resulting from the transmission of $\mathbf{U} = u_1, u_2, \ldots, u_n$ (one of 2^k n-tuples). We can therefore describe $\mathbf{r}$ as

$$\mathbf{r} = \mathbf{U} + \mathbf{e} \qquad (6.34)$$

where $\mathbf{e} = e_1, e_2, \ldots, e_n$ is an error vector or error pattern introduced by the channel. There are a total of $2^n - 1$ potential nonzero error patterns in the space of 2^n n-tuples. The *syndrome* of $\mathbf{r}$ is defined as

$$\mathbf{S} = \mathbf{rH}^T \qquad (6.35)$$

The syndrome is the result of a parity check performed on $\mathbf{r}$ to determine whether $\mathbf{r}$ is a valid member of the codeword set. If, in fact, $\mathbf{r}$ is a member, the syndrome $\mathbf{S}$ has a value $\mathbf{0}$. If $\mathbf{r}$ contains detectable errors, the syndrome has some nonzero value. If $\mathbf{r}$ contains correctable errors, the syndrome (like the symptom of an illness) has some nonzero value that can earmark the particular error pattern. The decoder, depending upon whether it has been implemented to perform FEC or ARQ, will then take actions to locate the errors and correct them (FEC), or else it will request a retransmission (ARQ). Combining Equations (6.34) and (6.35), the syndrome of $\mathbf{r}$ is seen to be

$$\begin{aligned} \mathbf{S} &= (\mathbf{U} + \mathbf{e})\mathbf{H}^T \\ &= \mathbf{UH}^T + \mathbf{eH}^T \end{aligned} \qquad (6.36)$$

However, $\mathbf{UH}^T = \mathbf{0}$ for all members of the codeword set. Therefore,

$$\mathbf{S} = \mathbf{eH}^T \qquad (6.37)$$

The foregoing development, starting with Equation (6.34) and terminating with Equation (6.37), is evidence that the syndrome test, whether performed on either a corrupted code vector or on the error pattern that caused it, yields the same syndrome. An important property of linear block codes, fundamental to the decoding process, is that the mapping between correctable error patterns and syndromes is one to one.

It is interesting to note the following two required properties of the parity-check matrix.

1. No column of **H** can be all zeros, or else an error in the corresponding codeword position would not affect the syndrome and would be undetectable.
2. All columns of **H** must be unique. If two columns of **H** were identical, errors in these two corresponding codeword positions would be indistinguishable.

Example 6.3   Syndrome Test

Suppose that codeword $\mathbf{U} = 1\ 0\ 1\ 1\ 1\ 0$ from the example in Section 6.4.3 is transmitted and the vector $\mathbf{r} = 0\ 0\ 1\ 1\ 1\ 0$ is received; that is, the leftmost bit is received in error. Find the syndrome vector value $\mathbf{S} = \mathbf{rH}^T$ and verify that it is equal to $\mathbf{eH}^T$.

Solution

$$\mathbf{S} = \mathbf{rH}^T$$

$$= [0\ 0\ 1\ 1\ 1\ 0] \begin{bmatrix} 1 & 0 & 0 \\ 0 & 1 & 0 \\ 0 & 0 & 1 \\ 1 & 1 & 0 \\ 0 & 1 & 1 \\ 1 & 0 & 1 \end{bmatrix}$$

$$= [1,\ 1+1,\ 1+1] = [1\ 0\ 0] \quad \text{(syndrome of corrupted code vector)}$$

Next, we verify that the syndrome of the corrupted code vector is the same as the syndrome of the error pattern that caused the error:

$$\mathbf{S} = \mathbf{eH}^T = [1\ 0\ 0\ 0\ 0\ 0]\mathbf{H}^T = [1\ 0\ 0] \quad \text{(syndrome of error pattern)}$$

6.4.8 Error Correction

We have detected a single error and have shown that the syndrome test performed on either the corrupted codeword, or on the error pattern that caused it, yields the same syndrome. This should be a clue that we not only can detect the error, but since there is a one-to-one correspondence between correctable error patterns and syndromes, we can correct such error patterns. Let us arrange the 2^n n-tuples that represent possible received vectors in an array, called the *standard array,* such that the first row contains all the codewords, starting with the all-zeros codeword, and the first column contains all the correctable error patterns. Recall from the basic properties of linear codes (see Section 6.4.2) that the all-zeros vector must be a member of the codeword set. Each row, called a *coset,* consists of an error pattern in the first column, called the *coset leader,* followed by the codewords perturbed by that error pattern. The standard array format for an (n, k) code is as follows:

$$
\begin{array}{ccccc}
\mathbf{U}_1 & \mathbf{U}_2 & \cdots & \mathbf{U}_i & \cdots & \mathbf{U}_{2^k} \\
\mathbf{e}_2 & \mathbf{U}_2 + \mathbf{e}_2 & \cdots & \mathbf{U}_i + \mathbf{e}_2 & \cdots & \mathbf{U}_{2^k} + \mathbf{e}_2 \\
\mathbf{e}_3 & \mathbf{U}_2 + \mathbf{e}_3 & \cdots & \mathbf{U}_i + \mathbf{e}_3 & \cdots & \mathbf{U}_{2^k} + \mathbf{e}_3 \\
\vdots & \vdots & & \vdots & & \\
\mathbf{e}_j & \mathbf{U}_2 + \mathbf{e}_j & \cdots & \mathbf{U}_i + \mathbf{e}_j & \cdots & \mathbf{U}_{2^k} + \mathbf{e}_j \\
\vdots & \vdots & & \vdots & & \\
\mathbf{e}_{2^{n-k}} & \mathbf{U}_2 + \mathbf{e}_{2^{n-k}} & \cdots & \mathbf{U}_i + \mathbf{e}_{2^{n-k}} & \cdots & \mathbf{U}_{2^k} + \mathbf{e}_{2^{n-k}}
\end{array}
\qquad (6.38)
$$

Note that codeword $\mathbf{U}_1$, the all-zeros codeword, plays two roles. It is one of the code-words, and it can also be thought of as the error pattern $\mathbf{e}_1$—the pattern that represents no error, such that $\mathbf{r} = \mathbf{U}$. The array contains all 2^n n-tuples in the space V_n. Each n-tuple appears in *only one* location—none are missing, and none are replicated. Each coset consists of 2^k n-tuples. Therefore, there are $(2^n/2^k) = 2^{n-k}$ cosets.

The decoding algorithm calls for replacing a corrupted vector (any n-tuple excluding those in the first row) with a valid codeword from the top of the column containing the corrupted vector. Suppose that a codeword $\mathbf{U}_i$ $(i = 1, \ldots, 2^k)$ is transmitted over a noisy channel, resulting in a received (corrupted) vector $\mathbf{U}_i + \mathbf{e}_j$. If the error pattern $\mathbf{e}_j$ caused by the channel is a coset leader, where the index $j = 1, \ldots, 2^{n-k}$, the received vector will be decoded correctly into the transmitted codeword $\mathbf{U}_i$. If the error pattern is not a coset leader, then an erroneous decoding will result.

6.4.8.1 The Syndrome of a Coset

If $\mathbf{e}_j$ is the coset leader or error pattern of the jth coset, then $\mathbf{U}_i + \mathbf{e}_j$ is an n-tuple in this coset. The syndrome of this n-tuple can be written

$$
\mathbf{S} = (\mathbf{U}_i + \mathbf{e}_j)\mathbf{H}^T = \mathbf{U}_i\mathbf{H}^T + \mathbf{e}_j\mathbf{H}^T
$$

Since $\mathbf{U}_i$ is a code vector, $\mathbf{U}_i\mathbf{H}^T = \mathbf{0}$, and we can write, as in Equation (6.37),

$$
\mathbf{S} = (\mathbf{U}_i + \mathbf{e}_j)\mathbf{H}^T = \mathbf{e}_j\mathbf{H}^T \qquad (6.39)
$$

The name *coset* is short for "a *set* of numbers having a *common* feature." What do the members of any given row (coset) have in common? From Equation (6.39) it is clear that each member of a coset has the *same syndrome*. The syndrome for each coset is different from that of any other coset in the code; it is the syndrome that is used to estimate the error pattern.

6.4.8.2 Error Correction Decoding

The procedure for error correction decoding proceeds as follows:

1. Calculate the syndrome of $\mathbf{r}$ using $\mathbf{S} = \mathbf{r}\mathbf{H}^T$.
2. Locate the coset leader (error pattern) $\mathbf{e}_j$, whose syndrome equals $\mathbf{r}\mathbf{H}^T$.

3. This error pattern is assumed to be the corruption caused by the channel.

4. The corrected received vector, or codeword, is identified as $\mathbf{U} = \mathbf{r} + \mathbf{e}_j$. We can say that we retrieve the valid codeword by subtracting out the identified error; in modulo-2 arithmetic, the operation of subtraction is identical to that of addition.

6.4.8.3 Locating the Error Pattern

Returning to the example of Section 6.4.3, we arrange the $2^6 =$ sixty-four 6-tuples in a standard array as shown in Figure 6.11. The valid codewords are the eight vectors in the first row, and the *correctable error patterns* are the seven nonzero *coset leaders* in the first column. Notice that all 1-bit error patterns are correctable. Also notice that after exhausting all 1-bit error patterns, there remains some error-correcting capability since we have not yet accounted for all sixty-four 6-tuples. There is one unassigned coset leader; therefore, there remains the capability of correcting one additional error pattern. We have the flexibility of choosing this error pattern to be any of the *n*-tuples in the remaining coset. In Figure 6.11 this final correctable error pattern is chosen, somewhat arbitrarily, to be the 2-bit error pattern 0 1 0 0 0 1. Decoding will be correct if, and only if, the error pattern caused by the channel is one of the coset leaders.

000000	110100	011010	101110	101001	011101	110011	000111
000001	110101	011011	101111	101000	011100	110010	000110
000010	110110	011000	101100	101011	011111	110001	000101
000100	110000	011110	101010	101101	011001	110111	000011
001000	111100	010010	100110	100001	010101	111011	001111
010000	100100	001010	111110	111001	001101	100011	010111
100000	010100	111010	001110	001001	111101	010011	100111
010001	100101	001011	111111	111000	001100	100010	010110

Figure 6.11   Example of a standard array for a (6, 3) code.

We now determine the syndrome corresponding to each of the correctable error sequences by computing $\mathbf{e}_j\mathbf{H}^T$ for each coset leader, as follows:

$$\mathbf{S} = \mathbf{e}_j \begin{bmatrix} 1 & 0 & 0 \\ 0 & 1 & 0 \\ 0 & 0 & 1 \\ 1 & 1 & 0 \\ 0 & 1 & 1 \\ 1 & 0 & 1 \end{bmatrix}$$

The results are listed in Table 6.2. Since each syndrome in the table is unique, the decoder can identify the error pattern **e** to which it corresponds.

TABLE 6.2   Syndrome Look-Up Table

Error pattern	Syndrome
0 0 0 0 0 0	0 0 0
0 0 0 0 0 1	1 0 1
0 0 0 0 1 0	0 1 1
0 0 0 1 0 0	1 1 0
0 0 1 0 0 0	0 0 1
0 1 0 0 0 0	0 1 0
1 0 0 0 0 0	1 0 0
0 1 0 0 0 1	1 1 1

6.4.8.4 Error Correction Example

As outlined in Section 6.4.8.2 we receive the vector **r** and calculate its syndrome using $\mathbf{S} = \mathbf{r}\mathbf{H}^T$. We then use the syndrome look-up table (Table 6.2), developed in the preceding section, to find the corresponding error pattern. This error pattern is an estimate of the error, and we denote it $\hat{\mathbf{e}}$. The decoder then adds $\hat{\mathbf{e}}$ to **r** to obtain an estimate of the transmitted codeword $\hat{\mathbf{U}}$:

$$\hat{\mathbf{U}} = \mathbf{r} + \hat{\mathbf{e}} = (\mathbf{U} + \mathbf{e}) + \hat{\mathbf{e}} = \mathbf{U} + (\mathbf{e} + \hat{\mathbf{e}}) \qquad (6.40)$$

If the estimated error pattern is the same as the actual error pattern, that is, if $\hat{\mathbf{e}} = \mathbf{e}$, then the estimate $\hat{\mathbf{U}}$ is equal to the transmitted codeword **U**. On the other hand, if the error estimate is incorrect, the decoder will estimate a codeword that was not transmitted, and we have an *undetectable decoding error*.

Example 6.4   Error Correction

Assume that codeword **U** = 1 0 1 1 1 0, from the Section 6.4.3 example, is transmitted, and the vector **r** = 0 0 1 1 1 0 is received. Show how a decoder, using the Table 6.2 syndrome look-up table, can correct the error.

Solution

The syndrome of **r** is computed:

$$\mathbf{S} = [0 \ 0 \ 1 \ 1 \ 1 \ 0]\mathbf{H}^T = [1 \ 0 \ 0]$$

Using Table 6.2, the error pattern corresponding to the syndrome above is estimated to be

$$\hat{\mathbf{e}} = 1 \ 0 \ 0 \ 0 \ 0 \ 0$$

The corrected vector is then estimated by

$$\hat{\mathbf{U}} = \mathbf{r} + \hat{\mathbf{e}}$$
$$= 0 \ 0 \ 1 \ 1 \ 1 \ 0 + 1 \ 0 \ 0 \ 0 \ 0 \ 0$$
$$= 1 \ 0 \ 1 \ 1 \ 1 \ 0$$

Since the estimated error pattern is the actual error pattern in this example, the error correction procedure yields $\hat{\mathbf{U}} = \mathbf{U}$.

Notice that the process of decoding a corrupted codeword by first detecting and then correcting an error can be compared to a familiar medical analogy. A patient (potentially corrupted codeword) enters a medical facility (decoder). The examining physician performs diagnostic testing (multiplies by $\mathbf{H}^T$) in order to find a symptom (syndrome). Imagine that the physician finds characteristic spots on the patient's x-rays. An experienced physician would immediately recognize the correspondence between the symptom and the disease (error pattern) tuberculosis. A novice physician might have to refer to a medical handbook (Table 6.2) to associate the symptom or syndrome with the disease or error pattern. The final step is to provide the proper medication that removes the disease, as seen in Equation (6.40). In the context of binary codes and the medical analogy, Equation (6.40) reveals an unusual type of medicine practiced here. The patient is cured by reapplying the original disease.

6.4.9 Decoder Implementation

When the code is short as in the case of the (6, 3) code described in the previous sections, the decoder can be implemented with simple circuitry. Consider the steps that the decoder must take: (1) calculate the syndrome, (2) locate the error pattern, and (3) perform modulo-2 addition of the error pattern and the received vector (which removes the error). In Example 6.4, we started with a corrupted vector and saw how these steps yielded the corrected codeword. Now, consider the circuit in Figure 6.12, made up of exclusive-OR gates and AND gates that can accomplish the same result for any *single-error pattern* in the (6, 3) code. From Table 6.2 and Equation 6.39, we can write an expression for each of the syndrome digits in terms of the received codeword digits as

$$\mathbf{S} = \mathbf{r}\mathbf{H}^T$$

$$\mathbf{S} = \begin{bmatrix} r_1 & r_2 & r_3 & r_4 & r_5 & r_6 \end{bmatrix} \begin{bmatrix} 1 & 0 & 0 \\ 0 & 1 & 0 \\ 0 & 0 & 1 \\ 1 & 1 & 0 \\ 0 & 1 & 1 \\ 1 & 0 & 1 \end{bmatrix}$$

and

$$s_1 = r_1 + r_4 + r_6$$
$$s_2 = r_2 + r_4 + r_5$$
$$s_3 = r_3 + r_5 + r_6$$

We use these syndrome expressions for wiring up the circuit in Figure 6.12. The exclusive-OR gate is the same operation as modulo-2 arithmetic and hence uses the same symbol. A small circle at the termination of any line entering the AND gate indicates the logic COMPLEMENT of the signal.

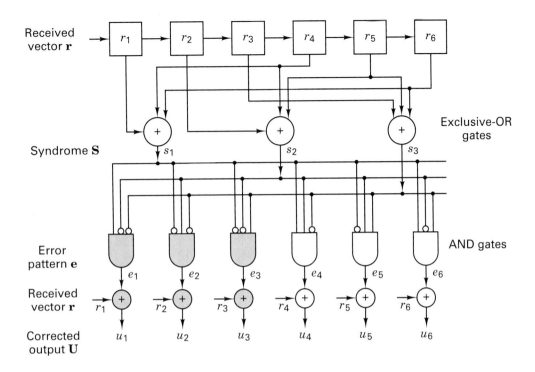

Figure 6.12 Implementation of the (6, 3) decoder.

The corrupted signal enters the decoder at two places simultaneously. At the upper part of the circuit, the syndrome is computed, and at the lower part, that syndrome is transformed to its corresponding error pattern. The error is removed by adding it back to the received vector yielding the corrected codeword.

Note that, for tutorial reasons, Figure 6.12 has been drawn to emphasize the algebraic decoding steps—calculation of syndrome, error pattern, and corrected output. In the real world, an (n, k) code is usually configured in systematic form. The decoder would not need to deliver the entire codeword; its output would consist of the data bits only. Hence, the Figure 6.12 circuitry becomes simplified by eliminating the gates that are shown with shading. For longer codes, such an implementation is very complex, and the preferred decoding techniques conserve circuitry by using a sequential approach instead of this parallel method [4]. It is also important to emphasize that Figure 6.12 has been configured to only detect and correct *single-error patterns* for the (6, 3) code. Error control for a double-error pattern would require additional circuitry.

6.4.9.1 Vector Notation

Codewords, error patterns, received vectors, and syndromes have been denoted by the vectors **U**, **e**, **r**, and **S** respectively. For notational simplicity, an index to denote a particular vector has generally been left off. However, to be precise, each of these vectors **U**, **e**, **r**, and **S** is one of a set having the general form

$$\mathbf{x}_j = \{x_1, x_2, \cdots, x_i, \cdots\}$$

Consider the range of the indices j and i in the context of the (6, 3) code in Table 6.1. For the codeword $\mathbf{U}_j$, the index $j = 1, \ldots, 2^k$ indicates that there are $2^3 = 8$ distinct codewords, and the index $i = 1, \ldots, n$ indicates that each codeword is made up of $n = 6$ bits. For a correctable error pattern $\mathbf{e}_j$, the index $j = 1, \ldots, 2^{n-k}$ indicates that there are $2^3 = 8$ coset leaders (7 nonzero correctable error patterns), and the index $i = 1, \ldots, n$ indicates that each error pattern is made up of $n = 6$ bits. For the received vector $\mathbf{r}_j$, the index $j = 1, \ldots, 2^n$ indicates that there are $2^6 = 64$ possible n-tuples that can be received, and the index $i = 1, \ldots, n$ indicates that each received n-tuple is made up of $n = 6$ bits. Finally, for the syndrome $\mathbf{S}_j$, the index $j = 1, \ldots, 2^{n-k}$ indicates that there are $2^3 = 8$ distinct syndrome vectors, and the index $i = 1, \ldots, n - k$ indicates that each syndrome is made up of $n - k = 3$ bits. In this chapter, the index is often dropped, and the vectors $\mathbf{U}_j$, $\mathbf{e}_j$, $\mathbf{r}_j$, and $\mathbf{S}_j$ are denoted as $\mathbf{U}$, $\mathbf{e}$, $\mathbf{r}$, and $\mathbf{S}$, respectively. The reader must be aware that for these vectors, an index is always inferred, even when it has been left off for notational simplicity.

6.5 ERROR-DETECTING AND CORRECTING CAPABILITY

6.5.1 Weight and Distance of Binary Vectors

It should be clear that not all error patterns can be correctly decoded. The error correction capability of a code will be investigated by first defining its structure. The *Hamming weight* $w(\mathbf{U})$ of a codeword $\mathbf{U}$ is defined to be the number of nonzero elements in $\mathbf{U}$. For a binary vector this is equivalent to the number of ones in the vector. For example, if $\mathbf{U} = 1\ 0\ 0\ 1\ 0\ 1\ 1\ 0\ 1$, then $w(\mathbf{U}) = 5$. The *Hamming distance* between two codewords $\mathbf{U}$ and $\mathbf{V}$, denoted $d(\mathbf{U}, \mathbf{V})$, is defined to be the number of elements in which they differ—for example,

$$\mathbf{U} = 1\ 0\ 0\ 1\ 0\ 1\ 1\ 0\ 1$$
$$\mathbf{V} = 0\ 1\ 1\ 1\ 1\ 0\ 1\ 0\ 0$$
$$d(\mathbf{U}, \mathbf{V}) = 6$$

By the properties of modulo-2 addition, we note that the sum of two binary vectors is another vector whose binary ones are located in those positions in which the two vectors differ—for example

$$\mathbf{U} + \mathbf{V} = 1\ 1\ 1\ 0\ 1\ 1\ 0\ 0\ 1$$

Thus, we observe that the Hamming distance between two codewords is equal to the Hamming weight of their sum: that is, $d(\mathbf{U}, \mathbf{V}) = w(\mathbf{U} + \mathbf{V})$. Also we see that the Hamming weight of a codeword is equal to its Hamming distance from the all-zeros vector.

6.5.2 Minimum Distance of a Linear Code

Consider the set of distances between all pairs of codewords in the space V_n. The smallest member of the set is the *minimum distance* of the code and is denoted d_{min}. Why do you suppose we have an interest in the minimum distance; why not the maximum distance? The minimum distance, like the weakest link in a chain, gives us a measure of the code's minimum capability and therefore characterizes the code's strength.

As discussed earlier, the sum of any two codewords yields another codeword member of the subspace. This property of linear codes is stated simply as: If $\mathbf{U}$ and $\mathbf{V}$ are codewords, then $\mathbf{W} = \mathbf{U} + \mathbf{V}$ must also be a codeword. Hence the distance between two codewords is equal to the weight of a third codeword; that is, $d(\mathbf{U}, \mathbf{V}) = w(\mathbf{U} + \mathbf{V}) = w(\mathbf{W})$. Thus the minimum distance of a linear code can be ascertained without examining the distance between all combinations of codeword pairs. We only need to examine the weight of each codeword (excluding the all-zeros codeword) in the subspace; the minimum weight corresponds to the minimum distance, d_{min}. Equivalently, d_{min} corresponds to the smallest of the set of distances between the all-zeros codeword and all the other codewords.

6.5.3 Error Detection and Correction

The task of the decoder, having received the vector $\mathbf{r}$, is to estimate the transmitted codeword $\mathbf{U}_i$. The optimal decoder strategy can be expressed in terms of the *maximum likelihood* algorithm (see Appendix B) as follows: Decide in favor of $\mathbf{U}_i$ if

$$P(\mathbf{r}|\mathbf{U}_i) = \max_{\text{over all } \mathbf{U}_j} P(\mathbf{r}|\mathbf{U}_j) \qquad (6.41)$$

Since for the binary symmetric channel (BSC), the likelihood of $\mathbf{U}_i$ with respect to $\mathbf{r}$ is inversely proportional to the distance between $\mathbf{r}$ and $\mathbf{U}_i$, we can write: Decide in favor of $\mathbf{U}_i$ if

$$d(\mathbf{r}, \mathbf{U}_i) = \min_{\text{over all } \mathbf{U}_j} d(\mathbf{r}, \mathbf{U}_j) \qquad (6.42)$$

In other words, the decoder determines the distance between $\mathbf{r}$ and each of the possible transmitted codewords $\mathbf{U}_j$, and selects as most likely a $\mathbf{U}_i$ for which

$$d(\mathbf{r}, \mathbf{U}_i) \le d(\mathbf{r}, \mathbf{U}_j) \qquad \text{for } i, j = 1, \dots, M \qquad \text{and} \qquad i \ne j \qquad (6.43)$$

where $M = 2^k$ is the size of the codeword set. If the minimum is not unique, the choice between minimum distance codewords is arbitrary. Distance metrics are treated further in Chapter 7.

In Figure 6.13 the distance between two codewords $\mathbf{U}$ and $\mathbf{V}$ is shown using a number line calibrated in *Hamming distance*. Each black dot represents a corrupted codeword. Figure 6.13a illustrates the reception of vector $\mathbf{r}_1$, which is distance 1 from $\mathbf{U}$ and distance 4 from $\mathbf{V}$. An error-correcting decoder, following the

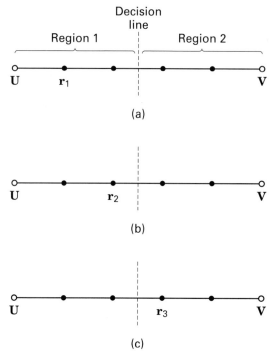

Region 1   Decision line   Region 2

$\mathbf{U}$ $\mathbf{r}_1$ $\mathbf{V}$

(a)

$\mathbf{U}$ $\mathbf{r}_2$ $\mathbf{V}$

(b)

$\mathbf{U}$ $\mathbf{r}_3$ $\mathbf{V}$

(c)

Figure 6.13 Error correction and detection capability. (a) Received vector $\mathbf{r}_1$. (b) Received vector $\mathbf{r}_2$. (c) Received vector $\mathbf{r}_3$.

maximum likelihood strategy, will select $\mathbf{U}$ upon receiving $\mathbf{r}_1$. If $\mathbf{r}_1$ had been the result of a 1-bit corruption to the transmitted code vector $\mathbf{U}$, the decoder has successfully corrected the error. But if $\mathbf{r}_1$ had been the result of a 4-bit corruption to the transmitted code vector $\mathbf{V}$, the result is a decoding error. Similarly, a double error in transmission of $\mathbf{U}$ might result in the received vector $\mathbf{r}_2$, which is distance 2 from $\mathbf{U}$ and distance 3 from $\mathbf{V}$, as shown in Figure 6.13b. Here, too, the decoder will select $\mathbf{U}$ upon receiving $\mathbf{r}_2$. A triple error in transmission of $\mathbf{U}$ might result in a received vector $\mathbf{r}_3$ that is distance 3 from $\mathbf{U}$ and distance 2 from $\mathbf{V}$, as shown in Figure 6.13c. Here the decoder will select $\mathbf{V}$ upon receiving $\mathbf{r}_3$, and will have made an error in decoding. From Figure 6.13 it should be clear that if error detection and not correction is the task, a corrupted vector—characterized by a black dot and representing a 1-bit, 2-bit, 3-bit, or 4-bit error—can be detected. However, five errors in transmission might result in codeword $\mathbf{V}$ being received when codeword $\mathbf{U}$ was actually transmitted; such an error would be *undetectable.*

From Figure 6.13 we can see that the error-detecting and error-correcting capabilities of a code are related to the *minimum distance* between codewords. The decision line in the figure serves the same purpose in the process of decoding as it does in demodulation, to define the decision regions. In the Figure 6.13 example, the decision criterion of choosing $\mathbf{U}$ if $\mathbf{r}$ falls in region 1, and choosing $\mathbf{V}$ if $\mathbf{r}$ falls in region 2, illustrates that such a code (with $d_{min} = 5$) can correct two errors. In general, the *error-correcting capability* t of a code is defined as the maximum number of guaranteed correctable errors per codeword, and is written [4]

$$t = \left\lfloor \frac{d_{\min} - 1}{2} \right\rfloor \tag{6.44}$$

where $\lfloor x \rfloor$ means the largest integer not to exceed x. Often, a code that corrects all possible sequences of t or fewer errors can also correct certain sequences of $t + 1$ errors. This can be seen in Figure 6.11. In this example $d_{\min} = 3$, and thus from Equation (6.44), we can see that *all* $t = 1$ bit-error patterns are correctable. Also, *a single* $t + 1$ or 2-bit error pattern is correctable. In general, a t-error-correcting (n, k) linear code is capable of correcting a total of 2^{n-k} error patterns. If a t-error-correcting block code is used strictly for error correction on a binary symmetric channel (BSC) with transition probability p, the message-error probability, P_M, that the decoder commits an erroneous decoding, and that the n-bit block is in error, can be calculated by using Equation (6.18) as an upper bound:

$$P_M \leq \sum_{j=t+1}^{n} \binom{n}{j} p^j (1-p)^{n-j} \tag{6.45}$$

The bound becomes an equality when the decoder corrects all combinations of errors up to and including t errors, but no combinations of errors greater than t. Such decoders are called *bounded distance decoders*. The decoded bit-error probability, P_B, depends on the particular code and decoder. It can be expressed [5] by the following approximation:

$$P_B \approx \frac{1}{n} \sum_{j=t+1}^{n} j \binom{n}{j} p^j (1-p)^{n-j} \tag{6.46}$$

A block code needs to detect errors prior to correcting them. Or, it may be used for error-detection only. It should be clear from Figure 6.13 that any received vector characterized by a black dot (a corrupted codeword) can be identified as an error. Therefore, the error-detecting capability is defined in terms of $d_{\min}$ as

$$e = d_{\min} - 1 \tag{6.47}$$

A block code with minimum distance $d_{\min}$ guarantees that all error patterns of $d_{\min} - 1$ or fewer errors can be detected. Such a code is also capable of detecting a large fraction of error patterns with $d_{\min}$ or more errors. In fact, an (n, k) code is capable of detecting $2^n - 2^k$ error patterns of length n. The reasoning is as follows. There are a total of $2^n - 1$ possible nonzero error patterns in the space of 2^n n-tuples. Even the bit pattern of a valid codeword represents a potential error pattern. Thus there are $2^k - 1$ error patterns that are identical to the $2^k - 1$ nonzero codewords. If any of these $2^k - 1$ error patterns occurs, it alters the transmitted codeword $\mathbf{U}_i$ into another codeword $\mathbf{U}_j$. Thus $\mathbf{U}_j$ will be received and its syndrome is zero. The decoder accepts $\mathbf{U}_j$ as the transmitted codeword and thereby commits an incorrect decoding. Therefore, there are $2^k - 1$ undetectable error patterns. If the error pattern is not identical to one of the 2^k codewords, the syndrome test on the received vector $\mathbf{r}$ yields a nonzero syndrome, and the error is detected. Therefore, there are exactly $2^n - 2^k$ detectable error patterns. For large n, where $2^k \ll 2^n$, only a small fraction of error patterns are undetected.

6.5.3.1 Codeword Weight Distribution

Let A_j be the number of codewords of weight j within an (n, k) linear code. The numbers $A_0, A_1, \ldots, A_n$ are called the *weight distribution* of the code. If the code is used only for error detection, on a BSC, the probability that the decoder does not detect an error can be computed from the weight distribution of the code [5] as

$$P_{\text{nd}} = \sum_{j=1}^{n} A_j p^j (1 - p)^{n-j} \qquad (6.48)$$

where p is the transition probability of the BSC. If the minimum distance of the code is $d_{\min}$, the values of A_1 to $A_{d_{\min}-1}$ are zero.

Example 6.5 Probability of an Undetected Error in an Error Detecting Code

Consider that the $(6, 3)$ code, given in Section 6.4.3, is used only for error detection. Calculate the probability of an undetected error if the channel is a BSC and the transition probability is 10^{-2}.

Solution

The weight distribution of this code is $A_0 = 1, A_1 = A_2 = 0, A_3 = 4, A_4 = 3, A_5 = 0, A_6 = 0$. Therefore, we can write, using Equation (6.48),

$$P_{\text{nd}} = 4p^3(1 - p)^3 + 3p^4(1 - p)^2$$

For $p = 10^{-2}$, the probability of an undetected error is 3.9×10^{-6}.

6.5.3.2 Simultaneous Error Correction and Detection

It is possible to trade correction capability from the maximum guaranteed (t), where t is defined in Equation (6.44), for the ability to simultaneously detect a class of errors. A code can be used for the simultaneous correction of α errors and detection of β errors, where $\beta \geq \alpha$, provided that its minimum distance is [4]

$$d_{\min} \geq \alpha + \beta + 1 \qquad (6.49)$$

When t or fewer errors occur, the code is capable of detecting and correcting them. When more than t but fewer than $e + 1$ errors occur, where e is defined in Equation (6.47), the code is capable of detecting their presence but correcting only a subset of them. For example, a code with $d_{\min} = 7$ can be used to simultaneously detect and correct in any one of the following ways:

Detect (β)	Correct (α)
3	3
4	2
5	1
6	0

Note that correction implies prior detection. For the above example, when there are three errors, all of them can be detected and corrected. When there are five errors, all of them can be detected but only a subset of them (one) can be corrected.

6.5.4 Visualization of a 6-Tuple Space

Figure 6.14 is a visualization of the eight codewords from the example of Section 6.4.3. The codewords are generated from linear combinations of the three independent 6-tuples in Equation (6.26); the codewords form a three-dimensional subspace. The figure shows such a subspace completely occupied by the eight codewords (large black circles); the coordinates of the subspace have purposely been drawn to emphasize their nonorthogonality. Figure 6.14 is an attempt to illustrate the entire space, containing sixty-four 6-tuples, even though there is no

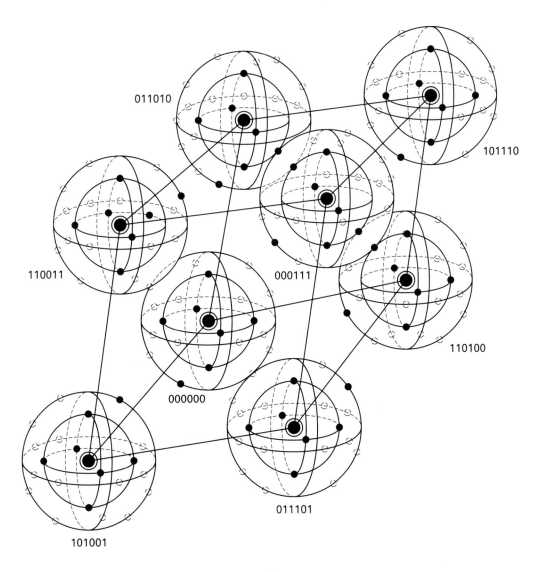

Figure 6.14 Example of eight codewords in a 6-tuple space.

precise way to draw or construct such a model. Spherical layers or shells are shown around each codeword. Each of the nonintersecting inner layers is a Hamming distance of 1 from its associated codeword; each outer layer is a Hamming distance of 2 from its codeword. Larger distances are not useful in this example. For each codeword, the two layers shown are occupied by perturbed codewords. There are six such points on each inner sphere (a total of 48 points), representing the six possible 1-bit error-perturbed vectors associated with each codeword. These 1-bit perturbed codewords are distinct in the sense that they can best be associated with only one codeword, and therefore can be corrected. As is seen from the standard array of Figure 6.11, there is also one 2-bit error pattern that can be corrected. There is a total of $\binom{6}{2} = 15$ different 2-bit error patterns that can be inflicted on each codeword, but only one of them, in our example the 0 1 0 0 0 1 error pattern, can be corrected. The other fourteen 2-bit error patterns yield vectors that cannot be uniquely identified with just one codeword; these noncorrectable error patterns yield vectors that are equivalent to the error-perturbed vectors of two or more codewords. In the figure, all correctable (fifty-six) 1- and 2-bit error-perturbed codewords are shown as small black circles. Perturbed codewords that cannot be corrected are shown as small clear circles.

Figure 6.14 is useful for visualizing the properties of a class of codes known as *perfect codes*. A *t*-error-correcting code is called a perfect code if its standard array has all the error patterns of *t* and fewer errors and no others as coset leaders (no residual error-correcting capacity). In terms of Figure 6.14, a *t*-error-correcting perfect code is one that can, with maximum likelihood decoding, correct all perturbed codewords occupying a shell at Hamming distance *t* or less from its originating codeword, and cannot correct any perturbed vectors occupying shells at distances greater than *t*.

Figure 6.14 is also useful for understanding the basic goal in the search for good codes. We would like for the space to be filled with as many codewords as possible (efficient utilization of the added redundancy), and we would also like these codewords to be as far away from one another as possible. Obviously, these goals conflict.

6.5.5 Erasure Correction

A receiver may be designed to declare a symbol *erased* when it is received ambiguously or when the receiver recognizes the presence of interference or a transient malfunction. Such a channel has an input alphabet of size Q and an output alphabet of size $Q + 1$; the extra output symbol is called an *erasure flag,* or simply an *erasure.* When a demodulator makes a symbol error, two parameters are needed to correct that error, its *location* and its *correct* symbol value. In the case of binary symbols, this reduces to needing only the error location. However, if the demodulator declares a symbol *erased,* although the correct symbol value is not known, the symbol location *is* known, and for this reason, the decoding of erased codewords can be simpler than error correcting. An error control code can be used to correct erasures or to correct errors and erasures simultaneously. If the code has minimum distance d_{min}, any pattern of ρ or fewer erasures can be corrected if [6]

$$d_{\min} \geq \rho + 1 \qquad (6.50)$$

Assume for the moment that no errors occur outside the erasure positions. The advantage of correcting by means of erasures is expressed quantitatively as follows: If a code has a minimum distance $d_{\min}$, then from Equation (6.50), $d_{\min} - 1$ erasures can be reconstituted. Since the number of errors that can be corrected without erasure information is $(d_{\min} - 1)/2$ at most, from Equation (6.44), the advantage of correcting by means of erasures is clear. Further, any pattern of α errors and γ erasures can be corrected simultaneously if [6]

$$d_{\min} \geq 2\alpha + \gamma + 1 \qquad (6.51)$$

Simultaneous erasure correction and error correction can be accomplished in the following way. First, the γ-erased positions are replaced with zeros and the resulting codeword is decoded normally. Next, the γ-erased positions are replaced with ones, and the decoding operation is repeated on this version of the codeword. Of the two codewords obtained (one with erasures replaced by zeros, and the other with erasures replaced by ones) the one corresponding to the smallest number of errors corrected outside the γ-erased positions is selected. This technique will always result in correct decoding if Equation (6.51) is satisfied.

Example 6.6   Erasure Correction

Consider the codeword set presented in Section 6.4.3:

000000   110100   011010   101110   101001   011101   110011   000111

Suppose that the codeword 110011 was transmitted and that the two leftmost digits were declared by the receiver to be erasures. Verify that the received flawed sequence xx0011 can be corrected.

Solution

Since $d_{\min} = \rho + 1 = 3$, the code can correct as many as $\rho = 2$ erasures. This is easily verified above or with Figure 6.11 by comparing the rightmost four digits of xx0011 with each of the allowable codewords. The codeword that was actually transmitted is closest in Hamming distance to the flawed sequence.

6.6 USEFULNESS OF THE STANDARD ARRAY

6.6.1 Estimating Code Capability

The standard array can be thought of as an organizational tool, a filing cabinet that contains all of the possible 2^n entries in the space of n-tuples—nothing missing, and nothing replicated. At first glance, the benefits of this tool seem limited to *small* block codes, because for code lengths beyond $n = 20$, there are millions of n-tuples in the space. However, even for large codes, the standard array allows visualization of important performance issues, such as possible trade-offs between error correction and detection, as well as bounds on error-correction capability. One such bound, called the *Hamming bound* [7], is described as

Number of parity bits: $\quad n - k \geq \log_2 \left[1 + \binom{n}{1} + \binom{n}{2} + \cdots + \binom{n}{t} \right]$ (6.52a)

or

Number of cosets: $\quad 2^{n-k} \geq \left[1 + \binom{n}{1} + \binom{n}{2} + \cdots + \binom{n}{t} \right]$ (6.52b)

where $\binom{n}{j}$, defined in Equation (6.16), represents the number of ways in which j bits out of n may be in error. Note that the sum of the terms within the square brackets of Equation (6.52) yields the minimum number of rows needed in the standard array to correct all combinations of errors through t-bit errors. The inequality gives a lower bound on $n - k$, the number of parity bits (or the number of 2^{n-k} cosets) as a function of the t-bit error-correction capability of the code. Similarly, the inequality can be described as giving an upper bound on the t-bit error correction capability as a function of the number of $n - k$ parity bits (or 2^{n-k} cosets). For any (n, k) linear block code to provide a t-bit error-correcting capability, it is a necessary condition that the Hamming bound be met.

To demonstrate how the standard array provides a visualization of this bound, let us choose the (127, 106) BCH code as an example. The array contains all $2^n = 2^{127} \approx 1.70 \times 10^{38}$ n-tuples in the space. The topmost row of the array contains the $2^k = 2^{106} \approx 8.11 \times 10^{31}$ codewords; hence, this is the number of columns in the array. The leftmost column contains the $2^{n-k} = 2^{21} = 2,097,152$ coset leaders; hence, this is the number of rows in the array. Although the number of n-tuples and codewords is enormous, the concern is not with any individual entry; the primary interest is in the number of cosets. There are 2,097,152 cosets, and hence there are at most 2,097,151 error patterns that can be corrected by this code. Next, it is shown how this number of cosets dictates an upper bound on the t-bit error correcting capability of the code.

Since each codeword contains 127 bits, there are 127 ways to make single errors. We next compute how many ways there are to make double errors, namely $\binom{127}{2} = 8,001$. We move on to triple errors because thus far only a small portion of the total 2,097,151 correctable error-patterns have been used. There are $\binom{127}{3} = 333,375$ ways to make triple errors. On Table 6.3 these computations are listed, indicating that the all-zeros error pattern requires the presence of the first coset, followed by single, double, and triple errors. Also shown are the number of cosets required for each error type and the cumulative number of cosets necessary through that error type. From this table it can be seen that a (127, 106) code can correct all single-, double-, and triple-error patterns, which only account for 341,504 of the available 2,097,152 cosets. The unused 1,755,648 rows are indicative of the fact that more error correction is possible. Indeed, it might be tempting to try fitting all possible 4-bit error patterns into the array. However, from Table 6.3 it can be seen that this is not possible, because the number of remaining cosets in the array is much smaller than the cumulative number of cosets required for correcting 4-bit errors, as indicated by the last line of the table. Therefore, for this (127, 106) example, the code has a Hamming bound that guarantees the correction of up to and including all 3-bit errors.

TABLE 6.3 Error-Correction Bound for the (127, 106) Code

Number of Bit Errors	Number of Cosets Required	Cumulative Number of Cosets Required
0	1	1
1	127	128
2	8,001	8,129
3	333,375	341,504
4	10,334,625	10,676,129

6.6.2 An (n, k) Example

The standard array provides insight into the trade-offs that are possible between error correction and detection. Consider a new (n, k) code example, and the factors that dictate what values of (n, k) should be chosen.

1. In order to perform a nontrivial trade-off between error correction and error detection, it is desired that the code have an error-correcting capability of at least $t = 2$. We use Equation (6.44) for finding the minimum distance, as follows: $d_{\min} = 2t + 1 = 5$.

2. For a nontrivial code system, it is desired that the number of data bits be at least $k = 2$. Thus, there will be $2^k = 4$ codewords. The code can now be designated as an (n, 2) code.

3. We look for the minimum value of n that will allow correcting all possible single and double errors. In this example, each of the 2^n n-tuples in the array will be tabulated. The minimum value of n is desired because whenever n is incremented by just a single integer, the number of n-tuples in the standard array doubles. It is, of course, desired that the list be of manageable size. For real world codes, we want the minimum n for different reasons—bandwidth efficiency and simplicity. If the Hamming bound is used in choosing n, then $n = 7$ could be selected. However, the dimensions of such a (7, 2) code will not meet our stated requirements of $t = 2$-bit error-correction capability and $d_{\min} = 5$. To see this, it is necessary to introduce another upper bound on the t-bit error correction capability (or $d_{\min}$). This bound, called the *Plotkin bound* [7], is described by

$$d_{\min} \leq \frac{n \times 2^{k-1}}{2^k - 1} \tag{6.53}$$

In general, a linear (n, k) code must meet all upper bounds involving error-correction capability (or minimum distance). For high-rate codes, if the Hamming bound is met, then the Plotkin bound will also be met; this was the case for the earlier (127, 106) code example. For low rate codes, it is the other way around [7]. Since this example entails a low-rate code, it is important to test error-correction capability via the Plotkin bound. Because $d_{\min} = 5$, it should be clear from Equation (6.53) that n must be 8, and therefore, the minimum dimensions of the code are (8, 2), in order to meet the requirements for this example. Would anyone use such a low-rate double-error correcting code as this (8, 2) code? No, it would be too bandwidth expansive compared with more efficient codes that are available. It is used here for tutorial purposes, because its standard array is of manageable size.

6.6.3 Designing the (8, 2) Code

A natural question to ask is, How does one select codewords out of the space of 2^8 8-tuples? There is not a single solution, but there are constraints in how choices are made. The following are the elements that help point to a solution:

1. The number of codewords is $2^k = 2^2 = 4$.
2. The all-zeros vector must be one of the codewords.
3. The property of closure must apply. This property dictates that the sum of any two codewords in the space must yield a valid codeword in the space.
4. Each codeword is 8 bits long.
5. Since $d_{min} = 5$, the weight of each codeword (except for the all-zeros codeword) must also be at least 5 (by virtue of the closure property). The weight of a vector is defined as the number of nonzero components in the vector.
6. Assume that the code is systematic, and thus the rightmost 2 bits of each codeword are the corresponding message bits.

The following is a candidate assignment of codewords to messages that meets all of the preceeding conditions:

Messages	Codewords
00	00000000
01	11110001
10	00111110
11	11001111

The design of the codeword set can begin in a very arbitrary way; it is only necessary to adhere to the properties of weight and systematic form of the code. The selection of the first few codewords is often simple. However, as the process continues, the selection routine becomes harder, and the choices become more constrained because of the need to adhere to the closure property.

6.6.4 Error Detection versus Error Correction Trade-offs

For the (8, 2) code system selected in the previous section, the $(k \times n) = (2 \times 8)$ generator matrix can be written as

$$\mathbf{G} = \begin{bmatrix} 0\,0\,1\,1\,1\,1\,1\,0 \\ 1\,1\,1\,1\,0\,0\,0\,1 \end{bmatrix}$$

Decoding starts with the computation of a syndrome, which can be thought of as learning the "symptom" of an error. For an (n, k) code, an $(n - k)$-bit syndrome $\mathbf{S}$ is the product of an n-bit received vector $\mathbf{r}$, and the transpose of an $(n-k) \times n$ parity-check matrix $\mathbf{H}$, where $\mathbf{H}$ is constructed so that the rows of $\mathbf{G}$ are orthogonal to the rows of $\mathbf{H}$; that is $\mathbf{GH}^T = \mathbf{0}$. For this $(8, 2)$ example, $\mathbf{S}$ is a 6-bit vector, and $\mathbf{H}$ is a 6×8 matrix, where

$$
\mathbf{H}^T = \begin{bmatrix}
1 & 0 & 0 & 0 & 0 & 0 \\
0 & 1 & 0 & 0 & 0 & 0 \\
0 & 0 & 1 & 0 & 0 & 0 \\
0 & 0 & 0 & 1 & 0 & 0 \\
0 & 0 & 0 & 0 & 1 & 0 \\
0 & 0 & 0 & 0 & 0 & 1 \\
0 & 0 & 1 & 1 & 1 & 1 \\
1 & 1 & 1 & 1 & 0 & 0
\end{bmatrix}
$$

The syndrome for each error pattern can be calculated using Equation (6.37), namely

$$ \mathbf{S}_i = \mathbf{e}_i\mathbf{H}^T \qquad i = 1, \cdots, 2^{n-k} $$

where $\mathbf{S}_i$ is one of the $2^{n-k} = 64$ syndromes, and $\mathbf{e}_i$ is one of the 64 coset leaders (error patterns) in the standard array. Figure 6.15 shows a tabulation of the standard array as well as all 64 syndromes for the $(8, 2)$ code. The set of syndromes were calculated by using Equation (6.37); the entries of any given row (coset) of the standard array have the same syndrome. The correction of a corrupted codeword proceeds by computing its syndrome and locating the error pattern that corresponds to that syndrome. Finally, the error pattern is modulo-2 added to the corrupted codeword yielding the corrected output. Equation (6.49), repeated below, indicates that error-detection and error-correction capabilities can be traded, provided that the distance relationship

$$ d_{\min} \geq \alpha + \beta + 1 $$

prevails. Here, α represents the number of bit errors to be corrected, β represents the number of bit errors to be detected, and $\beta \geq \alpha$. The trade-off choices available for the $(8, 2)$ code example are as follows:

Detect (β)	Correct (α)
2	2
3	1
4	0

This table shows that the $(8, 2)$ code can be implemented to perform only error correction, which means that it first detects as many as $\beta = 2$ errors and then corrects them. If some error correction is sacrificed so that the code will only correct single errors, then the detection capability is increased so that all $\beta = 3$ errors can be detected. Finally, if error correction is completely sacrificed, the decoder can be implemented so that all $\beta = 4$ errors can be detected. In the case of error detection

Syndromes		Standard array			
000000	1.	00000000	11110001	00111110	11001111
111100	2.	00000001	11110000	00111111	11001110
001111	3.	00000010	11110011	00111100	11001101
000001	4.	00000100	11110101	00111010	11001011
000010	5.	00001000	11111001	00110110	11000111
000100	6.	00010000	11100001	00101110	11011111
001000	7.	00100000	11010001	00011110	11101111
010000	8.	01000000	10110001	01111110	10001111
100000	9.	10000000	01110001	10111110	01001111
110011	10.	00000011	11110010	00111101	11001100
111101	11.	00000101	11110100	00111011	11001010
111110	12.	00001001	11111000	00110111	11000110
111000	13.	00010001	11100000	00101111	11011110
110100	14.	00100001	11010000	00011111	11101110
101100	15.	01000001	10110000	01111111	10001110
011100	16.	10000001	01110000	10111111	01001110
001110	17.	00000110	11110111	00111000	11001001
001101	18.	00001010	11111011	00110100	11000101
001011	19.	00010010	11100011	00101100	11011101
000111	20.	00100010	11010011	00011100	11101101
011111	21.	01000010	10110011	01111100	10001101
101111	22.	10000010	01110011	10111100	01001101
000011	23.	00001100	11111101	00110010	11000011
000101	24.	00010100	11100101	00101010	11011011
001001	25.	00100100	11010101	00011010	11101011
010001	26.	01000100	10110101	01111010	10001011
100001	27.	10000100	01110101	10111010	01001011
000110	28.	00011000	11101111	00100110	11010111
001010	29.	00101000	11011001	00010110	11100111
010010	30.	01001000	10111001	01110110	10000111
100010	31.	10001000	01111001	10110110	01000111
001100	32.	00110000	11000001	00001110	11111111
010100	33.	01010000	10100001	01101110	10011111
100100	34.	10010000	01100001	10101110	01011111
011000	35.	01100000	10010001	01011110	10101111
101000	36.	10100000	01010001	10011110	01101111
110000	37.	11000000	00110001	11111110	00001111
110010	38.	00000111	11110110	00111001	11001000
110111	39.	00010011	11100010	00101101	11011100
111011	40.	00100011	11010010	00011101	11101100
100011	41.	01000011	10110010	01111101	10001100
010011	42.	10000011	01110010	10111101	01001100
111111	43.	00001101	11111100	00110011	11000010
111001	44.	00010101	11100100	00101011	11011010
110101	45.	00100101	11010100	00011011	11101010
101101	46.	01000101	10110100	01111011	10001010
011101	47.	10000101	01110100	10111011	01001010
011110	48.	01000110	10110111	01111000	10001001
101110	49.	10000110	01110111	10111000	01001001
100101	50.	10010100	01100101	10101010	01011011
011001	51.	01100100	10010101	01011010	10101011
110001	52.	11000100	00110101	11111010	00001011
011010	53.	01101000	10011001	01010110	10100111
010110	54.	01011000	10101001	01100110	10010111
100110	55.	10011000	01101001	10100110	01010111
101010	56.	10101000	01011001	10010110	01100111
101001	57.	10100100	01010101	10011010	01101011
100111	58.	10100010	01010011	10011100	01101101
010111	59.	01100010	10010011	01011100	10101101
010101	60.	01010100	10100101	01101010	10011011
011011	61.	01010010	10100011	01101100	10011101
110110	62.	00101001	11011000	00010111	11100110
111010	63.	00011001	11101000	00100111	11010110
101011	64.	10010010	01100011	10101100	01011101

Figure 6.15 The syndromes and the standard array for the (8, 2) code.

only, the circuitry is very simple. The syndrome is computed and an error is detected whenever a nonzero syndrome occurs.

For correcting single errors, the decoder can be implemented with gates [4], similar to the circuitry in Figure 6.12, where a received code vector **r** enters at two places. In the top part of the figure, the received digits are connected to exclusive-OR gates, which yield the syndrome. For any given received vector, the syndrome is obtained from Equation (6.35) as

$$\mathbf{S}_i = \mathbf{r}_i \mathbf{H}^T \qquad i = 1, \cdots, 2^{n-k}$$

Using the $\mathbf{H}^T$ values for the (8, 2) code, the wiring between the received digits and the exclusive-OR gates in a circuit similar to the one in Figure 6.12, must be connected to yield

$$\mathbf{S}_i = \begin{bmatrix} r_1 & r_2 & r_3 & r_4 & r_5 & r_6 & r_7 & r_8 \end{bmatrix} \begin{bmatrix} 1 & 0 & 0 & 0 & 0 & 0 \\ 0 & 1 & 0 & 0 & 0 & 0 \\ 0 & 0 & 1 & 0 & 0 & 0 \\ 0 & 0 & 0 & 1 & 0 & 0 \\ 0 & 0 & 0 & 0 & 1 & 0 \\ 0 & 0 & 0 & 0 & 0 & 1 \\ 0 & 0 & 1 & 1 & 1 & 1 \\ 1 & 1 & 1 & 1 & 0 & 0 \end{bmatrix}$$

Each of the s_j digits ($j = 1, \ldots, 6$) making up syndrome $\mathbf{S}_i$ ($i = 1, \ldots, 64$) is related to the input-received code vector in the following way:

$$s_1 = r_1 + r_8 \qquad s_2 = r_2 + r_8 \qquad s_3 = r_3 + r_7 + r_8$$
$$s_4 = r_4 + r_7 + r_8 \qquad s_5 = r_5 + r_7 \qquad s_6 = r_6 + r_7$$

To implement a decoder circuit similar to Figure 6.12 for the (8, 2) code necessitates that the eight received digits be connected to six modulo-2 adders yielding the syndrome digits as described above. Additional modifications to the figure need to be made accordingly.

If the decoder is implemented to correct only single errors; that is $\alpha = 1$ and $\beta = 3$, then this is tantamount to drawing a line under coset 9 in Figure 6.15, and error correction takes place only when one of the eight syndromes associated with a single error appears. The decoding circuitry (similar to Figure 6.12) then transforms the syndrome to its corresponding error pattern. The error pattern is then modulo-2 added to the "potentially" corrupted received vector, yielding the corrected output. Additional gates are needed to test for the case in which the syndrome is nonzero and there is no correction designed to take place. For single-error correction, such an event happens for any of the syndromes numbered 10 through 64. This outcome is then used to indicate an error detection.

If the decoder is implemented to correct single and double errors, which means that $\beta = 2$ errors are detected and then corrected, then this is tantamount to drawing a line under coset 37 in the standard array of Figure 6.15. Even though this (8, 2) code is capable of correcting some combination of triple errors corresponding to the coset leaders 38 through 64, a decoder is most often implemented as a *bounded distance* decoder, which means that it corrects all combinations of errors up to and including t errors, but no combinations of errors greater than t. The unused error-correction capability can be applied toward some error-detection enhancement. As before, the decoder can be implemented with gates similar to those shown in Figure 6.12.

6.6.5 The Standard Array Provides Insight

In the context of Figure 6.15, the (8, 2) code satisfies the Hamming bound. That is, from the standard array it is recognizable that the (8, 2) code can correct all combinations of single and double errors. Consider the following question: Given that transmission takes place over a channel that always introduces errors in the form of a burst of 3-bit errors and thus there is no interest in correcting single or double errors, wouldn't it be possible to set up the coset leaders to correspond to only triple errors? It is simple to see that in a sequence of 8 bits there are $\binom{8}{3} = 56$ ways to make triple errors. If we only want to correct all these 56 combinations of triple errors, there is sufficient room (sufficient number of cosets) in the standard array, since there are 64 rows. Will that work? No, it will not. For any code, the overriding parameter for determining error-correcting capability is d_{min}. For the (8, 2) code, $d_{min} = 5$ dictates that only 2-bit error correction is possible.

How can the standard array provide some insight as to why this scheme won't work? In order for a group of x-bit error patterns to enable x-bit error correction, the entire group of weight-x vectors must be coset leaders; that is, they must only occupy the leftmost column. In figure 6.15, it can be seen that all weight-1 and weight-2 vectors appear in the leftmost column of the standard array, and nowhere else. Even if we forced all weight-3 vectors into row numbers 2 through 57, we would find that some of these vectors would have to reappear elsewhere in the array (which violates a basic property of the standard array). In Figure 6.15 a shaded box is drawn around every one of the 56 vectors having a weight of 3. Look at the coset leaders representing 3-bit error patterns, in rows 38, 41–43, 46–49, and 52 of the standard array. Now look at the entries of the same row numbers in the rightmost column, where shaded boxes indicate other weight-3 vectors. Do you see the ambiguity that exists for each of the rows listed above, and why it is not possible to correct all 3-bit error patterns with this (8, 2) code? Suppose the decoder receives the weight-3 vector 1 1 0 0 1 0 0 0, located at row 38 in the rightmost column. This flawed codeword could have arisen in one of two ways: One would be that codeword 1 1 0 0 1 1 1 1 was sent and the 3-bit error pattern 0 0 0 0 0 1 1 1 perturbed it; the other would be that codeword 0 0 0 0 0 0 0 0 was sent and the 3-bit error pattern 1 1 0 0 1 0 0 0 perturbed it.

6.7 CYCLIC CODES

Binary cyclic codes are an important subclass of linear block codes. The codes are easily implemented with feedback shift registers; the syndrome calculation is easily accomplished with similar feedback shift registers; and the underlying algebraic structure of a cyclic code lends itself to efficient decoding methods. An (n, k) linear code is called a *cyclic code* if it can be described by the following property. If the n-tuple $\mathbf{U} = (u_0, u_1, u_2, \ldots, u_{n-1})$ is a codeword in the subspace S, then $\mathbf{U}^{(1)} = (u_{n-1}, u_0, u_1, u_2, \ldots, u_{n-2})$ obtained by an end-around shift, is also a codeword in S. Or, in general, $\mathbf{U}^{(i)} = (u_{n-i}, u_{n-i+1}, \ldots u_{n-1}, u_0, u_1, \ldots, u_{n-i-1})$, obtained by i end-around or cyclic shifts, is also a codeword in S.

The components of a codeword $\mathbf{U} = (u_0, u_1, u_2, \ldots, u_{n-1})$ can be treated as the coefficients of a polynomial $\mathbf{U}(X)$ as follows:

$$\mathbf{U}(X) = u_0 + u_1 X + u_2 X^2 + \cdots + u_{n-1} X^{n-1} \tag{6.54}$$

The polynomial function $\mathbf{U}(X)$ can be thought of as a "placeholder" for the digits of the codeword $\mathbf{U}$; that is, an n-tuple vector is described by a polynomial of degree $n - 1$ or less. The presence or absence of each term in the polynomial indicates the presence of a 1 or 0 in the corresponding location of the n-tuple. If the u_{n-1} component is nonzero, the polynomial is of degree $n - 1$. The usefulness of this polynomial description of a codeword will become clear as we discuss the algebraic structure of the cyclic codes.

6.7.1 Algebraic Structure of Cyclic Codes

Expressing the codewords in polynomial form, the cyclic nature of the code manifests itself in the following way. If $\mathbf{U}(X)$ is an $(n - 1)$-degree codeword polynomial, then $\mathbf{U}^{(i)}(X)$, the remainder resulting from dividing $X^i \mathbf{U}(X)$ by $X^n + 1$, is also a codeword; that is,

$$\frac{X^i \mathbf{U}(X)}{X^n + 1} = \mathbf{q}(X) + \frac{\mathbf{U}^{(i)}(X)}{X^n + 1} \tag{6.55a}$$

or, multiplying through by $X^n + 1$,

$$X^i \mathbf{U}(X) = \mathbf{q}(X)(X^n + 1) + \underbrace{\mathbf{U}^{(i)}(X)}_{\text{remainder}} \tag{6.55b}$$

which can also be described in terms of modulo arithmetic as

$$\mathbf{U}^{(i)}(X) = X^i \mathbf{U}(X) \text{ modulo } (X^n + 1) \tag{6.56}$$

where x modulo y is defined as the remainder obtained from dividing x by y. Let us demonstrate the validity of Equation (6.56) for the case of $i = 1$:

$$\mathbf{U}(X) = u_0 + u_1 X + u_2 X^2 + \cdots + u_{n-2} X^{n-2} + u_{n-1} X^{n-1}$$
$$X\mathbf{U}(X) = u_0 X + u_1 X^2 + u_2 X^3 + \cdots + u_{n-2} X^{n-1} + u_{n-1} X^n$$

We now add and subtract u_{n-1}; or, since we are using modulo-2 arithmetic, we add u_{n-1} twice, as follows:

$$X\mathbf{U}(X) = \underbrace{u_{n-1} + u_0 X + u_1 X^2 + u_2 X^3 + \cdots + u_{n-2} X^{n-1}}_{\mathbf{U}^{(1)}(X)} + u_{n-1} X^n + u_{n-1}$$

$$= \mathbf{U}^{(1)}(X) + u_{n-1}(X^n + 1)$$

Since $\mathbf{U}^{(1)}(X)$ is of degree $n - 1$, it cannot be divided by $X^n + 1$. Thus, from Equation (6.55a), we can write

$$\mathbf{U}^{(1)}(X) = X\mathbf{U}(X) \text{ modulo } (X^n + 1)$$

By extension, we arrive at Equation (6.56):

$$\mathbf{U}^{(i)}(X) = X^i\mathbf{U}(X) \text{ modulo } (X^n + 1)$$

Example 6.7   Cyclic Shift of a Code Vector

Let $\mathbf{U} = 1\ 1\ 0\ 1$, for $n = 4$. Express the codeword in polynomial form, and using Equation (6.56), solve for the third end-around shift of the codeword.

Solution

$$\mathbf{U}(X) = 1 + X + X^3 \qquad \text{(polynomial is written low order to high order)};$$
$$X^i\mathbf{U}(X) = X^3 + X^4 + X^6, \quad \text{where } i = 3.$$

Divide $X^3\mathbf{U}(X)$ by $X^4 + 1$, and solve for the remainder using polynomial division:

$$
\begin{array}{r}
X^2 + 1 \\
X^4 + 1 \overline{)X^6 + X^4 + X^3 } \\
\underline{X^6 + X^2 } \\
X^4 + X^3 + X^2 \\
\underline{X^4 + 1} \\
X^3 + X^2 + 1 \qquad \text{remainder } \mathbf{U}^{(3)}(X)
\end{array}
$$

Writing the remainder low order to high order: $1 + X^2 + X^3$, the codeword $\mathbf{U}^{(3)} = 1\ 0\ 1\ 1$ is three cyclic shifts of $\mathbf{U} = 1\ 1\ 0\ 1$. Remember that for binary codes, the addition operation is performed modulo-2, so that $+ 1 = -1$, and we consequently do not show any minus signs in the computation.

6.7.2 Binary Cyclic Code Properties

We can generate a cyclic code using a *generator polynomial* in much the way that we generated a block code using a generator matrix. The generator polynomial $\mathbf{g}(X)$ for an (n, k) cyclic code is unique and is of the form

$$\mathbf{g}(X) = g_0 + g_1X + g_2X^2 + \cdots + g_pX^p \qquad (6.57)$$

where g_0 and g_p must equal 1. Every codeword polynomial in the subspace is of the form $\mathbf{U}(X) = \mathbf{m}(X)\mathbf{g}(X)$, where $\mathbf{U}(X)$ is a polynomial of degree $n - 1$ or less. Therefore, the message polynomial $\mathbf{m}(X)$ is written as

$$\mathbf{m}(X) = m_0 + m_1X + m_2X^2 + \cdots + m_{n-p-1}X^{n-p-1} \qquad (6.58)$$

There are 2^{n-p} codeword polynomials, and there are 2^k code vectors in an (n, k) code. Since there must be one codeword polynomial for each code vector,

$$n - p = k$$

or

$$p = n - k$$

Hence, $\mathbf{g}(X)$, as shown in Equation (6.57), must be of degree $n - k$, and every codeword polynomial in the (n, k) cyclic code can be expressed as

$$\mathbf{U}(X) = (m_0 + m_1 X + m_2 X^2 + \cdots + m_{k-1} X^{k-1}) \mathbf{g}(X) \tag{6.59}$$

$\mathbf{U}$ is said to be a valid codeword of the subspace S *if, and only if,* $\mathbf{g}(X)$ divides into $\mathbf{U}(X)$ without a remainder.

A generator polynomial $\mathbf{g}(X)$ of an (n, k) cyclic code is a factor of $X^n + 1$; that is, $X^n + 1 = \mathbf{g}(X)\mathbf{h}(X)$. For example,

$$X^7 + 1 = (1 + X + X^3)(1 + X + X^2 + X^4)$$

Using $\mathbf{g}(X) = 1 + X + X^3$ as a generator polynomial of degree $n - k = 3$, we can generate an $(n, k) = (7, 4)$ cyclic code. Or, using $\mathbf{g}(X) = 1 + X + X^2 + X^4$ where $n - k = 4$ we can generate a $(7, 3)$ cyclic code. In summary, if $\mathbf{g}(X)$ is a polynomial of degree $n - k$ and is a factor of $X^n + 1$, then $\mathbf{g}(X)$ uniquely generates an (n, k) cyclic code.

6.7.3 Encoding in Systematic Form

In Section 6.4.5 we introduced the *systematic* form and discussed the reduction in complexity that makes this encoding form attractive. Let us use some of the algebraic properties of the cyclic code to establish a systematic encoding procedure. We can express the message vector in polynomial form, as follows:

$$\mathbf{m}(X) = m_0 + m_1 X + m_2 X^2 + \cdots + m_{k-1} X^{k-1} \tag{6.60}$$

In systematic form, the message digits are utilized as part of the codeword. We can think of shifting the message digits into the rightmost k stages of a codeword register, and then appending the parity digits by placing them in the leftmost $n - k$ stages. Therefore, we manipulate the message polynomial algebraically so that it is right-shifted $n - k$ positions. If we multiply $\mathbf{m}(X)$ by X^{n-k}, we get the right-shifted message polynomial:

$$X^{n-k} \mathbf{m}(X) = m_0 X^{n-k} + m_1 X^{n-k+1} + \cdots + m_{k-1} X^{n-1} \tag{6.61}$$

If we next divide Equation (6.61) by $\mathbf{g}(X)$, the result can be expressed as

$$X^{n-k} \mathbf{m}(X) = \mathbf{q}(X)\mathbf{g}(X) + \mathbf{p}(X) \tag{6.62}$$

where the remainder $\mathbf{p}(X)$ can be expressed as

$$\mathbf{p}(X) = p_0 + p_1 X + p_2 X^2 + \cdots + p_{n-k-1} X^{n-k-1}$$

We can also say that

$$\mathbf{p}(X) = X^{n-k} \mathbf{m}(X) \text{ modulo } \mathbf{g}(X) \tag{6.63}$$

Adding $\mathbf{p}(X)$ to both sides of Equation (6.62), using modulo-2 arithmetic, we get

$$\mathbf{p}(X) + X^{n-k} \mathbf{m}(X) = \mathbf{q}(X)\mathbf{g}(X) = \mathbf{U}(X) \tag{6.64}$$

The left-hand side of Equation (6.64) is recognized as a valid codeword polynomial, since it is a polynomial of degree $n - 1$ or less, and when divided by $\mathbf{g}(X)$ there is a zero remainder. This codeword can be expanded into its polynomial terms as follows:

$$\mathbf{p}(X) + X^{n-k}\mathbf{m}(X) = p_0 + p_1 X + \cdots + p_{n-k-1}X^{n-k-1}$$
$$+ m_0 X^{n-k} + m_1 X^{n-k+1} + \cdots + m_{k-1}X^{n-1}$$

The codeword polynomial corresponds to the code vector

$$\mathbf{U} = \underbrace{(p_0, p_1, \ldots, p_{n-k-1},}_{(n-k) \text{ parity bits}} \underbrace{m_0, m_1, \ldots, m_{k-1})}_{k \text{ message bits}} \tag{6.65}$$

Example 6.8   Cyclic Code in Systematic Form

Using the generator polynomial $\mathbf{g}(X) = 1 + X + X^3$, generate a systematic codeword from the $(7, 4)$ codeword set for the message vector $\mathbf{m} = 1\,0\,1\,1$.

Solution

$$\mathbf{m}(X) = 1 + X^2 + X^3, \quad n = 7, \quad k = 4, \quad n - k = 3;$$
$$X^{n-k}\mathbf{m}(X) = X^3(1 + X^2 + X^3) = X^3 + X^5 + X^6$$

Dividing $X^{n-k}\mathbf{m}(X)$ by $\mathbf{g}(X)$ using polynomial division, we can write

$$X^3 + X^5 + X^6 = \underbrace{(1 + X + X^2 + X^3)}_{\substack{\text{quotient} \\ \mathbf{q}(X)}} \underbrace{(1 + X + X^3)}_{\substack{\text{generator} \\ \mathbf{g}(X)}} + \underbrace{1}_{\substack{\text{remainder} \\ \mathbf{p}(X)}}$$

Using Equation (6.64) yields

$$\mathbf{U}(X) = \mathbf{p}(X) + X^3\mathbf{m}(X) = 1 + X^3 + X^5 + X^6$$
$$\mathbf{U} = \underbrace{1\,0\,0}_{\substack{\text{parity} \\ \text{bits}}} \quad \underbrace{1\,0\,1\,1}_{\substack{\text{message} \\ \text{bits}}}$$

6.7.4 Circuit for Dividing Polynomials

We have seen that the cyclic shift of a codeword polynomial and that the encoding of a message polynomial involves the division of one polynomial by another. Such an operation is readily accomplished by a *dividing circuit* (feedback shift register). Given two polynomials $\mathbf{V}(X)$ and $\mathbf{g}(X)$, where

$$\mathbf{V}(X) = v_0 + v_1 X + v_2 X^2 + \cdots + v_m X^m$$

and

$$\mathbf{g}(X) = g_0 + g_1 X + g_2 X^2 + \cdots + g_p X^p$$

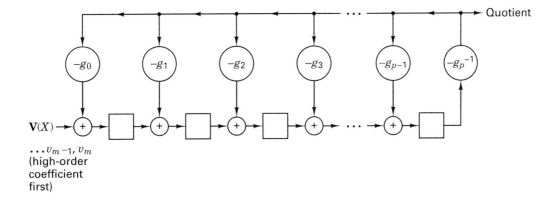

Quotient

$V(X) \rightarrow$

$\ldots v_{m-1}, v_m$
(high-order
coefficient
first)

Figure 6.16 Circuit for dividing polynomials.

such that $m \geq p$, the divider circuit of Figure 6.16 performs the polynomial division steps of dividing $\mathbf{V}(X)$ by $\mathbf{g}(X)$, thereby determining the quotient and remainder terms:

$$\frac{\mathbf{V}(X)}{\mathbf{g}(X)} = \mathbf{q}(X) + \frac{\mathbf{p}(X)}{\mathbf{g}(X)}$$

The stages of the register are first initialized by being filled with zeros. The first p shifts enter the most significant (higher-order) coefficients of $\mathbf{V}(X)$. After the pth shift, the quotient output is $g_p^{-1} v_m$; this is the highest-order term in the quotient. For each quotient coefficient q_i the polynomial $q_i \mathbf{g}(X)$ must be subtracted from the dividend. The feedback connections in Figure 6.16 perform this subtraction. The difference between the leftmost p terms remaining in the dividend and the feedback terms $q_i \mathbf{g}(X)$ is formed on each shift of the circuit and appears as the contents of the register. At each shift of the register, the difference is shifted one stage; the highest-order term (which by construction is zero) is shifted out, while the next significant coefficient of $\mathbf{V}(X)$ is shifted in. After $m + 1$ total shifts into the register, the quotient has been serially presented at the output and the remainder resides in the register.

Example 6.9   Dividing Circuit

Use a dividing circuit of the form shown in Figure 6.16 to divide $\mathbf{V}(X) = X^3 + X^5 + X^6$ ($\mathbf{V} = 0\ 0\ 0\ 1\ 0\ 1\ 1$) by $\mathbf{g}(X) = (1 + X + X^3)$. Find the quotient and remainder terms. Compare the circuit implementation to the polynomial division steps performed by hand.

Solution

The dividing circuit needs to perform the following operation:

$$\frac{X^3 + X^5 + X^6}{1 + X + X^3} = \mathbf{q}(X) + \frac{\mathbf{p}(X)}{1 + X + X^3}$$

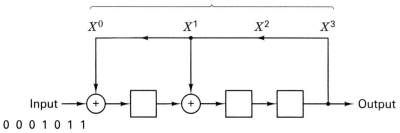

Feedback polynomial

Input → 0 0 0 1 0 1 1

Output

Figure 6.17   Dividing circuit for Example 6.9.

The required feedback shift register, following the general form of Figure 6.16, is shown in Figure 6.17. Assume that the register contents are initially zero. The operational steps of the circuit are as follows:

Input queue	Shift number	Register contents	Output and Feedback
0 0 0 1 0 1 1	0	0 0 0	–
0 0 0 1 0 1	1	1 0 0	0
0 0 0 1 0	2	1 1 0	0
0 0 0 1	3	0 1 1	0
0 0 0	4	0 1 1	1
0 0	5	1 1 1	1
0	6	1 0 1	1
–	7	1 0 0	1

After the fourth shift, the quotient coefficients $\{q_i\}$ serially presented at the output are seen to be 1 1 1 1, or the quotient polynomial is $\mathbf{q}(X) = 1 + X + X^2 + X^3$. The remainder coefficients $\{p_i\}$ are 1 0 0, or the remainder polynomial $\mathbf{p}(X) = 1$. In summary, the circuit computation $\mathbf{V}(X)/\mathbf{g}(X)$ is seen to be

$$\frac{X^3 + X^5 + X^6}{1 + X + X^3} = 1 + X + X^2 + X^3 + \frac{1}{1 + X + X^3}$$

The polynomial division steps are as follows:

Output after shift number:

$$
\begin{array}{c}
\ \ 4\quad\ 5\quad\ 6\quad\ 7\\
\ \ \downarrow\quad\downarrow\quad\downarrow\quad\downarrow
\end{array}
$$

$$
\begin{array}{r}
X^3 + X^2 + X + 1 \\
X^3 + X + 1\ \overline{)\ X^6 + X^5 + X^3}
\end{array}
$$

$X^6 + X^4 + X^3$ ←———— feedback after 4th shift

$X^5 + X^4$ ←———— register after 4th shift

$X^5 + X^3 + X^2$ ←———— feedback after 5th shift

$X^4 + X^3 + X^2$ ←———— register after 5th shift

$X^4 + X^2 + X$ ←———— feedback after 6th shift

$X^3 + X$ ←———— register after 6th shift

$X^3 + X + 1$ ←——— feedback after 7th shift

1 ←——— register after 7th shift

(remainder)

6.7.5 Systematic Encoding with an (n − k)-Stage Shift Register

The encoding of a cyclic code in systematic form has been shown, in Section 6.7.3, to involve the computation of parity bits as the result of the formation of $X^{n-k}\mathbf{m}(X)$ modulo $\mathbf{g}(X)$, in other words, the *division* of an *upshifted* (right shifted) message polynomial by a generator polynomial $\mathbf{g}(X)$. The need for upshifting is to make room for the parity bits, which are appended to the message bits, yielding the code vector in systematic form. Upshifting the message bits by $n - k$ positions is a trivial operation and is not really performed as part of the dividing circuit. Instead, only the parity bits are computed; they are then placed in the appropriate location alongside the message bits. The parity polynomial is the *remainder* after dividing by the generator polynomial; it is available in the register after n shifts through the $(n - k)$-stage feedback register shown in Figure 6.17. Notice that the first $n - k$ shifts through the register are simply filling the register. We cannot have any feedback until the rightmost stage has been filled; we therefore can shorten the shifting cycle by loading the input data to the output of the last stage, as shown in Figure 6.18. Further, the feedback term into the leftmost stage is the sum of the input and the rightmost stage. We guarantee that this sum is generated by ensuring that $g_0 = g_{n-k} = 1$ for any generator polynomial $\mathbf{g}(X)$. The circuit feedback connections correspond to the coefficients of the generator polynomial, which is written as

$$\mathbf{g}(X) = 1 + g_1 X + g_2 X^2 + \cdots + g_{n-k-1} X^{n-k-1} + X^{n-k} \qquad (6.66)$$

The following steps describe the encoding procedure used with the Figure 6.18 encoder:

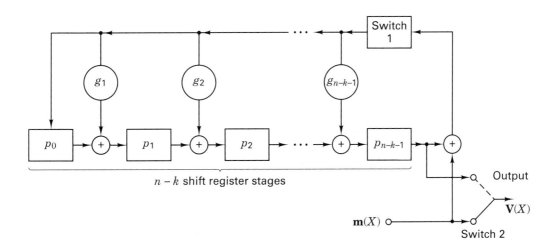

Figure 6.18 Encoding with an (n − k)-stage shift register.

1. Switch 1 is closed during the first k shifts, to allow transmission of the message bits into the $n - k$ stage encoding shift register.
2. Switch 2 is in the down position to allow transmission of the message bits directly to an output register during the first k shifts.
3. After transmission of the kth message bit, switch 1 is opened and switch 2 is moved to the up position.
4. The remaining $n - k$ shifts clear the encoding register by moving the parity bits to the output register.
5. The total number of shifts is equal to n, and the contents of the output register is the codeword polynomial $\mathbf{p}(X) + X^{n-k}\mathbf{m}(X)$.

Example 6.10   Systematic Encoding of a Cyclic Code

Use a feedback shift register of the form shown in Figure 6.18 to encode the message vector $\mathbf{m} = 1\ 0\ 1\ 1$ into a (7, 4) codeword using the generator polynomial $\mathbf{g}(X) = 1 + X + X^3$.

Solution

$$\mathbf{m} = 1\ 0\ 1\ 1$$

$$\mathbf{m}(X) = 1 + X^2 + X^3$$

$$X^{n-k}\mathbf{m}(X) = X^3\mathbf{m}(X) = X^3 + X^5 + X^6$$

$$X^{n-k}\mathbf{m}(X) = \mathbf{q}(X)\mathbf{g}(X) + \mathbf{p}(X)$$

$$\mathbf{p}(X) = (X^3 + X^5 + X^6)\ \text{modulo}\ (1 + X + X^3)$$

For the $(n - k) = 3$-stage encoding shift register shown in Figure 6.19, the operational steps are as follows:

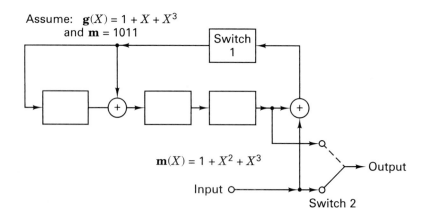

Figure 6.19 Example of encoding a (7, 4) cyclic code with an $(n - k)$-stage shift register.

Input queue	Shift number	Register contents	Output
1 0 1 1	0	0 0 0	–
1 0 1	1	1 1 0	1
1 0	2	1 0 1	1
1	3	1 0 0	0
–	4	1 0 0	1

After the fourth shift, switch 1 is opened, switch 2 is moved to the up position, and the parity bits contained in the register are shifted to the output. The output codeword is $\mathbf{U} = 1\,0\,0\,1\,0\,1\,1$, or in polynomial form, $\mathbf{U}(X) = 1 + X^3 + X^5 + X^6$.

6.7.6 Error Detection with an (n – k)-Stage Shift Register

A transmitted codeword may be perturbed by noise, and hence the vector received may be a corrupted version of the transmitted codeword. Let us assume that a codeword with polynomial representation $\mathbf{U}(X)$ is transmitted and that a vector with polynomial representation $\mathbf{Z}(X)$ is received. Since $\mathbf{U}(X)$ is a code polynomial, it must be a multiple of the generator polynomial $\mathbf{g}(X)$; that is,

$$\mathbf{U}(X) = \mathbf{m}(X)\,\mathbf{g}(X) \qquad (6.67)$$

and $\mathbf{Z}(X)$, the corrupted version of $\mathbf{U}(X)$, can be written as

$$\mathbf{Z}(X) = \mathbf{U}(X) + \mathbf{e}(X) \qquad (6.68)$$

where $\mathbf{e}(X)$ is the error pattern polynomial. The decoder tests whether $\mathbf{Z}(X)$ is a codeword polynomial, that is, whether it is divisible by $\mathbf{g}(X)$, with a zero remainder. This is accomplished by *calculating the syndrome* of the received polynomial. The syndrome $\mathbf{S}(X)$ is equal to the remainder resulting from dividing $\mathbf{Z}(X)$ by $\mathbf{g}(X)$; that is,

$$\mathbf{Z}(X) = \mathbf{q}(X)\,\mathbf{g}(X) + \mathbf{S}(X) \qquad (6.69)$$

where $\mathbf{S}(X)$ is a polynomial of degree $n - k - 1$ or less. Thus, the syndrome is an $(n - k)$-tuple. By combing Equations (6.67) to (6.69), we obtain

$$\mathbf{e}(X) = [\mathbf{m}(X) + \mathbf{q}(X)]\mathbf{g}(X) + \mathbf{S}(X) \qquad (6.70)$$

By comparing Equations (6.69) and (6.70), we see that the syndrome $\mathbf{S}(X)$, obtained as the remainder of $\mathbf{Z}(X)$ modulo $\mathbf{g}(X)$, is exactly the same polynomial obtained as the remainder of $\mathbf{e}(X)$ modulo $\mathbf{g}(X)$. Thus the syndrome of the received polynomial $\mathbf{Z}(X)$ contains the information needed for correction of the error pattern. The syndrome calculation is accomplished by a division circuit, almost identical to the encoding circuit used at the transmitter. An example of syndrome calculation with an $n - k$ shift register is shown in Figure 6.20 using the code vector generated in Example 6.10. Switch 1 is initially closed, and switch 2 is open. The received vector is shifted into the register input, with all stages initially set to zero. After the entire received vector has been entered into the shift register, the con-

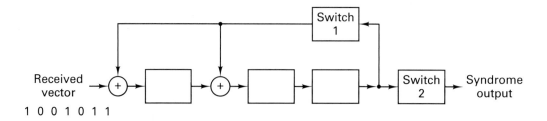

Figure 6.20 Example of syndrome calculation with an $(n - k)$-stage shift register.

tents of the register is the syndrome. Switch 1 is then opened and switch 2 is closed, so that the syndrome vector can be shifted out of the register. The operational steps of the decoder are as follows:

Input queue	Shift number	Register contents
1 0 0 1 0 1 1	0	0 0 0
1 0 0 1 0 1	1	1 0 0
1 0 0 1 0	2	1 1 0
1 0 0 1	3	0 1 1
1 0 0	4	0 1 1
1 0	5	1 1 1
1	6	1 0 1
–	7	0 0 0 Syndrome

If the syndrome is an all-zeros vector, the received vector is assumed to be a valid codeword. If the syndrome is a nonzero vector, the received vector is a perturbed codeword and errors have been detected; such errors can be corrected by adding the error vector (indicated by the syndrome) to the received vector, similar to the procedure described in Section 6.4.8. This method of decoding is useful for simple codes. More complex codes require the use of algebraic techniques to obtain practical decoders [6, 8].

6.8 WELL-KNOWN BLOCK CODES

6.8.1 Hamming Codes

Hamming codes are a simple class of block codes characterized by the structure

$$(n, k) = (2^m - 1, 2^m - 1 - m) \tag{6.71}$$

where $m = 2, 3, \ldots$. These codes have a minimum distance of 3 and thus, from Equations (6.44) and (6.47), they are capable of correcting all single errors or detecting all combinations of two or fewer errors within a block. Syndrome decoding is especially suited for Hamming codes. In fact, the syndrome can be formed to act as a binary pointer to identify the error location [5]. Although Hamming codes are not very powerful, they belong to a very limited class of block codes known as *perfect* codes, described in Section 6.5.4.

Assuming hard decision decoding, the bit error probability can be written, from Equation (6.46), as

$$P_B \approx \frac{1}{n} \sum_{j=2}^{n} j \binom{n}{j} p^j (1-p)^{n-j} \tag{6.72}$$

where p is the channel symbol error probability (transition probability on the binary symmetric channel). For single error-correcting codes (such as Hamming codes), in place of Equation (6.72) we can use the following equivalent equation. Its identity with Equation (6.72) is proven in Appendix D, Equation (D.16):

$$P_B \approx p - p(1-p)^{n-1} \tag{6.73}$$

Figure 6.21 is a plot of the decoded P_B versus channel-symbol error probability, illustrating the comparative performance for different types of block codes. For the

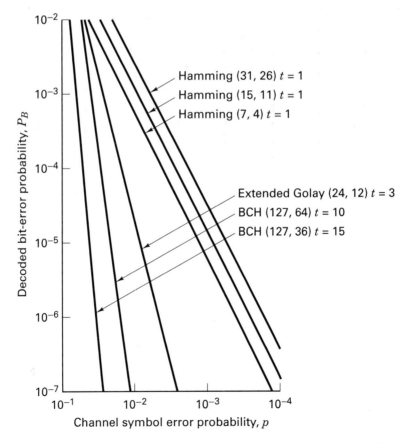

Figure 6.21 Bit error probability versus channel symbol error probability for several block codes.

Hamming codes, the plots are shown for $m = 3$, 4, and 5, or $(n, k) = (7, 4)$, $(15, 11)$, and $(31, 26)$. For performance over a Gaussian channel using coherently demodulated BPSK, we can express the channel symbol error probability in terms of E_c/N_0, similar to Equation (4.79), as

$$p = Q\left(\sqrt{\frac{2E_c}{N_0}}\right) \tag{6.74}$$

where E_c/N_0 is the code symbol energy per noise spectral density, and where $Q(x)$ is as defined in Equation (3.43). To relate E_c/N_0 to information bit energy per noise spectral density (E_b/N_0), we use

$$\frac{E_c}{N_0} = \left(\frac{k}{n}\right)\frac{E_b}{N_0} \tag{6.75}$$

For Hamming codes, Equation (6.75) becomes

$$\frac{E_c}{N_0} = \frac{2^m - 1 - m}{2^m - 1}\frac{E_b}{N_0} \tag{6.76}$$

Combining Equation (6.73), (6.74), and (6.76), P_B can be expressed as a function of E_b/N_0 for coherently demodulated BPSK over a Gaussian channel. The results are plotted in Figure 6.22 for different types of block codes. For the Hamming codes, plots are shown for $(n, k) = (7, 4)$, $(15, 11)$, and $(31, 26)$.

Example 6.11 Error Probability for Modulated and Coded Signals

A coded orthogonal BFSK modulated signal is transmitted over a Gaussian channel. The signal is noncoherently detected and hard-decision decoded. Find the decoded bit error probability if the coding is a Hamming (7, 4) block code and the received E_b/N_0 is equal to 20.

Solution

First we need to find E_c/N_0 using Equation (6.75):

$$\frac{E_c}{N_0} = \frac{4}{7}(20) = 11.43$$

Then, for coded noncoherent BFSK, we can relate the probability of a channel symbol error to E_c/N_0, similar to Equation (4.96), as follows

$$p = \frac{1}{2}\exp\left(-\frac{E_c}{2N_0}\right)$$
$$= \frac{1}{2}\exp\left(-\frac{11.43}{2}\right) = 1.6 \times 10^{-3}$$

Using this result in Equation (6.73), we solve for the probability of a decoded bit error, as follows:

$$P_B \approx p - p(1 - p)^6 \approx 1.6 \times 10^{-5}$$

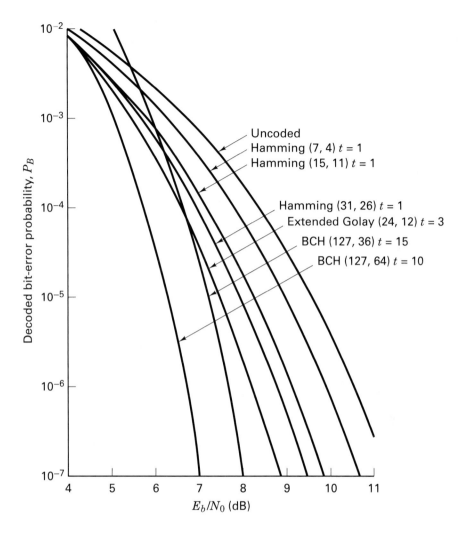

Figure 6.22 P_B versus E_b/N_0 for coherently demodulated BPSK over a Gaussian channel for several block codes.

6.8.2 Extended Golay Code

One of the more useful block codes is the binary (24, 12) *extended Golay code,* which is formed by adding an overall parity bit to the perfect (23, 12) code, known as the *Golay code.* This added parity bit increases the minimum distance d_{min} from 7 to 8 and produces a rate $\frac{1}{2}$ code, which is easier to implement (with regard to system clocks) than the rate 12/23 original Golay code. Extended Golay codes are considerably more powerful than the Hamming codes described in the preceding section. The price paid for the improved performance is a more complex decoder, a lower code rate, and hence a larger bandwidth expansion.

Since $d_{min} = 8$ for the extended Golay code, we see from Equation (6.44) that the code is guaranteed to correct all triple errors. The decoder can additionally be designed to correct *some but not all* four-error patterns. Since only 16.7% of the four-error patterns can be corrected, the decoder, for the sake of simplicity, is usually designed to only correct three-error patterns [5]. Assuming hard decision decoding, the bit error probability for the extended Golay code can be written as a function of the channel symbol error probability p from Equation (6.46), as follows:

$$P_B \approx \frac{1}{24} \sum_{j=4}^{24} j \binom{24}{j} p^j (1 - p)^{24 - j} \qquad (6.77)$$

The plot of Equation (6.77) is shown in Figure 6.21; the error performance of the extended Golay code is seen to be significantly better than that of the Hamming codes. Combining Equations (6.77), (6.74), and (6.75), we can relate P_B versus E_b/N_0 for coherently demodulated BPSK with extended Golay coding over a Gaussian channel. The result is plotted in Figure 6.22.

6.8.3 BCH Codes

Bose–Chadhuri–Hocquenghem (BCH) codes are a generalization of Hamming codes that allow multiple error correction. They are a *powerful class of cyclic codes* that provide a large selection of block lengths, code rates, alphabet sizes, and error-correcting capability. Table 6.4 lists some code generators $\mathbf{g}(x)$ commonly used for the construction of BCH codes [8] for various values of n, k, and t, up to a block length of 255. The coefficients of $\mathbf{g}(x)$ are presented as octal numbers arranged so that when they are converted to binary digits the rightmost digit corresponds to the zero-degree coefficient of $\mathbf{g}(x)$. From Table 6.4, one can easily verify a cyclic code property—the generator polynomial is of degree $n - k$. BCH codes are important because at block lengths of a few hundred, the BCH codes outperform all other block codes with the same block length and code rate. The most commonly used BCH codes employ a binary alphabet and a codeword block length of $n = 2^m - 1$, where $m = 3, 4, \ldots$.

The title of Table 6.4 indicates that the generators shown are for those BCH codes known as *primitive codes*. The term "primitive" is a number-theoretic concept requiring an algebraic development [7, 10–11], which is presented in Section 8.1.4. In Figures 6.21 and 6.22 are plotted error performance curves of two BCH codes (127, 64) and (127, 36), to illustrate comparative performance. Assuming hard decision decoding, the P_B versus channel error probability is shown in Figure 6.21. The P_B versus E_b/N_0 for coherently demodulated BPSK over a Gaussian channel is shown in Figure 6.22. The curves in Figure 6.22 seem to depart from our expectations. They each have the same block size, yet the more redundant (127, 36) code does not exhibit as much coding gain as does the less redundant (127, 64) code. It has been shown that a relatively broad maximum of coding gain versus code rate for fixed n occurs roughly between coding rates of $\frac{1}{3}$ and $\frac{3}{4}$ for BCH codes [12]. Performance over a Gaussian channel degrades substantially at very high or very low rates [11].

Figure 6.23 represents computed performance of BCH codes [13] using coherently demodulated BPSK with both *hard-* and *soft-decision decoding.* Soft-decision decoding is not usually used with block codes because of its complexity. However, whenever it is implemented, it offers an approximate 2-dB coding gain over hard-decision decoding. For a given code rate, the decoded error probability is known to improve with increasing block length n [4]. Thus, for a given code rate, it is

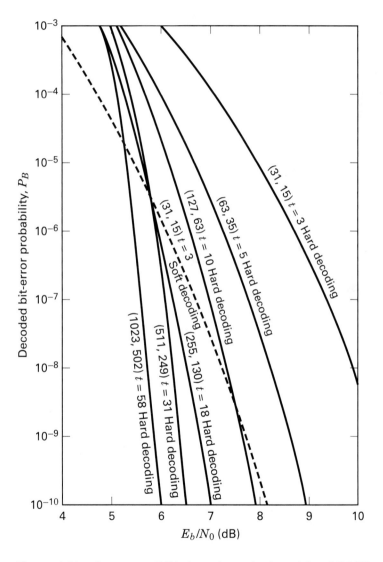

Figure 6.23 P_B versus E_b/N_0 for coherently demodulated BPSK over a Gaussian channel using BCH codes. (Reprinted with permission from L. J. Weng, "Soft and Hard Decoding Performance Comparisons for BCH Codes," *Proc. Int. Conf. Commun.,* 1979, Fig. 3, p. 25.5.5. © 1979 IEEE.)

TABLE 6.4 Generators of Primitive BCH Codes

n	k	t	$g(x)$	n	k	t	$g(x)$
7	4	1	13	255	171	11	15416214212342356077061630637
15	11	1	23		163	12	7500415510075602551574724514601
	7	2	721		155	13	3757513005407665015722506464677633
	5	3	2467		147	14	1642130173557165525304165305441011711
31	26	1	45		139	15	4614017320601755615707227302474535567445
	21	2	3551		131	18	2157133314715101512612502774421420241 65471
	16	3	107657		123	19	120614505224206600371721032651614122262 72506267
	11	5	5423325		115	21	605266655721002472636404060027635255 6313472737
	6	7	313365047		107	22	22205772322066256312417300235334742017 6574750154441
63	57	1	103		99	23	1065666725347317422274141620157433225 2411076432303431
	51	2	12471		91	25	6750265030327444172723631724732511075 5507627072434561
	45	3	1701317		87	26	1101367634147432364352316343071720462 0672254527311721317
	39	4	166623567		79	27	667000356376575000202703442073617462 10155326711766541342355
	36	5	1033500423				
	30	6	157464165547				
	24	7	17323260404441				
	18	10	1363026512351725				
	16	11	6331141367235453				
	10	13	472622305527250155				
	7	15	5231045543503271737				

n	k	t	Generator (octal)
127	120	1	211
	113	2	41567
	106	3	11554743
	99	4	3447023271
	92	5	624730022327
	85	6	130704476322273
	78	7	26230002166130115
	71	9	6255010713253127753
	64	10	1206534025570773100045
	57	11	335265252505705053517721
	50	13	54446512523314012421501421
	43	14	17721772213651227521220574343
	36	15	3146074666522075044764574721735
	29	21	403114461367670603667530141176155
	22	23	123376070404722522435445626637647043
	15	27	22057042445604554770523013762217604353
	8	31	70472640527510306514762242715677331310217
	71	29	24024710520644321515541721123311632054442503625576432221706035
	63	30	1075447505516354432531521735770700366611172645267613656702543301
	55	31	7315425203501100133015275306032054325414326755010557044426035473617
	47	42	25335420170626465630330413774062331751233341454460450050660245552 43173
	45	43	15202056055234161131101346376423701563670024470762373033202157025051541
	37	45	51363302550670074141774472454375304207357061743234323476443547374030 44003
	29	47	302571553667307146552706401236137711534224324201174114060254757410403565037
	21	55	1256215257060332656001773153607612103227341405653074542511531211614466513473725
	13	59	4641732005052564544426573714250066004330677445476561403174677213570261344605 00547
	9	63	15726025217472463201031043255351346141623672120440754511276611554770556177516057
255	247	1	435
	239	2	267543
	231	3	156720665
	223	4	75626641375
	215	5	23157564726421
	207	6	161765605676636227
	199	7	7633031270420722341
	191	8	266347017611533714567
	187	9	52755313540001322236351
	179	10	22624710717340432416300455

Source: Reprinted with permission from "Table of Generators for BCH Codes," *IEEE Trans. Inf. Theory*, vol. IT10, no. 4, Oct. 1964, p. 391. © 1964 IEEE.

interesting to consider the block length that would be required for the hard-decision-decoding performance to be comparable to the soft-decision-decoding performance. In Figure 6.23, the BCH codes shown all have code rates of approximately $\frac{1}{2}$. From the figure [13] it appears that for a fixed code rate, the hard-decision-decoded BCH code of length 8 times n or longer has a better performance (for P_B of about 10^{-6} or less) than that of a soft-decision-decoded BCH code of length n. One special subclass of the BCH codes (the discovery of which preceded the BCH codes) is the particularly useful *nonbinary* set called *Reed-Solomon* codes. They are described in Section 8.1.

6.9 CONCLUSION

In this chapter we have explored the general goals of channel coding, leading to improved performance (error probability, E_b/N_0, or capacity) at a cost in bandwidth. We partitioned channel coding into two study groups: waveform coding and structured sequences. Waveform coding represents a transformation of waveforms into improved waveforms, such that the distance properties are improved over those of the original waveforms. Structured sequences involve the addition of redundant digits to the data, such that the redundant digits can then be employed for detecting and/or correcting specific error patterns.

We also closely examined linear block codes. Geometric analogies can be drawn between the coding and modulation disciplines. They both seek to pack the signal space efficiently and to maximize the distance between signals in the signaling set. Within block codes, we looked at cyclic codes, which are relatively easy to implement using modern integrated circuit techniques. We considered the polynomial representation of codes and the correspondence between the polynomial structure, the necessary algebraic operations, and the hardware implementation. Finally, we looked at performance details of some of the well-known block codes. Other coding subjects are treated in later chapters. In Chapter 7 we study the large class of convolutional codes; in Chapter 8 we discuss Reed–Solomon codes, concatenated codes, and turbo codes; and in Chapter 9 we examine trellis-coded modulation.

REFERENCES

1. Viterbi, A. J., "On Coded Phase-Coherent Communications," *IRE Trans. Space Electron. Telem.,* vol. SET7, Mar. 1961, pp. 3–14.

2. Lindsey, W. C., and Simon, M. K., *Telecommunication Systems Engineering,* Prentice-Hall, Inc., Englewood Cliffs, N.J., 1973.

3. Proakis, J. G., *Digital Communications,* McGraw-Hill Book Company, New York, 1983.

4. Lin, S., and Costello, D. J., Jr., *Error Control Coding: Fundamentals and Applications,* Prentice-Hall, Inc., Englewood Cliffs, N.J., 1983.

5. Odenwalder, J. P., *Error Control Coding Handbook,* Linkabit Corporation, San Diego, Calif., July 15, 1976.

6. Blahut, R. E., *Theory and Practice of Error Control Codes,* Addison-Wesley Publishing Company, Inc., Reading, Mass, 1983.

7. Peterson, W. W., and Weldon, E. J., *Error Correcting Codes,* 2nd ed., The MIT Press, Cambridge, Mass., 1972.

8. Blahut, R. E., "Algebraic Fields, Signal Processing, and Error Control," *Proc. IEEE,* vol. 73, May 1985, pp. 874–893.

9. Stenbit, J. P., "Table of Generators for Bose–Chadhuri Codes, *IEEE Trans. Inf. Theory,* vol. IT10, no. 4, Oct. 1964, pp. 390–391.

10. Berlekamp, E. R., *Algebraic Coding Theory,* McGraw-Hill Book Company, New York, 1968.

11. Clark, G. C., Jr., and Cain, J. B. *Error-Correction Coding for Digital Communications,* Plenum Press, New York, 1981.

12. Wozencraft, J. M., and Jacobs, I. M., *Principles of Communication Engineering,* John Wiley & Sons, Inc., New York, 1965.

13. Weng, L. J., "Soft and Hard Decoding Performance Comparisons for BCH Codes," *Proc. Int. Conf. Commun.,* 1979, pp. 25.5.1–25.5.5

PROBLEMS

6.1. Design an (n, k) single-parity code that will detect all 1-, 3-, 5-, and 7-error patterns in a block. Show the values of n and $k,$ and find the probability of an undetected block error if the probability of channel symbol error is 10^{-2}.

6.2. Calculate the probability of message error for a 12-bit data sequence encoded with a (24, 12) linear block code. Assume that the code corrects all 1-bit and 2-bit error patterns and assume that it corrects no error patterns with more than two errors. Also, assume that the probability of a channel symbol error is 10^{-3}.

6.3. Consider a (127, 92) linear block code capable of triple error corrections.
   (a) What is the probability of message error for an uncoded block of 92 bits if the channel symbol error probability is 10^{-3}?
   (b) What is the probability of message error when using the (127, 92) block code if the channel symbol error probability of 10^{-3}?

6.4. Calculate the improvement in probability of message error relative to an uncoded transmission for a (24, 12) double-error-correcting linear block code. Assume that coherent BPSK modulation is used and that the received $E_b/N_0 = 10$ dB.

6.5. Consider a (24, 12) linear block code capable of double-error corrections. Assume that a noncoherently detected binary orthogonal frequency-shift keying (BFSK) modulation format is used and that the received $E_b/N_0 = 14$ dB.
   (a) Does the code provide any improvement in probability of message error? If it does, how much? If it does not, explain why not.
   (b) Repeat part (a) with $E_b/N_0 = 10$ dB.

6.6. The telephone company uses a "best-of-five" encoder for some of its digital data channels. In this system every data bit is repeated five times, and at the receiver, a majority vote decides the value of each data bit. If the uncoded probability of bit error is 10^{-3}, calculate the decoded bit-error probability when using such a best-of-five code.

6.7. The minimum distance for a particular linear block code is 11. Find the maximum error-correcting capability, the maximum error-detecting capability, and the maximum erasure-correcting capability in a block length.

6.8. Consider a $(7, 4)$ code whose generator matrix is

$$\mathbf{G} = \begin{bmatrix} 1 & 1 & 1 & 1 & 0 & 0 & 0 \\ 1 & 0 & 1 & 0 & 1 & 0 & 0 \\ 0 & 1 & 1 & 0 & 0 & 1 & 0 \\ 1 & 1 & 0 & 0 & 0 & 0 & 1 \end{bmatrix}$$

(a) Find all the codewords of the code.
(b) Find $\mathbf{H}$, the parity-check matrix of the code.
(c) Compute the syndrome for the received vector 1 1 0 1 1 0 1. Is this a valid code vector?
(d) What is the error-correcting capability of the code?
(e) What is the error-detecting capability of the code?

6.9. Consider a systematic block code whose parity-check equations are

$$p_1 = m_1 + m_2 + m_4$$
$$p_2 = m_1 + m_3 + m_4$$
$$p_3 = m_1 + m_2 + m_3$$
$$p_4 = m_2 + m_3 + m_4$$

where m_i are message digits and p_i are check digits.

(a) Find the generator matrix and the parity-check matrix for this code.
(b) How many errors can the code correct?
(c) Is the vector 10101010 a codeword?
(d) Is the vector 01011100 a codeword?

6.10. Consider the linear block code with the codeword defined by

$$\mathbf{U} = m_1 + m_2 + m_4 + m_5, \ m_1 + m_3 + m_4 + m_5, \ m_1 + m_2 + m_3 + m_5,$$
$$m_1 + m_2 + m_3 + m_4, \ m_1, \ m_2, \ m_3, \ m_4, \ m_5$$

(a) Show the generator matrix.
(b) Show the parity-check matrix.
(c) Find n, k, and d_{min}.

6.11. Design an $(n, k) = (5, 2)$ linear block code.

(a) Choose the codewords to be in systematic form, and choose them with the goal of maximizing d_{min}.
(b) Find the generator matrix for the codeword set.
(c) Calculate the parity-check matrix.
(d) Enter all of the n-tuples into a standard array.
(e) What are the error-correcting and error-detecting capabilities of the code?
(f) Make a syndrome table for the correctable error patterns.

6.12. Consider the $(5, 1)$ repetition code, which consists of the two codewords 00000 and 11111, corresponding to message 0 and 1, respectively. Derive the standard array for this code. Is this a perfect code?

6.13. Design a $(3, 1)$ code that will correct all single-error patterns. Choose the codeword set and show the standard array.

6.14. Is a (7, 3) code a perfect code? Is a (7, 4) code a perfect code? Is a (15, 11) code a perfect code? Justify your answers.

6.15. A (15, 11) linear block code can be defined by the following parity array:

$$\mathbf{P} = \begin{bmatrix} 0 & 0 & 1 & 1 \\ 0 & 1 & 0 & 1 \\ 1 & 0 & 0 & 1 \\ 0 & 1 & 1 & 0 \\ 1 & 0 & 1 & 0 \\ 1 & 1 & 0 & 0 \\ 0 & 1 & 1 & 1 \\ 1 & 1 & 1 & 0 \\ 1 & 1 & 0 & 1 \\ 1 & 0 & 1 & 1 \\ 1 & 1 & 1 & 1 \end{bmatrix}$$

(a) Show the parity-check matrix for this code.

(b) List the coset leaders from the standard array. Is this code a perfect code? Justify your answer.

(c) A received vector is $\mathbf{V} = 0\ 1\ 1\ 1\ 1\ 1\ 0\ 0\ 1\ 0\ 1\ 1\ 0\ 1\ 1$. Compute the syndrome. Assuming that a single bit error has been made, find the correct codeword.

(d) How many erasures can this code correct? Explain.

6.16. Is it possible that a nonzero error pattern can produce a syndrome of $\mathbf{S} = \mathbf{0}$? If yes, how many such error patterns can give this result for an (n, k) code? Use Figure 6.11 to justify your answer.

6.17. Determine which, if any, of the following polynomials can generate a cyclic code with codeword length $n \leq 7$. Find the (n, k) values of any such codes that can be generated.

(a) $1 + X^3 + X^4$

(b) $1 + X^2 + X^4$

(c) $1 + X + X^3 + X^4$

(d) $1 + X + X^2 + X^4$

(e) $1 + X^3 + X^5$

6.18. Encode the message 1 0 1 in systematic form using polynomial division and the generator $\mathbf{g}(X) = 1 + X + X^2 + X^4$.

6.19. Design a feedback shift register encoder for an (8, 5) cyclic code with a generator $\mathbf{g}(x) = 1 + X + X^2 + X^3$. Use the encoder to find the codeword for the message 1 0 1 0 1 in systematic form.

6.20. In Figure P6.1 the signal is differentially coherent PSK (DPSK), the encoded symbol rate is 10,000 code symbols per second, and the decoder is a single-error-correcting (7, 4) decoder. Is a predetection signal-to-noise spectral density ratio of $P_r/N_0 = 48$ dB-Hz sufficient to provide a probability of message error of 10^{-3} at the output?

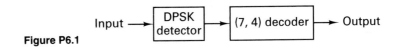

Input $\longrightarrow$ [DPSK detector] $\longrightarrow$ [(7, 4) decoder] $\longrightarrow$ Output

Figure P6.1

Justify your answer. Assume that a message block contains 4 data bits and that any single-error pattern in a block length of 7 bits can be corrected.

6.21. A $(15, 5)$ cyclic code has a generator polynomial as follows:

$$\mathbf{g}(X) = 1 + X + X^2 + X^5 + X^8 + X^{10}$$

(a) Draw a diagram of an encoder for this code.

(b) Find the code polynomial (in systematic form) for the message $\mathbf{m}(X) = 1 + X^2 + X^4$.

(c) Is $\mathbf{V}(X) = 1 + X^4 + X^6 + X^8 + X^{14}$ a code polynomial in this system? Justify your answer.

6.22. Consider the $(15, 11)$ cyclic code generated by $\mathbf{g}(X) = 1 + X + X^4$.

(a) Devise a feedback register encoder and decoder for this code.

(b) Illustrate the encoding procedure with the message vector 11001101011 by listing the states of the register (the rightmost bit is the earliest bit).

(c) Repeat part (b) for the decoding procedure.

6.23. For a fixed probability of channel symbol error, the probability of bit error for a Hamming $(15, 11)$ code is worse than that for a Hamming $(7, 4)$ code. Explain why. What, then, is the advantage of the $(15, 11)$ code? What basic trade-off is involved?

6.24. A $(63, 36)$ BCH code can correct five errors. Nine blocks of a $(7, 4)$ code can correct nine errors. Both codes have the same code rate.

(a) The $(7, 4)$ code can correct more errors. Is it more powerful? Explain.

(b) Compare the two codes when five errors occur randomly in 63 bits.

6.25. Information from a source is organized in 36-bit messages that are to be transmitted over an AWGN channel using noncoherently detected BFSK modulation.

(a) If no error control coding is used, compute the E_b/N_0 required to provide a message error probability of 10^{-3}.

(b) Consider the use of a $(127, 36)$ linear block code (minimum distance is 31) in the transmission of these messages. Compute the coding gain for this code for a message error probability of 10^{-3}. (*Hint:* The coding gain is defined as the difference between the E_b/N_0 required without coding and the E_b/N_0 required with coding.)

6.26. (a) Consider a data sequence encoded with a $(127, 64)$ BCH code and then modulated using coherent 16-ary PSK. If the received E_b/N_0 is 10 dB, find the MPSK probability of symbol error, the probability of code-bit error (assuming that a Gray code is used for symbol-to-bit assignment), and the probability of information-bit error.

(b) For the same probability of information-bit error found in part (a), determine the value of E_b/N_0 required if the modulation in part (a) is changed to coherent orthogonal 16-ary FSK. Explain the difference.

6.27. A message consists of English text (assume that each word in the message contains six letters). Each letter is encoded using the 7-bit ASCII character code. Thus, each word of text consists of a 42-bit sequence. The message is to be transmitted over a channel having a symbol error probability of 10^{-3}.

(a) What is the probability that a word will be received in error?

(b) If a repetition code is used such that each letter in each word is repeated three times, and at the receiver, majority voting is used to decode the message, what is the probability that a decoded word will be in error?

(c) If a $(126, 42)$ BCH code with error-correcting capability of $t = 14$ is used to encode each 42-bit word, what is the probability that a decoded word will be in error?

(d) For a real system, it is not fair to compare uncoded versus coded message error performance on the basis of a fixed probability of channel symbol error, since this implies a fixed level of received E_c/N_0 for all choices of coding (or lack of coding). Therefore, repeat parts (a), (b), and (c) under the condition that the channel symbol error probability is determined by a received E_b/N_0 of 12 dB, where E_b/N_0 is the information bit energy per noise spectral density. Assume that the information rate must be the same for all choices of coding or lack of coding. Also assume that noncoherent orthogonal binary FSK modulation is used over an AWGN channel.

(e) Discuss the relative error performance capabilities of the above coding schemes under the two postulated conditions—fixed channel symbol error probability, and fixed E_b/N_0. Under what circumstances can a repetition code offer error performance improvement? When will it cause performance degradation?

6.28. A 5-bit data sequence is transformed into an orthogonal coded sequence using a Hadamard matrix. Coherent detection is performed over a codeword interval of time, as shown in Figure 6.5. Using a $P_B = 10^{-5}$ reference, compute the coding gain relative to transmitting the data one bit at a time using BPSK.

6.29. For the (8, 2) code described in Section 6.6.3, verify that the values given for the generator matrix, the parity-check matrix, and the syndrome vectors for each of the cosets from 1 through 10, are valid.

6.30. Using exclusive-OR gates and AND gates, implement a decoder circuit, similar to the one shown in Figure 6.12, that will perform error correction for all single-error patterns of the (8, 2) code described by the coset leaders 2 through 9 in Figure 6.15.

6.31. Explain in detail how you could use exclusive-OR gates and AND gates to implement a decoder circuit, similar to the one shown in Figure 6.12, that performs error correction for all single and double-error patterns of the (8, 2) code, and performs error detection for the triple-error patterns (coset leaders or rows 38 through 64).

6.32. Verify that all of the BCH codes of length $n = 31$, shown in Table 6.4, meet the Hamming bound and the Plotkin bound.

6.33. When encoding an all-zeros message block, the result is an all-zeros codeword. It is generally undesirable to transmit long runs of such zeros. One cyclic encoding technique that avoids such transmissions involves preloading the shift-register stages with ones instead of zeros, prior to encoding. The resulting "pseudoparity" is then guaranteed to contain some ones. At the decoder, there needs to be a reversal of the pseudoparity before the decoding operation starts. Devise a general scheme for reversing the pseudo-parity at any such cyclic decoder. Use a (7, 4) BCH encoder, preloaded with ones, to encode the message 1 0 1 1 (right-most bit is earliest). Then demonstrate that your reversal scheme, which is applied prior to decoding, yields the correct decoded message.

6.34. (a) Using the generator polynomial for the (15, 5) cyclic code in Problem 6.21, encode the message sequence 1 1 0 1 1 in systematic form. Show the resulting codeword polynomial. What property characterizes the degree of the generator polynomial?

(b) Consider that the received codeword is corrupted by an error pattern $\mathbf{e}(X) = X^8 + X^{10} + X^{13}$. Show the corrupted codeword polynomial.

(c) Form the syndrome polynomial by using the generator and received-codeword polynomials.

(d) Form the syndrome polynomial by using the generator and error-pattern polynomials, and verify that this is the same syndrome computed in part (c).

(e) Explain why the syndrome computations in parts (c) and (d) must yield identical results.

(f) Using the properties of the standard array of a (15, 5) linear block code, find the maximum amount of error correction possible for a code with these parameters. Is a (15, 5) code a perfect code?

(g) If we want to implement the (15, 5) cyclic code to simultaneously correct two erasures and still perform error correction, how much error correction would have to be sacrificed?

QUESTIONS

6.1. Describe four types of trade-offs that can be accomplished by using an error-correcting code. (See Section 6.3.4.)

6.2. In a *real-time* communication system, achieving coding gain by adding redundancy costs *bandwidth*. What is the usual cost for achieving coding gain in a *non-real-time* communication system? (See Section 6.3.4.2.)

6.3. In a real-time communication system, added redundancy means faster signaling, less energy per channel symbol, and more errors out of the demodulator. In the face of such degradation, explain how coding gain is achieved. (See Example 6.2.)

6.4. Why do error-correcting codes typically yield error-performance degradation at low values of E_b/N_0? (See Section 6.3.4.6.)

6.5. Describe the process of syndrome testing, error detection and correction in the context of a medical analogy. (See Section 6.4.8.4.)

6.6. Of what use is the *standard array* in understanding a block code, and in evaluating its capability? (See Section 6.6.5.)

EXERCISES

Using the Companion CD, run the exercises associated with Chapter 6.

CHAPTER 7

Channel Coding: Part 2

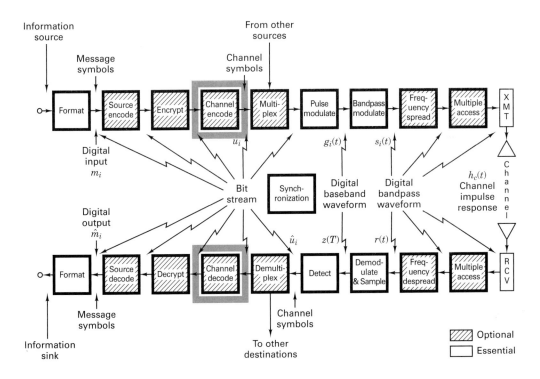

This chapter deals with convolutional coding. Chapter 6 presented the fundamentals of linear block codes, which are described by two integers, n and k, and a generator matrix or polynomial. The integer k is the number of data bits that form an input to a block encoder. The integer n is the total number of bits in the associated codeword out of the encoder. A characteristic of linear block codes is that each codeword n-tuple is uniquely determined by the input message k-tuple. The ratio k/n is called the *rate* of the code—a measure of the amount of added redundancy. A *convolutional code* is described by three integers, n, k, and K, where the ratio k/n has the same code rate significance (information per coded bit) that it has for block codes; however, n does *not* define a block or codeword length as it does for block codes. The integer K is a parameter known as the *constraint length*; it represents the number of k-tuple stages in the encoding shift register. An important characteristic of convolutional codes, different from block codes, is that the encoder has memory—the n-tuple emitted by the convolutional encoding procedure is not only a function of an input k-tuple, but is also a function of the previous $K - 1$ input k-tuples. In practice, n and k are small integers and K is varied to control the capability and complexity of the code.

7.1 CONVOLUTIONAL ENCODING

In Figure 1.2 we presented a typical block diagram of a digital communication system. A version of this functional diagram, focusing primarily on the convolutional encode/decode and modulate/demodulate portions of the communication link, is shown in Figure 7.1. The input message source is denoted by the sequence $\mathbf{m} = m_1,$ $m_2, \ldots, m_i, \ldots$, where each m_i represents a binary digit (bit), and i is a time index. To be precise, one should denote the elements of $\mathbf{m}$ with an index for class membership (e.g., for binary codes, 1 or 0) and an index for time. However, in this chapter, for simplicity, indexing is only used to indicate time (or location within a sequence). We shall assume that each m_i is equally likely to be a one or a zero, and independent from digit to digit. Being independent, the bit sequence lacks any redundancy; that is, knowledge about bit m_i gives no information about m_j $(i \neq j)$. The encoder transforms each sequence $\mathbf{m}$ into a unique codeword sequence $\mathbf{U} = G(\mathbf{m})$. Even though the sequence $\mathbf{m}$ uniquely defines the sequence $\mathbf{U}$, a key feature of convolutional codes is that a given k-tuple within $\mathbf{m}$ does *not* uniquely define its associated n-tuple within $\mathbf{U}$ since the encoding of each k-tuple is *not only* a function of that k-tuple but is also a function of the $K - 1$ input k-tuples that precede it. The sequence $\mathbf{U}$ can be partitioned into a sequence of branch words: $\mathbf{U} = U_1, U_2, \ldots,$ $U_i, \ldots$. Each branch word U_i is made up of binary *code symbols,* often called *channel symbols, channel bits,* or *code bits;* unlike the input message bits the code symbols are not independent.

In a typical communication application, the codeword sequence $\mathbf{U}$ modulates a waveform $s(t)$. During transmission, the waveform $s(t)$ is corrupted by noise, resulting in a received waveform $\hat{s}(t)$ and a demodulated sequence $\mathbf{Z} = Z_1, Z_2, \ldots,$ $Z_i, \ldots$, as indicated in Figure 7.1. The task of the decoder is to produce an estimate $\hat{\mathbf{m}} = \hat{m}_1, \hat{m}_2, \ldots, \hat{m}_i, \ldots$ of the original message sequence, using the received sequence $\mathbf{Z}$ together with a priori knowledge of the encoding procedure.

A general convolutional encoder, shown in Figure 7.2, is mechanized with a kK-stage shift register and n modulo-2 adders, where K is the constraint length.

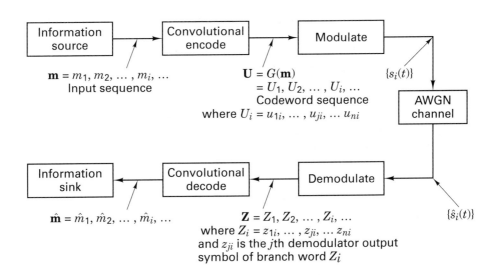

Figure 7.1 Encode/decode and modulate/demodulate portions of a communication link.

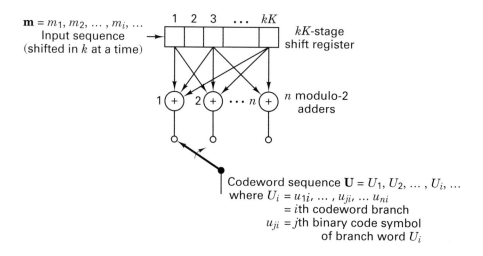

$\mathbf{m} = m_1, m_2, \ldots, m_i, \ldots$
Input sequence
(shifted in k at a time)

1 2 3 $\ldots$ kK

kK-stage
shift register

1 $\oplus$ 2 $\oplus$ $\ldots n$ $\oplus$

n modulo-2
adders

Codeword sequence $\mathbf{U} = U_1, U_2, \ldots, U_i, \ldots$
where $U_i = u_{1i}, \ldots, u_{ji}, \ldots u_{ni}$
$= i$th codeword branch
$u_{ji} = j$th binary code symbol
of branch word U_i

Figure 7.2 Convolutional encoder with constraint length K and rate k/n.

The constraint length represents the number of k-bit shifts over which a single information bit can influence the encoder output. At each unit of time, k bits are shifted into the first k stages of the register; all bits in the register are shifted k stages to the right, and the outputs of the n adders are sequentially sampled to yield the binary code symbols or code bits. These code symbols are then used by the modulator to specify the waveforms to be transmitted over the channel. Since there are n code bits for each input group of k message bits, the code rate is k/n message bit per code bit, where $k < n$.

We shall consider only the most commonly used binary convolutional encoders for which $k = 1$—that is, those encoders in which the message bits are shifted into the encoder one bit at a time, although generalization to higher order alphabets is straightforward [1, 2]. For the $k = 1$ encoder, at the ith unit of time, message bit m_i is shifted into the first shift register stage; all previous bits in the register are shifted one stage to the right, and as in the more general case, the outputs of the n adders are sequentially sampled and transmitted. Since there are n code bits for each message bit, the code rate is $1/n$. The n code symbols occurring at time t_i comprise the ith branch word, $U_i = u_{1i}, u_{2i}, \ldots, u_{ni}$, where u_{ji} ($j = 1, 2, \ldots, n$) is the jth code symbol belonging to the ith branch word. Note that for the rate $1/n$ encoder, the kK-stage shift register can be referred to simply as a K-stage register, and the constraint length K, which was expressed in units of k-tuple stages, can be referred to as constraint length in units of bits.

7.2 CONVOLUTIONAL ENCODER REPRESENTATION

To describe a convolutional code, one needs to characterize the encoding function $G(\mathbf{m})$, so that given an input sequence $\mathbf{m}$, one can readily compute the output sequence $\mathbf{U}$. Several methods are used for representing a convolutional encoder, the most popular being the *connection pictorial, connection vectors or polynomials*, the *state diagram*, the *tree diagram*, and the *trellis diagram*. They are each described below.

7.2.1 Connection Representation

We shall use the convolutional encoder, shown in Figure 7.3, as a model for discussing convolutional encoders. The figure illustrates a (2, 1) convolutional encoder with constraint length $K = 3$. There are $n = 2$ modulo-2 adders; thus the code rate k/n is $\frac{1}{2}$. At each input bit time, a bit is shifted into the leftmost stage and the bits in the register are shifted one position to the right. Next, the output switch samples the output of each modulo-2 adder (i.e., first the upper adder, then the lower adder), thus forming the code symbol pair making up the branch word associated with the bit just inputted. The sampling is repeated for each inputted bit. The choice of connections between the adders and the stages of the register gives rise to the characteristics of the code. Any change in the choice of connections results in a different code. The connections are, of course, *not* chosen or changed arbitrarily. The problem of choosing connections to yield good distance properties is complicated and has not been solved in general; however, good codes have been found by computer search for all constraint lengths less than about 20 [3–5].

Unlike a block code that has a fixed word length n, a convolutional code has no particular block size. However, convolutional codes are often forced into a block structure by *periodic truncation*. This requires a number of zero bits to be appended to the end of the input data sequence, for the purpose of clearing or *flushing* the encoding shift register of the data bits. Since the added zeros carry no information, the *effective code rate* falls below k/n. To keep the code rate close to k/n, the truncation period is generally made as long as practical.

One way to represent the encoder is to specify a set of n *connection vectors*, one for each of the n modulo-2 adders. Each vector has dimension K and describes the connection of the encoding shift register to that modulo-2 adder. A one in the ith position of the vector indicates that the corresponding stage in the shift register

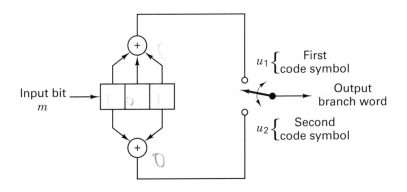

Figure 7.3 Convolutional encoder (rate $\frac{1}{2}$, $K = 3$).

is connected to the modulo-2 adder, and a zero in a given position indicates that no connection exists between the stage and the modulo-2 adder. For the encoder example in Figure 6.3, we can write the connection vector $\mathbf{g}_1$ for the upper connections and $\mathbf{g}_2$ for the lower connections as follows:

$$\mathbf{g}_1 = 1\ 1\ 1$$
$$\mathbf{g}_2 = 1\ 0\ 1$$

Now consider that a message vector $\mathbf{m} = 1\ 0\ 1$ is convolutionally encoded with the encoder shown in Figure 7.3. The three message bits are inputted, one at a time, at times t_1, t_2, and t_3, as shown in Figure 7.4. Subsequently, $(K - 1) = 2$ zeros are inputted at times t_4 and t_5 to flush the register and thus ensure that the tail end of the message is shifted the full length of the register. The output sequence is seen to be $1\ 1\ 1\ 0\ 0\ 0\ 1\ 0\ 1\ 1$, where the leftmost symbol represents the earliest transmission. The entire output sequence, including the code symbols as a result of flushing, are needed to decode the message. To flush the message from the encoder requires one less zero than the number of stages in the register, or $K - 1$ flush bits. Another zero input is shown at time t_6, for the reader to verify that the flushing is completed at time t_5. Thus, a new message can be entered at time t_6.

7.2.1.1 Impulse Response of the Encoder

We can approach the encoder in terms of its *impulse response*—that is, the response of the encoder to a single "one" bit that moves through it. Consider the contents of the register in Figure 7.3 as a one moves through it:

Register contents	Branch word	
	u_1	u_2
1 0 0	1	1
0 1 0	1	0
0 0 1	1	1

Input sequence: 1 0 0
Output sequence: 1 1   1 0   1 1

The output sequence for the input "one" is called the impulse response of the encoder. Then, for the input sequence $\mathbf{m} = 1\ 0\ 1$, the output may be found by the *superposition* or the *linear addition* of the time-shifted input "impulses" as follows:

Input $\mathbf{m}$	Output				
1	1 1	1 0	1 1		
0		0 0	0 0	0 0	
1			1 1	1 0	1 1
Modulo-2 sum:	1 1	1 0	0 0	1 0	1 1

Observe that this is the same output as that obtained in Figure 7.4, demonstrating that *convolutional codes are linear*—just like the linear block codes of Chapter 6. It

Channel Coding: Part 2   Chap. 7

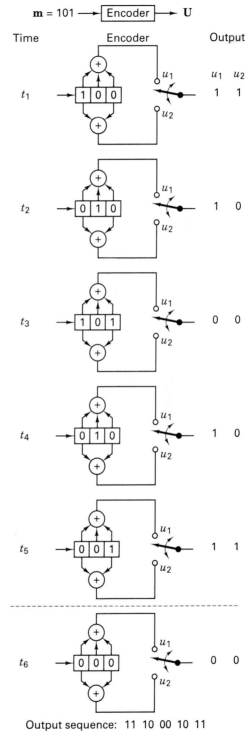

Figure 7.4 Convolutionally encoding a message sequence with a rate $\frac{1}{2}$, $K = 3$ encoder.

Output sequence: 11 10 00 10 11

is from this property of generating the output by the linear addition of time-shifted impulses, or the convolution of the input sequence with the impulse response of the encoder, that we derive the name *convolutional encoder*. Often, this encoder characterization is presented in terms of an infinite-order generator matrix [6].

Notice that the *effective code rate* for the foregoing example with 3-bit input sequence and 10-bit output sequence is $k/n = \frac{3}{10}$—quite a bit less than the rate $\frac{1}{2}$ that might have been expected from the knowledge that each input data bit yields a pair of output channel bits. The reason for the disparity is that the final data bit into the encoder needs to be shifted through the encoder. All of the output channel bits are needed in the decoding process. If the message had been longer, say 300 bits, the output codeword sequence would contain 604 bits, resulting in a code rate of 300/604—much closer to $\frac{1}{2}$.

7.2.1.2 Polynomial Representation

Sometimes, the encoder connections are characterized by *generator polynomials,* similar to those used in Chapter 6 for describing the feedback shift register implementation of cyclic codes. We can represent a convolutional encoder with a set of n generator polynomials, one for each of the n modulo-2 adders. Each polynomial is of degree $K - 1$ or less and describes the connection of the encoding shift register to that modulo-2 adder, much the same way that a connection vector does. The coefficient of each term in the $(K - 1)$-degree polynomial is either 1 or 0, depending on whether a connection exists or does not exist between the shift register and the modulo-2 adder in question. For the encoder example in Figure 7.3, we can write the generator polynomial $\mathbf{g}_1(X)$ for the upper connections and $\mathbf{g}_2(X)$ for the lower connections as follows:

$$\mathbf{g}_1(X) = 1 + X + X^2$$
$$\mathbf{g}_2(X) = 1 + X^2$$

where the lowest order term in the polynomial corresponds to the input stage of the register. The output sequence is found as follows:

$$\mathbf{U}(X) = \mathbf{m}(X)\mathbf{g}_1(X) \text{ interlaced with } \mathbf{m}(X)\mathbf{g}_2(X)$$

First, express the message vector $\mathbf{m} = 1\ 0\ 1$ as a polynomial—that is, $\mathbf{m}(X) = 1 + X^2$. We shall again assume the use of zeros following the message bits, to flush the register. Then the output polynomial $\mathbf{U}(X)$, or the output sequence $\mathbf{U}$, of the Figure 7.3 encoder can be found for the input message $\mathbf{m}$ as follows:

$$
\begin{aligned}
\mathbf{m}(X)\mathbf{g}_1(X) &= (1 + X^2)(1 + X + X^2) = 1 + X + X^3 + X^4 \\
\mathbf{m}(X)\mathbf{g}_2(X) &= (1 + X^2)(1 + X^2) = 1 + X^4 \\
\hline
\mathbf{m}(X)\mathbf{g}_1(X) &= 1 + \quad X + 0X^2 + \quad X^3 + X^4 \\
\mathbf{m}(X)\mathbf{g}_2(X) &= 1 + 0X + 0X^2 + 0X^3 + X^4
\end{aligned}
$$

$$
\begin{aligned}
\mathbf{U}(X) &= (1, 1) + (1, 0)X + (0, 0)X^2 + (1, 0)X^3 + (1, 1)X^4 \\
\mathbf{U} &= 1\ 1 \quad\quad 1\ 0 \quad\quad 0\ 0 \quad\quad 1\ 0 \quad\quad 1\ 1
\end{aligned}
$$

In this example we started with another point of view—namely, that the convolutional encoder can be treated as a set of *cyclic code shift registers.* We represented the encoder with *polynomial generators* as used for describing cyclic codes. However, we arrived at the same output sequence as in Figure 7.4 and at the same output sequence as the impulse response treatment of the preceding section. (For a good presentation of convolutional code structure in the context of linear sequential circuits, see Reference [7].)

7.2.2 State Representation and the State Diagram

A convolutional encoder belongs to a class of devices known as *finite-state machines,* which is the general name given to machines that have a memory of past signals. The adjective *finite* refers to the fact that there are only a finite number of unique states that the machine can encounter. What is meant by the *state* of a finite-state machine? In the most general sense, the state consists of the smallest amount of information that, together with a current input to the machine, can predict the output of the machine. The state provides some knowledge of the past signaling events and the restricted set of possible outputs in the future. A future state is restricted by the past state. For a rate $1/n$ convolutional encoder, the state is represented by the contents of the rightmost $K - 1$ stages (see Figure 7.3). Knowledge of the state together with knowledge of the next input is necessary and sufficient to determine the next output. Let the state of the encoder at time t_i be defined as $X_i = m_{i-1}, m_{i-2}, \ldots, m_{i-K+1}$. The ith codeword branch U_i is completely determined by state X_i and the present input bit m_i; thus the state X_i represents the past history of the encoder in determining the encoder output. The encoder state is said to be *Markov,* in the sense that the probability $P(X_{i+1}|X_i, X_{i-1}, \ldots, X_0)$ of being in state X_{i+1}, given all previous states, depends only on the most recent state X_i; that is, the probability is equal to $P(X_{i+1}|X_i)$.

One way to represent simple encoders is with a *state diagram;* such a representation for the encoder in Figure 7.3 is shown in Figure 7.5. The states, shown in the boxes of the diagram, represent the possible contents of the rightmost $K - 1$ stages of the register, and the paths between the states represent the output branch words resulting from such state transitions. The states of the register are designated $a = 00$, $b = 10$, $c = 01$, and $d = 11$; the diagram shown in Figure 7.5 illustrates all the state transitions that are possible for the encoder in Figure 7.3. There are *only two transitions* emanating from each state, corresponding to the two possible input bits. Next to each path between states is written the output branch word associated with the state transition. In drawing the path, we use the convention that a solid line denotes a path associated with an input bit, zero, and a dashed line denotes a path associated with an input bit, one. Notice that it is *not possible* in a single transition to move from a given state to *any arbitrary state.* As a consequence of shifting-in one bit at a time, there are only two possible state transitions that the register can make at each bit time. For example, if the present encoder state is 00, the *only possibilities* for the state at the next shift are 00 or 10.

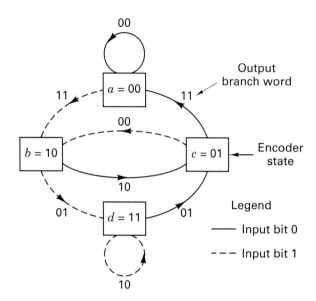

00

Output
branch word

11 - - - $a = 00$ 11

00

$b = 10$ $c = 01$ ← Encoder
state

10

01 - - - $d = 11$ 01

Legend

— Input bit 0

- - - Input bit 1

10

Figure 7.5 Encoder state diagram
(rate $\frac{1}{2}$, $K = 3$).

Example 7.1   Convolutional Encoding

For the encoder shown in Figure 7.3, show the state changes and the resulting output
codeword sequence **U** for the message sequence **m** = 1 1 0 1 1, followed by $K - 1 = 2$
zeros to flush the register. Assume that the initial contents of the register are all zeros.

Solution

Input bit m_i	Register contents	State at time t_i	State at time t_{i+1}	Branch word at time t_i	
				u_1	u_2
—	0 0 0	0 0	0 0	—	
1	1 0 0	0 0	1 0	1	1
1	1 1 0	1 0	1 1	0	1
0	0 1 1	1 1	0 1	0	1
1	1 0 1	0 1	1 0	0	0
1	1 1 0	1 0	1 1	0	1
0	0 1 1	1 1	0 1	0	1
0	0 0 1	0 1	0 0	1	1

state t_i

state
t_{i+1}

Output sequence: **U** = 1 1   0 1   0 1   0 0   0 1   0 1   1 1

Example 7.2   Convolutional Encoding

In Example 7.1 the initial contents of the register are all zeros. This is equivalent to the condition that the given input sequence is preceded by two zero bits (the encoding is a function of the present bit and the $K - 1$ prior bits). Repeat Example 7.1 with the assumption that the given input sequence is preceded by two one bits, and verify that now the codeword sequence **U** for input sequence **m** = 1 1 0 1 1 is different than the codeword found in Example 7.1.

Solution

The entry "×" signifies "don't know."

Input bit m_i	Register contents	State at time t_i	State at time t_{i+1}	Branch word at time t_i u_1	u_2
—	1 1 ×	1 ×	1 1		—
1	1 1 1	1 1	1 1	1	0
1	1 1 1	1 1	1 1	1	0
0	0 1 1	1 1	0 1	0	1
1	1 0 1	0 1	1 0	0	0
1	1 1 0	1 0	1 1	0	1
0	0 1 1	1 1	0 1	0	1
0	0 0 1	0 1	0 0	1	1

state t_i

state t_{i+1}

Output sequence: **U** = 1 0   1 0   0 1   0 0   0 1   0 1   1 1

By comparing this result with that of Example 7.1, we can see that each branch word of the output sequence **U** is *not only* a function of the input bit, but is also a function of the $K - 1$ prior bits.

7.2.3 The Tree Diagram

Although the state diagram completely characterizes the encoder, one cannot easily use it for tracking the encoder transitions as a function of time since the diagram cannot represent time history. The tree diagram adds the *dimension of time* to the state diagram. The tree diagram for the convolutional encoder shown in Figure 7.3 is illustrated in Figure 7.6. At each successive input bit time the encoding procedure can be described by traversing the diagram from left to right, each tree branch describing an output branch word. The branching rule for finding a codeword sequence is as follows: If the input bit is a zero, its associated branch word is found by moving to the next rightmost branch in the upward direction. If the input bit is a one, its branch word is found by moving to the next rightmost branch in the downward direction. Assuming that the initial contents of the encoder is all zeros, the

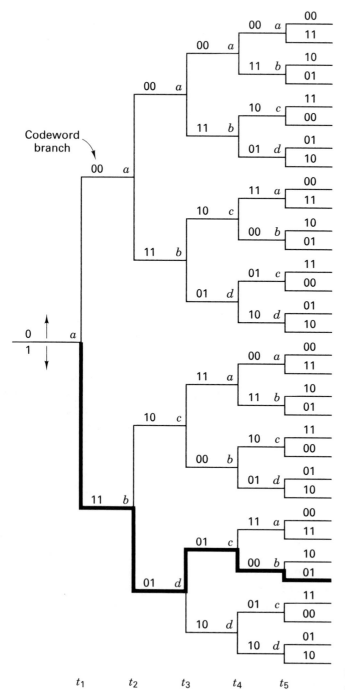

Figure 7.6 Tree representation of encoder (rate $\frac{1}{2}$, $K = 3$).

diagram shows that if the first input bit is a zero, the output branch word is 00 and, if the first input bit is a one, the output branch word is 11. Similarly, if the first input bit is a one and the second input bit is a zero, the second output branch word is 10. Or, if the first input bit is a one and the second input bit is a one, the second output branch word is 01. Following this procedure we see that the input sequence 1 1 0 1 1 traces the heavy line drawn on the tree diagram in Figure 7.6. This path corresponds to the output codeword sequence 1 1 0 1 0 1 0 0 0 1.

The added dimension of time in the tree diagram (compared to the state diagram) allows one to dynamically describe the encoder as a function of a particular input sequence. However, can you see one problem in trying to use a tree diagram for describing a sequence of any length? The number of branches increases as a function of 2^L, where L is the number of branch words in the sequence. You would quickly run out of paper, and patience.

7.2.4 The Trellis Diagram

Observation of the Figure 7.6 tree diagram shows that for this example, the structure repeats itself at time t_4, after the third branching (in general, the tree structure *repeats after K branchings*, where K is the constraint length). We label each node in the tree of Figure 7.6 to correspond to the four possible states in the shift register, as follows: $a = 00$, $b = 10$, $c = 01$, and $d = 11$. The first branching of the tree structure, at time t_1, produces a pair of nodes labeled a and b. At each successive branching the number of nodes double. The second branching, at time t_2, results in four nodes labeled a, b, c, and d. After the *third* branching, there are a total of eight nodes: two are labeled a, two are labeled b, two are labeled c, and two are labeled d. We can see that all branches emanating from two nodes of the same state generate identical branch word sequences. From this point on, the upper and the lower halves of the tree are identical. The reason for this should be obvious from examination of the encoder in Figure 7.3. As the fourth input bit enters the encoder on the left, the first input bit is ejected on the right and no longer influences the output branch words. Consequently, the input sequences 1 0 0 x y . . . and 0 0 0 x y . . . , where the leftmost bit is the earliest bit, generate the same branch words after the $(K = 3)$rd branching. This means that any two nodes having the same state label at the same time t_i can be merged, since all succeeding paths will be indistinguishable. If we do this to the tree structure of Figure 7.6, we obtain another diagram, called the trellis diagram. The *trellis diagram,* by exploiting the repetitive structure, provides a more manageable encoder description than does the tree diagram. The trellis diagram for the convolutional encoder of Figure 7.3 is shown in Figure 7.7.

In drawing the trellis diagram, we use the same convention that we introduced with the state diagram—a solid line denotes the output generated by an input bit zero, and a dashed line denotes the output generated by an input bit one. The nodes of the trellis characterize the encoder states; the first row nodes correspond to the state $a = 00$, the second and subsequent rows correspond to the states $b = 10$, $c = 01$, and $d = 11$. At each unit of time, the trellis requires 2^{K-1} nodes to represent the 2^{K-1} possible encoder states. The trellis in our example assumes a

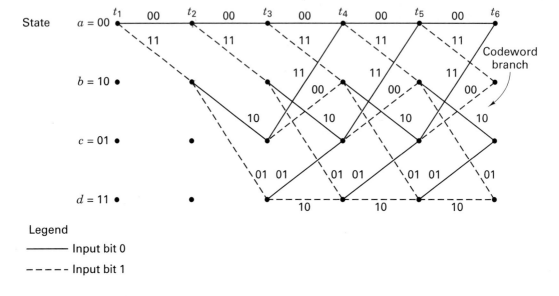

Figure 7.7 Encoder trellis diagram (rate $\frac{1}{2}$, $K = 3$).

fixed periodic structure after trellis depth 3 is reached (at time t_4). In the general case, the fixed structure prevails after depth K is reached. At this point and thereafter, each of the states can be entered from either of two preceding states. Also, each of the states can transition to one of two states. Of the two outgoing branches, one corresponds to an input bit zero and the other corresponds to an input bit one. On Figure 7.7 the output branch words corresponding to the state transitions appear as labels on the trellis branches.

One time-interval section of a fully-formed encoding trellis structure completely defines the code. The only reason for showing several sections is for viewing a code-symbol sequence as a function of time. The state of the convolutional encoder is represented by the contents of the rightmost $K - 1$ stages in the encoder register. Some authors describe the state as the contents of the leftmost $K - 1$ stages. Which description is correct? They are both correct in the following sense. Every transition has a starting state and a terminating state. The rightmost $K - 1$ stages describe the starting state for the current input, which is in the leftmost stage (assuming a rate $1/n$ encoder). The leftmost $K - 1$ stages represent the terminating state for that transition. A code-symbol sequence is characterized by N branches (representing N data bits) occupying N intervals of time and associated with a particular state at each of $N + 1$ times (from start to finish). Thus, we launch bits at times $t_1, t_2, \ldots, t_N$, and are interested in state metrics at times $t_1, t_2, \ldots, t_{N+1}$. The convention used here is that the current bit is located in the leftmost stage (not on a wire leading to that stage), and the rightmost $K - 1$ stages start in the all-zeros state. We refer to this time as the *start time* and label it t_1. We refer to the concluding time of the last transition as the *terminating time* and label it t_{N+1}.

7.3 FORMULATION OF THE CONVOLUTIONAL DECODING PROBLEM

7.3.1 Maximum Likelihood Decoding

If all input message sequences are equally likely, a decoder that achieves the minimum probability of error is one that compares the conditional probabilities, also called the *likelihood functions* $P(\mathbf{Z}|\mathbf{U}^{(m)})$, where $\mathbf{Z}$ is the received sequence and $\mathbf{U}^{(m)}$ is one of the possible transmitted sequences, and chooses the maximum. The decoder chooses $\mathbf{U}^{(m')}$ if

$$P(\mathbf{Z}|\mathbf{U}^{(m')}) = \max_{\text{over all } \mathbf{U}^{(m)}} P(\mathbf{Z}|\mathbf{U}^{(m)}) \tag{7.1}$$

The *maximum likelihood* concept, as stated in Equation (7.1), is a fundamental development of decision theory (see Appendix B); it is the formalization of a "common-sense" way to make decisions when there is statistical knowledge of the possibilities. In the binary demodulation treatment in Chapters 3 and 4 there were *only two* equally likely possible signals, $s_1(t)$ or $s_2(t)$, that might have been transmitted. Therefore, to make the binary maximum likelihood decision, given a received signal, meant only to decide that $s_1(t)$ was transmitted if

$$p(z|s_1) > p(z|s_2)$$

otherwise, to decide that $s_2(t)$ was transmitted. The parameter z represents $z(T)$, the receiver predetection value at the end of each symbol duration time $t = T$. However, when applying maximum likelihood to the convolutional decoding problem, we observe that the convolutional code has memory (the received sequence represents the superposition of current bits and prior bits). Thus, applying maximum likelihood to the decoding of convolutionally encoded bits is performed in the context of choosing the *most likely sequence,* as shown in Equation (7.1). There are typically a *multitude* of possible codeword sequences that might have been transmitted. To be specific, for a binary code, a sequence of L branch words is a member of a set of 2^L possible sequences. Therefore, in the maximum likelihood context, we can say that the decoder chooses a particular $\mathbf{U}^{(m')}$ as the transmitted sequence if the likelihood $P(\mathbf{Z}|\mathbf{U}^{(m')})$ is greater than the likelihoods of all the other possible transmitted sequences. Such an optimal decoder, which minimizes the error probability (for the case where all transmitted sequences are equally likely), is known as a *maximum likelihood decoder.* The likelihood functions are given or computed from the specifications of the channel.

We will assume that the noise is additive white Gaussian with zero mean and thus the channel is *memoryless,* which means that the noise affects each code symbol *independently* of all the other symbols. For a convolutional code of rate $1/n$, we can therefore express the likelihood as

$$P(\mathbf{Z}|\mathbf{U}^{(m)}) = \prod_{i=1}^{\infty} P(Z_i|U_i^{(m)}) = \prod_{i=1}^{\infty} \prod_{j=1}^{n} P(z_{ji}|u_{ji}^{(m)}) \tag{7.2}$$

where Z_i is the ith branch of the received sequence $\mathbf{Z}$, $U_i^{(m)}$ is the ith branch of a particular codeword sequence $\mathbf{U}^{(m)}$, z_{ji} is the jth code symbol of Z_i, and $u_{ji}^{(m)}$ is the jth code symbol of $U_i^{(m)}$, and each branch comprises n code symbols. The decoder problem consists of choosing a path through the trellis of Figure 7.7 (each possible path defines a codeword sequence) such that

$$\prod_{i=1}^{\infty} \prod_{j=1}^{n} P(z_{ji}|u_{ji}^{(m)}) \text{ is maximized} \tag{7.3}$$

Generally, it is computationally more convenient to use the logarithm of the likelihood function since this permits the summation, instead of the multiplication, of terms. We are able to use this transformation because the logarithm is a monotonically increasing function and thus will not alter the final result in our codeword selection. We can define the log-likelihood function as

$$\gamma_{\mathbf{U}}(m) = \log P(\mathbf{Z}|\mathbf{U}^{(m)}) = \sum_{i=1}^{\infty} \log P(Z_i|U_i^{(m)}) = \sum_{i=1}^{\infty} \sum_{j=1}^{n} \log P(z_{ji}|u_{ji}^{(m)}) \tag{7.4}$$

The decoder problem now consists of choosing a path through the tree of Figure 7.6 or the trellis of Figure 7.7 such that $\gamma_{\mathbf{U}}(m)$ is maximized. For the decoding of convolutional codes, either the tree or the trellis structure can be used. In the tree representation of the code, the fact that the paths remerge is ignored. Since for a binary code, the number of possible sequences made up of L branch words is 2^L, maximum likelihood decoding of such a received sequence, using a tree diagram, requires the "brute force" or exhaustive comparison of 2^L accumulated log-likelihood metrics, representing all the possible different codeword sequences that could have been transmitted. Hence it is not practical to consider maximum likelihood decoding with a tree structure. It is shown in a later section that with the use of the trellis representation of the code, it is possible to configure a decoder which can discard the paths that could not possibly be candidates for the maximum likelihood sequence. The decoded path is chosen from some reduced set of *surviving paths*. Such a decoder is still optimum in the sense that the decoded path is the same as the decoded path obtained from a "brute force" maximum likelihood decoder, but the early rejection of unlikely paths reduces the decoding complexity.

For an excellent tutorial on the structure of convolutional codes, maximum likelihood decoding, and code performance, see Reference [8]. There are several algorithms that yield *approximate* solutions to the maximum likelihood decoding problem, including sequential [9, 10] and threshold [11]. Each of these algorithms is suited to certain special applications, but are all suboptimal. In contrast, the *Viterbi decoding algorithm* performs maximum likelihood decoding and is therefore optimal. This does not imply that the Viterbi algorithm is best for every application; there are severe constraints imposed by hardware complexity. The Viterbi algorithm is considered in Sections 7.3.3 and 7.3.4.

7.3.2 Channel Models: Hard versus Soft Decisions

Before specifying an algorithm that will determine the maximum likelihood decision, let us describe the channel. The codeword sequence $\mathbf{U}^{(m)}$, made up of branch words, with each branch word comprised of n code symbols, can be considered to be an endless stream, as opposed to a block code, in which the source data and their codewords are partitioned into precise block sizes. The codeword sequence shown in Figure 7.1 emanates from the convolutional encoder and enters the modulator, where the code symbols are transformed into signal waveforms. The modulation may be baseband (e.g., pulse waveforms) or bandpass (e.g., PSK or FSK). In general, ℓ symbols at a time, where ℓ is an integer, are mapped into signal waveforms $s_i(t)$, where $i = 1, 2, \ldots, M = 2^\ell$. When $\ell = 1$, the modulator maps each code symbol into a binary waveform. The channel over which the waveform is transmitted is assumed to corrupt the signal with Gaussian noise. When the corrupted signal is received, it is first processed by the demodulator and then by the decoder.

Consider that a binary signal transmitted over a symbol interval $(0, T)$ is represented by $s_1(t)$ for a binary one and $s_2(t)$ for a binary zero. The received signal is $r(t) = s_i(t) + n(t)$, where $n(t)$ is a zero-mean Gaussian noise process. In Chapter 3 we described the detection of $r(t)$ in terms of two basic steps. In the first step, the received waveform is reduced to a single number, $z(T) = a_i + n_0$, where a_i is the signal component of $z(T)$ and n_0 is the noise component. The noise component, n_0, is a zero-mean *Gaussian random variable,* and thus $z(T)$ is a *Gaussian random variable* with a mean of either a_1 or a_2 depending on whether a binary one or binary zero was sent. In the second step of the detection process a decision was made as to which signal was transmitted, on the basis of comparing $z(T)$ to a threshold. The conditional probabilities of $z(T)$, $p(z|s_1)$, and $p(z|s_2)$ are shown in Figure 7.8, labeled likelihood of s_1 and likelihood of s_2. The demodulator in Figure 7.1, converts the set of time-ordered random variables $\{z(T)\}$ into a code sequence **Z**, and passes it on to the decoder. The demodulator output can be configured in a variety of ways. It can be implemented to make a *firm or hard decision* as to whether $z(T)$ represents a zero or a one. In this case, the output of the demodulator is quantized to two levels, zero and one, and fed into the decoder (this is exactly the same threshold decision that was made in Chapters 3

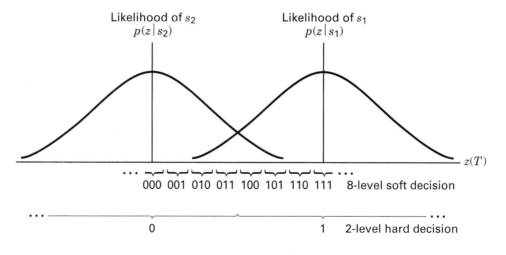

Figure 7.8 Hard and soft decoding decisions.

and 4). Since the decoder operates on the hard decisions made by the demodulator, the decoding is called *hard-decision decoding.*

The demodulator can also be configured to feed the decoder with a *quantized value* of $z(T)$ *greater than two levels.* Such an implementation furnishes the decoder with more information than is provided in the hard-decision case. When the quantization level of the demodulator output is greater than two, the decoding is called *soft-decision decoding.* Eight levels (3-bits) of quantization are illustrated on the abscissa of Figure 7.8. When the demodulator sends a hard binary decision to the decoder, it sends it a single binary symbol. When the demodulator sends a soft binary decision, quantized to eight levels, it sends the decoder a 3-bit word describing an interval along $z(T)$. In effect, sending such a 3-bit word in place of a single binary symbol is equivalent to sending the decoder a *measure of confidence* along with the code-symbol decision. Referring to Figure 7.8, if the demodulator sends 1 1 1 to the decoder, this is tantamount to declaring the code symbol to be a one with very high confidence, while sending a 1 0 0 is tantamount to declaring the code symbol to be a one with very low confidence. It should be clear that ultimately, every message decision out of the decoder must be a hard decision; otherwise, one might see computer printouts that read: "think it's a 1," "think it's a 0," and so on. The idea behind the demodulator *not making hard decisions* and sending more data (soft decisions) to the decoder can be thought of as an interim step to provide the decoder with more information, which the decoder then uses for recovering the message sequence (with better error performance than it could in the case of hard-decision decoding). In Figure 7.8, the 8-level soft-decision metric is often shown as −7, −5, −3, −1, 1, 3, 5, 7. Such a designation lends itself to a simple interpretation of the soft decision: The sign of the metric represents a decision (e.g., choose s_1 if positive, choose s_2 if negative), and the magnitude of the metric represents the confidence level of that decision. The only advantage for the metric shown in Figure 7.8 is that it avoids the use of negative numbers.

For a Gaussian channel, eight-level quantization results in a performance improvement of approximately 2 dB in required signal-to-noise ratio compared to two-level quantization. This means that eight-level soft-decision decoding can provide the same probability of bit error as that of hard-decision decoding, but requires 2 dB *less* E_b/N_0 for the same performance. Analog (or infinite-level quantization) results in a 2.2-dB performance improvement over two-level quantization; therefore, *eight-level quantization* results in a loss of approximately 0.2 dB compared to infinitely fine quantization. For this reason, quantization to more than eight levels can yield little performance improvement [12]. What price is paid for such improved soft-decision-decoder performance? In the case of hard-decision decoding, a single bit is used to describe each code symbol, while for eight-level quantized soft-decision decoding 3 bits are used to describe each code symbol; therefore, three times the amount of data must be handled during the decoding process. Hence the price paid for soft-decision decoding is an increase in required memory size at the decoder (and possibly a speed penalty).

Block decoding algorithms and convolutional decoding algorithms have been devised to operate with hard *or* soft decisions. However, soft-decision decoding is generally not used with block codes because it is considerably more difficult than hard-decision decoding to implement. The most prevalent use of soft-decision decoding is with the *Viterbi convolutional decoding algorithm,* since with Viterbi decoding, soft decisions represent only a trivial increase in computation.

7.3.2.1 Binary Symmetric Channel

A binary symmetric channel (BSC) is a discrete memoryless channel (see Section 6.3.1) that has binary input and output alphabets and symmetric transition probabilities. It can be described by the conditional probabilities

$$P(0|1) = P(1|0) = p$$

$$P(1|1) = P(0|0) = 1 - p$$

(7.5)

as illustrated in Figure 7.9. The probability that an output symbol will differ from the input symbol is p, and the probability that the output symbol will be identical to the input symbol is $(1 - p)$. The BSC is an example of a *hard-decision channel,* which means that, even though continuous-valued signals may be received by the demodulator, a BSC allows only firm decisions such that each demodulator output symbol, z_{ji}, as shown in Figure 7.1, consists of one of two binary values. The indexing of z_{ji} pertains to the jth code symbol of the ith branch word, Z_i. The demodulator then feeds the sequence $\mathbf{Z} = \{Z_i\}$ to the decoder.

Let $\mathbf{U}^{(m)}$ be a transmitted codeword over a BSC with symbol error probability p, and let $\mathbf{Z}$ be the corresponding received decoder sequence. As noted previously, a maximum likelihood decoder chooses the codeword $\mathbf{U}^{(m')}$ that maximizes the likelihood $P(\mathbf{Z}|\mathbf{U}^{(m)})$ or its logarithm. For a BSC, this is equivalent to choosing the codeword $\mathbf{U}^{(m')}$ that is closest in *Hamming distance* to $\mathbf{Z}$ [8]. Thus Hamming distance is an appropriate metric to describe the distance or closeness of fit between $\mathbf{U}^{(m)}$ and $\mathbf{Z}$. From all the possible transmitted sequences $\mathbf{U}^{(m)}$, the decoder chooses the $\mathbf{U}^{(m')}$ sequence for which the distance to $\mathbf{Z}$ is minimum.

Suppose that $\mathbf{U}^{(m)}$ and $\mathbf{Z}$ are each L-bit-long sequences and that they differ in d_m positions [i.e., the Hamming distance between $\mathbf{U}^{(m)}$ and $\mathbf{Z}$ is d_m]. Then, since the

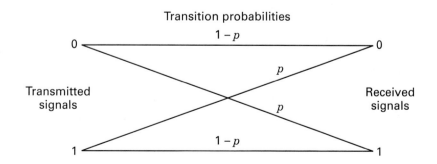

Figure 7.9 Binary symmetric channel (hard-decision channel).

channel is assumed to be memoryless, the probability that this $\mathbf{U}^{(m)}$ was transformed to the specific received $\mathbf{Z}$ at distance d_m from it can be written as

$$P(\mathbf{Z}|\mathbf{U}^{(m)}) = p^{d_m}(1-p)^{L-d_m} \tag{7.6}$$

and the log-likelihood function is

$$\log P(\mathbf{Z}|\mathbf{U}^{(m)}) = -d_m \log\left(\frac{1-p}{p}\right) + L\log(1-p) \tag{7.7}$$

If we compute this quantity for each possible transmitted sequence, the last term in the equation will be constant in each case. Assuming that $p < 0.5$, we can express Equation (7.7) as

$$\log P(\mathbf{Z}|\mathbf{U}^{(m)}) = -Ad_m - B \tag{7.8}$$

where A and B are positive constants. Therefore, choosing the codeword $\mathbf{U}^{(m')}$, such that the Hamming distance d_m to the received sequence $\mathbf{Z}$ is minimized, corresponds to *maximizing the likelihood or log-likelihood metric.* Consequently, over a BSC, the log-likelihood metric is conveniently replaced by the Hamming distance, and a maximum likelihood decoder will choose, in the tree or trellis diagram, the path whose corresponding sequence $\mathbf{U}^{(m')}$ is at the *minimum Hamming distance* to the received sequence $\mathbf{Z}$.

7.3.2.2 Gaussian Channel

For a Gaussian channel, each demodulator output symbol z_{ji}, as shown in Figure 7.1, is a value from a continuous alphabet. The symbol z_{ji} cannot be labeled as a correct or incorrect detection decision. Sending the decoder such soft decisions can be viewed as sending a family of conditional probabilities of the different symbols (see Section 6.3.1). It can be shown [8] that maximizing $P(\mathbf{Z}|\mathbf{U}^{(m)})$ is equivalent to maximizing the inner product between the codeword sequence $\mathbf{U}^{(m)}$ (consisting of binary symbols represented as bipolar values) and the analog-valued received sequence $\mathbf{Z}$. Thus, the decoder chooses the codeword $\mathbf{U}^{(m')}$ if it maximizes

$$\sum_{i=1}^{\infty} \sum_{j=1}^{n} z_{ji} u_{ji}^{(m)} \tag{7.9}$$

This is equivalent to choosing the codeword $\mathbf{U}^{(m')}$ that is closest in *Euclidean distance* to $\mathbf{Z}$. Even though the hard- and soft-decision channels require different metrics, the concept of choosing the codeword $\mathbf{U}^{(m')}$ that is closest to the received sequence, $\mathbf{Z}$, is the same in both cases. To implement the maximization of Equation (7.9) exactly, the decoder would have to be able to handle analog-valued arithmetic operations. This is impractical because the decoder is generally implemented digitally. Thus it is necessary to quantize the received symbols z_{ji}. Does Equation (7.9) remind you of the demodulation treatment in Chapters 3 and 4? Equation (7.9) is the discrete version of correlating an input received waveform, $r(t)$, with a reference waveform, $s_i(t)$, as expressed in Equation (4.15). The quantized Gaussian channel, typically referred to as a *soft-decision channel,* is the channel model assumed for the soft-decision decoding described earlier.

7.3.3 The Viterbi Convolutional Decoding Algorithm

The Viterbi decoding algorithm was discovered and analyzed by Viterbi [13] in 1967. The Viterbi algorithm essentially performs maximum likelihood decoding; however, it reduces the computational load by taking advantage of the special structure in the code trellis. The advantage of Viterbi decoding, compared with brute-force decoding, is that the complexity of a Viterbi decoder is not a function of the number of symbols in the codeword sequence. The algorithm involves calculating a *measure of similarity, or distance,* between the received signal, at time t_i, and all the trellis paths entering each state at time t_i. The Viterbi algorithm removes from consideration those trellis paths that could not possibly be candidates for the maximum likelihood choice. When two paths enter the same state, the one having the best metric is chosen; this path is called the *surviving path.* This selection of surviving paths is performed for all the states. The decoder continues in this way to advance deeper into the trellis, making decisions by eliminating the least likely paths. The early rejection of the unlikely paths reduces the decoding complexity. In 1969, Omura [14] demonstrated that the Viterbi algorithm is, in fact, maximum likelihood. Note that the goal of selecting the optimum path can be expressed, equivalently, as choosing the codeword with the *maximum likelihood metric,* or as choosing the codeword with the *minimum distance metric.*

7.3.4 An Example of Viterbi Convolutional Decoding

For simplicity, a BSC is assumed; thus Hamming distance is a proper distance measure. The encoder for this example is shown in Figure 7.3, and the encoder trellis diagram is shown in Figure 7.7. A similar trellis can be used to represent the decoder, as shown in Figure 7.10. We start at time t_1 in the 00 state (flushing the encoder between messages provides the decoder with starting-state knowledge). Since in this example, there are only two possible transitions leaving any state, not all branches need be shown initially. The full trellis structure evolves after time t_3. The basic idea behind the decoding procedure can best be understood by examining the Figure 7.7 encoder trellis in concert with the Figure 7.10 decoder trellis. For the decoder trellis it is convenient at each time interval, to label each branch with the *Hamming distance* between the received code symbols and the branch word corresponding to the same branch from the encoder trellis. The example in Figure 7.10 shows a message sequence **m**, the corresponding codeword sequence **U**, and a noise corrupted received sequence **Z** = 11 01 01 10 01 . . . . The branch words seen on the *encoder trellis* branches characterize the encoder in Figure 7.3, and are known a priori to both the encoder and the decoder. These encoder branch words are the code symbols that would be expected to come from the encoder output as a result of each of the state transitions. The labels on the *decoder trellis* branches are accumulated by the decoder *on the fly.* That is, as the code symbols are received,

Input data sequence	**m:**	1	1	0	1	1	$\cdots$
Transmitted codeword	**U:**	11	01	01	00	01	$\cdots$
Received sequence	**Z:**	11	01	01	10	01	$\cdots$

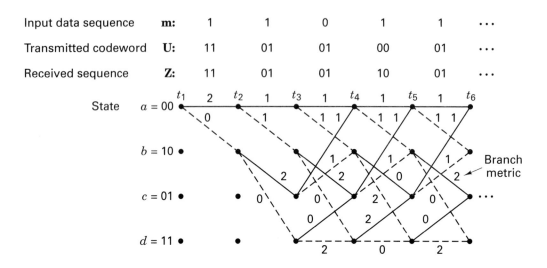

Figure 7.10 Decoder trellis diagram (rate $\frac{1}{2}$, $K = 3$).

each branch of the decoder trellis is labeled with a metric of similarity (Hamming distance) between the received code symbols and each of the branch words for that time interval. From the received sequence **Z**, shown in Figure 7.10, we see that the code symbols received at (following) time t_1 are 11. In order to label the decoder branches at (departing) time t_1 with the appropriate Hamming distance metric, we look at the Figure 7.7 encoder trellis. Here we see that a state $00 \rightarrow 00$ transition yields an output branch word of 00. But we received 11. Therefore, on the decoder trellis we label the state $00 \rightarrow 00$ transition with Hamming distance between them, namely 2. Looking at the encoder trellis again, we see that a state $00 \rightarrow 10$ transition yields an output branch word of 11, which corresponds exactly with the code symbols we received at time t_1. Therefore, on the decoder trellis, we label the state $00 \rightarrow 10$ transition with a Hamming distance of 0. In summary, the metric entered on a decoder trellis branch represents the difference (distance) between what was received and what "should have been" received had the branch word associated with that branch been transmitted. In effect, these metrics describe a correlation-like measure between a received branch word and each of the candidate branch words. We continue labeling the decoder trellis branches in this way as the symbols are received at each time t_i. The decoding algorithm uses these Hamming distance metrics to find the *most likely* (minimum distance) path through the trellis.

The basis of *Viterbi decoding* is the following observation: If any two paths in the trellis merge to a single state, one of them can always be eliminated in the search for an optimum path. For example, Figure 7.11 shows two paths merging at time t_5 to state 00. Let us define the *cumulative Hamming path metric* of a given path at time t_i as the sum of the branch Hamming distance metrics along that path up to time t_i. In Figure 7.11 the upper path has metric 4; the lower has metric 1. The upper path cannot be a portion of the optimum path because the lower path, which

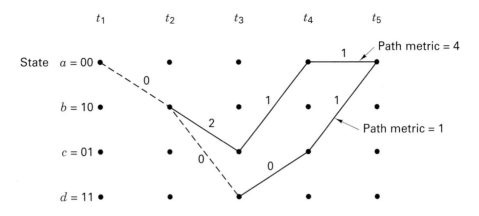

Figure 7.11 Path metrics for two merging paths.

enters the same state, has a lower metric. This observation holds because of the Markov nature of the encoder state: The present state summarizes the encoder history in the sense that previous states cannot affect future states or future output branches.

At each time t_i there are 2^{K-1} states in the trellis, where K is the constraint length, and each state can be *entered by means of two paths*. Viterbi decoding consists of computing the metrics for the two paths entering each state and *eliminating one of them*. This computation is done for each of the 2^{K-1} states or nodes at time t_i; then the decoder moves to time t_{i+1} and repeats the process. At a given time, the winning path metric for each state is designated as the *state metric* for that state at that time. The first few steps in our decoding example are as follows (see Figure 7.12). Assume that the input data sequence **m**, codeword **U**, and received sequence **Z** are as shown in Figure 7.10. Assume that the decoder knows the correct initial state of the trellis. (This assumption is not necessary in practice, but simplifies the explanation.) At time t_1 the received code symbols are 11. From state 00 the only possible transitions are to state 00 or state 10, as shown in Figure 7.12a. State $00 \rightarrow 00$ transition has branch metric 2; state $00 \rightarrow 10$ transition has branch metric 0. At time t_2 there are two possible branches leaving each state, as shown in Figure 7.12b. The cumulative metrics of these branches are labeled state metrics Γ_a, Γ_b, Γ_c, and Γ_d, corresponding to the terminating state. At time t_3 in Figure 7.12c there are again two branches diverging from each state. As a result, there are two paths entering each state at time t_4. One path entering each state can be eliminated, namely, the one having the larger cumulative path metric. Should metrics of the two entering paths be of equal value, one path is chosen for elimination by using an arbitrary rule. The surviving path into each state is shown in Figure 7.12d. At this point in the decoding process, there is only a single surviving path, termed the *common stem*, between times t_1 and t_2. Therefore, the decoder can now decide that the state transition which occurred between t_1 and t_2 was $00 \rightarrow 10$. Since this transition is produced by an input bit one, the decoder outputs a one as the first decoded bit.

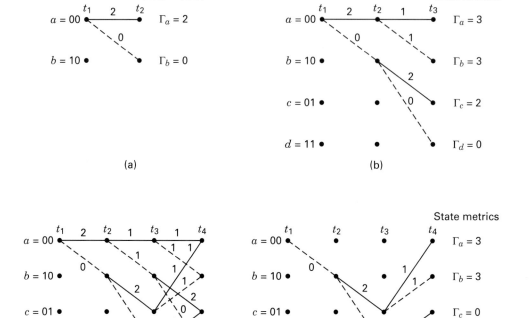

Figure 7.12 Selection of survivor paths, (a) Survivors at t_2. (b) Survivors at t_3. (c) Metric comparisons at t_4. (d) Survivors at t_4. (e) Metric comparisons at t_5. (f) Survivors at t_5. (g) Metric comparisons at t_6. (h) Survivors at t_6.

Here we can see how the decoding of the surviving branch is facilitated by having drawn the trellis branches with solid lines for input zeros and dashed lines for input ones. Note that the first bit was not decoded until the path metric computation had proceeded to a much greater depth into the trellis. For a typical decoder implementation, this represents a decoding delay which can be as much as five times the constraint length in bits.

At each succeeding step in the decoding process, there will always be two possible paths entering each state; one of the two will be eliminated by comparing the path metrics. Figure 7.12e shows the next step in the decoding process. Again, at time t_5 there are two paths entering each state, and one of each pair can be eliminated. Figure 7.12f shows the survivors at time t_5. Notice that in our example we cannot yet make a decision on the second input data bit because there still are two paths leaving the state 10 node at time t_2. At time t_6 in Figure 7.12g we again see the pattern of remerging paths, and in Figure 7.12h we see the survivors at time t_6. Also, in Figure 7.12h the decoder outputs one as the second decoded bit, corre-

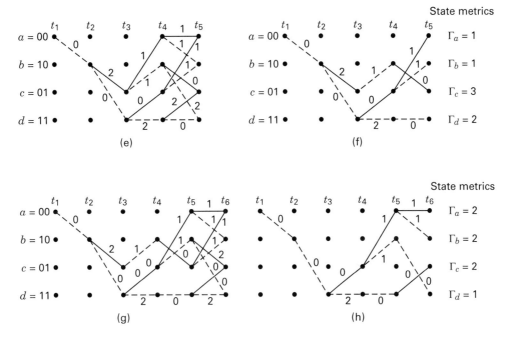

Figure 7.12 (*Continued*)

sponding to the single surviving path between t_2 and t_3. The decoder continues in this way to advance deeper into the trellis and to make decisions on the input data bits by eliminating all paths but one.

Pruning the trellis (as paths remerge) guarantees that there are never more paths than there are states. For this example, verify that after each pruning in Figures 7.12b, d, f, and h, there are only 4 paths. Compare this to attempting a "brute force" maximum-likelihood sequence estimation without using the Viterbi algorithm. In that case, the number of possible paths (representing possible sequences) is an exponential function of sequence length. For a binary codeword sequence that has a length of L branch words, there are 2^L possible sequences.

7.3.5 Decoder Implementation

In the context of the trellis diagram of Figure 7.10, transitions during any one time interval can be grouped into 2^{v-1} disjoint cells, each cell depicting four possible transitions, where $v = K - 1$ is called the *encoder memory*. For the $K = 3$ example, $v = 2$ and $2^{v-1} = 2$ cells. These cells are shown in Figure 7.13, where a, b, c, and d refer to the states at time t_i, and a', b', c', and d' refer to the states at time t_{i+1}. Shown on each transition is the *branch metric* δ_{xy}, where the subscript indicates that the metric corresponds to the transition from state x to state y. These cells and the associated logic units that update the *state metrics* $\{\Gamma_x\}$, where x designates a particular state, represent the basic building blocks of the decoder.

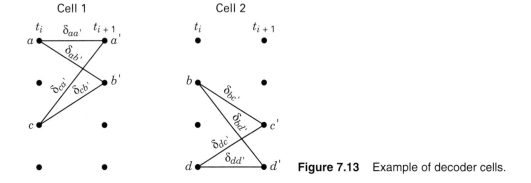

Figure 7.13 Example of decoder cells.

7.3.5.1 Add-Compare-Select Computation

Continuing with the $K = 3$, 2-cell example, Figure 7.14 illustrates the logic unit that corresponds to cell 1. The logic executes the special purpose computation called *add-compare-select* (ACS). The state metric $\Gamma_{a'}$ is calculated by adding the previous-time state metric of state a, Γ_a, to the branch metric $\delta_{aa'}$ and the previous-time state metric of state c, Γ_c, to the branch metric $\delta_{ca'}$. This results in two possible path metrics as candidates for the new state metric $\Gamma_{a'}$. The two candidates are compared in the logic unit of Figure 7.14. The largest likelihood (smallest distance) of the two path metrics is stored as the new state metric $\Gamma_{a'}$ for state a. Also stored is the *new* path history $\hat{m}_{a'}$ for state a, where $\hat{m}_{a'}$ is the message-path history of the state augmented by the data of the winning path.

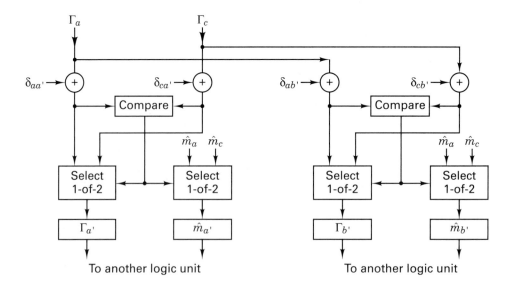

Figure 7.14 Logic unit that implements the add-compare-select functions corresponding to cell #1.

Channel Coding: Part 2   Chap. 7

Also shown in Figure 7.14 is the cell-1 ACS logic that yields the new state metric $\Gamma_{b'}$ and the new path history $\hat{m}_{b'}$. This ACS operation is similarly performed for the paths in other cells. The oldest bit on the path with the smallest state metric forms the decoder output.

7.3.5.2 Add-Compare-Select as seen on the Trellis

Consider the same example that was used for describing Viterbi decoding in Section 7.3.4. The message sequence was $\mathbf{m} = 1\,1\,0\,1\,1$, the codeword sequence was $\mathbf{U} = 11\ 01\ 01\ 00\ 01$, and the received sequence was $\mathbf{Z} = 11\ 01\ 01\ 10\ 01$. Figure 7.15 depicts a decoding trellis diagram similar to Figure 7.10. A branch metric that labels each branch is the Hamming distance between the received code symbols and the corresponding branch word from the encoder trellis. Additionally, the Figure 7.15 trellis indicates a value at each state x, and for each time from time t_2 to t_6, which is a state metric Γ_x. We perform the add-compare-select (ACS) operation when there are two transitions entering a state, as there are for times t_4 and later. For example at time t_4, the value of the state metric for state a is obtained by incrementing the state metric $\Gamma_a = 3$ at time t_3 with the branch metric $\delta_{aa'} = 1$ yielding a candidate value of 4. Simultaneously, the state metric $\Gamma_c = 2$ at time t_3 is incremented with branch metric $\delta_{ca'} = 1$ yielding a candidate value of 3. The select operation of the ACS process selects the largest-likelihood (minimum distance) path metric as the new state metric; hence, for state a at time t_4, the new state metric is $\Gamma_{a'} = 3$. The winning path is shown with a heavy line and the path that has been dropped is shown with a lighter line. On the trellis of Figure 7.15, observe the state metrics from left to right. Verify that at each time, the value of each state metric is obtained by incrementing the connected state metric from the previous time along the winning path (heavy line) with the branch metric between them. At a given time, starting with the

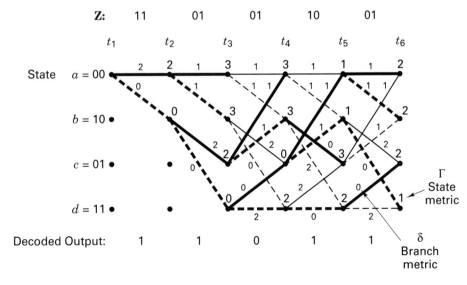

Figure 7.15 Add-compare-select computations in Viterbi decoding.

state having the smallest Γ_x, the decoder output is formed by tracing the winning branches back to the oldest bit on the path. As an example, looking at time t_6 in Figure 7.15, we see that the minimum-distance state metric has a value of 1. From this state d, the winning path can be traced back to time t_1, and one can verify that the decoded message is the same as the original message, by the convention that dashed and solid lines represent binary ones and zeros respectively.

7.3.6 Path Memory and Synchronization

The storage requirements of the Viterbi decoder grow exponentially with constraint length K. For a code with rate $1/n$, the decoder retains a set of 2^{K-1} paths after each decoding step. With high probability, these paths will not be mutually disjoint very far back from the present decoding depth [12]. All of the 2^{K-1} paths tend to have a common stem which eventually branches to the various states. Thus if the decoder stores enough of the history of the 2^{K-1} paths, the oldest bits on all paths will be the same. A simple decoder implementation, then, contains a *fixed amount of path history* and outputs the oldest bit on an arbitrary path each time it steps one level deeper into the trellis. The amount of path storage required is [12]

$$u = h2^{K-1} \qquad (7.10)$$

where h is the length of the information bit path history per state. A refinement, which minimizes the value of h, uses the oldest bit on the most likely path as the decoder output, instead of the oldest bit on an arbitrary path. It has been demonstrated [12] that a value of h of 4 or 5 times the code constraint length is sufficient for near-optimum decoder performance. The storage requirement u is the basic limitation on the implementation of Viterbi decoders. Commercial decoders are limited to a constraint length of about $K = 10$. Efforts to increase coding gain by further increasing constraint length are met by the exponential increase in memory requirements (and complexity) that follows from Equation (7.10).

Branch word synchronization is the process of determining the beginning of a branch word in the received sequence. Such synchronization can take place without new information being added to the transmitted symbol stream because the received data appear to have an excessive error rate when not synchronized. Therefore, a simple way of accomplishing synchronization is to monitor some concomitant indication of this large error rate, that is, the rate at which the state metrics are increasing or the rate at which the surviving paths in the trellis merge. The monitored parameters are compared to a threshold, and synchronization is then adjusted accordingly.

7.4 PROPERTIES OF CONVOLUTIONAL CODES

7.4.1 Distance Properties of Convolutional Codes

Consider the distance properties of convolutional codes in the context of the simple encoder in Figure 7.3 and its trellis diagram in Figure 7.7. We want to evaluate the distance between all possible pairs of codeword sequences. As in the case of block codes (see Section 6.5.2), we are interested in the *minimum distance* between all pairs of such codeword sequences in the code, since the minimum distance is related to the error-correcting capability of the code. Because a convolutional code is a group or *linear code* [6], there is no loss in generality in simply finding the minimum distance between each of the codeword sequences and the all-zeros sequence. In other words, for a linear code, any test message is just as "good" as any other test message. So, why not choose one that is easy to keep track of—namely, the all-zeros sequence? Assuming that the all-zeros input sequence was transmitted, the paths of interest are those that start and end in the 00 state and do not return to the 00 state anywhere in between. An error will occur whenever the distance of any other path that merges with the $a = 00$ state at time t_i is less than that of the all-zeros path up to time t_i, causing the all-zeros path to be discarded in the decoding process. In other words, given the all-zeros transmission, an error occurs whenever the *all-zeros path does not survive.* Thus, an error of interest is associated with a surviving path that diverges from and then remerges to the all-zeros path. One might ask, Why is it necessary for the path to remerge? Isn't the divergence enough to indicate an error? Yes, of course, but an error characterized by *only* a divergence means that the decoder, from that point on, will be outputting "garbage" for the rest of the message duration. We want to quantify the decoder's capability in terms of errors that will usually take place—that is, we want to learn the "easiest" way for the decoder to make an error. The minimum distance for making such an error can be found by exhaustively examining every path from the 00 state to the 00 state. First, let us redraw the trellis diagram, shown in Figure 7.16, labeling each branch with its Hamming distance from the all-zeros codeword instead of with its branch word symbols. The Hamming distance between two unequal-length sequences will be found by first appending the necessary number of zeros to the shorter sequence to make the two sequences equal in length. Consider all the paths that diverge from the all-zeros path and then remerge for the first time at some arbitrary node. From Figure 7.16 we can compute the distances of these paths from the all-zeros path. There is one path at distance 5 from the all-zeros path; this path departs from the all-zeros path at time t_1 and merges with it at time t_4. Similarly, there are two paths at distance 6, one which departs at time t_1 and merges at time t_5, and the other which departs at time t_1 and merges at time t_6, and so on. We can also see from the dashed and solid lines of the diagram that the input bits for the distance 5 path are 1 0 0; it differs in only one input bit from the all-zeros input sequence. Similarly, the input bits for the distance 6 paths are 1 1 0 0 and 1 0 1 0 0; each differs in two positions from the all-zeros path. The minimum distance in the set of all arbitrarily long paths that diverge and remerge, called the *minimum free distance,* or simply the *free distance,* is seen to be 5 in this example, as shown with the heavy lines in Figure 7.16. For calculating the error-correcting capability of the code, we repeat Equation (6.44) with the minimum distance d_{min} replaced by the free distance d_f as

$$t = \left\lfloor \frac{d_f - 1}{2} \right\rfloor \tag{7.11}$$

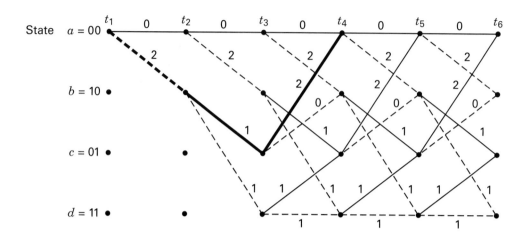

State
a = 00
b = 10
c = 01
d = 11

t_1 0 t_2 0 t_3 0 t_4 0 t_5 0 t_6

2 2 2 2 2

2 2 2

0 0 0

1 1 1 1 1

1 1 1 1 1 1

1 1 1

Figure 7.16 Trellis diagram, labeled with distances from the all-zeros path.

where $\lfloor x \rfloor$ means the largest integer no greater than x. Setting $d_f = 5$, we see that the code, characterized by the Figure 7.3 encoder, can correct any two channel errors. (See Section 7.4.1.1.)

A trellis diagram represents "the rules of the game." It is a shorthand description of all the possible transitions and their corresponding start and finish states associated with a particular finite-state machine. The trellis diagram offers some insight into the benefit (coding gain) when using error-correction coding. Consider Figure 7.16 and the possible divergence-remergence error paths. From this picture one sees that the decoder cannot make an error in any *arbitrary way*. The error path must follow one of the allowable transitions. The trellis pinpoints all such allowable paths. By having encoded the data in this way, we have placed constraints on the transmitted signal. The decoder knows these constraints, and this knowledge enables the system to more easily (using less E_b/N_0) meet some error performance requirements.

Although Figure 7.16 presents the computation of free distance in a straightforward way, a more direct closed-form expression can be obtained by starting with the state diagram in Figure 7.5. First, we label the branches of the state diagram as either $D^0 = 1$, D^1, or D^2, shown in Figure 7.17, where the exponent of D denotes the Hamming distance from the branch word of that branch to the all-zeros branch. The self-loop at node a can be eliminated since it contributes nothing to the distance properties of a codeword sequence relative to the all-zeros sequence. Furthermore, node a can be split into two nodes (labeled a and e), one of which represents the input and the other the output of the state diagram. All paths originating at $a = 00$ and terminating at $e = 00$ can be traced on the modified state diagram of Figure 7.17. We can calculate the transfer function of path $a\ b\ c\ e$ (starting and ending at state 00) in terms of the indeterminate "placeholder" D, as $D^2\ D\ D^2 = D^5$. The exponent of D represents the cumulative tally of the number of ones in the path, and hence the Hamming distance from the all-zeros path. Similarly, the

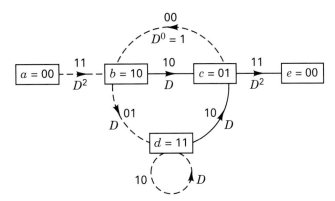

Figure 7.17 State diagram, labeled according to distance from the all-zeros path.

paths $a\,b\,d\,c\,e$ and $a\,b\,c\,b\,c\,e$ each have the transfer function D^6 and thus a Hamming distance of 6 from the all-zeros path. We now write the state equations as

$$X_b = D^2 X_a + X_c$$
$$X_c = DX_b + DX_d$$
$$X_d = DX_b + DX_d$$
$$X_e = D^2 X_c$$

(7.12)

where $X_a, \ldots, X_e$ are dummy variables for the partial paths to the intermediate nodes. The *transfer function, $T(D)$*, sometimes called the *generating function* of the code can be expressed as $T(D) = X_e/X_a$. By solving the state equations shown in Equation (7.12), we obtain [15, 16]

$$T(D) = \frac{D^5}{1 - 2D}$$

(7.13)

$$= D^5 + 2D^6 + 4D^7 + \cdots + 2^\ell D^{\ell+5} + \cdots$$

The transfer function for this code indicates that there is a single path of distance 5 from the all-zeros path, two of distance 6, four of distance 7, and in general, there are 2^ℓ paths of distance $\ell + 5$ from the all-zeros path, where $\ell = 0, 1, 2, \ldots$. The free distance d_f of the code is the Hamming weight of the lowest-order term in the expansion of $T(D)$. In this example $d_f = 5$. In evaluating distance properties, the transfer function, $T(D)$, cannot be used for long constraint lengths since the complexity of $T(D)$ increases exponentially with constraint length.

The transfer function can be used to provide more detailed information than just the distance of the various paths. Let us introduce a factor L into each branch of the state diagram so that the exponent of L can serve as a counter to indicate the number of branches in any given path from state $a = 00$ to state $e = 00$. Furthermore, we can introduce a factor N into all branch transitions caused by the input bit one. Thus, as each branch is traversed, the cumulative exponent on N increases by one, only if that branch transition is due to an input bit one. For the convolutional code characterized in the Figure 7.3 example, the additional factors L and N are shown on the modified state diagram of Figure 7.18. Equations (7.12) can now be modified as follows:

$$\begin{aligned}
X_b &= D^2 L N X_a + L N X_c \\
X_c &= D L X_b + D L X_d \\
X_d &= D L N X_b + D L N X_d \\
X_e &= D^2 L X_c
\end{aligned} \qquad (7.14)$$

The transfer function of this augmented state diagram is

$$\begin{aligned}
T(D, L, N) &= \frac{D^5 L^3 N}{1 - D L (1 + L) N} \\
&= D^5 L^3 N + D^6 L^4 (1 + L) N^2 + D^7 L^5 (1 + L)^2 N^3 \qquad (7.15) \\
&\quad + \cdots + D^{\ell+5} L^{\ell+3} N^{\ell+1} + \cdots
\end{aligned}$$

Thus, we can verify some of the path properties displayed in Figure 7.16. There is one path of distance 5, length 3, which differs in one input bit from the all-zeros path. There are two paths of distance 6, one of which is length 4, the other length 5, and both differ in two input bits from the all-zeros path. Also, of the distance 7 paths, one is of length 5, two are of length 6, and one is of length 7; all four paths correspond to input sequences that differ in three input bits from the all-zeros path. Thus if the all-zeros path is the correct path and the noise causes us to choose one of the incorrect paths of distance 7, three bit errors will be made.

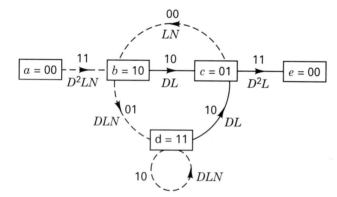

Figure 7.18 State diagram, labeled according to distance, length, and number of input ones.

7.4.1.1 Error-Correcting Capability of Convolutional Codes

In the study of block codes in Chapter 6, we saw that the error-correcting capability, t, represented the number of code symbol errors that could, with maximum likelihood decoding, be corrected in each block length of the code. However, when decoding convolutional codes, the error-correcting capability cannot be stated so succinctly. With regard to Equation (7.11), we can say that the code can, with maximum likelihood decoding, correct t errors within a few constraint lengths, where "few" here means 3 to 5. The exact length depends on how the errors are distributed. For a particular code and error pattern, the length can be bounded using transfer function methods. Such bounds are described later.

7.4.2 Systematic and Nonsystematic Convolutional Codes

A *systematic* convolutional code is one in which the input k-tuple appears as part of the output branch word n-tuple associated with that k-tuple. Figure 7.19 shows a binary, rate $\frac{1}{2}$, $K = 3$ systematic encoder. For linear block codes, any nonsystematic code can be transformed into a systematic code with the same block distance properties. This is not the case for convolutional codes. The reason for this is that convolutional codes depend largely on *free distance*; making the convolutional code systematic, in general, *reduces* the maximum possible free distance for a given constraint length and rate.

Table 7.1 shows the maximum free distance for rate $\frac{1}{2}$ systematic and nonsystematic codes for $K = 2$ through 8. For large constraint lengths the results are even more widely separated [17].

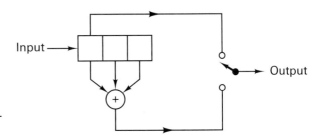

Figure 7.19 Systematic convolutional encoder, rate $\frac{1}{2}$, $K = 3$.

TABLE 7.1 Comparison of Systematic and Nonsystematic Free Distance, Rate $\frac{1}{2}$

Constraint Length	Free Distance Systematic	Free Distance Nonsystematic
2	3	3
3	4	5
4	4	6
5	5	7
6	6	8
7	6	10
8	7	10

Source: A. J. Viterbi and J. K. Omura, *Principles of Digital Communication and Coding,* McGraw-Hill Book Company, New York, 1979, p. 251.

7.4.3 Catastrophic Error Propagation in Convolutional Codes

A *catastrophic error* is defined as an event whereby a finite number of code symbol errors cause an infinite number of decoded data bit errors. Massey and Sain [18] have derived a necessary and sufficient condition for convolutional codes to display catastrophic error propagation. For rate $1/n$ codes with register taps designated by polynomial generators, as described in Section 7.2.1, the condition for catastrophic error propagation is that the generators have a *common polynomial factor* (of degree at least one). For example, Figure 7.20a illustrates a rate $\frac{1}{2}$, $K = 3$ encoder with upper polynomial $\mathbf{g}_1(X)$ and lower polynomial $\mathbf{g}_2(X)$, as follows:

$$\mathbf{g}_1(X) = 1 + X \qquad\qquad (7.16)$$
$$\mathbf{g}_2(X) = 1 + X^2$$

The generators $\mathbf{g}_1(X)$ and $\mathbf{g}_2(X)$ have in common the polynomial factor $1 + X$, since

$$1 + X^2 = (1 + X)(1 + X)$$

Therefore, the encoder in Figure 7.20a can manifest *catastrophic error propagation*.

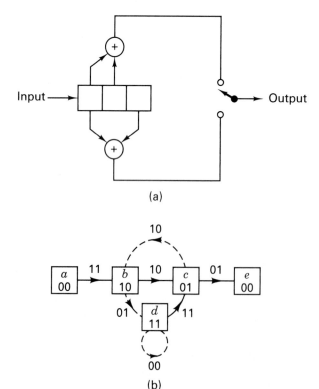

(a)

(b)

Figure 7.20 Encoder displaying cata-strophic error propagation. (a) Encoder. (b) State diagram.

In terms of the state diagram for any-rate code, catastrophic errors can occur if, and only if, any closed-loop path in the diagram has zero weight (zero distance from the all-zeros path). To illustrate this, consider the example of Figure 7.20. The state diagram in Figure 7.20b is drawn with the state $a = 00$ node split into two nodes, a and e, as before. Assuming that the all-zeros path is the correct path, the incorrect path $a\,b\,d\,d \ldots d\,c\,e$ has exactly 6 ones, no matter how many times we go around the self-loop at node d. Thus for a BSC, for example, three channel errors may cause us to choose this incorrect path. An arbitrarily large number of errors (two plus the number of times the self-loop is traversed) can be made on such a path. We observe that for rate $1/n$ codes, if each adder in the encoder has an even number of connections, the self-loop corresponding to the all-ones data state will have zero weight, and consequently, *the code will be catastrophic.*

The only advantage of a systematic code, described earlier, is that it can never be catastrophic, since each closed loop must contain at least one branch generated by a nonzero input bit, and thus each closed loop must have a nonzero code symbol. However, it can be shown [19] that only a small fraction of nonsystematic codes (excluding those where all adders have an even number of taps) are catastrophic.

7.4.4 Performance Bounds for Convolutional Codes

The probability of bit error, P_B, for a binary convolutional code using hard-decision decoding can be shown [8] to be upper bounded as follows:

$$P_B \le \left. \frac{dT(D, N)}{dN} \right|_{N=1, D=2\sqrt{p(1-p)}} \tag{7.17}$$

where p is the probability of channel symbol error. For the example of Figure 7.3, $T(D, N)$ is obtained from $T(D, L, N)$ by setting $L = 1$ in Equation (7.15).

$$T(D, N) = \frac{D^5 N}{1 - 2DN} \tag{7.18}$$

and

$$\left. \frac{dT(D, N)}{dN} \right|_{N=1} = \frac{D^5}{(1 - 2D)^2} \tag{7.19}$$

Combining Equations (7.17) and (7.19), we can write

$$P_B \le \frac{\{2[p(1-p)]^{1/2}\}^5}{\{1 - 4[p(1-p)]^{1/2}\}^2} \tag{7.20}$$

For coherent BPSK modulation over an additive white Gaussian noise (AWGN) channel, it can be shown [8] that the bit error probability is bounded by

$$P_B \le Q\left(\sqrt{2d_f \frac{E_c}{N_0}}\right) \exp\left(d_f \frac{E_c}{N_0}\right) \left. \frac{dT(D, N)}{dN} \right|_{N=1, D=\exp(-E_c/N_0)} \tag{7.21}$$

where

$E_c/N_0 = rE_b/N_0$

E_b/N_0 = ratio of information bit energy to noise power spectral density

E_c/N_0 = ratio of channel symbol energy to noise power spectral density

$r = k/n$ = rate of the code

and $Q(x)$ is defined in Equations (3.43) and (3.44) and tabulated in Table B.1. Therefore, for the rate $\frac{1}{2}$ code with free distance $d_f = 5$, in conjunction with coherent BPSK and hard-decision decoding, we can write

$$P_B \le Q\left(\sqrt{\frac{5E_b}{N_0}}\right) \exp\left(\frac{5E_b}{2N_0}\right) \frac{\exp(-5E_b/2N_0)}{[1 - 2\exp(-E_b/2N_0)]^2} \qquad (7.22)$$

$$\le \frac{Q(\sqrt{5E_b/N_0})}{[1 - 2\exp(-E_b/2N_0)]^2}$$

7.4.5 Coding Gain

Coding gain, as presented in Equation (6.19), is defined as the reduction, usually expressed in decibels, in the required E_b/N_0 to achieve a specified error probability of the coded system over an uncoded system with the same modulation and channel characteristics. Table 7.2 lists an upper bound on the coding gains, compared to uncoded coherent BPSK, for several maximum free distance convolutional codes with constraint lengths varying from 3 to 9 over a Gaussian channel with hard-decision decoding. The table illustrates that it is possible to achieve significant coding gain even with a simple convolutional code. The actual coding gain will vary with the required bit error probability [20].

Table 7.3 lists the measured coding gains, compared to uncoded coherent BPSK, achieved with hardware implementation or computer simulation over a Gaussian channel with soft-decision decoding [21]. The uncoded E_b/N_0 is given in the leftmost column. From Table 7.3 we can see that coding gain increases as the bit error probability is decreased. However, the coding gain cannot increase indefinitely; it has an upper bound as shown in the table. This bound in decibels can be shown [21] to be

$$\text{coding gain} \le 10\log_{10}(rd_f) \qquad (7.23)$$

where r is the code rate and d_f is the free distance. Examination of Table 7.3 also reveals that at $P_B = 10^{-7}$, for code rates of $\frac{1}{2}$ and $\frac{2}{3}$, the weaker codes tend to be closer to the upper bound than are the more powerful codes.

Typically, Viterbi decoding is used over binary input channels with either hard or 3-bit soft quantized outputs. The constraint lengths vary between 3 and 9, the code rate is rarely smaller than $\frac{1}{3}$, and the path memory is usually a few con-

TABLE 7.2 Coding Gain Upper Bounds for Some Convolutional Codes

Rate $\frac{1}{2}$ Codes			Rate $\frac{1}{2}$ Codes		
K	d_f	Upper Bound (dB)	K	d_f	Upper Bound (dB)
3	5	3.97	3	8	4.26
4	6	4.76	4	10	5.23
5	7	5.43	5	12	6.02
6	8	6.00	6	13	6.37
7	10	6.99	7	15	6.99
8	10	6.99	8	16	7.27
9	12	7.78	9	18	7.78

Source: V. K. Bhargava, D. Haccoun, R. Matyas, and P. Nuspl, *Digital Communications by Satellite,* John Wiley & Sons, Inc., New York, 1981.

straint lengths [12]. The path memory refers to the depth of the input bit history stored by the decoder. From the Viterbi decoding example in Section 7.3.4, one might question the notion of a fixed path memory. It seems from the example that the decoding of a branch word, at any arbitrary node, can take place as soon as there is only a single surviving branch at that node. That is true; however, to actually implement the decoder in this way would entail an extensive amount of processing to continually check when the branch word can be decoded. Instead, *a fixed delay is provided,* after which the branch word is decoded. It has been shown [12, 22] that tracing back from the state with the lowest state metric, over a fixed amount of path history (about 4 or 5 times the constraint length), is sufficient to limit the degradation from the optimum decoder performance to about 0.1 dB for the BSC and Gaussian channels. Typical error performance simulation results are shown in Figure 7.21 for Viterbi decoding with hard decision quantization [12]. Notice that each increment in constraint length improves the required E_b/N_0 by a factor of approximately 0.5 dB at $P_B = 10^{-5}$.

TABLE 7.3 Basic Coding Gain (dB) for Soft Decision Viterbi Decoding

Uncoded E_b/N_0 (dB)	Code Rate		$\frac{1}{3}$		$\frac{1}{2}$			$\frac{2}{3}$		$\frac{3}{4}$	
	P_B	K	7	8	5	6	7	6	8	6	9
6.8	10^{-3}		4.2	4.4	3.3	3.5	3.8	2.9	3.1	2.6	2.6
9.6	10^{-5}		5.7	5.9	4.3	4.6	5.1	4.2	4.6	3.6	4.2
11.3	10^{-7}		6.2	6.5	4.9	5.3	5.8	4.7	5.2	3.9	4.8
Upper bound			7.0	7.3	5.4	6.0	7.0	5.2	6.7	4.8	5.7

Source: I. M. Jacobs, "Practical Applications of Coding," *IEEE Trans. Inf. Theory,* vol. IT20, May 1974, pp. 305–310.

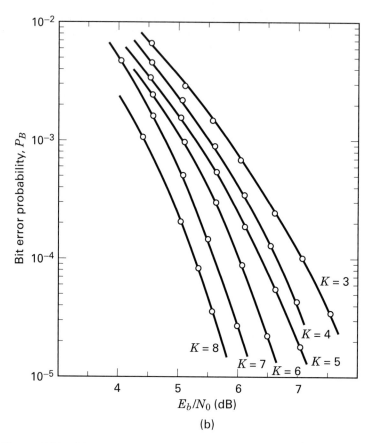

(b)

Figure 7.21 Bit error probability versus E_b/N_0 for rate $\frac{1}{2}$ codes using coherent BPSK over a BSC, Viterbi decoding, and a 32-bit path memory. (Reprinted with permission from J. A. Heller and I. M. Jacobs, "Viterbi Decoding for Satellite and Space Communication," *IEEE Trans. Commun. Technol.*, vol. COM19, no. 5, October 1971, Fig. 7, p. 84. © 1971 IEEE.)

7.4.6 Best Known Convolutional Codes

The connection vectors or polynomial generators of a convolutional code are usually selected based on the code's free distance properties. The first criterion is to select a code that does not have catastrophic error propagation and that has the maximum free distance for the given rate and constraint length. Then the number of paths at the free distance d_f, or the number of data bit errors the paths represent, should be minimized. The selection procedure can be further refined by considering the number of paths or bit errors at $d_f + 1$, at $d_f + 2$, and so on, until only one code or class of codes remains. A list of the best known codes of rate $\frac{1}{2}$, $K = 3$ to 9, and rate $\frac{1}{3}$, $K = 3$ to 8, based on this criterion was compiled by Odenwalder [3, 23] and is given in Table 7.4. The connection vectors in this table represent the presence or absence (1 or 0) of a tap connection on the corresponding stage of the convolutional encoder, the leftmost term corresponding to the leftmost stage of the encoder register. It is interesting to note that these connections can be inverted (leftmost and rightmost can be interchanged in the above description). Under the condition of Viterbi decoding, the inverted connections give rise to codes with identical distance properties, and hence identical performance, as those in Table 7.4.

TABLE 7.4 Optimum Short Constraint Length Convolutional Codes
(Rate $\frac{1}{2}$ and Rate $\frac{1}{3}$)

Rate	Constraint Length	Free Distance	Code Vector
$\frac{1}{2}$	3	5	111 101
$\frac{1}{2}$	4	6	1111 1011
$\frac{1}{2}$	5	7	10111 11001
$\frac{1}{2}$	6	8	101111 110101
$\frac{1}{2}$	7	10	1001111 1101101
$\frac{1}{2}$	8	10	10011111 11100101
$\frac{1}{2}$	9	12	110101111 100011101
$\frac{1}{3}$	3	8	111 111 101
$\frac{1}{3}$	4	10	1111 1011 1101
$\frac{1}{3}$	5	12	11111 11011 10101
$\frac{1}{3}$	6	13	101111 110101 111001
$\frac{1}{3}$	7	15	1001111 1010111 1101101
$\frac{1}{3}$	8	16	11101111 10011011 10101001

Source: J. P. Odenwalder, *Error Control Coding Handbook,* Linkabit Corp., San Diego, Calif., July 15, 1976.

7.4.7 Convolutional Code Rate Trace-Off

7.4.7.1 Performance with Coherent PSK Signaling

The error-correcting capability of a coding scheme increases as the number of channel symbols n per information bit k increases, or the rate k/n decreases. However, the channel bandwidth and the decoder complexity both increase with n. The advantage of lower code rates when using convolutional codes with coherent PSK, is that the required E_b/N_0 is decreased (for a large range of code rates), permitting the transmission of higher data rates for a given amount of power, or permitting reduced power for a given data rate. Simulation studies have shown [16, 22] that for a fixed constraint length, a decrease in the code rate from $\frac{1}{2}$ to $\frac{1}{3}$ results in a reduction of the required E_b/N_0 of roughly 0.4 dB. However, the corresponding increase in decoder complexity is about 17%. For smaller values of code rate, the improvement in performance relative to the increased decoding complexity diminishes rapidly [22]. Eventually, a point is reached where further decrease in code rate is characterized by a reduction in coding gain. (See Section 9.7.7.2.)

7.4.7.2 Performance with Noncoherent Orthogonal Signaling

In contrast to PSK, there is an optimum code rate of about $\frac{1}{2}$ for noncoherent orthogonal signaling. Error performance at rates of $\frac{1}{3}$, $\frac{2}{3}$, and $\frac{3}{4}$ are each worse than those for rate $\frac{1}{2}$. For a fixed constraint length, the rate $\frac{1}{3}$, $\frac{2}{3}$, and $\frac{3}{4}$ codes typically degrade by about 0.25, 0.5, and 0.3 dB, respectively, relative to the rate $\frac{1}{2}$ performance [16].

7.4.8 Soft-Decision Viterbi Decoding

For a rate $\frac{1}{2}$ binary convolutional code system, the demodulator delivers two code symbols at a time to the decoder. For hard-decision (2-level) decoding, each pair of received code symbols can be depicted on a plane, as one of the corners of a square, as shown in Figure 7.22a. The corners are labeled with the binary numbers (0,0), (0,1), (1,0), and (1,1), representing the four possible hard-decision values that the two code symbols might have. For 8-level soft-decision decoding, each pair of code symbols can be similarly represented on an equally spaced 8-level by 8-level plane, as a point from the set of 64 points shown in Figure 7.22b. In this soft-decision case, the demodulator no longer delivers firm decisions; it delivers quantized noisy signals (soft decisions).

The primary difference between hard-decision and soft-decision Viterbi decoding, is that the soft-decision algorithm cannot use a Hamming distance metric because of its limited resolution. A distance metric with the needed resolution is Euclidean distance, and to facilitate its use, the binary numbers of 1 and 0 are transformed to the octal numbers 7 and 0, respectively. This can be seen in Figure 7.22c, where the corners of the square have been re-labeled accordingly; this allows us to use a pair of integers, each in the range of 0 to 7, for describing any point in the 64-point set. Also shown in Figure 7.22c is the point 5,4, representing an example of a pair of noisy code-symbol values that might stem from a

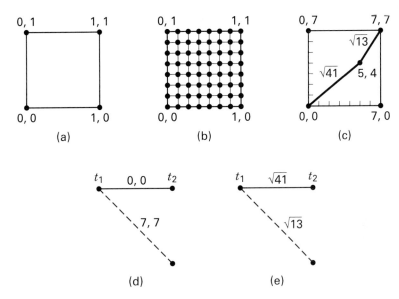

Figure 7.22 (a) Hard-decision plane (b) 8-level by 8-level soft-decision plane (c) Example of soft code symbols (d) Encoding trellis section (e) Decoding trellis section.

demodulator. Imagine that the square in Figure 7.22c has coordinates x and y. Then, what is the Euclidean distance between the noisy point 5,4 and the noiseless point 0,0? It is $\sqrt{(5-0)^2 + (4-0)^2} = \sqrt{41}$. Similarly, if we ask what is the Euclidean distance between the noisy point 5,4 and the noiseless point 7,7? It is $\sqrt{(5-7)^2 + (4-7)^2} = \sqrt{13}$.

Soft-decision Viterbi decoding, for the most part, proceeds in the same way as hard-decision decoding (as described in Sections 7.3.4 and 7.3.5). The only difference is that Hamming distances are not used. Consider how soft-decision decoding is performed with the use of Euclidean distances. Figure 7.22d shows the first section of an encoding trellis, originally presented in Figure 7.7, with the branch words transformed from binary to octal. Suppose that a pair of soft-decision code symbols with values 5,4 arrives at a decoder during the first transition interval. Figure 7.22e shows the first section of a decoding trellis. The metric ($\sqrt{41}$), representing the Euclidean distance between the arriving 5,4 and the 0,0 branch word, is placed on the solid line. Similarly, the metric ($\sqrt{13}$), representing the Euclidean distance between the arriving 5,4 and the 7,7 code symbols, is placed on the dashed line. The rest of the task, pruning the trellis in search of a common stem, proceeds in the same way as hard-decision decoding. Note that in a real convolutional decoding chip, the Euclidean distance is not actually used for a soft-decision metric; instead, a monotonic metric that has similar properties and is easier to implement is used. An example of such a metric is the Euclidean distance-squared, in which case the square-root operation shown above is eliminated. Further, if the binary code symbols are represented with bipolar values, then the inner-product metric in Equation (7.9) can be used. With such a metric, we would seek maximum correlation rather than minimum distance.

7.5 OTHER CONVOLUTIONAL DECODING ALGORITHMS

7.5.1 Sequential Decoding

Prior to the discovery of an optimum algorithm by Viterbi, other algorithms had been proposed for decoding convolutional codes. The earliest was the *sequential decoding algorithm*, originally proposed by Wozencraft [24, 25] and subsequently modified by Fano [2]. A sequential decoder works by generating hypotheses about the transmitted codeword sequence; it computes a metric between these hypotheses and the received signal. It goes forward as long as the metric indicates that its choices are likely; otherwise, it goes backward, changing hypotheses until, through a systematic trial-and-error search, it finds a likely hypothesis. Sequential decoders can be implemented to work with hard or soft decisions, but soft decisions are usually avoided because they greatly increase the amount of the required storage and the complexity of the computations.

Consider that using the encoder shown in Figure 7.3, a sequence $\mathbf{m} = 1\ 1\ 0\ 1\ 1$ is encoded into the codeword sequence $\mathbf{U} = 1\ 1\ 0\ 1\ 0\ 1\ 0\ 0\ 0\ 1$, as shown in Example 7.1. Assume that the received sequence $\mathbf{Z}$ is, in fact, a *correct* rendition of $\mathbf{U}$. The decoder has available a replica of the encoder code tree, shown in Figure 7.6, and can use the received sequence $\mathbf{Z}$ to penetrate the tree. The decoder starts at the time t_1 node of the tree and generates both paths leaving that node. The decoder follows that path which agrees with the received n code symbols. At the next level in the tree, the decoder again generates both paths leaving that node, and follows the path agreeing with the second group of n code symbols. Proceeding in this manner, the decoder quickly penetrates the tree.

Suppose, however, that the received sequence $\mathbf{Z}$ is a *corrupted* version of $\mathbf{U}$. The decoder starts at the time t_1 node of the code tree and generates both paths leading from that node. If the received n code symbols coincide with one of the generated paths, the decoder follows that path. If there is not agreement, the decoder follows the *most likely path* but keeps a cumulative count on the number of disagreements between the received symbols and the branch words on the path being followed. If two branches appear equally likely, the receiver uses an arbitrary rule, such as following the zero input path. At each new level in the tree, the decoder generates new branches and compares them with the next set of n received code symbols. The search continues to penetrate the tree along the most likely path and maintains the cumulative disagreement count.

If the disagreement count exceeds a certain number (which may increase as we penetrate the tree), the decoder decides that it is on an incorrect path, backs out of the path, and tries another. The decoder keeps track of the discarded pathways to avoid repeating any path excursions. For example, assume that the encoder in Figure 7.3 is used to encode the message sequence $\mathbf{m} = 1\ 1\ 0\ 1\ 1$ into the codeword sequence $\mathbf{U}$ as shown in Example 7.1. Suppose that the fourth and seventh bits of the transmitted sequence $\mathbf{U}$ are received in error, such that:

Time:		t_1	t_2	t_3	t_4	t_5
Message sequence:	$\mathbf{m} =$	1	1	0	1	1
Transmitted sequence:	$\mathbf{U} =$	1 1	0 1	0 1	0 0	0 1
Received sequence:	$\mathbf{Z} =$	1 1	0 0	0 1	1 0	0 1

Let us follow the decoder path trajectory with the aid of Figure 7.23. Assume that a cumulative path disagreement count of 3 is the criterion for backing up and trying an alternative path. On Figure 7.23 the numbers along the path trajectory represent the current disagreement count.

1. At time t_1 we receive symbols 11 and compare them with the branch words leaving the first node.
2. The most likely branch is the one with branch word 11 (corresponding to an input bit one or downward branching), so the decoder decides that input bit one is the correct decoding, and moves to the next level.
3. At time t_2, the decoder receives symbols 00 and compares them with the available branch words 10 and 01 at this second level.
4. There is no "best" path, so the decoder arbitrarily takes the input bit zero (or branch word 10) path, and the disagreement count registers a disagreement of 1.
5. At time t_3, the decoder receives symbols 01 and compares them with the available branch words 11 and 00 at this third level.
6. Again, there is no best path, so the decoder arbitrarily takes the input zero (or branch word 11) path, and the disagreement count is increased to 2.
7. At time t_4, the decoder receives symbols 10 and compares them with the available branch words 00 and 11 at this fourth level.
8. Again, there is no best path, so the decoder takes the input bit zero (or branch word 00) path, and the disagreement count is increased to 3.
9. But a disagreement count of 3 is the turnaround criterion, so the decoder "backs out" and tries the alternative path. The disagreement counter is reset to 2.
10. The alternative path is the input bit one (or branch word 11) path at the t_4 level. The decoder tries this, but compared to the received symbols 10, there is still a disagreement of 1, and the counter is reset to 3.
11. But, 3 being the turnaround criterion, the decoder backs out of this path, and the counter is reset to 2. All of the alternatives have now been traversed at this t_4 level, so the decoder returns to the node at t_3, and resets the counter to 1.
12. At the t_3 node, the decoder compares the symbols received at time t_3, namely 01, with the untried 00 path. There is a disagreement of 1, and the counter is increased to 2.

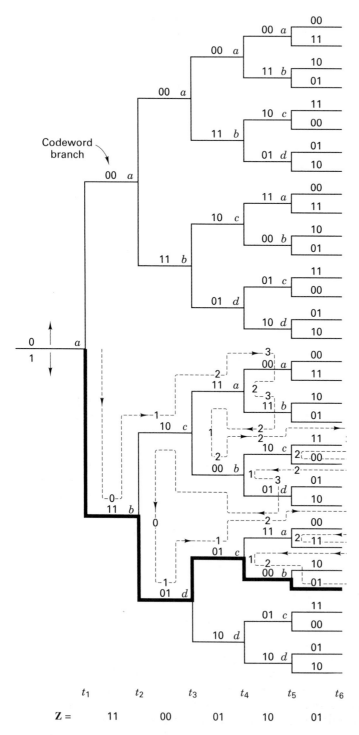

Codeword
branch

t_1 t_2 t_3 t_4 t_5 t_6

$\mathbf{Z} =$ 11   00   01   10   01

Figure 7.23 Sequential decoding example.

13. At the t_4 node, the decoder follows the branch word 10 that matches its t_4 code symbols of 10. The counter remains unchanged at 2.

14. At the t_5 node, there is no best path, so the decoder follows the upper branch, as is the rule, and the counter is increased to 3.

15. At this count, the decoder backs up, resets the counter to 2, and tries the alternative path at node t_5. Since the alternate branch word is 00, there is a disagreement of 1 with the received code symbols 01 at time t_5, and the counter is again increased to 3.

16. The decoder backs out of this path, and the counter is reset to 2. All of the alternatives have now been traversed at this t_5 level, so the decoder returns to the node at t_4 and resets the counter to 1.

17. The decoder tries the alternative path at t_4, which raises the metric to 3 since there is a disagreement in two positions of the branch word. This time the decoder must back up all the way to the time t_2 node because all of the other paths at higher levels have been tried. The counter is now decremented to zero.

18. At the t_2 node, the decoder now follows the branch word 01, and because there is a disagreement of 1 with the received code symbols 00 at time t_2, the counter is increased to 1.

The decoder continues in this way. As shown in Figure 7.23, the final path, which has not increased the counter to its turnaround criterion, yields the correctly decoded message sequence, 1 1 0 1 1. Sequential decoding can be viewed as a trial-and-error technique for searching out the correct path in the code tree. It performs the search in a sequential manner, always operating on just a single path at a time. If an incorrect decision is made, subsequent extensions of the path will be wrong. The decoder can eventually recognize its error by monitoring the path metric. The algorithm is similar to the case of an automobile traveler following a road map. As long as the traveler recognizes that the passing landmarks correspond to those on the map, he continues on the path. When he notices strange landmarks (an increase in his dissimilarity metric) the traveler eventually assumes that he is on an incorrect road, and he backs up to a point where he can now recognize the landmarks (his metric returns to an acceptable range). He then tries an alternative road.

7.5.2 Comparisons and Limitations of Viterbi and Sequential Decoding

The major drawback of the Viterbi algorithm is that while error probability decreases exponentially with constraint length, the number of code states, and consequently decoder complexity, *grows exponentially with constraint length*. On the other hand, the computational complexity of the Viterbi algorithm is independent of channel characteristics (compared to hard-decision decoding, soft-decision decoding requires only a trivial increase in the number of computations). Sequential decoding achieves asymptotically the same error probability as maximum likelihood decoding but without searching all possible states. In fact, with sequential de-

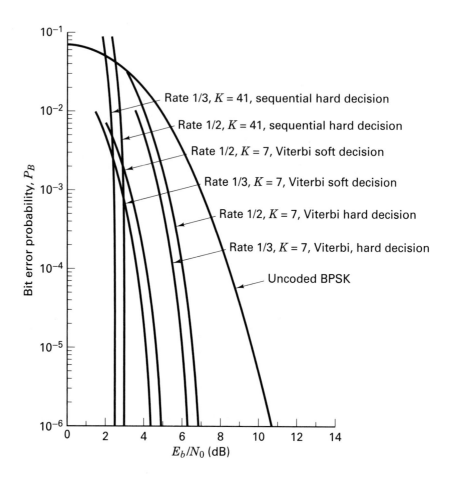

Rate 1/3, $K = 41$, sequential hard decision

Rate 1/2, $K = 41$, sequential hard decision

Rate 1/2, $K = 7$, Viterbi soft decision

Rate 1/3, $K = 7$, Viterbi soft decision

Rate 1/2, $K = 7$, Viterbi hard decision

Rate 1/3, $K = 7$, Viterbi, hard decision

Uncoded BPSK

Figure 7.24 Bit error performance for various Viterbi and sequential decoding schemes using coherent BPSK over an AWGN channel. (Reprinted with permission from J. K. Omura and B. K. Levitt, "Coded Error Probability Evaluation for Antijam Communication Systems," *IEEE Trans. Commun.*, vol. COM30, no. 5, May 1982, Fig. 4, p. 900. © 1982 IEEE.)

coding the number of states searched is essentially *independent of constraint length,* thus making it possible to use very large ($K = 41$) constraint lengths. This is an important factor in providing such low error probabilities. The major drawback of sequential decoding is that the number of state metrics searched is a random variable. For sequential decoding, the expected number of poor hypotheses and backward searches is a function of the channel SNR. With a low SNR, more hypotheses must be tried than with a high SNR. Because of this variability in computational load, buffers must be provided to store the arriving sequences. Under low SNR, the received sequences must be buffered while the decoder is laboring to find a likely hypothesis. If the average symbol arrival rate exceeds the average symbol decode rate, the buffer will overflow, no matter how large it is, causing a loss of data. The sequential decoder typically puts out error-free data until the buffer overflows, at which time the decoder has to go through a recovery procedure. The buffer overflow threshold is a very sensitive function of SNR. Therefore, an important part of a sequential decoder specification is the *probability of buffer overflow*.

In Figure 7.24, some typical P_B versus E_b/N_0 curves for these two popular solutions to the convolutional decoding problem, Viterbi decoding and sequential decoding, illustrate their comparative performance using coherent BPSK over an AWGN channel. The curves compare Viterbi decoding (rates $\frac{1}{2}$ and $\frac{1}{3}$ hard decision, $K = 7$) versus Viterbi decoding (rates $\frac{1}{2}$ and $\frac{1}{3}$ soft decision, $K = 7$) versus sequential decoding (rates $\frac{1}{2}$ and $\frac{1}{3}$ hard decision, $K = 41$). One can see from Figure 7.24 that coding gains of approximately 8 dB at $P_B = 10^{-6}$ can be achieved with sequential decoders. Since the work of Shannon [26] foretold the potential of approximately 11 dB of coding gain compared to uncoded BPSK, it appears that the major portion of _____ published.

7.5

A _____ bit at stage j based on metrics _____ a preselected positive integer _____ number of received code
sym _____ er of encoder input bits that
are _____ whether the data bit is zero
or _____ g distance path traverses in
the _____ e detailed operation is best
und _____ er the use of a feedback decode _____ 7.3. Figure 7.25 illustrates
the _____ coder for $L = 3$. That is, in
deco _____ paths at branches j, $j + 1$,
and j _____

Ham _____ ates 2^L or eight cumulative
minin _____ first branch is zero if the
if the _____ the tree, and decides one
ceivec _____ tree. Assume that the re-
time l _____ nine the eight paths from
comp _____ .24, and compute metrics
branch _____ ed code symbols (three
path n _____ he Hamming cumulative
_____ are

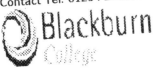
We see _____ part of the tree. There-
fore, th _____ ward movement on the
tree). T _____ (the part that survived)

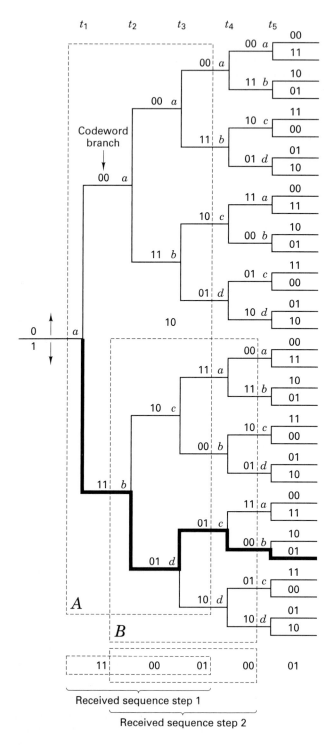

Figure 7.25 Feedback decoding example.

one stage deeper, and again compute eight metrics, this time from t_2 through t_4. Having decoded the first two code symbols, we now slide over two code symbols to the right and again compute the path metrics for six code symbols. This takes place in the block marked B in Figure 7.25. Again, listing the metrics from top path to bottom path, we find that they are

$$\text{Upper-half metrics:} \quad 2, 4, 3, 3$$

$$\text{Lower-half metrics:} \quad 3, 1, 4, 4$$

For the assumed received sequence, the minimum metric is found in the lower half of block B. Therefore, the second decoded bit is one.

The same procedure continues until the entire message is decoded. The decoder is called a *feedback decoder* because the detection decisions are *fed back* to the decoder in determining the subset of code paths that are to be considered next. On the BSC, the feedback decoder can perform nearly as well as the Viterbi decoder [17] in that it can correct all the more probable error patterns, namely all those of weight $(d_f - 1)/2$ or less, where d_f is the free distance of the code. An important design parameter for feedback convolutional decoders is L, the look-ahead length. Increasing L increases the coding gain but also increases the decoder implementation complexity.

7.6 CONCLUSION

In the last decade, coding emphasis has been in the area of convolutional codes since in almost every application, convolutional codes outperform block codes for the same implementation complexity of the encoder–decoder. For satellite communication channels, forward error correction techniques can easily reduce the required SNR for a specified error performance by 5 to 6 dB. This coding gain can translate directly into an equivalent reduction in required satellite effective radiated power (EIRP), with consequently reduced satellite weight and cost.

In this chapter we have outlined the essential structural difference between block codes and convolutional codes—the fact that rate $1/n$ convolutional codes have a memory of the prior $K - 1$ bits, where K is the encoder constraint length. With such memory, the encoding of each input data bit not only depends on the value of that bit but on the values of the $K - 1$ input bits that precede it. We presented the decoding problem in the context of the maximum likelihood algorithm, examining all the candidate codeword sequences which could possibly be created by the encoder, and selecting the one that appears statistically most likely; the decision is based on a distance metric for the received code symbols. The error performance analysis of convolutional codes is more complicated than the simple binomial expansion describing the error performance of many block codes. We laid out the concept of free distance, and we presented the relationship between free distance and error performance in terms of bounds. We also described the basic idea behind sequential decoding and feedback decoding and showed some comparative performance curves and tables for various coding schemes.

REFERENCES

1. Gallager, R. G., *Information Theory and Reliable Communication,* John Wiley & Sons, Inc., New York, 1968.

2. Fano, R. M., "A Heuristic Discussion of Probabilistic Decoding," IRE *Trans. Inf. Theory,* vol. IT9, no. 2, 1963, pp. 64–74.

3. Odenwalder, J. P., *Optimal Decoding of Convolutional Codes,* Ph.D. dissertation, University of California, Los Angeles, 1970.

4. Curry, S. J., *Selection of Convolutional Codes Having Large Free Distance,* Ph.D. dissertation, University of California, Los Angeles, 1971.

5. Larsen, K. J., "Short Convolutional Codes with Maximal Free Distance for Rates $\frac{1}{2}$, $\frac{1}{3}$, and $\frac{1}{4}$," *IEEE Trans. Inf. Theory,* vol. IT19, no. 3, 1973, pp. 371–372.

6. Lin, S., and Costello, D. J., Jr., *Error Control Coding: Fundamentals and Applications,* Prentice-Hall, Inc., Englewood Cliffs, N.J., 1983.

7. Forney, G. D., Jr., "Convolutional Codes: I. Algebraic Structure," *IEEE Trans. Inf. Theory,* vol. IT16, no. 6, Nov. 1970, pp. 720–738.

8. Viterbi, A., "Convolutional Codes and Their Performance in Communication Systems," *IEEE Trans. Commun. Technol.,* vol. COM 19, no. 5, Oct. 1971, pp. 751–772.

9. Forney, G. D., Jr., and Bower, E. K., "A High Speed Sequential Decoder: Prototype Design and Test," *IEEE Trans. Commun. Technol.,* vol. COM19, no. 5, Oct. 1971, pp. 821–835.

10. Jelinek, F., "Fast Sequential Decoding Algorithm Using a Stack," *IBM J. Res. Dev.,* vol. 13, Nov. 1969, pp. 675–685.

11. Massey, J. L., *Threshold Decoding,* The MIT Press, Cambridge, Mass., 1963.

12. Heller, J. A., and Jacobs, I. W., "Viterbi Decoding for Satellite and Space Communication," *IEEE Trans. Commun. Technol.,* vol. COM19, no. 5, October 1971, pp. 835–848.

13. Viterbi, A. J., "Error Bounds for Convolutional Codes and an Asymptotically Optimum Decoding Algorithm," *IEEE Trans. Inf. Theory,* vol. IT13, April 1967, pp. 260–269.

14. Omura, J. K., "On the Viterbi Decoding Algorithm" (correspondence), *IEEE Trans. Inf. Theory,* vol. IT15, Jan. 1969, pp. 177–179.

15. Mason, S. J., and Zimmerman, H. J., *Electronic Circuits, Signals, and Systems,* John Wiley & Sons, Inc., New York, 1960.

16. Clark, G. C., Jr., and Cain, J. B., *Error-Correction Coding for Digital Communications,* Plenum Press, New York, 1981.

17. Viterbi, A. J., and Omura, J. K., *Principles of Digital Communication and Coding,* McGraw-Hill Book Company, New York, 1979.

18. Massey, J. L., and Sain, M. K., "Inverse of Linear Sequential Circuits," *IEEE Trans. Comput.,* vol. C17, Apr. 1968, pp. 330–337.

19. Rosenberg, W. J., *Structural Properties of Convolutional Codes,* Ph.D. dissertation, University of California, Los Angeles, 1971.

20. Bhargava, V. K., Haccoun, D., Matyas, R., and Nuspl, P., *Digital Communications by Satellite*, John Wiley & Sons, Inc., New York, 1981.

21. Jacobs, I. M., "Practical Applications of Coding," *IEEE Trans. Inf. Theory*, vol. IT20, May 1974, pp. 305–310.

22. Linkabit Corporation, "Coding Systems Study for High Data Rate Telemetry Links," *NASA Ames Res. Center, Final Rep. CR-114278*, Contract NAS-2-6-24, Moffett Field, Calif., 1970.

23. Odenwalder, J. P., *Error Control Coding Handbook*, Linkabit Corporation, San Diego, Calif., July 15, 1976.

24. Wozencraft, J. M., "Sequential Decoding for Reliable Communication," *IRE Natl. Conv. Rec.*, vol. 5, pt. 2, 1957, pp. 11–25.

25. Wozencraft, J. M., and Reiffen, B., *Sequential Decoding*, The MIT Press, Cambridge, Mass., 1961.

26. Shannon, C. E., "A Mathematical Theory of Communication," *Bell Syst. Tech. J.*, vol. 27, 1948, pp. 379–423, 623–656.

PROBLEMS

7.1. Draw the state diagram, tree diagram, and trellis diagram for the $K = 3$, rate $\frac{1}{3}$ code generated by

$$\mathbf{g}_1(X) = X + X^2$$

$$\mathbf{g}_2(X) = 1 + X$$

$$\mathbf{g}_3(X) = 1 + X + X^2$$

7.2. Given a $K = 3$, rate $\frac{1}{2}$, binary convolutional code with the partially completed state diagram shown in Figure P7.1, find the complete state diagram and sketch a diagram for the encoder.

7.3. Draw the state diagram, tree diagram, and trellis diagram for the convolutional encoder characterized by the block diagram in Figure P7.2.

7.4. Suppose that you were trying to find the quickest way to get from London to Vienna by boat or train. The diagram in Figure P7.3 was constructed from various schedules. The labels on each path are travel times. Using the Viterbi algorithm, find the fastest

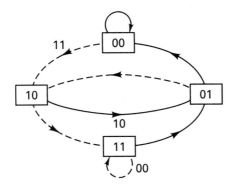

Figure P7.1

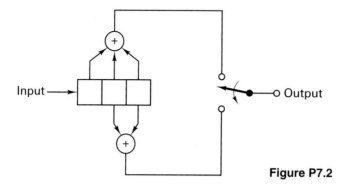

Figure P7.2

route from London to Vienna. In a general sense, explain how the algorithm works, what calculations must be made, and what information must be retained in the memory used by the algorithm.

7.5. Consider the convolutional encoder shown in Figure P7.4.
   (a) Write the connection vectors and polynomials for this encoder.
   (b) Draw the state diagram, tree diagram, and trellis diagram.

7.6. What is the impulse response of the encoder of Problem 7.5? Using the impulse response, determine the output sequence when the input is 1 0 1. Verify by using the generator polynomials.

7.7. Does the encoder of Problem 7.5 exhibit the properties of catastrophic error propagation? Justify your answer with an example.

7.8. Find the free distance of the encoder of Problem 7.3 by the transfer function method.

7.9. Let the codewords of a coding scheme be

$$a = 0\,0\,0\,0\,0\,0$$

$$b = 1\,0\,1\,0\,1\,0$$

$$c = 0\,1\,0\,1\,0\,1$$

$$d = 1\,1\,1\,1\,1\,1$$

If the received sequence over a binary symmetric channel is 1 1 1 0 1 0 and a maximum likelihood decoder is used, what will be the decoded symbol?

7.10. Consider that the $K = 3$, rate $\frac{1}{2}$ encoder of Figure 7.3 is used over a binary symmetric channel (BSC). Assume that the initial encoder state is the 00 state. At the output of the BSC, the sequence $\mathbf{Z} = (1\,1\,0\,0\,0\,0\,1\,0\,1\,1$ rest all "0") is received.

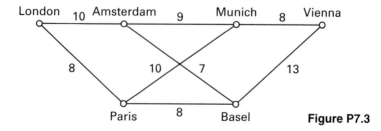

Figure P7.3

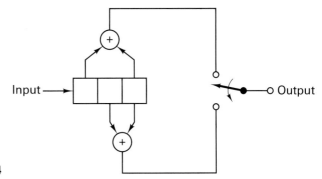

Figure P7.4

(a) Find the maximum likelihood path through the trellis diagram, and determine the first 5 decoded information bits. If a tie occurs between any two merged paths, choose the upper branch entering the particular state.

(b) Identify any channel bits in **Z** that were inverted by the channel during transmission.

7.11. Determine which of the following rate $\frac{1}{2}$ codes are catastrophic.

    (a) $g_1(X) = X^2$, $g_2(X) = 1 + X + X^3$

    (b) $g_1(X) = 1 + X^2$, $g_2(X) = 1 + X^3$

    (c) $g_1(X) = 1 + X + X^2$, $g_2(X) = 1 + X + X^3 + X^4$

    (d) $g_1(X) = 1 + X + X^3 + X^4$, $g_2(X) = 1 + X^2 + X^4$

    (e) $g_1(X) = 1 + X^4 + X^6 + X^7$, $g_2(X) = 1 + X^3 + X^4$

    (f) $g_1(X) = 1 + X^3 + X^4$, $g_2(X) = 1 + X + X^2 + X^4$

7.12. (a) Consider a coherently detected BPSK signal encoded with the encoder shown in Figure 7.3. Find an upper bound on the bit error probability, P_B, if the available E_b/N_0 is 6 dB. Assume hard decision decoding.

    (b) Compare P_B with the uncoded case and calculate the improvement factor.

7.13. Using sequential decoding, illustrate the path along the tree diagram shown in Figure 7.22 when the received sequence is 0 1 1 1 0 0 0 1 1 1. The backup criterion is three disagreements.

7.14. Repeat the decoding example of Problem 7.13 using feedback decoding, with a look-ahead length of 3. In the event of a tie, select the upper half of the tree.

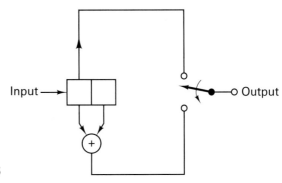

Figure P7.5

7.15. Figure P7.5 depicts a constraint length 2 convolutional encoder.
  (a) Draw the state diagram, tree diagram, and trellis diagram.
  (b) Assume that a received message from this encoder is 1 1 0 0 1 0. Use a feedback decoding algorithm with a look-ahead length of 2 to decode the coded message sequence.

7.16. Using the branch word information on the encoder trellis of Figure 7.7, decode the sequence $\mathbf{Z} = (01\ 11\ 00\ 01\ 11$ rest all "0"), using hard-decision Viterbi decoding.

7.17. Consider the rate $\frac{2}{3}$ convolutional encoder shown in Figure P7.6. In this encoder, $k = 2$ bits at a time are shifted into the encoder and $n = 3$ bits are generated at the encoder output. There are $kK = 4$ stages in the register, and the constraint length is $K = 2$ in units of 2-bit bytes. The state of the encoder is defined as the contents of the rightmost $K - 1$ k-tuple stages. Draw the state diagram, the tree diagram, and the trellis diagram.

7.18. Find the ratio of the predetection signal-to-noise spectral density, P_r/N_0, in decibels, required to yield a decoded data rate of 1 Mbit/s with error probability of 10^{-5}. Assume binary noncoherent FSK modulation. Also, assume convolutional encoding with the decoder relationship

$$P_B = 2000\, p_c^4$$

where p_c and P_B are bit error probabilities into and out of the decoder, respectively.

7.19. Using Table 7.4, devise a $K = 4$, rate $\frac{1}{2}$ binary convolutional encoder.
  (a) Draw the circuit.
  (b) Draw the encoding trellis showing its states and branch words.
  (c) Configure the cells that would be implemented in an ACS algorithm.

7.20. For the $K = 3$, rate $\frac{1}{2}$ code described by the encoder circuit of Figure 7.3, perform soft-decision decoding for the following demodulated sequence. The signals are 8-level quantized integers in the range of 0 to 7. The level 0 represents the perfect binary 0, and the level 7 represents the perfect binary 1. If the digits into the decoder are: 6, 7, 5, 3, 1, 0, 1, 1, 2, 0, where the leftmost digit is the earliest, use a decoding

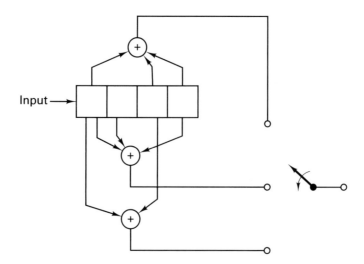

Figure P7.6

trellis diagram to decode the first two data bits. Assume that the encoder had started in the 00 state, and that the decoding process is perfectly synchronized.

QUESTIONS

7.1. In convolutional encoding, why is *flushing* of the register periodically performed? (See Sections 7.2.1 and 7.3.4.)

7.2. Define what is meant by the *state* of a machine. (See Section 7.2.2.)

7.3. What is a *finite-state machine*? (See Section 7.2.2.)

7.4. What are *soft decisions,* and how much *greater complexity* is there in the process of soft-decision Viterbi decoding as compared with hard decision decoding? (See Sections 7.3.2 and 7.4.8.)

7.5. What is another (descriptive) name for a binary symmetric channel (BSC)? (See Section 7.3.2.1.)

7.6. Describe the *Add-Compare-Select* (ACS) computations performed in the process of Viterbi decoding. (See Section 7.3.5.)

7.7. On a trellis diagram, an *error* is associated with a surviving path that *diverges from,* and then *remerges to* the correct path. Why is it necessary for the path to remerge? (See Section 7.4.1.)

EXERCISES

Using the Companion CD, run the exercises associated with Chapter 7.

CHAPTER 8

Channel Coding: Part 3

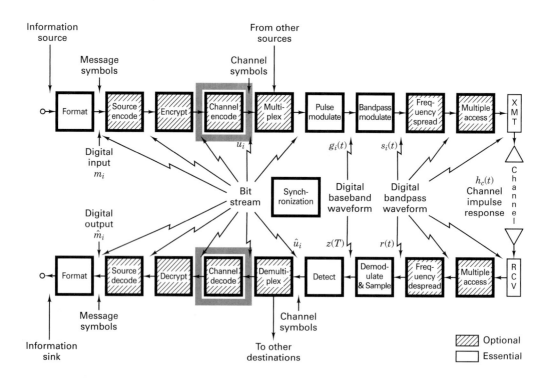

8.1 REED–SOLOMON CODES

Reed–Solomon (R–S) codes are *nonbinary cyclic* codes with symbols made up of *m*-bit sequences, where *m* is any positive integer having a value greater than 1. R–S (n, k) codes on *m*-bit symbols exist for all *n* and *k* for which

$$0 < k < n < 2^m + 2 \qquad (8.1)$$

where *k* is the number of data symbols being encoded, and *n* is the total number of code symbols in the encoded block. For the most conventional R–S (n, k) code,

$$(n, k) = (2^m - 1, \ 2^m - 1 - 2t) \qquad (8.2)$$

where *t* is the symbol-error correcting capability of the code, and $n - k = 2t$ is the number of parity symbols. An extended R–S code can be made up with $n = 2^m$ or $n = 2^m + 1$, but not any further.

Reed–Solomon (R–S) codes achieve the *largest possible* code minimum distance for any linear code with the same encoder input and output block lengths. For nonbinary codes, the distance between two codewords is defined (analogous to Hamming distance) as the number of symbols in which the sequences differ. For Reed–Solomon codes the code minimum distance is given by [1]

$$d_{\min} = n - k + 1 \qquad (8.3)$$

The code is capable of correcting any combination of *t* or fewer errors, where *t* obtained from Equation (6.44), can be expressed as

$$t = \left\lfloor \frac{d_{min} - 1}{2} \right\rfloor = \left\lfloor \frac{n - k}{2} \right\rfloor \qquad (8.4)$$

where $\lfloor x \rfloor$ means the largest integer not to exceed x. Equation (8.4) illustrates that for the case of R–S codes, correcting t symbol errors requires no more than $2t$ parity symbols. Equation (8.4) lends itself to the following intuitive reasoning. One can say that the decoder has $n - k$ redundant symbols "to spend," which is twice the amount of correctable errors. For each error, one redundant symbol is used to locate the error, and another redundant symbol is used to find its correct value.

The erasure-correcting capability of the code is

$$\rho = d_{min} - 1 = n - k \qquad (8.5)$$

Simultaneous error-correction and erasure-correction capability can be expressed by the requirement that

$$2\alpha + \gamma < d_{min} < n - k \qquad (8.6)$$

where α is the number of symbol error patterns that can be corrected, and γ is the number of symbol erasure patterns that can be corrected. An advantage of nonbinary codes such as a Reed–Solomon code can be seen by the following comparison. Consider a binary $(n, k) = (7, 3)$ code. The entire n-tuple space contains $2^n = 2^7 = 128$ n-tuples, of which $2^k = 2^3 = 8$ (or 1/16 of the n-tuples) are codewords. Next consider a nonbinary $(n, k) = (7, 3)$ code where each symbol comprises $m = 3$ bits. The n-tuple space amounts to $2^{nm} = 2^{21} = 2,097,152$ n-tuples, of which $2^{km} = 2^9 = 512$ (or 1/4096 of the n-tuples) are codewords. When dealing with nonbinary symbols, each made up of m bits, only a small fraction (i.e., 2^{km} of the large number 2^{nm}) of possible n-tuples are codewords. This fraction decreases with increasing values of m. The important point here is that, when a small fraction of the n-tuple space is used for codewords, a large d_{min} can be created.

Any linear code is capable of correcting $n - k$ symbol erasure patterns if the $n - k$ erased symbols all happen to lie on the parity symbols. However, R–S codes have the remarkable property that they are able to correct *any* set of $n - k$ symbol erasures within the block. R–S codes can be designed to have any redundancy. However, the complexity of a high speed implementation increases with redundancy. Thus, the most attractive R–S codes have high code rates (low redundancy).

8.1.1 Reed-Solomon Error Probability

The Reed–Solomon (R–S) codes are particularly useful for *burst-error correction*; that is, they are effective for channels that have memory. Also, they can be used efficiently on channels where the set of input symbols is large. An interesting feature of the R–S code is that as many as two information symbols can be added to an R–S code of length n without reducing its minimum distance. This extended R–S code has length $n + 2$ and the same number of parity check symbols as the original code. From Equation (6.46), the R–S decoded symbol error probability, P_E, in terms of the channel symbol error probability, p, can be written as follows [2]:

$$P_E \approx \frac{1}{2^m - 1} \sum_{j=t+1}^{2^m-1} j \binom{2^m - 1}{j} p^j (1 - p)^{2^m - 1 - j} \tag{8.7}$$

where t is the symbol-error correcting capability of the code, and the symbols are made up of m bits each.

The bit error probability can be upper bounded by the symbol error probability for specific modulation types. For MFSK modulation with $M = 2^m$, the relationship between P_B and P_E as given in Equation (4.112) is repeated here:

$$\frac{P_B}{P_E} = \frac{2^{m-1}}{2^m - 1} \tag{8.8}$$

Figure 8.1 shows P_B versus the channel symbol error probability p, plotted from Equations (8.7) and (8.8) for various t-error-correcting 32-ary orthogonal Reed–Solomon codes with $n = 31$ (thirty-one 5-bit symbols per code block).

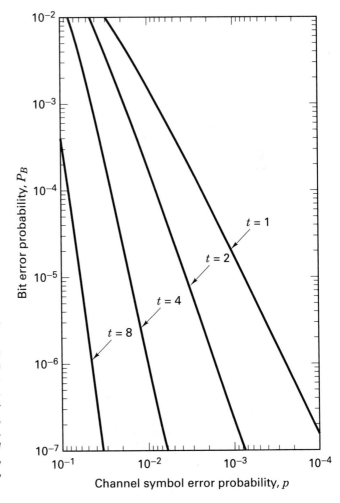

Figure 8.1 P_B versus p for 32-ary orthogonal signaling and $n = 31$, t-error-correcting Reed–Solomon coding. (Reprinted with permission from *Data Communications, Networks and Systems,* ed. Thomas C. Bartee, Howard W. Sams Company, Indianapolis, Ind., 1985, p. 311. Originally published in J. P. Odenwalder, *Error Control Coding Handbook,* M/A-COM LINKABIT, Inc., San Diego, Calif., July 15, 1976, p. 91.)

8.1   Reed–Solomon Codes

Figure 8.2 shows P_B versus E_b/N_0 for such a coded system using 32-ary MFSK modulation and noncoherent demodulation over an AWGN channel [2]. For R–S codes, error probability is an exponentially decreasing function of block length, n, and decoding complexity is proportional to a small power of the block length [1]. The R–S codes are sometimes used in a concatenated arrangement. In such a

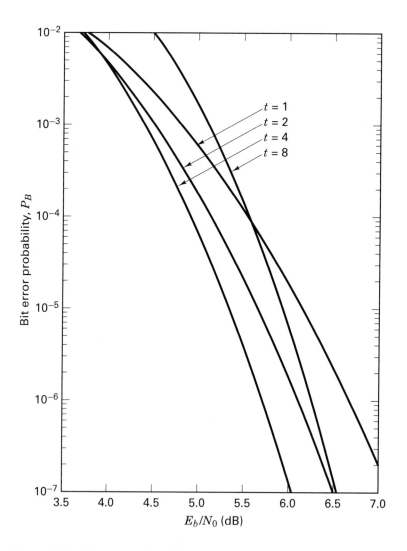

Figure 8.2 Bit error probability versus E_b/N_0 performance of several $n = 31$, t-error correcting Reed–Solomon coding systems with 32-ary MFSK modulation over an AWGN channel. (Reprinted with permission from *Data Communications, Networks, and Systems,* ed. Thomas C. Bartee, Howard W. Sams Company, Indianapolis, Ind., 1985, p. 312. Originally published in J. P. Odenwalder, *Error Control Coding Handbook,* M/A-COM LINKABIT, Inc. San Diego, Calif., July 15, 1976, p. 92.)

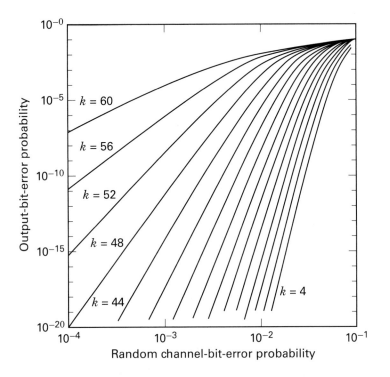

Figure 8.5 Reed–Solomon (64, *k*) decoder performance as a function of redundancy.

in Figure 8.6. Here, the performance curves are plotted for BPSK modulation and an R–S (31, *k*) code for various channel types. Figure 8.6 can reflect a real-time communication system, where the price paid for error-correction coding is bandwidth expansion by a factor equal to the inverse of the code rate. The curves plotted show clear optimum code rates which minimize the required E_b/N_0 [4]. The optimum code rate is about 0.6 to 0.7 for a Gaussian channel, 0.5 for a Rician-fading channel (with the ratio of direct to reflected received signal power, $K = 7$ dB), and 0.3 for a Rayleigh-fading channel. (Fading channels are treated in Chapter 15.) Why is there an error-performance degradation for very large rates (small redundancy) and very low rates (large redundancy)? It is easy to explain the degradation at high rates compared with the optimum rate. Any code generally provides a coding gain benefit; thus, as the code rate approaches unity (no coding), the system will suffer worse error performance. The degradation at low code rates is more subtle because when E_b/N_0 is fixed, there are two mechanisms at work. One mechanism works to improve error performance, and the other works to degrade it. The improving mechanism is the coding; the greater the redundancy, the greater will be the error-correcting capability of the code. The degrading mechanism is the energy reduction per channel symbol (compared with the data symbol) which stems from the increased redundancy causing faster signaling (in the case of a real-time communication system). The reduced channel-symbol energy causes the

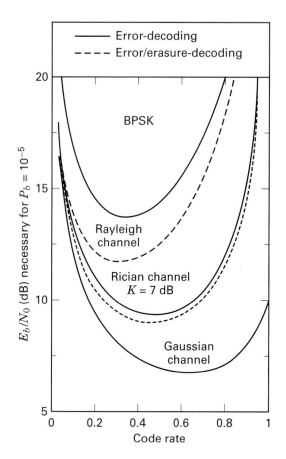

E_b/N_0 (dB) necessary for $P_b = 10^{-5}$

BPSK

Rayleigh
channel

Rician channel
$K = 7$ dB

Gaussian
channel

Code rate

Figure 8.6 BPSK plus Reed–Solomon (31, k) decoder performance as a function of code rate.

demodulator to make more errors. Eventually, the second mechanism wins out, and thus, at very low code rates the system experiences error-performance degradation.

Let us see if we can corroborate the error performance versus code rate in Figure 8.6 with the curves in Figure 8.2. The figures are really not directly comparable because the modulation is BPSK in Figure 8.6, while it is 32-ary MFSK in Figure 8.2. However, perhaps we can verify that R-S error performance-versus-code rate exhibits the same general curvature with MFSK modulation as it does with BPSK. In Figure 8.2, the error performance over an AWGN channel, improves as the symbol error-correcting capability t increases from $t = 1$ to $t = 4$; the $t = 1$ and $t = 4$ cases correspond to R-S (31, 29) and R-S (31, 23) with code rates of 0.94 and 0.74, respectively. However at $t = 8$, which corresponds to R-S (31, 15) with code rate equal to 0.48, the error performance at $P_B = 10^{-5}$ degrades by about 0.5 dB of E_b/N_0, compared with the $t = 4$ case. From Figure 8.2, we can conclude that if we were to plot error performance versus code rate, the curve would have the same general shape as it does in Figure 8.6. Note that this manifestation cannot be gleaned from Figure 8.1, since that figure represents a decoder transfer function, which provides no information about the channel and the demodulation. Therefore, of the two mechanisms at work in the channel, the Figure 8.1 transfer function only presents the output-versus-input benefits of the decoder, and displays nothing about the loss of channel-symbol energy as a function of lower code rate. More is said about choosing a code in concert with a modulation type in Section 9.7.7.

8.1.4 Finite Fields

In order to understand the encoding and decoding principles of nonbinary codes, such as a Reed–Solomon (R-S) codes, it is necessary to venture into the area of finite fields known as *Galois Fields* (GF). For any prime number p there exists a finite field denoted GF(p), containing p elements. It is possible to extend GF(p) to a field of p^m elements, called an extension field of GF(p), and denoted by GF(p^m), where m is a nonzero positive integer. Note that GF(p^m) contains as a subset the elements of GF(p). Symbols from the extension field GF(2^m) are used in the construction of Reed–Solomon (R–S) codes.

The binary field GF(2) is a subfield of the extension field GF(2^m), much the same way as the real number field is a subfield of the complex number field. Besides the numbers 0 and 1, there are additional unique elements in the extension field that will be represented with a new symbol α. Each nonzero element in GF(2^m) can be represented by a power of α. An *infinite* set of elements, F, is formed by starting with the elements $\{0, 1, \alpha\}$ and generating additional elements by progressively multiplying the last entry by α which yields

$$F = \{0, 1, \alpha, \alpha^2, \cdots, \alpha^j, \cdots\} = \{0, \alpha^0, \alpha^1, \alpha^2, \cdots, \alpha^j, \cdots\} \tag{8.9}$$

To obtain the *finite* set of elements of GF(2^m) from F, a condition must be imposed on F so that it may contain only 2^m elements and is closed under multiplication. The condition that closes the set of field elements under multiplication is characterized by the irreducible polynomial

$$\alpha^{(2^m - 1)} + 1 = 0$$

or equivalently,

$$\alpha^{(2^m - 1)} = 1 = \alpha^0 \tag{8.10}$$

Using this polynomial constraint, any field element that has a power equal to or greater than $2^m - 1$ can be reduced to an element with a power less than $2^m - 1$ as follows:

$$\alpha^{(2^m + n)} = \alpha^{(2^m - 1)} \alpha^{n+1} = \alpha^{n+1} \tag{8.11}$$

Thus, Equation (8.10) can be used to form the finite sequence F^* from the infinite sequence F, as follows:

$$F^* = \{0, 1, \alpha, \alpha^2, \cdots, \alpha^{2^m - 2}, \alpha^{2^m - 1}, \alpha^{2^m}, \cdots\} \tag{8.12}$$

$$= \{0, \alpha^0, \alpha^1, \alpha^2, \cdots, \alpha^{2^m - 2}, \alpha^0, \alpha^1, \alpha^2, \cdots\}$$

Therefore, it can be seen from Equation (8.12) that the elements of the finite field GF(2)^m are given by

$$GF(2^m) = \{0, \alpha^0, \alpha^1, \alpha^2, \cdots, \alpha^{2^m-2}\} \qquad (8.13)$$

8.1.4.1 Addition in the Extension Field GF(2^m)

Each of the 2^m elements of the finite field $GF(2^m)$ can be represented as a distinct polynomial of *degree* $m-1$ or less. The degree of a polynomial is the value of its highest order exponent. We denote each of the nonzero elements of $GF(2^m)$ as a polynomial $a_i(X)$, where at least one of the m coefficients of $a_i(X)$ is nonzero. For $i = 0, 1, 2, \ldots, 2^m - 2$,

$$\alpha^i = a_i(X) = a_{i,0} + a_{i,1}X + a_{i,2}X^2 + \cdots + a_{i,m-1}X^{m-1} \qquad (8.14)$$

Consider the case of $m = 3$, where the finite field is denoted $GF(2^3)$. Figure 8.7 shows the mapping (developed later) of the seven elements $\{\alpha^i\}$ and the zero element, in terms of the basis elements $\{X^0, X^1, X^2\}$ described by Equation (8.14). Since Equation (8.10) indicates that $\alpha^0 = \alpha^7$, there are seven nonzero elements or a total of eight elements in this field. Each row in the Figure 8.7 mapping comprises a sequence of binary values representing the coefficients $\alpha_{i,0}$, $\alpha_{i,1}$, and $\alpha_{i,2}$ in Equation (8.14). One of the benefits of using extension field elements $\{\alpha^i\}$ in place of binary elements is the compact notation that facilitates the mathematical representation of nonbinary encoding and decoding processes. Addition of two elements of the finite field is then defined as the modulo-2 sum of each of the polynomial coefficients of like powers, i.e.,

$$\alpha^i + \alpha^j = (a_{i,0} + a_{j,0}) + (a_{i,1} + a_{j,1})X + \cdots + (a_{i,m-1} + a_{j,m-1})X^{m-1} \qquad (8.15)$$

8.1.4.2 A Primitive Polynomial is Used to Define the Finite Field

A class of polynomials called *primitive polynomials,* is of interest because such functions define the finite fields of $GF(2^m)$ which in turn are needed to define R-S codes. The following condition is necessary and sufficient to guarantee that a

Basis elements

X^0 X^1 X^2

Field elements	X^0	X^1	X^2
	0	0	0	0
α^0	1	0	0
α^1	0	1	0
α^2	0	0	1
α^3	1	1	0
α^4	0	1	1
α^5	1	1	1
α^6	1	0	1
α^7	1	0	0

Figure 8.7 Mapping field elements in terms of basis elements for GF(8) with $f(X) = 1 + X + X^3$.

polynomial is primitive. An irreducible polynomial, $f(X)$, of degree m is said to be primitive, if the smallest positive integer n for which $f(X)$ divides $X^n + 1$ is $n = 2^m - 1$. Note that an irreducible polynomial is one that cannot be factored to yield lower order polynomials, and that the statement A divides B means that A divided into B yields a nonzero quotient and a zero remainder. Polynomials will usually be shown low order-to-high order. Sometimes, it is convenient to follow the reverse format (e.g., when performing polynomial division).

Example 8.1 Recognizing a Primitive Polynomial

Based on the foregoing definition of a primitive polynomial, determine whether the following irreducible polynomials are primitive:

(a) $1 + X + X^4$

(b) $1 + X + X^2 + X^3 + X^4$

Solution

(a) We can verify whether or not this degree $m = 4$ polynomial is primitive by determining if it divides $X^n + 1 = X^{(2^m - 1)} + 1 = X^{15} + 1$, but does not divide $X^n + 1$, for values of n in the range of $1 \leq n < 15$. It is easy to verify that $1 + X + X^4$ divides $X^{15} + 1$, and after repeated computations it can be verified that $1 + X + X^4$ will not divide $X^n + 1$ for any n in the range of $1 \leq n < 15$. Therefore, $1 + X + X^4$ is a primitive polynomial.

(b) It is simple to verify that the polynomial $1 + X + X^2 + X^3 + X^4$ divides $X^{15} + 1$. Testing to see if it will divide $X^n + 1$ for some n that is less than 15, yields the fact that it also divides $X^5 + 1$. Thus, although $1 + X + X^2 + X^3 + X^4$ is irreducible, it is not primitive.

8.1.4.3 The Extension Field GF(2^3)

Consider an example involving a primitive polynomial and the finite field that it defines. Table 8.1 contains a listing of some primitive polynomials. We choose the first one shown, $f(X) = 1 + X + X^3$ which defines a finite field GF(2^m), where the degree of the polynomial is $m = 3$. Thus, there are $2^m = 2^3 = 8$ elements in the field defined by $f(X)$. Solving for the roots of $f(X)$ means that the values of X that correspond to $f(X) = 0$ must be found. The familiar binary elements 1 and 0 do not satisfy (are not roots of) the polynomial $f(X) = 1 + X + X^3$, since $f(1) = 1$ and $f(0) = 1$ (using modulo-2 arithmetic). Yet, a fundamental theorem of algebra states that a polynomial of degree m must have precisely m roots. Therefore for this example, $f(X) = 0$ must yield 3 roots. Clearly a dilemma arises, since the 3 roots do not lie in the same finite field as the coefficients of $f(X)$. Therefore, they must lie somewhere else; the roots lie in the extension field GF(2^3). Let α, an element of the extension field, be defined as a root of the polynomial $f(X)$. Therefore, it is possible to write

$$f(\alpha) = 0$$

$$1 + \alpha + \alpha^3 = 0 \tag{8.16}$$

$$\alpha^3 = -1 - \alpha$$

TABLE 8.1   Some Primitive Polynomials

m		m	
3	$1 + X + X^3$	14	$1 + X + X^6 + X^{10} + X^{14}$
4	$1 + X + X^4$	15	$1 + X + X^{15}$
5	$1 + X^2 + X^5$	16	$1 + X + X^3 + X^{12} + X^{16}$
6	$1 + X + X^6$	17	$1 + X^3 + X^{17}$
7	$1 + X^3 + X^7$	18	$1 + X^7 + X^{18}$
8	$1 + X^2 + X^3 + X^4 + X^8$	19	$1 + X + X^2 + X^5 + X^{19}$
9	$1 + X^4 + X^9$	20	$1 + X^3 + X^{20}$
10	$1 + X^3 + X^{10}$	21	$1 + X^2 + X^{21}$
11	$1 + X^2 + X^{11}$	22	$1 + X + X^{22}$
12	$1 + X + X^4 + X^6 + X^{12}$	23	$1 + X^5 + X^{23}$
13	$1 + X + X^3 + X^4 + X^{13}$	24	$1 + X + X^2 + X^7 + X^{24}$

Since in the binary field $+1 = -1$, then α^3 can be represented as

$$\alpha^3 = 1 + \alpha \tag{8.17}$$

Thus, α^3 is expressed as a weighted sum of α-terms having lower orders. In fact, all powers of α can be so expressed. For example, consider

$$\alpha^4 = \alpha \cdot \alpha^3 = \alpha \cdot (1 + \alpha) = \alpha + \alpha^2 \tag{8.18a}$$

Now consider

$$\alpha^5 = \alpha \cdot \alpha^4 = \alpha \cdot (\alpha + \alpha^2) = \alpha^2 + \alpha^3 \tag{8.18b}$$

From Equations (8.17) and (8.18b), we obtain

$$\alpha^5 = 1 + \alpha + \alpha^2 \tag{8.18c}$$

Now, using Equation (8.18c), we obtain

$$\alpha^6 = \alpha \cdot \alpha^5 = \alpha \cdot (1 + \alpha + \alpha^2) = \alpha + \alpha^2 + \alpha^3 = 1 + \alpha^2 \tag{8.18d}$$

And using Equation (8.18d), we obtain

$$\alpha^7 = \alpha \cdot \alpha^6 = \alpha \cdot (1 + \alpha^2) = \alpha + \alpha^3 = 1 = \alpha^0 \tag{8.18e}$$

Note that $\alpha^7 = \alpha^0$, and therefore, the eight finite field elements of $GF(2^3)$ are

$$\{0, \alpha^0, \alpha^1, \alpha^2, \alpha^3, \alpha^4, \alpha^5, \alpha^6\} \tag{8.19}$$

The mapping of field elements in terms of basis elements, described by Equation (8.14) can be demonstrated with the linear feedback shift register (LFSR) circuit shown in Figure 8.8. The circuit generates (with $m = 3$) the $2^m - 1$ nonzero elements of the field, and thus summarizes the findings of Figure 8.7 and Equations (8.17) through (8.19). Note that in Figure 8.8, the circuit feedback connections correspond to the coefficients of the polynomial $f(X) = 1 + X + X^3$, just like for binary cyclic codes. (See Section 6.7.5.) By starting the circuit in any nonzero state, say 1 0 0, and performing a right-shift at each clock time, it is possible to verify that each of

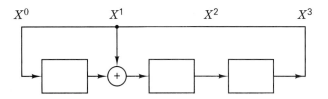

X^0 X^1 X^2 X^3

Figure 8.8 Extension field elements can be represented by the contents of a binary linear feedback shift register (LFSR) formed from a primitive polynomial.

the field elements shown in Figure 8.7 (except the all-zeros element) will cyclicly appear in the stages of the shift register. Two arithmetic operations, addition and multiplication, can be defined for this $GF(2^3)$ finite field. Addition is shown in Table 8.2, and multiplication is shown in Table 8.3 for the nonzero elements only. The rules of addition follow from Equations (8.17) through (8.18e), and can be verified by noticing in Figure 8.7 that the sum of any field elements can be obtained by adding (modulo-2) the respective coefficients of their basis elements. The multiplication rules in Table 8.3 follow the usual procedure in which the product of the field elements is obtained by adding their exponents modulo-$(2^m - 1)$, or for this case, modulo-7.

TABLE 8.2 Addition Table for GF(8) with $f(X) = 1 + X + X^3$

	α^0	α^1	α^2	α^3	α^4	α^5	α^6
α^0	0	α^3	α^6	α^1	α^5	α^4	α^2
α^1	α^3	0	α^4	α^0	α^2	α^6	α^5
α^2	α^6	α^4	0	α^5	α^1	α^3	α^0
α^3	α^1	α^0	α^5	0	α^6	α^2	α^4
α^4	α^5	α^2	α^1	α^6	0	α^0	α^3
α^5	α^4	α^6	α^3	α^2	α^0	0	α^1
α^6	α^2	α^5	α^0	α^4	α^3	α^1	0

TABLE 8.3 Multiplication Table for GF(8) with $f(X) = 1 + X + X^3$

	α^0	α^1	α^2	α^3	α^4	α^5	α^6
α^0	α^0	α^1	α^2	α^3	α^4	α^5	α^6
α^1	α^1	α^2	α^3	α^4	α^5	α^6	α^0
α^2	α^2	α^3	α^4	α^5	α^6	α^0	α^1
α^3	α^3	α^4	α^5	α^6	α^0	α^1	α^2
α^4	α^4	α^5	α^6	α^0	α^1	α^2	α^3
α^5	α^5	α^6	α^0	α^1	α^2	α^3	α^4
α^6	α^6	α^0	α^1	α^2	α^3	α^4	α^5

8.1.4.4 A Simple Test to Determine if a Polynomial is Primitive

There is another way of defining a primitive polynomial that makes its verification relatively easy. For an irreducible polynomial to be a primitive polynomial, at least one of its roots must be a primitive element. A primitive element is one that when raised to higher order exponents will yield all the nonzero elements in the field. Since the field is a finite field, the number of such elements is finite.

Example 8.2   A Primitive Polynomial Must Have at Least one Primitive Element

Find the $m = 3$ roots of $f(X) = 1 + X + X^3$, and verify that the polynomial is primitive by checking that at least one of the roots is a primitive element. What are the roots? Which ones are primitive?

Solution

The roots will be found by enumeration. Clearly, $\alpha^0 = 1$ is not a root because $f(\alpha^0) = 1$. Now use Table 8.2 to check if a α^1 is a root. Since $f(\alpha) = 1 + \alpha + \alpha^3 = 1 + \alpha^0 = 0$, then α is a root. Now check if α^2 is a root. $f(\alpha^2) = 1 + \alpha^2 + \alpha^6 = 1 + \alpha^0 = 0$. Hence, α^2 is a root. Now check if α^3 is a root. $f(\alpha^3) = 1 + \alpha^3 + \alpha^9 = 1 + \alpha^3 + \alpha^2 = 1 + \alpha^5 = \alpha^4 \neq 0$. Hence, α^3 is *not* a root. Is α^4 a root? $f(\alpha^4) = \alpha^{12} + \alpha^4 + 1 = \alpha^5 + \alpha^4 + 1 = 1 + \alpha^0 = 0$. Yes, it is a root. Hence, the roots of $f(X) = 1 + X + X^3$, are α, α^2, and α^4. It is not difficult to verify that starting with any one of these roots and generating higher order exponents yields all of the 7 nonzero elements in the field. Hence, each of the roots is a primitive element. Since our verification requires that at least one root be a primitive element, the polynomial is primitive.

A relatively simple method to verify if a polynomial is primitive can be described in a manner that is related to this example. For any given polynomial under test, draw the LFSR, with the feedback connections corresponding to the polynomial coefficients as shown by the example of Figure 8.8. Load into the circuit-registers any nonzero setting, and perform a right shift with each clock pulse. If the circuit generates each of the nonzero field elements within one period, then the polynomial that defines this GF(2^m) field is a primitive polynomial.

8.1.5 Reed–Solomon Encoding

Equation (8.2) expresses the most conventional form of Reed–Solomon (R–S) codes in terms of the parameters n, k, t, and any positive integer $m > 2$. Repeated here, that equation is

$$(n, k) = (2^m - 1, \ 2^m - 1 - 2t) \tag{8.20}$$

where $n - k = 2t$ is the number of parity symbols, and t is the symbol-error correcting capability of the code. The generating polynomial for an R-S code takes the following form:

$$\mathbf{g}(X) = g_0 + g_1 X + g_2 X^2 + \cdots + g_{2t-1} X^{2t-1} + X^{2t} \tag{8.21}$$

The degree of the generator polynomial is equal to the number of parity symbols. R-S codes are a subset of the BCH codes described in Section 6.8.3 and Table 6.4. Hence, it should be no surprise that this relationship between the degree of the generator polynomial and the number of parity symbols holds just as it does for BCH codes. This can be verified by checking any of the generator polynomials in Table 6.4. Since the generator polynomial is of degree $2t$, there must be precisely $2t$ successive powers of α that are roots of the polynomial. We designate the roots of $\mathbf{g}(X)$ as: α, α^2, ... , α^{2t}. It is not necessary to start with the root α; starting with any power of α is possible. Consider as an example, the (7, 3) double-symbol error correcting R-S code. We describe the generator polynomial in terms of its $2t = n - k = 4$ roots, as follows:

$$\mathbf{g}(X) = (X - \alpha)(X - \alpha^2)(X - \alpha^3)(X - \alpha^4)$$
$$= (X^2 - (\alpha + \alpha^2)X + \alpha^3)(X^2 - (\alpha^3 + \alpha^4)X + \alpha^7)$$
$$= (X^2 - \alpha^4 X + \alpha^3)(X^2 - \alpha^6 X + \alpha^0)$$
$$= X^4 - (\alpha^4 + \alpha^6)X^3 + (\alpha^3 + \alpha^{10} + \alpha^0)X^2 - (\alpha^4 + \alpha^9)X + \alpha^3$$
$$= X^4 - \alpha^3 X^3 + \alpha^0 X^2 - \alpha^1 X + \alpha^3$$

Following the format of low order to high order, and changing negative signs to positive, since in the binary field $+1 = -1$, the generator $\mathbf{g}(X)$ can be expressed as

$$\mathbf{g}(X) = \alpha^3 + \alpha^1 X + \alpha^0 X^2 + \alpha^3 X^3 + X^4 \qquad (8.22)$$

8.1.5.1 Encoding in Systematic Form

Since R-S codes are cyclic codes, encoding in systematic form is analogous to the binary encoding procedure established in Section 6.7.3. We can think of shifting a message polynomial $\mathbf{m}(X)$ into the rightmost k stages of a codeword register and then appending a parity polynomial $\mathbf{p}(X)$ by placing it in the leftmost $n - k$ stages. Therefore we multiply $\mathbf{m}(X)$ by X^{n-k}, thereby manipulating the message polynomial algebraically so that it is right-shifted $n - k$ positions. In Chapter 6, this is shown in Equation (6.61) in the context of binary encoding. Next, we divide $X^{n-k} \mathbf{m}(X)$ by the generator polynomial $\mathbf{g}(X)$, which is written as

$$X^{n-k} \mathbf{m}(X) = \mathbf{q}(X) \mathbf{g}(X) + \mathbf{p}(X) \qquad (8.23)$$

where $\mathbf{q}(X)$ and $\mathbf{p}(X)$ are quotient and remainder polynomials, respectively. As in the binary case, the remainder is the parity. Equation (8.23) can also be expressed as

$$\mathbf{p}(X) = X^{n-k} \mathbf{m}(X) \text{ modulo } \mathbf{g}(X) \qquad (8.24)$$

The resulting codeword polynomial $\mathbf{U}(X)$, shown in Equation (6.64), is rewritten as

$$\mathbf{U}(X) = \mathbf{p}(X) + X^{n-k} \mathbf{m}(X) \qquad (8.25)$$

We demonstrate the steps implied by Equations (8.24) and (8.25) by encoding the three-symbol message

$$\underbrace{010}_{\alpha^1} \quad \underbrace{110}_{\alpha^3} \quad \underbrace{111}_{\alpha^5}$$

with the (7, 3) R-S code whose generator polynomial is given in Equation (8.22). We first multiply (upshift) the message polynomial $\alpha^1 + \alpha^3 X + \alpha^5 X^2$ by $X^{n-k} = X^4$, yielding $\alpha^1 X^4 + \alpha^3 X^5 + \alpha^5 X^6$. We next divide this upshifted message polynomial by the generator polynomial in Equation (8.22), $\alpha^3 + \alpha^1 X + \alpha^0 X^2 + \alpha^3 X^3 + X^4$. Polynomial division with nonbinary coefficients is more tedious than its binary counterpart (see Example 6.9), because the required operations of addition (subtraction) and multiplication (division) must follow the rules in Tables 8.2 and 8.3, respectively. It is left as an exercise for the reader to verify that this polynomial division results in the following remainder (parity) polynomial:

$$\mathbf{p}(X) = \alpha^0 + \alpha^2 X + \alpha^4 X^2 + \alpha^6 X^3$$

Then, from Equation (8.25), the codeword polynomial can be written as

$$\mathbf{U}(X) = \alpha^0 + \alpha^2 X + \alpha^4 X^2 + \alpha^6 X^3 + \alpha^1 X^4 + \alpha^3 X^5 + \alpha^5 X^6$$

8.1.5.2 Systematic Encoding with an $(n - k)$-Stage Shift Register

Using circuitry to encode a 3-symbol sequence in systematic form with the $(7, 3)$ R-S code described by $\mathbf{g}(X)$ in Equation (8.22) requires the implementation of a LFSR, as shown in Figure 8.9. It can be easily verified that the multiplier terms in Figure 8.9 taken from left to right correspond to the coefficients of the polynomial in Equation (8.22) (low order to high order). This encoding process is the nonbinary equivalent of the cyclic encoding that was described in Section 6.7.5. Here, corresponding to Equation (8.20), the $(7, 3)$ R-S nonzero codewords are made up of $2^m - 1 = 7$ symbols, and each symbol is made of $m = 3$ bits.

Notice the similarity amongst Figures 8.9, 6.18, and 6.19. In all three cases the number of stages in the shift register is $n - k$. The figures in Chapter 6 illustrate binary examples where each shift-register stage holds 1 bit. Here the example is nonbinary, so that each stage in the shift register of Figure 8.9 holds a 3-bit symbol. In Figure 6.18, the coefficients labeled $g_1, g_2, \ldots$ are binary. Therefore, they take on values of 1 or 0, simply dictating the presence or absence of a connection in the LFSR. However in Figure 8.9, since each coefficient is specified by 3-bits, it can take on one of 8 values.

The nonbinary operation implemented by the encoder of Figure 8.9, forming codewords in a systematic format, proceeds in the same way as the binary one. The steps can be described as follows:

1. Switch 1 is closed during the first k clock cycles to allow shifting the message symbols into the $(n - k)$-stage shift register.

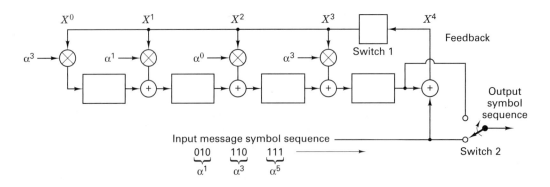

Figure 8.9 LFSR Encoder for a (7,3) R–S code.

2. Switch 2 is in the down position during the first k clock cycles in order to allow simultaneous transfer of the message symbols directly to an output register (not shown in Figure 8.9).

3. After transfer of the kth message symbol to the output register, switch 1 is opened and switch 2 is moved to the up position.

4. The remaining $(n - k)$ clock cycles clear the parity symbols contained in the shift register by moving them to the output register.

5. The total number of clock cycles is equal to n, and the contents of the output register is the codeword polynomial $\mathbf{p}(X) + X^{n-k}\mathbf{m}(X)$, where $\mathbf{p}(X)$ represents the parity symbols, and $\mathbf{m}(X)$ the message symbols in polynomial form.

We use the same symbol sequence that was chosen as a test message in Section 8.1.5.1, and we write

$$
\underbrace{010}_{\alpha^1} \quad \underbrace{110}_{\alpha^3} \quad \underbrace{111}_{\alpha^5}
$$

where the rightmost symbol is the earliest symbol, and the rightmost bit is the earliest bit. The operational steps during the first $k = 3$ shifts of the encoding circuit of Figure 8.9 are as follows:

INPUT QUEUE			CLOCK CYCLE	REGISTER CONTENTS				FEEDBACK
α^1	α^3	α^5	0	0	0	0	0	α^5
	α^1	α^3	1	α^1	α^6	α^5	α^1	α^0
		α^1	2	α^3	0	α^2	α^2	α^4
		—	3	α^0	α^2	α^4	α^6	—

After the third clock cycle, the register contents are the 4 parity symbols, α^0, α^2, α^4, and α^6, as shown. Then, switch 1 of the circuit is opened, switch 2 is toggled to the up position, and the parity symbols contained in the register are shifted to the output. Therefore the output codeword, written in polynomial form, can be expressed as

$$
\mathbf{U}(X) = \sum_{n=0}^{6} u_n X^n
$$

$$
\mathbf{U}(X) = \alpha^0 + \alpha^2 X + \alpha^4 X^2 + \alpha^6 X^3 + \alpha^1 X^4 + \alpha^3 X^5 + \alpha^5 X^6 \qquad (8.26)
$$

$$
= (100) + (001) X + (011) X^2 + (101) X^3 + (010) X^4 + (110) X^5 + (111) X^6
$$

The process of verifying the contents of the register at various clock cycles is somewhat more tedious than in the binary case. Here, the field elements must be added and multiplied by using Table 8.2 and Table 8.3, respectively.

The roots of a generator polynomial $\mathbf{g}(X)$ must also be the roots of the codeword generated by $\mathbf{g}(X)$, because a valid codeword is of the form

$$
\mathbf{U}(X) = \mathbf{m}(X)\,\mathbf{g}(X) \qquad (8.27)
$$

Therefore an arbitrary codeword, when evaluated at any root of $\mathbf{g}(X)$, must yield zero. It is of interest to verify that the codeword polynomial in Equation (8.26) does indeed yield zero when evaluated at the 4 roots of $\mathbf{g}(X)$. In other words, this means checking that

$$\mathbf{U}(\alpha) = \mathbf{U}(\alpha^2) = \mathbf{U}(\alpha^3) = \mathbf{U}(\alpha^4) = \mathbf{0}$$

Evaluating each term independently yields

$$\begin{aligned}
\mathbf{U}(\alpha) &= \alpha^0 + \alpha^3 + \alpha^6 + \alpha^9 + \alpha^5 + \alpha^8 + \alpha^{11} \\
&= \alpha^0 + \alpha^3 + \alpha^6 + \alpha^2 + \alpha^5 + \alpha^1 + \alpha^4 \\
&= \alpha^1 + \alpha^0 + \alpha^6 + \alpha^4 \\
&= \alpha^3 + \alpha^3 = \mathbf{0}
\end{aligned}$$

$$\begin{aligned}
\mathbf{U}(\alpha^2) &= \alpha^0 + \alpha^4 + \alpha^8 + \alpha^{12} + \alpha^9 + \alpha^{13} + \alpha^{17} \\
&= \alpha^0 + \alpha^4 + \alpha^1 + \alpha^5 + \alpha^2 + \alpha^6 + \alpha^3 \\
&= \alpha^5 + \alpha^6 + \alpha^0 + \alpha^3 \\
&= \alpha^1 + \alpha^1 = \mathbf{0}
\end{aligned}$$

$$\begin{aligned}
\mathbf{U}(\alpha^3) &= \alpha^0 + \alpha^5 + \alpha^{10} + \alpha^{15} + \alpha^{13} + \alpha^{18} + \alpha^{23} \\
&= \alpha^0 + \alpha^5 + \alpha^3 + \alpha^1 + \alpha^6 + \alpha^4 + \alpha^2 \\
&= \alpha^4 + \alpha^0 + \alpha^3 + \alpha^2 \\
&= \alpha^5 + \alpha^5 = \mathbf{0}
\end{aligned}$$

$$\begin{aligned}
\mathbf{U}(\alpha^4) &= \alpha^0 + \alpha^6 + \alpha^{12} + \alpha^{18} + \alpha^{17} + \alpha^{23} + \alpha^{29} \\
&= \alpha^0 + \alpha^6 + \alpha^5 + \alpha^4 + \alpha^3 + \alpha^2 + \alpha^1 \\
&= \alpha^2 + \alpha^0 + \alpha^5 + \alpha^1 \\
&= \alpha^6 + \alpha^6 = \mathbf{0}
\end{aligned}$$

This demonstrates the expected results that a codeword evaluated at any root of $\mathbf{g}(X)$ must yield zero.

8.1.6 Reed–Solomon Decoding

In Section 8.1.5, a test message encoded in systematic form using a (7, 3) R–S code, resulted in a codeword polynomial described by Equation (8.26). Now, assume that during transmission, this codeword becomes corrupted so that 2 symbols are received in error. (This number of errors corresponds to the maximum error-correcting capability of the code.) For this 7-symbol codeword example, the error pattern can be described in polynomial form as

$$\mathbf{e}(X) = \sum_{n=0}^{6} e_n X^n \tag{8.28}$$

For this example, let the double-symbol error be such that

$$\mathbf{e}(X) = 0 + 0X + 0X^2 + \alpha^2 X^3 + \alpha^5 X^4 + 0X^5 + 0X^6 \tag{8.29}$$

$$= (000) + (000)X + (000)X^2 + (001)X^3 + (111)X^4 + (000)X^5 + (000)X^6$$

In other words, one parity symbol has been corrupted with a 1-bit error (seen as α^2), and one data symbol has been corrupted with a 3-bit error (seen as α^5). The received corrupted-codeword polynomial $\mathbf{r}(X)$ is then represented by the sum of the transmitted-codeword polynomial and the error-pattern polynomial as follows:

$$\mathbf{r}(X) = \mathbf{U}(X) + \mathbf{e}(X) \tag{8.30}$$

Following Equation (8.30), we add $\mathbf{U}(X)$ from Equation (8.26) to $\mathbf{e}(X)$ from Equation (8.29) to yield

$$\mathbf{r}(X) = (100) + (001)X + (011)X^2 + (100)X^3 + (101)X^4 + (110)X^5 + (111)X^6$$

$$= \alpha^0 + \alpha^2 X + \alpha^4 X^2 + \alpha^0 X^3 + \alpha^6 X^4 + \alpha^3 X^5 + \alpha^5 X^6 \tag{8.31}$$

In this 2-symbol error-correction example, there are four unknowns—two error locations and two error values. Notice an important difference between the nonbinary decoding of $\mathbf{r}(X)$ that we are faced with in Equation (8.31) and the binary decoding that was described in Chapter 6. In binary decoding, the decoder only needs to find the error locations. Knowledge that there is an error at a particular location dictates that the bit must be "flipped" from a 1 to a 0, or vice versa. But here, the nonbinary symbols require that we not only learn the error locations, but that we also determine the correct symbol values at those locations. Since there are four unknowns in this example, four equations are required for their solution.

8.1.6.1 Syndrome Computation

Recall from Section 6.4.7, that the *syndrome* is the result of a parity check performed on $\mathbf{r}$ to determine whether $\mathbf{r}$ is a valid member of the codeword set. If in fact $\mathbf{r}$ is a member, then the syndrome $\mathbf{S}$ has value $\mathbf{0}$. Any nonzero value of $\mathbf{S}$ indicates the presence of errors. Similar to the binary case, the syndrome $\mathbf{S}$ is made up of $n - k$ symbols, $\{S_i\}$ $(i = 1, \ldots, n - k)$. Thus, for this (7, 3) R–S code, there are four symbols in every syndrome vector; their values can be computed from the received polynomial $\mathbf{r}(X)$. Note how the computation is facilitated by the structure of the code, given by Equation (8.27) and rewritten as

$$\mathbf{U}(X) = \mathbf{m}(X)\,\mathbf{g}(X)$$

From this structure it can be seen that every valid codeword polynomial $\mathbf{U}(X)$ is a multiple of the generator polynomial $\mathbf{g}(X)$. Therefore, the roots of $\mathbf{g}(X)$ must also be the roots of $\mathbf{U}(X)$. Since $\mathbf{r}(X) = \mathbf{U}(X) + \mathbf{e}(X)$, then $\mathbf{r}(X)$ evaluated at each of the roots of $\mathbf{g}(X)$ should yield zero only when it is a valid codeword. Any nonzero result is an indication that an error is present. The computation of a syndrome symbol can be described as

$$S_i = \mathbf{r}(X)\Big|_{X=\alpha^i} = \mathbf{r}(\alpha^i) \quad i = 1, \cdots, n - k \tag{8.32}$$

where $\mathbf{r}(X)$ contains the postulated 2-symbol errors as shown in Equation (8.29). If $\mathbf{r}(X)$ were a valid codeword, it would cause each syndrome symbol S_i to equal 0. For this example, the four syndrome symbols are found as follows:

$$S_1 = \mathbf{r}(\alpha) = \alpha^0 + \alpha^3 + \alpha^6 + \alpha^3 + \alpha^{10} + \alpha^8 + \alpha^{11}$$
$$= \alpha^0 + \alpha^3 + \alpha^6 + \alpha^3 + \alpha^2 + \alpha^1 + \alpha^4 \qquad (8.33)$$
$$= \alpha^3$$

$$S_2 = \mathbf{r}(\alpha^2) = \alpha^0 + \alpha^4 + \alpha^8 + \alpha^6 + \alpha^{14} + \alpha^{13} + \alpha^{17}$$
$$= \alpha^0 + \alpha^4 + \alpha^1 + \alpha^6 + \alpha^0 + \alpha^6 + \alpha^3 \qquad (8.34)$$
$$= \alpha^5$$

$$S_3 = \mathbf{r}(\alpha^3) = \alpha^0 + \alpha^5 + \alpha^{10} + \alpha^9 + \alpha^{18} + \alpha^{18} + \alpha^{23}$$
$$= \alpha^0 + \alpha^5 + \alpha^3 + \alpha^2 + \alpha^4 + \alpha^4 + \alpha^2 \qquad (8.35)$$
$$= \alpha^6$$

$$S_4 = \mathbf{r}(\alpha^4) = \alpha^0 + \alpha^6 + \alpha^{12} + \alpha^{12} + \alpha^{22} + \alpha^{23} + \alpha^{29}$$
$$= \alpha^0 + \alpha^6 + \alpha^5 + \alpha^5 + \alpha^1 + \alpha^2 + \alpha^1 \qquad (8.36)$$
$$= 0$$

The results confirm that the received codeword contains an error (which we inserted) since $\mathbf{S} \neq \mathbf{0}$.

Example 8.3 A Secondary Check on the Syndrome Values

For the (7, 3) R–S code example under consideration, the error pattern is known since it was chosen earlier. Recall the property of codes presented in Section 6.4.8.1 when describing the standard array. Each element of a coset (row) in the standard array has the same syndrome. Show that this property is also true for the R-S code by evaluating the error polynomial $\mathbf{e}(X)$ at the roots of $\mathbf{g}(X)$ to demonstrate that it must yield the same syndrome values as when $\mathbf{r}(X)$ is evaluated at the roots of $\mathbf{g}(X)$. In other words, it must yield the same values obtained in Equations (8.33) through (8.36).

Solution

$$S_i = \mathbf{r}(X)\Big|_{X=\alpha^i} = \mathbf{r}(\alpha^i) \quad i = 1, 2, \cdots, n - k$$

$$S_i = [\mathbf{U}(X) + \mathbf{e}(X)]\Big|_{X=\alpha^i} = \mathbf{U}(\alpha^i) + \mathbf{e}(\alpha^i)$$

$$S_i = \mathbf{r}(\alpha^i) = \mathbf{U}(\alpha^i) + \mathbf{e}(\alpha^i) = 0 + \mathbf{e}(\alpha^i)$$

From Equation (8.29), $\mathbf{e}(X) = \alpha^2 X^3 + \alpha^5 X^4$; therefore,

$$S_1 = \mathbf{e}(\alpha^1) = \alpha^5 + \alpha^9$$
$$= \alpha^5 + \alpha^2$$
$$= \alpha^3$$

$$S_2 = \mathbf{e}(\alpha^2) = \alpha^8 + \alpha^{13}$$
$$= \alpha^1 + \alpha^6$$
$$= \alpha^5$$

$$S_3 = \mathbf{e}(\alpha^3) = \alpha^{11} + \alpha^{17}$$
$$= \alpha^4 + \alpha^3$$
$$= \alpha^6$$

$$S_4 = \mathbf{e}(\alpha^4) = \alpha^{14} + \alpha^{21}$$
$$= \alpha^0 + \alpha^0$$
$$= 0$$

These results confirm that the syndrome values are the same, whether obtained by evaluating $\mathbf{e}(X)$ at the roots of $\mathbf{g}(X)$, or $\mathbf{r}(X)$ at the roots of $\mathbf{g}(X)$.

8.1.6.2 Error Location

Suppose there are v errors in the codeword at location $X^{j_1}, X^{j_2}, \ldots X^{j_v}$. Then, the error polynomial shown in Equations (8.28) and (8.29) can be written as

$$\mathbf{e}(X) = e_{j_1} X^{j_1} + e_{j_2} X^{j_2} + \cdots + e_{j_v} X^{j_v} \tag{8.37}$$

The indices $1, 2, \ldots, v$ refer to the $1^{\text{st}}, 2^{\text{nd}}, \ldots, v^{\text{th}}$ errors, and the index j refers to the error location. To correct the corrupted codeword, each error value e_{j_ℓ} and its location X^{j_ℓ}, where $\ell = 1, 2, \ldots v$ must be determined. We define an error locator number as $\beta_\ell = \alpha^{j_\ell}$. Next, we obtain the $n - k = 2t$ syndrome symbols by substituting α^i into the received polynomial for $i = 1, 2, \ldots, 2t$:

$$S_1 = \mathbf{r}(\alpha) = e_{j_1} \beta_1 + e_{j_2} \beta_2 + \cdots + e_{j_v} \beta_v$$
$$S_2 = \mathbf{r}(\alpha^2) = e_{j_1} \beta_1^2 + e_{j_2} \beta_2^2 + \cdots + e_{j_v} \beta_v^2 \tag{8.38}$$
$$\vdots$$
$$S_{2t} = \mathbf{r}(\alpha^{2t}) = e_{j_1} \beta_1^{2t} + e_{j_2} \beta_2^{2t} + \cdots + e_{j_v} \beta_v^{2t}$$

There are $2t$ unknowns (t error values and t locations), and $2t$ simultaneous equations. However, these $2t$ simultaneous equations cannot be solved in the usual way because they are nonlinear (as some of the unknowns have exponents). Any technique that solves this system of equations is known as a Reed–Solomon decoding algorithm.

When a nonzero syndrome vector (one or more of its symbols are nonzero) has been computed, it signifies that an error has been received. Next, it is necessary to learn the location of the error or errors. An error-locator polynomial can be defined as

$$\boldsymbol{\sigma}(X) = (1 + \beta_1 X)(1 + \beta_2 X) \cdots (1 + \beta_v X) \tag{8.39}$$
$$= 1 + \sigma_1 X + \sigma_2 X^2 + \cdots + \sigma_v X^v$$

The roots of $\boldsymbol{\sigma}(X)$ are $1/\beta_1, 1/\beta_2, \ldots, 1/\beta_v$. The reciprocal of the roots of $\boldsymbol{\sigma}(X)$ are the error-location numbers of the error pattern $\mathbf{e}(X)$. Then using autoregressive modeling techniques [5], we form a matrix from the syndromes, where the first t syndromes are used to predict the next syndrome. That is,

$$\begin{bmatrix} S_1 & S_2 & S_3 & \cdots & S_{t-1} & S_t \\ S_2 & S_3 & S_4 & \cdots & S_t & S_{t+1} \\ & & & \vdots & & \\ S_{t-1} & S_t & S_{t+1} & \cdots & S_{2t-3} & S_{2t-2} \\ S_t & S_{t+1} & S_{t+2} & \cdots & S_{2t-2} & S_{2t-1} \end{bmatrix} \begin{bmatrix} \sigma_t \\ \sigma_{t-1} \\ \vdots \\ \sigma_2 \\ \sigma_1 \end{bmatrix} = \begin{bmatrix} -S_{t+1} \\ -S_{t+2} \\ \vdots \\ -S_{2t-1} \\ -S_{2t} \end{bmatrix} \tag{8.40}$$

We apply the autoregressive model of Equation (8.40) by using the largest dimensioned matrix that has a nonzero determinant. For the $(7, 3)$ double symbol error-correcting R-S code, the matrix size is 2×2, and the model is written as

$$\begin{bmatrix} S_1 & S_2 \\ S_2 & S_3 \end{bmatrix} \begin{bmatrix} \sigma_2 \\ \sigma_1 \end{bmatrix} = \begin{bmatrix} S_3 \\ S_4 \end{bmatrix} \tag{8.41}$$

$$\begin{bmatrix} \alpha^3 & \alpha^5 \\ \alpha^5 & \alpha^6 \end{bmatrix} \begin{bmatrix} \sigma_2 \\ \sigma_1 \end{bmatrix} = \begin{bmatrix} \alpha^6 \\ 0 \end{bmatrix} \tag{8.42}$$

To solve for the coefficients σ_1 and σ_2 of the error-locator polynomial $\boldsymbol{\sigma}(X)$, we first take the inverse of the matrix in Equation (8.42). The inverse of a matrix $[A]$ is found as follows:

$$\mathrm{Inv}\,[A] = \frac{\mathrm{cofactor}\,[A]}{\det\,[A]}$$

Therefore,

$$\det \begin{bmatrix} \alpha^3 & \alpha^5 \\ \alpha^5 & \alpha^6 \end{bmatrix} = \alpha^3 \alpha^6 - \alpha^5 \alpha^5 = \alpha^9 + \alpha^{10} \tag{8.43}$$

$$= \alpha^2 + \alpha^3 = \alpha^5$$

$$\mathrm{cofactor} \begin{bmatrix} \alpha^3 & \alpha^5 \\ \alpha^5 & \alpha^6 \end{bmatrix} = \begin{bmatrix} \alpha^6 & \alpha^5 \\ \alpha^5 & \alpha^3 \end{bmatrix} \tag{8.44}$$

and

$$\mathrm{Inv} \begin{bmatrix} \alpha^3 & \alpha^5 \\ \alpha^5 & \alpha^6 \end{bmatrix} = \frac{\begin{bmatrix} \alpha^6 & \alpha^5 \\ \alpha^5 & \alpha^3 \end{bmatrix}}{\alpha^5} = \alpha^{-5} \begin{bmatrix} \alpha^6 & \alpha^5 \\ \alpha^5 & \alpha^3 \end{bmatrix} \tag{8.45}$$

$$= \alpha^2 \begin{bmatrix} \alpha^6 & \alpha^5 \\ \alpha^5 & \alpha^3 \end{bmatrix} = \begin{bmatrix} \alpha^8 & \alpha^7 \\ \alpha^7 & \alpha^5 \end{bmatrix} = \begin{bmatrix} \alpha^1 & \alpha^0 \\ \alpha^0 & \alpha^5 \end{bmatrix}$$

Safety Check. If the inversion was performed correctly, then the multiplication of the original matrix by the inverted matrix should yield an identity matrix:

$$\begin{bmatrix} \alpha^3 & \alpha^5 \\ \alpha^5 & \alpha^6 \end{bmatrix} \begin{bmatrix} \alpha^1 & \alpha^0 \\ \alpha^0 & \alpha^5 \end{bmatrix} = \begin{bmatrix} \alpha^4 + \alpha^5 & \alpha^3 + \alpha^{10} \\ \alpha^6 + \alpha^6 & \alpha^5 + \alpha^{11} \end{bmatrix} = \begin{bmatrix} 1 & 0 \\ 0 & 1 \end{bmatrix} \tag{8.46}$$

Continuing from Equation (8.42), we begin our search for the error locations by solving for the coefficients of the error-locator polynomial $\boldsymbol{\sigma}(X)$, as follows:

$$\begin{bmatrix} \sigma_2 \\ \sigma_1 \end{bmatrix} = \begin{bmatrix} \alpha^3 & \alpha^5 \\ \alpha^5 & \alpha^6 \end{bmatrix}^{-1} \begin{bmatrix} \alpha^6 \\ 0 \end{bmatrix} = \begin{bmatrix} \alpha^1 & \alpha^0 \\ \alpha^0 & \alpha^5 \end{bmatrix} \begin{bmatrix} \alpha^6 \\ 0 \end{bmatrix} = \begin{bmatrix} \alpha^7 \\ \alpha^6 \end{bmatrix} = \begin{bmatrix} \alpha^0 \\ \alpha^6 \end{bmatrix} \tag{8.47}$$

From Equations (8.39) and (8.47),

$$\boldsymbol{\sigma}(X) = \alpha^0 + \sigma_1 X + \sigma_2 X^2 \tag{8.48}$$

$$= \alpha^0 + \alpha^6 X + \alpha^0 X^2$$

The roots of $\boldsymbol{\sigma}(X)$ are the reciprocals of the error locations. Once these roots are located, the error locations will be known. In general, the roots of $\boldsymbol{\sigma}(X)$ may be one or more of the elements of the field. We determine these roots by exhaustive testing of the $\boldsymbol{\sigma}(X)$ polynomial with each of the field elements, as shown below. Any element X that yields $\boldsymbol{\sigma}(X) = \mathbf{0}$ is a root, and allows us to locate an error:

$$\boldsymbol{\sigma}(\alpha^0) = \alpha^0 + \alpha^6 + \alpha^0 = \alpha^6 \neq \mathbf{0}$$

$$\boldsymbol{\sigma}(\alpha^1) = \alpha^0 + \alpha^7 + \alpha^2 = \alpha^2 \neq \mathbf{0}$$

$$\boldsymbol{\sigma}(\alpha^2) = \alpha^0 + \alpha^8 + \alpha^4 = \alpha^6 \neq \mathbf{0}$$

$$\boldsymbol{\sigma}(\alpha^3) = \alpha^0 + \alpha^9 + \alpha^6 = \mathbf{0} \Rightarrow \text{ERROR}$$

$$\boldsymbol{\sigma}(\alpha^4) = \alpha^0 + \alpha^{10} + \alpha^8 = \mathbf{0} \Rightarrow \text{ERROR}$$

$$\boldsymbol{\sigma}(\alpha^5) = \alpha^0 + \alpha^{11} + \alpha^{10} = \alpha^2 \neq \mathbf{0}$$

$$\boldsymbol{\sigma}(\alpha^6) = \alpha^0 + \alpha^{12} + \alpha^{12} = \alpha^0 \neq \mathbf{0}$$

As seen in Equation (8.39), the error locations are at the inverse of the roots of the polynomial. Therefore $\boldsymbol{\sigma}(\alpha^3) = \mathbf{0}$ indicates that one root exits at $1/\beta_\ell = \alpha^3$. Thus, $\beta_\ell = 1/\alpha^3 = \alpha^4$. Similarly, $\boldsymbol{\sigma}(\alpha^4) = \mathbf{0}$ indicates that another root exits at $1/\beta_{\ell'} = 1/\alpha^4 = \alpha^3$, where (for this example) ℓ and ℓ' refer to the 1st and 2nd error respectively. Since there are 2-symbol errors here, the error polynomial is of the form

$$\mathbf{e}(X) = e_{j_1} X^{j_1} + e_{j_2} X^{j_2} \tag{8.49}$$

The two errors were found at locations α^3 and α^4. Note that the indexing of the error-location numbers is completely arbitrary. Thus, for this example, we can designate the $\beta_\ell = \alpha^{j_\ell}$ values as $\beta_1 = \alpha^{j_1} = \alpha^3$ and $\beta_2 = \alpha^{j_2} = \alpha^4$.

8.1.6.3 Error Values

An error had been denoted e_{j_ℓ}, where the index j refers to the error location and the index ℓ identifies the ℓth error. Since each error value is coupled to a particular location, the notation can be simplified by denoting e_{j_ℓ} simply as e_ℓ. Now, preparing to determine the error values e_1 and e_2, associated with locations $\beta_1 = \alpha^3$ and $\beta_2 = \alpha^4$, any of the four syndrome equations can be used. From Equation (8.38), let us use S_1 and S_2:

$$S_1 = \mathbf{r}(\alpha) = e_1\beta_1 + e_2\beta_2 \tag{8.50}$$

$$S_2 = \mathbf{r}(\alpha^2) = e_1\beta_1^2 + e_2\beta_2^2$$

We can write these equations in matrix form as follows:

$$\begin{bmatrix} \beta_1 & \beta_2 \\ \beta_1^2 & \beta_2^2 \end{bmatrix} \begin{bmatrix} e_1 \\ e_2 \end{bmatrix} = \begin{bmatrix} S_1 \\ S_2 \end{bmatrix} \tag{8.51}$$

$$\begin{bmatrix} \alpha^3 & \alpha^4 \\ \alpha^6 & \alpha^8 \end{bmatrix} \begin{bmatrix} e_1 \\ e_2 \end{bmatrix} = \begin{bmatrix} \alpha^3 \\ \alpha^5 \end{bmatrix} \tag{8.52}$$

To solve for the error values e_1 and e_2, the matrix in Equation (8.52) is inverted in the usual way, yielding

$$\text{Inv}\begin{bmatrix} \alpha^3 & \alpha^4 \\ \alpha^6 & \alpha^1 \end{bmatrix} = \frac{\begin{bmatrix} \alpha^1 & \alpha^4 \\ \alpha^6 & \alpha^3 \end{bmatrix}}{\alpha^3\alpha^1 - \alpha^6\alpha^4}$$

$$= \frac{\begin{bmatrix} \alpha^1 & \alpha^4 \\ \alpha^6 & \alpha^3 \end{bmatrix}}{\alpha^4 + \alpha^3} = \alpha^{-6}\begin{bmatrix} \alpha^1 & \alpha^4 \\ \alpha^6 & \alpha^3 \end{bmatrix} = \alpha^1\begin{bmatrix} \alpha^1 & \alpha^4 \\ \alpha^6 & \alpha^3 \end{bmatrix} \qquad (8.53)$$

$$= \begin{bmatrix} \alpha^2 & \alpha^5 \\ \alpha^7 & \alpha^4 \end{bmatrix} = \begin{bmatrix} \alpha^2 & \alpha^5 \\ \alpha^0 & \alpha^4 \end{bmatrix}$$

Now, we solve Equation (8.52) for the error values, as follows:

$$\begin{bmatrix} e_1 \\ e_2 \end{bmatrix} = \begin{bmatrix} \alpha^2 & \alpha^5 \\ \alpha^0 & \alpha^4 \end{bmatrix}\begin{bmatrix} \alpha^3 \\ \alpha^5 \end{bmatrix} = \begin{bmatrix} \alpha^5 + \alpha^{10} \\ \alpha^3 + \alpha^9 \end{bmatrix} = \begin{bmatrix} \alpha^5 + \alpha^3 \\ \alpha^3 + \alpha^2 \end{bmatrix} = \begin{bmatrix} \alpha^2 \\ \alpha^5 \end{bmatrix} \qquad (8.54)$$

8.1.6.4 Correcting the Received Polynomial with Estimates of the Error Polynomial

From Equation (8.49) and (8.54), the estimated error polynomial is formed, to yield

$$\hat{e}(X) = e_1 X^{j_1} + e_2 X^{j_2} \qquad (8.55)$$

$$= \alpha^2 X^3 + \alpha^5 X^4$$

The demonstrated algorithm repairs the received polynomial yielding an estimate of the transmitted codeword, and ultimately delivers a decoded message. That is,

$$\hat{U}(X) = r(X) + \hat{e}(X) = U(X) + e(X) + \hat{e}(X) \qquad (8.56)$$

$$r(X) = (100) + (001)X + (011)X^2 + (100)X^3 + (101)X^4 + (110)X^5 + (111)X^6$$

$$\hat{e}(X) = (000) + (000)X + (000)X^2 + (001)X^3 + (111)X^4 + (000)X^5 + (000)X^6$$

$$\hat{U}(X) = (100) + (001)X + (011)X^2 + (101)X^3 + (010)X^4 + (110)X^5 + (111)X^6$$

$$= \alpha^0 + \alpha^2 X + \alpha^4 X^2 + \alpha^6 X^3 + \alpha^1 X^4 + \alpha^3 X^5 + \alpha^5 X^6 \qquad (8.57)$$

Since the message symbols constitute the rightmost $k = 3$ symbols, the decoded message is

$$\underbrace{010}_{\alpha^1} \quad \underbrace{110}_{\alpha^3} \quad \underbrace{111}_{\alpha^5}$$

which is exactly the test message that was chosen in Section 8.1.5 for this example. (For further reading on R-S coding, see the collection of papers in reference [6].)

8.2 INTERLEAVING AND CONCATENATED CODES

Throughout this and earlier chapters we have assumed that the channel is *memoryless,* since we have considered codes that are designed to combat random independent errors. A channel that has *memory* is one that exhibits mutually dependent signal transmission impairments. A channel that exhibits *multipath fading,* where signals arrive at the receiver over two or more paths of different lengths, is an example of a channel with memory. The effect is that the signals can arrive out of phase with each other, and the cumulative received signal is distorted. Wireless mobile communication channels, as well as ionospheric and tropospheric propagation channels, suffer from such phenomena. (See Chapter 15 for details on fading channels.) Also, some channels suffer from switching noise and other burst noise (e.g., telephone channels or channels disturbed by pulse jamming). All of these time-correlated impairments result in statistical dependence among successive symbol transmissions. That is, the disturbances tend to cause errors that occur in bursts, instead of as isolated events.

Under the assumption that the channel has memory, the errors no longer can be characterized as single randomly distributed bit errors whose occurrence is independent from bit to bit. Most block or convolutional codes are designed to combat random independent errors. The result of a channel having memory on such coded signals is to cause degradation in error performance. Coding techniques for channels with memory have been proposed, but the greatest problem with such coding is the difficulty in obtaining accurate models of the often time-varying statistics of such channels. One technique, which only requires a knowledge of the *duration or span* of the channel memory, *not* its exact statistical characterization, is the use of time diversity or *interleaving.*

Interleaving the coded message before transmission and deinterleaving after reception causes bursts of channel errors to be spread out in time and thus to be handled by the decoder as if they were random errors. Since, in all practical cases, the channel memory decreases with time separation, the idea behind interleaving is to separate the codeword symbols in time. The intervening times are similarly filled by the symbols of other codewords. Separating the symbols in time effectively transforms a channel with memory to a *memoryless* one, and thereby enables the random-error-correcting codes to be useful in a burst-noise channel.

The interleaver shuffles the code symbols over a span of several block lengths (for block codes) or several constraint lengths (for convolutional codes). The span required is determined by the burst duration. The details of the bit redistribution pattern must be known to the receiver in order for the symbol stream to be deinterleaved before being decoded. Figure 8.10 illustrates a simple interleaving example. In Figure 8.10a we see seven uninterleaved codewords, *A* through *G.* Each codeword is comprised of seven code symbols. Let us assume that the code has a single-error-correcting capability within each seven-symbol sequence. If the memory span of the channel is one codeword in duration, such a seven-symbol-time noise burst could destroy the information contained in one or two codewords. However, suppose that, after having encoded the data, the code symbols were then *interleaved* or shuffled, as shown in Figure 8.10b. That is, each code symbol of each codeword is

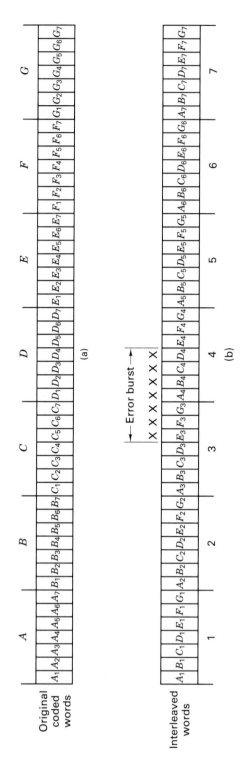

Figure 8.10 Interleaving example. (a) Original uninterleaved codewords, each comprised of seven code symbols. (b) Interleaved code symbols.

separated from its preinterleaved neighbors by a span of seven symbol times. The interleaved stream is then used to modulate a waveform that is transmitted over the channel. A contiguous channel noise burst occupying seven symbol times is seen in Figure 8.10b, to affect one code symbol from each of the original seven codewords. Upon reception, the stream is first deinterleaved so that it resembles the original coded sequence in Figure 8.10a. Then the stream is decoded. Since each codeword possesses a single-error-correcting capability, the burst noise has no degrading effect on the final sequence.

Interleaving techniques have proven useful for all the block and convolutional codes described here and in earlier chapters. Two types of interleavers are commonly used, *block interleavers* and *convolutional interleavers*. They are each described below.

8.2.1 Block Interleaving

A block interleaver accepts the coded symbols in blocks from the encoder, permutes the symbols, and then feeds the rearranged symbols to the modulator. The usual permutation of the block is accomplished by *filling the columns* of an M-row-by N-column ($M \times N$) array with the encoded sequence. After the array is completely filled, the symbols are then fed to the modulator *one row at a time* and transmitted over the channel. At the receiver, the deinterleaver performs the inverse operation; it accepts the symbols from the demodulator, deinterleaves them, and feeds them to the decoder. Symbols are entered into the deinterleaver array by rows, and removed by columns. Figure 8.11a illustrates an example of an interleaver with $M = 4$ rows and $N = 6$ columns. The entries in the array illustrate the order in which the 24 code symbols are placed into the interleaver. The output sequence to the transmitter consists of code symbols removed from the array by rows, as shown in the figure. The most important characteristics of such a block interleaver are as follows:

1. Any burst of less than N contiguous channel symbol errors results in isolated errors at the deinterlever output that are separated from each other by at least M symbols.

2. Any bN burst of errors, where $b > 1$, results in output bursts from the deinterleaver of no more than $\lceil b \rceil$ symbol errors. Each output burst is separated from the other bursts by no less than $M - \lfloor b \rfloor$ symbols. The notation $\lceil x \rceil$ means the smallest integer no less than x, and $\lfloor x \rfloor$ means the largest integer no greater than x.

3. A periodic sequence of single errors spaced N symbols apart results in a single burst of errors of length M at the deinterleaver output.

4. The interleaver/deinterleaver end-to-end delay is approximately $2MN$ symbol times. To be precise, only $M(N - 1) + 1$ memory cells need to be filled before transmission can begin (as soon as the first symbol of the last column of the $M \times N$ array is filled). A corresponding number needs to be filled at the receiver before decoding begins. Thus the minimum end-to-end delay is $(2MN - 2M + 2)$ symbol times, not including any channel propagation delay.

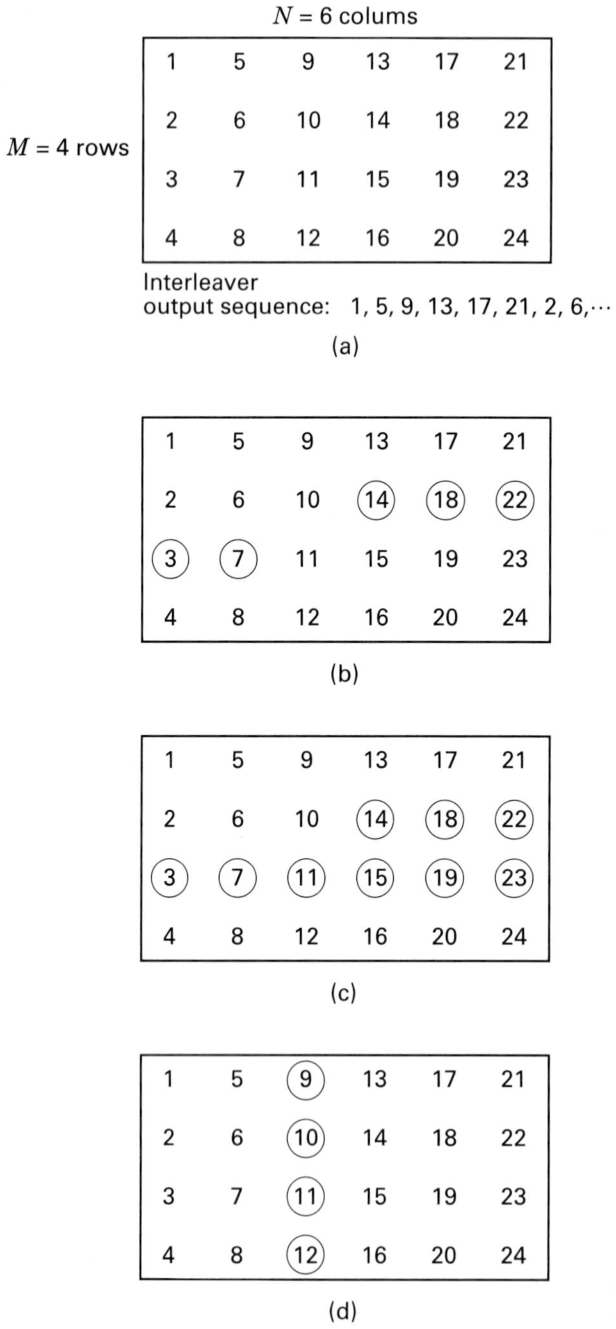

$N = 6$ colums

1	5	9	13	17	21
2	6	10	14	18	22
3	7	11	15	19	23
4	8	12	16	20	24

$M = 4$ rows

Interleaver
output sequence: 1, 5, 9, 13, 17, 21, 2, 6,···

(a)

(b)

(c)

(d)

Figure 8.11 Block interleaver example. (a) $M \times N$ block interleaver. (b) Five-symbol error burst. (c) Nine-symbol error burst. (d) Periodic single-error sequence spaced $N = 6$ symbols apart.

5. The memory requirement is MN symbols for each location (interleaver and deinterleaver). However, since the $M \times N$ array needs to be (mostly) filled before it can be read out, a memory of $2MN$ symbols is generally implemented at each location to allow the emptying of one $M \times N$ array while the other is being filled, and vice versa.

Example 8.4   Interleaver Characteristics

Using the $M = 4$, $N = 6$ interleaver structure of Figure 8.11a, verify each of the block interleaver characteristics described above.

Solution

1. Let there be a noise burst of five symbol times, such that the symbols shown encircled in Figure 8.11b experience errors in transmission. After deinterleaving at the receiver, the sequence is

   1   2   ③   4   5   6   ⑦   8   9   10   11   12

                    13   ⑭   15   16   17   ⑱   19   20   21   ㉒   23   24

   where the encircled symbols are in error. It is seen that the smallest separation between symbols in error is $M = 4$.

2. Let $b = 1.5$ so that $bN = 9$. Figure 8.11c illustrates an example of nine-symbol error burst. After deinterleaving at the receiver, the sequence is

   1   2   ③   4   5   6   ⑦   8   9   10   ⑪   12

                    13   ⑭   ⑮   16   17   ⑱   ⑲   20   21   ㉒   ㉓   24

   Again, the encircled symbols are in error. It is seen that the bursts consist of no more than $\lceil 1.5 \rceil = 2$ contiguous symbols and that they are separated by at least $M - \lfloor 1.5 \rfloor = 4 - 1 = 3$ symbols.

3. Figure 8.11d illustrates a sequence of single errors spaced by $N = 6$ symbols apart. After deinterleaving at the receiver, the sequence is

   1   2   3   4   5   6   7   8   ⑨   ⑩   ⑪   ⑫

                    13   14   15   16   17   18   19   20   21   22   23   24

   It is seen that the deinterleaved sequence has a singe error burst of length $M = 4$ symbols.

4. End-to-end delay: The minimum end-to-end delay due to the interleaver and deinterleaver is $(2MN - 2M + 2) = 42$ symbol times.

5. Memory requirement: The interleaver and the deinterleaver arrays are each of size $M \times N$. Therefore, storage for $MN = 24$ symbols is required at each end of the channel. As mentioned earlier, storage for $2MN = 48$ symbols would generally be implemented.

Typically, for use with a single-error-correcting code the interleaver parameters are selected such that the number of columns N overbounds the *expected burst length*. The choice of the number of rows M is dependent on the coding scheme used. For block codes, M should be larger than the code block length, while for convolutional codes, M should be larger than the constraint length. Thus a burst of length N can cause at most a single error in any block codeword; similarly, with convolutional codes, there will be at most a single error in any decoding constraint length. For t-error-correcting codes, the choice of N need only overbound the expected burst length divided by t.

8.2.2 Convolutional Interleaving

Convolutional interleavers have been proposed by Ramsey [7] and Forney [8]. The structure proposed by Forney appears in Figure 8.12. The code symbols are sequentially shifted into the bank of N registers; each successive register provides J symbols more storage than did the preceding one. The zeroth register provides no storage (the symbol is transmitted immediately). With each new code symbol the commutator switches to a new register, and the new code symbol is shifted in while the oldest code symbol in that register is shifted out to the modulator/transmitter. After the $(N-1)$th register, the commutator returns to the zeroth register and starts again. The deinterleaver performs the inverse operation, and the input and output commutators for both interleaving and deinterleaving must be synchronized.

Figure 8.13 illustrates an example of a simple convolutional four-register ($J = 1$) interleaver being loaded by a sequence of code symbols. The synchronized deinterleaver is shown simultaneously feeding the deinterleaved symbols to the decoder. Figure 8.13a shows symbols 1 to 4 being loaded; the ×s represent unknown states. Figure 8.13b shows the first four symbols shifted within the registers and the entry of symbols 5 to 8 to the interleaver input. Figure 8.13c shows symbols 9 to 12 entering the interleaver. The deinterleaver is now filled with message symbols, but nothing useful is being fed to the decoder yet. Finally, Figure 8.13d shows symbols

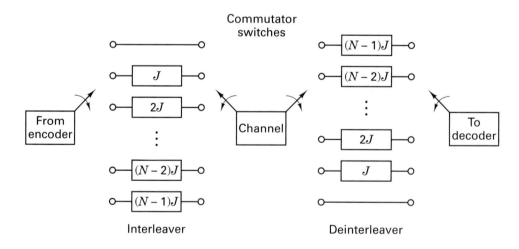

Figure 8.12 Shift register implementation of a convolutional interleaver/deinterleaver.

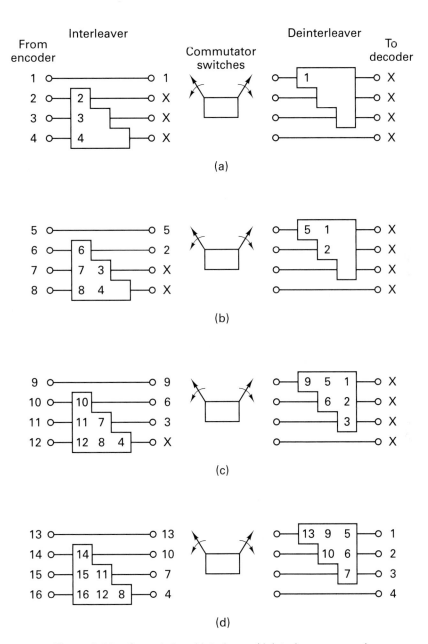

Interleaver

From encoder

Deinterleaver

Commutator switches

To decoder

(a)

(b)

(c)

(d)

Figure 8.13   Convolutional interleaver/deinterleaver example.

13 to 16 entering the interleaver, and at the output of the deinterleaver, symbols 1 to 4 are being passed to the decoder. The process continues in this way until the entire codeword sequence, in its original preinterleaved form, is presented to the decoder.

The performance of a convolutional interleaver is very similar to that of a block interleaver. The important advantage of convolutional over block interleaving is that with convolutional interleaving the end-to-end delay is $M(N-1)$ symbols, where $M = NJ$, and the memory required is $M(N-1)/2$ at both ends of the channel. Therefore, there is a reduction of one-half in delay and memory over the block interleaving requirements [9].

8.2.3 Concatenated Codes

A concatenated code is one that uses two levels of coding, an inner code and an outer code, to achieve the desired error performance. Figure 8.14 illustrates the order of encoding and decoding. The inner code, the one that interfaces with the modulator/demodulator and channel, is usually configured to correct most of the channel errors. The outer code, usually a higher-rate (lower-redundancy) code, then reduces the probability of error to the specified level. The primary reason for using a concatenated code is to achieve a low error rate with an overall implementation complexity which is less than that which would be required by a single coding operation. In Figure 8.14 an interleaver is shown between the two coding steps. This is usually required to spread any error bursts that may appear at the output of the inner coding operation.

One of the most popular concatenated coding systems uses a Viterbi-decoded convolutional inner code and a Reed–Solomon (R–S) outer code, with interleaving between the two coding steps [2]. Operation of such systems with E_b/N_0 in the range 2.0 to 2.5 dB to achieve $P_B = 10^{-5}$ is feasible with practical hardware [9]. In this system, the demodulator outputs soft quantized code symbols to the inner convolutional decoder, which in turn outputs hard quantized code symbols with bursty errors to the R–S decoder. (In a Viterbi-decoded system, the output errors tend to

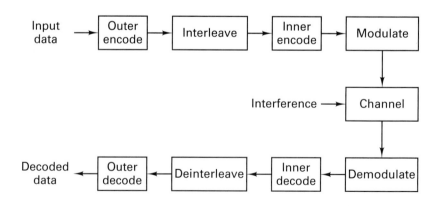

Figure 8.14 Block diagram of a concatenated coding system.

occur in bursts.) The outer R–S code is formed from m-bit segments of the binary data stream. The performance of such a (nonbinary) R–S code depends only on the number of *symbol errors* in the block. The code is undisturbed by burst errors within an m-bit symbol. That is, for a given symbol error, the R–S code performance is the same whether the symbol error is due to one bit being in error or m bits being in error. However, the concatenated system performance is severely degraded by correlated errors among successive symbols. Hence interleaving between codes at the symbol level (not at the bit level) needs to be provided. Reference [10] presents a review of concatenated codes that have been investigated for deep-space communications. In the next section we consider a popular consumer application of symbol interleaving in a concatenated system.

8.3 CODING AND INTERLEAVING APPLIED TO THE COMPACT DISC DIGITAL AUDIO SYSTEM

In 1979, Philips Corp. of the Netherlands and Sony Corp. of Japan defined a standard for the digital storage and reproduction of audio signals, known as the *compact disc (CD) digital audio system.* This CD system has become the world standard for achieving fidelity of sound reproduction that far surpasses any other available technique. A plastic disc 120 mm in diameter is used to store the digitized audio waveform. The waveform is sampled at 44.1 kilosamples/s to provide a recorded bandwidth of 20 kHZ; each audio sample is uniformly quantized to one of 2^{16} levels (16 bits/sample), resulting in a dynamic range of 96 dB and a total harmonic distortion of 0.005%. A single disc (playing time approximately 70 minutes) stores about 10^{10} bits in the form of minute *pits* that are optically scanned by a laser.

There are several sources of channel errors: (1) small unwanted particles or air bubbles in the plastic material or pit inaccuracies arising in manufacturing, and (2) fingerprints or scratches during handling. It is difficult to predict how, on the average, a CD will get damaged; but in the absence of an accurate channel model, it is safe to assume that the channel mainly has a *burstlike* error behavior, since a scratch or fingerprint will cause *several* consecutive data samples to be in error. An important aspect of the system design contributing to the high-fidelity performance is a concatenated error-control scheme called the *cross-interleave Reed–Solomon code (CIRC).* The data are rearranged in time so that digits stemming from contiguous samples of the waveform are *spread out in time.* In this way, error bursts are made to appear as single random events (see the earlier sections on interleaving). The digital information is protected by adding parity bytes derived in two Reed–Solomon (R–S) encoders. Error control applied to the compact disc depends mostly on R–S coding and multiple layers of interleaving.

In digital audio applications, an undetected decoding error is very serious since it results in clicks, while occasional *detected* failures are not so serious because they can be concealed. The CIRC error-control scheme in the CD system involves both *correction* and *concealment* of errors. The performance specifications for the CIRC are given in Table 8.4. From the specifications in the table it would appear

TABLE 8.4   Specifications for the CD Cross-Interleave Reed–Solomon Code

Maximum correctable burst length	≈ 4000 bits (2.5-mm track length on the disc)
Maximum interpolatable burst length	$\approx 12,000$ bits (8 mm)
Sample interpolation rate	One sample every 10 hours at $P_B = 10^{-4}$
	1000 samples/min at $P_B = 10^{-3}$
Undetected error samples (clicks)	Less than one every 750 hours at $P_B = 10^{-3}$
	Negligible at $P_B \leq 10^{-4}$
New discs are characterized by	$P_B \approx 10^{-4}$

that the CD can endure much damage (e.g., 8-mm holes punched in the disc) without any noticeable effect on the sound quality.

The CIRC system achieves its error control by a hierarchy of the following techniques:

1. The decoder provides a level of error correction.
2. If the error correction capability is exceeded, the decoder provides a level of erasure correction (see Section 6.5.5).
3. If the erasure correction capability is exceeded, the decoder attempts to conceal unreliable data samples by *interpolating* between reliable neighboring samples.
4. If the interpolation capability is exceeded, the decoder blanks out or *mutes* the system for the duration of the unreliable samples.

8.3.1 Circ Encoding

Figure 8.15 illustrates the basic CIRC encoder block diagram (within the CD recording equipment) and the decoder block diagram (within the CD player equipment). Encoding consists of the encoding and interleaving steps designated as Δ interleave, C_2 encode, D^* interleave, C_1 encode, and D interleave. The decoder steps, consisting of deinterleaving and decoding, are preformed in the *reverse* order of the encoding steps and are designated as D deinterleave, C_1 decode, D^* deinterleave, C_2 decode, and Δ deinterleave.

Figure 8.16 illustrates the basic system frame time, comprising six sampling periods, each made up of a stereo sample pair (16-bit left sample and 16-bit right sample). The bits are organized into symbols or bytes of 8 bits each. Therefore, each sample pair contains 4 bytes, and the uncoded frame contains $k = 24$ bytes. Figure 8.16a–e summarizes the *five encoding steps* that characterize the CIRC system. The function of each of these steps will best be understood when we consider the decoding operation. The steps are as follows:

(a) Δ *interleave.* Even-numbered samples are separated from odd-numbered samples by two frame times in order to scramble uncorrectable but detectable byte errors. This facilitates the interpolation process.

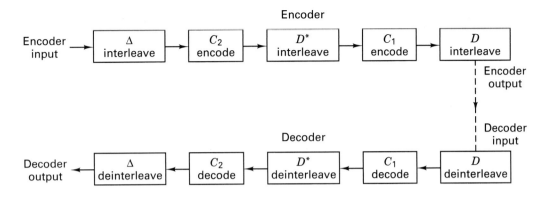

Figure 8.15 CIRC encoder and decoder.

(b) C_2 *encode.* Four Reed–Solomon (R–S) parity bytes are added to the Δ-interleaved 24-byte frame, resulting in a total of $n = 28$ bytes. This (28, 24) code is called the *outer code.*

(c) D^* *interleave.* Here each byte is delayed a different length, thereby spreading errors over several codewords. C_2 encoding together with D^* interleaving have the function of providing for the correction of burst errors and error patterns that the C_1 decoder cannot correct.

(d) C_1 *encode.* Four R-S parity bytes are added to the $k = 28$ bytes of the D^*-interleaved frame, resulting in a total of $n = 32$ bytes. This (32, 28) code is called the *inner code.*

(e) D *interleave.* The purpose is to *cross-interleave* the *even bytes* of a frame with the *odd bytes* of the next frame. By this procedure, two consecutive bytes on the disc will always end up in two different codewords. Upon decoding, this interleaving, together with the C_1 decoding, results in the correction of most random single errors and the detection of longer burst errors.

8.3.1.1 Shortening the R-S Code

In Section 8.1 an (n, k) R–S code is expressed in terms of $n = 2^m - 1$ total symbols and $k = 2^m - 1 - 2t$ data symbols, where m is the number of bits per symbol and t is the error-correcting capability of the code in symbols. For the CD system, where a symbol is made up of 8 bits, a 2-symbol error-correcting code can be configured as a (255, 251) code. However, the CD system uses a considerably shorter block length. Any block code (in systematic form) can be shortened without affecting the number of errors that can be corrected within a block length. In terms of the (255, 251) R–S code, imagine that 227 of the 251 data symbols are a set of all-zero symbols (which are not actually transmitted and hence are not subject to any errors). Then the code is really a (28, 24) code with the same 2-symbol error-correcting capability. This is what is done in the C_2 encoder of the CD system.

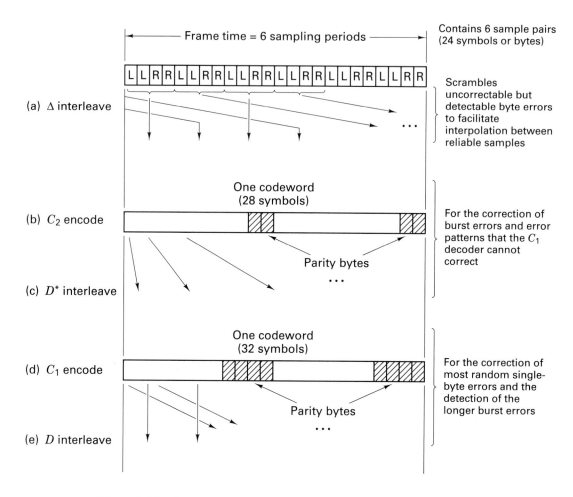

Figure 8.16 Compact disc encoder. (a) Δ interleave. (b) C_2 encode. (c) D^* interleave. (d) C_1 encode. (e) D interleave.

We can think of the 28 total symbols out of the C_2 encoder as the data symbols into the C_1 encoder. Again, we can configure a shortened 2-symbol error-correcting (255, 251) code by throwing away 223 data symbols—the result being a (32, 28) code.

8.3.2 CIRC Decoding

The inner and outer R–S codes with (n, k) values (32, 28) and 28, 24) each use four parity bytes. The code rate of the CIRC is $(k_1/n_1)(k_2/n_2) = 24/32 = 3/4$. From Equation (8.3) the minimum distance of the C_1 and C_2 R-S codes is $d_{\min} = n - k + 1 = 5$. From Equations (8.4) and (8.5)

$$t \leq \left\lfloor \frac{d_{\min} - 1}{2} \right\rfloor = \left\lfloor \frac{n - k}{2} \right\rfloor \tag{8.58}$$

and

$$\rho \leq d_{\min} - 1 \tag{8.59}$$

where t is the error-correcting capability and ρ is the erasure-correcting capability, it is seen that the C_1 or C_2 decoder can correct a maximum of 2 symbol errors or 4 symbol erasures per codeword. Or, as described by Equation (8.6), it is possible to correct any pattern of α errors and γ erasures simultaneously, provided that

$$2\alpha + \gamma < d_{\min} < n - k \tag{8.60}$$

There is a trade-off between error correction and erasure correction; the larger the error correcting capability used, the smaller will be the erasure correcting capability.

The benefits of CIRC are best seen at the *decoder,* where the processing steps, shown in Figure 8.17 are in the reverse order of the encoder steps. The decoder steps are as follows:

1. *D deinterleave.* This function is performed by the alternating delay lines marked D. The 32 bytes $(B_{i1}, \ldots, B_{i32})$ of an encoded frame are applied in

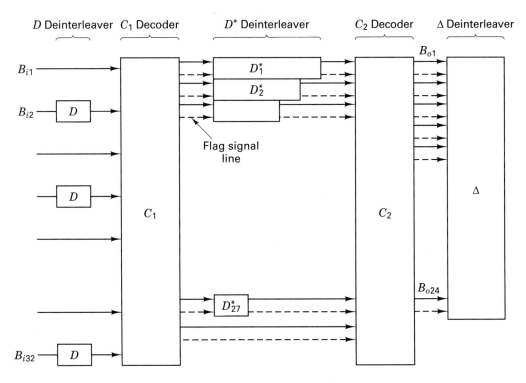

Figure 8.17 Compact disc decoder.

parallel to the 32 inputs of the D deinterleaver. Each delay is equal to the duration of 1 byte, so that the information of the *even bytes* of a frame is cross-deinterleaved with that of the *odd bytes* of the next frame.

2. **C_1 decode.** The D deinterleaver and the C_1 decoder are designed to correct a single byte error in the block of 32 bytes and to detect larger burst errors. If multiple errors occur, the C_1 decoder passes them on unchanged, attaching to all 28 remaining bytes an erasure flag, sent via the dashed lines (the four parity bytes used in the C_1 decoder are no longer retained).

3. **D^* deinterleave.** Due to the different lengths of the deinterleaving delay lines $D^*(1, \ldots, 27)$, errors that occur in one word at the output of the C_1 decoder are *spread over a number of words* at the input of the C_2 decoder. This results in reducing the number of errors per input word of the C_2 decoder, enabling the C_2 decoder to correct these errors.

4. **C_2 decode.** The C_2 decoder is intended for the correction of burst errors that the C_1 decoder could not correct. If the C_2 decoder cannot correct these errors, the 24-byte codeword is passed on unchanged to the Δ deinterleaver and the associated positions are given an *erasure flag* via the dashed output lines, $B_{o1}, \ldots, B_{o24}$.

5. **Δ deinterleave.** The final operation deinterleaves uncorrectable but detected byte errors in such a way that *interpolation* can be used between reliable neighboring samples.

Figure 8.18 highlights the decoder steps 2, 3, and 4. At the output of the C_1 decoder is seen a sequence of four 28-byte codewords that have exceeded the 1 byte per codeword error correction design. Therefore, each of the symbols in these codewords is tagged with an erasure flag (shown with circles). The D^* deinterleaver provides a staggered delay for each byte of a codeword, so that the bytes of a given codeword arrive in different codewords at the input to the C_2 decoder. If we assume that the delay increments of the D^* deinterleaver in Figure 8.18 are 1 byte, it would be possible to correct error bursts of as many as four consecutive C_1 codewords (since the C_2 decoder is capable of four erasure corrections per codeword). In the actual CD system, the delay increments are 4 bytes; therefore, the maximum burst error correction capability consists of 16 consecutive uncorrectable C_1 words.

8.3.3 Interpolation and Muting

Samples that cannot be corrected by the C_2 decoder could cause audible disturbances. The function of the *interpolation* process is to insert new samples, estimated from reliable neighbors, in place of the unreliable ones. If an entire C_2 word is detected as unreliable, this would make it impossible to apply interpolation without additional interleaving, since both even- and odd-numbered samples are unreliable. This can happen if the C_1 decoder fails to detect an error but the C_2 decoder detects it. It is the purpose of Δ deinterleaving (over a span of two frame times) to obtain a pattern where even-numbered samples can be interpolated from reliable odd-numbered samples, or vice versa.

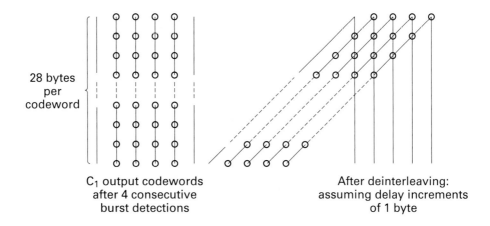

28 bytes
per
codeword

C$_1$ output codewords
after 4 consecutive
burst detections

After deinterleaving:
assuming delay increments
of 1 byte

Figure 8.18 Example of 4-byte erasure capability. (Rightmost event is at the earliest time.)

Two successive unreliable words consisting of 12 sample pairs are shown in Figure 8.19. A sample pair consists of a sample (2 bytes) from the right audio channel and a sample from the left audio channel. The numbers indicate the ordering of the sets of samples. An encircled sample set denotes an *erasure* flag. After Δ deinterleaving, the unreliable samples shown in the figure are estimated by a first-order linear interpolation between neighboring samples that stem from a different location on the disc.

In CD players, another level of error control is provided in case a burst length of 48 frames is exceeded and 2 or more consecutive unreliable samples result. In this case the system is *muted* (audio is softly blanked out), which is not discernible to the human ear if the muting time does not exceed a few milliseconds. For a more detailed treatment of the CIRC coding scheme in the CD system, see References [11–15].

8.4 TURBO CODES

Concatenated coding schemes were first proposed by Forney [16} as a method for achieving large coding gains by combining two or more relatively simple building-block or *component codes* (sometimes called *constituent codes*). The resulting codes had the error-correction capability of much longer codes, and they were endowed with a structure that permitted relatively easy to moderately complex decoding. A serial concatenation of codes is most often used for power-limited systems such as transmitters on deep-space probes. The most popular of these schemes consists of a Reed–Solomon outer (applied first, removed last) code followed by a convolutional inner (applied last, removed first) code [10]. A turbo code can be thought of as a refinement of the concatenated encoding structure plus an iterative algorithm for decoding the associated code sequence. Because of

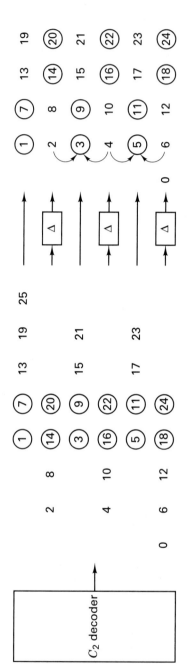

Figure 8.19 Effect of interleaving. (Rightmost event is at the earliest time.)

Channel Coding: Part 3   Chap. 8

its unique iterative form, we choose to list *turbo* as a separate category under structured sequences in Figure 1.3.

Turbo codes were first introduced in 1993 by Berrou, Glavieux, and Thitimajshima, and reported in [17, 18], where a scheme is described that achieves a bit-error-probability of 10^{-5}, using a rate 1/2 code over an additive white Gaussian noise (AWGN) channel and BPSK modulation at an E_b/N_0 of 0.7 dB. The codes are constructed by using two or more component codes on different interleaved versions of the same information sequence. Whereas, for conventional codes, the final step at the decoder yields hard-decision decoded bits (or more generally, decoded symbols), for a concatenated scheme, such as a turbo code, to work properly, the decoding algorithm should not limit itself to passing hard-decisions among the decoders. To best exploit the information learned from each decoder, the decoding algorithm must effect an exchange of soft decisions rather than hard decisions. For a system with two component codes, the concept behind turbo decoding is to pass soft decisions from the output of one decoder to the input of the other decoder, and to iterate this process several times so as to produce more reliable decisions.

8.4.1 Turbo Code Concepts

8.4.1.1 Likelihood Functions

The mathematical foundations of hypothesis testing rests on Bayes' theorem, which is developed in Appendix B. For communications engineering, where applications involving an AWGN channel are of great interest, the most useful form of Bayes' theorem expresses the a posteriori probability (APP) of a decision in terms of a continuous-valued random variable x as

$$P(d = i|x) = \frac{p(x|d = i)\,P(d = i)}{p(x)} \quad i = 1, \cdots, M \qquad (8.61)$$

and

$$p(x) = \sum_{i=1}^{M} p(x|d = i)\,P(d = i) \qquad (8.62)$$

where $P(d = i|x)$ is the APP, and $d = i$ represents data d belonging to the ith signal class from a set of M classes. Further, $p(x|d = i)$ represents the probability density function (pdf) of a received continuous-valued data-plus-noise signal x, conditioned on the signal class $d = i$. Also, $p(d = i)$, called the a priori probability, is the probability of occurrence of the ith signal class. Typically x is an "observable" random variable or a test statistic that is obtained at the output of a demodulator or some other signal processor. Therefore, $p(x)$ is the pdf of the received signal x, yielding the test statistic over the entire space of signal classes. In Equation (8.61), for a particular observation, $p(x)$ is a scaling factor since it is obtained by averaging over all the classes in the space. Lower case p is used to designate the pdf of a continuous-valued random variable, and upper case P is used to designate probability (a priori and APP). Determining the APP of a received signal from Equation (8.61) can be thought of as the result of an experiment. Before the experiment, there generally exists (or one can estimate) an a priori probability $P(d = i)$. The experiment consists of using Equation (8.61) for computing the APP, $P(d = i|x)$, which can be thought of as a "refinement" of the prior knowledge about the data, brought about by examining the received signal x.

8.4.1.2 The Two-Signal Class Case

Let the binary logical elements 1 and 0 be represented electronically by voltages +1 and −1, respectively. The variable d is used to represent the transmitted data bit, whether it appears as a voltage or as a logical element. Sometimes one format is more convenient than the other; the reader should be able to recognize the difference from the context. Let the binary 0 (or the voltage value −1) be the null element under addition. For signal transmission over an AWGN channel, Figure 8.20 shows the conditional pdfs, referred to as likelihood functions. The rightmost function $p(x|d = +1)$ shows the pdf of the random variable x conditioned on $d = +1$ being transmitted. The leftmost function $p(x|d = -1)$ illustrates a similar pdf conditioned on $d = -1$ being transmitted. The abscissa represents the full range of possible values of the test statistic x generated at the receiver. In Figure 8.20, one such arbitrary value x_k is shown, where the index denotes an observation in the kth time interval. A line subtended from x_k intercepts the two likelihood functions yielding two likelihood values $\ell_1 = p(x_k|d_k = +1)$ and $\ell_2 = p(x_k|d_k = -1)$. A well-known hard-decision rule, known as *maximum likelihood*, is to choose the data $d_k = +1$ or $d_k = -1$ associated with the larger of the two intercept values ℓ_1 or ℓ_2, respectively. For each data bit at time k, this is tantamount to deciding that $d_k = +1$ if x_k falls on the right side of the decision line labeled γ_0, otherwise deciding that $d_k = -1$.

A similar decision rule, known as *maximum a posteriori* (MAP), which can be shown to be a *minimum-probability-of-error rule,* takes into account the a priori probabilities of the data. The general expression for the MAP rule in terms of APPs is

$$P(d = +1|x) \underset{H_2}{\overset{H_1}{\gtrless}} P(d = -1|x) \tag{8.63}$$

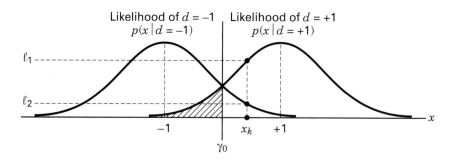

Figure 8.20 Likelihood functions.

Equation (8.63) states that one should choose the hypothesis H_1, $(d = +1)$ if the APP, $P(d = +1 | x)$, is greater than the APP, $P(d = -1 | x)$. Otherwise, one should choose hypothesis H_2, $(d = -1)$. Using the Bayes' theorem of Equation (8.61), the APPs in Equation (8.63) can be replaced by their equivalent expressions, yielding

$$p(x | d = +1) \, P(d = +1) \underset{H_2}{\overset{H_1}{\gtrless}} p(x | d = -1) \, P(d = -1) \qquad (8.64)$$

where the pdf $p(x)$ appearing on both sides of the inequality in Equation (8.61) has been canceled. Equation (8.64) is generally expressed in terms of a ratio, yielding the so-called *likelihood ratio test,* as follows:

$$\frac{p(x | d = +1)}{p(x | d = -1)} \underset{H_2}{\overset{H_1}{\gtrless}} \frac{P(d = -1)}{P(d = +1)} \quad \text{or} \quad \frac{p(x | d = +1) P(d = +1)}{p(x | d = -1) P(d = -1)} \underset{H_2}{\overset{H_1}{\gtrless}} 1 \qquad (8.65)$$

8.4.1.3 Log-Likelihood Ratio

By taking the logarithm of the likelihood ratio developed in Equations (8.63) through (8.65), we obtain a useful metric called the log-likelihood ratio (LLR). It is a real number representing a soft decision out of a detector, designated by

$$L(d | x) = \log \left[\frac{P(d = +1 | x)}{P(d = -1 | x)} \right] = \log \left[\frac{p(x | d = +1) \, P(d = +1)}{p(x | d = -1) \, P(d = -1)} \right] \qquad (8.66)$$

so that

$$L(d | x) = \log \left[\frac{p(x | d = +1)}{p(x | d = -1)} \right] + \log \left[\frac{P(d = +1)}{P(d = -1)} \right] \qquad (8.67)$$

or

$$L(d | x) = L(x | d) + L(d) \qquad (8.68)$$

where $L(x | d)$ is the LLR of the test statistic x obtained by measurements of the channel output x under the alternate conditions that $d = +1$ or $d = -1$ may have been transmitted, and $L(d)$ is the a priori LLR of the data bit d. To simplify the notation, Equation (8.68) is rewritten as

$$L'(\hat{d}) = L_c(x) + L(d) \qquad (8.69)$$

where the notation $L_c(x)$ emphasizes that this LLR term is the result of a channel measurement made at the receiver. Equations (8.61) through (8.69) were developed with only a data detector in mind. Next, the introduction of a decoder will typically yield decision-making benefits. For a systematic code, it can be shown [17] that the LLR (soft output) out of the decoder is equal to

$$L(\hat{d}) = L'(\hat{d}) + L_e(\hat{d}) \qquad (8.70)$$

where $L'(\hat{d})$ is the LLR of a data bit out of the demodulator (input to the decoder), and $L_e(\hat{d})$, called the *extrinsic* LLR, represents extra knowledge that is gleaned from the decoding process. The output sequence of a systematic decoder is made up of values representing data bits and parity bits. From Equations (8.69) and (8.70), the output LLR of the decoder is now written as

$$L(\hat{d}) = L_c(x) + L(d) + L_e(\hat{d}) \qquad (8.71)$$

Equation (8.71) shows that the output LLR of a systematic decoder can be represented as having three LLR elements—a channel measurement, a priori knowledge of the data, and an extrinsic LLR stemming solely from the decoder. To yield the final $L(\hat{d})$, each of the individual LLRs can be added as shown in Equation (8.71), because the three terms are statistically independent [17, 19]. The proof is left as an exercise for the reader. (See Problem 8.18.) This soft decoder output $L(\hat{d})$ is a real number that provides a hard decision as well as the reliability of that decision. The sign of $L(\hat{d})$ denotes the hard decision—that is, for positive values of $L(\hat{d})$ decide that $d = +1$, and for negative values that $d = -1$. The magnitude of $L(\hat{d})$ denotes the reliability of that decision. Often the value of $L_e(\hat{d})$ due to the decoding has the same sign as $L_c(x) + L(d)$ and therefore acts to improve the reliability of $L(\hat{d})$.

8.4.1.4 Principles of Iterative (Turbo) Decoding

In a typical communications receiver, a demodulator is often designed to produce soft decisions which are then transferred to a decoder. In Chapter 7, the error-performance improvement of systems utilizing such soft decisions compared with hard decisions were quantified as being approximately 2 dB in AWGN. Such a decoder could be called a soft-input/hard-output decoder, because the final decoding process out of the decoder must terminate in bits (hard decisions). With turbo codes, where two or more component codes are used, and decoding involves feeding outputs from one decoder to the inputs of other decoders in an iterative fashion, a hard-output decoder would not be suitable. That is because hard decisions into a decoder degrades system performance (compared with soft decisions). Hence, what is needed for the decoding of turbo codes is a *soft-input/soft-output* decoder. For the first decoding iteration of such a soft-input/soft-output decoder, illustrated in Figure 8.21, one generally assumes the binary data to be equally likely, yielding an initial a priori LLR value of $L(d) = 0$ for the third term in Equation (8.67). The channel LLR value $L_c(x)$ is measured by forming the logarithm of the ratio of the values of ℓ_1 and ℓ_2 for a particular observation of x (see Figure 8.20), which appears as the second term in Equation (8.67). The output $L(\hat{d})$ of the decoder in Figure 8.21 is made up of the LLR from the detector $L'(\hat{d})$ and the extrinsic LLR output $L_e(\hat{d})$, representing knowledge gleaned from the decoding process. As illustrated in Figure 8.21, for iterative decoding, the extrinsic likelihood is fed back to the input (of another component decoder) to serve as a refinement of the a-priori probability of the data for the next iteration.

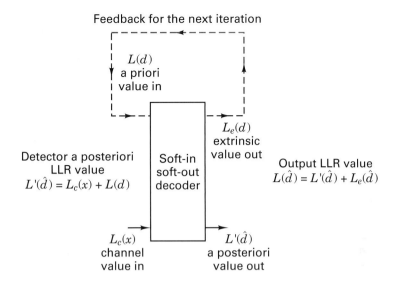

Feedback for the next iteration

$L(d)$
a priori
value in

$L_e(d)$
extrinsic
value out

Detector a posteriori
LLR value
$L'(\hat{d}) = L_c(x) + L(d)$

Soft-in
soft-out
decoder

Output LLR value
$L(\hat{d}) = L'(\hat{d}) + L_e(\hat{d})$

$L_c(x)$
channel
value in

$L'(\hat{d})$
a posteriori
value out

Figure 8.21   Soft input/soft output decoder (for a systematic code).

8.4.2 Log-Likelihood Algebra

To best explain the iterative feedback of soft decoder outputs, the concept of a log-likelihood algebra [19] is introduced. For statistically independent data d, the sum of two log likelihood ratios (LLRs) is defined as

$$L(d_1) \boxplus L(d_2) \triangleq L(d_1 \oplus d_2) = \log_e \left[\frac{e^{L(d_1)} + e^{L(d_2)}}{1 + e^{L(d_1)} e^{L(d_2)}} \right] \qquad (8.72)$$

$$\approx (-1) \times \mathrm{sgn}\,[L(d_1)] \times \mathrm{sgn}\,[L(d_2)] \times \min\,(|L(d_1)|, |L(d_2)|) \qquad (8.73)$$

where the natural logarithm is used, and the function sgn $(\cdot)$ represents the "polarity of." There are three addition operations in Equation (8.72). The + sign is used for ordinary addition. The $\oplus$ sign is used to denote the modulo-2 sum of data expressed as binary digits. The $\boxplus$ sign denotes log-likelihood addition, or equivalently, the mathematical operation described by Equation (8.72). The sum of two LLRs denoted by the operator $\boxplus$ is defined as the LLR of the modulo-2 sum of the underlying statistically independent data bits. The development of Equation (8.72) is shown in Appendix 8A. Equation (8.73) is an approximation of Equation (8.72) that will prove useful later in a numerical example. The sum of LLRs, as described by Equations (8.72) or (8.73), yields the following interesting results when one of the LLRs is very large or very small:

$$L(d) \boxplus \infty = -L(d)$$

and

$$L(d) \boxplus 0 = 0$$

Note that the log-likelihood algebra described here differs slightly from that used in [19] because of a different choice of the null element. In this treatment, the null element of the binary set $(1, 0)$ has been chosen to be 0.

8.4.3 Product Code Example

Consider the 2-dimensional code (product code) depicted in Figure 8.22. The configuration can be described as a data array made up of k_1 rows and k_2 columns. The k_1 rows contain codewords made up of k_2 data bits and $n_2 - k_2$ parity bits. Thus each of the k_1 rows represents a codeword from an (n_2, k_2) code. Similarly, the k_2 columns contain codewords made up of k_1 data bits and $n_1 - k_1$ parity bits. Thus, each of the k_2 columns represents a codeword from an (n_1, k_1) code. The various portions of the structure are labeled d for data, p_h for horizontal parity (along the rows), and p_v for vertical parity (along the columns). In effect, the block of $k_1 \times k_2$ data bits is encoded with two codes—a horizontal code, and a vertical code.

Additionally, in Figure 8.22, there are blocks labeled L_{eh} and L_{ev} containing the extrinsic LLR values learned from the horizontal and vertical decoding steps, respectively. Error-correction codes generally provide some improved performance. We will see that the extrinsic LLRs represent a measure of that improvement. Notice that this product code is a simple example of a concatenated code. Its structure encompasses two separate encoding steps—horizontal and vertical.

Recall that the final decoding decision for each bit and its reliability hinges on the value of $L(\hat{d})$, as shown in Equation (8.71). With this equation in mind, an algorithm yielding the extrinsic LLRs (horizontal and vertical) and a final $L(\hat{d})$ can be described. For the product code, this iterative decoding algorithm proceeds as follows:

1. Set the a-priori LLR $L(d) = 0$ (unless the a priori probabilities of the data bits are other than equally likely).

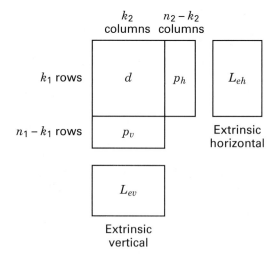

Figure 8.22   Product code.

2. Decode horizontally, and, using Equation (8.71), obtain the horizontal extrinsic LLR

$$L_{eh}(\hat{d}) = L(\hat{d}) - L_c(x) - L(d)$$

3. Set $L(d) = L_{eh}(\hat{d})$ for the vertical decoding of step 4.

4. Decode vertically, and, using Equation (8.71), obtain the vertical extrinsic LLR

$$L_{ev}(\hat{d}) = L(\hat{d}) - L_c(x) - L(d)$$

5. Set $L(d) = L_{ev}(\hat{d})$ for the step 2 horizontal decoding. Then repeat steps 2 through 5.

6. After enough iterations (i.e., repetitions of steps 2 through 5) to yield a reliable decision, go to step 7.

7. The soft output is

$$L(\hat{d}) = L_c(x) + L_{eh}(\hat{d}) + L_{ev}(\hat{d}) \qquad (8.74)$$

An example is next used to demonstrate the application of this algorithm to a very simple product code.

8.4.3.1 Two-Dimensional Single-Parity Code Example

At the encoder, let the data bits and parity bits take on the values shown in Figure 8.23a, where the relationships between data and parity bits within a particular row (or column) expressed as the binary digits (1, 0) are

$$d_i \oplus d_j = p_{ij} \qquad (8.75)$$

and

$$d_i = d_j \oplus p_{ij} \qquad i,j \in \{(1, 2), \quad (3, 4), \quad (1, 3), \quad (2, 4)\} \qquad (8.76)$$

in which $\oplus$ denotes modulo-2 addition. The transmitted bits are represented by the sequence d_1, d_2, d_3, d_4, p_{12}, p_{34}, p_{13}, p_{24}. At the receiver input, the noise-corrupted bits are represented by the sequence $\{x_i\}$, $\{x_{ij}\}$, where $x_i = d_i + n$ for each received data bit, $x_{ij} = p_{ij} + n$ for each received parity bit, and n represents the noise contribution that is statistically independent for both d_i and p_{ij}. The indices i and j represent position in the encoder output array shown in Figure 8.23a. However, it is often more useful to denote the received sequence as $\{x_k\}$, where k is a time index. Both conventions will be followed below—using i and j when focusing on the positional relationships within the product code, and using k when focusing on the more general aspect of a time-related signal. The distinction as to which convention is being used should be clear from the context. Using the relationships developed in Equations (8.67) through (8.69), and assuming an AWGN interference model, the LLR for the channel measurement of a signal x_k received at time k, is written

$$L_c(x_k) = \log_e \left[\frac{p(x_k \mid d_k = +1)}{p(x_k \mid d_k = -1)} \right] \qquad (8.77a)$$

$d_1 = 1$	$d_2 = 0$	$p_{12} = 1$
$d_3 = 0$	$d_4 = 1$	$p_{34} = 1$
$p_{13} = 1$	$p_{24} = 1$	

(a) Encoder output binary digits

$L_c(x_1) = 1.5$	$L_c(x_2) = 0.1$	$L_c(x_{12}) = 2.5$
$L_c(x_3) = 0.2$	$L_c(x_4) = 0.3$	$L_c(x_{34}) = 2.0$
$L_c(x_{13}) = 6.0$	$L_c(x_{24}) = 1.0$	

(b) Decoder input log-likelihood ratios $L_c(x)$ **Figure 8.23**   Product code example.

$$= \log_e \left(\frac{\frac{1}{\sigma\sqrt{2\pi}} \exp\left[-\frac{1}{2}\left(\frac{x_k - 1}{\sigma}\right)^2\right]}{\frac{1}{\sigma\sqrt{2\pi}} \exp\left[-\frac{1}{2}\left(\frac{x_k + 1}{\sigma}\right)^2\right]} \right) \tag{8.77b}$$

$$= -\frac{1}{2}\left(\frac{x_k - 1}{\sigma}\right)^2 + \frac{1}{2}\left(\frac{x_k + 1}{\sigma}\right)^2 = \frac{2}{\sigma^2} x_k \tag{8.77c}$$

where the natural logarithm is used. If a further simplifying assumption is made that the noise variance σ^2 is unity, then

$$L_c(x_k) = 2x_k \tag{8.78}$$

Consider the following example, where the data sequence d_1, d_2, d_3, d_4 is made up of the binary digits 1 0 0 1, as shown in Figure 8.23a. By the use of Equation (8.75), it is seen that the parity sequence $p_{12}, p_{34}, p_{13}, p_{24}$ must be equal to the digits 1 1 1 1. Thus, the transmitted sequence is

$$\{d_i\}, \{p_{ij}\} = 1\ 0\ 0\ 1\ 1\ 1\ 1\ 1 \tag{8.79}$$

When the data bits are expressed as bipolar voltage values of +1 and −1 corresponding to the binary logic levels 1 and 0, the transmitted sequence is

$$\{d_i\}, \{p_{ij}\} = +1\ -1\ -1\ +1\ +1\ +1\ +1\ +1$$

Assume now that the noise transforms this data-plus-parity sequence into the received sequence

$$\{x_i\}, \{x_{ij}\} = 0.75,\ 0.05,\ 0.10,\ 0.15,\ 1.25,\ 1.0,\ 3.0,\ 0.5 \tag{8.80}$$

where the members of $\{x_i\}, \{x_{ij}\}$ positionally correspond to the data and parity $\{d_i\}$, $\{p_{ij}\}$ that was transmitted. Thus, in terms of the positional subscripts, the received sequence can be denoted as

$$\{x_i\}, \{x_{ij}\} = x_1, x_2, x_3, x_4, x_{12}, x_{34}, x_{13}, x_{24}$$

From Equation (8.78), the assumed channel measurements yield the LLR values

$$\{L_c(x_i)\}, \{L_c(x_{ij})\} = 1.5, 0.1, 0.20, 0.3, 2.5, 2.0, 6.0, 1.0 \qquad (8.81)$$

These values are shown in Figure 8.23b as the decoder input measurements. It should be noted that, given equal prior probabilities for the transmitted data, if hard decisions are made based on the $\{x_k\}$ or the $\{L_c(x_k)\}$ values shown above, such a process would result in two errors, since d_2 and d_3 would each be incorrectly classified as binary 1.

8.4.3.2 Extrinsic Likelihoods

For the product-code example in Figure 8.23, we use Equation (8.71) to express the soft output for the received signal corresponding to data d_1 as

$$L(\hat{d}_1) = L_c(x_1) + L(d_1) + \{[L_c(x_2) + L(d_2)] \boxplus L_c(x_{12})\} \qquad (8.82)$$

where the terms $\{[L_c(x_2) + L(d_2)] \boxplus L_c(x_{12})\}$ represent the extrinsic LLR contributed by the code (i.e., the reception corresponding to data d_2 and its a priori probability, in conjunction with the reception corresponding to parity p_{12}). In general the soft output $L(\hat{d}_i)$ for the received signal corresponding to data d_i is

$$L(\hat{d}_i) = L_c(x_i) + L(d_i) + \{[L_c(x_j) + L(d_j)] \boxplus L_c(x_{ij})\} \qquad (8.83)$$

where $L_c(x_i)$, $L_c(x_j)$, and $L_c(x_{ij})$ are the channel LLR measurements of the reception corresponding to d_i, d_j, and p_{ij}, respectively. $L(d_i)$ and $L(d_j)$ are the LLRs of the a priori probabilities of d_i and d_j respectively, and $\{[L_c(x_j) + L(d_j)] \boxplus L_c(x_{ij})\}$ is the extrinsic LLR contribution from the code. Equations (8.82) and (8.83) can best be understood in the context of Figure 8.23b. For this example, assuming equally likely signaling, the soft output $L(\hat{d}_1)$ is represented by the detector LLR measurement of $L_c(x_1) = 1.5$ for the reception corresponding to data d_1, plus the extrinsic LLR of $[L_c(x_2) = 0.1] \boxplus [L_c(x_{12}) = 2.5]$ gleaned from the fact that the data d_2 and the parity p_{12} also provide knowledge about the data d_1 as seen from Equations (8.75) and (8.76).

8.4.3.3 Computing the Extrinsic Likelihoods

For the example in Figure 8.23, the horizontal calculations for $L_{eh}(\hat{d})$ and the vertical calculations for $L_{ev}(\hat{d})$ are expressed as follows:

$$L_{eh}(\hat{d}_1) = [L_c(x_2) + L(d_2)] \boxplus L_c(x_{12}) \qquad (8.84a)$$

$$L_{ev}(\hat{d}_1) = [L_c(x_3) + L(d_3)] \boxplus L_c(x_{13}) \qquad (8.84b)$$

$$L_{eh}(\hat{d}_2) = [L_c(x_1) + L(d_1)] \boxplus L_c(x_{12}) \qquad (8.85a)$$

$$L_{ev}(\hat{d}_2) = [L_c(x_4) + L(d_4)] \boxplus L_c(x_{24}) \qquad (8.85b)$$

$$L_{eh}(\hat{d}_3) = [L_c(x_4) + L(d_4)] \boxplus L_c(x_{34}) \qquad (8.86a)$$

$$L_{ev}(\hat{d}_3) = [L_c(x_1) + L(d_1)] \boxplus L_c(x_{13}) \qquad (8.86b)$$

$$L_{eh}(\hat{d}_4) = [L_c(x_3) + L(d_3)] \boxplus L_c(x_{34}) \qquad (8.87a)$$

$$L_{ev}(\hat{d}_4) = [L_c(x_2) + L(d_2)] \boxplus L_c(x_{24}) \qquad (8.87b)$$

The LLR values shown in Figure 8.23 are entered into the $L_{eh}(\hat{d})$ expressions in Equations (8.84) through (8.87), and, assuming equally likely signaling, the $L(d)$ values are initially set equal to zero, yielding

$$L_{eh}(\hat{d}_1) = (0.1 + 0) \boxplus 2.5 \approx -0.1 = \text{new } L(d_1) \qquad (8.88)$$

$$L_{eh}(\hat{d}_2) = (1.5 + 0) \boxplus 2.5 \approx -1.5 = \text{new } L(d_2) \qquad (8.89)$$

$$L_{eh}(\hat{d}_3) = (0.3 + 0) \boxplus 2.0 \approx -0.3 = \text{new } L(d_3) \qquad (8.90)$$

$$L_{eh}(\hat{d}_4) = (0.2 + 0) \boxplus 2.0 \approx -0.2 = \text{new } L(d_4) \qquad (8.91)$$

where the log-likelihood addition has been calculated using the approximation in Equation (8.73). Next, we proceed to obtain the first vertical calculations, using the $L_{ev}(\hat{d})$ expressions in Equations (8.84) through (8.87). Now, the values of $L(d)$ can be refined by using the new $L(d)$ values gleaned from the first horizontal calculations, shown in Equations (8.88) through (8.91). That is,

$$L_{ev}(\hat{d}_1) = (0.2 - 0.3) \boxplus 6.0 \approx 0.1 = \text{new } L(d_1) \qquad (8.92)$$

$$L_{ev}(\hat{d}_2) = (0.3 - 0.2) \boxplus 1.0 \approx -0.1 = \text{new } L(d_2) \qquad (8.93)$$

$$L_{ev}(\hat{d}_3) = (1.5 - 0.1) \boxplus 6.0 \approx -1.4 = \text{new } L(d_3) \qquad (8.94)$$

$$L_{ev}(\hat{d}_4) = (0.1 - 1.5) \boxplus 1.0 \approx 1.0 = \text{new } L(d_4) \qquad (8.95)$$

The results of the first full iteration of the two decoding steps (horizontal and vertical) are as follows:

Original $L_c(x_k)$ measurements

1.5	0.1
0.2	0.3

-0.1	-1.5
-0.3	-0.2

$L_{eh}(\hat{d})$ after first horizontal decoding

0.1	-0.1
-1.4	1.0

$L_{ev}(\hat{d})$ after first vertical decoding

Each decoding step improves the original LLRs that are based on channel measurements only. This is seen by calculating the decoder output LLR, using Equation (8.74). The original LLR plus the horizontal extrinsic LLRs yield the following improvement (the extrinsic vertical terms are not yet being considered):

Improved LLRs due to $L_{eh}(\hat{d})$

1.4	−1.4
−0.1	0.1

The original LLR plus both the horizontal and vertical extrinsic LLRs yield the following improvement:

Improved LLRs due to $L_{eh}(\hat{d}) + L_{ev}(\hat{d})$

1.5	−1.5
−1.5	1.1

For this example, it is seen that the knowledge gained from horizontal decoding alone is sufficient to yield the correct hard decisions out of the decoder, but with very low confidence for data bits d_3 and d_4. After incorporating the vertical extrinsic LLRs into the decoder, the new LLR values exhibit a higher level of reliability or confidence. Let us pursue one additional horizontal and vertical decoding iteration to determine if there are any significant changes in the results. We again use the relationships shown in Equations (8.84) through (8.87) and proceed with the second horizontal calculations for $L_{eh}(\hat{d})$, using the new $L(d)$ from the first vertical calculations, shown in Equations (8.92) through (8.95), so that

$$L_{eh}(\hat{d}_1) = (0.1 - 0.1) \boxplus 2.5 \approx \quad 0 \ = \text{new } L(d_1) \tag{8.96}$$

$$L_{eh}(\hat{d}_2) = (1.5 + 0.1) \boxplus 2.5 \approx -1.6 = \text{new } L(d_2) \tag{8.97}$$

$$L_{eh}(\hat{d}_3) = (0.3 + 1.0) \boxplus 2.0 \approx -1.3 = \text{new } L(d_3) \tag{8.98}$$

$$L_{eh}(\hat{d}_4) = (0.2 - 1.4) \boxplus 2.0 \approx \quad 1.2 = \text{new } L(d_4) \tag{8.99}$$

Next, we proceed with the second vertical calculations for $L_{ev}(\hat{d})$, using the new $L(d)$ from the second horizontal calculations, shown in Equations (8.96) through (8.99). This yields

$$L_{ev}(\hat{d}_1) = (0.2 - 1.3) \boxplus 6.0 \approx \quad 1.1 = \text{new } L(d_1) \tag{8.100}$$

$$L_{ev}(\hat{d}_2) = (0.3 + 1.2) \boxplus 1.0 \approx -1.0 = \text{new } L(d_2) \tag{8.101}$$

$$L_{ev}(\hat{d}_3) = (1.5 + \quad 0) \boxplus 6.0 \approx -1.5 = \text{new } L(d_3) \tag{8.102}$$

$$L_{ev}(\hat{d}_4) = (0.1 - 1.6) \boxplus 1.0 \approx \quad 1.0 = \text{new } L(d_4) \tag{8.103}$$

The second iteration of horizontal and vertical decoding, yielding the preceding values, results in soft-output LLRs that are again calculated from Equation (8.74), which is rewritten below:

$$L(\hat{d}) = L_c(x) + L_{eh}(\hat{d}) + L_{ev}(\hat{d}) \tag{8.104}$$

The horizontal and vertical extrinsic LLRs of Equations (8.96) through (8.103) and the resulting decoder LLRs are displayed below. For this example, the second horizontal and vertical iteration (yielding a total of four iterations) suggests a modest improvement over a single horizontal and vertical iteration. The results show a balancing of the confidence values amongst each of the four data decisions:

Original $L_c(x)$ Measurements

1.5	0.1
0.2	0.3

0	−1.6
−1.3	1.2

$L_{eh}(\hat{d})$ after second horizontal decoding

1.1	−1.0
−1.5	1.0

$L_{ev}(\hat{d})$ after second vertical decoding

The soft output is $L(\hat{d}) = L_c(x) + L_{eh}(\hat{d}) + L_{ev}(\hat{d})$, which, after a total of four iterations, yields the values for $L(\hat{d})$ of

2.6	−2.5
−2.6	2.5

Observe that correct decisions about the four data bits will result, and the level of confidence about these decisions is high. The iterative decoding of turbo codes is similar to the process used when solving a crossword puzzle. The first pass through the puzzle is likely to contain a few errors. Some words seem to fit, but when the letters intersecting a row and column do not match, it is necessary to go back and correct the first-pass answers.

8.4.4 Encoding with Recursive Systematic Codes

The basic concepts of concatenation, iteration, and soft decision decoding using a simple product-code example have been described. These ideas are next applied to the implementation of turbo codes that are formed by the parallel concatenation of component convolutional codes [17, 20].

A short review of simple binary rate 1/2 convolutional encoders with constraint length K and memory $K − 1$ is in order. The input to the encoder at time k is a bit d_k, and the corresponding codeword is the bit pair (u_k, v_k), where

$$u_k = \sum_{i=0}^{K-1} g_{1i} d_{k-i} \quad \text{modulo-2}, \quad g_{1i} = 0, 1 \tag{8.105}$$

and

$$v_k = \sum_{i=0}^{K-1} g_{2i} d_{k-i} \quad \text{modulo-2}, \quad g_{2i} = 0, 1 \tag{8.106}$$

$\mathbf{G}_1 = \{g_{1i}\}$ and $\mathbf{G}_2 = \{g_{2i}\}$ are the code generators, and d_k is represented as a binary digit. This encoder can be visualized as a discrete-time finite impulse response (FIR) linear system, giving rise to the familiar nonsystematic convolutional (NSC) code, an example of which is shown in Figure 8.24. Its trellis structure can be seen in Figure 7.7. In this example, the constraint length is $K = 3$, and the two code generators are described by $\mathbf{G}_1 = \{111\}$ and $\mathbf{G}_2 = \{101\}$. It is well known that at large E_b/N_0 values, the error performance of a NSC is better than that of a systematic code having the same memory. At small E_b/N_0 values, it is generally the other way around [17]. A class of infinite impulse response (IIR) convolutional codes [17] has been proposed as building blocks for a turbo code. Such building blocks are also referred to as recursive systematic convolutional (RSC) codes because previously encoded information bits are continually fed back to the encoder's input. For high code rates, RSC codes result in better error performance than the best NSC codes at any value of E_b/N_0. A binary rate 1/2 RSC code is obtained from a NSC code by using a feedback loop, and setting one of the two outputs (u_k or v_k) equal to d_k. Figure 8.25a illustrates an example of such an RSC code, with $K = 3$, where a_k is recursively calculated as

$$a_k = d_k + \sum_{i=1}^{K-1} g_i' a_{k-i} \quad \text{modulo-2} \tag{8.107}$$

and g_i' is equal to g_{1i} if $u_k = d_k$, and to g_{2i} if $v_k = d_k$. Figure 8.25b shows the trellis structure for the RSC code in Figure 8.25a.

It is assumed that an input bit d_k takes on values of 1 or 0 with equal probability. Furthermore, $\{a_k\}$ exhibits the same statistical properties as $\{d_k\}$ [17]. The free distance is identical for the RSC code of Figure 8.25a and the NSC code of Figure 8.24. Similarly, their trellis structures are identical with respect to state transitions and their corresponding output bits. However, the two output sequences $\{u_k\}$ and $\{v_k\}$ do not correspond to the same input sequence $\{d_k\}$ for RSC and NSC codes. For the same code generators, it can be said that the weight distribution of the output codewords from an RSC encoder is not modified compared with the weight distribution from the NSC counterpart. The only change is the mapping between input data sequences and output codeword sequences.

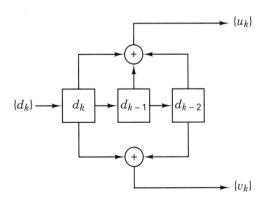

Figure 8.24 Nonsystematic convolutional (NSC) code.

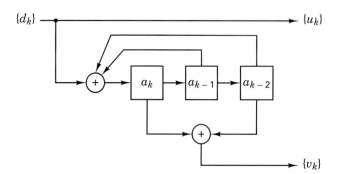

Figure 8.25a Recursive systematic convolutional (RSC) code.

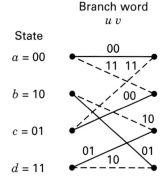

Figure 8.25b Trellis structure for the RSC code in part a).

Example 8.5 Recursive Encoders and their Trellis Diagrams

a) Using the RSC encoder in Figure 8.25a, verify the section of the trellis structure (diagram) shown in Figure 8.25b.

b) For the encoder in part a), start with the input data sequence $\{d_k\} = 1\ 1\ 1\ 0$, and show the step-by-step encoder procedure for finding the output codeword.

Solution

a) For NSC encoders, keeping track of the register contents and state transitions is a straightforward procedure. However, when the encoders are recursive, more care must be taken. Table 8.5 is made up of 8 rows corresponding to the 8 possible transitions in this 4-state machine. The first four rows represent transitions when the input data bit, d_k, is a binary zero, and the last four rows represent transitions when d_k is a one. For this example, the step-by-step encoding procedure can be described with reference to Table 8.5 and Figure 8.25 as follows:

1. At any input-bit time, k, the (starting) state of a transition, is denoted by the contents of the two rightmost stages in the register, namely a_{k-1} and a_{k-2}.

2. For any row on the table (transition on the trellis), the contents of the a_k stage is found by the modulo-2 addition of bits d_k, a_{k-1} and a_{k-2} on that row.

3. The output code-bit sequence, $u_k v_k$, for each possible starting state (i.e., $a = 00$, $b = 10$, $c = 01$, and $d = 11$) is found by appending the modulo-2 addition of a_k and a_{k-2} to the current data bit $d_k = u_k$.

TABLE 8.5   Validation of the Figure 8.25b Trellis Section

Input bit	Current bit	Starting state		Code bits		Ending state	
$d_k = u_k$	a_k	a_{k-1}	a_{k-2}	u_k	v_k	a_k	a_{k-1}
	0	0	0	0	0	0	0
	1	1	0	0	1	1	1
0	1	0	1	0	0	1	0
	0	1	1	0	1	0	1
	1	0	0	1	1	1	0
	0	1	0	1	0	0	1
1	0	0	1	1	1	0	0
	1	1	1	1	0	1	1

It is easy to verify that the details in Table 8.5 correspond to the trellis section of Figure 8.25b. An interesting property of the most useful recursive shift registers used as component codes for turbo encoders is that the two transitions entering a state *should not* correspond to the same input bit value (i.e., two solid lines or two dashed lines should not enter a given state). This property is assured if the polynomial describing the feedback in the shift register is of full degree, which means one of the feedback lines must emanate from the highest-order stage, in this example, stage a_{k-2}.

b) There are two ways to proceed with encoding the input data sequence $\{d_k\}$ = 1 1 1 0. One way uses the trellis diagram, and the other way uses the encoder circuit. Using the trellis section in Figure 8.25b, we choose the dashed-line transition (representing input bit binary one) from the state $a = 00$ (a natural choice for the starting state) to the next state $b = 10$ (which becomes the starting state for the next input bit). We denote the bits shown on that transition as the output coded-bit sequence 11. This procedure is repeated for each input bit. Another way to proceed is to build a table, such as Table 8.6, based on the encoder circuit in Figure 8.25a. Here, time, k, is shown from start to finish (5 time instances and 4 time intervals). Table 8.6 is read as follows:

1. At any instant of time, a data bit d_k becomes transformed to a_k by summing it (modulo-2) to the bits a_{k-1} and a_{k-2} on the same row.
2. For example, at time $k = 2$, the data bit $d_k = 1$ is tranformed to $a_k = 0$ by summing it to the bits a_{k-1} and a_{k-2} on the same $k = 2$ row.
3. The resulting output, $u_k v_k = 10$ dictated by the encoder logic circuitry, is the coded-bit sequence associated with time $k = 2$ (actually the time interval between times $k = 2$ and $k = 3$).

TABLE 8.6   Encoding a Bit Sequence with the Figure 8.25a Encoder

Time	Input bit	First stage	State at time k		Code bits	
k	$d_k = u_k$	a_k	a_{k-1}	a_{k-2}	u_k	v_k
1	1	1	0	0	1	1
2	1	0	1	0	1	0
3	1	0	0	1	1	1
4	0	0	0	0	0	0
5			0	0		

4. At time $k = 2$, the contents, 10, of the rightmost two stages, $a_{k-1}\, a_{k-2}$, represents the state of the machine at the start of that transition.
5. The state at the end of that transition is seen as the contents, 01, in the two leftmost stages, $a_k\, a_{k-1}$, on that same row. Since the bits shift from left to right, this transition-terminating state reappears as the starting state for time $k = 3$ on the next row.
6. Each row can be described in the same way. Thus, the encoded sequence seen in the final column of Table 8.6 is 1 1 1 0 1 1 0 0.

8.4.4.1 Concatenation of RSC Codes

Consider the parallel concatenation of two RSC encoders of the type shown in Figure 8.25. Good turbo codes have been constructed from component codes having short constraint lengths ($K = 3$ to 5). An example of such a turbo encoder is shown in Figure 8.26, where the switch yielding v_k provides puncturing, making the overall code rate 1/2. Without the switch, the code rate would be 1/3. There is no limit to the number of encoders that may be concatenated, and, in general, the component codes need not be identical with regard to constraint length and rate. The goal in designing turbo codes is to choose the best component codes by maximizing the effective free distance of the code [21]. At large values of E_b/N_0, this is tantamount to maximizing the minimum weight codeword. However, at low values of E_b/N_0 (the region of greatest interest), optimizing the weight distribution of the codewords is more important than maximizing the minimum weight codeword [20].

The turbo encoder in Figure 8.26 produces codewords from each of two component encoders. The weight distribution for the codewords out of this parallel concatenation depends on how the codewords from one of the component encoders are combined with codewords from the other encoder.

Intuitively, we should avoid pairing low-weight codewords from one encoder with low-weight codewords from the other encoder. Many such pairings can be avoided by proper design of the interleaver. An interleaver that permutes the data in a random fashion provides better performance than the familiar block interleaver [22].

If the component encoders are not recursive, the unit weight input sequence (0 0 ... 0 0 1 0 0 ... 0 0) will always generate a low weight codeword at the input of a second encoder for any interleaver design. In other words, the interleaver would not influence the output codeword weight distribution if the component codes were not recursive. However, if the component codes are recursive, a weight-1 input sequence generates an infinite impulse response (infinite-weight output). Therefore, for the case of recursive codes, the weight-1 input sequence does not yield the minimum weight codeword out of the encoder. The encoded output weight is kept finite only by trellis termination, a process that forces the coded sequence to terminate in such a way that the encoder returns to the zero state. In effect, the convolutional code is converted to a block code.

For the encoder of Figure 8.26, the minimum weight codeword for each component encoder is generated by the weight-3 input sequence (0 0 ... 0 0 1 1 1 0 0 0 ... 0 0), with three consecutive 1's. Another input that produces fairly low weight codewords is the weight-2 sequence (0 0 ... 0 0 1 0 0 1 0 0 ... 0 0). However, after the permutations introduced by an interleaver, either of these deleterious input

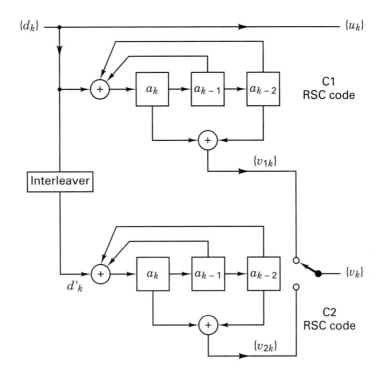

Figure 8.26 Parallel concatenation of two RSC encoders.

patterns is not likely to appear again at the input to another encoder, making it unlikely that a minimum weight codeword will be combined with another minimum weight codeword.

The important aspect of the building blocks used in turbo codes is that they are recursive (the systematic aspect is merely incidental). It is the RSC code's IIR property that protects against the generation of low-weight codewords that cannot be remedied by an interleaver. One can argue that turbo code performance is largely influenced by minimum weight codewords that result from the weight-2 input sequence. The argument is that weight-1 inputs can be ignored since they yield large codeword weights due to the IIR encoder structure. For input sequences having weight-3 and larger, a properly designed interleaver makes the occurrence of low weight output codewords relatively rare [21–25].

8.4.5 A Feedback Decoder

The Viterbi algorithm (VA) is an optimal decoding method for minimizing the probability of sequence error. Unfortunately, the (hard-decision output) VA is not suited to generate the a posteriori probability (APP) or soft-decision output for each decoded bit. A relevant algorithm for doing this has been proposed by Bahl et. al. [26]. The Bahl algorithm was modified by Berrou, et. al. [17] for use in decoding RSC codes. The APP that a decoded data bit $d_k = i$ can be derived from the joint probability $\lambda_k^{i,m}$ defined by

$$\lambda_k^{i,m} = P\{d_k = i, S_k = m \mid R_1^N\} \qquad (8.108)$$

where $S_k = m$ is the encoder state at time k, and R_1^N is a received binary sequence from time $k = 1$ through some time N.

Thus, the APP that a decoded data bit $d_k = i$, represented as a binary digit, is obtained by summing the joint probability over all states, as follows:

$$P\{d_k = i \mid R_1^N\} = \sum_m \lambda_k^{i,m} \qquad i = 0, 1 \qquad (8.109)$$

Next, the log-likelihood ratio (LLR) is written as the logarithm of the ratio of APPs, as

$$L(\hat{d}_k) = \log \left[\frac{\sum_m \lambda_k^{1,m}}{\sum_m \lambda_k^{0,m}} \right] \qquad (8.110)$$

The decoder makes a decision, known as the *maximum a posteriori* (MAP) decision rule, by comparing $L(\hat{d}_k)$ to a zero threshold. That is,

$$\hat{d}_k = 1 \quad \text{if} \quad L(\hat{d}_k) > 0$$
$$\hspace{7cm} (8.111)$$
$$\hat{d}_k = 0 \quad \text{if} \quad L(\hat{d}_k) < 0$$

For a systematic code, the LLR $L(\hat{d}_k)$ associated with each decoded bit $\hat{d}_k$ can be described as the sum of the LLR of $\hat{d}_k$, out of the demodulator and of other LLRs generated by the decoder (extrinsic information), as was expressed in Equations (8.72) and (8.73). Consider the detection of a noisy data sequence that stems from the encoder of Figure 8.26, with the use of a decoder shown in Figure 8.27.

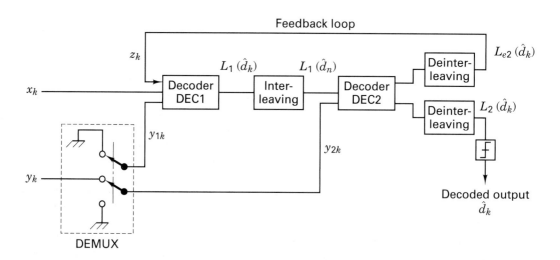

Figure 8.27 Feedback decoder.

Assume binary modulation and a discrete memoryless Gaussian channel. The decoder input is made up of a set R_k of two random variables x_k and y_k. For the bits d_k and v_k at time k, expressed as binary numbers (1, 0), the conversion to received bipolar (+1, −1) pulses can be expressed as

$$x_k = (2d_k - 1) + i_k \qquad (8.112)$$

and

$$y_k = (2v_k - 1) + q_k \qquad (8.113)$$

where i_k and q_k are two statistically independent random variables with the same variance σ^2, accounting for the noise contribution. The redundant information y_k is demultiplexed and sent to decoder DEC1 as y_{1k}, when $v_k = v_{1k}$, and to decoder DEC2 as y_{2k}, when $v_k = v_{2k}$. When the redundant information of a given encoder (C1 or C2) is not emitted, the corresponding decoder input is set to zero. Note that the output of DEC1 has an interleaver structure identical to the one used at the transmitter between the two component encoders. This is because the information processed by DEC1 is the noninterleaved output of C1 (corrupted by channel noise). Conversely, the information processed by DEC2 is the noisy output of C2 whose input is the same data going into C1, however permuted by the interleaver. DEC2 makes use of the DEC1 output, provided this output is time ordered in the same way as the input to C2 (i.e., the two sequences into DEC2 must appear "in step" with respect to the positional arrangement of the signals in each sequence).

8.4.5.1 Decoding with a Feedback Loop

We rewrite Equation (8.71) for the soft-decision output at time k, with the a priori LLR $L(d_k)$ initially set to zero. This follows from the assumption that the data bits are equally likely. Therefore,

$$L(\hat{d}_k) = L_c(x_k) + L_e(\hat{d}_k) \qquad (8.114)$$

$$= \log \left[\frac{p(x_k | d_k = 1)}{p(x_k | d_k = 0)} \right] + L_e(\hat{d}_k)$$

where $L(\hat{d}_k)$ is the soft-decision output at the decoder, and $L_c(x_k)$ is the LLR channel measurement, stemming from the ratio of likelihood functions $p(x_k | d_k = i)$ associated with the discrete memoryless channel model. $L_e(\hat{d}_k) = L(\hat{d}_k)|_{x_k = 0}$ is a function of the redundant information. It is the extrinsic information supplied by the decoder and does not depend on the decoder input x_k. Ideally $L_c(x_k)$ and $L_e(\hat{d}_k)$ are corrupted by uncorrelated noise, and thus $L_e(\hat{d}_k)$ may be used as a new observation of d_k by another decoder to form an iterative process. The fundamental principle for feeding back information to another decoder is that a decoder should never be supplied with information that stems from its own input (because the input and output corruption will be highly correlated).

For the Gaussian channel, the natural logarithm in Equation (8.114) is used to describe the channel LLR $L_c(x_k)$, as was done in Equations (8.77). We rewrite the Equation (8.77c) LLR result as

$$L_c(x_k) = -\frac{1}{2}\left(\frac{x_k - 1}{\sigma}\right)^2 + \frac{1}{2}\left(\frac{x_k + 1}{\sigma}\right)^2 = \frac{2}{\sigma^2}x_k \qquad (8.115)$$

Both decoders, DEC1 and DEC2 use the modified Bahl algorithm [26]. If the inputs $L_1(\hat{d}_k)$ and y_{2k} to decoder DEC2 (see Figure 8.27) are statistically independent, then the LLR $L_2(\hat{d}_k)$ at the output of DEC2 can be written as

$$L_2(\hat{d}_k) = f[L_1(\hat{d}_k)] + L_{e2}(\hat{d}_k) \qquad (8.116)$$

with

$$L_1(\hat{d}_k) = \frac{2}{\sigma_0^2}x_k + L_{e1}(\hat{d}_k) \qquad (8.117)$$

where $f[\cdot]$ indicates a functional relationship. The extrinsic information $L_{e2}(\hat{d}_k)$ out of DEC2 is a function of the sequence $\{L_1(\hat{d}_k)\}_{n \neq k}$. Since $L_1(\hat{d}_n)$ depends on the observation R_1^N, then the extrinsic information $L_{e2}(\hat{d}_k)$ is correlated with observations x_k and y_{1k}. Nevertheless, the greater $|n - k|$ is, the less correlated are $L_1(\hat{d}_n)$ and the observations x_k, y_k. Thus, due to the interleaving between DEC1 and DEC2, the extrinsic information $L_{e2}(\hat{d}_k)$ and the observations x_k, y_{1k} are weakly correlated. Therefore, they can be jointly used for the decoding of bit d_k [17]. In Figure 8.27, the parameter $z_k = L_{e2}(\hat{d}_k)$ feeding into DEC1 acts as a diversity effect in an iterative process. In general, $L_{e2}(\hat{d}_k)$ will have the same sign as d_k. Therefore, $L_{e2}(\hat{d}_k)$ may increase the associated LLR and thereby improve the reliability of each decoded data bit.

The algorithmic details for computing the LLR $L(\hat{d}_k)$ of the a posteriori probability (APP) for each data bit has been described by several authors [17–18, 30]. Suggestions for decreasing the implementational complexity of the algorithms can be found in [27–31]. A reasonable way to think of the process that produces APP values for each data bit is to imagine implementing a maximum likelihood sequence estimation or Viterbi algorithm (VA) and computing it in two directions over a block of code bits. Proceeding with this bi-directional VA in a sliding-window fashion—and thereby obtaining metrics associated with states in the forward and backward direction—allows computing the APP for each data bit represented in the block. With this view in mind, the decoding of turbo codes can be estimated to be at least two times more complex than decoding one of its component codes using the VA.

8.4.5.2 Turbo Code Error-Performance Example

Performance results using Monte Carlo simulations have been presented in [17] for a rate 1/2, $K = 5$ encoder implemented with generators $\mathbf{G}_1 = \{1\ 1\ 1\ 1\ 1\}$ and $\mathbf{G}_2 = \{1\ 0\ 0\ 0\ 1\}$, using parallel concatenation and a 256×256 array interleaver. The modified Bahl algorithm was used with a data block length of 65,536 bits. After 18 decoder iterations, the bit-error probability P_B was less than 10^{-5} at $E_b/N_0 = 0.7$ dB. The error-performance improvement as a function of the number of decoder iterations is seen in Figure 8.28. Note that as the Shannon limit of −1.6 dB is approached, the required system bandwidth approaches infinity, and the capacity

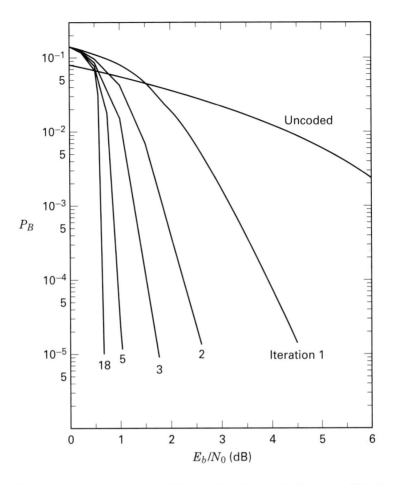

Figure 8.28 Bit-error probability as a function of E_b/N_o and multiple iterations.

REF: Berrou, C., Glavieux, A., and Thitimajshima, P, "Near Shannon Limit Error-Correcting Coding and Decoding: Turbo Codes," IEEE Proc. of Int'l. Conf. on Communications, Geneva, Switzerland, May 1993 (ICC '93), pp. 1064-1070.

(code rate) approaches zero. Therefore, the Shannon limit represents an interesting theoretical bound, but it is not a practical goal. For binary modulation, several authors use $P_B = 10^{-5}$ and $E_b/N_0 = 0.2$ dB as a *pragmatic* Shannon limit reference for a rate 1/2 code. Thus, with parallel concatenation of RSC convolutional codes and feedback decoding, the error performance of a turbo code at $P_B = 10^{-5}$ is within 0.5 dB of the (pragmatic) Shannon limit. A class of codes that use serial instead of parallel concatenation of the interleaved building blocks has been proposed. It has been suggested that serial concatenation of codes may have superior performance [28] to those that use parallel concatenation.

8.4.6 The MAP Algorithm

The process of turbo-code decoding starts with the formation of *a posteriori proba-bilities* (APPs) for each data bit, which is followed by choosing the data-bit value that corresponds to the *maximum a posteriori* (MAP) probability for that data bit. Upon reception of a corrupted code-bit sequence, the process of decision making with APPs, allows the MAP algorithm to determine the most likely information bit to have been transmitted at each bit time. This is unlike the Viterbi algorithm (VA), where the APP for each data bit is not available. Instead, the VA finds the most likely sequence to have been transmitted. There are, however, similarities in the implementation of the two algorithms. (See Section 8.4.6.3.) When the decoded P_B is small, there is very little performance difference between the MAP and a soft-output Viterbi algorithm called SOVA. However, at low E_b/N_0 and high P_B values, the MAP algorithm can outperform SOVA decoding by 0.5 dB or more [30, 31]. For turbo codes, this can be very important, since the first decoding iterations can yield poor error performance. The implementation of the MAP algorithm proceeds somewhat like performing a Viterbi algorithm in two directions over a block of code bits. Once this bi-directional computation yields state and branch metrics for the block, the APPs and the MAP can be obtained for each data bit represented within the block. We describe here a derivation of the MAP decoding algorithm for systematic convolutional codes assuming an AWGN channel model, as presented by Pietrobon [30]. We start with the ratio of the APPs, known as the likelihood ratio $\Lambda(\hat{d}_k)$, or its logarithm, $L(\hat{d}_k)$ called the log-likelihood ratio (LLR), as shown earlier in Equation (8.110):

$$\Lambda(\hat{d}_k) = \frac{\sum_m \lambda_k^{1,m}}{\sum_m \lambda_k^{0,m}} \qquad (8.118a)$$

and

$$L(\hat{d}_k) = \log \left[\frac{\sum_m \lambda_k^{1,m}}{\sum_m \lambda_k^{0,m}} \right] \qquad (8.118b)$$

Here $\lambda_k^{i,m}$—the joint probability that data $d_k = i$ and state $S_k = m$, conditioned on the received binary sequence R_1^N, observed from time $k = 1$ through some time N—is described by Equation (8.108), and rewritten below:

$$\lambda_k^{i,m} = P(d_k = i, S_k = m \mid R_1^N) \qquad (8.119)$$

R_1^N represents a corrupted code-bit sequence after it has been transmitted through the channel, demodulated, and presented to the decoder in soft-decision form. In effect, the MAP algorithm requires that the output sequence from the demodulator be presented to the decoder as a block of N bits at a time. Let R_1^N be written as follows:

$$R_1^N = \{R_1^{k-1}, R_k, R_{k+1}^N\} \tag{8.120}$$

To facilitate the use of Bayes' rule, Equation (8.119) is partitioned using the letters A, B, C, D, and Equation (8.120). Thus, Equation (8.119) can be written in the form

$$\lambda_k^{i,m} = P\,(\underbrace{d_k = i, S_k = m}_{A}\,|\,\underbrace{R_1^{k-1}}_{B},\ \underbrace{R_k}_{C},\ \underbrace{R_{k+1}^N}_{D}) \tag{8.121}$$

Recall from Bayes' rule that

$$
\begin{aligned}
P(A|B,\ C,\ D) &= \frac{P(A,\ B,\ C,\ D)}{P(B,\ C,\ D)} = \frac{P(B|A,\ C,\ D)\,P(A,\ C,\ D)}{P(B,\ C,\ D)} \\
&= \frac{P(B|A,\ C,\ D)\,P(D|A,\ C)\,P(A,\ C)}{P(B,\ C,\ D)}
\end{aligned} \tag{8.122}
$$

Hence, application of this rule to Equation (8.121) yields

$$
\begin{aligned}
\lambda_k^{i,m} &= P(R_1^{k-1}|d_k = i, S_k = m, R_k^N)\,P(R_{k+1}^N|d_k = i, S_k = m, R_k) \\
&\quad \times P(d_k = i, S_k = m, R_k)/P(R_1^N)
\end{aligned} \tag{8.123}
$$

where $R_k^N = \{R_k, R_{k+1}^N\}$. Equation (8.123) can be expressed in a way that gives greater meaning to the probability terms contributing to $\lambda_k^{i,m}$. In the sections that follow, the three numerator factors on the right side of Equation (8.123) will be defined and developed as the forward state metric, the reverse state metric, and the branch metric.

8.4.6.1 The State Metrics and the Branch Metric

We define the first numerator factor on the right side of Equation (8.123) as the forward state metric at time k and state m, and denote it as α_k^m. Thus, for $i = 1, 0$

$$P(R_1^{k-1}\,|\,\overbrace{d_k = i}^{\text{IRRELEVANT}}, S_k = m, \overbrace{R_k^N}^{\text{IRRELEVANT}}) = P(R_1^{k-1}|S_k = m) \triangleq \alpha_k^m \tag{8.124}$$

Note that $d_k = i$ and R_k^N are designated as irrelevant, since the assumption that $S_k = m$ implies that events before time k are not influenced by observations after time k. In other words, the past is not affected by the future; hence, $P(R_1^{k-1})$ is independent of the fact that $d_k = i$ and the sequence R_k^N. However, since the encoder has memory, the encoder state $S_k = m$ is based on the past, so this term is relevant and must be left in the expression. The form of Equation (8.124) is intuitively satisfying since it presents the forward state metric α_k^m, at time k, as being a probability of the past sequence that is only dependent on the current state induced by this sequence and nothing more. This should be familiar to us from the convolutional encoder and its state representation as a Markov process in Chapter 7.

Similarly, the second numerator factor on the right side of Equation (8.123) represents a reverse state metric β_k^m, at time k and state m, described by

$$P(R_{k+1}^N \,|\, d_k = i, S_k = m, R_k) = P(R_{k+1}^N \,|\, S_{k+1} = f(i, m)) \underline{\triangle} \beta_{k+1}^{f(i,\,m)} \quad (8.125)$$

where $f(i, m)$ is the next state, given an input i and state m, and $\beta_{k+1}^{f(i,\,m)}$ is the reverse state metric at time $k + 1$ and state $f(i, m)$. The form of Equation (8.125) is intuitively satisfying since it presents the reverse state metric β_{k+1}^m, at future time $k + 1$, as being a probability of the future sequence, which depends on the state (at future time $k + 1$), which in turn is a function of the input bit and the state (at current time k). This should be familiar to us because it engenders the basic definition of a finite-state machine (see Section 7.2.2).

We define the third numerator factor on the right side of Equation (8.123) as the branch metric at time k and state m, which we denote $\delta_k^{i,m}$. Thus, we write

$$P(d_k = i, S_k = m, R_k) \underline{\triangle} \delta_k^{i,m} \quad (8.126)$$

Substituting Equations (8.124) through (8.126) into Equation (8.123) yields the following more compact expression for the joint probability, as follows:

$$\lambda_k^{i,\,m} = \frac{\alpha_k^m \, \delta_k^{i,\,m} \, \beta_{k+1}^{f(i,\,m)}}{P(R_1^N)} \quad (8.127)$$

Equation (8.127) can be used to express Equation (8.118) as

$$\Lambda(\hat{d}_k) = \frac{\displaystyle\sum_m \alpha_k^m \, \delta_k^{1,\,m} \, \beta_{k+1}^{f(1,\,m)}}{\displaystyle\sum_m \alpha_k^m \, \delta_k^{0,\,m} \, \beta_{k+1}^{f(0,\,m)}} \quad (8.128a)$$

and

$$L(\hat{d}_k) = \log \left[\frac{\displaystyle\sum_m \alpha_k^m \, \delta_k^{1,\,m} \, \beta_{k+1}^{f(1,\,m)}}{\displaystyle\sum_m \alpha_k^m \, \delta_k^{0,\,m} \, \beta_{k+1}^{f(0,\,m)}} \right] \quad (8.128b)$$

where $\Lambda(\hat{d}_k)$ is the likelihood ratio of the k-th data bit, and $L(\hat{d}_k)$ the logarithm of $\Lambda(\hat{d}_k)$, is the LLR of the k-th data bit, where the logarithm is generally taken to the base e.

8.4.6.2 Calculating the Forward State Metric

Starting from Equation (8.124), α_k^m can be expressed as the summation of all possible transition probabilities from time $k - 1$, as follows:

$$\alpha_k^m = \sum_{m'} \sum_{j=0}^{1} P(d_{k-1} = j, S_{k-1} = m', R_1^{k-1} \,|\, S_k = m) \quad (8.129)$$

We can rewrite R_1^{k-1} as $\{R_1^{k-2}, R_{k-1}\}$, and from Bayes' Rule,

$$\alpha_k^m = \sum_{m'} \sum_{j=0}^{1} P(R_1^{k-2} \,|\, S_k = m, d_{k-1} = j, S_{k-1} = m', R_{k-1})$$

$$\times \, P(d_{k-1} = j, S_{k-1} = m', R_{k-1} \,|\, S_k = m) \tag{8.130a}$$

$$= \sum_{j=0}^{1} P(R_1^{k-2} \,|\, S_{k-1} = b(j, m)) \, P(d_{k-1} = j, S_{k-1} = b(j, m), R_{k-1}) \tag{8.130b}$$

where $b(j, m)$ is the state going backwards in time from state m, via the previous branch corresponding to input j. Equation (8.130b) can replace Equation (8.130a), since knowledge about the state m' and the input j, at time $k - 1$, completely defines the path resulting in state $S_k = m$. Using Equations (8.124) and (8.126) to simplify the notation of Equation (8.130b) yields

$$\alpha_k^m = \sum_{j=0}^{1} \alpha_{k-1}^{b(j,m)} \, \delta_{k-1}^{j,\, b(j,m)} \tag{8.131}$$

Equation (8.131) indicates that a new forward state metric at time k and state m is obtained by summing two weighted state metrics from time $k - 1$. The weighting consists of the branch metrics associated with the transitions corresponding to data bits 0 and 1. Figure 8.29a illustrates the use of two different types of notation for the parameter alpha. We use $\alpha_{k-1}^{b(j,m)}$ for the forward state metric at time $k - 1$, when there are two possible underlying states (depending upon whether $j = 0$ or 1). And we use α_k^m for the forward state metric at time k, when the two possible transitions from the previous time terminate on the same state m at time k.

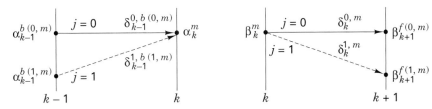

(a) Forward state metric:	(b) Reverse state metric:

$$\alpha_k^m = \alpha_{k-1}^{b(0,m)} \delta_{k-1}^{0,\,b(0,m)} + \alpha_{k-1}^{b(1,m)} \delta_{k-1}^{1,\,b(1,m)} \qquad \beta_k^m = \beta_{k+1}^{f(0,m)} \delta_k^{0,\,m} + \beta_{k+1}^{f(1,m)} \delta_k^{1,\,m}$$

Where $b(j, m)$ is the state going Where $f(j, m)$ is the next state
backwards in time corresponding given an input j and state m
to an input j

Branch metric:

$$\delta_k^{i,\,m} = \pi_k^i \exp\left(x_k \, u_k^i + y_k \, v_k^{i,\,m} \right)$$

Figure 8.29 Graphical representation for calculating α_k^m and β_k^m.
REF: Pietrobon, S.S., "Implementation and Performance of a Turbo/Map Decoder," Int'l. J. of Satellite Communications, vol. 16, Jan.-Feb. 1998, pp. 23–46.

8.4.6.3 Calculating the Reverse State Metric

Starting from Equation (8.125), where $\beta_{k+1}^{f(i,\,m)} = P[R_{k+1}^N \mid S_{k+1} = f(i,\,m)]$, we have

$$\beta_k^m = P(R_k^N \mid S_k = m) = P(R_k, R_{k+1}^N \mid S_k = m) \tag{8.132}$$

We can express β_k^m as the summation of all possible transition probabilities to time $k + 1$, as follows:

$$\beta_k^m = \sum_{m'} \sum_{j=0}^{1} P(d_k = j, S_{k+1} = m', R_k, R_{k+1}^N \mid S_k = m) \tag{8.133}$$

Using Bayes' Rule,

$$\beta_k^m = \sum_{m'} \sum_{j=0}^{1} P(R_{k+1}^N \mid S_k = m, d_k = j, S_{k+1} = m', R_k)$$
$$\times P(d_k = j, S_{k+1} = m', R_k \mid S_k = m) \tag{8.134}$$

$S_k = m$ and $d_k = j$ in the first term on the right side of Equation (8.134) completely defines the path resulting in $S_{k+1} = f(j,\,m)$, the next state given an input j and state m. Thus, these conditions allow replacing $S_{k+1} = m'$ with $S_k = m$ in the second term of Equation (8.134), yielding

$$\beta_k^m = \sum_{j=0}^{1} P(R_{k+1}^N \mid S_{k+1} = f(j,m))\, P(d_k = j, S_k = m, R_k)$$
$$= \sum_{j=0}^{1} \delta_k^{j,\,m}\, \beta_{k+1}^{f(j,\,m)} \tag{8.135}$$

Equation (8.135) indicates that a new reverse state metric at time k and state m is obtained by summing two weighted state metrics from time $k + 1$. The weighting consists of the branch metrics associated with the transitions corresponding to data bits 0 and 1. Figure 8.29b illustrates the use of two different types of notation for the parameter beta. First, we use $\beta_{k+1}^{f(j,\,m)}$ for the reverse state metric at time $k + 1$, when there are two possible underlying states (depending on whether $j = 0$ or 1). Second, we use β_k^m for the reverse state metric at time k, where the two possible transitions arriving at time $k + 1$ stem from the same state m at time k. Figure 8.29 presents a graphical illustration for calculating the forward and reverse state metrics.

Implementing the MAP decoding algorithm has some similarities to implementing the Viterbi decoding algorithm (see Section 7.3). In the Viterbi algorithm, we add branch metrics to state metrics. Then we compare and select the minimum distance (maximum likelihood) in order to form the next state metric. The process is called Add-Compare-Select (ACS). In the MAP algorithm, we multiply (add, in the logarithmic domain) state metrics by branch metrics. Then, instead of comparing them, we sum them to form the next forward (or reverse) state metric, as seen in Figure 8.29. The differences should make intuitive sense. With the Viterbi algorithm, the most likely sequence (path) is being sought; hence, there is a continual comparison and selection to find the best path. With the MAP algorithm, a soft number (likelihood or log-likelihood) is being sought; hence, the process uses all the metrics from all the possible transitions within a time interval, in order to come up with the best overall statistic regarding the data bit associated with that time interval.

8.4.6.4 Calculating the Branch Metric

We start with Equation (8.126),

$$\delta_k^{i,m} = P(d_k = i, S_k = m, R_k) \tag{8.136}$$

$$= P(R_k|d_k = i, S_k = m) \, P(S_k = m|d_k = i) \, P(d_k = i)$$

where R_k represents the sequence $\{x_k, y_k\}$, x_k is the noisy received data bit, and y_k is the corresponding noisy received parity bit. Since the noise affecting the data and the parity are independent, the current state is independent of the current input and can therefore be any one of the 2^v states, where v is the number of memory elements in the convolutional code system. That is, the constraint length K of the code is equal to $v + 1$. Hence,

$$P(S_k = m|d_k = i) = \frac{1}{2^v}$$

and

$$\delta_k^{i,m} = P(x_k|d_k = i, S_k = m) \, P(y_k|d_k = i, S_k = m) \, \frac{\pi_k^i}{2^v} \tag{8.137}$$

where π_k^i is defined as $P(d_k = i)$, the a priori probability of d_k.

From Equation (1.25d) in Chapter 1, the probability $P(X_k = x_k)$ of a random variable, X_k taking on the value x_k, is related to the probability density function (pdf) $p_{x_k}(x_k)$, as follows:

$$P(X_k = x_k) = p_{x_k}(x_k) \, dx_k \tag{8.138}$$

For notational convenience, the random variable X_k, which takes on values x_k, is often termed "the random variable x_k", which represents the meanings of x_k and y_k in Equation (8.137). Thus, for an AWGN channel, where the noise has zero mean and variance σ^2, we use Equation (8.138) in order to replace the probability terms in Equation (8.137) with their pdf equivalents, and we write

$$\delta_k^{i,m} = \frac{\pi_k^i}{2^v \sqrt{2\pi}\sigma} \exp\left[-\frac{1}{2}\left(\frac{x_k - u_k^i}{\sigma}\right)^2\right] dx_k \frac{1}{\sqrt{2\pi}\sigma} \exp\left[-\frac{1}{2}\left(\frac{y_k - v_k^{i,m}}{\sigma}\right)^2\right] dy_k \tag{8.139}$$

where u_k and v_k represent the transmitted data bits and parity bits, respectively (in bipolar form), and dx_k and dy_k are the differentials of x_k and y_k, and get absorbed into the constant A_k, below. Note that the parameter u_k^i represents data that have no dependence on the state m. However, the parameter $v_k^{i,m}$ represents parity that does depend on the state m, since the code has memory. Simplifying the notation by eliminating all terms that will appear in both the numerator and denominator of the likelihood ratio, resulting in cancellation, we can write

8.4   Turbo Codes

$$\delta_k^{i,m} = A_k \, \pi_k^i \exp\left[\frac{1}{\sigma^2}\left(x_k \, u_k^i + y_k \, v_k^{i,m}\right)\right] \qquad (8.140)$$

If we substitute Equation (8.140) into Equation (8.128a), we obtain

$$\Lambda(\hat{d}_k) = \pi_k \exp\left(\frac{2x_k}{\sigma^2}\right) \frac{\sum_m \alpha_k^m \exp\left(\frac{y_k v_k^{1,m}}{\sigma^2}\right) \beta_{k+1}^{f(1,m)}}{\sum_m \alpha_k^m \exp\left(\frac{y_k v_k^{0,m}}{\sigma^2}\right) \beta_{k+1}^{f(0,m)}} \qquad (8.141a)$$

$$= \pi_k \exp\left(\frac{2x_k}{\sigma^2}\right) \pi_k^e \qquad (8.141b)$$

and

$$L(\hat{d}_k) = L(d_k) + L_c(x_k) + L_e(\hat{d}_k) \qquad (8.141c)$$

where $\pi_k = \pi_k^1/\pi_k^0$ is the input a priori probability ratio (prior likelihood), and π_k^e is the output extrinsic likelihood, each at time k. In Equation (8.141b), one can think of π_k^e as a correction term (due to the coding) that changes the input prior knowledge about a data bit. In a turbo code, such correction terms are passed from one decoder to the next, in order to improve the likelihood ratio for each data bit, and thus minimize the probability of decoding error. Thus, the decoding process entails the use of Equation (8.141b) to compute $\Lambda(\hat{d}_k)$ for several iterations. The extrinsic likelihood π_k^e, resulting from a particular iteration replaces the a priori likelihood ratio π_{k+1} for the next iteration. Taking the logarithm of $\Lambda(\hat{d}_k)$ in Equation (8.141b) yields Equation (8.141c), which is the same result provided by Equation (8.71) showing that the final soft number $L(\hat{d}_k)$ is made up of three LLR terms—the a priori LLR, the channel-measurement LLR, and the extrinsic LLR.

The MAP algorithm can be implemented in terms of a likelihood ratio $\Lambda(\hat{d}_k)$, as shown in Equation (8.128a) or (8.141a,b). However, implementation using likelihood ratios is very complex because of the multiply operations that are required. By operating the MAP algorithm in the logarithmic domain [30, 31], as described by the LLR in Equation (8.128b) or (8.141c), the complexity can be greatly reduced by eliminating the multiply operations.

8.4.7 MAP Decoding Example

Figure 8.30 illustrates a MAP decoding example. Figure 8.30a shows a simple systematic convolutional encoder, with constraint length, $K = 3$, and rate $\frac{1}{2}$. The input data consists of the sequence $\mathbf{d} = \{1, 0, 0\}$, corresponding to the times $k = 1, 2, 3$. The output code-bit sequence, being systematic, is formed by consecutively taking one bit from the sequence $\mathbf{u} = \{1, 0, 0\}$, followed by one bit from the parity-bit sequence $\mathbf{v} = \{1, 0, 1\}$. In each case, the leftmost bit is the earliest bit. Thus, the output sequence is 1 1 0 0 0 1, or in bipolar form the sequence is +1 +1 −1 −1 −1 +1. Figure 8.30b shows the results of some postulated noise vectors $\mathbf{n}_x$ and $\mathbf{n}_y$, having cor-

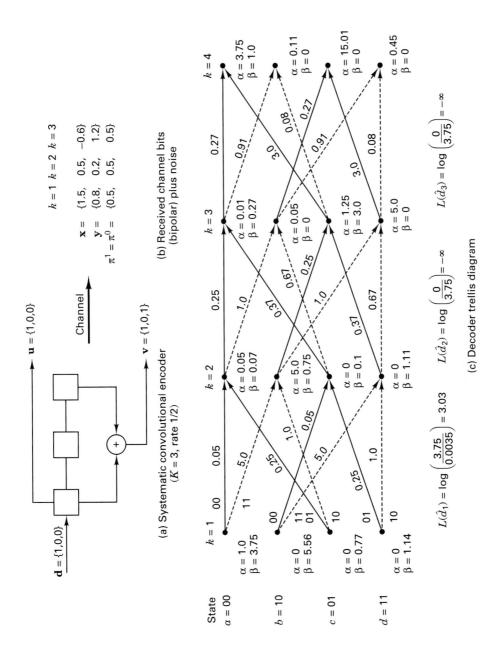

Figure 8.30 Example of MAP decoding ($K = 3$, rate ½, systematic).

rupted sequences $\mathbf{u}$ and $\mathbf{v}$, so they are now designated as $\mathbf{x} = \mathbf{u} + \mathbf{n}_x$ and $\mathbf{y} = \mathbf{v} + \mathbf{n}_y$. As shown in Figure 8.30b, the demodulator outputs arriving at the decoder corresponding to the times $k = 1, 2, 3$, have values of 1.5, 0.8, 0.5, 0.2, −0.6, 1.2. Also shown are the a priori probabilities of a data bit being 1 or 0, designated as π^1 and π^0, respectively, and assumed to be equally likely for all k times. For this example, all information is now available to calculate the branch metrics and the state metrics, and enter their values onto the decoder trellis diagram of Figure 8.30c. On the trellis diagram, each transition occurring between times k and $k + 1$ corresponds to a data bit d_k that appears at the encoder input at the transition-start time k. At time k, the encoder will be in some state m, and at time $k + 1$ it transitions to a new state (possibly the same state). When such a trellis diagram is used to depict a sequence of code bits (representing N data bits), the sequence is characterized by N transition intervals and $N + 1$ states (from start to finish).

8.4.7.1 Calculating the Branch Metrics

We start with Equation (8.140), with $\pi^i_k = 0.5$ (for this exercise, data bits are assumed equally likely for all time), and for simplicity assume that $A_k = 1$ for all time and that $\sigma^2 = 1$. Thus, $\delta^{i,m}_k$ becomes

$$\delta^{i,m}_k = 0.5 \exp\left(x_k u^i_k + y_k v^{i,m}_k\right) \qquad (8.142)$$

What basic receiver function does Equation (8.142) resemble? The expression looks somewhat like a correlation process. At the decoder, a pair of receptions (data-bit related x_k, and parity-bit related y_k) arrive at each time k. The branch metric is calculated by taking the product of the received x_k with each of the prototype signals u_k, and similarly the product of the received y_k with each of the prototype signals v_k. For each trellis transition, the magnitude of the branch metric will be a function of how good a match there is between the pair of noisy receptions and the code-bit meaning of that trellis transition. For $k = 1$, Equation (8.142) is used with the data in Figure 8.30b for evaluating eight branch metrics (a transition from each state m and for each data value i), as shown below. For notational simplicity, we designate the trellis states as follows: $a = 00, b = 10, c = 01, d = 11$. Note that the code-bit meaning, u_k, v_k, of each trellis transition is written on the transition in Figure 8.30c (for $k = 1$ only) and was obtained from the encoder structure in the usual way. (See Section 7.2.4.) Also, for the trellis transitions of Figure 8.30c, the convention that dashed lines and solid lines correspond to the underlying data bits 1 and 0, respectively, is used:

$$\delta^{1,m=a}_{k=1} = \delta^{1,m=b}_{k=1} = 0.5 \exp\left[(1.5)(1) + (0.8)(1)\right] = 5.0$$

$$\delta^{0,m=a}_{k=1} = \delta^{0,m=b}_{k=1} = 0.5 \exp\left[(1.5)(-1) + (0.8)(-1)\right] = 0.05$$

$$\delta^{1,m=c}_{k=1} = \delta^{1,m=d}_{k=1} = 0.5 \exp\left[(1.5)(1) + (0.8)(-1)\right] = 1.0$$

$$\delta^{0,m=c}_{k=1} = \delta^{0,m=d}_{k=1} = 0.5 \exp\left[(1.5)(-1) + (0.8)(1)\right] = 0.25$$

Next, we repeat these calculations using Equation (8.142) for the eight branch metric values at time $k = 2$:

$$\delta_{k=2}^{1, m=a} = \delta_{k=2}^{1, m=b} = 0.5 \exp\left[(0.5)(1) + (0.2)(1)\right] = 1.0$$

$$\delta_{k=2}^{0, m=a} = \delta_{k=2}^{0, m=b} = 0.5 \exp\left[(0.5)(-1) + (0.2)(-1)\right] = 0.25$$

$$\delta_{k=2}^{1, m=c} = \delta_{k=2}^{1, m=d} = 0.5 \exp\left[(0.5)(1) + (0.2)(-1)\right] = 0.67$$

$$\delta_{k=2}^{0, m=c} = \delta_{k=2}^{0, m=d} = 0.5 \exp\left[(0.5)(-1) + (0.2)(1)\right] = 0.37$$

Again, we repeat the calculations for the eight branch metric values at time $k = 3$:

$$\delta_{k=3}^{1, m=a} = \delta_{k=3}^{1, m=b} = 0.5 \exp\left[(-0.6)(1) + (1.2)(1)\right] = 0.91$$

$$\delta_{k=3}^{0, m=a} = \delta_{k=3}^{0, m=b} = 0.5 \exp\left[(-0.6)(-1) + (1.2)(-1)\right] = 0.27$$

$$\delta_{k=3}^{1, m=c} = \delta_{k=3}^{1, m=d} = 0.5 \exp\left[(-0.6)(1) + (1.2)(-1)\right] = 0.08$$

$$\delta_{k=3}^{0, m=c} = \delta_{k=3}^{0, m=d} = 0.5 \exp\left[(-0.6)(-1) + (1.2)(1)\right] = 3.0$$

8.4.7.2 Calculating the State Metrics

Once the eight values of $\delta_k^{i,m}$ are computed for each k, the forward state metrics α_k^m can be calculated with the help of Figures 8.29, 8.30c, and Equation (8.131), rewritten below

$$\alpha_{k+1}^m = \sum_{j=0}^{1} \delta_k^{j, b(j, m)} \alpha_k^{b(j, m)}$$

Assume that the encoder starting state is $a = 00$. Then,

$$\alpha_{k=1}^{m=a} = 1.0 \quad \text{and} \quad \alpha_{k=1}^{m=b} = \alpha_{k=1}^{m=c} = \alpha_{k=1}^{m=d} = 0$$

$$\alpha_{k=2}^{m=a} = (0.05)(1.0) + (0.25)(0) = 0.05$$

$$\alpha_{k=2}^{m=b} = (5.0)(1.0) + (1.0)(0) = 5.0$$

$$\alpha_{k=2}^{m=c} = \alpha_{k=2}^{m=d} = 0$$

and so forth, as shown on the trellis diagram of Figure 8.30c. Similarly, the reverse state metric β_k^m can be calculated with the help of Figures 8.29, 8.30c, and Equation (8.135), rewritten below

$$\beta_k^m = \sum_{j=0}^{1} \delta_k^{j, m} \beta_{k+1}^{f(j, m)}$$

The data sequence and the code in this example were purposely chosen so that the final state of the trellis at time $k = 4$ is the $a = 00$ state. Otherwise, it would be necessary to use tail bits to force the final state into such a known state. Thus, for this example, illustrated in Figure 8.30, knowing that the final state is $a = 00$, the reverse state metrics can be calculated as follows:

$$\beta_{k=4}^{m=a} = 1.0 \quad \text{and} \quad \beta_{k=4}^{m=b} = \beta_{k=4}^{m=c} = \beta_{k=4}^{m=d} = 0$$

$$\beta_{k=3}^{m=a} = (0.27)(1.0) + (0.91)(0) = 0.27$$

$$\beta_{k=3}^{m=b} = \beta_{k=3}^{m=d} = 0$$

$$\beta_{k=3}^{m=c} = (3.0)(1.0) + (0.08)(0) = 3.0$$

and so forth. All the reverse state metric values are shown on the trellis of Figure 8.30c.

8.4.7.3 Calculating the Log-Likelihood Ratio

Now that the metrics δ, α, and β have all been computed for the code-bit sequence in this example, the turbo decoding process can use Equation (8.128) or (8.141) for finding a soft decision, $\Lambda(\hat{d}_k)$ or $L(\hat{d}_k)$, for each data bit. When using turbo codes, this process can be iterated several times to improve the reliability of that decision. This is generally accomplished by using the extrinsic likelihood parameter of Equation (8.141b) to compute and re-compute the likelihood ratio $\Lambda(\hat{d}_k)$ for several iterations. The extrinsic likelihood π_k^e of any iteration is used to replace the a priori likelihood ratio π_{k+1} for the next iteration.

For this example, let us now use the metrics calculated above (with a single pass through the decoder). We choose Equation (8.128b) to compute the LLR for each data bit in the sequence $\{d_k\}$, and then use the decision rules of Equation (8.111) to transform the resulting soft numbers into hard decisions. For $k = 1$, omitting some of the zero factors, we obtain

$$L(\hat{d}_k) = \log\left(\frac{1.0 \times 5.0 \times 0.75}{1.0 \times 0.05 \times 0.07}\right) = \log\left(\frac{3.75}{0.0035}\right) = 3.03$$

For $k = 2$, again omitting some of the zero factors, we obtain

$$L(\hat{d}_2) = \log\left[\frac{(0.05 \times 1.0 \times 0) + (5.0 \times 1.0 \times 0)}{(0.05 \times 0.25 \times 0.27) + (5.0 \times 0.25 \times 3.0)}\right] = \log\left(\frac{0}{3.75}\right) = -\infty$$

For $k = 3$, we obtain

$$L(\hat{d}_3) = \log\left[\frac{(0.01 \times 0.91 \times 0) + (0.05 \times 0.91 \times 0)}{(0.01 \times 0.27 \times 1.0) + (0.05 \times 0.27 \times 0)}\right.$$

$$\left.\frac{+ (1.25 \times 0.08 \times 0) + (5.0 \times 0.08 \times 0)}{+ (1.25 \times 3.0 \times 1.0) + (5.0 \times 3.0 \times 0)}\right]$$

$$= \log\left(\frac{0}{3.75}\right) = -\infty$$

Using Equation (8.111) to make the final decisions about the bits at times $k = 1, 2, 3$, the sequence is decoded as $\{1\ 0\ 0\}$. This is clearly correct, given the specified input to the encoder.

8.4.7.4 Shift Register Representation for Finite State Machines

The shift registers used throughout this book, whether feed-forward or feedback, are mostly represented with storage stages and connecting lines. It is important to point out that it is often useful to represent an encoder shift register, particularly a recursive encoder, in a slightly different way. Some authors use blocks labeled with the letter D or T to denote time delays (typically 1-bit delays). The junctions outside the blocks, carrying voltage or logic levels, represent the storage in the encoder between clock times. The two formats—storage blocks versus delay blocks—do not change the characteristics or the operation of the underlying process in any way. For some finite-state machines, with many recursive connections, it may be somewhat easier to track the signal when the delay-block format is used. Problems 8.23 and 8.24 employ such encoders in Figures P8.2 and P8.3, respectively. For the storage-stage format, the current state of a machine is described by the contents of the rightmost $K - 1$ stages. For the delay-block format, the current state is similarly described by the logic levels at the outputs of the rightmost $K - 1$ delay blocks. For both formats, the relationship between memory v and constraint length K is the same, that is $v = K - 1$. Thus, in Figure P8.2, three delay blocks means that $v = 3$ and $K = 4$. Similarly, in Figure P8.3, two delay blocks means that $v = 2$ and $K = 3$.

8.5 CONCLUSION

In this chapter, we examined Reed–Solomon (R–S) codes, an important class of non-binary block codes, particularly useful for correcting burst errors. Because coding efficiency increases with code length, R–S codes have a special attraction. They can be configured with long block lengths (in bits) with less decoding time than other codes of similar lengths. That is because the decoder logic works with symbol-based, not bit-based, arithmetic. Hence, for 8-bit symbols, the arithmetic operations would all be at the byte level. This increases the complexity of the logic, compared with binary codes of the same length, but it also increases the throughput.

We next described a technique called interleaving, which allows the popular block and convolutional coding schemes to be used over channels that exhibit bursty noise or periodic fading without suffering degradation. We used the CD digital audio system as an example of how both R–S coding and interleaving play an important role in ameliorating the effects of burst noise.

We described concatenated codes and the concept of turbo coding, whose basic configuration depends on the concatenation of two or more component codes. Basic statistical measures, such as a posteriori probability and likelihood also were reviewed. We then used these measures for describing the error performance of a soft-input/soft-output decoder. We showed how performance is improved when soft outputs from concatenated decoders are used in an iterative decoding process. We then proceeded to apply these concepts to the parallel concatenation of recursive systematic convolutional (RSC) codes, and we explained why such codes are the preferred building blocks in turbo codes. A feedback decoder was described in general ways, and its remarkable performance was presented. We next developed the mathematics of a maximum a posteriori (MAP) decoder, and used a numerical example (traversing a trellis diagram in two directions) that resulted in soft-decision outputs.

APPENDIX 8A THE SUM OF LOG-LIKELIHOOD RATIOS

Following are the algebraic details yielding the results shown in Equation (8.72), rewritten below:

$$L(d_1) \boxplus L(d_2) \triangleq L(d_1 \oplus d_2) = \log_e \left(\frac{e^{L(d_1)} + e^{L(d_2)}}{1 + e^{L(d_1)} e^{L(d_1)}} \right) \qquad (8A.1)$$

We start with a likelihood ratio of the APP that a data bit equals +1 compared to the APP that it equals −1. Since the logarithm of this likelihood ratio, denoted $L(d)$, has been conveniently taken to the base e, it can be expressed as

$$L(d) = \log_e \left[\frac{P(d = +1)}{P(d = -1)} \right] = \log_e \left[\frac{P(d = +1)}{1 - P(d = +1)} \right] \qquad (8A.2)$$

so that

$$e^{L(d)} = \left[\frac{P(d = +1)}{1 - P(d = +1)} \right] \qquad (8A.3)$$

Solving for $P(d + 1)$, we obtain

$$e^{L(d)} - e^{L(d)} \times P(d = +1) = P(d = +1) \qquad (8A.4)$$

$$e^{L(d)} = P(d = +1) \times [1 + e^{L(d)}] \qquad (8A.5)$$

and

$$P(d = +1) = \frac{e^{L(d)}}{1 + e^{L(d)}} \qquad 8A.6)$$

Observe from Equation 8A.6 that

$$P(d = -1) = 1 - P(d = +1) = 1 - \frac{e^{L(d)}}{1 + e^{L(d)}} = \frac{1}{1 + e^{L(d)}} \qquad (8A.7)$$

Let d_1 and d_2 be two statistically independent data bits taking on voltage values of +1 and −1 corresponding to logic levels 1 and 0 respectively.

When formatted in this way, the modulo-2 summation of d_1 and d_2 yields −1 whenever d_1 and d_2 have identical values (both +1 or both −1), and the summation yields +1 whenever d_1 and d_2 have different values. Then

$$L(d_1 \oplus d_2) = \log_e \left[\frac{P(d_1 \oplus d_2 = 1)}{P(d_1 \oplus d_2 = -1)} \right]$$

$$= \log_e \left[\frac{P(d_1 = +1) \times P(d_2 = -1) + [1 - P(d_1 = +1)][1 - P(d_2 = -1)]}{P(d_1 = +1) \times P(d_2 = +1) + [1 - P(d_1 = +1)][1 - P(d_2 = +1)]} \right]$$

$$(8A.8)$$

Using Equations (8A.6) and 8A.7) to replace the probability terms of Equation (8A.8), we obtain

$$L(d_1 \oplus d_2) = \log_e \left[\frac{\left(\dfrac{e^{L(d_1)}}{1 + e^{L(d_1)}} \right) \left(\dfrac{1}{1 + e^{L(d_2)}} \right) + \left(\dfrac{1}{1 + e^{L(d_1)}} \right) \left(\dfrac{e^{L(d_2)}}{1 + e^{L(d_2)}} \right)}{\left(\dfrac{e^{L(d_1)}}{1 + e^{L(d_1)}} \right) \left(\dfrac{e^{L(d_2)}}{1 + e^{L(d_2)}} \right) + \left(\dfrac{1}{1 + e^{L(d_1)}} \right) \left(\dfrac{1}{1 + e^{L(d_2)}} \right)} \right] \quad (8A.9)$$

$$= \log_e \left[\frac{\left(\dfrac{e^{L(d_1)} + e^{L(d_2)}}{[1 + e^{L(d_1)}] [1 + e^{L(d_2)}]} \right)}{\left(\dfrac{e^{L(d_1)} \, e^{L(d_2)} + 1}{[1 + e^{L(d_1)}] [1 + e^{L(d_2)}]} \right)} \right] \quad (8A.10)$$

$$= \log_e \left[\frac{e^{L(d_1)} + e^{L(d_2)}}{1 + e^{L(d_1)} \, e^{L(d_2)}} \right] \quad (8A.11)$$

REFERENCES

1. Gallager, R. G., *Information Theory and Reliable Communication,* John Wiley and Sons, New York, 1968.
2. Odenwalder, J. P., *Error Control Coding Handbook,* Linkabit Corporation, San Diego, CA, July 15, 1976.
3. Berlekamp, E. R., Peile, R. E., and Pope, S. P., "The Application of Error Control to Communications," *IEEE Communications Magazine,* vol. 25, no. 4, April 1987, pp. 44–57.
4. Hagenauer, J., and Lutz, E., "Forward Error Correction Coding for Fading Compensation in Mobile Satellite Channels," *IEEE J. on Selected Areas in Comm.,* vol. SAC-5, no. 2, February 1987, pp. 215–225.
5. Blahut, R. E., *Theory and Practice of Error Control Codes,* Addison-Wesley Publishing Co., Reading, Massachusetts, 1983.
6. *Reed–Solomon Codes and Their Applications,* ed. Wicker, S. B., and Bhargava, V. K., IEEE Press, Piscataway, New Jersey, 1983.
7. Ramsey, J. L., "Realization of Optimum Interleavers, *IEEE Trans. Inform. Theory,* vol. IT-16, no. 3, May 1970, pp 338–345.
8. Forney, G. D., "Burst-Correcting Codes for the Classic Bursty Channel,"*IEEE Trans. Commun. Technol., vol. COM-19, Oct. 1971, pp. 772–781.*
9. Clark, G. C., Jr., and Cain, J. B., *Error-Correction Coding for Digital Communications,* Plenum Press, New York, 1981.
10. J. H. Yuen, et. al., "Modulation and Coding for Satellite and Space Communications," *Proc. IEEE,* vol. 78, no. 7, July 1990, pp. 1250–1265.
11. Peek, J. B. H., "Communications Aspects of the Compact Disc Digital Audio System," *IEEE Communications Magazine,* vol. 23, no. 2, February 1985, pp. 7–20.
12. Berkhout, P. J., and Eggermont, L. D. J., "Digital Audio Systems," *IEEE ASSP Magazine,* October 1985, pp. 45–67.

13. Driessen, L. M. H. E., and Vries, L. B., "Performance Calculations of the Compact Disc Error Correcting Code on Memoryless Channel," *Fourth Int'l. Conf. Video and Data Recording,* Southampton, England, April 20–23, 1982, IERE Conference Proc #54, pp. 385–395.

14. Hoeve, H., Timmermans, J., and Vries, L. B., "Error Correction in the Compact Disc System," *Philips Tech. Rev.,* vol. 40, no. 6, 1982, pp. 166–172.

15. Pohlmann, K. C., *The Compact Disc Handbook,* A-R Editions, Inc., Madison, Wisconsin, 1992.

16. Forney, G. D., Jr., *Concatenated Codes,* Cambridge, Massachusetts: M. I. T. Press, 1966.

17. Berrou, C., Glavieux, A., and Thitimajshima, P. "Near Shannon Limit Error-Correcting Coding and Decoding: Turbo Codes," *IEEE Proceedings of the Int. Conf. on Communications,* Geneva, Switzerland, May 1993 (ICC '93), pp. 1064–1070.

18. Berrou, C. and Glavieux, A. "Near Optimum Error Correcting Coding and Decoding: Turbo-Codes," *IEEE Trans. On Communications,* vol. 44, no. 10, October 1996, pp. 1261–1271.

19. Hagenauer, J. "Iterative Decoding of Binary Block and Convolutional Codes," *IEEE Trans. On Information Theory,* vol. 42, no. 2, March 1996, pp. 429–445.

20. Divsalar, D. and Pollara, F. "On the Design of Turbo Codes," *TDA Progress Report 42–123,* Jet Propulsion Laboratory, Pasadena, California, November 15, 1995, pp. 99–121.

21. Divsalar, D. and McEliece, R. J. "Effective Free Distance of Turbo Codes," *Electronic Letters,* vol. 32, no. 5, Feb. 29, 1996, pp. 445–446.

22. Dolinar, S. and Divsalar, D. "Weight distributions for Turbo Codes Using Random and Nonrandom Permutations," *TDA Progress Report 42–122,* Jet Propulsion Laboratory, Pasadena, California, August 15, 1995, pp. 56–65.

23. Divsalar, D. and Pollara, F. "Turbo Codes for Deep-Space Communications," *TDA Progress Report 42–120,* Jet Propulsion Laboratory, Pasadena, California, February 15, 1995, pp. 29–39.

24. Divsalar, D. and Pollara, F. "Multiple Turbo Codes for Deep-Space Communications," *TDA Progress Report 42–121,* Jet Propulsion Laboratory, Pasadena, California, May 15, 1995, pp. 66–77.

25. Divsalar, D. and Pollara, F. "Turbo Codes for PCS Applications," *Proc. ICC '95,* Seattle, Washington, June 18–22, 1995.

26. Bahl, L. R., Cocke, J., Jelinek, F. and Raviv, J. "Optimal Decoding of Linear Codes for Minimizing Symbol Error Rate," *Trans. Inform. Theory,* vol. IT-20, March 1974, pp. 248–287.

27. Benedetto, S. et. al., "Soft Output Decoding Algorithm in Iterative Decoding of Turbo Codes," *TDA Progress Report 42–124,* Jet Propulsion Laboratory, Pasadena, California, February 15, 1996, pp. 63–87.

28. Benedetto, S. et. al., "A Soft-Input Soft-Output Maximum A Posteriori (MAP) Module to Decode Parallel and Serial Concatenated Codes," *TDA Progress Report 42–127,* Jet Propulsion Laboratory, Pasadena, California, November 15, 1996, pp. 63–87.

29. Benedetto, S. et. al., "A Soft-Input Soft-Output APP Module for Iterative Decoding of Concatenated Codes," *IEEE Communications Letters,* vol. 1, no. 1, January 1997, pp. 22–24.

30. Pietrobon, S., "Implementation and Performance of a Turbo/MAP Decoder," *Int'l. J. Satellite Commun.,* vol. 16, Jan–Feb 1998, pp. 23–46.

31. Robertson, P., Villebrun, E., and Hoeher, P., "A Comparison of Optimal and Sub-Optimal MAP Decoding Algorithms Operating in the Log Domain," *Proc. of ICC '95,* Seattle, Washington, June 1995, pp. 1009–1013.

PROBLEMS

8.1. Determine which if any of the following polynomials are primitive. Hint: One of the easiest way is with the use of an LFSR, similar to the one shown in Figure 8.8.
   a) $1 + X^2 + X^3$
   b) $1 + X + X^2 + X^3$
   c) $1 + X^2 + X^4$
   d) $1 + X^3 + X^4$
   e) $1 + X + X^2 + X^3 + X^4$
   f) $1 + X + X^5$
   g) $1 + X^2 + X^5$
   h) $1 + X^3 + X^5$
   i) $1 + X^4 + X^5$

8.2. a) What is the symbol-error correcting capability of a (7, 3) R–S code? How many bits are there per symbol?
   b) Compute the number of rows and columns in the standard array (see Section 6.6) required to represent the (7, 3) R–S code in part a).
   c) Use the dimensions of the standard array in part b) to corroborate the symbol-error correcting capability found in part a).
   d) Is the (7, 3) R–S code a perfect code? If not, how much residual symbol-error correcting capability does it have?

8.3. a) Define a set of elements $\{0, \alpha^0, \alpha^1, \alpha^2, \ldots, \alpha^{2^m - 2}\}$ in terms of basis elements from the finite field GF (2^m), where $m = 4$.
   b) For the finite field defined in part a), develop an addition table similar to Table 8.2.
   c) Develop a multiplication table similar to Table 8.3.
   d) Find the generator polynomial for the (31, 27) R–S code.
   e) Encode the message {96 leading zeros followed by 110010001111} (rightmost bit is earliest) with the (31, 27) R–S code in systematic form. Why do you suppose the message was configured with so many leading zeros?

8.4. Use the generator polynomial for the (7, 3) R–S code to encode the message 010110111 (rightmost bit is earliest bit) in systematic form. Use polynomial division to find the parity polynomial, and show the resulting codeword in polynomial form and in binary form.

8.5. a) Use a LFSR to encode the symbols {6, 5, 1} (rightmost symbol is the earliest) with a (7, 3) R–S code in systematic form. Show the resulting codeword in binary form.
   b) Verify the encoding results from part a) by evaluating the codeword polynomial at the roots of the (7, 3) R–S generator polynomial, $\mathbf{g}(X)$.

8.6. a) Suppose that the codeword found in Problem 8.5 was degraded during transmission, so that its rightmost 6 bits are inverted. Find the value of each syndrome by evaluating the flawed codeword polynomial at the roots of the generator polynomial $\mathbf{g}(X)$.
   b) Verify that the same syndrome values found in part a) can be found by evaluating the error polynomial, $\mathbf{e}(X)$, at the roots of $\mathbf{g}(X)$.

8.7. a) Use the autoregressive model in Equation (8.40) with the flawed codeword from Problem 8.6 to find the location of each symbol error.

b) Find the value of each symbol error.

c) Use the information found in parts a) and b) to correct the flawed codeword.

8.8. The sequence 1011011000101100 is the input to a 4×4 block interleaver. What is the output sequence? The same input sequence is applied to the convolutional interleaver of Figure 8.13. What is the output sequence?

8.9. For each of the following conditions, design an interleaver for a communication system operating over a bursty noise channel at a transmission rate of 19,200 code symbols/s.

a) A contiguous noise burst typically lasts for 250 ms. The system code consists of a (127, 36) BCH code with $d_{min} = 31$. The end-to-end delay is not to exceed 5 s.

b) A contiguous noise burst typically lasts for 20 ms. The system code consists of a rate $\frac{1}{2}$ convolutional code with a feedback decoding algorithm that corrects an average of 3 symbols in a sequence of 21 symbols. The end-to-end delay is not to exceed 160 ms.

8.10. a) Calculate the probability of a byte (symbol) error after decoding the data stored on a compact disc (CD) as described in Section 8.3. Assume that the probability of a channel-symbol error for the disc is 10^{-3}. Also assume that the inner and outer R–S decoders are each configured to correct all 2-symbol errors, and that the interleaving process results in channel symbol errors being uncorrelated from one another.

b) Repeat part a) for a disc that has a probability of channel-symbol error equal to 10^{-2}.

8.11. A BPSK system receives equiprobable bipolar symbols (+1 or –1) plus AWGN. Assume unity noise variance. At time k, the value of the received signal x_k is equal to 0.11.

a) Calculate the two likelihood values for this received signal.

b) What would be the maximum a posteriori decision, +1 or –1?

c) The a priori probability that the transmitted symbol was +1 is equal to 0.3. What would be the maximum a posteriori decision, +1 or –1?

d) Assuming the a priori probabilities from part c), calculate the log-likelihood ratio $L(d_k | x_k)$.

8.12. Consider the two-dimensional parity-check code example described in Section 8.4.3. As outlined there, the transmitted symbols are represented by the sequence $d_1, d_2, d_3, d_4, p_{12}, p_{34}, p_{13}, p_{24}$, resulting in a code rate of 1/2. A particular application requiring a higher data rate allows the output sequence from this code to be *punctured* by discarding every other parity bit, resulting in an overall code rate of 2/3. The transmitted output is now given by the sequence $d_1, d_2, d_3, d_4, p_{12}, \_, p_{13}, \_$ (parity bits p_{34} and p_{24} are not transmitted). The transmitted sequence is $\{d_i\}, \{p_{ij}\} = +1 -1 -1 +1 +1 +1$, where i and j are location indices. The noise transforms this data plus parity sequence into the received sequence $\{x_k\} = 0.75, 0.05, 0.10, 0.15, 1.25, 3.0$, where k is a time index. Calculate the values of the soft outputs for the data bits after two horizontal and two vertical decoding iterations. Assume unity noise variance.

8.13. Consider the parallel concatenation of two RSC component encoders as shown in Figure 8.26. An interleaver of block size 10, maps a sequence of input bits $\{d_k\}$ to bits $\{d'_k\}$ where the interleaver permutation is given by [6, 3, 8, 9, 5, 7, 1, 4, 10, 2], i.e., the 1st bit of the incoming data block is mapped to position 6, the 2nd bit is mapped to position 3 etc. The input sequence is given by (0, 1, 1, 0, 0, 1, 0, 1, 1, 0). Assume that the component encoders start in the all-zeros state and that no termination bits are added to force them back to the all-zeros state.

a) Calculate the 10-bit parity sequence $\{v_{1k}\}$.

b) Calculate the 10-bit parity sequence $\{v_{2k}\}$.

c) The switch yielding the sequence $\{v_k\}$ performs puncturing such that $\{v_k\}$ is given by the following: $v_{1k}, v_{2(k+1)}, v_{1(k+2)}, v_{2(k+3)}, \ldots$, and the code rate is $\frac{1}{2}$. Calculate the weight of the output codeword.

d) When decoding with the MAP algorithm, what changes do you think need to be made with regard to initializing the state metrics and branch metrics, if the encoders are left unterminated?

8.14. a) For the nonrecursive encoder shown in Figure P8.1, calculate the minimum distance of the overall code.

b) For the recursive encoder shown in Figure 8.26, calculate the minimum distance of the overall code. Assume that there is no puncturing, so that the code rate is $\frac{1}{3}$.

c) For the encoder shown in Figure 8.26, discuss the effect on the output code weight if the input to each component encoder is given by the weight-2 sequence $(00\ldots00100100\ldots00)$ (Assume no puncturing).

d) Repeat part c) for the case where the weight-2 sequence is given by $(00\ldots0010100\ldots00)$.

8.15. Consider that the encoder in Figure 8.25a is used as a component code within a turbo code. Its 4-state trellis structure is shown in Figure 8.25b. The code rate is $\frac{1}{2}$ and the branch labeling, $u\,v$, represents the output branch word (code bits) for that branch, where u is a data bit (systematic code) and v is a parity bit, and at each time k, a data bit and parity bit are transmitted. Signals received from a demodulator have the noise-disturbed u, v values of 1.9, 0.7 at time $k = 1$, and -0.4, 0.8 at time $k = 2$. Assume that the a priori probability for the data bit being a 1 or 0 is equally likely and that the encoder begins in the all-zeros state at starting time $k = 1$. Also assume that the noise variance is equal to 1.3. Recall that a data sequence of N bits is character-

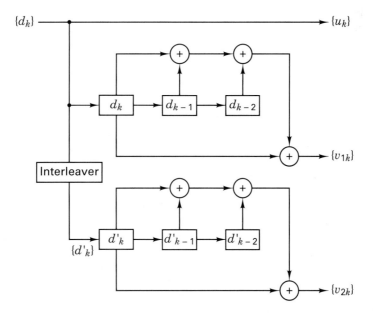

Figure P8.1 Encoder with non-recursive component codes.

ized by N transition-time intervals and $N + 1$ states (from start to finish). Thus, for this example, bits are launched at times $k = 1, 2$, and we are interested in the state metrics at times $k = 1, 2, 3$.

a) Calculate the branch metrics for times $k = 1$ and $k = 2$, that are needed for using the MAP algorithm.

b) Calculate the forward state metrics for times $k = 1, 2$, and 3.

c) The values of the reverse state metrics at times $k = 2$ and 3 are given below in Table P8.1 for each valid state. Based on the values in the table and the values calculated in parts a) and b) calculate the values for the log-likelihood ratio associated with the data bits at time $k = 1$ and $k = 2$. Use the MAP decision rule to find the most likely data bit sequence that was transmitted.

TABLE P8.1

β_k^m	$k = 2$	$k = 3$
$m = a$	4.6	2.1
$m = b$	2.4	11.5
$m = c$	5.7	3.4
$m = d$	4.3	0.9

8.16. Suppose the received sequence obtained in Problem 8.15 is in fact for a rate $\frac{2}{3}$ code which is obtained by puncturing the rate $\frac{1}{2}$ code (defined by the trellis in Figure 8.25b). The puncturing is such that only every second parity bit generated is transmitted. Therefore the four-signal sequence received represents data symbol, parity symbol, data symbol, data symbol. Calculate the branch metrics and forward state metrics for times $k = 1$ and $k = 2$ that would be needed for using the MAP algorithm.

8.17. The trellis for a four-state code used as a component code within a turbo code is shown in Figure 8.25b. The code rate is $\frac{1}{2}$ and the branch labeling, $u\ v$ represents the output, branch word (code bits) for that branch, where u is a data bit (systematic code) and v is a parity bit. A block of $N = 1024$ samples are received from a demodulator. Assume that the first signals in the block arrive at time $k = 1$, and at each time k, a noisy data bit and parity bit is received. At time $k = 1023$, the received signals have noisy u, v values of 1.3, –0.8, and at time $k = 1024$, the values are –1.4, –0.9. Assume that the a priori probability for the data bit being a 1 or 0 is equally likely and the encoder ends in a state $a = 00$ at termination time $k = 1025$. Also, assume that the noise variance is equal to 2.5.

a) Calculate the branch metrics for time $k = 1023$ and $k = 1024$.

b) Calculate the reverse state metrics for time $k = 1023, 1024$, and 1025.

c) The values of the forward state metrics at time $k = 1023$ and $k = 1024$ are given below in Table P8.2 for each valid state. Based on the values in the table and the values calculated in parts a) and b) calculate the values for the likelihood ratio associated with the data bits at time $k = 1023$ and $k = 1024$, and using the MAP decision rule, find the most likely data bit sequence that was transmitted.

TABLE P8.2

α_k^m	$k = 1023$	$k = 1024$
$m = a$	6.6	12.1
$m = b$	7.0	1.5
$m = c$	4.2	13.4
$m = d$	4.0	5.9

8.18. Given two statistically independent observations of a noisy signal x_1 and x_2, prove that the log-likelihood ratio (LLR) $L(d \mid x_1, x_2)$ can be expressed in terms of individual LLRs as

$$L(d \mid x_1, x_2) = L(x_1 \mid d) + L(x_2 \mid d) + L(d)$$

where $L(d)$ is the a priori LLR of the underlying data bit d.

8.19. a) Using Bayes' theorem, show the detailed steps that transform α_k^m in Equation (8.129) to Equation (8.130b). Hint: Use a simple lettering scheme as is used in Equations (8.121) and (8.122).

b) Explain how the summation over the states m' in Equation (8.130a) results in the expression seen in Equation (130b).

c) Repeat part a) to show in detail how Equation (8.133) evolves to Equation (8.135). Also explain how the summation over the states m' at the future time $k + 1$ results in the form of Equation (8.135).

8.20. Starting with Equation (8.139) for the branch metric $\delta_k^{i,m}$, show the detailed development resulting in Equation (8.140), and indicate which terms can be identified as the constant A_k in Equation (8.140). Why does the A_k term disappear in Equation (8.141a)?

8.21. The interleaver in Figure 8.27 (identical to the interleaver in the corresponding encoder) is needed to insure that the sequence out of DEC1 is time ordered in the same way as the sequence $\{y_{2k}\}$ Can this be implemented in a simpler way? What about using a deinterleaver in the lower line? Wouldn't that accomplish the same time ordering more simply? If we did that, then the two deinterleavers, just prior to the output, could be eliminated. Explain why that would not work.

8.22. In the implementation of the Viterbi decoding algorithm, an add-compare-select (ACS) processor is used. But, in performing the maximum a posteriori (MAP) algorithm in turbo decoding, there is no such concept as comparing and selecting one transition over another. Instead the MAP algorithm incorporates all of the branch and state metrics at each time interval. Explain the reason for this fundamental difference between the two algorithms.

8.23. Figure P8.2 illustrates a recursive systematic convolutional (RSC) rate $\frac{1}{2}$, $K = 4$, encoder. Note that the figure uses the format of 1-bit delay blocks rather than storage stages (see Section 8.4.7.4). Thus, the current state of this circuit can be described by

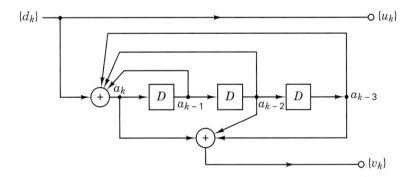

Figure P8.2 Recursive systematic convolutional (RSC) encoder, rate ½, $K = 4$.

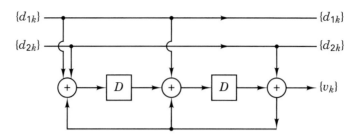

Figure P8.3 Recursive systematic convolutional (RSC) encoder, rate ⅔, $K = 3$.

the signal levels at points a_{k-1}, a_{k-2}, and a_{k-3}, similar to the way a state is described in the format using storage stages. Form a table, similar to Table 8.5 that describes all possible transitions in this circuit, and use the table to draw a trellis section.

8.24. Figure P8.3 illustrates a recursive systematic convolutional (RSC) rate $\frac{2}{3}$, $K = 3$, encoder. Note that the figure uses the format of 1-bit delay blocks rather than storage stages (see Section 8.4.7.4). Form a table, similar to Table 8.5 that describes all possible transitions in this circuit, and use the table to draw a trellis section. Use a table, similar to Table 8.6 to find the output codeword for the message sequence 1100110011. At each clock time, data bits enter the circuit in pairs $\{d_{1k}, d_{2k}\}$, and each output branch word $\{d_{1k}, d_{2k}, v_k\}$ is made up of that pair of data bits plus one parity bit, v_k.

8.25. Consider a turbo code consisting of two, four state convolutional codes as the component codes, both of which are described by the same trellis as shown in Figure 8.25b. The code rate is equal to $\frac{1}{2}$ and the block length is equal to 12. The second encoder is left unterminated. The branch metrics, forward state metrics, and reverse state metrics for the data bits associated with the terminated encoder are described by the matrices that follow. The received 12-signal vector is of the form data signal, parity signal, data signal, parity signal, and so forth, and has the following values:

$$1.2 \quad 1.3 \quad -1.2 \quad 0.6 \quad -0.4 \quad 1.9 \quad -0.7 \quad -1.9 \quad -2.2 \quad 0.2 \quad -0.1 \quad 0.6$$

Branch $\delta_k^{i,m}$ matrix

$$\delta_k^{i,m} = \begin{bmatrix} \delta_1^{0,a} & \delta_2^{0,a} & \cdots & \delta_6^{0,a} \\ \delta_1^{1,a} & \ddots & \cdots & \delta_6^{1,a} \\ \delta_1^{0,b} & \ddots & \ddots & \vdots \\ \delta_1^{1,b} & \ddots & \ddots & \vdots \\ \delta_1^{0,c} & \ddots & \ddots & \vdots \\ \delta_1^{1,c} & \ddots & \ddots & \vdots \\ \delta_1^{0,d} & \ddots & \ddots & \vdots \\ \delta_1^{1,d} & \cdots & \cdots & \delta_6^{1,d} \end{bmatrix} = \begin{bmatrix} 1.00 & 1.00 & 1.00 & 1.00 & 1.00 & 1.00 \\ 3.49 & 0.74 & 2.12 & 0.27 & 0.37 & 1.28 \\ 1.00 & 1.00 & 1.00 & 1.00 & 1.00 & 1.00 \\ 3.49 & 0.74 & 2.12 & 0.27 & 0.37 & 1.28 \\ 1.92 & 1.35 & 2.59 & 0.39 & 1.11 & 1.35 \\ 1.82 & 0.55 & 0.82 & 0.70 & 0.33 & 0.95 \\ 1.92 & 1.35 & 2.59 & 0.39 & 1.11 & 1.35 \\ 1.82 & 0.55 & 0.82 & 0.70 & 0.33 & 0.95 \end{bmatrix}$$

Alpha (α_k^m) *matrix*

$$
\alpha_k^m = \begin{bmatrix} \alpha_1^u & \alpha_2^a & \cdots & \alpha_7^a \\ \alpha_1^b & \ddots & \cdots & \alpha_7^b \\ \alpha_1^c & \cdots & \ddots & \vdots \\ \alpha_1^d & \cdots & \cdots & \alpha_7^d \end{bmatrix} = \begin{bmatrix} 1.00 & 1.00 & 1.00 & 5.05 & 8.54 & 10.41 & 24.45 \\ 0.00 & 0.00 & 1.92 & 12.79 & 5.07 & 10.93 & 31.48 \\ 0.00 & 3.49 & 0.74 & 4.03 & 14.16 & 8.22 & 24.30 \\ 0.00 & 0.00 & 4.71 & 5.77 & 5.63 & 17.53 & 27.76 \end{bmatrix}
$$

Beta (β_k^m) *matrix*

$$
\beta_k^m = \begin{bmatrix} \beta_1^a & \beta_2^a & \cdots & \beta_7^a \\ \beta_1^b & \ddots & \cdots & \beta_7^b \\ \beta_1^c & \cdots & \ddots & \vdots \\ \beta_1^d & \cdots & \cdots & \beta_7^d \end{bmatrix} = \begin{bmatrix} 24.45 & 5.44 & 2.83 & 1.12 & 1.00 & 1.00 & 1.00 \\ 24.43 & 5.62 & 3.17 & 0.70 & 0.37 & 1.28 & 0.00 \\ 21.32 & 5.45 & 3.53 & 0.81 & 0.43 & 0.00 & 0.00 \\ 21.31 & 5.79 & 2.75 & 1.14 & 1.42 & 0.00 & 0.00 \end{bmatrix}
$$

Calculate the log-likelihood ratio for each of the six data bits $\{d_k\}$, and by using the MAP decision rule, find the most likely data-bit sequence that was transmitted.

QUESTIONS

8.1. Explain why R–S codes perform so well in a *bursty-noise* environment. (See Section 8.1.2.)

8.2. Explain why the curves in Figure 8.6 of the text show error-performance *degradation* at low values of code rate. (See Section 8.1.3.)

8.3. Considering all the ways that there are to determine whether a polynomial is *primitive,* the method involving a linear feedback shift register (LFSR) is one of the simplest. Explain the procedure. (See Example 8.2.)

8.4. Explain why a *syndrome* can be calculated by evaluating the received polynomial at each of the roots of the code's generator polynomial. (See Section 8.1.6.1.)

8.5. What key transformation does an interleaver/deinterleaver system perform on *bursty noise?* (See Section 8.2.1.)

8.6. Why is the *Shannon limit* of −1.6 dB not a useful goal in the design of real systems? (See Section 8.4.5.2.)

8.7. What are the consequences of the *Viterbi decoding algorithm* not yielding *a posteriori* probabilities? (See Section 8.4.6.)

8.8. What is a more descriptive name for the Viterbi algorithm? (See Section 8.4.6.)

8.9. Describe the similarities and differences between implementing a Viterbi decoding algorithm and implementing a *maximum a posteriori* (MAP) decoding algorithm? (See Section 8.4.6.)

EXERCISES

Using the Companion CD, run the exercises associated with Chapter 8.

CHAPTER 9

Modulation and Coding Trade-Offs

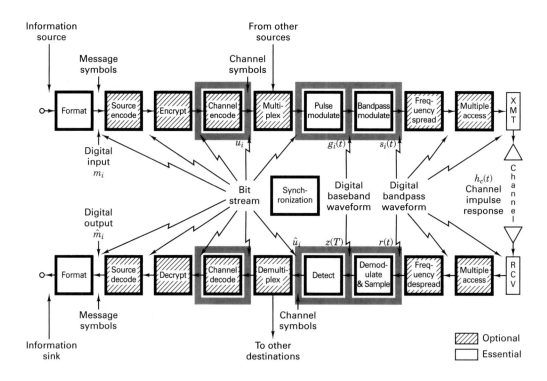

9.1 GOALS OF THE COMMUNICATIONS SYSTEM DESIGNER

System trade-offs are fundamental to all digital communication designs. The goals of the designer may include any of the following (1) to maximize transmission bit rate R; (2) to minimize probability of bit error P_B; (3) to minimize required power, or equivalently, to minimize required bit energy to noise power spectral density E_b/N_0; (4) to minimize required system bandwidth W; (5) to maximize system utilization, that is, to provide reliable service for a maximum number of users with minimum delay and with maximum resistance to interference; and (6) to minimize system complexity, computational load, and system cost. A system designer may seek to achieve all these goals simultaneously. However, goals 1 and 2 are clearly in conflict with goals 3 and 4; they call for simultaneously maximizing R, while minimizing P_B, E_b/N_0, and W. There are several constraints and theoretical limitations that necessitate the trading off of any one system requirement with each of the others:

The Nyquist theoretical minimum bandwidth requirement
The Shannon-Hartley capacity theorem (and the Shannon limit)
Government regulations (e.g., frequency allocations)
Technological limitations (e.g., state-of-the-art components)
Other system requirements (e.g., satellite orbits)

Some of the realizable modulation and coding trade-offs can best be viewed as a change in operating point on one of two performance planes. These planes will be referred to as the error probability plane and the bandwidth efficiency plane, and they are described in the following sections.

9.2 ERROR PROBABILITY PLANE

Figure 9.1 illustrates the family of P_B versus E_b/N_0 curves for the coherent detection of orthogonal signaling (Figure 9.1a) and multiple phase signaling (Figure 9.1b). The modulator uses one of its $M = 2^k$ waveforms to represent each k-bit sequence, where M is the size of the symbol set. Figure 9.1a illustrates the potential bit error improvement with orthogonal signaling as k (or M) is increased. For orthogonal signal sets, such as orthogonal frequency shift keying (FSK) modulation, increasing the size of the symbol set can provide an improvement in P_B, or a reduction in the E_b/N_0 required, at the cost of increased bandwidth. Figure 9.1b illustrates potential bit error degradation with nonorthogonal signaling as k (or M) increases. For nonorthogonal signal sets, such as multiple phase shift keying (MPSK) modulation, increasing the size of the symbol set can reduce the bandwidth requirement, but at the cost of a degraded P_B, or an increased E_b/N_0 requirement. We shall refer to these families of curves (Figure 9.1a or b) as *error probability performance curves,* and to the plane on which they are plotted as an *error probability plane.* Such a plane describes the locus of operating points available for a particular type of modulation and coding. For a given system information rate, each curve in the plane can be associated with a different fixed minimum required bandwidth; therefore, the set of curves can be termed *equibandwidth curves.* As the curves move in the direction of the ordinate, the required transmission bandwidth increases; as the curves move in the opposite direction, the required bandwidth decreases. Once a modulation and coding scheme and an available E_b/N_0 are determined, system operation is characterized by a particular point in the error probability plane. Possible trade-offs can be viewed as changes in the operating point on one of the curves or as changes in the operating point from one curve to another curve of the family. These trade-offs are seen in Figure 9.1a and b as changes in the system operating point in the direction shown by the arrows. Movement of the operating point along line 1, between points *a* and *b,* can be viewed as trading off between P_B and E_b/N_0 performance (with W fixed). Similarly, movement along line 2, between points *c* and *d,* is seen as trading P_B versus W (with E_b/N_0 fixed). Finally, movement along line 3, between points *e* and *f,* illustrates trading W versus E_b/N_0 (with P_B fixed). Movement along line 1 is effected by increasing or decreasing the available E_b/N_0. This can be achieved, for example, by increasing transmitter power, which means that the trade-off might be accomplished simply by "turning a knob," even after the system is configured. However, the other trade-offs (movement along line 2 or line 3) involve some changes in the system modulation or coding scheme, and therefore need to be accomplished during the system design phase. The advent of *software radios* [1] will even allow changes to a system's modulation and coding by programmable means.

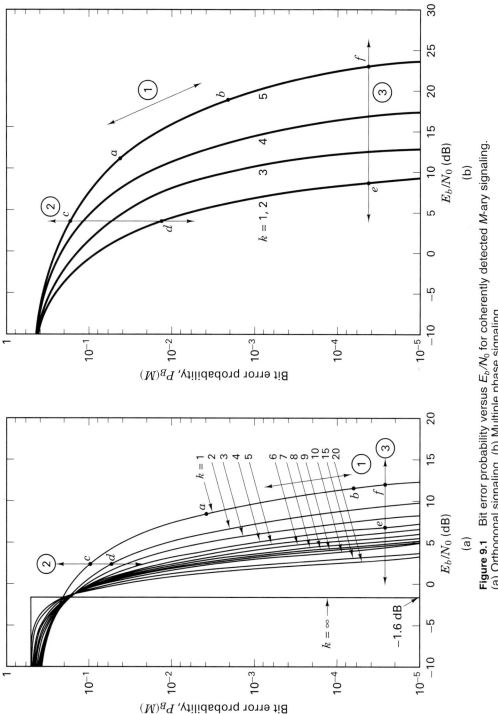

Figure 9.1 Bit error probability versus E_b/N_0 for coherently detected M-ary signaling. (a) Orthogonal signaling. (b) Multiple phase signaling.

9.3 NYQUIST MINIMUM BANDWIDTH

Every realizable system having some nonideal filtering will suffer from intersymbol interference (ISI)—the tail of one pulse spilling over into adjacent symbol intervals so as to interfere with correct detection. Nyquist [2] showed that the theoretical minimum bandwidth (Nyquist bandwidth) needed for the baseband transmission of R_s symbols per second without ISI is $R_s/2$ hertz. This is a basic theoretical constraint, limiting the designer's goal to expend as little bandwidth as possible. (See Section 3.3.) In practice, the Nyquist minimum bandwidth is expanded by about 10% to 40%, because of the constraints of real filters. Thus, *typical* baseband digital communication throughput is reduced from the ideal 2 symbols/s/Hz to the range of about 1.8 to 1.4 symbols/s/Hz. From its set of M symbols, the modulation or coding system assigns to each symbol a k-bit meaning, where $M = 2^k$. Thus, the number of bits per symbol can be expressed as $k = \log_2 M$, and the data rate or bit rate R must be k times faster than the symbol rate R_s, as expressed by the basic relationship

$$R = kR_s \quad \text{or} \quad R_s = \frac{R}{k} = \frac{R}{\log_2 M} \tag{9.1}$$

For signaling at a fixed symbol rate, Equation (9.1) shows that, as k is increased, the data rate R is increased. In the case of MPSK, increasing k, thereby results in an increased bandwidth efficiency R/W measured in bits/s/Hz. For example, movement along line 3, from point e to point f in Figure 9.1b, represents trading E_b/N_0 for a reduced bandwidth requirement. In other words, with the same system bandwidth, one can transmit MPSK signals at an increased date rate and hence at an increased R/W.

Example 9.1   Digital Modulation Schemes Fall into One of Two Classes

In some sense, all digital modulation schemes fall into one of two classes with opposite behavior characteristics. The first class constitutes orthogonal signaling, and its error performance follows the curves shown in Figure 9.1a. The second class constitutes nonorthogonal signaling (the constellation of signal phasors can be depicted on a plane). Figure 9.1b illustrates an MPSK example of such nonorthogonal signaling. However, any phase/amplitude modulation (e.g., QAM) falls into this second class. In the context of Figure 9.1, answer the following questions:

(a) Does error-performance improve or degrade with increasing M, for M-ary signaling?
(b) The choices available in digital communications almost always involves a trade-off. If error-performance improves, what price must we pay?
(c) If error-performance degrades, what benefit is exhibited?

Solution

(a) When examining Figure 9.1, we see that error-performance improvement or degradation depends upon the class of signaling in question. Consider the orthogonal signaling in Figure 9.1a, where error-performance improves with increased k or M. Recall that there are only two fair ways to compare error-performance with such curves. A vertical line can be drawn through some fixed value of E_b/N_0, and as k or M is increased, it is seen that P_B is reduced. Or, a horizontal line can be drawn through some fixed P_B requirement, and as k or M is increased, it is seen that the E_b/N_0 requirement is reduced. Similarly, it can be seen that the curves in Figure 9.1b, for nonorthogonal signaling such as MPSK, behave in the opposite fashion. Error-performance degrades as k or M is increased.

(b) In the case of orthogonal signaling, where error performance improves with increasing k of M, what is the cost? In terms of the orthogonal signaling we are most familiar with, MFSK, when $k = 1$ and $M = 2$ there are two tones in the signaling set. When $k = 2$ and $M = 4$, there are four tones in the set. When $k = 3$ and $M = 8$, there are eight tones, and so forth. With MFSK, only one tone is sent during each symbol time, but the available transmission bandwidth consists of the entire set of tones. Hence, as k or M is increased, it should be clear that the cost of improved error-performance is an expansion of required bandwidth.

(c) In the case of nonorthogonal signaling, such as MPSK or QAM, where error-performance degrades as k or M is increased, one might rightfully guess that the tradeoff will entail a reduction in the required bandwidth. Consider the following example. Suppose we require a data rate of $R = 9600$ bit/s. And, suppose that the modulation chosen is 8-ary PSK. Then, using Equation (9.1), we find that the symbol rate is

$$R_s = \frac{R}{\log_2 M} = \frac{9600 \text{ bit/s}}{3 \text{ bit/symbol}} = 3200 \text{ symbol/s}$$

If we decide to use 16-ary PSK for this example, the symbol rate would then be

$$R_s = = \frac{9600 \text{ bit/s}}{4 \text{ bit/symbol}} = 2400 \text{ symbol/s}$$

If we continue in this direction and use 32-ary PSK, the symbol rate becomes

$$R_s = = \frac{9600 \text{ bit/s}}{5 \text{ bit/symbol}} = 1920 \text{ symbol/s}$$

Do you see what happens as the operating point in Figure 9.1b is moved along a horizontal line from the $k = 3$ curve to the $k = 4$ curve, and finally to the $k = 5$ curve? For a given data rate and bit-error probability, each such movement allows us to signal at a slower rate. Whenever you hear the words, "slower signaling rate," that is tantamount to saying that the transmission bandwidth can be reduced. Similarly, any case of increasing the signaling rate, corresponds to a need for increasing the transmission bandwidth.

9.4 SHANNON–HARTLEY CAPACITY THEOREM

Shannon [3] showed that the system capacity C of a channel perturbed by additive white Gaussian noise (AWGN) is a function of the average received signal power S, the average noise power N, and the bandwidth W. The capacity relationship (Shannon–Hartley theorem) can be stated as

$$C = W \log_2 \left(1 + \frac{S}{N} \right) \tag{9.2}$$

When W is in hertz and the logarithm is taken to the base 2, as shown, the capacity is given in bits/s. It is theoretically possible to transmit information over such a channel at any rate R, where $R \leq C$, with an *arbitrarily small* error probability by using a sufficiently complicated coding scheme. For an information rate $R > C$, it is not possible to find a code that can achieve an arbitrarily small error probability. Shannon's work showed that the values of S, N, and W *set a limit on transmission rate, not on error probability.* Shannon [4] used Equation (9.2) to graphically exhibit a bound for the achievable performance of practical systems. This plot, shown in Figure 9.2, gives the normalized channel capacity C/W in bits/s/Hz as a function of the channel signal-to-noise ratio (SNR). A related plot, shown in Figure 9.3, indicates the normalized channel bandwidth W/C in Hz/bits/s as a function of SNR in the channel. Figure 9.3 is sometimes used to illustrate the power-bandwidth trade-off inherent in the ideal channel. However, it is not a pure trade-off [5] because the detected noise power is proportional to bandwidth:

$$N = N_0 W \tag{9.3}$$

Substituting Equation 9.3 into Equation 9.2 and rearranging terms yields

$$\frac{C}{W} = \log_2\left(1 + \frac{S}{N_0 W}\right) \tag{9.4}$$

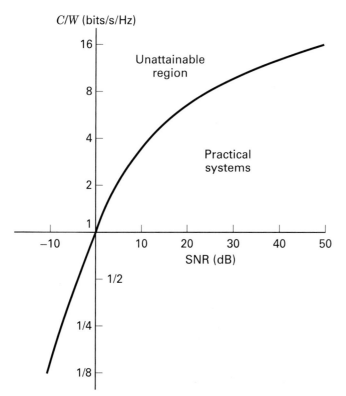

Figure 9.2 Normalized channel capacity versus channel SNR.

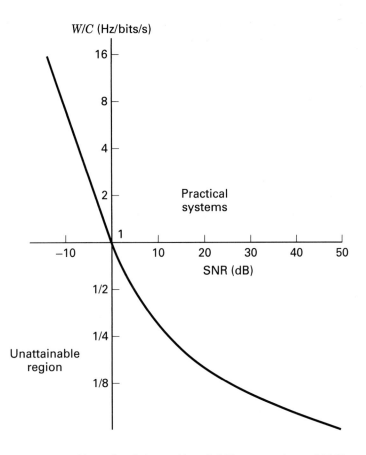

Figure 9.3 Normalized channel bandwidth versus channel SNR.

For the case where transmission bit rate is equal to channel capacity, $R = C$, we can use the identity presented in Equation 3.30 to write

$$\frac{S}{N_0 C} = \frac{E_b}{N_0} \qquad (9.5)$$

Hence, we can modify Equation 9.4 as follows:

$$\frac{C}{W} = \log_2 \left[1 + \frac{E_b}{N_0} \left(\frac{C}{W} \right) \right] \qquad (9.6a)$$

$$2^{C/W} = 1 + \frac{E_b}{N_0} \left(\frac{C}{W} \right) \qquad (9.6b)$$

$$\frac{E_b}{N_0} = \frac{W}{C} (2^{C/W} - 1) \qquad (9.6c)$$

9.4   Shannon–Hartley Capacity Theorem   **527**

Figure 9.4 is a plot of W/C versus E_b/N_0 in accordance with Equation (9.6c). The asymptotic behavior of this curve as $C/W \to 0$ (or $W/C \to \infty$) is discussed in the next section.

9.4.1 Shannon Limit

There exists a limiting value of E_b/N_0 below which there can be no error-free communication at any information rate. Using the identity

$$\lim_{x \to 0} (1 + x)^{1/x} = e$$

we can calculate the limiting value of E_b/N_0 as follows: Let

$$x = \frac{E_b}{N_0} \left(\frac{C}{W} \right)$$

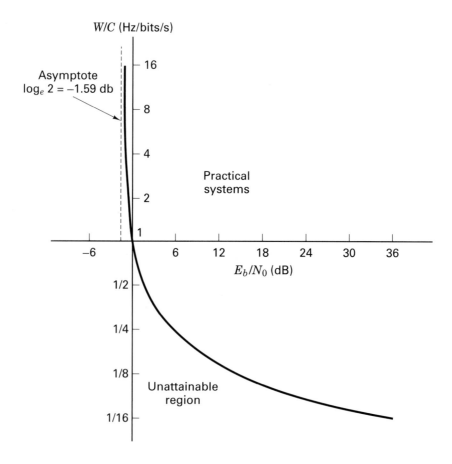

Figure 9.4 Normalized channel bandwidth versus channel E_b/N_0.

Then, from Equation (9.6a),

$$\frac{C}{W} = x \log_2 (1 + x)^{1/x}$$

and

$$1 = \frac{E_b}{N_0} \log_2 (1 + x)^{1/x}$$

In the limit, as $C/W \to 0$, we get

$$\frac{E_b}{N_0} = \frac{1}{\log_2 e} = 0.693 \tag{9.7}$$

or, in decibels,

$$\frac{E_b}{N_0} = -1.6 \text{ dB}$$

This value of E_b/N_0 is called the *Shannon limit*. On Figure 9.1a the Shannon limit is the P_B versus E_b/N_0 curve corresponding to $k \to \infty$. The curve is discontinuous, going from a value of $P_B = \frac{1}{2}$ to $P_B = 0$ at $E_b/N_0 = -1.6$ dB. It is not possible in practice to reach the Shannon limit, because as k increases without bound, the bandwidth requirement and the implementation complexity increases without bound. Shannon's work provided a theoretical proof for the existence of codes that could improve the P_B performance, or reduce the E_b/N_0 required, from the levels of the uncoded binary modulation schemes to levels approaching the limiting curve. For a bit error probability of 10^{-5}, binary phase-shift-keying (BPSK) modulation requires an E_b/N_0 of 9.6 dB (the optimum uncoded binary modulation). Therefore, for this case, Shannon's work promised the existence of a theoretical performance improvement of 11.2 dB over the performance of optimum uncoded binary modulation, through the use of coding techniques. Today, most of that promised improvement (as much as 10 dB) is realizable with turbo codes (see Section 8.4). Optimum system design can best be described as a search for rational compromises or trade-offs among the various constraints and conflicting goals. The modulation and coding trade-off, that is, the selection of modulation and coding techniques to make the best use of transmitter power and channel bandwidth, is important, since there are strong incentives to reduce the cost of generating power and to conserve the radio spectrum.

9.4.2 Entropy

To design a communications system with a specified message handling capability, we need a metric for measuring the information content to be transmitted. Shannon [3] developed such a metric, H, called the entropy of the message source (having n possible outputs). *Entropy* is defined as the average amount of information per source output and is expressed by

$$H = -\sum_{i=1}^{n} p_i \log_2 p_i \quad \text{bits/source output} \tag{9.8}$$

where p_i is the probability of the ith output and $\Sigma p_i = 1$. In the case of a binary message or a source having only two possible outputs, with probabilities p and $q = (1 - p)$, the entropy is written

$$H = -(p \log_2 p + q \log_2 q) \tag{9.9}$$

and is plotted versus p in Figure 9.5.

The quantity H has a number of interesting properties, including the following:

1. When the logarithm in Equation (9.8) is taken to the base 2, as shown, the unit for H is average bits per event. The unit *bit,* here, is a measure of *information content* and is not to be confused with the term "bit," meaning "binary digit."

2. Ther term "entropy" has the same uncertainty connotation as it does in certain formulations of statistical mechanics. For the information source with two equally likely possibilities (e.g., the flipping of a fair coin), it can be seen from Figure 9.5 that the uncertainty in the event, and hence the average infor-

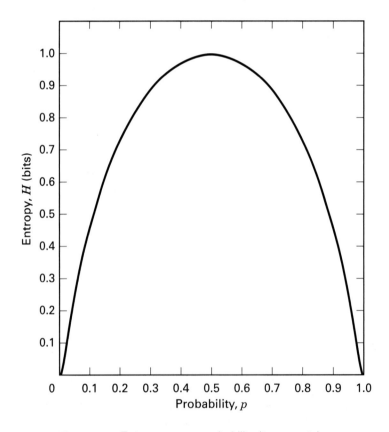

Figure 9.5 Entropy versus probability (two events).

Modulation and Coding Trade-Offs   Chap. 9

mation content, is maximum. As the probabilities depart from the equally likely case, the average information content decreases. In the limit, when one of the probabilities goes to zero, H also goes to zero. We know the result before the event happens, so the result conveys no additional information.

3. To illustrate that information content is related to a priori probability (if the a priori message probability at the receiver is zero or one, we need not send the message), consider the following example: At the end of her nine-month pregnancy, a woman enters the delivery room of a local hospital to give birth. Her husband waits anxiously in the waiting room. After some time, a physician approaches the husband and says: "Congratulations, you are the father of a child." How much information has the physician given the father *beyond the medical outcome*? Almost none; the father has known with virtual certainty that a child was forthcoming. Had the physician said, "you are the father of a boy" or "you are the father of a girl," he would have transmitted 1 bit of information, since there was a 50% chance that the child could have been a boy or a girl.

Example 9.2   Average Information Content in the English Language

   (a) Calculate the average information in bits/character for the English language, assuming that each of the 26 characters in the alphabet occurs with equal likelihood. Neglect spaces and punctuation.
   (b) Since the alphabetic characters do not appear with equal frequency in the English language (or any other language), the answer to part (a) will represent an upper bound on average information content per character. Repeat part (a) under the assumption that the alphabetic characters occur with the following probabilities:

$$p = 0.10: \quad \text{for the letters a, e, o, t}$$

$$p = 0.07: \quad \text{for the letters h, i, n, r, s}$$

$$p = 0.02: \quad \text{for the letters c, d, f, l, m, p, u, y}$$

$$p = 0.01: \quad \text{for the letters b, g, j, k, q, v, w, x, z}$$

Solution

(a) $H = -\displaystyle\sum_{i=1}^{26} \frac{1}{26} \log_2 \left(\frac{1}{26} \right)$

   $= 4.7$ bits/character

(b) $H = -(4 \times 0.1 \log_2 0.1 + 5 \times 0.07 \log_2 0.07$
   $+ 8 \times 0.02 \log_2 0.02 + 9 \times 0.01 \log_2 0.01)$
   $= 4.17$ bits/character

If we want to express the 26 letters of the alphabet with some binary-digit coding scheme, we generally need five binary digits for each character. Example 9.2 demonstrates that there may be a way to encode the English language with a fewer number of binary digits per character, *on the average,* by exploiting the fact that the average amount of information contained within each character is less than

5 bits. The subject of source coding, which deals with this exploitation, is treated in Chapter 13.

9.4.3 Equivocation and Effective Transmission Rate

Suppose that we are transmitting information at a rate of 1000 binary symbols/s over a binary symmetric channel (defined in Section 6.3.1), and that the a priori probability of transmitting either a one or a zero is equally likely. Suppose also that the noise in the channel is so great that the probability of receiving a one is $\frac{1}{2}$, whatever was transmitted, and similarly for receiving a zero. In such a case, half the received symbols would be correct *due to chance alone,* and the system might appear to be providing 500 bits/s while actually no information is being received at all. Equally "good" reception could be obtained by dispensing with the channel entirely and "flipping a coin" within the receiver. The proper correction to apply to the amount of information transmitted is the amount of information that is lost in the channel. Shannon [3] uses a correction factor called *equivocation* to account for the uncertainty in the received signal. Equivocation is defined as the *conditional entropy* of the message X, given Y, or

$$H(X|Y) = -\sum_{X,Y} P(X, Y) \log_2 P(X|Y)$$

$$= -\sum_{Y} P(Y) \sum_{X} P(X|Y) \log_2 P(X|Y)$$

(9.10)

where X is the transmitted source message, Y is the received signal, $P(X, Y)$ is the joint probability of X and Y, and $P(X|Y)$ is the conditional probability of X given Y. Equivocation can be thought of as the uncertainty that message X was sent, having received Y. For an *error-free channel,* $H(X|Y) = 0$, because having received Y, there is complete certainty about the message X. However, for a channel with a nonzero probability of symbol error, $H(X|Y) > 0$, because the channel introduces uncertainty. Consider a binary sequence, X, where the a priori source probabilities are $P(X = 1) = P(X = 0) = \frac{1}{2}$, and where, on the average, the channel produces one error in a received sequence of 100 bits ($P_B = 0.01$). Using Equation (9.10), the equivocation $H(X|Y)$ is expressed as

$$H(X|Y) = -[(1 - P_B) \log_2 (1 - P_B) + P_B \log_2 P_B]$$

$$= -(0.99 \log_2 0.99 + 0.01 \log_2 0.01)$$

$$= 0.081 \text{ bit/received symbol}$$

Thus, the channel introduces 0.081 bit of uncertainty to each received symbol.

Shannon showed that the average effective information content H_{eff} at the receiver is obtained by subtracting the equivocation from the entropy of the source. Therefore,

$$H_{\text{eff}} = H(X) - H(X|Y)$$

(9.11)

For a system transmitting equally likely binary symbols, the entropy $H(X)$ is 1 bit/symbol. When the symbols are received with $P_B = 0.01$, the equivocation is 0.081 bit/received symbol as was calculated above. Then, using Equation (9.11), the effective entropy of the received signal H_{eff} is

$$H_{eff} = 1 - 0.081 = 0.919 \text{ bit/received symbol}$$

Thus, if $R = 1000$ binary symbols transmitted per second, for example, the effective information bit rate R_{eff} can then be expressed as

$$\begin{aligned} R_{eff} &= RH_{eff} \\ &= 1000 \text{ symbols/s} \times 0.919 \text{ bit/symbol} = 919 \text{ bits/s} \end{aligned} \qquad (9.12)$$

Notice that in the extreme case, where $P_B = 0.5$,

$$H(X|Y) = -(0.5 \log_2 0.5 + 0.5 \log_2 0.5)$$

$$= 1 \text{ bit/symbol}$$

and, applying Equations (9.12) and (9.11) to the $R = 1000$ symbols/s example, yields

$$R_{eff} = 1000 \text{ symbols/s} (1 - 1) = 0 \text{ bit/s}$$

as should be expected.

Example 9.3   Apparent Contradiction in the Shannon Limit

Plots of P_B versus E_b/N_0 typically display a smooth increase of P_B as E_b/N_0 is decreased. For example, the bit error probability for the curves in Figure 9.1 shows P_B *tending* to 0.5 in the limit as E_b/N_0 approaches zero. Thus there is apparently always a nonvanishing information rate, regardless of how small E_b/N_0 becomes. This *appears to contradict* the Shannon limit of $E_b/N_0 = -1.6$ dB, below which no error-free information rate can be supported per unit bandwidth, or below which even an infinite bandwidth cannot support a finite information rate (see Figure 9.4).

(a) Suggest a way of resolving the apparent contradiction.
(b) Show how Shannon's equivocation correction can resolve it for a binary PSK system where the source has an entropy of 1 bit/symbol. Consider that the operating point on Figure 9.1b corresponds to $E_b/N_0 = 0.1$ (−10 dB).

Solution

(a) The value of E_b, traditionally used in link calculations for practical systems, is invariably the received signal energy per *transmitted symbol*. However, the meaning of E_b in Equation (9.6) is the signal energy per bit of *received information*. The information loss caused by the noisy channel must be taken into account to resolve the apparent contradiction.
(b) Following Equation (4.79) for BPSK, we write

$$P_B = Q(\sqrt{2E_b/N_0}) = Q(0.447)$$

where Q is defined in Equation (3.43) and tabulated in Table B.1. From the tabulation, P_B is found to be 0.33. Next, we solve for the equivocation and effective entropy:

$$H(X|Y) = -[(1 - P_B) \log_2 (1 - P_B) + P_B \log_2 P_B]$$

$$= -(0.67 \log_2 0.67 + 0.33 \log_2 0.33)$$
$$= 0.915 \text{ bit/symbol}$$

$$H_{\text{eff}} = H(X) - H(X|Y)$$
$$= 1 - 0.915$$
$$= 0.085 \text{ bit/symbol}$$

Hence,

$$\left(\frac{E_b}{N_0}\right)_{\text{eff}} = \frac{(E_b/N_0) \text{ joules per symbol/watts per hertz}}{H_{\text{eff}} \text{ bits/symbol}}$$

$$= \frac{0.1}{0.085} = 1.176 \frac{\text{joules per bit}}{\text{watts/Hz}}$$

$$= 0.7 \text{ dB}$$

Thus, the effective value of E_b/N_0 is equal to 0.7 dB per received information bit, which is well above Shannon's limit of -1.6 dB.

9.5 BANDWIDTH-EFFICIENCY PLANE

Using Equation (9.6), we can plot normalized channel bandwidth W/C in Hz/bits/s versus E_b/N_0, as shown in Figure 9.4. Here, with the abscissa taken as E_b/N_0, we see the *true power-bandwidth trade-off* at work. It can be shown [5] that well-designed systems tend to operate near the "knee" of this power-bandwidth trade-off curve for the ideal $(R = C)$ channel. Actual systems are frequently within 10 dB or less of the performance of the ideal. The existence of the knee means that systems seeking to reduce the channel bandwidth they occupy or to reduce the signal power they require must make an increasingly unfavorable exchange in the other parameter. For example, from Figure 9.4, an ideal system operating at an E_b/N_0 of 1.8 dB and using a normalized bandwidth of 0.5 Hz/bits/s would have to increase E_b/N_0 to 20 dB to reduce the bandwidth occupancy to 0.1 Hz/bits/s. Trade-offs in the other direction are similarly inequitable.

Using Equation (9.6c), we an also plot C/W versus E_b/N_0. This relationship is shown plotted on the R/W versus E_b/N_0 plane in Figure 9.6. We shall denote this plane as the *bandwidth-efficiency plane*. The ordinate R/W is a measure of how much data can be communicated in a specified bandwidth within a given time; it therefore reflects how efficiently the bandwidth resource is utilized. The abscissa is E_b/N_0, in units of decibels. For the case in which $R = C$ in Figure 9.6, the curve represents a boundary that separates a region characterizing practical communication systems from a region where such communication systems are not theoretically possible. Like Figure 9.2, the bandwidth-efficiency plane in Figure 9.6 sets the limiting performance that can be achieved by practical systems. Since the abscissa in Figure 9.6 is E_b/N_0 rather than SNR, Figure 9.6 is more useful for comparing digital communication modulation and coding trade-offs than is Figure 9.2. Note that Figure 9.6 illustrates bandwidth efficiency versus E_b/N_0 for single-carrier systems. For multiple-carrier systems, bandwidth efficiency is also a function of carrier spacing (which depends on the modulation type). The trade-off becomes how closely can

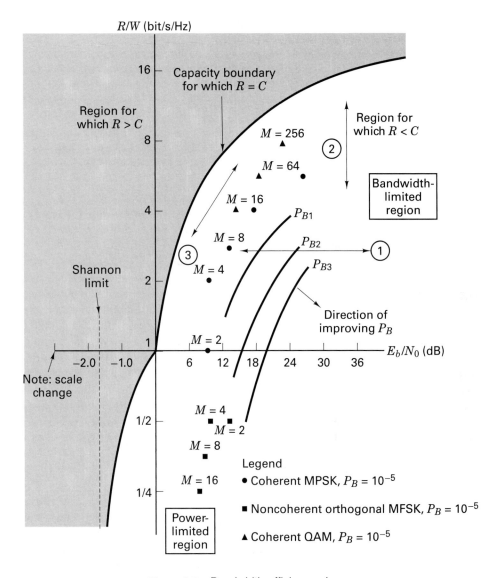

Figure 9.6 Bandwidth-efficiency plane.

the carriers be spaced (thereby improving bandwidth efficiency) without suffering an unacceptable amount of adjacent channel interference (ACI).

9.5.1 Bandwidth Efficiency of MPSK and MFSK Modulation

On the bandwidth-efficiency plane of Figure 9.6 are plotted the operating points for coherent MPSK modulation at a bit error probability of 10^{-5}. We assume Nyquist (ideal rectangular) filtering at baseband, so that the minimum double-sideband (DSB) bandwidth at an intermediate frequency (IF) is $W_{IF} = 1/T$, where T is the symbol duration. Thus using (Equation (9.1), the bandwidth efficiency is $R/W = \log_2 M$, where M is the symbol set size. For realistic channels and waveforms, the performance must be reduced to account for the bandwidth increase required to implement realizable filters. Notice that for MPSK modulation, R/W increases with increasing M. Notice also that the location of the MPSK points indicates that BPSK ($M = 2$) and quaternary PSK or QPSK ($M = 4$) require the same E_b/N_0. That is, for the same value of E_b/N_0, QPSK has a bandwidth efficiency of 2 bits/s/Hz, compared to 1 bit/s/Hz for BPSK. This unique features stems from the fact that QPSK is effectively a composite of two BPSK signals transmitted on orthogonal components of the carrier.

Also plotted on the bandwidth-efficiency plane of Figure 9.6 are the operating points for noncoherent orthogonal MFSK modulation, at a bit error probability of 10^{-5}. We assume that the IF transmission bandwidth is $W_{IF} = M/T$ (see Section 4.5.4.1), and thus using Equation (9.1), the bandwidth efficiency is $R/W = (\log_2 M)/M$. Notice that for MFSK modulation, R/W decreases with increasing M. Notice also that the position of the MFSK points indicates that BFSK ($M = 2$) and quaternary FSK ($M = 4$) have the same bandwidth efficiency, even though the former requires greater E_b/N_0 for the same error probability. The bandwidth efficiency varies with the modulation index (tone spacing in hertz divided by bit rate). Under the assumption that an equal increment of bandwidth is required for each MFSK tone the system uses, it can be seen that for $M = 2$, the bandwidth efficiency is 1 bit/s/2 Hz or $\frac{1}{2}$, and for $M = 4$, similarly, the R/W is 2 bits/s/4 Hz or $\frac{1}{2}$. Thus binary and 4-ary orthogonal FSK are curiously characterized by the same value of R/W.

Operating points for coherent quadrature amplitude modulation (QAM) are also plotted in Figure 9.6. Of the modulations shown, QAM is clearly the most bandwidth efficient; it is treated in greater detail in Section 9.8.3.

9.5.2 Analogies Between Bandwidth-Efficiency and Error-Probability Planes

The bandwidth-efficiency plane in Figure 9.6 is analogous to the error-probability plane in Figure 9.1. The Shannon limit of the Figure 9.1 plane is analogous to the capacity boundary of the Figure 9.6 plane. The curves in Figure 9.1 were referred to as equibandwidth curves. In Figure 9.6, we can analogously describe equi-error-probability curves for various modulation and coding schemes. The curves, labeled P_{B1}, P_{B2}, and P_{B3}, are hypothetical constructions for some arbitrary modulation and coding scheme; the P_{B1} curve represents the largest error probability of the three curves, and the P_{B3} curve represents the smallest. The general direction in which the curves move for improved P_B is indicated on the figure.

Just as potential trade-offs among P_B, E_b/N_0, and W were considered for the error-probability plane, the same trade-offs can be considered on the bandwidth efficiency plane. The potential trade-offs are seen in Figure 9.6 as changes in operating point in the direction shown by the arrows. Movement of the operating point along line 1 can be viewed as trading P_B versus E_b/N_0, with R/W fixed. Similarly, movement along line 2 is seen as trading P_B versus W (or R/W), with E_b/N_0 fixed.

Finally, movement along line 3 illustrates trading W (or R/W) versus E_b/N_0, with P_B fixed. In Figure 9.6, as in Figure 9.1, movement along line 1 can be effected by increasing or decreasing the available E_b/N_0. However, movement along line 2 or line 3 requires changes in the system modulation or coding scheme.

The two primary communications resources are the transmitted power and the channel bandwidth. In many communication systems, one of these resources may be more precious than the other, and hence most systems can be classified as either power limited or bandwidth limited. In *power-limited systems,* coding schemes can be used to save power at the expense of bandwidth, whereas in *bandwidth-limited systems,* spectrally efficient modulation techniques can be used to save bandwidth at the expense of power.

9.6 MODULATION AND CODING TRADE-OFFS

Figure 9.7 is useful in pointing out analogies between the two performance planes, the error-probability plane of Figure 9.1 and the bandwidth-efficiency plane of Figure 9.6. Figure 9.7a and b represent the same planes as Figures 9.1 and 9.6, respectively. They have been redrawn as symmetrical by choosing appropriate scales. In each case the arrows and their labels describe the general effect of moving an operating point in the direction of the arrow by means of appropriate modulation and coding techniques. The notations G, C, and F stand for the trade-off considerations "*G*ained or achieved," "*C*ost or expended," and "*F*ixed or unchanged," respectively. The parameters being traded are P_B, W, R/W, and P (power or S/N). Just as the movement of an operating point toward the Shannon limit in Figure 9.7a can

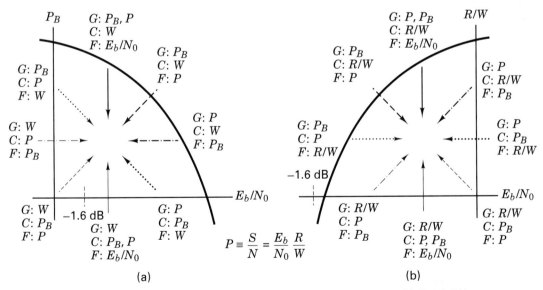

Figure 9.7 Modulation/coding trade-offs. (a) Error probability plane. (b) Bandwidth-efficiency plane.

achieve improved P_B or reduced required transmitter power at the cost of bandwidth, so too movement toward the capacity boundary in Figure 9.7b can improve bandwidth efficiency at the cost of increased required power or degraded P_B.

Most often, these trade-offs are examined with a fixed P_B (constrained by the system requirement) in mind. Therefore, the most interesting arrows are those having bit error probability (marked F: P_B). There are four such arrows on Figure 9.7, two on the error probability plane and two on the bandwidth-efficiency plane. Arrows marked with the same pattern indicate correspondence between the two planes. System operation can be characterized by either of these two planes. The planes represent two ways of looking at some of the key system parameters; each plane highlights slightly different aspects of the overall design problem. The error probability plane tends to be most useful with *power-limited systems*, whereas when we move from curve to curve, the bandwidth requirements are only inferred, while the bit error probability is clearly displayed. The bandwidth-efficiency plane is generally more useful for examining *bandwidth-limited systems;* here, as we move from curve to curve, the bit-error probability is only inferred, but the bandwidth requirements are explicit.

The two system trade-off planes, error probability and bandwidth efficiency, have been presented *heuristically* with simple examples (orthogonal and multiple phase signaling) to provide some insight into the design issues of trading-off error probability, bandwidth, and power. The ideas are useful for *most modulation and coding schemes,* with the following caveat. For *some* codes or combined modulation and coding schemes, the performance curves *do not move as predictably* as those for the examples chosen here. The reason has to do with the error-correcting capability and bandwidth expansion features of the particular code. For example, the performance of coherent PSK combined with several codes was illustrated in Figure 6.22. Examine the curves characterizing the two BCH codes, (127, 64) and (127, 36). It should be clear from their relative positions that the (127, 64) code manifests *greater coding gain* than the (127, 36) code. This violates our expectations since, within the same block size, the latter code has greater redundancy (requires more bandwidth expansion) than the former. Also, in the area of trellis-coded modulation covered in Section 9.10, we consider codes that provide coding gain without any bandwidth expansion. Performance curves for such coding schemes will also behave differently from the curves of most modulation and coding schemes discussed so far.

9.7 DEFINING, DESIGNING, AND EVALUATING DIGITAL COMMUNICATION SYSTEMS

This section is intended to serve as a "road map" for outlining typical steps that need to be considered in meeting the bandwidth, power, and error-performance requirements of a digital communication system. The criteria for choosing modulation and coding schemes, based on whether a system is bandwidth limited or power limited, are reviewed for several system examples. We will emphasize the subtle but straightforward relationships that exist when transforming from data-bits to channel-bits to symbols to chips.

The design of any digital communication system begins with a description of the channel (received power, available bandwidth, noise statistics and other impairments, such as fading), and a definition of the system requirements (data rate and error performance). Given the channel description, we need to determine design choices that best match the channel and meet the performance requirements. An orderly set of transformations and computations has evolved to aid in characterizing a system's performance. Once this approach is understood, it can serve as the format for evaluating most communication systems. In subsequent sections, we examine three system examples, chosen to provide a representative assortment: a bandwidth-limited uncoded system, a power-limited uncoded system, and a bandwidth-limited and power-limited coded system. In this section, we deal with real-time communication systems, where the term *coded* (or *uncoded*) refers to the presence (or absence) of error-correction coding schemes involving the use of *redundant bits* and expanded bandwidth.

Two primary communications resources are the *received power* and the *available transmission bandwidth*. In many communication systems, one of these resources may be more precious than the other, and hence most systems can be classified as either bandwidth limited or power limited. In bandwidth-limited systems, spectrally efficient modulation techniques can be used to save bandwidth at the expense of power, whereas in power-limited systems, power-efficient modulation techniques can be used to save power at the expanse of bandwidth. In both bandwidth- and power-limited systems, error-correction coding (often called *channel coding*) can be used to save power or to improve error performance at the expense of bandwidth. Trellis-coded modulation (TCM) schemes have been used to improve the error performance of bandwidth-limited channels without *any* increase in bandwidth [6]. These methods are considered in Section 9.10.

9.7.1 *M*-ary Signaling

For signaling schemes that process k bits at a time, the signaling is called *M*-ary (see Section 3.8). Each symbol in an *M*-ary alphabet can be related to a unique sequence of k bits, where

$$M = 2^k \quad \text{or} \quad k = \log_2 M \tag{9.13}$$

and where M is the size of the alphabet. In the case of digital transmission, the term *symbol* refers to the member of the *M*-ary alphabet that is transmitted during each symbol duration T_s. In order to transmit the symbol, it must be mapped onto an electrical voltage or current waveform. Because the waveform represents the symbol, the terms *symbol* and *waveform* are sometimes used interchangeably. Since one of M symbols or waveforms is transmitted during each symbol duration T_s, the date rate R can be expressed as

$$R = \frac{k}{T_s} = \frac{\log_2 M}{T_s} \quad \text{bit/s} \tag{9.14}$$

From Equation (9.14), we write that the *effective* duration T_b of each bit in terms of the symbol duration T_s or the symbol rate R_s is

$$T_b = \frac{1}{R} = \frac{T_s}{k} = \frac{1}{kR_s} \qquad (9.15)$$

Then, using Equations (9.13) and (9.15), we can express the symbol rate R_s in terms of the bit rate R, as was presented earlier:

$$R_s = \frac{R}{\log_2 M} \qquad (9.16)$$

From Equations (9.14) and (9.15), it is seen that any digital scheme that transmits $k = (\log_2 M)$ bits in T_s seconds, using a bandwidth of W Hz, operates at a bandwidth efficiency of

$$\frac{R}{W} = \frac{\log_2 M}{WT_s} = \frac{1}{WT_b} \quad \text{bits/s/Hz} \qquad (9.17)$$

where T_b is the effective time duration of each data bit.

9.7.2 Bandwidth-Limited Systems

From Equation (9.17), it can be seen that any digital communication system will become more bandwidth efficient as its WT_b product is decreased. Thus, signals with small WT_b products are often used with bandwidth-limited systems. For example, the Global System for Mobile (GSM) Communication uses Gaussian minimum shift keying (GMSK) modulation having a WT_b product equal to 0.3 Hz/bit/s [7], where W is the 3-dB bandwidth of a Gaussian filter.

For uncoded bandwidth-limited systems, the objective is to maximize the transmitted information rate within the allowable bandwidth, at the expense of E_b/N_0 (while maintaining a specified value of bit-error probability P_B). On the bandwidth-efficiency plane of Figure 9.6 are plotted the operating points for coherent M-ary PSK (MPSK) at $P_B = 10^{-5}$. We shall assume Nyquist (ideal rectangular) filtering at baseband [2], so that, for MPSK, the required double-sideband (DSB) bandwidth at an intermediate frequency (IF) is related to the symbol rate by

$$W = \frac{1}{T_s} = R_s \qquad (9.18)$$

where T_s is the symbol duration and R_s is the symbol rate. The use of Nyquist filtering results in the *minimum* required transmission bandwidth that yields zero intersymbol interference; such ideal filtering gives rise to the name *Nyquist minimum bandwidth.* Note that the bandwidth of nonorthogonal signaling, such as MPSK or MQAM, does not depend on the density of the signaling points in the constellation but only on the speed of signaling. When a phasor is transmitted, the system cannot distinguish as to whether that signal arose from a sparse alphabet set or a dense alphabet set. It is this aspect of nonorthogonal signals that allows us to pack the signaling space densely and thus achieve improved bandwidth efficiency at the expense of power. From Equations (9.17) and (9.18), the bandwidth efficiency of MPSK modulated signals using Nyquist filtering can be expressed as

$$\frac{R}{W} = \log_2 M \quad \text{bits/s/Hz} \tag{9.19}$$

The MPSK points plotted in Figure 9.6 confirm the relationship shown in Equation (9.19). Note that MPSK modulation is a bandwidth-efficient scheme. As M increases in value, R/W also increases. From Figure 9.6, it can be verified that MPSK modulation can achieve improved bandwidth efficiency at the cost of increased E_b/N_0. Many highly bandwidth-efficient modulation schemes have been investigated [8], but such schemes are beyond the scope of this book.

Two regions, the bandwidth-limited region and the power-limited region, are shown on the bandwidth-efficiency plane of Figure 9.6. Notice that the desirable trade-offs associated with each of these regions are not equitable. For the bandwidth-limited region, large R/W is desired; however, as E_b/N_0 is increased, the capacity boundary curve flattens out and ever-increasing amounts of additional E_b/N_0 are required to achieve improvement in R/W. A similar relationship is at work in the power-limited region. Here a savings in E_b/N_0 is desired, but the capacity boundary curve is steep; to achieve a small reduction in required E_b/N_0, requires a large reduction in R/W.

9.7.3 Power-Limited Systems

For the case of power-limited systems in which power is scarce but system bandwidth is available (e.g., a space communication link), the following trade-offs, which can be seen in Figure 9.1a, are possible: (1) improved P_B at the expense of bandwidth for a fixed E_b/N_0; or (2) reduction in E_b/N_0 at the expense of bandwidth for a fixed P_B. A "natural" modulation choice for a power-limited system is M-ary FSK (MFSK). Plotted on Figure 9.6 are the operating points for noncoherent orthogonal MFSK modulation at $P_B = 10^{-5}$. For such MFSK, the IF minimum bandwidth, assuming minimum tone spacing, is given by (see Section 4.5.4.1)

$$W = \frac{M}{T_s} = MR_s \tag{9.20}$$

where T_s is the symbol duration, and R_s is the symbol rate. With M-ary FSK, the required transmission bandwidth is expanded M-fold over binary FSK since there are M different orthogonal waveforms, each requiring a bandwidth of $1/T_s$. Thus, from Equations (9.17) and (9.20), the bandwidth efficiency of noncoherent MFSK signals can be expressed as

$$\frac{R}{W} = \frac{\log_2 M}{M} \quad \text{bits/s/Hz} \tag{9.21}$$

Notice the important difference between the *bandwidth efficiency* (R/W) of MPSK expressed in Equation (9.19) and that of MFSK expressed in Equation (9.21). With MPSK, R/W increases as the signal dimensionality M increases. With

MFSK there are two mechanisms at work. The numerator shows the same increase in R/W with larger M, as in the case of MPSK. But the denominator indicates a decrease in R/W with larger M. As M grows larger, the denominator grows faster than the numerator, and thus R/W decreases. The MFSK points plotted in Figure 9.6 confirm the relationship shown in Equation (9.21), that orthogonal signaling such as MFSK is a bandwidth-expansive scheme. From Figure 9.6, it can be seen that MFSK modulation can be used for realizing a reduction in required E_b/N_0, at the cost of increased bandwidth.

It is important to emphasize that in Equations (9.18) and (9.19) for MPSK, and for all the MPSK points plotted in Figure 9.6, Nyquist (ideal rectangular) filtering has been assumed. Such filters are not realizable. For *realistic* channels and waveforms, the required transmission bandwidth must be *increased* in order to account for *realizable* filters.

In each of the examples that follow, we consider radio channels, disturbed *only* by additive white Gaussian noise (AWGN) and having no other impairments. For simplicity, the modulation choice is limited to *constant-envelope types*—either MPSK or noncoherent orthogonal MFSK. Thus, for an *uncoded* system, if the channel is bandwidth limited, MPSK is selected, and if the channel is power limited, MFSK is selected. Note that, *when error-correction coding is considered,* modulation selection is not so simple, because there exist coding techniques [9] that can provide power-bandwidth trade-offs more effectively than would be possible through the use of any M-ary modulation scheme.

Note that in the most general sense, M-ary signaling can be regarded as a *waveform-coding* procedure. That is, whenever we select an M-ary modulation technique instead of a binary one, we *in effect* have replaced the binary waveforms with *better* waveforms—either better for bandwidth performance (MPSK), or better for power performance (MFSK). Even though orthogonal MFSK signaling can be thought of as being a coded system (it can be described as a first-order Reed–Muller code [10]), we shall here restrict our use of the term *coded system* to refer only to those traditional error-correction codes using redundancies, such as block codes or convolutional codes.

9.7.4 Requirements for MPSK and MFSK Signaling

The basic relationship between the symbol (or waveform) transmission rate R_s and the data rate R was shown in Equation (9.16) to be

$$ R_s = \frac{R}{\log_2 M} $$

Using this relationship together with Equations (9.18) through (9.21), and a given data rate of $R = 9600$ bit/s, Table 9.1 has been compiled [11]. The table is a summary of symbol rate, minimum bandwidth, and bandwidth efficiency for MPSK and noncoherent orthogonal MFSK, for the values of $M = 2, 4, 8, 16$, and 32. Also included in Table 9.1 are the required values of E_b/N_0 to achieve a bit-error probability of 10^{-5} for MPSK and MFSK for each value of M shown. These E_b/N_0

M	k	R (bit/s)	R_s (symb/s)	MPSK Minimum Bandwidth (Hz)	MPSK R/W	MPSK E_b/N_0 (dB) $P_B = 10^{-5}$	Noncoherent Orthog MFSK Min Bandwidth (Hz)	MFSK R/W	MFSK E_b/N_0 (dB) $P_B = 10^{-5}$
2	1	9600	9600	9600	1	9.6	19,200	1/2	13.4
4	2	9600	4800	4800	2	9.6	19,200	1/2	10.6
8	3	9600	3200	3200	3	13.0	25,600	1/3	9.1
16	4	9600	2400	2400	4	17.5	38,400	1/4	8.1
32	5	9600	1920	1920	5	22.4	61,440	5/32	7.4

entries were computed using relationships that are presented later. The E_b/N_0 entries corroborate the trade-offs shown in Figure 9.6. As M increases, MPSK signaling provides more bandwidth efficiency at the cost of increased E_b/N_0, while MFSK signaling allows for a reduction in E_b/N_0 at the cost of increased bandwidth. The next three sections are presented in the context of examples taken from Table 9.1.

9.7.5 Bandwidth-Limited Uncoded System Example

Suppose we are given a bandwidth-limited AWGN radio channel with an available bandwidth of $W = 4000$ Hz. Also, consider that the link constraints (transmitter power, antenna gains, path loss, etc.) result in the ratio of received signal power to noise-power spectral density (P_r/N_0) being equal to 53 dB-Hz. Let the required data rate R be equal to 9600 bits/s, and let the required bit-error performance P_B be *at most* 10^{-5}. The goal is to choose a modulation scheme that meets the required performance. In general, an error-correction coding scheme may be needed if none of the allowable modulation schemes can meet the requirements. However, in this example, we will see that the use of error-correction coding is not necessary.

For any digital communication system, the relationship between received power to noise-power spectral density (P_r/N_0) and received bit-energy to noise-power spectral density (E_b/N_0) was shown in Equation (5.20c) to be

$$\frac{P_r}{N_0} = \frac{E_b}{N_0} R \qquad (9.22)$$

Solving for E_b/N_0, in decibels, we obtain

$$\frac{E_b}{N_0} \text{ (dB)} = \frac{P_r}{N_0} \text{ (dB-Hz)} - R \text{ (dB-bit/s)} \qquad (9.23)$$

$$= 53 \text{ dB-Hz} - (10 \times \log_{10} 9600) \text{ dB-bit/s} = 13.2 \text{ dB (or 20.89)}$$

Since the required data rate of 9600 bits/s is much larger than the available bandwidth of 4000 Hz, the channel can be described as *bandwidth limited*. We therefore select MPSK as our modulation scheme. Recall that we have confined the possible modulation choices to be constant-envelope types; without such a restriction, it would be possible to select a modulation type with greater bandwidth efficiency. In an effort to conserve power, we next compute the *smallest possible* value of M, such that the symbol rate is *at most* equal to the available bandwidth of 4000 Hz. From Table 9.1, it is clear that the smallest value of M meeting this requirement is $M = 8$. Our next task is to determine whether the required bit-error performance of $P_B \leq 10^{-5}$ can be met by using 8-PSK modulation alone, or whether it is necessary to additionally use an error-correction coding scheme. It can be seen from Table 9.1, that 8-PSK *alone* will meet the requirements, since the required E_b/N_0 listed for 8-PSK is less then the received E_b/N_0 that was derived in Equation (9.23). However, imagine that we do not have Table 9.1. Let us demonstrate how to evaluate whether or not error-correction coding is necessary.

Figure 9.8 shows the basic modulator/demodulator (MODEM) block diagram summarizing the functional details of this design. At the modulator, the transformation from data bits to symbols yields an output symbol rate R_s that is a factor $(\log_2 M)$ smaller than the input data-bit rate R, as can be seen in Equation (9.16). Similarly, at the input to the demodulator, the symbol-energy to noise-power spectral density E_s/N_0 is a factor $(\log_2 M)$ larger than E_b/N_0, since each symbol is made up of $(\log_2 M)$ bits. Because E_s/N_0 is larger than E_b/N_0 by the same factor that R_s is smaller than R, we can expand Equation (9.22), as follows:

$$\frac{P_r}{N_0} = \frac{E_b}{N_0} R = \frac{E_s}{N_0} R_s \tag{9.24}$$

The demodulator receives a waveform (in this example, one of $M = 8$ possible phase shifts) during each time interval T_s. The probability that the demodulator makes a symbol error $P_E(M)$ is well approximated by [12], and we write

$$P_E(M) \approx 2Q\left[\sqrt{\frac{2E_s}{N_0}} \sin\left(\frac{\pi}{M}\right)\right] \quad \text{for } M > 2 \tag{9.25}$$

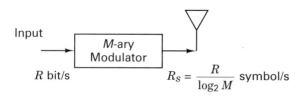

Input →
R bit/s

M-ary Modulator

$R_s = \dfrac{R}{\log_2 M}$ symbol/s

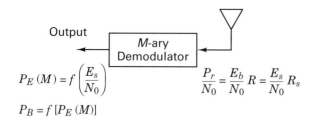

Output ←

M-ary Demodulator

$P_E(M) = f\left(\dfrac{E_s}{N_0}\right)$

$P_B = f[P_E(M)]$

$\dfrac{P_r}{N_0} = \dfrac{E_b}{N_0} R = \dfrac{E_s}{N_0} R_s$

Figure 9.8: Basic modulator/ demodulator (MODEM) without channel coding.

where $Q(x)$, the *complementary error function*, was defined in Equation (3.43) as

$$Q(x) = \frac{1}{\sqrt{2\pi}} \int_x^\infty \exp\left(-\frac{u^2}{2}\right) du$$

In Figure 9.8 and all the figures that follow, rather than show explicit probability relationships, the generalized notation $f(x)$ has been used to indicate some functional dependence on x.

A traditional way of characterizing communication (power) efficiency or error performance in digital systems is in terms of the received E_b/N_0 in decibels. This E_b/N_0 description has become standard practice. However, recall that at the input to the demodulator/detector, there are no bits; there are only waveforms that have been assigned bit meanings. Thus, the received E_b/N_0 value represents a bit-apportionment of the arriving waveform energy. A more precise (but unwieldy) name would be the energy per *effective bit* versus N_0. To solve for $P_E(M)$ in Equation (9.25), we first need to compute the ratio of received symbol-energy to noise-power spectral density, E_s/N_0. Since, from Equation (9.23), $E_b/N_0 = 13.2$ dB (or 20.89), and because each symbol is made up of $(\log_2 M)$ bits, we compute, with $M = 8$,

$$\frac{E_s}{N_0} = (\log_2 M) \frac{E_b}{N_0} = 3 \times 20.89 = 62.67 \qquad (9.26)$$

Using the results of Equation (9.26) in Equation (9.25), yields the symbol-error probability, $P_E = 2.2 \times 10^{-5}$. To transform this to bit-error probability, we need to use the relationship between bit-error probability P_B and symbol-error probability P_E for multiple-phase signaling [10]. We write

$$P_B \approx \frac{P_E}{\log_2 M} \quad (\text{for } P_E \ll 1) \qquad (9.27)$$

which is a good approximation, when Gray coding [12] is used for the bit-to-symbol assignment. This last computation yields $P_B = 7.3 \times 10^{-6}$, which meets the required bit-error performance. Thus, in this example, no error-correction coding is necessary and 8-PSK modulation represents the design choice to meet the requirements of the bandwidth-limited channel (which we had predicted by examining the required E_b/N_0 values in Table 9.1).

9.7.6 Power-Limited Uncoded System Example

Now, suppose that we have exactly the same data rate and bit-error probability requirements as in the example of Section 9.7.5. However, in this example, let the available bandwidth W be equal to 45 kHz, and let the available P_r/N_0 be equal to 48 dB-Hz. As before, the goal is to choose a modulation or modulation/coding scheme that yields the required performance. In this example, we shall again find that error-correction coding is not required.

The channel in this example is clearly not bandwidth limited since the available bandwidth of 45 kHz is more than adequate for supporting the required data rate of 9600 bits/s. The received E_b/N_0 is found from Equation (9.23), as follows:

$$\frac{E_b}{N_0} \text{(dB)} = 48 \text{ db-Hz} - (10 \times \log_{10} 9600) \text{ dB-bit/s} = 8.2 \text{ dB (or 6.61)} \quad (9.28)$$

Since there is abundant bandwidth but a relatively small amount of E_b/N_0 for the required bit-error probability, this channel may be referred to as *power limited.* We therefore choose MFSK as the modulation scheme. In an effort to conserve power, we next search for the *largest possible M* such that the MFSK minimum bandwidth is not expanded beyond our available bandwidth of 45 kHz. From Table 9.1, we see that such a search results in the choice of $M = 16$. Out next task is to determine whether the required error performance of $P_B \leq 10^{-5}$ can be met by using 16-FSK alone, without the use of any error-correction coding. Similar to the previous example, it can be seen from Table 9.1, that 16-FSK *alone* will meet the requirements, since the required E_b/N_0 listed for 16-FSK is less than the received E_b/N_0 that was derived in Equation (9.28). However, imagine again that we do not have Table 9.1. Let us demonstrate how to evaluate whether or not error-correction coding is necessary.

As before, the block diagram in Figure 9.8 summarizes the relationship between symbol rate R_s and bit rate R, and between E_s/N_0 and E_b/N_0, which is identical to each of the respective relationships in the previous bandwidth-limited example. In this example, the 16-FSK demodulator receives a waveform (one of 16 possible frequencies) during each symbol time interval T_s. For noncoherent MFSK, the probability that the demodulator makes a symbol error is approximated by [13]

$$P_E(M) \leq \frac{M-1}{2} \exp\left(-\frac{E_s}{2N_0}\right) \quad (9.29)$$

To solve for $P_E(M)$ in Equation (9.29), we need to compute E_s/N_0, as we did in Example 1. Using the results of Equation (9.28) in Equation (9.26), with $M = 16$, we get

$$\frac{E_s}{N_0} = (\log_2 M) \frac{E_b}{N_0} = 4 \times 6.61 = 26.44 \quad (9.30)$$

Next, we combine the results of Equation (9.30) in Equation (9.29) to yield the symbol-error probability $P_E = 1.4 \times 10^{-5}$. To transform this to bit-error probability P_B, we need to use the relationship between P_B and P_E for orthogonal signaling [13], given by

$$P_B = \frac{2^{k-1}}{2^k - 1} P_E \quad (9.31)$$

This last computation yields $P_B = 7.3 \times 10^{-6}$, which meets the required bit-error performance. Thus, we can meet the given specifications for this power-limited channel by using 16-FSK modulation, without any need for error-correction coding (which we had predicted by examining the required E_b/N_0 values in Table 9.1).

9.7.7 Bandwidth-Limited and Power-Limited Coded System Example

In this example, we start with the same channel parameters as in the bandwidth-limited example of Section 9.7.5, namely, $W = 4000$ Hz, $P_r/N_0 = 53$ dB-Hz, and $R = 9600$ bits/s, with one exception. In the present example, we specify that the bit-error probability must be *at most* 10^{-9}. Since the available bandwidth is 4000 Hz, and from Equation (9.23) the available E_b/N_0 is 13.2 dB, it should be clear from Table 9.1, that the system is both bandwidth limited *and* power limited (8-PSK is the only possible choice to meet the bandwidth constraint; however, the available E_b/N_0 of 13.2 dB is certainly insufficient to meet the required bit-error probability of 10^{-9}). For such a small value of P_B, the system shown in Figure 9.8 will obviously be inadequate, and we need to consider the performance improvement that error-correction coding (within the available bandwidth) can provide. In general, one can use convolutional codes or block codes. To simplify the explanation, we shall choose a block code. The Bose, Chaudhuri, and Hocquenghem (BCH) codes form a large class of powerful error-correcting cyclic (block) codes [14]. For this example, let us select one of the codes from this family of codes. Table 9.2

TABLE 9.2 BCH Codes (Partial Catalog)

n	k	t
7	4	1
15	11	1
	7	2
	5	3
31	26	1
	21	2
	16	3
	11	5
63	57	1
	51	2
	45	3
	39	4
	36	5
	30	6
127	120	1
	113	2
	106	3
	99	4
	92	5
	85	6
	78	7
	71	9
	64	10
	57	11
	50	13
	43	14
	36	15
	29	21
	22	23
	15	27
	8	31

presents a partial catalog of the available BCH codes in terms of n, k, and t, where k represents the number of information or data bits that the codes transforms into a longer block or n code bits (also called *channel bits* or *channel symbols*), and t represents the largest number of incorrect channel bits that the code can correct within each n-sized block. The *rate* of a code is defined as the ratio k/n; its inverse represents a measure of the code's redundancy.

Since this example is represented by the same bandwidth-limited parameters that were given in Section 9.7.5, we start with the same 8-PSK modulation as before in order to meet the stated bandwidth constraint. However, we now additionally need to employ error-correction coding so that the bit-error probability can be lowered to $P_B \leq 10^{-9}$. To make the optimum code selection from Table 9.2, we are guided by the following goals:

1. The output bit-error probability of the combined modulation/coding system must meet the system error requirement.
2. The rate of the code must not expand the required transmission bandwidth beyond the available channel bandwidth.
3. The code should be as simple as possible. Generally, the shorter the code, the simpler will be its implementation.

The uncoded 8-PSK minimum bandwidth requirement is 3200 Hz (see Table 9.1), and the allowable channel bandwidth is specified as 4000 Hz. Therefore, the uncoded signal bandwidth may be increased by *no more than* a factor of 1.25 (or an expansion of 25%). Thus, the very first step in this (simplified) code selection example is to eliminate the candidates from Table 9.2 that would expand the bandwidth by more than 25%. The remaining entries in Table 9.2 form a much reduced set of "bandwidth-compatible" codes, which have been listed in Table 9.3. In Table 9.3, two columns designated Coding Gain, G, have been added, where coding gain in decibels is defined as

$$G(\text{dB}) = \left(\frac{E_b}{N_0}\right)_{\text{uncoded}} (\text{dB}) - \left(\frac{E_b}{N_0}\right)_{\text{coded}} (\text{dB}) \qquad (9.32)$$

From Equation (9.32), coding gain can be described as a measure of the *reduction* in the required E_b/N_0 (in decibels) that needs to be provided, due to the error-

TABLE 9.3   Bandwidth-Compatible BCH Codes

n	k	t	Coding Gain, G (dB), with MPSK	
			$P_B = 10^{-5}$	$P_B = 10^{-9}$
31	26	1	1.8	2.0
63	57	1	1.8	2.2
	51	2	2.6	3.2
127	120	1	1.7	2.2
	113	2	2.6	3.4
	106	3	3.1	4.0

performance properties of the channel coding. Coding gain is a function of the modulation type and bit-error probability. In Table 9.3, the coding gain G has been computed for MPSK at $P_B = 10^{-5}$ and 10^{-9}. For MPSK modulation, G is relatively independent of the value of M. Thus, for a particular bit-error probability, a given code will provide approximately the same coding gain when used with any of the MPSK modulation schemes. The coding gains in Table 9.3 were calculated using a procedure outlined under the section below, entitled "Calculating Coding Gain."

Figure 9.9 illustrates a block diagram that summarizes the details of this system containing a modulator/demodulator (MODEM) and coding. When comparing Figure 9.9 with Figure 9.8, we see that the introduction of the encoder/decoder blocks has brought about additional transformations. In Figure 9.9, at the encoder/modulator are shown the relationships that exist when transforming from R bit/s to R_c channel-bit/s to R_s symbol/s.

We assume that our communication system is a real-time system and thus cannot tolerate any message delay. Therefore, the channel-bit rate R_c must *exceed* the data-bit rate R by the factor n/k. Further, each transmission symbol is made up of $(\log_2 M)$ channel bits, so the symbol rate R_s is *less* than R_c by the factor $(\log_2 M)$. For a system containing both modulation and coding, we summarize the rate transformations, as follows:

$$R_c = \left(\frac{n}{k}\right) R \qquad (9.33)$$

$$R_s = \frac{R_c}{\log_2 M} \qquad (9.34)$$

At the demodulator/decoder shown in Figure 9.9, the transformations amongst data-bit energy, channel-bit energy, and symbol energy are related (in a reciprocal

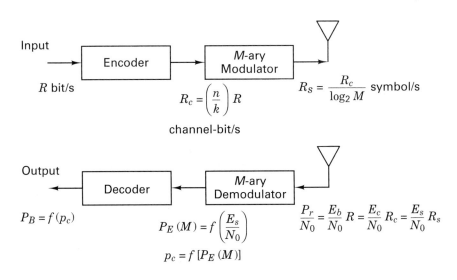

Figure 9.9: MODEM with channel coding.

fashion) by the same factors as shown amongst the rate transformations in Equations (9.33) and (9.34). Since the encoding transformation has replaced k data bits with n channel bits, then the ratio of channel-bit energy to noise-power spectral density, E_c/N_0, is computed by decrementing the value of E_b/N_0 by the factor k/n. Also, since each transmission symbol is made up of $(\log_2 M)$ channel bits, then E_s/N_0, which is needed in Equation (9.25) to solve for P_E, is computed by incrementing E_c/N_0 by the factor $(\log_2 M)$. For a system containing both modulation and coding, we summarize the energy-to-noise-power spectral density transformations, as follows:

$$\frac{E_c}{N_0} = \left(\frac{k}{n}\right)\frac{E_b}{N_0} \tag{9.35}$$

$$\frac{E_s}{N_0} = (\log_2 M)\frac{E_c}{N_0} \tag{9.36}$$

Therefore, using Equations (9.33) through (9.36), we can now expand the expression for P_r/N_0 in Equation (9.24), as follows:

$$\frac{P_r}{N_0} = \frac{E_b}{N_0}R = \frac{E_c}{N_0}R_c = \frac{E_s}{N_0}R_s \tag{9.37}$$

As before, a standard way of describing the link is in terms of the received E_b/N_0 in decibels. However, *there are no data bits at the input to the demodulator/detector; neither are there any channel bits.* There are *only* waveforms (transmission symbols) that have bit meanings, and thus the waveforms can be described in terms of bit-energy apportionments. Equation (9.37) illustrates that the predetection point in the receiver is a useful reference point at which we can relate the *effective* energy and the *effective* speed of various parameters of interest. We use the word "effective" because the only type of signals that actually appear at the predetection point are waveforms (transformed to baseband pulses) that we call symbols. Of course, these symbols are related to channel bits, which in turn are related to data bits. To emphasize the point that Equation (9.37) represents a useful kind of "bookkeeping," consider a system wherein a stream of some number of bits, say 273 bits, appears so repeatedly as a module, that we give this group 273 bits a name; we call it a "chunk." Engineers do that all the time—e.g., eight bits are referred to as a byte. The moment we identify this new entity, the chunk, it can immediately be related to the parameters in Equation (9.37), since P_r/N_0 will now also equal the energy in a chunk over N_0 times the chunk rate. In Chapter 12, we will in fact do something similar when we extend Equation (9.37) to include spread-spectrum *chips.*

Since P_r/N_0 and R were given as 53 dB-Hz and 9600 bits/s, respectively, we find as before from Equation (9.23) that the received $E_b/N_0 = 13.2$ dB. Note that the received E_b/N_0 is fixed and independent of the code parameters n and k, and the modulation parameter M. As we search, in Table 9.3, for the ideal code that will meet the specifications, we can iteratively repeat the computations that are summarized in Figure 9.9. It might be useful to program on a PC (or calculator) the following four steps as a function of n, k, and t. Step 1 starts by combining Equations (9.35) and (9.36), as follows:

Step 1:
$$\frac{E_s}{N_0} = (\log_2 M) \frac{E_c}{N_0} = (\log_2 M) \left(\frac{k}{n}\right) \frac{E_b}{N_0} \qquad (9.38)$$

Step 2:
$$P_E(M) \approx 2Q\left[\sqrt{\frac{2 E_s}{N_0}} \sin\left(\frac{\pi}{M}\right)\right] \qquad (9.39)$$

The expression in step 2 is the approximation (for *M*-ary PSK) for symbol-error probability P_E, rewritten from Equation (9.25). At each symbol-time interval, the demodulator makes a symbol decision, but it delivers to the decoder a channel-bit sequence representing that symbol. When the channel-bit output of the demodulator is quantized to two levels, denoted by 1 and 0, the demodulator is said to make *hard decisions*. When the output is quantized to more than two levels, the demodulator is said to make *soft decisions*. Throughout this section, hard-decision demodulation is assumed.

Now that a decoder block is present in the system, we designate the channel-bit-error probability out of the demodulator and into the decoder as p_c, and reserve the notation P_B for the bit-error probability *out of the decoder* (the decoded bit-error probability). Equation (9.27) is rewritten in terms of p_c as follows:

Step 3:
$$p_c \approx \frac{P_E}{\log_2 M} \quad \text{(for } P_E \ll 1) \qquad (9.40)$$

Step 3 relates the channel-bit-error probability to the symbol-error probability out of the demodulator, assuming Gray coding, as referenced in Equation (9.27).

For a real-time communication system, using traditional channel-coding schemes, and a given value of received P_r/N_0, the value of E_s/N_0 with coding will *always be less* than the value of E_s/N_0 without coding. Since the demodulator, with coding, receives less E_s/N_0 it makes more errors! However, when coding is used, the system error-performance doesn't only depend on the performance of the demodulator, it also depends on the performance of the decoder. Thus, for error-performance improvement due to coding, we require that the decoder provides enough error correction to *more than compensate* for the poor performance of the demodulator. The final output decoded bit-error probability P_B depends on the particular code, the decoder, and the channel-bit-error probability p_c. It can be expressed [15] by the following approximation:

Step 4:
$$P_B \approx \frac{1}{n} \sum_{j=t+1}^{n} j \binom{n}{j} p_c^j (1 - p_c)^{n-j} \qquad (9.41)$$

In Step 4, *t* is the largest number of channel bits that the code can correct within each block of *n* bits. Using Equations (9.38) through (9.41) in the above four steps, the decoded bit-error probability P_B can be computed as a function of *n*, *k*, and *t* for each of the codes listed in Table 9.3. The entry that meets the stated error requirement with the *largest possible* code rate and the *smallest* value of *n* is the double-error correcting (63, 51) code. The computations are as follows:

Step 1:
$$\frac{E_s}{N_0} = 3\left(\frac{51}{63}\right) 20.89 = 50.73$$

where $M = 8$, and the received $E_b/N_0 = 13.2$ dB (or 20.89)

Step 2: $P_E \approx 2Q\left[\sqrt{101.5} \times \sin\left(\dfrac{\pi}{8}\right)\right] = 2Q\,(3.86) = 1.2 \times 10^{-4}$

Step 3: $p_c \approx \dfrac{1.2 \times 10^{-4}}{3} = 4 \times 10^{-5}$

Step 4: $P_B \approx \dfrac{3}{63}\begin{pmatrix}63\\3\end{pmatrix}(4 \times 10^{-5})^3\,(1 - 4 \times 10^{-5})^{60}$

$$+ \dfrac{4}{63}\begin{pmatrix}63\\4\end{pmatrix}(4 \times 10^{-5})^4\,(1 - 4 \times 10^{-5})^{59} + \cdots$$

$$= 1.2 \times 10^{-10}$$

In Step 4, the bit-error-correcting capability of the code is $t = 2$. For the computation of P_B in Step 4, only the first two terms in the summation of Equation (9.41) have been used, since the other terms have a vanishingly small effect on the result whenever p_c is small or E_b/N_0 reasonably large. It is important to note that when performing this computation with a computer, it is advised (for being safe) to *always* include all of the summation terms in Equation (9.41), since a truncated solution can be very erroneous whenever E_b/N_0 is small. Now that we have selected the (63, 51) code, the values of channel-bit rate R_c and symbol rate R_s are computed using Equations (9.33) and (9.34), with $M = 8$:

$$R_c = \left(\dfrac{n}{k}\right)R = \left(\dfrac{63}{51}\right)9600 \approx 11{,}859 \text{ channel-bits/s}$$

$$R_s = \dfrac{R_c}{\log_2 M} = \dfrac{11{,}859}{3} = 3953 \text{ symbols/s}$$

9.7.7.1 Calculating Coding Gain

A more direct way to find the simplest code that meets the specified error performance for the example in Section 9.7.7 is to first compute, for the *uncoded* 8-PSK, how much more E_b/N_0 beyond the available 13.2 dB would be required to yield $P_B = 10^{-9}$. This additional E_b/N_0 is the required coding gain. Next, we simply choose, from Table 9.3, the code that provides this coding gain. The *uncoded* E_s/N_0 that will yield an error probability of $P_B = 10^{-9}$ is found by writing, from Equations (9.27) and (9.39),

$$P_B \approx \dfrac{P_E}{\log_2 M} \approx \dfrac{2Q\left[\sqrt{\dfrac{2E_s}{N_0}}\sin\left(\dfrac{\pi}{M}\right)\right]}{\log_2 M} = 10^{-9} \qquad (9.42)$$

At this low value of bit-error probability, it is valid to use Equation (3.44) to approximate $Q(x)$ in Equation (9.42). By trial-and-error (on a programmable calculator), we find that the *uncoded* $E_s/N_0 = 120.67 = 20.8$ dB, and since each symbol is made up of $(\log_2 8) = 3$ bits, the required $(E_b/N_0)_{\text{uncoded}} = 120.67/3 = 40.22 = 16$ dB.

We know from the given parameters in this example and Equation (9.23), that the received $(E_b/N_0)_{\text{coded}} = 13.2$ dB. Therefore, using Equation (9.32), we see that the required coding gain to meet the bit-error performance of $P_B = 10^{-9}$ is

$$G(\text{dB}) = \left(\frac{E_b}{N_0}\right)_{\text{uncoded}} (\text{dB}) - \left(\frac{E_b}{N_0}\right)_{\text{coded}} (\text{dB}) = 16 \text{ dB} - 13.2 \text{ dB} = 2.8 \text{ dB}$$

To be precise in this computation, each of the E_b/N_0 values in the above computation must correspond to exactly the same value of bit-error probability (which they do not). They correspond to $P_B = 10^{-9}$ and $P_B = 1.2 \times 10^{-10}$, respectively. However, at these low probability values, even with such a discrepancy, this computation still provides a good approximation of the required coding gain. In searching Table 9.3 for the simplest code that will yield a coding gain of *at least* 2.8 dB, we see that the choice is the (63, 51) code, which corresponds to the same code choice that was made earlier. Note that coding gain must always be specified for a particular error probability and modulation type, as it is in Table 9.3.

9.7.7.2 Code Selection

Consider a real-time communication system, where the specifications cause it to be power-limited, but there is ample available bandwidth, and the users require a very small bit-error probability. The situation calls for error-correction coding. Suppose that we were asked to select one of the BCH codes listed in Table 9.2. Since the system is not bandwidth-limited, and it requires very good error performance, one might be tempted to simply choose the most powerful code in Table 9.2, that is the (127, 8) code, capable of correcting any combination of up to 31 flawed bits within a block of 127 code bits. Would anyone use such a code in a real-time communication system? No, they would not. Let us explain why such a choice would be *unwise*.

Whenever error-correction coding is used and E_b/N_0 is fixed, there are two mechanisms at work that influence error performance. One mechanism works to improve the performance, and the other works to degrade it. The improving mechanism is the coding; the greater redundancy, the greater will be the error-correcting capability of the code. The degrading mechanism is the energy reduction per channel symbol or code bit (compared with the data bit). This reduced energy stems from the increased redundancy (giving rise to faster signaling in a real-time communication system). The reduced symbol energy causes the demodulator to make more errors. Eventually, the second mechanism wins out, and thus at very low code rates we see degradation. This is demonstrated in Example 9.4 below. Note that the degrading mechanism applies for coding in a real-time system (where messages cannot be delayed). For systems with fixed power and extended transmission time (i.e., delay), there is no degradation with reduced code rate since there is no reduction in channel-symbol energy.

Example 9.4   Choosing a Code to Meet Performance Requirements

A system is specified with the following parameters: $P_r/N_0 = 67$ dB-Hz, data rate $R = 10^6$ bits/s, available bandwidth $W = 20$ MHz, decoded bit-error probability $P_B \leq 10^{-7}$, and the modulation is BPSK. Choose a code from Table 9.2 that will fulfill these requirements. Start by considering the (127, 8) code. It appears attractive because it has the greatest bit-error correcting capability on the list.

Solution

The (127, 8) code expands the transmission bandwidth by a factor of 127/8 = 15.875. Hence, the signaling rate of 1 Mbit/s (giving rise to a nominal bandwidth of 1 MHz) will be expanded by using this code to 15.875 MHz. The transmission signal is within the available bandwidth of 20 MHz, even after allowing another 25% bandwidth expansion for filtering. After choosing this code, we next evaluate the error performance, by following the steps outlined in Section 9.7.7, which yields

$$\frac{E_b}{N_0} = \frac{P_r}{N_0}\left(\frac{1}{R}\right) = 67 \text{ dB} - 60 \text{ dB} = 7 \text{ dB (or 5)}$$

$$\frac{E_s}{N_0} = \frac{E_c}{N_0} = \left(\frac{k}{n}\right)\frac{E_b}{N_0} = \left(\frac{8}{127}\right)5 = 0.314$$

Since the modulation is binary, then $p_c = P_E$, so that

$$p_c = P_E \approx Q\left(\sqrt{\frac{2E_s}{N_0}}\right) = Q(\sqrt{0.628}) = Q(0.7936) = 0.2156$$

Since the (127, 8) code is a $t = 31$ error-correcting code, we next use Equation (9.41) to find the decoded bit-error probability, as follows:

$$P_B \approx \frac{1}{n} \sum_{j=t+1}^{n} j\binom{n}{j} p_c^j (1 - p_c)^{n-j} = \frac{1}{127} \sum_{j=32}^{127} j\binom{127}{j}(0.2156)^j (1 - 0.2156)^{127-j}$$

Whenever p_c is very small, it suffices to only use the first term, or the first few terms in the summation. But when p_c is large, as it is here, computer assistance is helpful. Solving the above with $p_c = 0.2156$ yields a decoded bit-error probability of $P_B = 0.05$, which is a far cry from the system requirement of 10^{-7}. Let us next select a code whose code rate is close to the popular rate $\frac{1}{2}$—that is, the (127, 64) code. It is not as capable as the first choice because it only corrects 10 flawed bits in a block of 127 code bits. However, watch what happens. Using the same steps as before yields

$$\frac{E_s}{N_0} = \frac{E_c}{N_0} = \left(\frac{k}{n}\right)\frac{E_b}{N_0} = \left(\frac{64}{127}\right)5 = 2.519$$

Notice how much larger the E_s/N_0 is here, compared with the case using (127, 8) coding:

$$p_c = Q(\sqrt{2 \times 2.519}) = Q(2.245) = 0.0124$$

$$P_B \approx \frac{1}{127} \sum_{j=11}^{127} j\binom{127}{j}(0.0124)^j (1 - 0.0124)^{127-j}$$

And the result yields $P_B = 5.6 \times 10^{-8}$, which meets the system requirements. From this example, one should see that the selection of a code needs to be made in concert with the modulation choice and the available E_b/N_0. One can be guided by the fact that very high rates and very low rates generally perform poorly in a real-time communication system, as evidenced by the Figure 8.6 curves presented in Chapter 8.

9.8 BANDWIDTH-EFFICIENT MODULATION

The primary objective of spectrally efficient modulation techniques is to maximize bandwidth efficiency. The increasing demand for digital transmission channels has led to the investigation of spectrally efficient modulation techniques [8, 16] to maximize bandwidth efficiency and thus help ameliorate the spectral congestion problem.

Some systems have additional modulation requirements besides spectral efficiency. For example, satellite systems with highly nonlinear transponders require a constant envelope modulation. This is because the nonlinear transponder produces extraneous sidebands when passing a signal with amplitude fluctuations (due to a mechanism called AM-to-PM conversion). These sidebands deprive the information signals of some of their portion of transponder power, and also can interfere with nearby channels (adjacent channel interference) or with other communication systems (co-channel interference). *Offset QPSK* (OQPSK) and *Minimum shift keying* (MSK) are two examples of constant envelope modulation schemes that are attractive for systems using nonlinear transponders.

9.8.1 QPSK and Offset QPSK Signaling

Figure 9.10 illustrates the partitioning of a typical pulse stream for QPSK modulation. Figure 9.10a shows the original data stream $d_k(t) = d_0, d_1, d_2, \ldots$ consisting of bipolar pulses; that is, the values of $d_k(t)$ are $+1$ or -1, representing binary one and zero, respectively. This pulse stream is divided into an in-phase stream, $d_I(t)$, and a quadrature stream, $d_Q(t)$, illustrated in Figure 9.10b, as follows:

$$d_I(t) = d_0, d_2, d_4, \ldots \text{ (even bits)} \tag{9.43}$$

$$d_Q(t) = d_1, d_3, d_5, \ldots \text{ (odd bits)}$$

Note that $d_I(t)$ and $d_Q(t)$ each have half the bit rate of $d_k(t)$. A convenient orthogonal realization of a QPSK waveform, $s(t)$, is achieved by amplitude modulating the in-phase and quadrature data streams onto the cosine and sine functions of a carrier wave, as follows:

$$s(t) = \frac{1}{\sqrt{2}} d_I(t) \cos\left(2\pi f_0 t + \frac{\pi}{4}\right) + \frac{1}{\sqrt{2}} d_Q(t) \sin\left(2\pi f_0 t + \frac{\pi}{4}\right) \tag{9.44}$$

Using the trigonometric identities shown in Equations (D.5) and (D.6), Equation (9.44) can also be written as

$$s(t) = \cos\left[2\pi f_0 t + \theta(t)\right] \tag{9.45}$$

The QPSK modulator shown in Figure 9.10c uses the sum of cosine and sine terms, while a similar device, described in Section 4.6, uses the difference of such terms. The treatment in this section corresponds to that of Pasupathy [17]. Because a coherent receiver needs to resolve any phase ambiguities, then the use of a different phase format at the transmitter can be handled as part of that ambiguity. The pulse stream $d_I(t)$ amplitude-modulates the cosine function with an amplitude of $+1$ or

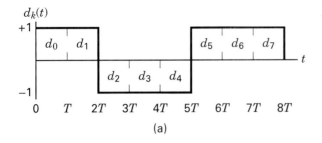

$$d_k(t)$$

(a)

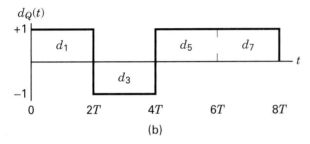

$$d_I(t)$$

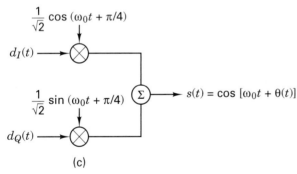

$$d_Q(t)$$

(b)

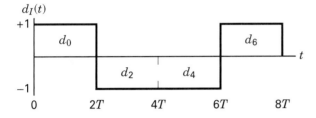

$$\frac{1}{\sqrt{2}} \cos (\omega_0 t + \pi/4)$$

$$d_I(t)$$

$$\frac{1}{\sqrt{2}} \sin (\omega_0 t + \pi/4)$$

$$d_Q(t)$$

$$s(t) = \cos [\omega_0 t + \theta(t)]$$

(c)

Figure 9.10   QPSK modulation.

−1. This is equivalent to shifting the phase of the cosine function by 0 or π; consequently, this produces a BPSK waveform. Similarly, the pulse stream $d_Q(t)$ modulates the sine function, yielding a BPSK waveform orthogonal to the cosine function. The summation of these two orthogonal components of the carrier yields the QPSK waveform. The value of $\theta(t)$ will correspond to one of the four possible combinations of $d_I(t)$ and $d_Q(t)$ in Equation 9.44: $\theta(t) = 0°, \pm 90°,$ or $180°$; the resulting signal vectors are seen in the signal space illustrated in Figure 9.11. Because

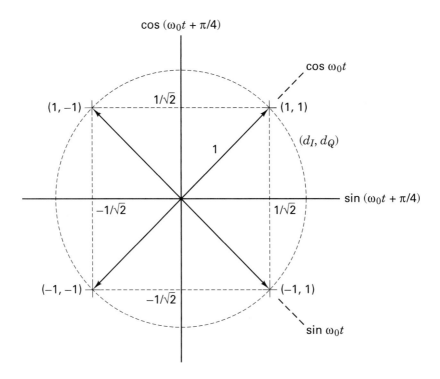

Figure 9.11 Signal space for QPSK and OQPSK.

cos $(2\pi f_0 t + \pi/4)$ and sin $(2\pi f_0 t + \pi/4)$ are orthogonal, the two BPSK signals can be detected separately.

Offset QPSK (OQPSK) signaling can also be represented by Equations (9.44) and (9.45); the difference between the two modulation schemes, QPSK and OQPSK, is only in the *alignment* of the two baseband waveforms. As shown in Figure 9.10, the duration of each original pulse is T (Figure 9.10a), and hence in the partitioned streams of Figure 9.10b, the duration of each pulse is $2T$. In standard QPSK, the odd and even pulse streams are both transmitted at the rate of $1/2T$ bit/s and are synchronously aligned, such that their transitions coincide, as shown in Figure 9.10b. In OQPSK, sometimes called *staggered QPSK* (SQPSK), there is the same data stream partitioning and orthogonal transmission; the difference is that the timing of the pulse stream $d_I(t)$ and $d_Q(t)$ is shifted such that the alignment of the two streams is offset by T. Figure 9.12 illustrates this offset.

In standard QPSK, due to the coincident alignment of $d_I(t)$ and $d_Q(t)$, the carrier phase can change only once every $2T$. The carrier phase during any $2T$ interval can be any one of the four phases shown in Figure 9.11, depending on the values of $d_I(t)$ and $d_Q(t)$ during that interval. During the next $2T$ interval, if neither pulse stream changes sign, the carrier phase remains the same. If only one of the pulse streams change sign, a phase shift of $\pm 90°$ occurs. A change in both streams results in a carrier phase shift of $180°$. Figure 9.13a shows a typical QPSK waveform for the sample sequence $d_I(t)$ and $d_Q(t)$ shown in Figure 9.10.

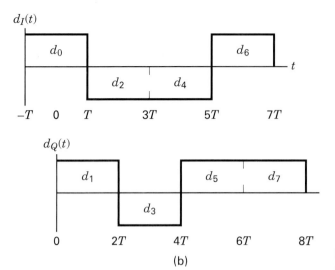

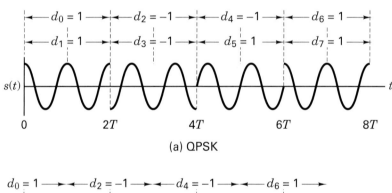

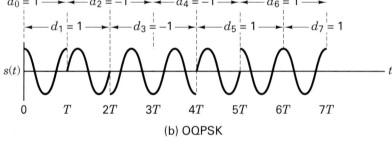

Figure 9.12 Offset QPSK (OQPSK) data streams.

(a) QPSK

(b) OQPSK

Figure 9.13 (a) QPSK and (b) OQPSK waveforms. (Reprinted with permission from S. Pasupathy, "Minimum Shift Keying: A Spectrally Efficient Modulation," *IEEE Commun. Mag.*, July 1979, Fig. 4, p. 17. © 1979 IEEE.)

If a QPSK modulated signal undergoes filtering to reduce the spectral side-lobes, the resulting waveform will no longer have a constant envelope and in fact, the occasional 180° phase shifts will cause the envelope to go to zero momentarily (see Figure 9.13a). When these signals are used in satellite channels employing highly nonlinear amplifiers, the constant envelope will tend to be restored. However, at the same time, all of the *undesirable* frequency side-lobes, which can interfere with nearby channels and other communication systems, are also restored.

In OQPSK, the pulse streams $d_I(t)$ and $d_Q(t)$ are staggered and thus do not change states simultaneously. The possibility of the carrier changing phase by 180° is eliminated, since only one component can make a transition at one time. Changes are limited to 0° and ± 90° every T seconds. Figure 9.13b shows a typical OQPSK waveform for the sample sequence in Figure 9.12. When an OQPSK signal undergoes bandlimiting, the resulting intersymbol interference causes the envelope to droop slightly in the region of ± 90° phase transition, but since the phase transitions of 180° have been avoided in OQPSK, the envelope will not go to zero as it does with QPSK. When the bandlimited OQPSK goes through a nonlinear transponder, the envelope droop is removed; however, the high-frequency components associated with the collapse of the envelope are not reinforced. Thus out-of-band interference is avoided [17].

9.8.2 Minimum Shift Keying

The main advantage of OQPSK over QPSK, that of suppressing out-of-band interference, suggests that further improvement is possible if the OQPSK format is modified to avoid discontinuous phase transitions. This was the motivation for designing continuous phase modulation (CPM) schemes. *Minimum shift keying* (MSK) is one such scheme [17–20]. MSK can be viewed as either a special case of *continuous-phase frequency shift keying* (CPFSK), or a special case of OQPSK with sinusoidal symbol weighting. When viewed as CPFSK, the MSK waveform can be expressed as [18]

$$s(t) = \cos\left[2\pi\left(f_0 + \frac{d_k}{4T}\right)t + x_k\right] \quad kT < t < (k+1)T \quad (9.46)$$

where f_0 is the carrier frequency, $d_k = \pm 1$ represents the bipolar data being transmitted at a rate $R = 1/T$, and x_k is a phase constant which is valid over the kth binary data interval. Notice that for $d_k = 1$, the frequency transmitted is $f_0 + 1/4T$, and for $d_k = -1$, the frequency transmitted is $f_0 - 1/4T$. The tone spacing in MSK is thus one-half that employed for noncoherently demodulated orthogonal FSK, giving rise to the name *minimum* shift keying. During each T-second data interval, the value of x_k is a constant, that is, $x_k = 0$ or π, determined by the requirement that the phase of the waveform be continuous at $t = kT$. This requirement results in the following recursive phase constraint for x_k:

$$x_k = \left[x_{k-1} + \frac{\pi k}{2}(d_{k-1} - d_k)\right] \text{ modulo } 2\pi \quad (9.47)$$

Equation (9.46) can be expressed in a quadrature representation, using the identities in Equation (D.5) and (D.6), and we write

$$s(t) = a_k \cos \frac{\pi t}{2T} \cos 2\pi f_0 t - b_k \sin \frac{\pi t}{2T} \sin 2\pi f_0 t$$

$$kT < t < (k+1)T \qquad (9.48)$$

where

$$a_k = \cos x_k = \pm 1$$

$$(9.49)$$

$$b_k = d_k \cos x_k = \pm 1$$

The in-phase (*I*) component is identified as $a_k \cos (\pi t/2T) \cos 2\pi f_0 t$, where $\cos 2\pi f_0 t$ is the carrier, $\cos (\pi t/2T)$ can be regarded as a *sinusoidal symbol weighting*, and a_k is a data-dependent term. Similarly, the quadrature (*Q*) component is identified as $b_k \sin (\pi t/2T) \sin 2\pi f_0 t$, where $\sin 2\pi f_0 t$ is the quadrature carrier term, $\sin (\pi t/2T)$ can be regarded as a sinusoidal symbol weighting, and b_k is a data-dependent term. It might appear that the a_k and b_k terms can change every *T* seconds, since the source data d_k can change every *T* seconds. However, because of the continuous phase constraint, the a_k term can only change value at the zero crossings of cos $(\pi t/2T)$ and the b_k term can only change value at the zero crossings of sin $(\pi t/2T)$. Thus, the symbol weighting in either the *I*- or *Q*-channel is a half-cycle sinusoidal pulse of duration 2*T* seconds with alternating sign. As in the case of OQPSK, the *I* and *Q* components are offset *T* seconds with respect to one another.

Notice that x_k in Equation (9.46) is a function of the difference between the prior data bit and the present data bit (differential encoding). Hence the a_k and b_k terms in Equation (9.48) can be viewed as *differentially encoded* components of the d_k source data. However, for bit-to-bit independent data d_k, the signs of successive *I*- or *Q*-channel pulses are also random from one 2*T*-second pulse interval to the next. Thus when viewed as a special case of OQPSK, Equation (9.48) can be rewritten with more straightforward (nondifferential) data encoding [18] as follows:

$$s(t) = d_I(t) \cos \frac{\pi t}{2T} \cos 2\pi f_0 t + d_Q(t) \sin \frac{\pi t}{2T} \sin 2\pi f_0 t \qquad (9.50)$$

where $d_I(t)$ and $d_Q(t)$ have the same in-phase and quadrature data stream interpretation as in Equation (9.43). This MSK format in Equation (9.50) is sometimes referred to as *precoded MSK*. Figure 9.14 illustrates Equation (9.50) pictorially. Figure 9.14a and c show the sinusoidal weighting of the *I*- and *Q*-channel pulses. These sequences represent the same data sequences as in Figure 9.12, but here, multiplication by a sinusoid results in more gradual phase transitions compared to those of the original data representation. Figure 9.14b and d illustrate the modulation of the orthogonal components cos $2\pi f_0 t$ and sin $2\pi f_0 t$, respectively, by the sinusoidally shaped data streams. Figure 9.14e illustrates the summation of the orthogonal components from Figure 9.14b and d. In summary, the following properties of MSK modulation can be deduced from Equation (9.50) and Figure 9.14:

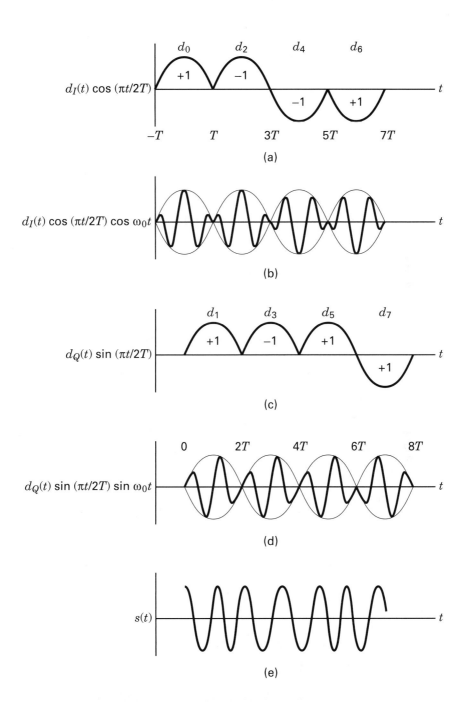

Figure 9.14 Minimum shift keying (MSK). (a) Modified *I* bit stream. (b) *I* bit stream times carrier. (c) Modified *Q* bit stream. (d) *Q* bit stream times carrier. (e) MSK waveform. (Reprinted with permission from S. Pasupathy, "Minimum Shift Keying: A Spectrally Efficient Modulation," *IEEE Commun. Mag.,* July 1979, Fig. 5, p. 18. © 1979 IEEE.)

(1) the waveform $s(t)$ has constant envelope; (2) there is phase continuity in the RF carrier at the bit transitions; and (3) the waveform $s(t)$ can be regarded as an FSK waveform with signaling frequencies $f_0 + 1/4T$ and $f_0 - 1/4T$. Therefore, the minimum tone separation required for MSK modulation is

$$\left(f_0 + \frac{1}{4T} \right) - \left(f_0 - \frac{1}{4T} \right) = \frac{1}{2T} \qquad (9.51)$$

which is equal to half the bit rate. Notice that the required tone spacing for MSK is one-half the spacing, $1/T$, required for the noncoherent detection of FSK signals (see Section 4.5.4). This is because the carrier phase is known and continuous, enabling the signal to be coherently demodulated.

The power spectral density $G(f)$ for QPSK and OQPSK is given by [18]

$$G(f) = 2PT \left(\frac{\sin 2\pi fT}{2\pi fT} \right)^2 \qquad (9.52)$$

where P is the average power in the modulated waveform. For MSK, $G(f)$ is given by [18]

$$G(f) = \frac{16PT}{\pi^2} \left(\frac{\cos 2\pi fT}{1 - 16f^2T^2} \right)^2 \qquad (9.53)$$

The normalized power spectral density ($P = 1$ W) for QPSK, OQPSK, and MSK are sketched in Figure 9.15. A spectral plot of BPSK is included for comparison. The fact that BPSK requires more bandwidth than the others for a given level of spectral density should come as no surprise. In Section 9.5.1 and Figure 9.6 we saw that the theoretical bandwidth efficiency of BPSK is half that of QPSK. It is seen from Figure 9.15 that MSK has lower sidelobes than QPSK or OQPSK. This is a consequence of multiplying the data stream with a sinusoid, yielding more *gradual phase transitions*. The more gradual the transition, the faster the spectral tails drop to zero. MSK is *spectrally more efficient* than QPSK or OQPSK; however, as can be seen from Figure 9.15, the MSK spectrum has a wider mainlobe than QPSK and OQPSK. Therefore, MSK may not be the preferred method for narrowband links. However, MSK might be the preferred choice for multiple-carrier systems, because its relatively low spectral sidelobes help to avoid excessive adjacent channel interference (ACI). The reason for the QPSK spectrum having a narrower mainlobe than MSK is that, for a given bit rate, the QPSK symbol rate is half the MSK symbol rate.

9.8.2.1 Error Performance of OQPSK and MSK

We have seen that BPSK and QPSK have the same bit-error probability because QPSK is configured as two BPSK signals modulating orthogonal components of the carrier. Since staggering the bit streams does not change the orthogonality of the carriers, OQPSK has the same theoretical bit error performance as BPSK and QPSK.

Minimum shift keying uses antipodal symbol shapes, $\pm \cos (\pi t/2T)$ and $\pm \sin (\pi t/2T)$, over $2T$ to modulate the two quadrature components of the carrier. Thus

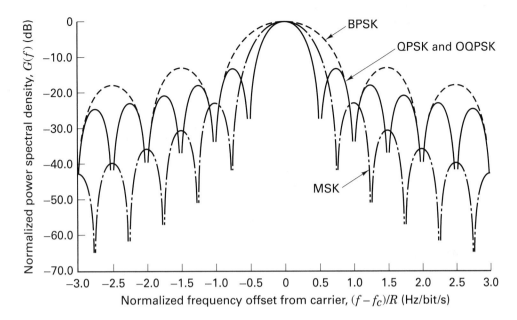

Figure 9.15 Normalized power spectral density for BPSK, QPSK, OQPSK, and MSK. (Reprinted with permission from F. Amoroso, "The Bandwidth of Digital Data Signals," *IEEE Commun. Mag.*, vol. 18, no. 6, Nov. 1980, Fig. 2A, p. 16. © 1980 IEEE.)

when a matched filter is used to recover the data from each of the quadrature components independently, MSK, as defined in Equation (9.50), has the same error performance properties as BPSK, QPSK, and OQPSK [17]. However, if MSK is coherently detected as an FSK signal over an observation interval of T seconds, it would be poorer than BPSK by 3 dB [17]. MSK, with differentially encoded data, as defined in Equation (9.46), has the same error probability performance as the coherent detection of differentially encoded PSK. MSK can also be noncoherently detected [19]. This permits inexpensive demodulation when the value of received E_b/N_0 permits.

9.8.3 Quadrature Amplitude Modulation

Coherent M-ary phase shift keying (MPSK) modulation is a well-known technique for achieving bandwidth reduction. Instead of using a binary alphabet with 1 bit of information per channel symbol period, an alphabet with M symbols is used, permitting the transmission of $k = \log_2 M$ bits during each symbol period. Since the use of M-ary symbols allows a k-fold increase in the data rate within the same bandwidth, then for a fixed data rate, use of M-ary PSK reduces the required bandwidth by a factor k. (See Section 4.8.3.)

From Equation (9.44) it can be seen that QPSK modulation consists of two independent streams. One stream amplitude-modulates the cosine function of a carrier wave with levels +1 and −1, and the other stream similarly amplitude-modulates the sine function. The resultant waveform is termed a double-sideband suppressed-carrier (DSB-SC) wave, since the RF bandwidth is twice the baseband bandwidth (see Section 1.7.1) and there is no isolated carrier term. *Quadrature amplitude modulation* (QAM) can be considered a logical extension of QPSK, since QAM also consists of two independently amplitude-modulated carriers in quadrature. Each block of k bits (k assumed even) can be split into two $(k/2)$-bit blocks which use $(k/2)$-bit digital-to-analog (D/A) converters to provide the required modulating voltages for the carriers. At the receiver, each of the two signals is independently detected using matched filters. QAM signaling can also be viewed as a combination of amplitude shift keying (ASK) and phase shift keying (PSK), giving rise to the alternative name, *amplitude phase keying* (APK). Finally, it can also be viewed as amplitude shift keying in two dimensions, giving rise to the name *quadrature amplitude shift keying* (QASK).

Figure 9.16a illustrates a two-dimensional signal space and a set of 16-ary QAM signal vectors or points arranged in a rectangular constellation. A canonical QAM modulator is shown in Figure 9.16b. Assuming that Gaussian noise is the only channel disturbance, the simple channel model of Figure 9.16c applies. Signals are sent in pairs (x, y). The model indicates that the signal point coordinates (x, y) are transmitted over separate channels and independently perturbed by Gaussian

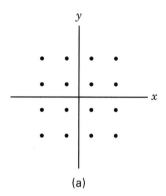

(a)

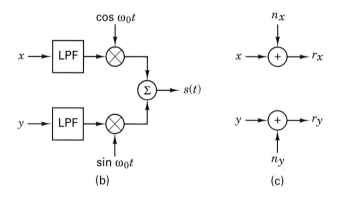

(b)   (c)

Figure 9.16 QAM modulation. (a) 16-ary signal space. (b) Canonical QAM modulator. (c) QAM channel model.

noise variables (n_x, n_y), each with zero mean and variance N. Or we can say that the two-dimensional signal point is perturbed by a two dimensional Gaussian noise variable. If the average signal energy (mean-square value of the signal coordinates) is S, then the signal-to-noise ratio is S/N. The simplest method of digital signaling through such a system is to use one-dimensional pulse amplitude modulation (PAM) independently for each signal coordinate. In PAM, to send k bits/dimension over a Gaussian channel, each signal point coordinate takes on one of 2^k equally likely equispaced amplitudes. By convention, the signal points are grouped about the center of the space at amplitudes $\pm 1, \pm 3, \ldots, \pm(2^k - 1)$.

9.8.3.1 QAM Probability of Bit Error

For a rectangular constellation, a Gaussian channel, and matched filter reception, the bit-error probability for M–QAM, where $M = 2^k$ and k is even, is [12]

$$P_B \approx \frac{2(1 - L^{-1})}{\log_2 L} Q\left[\sqrt{\left(\frac{3\log_2 L}{L^2 - 1}\right)\frac{2E_b}{N_0}}\right] \tag{9.54}$$

where $Q(x)$ is as defined in Equation (3.43) and $L = \sqrt{M}$ represents the number of amplitude levels in one dimension. In the context of L–PAM, a sequence of $k/2 = \log_2 L$ bits are assigned to an L-ary symbol using a Gray code (defined in Section 4.9.4).

9.8.3.2 Bandwidth–Power Trade-Off

The bandwidth–power trade-off of M-ary QAM at a bit error probability of 10^{-5} is displayed on the bandwidth-efficiency plane in Figure 9.6, with the abscissa measured in average E_b/N_0. We assume Nyquist filtering of the baseband pluses so that the DSB transmission bandwidth at IF is $W_{IF} = 1/T$, where T is the symbol duration. Thus the bandwidth efficiency is $R/W = \log_2 M$, where M is the symbol set size. For realistic channels and waveforms, the performance must be reduced to account for the increased bandwidth necessary to implement realizable filters. From Figure 9.6 it can be seen that QAM represents a method of reducing the bandwidth required for the transmission of digital data. As with M-ary PSK, bandwidth efficiency can be exchanged for power or E_b/N_0; however, in the case of QAM, a *much more efficient exchange* is possible than in the case of M-ary PSK.

Example 9.5   Waveform Design

Assume that a data stream with data rate $R = 144$ Mbits/s is to be transmitted on an RF channel using a DSB modulation scheme. Assume Nyquist filtering and an allowable DSB bandwidth of 36 MHz. Which modulation technique would you choose for this requirement? If the available E_b/N_0 is 20, what would be the resulting probability of bit error?

Solution

The required spectral efficiency is

$$\frac{R}{W} = \frac{144 \text{ Mbits/s}}{36 \text{ MHz}} = 4 \text{ bits/s/Hz}$$

From Figure 9.6 we note that 16-ary QAM, with a theoretical spectral efficiency of 4 bits/s/Hz, requires a lower E_b/N_0 than that of 16-ary PSK for the same P_B. Based on these considerations we choose a 16-ary QAM modem.

With the available E_b/N_0 given as 20, we use Equation (9.54) to calculate the expected bit error probability as

$$P_B \approx \frac{3}{4} Q \left(\sqrt{\frac{4}{5} \frac{E_b}{N_0}} \right) = 2.5 \times 10^{-5}$$

Example 9.6 Spectral Efficiency

(a) Explain the computation of the QAM spectral efficiency in Example 9.5, considering that QAM is transmitted on orthogonal components of a carrier wave.
(b) Since the DSB bandwidth is 36 MHz in Example 9.5, consider using half that amount at baseband to transmit the 144-Mbits/s data stream, using multilevel PAM. What is the spectral efficiency needed to accomplish this, and how many levels of PAM would be required? Assume Nyquist filtering.

Solution

(a) *Bandpass channel using QAM:* The 144-Mbits/s data stream is partitioned into a 72-Mbits/s in-phase and a 72-Mbits/s quadrature stream; one stream amplitude-modulates the cosine component of a carrier over a bandwidth of 36 MHz, and the other stream amplitude-modulates the sine component of the carrier wave over the same 36-MHz bandwidth. Since each 72-Mbits/s stream modulates an orthogonal component of the carrier, the 36 MHz suffices for both streams, or for the full 144 Mbits/s. Thus the spectral efficiency is (144 Mbits/s)/36 MHz = 4 bits/s/Hz.
(b) *Required spectral efficiency at baseband*

$$\frac{R}{W} = \frac{144 \text{ Mbits/s}}{18 \text{ MHz}} = 8 \text{ bits/s/Hz}$$

Assuming Nyquist filtering, a bandwidth of 18 MHz can support a maximum symbol rate of $R_s = 2W = 36$ megasymbols/s [see Equation (3.80)]. Each PAM pulse must therefore have an ℓ-bit meaning, such that

$$R = \ell R_s$$

Hence,

$$\ell = \frac{144 \text{ Mbits/s}}{36 \text{ megapulses/s}} = 4 \text{ bits/pulse}$$

where $\ell = \log_2 L$, and $L = 16$ levels.

9.9 MODULATION AND CODING FOR BANDLIMITED CHANNELS

The channel coding techniques of Chapters 6–8 have generally *not* been associated with voice-grade telephone channels (although the first field test of sequential decoding of convolutional codes was on a telephone line). Recently, however, there has been considerable interest in techniques that can provide coding gain for bandlimited channels. The motivation is to enable the reliable transmission of *higher data rates* over voice-grade channels. The potential gain is about 3 bits/symbol (for a given signal-to-noise ratio) [21] or, alternatively, a given error performance could be achieved with a power savings of 9 dB [21].

The greatest interest is in the following three separate coding research areas:

1. Optimum signal constellation boundaries (choosing a closely packed signal subset from any regular array or lattice of candidate points)
2. Higher-density lattice structures (adding improvement to the signal subset choice by starting with the densest possible lattice for the space)
3. Trellis-coded modulation (combined modulation and coding techniques for obtaining coding gain for bandlimited channels)

The first two areas are not "true" error control coding schemes. By "true error control coding" we refer to those techniques that employ some structured redundancy to improve the error performance. Only the third technique, trellis-coded modulation, involves redundancy. Each of these coding research areas and their expected performance improvements are discussed below.

9.9.1 Commercial Telephone Modems

The use of efficient modulation techniques has traditionally been spearheaded by the telecommunications industry, since the telephone company's foremost resource consists of sharply bandlimited voice-grade channels. The typical telephone channel is characterized by a high signal-to-noise ratio (SNR) of approximately 30 dB and a bandwidth of approximately 3 kHz. Table 9.4 lists the evolution of leased-line telephone modems, and Table 9.5 lists the evolution of dial-line telephone modem standards.

TABLE 9.4 Evolution of Leased-Line Telephone Modems

Year	Name	Maximum Bit Rate (bits/s)	Signaling Rate (symbols/s)	Modulation Technique	Signaling Efficiency (bits/symbol)
1962	Bell 201	2400	1200	4-PSK	2
1967	Milgo 4400/48	4800	1600	8-PSK	3
1971	Codex 9600C	9600	2400	16-QAM	4
1980	Paradyne MP14400	14,400	2400	64-QAM	6
1981	Codex SP14.4	14,400	2400	64-QAM	6
1984	Codex 2660	16,800	2400	Trellis-coded 256-QAM	7
1985	Codex 2680	19,200	2743	8-D Trellis-coded 160-QAM	7

TABLE 9.5   Evolution of Dial-Line Telephone Modem Standards

Year	Name	Maximum Bit Rate (bits/s)	Signaling Rate (symbols/s)	Modulation Technique	Signaling Efficiency (bits/symbol)
1984	V.32	9600	2400	2-D Trellis Coded 32-QAM	4
1991	V.32bis	14,400	2400	2-D Trellis Coded 128-QAM	6
1994	V.34	28,800	2400, 2743, 2800, 3000, 3200, 3429	4-D Trellis Coded 960-QAM	≈ 9
1996	V.34	33,600	2400, 2743, 2800, 3000, 3200, 3429	4-D Trellis Coded 1664-QAM	≈ 10
1998	4.90	downstream: 56,000 upstream: 33,600	8000 as in V.34	PCM* (M-PAM) as in V.34	7 ≈ 10
2000	V.92	downstream: 56,000 upstream: 48,000	8000 8000	PCM* (M-PAM) Trellis Coded PCM*	7 6

*In the G.711 ITU-T Recommendation, PCM is the term used for M-ary PAM signaling.

9.9.2 Signal Constellation Boundaries

Several researchers [22–26] have examined large numbers of possible QAM signal constellations in a search for designs that result in the best error performance for a given average signal-to-noise ratio. Figure 9.17 illustrates some examples of symbol constellations for $M = 4, 8,$ and 16 that have been considered [22]. The circular sets are designated by the notation $(a, b, \ldots)$, where there are a quantity of a signals on the inner circle, b signals on the next circle, and so on. In general, the constellation rule, known as the Campopiano-Glazer construction rule [24], that yields optimum signal set performance can be summarized as follows: From an infinite array of points closely packed in a *regular array or lattice*, select a closely packed subset of 2^k points as a signal constellation. In this case "optimum" means minimum average or peak power for a given error probability. In a two-dimensional signal space the optimum boundary surrounding an array of points tends toward a circle. Figure 9.18 illustrates examples of 64-ary ($k = 6$) and 128-ary ($k = 7$) signal sets from a rectangular array. The cross-shaped boundaries are a compromise to the optimum circle. The $k = 6$ constellation was used in the Paradyne 14.4-kbits/s modem. Compared with a square, the performance improvement resulting from a circular boundary is only a modest 0.2 dB [21].

9.9.3 Higher-Demensional Signal Constellations

For any particular information rate and channel-noise process that is independent and identically distributed in two dimensions, signaling in a two-dimensional space can provide the same error performance with less average (or peak) power than signaling in a one-dimensional pulse-amplitude (PAM) space. This is accomplished

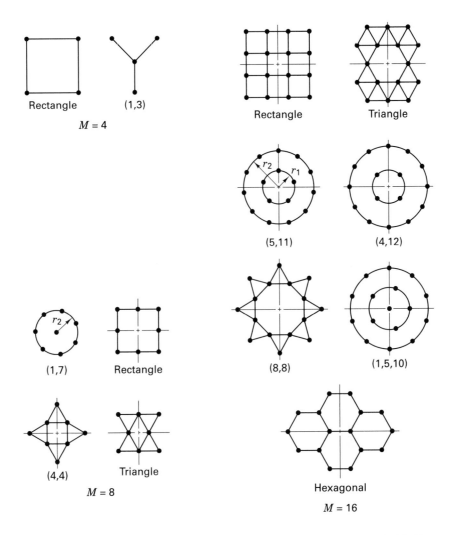

Figure 9.17 *M*-ary symbol constellations. (Reprinted with permission from C. M. Thomas, M. Y. Weidner, and S. H. Durrani, "Digital Amplitude-Phase Keying with *M*-ary Alphabets," *IEEE Trans. Commun.*, vol. COM22, no. 2, Feb. 1974, Figs. 2 and 3, p. 170. © 1974 IEEE.)

by choosing signaling points on a two-dimensional lattice from within a circular rather than a rectangular boundary. In the same way, by going to a higher number, N dimensions, and choosing points on an n-dimensional lattice from within an N-sphere rather than an N-cube, further energy savings are possible [27–30]. The goal of this constellation shaping is to make the required average energy of signal points from the N-sphere less than that from the N-cube; such reduction in required energy for a given error performance is referred to as a *shaping gain* [16]. Table 9.6 gives the energy savings possible in N dimensions. As N goes to infinity, the gain goes to 1.53 dB; it is not difficult to achieve shaping gain on the order of 1 dB [16, 21].

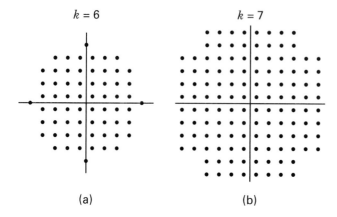

$k = 6$ $k = 7$

(a)   (b)

Figure 9.18 Examples of *M*-ary constellations using a rectangular array.

The channel is essentially two-dimensional, since symbols represented as points on a two-dimensional plane are transmitted in quadrature fashion. Multidimensional signaling is generally taken to mean signaling with two or more such planes. For transmitting n bits/symbol with N-dimensional (N even and greater than 2) signaling, the incoming bits are grouped into blocks of $nN/2$. A mapping must then be made that assigns data bits to $2^{nN/2}$ N-dimensional vectors that have the least energy among all such vectors. A corresponding inverse mapping must be made at the receiver.

Consider an example of mapping signals from a two-dimensional to a four-dimensional space. We start with a two-dimensional *M*-ary constellation, such as *M*-QAM with $M = 16$. Here the transmitted symbol, viewed as a point on a plane, is represented by $n = 4$ bits (two 4-ary amplitudes, and two bits per amplitude). Each symbol transmission consists of sending a vector from a space of 16 possible vectors. With four-dimensional signaling, the transmitted symbol, viewed as two points, one from each of two planes, is represented by 8 bits. Then, each (two-point) transmission consists of sending a vector from a space of $16 \times 16 = 256$ vec-

TABLE 9.6 Energy Savings from *N*-Sphere Mapping versus *N*-Cube Mapping (Shaping Gain)

Dimensions (N)	*N*-Sphere Mapping Gain (dB)
2	0.20
4	0.45
8	0.73
16	0.98
24	1.10
32	1.17
48	1.26
64	1.31

Source: G. D. Forney, Jr., et al., "Efficient Modulation for Bandlimited Channels," *IEEE J. Sel. Areas Commun.*, vol. SAC2, no. 5, September 1984, pp. 632–647.

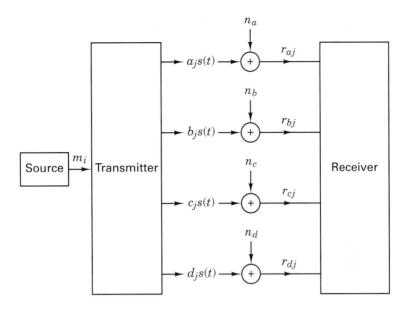

Figure 9.19 Four-dimensional system configuration.

tors. In general, the source-data bits are grouped into blocks of $nN/2$. In this example for four-dimensional signaling, we group the source-data bits into blocks of 8 bits (2 planes × $n = 4$ bits/plane). Such an 8-bit transmission can be viewed as a mapping from a space of 2^n two-dimensional vectors to a space of $2^{nN/2}$ four-dimensional vectors. For the four-dimensional system depicted in Figure 9.19, a given source produces one of 256 four-dimensional vectors m_i ($i = 1, 2, \ldots, 256$) by grouping two 16-ary symbols (two planes) at a time and transmitting waveforms, $a_js(t)$, $b_js(t)$, $c_js(t)$, $d_js(t)$, where $j = 1, \ldots, 4$ represents one of 4-ary amplitude values. These baseband or bandpass waveforms are transmitted on separate noninterfering channels. In each channel, the waveforms are distorted by independent AWGN, and at the receiver they are demodulated with matched filters. We may choose to transmit the N-dimensional signal in a number of ways:

1. Using four separate wires representing four baseband channels.
2. Using two bandpass channels, each with separately modulated inphase and quadrature components.
3. Using time- or frequency-division multiplexing to carry the baseband or bandpass channels on a common transmission line.
4. Using orthogonal electromagnetic wave polarization.

Thus, if the Figure 9.19 example represents a radio system, we could follow method 2 above and modulate waveforms $a_js(t)$, and $b_js(t)$ in quadrature fashion onto a particular carrier wave, while modulating waveforms $c_js(t)$, and $d_js(t)$ onto a second carrier wave. Thus, during each $2T$ second interval, one would transmit four 4-ary numbers representing 8 bits or a vector from a 256-ary space. Further shaping gain can be similarly achieved for the delivery of 16-ary symbols per plane with six-dimensional signaling, where every $3T$ seconds, a 16-ary symbol from each of three planes is transmitted. Thus, each six-dimensional signal constitutes three 16-ary values representing 12 bits or a point on a 4096-ary signal space. It is important to emphasize that it is not the mere grouping of the 16-ary symbols that brings about the shaping gain. The gain comes about because detection performed over a larger signal space can achieve a given error performance with a smaller E_b/N_0. In the case of sending 16-ary symbols with six-dimensional signaling, a 12-bit sequence is detected every $3T$ seconds (not a 4-bit sequence every T seconds). Detection in a higher dimensional space entails greater (signal-mapping) complexity. Compromises are generally used to simplify the mapping complexity at the cost of some suboptimality in energy efficiency.

9.9.4 Higher-Density Lattice Structures

In Section 9.9.3, we discussed the selection of a closely packed subset of points from any regular array or lattice. Here we consider the added improvement by starting with the *densest possible lattice* in the space. In a two-dimensional signal space, the densest lattice is the hexagonal lattice (try penny packing). The result of employing a hexagonal lattice instead of a rectangular one, such as those shown in Figure 9.18, can be a 0.6-dB savings in average energy. Figure 9.20 illustrates some examples of hexagonal packing. The strange-looking $k = 4$ constellation in Figure 9.20a was discovered by Foschini et al. [26] and is still the best 16-ary constellation known. The $k = 6$ constellation in Figure 9.20b was used in the Codex SP14.4 modem.

The hexagonal lattice is optimum for two dimensions. For higher dimensions there are other lattice structures that provide the densest packing. Table 9.7 gives the gain over the rectangular lattice, in decibels, due to the densest packings currently known for various dimensions.

9.9.5 Combined Gain: *N*-Sphere Mapping and Dense Lattice

It is possible to combine the benefits of the Campopiano–Glazer boundary construction in N dimensions with the gain from the densest lattice in N-space. The

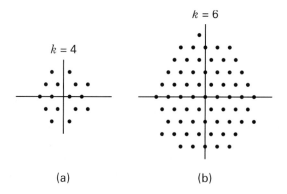

(a)   (b)

Figure 9.20 Examples of *M*-ary constellations using a hexagonal array.

TABLE 9.7 Energy Savings from Dense Lattices versus the Rectangular Lattice

Dimensions (N)	Dense Lattice Gain (dB)
2	0.62
4	1.51
8	3.01
16	4.52
24	6.02
32	6.02
48	7.78
64	8.09

Source: G. D. Forney, Jr., et al., "Efficient Modulation for Bandlimited Channels," *IEEE J. Sel. Areas Commun.,* vol. SAC2, no. 5, September 1984, pp. 632–647.

resulting gain is a combination of N-sphere versus N-cube boundary gain of Table 9.6 and the lattice packing density gain of Table 9.7. The combined energy savings are shown in Table 9.8.

TABLE 9.8 Combined Energy Savings from N-Sphere Mapping and Dense Lattices

Dimensions (N)	Combined Savings Gain (dB)
2	0.82
4	1.96
8	3.74
16	5.50
24	7.12
32	7.19
48	9.04
64	9.40

Source: G. D. Forney, Jr., et al., "Efficient Modulation for Bandlimited Channels," *IEEE J. Sel. Areas Commun.,* vol. SAC2, no. 5, September 1984, pp. 632–647.

9.10 TRELLIS-CODED MODULATION

The error-correction codes described in Chapters 6–8, when used in real-time communication systems, provide improvements in error performance at the cost of bandwidth expansion. For both block codes and convolutional codes, transforming each input data k-tuple into a larger output codeword n-tuple, requires additional transmission bandwidth. Therefore, in the past, coding generally was not popular for bandlimited channels such as telephone channels, where signal bandwidth expansion is not practical. Since about 1984, however, there has been active interest in combined modulation and coding schemes, called *trellis-coded modulation* (TCM), that achieve error-performance improvements without expansion of signal bandwidth. TCM schemes use redundant nonbinary modulation in combination with a *finite-state machine* (the encoder). What is a finite-state machine, and what is meant by its state? Finite-state machine is the general name given to a device that has a memory of past signals; the adjective *finite* refers to the fact that there are only a finite number of unique states that the machine can encounter. What is meant by the *state* of a finite-state machine? In the most general sense, the state consists of the smallest amount of information that, together with a current input to the machine, can predict the output of the machine. The state provides some knowledge of the past signaling events and the restricted set of possible outputs in the future. A future state is restricted by the past state.

For each symbol interval, a TCM finite-state encoder selects one of a set of waveforms, thereby generating a sequence of coded waveforms to be transmitted. The noisy received signals are detected and decoded by a soft-decision maximum-likelihood detector/decoder. In conventional systems involving modulation and coding, it is common to separately describe and implement the detector and the decoder. With TCM systems, however, these functions must be treated jointly. Coding gain can be achieved without sacrificing data rate or without increasing either bandwidth or power [6, 31]. At first, it may seem that this statement violates some basic principle of power-bandwidth, error-probability trade-off. However, there is still a trade-off involved, since TCM achieves coding gain at the expense of decoder complexity.

Trellis-coded modulation combines a multilevel/phase modulation signaling set with a *trellis-coding scheme.* The term "trellis-coding scheme" refers to any code system that has memory (a finite state machine), such as a convolutional code. Multilevel/phase signals have constellations involving multiple amplitudes, multiple phases, or combinations of multiple amplitudes and multiple phases. In other words, a TCM signal set is best represented by any signal set (greater than binary) whose vector representations can be depicted on a plane, such as that shown in Figure 9.16a for QAM signals. A trellis-coding scheme is one that can be characterized with a state-transition (trellis) diagram, similar to the trellis diagrams describing convolutional codes. The convolutional codes presented in Chapter 7 are linear, although trellis codes are not constrained to be linear. Coding gains can be realized with block codes or trellis codes, but only trellis codes will be considered because the availability of the *Viterbi decoding algorithm* makes trellis decoding simple and efficient. Ungerboeck showed that in the presence of AWGN, TCM schemes can yield net coding gains of about 3-dB relative to uncoded systems with relative ease, while gains of about 6-dB can be achieved with greater complexity.

9.10.1 The Idea Behind Trellis-Coded Modulation

In TCM, channel coding and modulation are performed together; where one begins and the other ends cannot be easily established. Can you speculate what ideas might have prompted the development of TCM? Perhaps it started with the notion that "not all signal subsets (in a constellation) have equal distance properties." That is, for a nonorthogonal signal set, such as MPSK, antipodal signals have the best distance properties for easily discriminating one signal from the other, while nearest neighbor signals have relatively poor distance properties. It may be that the initial idea of coded modulation came about by trying to exploit these differences.

A simple analogy might be helpful in understanding the overall goals in TCM. Imagine that there is an all-knowing wizard at the transmitter. As the message bits enter the system, the wizard recognizes that some of the bits are most vulnerable to the degradation effects of channel impairments; hence, they are assigned modulation waveforms associated with the best distance properties. Similarly, other bits are judged to be very robust, and hence, they are assigned waveforms with poorer distance properties. Modulation and coding take place together. The wizard is assigning waveforms to bits (modulation), but, the assignment is being performed according to the criterion of better or worse distance properties (channel coding).

9.10.1.1 Increasing Signal Redundancy

TCM may be implemented with a convolutional encoder, wherein k current bits and $K - 1$ prior bits are used to produce $n = k + p$ code bits, where K is the encoder constraint length (see Chapter 7) and p is the number of parity bits. Notice that encoding increases the signal set size from 2^k to 2^{k+p}. Ungerboeck [31] investigated the increase in channel capacity achievable by signal set expansion, and concluded that most of the achievable coding gain over conventional uncoded multilevel modulation could be realized by expanding the uncoded signaling set by the factor of two ($p = 1$). This can be accomplished by encoding with a rate $k/(k + 1)$ code and subsequently mapping groups of $(k + 1)$ bits into the set of 2^{k+1} waveforms. Figure 9.21a illustrates an uncoded 4-ary PAM signal set, before and after being rate 2/3 encoded into an 8-ary PAM signal set. Similarly, Figure 9.21b illustrates an uncoded 4-ary PSK (QPSK) signal set, before and after being rate 2/3 encoded into an 8-ary PSK signal set. Similarly, Figure 9.21c illustrates an uncoded 16-ary QAM signal set, before and after being rate 4/5 encoded into a 32-ary QAM signal set. In each of the cases shown in Figure 9.21, the system is configured to use the same average signal power before and after coding. Also, to provide the needed redundancy for coding, the signaling set is increased from $M = 2^k$ to $M' = 2^{k+1}$. Thus, $M' = 2M$; however, the increase in the alphabet size does not result in an increase in required bandwidth. Recall from Section 9.7.2 that the transmission bandwidth of nonorthogonal signaling does not depend on the density of signaling points in the constellation; it depends only on the rate of signaling. The expanded signal set *does* result in a reduced distance between adjacent symbol points (for signal sets with a constant average power), as seen in Figure 9.21. In an *uncoded* system, such a reduced distance degrades the error performance. However, because of the redundancy introduced by the code, this reduced distance no longer plays a critical role in determining the error performance. Instead, the *free distance,* which is the minimum distance between members of the set of *allowed* code sequences, determines the error performance. The free distance characterizes the "easiest

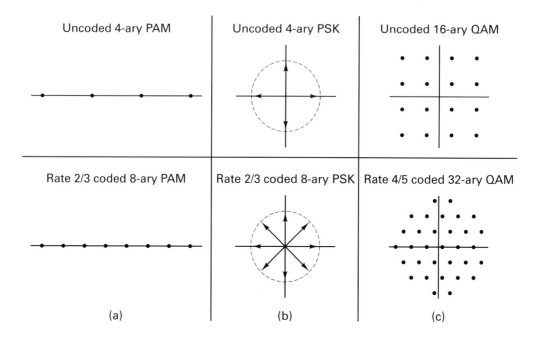

Uncoded 4-ary PAM   Uncoded 4-ary PSK   Uncoded 16-ary QAM

Rate 2/3 coded 8-ary PAM   Rate 2/3 coded 8-ary PSK   Rate 4/5 coded 32-ary QAM

(a)   (b)   (c)

Figure 9.21   Increase of signal set size for trellis-coded modulation.

way" for the decoder to make an error. (See Section 9.10.3.1.) Whenever there is a code at work, the signaling space is not the proper place for examining the error-performance advantage that coding can achieve. This is because the code is defined by rules and constraints that are not visible in the signaling space. When two signals are in close proximity in the signaling space of a coded system, their closeness may not have much performance significance because the rules of the code may not allow for the transitioning between two such vulnerable signal points. Where is the proper place for evaluating allowed code sequences and distance properties? It is the trellis diagram. Using this diagram, the objective of TCM is to assign waveforms to trellis transitions so as to increase the free distance between the waveforms that are the most likely to be confused.

9.10.2 TCM Encoding

9.10.2.1 Ungerboeck Partitioning

Assume that the receiver uses soft decision, so that the appropriate distance metric is Euclidean. In order to maximize the free Euclidean distance (ED), the code-to-signal mapping pioneered by Ungerboeck [31] follows from the successive partitioning of a modulation-signal constellation into subsets having increasing minimum distances $d_0 < d_1 < d_2 \ldots$ between the elements of the subsets. The concept is illustrated in Figure 9.22 for the 8-PSK signal set. In Figure 9.22, the original constellation of signals is labeled A_0, and the individual signals are numbered sequentially from 0 to 7. If the average signal power (amplitude squared) is chosen

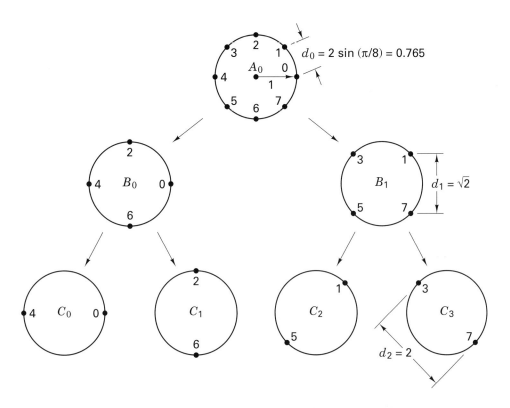

Figure 9.22 Ungerboeck partitioning of an 8-PSK signal set.

equal to unity, then the distance d_0 between any two adjacent signals is seen to be $2 \sin(\pi/8) = 0.765$. The first level of partitioning results in subsets B_0 and B_1, where the distance between adjacent signals is $d_1 = \sqrt{2}$. The next level of partitioning results in subsets C_0 to C_3, where the distance between adjacent signals is $d_2 = 2$. The construction of simple codes (up to eight states) can be determined heuristically. First a suitable trellis structure is selected, which can be done without any particular encoder in mind. TCM is classified as a waveform-coding technique because in order to describe the concept, only a suitable trellis and a set of modulation waveforms are required; it is not yet necessary to introduce bits. Waveforms from the extended set of $M' = 2^{k+1}$ waveforms are assigned to the trellis transitions so as to maximize the free ED. In the treatment of convolutional coding in Chapter 7, the encoder-trellis transitions (reflecting the behavior of an encoding circuit) were labeled with code bits. For TCM, the trellis transitions are labeled with modulation waveforms. The uncoded 4-PSK waveform set will serve as a reference for the 8-PSK coded set. This reference set, as seen in Figure 9.23, has a trivial one-state trellis diagram with four parallel transitions. This is a trivial trellis because a one-state trellis means that the system has no memory. There are no constraints or restrictions as to which of the 4-PSK signals can be transmitted during any time interval; hence, for this uncoded case, the optimum detector simply makes independent, nearest-neighbor decisions for each noisy 4-PSK signal received.

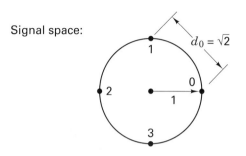

Signal space:

$d_0 = \sqrt{2}$

Trellis diagram:

Waveform number

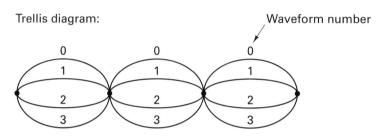

Figure 9.23 Uncoded 4-PSK and its one-state trellis diagram.

9.10.2.2 Mapping of Waveforms to Trellis Transitions

Ungerboeck devised a heuristic set of rules [31] for assigning trellis-transition branches to waveforms so that coding gain is assured, given an adequate choice of trellis states. The rules for forming the trellis structure and for partitioning the waveform set (in the context of coded 8-ary PSK modulation) can be summarized as follows:

1. If k bits are to be encoded per modulation interval, the trellis must allow for 2^k possible transitions from each state to a successor state.
2. More than one transition may occur between pairs of states.
3. All waveforms should occur with equal frequency and with a fair amount of regularity and symmetry.
4. Transitions originating from the same state are assigned waveforms either from subset B_0 or B_1—never a mixture between them.
5. Transitions joining into the same state are assigned waveforms either from subset B_0 or B_1—never a mixture between them.
6. Parallel transitions are assigned waveforms either from subset C_0 or C_1 or C_2 or C_3—never a mixture between them.

The rules guarantee that codes constructed in this way will have a regular structure and a free ED that will always exceed the minimum distance between signal points of the uncoded reference modulation. Figure 9.24 illustrates a possible code-to-signal mapping, using a four-state trellis with parallel paths. The code-to-signal assignments are made by examining the partitioned signal space in Figure 9.22 in con-

State

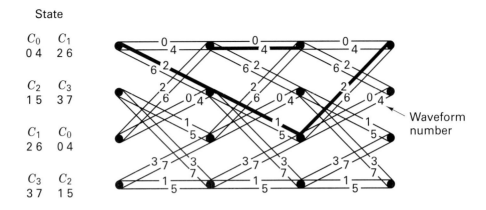

| C_0 | C_1 |
| 0 4 | 2 6 |

| C_2 | C_3 |
| 1 5 | 3 7 |

| C_1 | C_0 |
| 2 6 | 0 4 |

| C_3 | C_2 |
| 3 7 | 1 5 |

Waveform
number

Figure 9.24 Four-state trellis with parallel paths.

cert with the trellis diagram in Figure 9.24 and the rules listed above. Written on the trellis transitions are the waveform numbers that have been assigned to those transitions by following the partitioning rules. Notice that for 8-PSK, the waveform assignments comply with rule 1: There are $k + 1 = 3$ code bits and thus $k = 2$ information bits, and there are $2^2 = 4$ transitions into and out of each state. The waveform assignments comply with rule 6 because each parallel pair of transitions have been assigned waveforms from subset C_0 or C_1 or C_2 or C_3. Also, the assignments comply with rules 4 and 5, because the four branches leaving (or entering) a state have been assigned waveforms from subset B_0 or B_1. In Figure 9.24, the states of the trellis have been designated according to the waveform types that may appear on the transitions leaving that state. Thus, we can refer to the states in terms of the signal subsets as the $C_0\,C_1$ state or the $C_2\,C_3$ state, and so forth. Or, another possible designation in terms of waveform numbers would be the 0426 state or the 1537 state, and so forth. In Figure 9.24, both designations are shown. This assignment of modulation waveforms to trellis transitions by observing the set-partitioning rules concludes the specification of the trellis *encoder*. Note that the final assignment of the code bits to the waveform signals (codeword-to-transition mapping) can now be made in an arbitrary fashion. Although you might think it surprising that bits can now be assigned to trellis transitions and waveforms with impunity, recall that a circuit does not yet exist. Hence, there are no bits, and the trellis transitions can mean anything that we choose for them to mean. What are the consequences of such an *arbitrary* assignment? Different codeword-to-transition choices will reflect upon the encoder design. Hence, if we are lucky, designing the encoder circuit whose output bits correspond to the way that we assigned them to state transitions will be easily accomplished. If however, we are not lucky, such a circuit design will be difficult to accomplish. Some choices of codeword assignments will give rise to simple encoder designs, while other choices may dictate unwieldy designs.

The same trellis structure of Figure 9.24 will soon be examined in the context of detection and decoding to verify that a coding gain is assured by the fact that Ungerboeck's rules have been followed during the encoding process.

9.10.3 TCM Decoding

9.10.3.1 An Error Event and Free Distance

The task of a convolutional decoder is to estimate the path that the message had traversed through the encoding trellis. If all input message sequences are equally likely, a decoder that will achieve the minimum probability of error is one that compares the conditional probabilities $P(\mathbf{Z}|\mathbf{U}^{(m)})$—where $\mathbf{Z}$ is the received sequence of waveforms, and $\mathbf{U}^{(m)}$ is one of the possible transmitted sequences of waveforms—and chooses the maximum. This decision-making criterion, known as *maximum likelihood*, is described in Section 7.3.1. Finding the sequence $\mathbf{U}^{(m')}$ which maximizes $P(\mathbf{Z}|\mathbf{U}^{(m)})$ is equivalent to finding the sequence $\mathbf{U}^{(m')}$ that has the greatest similarity to $\mathbf{Z}$. Since a maximum-likelihood decoder will choose the trellis path whose corresponding sequence $\mathbf{U}^{(m')}$ is at the minimum distance to the received sequence $\mathbf{Z}$, the maximum-likelihood problem is identical to the problem of finding the shortest distance through the trellis diagram.

Because a convolutional code is a group (or linear) code, the set of distances that must be examined is independent of which sequence is selected as a test sequence. Therefore, there is no loss in generality in selecting an all-zeros test sequence, which is shown as a dashed line in Figure 9.25. Assuming that the all-zeros sequence is transmitted, an error event is identified as a divergence from the all-zeros path followed by a remergence with the all-zeros path. Error events both start and end in state *a,* and they do not return to state *a* anywhere in between.

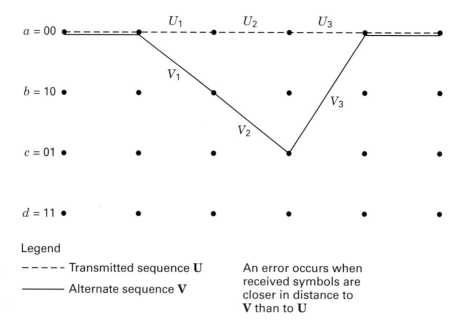

Legend

– – – – – Transmitted sequence **U**

———— Alternate sequence **V**

An error occurs when received symbols are closer in distance to **V** than to **U**

Figure 9.25 Illustration of an error event.

Figure 9.25 illustrates an error event in a trellis code; that is, the figure illustrates a transmitted all-zeros sequence marked $\mathbf{U} = \ldots, U_1, U_2, U_3, \ldots$ and an alternate sequence marked $\mathbf{V} = \ldots, V_1, V_2, V_3, \ldots$ The alternate sequence is seen to diverge and then remerge with the transmitted sequence. Assuming soft-decision decoding, an error event occurs whenever the received symbols are closer in Euclidean distance to some alternate sequence $\mathbf{V}$ than to the actually transmitted sequence $\mathbf{U}$. This implies that codes for multilevel/phase signals should be designed to achieve maximum free Euclidean distance (ED); the larger the Euclidean distance, the lower the probability of error. Therefore, assigning waveforms to trellis transitions at the encoder in a way that maximizes this free ED (see Section 9.10.2) is the key to optimizing the trellis codes.

9.10.3.2 Coding Gain

Using soft-decision, maximum-likelihood decoding, and assuming unit average signal power and Gaussian noise with variance σ^2 per dimension, a lower bound on the error-event probability can be expressed in terms of the free distance d_f [32] as

$$P_e \geq Q\left(\frac{d_f}{2\sigma}\right) \tag{9.55}$$

where $Q(\cdot)$ is the complementary error function defined in Equation (3.43). The term "error event" is used, rather than "bit-error," because the error might entail more than one flawed bit. At high signal-to-noise ratios (SNR), the bound in Equation (9.55) is asymptotically exact. The *asymptotic coding gain G*, in dB, compared with some uncoded reference system with the same average signal power and noise variance, is therefore expressed as a ratio of distances or distances squared, and we write

$$G\,(\text{dB}) = 20 \times \log_{10}\left(\frac{d_f}{d_{\text{ref}}}\right) \quad \text{or} \quad G\,(\text{dB}) = 10 \times \log_{10}\left(\frac{d_f^2}{d_{\text{ref}}^2}\right) \tag{9.56}$$

where d_f and d_{ref} represent the free ED of the coded system and the free ED of an uncoded reference system, respectively. Note that for high SNR and a given error probability, Equation (9.56) yields the same result as the coding gain expressed in Equation (6.19), which is rewritten as

$$G\,(\text{dB}) = \left(\frac{E_b}{N_0}\right)_u (\text{dB}) - \left(\frac{E_b}{N_0}\right)_c (\text{dB}) \tag{9.57}$$

where $(E_b/N_0)_u$ and $(E_b/N_0)_c$ represent the required E_b/N_0 (given in dB) for the uncoded system and the coded system, respectively. Keep in mind that the coding gain expressed in Equation (9.56) provides the same information (at high SNR) as the more familiar error-performance improvement expression in Equation (9.57). Equation (9.56), in effect, summarizes the primary goal of TCM code construction. It is to achieve a free distance that exceeds the minimum distance between uncoded modulation signals (at the same information rate, bandwidth, and power).

9.10.3.3 Coding Gain for 8-PSK with a 4-State Trellis

We now calculate the coding gain for 8-PSK with the 4-state trellis that was designed by following the encoding rules in Section 9.10.2.2. The trellis in Figure 9.24 will now be examined in the context of a decoding procedure. First the all-zeros sequence is chosen as the test vehicle. In other words, assume that the transmitter sent a sequence containing only copies of the waveform number 0. To demonstrate the benefits of this TCM system (using the Viterbi decoding algorithm), it must be shown that the easiest way for the coded system to make an error is still more difficult than the easiest way for the uncoded system to make an error. It is necessary to examine every possible divergence-from and remergence-to the correct path (all-zeros sequence), and find the one with the minimum ED from the correct path. First, look at the candidate error-event path in Figure 9.24 that is shown darkened and labeled with waveform numbers 2, 1, 2. We compute the squared distance from the all zeros path as the sum of the individual squared distances of waveform 2 from waveform 0, waveform 1 from waveform 0, and waveform 2 from waveform 0. The individual distances are taken from the partitioning diagram of Figure 9.22, and the computation results in

$$d^2 = d_1^2 + d_0^2 + d_1^2 = 2 + 0.585 + 2 = 4.585$$

or, (9.58)

$$d = \sqrt{4.585} = 2.2$$

In Equation (9.58), Euclidean distance d is formed in the same way that a resultant vector in a Euclidean space is constructed—that is, as the square root of the sum of the squares of the individual components (distances). In Figure 9.24, there is a divergence/remergence path that actually has a smaller ED than $d = 2.2$. It is the darkened error event (labeled waveform 4) that results if the parallel path survives (using Viterbi decoding), rather than the correct path associated with waveform 0. One might ask: If the decoder chooses the parallel path,—that is, the successor state is the same state in either case—is that really a serious error? If the parallel path is incorrectly chosen (it is still a divergence/remergence path, even though it occupies only one interval of time), then when we later introduce circuits and bits, a surviving waveform 4 will yield an incorrect bit meaning. The distance of the waveform 4 parallel path from waveform 0 is seen in Figure 9.22 to be $d = 2$. This distance is smaller than that of any other potential error event (try them all to be convinced); hence, the free ED for this coded system is $d_f = 2$. The minimum ED for the uncoded reference signal set in Figure 9.23 is $d_{ref} = \sqrt{2}$. Equation (9.56) is now used to compute the asymptotic gain as follows:

$$G(\text{dB}) = 10 \log_{10} \left(\frac{d_f^2}{d_{ref}^2} \right) = 10 \log_{10} \left(\frac{4}{2} \right) = 3 \text{ dB} \qquad (9.59)$$

9.10.4 Other Trellis Codes

9.10.4.1 Parallel Paths

The trellis diagram requires parallel paths whenever the number of states is less than the size of the coded waveform set M'. Hence the 4-state trellis for 8-PSK requires parallel paths. To understand why this is the case, we first state again rule 1 from Ungerboeck: If k bits are to be encoded per modulation interval, the trellis must allow for 2^k possible transitions from each state to a successor state. Each waveform in the case of coded 8-PSK represents $k + 1 = 3$ code bits or $k = 2$ data bits. Hence, this first rule calls for $2^k = 2^2 = 4$ transitions to each successor state. At first glance, the 4-state trellis without parallel paths can meet this provision, since a fully connected trellis (each state connected to each of the other succeeding states) complies. However, try drawing a 4-state, fully connected trellis without parallel paths, and try meeting the Ungerboeck rules 4 and 5 in the context of a coded 8-PSK system. It is not possible. Violating the rules will yield suboptimum results. In the next section, an 8-state trellis for 8-PSK (the number of states is no longer less than M') is shown, where all of the partitioning rules can be observed without the need for parallel paths.

9.10.4.2 8-State Trellis

After experimentation with various trellis structures and channel-signal assignments, the 8-PSK code depicted in Figure 9.26 was chosen as optimum for eight states [31]. The error-event path with minimum distance from the all-zeros path is labeled with waveform numbers 6, 7, 6. Here, since there are no parallel paths to limit the free ED, the squared free ED is $d_f^2 = d_1^2 + d_0^2 + d_1^2 = 4.585$, where the distances d_0 and d_1 are obtained from Figure 9.22. The asymptotic coding gain of the 8-state TCM system compared with the reference 4-PSK system is

$$G(\text{dB}) = 10 \times \log_{10} \frac{(d_1^2 + d_0^2 + d_1^2)_{\text{coded 8-PSK}}}{(d_{\text{ref}}^2)_{\text{uncoded 4-PSK}}} = 10 \times \log_{10}\left(\frac{4.585}{2}\right) = 3.6 \text{ dB} \qquad (9.60)$$

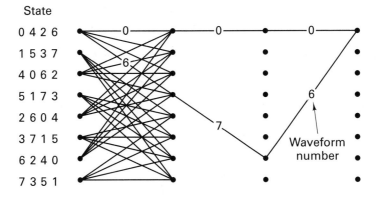

Figure 9.26 Eight-state trellis diagram for coded 8-PSK.

Similarly, a 16-state trellis structure for the coded 8-PSK constellation can be shown to yield a coding gain of 4.1 dB over uncoded 4-PSK [31]. For less than eight states, further coding gain may be realized by introducing asymmetry into the constellation of modulation signals [33].

9.10.4.3 Trellis Coding for QAM

The set-partitioning technique can be applied to other types of modulation waveforms. Consider the use of coded 16-QAM, with three information bits per modulation interval, where uncoded 8-PSK is designated as the reference system. For normalizing the 16-QAM space, we choose the average value of the signal-set squared-amplitude equal to unity, resulting in $d_0 = 2/\sqrt{10}$. Figure 9.27 illustrates a partitioning of 16-QAM waveforms into subsets with increasing subset distances ($d_0 < d_1 < d_2 < d_3$). An 8-state 16-QAM code system, obtained by using set-partitioning similar to the procedure outlined earlier, is shown in Figure 9.28 [31]. The minimum-distance error-event path is labeled D_6, D_5, D_2. Despite the coding gain that can be achieved with TCM schemes, potential phase ambiguities exist in decoding the expanded signal space that can seriously degrade performance. Wei [34] has applied the concepts of differential encoding to TCM techniques, so that the resultant codes are transparent to signal element rotations of 90-, 180-, and 270-degrees.

In summary, trellis coding for bandlimited channels employs larger signal alphabets (i.e., M-ary PAM, PSK, or QAM) to accommodate the redundancy introduced by coding, so that channel bandwidth is not increased. Even though an increase in the modulation signal-set size reduces the minimum distance between signals, the free ED between *valid* code sequences more than compensates for this reduction. The result is a net coding gain of 3 to 6 dB, without any bandwidth

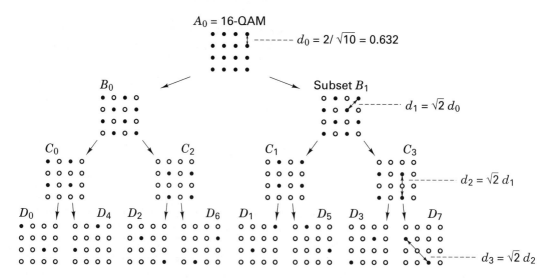

Figure 9.27   Ungerboeck partitioning of 16-QAM signals.

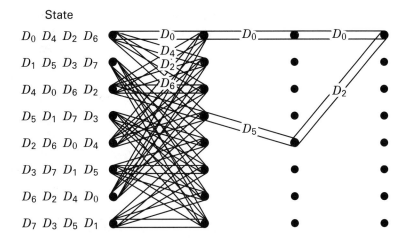

State

$D_0\ D_4\ D_2\ D_6$

$D_1\ D_5\ D_3\ D_7$

$D_4\ D_0\ D_6\ D_2$

$D_5\ D_1\ D_7\ D_3$

$D_2\ D_6\ D_0\ D_4$

$D_3\ D_7\ D_1\ D_5$

$D_6\ D_2\ D_4\ D_0$

$D_7\ D_3\ D_5\ D_1$

Figure 9.28   Eight-state trellis diagram for 16-QAM signaling.

expansion [6, 31]. In the following section, these ideas are further illustrated with an example.

9.10.5 Trellis-Coded Modulation Example

The previous sections dealt with mapping waveforms to trellis transitions, without regard to the final mapping of channel symbols (code bits or codewords) to the trellis transitions. In this section, a TCM example is approached by first specifying an encoder design. Once the encoder circuit is specified, a trellis diagram and the assignment of codewords to trellis transitions are also specified (defined by the circuitry). Thus, in this example, when waveforms are assigned to trellis transitions (and hence, to the underlying codewords), it will no longer be possible to arbitrarily assign codewords to waveforms, as was done earlier when there was no encoder circuit present.

Consider an encoder that uses a rate 2/3 convolutional code to transmit two information bits per modulation interval. A candidate encoder is shown in Figure 9.29. The rate 2/3 encoding is accomplished by transmitting one bit from each pair of bits in the input sequence unmodified, and encoding the other bit into two code bits using a rate 1/2, constraint length $K = 3$ encoder. As the figure shows, only every other bit $(m_2, m_4, \ldots)$ from the input sequence enters the shift register. One might ask: How good can such a system be, since only 50% of the input bits are receiving the benefits of redundancy? Recall the analogy of the wizard who recognizes that some bits are quite vulnerable and thus they are assigned modulation waveforms with the best distance properties, while other bits are judged to be robust and are assigned waveforms with poorer distance properties. Modulation and coding take place together; the seemingly "uncoded" bits will not be forgotten, but will be receiving the benefits of the best waveform assignments. We emphasize that the encoding and decoding of TCM take place primarily at the waveform level (our

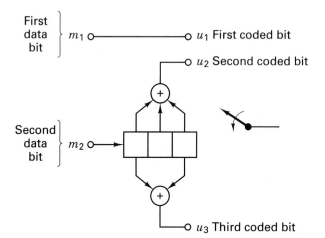

First data bit } m_1 ○————————○ u_1 First coded bit

○ u_2 Second coded bit

Second data bit } m_2 ○——▶

○ u_3 Third coded bit

Figure 9.29 Rate $\frac{2}{3}$ convolutional encoder.

first description of TCM was devoid of any encoder circuit) as compared to traditional error-correction codes where encoding and decoding take place only at the bit level.

The trellis diagram in Figure 9.30 describes the encoder circuit in Figure 9.29. As in Chapter 7, the state names were chosen to correspond to the contents of the

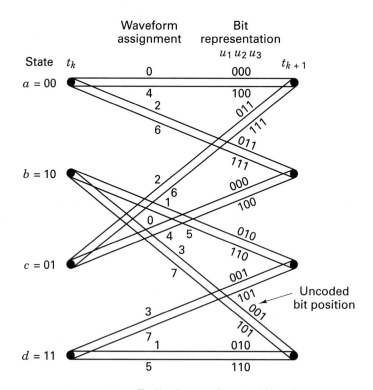

Figure 9.30 Trellis diagram for rate 2/3 code.

Modulation and Coding Trade-Offs   Chap. 9

rightmost $K - 1 = 2$ stages in the shift register. The parallel transitions on the trellis of Figure 9.30 are due to the uncoded bits; an uncoded bit is represented as the leftmost bit on each trellis transition. Four transitions emerge from each state. For each state, the two upper transitions stem from a pair of input information bits (such as m_1 m_2 being 00 and 10, respectively); the two lower transitions stem from the pair being 01 and 11, respectively. Figure 9.30 illustrates the same trellis structure as the trellis in Figure 9.24, except that each transition in Figure 9.30 is labeled with its codeword designation. It is worth repeating that the encoder circuit determines which codewords appear on the trellis transitions; the system designer only assigns waveforms to the transitions. Therefore, once there is a circuit (whose behavior is described by the trellis), any waveform assigned to a trellis transition automatically carries the codeword meaning corresponding to that transition.

Assume that the coded modulation is 8-ary pulse amplitude modulation (8-PAM), as shown in Figure 9.31. Part (a) of the figure illustrates the coded signal set, where the ED of each signal from the center of the signal space is shown in arbitrary units that are equally spaced and symmetrical about zero. Figure 9.31b illustrates the 4-ary PAM reference set, with signal points and distances similarly labeled. The important step in the encoder design is to assign the 8-ary PAM waveforms to the trellis transitions according to the Ungerboeck partitioning rules shown in Figure 9.32. Observing these rules can lead to the same assignment of waveform numbers to trellis transitions as seen in the trellis of Figure 9.24. These waveform assignments, as well as the codewords assigned by the circuit, are shown in Figure 9.30. The most disparate waveform pairs (with distance $d_2 = 8$) have been assigned to the most vulnerable error events—the parallel transitions. Also, as directed by the Ungerboeck rules, the next most distant ($d_1 = 4$) waveforms have been assigned to transitions originating from or joining into the same state. For easy reference, the codeword-to-waveform assignment (the result of waveforms mapped to trellis transitions) is also shown in Figure 9.31a.

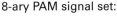

8-ary PAM signal set:

101	111	110	100	001	011	010	000	◄— Codeword assignment
7	6	5	4	3	2	1	0	◄— Waveform number
−7	−5	−3	−1	1	3	5	7	◄— Euclidean distance

(a)

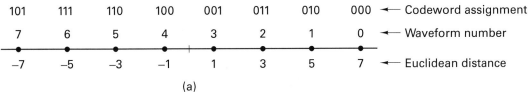

4-ary PAM signal set:

3	2	1	0	◄— Waveform number
−3	−1	1	3	◄— Euclidean distance

(b)

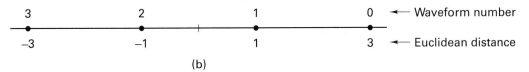

Figure 9.31 Coded 8-ary and uncoded 4-ary PAM signal sets.

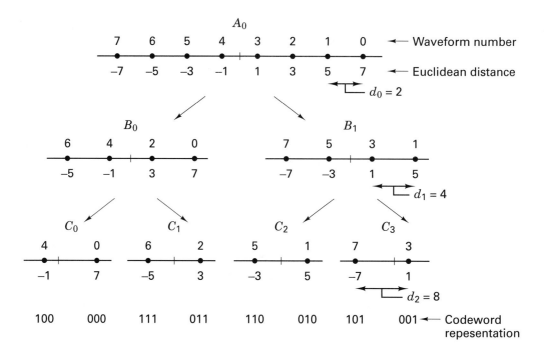

Figure 9.32 Ungerboeck partitioning of 8-PAM signals.

In Figure 9.24, the error event path labeled with waveform numbers 2, 1, 2, is the minimum distance path for this 8-PAM example. The distance from the all-zeros path is computed in the same way as was done in Equation (9.58). In this example, taking the individual distances from Figure 9.32, the computation for d_f yields

$$d_f^2 = d_1^2 + d_0^2 + d_1^2 = 16 + 4 + 16 = 36$$

or,

$$d_f = 6$$

(9.61)

For this modulation type, one can easily verify that the parallel path (with $d = 8$) is not the minimum distance error path (as it was for the 8-PSK case). Next, for finding the 4-PAM reference distance, observe from the arbitrary metric used in Figure 9.31b that $d_{\text{ref}} = 2$. We can now compute the asymptotic coding gain for this example by comparing the squared free ED of the coded system to that of the reference system. But here it is necessary to insure that the distances in the coded and uncoded systems are on an equal footing, meaning that the average power of the signals in each set is the same. In the earlier 8-PSK example, the choice of a unit circle for both the coded and uncoded systems meant that the average signal power was the same for both sets. But that is not the case in this example. Therefore, a modified version of Equation (9.56) for computing the asymptotic gain is necessary to normalize the effect of unequal signal-set average power [35]. Thus, we write

$$G(\text{dB}) = 10 \times \log_{10} \left(\frac{d_f^2/S_{av}}{d_{\text{ref}}^2/S_{av}'} \right) \tag{9.62}$$

where S_{av} and S_{av}' represent average signal power in the coded set and in the reference set, respectively. Distance corresponds to a signal amplitude or voltage; thus, distance squared corresponds to voltage squared, or power. The average signal power of the signal constellation is then computed as

$$S_{av} = \frac{d_1^2 + d_2^2 + \cdots + d_M^2}{M} \tag{9.63}$$

where d_i is the Euclidean distance of the ith signal from the center of the space, and M is the number of codeword symbols in that set. For the 8-PAM coded signal set, illustrated in Figure 9.31a, use of Equation (9.63) yields $S_{av} = 21$. For the 4-ary PAM reference signal set, illustrated in Figure 9.31b, use of Equation (9.63) yields $S_{av}' = 5$.

Using Equation (9.62), the asymptotic coding gain for an 8-PAM coded system with a 4-state trellis, is given by

$$G(\text{dB}) = 10 \times \log_{10} \left(\frac{36/21}{4/5} \right) = 3.3 \text{ dB} \tag{9.64}$$

Larger coding gains can be achieved with an increased number of trellis states (larger constraint length) at the expense of increased decoding complexity. For the rate 2/3 coded 8-ary PAM example discussed here, a 256-state trellis yields a coding gain of 5.83 dB, relative to an uncoded 4-ary PAM signal set [9]. There is only a slight increase in complexity at the transmitter, due to the trellis coding. The decoding problem at the receiver is made more complex, but the use of large scale integrated (LSI) circuits and very high-speed integrated circuits (VHSIC) makes these coding techniques extremely attractive for achieving significant coding gain without bandwidth expansion.

9.10.6 Multi-Dimensional Trellis-Coded Modulation

It was pointed out in Section 9.9.3 that for a given data rate, signaling in a two-dimensional space can provide the same error performance with less average power than signaling in a one-dimensional PAM space. This is accomplished by choosing signal points on a two dimensional lattice from within a circular-shaped boundary rather than a rectangular-shaped one. By performing similar shaping at higher dimensions, the potential energy savings approaches 1.53 dB as N approaches infinity. For practical systems, such multidimensional signaling can achieve an energy savings, called the *shaping gain,* of approximately 1-dB over one-dimensional signaling [21, 36, 37]. In the V.34 High-Speed Modem Standard, 16-dimensional QAM is specified, where the technique for mapping bits to higher-dimensional signal-constellation points is called *shell mapping,* and can provide a shaping gain of 0.8 dB [16]. Using four-, eight-, or 16-dimensional signal constellations can have some advantages over the usual two-dimensional schemes. These advantages include smaller, two-dimensional constellation building blocks, better tolerance-to-phase ambiguities, and a better trade-off between coding gain and complexity. Wei [36] has presented and evaluated a number of such schemes. (For readers who may be interested in reading further in the area of coded modulation, particularly trellis-coded modulation, references [38–46] are recommended.)

9.11 CONCLUSION

In this chapter we have integrated some of the ideas in earlier Chapters dealing with modulation and coding. We have reviewed the basic system design goals: to maximize data rate while simultaneously minimizing error probability, bandwidth, E_b/N_0, and complexity. We examined the trade-offs heuristically on two performance planes: the error probability plane and the bandwidth efficiency plane. The former explicitly illustrates the P_B versus E_b/N_0 trade-offs while only implicitly displaying the bandwidth expenditure. The latter explicitly illustrates the R/W versus E_b/N_0 trade-offs while only implicitly displaying the P_B performance. We outlined typical steps that need to be considered in meeting bandwidth power, and error-performance requirements of a digital communication system, and we discussed some of the basic constraints to improvement without limit. The Nyquist criterion establishes that we cannot continue to reduce system bandwidth indefinitely. There is a theoretical limitation; in order to transmit R_s symbols/second without intersymbol interference, we must utilize a minimum of $R_s/2$ hertz of bandwidth. The Shannon–Hartley theorem relates to the power–bandwidth trade-off and results in another important limitation, the Shannon limit. The Shannon limit of –1.6 dB is the theoretical minimum amount of E_b/N_0 that is necessary (in concert with channel coding) to achieve an arbitrarily low error probability over an AWGN channel. The more general limitation is the channel capacity, above which there cannot be error-free signaling. We have also examined some of the bandwidth-efficient modulation schemes, such as minimum shift keying (MSK), quadrature amplitude modulation (QAM), and trellis-coded modulation. The latter technique offers an attractive way to obtain coding gain without paying the price of additional bandwidth.

REFERENCES

1. *IEEE Personal Communications,* Special Issue on Software Radio, vol. 6, no. 4, August 1999.
2. Nyquist, H., "Certain Topics on Telegraph Transmission Theory," *Trans. AIEE,* vol. 47, April 1928, pp. 617–644.

3. Shanon, C. E., "A Mathematical Theory of Communication," *BSTJ,* vol. 27, 1948, pp. 379–423, 623–657.

4. Shannon, C. E., "Communication in the presence of Noise," *Proc. IRE,* vol. 37, no. 1, January 1949, pp. 10–21.

5. Bedrosian, E., "Spectrum Conservation by Efficient Channel Utilization," *Rand Corp., Report* WN-9275-ARPA, Contract DAHC-15–73–C–0181, Santa Monica, California, October, 1975.

6. Ungerboeck, G., "Trellis-Coded Modulation with Redundant Signal Sets," Part I and Part II, *IEEE Communications Magazine,* vol. 25, February 1987, p. 5–21.

7. Hodges, M.R.L., "The GSM Radio Interface," *British Telecom Tech. J.,* vol. 8, no. 1, January 1990, pp. 31–43.

8. Anderson, J. B., and Sundberg, C-E. W., "Advances in Constant Envelope Coded Modulation," *IEEE Commun., Mag.,* vol. 29, no. 12, Dec. 1991, pp. 36–45.

9. Clark, G. C. Jr., and Cain, J. B., *Error Correction Coding for Digital Communications,* Plenum Press, New York, 1981.

10. Lindsey, W. C., and Simon, M. K., *Telecommunication Systems Engineering,* Prentice-Hall, Englewood Cliffs, NJ, 1973.

11. Sklar, B., "Defining, Designing, and Evaluating Digital Communication Systems," *IEEE Commun. Mag.,* Vol. 31, no. 11, Nov. 1993, pp. 92–101

12. Korn, I., *Digital Communications,* Van Nostrand Reinhold Co., New York, 1985.

13. Viterbi, A. J., *Principles of Coherent Communications,* McGraw-Hill Book Co., New York, 1966.

14. Lin, S., and Costello, D. J., Jr., *Error Control Coding: Fundamentals and Applications,* Prentice-Hall, Englewood Cliffs, NJ, 1983.

15. Odenwalder, J. P., *Error Control Coding Handbook,* Linkabit Corporation, San Diego, California, July 15, 1976.

16. Forney, G. D., Jr., et. al., "The V.34 High-Speed Modem Standard," *IEEE Communications Magazine,* December 1996.

17. Pasupathy S., "Minimum Shift Keying: A Spectrally Efficient Modulation," *IEEE Commun. Mag.,* July 1979, pp. 14–22.

18. Gronemeyer, S. A., and McBride, A. L., "MSK and Offset QPSK Modulation," *IEEE Trans. Commun.,* vol. COM-24, August 1976, pp. 809–820.

19. Simon, M. K. "A Generalization of Minimum Shift Keying (MSK) Type Signaling Based Upon Input Data Symbol Pulse Shaping," *IEEE Trans. Commun.,* vol. COM-24, August 1976, pp. 845–857.

20. Leib, H., and Pasupathy, S., "Inherent Error Control Properties of Minimum Shift Keying," *IEEE Communications Mag.,* vol. 31, no. 1, January 1993, pp. 52–61.

21. Forney, G. D. Jr. et. al., "Efficient Modulation for Bandlimited Channels," *IEEE J. Selected Areas in Commun.,* vol. SAC-2, no. 5, September 1984, pp. 632–647.

22. Thomas, C. M., Weidner, M. Y., and Durrani, S. H., "Digital Amplitude-Phase Keying with M-ary Alphabets," *IEEE Trans. Commun.,* vol. COM-22, no. 2, February 1974, pp. 168–180.

23. Lucky, R. W., and Hancock, J. C., "On the Optimum Performance of N-ary Systems Having Two Degrees of Freedom," *IRE Trans. on Commun. Sys.,* vol. CS-10, June 1962, pp. 185–192.

24. Campopiano, C. N., and Glazer, B. G., "A Coherent Digital Amplitude and Phase Modulation Scheme," *IRE Trans. on Commun. Sys.,* vol. CS-10, June 1962, pp. 90–95.

25. Cahn, C. R., "Combined Digital Phase and Amplitude Modulation Communication Systems," *IRE Trans. on Commun. Tech.,* Sept. 1960.

26. Foschini, G. J., and Gitlin, R. D., "Optimization of Two Dimensional Signal Constellations in the Presence of Gaussian Noise," *IEEE Trans. Commun.,* vol. COM-22, no. 1, January 1974, pp. 23–38.

27. Welti, G. R., and Jhong, S. L., "Digital Transmission with Coherent Four-Dimensional Modulation," *IEEE Trans. Inform. Theory,* vol. IT-20, no. 4, July 1974, pp. 497–502.

28. Gersho, A., and Lawrence, V. B., "Multidimensional Signal Constellations for Voiceband Data Transmission," *IEEE J. Selected Areas in Commun.,* vol. SAC-2, no. 5, September 1984, pp. 687–702.

29. Zetterberg, L. H., and Brandstrom, H., "Codes for Combined Phase and Amplitude Modulated Signals in a Four-Dimensional Space," *IEEE Trans. Commun.,* vol. COM-25, no. 9, September 1977, pp. 943–950.

30. Wilson, S. G., Sleeper, H. A., and Srinath, N. K., "Four-Dimensional Modulation and Coding: An Alternative to Frequency Reuse," *IEEE 1984 Int'l. Commun. Conf.,* pp. 919–923.

31. Ungerboeck, G., "Channel Coding with Multilevel/Phase Signals," *IEEE Trans. Inform. Theory,* vol. IT-28, January 1982, pp. 55–67.

32. Forney, G. D., "The Viterbi Algorithm," *Proceedings of the IEEE,* vol. 61, no. 3, March 1978, pp. 268–278.

33. Divsalar, D., Simon, M. K., and Yuen, J. H., "Trellis Coding with Asymmetric Modulations," *IEEE Trans. Commun.* vol. COM-35, no. 2, February 1987.

34. Wei, J.-F., "Rotationally Invariant Convolutional Channel Coding with Expanded Signal Space—Parts I and II," *IEEE J. Sel. Areas Commun.,* vol. SAC-2, no. 5, Sept. 1984, pp. 659–686.

35. Thapar, H. K., "Real-Time Application of Trellis Coding to Highspeed Voiceband Data Transmission," *IEEE J. Sel. Areas Commun.,* vol. SAC-2, no. 5, Sept. 1984, pp. 648–658.

36. Wei, L.-F., "Trellis-Coded Modulation with Multidimensional Constellations," *IEEE Trans. Information Theory,* vol. IT-33, no. 4, July 1987, pp. 483–501.

37. Tretter, S. A., "An Eight-Dimensional 64-State Trellis Code for Transmitting 4 Bits Per 2-D Symbol," *IEEE J. on Sel. Areas of Commun.,* vol. 7, no. 9, December 1989, pp. 1392–1395.

38. Kato, S., Morikura, M., and Kubota, S., "Implementation of Coded Modems," *IEEE Communications Magazine,* vol. 29, no. 12, December, 1991, pp. 88–97.

39. Special Issue on Coded Modulation, *IEEE Communications Magazine,* vol. 29, no. 12, December 1991.

40. Biglieri, E., et. al., *Introduction to Trellis-Coded Modulation with Application,* MacMillan, New York, NY, 1991.

41. Edbauer, F., "Performance of Interleaved Trellis-Code Differential 8-PSK Modulation over Fading Channels," *IEEE J. on Selected Areas in Commun.,* vol. 7, no. 9, December 1989, pp. 1340–1346.

42. Rimoldi, B., "Design of Coded CPFSK Modulation Systems for Bandwidth and Energy Efficiency," *IEEE Transactions on Communications,* vol. 37, no. 9, September 1989, pp. 897–905.

43. Viterbi, A. J., et. al., "A Pragmatic Approach to Trellis-Coded Modulation," *IEEE Communications Magazine,* vol. 27, no. 7, July 1989, pp 11–19.

44. Divsalar, D., and Simon, M. K., "The Design of Trellis Coded MPSK for Fading Channels: Performance Criteria," *IEEE Trans. on Comm.,* vol. 36, no. 9, September 1988, pp. 1004–1012.

45. Divsalar, D., and Simon, M. K., "The Design of Trellis Coded MPSK for Fading Channels: Set Partitioning for Optimum Code Design," *IEEE Trans. on Comm.,* vol. 36, no. 9, September 1988, pp. 1013–1021.

46. Divsalar, D., and Simon, M. K., "Multiple Trellis Coded Modulation (MTCM)," *IEEE Trans. on Commun.,* vol. 36, no. 4, April 1988, pp. 410–419.

PROBLEMS

9.1. Consider a voice-grade telephone circuit with a bandwidth of 3 kHz. Assume that the circuit can be modeled as an AWGN channel.
   (a) What is the capacity of such a circuit if the SNR is 30 dB?
   (b) What is the minimum SNR required for a data rate of 4800 bits/s on such a voice-grade circuit?
   (c) Repeat part (b) for a data rate of 19,200 bits/s.

9.2. Consider that a 100-kbits/s data stream is to be transmitted on a voice-grade telephone circuit (with a bandwidth of 3 kHz). Is it possible to approach error-free transmission with a SNR of 10 dB? Justify your answer. If it is not possible, suggest system modifications that might be made.

9.3. Consider a source that produces six messages with probabilities $\frac{1}{2}, \frac{1}{4}, \frac{1}{8}, \frac{1}{16}, \frac{1}{32},$ and $\frac{1}{32}$. Determine the average information content in bits, of a message.

9.4. A given source alphabet consists of 300 words, of which 15 occur with probability 0.06 each and the remaining 285 words occur with probability 0.00035 each. If 1000 words are transmitted each second, what is the average rate of information transmission?

9.5. **(a)** Find the average capacity in bits per second that would be required to transmit a high-resolution black-and-white TV signal at the rate of 32 pictures per second if each picture is made up of 2×10^6 picture elements and 16 different brightness levels. All picture elements are assumed to be independent and all levels have equal likelihood occurrence.
   (b) For color TV, this system additionally provides for 64 different shades of color. How much more system capacity is required for a color system compared to the black and white system?
   (c) Find the required capacity if 100 of the possible brightness–color combinations occur with a probability of 0.003 each, 300 of the combinations occur with a probability of 0.001, and 624 of the combinations occur with a probability of 0.00064.

9.6. Prove that entropy is maximized when all source outputs have equal probability.

9.7. Compute the equivocation or message uncertainty in bits per character for a textual transmission using 7-bit ASCII coding. Assume that each character is equally likely and that the noise on the channel results in a bit error probability of 0.01.

9.8. Suppose a binary noncoherent FSK link has a maximum data rate of 2.4 kbits/s without ISI over a channel whose nominal bandwidth is 2.4 kHz. Suggest ways of increasing the data rate under the following system constraints.

(a) The system is power limited.

(b) The system is bandwidth limited.

(c) The system is both power and bandwidth limited.

9.9. Table P9.1 characterizes four different satellite-to-earth-terminal links. For each link assume that the space loss is 196 dB, the margin is 0 dB, and there are no other incidental losses. For each link, plot an operating point on the bandwidth efficiency plane, R/W versus E_b/N_0, and characterize the link according to one of the following descriptions: bandwidth limited, severely bandwidth limited, power limited, and severely power limited. Justify your answers.

TABLE P9.1 Downlink Capacity for Four Satellite Links

Satellite	Receive Terminal	Maximum Data Rate
INTELSAT IV EIRP = 22.5 dBW Bandwidth = 36 MHz	Large fixed antenna diameter = 30 m G/T = 40.7 dB/K	165 Mbits/s
DSCS II EIRP = 28 dBW Bandwidth = 50 MHz	Shipboard antenna diameter = 4 ft G/T = 10 dB/K	100 kbits/s
DSCS II EIRP = 28 dBW Bandwidth = 50 MHz	Large fixed antenna diameter = 60 ft G/T = 39 dB/K	72 Mbits/s
GAPSAT/MARISAT EIRP = 28 dBW Bandwidth = 500 kHz	Aircraft antenna gain = 0 dB G/T = −30 dB/K	500 bits/s

9.10. You are required to make modulation and error-correction code choices for a real-time communication system operating over an AWGN channel with an available bandwidth of 2400 Hz. The available E_b/N_0 is 14 dB. The required data rate and bit-error probability are 9600 bits/s and 10^{-5}, respectively. You are allowed to choose one of two modulation types—either noncoherent orthogonal 8-FSK or 16-QAM with matched filter detection. You are allowed to choose one of two codes—either the (127, 92) BCH code or a rate $\frac{1}{2}$ convolutional code that provides 5 dB of coding gain at a bit-error probability of 10^{-5}. Assuming ideal filtering, verify that your choices achieve the desired bandwidth- and error-performance requirements.

9.11. You are required to make the same kind of design choices of modulation and coding as in Problem 9.10. The requirements and assumptions in this problem are the same as in Problem 9.10, except that the available bandwidth is now increased to 40 kHz, and the available E_b/N_0 is now reduced to 7.3 dB. Verify that your choices achieve the desired bandwidth-and error-performance requirements.

9.12. You are required to make the same kind of design choices as in Problem 9.10. The requirements and assumptions in this problem are the same as in Problem 9.10, except that the AWGN channel now manifests deep fades lasting up to 100 ms in duration. The available bandwidth is now 3400 Hz, and the available E_b/N_0 is now 10 dB. In addition to making modulation and coding choices you are asked to select an interleaver design (see Section 8.2) to combat the worst-case fading problem. You are allowed to choose one of two types—either a 16×32 block interleaver or a 150×300 convolutional interleaver. Verify that your choices achieve the desired bandwidth- and error-performance requirements, as well as providing the needed mitigation against the longest fade duration.

9.13. (a) Consider a real-time communication system operating over an AWGN channel that uses 8-PSK modulation with Gray coding. Select an error-correcting code that provides a decoded bit-error probability of at most 10^{-7} when the received P_r/N_0 is 70 dB-Hz, and the data rate is 1 Mbit/s. You are permitted to choose one of the following codes: (24, 12) Extended Golay, or (127, 64) BCH, or (127, 36) BCH. The input-output transfer functions for these codes appear in Figure 6.21. To guide you in making a selection, consider that at $P_B = 10^{-7}$, the transfer functions intercept the abscissa at the following points: the (24, 12) code intercepts at 3×10^{-3}, the (127, 64) code intercepts at 1.3×10^{-2}, and the (127, 36) BCH code intercepts at 3×10^{-2}.

(b) The general appearance of the code transfer functions may have caused you to form an initial judgement as to which code is the best choice for the stated design requirements. Does your final choice agree with your initial guess? Do the results of part a) surprise you? Explain your results in the context of the two mechanisms at work when using error-correction coding in a real-time communication system.

(c) How much coding gain in decibels does your code choice in part a) provide?

9.14. Consider a real-time communication satellite system, operating over an AWGN channel (disturbed by periodic fades). The overall link is described by the following specifications from a mobile transmitter to a low-earth-orbit satellite receiver:

Data rate $R = 9600$ bits/s
Available bandwidth $W = 3000$ Hz
Link margin $M = 0$ dB (see Section 5.6)
Carrier frequency $f_c = 1.5$ GHz
EIRP = 6 dBW
Distance between transmitter and receiver $d = 1000$ km
Satellite receiver figure of merit $G/T = 30$ dBI
Receiver antenna temperature $T_A^\circ = 290$ K
Line loss from the receiver antenna to the receiver, $L = 3$ dB
Receiver noise figure $F = 10$ dB
Losses due to fading $L_f = 20$ dB
Other losses $L_o = 6$ dB

You are allowed to choose one of two modulation schemes—MPSK with Gray coding, or noncoherent orthogonal MFSK—such that the available bandwidth is not exceeded and power is conserved. For error-correction coding, you are to choose one of the (127, k) BCH codes from Table 9.2 that provides the most redundancy, but still meets the bandwidth constraints. Calculate the output decoded bit-error probability. How much coding gain, if any, characterizes your choices. Hint: Proceed by calculating parameters in the following order, E_b/N_0, E_s/N_0, $P_E(M)$, p_c, P_B. When using Equation (9.41) for computing decoded bit-error probability, a small E_b/N_0 necessitates using many terms in the summation. Hence, computer assistance is helpful here.

9.15. You are required to provide a real-time communication system to support 9600 bits/s with a required bit-error probability of at most 10^{-5} within an available bandwidth of

2700 Hz. The predetection P_r/N_0 is 54.8 dB-Hz. Choose one of two modulation schemes—either MPSK with Gray coding or noncoherent orthogonal MFSK, such that the available bandwidth is not exceeded and power is conserved. If error-correction coding is needed, choose the simplest (shortest) code in Table 9.3 that provides the needed error performance, but does not exceed the available bandwidth. Verify that your design choices result in meeting the requirements.

9.16. (a) For a fixed error probability, show that the relationship between alphabet size M and required average power for MPSK versus QAM can be expressed as

$$\frac{\text{average power for MPSK}}{\text{aveage power for QAM}} \simeq \frac{3M^2}{2(M-1)\pi^2}$$

(b) Discuss the advantage of one type of signaling over the other.

9.17. Consider a telephone modem operating at 28.8 kbits/s that uses trellis-coded QAM modulation.

(a) Calculate the bandwidth efficiency of such a modem, assuming that the usable channel bandwidth is 3429 Hz.

(b) Assuming AWGN and an available $E_b/N_0 = 10$ dB, calculate the theoretically available capacity in the 3429-Hz bandwidth.

(c) What is the required E_b/N_0 that will enable a 3429-Hz bandwidth to have a capacity of 28.8 kbits/s?

9.18. Figure 9.17 shows several 16-ary symbol constellations.

(a) For the (5, 11) circular constellations, compute the minimum radial distances r_1 and r_2 if the minimum distance between each symbol must be 1 unit.

(b) Compute the average signal power for the (5, 11) circular constellation, and compare it to the average signal power for the 4×4 ($M = 16$) square constellation (with the same minimum distance between symbols.)

(c) Why might the square constellation be more practical?

9.19. Consider that the rate $\frac{2}{3}$ trellis-coded system of Section 9.10.5 is used over a binary symmetric channel (BSC). Assume that the initial encoder state is the 00 state. At the output of the BSC, the sequence $\mathbf{Z} = (1\ 1\ 1\ 0\ 0\ 1\ 1\ 0\ 1\ 0\ 1\ 1$ rest all "0") is received.

(a) Find the maximum likelihood path through the trellis diagram and determine the first 6 decoded information bits. If a tie occurs between any two merged paths, choose the upper branch entering the particular state.

(b) Determine if any channel bits in $\mathbf{Z}$ had been inverted by the channel during transmission, and if so, identify them.

(c) Explain how you would proceed with the problem if the channel were specified as a Gaussian channel instead of a BSC.

9.20. Find the asymptotic coding gain for a 4-state trellis-coded modulation (TCM) scheme. A code rate of $\frac{2}{3}$ is achieved by using the encoder configuration shown in Figure 9.29, where 50% of the data bits are inputted to a rate $\frac{1}{2}$ convolutional encoder, and the remaining 50% are passed directly to the output. The coded modulation is 8-PAM as seen in the upper part of Figure 9.31. Use a 4-PAM signal set with amplitudes of $-16, -1, +1, +16$ to serve as the reference set. Does your answer appear to violate Shannon's theorem that predicts a coding-gain limit of about 11–12 dB? Would anyone ever use a reference set, such as the one proposed here? Perhaps the notion of coding gain for combined modulation/coding schemes is slightly different than that for coding alone. In this context explain your results.

9.21. Find the asymptotic coding gain for an 8-state trellis-coded modulation scheme having the following features. The coded modulation is 8-PSK, and the uncoded reference is 4-PSK. The trellis structure is formed from time t_k to time t_{k+1} in the following way: Number the states (from top to bottom) with the arbitrary designations 1–8. Then, connect each of the states 1, 3, 5, and 7 at time t_k to states 1–4 at time t_{k+1}. Similarly, connect each of the states 2, 4, 6, and 8, at time t_k, to states 5–8, at time t_{k+1}. Next, draw three sections (time intervals) of the trellis structure. Then, make waveform assignments to branches, and search for the shortest error path.

QUESTIONS

9.1. Why do binary and 4-ary orthogonal frequency shift keying (4-FSK) manifest the same *bandwidth-efficiency* relationship? (See Section 9.5.1.)

9.2. For MPSK modulation, *bandwidth efficiency* increases with higher-dimensional signaling, but for MFSK, it decreases. Explain why this is the case. (See Sections 9.7.2 and 9.7.3.)

9.3. Consider the assortment of signaling elements that flow through a typical system, and describe the subtle energy and rate transformations among them: from data-bits to channel-bits to symbols to chips. (See Section 9.7.7.)

9.4. The steep decrease in MSK *spectral sidelopes* in Figure 9.15 illustrates why MSK is considered to be more spectrally efficient then QPSK. How do you explain then that the QPSK spectrum has a *narrower mainlobe* than the MSK spectrum? (See Section 9.8.2.)

9.5. In Chapter 4, we presented binary phase shift keying (BPSK) and quaternary phase shift keying (QPSK) as manifesting the same bit-error-probability relationship. (See Section 4.8.4.) Does the same hold true for *M*-ary pulse amplitude modulation (*M*-PAM) and M^2-ary quadrature amplitude modulation (M^2-QAM); that is, do these schemes also manifest the same bit-error probability? (See Section 9.8.3.1.)

9.6 Although trellis-coded modulation schemes do not require additional bandwidth or power, their use still involves a *trade-off*. What is the cost of achieving coding gain with TCM? (See Section 9.10.)

9.7 What is meant by the *state* of a finite-state machine? (See Section 9.10.)

9.8 When using TCM, how much *waveform redundancy* is sufficient to realize the benefits of coding (improved error performance or increased capacity)? (See Section 9.10.1.1.)

9.9 For TCM schemes, define what is meant by *asymptotic coding gain,* and from that definition, explain what goals one should strive for in the construction of TCM codes. (See Section 9.10.3.2.)

9.10. In a TCM trellis diagram, when are *parallel paths* required in order to comply with the Ungerboeck partitioning rules? What is the result of violating the rules? (See Section 9.10.4.1.)

EXERCISES

Using the Companion CD, run the exercises associated with Chapter 9.

CHAPTER 10

Synchronization

Maurice A. King, Jr.
The Aerospace Corporation
El Segundo, California

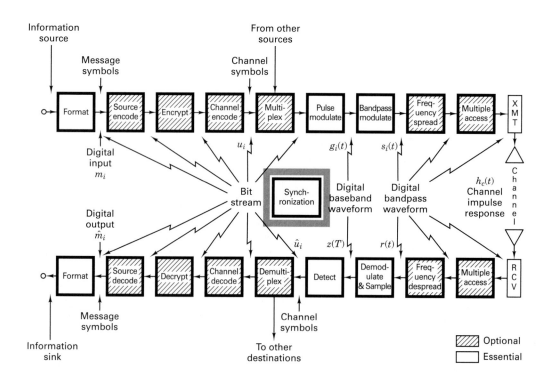

10.1 INTRODUCTION

10.1.1 Synchronization Defined

In almost every discussion of receiver or demodulator performance, some level of signal synchronization is assumed, although this assumption is often not explicitly stated. For example, in the case of coherent phase demodulation (PSK), the receiver is assumed to be able to generate reference signals whose phases are identical (except perhaps for a constant offset) to those of the signaling alphabet at the transmitter. These reference signals are compared with the incoming signals in the process of making maximum-likelihood symbol decisions.

In order to be able to generate these reference signals, the receiver has to be in synchronization with the received carrier. This means that there has to be phase concurrence between the incoming carrier and a replica of it in the receiver. In other words, if there were no information modulated on the incoming carrier, the incoming carrier and the replica in the receiver would pass through zero simultaneously. This is what is known as being in *phase lock* and is a condition that must be closely approximated if coherently modulated signals are going to be accurately demodulated at the receiver. Being in phase lock means that the receiver's local oscillator is synchronized in both frequency and phase with the received signal. If the information-bearing signal is not modulated directly on the carrier but indirectly through the use of a subcarrier, both the phase of the carrier and that of the subcarrier must be determined. If the carrier and subcarrier are not kept in phase synchronism by the transmitter (they typically are not), this will require the generation of a replica of the subcarrier by the receiver, where the phase of the subcarrier replica is controlled separately from that of the carrier replica. This will enable the receiver to achieve phase lock on both the carrier and subcarrier.

It is also assumed that the receiver has accurate knowledge of when an incoming symbol started and when it is over. This knowledge is required in order to know the proper symbol integration interval—the interval over which energy is integrated prior to making symbol decisions. Clearly if the receiver integrates over an interval of an inappropriate length, or over an interval that spans two symbols, the ability to make accurate symbol decisions will be degraded.

It can be seen that symbol synchronization and phase synchronization are similar in that they both involve producing in the receiver a replica of a portion of the transmitted signal. For phase synchronization, it was an accurate replica of the carrier. For symbol synchronization, it is a square wave at the symbol transition rate. The receiver must, in effect, be able to produce a square wave that will transition through zero simultaneously with the incoming signal's transitions between symbols. A receiver that is able to do this can be said to have symbol synchronization, or to be in *symbol lock.* Since there are typically a very large number of carrier cycles per symbol period, this second level of synchronization is much coarser than phase synchronization and is usually done with different circuitry than that used for phase synchronization.

In many communication systems an even higher level of synchronization is required. This is usually called *frame synchronization.* Frame synchronization is required when the information is organized in blocks, or messages of some uniform number of symbols. This will occur, for example, if a block code is used for forward error control, or if the communications channel is being time-shared, on a regular basis, by several users (TDMA). In the case of block coding, the decoder needs to know the location of boundaries between code words in order to decode the message correctly. In the case of a time-shared channel, it is necessary to know where the location of boundaries between channel users are, in order to route the information appropriately. Similar to symbol synchronization, frame synchronization is equivalent to being able to generate a square wave at the frame rate, with the zero crossings coincident with the transitions from one frame to the next.

Most digital communications systems using coherent modulation require all three levels of synchronization: phase, symbol, and frame. Systems using noncoherent modulation techniques will typically require symbol and frame synchronization, but since the modulation is not coherent, accurate phase lock is not required. Instead, noncoherent systems require *frequency synchronization.* Frequency synchronization differs from phase synchronization in that the replica of the carrier that is generated by the receiver is allowed to have an arbitrary constant phase offset from the received carrier. Receiver designs can be simplified by removing the requirement to determine the exact value of the incoming carrier phase. Unfortunately, as is shown in the discussion of modulation techniques, this simplification carries a penalty in terms of degraded performance versus signal-to-noise ratio. The relative trade-offs of synchronization levels versus performance and system versatility are discussed further in the next section.

All of the discussion thus far had been oriented toward the receiving end of a communication link. There are instances, however, when the transmitter assumes the more active role in synchronization—by varying the timing and frequency of its transmissions to correspond to the expectations of the receiver. An example of this situation is a satellite communication network, where many terrestrial terminals are beaming signals toward a single satellite receiver. In most of these cases the transmitter relies on a return path from the receiver to determine the accuracy of its synchronization. Thus, transmitter synchronization often implies two-way communications or a network in order to be successful. Thus, transmitter synchronization is often called *network synchronization*. Transmitter or network synchronization is discussed later in this chapter.

10.1.2 Costs versus Benefits

There is a cost associated with the need for receiver synchronization. Each additional level of synchronization implies more cost. The most obvious cost is in the need for additional hardware or software in the receiver for acquisition and tracking. Possibly less obvious costs lie in the extra time required to achieve synchronization before commencing communications, or in the energy expended by the transmitter on signals to be used at the receiver as acquisition or tracking aids. In the face of these costs to the system, one might question why a communications system designer would consider a system design requiring a high degree of synchronization. The answer: improved performance and versatility.

Consider a standard commercial analog AM radio. This radio may be considered part of a broadcast communication system involving a central transmitter and many receivers. This communication system involves no synchronization. However, the receiver passband must be wide enough to accommodate not only the information-bearing signal, but also any fluctuations in the carrier, due perhaps to Doppler shift* or drift in the transmitter's frequency reference. This requirement in the receiver passband means that additional noise energy is passed to the detector, over and above the amount theoretically required by the bandwidth of the information. A somewhat more complicated receiver that employs a carrier frequency tracking loop would be able to keep a narrow passband filter centered about the carrier, thereby substantially reducing the detected noise energy and improving the received signal-to-noise ratio. Thus, although a standard radio may be perfectly adequate for reception of signals from large transmitters a few tens of kilometers distant, it may prove totally inadequate under less benign conditions.

For digital communications, examples of the trade-off between performance and receiver complexity are often seen in the choice of modulation. Among the

---

*An offset in frequency as perceived by a receiver, from the nominally transmitted frequency caused by relative motion of the transmitter and receiver. Ignoring second- and higher-order effects the value of the frequency offset, Δf, is given by $V f_0/c$, where V is the relative velocity (positive when the relative distance between transmitter and receiver is being reduced), f_0 the nominal frequency and c the speed of light.

simplest digital receivers are those designed to be used with noncoherently detected binary FSK. The only synchronization requirements are bit timing and frequency tracking. however, the same bit error probability could be achieved with approximately 4 dB less signal-to-noise ratio if the modulation is coherent BPSK. The disadvantage of BPSK is that the receiver requires accurate phase tracking, which can present a complex design problem if the signals experience high Doppler rates** or fading. (See Chapter 15.)

A third cost-versus-performance trade-off involves the use of error-control coding. As was established in earlier chapters, there are substantial performance advantages in the use of appropriate error-control coding techniques. The cost, however, measured in receiver complexity, can be high. For a block decoder to operate properly requires the receiver to achieve block, message, or frame synchronism. This is a procedure over and above the usual decoding procedure, although some error-correcting codes have been designed with block synchronization aids built in [1]. Convolutional codes also require some degree of additional synchronization in order to provide optimum performance. Although the performance analysis of convolutional codes often makes the assumption that the input data sequence is infinitely long, in practice it is not. In order to provide the minimum error probability, the decoder must know the beginning state (usually all zeros) when the data sequence will begin, the eventual ending state, and when the ending state is to be reached. Knowing when the beginning state was left and when the ending state is to be reached, however, is equivalent to having frame synchronization. In addition, the decoder will have to know how to group the channel symbols in order to make branch decisions. This is also a synchronization requirement.

The trade-offs discussed thus far have been in terms of the performance versus complexity of individual links and receivers. The ability to synchronize has a large potential consequence in terms of system efficiency and versatility as well. Frame synchronization allows the use of advanced, versatile, multiple-access techniques, such as the variety of demand-assignment-multiple-access (DAMA) schemes, which have become increasingly popular as communication channel resources become increasingly scarce. In addition, the use of spread-spectrum techniques—both as multiple access schemes and for interference rejection—requires a high level of system synchronization. (Spread-spectrum techniques are treated in Chapter 12.) It will be seen that these techniques provide the potential for a great deal of system versatility, which is a very valuable feature if the system encounters changing or unstable conditions, such as the effects of intentional and unintentional interference from external sources.

10.1.3 Approach and Assumptions

There have been at least two substantial developments in the general area of synchronization since the first edition of this text. One has been the emergence, and then near total dominance, of signal processing (including synchronizers) by sam-

---

**The rate of change of the Doppler shift. This rate sets requirements on the tracking ability of the phase tracking loop.

pled data techniques. The other has been the publication of several book-length treatments of synchronization [2–4]. This single chapter will not attempt the treatment of a full-length text. The goal here is to provide a broad intuitive understanding of the issues, rather than attempt to describe a catalog of synchronizer-design methods. Thus, we will generally follow a traditional analog development knowing that these principles apply equally well to sampled-data systems, even if the implementation of the synchronizers differ. Phase-locked-loops are commercially available as relatively small gate-count chips, or as one part of a larger signal-processing device. It is assumed that the reader interested in modern design implementations will be able to make the reasonable straightforward transition from the principles presented here to the basic sampled data representations.

10.2 RECEIVER SYNCHRONIZATION

All digital communication systems require some degree of synchronization to incoming signals by the receivers. In this section the fundamentals of the various levels of receiver synchronization are discussed. The discussion begins with the basic levels of synchronization required for coherent reception—frequency and phase synchronization—and a brief discussion of the principles of phase-locked-loop (PLL) operation and design. The discussion then broadens into the topic of symbol synchronization. Some degree of symbol synchronization is required for all digital communications reception, either coherent or noncoherent. The final topics in the section are receiver frame synchronization and techniques for achieving and maintaining it.

10.2.1 Frequency and Phase Synchronization

At the heart of nearly all synchronization circuits is some version of a phase-locked-loop (PLL). In modern digital receivers this loop may be difficult to recognize, but the functional equivalent is essentially always present. A schematic diagram of the basic PLL is given in Figure 10.1. Phase-locked loops are servo-control loops, whose controlled parameter is the phase of a locally generated replica of the incoming carrier signal. Phase-locked loops have three basic components: a phase detector, a loop filter, and a voltage-controlled oscillator (VCO). The phase detector is a device that produces a measure of the difference in phase between an incoming signal and the local replica. As the incoming signal and the local replica change with respect to each other, the phase difference (or phase error) becomes a time-varying signal into the loop filter. The loop filter governs the PLL's response to these variations in the error signal. A well-designed loop should be able to track changes in the incoming signal's phase but not be overly responsive to receiver noise. The VCO is the device that produces the carrier replica. The VCO, as the name implies, is a sinusoidal oscillator whose frequency is controlled by a voltage level at the device input. In Figure 10.1, the phase detector is shown as a multiplier, the loop filter is described by its impulse response function $f(t)$, with Fourier transform $F(\omega)$, and the VCO is so indicated.

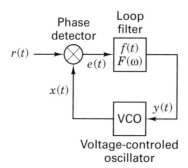

Voltage-controled
oscillator

Figure 10.1 Schematic of the basic phase-locked loop.

A VCO is an oscillator whose output frequency is a linear function of its input voltage over some range of input and output. A positive input voltage will cause the VCO output frequency to be greater than its uncontrolled value, ω_0, while a negative voltage will cause it to be less. Phase lock is achieved by feeding a filtered version of the phase difference (i.e., the phase error) between the incoming signal $r(t)$ and the output of the VCO, $x(t)$, back to the input of the VCO, $y(t)$.

In the case of modern digital receivers, the error detector may be mathematically much more complicated than the simple multiplier shown in Figure 10.1. For example, the error detector might be a set of matched-filter correlators, each matched to a slightly different phase offset feeding a weighting or decision function. The output of the weighting function would be the phase error estimate. Such a function might be mathematically very complex, but it would be easily approximated using modern digital technology. The VCO may not appear to be a sinusoidal oscillator, but it may be implemented as a read-only memory whose pointers are controlled by a combination of a clock and the output of the error estimator. The feedback path may not be continuous (as shown in Figure 10.1), but phase corrections may only be applied once per frame, or once per packet, depending on the signal structure. A special header or known sequence of symbols may be inserted into the information stream for the expressed purpose of aiding the estimation process. These obvious differences notwithstanding, the basic principles are still illuminated with the simple model of Figure 10.1.

Consider a normalized input signal of the form

$$r(t) = \cos\left[\omega_0 t + \theta(t)\right] \tag{10.1}$$

where ω_0 is the nominal carrier frequency and $\theta(t)$ is a slowly varying phase. Similarly, consider a normalized VCO output of the form

$$x(t) = -2\sin\left[\omega_0 t + \hat{\theta}(t)\right] \tag{10.2}$$

These signals will produce an output error signal at the phase detector output of the form

$$
\begin{aligned}
e(t) &= x(t)r(t) = 2\sin\left[\omega_0 t + \hat{\theta}(t)\right]\cos\left[\omega_0 t + \theta(t)\right] \\
&= \sin\left[\theta(t) - \hat{\theta}(t)\right] + \sin\left[2\omega_0 t + \theta(t) + \hat{\theta}(t)\right]
\end{aligned} \tag{10.3}
$$

Synchronization   Chap. 10

Assuming that the loop filter is low pass, the second term on the right hand side of Equation (10.3) will be filtered out and can be ignored. This low-pass assumption is a reasonable loop design decision. A low-pass filter provides an error signal that is solely a function of the difference in phases between the input [Equation (10.1)] and the VCO output [Equation (10.2)]. This is exactly the error signal that is needed. The VCO output frequency is the time derivative of the argument of the sine function in Equation (10.2). If we make the assumption that ω_0 is the uncontrolled frequency of the VCO (the output frequency when the input voltage is zero), we can express the difference in the VCO output frequency from ω_0 as the time differential of the phase term $\hat{\theta}(t)$. The output frequency of the VCO is a linear function of the input voltage. Therefore, since an input voltage of zero produces an output frequency of ω_0, the difference in the output frequency from ω_0 will be proportional to the value of the input voltage $y(t)$, or

$$\Delta\omega(t) = \frac{d}{dt}[\hat{\theta}(t)] = K_0 y(t)$$

$$= K_0 e(t) * f(t) \tag{10.4}$$

$$\approx K_0 [\theta(t) - \hat{\theta}(t)] * f(t)$$

where $\Delta\omega(t)$ denotes the frequency difference, the notation $*$ indicates the convolution operation (see Appendix A), and the small-angle approximation [i.e., $e(t) = \sin[\theta(t) - \hat{\theta}(t)] \cong \theta(t) - \hat{\theta}(t)$] has been used in the last line of Equation (10.4). The small-angle approximation will be accurate when the output phase error is small (the loop is close to phase lock). This will be the situation when the loop is operating normally. The factor K_0 is the gain of the VCO, and $f(t)$ is the loop-filter impulse response. This linear differential equation in $\hat{\theta}(t)$ (utilizing the small-angle approximation) is known as the linearized loop equation. It is the single most useful relationship in determining loop behavior during normal operation (where the phase error is small).

Example 10.1 Linearized Loop Equation

Show that for appropriately chosen K_0 and $f(t)$ the linearized loop equation [Equation (10.4)] demonstrates a tendency toward phase lock—that is, the phase difference between the incoming signal and the VCO output tends to decrease.

Solution

Consider the case where the phase of the input signal, $\theta(t)$, is slowly varying with time. It can be seen that if the phase difference on the right-hand side of Equation (10.4) is positive [i.e., $\theta(t) > \hat{\theta}(t)$], then by appropriate choice of K_0 and $f(t)$, the time derivative of $\hat{\theta}(t)$ will be positive, so that $\hat{\theta}(t)$ will increase with time, which will tend to reduce the magnitude of the difference $|\theta(t) - \hat{\theta}(t)|$. On the other hand, if the phase difference is negative, $\hat{\theta}(t)$ will decrease with time, which will also reduce the magnitude of the phase difference. Finally, if $\theta(t) = \hat{\theta}(t)$, then Equation (10.4) indicates that $\hat{\theta}(t)$ will not change with time, and the equality will be maintained.

Consider the Fourier transform of Equation (10.4),

$$j\omega\hat{\Theta}(\omega) = K_0[\Theta(\omega) - \hat{\Theta}(\omega)]F(\omega) \tag{10.5}$$

where the capitalized functions of ω are the Fourier transforms of the lowercase functions of t in Equation (10.4). That is, $\hat{\Theta}(\omega) \leftrightarrow \hat{\theta}(t)$, $\Theta(\omega) \leftrightarrow \theta(t)$, and $F(\omega) \leftrightarrow f(t)$. Reorganizing Equation (10.5) provides

$$\frac{\hat{\Theta}(\omega)}{\Theta(\omega)} = \frac{K_0 F(\omega)}{j\omega + K_0 F(\omega)} = H(\omega) \tag{10.6}$$

The term $H(\omega)$ is known as the closed-loop transfer function of the PLL. This term is very useful in characterizing the transient response of a PLL. The order of a PLL is defined to be the order of the highest-order term in $j\omega$ in the denominator of $H(\omega)$. Equation (10.6) indicates that this is always one more than the order of the loop filter $F(\omega)$. This is because when $F(\omega)$ is expressed analytically as $F(\omega) = N(\omega)/D(\omega)$, the denominator of $H(\omega)$ when expressed as a polynomial in $j\omega$ will have the term $j\omega D(\omega)$, which must have a term in $j\omega$ that is one order higher than the highest-order term in $D(\omega)$ alone. The order of a PLL is critical for determining the loop's steady-state response to a steady-state input. This is discussed in the next section.

10.2.1.1 Steady-State Tracking Characteristics

By reorganizing Equation (10.6), we can obtain an expression for the Fourier transform of the phase error:

$$E(\omega) = \mathcal{F}\{e(t)\}$$

$$= \Theta(\omega) - \hat{\Theta}(\omega)$$

$$= [1 - H(\omega)]\Theta(\omega)$$

$$= \frac{j\omega\Theta(\omega)}{j\omega + K_0 F(\omega)} \tag{10.7}$$

Equation (10.7) can be used in conjunction with the final value theorem of Fourier transforms to determine the steady-state error response of a loop to a variety of possible input characteristics. The steady-state error is the residual error after all transients have died away, and thus provides a measure of a loop's ability to cope with various types of changes in the input. The final value theorem states that

$$\lim_{t \to \infty} e(t) = \lim_{j\omega \to 0} j\omega E(\omega) \tag{10.8}$$

Combining Equations (10.7) and (10.8) yields

$$\lim_{t \to \infty} e(t) = \lim_{j\omega \to 0} \frac{(j\omega)^2 \Theta(\omega)}{j\omega + K_0 F(\omega)} \tag{10.9}$$

Example 10.2   Response to a Phase Step

Consider a loop's steady-state response to a phase step at the loop input.

Solution

Assuming that the PLL was originally in phase lock, a phase step will throw the loop out of lock. Having abruptly changed, however, the input phase again becomes stable.

This should be the easiest type of phase disturbance for a PLL to deal with. The Fourier transform of a phase step will be taken to be

$$\Theta(\omega) = \mathcal{F}\{\Delta\phi\, u(t)\}$$

$$= \frac{\Delta\phi}{j\omega} \tag{10.10}$$

where $\Delta\phi$ is the magnitude of the step and $u(t)$ is the unit step function

$$u(t) = \begin{cases} 1 & \text{for } t > 0 \\ 0 & \text{for } t < 0 \end{cases}$$

$$= \int_{-\infty}^{t} \delta(\tau)\, d\tau$$

in which $\delta(\tau)$ is the Dirac delta function. From Equation (10.9) and (10.10),

$$\lim_{t\to\infty} e(t) = \lim_{j\omega\to 0} \frac{j\omega\Delta\phi}{j\omega + K_0 F(\omega)} = 0$$

assuming that $F(0) \neq 0$. Thus the loop will eventually track out any phase step that appears at the input if the loop filter has a nonzero dc response. This means that for any loop filter with the property that $F(\omega) = N(\omega)/D(\omega)$ and $N(0) \neq 0$, the PLL will automatically tend to recover phase lock if the input is displaced by a constant phase. This is clearly a very desirable loop characteristic.

Example 10.3   Response to a Frequency Step

Next, consider a loop's steady-state response to a frequency step at the input.

Solution

A frequency step can approximate the effect of a Doppler shift in the incoming signal frequency due to relative motion between the transmitter and the receiver. Thus, this is an important example for systems with mobile terminals. Since phase is the integral of frequency, the input phase will change linearly as a function of time for a constant input-frequency offset. The Fourier transform of the phase characteristic will be the transform of the integral of the frequency characteristic. Since the frequency characteristic is a step, and the transform of an integral is the transform of the integrand divided by the parameter $j\omega$, it follows that

$$\Theta(\omega) = \frac{\Delta\omega}{(j\omega)^2} \tag{10.11}$$

where $\Delta\omega$ is the magnitude of the frequency step. Substituting Equation (10.11) into Equation (10.9) yields

$$\lim_{t\to\infty} e(t) = \lim_{j\omega\to 0} \frac{\Delta\omega}{j\omega + K_0 F(\omega)} = \frac{\Delta\omega}{K_0 F(0)} \tag{10.12}$$

The steady-state result in this case depends on more properties of the loop filter than merely a nonzero dc response. If the filter is "all-pass," then

$$F_{ap}(\omega) = 1 \tag{10.13}$$

If it is low-pass, then

$$F_{\ell p}(\omega) = \frac{\omega_1}{j\omega + \omega_1} \tag{10.14}$$

or if it is a lead-lag, then

$$F_{\ell\ell}(\omega) = \left(\frac{\omega_1}{\omega_2}\right)\frac{j\omega + \omega_2}{j\omega + \omega_1} \tag{10.15}$$

Equation (10.12) indicates that the loop will track the input phase ramp with a constant steady-state error whose value will depend on the gain term K_0 and the magnitude of the frequency step. Using any of $F_{ap}(\omega)$, $F_{\ell p}(\omega)$, or $F_{\ell\ell}(\omega)$ for $F(\omega)$ in Equation (10.12) yields

$$\lim_{t \to \infty} e(t) = \frac{\Delta\omega}{K_0}$$

Notice that a product of several filters with filter characteristics of the form of Equation (10.13), (10.14), or (10.15) would still produce this result. This steady-state error which is called the *velocity error,* will exist regardless of the order of the filter, unless the denominator of $F(\omega)$, contains $j\omega$ as a factor [$\omega_1 = 0$ in the denominator of Equation (10.14) or (10.15) with the appropriate renormalization in the numerators]. Having $j\omega$ as a factor of $D(\omega)$ is equivalent to having a perfect integrator in the loop filter. It is not possible to build a perfect integrator, but one may be closely approximated either digitally or by using active integrated circuits [5]. Thus if the system design requires the tracking of Doppler shifts with zero steady-state error the loop filter design must contain an approximation to a perfect integrator. It should be noted that even with a nonzero velocity error, the frequency is still being tracked: there are important applications where tracking to zero phase error is not important. Noncoherent signaling, such as the standard use of FSK modulation, is an example. For noncoherent signaling it is actually frequency tracking that is required, and the absolute value of phase is unimportant.

Example 10.4   Response to a Frequency Ramp

Consider a loop's steady state response when the input frequency is changing linearly with time (a frequency ramp function).

Solution

This example corresponds to the effect of a step change in the time derivative of the input frequency. This would approximate a change in the Doppler rate, which could model acceleration in the motion between a satellite or an aircraft and a ground receiver. In this case, the Fourier transform of the phase characteristic is given by

$$\Theta(\omega) = \frac{\Delta\dot{\omega}}{(j\omega)^3} \tag{10.16}$$

where $\dot{\omega}$ is the magnitude of the rate of frequency change. In this case, Equation (10.9) yields

$$\lim_{t \to \infty} e(t) = \lim_{j\omega \to 0} \frac{\Delta\dot{\omega}/j\omega}{j\omega + K_0 F(\omega)} = \lim_{j\omega \to 0} \frac{\Delta\dot{\omega}}{j\omega K_0 F(\omega)} \tag{10.17}$$

If the loop has a nonzero velocity error—that is, if the right-hand side of Equation (10.12) is not equal to zero—Equation (10.17) shows the steady-state phase error to be unbounded due to a frequency ramp. This says that a PLL with loop filters given by any of Equations (10.13) to (10.15) will not be able to track a frequency ramp. In order to track a frequency ramp, the denominator of the loop filter transform $D(\omega)$ must have $j\omega$ as a factor. From Equation (10.17) it can be seen that a loop filter with a transfer function of the type $F(\omega) = N(\omega)/j\omega D_1(\omega)$ will allow the PLL to track a frequency ramp with a constant phase error. This implies that in order to track a signal with a linearly changing Doppler shift (constant relative acceleration), the receiver must have a PLL that is second order or higher. To track a frequency ramp with zero phase error, the loop filter would be required to have a transfer function with $(j\omega)^2$ as a factor of the denominator, $F(\omega) = N(\omega)/(j\omega)^2 D_2(\omega)$. This implies a PLL that is third order or higher. Thus high-performance aircraft that need to track phase accurately through violent maneuvers may require third- or higher-order PLLs. In all cases, frequency lock is available with a loop of one order less than that required for phase lock. Steady-state error analysis is therefore a useful indicator of the required complexity of the loop filters.

In practice, the vast majority of PLL designs are second order. This is because a second-order loop can be made to be unconditionally stable [5]. Unconditionally, stable loops will always try to track the input. No set of input conditions—regardless of how extreme—will cause the loop to respond in the inappropriate direction to changes in the input. Second-order loops will track out the effect of a frequency step (Doppler shift), and they are relatively easy to analyze, since the closed-form results obtained for first-order loops are good approximations for second-order loop performance. Third-order loops are used for some special applications [e.g., some Global Positioning System (GPS) navigation receivers and some airborne receivers have third-order PLLs], but loop performance for third-order loops is relatively difficult to determine, and third- and higher-order loops are only conditionally stable. Often if the signal dynamics are expected to be such that high-order loops would be required for coherent demodulation, noncoherent demodulation is used instead.

10.2.1.2 Performance in Noise

The steady-state analysis of the preceding section tacitly assumed that the input signal was noise free. In some situations this may be approximately correct, but as in other parts of communication analysis, the more general case would include the effects of noise.

Reconsider the normalized loop input signal of Equation (10.1) and Figure 10.1. With the inclusion of normalized narrowband additive Gaussian noise $n(t)$, the expression for the input becomes

$$r(t) = \cos(\omega_0 t + \theta) + n(t) \tag{10.18}$$

where, for the moment, we consider the input phase offset, θ to be a constant. The noise process $n(t)$, assumed to be a zero-mean narrowband Gaussian process, can be expanded into quadrature components about the carrier frequency as [6]

$$n(t) = n_c(t) \cos \omega_0 t + n_s(t) \sin \omega_0 t \tag{10.19}$$

where both $n_c(t)$ and $n_s(t)$ are zero-mean Gaussian random processes and are statistically independent. Now the output of the phase detector can be written as [see Equation (10.3)]

$$e(t) = x(t)r(t) \tag{10.20}$$

$$= \sin(\theta - \hat{\theta}) + n_c(t)\cos\hat{\theta} + n_s(t)\sin\hat{\theta} + (\text{terms at twice the carrier frequency})$$

As before, the loop filter eliminates the twice-carrier-frequency terms. Denoting the second and third terms of Equation (10.20) as

$$n'(t) = n_c(t)\cos\hat{\theta} + n_s(t)\sin\hat{\theta} \tag{10.21}$$

we see that it is easy to verify that the variance of $n'(t)$ is identical to the variance of $n(t)$. This variance will be denoted by σ_n^2.

Consider the autocorrelation function of $n'(t)$,

$$R(t_1, t_2) = \mathbf{E}\{n'(t_1)n'(t_2)\}$$

$$= \mathbf{E}\{n_c(t_1)n_c(t_2)\}\cos^2\hat{\theta} + \mathbf{E}\{n_s(t_1)n_s(t_2)\}\sin^2\hat{\theta} \tag{10.22}$$

$$+ [\mathbf{E}\{n_c(t_1)n_s(t_2)\} + \mathbf{E}\{n_s(t_1)n_c(t_2)\}]\sin\hat{\theta}\cos\hat{\theta}$$

where $\mathbf{E}\{\cdot\}$ denotes the expected value. The cross-terms on the right-hand side of Equation (10.22) are equal to zero because n_c and n_s are mutually independent and have zero means [6]. With the assumption of wide-sense stationarity [7], we have

$$R(\tau) = R_c(\tau)\cos^2\hat{\theta} + R_s(\tau)\sin^2\hat{\theta} \tag{10.23}$$

where $\tau = t_1 - t_2$. Taking Fourier transforms, the power spectral density of $n'(t)$ is seen to be

$$G(\omega) = \mathscr{F}[R(t)]$$

$$= G_c(\omega)\cos^2\hat{\theta} + G_s(\omega)\sin^2\hat{\theta} \tag{10.24}$$

where G_c and G_s are the Fourier transforms of R_c and R_s, respectively. But from Equation (10.19), it can be seen that the spectra G_c and G_s are made of shifted versions of the spectra of the original noise process $n(t)$. Therefore, because of our construction [8],

$$G_s(\omega) = G_c(\omega) = G_n(\omega_0 - \omega) + G_n(\omega_0 + \omega)$$

where $G_n(\omega)$ is the spectral density of the original bandpass noise process $n(t)$. Equation (10.24) can be rewritten as

$$G(\omega) = G_n(\omega_0 - \omega) + G_n(\omega_0 + \omega) \tag{10.25}$$

For the special case of white noise, we have $G_n(\omega) = N_0/2$ Watts/Hertz, where N_0 is the single-sided spectral density of the white noise. Thus, from Equation (10.25), for this important special case,

$$G(\omega) = N_0 \tag{10.26}$$

The value in this development is that for the same small-angle approximations that were made in the preceding section, the spectral density of the VCO phase, $G_{\hat{\theta}}$, is related to the spectral density of the noise process through the loop transfer function [Equation (10.6)]. That is,

$$G_{\hat{\theta}}(\omega) = G(\omega)\,|H(\omega)|^2 \tag{10.27}$$

where $G(\omega)$ is as given in Equation (10.25) and $H(\omega)$ as defined in Equation (10.6). The variance of the output phase is then

$$\sigma_{\hat{\theta}}^2 = \frac{1}{2\pi} \int_{-\infty}^{\infty} G(\omega)|H(\omega)|^2\,d\omega \tag{10.28}$$

For the special case of white noise,

$$\sigma_{\hat{\theta}}^2 = \frac{N_0}{2\pi} \int_{-\infty}^{\infty} |H(\omega)|^2\,d\omega \tag{10.29}$$

The integral in Equation (10.29) (renormalized to natural frequency) is called the *two-sided loop bandwidth* W_L. The *single-sided loop bandwidth* is termed B_L. The definitions of these terms are

$$W_L = 2B_L = \frac{1}{2\pi} \int_{-\infty}^{\infty} |H(\omega)|^2\,d\omega \text{ Hertz} \tag{10.30}$$

Thus, if the noise process is white and the small-angle approximation holds (in other words, the loop is successfully tracking the input phase), the phase variance is given by

$$\sigma_{\hat{\theta}}^2 = 2N_0 B_L \tag{10.31}$$

The phase variance is a measure of the amount of jitter or wobble in the VCO output due to noise at the input. Equations (10.31) and (10.7) highlight one of the many trade-offs in communication theory. Clearly, one would wish $\sigma_{\hat{\theta}}^2$ to be small, which for a given noise level implies a small loop bandwidth, B_L, which from Equation (10.30) implies a narrow $H(\omega)$. However, it can be inferred from Equation (10.7) that the narrower the effective bandwidth of $H(\omega)$, the poorer will be the loop's ability to track incoming signal phase changes $\Theta(\omega)$. Thus a loop design must balance noise response with desired input phase response. The designer's dilemma is to design a loop that responds appropriately to the changes in the input signal, while not being overly responsive to the apparent changes, which are actually only artifacts of the noise process.

10.2.1.3 Nonlinear Loop Analysis

All of the PLL discussion in the previous sections has utilized what is called the linearized PLL model. This model is shown schematically in Figure 10.2. The model makes use of the small-angle approximation

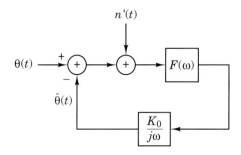

Figure 10.2 Schematic of linearized PLL model.

$$\sin{(\theta - \hat{\theta})} \approx \theta - \hat{\theta} \tag{10.32}$$

which is accurate when the loop is "in lock" and performing as desired (i.e., with small phase errors). Clearly, these conditions form only part of the picture. A complete analysis of PLL performance must allow for the times when Equation (10.32) is not accurate. When the small-angle approximation is inaccurate, an appropriate model is the one shown schematically in Figure 10.3. From Equations (10.4), (10.20), and (10.21) and Figure 10.3, the model can be described by the differential equation

$$\frac{d}{dt}\left[\hat{\theta}\,(t)\right] = K_0 f(t) * \sin{\left[\theta(t) - \hat{\theta}\,(t)\right]} + K_0 f(t) * n'(t) \tag{10.33}$$

where, as before, * denotes the convolution operation. In spite of the best efforts of many researchers, this differential equation has resisted general solution for many years. However, Viterbi [8] derived a closed-form solution for an important special case.

Consider the case where $\theta(t)$, the input phase as a function of time, is a constant θ. We can now define a new phase variable

$$\phi(t) = \left[\theta - \hat{\theta}(t)\right] \text{ modulo } 2\pi \tag{10.34}$$

Because θ is constant, Equation (10.33) can be rewritten as

$$\frac{d}{dt}\left[\phi(t)\right] = K_0 f(t) * \sin{\phi(t)} + K_0 f(t) * n'(t) \tag{10.35}$$

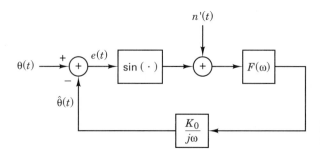

Figure 10.3 Schematic of nonlinearized PLL model.

Synchronization   Chap. 10

Since, from Equation (10.35), $\phi(t)$ is a function of the random process $n'(t)$, $\phi(t)$ it-self is a random process. Because $\phi(t)$ is defined modulo 2π, it can be shown [5] that $\phi(t)$ is stationary in the limit when all transient effects have died down (i.e., θ is a constant). Viterbi [8] determined that for a first-order PLL (i.e., the loop filter is a short circuit, or equivalently $f(t) = \delta(t)$), the probability density function of ϕ is of the form

$$p(\phi) = \frac{\exp(\rho \cos \phi)}{2\pi I_0(\rho)} \quad \text{for } |\phi| \leq \pi \tag{10.36}$$

where $\rho = 1/\sigma_\theta^2$ [see Equation (10.31)] is the normalized (to unit signal energy) loop signal-to-noise ratio, and $I_0(\rho)$ is the zeroth-order modified Bessel function of the first kind, evaluated at ρ. The phase variance, modulo 2π, can now be computed using Equation (10.36). The resulting value of the phase variance will be exact for first-order loops, and is an extremely useful approximation for the behavior of many second-order loops [5]. It has also been shown to be an exact form for higher-order loops under a modified definition of ρ [9].

The change of variable from a phase that can take any real value to a phase that is modulo 2π results in the concept of loop cycle slips. A cycle slip occurs when the magnitude of the original phase error, $|\theta - \hat{\theta}(t)|$, exceeds 2π radians. This will cause the value of ϕ [Equation (10.34)] to abruptly change from about 2π to about 0. This event can be thought of as a momentary loss of lock with an almost immediate reacquisition. The statistics of cycle slips can be as important an indicator of PLL performance as phase variance—especially at low-loop signal-to-noise ratios, when cycle slips may occur frequently.

By manipulating his phase-distribution results, Viterbi [8] derived an expression for the mean time to the first cycle slip, T_m, beginning at some arbitrary reference time:

$$T_m = \frac{\pi^2 \rho I_0^2(\rho)}{2B_L} \tag{10.37}$$

For large ρ, this expression can be approximated by

$$T_m \approx \frac{\pi \exp(2\rho)}{4B_L} \tag{10.38}$$

As was true with the probability density function of Equation (10.36), these results were derived for first-order loops, but they are useful approximations for the behavior of second-order loops, and they provide an upper bound to second-order loop performance at medium and large loop signal-to-noise ratios. In addition, computer simulations and laboratory measurements [5] indicate that the time T between cycle slips is exponentially distributed:

$$P(T) = 1 - \exp\left(-\frac{T}{T_m}\right) \tag{10.39}$$

This is to say that the probability that a loop will cycle-slip within time T, starting from zero phase error, is given by Equation (10.39).

10.2 Receiver Synchronization   **613**

10.2.1.4 Suppressed Carrier Loops

The discussion of PLLs to this point has presumed that the carrier input is a fairly stable sinusoid with some known positive average energy. In the case of a phase modulated communication system, if the carrier phase variation due to the modulation is less than $\pi/2$ radians, there will be positive energy at the carrier frequency. This is called a system design that has a residual carrier component, and all of the discussion of PLL development to this point would apply directly to this residual component. A diagram of the signal space for a binary phase modulated system with a residual carrier component is given in Figure 10.4, for a modulating angle of $\gamma \leq \pi/2$. At one time, most phase modulated systems were designed in this way. However, the residual carrier component is, in a sense, wasted energy—in the sense that the energy in the residual carrier is not being used to transmit the information, only to transmit the carrier. Thus most modern phase modulated systems are suppressed carrier systems. This means that there is no average energy transmitted at the carrier frequency. All of the transmitted energy goes into the modulation. Unfortunately, this means that there is no longer any signal for the basic PLL of Figure 10.1 to track.

Consider, as an example, a BPSK signal

$$r(t) = m(t) \sin(\omega_0 t + \theta) + n(t) \tag{10.40}$$

where $m(t) = \pm 1$ with equal probability. This is a suppressed carrier transmission—the average energy at radian frequency ω_0 is zero. This situation is represented graphically in Figure 10.4, when $\gamma = \pi/2$. The figure indicates that for this case the vertical carrier component will vanish. To acquire and track the phase of the carrier, the effects of the modulation must be eliminated. One way to eliminate the modulation is to square the signal:

$$\begin{aligned} r^2(t) &= m^2(t) \sin^2(\omega_0 t + \theta) + n^2(t) + 2n(t)m(t) \sin(\omega_0 t + \theta) \\ &= \tfrac{1}{2} - \tfrac{1}{2} \cos(2\omega_0 t + 2\theta) + n^2(t) + 2n(t)m(t) \sin(\omega_0 t + \theta) \end{aligned} \tag{10.41}$$

Here, we have made of the fact that $m^2(t) = 1$. The second term on the right-hand side of Equation (10.41) is a carrier-related term (at twice the original carrier

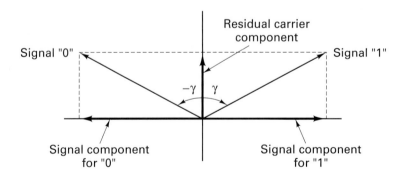

Figure 10.4 Residual carrier binary phase modulation.

frequency) that can be acquired and tracked with a basic PLL of the type illustrated in Figure 10.1. Such an arrangement is illustrated in Figure 10.5. When the incoming suppressed-carrier waveform is squared, the resulting twice-carrier component can be acquired and tracked by a PLL of standard design. Some of the problems with this procedure can be inferred from Equation (10.41). The first problem is simply that all phase angles have been doubled. Thus, the phase noise and phase jitter has been doubled, and the phase error variance (related to the phase noise squared) is larger by a factor of 4 than that of the original signal. This angle doubling is offset by the divide-by-2 circuit at the VCO output, and, therefore, does not directly affect the accuracy of the loop's output signal that is used by the data demodulator. However, this larger internal variation will cause the PLL to require a 6-dB-larger carrier signal-to-noise ratio than a residual carrier system in order to maintain phase lock. In addition, now there are two effective noise terms interfering with loop operation, because of the cross-correlation term between noise and signal in Equation (10.41). For cases of medium or low loop signal-to-noise ratio, these two noise terms will reduce the available signal-to-noise ratio even further relative to the original unmodulated carrier. This additional loss due to signal-times-noise and noise-times-noise terms is called the *loop squaring loss* S_L. Gardner [5] shows that if the input noise process $n(t)$ is a narrowband Gaussian noise of bandwidth B_i, the squaring loss is upper bounded by

$$S_L \leq 1 + N_0 B_i \qquad (10.42)$$

where, as before, N_0 is the single-sided power spectral density of the prefiltered, normalized white Gaussian noise process. Equation (10.42) is an upper bound because the filter bandwidth B_i is tacitly assumed to be wide enough to pass the signal undistorted. In an actual design, signal distortion can be traded for squaring loss, as is shown in [10].

Since the normalization in Equation (10.42) is with respect to the signal powers, the second term is proportional to a signal-to-noise ratio

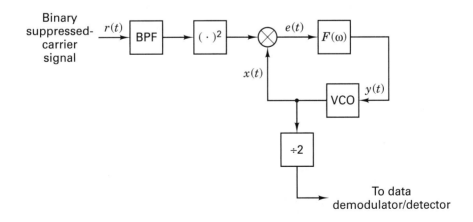

Figure 10.5   Basic squaring loop schematic.

$$\rho_i = \frac{1}{2 N_0 B_i} \tag{10.43}$$

where ρ_i is the signal-to-noise ratio in the input filter bandwidth. For large loop signal-to-noise ratios, the output phase variance can now be expressed as

$$\sigma_{\hat\theta}^2 = 2 N_0 B_L S_L = 2 N_0 B_L \left(1 + \frac{1}{2\rho_i} \right) \tag{10.44}$$

The leading term on the right-hand side of Equation (10.44) can be seen to be identical to that of Equation (10.31), the phase variance of the standard PLL. It can also be seen that for large input signal-to-noise ratios, the second term in the squaring loss will vanish, and we are left with the phase variance of the standard PLL.

Another potentially serious problem, associated mainly with suppressed carrier loops, is that of *false lock* [5, 11–13]. This can be a problem especially during acquisition or reacquisition of carrier phase. The interaction of the data stream with the loop nonlinearities (especially the squaring circuit) and loop filters will produce sidebands in the spectrum that is input to the phase detector. These sidebands can contain stable frequency components. Care must be taken that these stable components are not allowed to capture the tracking loop. If the loop is captured, it will appear to be operating correctly; the VCO control signal $y(t)$ will be small but the VCO output will be offset in frequency from the correct carrier component. This is false lock. The loop is tracking a sideband frequency component, and the loop filter is filtering out the real carrier. False lock is a hardware-implementation problem that typically sets an effective lower limit on the bandwidth of the loop filters. Because they have fewer nonlinear elements, false locking is not usually a problem with residual carrier loops.

10.2.1.5 Costas Loops

An important form of a suppressed carrier loop is the Costas loop, shown schematically in Figure 10.6. This loop design is important because it eliminates the square-law device, which can be difficult to implement at carrier frequencies, and replaces it with a multiplier and relatively simple low-pass filters. Although the appearance of the circuits in Figures 10.5 and 10.6 is quite different, their theoretical performance can be shown to be the same [5]. The main remaining implementation problem with Costas loops is that to achieve the theoretically optimum performance, the two low-pass arm filters must be perfectly matched. This can only be approximated in any analog hardware implementation. If the arm filters are implemented digitally, there will be no problem keeping them matched, but the designer will confront the usual sampled data design issues. Thus the decision as to whether to implement a Costas loop or the classical design of Figure 10.5 amounts to a design decision between the difficulty of implementing the squaring device and the difficulty of implementing closely matched arm filters. This design decision will depend on the parameters and requirements of the particular receiving system, and cannot be generalized here.

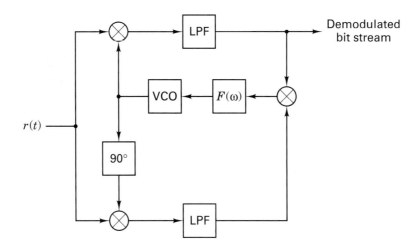

Figure 10.6 Costas loop.

10.2.1.6 High-Order Suppressed-Carrier Loops

Binary phase-shift keying is not the only type of suppressed-carrier modulation. In fact, assuming that all signals are equally likely *a priori,* any modulation scheme whose average amplitude, averaged over the signal set, is zero will have no average energy in the transmitted carrier. Perhaps the most common nonbinary suppressed-carrier modulation is quadrature phase-shift keying, or QPSK (4-ary PSK). If a QPSK signal is squared, the result "looks like" a BPSK signal. Thus, for equally likely QPSK signals, the carrier is still suppressed. However, squaring the signal a second time—equivalent to taking the original signal to the fourth power— can be seen to produce a term with a carrier component at four times the transmitted carrier's frequency. As in the binary case, operating on the incoming signal with a power-law device produces cross products among the noise terms and signal terms, and introduces the equivalent of a "squaring loss." Under the assumption that the noise bandwidth will pass the signal undistorted, the loss for fourth-power loops is upper bounded by [5]

$$S_L \leq 1 + \frac{9}{\rho_i} + \frac{6}{\rho_i^2} + \frac{3}{2\rho_i^3} \tag{10.45}$$

As was the case with the squaring loop, for sufficiently high input signal-to-noise ratios, ρ_i, Equation (10.45) indicates that the additional loss terms vanish, and the loop performance approaches that of the basic loop. As was also the case for the squaring loop, there are Costas loop designs equivalent to fourth-order loops [5, 14, 15] that may exhibit hardware implementation advantages. Their theoretical performance, however, is the same as that of the straightforward fourth-power design.

Example 10.5   Squaring Loss Bounds

Compare the upper bounds on squaring loss S_L given by Equations (10.42) and (10.45) for second- and fourth-power loops, respectively, for an input loop signal-to-noise ratio ρ_i of 10 dB.

Solution

A 10dB signal-to-noise ratio is also 10 in terms of its power ratio. Therefore, from Equations (10.42) to (10.44), for the squaring loop,

$$S_L = 1 + \frac{1}{2\rho_i} = 1.05 = 0.2 \text{ dB}$$

From Equation (10.45), for the fourth-power loop,

$$S_L = 1 + 0.9 + 0.06 + 0.0015 = 1.9615 = 2.9 \text{ dB}$$

Thus, while an input signal-to-noise ratio of 10 dB is adequate to keep losses small for the squaring loop, the same signal-to-noise ratio may allow significant losses for the fourth-power loop.

10.2.1.7 Acquisition

In most of the discussion thus far, the assumption has been that the PLL is in lock. This was the justification for assuming that the phase error $|\theta - \hat{\theta}|$ was small. At one time or another, however, every loop must acquire lock—that is, it must be brought into lock. Acquisition can be accomplished with the aid of external circuits or signals (aided acquisition) or in some cases by an unaided PLL (self-acquisition) [5].

Acquisition is an inherently nonlinear operation and therefore is difficult to analyze in general. However, some intuition may be obtained by considering a noise-free, first-order loop. Such a loop is shown schematically in Figure 10.3, where $n'(t) = 0$ (noise-free) and $F(\omega) = 1$ (first-order). Denote the input phase as

$$\theta(t) = \omega_i t$$

and the output phase as

$$\hat{\theta}(t) = \omega_0 t + \int_0^t K_0 \sin e(t) \, dt + \hat{\theta}(0) \tag{10.46}$$

where ω_i and ω_0 are the radian frequencies of the input and output signals, respectively. Thus the phase error is given by

$$e(t) = \theta(t) - \hat{\theta}(t)$$

$$\tag{10.47}$$

$$= (\omega_i - \omega_0)t - \int_0^t K_0 \sin e(t) \, dt - \hat{\theta}(0)$$

Differentiating both sides and letting $\Delta\omega = \omega_i - \omega_0$ provides

$$\frac{de}{dt} = \Delta\omega - K_0 \sin e \qquad (10.48)$$

where the time dependence of the function $e(t)$ has been suppressed to ease notation. This differential equation describes the behavior of the first-order noise-free PLL. The loop being in lock requires that

$$\frac{de}{dt} = 0 \qquad (10.49)$$

Equation (10.49) is a necessary, but not a sufficient, condition for phase lock. This can be verified by observing the phase plane diagram of Figure 10.7. This figure is obtained by dividing both sides of Equation (10.48) by the gain term K_0, and plotting the results. First observe point a. If the phase error is displaced a little to the left or right of point a, the sign of the derivative term is such that the phase error e, will be driven back toward a. Thus, point a is a stable point of the system, a point where phase lock can be obtained and will be maintained. Now consider the case of point b. If the phase error is exactly at b, Equation (10.49) will be satisfied. However, if there is any slight offset from b, the sign of the derivative term will be such that the error will be driven away from b. Thus b is a point of marginal stability for the loop, a point where Equation (10.49) is satisfied, but not a stable lock point.

The amount of time required for a loop to come into lock can be a very important system design consideration. By observing Equation (10.48), we can see that the requirement of Equation (10.49) for phase lock cannot be met unless

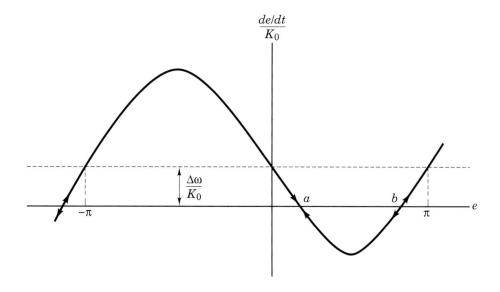

Figure 10.7 Phase-plane plot of first-order loop.

$$\frac{|\Delta\omega|}{|K_0|} \leq 1 \qquad (10.50)$$

This is because sinusoidal functions have a maximum amplitude of unity. This range of frequency difference, $-K_0 < \Delta\omega < K_0$, is sometimes called the lock-in range of the loop. Assuming that Equation (10.50) holds, Gardner [5] gives a rule of thumb of $3/K_0$ seconds for the time required for loop acquisition. Actual values can be obtained from Equation (10.47) for well-defined sets of initial conditions, or by extensive computer simulation. It can be seen from the phase plot of Figure 10.7 that the time required will vary widely as a function of the initial phase error. For phase errors very close to point b, the driving force $(de/dt)/K_0$ will be very small. Thus, for this worst-case phase error, the error could "linger" in the vicinity of b for a long time. This phenomenon is called terminal loop hang-up [16] and can be a serious problem for system designs that depend on self-acquisition.

Perhaps the most important operational difference between first-order and higher-order loops is the higher-order loop's ability to "pull in" from frequency differences that are larger than the lock-in range. A first-order loop with a frequency error larger than the lock-in range will drift toward lock but never quite lock in. Why? Second- and higher-order loops can pull in and achieve phase lock because of their more complicated phase-plane characteristics. (Interested readers should consult Viterbi [8] and other texts on PLLs for more details [5, 9, 17–19].)

The study of self-acquisition for phase-locked loops is mostly of academic interest. Gardner [5] states that loops using self-acquisition can be guaranteed to acquire in reasonable time only under very benign circumstances. This, unfortunately, is rarely the case in practice.

Acquisition aiding drives the loop through the region of phase space expected to contain the lock-in region by means of some external driving signal. This is the most common means of achieving acquisition. Aiding can be implemented by simply applying a voltage ramp to the input of the VCO. This driving signal will cause the VCO output frequency to vary linearly with time. As was shown earlier [Equation (10.17)], loops with loop filters that do not contain $j\omega$ as a factor of their transfer function's denominator cannot track a frequency ramp with finite phase error. Therefore, if frequency sweeping is to be employed with a first-order loop or a second-order loop without this transfer function characteristic, the rate of frequency sweep must be slow enough so that when the loop achieves lock, the presence of phase lock can be detected and the sweeping signal removed before it drives the loop back out of lock. With loops that contain $j\omega$ as a factor of $D(\omega)$, it may not be necessary to remove the sweeping signal at all, because, at least in theory, the loop will be able to track out the frequency ramp. In any case, the sweep rate must not be too large, or the loop will be driven through the lock point so fast that it will fail to acquire. For a second-order loop with loop transfer function [see Equation (10.6)],

$$H(\omega) = \frac{1}{-(j\omega/\omega_n)^2 + 2\zeta(j\omega/\omega_n) + 1} \qquad (10.51)$$

Gardner [5] indicates that the maximum sweep rate, $\Delta\dot{\omega}$, must be in the vicinity of

$$\Delta\dot{\omega} \approx \frac{1}{2}\,\omega_n^2\,(1 - 2\sigma_{\hat{\theta}}) \qquad (10.52)$$

where $\sigma_{\hat{\theta}}$ is as defined in Equation (10.31) and ω_n, implicitly defined in Equation (10.51), is called the *natural frequency* of a second-order PLL and is related to the loop bandwidth B_L and loop damping factor ζ [9] by

$$\omega_n = \frac{8\zeta}{4\zeta^2 + 1}\,B_L$$

Blanchard [17] gives more detailed results for aided phase acquisition.

10.2.1.8 Phase Tracking Errors and Link Performance

If a loop is unable to track out all phase errors, the received symbol-error probability will be degraded relative to what is theoretically achievable. The analysis required to determine the amount of the degradation is very involved, but for most of the standard coherent signaling systems, curves are available [14, 15, 20]. Figure 10.8 is an example of such a performance curve for a residual carrier-phase tracking loop operating on a signal with BPSK modulation in additive Gaussian noise. It can be seen that for signal-to-noise ratios of moderate value, small phase errors produce very little degradation. It is only when the standard deviation of phase error exceeds 0.3 that the degradations become significant. This means that the inherent degradation in performance caused by a well-designed loop operating in benign conditions can generally be ignored. The curve also indicates that if conditions are such that the phase variance is large, increasing the data signal-to-Gaussian noise ratio may not be effective in reducing the detected error probability. It should be noted that the presence of an irreducible error in these situations is a characteristic of residual carrier designs with constant loop signal-to-noise ratios ρ_i. Suppressed carrier tracking loops tend not to have irreducible errors, because an increase in the data signal-to-noise ratio will increase the signal-to-noise ratio of the suppressed carrier tracking loop, reducing the tracking error.

Example 10.6 PLL Signal-to-Noise Ratio

Develop an integral expression for the effect on link bit error probability of slowly varying phase tracking errors for a residual carrier BPSK link. Compare the effect of a normalized loop signal-to-noise ratio ($\rho = 1/\sigma_{\hat{\theta}}^2$) of 20 dB with one of 10 dB on error performance at a desired bit error probability of 10^{-5} using Figure 10.8.

Solution

From Chapter 4, the theoretically possible bit error probability for a BPSK link in additive white Gaussian noise of single-sided spectral density N_0 Watts/Hertz is given by

$$P_B = Q\left(\sqrt{\frac{2E_b}{N_0}}\right)$$

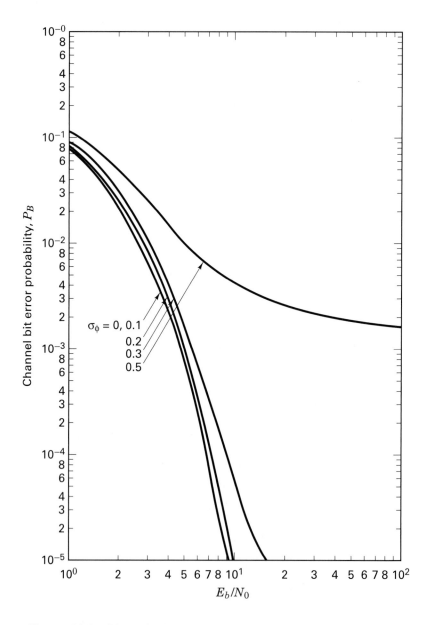

Figure 10.8 Channel bit error probability versus E_b/N_0 for BPSK with imperfect carrier synchronization. (Reprinted with permission from J. J. Stiffler, *Theory of Synchronous Communications,* Prentice-Hall, Inc., Englewood Cliffs, N.J., Fig. 9.1, p. 270.)

where E_b is the received energy per bit time. From the derivation of this expression for error probability, it can be shown that if there is a slowly varying (with respect to the data rate) phase-tracking error of β radians, the resulting probability of error will be given by

$$P_B(\beta) = Q\left(\sqrt{\frac{2E_b \cos \beta}{N_0}}\right)$$

Now if the phase error β is the result of tracking errors caused by system noise, β will be described stochastically by some probability density function $p(\beta)$. Then the expected bit-error probability is given by

$$P_B = \int_0^{2\pi} P_B(\beta)\, p(\beta)\, d\beta$$

For the special case of a first-order loop, the probability density function is given by Equation (10.36). Then the final expression for the expected bit-error probability is given by

$$P_B = \int_0^{2\pi} Q\left(\sqrt{\frac{2E_b \cos \beta}{N_0}}\right) \frac{\exp(\rho \cos \beta)}{2\pi I_0(\rho)}\, d\beta$$

A loop signal-to-noise ratio (ρ_i) of 20 dB will correspond to a standard deviation of phase noise of $\sigma_{\hat{\theta}} = 0.1$ rad. From Figure 10.8, this small amount of phase noise produces no appreciable degradation in the bit-error probability. A loop ρ_i of 10 dB, however, corresponds to a phase noise standard deviation of $\sigma_{\hat{\theta}} = 0.32$ rad. It can be seen from Figure 10.8 that for a bit-error probability of 10^{-5}, this phase noise standard deviation will require a data SNR of somewhat more than 11 (10.4 dB), rather than a data SNR of 9.1 (9.6 dB) for perfect phase tracking. Thus, this loop signal-to-noise will cause an error-performance degradation of somewhat more than 0.8 dB, at an error probability of 10^{-5}. It should be noted that for loop SNRs less than about 10 dB, the degradation in performance increases very rapidly. Thus, 10 dB is something of a threshold for reasonable system performance for residual carrier designs. Suppressed carrier designs, having no problem with irreducible error, may do better.

10.2.1.9 Spectrum Analysis Techniques

The techniques we have considered thus far belong to a class of synchronizers sometimes called *spectral line techniques*. These techniques all either use an existing spectral line at the carrier frequency or produce such a line at the carrier frequency or a multiple thereof, as a crucial part of the error determination. There is another set of techniques that are especially useful in carrier-frequency estimation or tracking that utilize the shape of the signal's passband spectrum. These techniques have solid roots in maximum-likelihood estimation theory [4], but they also are intuitively appealing and will be approached here from the intuitive direction.

Possibly the most intuitive technique in this class is a simple bank of matched filters, with each filter matched to the expected signal with a different carrier frequency offset. Such a bank of filters could be implemented directly, or by performing a weighting and combining operation on the output of a Fast Fourier

Transform. In any case, the filter with the maximum output would be associated with the signal's frequency offset. Such a frequency detector is shown notionally in Figure 10.9. Depending on the signal design and its sensitivity to frequency errors, and on the density of the frequency offsets, either the largest output could be taken as the frequency estimate directly, or additional processing to refine the estimate could be performed. In either case, it is clear that a filter bank that spans the range of possible frequency offsets could be designed in concept, and that such a design would provide a quick and reliable estimate of the carrier frequency offset.

An advantage of the filter bank approach discussed previously is the width of the frequency uncertainty that can be easily accommodated. A disadvantage is the granularity of the initial estimate. A second spectral technique, sometimes called *band-edge filtering*, can provide a mush more accurate estimate, at the cost of a reduction in the initial frequency uncertainty that can be accommodated. The idea can be easily seen through a graphical example.

In the upper graph shown in Figure 10.10, the signal-bandpass spectrum is shown as the wide shaded region, centered on a nominal carrier frequency ω_0. Also shown in this graph are two narrower passband filters at the edges, or roll-off regions of the signal spectrum. If, as is shown in the second graph, the detected signal in both of the two band-edge filters is equal, the signal spectrum must be centered between them, and the nominal carrier frequency error is zero. However, if, as is shown in the third and fourth graphs, the input signal spectrum is shifted relative to the band-edge filters, one of the filters will have more detectable signal, and an

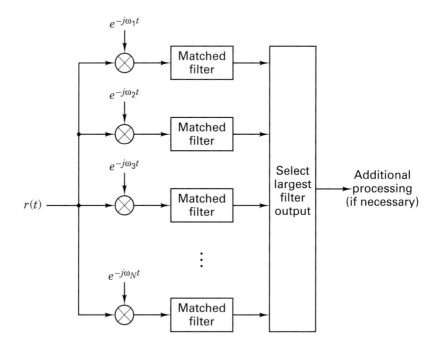

Figure 10.9 Matched filter-bank frequency estimator.

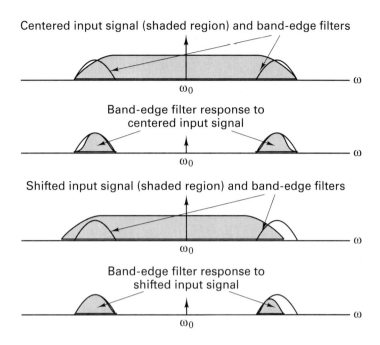

Centered input signal (shaded region) and band-edge filters

ω_0

ω

Band-edge filter response to centered input signal

ω_0

ω

Shifted input signal (shaded region) and band-edge filters

ω_0

ω

Band-edge filter response to shifted input signal

ω_0

ω

Figure 10.10   Band edge filter example.

error measure can be formed. This error measure could be used to drive a control loop, or it could be used to compute a frequency correction directly. The primary advantage of this kind of technique is that no noise-enhancing nonlinearity is required. The disadvantages are that it requires more knowledge of the signal spectrum, and the implementation of two narrow filters with well-matched passband characteristics. Narrow well-matched filters such as these could represent a design challenge if done with analog circuitry, but, in concept, it could be easily accomplished with digital techniques.

10.2.2 Symbol Synchronization—Discrete Symbol Modulations

All digital receivers need to be synchronized to the incoming digital symbol transitions in order to achieve optimum demodulation. In the discussion that follows, we will consider several of the basic types of designs of symbol or data synchronizers. The discussion will center on a random binary baseband signal, for ease of terminology and notation, but the extension to nonbinary baseband signals should be apparent.

The presentation in this section assumes that nothing is known about the actual data sequence. this class of synchronizers is called non-data-aided (NDA) synchronizers. There is another class of symbol synchronizers that use known information about the data stream. This knowledge may be obtained by feeding back decisions on received data, or because a known sequence has been injected into the data stream. Data-aided (DA) techniques have become more important and prevalent with the increasing use of bandwidth-efficient modulation. This is especially true with the class of continuous phase modulations. Data-aided techniques will be considered in somewhat more detail in the succeeding section.

The symbol synchronizers that will be considered here can be classified into two basic groups. The first group consists of the open-loop synchronizers. These circuits recover a replica of the transmitter data clock output directly from operations on the incoming data stream. The second group comprises the closed-loop synchronizers. Closed-loop data synchronizers attempt to lock a local data clock to the incoming signal by use of comparative measurements on the local and incoming signals. Closed-loop methods tend to be more accurate, but they are much more costly and complex.

10.2.2.1 Open-Loop Symbol Synchronizers

Open-loop symbol synchronizers are also occasionally called nonlinear filter synchronizers [20], a very descriptive title. This class of synchronizers generates a frequency component at the symbol rate by operating on the incoming baseband sequence with a combination of filtering and a nonlinear device. The operation is analogous to carrier recovery in a suppressed carrier-tracking loop. In the present case, the desired frequency component, at the data symbol rate, is isolated with a bandpass filter, and "shaped" with a high-gain saturating amplifier. The shaping recovers the square-wave appearance of the data clock signal.

Three examples of open-loop bit synchronizers are shown in Figure 10.11. In the first example (Figure 10.11a), the incoming signal $s(t)$ is filtered with a matched filter. The output of this filter will be the autocorrelation function of the input signal shape. For square-wave signaling, for example, the output will be the familiar isosceles-triangular waveshape. The sequence of bit autocorrelation waveshapes is then "rectified" by some type of memoryless even-law nonlinearity—a square-law device, for example. The resulting waveform will have positive amplitude peaks that correspond, to within a time delay, with the input symbol transitions. This sequence of processes is illustrated in Figure 10.12. Thus the output waveform from the even-law device will contain a Fourier component at the fundamental frequency of the data clock. This frequency component is isolated from its harmonics with a bandpass filter (BPF) and shaped with an ideal saturating amplifier, with transfer function

$$\text{sgn } x = \begin{cases} 1 & \text{for } x > 0 \\ -1 & \text{otherwise} \end{cases} \qquad (10.53)$$

The second example in Figure 10.11 produces a Fourier component at the data clock frequency by means of a delay and multiply. The delay shown in Figure 10.11b is half a bit period, which is the best value because it provides the strongest Fourier component [20]. The waveform $m(t)$ will always be positive in the second half of every bit period, but will have a negative first half if there has been a state change in the incoming bit stream, $s(t)$. This produces a square-wave signal with

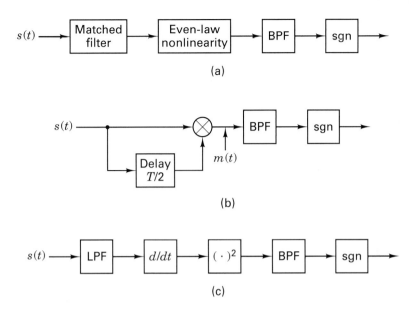

(a)

(b)

(c)

Figure 10.11 Three types of open-loop bit synchronizers.

spectral components at the data rate and all harmonics, as in Figure 10.11a. As before, the appropriate spectral component can be isolated with a BPF and shaped.

The final example (Figure 10.11c) amounts to an edge detector. The main operations are those of differentiation and rectification (by use of a square-law device). For a square-wave input, the differentiator will produce positive or negative spikes at all symbol transitions. When rectified, the resulting sequence of positive spikes will have a Fourier component at the data symbol rate. A potential problem with this particular scheme is that differentiators are typically very sensitive to wideband noise. This necessitates the low-pass filter (LPF) that precedes the differentiator in Figure 10.11c. The LPF, however, will also remove the high frequency components of the data symbols, causing them to lose their original rectangular wave shape. This will cause the resulting differential signal to have some finite rise and fall time, rather than being a set of impulses.

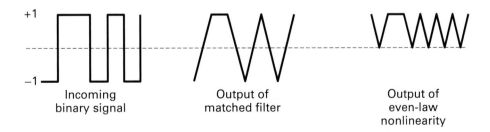

Figure 10.12 Open-loop bit synchronizer illustration.

Clearly, there will be some hardware delay associated with the signal processing steps illustrated in Figure 10.11. Wintz and Luecke [21] have shown that for a BPF that effectively averages K input symbols (bandwidth = $1/KT$), the magnitude of the fractional mean time error (delay) is approximated by

$$\frac{|\overline{\varepsilon}|}{T} \approx \frac{0.33}{\sqrt{KE_b/N_0}} \quad \text{for } \frac{E_b}{N_0} > 5, \quad K \geq 18 \tag{10.54}$$

where T is the bit period, E_b the detected energy per bit, and N_0 the single-sided received noise spectral density. Wintz and Luecke have also shown that at high signal-to-noise ratios the fractional standard deviation of the fractional timing error is given by

$$\frac{\sigma_\varepsilon}{T} \approx \frac{0.411}{\sqrt{KE_b/N_0}} \quad \text{for } \frac{E_b}{N_0} > 1 \tag{10.55}$$

Thus, for a given BPF, when the received signal-to-noise ratio is sufficiently large all of the techniques shown in Figure 10.11 will provide accurate bit timing.

10.2.2.2 Closed-Loop Symbol synchronizers

The primary disadvantage of open-loop symbol synchronization methods is that there is an unavoidable non-zero-mean tracking error. This error can be made small for large signal-to-noise ratios, but since the synchronization signal waveform depends directly on the incoming signal, the error will never vanish.

Closed-loop, symbol-data synchronizers use comparative measurements on the incoming signal and a locally generated data-clock signal to bring the locally generated signal into synchronism with the incoming data transitions. The procedure is essentially the same as that used for closed-loop carrier tracking.

Among the most popular of the closed-loop symbol synchronizers is the early/late-gate synchronizer. An example of such a synchronizer is shown schematically in Figure 10.13. The synchronizer operates by performing two separate integrations of the incoming signal energy over two different $(T - d)$ second portions of a symbol interval. The first integration (the early gate) begins integration at the loop's best estimate of the beginning of a symbol period (the nominal time zero) and integrates for the next $(T - d)$ seconds. The second integral (the late gate) delays the start of its integration for d seconds, and then integrates to the end of the symbol period (the nominal time T). The difference in the absolute values of the outputs of these two integrations, y_1 and y_2, is a measure of the receiver's symbol-timing error, and it can be fed back to the loop's timing reference to correct loop timing.

The action of the early/late-gate synchronizer can be understood by referring to Figure 10.14. In the case of perfect synchronization, Figure 10.14a shows that both gates are entirely within a signal symbol interval. In this case, both integrators will accumulate the same amount of signal, and their difference (the error signal e, in Figure 10.13) is zero. Thus, when the device is synchronized, it is stable—there is no tendency to drive itself away from synchronization. The case shown in Figure 10.14b is for a receiver whose data clock is early relative to the incoming data. In

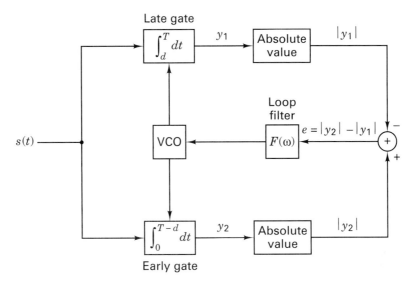

Figure 10.13 Early/late-gate data synchronizer.

this case the first portion of the early gate falls in the previous bit interval, while the late gate is still entirely inside the current symbol. The late-gate integrator will accumulate signal over its entire $(T - d)$ integration interval, as in the case in Figure 10.14a; but the early-gate integrator will end up with energy accumulated only over $[(T - d) - 2\Delta)]$, where Δ is the portion of the early-gate interval falling in the previous bit interval. Thus, for this case, the error signal will be $e = -2\Delta$, which will lower the input voltage to the VCO in Figure 10.13. This will reduce the VCO output frequency and retard the receiver's timing to bring it back toward the incoming signal's bit timing. Using Figure 10.14 as a guide, it can be seen that if the receiver's timing had been late, the amounts of energy integrated in the early gate and late gate would be reversed, as would the sign of the error signal. Thus, late receiver timing produces an increase in the VCO input voltage, increasing the output frequency and advancing the receiver's timing toward that of the incoming signal.

The example illustrated in Figure 10.14 tacitly assumes that there will be data state changes before and after the channel symbol of interest. If there are no transitions, it can be seen that the early gate and late gate will have the same integrated energy. Thus, there will be no error signal generated for cases where there is no data-state change. This is a practical implementation consideration in the use of all symbol synchronizers. Reconsider Figure 10.13. It is not possible to build two integrators that are exactly the same. Thus, the signals from the two arms of the early/late-gate loop will contain an offset with respect to each other, even when they should be identical. This offset will be small for well-designed integrators but will cause the loop to drift out of synchronism if there are long sequences of identical data symbols. There are two common responses to this problem. The first, and perhaps most obvious, is to format the data in a manner which ensures that there will be no transitionless intervals that are long enough to allow the loop to break lock. The second response is to modify the loop design so that it contains a single integrator. An example of this type of modified design is the tau-dither loop, considered in conjunction with the synchronization of spread-spectrum systems in Chapter 12.

Another loop design issue is the integration interval of the two gates. The example illustrated in Figure 10.14 shows the gates to occupy about three-fourths of a symbol period. Actually, this interval can vary from half a symbol interval to nearly a whole symbol interval. Why not less than half? The trade-off is between the amount of integrated noise and interference in a gate versus the amount of signal. As was true with the nonlinear model of phase-locked loops, loops of this type are difficult to analyze; the determination of performance is usually via computer simulation. This will be especially true for overlapping gates, as in Figure 10.14, because the noise samples in the two gates will be correlated. Gardner [5] has shown that for a normalized incoming signal of one volt, additive white Gaussian noise, random data (the probability of a transition is $\frac{1}{2}$), and early and late gates that are half a bit interval in duration, for large-loop signal-to-noise ratios, the fractional timing jitter is approximated by

$$\frac{\sigma_e^2}{T^2} = 2 N_0 B_L \tag{10.56}$$

where N_0 is the (normalized) noise power spectral density, T is the symbol interval, and B_L is the loop bandwidth.

10.2.2.3 Symbol Synchronization Errors and Symbol Error Performance

The effect of symbol-synchronization error on bit-error probability for a BPSK signal in additive white Gaussian noise is shown in Figure 10.15. It can be seen from the figure that the degradation is less than about 1 dB in signal-to-noise

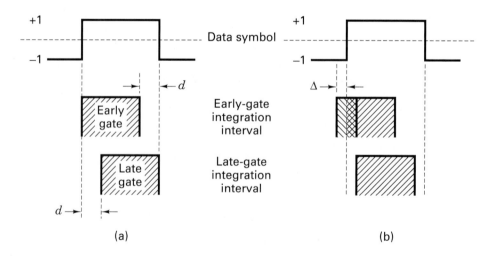

Figure 10.14   (a) Correct receiver timing. (b) Early receiver timing.

ratio for a fractional timing jitter of less than 5%. Comparing symbol timing error effects with the effect of phase noise (see Figure 10.8), it can be seen that the symbol synchronization error, taken as a fraction of the symbol interval, does not affect system performance as strongly as does phase noise taken as a fraction of a cycle. In both cases, however, the degradation increases with increases in error.

Example 10.7   Effect of Timing Jitter

Through the use of Figure 10.15, determine the effect of a 10% symbol-fractional timing jitter on a system required to maintain a 10^{-3} bit error probability.

Solution

It can be seen from Figure 10.15 that a 10^{-3}-bit-error probability will require a SNR of about 6.7 dB in the absence of all timing jitter. The same figure indicates that for a fractional timing jitter of 10% ($\sigma_e/T = 0.1$), a SNR of about 12.9 dB is required. Thus, the ability to accommodate this large timing jitter would require a 6.2-dB higher signal-to-noise ratio than that needed to maintain a 10^{-3}-bit-eror probability without jitter. This illustrates a use to which Figure 10.15 can be put; however, this example is clearly extreme. No communication system would be designed with over four times the nominally required power level in order to accommodate a large symbol-synchronization error. Some other answer would be found, such as redesigning the system filtering to increase the value of K in Equation (10.55), which will reduce the symbol timing jitter.

10.2.3 Synchronization with Continuous-Phase Modulations (CPM)

10.2.3.1 Background

Continuous-Phase Modulations (CPM) have grown from a research topic to an increasingly important signaling technique because of their bandwidth efficiency. As bandwidth becomes increasingly dear, their importance will continue to increase. These modulations raise new issues in synchronization, especially symbol synchronization. The bandwidth efficiency of CPM is obtained by increasing the smoothness of the waveforms in the time domain. If done properly, this smoothness will concentrate the signal's energy in a narrower bandwidth, reducing the amount of bandwidth required to pass the signal, and allowing adjacent signals to be packed closer together. However, this smoothness in the time domain also tends to eliminate the symbol transition features upon which many symbol synchronization schemes depend. A related problem is that, in general, it is difficult with CPM to separate the effects of carrier-phase error from symbol-timing error, making the phase and timing tasks interrelated. A mitigating feature of the smoothed waveforms is that in most cases of practical interest, the performance of the receivers is relatively insensitive to moderate timing errors [3].

A normalized CPM signal can be represented in complex notation as

$$s(t) = \exp\{j[\omega_0 t + \theta + \psi(t - \tau, \alpha)]\} \qquad (10.57)$$

where ω_0 is the carrier frequency, θ is the carrier phase (measured relative to the phase of the receiver), and $\psi(t, \alpha)$ is called the *excess phase* of $s(t)$. It is $\psi(t, \alpha)$ that carries the information in the signal. It is also $\psi(t, \alpha)$ that determines the amount

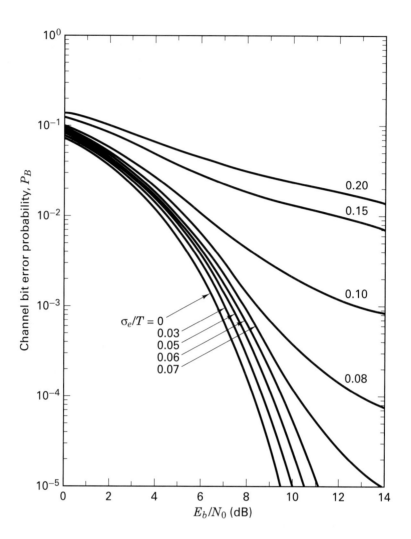

Figure 10.15 Channel bit error probability versus E_b/N_0 with the standard deviation of the symbol sync error σ_e as a parameter. (Reprinted from W. C. Lindsey and M. K. Simon, *Telecommunication Systems Engineering,* Prentice-Hall, Inc., Englewood Cliffs, N.J., 1973, courtesy of W. C. Lindsey and Marvin K. Simon.)

of bandwidth the signal will need. The required bandwidth is sometimes referred to as the *bandwidth occupancy* of the signal. When considering the goal of reducing or minimizing the required bandwidth from the standpoint of Fourier Theory, it can be seen that relatively high frequency components are associated with relatively abrupt changes in the time domain signal [22]. Thus, in order to reduce or eliminate the high frequency components, one must smooth out all the rough edges or abrupt changes in the time domain signal. In CPM signaling, this is accomplished through a combination of three techniques:

1. Use signal pulses that have several orders of continuous derivatives.
2. Allow individual signal pulses to occupy multiple signal time intervals (i.e., intentionally inject some intersymbol interference between symbols).
3. Reduce the maximum allowed phase change per symbol interval.

Not all CPM schemes use all of these techniques, but they all use some. For CPM schemes, it should be noted that at the beginning of each symbol interval, the excess phase, $\psi(t, \alpha)$, is a Markov Process [4], in that it depends only on the phase at the beginning of the symbol and the current symbol value. The phase value at the beginning of the symbol is a consequence of some number of previous symbols. Therefore, for the practical case where there are a finite number of possible phase states, the result is a finite state channel. Thus, the excess phase can be defined as

$$\psi(t, \alpha) = \eta(t, \mathbf{C}_k, \alpha_k) + \Phi_k \quad kT \le t \le (k+1)T \tag{10.58}$$

where

$$\eta(t, \mathbf{C}_k, \alpha_k) = 2\pi h \sum_{i=k-L+1}^{k} \alpha_i q(t - iT) \tag{10.59}$$

$\mathbf{C}_k$ is called the correlative state, k is a time index, and α_k is the kth information symbol drawn from the alphabet $\{\alpha_k\} = \{\pm 1, \pm 3, \ldots, \pm(M-1)\}$. This alphabet allows for the general case of M-ary (rather than just binary) signaling. The parameter h is the modulation index, and $q(t)$ is called the *modulation phase response,* defined outside of the region $0 < t < LT$, as

$$q(t) = \begin{cases} 0 & \text{for } t \le 0 \\ 1/2 & \text{for } t \ge LT \end{cases} \tag{10.60}$$

where L is called the correlation length. The correlation length is the number of information-symbol periods of length T seconds that are affected by a single information symbol. This is a measure of the amount of intentional intersymbol interference. When $L = 1$, we refer to the signaling as *full response.* This is the condition that was assumed in earlier chapters dealing with modulation. In full-response signaling, each pulse is confined to its own time boundaries. However, when $L > 1$, the signaling is called *partial response,* which means that each pulse is not restricted to its own symbol interval, but rather it is "smeared" into $L - 1$ neighboring symbol intervals. This is the vehicle used in many CPM schemes for purposely injecting controlled intersymbol interference between symbols and thereby increasing bandwidth efficiency. Classical minimum-shift-keying (MSK), one of the earliest examples of CPM (see Chapter 9), does not use multiple symbol intervals per pulse. Thus, classical MSK is an example of full-response signaling. Observing Equations (10.60), it can be seen that the consequence of $q(LT) = \frac{1}{2}$ is that the maximum possible phase change over an LT interval is $(M-1)\pi h$, as can be seen from Equation (10.58) and (10.59).

The vector $\mathbf{C}_k$, which is called the *correlative state,* is a sequence of information symbols $\{\alpha_k\}$, starting from the earliest time that can affect the signal's phase at the current time k, as follows:

$$\mathbf{C}_k = (\alpha_{k-L+1}, \cdots, \alpha_{k-2}, \alpha_{k-1})$$

The term Φ_k in Equation (10.58), expressed as

$$\Phi_k = \pi h \sum_{i=0}^{k-L} \alpha_i \quad \text{mod } 2\pi \tag{10.61}$$

is called the *phase state*. The phase state is one of a set of discrete phases that the signal can take as a consequence of the values of past symbols. Starting the phase transition for the next symbol from this phase state is a necessary condition for continuous phase. In the context of a trellis diagram, Φ_k can be viewed as an initial state or node, and $\mathbf{C}_k$ as defining a path to one of the other nodes. The definition of $q(t)$ in the interval $(0 < t < LT)$ is what gives the modulation its particular characteristics. MSK has the parameters $h = \frac{1}{2}$, $L = 1$, $M = 2$ and $q(t) = t/(2T)$ in the interval $(0 < t < T)$. The frequency response, defined as $g(t) \triangleq dq(t)/dt$, is clearly rectangular for MSK:

$$g(t) = \begin{cases} 1/(2T) & 0 \leq t \leq T \\ 0 & t < 0, \ t > T \end{cases} \tag{10.62}$$

Gaussian MSK (GMSK), another example of CPM, is defined as having a frequency response that is the convolution of this rectangle with a Gaussian shaped pulse.

Many of the synchronization techniques we have described in previous sections are based on *ad hoc* methods. People who had gained a fair amount of intuition in synchronization invented them by trying things that seemed to make sense. With a few exceptions, intuition has been less successful with CPM. Most techniques are based on classical estimation theory principles, the most popular being maximum likelihood estimation. The principles involved are the same as those developed for maximum likelihood signal detection.

Maximum likelihood estimation involves the maximization of conditional probabilities, and it is based on Baysian Theory [7]. Let $s(t, \gamma)$ represent a signal with some set of unknown parameters, γ. The parameters could be carrier phase, or symbol-timing offset, or the values of the transmitted information symbols, or possibly other parameters. Let

$$r(t) = s(t, \gamma) + n(t) \tag{10.63}$$

represent a received signal, where $n(t)$ is some additive receiver noise process. Let $R(t)$ represent a realization of the process $r(t)$. Then the maximum-likelihood estimate for the unknown parameter set γ is the value of γ that maximizes the likelihood $p[r(t) = R(t) | \gamma]$ over all γ. As has been developed in Chapter 3, a realization of a maximum-likelihood detector for a known signal is a filter matched to that signal. For the case of CPM, this leads to a receiver structure as shown in Figure 10.16.

In the primary signal detection process, the carrier frequency ω_0, carrier phase θ, and symbol timing offset τ are assumed known. The receiver structure is effectively a bank of matched filters, each filter matched to an L-symbol signal realization, which feeds a Viterbi Algorithm. The number of filters is M^L, and the number of nodes in the branch metric computation is PM^{L-1}, where P is the num-

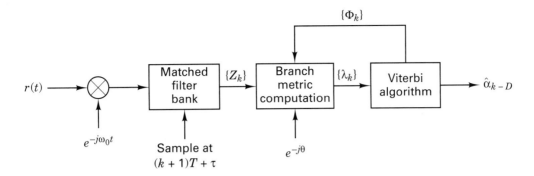

Figure 10.16 CPM receiver structure.
(Note: $\hat{\alpha}_{k-D}$ is the k-th output symbol with processing delay, D)

ber of phase states $\{\Phi_k\}$. Both of these numbers can be awkwardly large, and so simpler receiver structures are often used in practice [3, 4, 22]; however, the structure is still useful as a basis for synchronization.

From the preceding CPM description, the individual filters in the filter bank will have impulse responses given by

$$h^{(\ell)}(t) \triangleq \begin{cases} e^{-j\,\eta_\ell(T-t,\,\mathbf{C}_0^{(\ell)}\alpha_0^{(\ell)})} & 0 \le t \le T \\ 0 & \text{elsewhere} \end{cases} \tag{10.64}$$

with $(\ell = 1, 2, \ldots, M^L)$ denoting the generic L-symbol string $(\mathbf{C}_0^{(\ell)}, \alpha_0^{(\ell)}) = (\alpha_{-L+1}^{(\ell)}, \ldots, \alpha_{-1}^{(\ell)}, \alpha_0^{(\ell)})$, where each $\alpha_k^{(\ell)}$ is selected from the signal alphabet and ℓ denotes a particular path (symbol sequence) in the set of M^L possible paths. Similar to the earlier notation,

$$\eta_\ell(t, \mathbf{C}_0^{(\ell)}, \alpha_0^{(\ell)}) = 2\pi h \sum_{i=-L+1}^{0} \alpha_i^{(\ell)} q(t - iT) \tag{10.65}$$

From Figure 10.16, it can be seen that the individual filter outputs are given by

$$Z_k^{(\ell)}(\mathbf{C}_k, \alpha_k, \tau) \triangleq \int_{\tau+kT}^{\tau+(k+1)T} r(t)\, h^{(\ell)}(t - \tau - kT) e^{-j\omega_0 t}\, dt \tag{10.66}$$

This set of outputs $\{Z_k\}$, along with the carrier phase estimate $\hat{\theta}$ and the phase state $\{\Phi_k\}$, are used to compute the path metrics and, ultimately, to determine the Viterbi Algorithm's decisions.

10.2.3.2 Data-Aided Synchronization

Techniques for synchronizing CPM receivers can be divided into those that rely on knowledge of the information symbols, and those that do not. Those that rely on such knowledge are called data-aided techniques. Those that do not are termed non-data-aided (NDA). Clearly this is a distinction that could be applied to receiver synchronization for all modulation techniques, but data aided techniques appear especially useful and popular with CPM. There are two ways in which the data symbols could be known: either the symbols under consideration are a part of a known header or training sequence inserted into the data stream, or decisions from the Viterbi Algorithm are being fed back into the synchronization process. If decision feedback is to be employed, clearly the decisions must be fairly reliable, which implies that the receiver must be close to lock. Initial acquisition based on decision feedback is not likely to be a practical approach.

Given that the transmitted symbol string is known over some observation interval, L_0, the index (ℓ) in Equation (10.66) can be dropped. With the usual assumptions of Gaussian noise processes and equal energy signals, the likelihood function, $\Lambda(R|\hat{\theta}, \hat{\tau})$, associated with θ and τ, the unknown phase and time offsets, respectively, is given by [3]

$$\Lambda(R|\hat{\theta}, \hat{\tau}) = \exp\left\{ \sum_{k=0}^{L_0-1} \text{Re}\left[Z_k(\mathbf{C}_k, \alpha_k, \hat{\tau}) e^{-j(\hat{\theta} + \Phi_k)}\right]\right\} \qquad (10.67)$$

where unimportant constant factors have been dropped and Re{·} refers to the real part of the complex argument. It is clear that the right-hand side of Equation (10.67) will be maximized when the summation is maximized. Therefore, taking the partial derivatives of the summation with respect to $\hat{\theta}$ and $\hat{\tau}$ and then setting the results to zero yields

$$\sum_{k=0}^{L_0-1} \text{Im}\left[Z_k(\mathbf{C}_k, \alpha_k, \hat{\tau}) e^{-j(\hat{\theta} + \Phi_k)}\right] = 0 \qquad (10.68)$$

and

$$\sum_{k=0}^{L_0-1} \text{Re}\left[Y_k(\mathbf{C}_k, \alpha_k, \hat{\tau}) e^{-j(\hat{\theta} + \Phi_k)}\right] = 0 \qquad (10.69)$$

where $Y_k = \partial Z_k / \partial \hat{\tau}$, and Im{·} refers to the imaginary part of the complex argument. Mengali [3] points out that the left-hand side of Equation (10.69) can be obtained in two ways: either by taking the derivative in the straightforward manner, or by implementing a set of "derivative filters." Which is the preferred implementation would depend on the exact case at hand.

Unfortunately, Equations (10.68) and (10.69) do not provide much in the way of intuition or insight, and there are no known closed-form solutions. The equations must be solved numerically through some jointly iterative procedure for $\hat{\theta}$ and $\hat{\tau}$. Mengali suggests an iterative procedure, where the successive terms in each summation are used to form error terms for a successive approximation. That is,

$$\hat{\theta}_{k+1} = \hat{\theta}_k + \gamma_P e_P(k-1) \qquad (10.70)$$

$$\hat{\tau}_{k+1} = \hat{\tau}_k + \gamma_T e_T(k-1) \qquad (10.71)$$

where e_P and e_T are higher order terms from the left-hand sides of Equations (10.68) and (10.69), respectively, and γ_P and γ_T are "gains," selected to assure that the process converges. Clearly, this iterative procedure is more appropriate in a decision-feedback operation than with a fixed-length training sequence.

10.2.3.3 Non-data-Aided Synchronization

One of the first tenets of Information Theory is that having more information is better than having less. In the current context this means that knowing the symbol sequence will allow better estimates of carrier phase and symbol timing than not knowing. There may be cases, however, where training sequences are impractical or inconvenient, and the decision process is not sufficiently reliable for decision feedback. In these cases, non-data-aided (NDA) synchronization processes are called for. Two techniques of general applicability will be discussed, and one power-law technique that can be applied in a large number of cases.

The first technique is a direct extension of the development in the previous section. Clearly, if the symbol sequence ($\mathbf{C}_k$, α_k) is not known, a new likelihood function similar to Equation (10.67) can be written to accommodate that fact:

$$\Lambda(R \mid \hat{\mathbf{C}}_k, \hat{\alpha}_k, \hat{\theta}, \hat{\tau}) = \exp\left\{ \sum_{k=0}^{L_0-1} \text{Re}\left[Z_k(\hat{\mathbf{C}}_k, \hat{\alpha}_k, \hat{\tau}) e^{-j(\hat{\theta} + \Phi_k)} \right] \right\} \tag{10.72}$$

Since the likelihood function is proportional to a conditional probability, the chain rule of conditional probabilities can be applied to get back to a likelihood function dependent on $\hat{\theta}$ and $\hat{\tau}$ alone. The chain rule states that [7]

$$p(r(t) = R(t) \mid \gamma) = \int_{\text{all } \beta} p[r(t) = R(t) \mid \gamma, \beta] \, p(\beta) d\beta \tag{10.73}$$

which implies that the desired likelihood function is given by

$$\Lambda'(R \mid \hat{\theta}, \hat{\tau}) = \frac{1}{M^L} \sum_{\text{all } (\hat{C}_k, \hat{\alpha}_k)} \Lambda(R \mid \hat{\mathbf{C}}_k, \hat{\alpha}_k, \hat{\theta}, \hat{\tau}) \tag{10.74}$$

where the assumption has been made that all symbol sequences are equally likely. The likelihood function on the right-hand side of Equation (10.74) can now be differentiated to produce two equations analogous to Equations (10.68) and (10.69). Clearly, the result is considerably more computationally complicated than the results in Equations (10.68) and (10.69), and simpler if sub-optimal techniques may be desired. Mengali [3] discusses some approximations which lead to a somewhat simplified estimator for $\hat{\tau}$.

Another approach is indicated by the form of a sub-optimal receiver structure using Laurent filters [23, 24]. This approach approximates the CPM signal by a set of superimposed pulse amplitude modulated (PAM) waveforms. Considering only the first in the series, the expression is

$$e^{j\psi(t, \alpha)} \approx \sum_i a_{0,i} h_0 (t - iT) \tag{10.75}$$

where $\psi(t, \alpha)$ is defined in Equation (10.58) and the coefficients $a_{0,i}$ are called *pseudo-symbols*. The pseudo-symbols, whose values depend on the past and present data symbols, are defined by

$$a_{0,i} = \exp\left(j\,\pi h \sum_{\ell=0}^{i} \alpha_l \right) \qquad (10.76)$$

where the modulation index h can take on any noninteger value. For the important special case of MSK, where $h = \frac{1}{2}$, the expression in Equation (10.75) is exact for a filter function of the form

$$h_0(t) = \begin{cases} \sin\left(\dfrac{\pi t}{2T}\right) & 0 \le t \le 2T \\ 0 & \text{elsewhere} \end{cases} \qquad (10.77)$$

For other modulations, the approximation will be more or less accurate, and the form of $h_0(t)$ will vary [23]. In any case, ignoring the noise process momentarily, the normalized signal can now be rewritten in the form

$$s(t) \approx e^{\,j(\omega_0 t + \theta)} \sum_i a_{0,i} h_0(t - iT - \tau) \qquad (10.78)$$

From this expression, it is clear that the standard techniques for phase and symbol timing that were developed in the previous sections for linear modulations could be applied to this approximation. Mengali [3] points out that care must be taken in following this approach, however, because a filter actually matched to $h_0(t)$ might produce a very poorly shaped pulse. The issue is discussed by Kaleh [25].

Finally, in the special cases where the modulation index is rational, $h = k_1/k_2$, with (k_1, k_2) integers, power-law techniques can be applied [22]. For this case, Equation (10.57) can be rewritten as

$$s(t) = \exp\left\{ j\left[\omega_0 t + \theta + 2\pi \frac{k_1}{k_2} \sum_{i=k-L+1}^{k} \alpha_i q(t - iT) \right] \right\} \qquad (10.79)$$

where, for simplicity, Φ_k from Equation (10.58) has been absorbed into θ. Taking the k_2^{th} power of $s(t)$ gives

$$[s(t)]^{k_2} = \exp\left\{ j\left[k_2(\omega_0 t + \theta) + 2\pi k_1 \sum_{i=k-L+1}^{k} \alpha_i q(t - iT) \right] \right\} \qquad (10.80)$$

The $\omega_0 t + \theta$ term on the right-hand side is clearly a high-frequency term that can be filtered out. The right-most term is the k_1^{th} power of the information portion of the signal. From the development in Equations (10.57) to (10.60), this last term is repetitive with a period of at most LT. Depending on the exact nature of the phase response $q(t)$, Fourier series components at multiples of $2\pi k_1/(LT)$ radians may be produced. At least in theory, these components could be isolated and tracked. Even if spectral lines are not available, if the multiple of the signaling spectrum can be isolated, band-edge filtering techniques (described in Section 10.2.1.9) could be used to estimate a multiple of the symbol rate. The phase term θk_2 can also be isolated. There are several practical problems with this procedure. The symbol period will have a (k_1/L)-way ambiguity, and the phase estimate will have a k_2-way ambiguity that must be resolved in some fashion. Depending on the nature of $q(t)$, the

Fourier components may be fairly weak and potentially close together, making them hard to isolate. And finally, as is the case with all power-law techniques, the receiver noise is disproportionately increased, possibly decreasing the effective detector signal-to-noise ratio to unusable levels. The method does have the advantage of offering some degree of intuition into the process. It offers a direct connection to the spectral-line techniques considered earlier. These techniques used nonlinearities, usually power-law devices, to recover a pure spectral line at the frequency of interest, or a known multiple of the frequency. This is exactly what is going on here. The assumed rational nature of the modulation index h is utilized to produce potential spectral lines at multiples of the symbol rate and the carrier frequency. These lines can be used to acquire and track the symbol timing and carrier frequency and phase.

10.2.4 Frame Synchronization

Almost all digital data streams have some sort of frame structure. This is to say that the data stream is organized into uniformly sized groups of bits. If the data stream is digitized TV, each pixel is represented by a word having several bits, which is further organized into horizontal raster scans, which is further organized in terms of vertical raster scans. Computer data are typically organized into words of some number of 8-bit bytes, and these, in turn, are organized into card images, packets, frames, or files. Any system that uses block-error control coding must be organized around the codeword length. Digital speech is typically transmitted in packets or frames which are indistinguishable from other digital data.

For a receiver to make sense of the incoming data stream, the receiver needs to be synchronized with the data stream's frame structure. Frame synchronization is usually accomplished with the aid of some special signaling procedure from the transmitter. This procedure may be very simple, or fairly involved, depending on the environment in which the system is required to operate.

Probably the simplest frame synchronization aid is the frame marker, illustrated in Figure 10.17. The frame marker is a single bit, or a short pattern of bits that the transmitter injects periodically into the data stream. The receiver must know the pattern and the injection interval. The receiver, having achieved data synchronization, correlates the known pattern with the incoming data stream at the known injection interval. If the receiver is not in synchronization with the framing pattern, the accumulated correlation will be low. When the receiver comes into frame synch, however, the correlation should be nearly perfect, blemished only by an occasional detection error.

The advantage of the frame marker is its simplicity. Even a single bit can suffice as a frame marker if a sufficient number of correlations are accumulated before deciding whether or not the system has achieved synchronization. The major drawback is that the sufficient number may be very large, and thus the expected time required to acquire synchronization would be long. Therefore, frame markers are most useful in systems that transmit data continuously, like many telephony and computer links, and would be inappropriate for systems that transmit in iso-

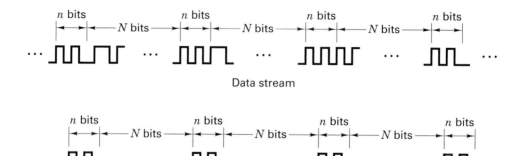

Data stream

Receiver generated frame marker replica

Figure 10.17 Frame marker illustration.

lated bursts or systems that require rapid frame acquisition. A secondary drawback is that the inserted bit(s) may make the organization of the data stream awkward.

An example is the T1 carrier stream developed by Bell Labs and in general use in North American telephony systems. The T1 carrier structure includes a single bit frame marker after each set of 24 8-bit bytes, each byte representing one of 24 possible voice data streams. This yields a data structure that is an integer multiple of 193 bits—an unhandy number from the standpoint of most integrated circuits.

An approach for systems with inconsistent or bursty transmissions, or systems with rapid acquisition requirements, is a synchronization codeword. A synchronization codeword would typically be sent as part of a message header. The receiver must know the codeword and be constantly searching for it in the data stream, possibly with a matched filter correlator. Detection of the codeword would indicate a known position (typically the beginning) in the data frame. The advantage of this system is that frame acquisition can be essentially immediate. The only delay would be that required to process the incoming codeword. The disadvantage is that the codeword must be long relative to the frame marker, to keep the probability of false detections low. The complexity of the correlation operation is proportional to the length of the sequence, so the correlator may be relatively complicated.

A good synchronization codeword is one that has the property that the absolute value of its "correlation sidelobes" is small. A correlation sidelobe is the value of the correlation of a codeword with a time-shifted version of itself. Thus, the correlation sidelobe value for a k-symbol shift of an N-bit code sequence $\{X_i\}$ is given by

$$C_k = \sum_{j=1}^{N-k} X_j X_{j+k} \qquad (10.81)$$

where X_i ($1 \le i \le N$) is an individual code symbol taking values ± 1 and the adjacent data symbols (associated with index value $i > N$) are assumed to be zero. An example of correlation sidelobe computation is shown in Figure 10.18. The 5-bit

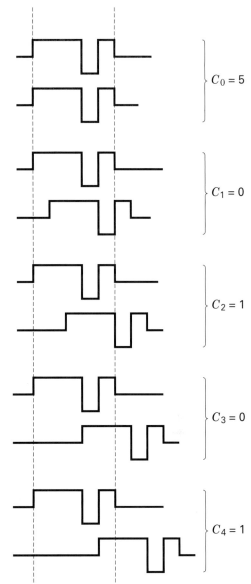

$C_0 = 5$

$C_1 = 0$

$C_2 = 1$

$C_3 = 0$

$C_4 = 1$

Figure 10.18 Correlation sidelobe example.

sequence in the example is seen to have good correlation properties, in that the largest sidelobe is one-fifth of the main lobe, C_0. Sequences like the example in Figure 10.18, with the property that their largest sidelobe has a magnitude of unity, are known as Barker sequences or Barker words [26]. There is no known constructive method for finding Barker words, and only 10 unique words are known, the longest of which has 13 symbols. The known unique Barker words are given in Table 10.1. Some thought should make it clear that a completely exhaustive list of known

Barker sequences would include those sequences produced by inverting the sign of the symbols and those produced by reversing the time ordering of the symbols in the sequences of Table 10.1.

The sidelobe correlation properties of Barker codes are based on the assumption that the adjacent symbols have zero value. This is an approximation to the effect of equally likely random binary data adjacent to the Barker word, taking the values ±1. Unfortunately, the Barker sequences are too short for this approximation to provide the best codeword in random binary data in all cases. Willard [27] found the best sequences in terms of the minimum probability of false synchronization for random adjacent symbols for the Barker word lengths by the use of a computer simulation. The Willard sequences are shown in Table 10.2.

Two probabilities characterize the performance of a system using a synchronization word. These are the probability of a missed detection and the probability of false alarm. Clearly, the system designer would wish both probabilities to be as small as possible. These are conflicting desires. In order to decrease the probability of a miss, the system design may allow less-than-perfect correlation of an incoming synchronization word. That is, a word may be accepted even if it contains a small number of errors. This, however, enlarges the number of symbol patterns that will be accepted and thereby increases the probability of a false alarm. The probability of a miss for an N-bit word where k or fewer errors are accepted is given by

$$P_m = \sum_{j=k+1}^{N} \binom{N}{j} p^j (1 - p)^{N-j} \tag{10.82}$$

where p is the probability of a detector bit error. The probability of a false alarm generated by N bits of random data is given by

$$P_{FA} = \sum_{j=0}^{k} \frac{\binom{N}{j}}{2^N} \tag{10.83}$$

It can be seen that for small p, P_m will decrease roughly exponentially with increasing k. Unfortunately, P_{FA} increases roughly exponentially with increasing k. To obtain acceptable values of both P_m and P_{FA} for a given value of p, the system designer often needs values of N larger than those provided by the Barker and

TABLE 10.1 Barker Synchronization Codewords

N	Barker Sequence
1	+
2	+ + or + −
3	+ + −
4	+ + + − or + + − +
5	+ + + − +
7	+ + + − − + −
11	+ + + − − − + − − + −
13	+ + + + + − − + + − + − +

TABLE 10.2 Willard Synchronization Codewords

N	Willard Sequences
1	$+$
2	$+-$
3	$++-$
4	$++--$
5	$++-+-$
7	$+++-+--$
11	$+++-++-+---$
13	$+++++--+-+---$

Willard sequences. fortunately, there is a fairly large body of literature dealing with longer sequences. Most of these sequences were discovered through exhaustive computer searches. Spilker [20] lists sequences of up to $N = 24$ found by Newman and Hofman [28] and mentions that their original paper had sequences to $N = 100$. Wu [29] provides a list of Maury-Styles sequences to length $N = 30$, a list of Linder sequences to length 40. He also provides a fairly complete discussion of the topic of synchronization sequences, including constructive techniques for reasonable but non-optimum sequences, and insight into the frame synchronization procedures of some operational satellite digital communication systems.

10.3 NETWORK SYNCHRONIZATION

For systems using coherent modulation techniques, one-direction communications such as broadcast channels, or single-link communications, such as most microwave links, land-line links, or fiber-optics links, the synchronization architecture that makes the most sense is to make synchronization totally a receiver function. For communications systems using noncoherent modulation techniques, or that involve many users accessing a central communication node, such as many satellite communication systems, it often makes sense for synchronization to be mostly or entirely a terminal function. This means that the terminal transmitter parameters are modified to achieve synchronization, rather than modifying the central node's receiver parameters. This must be the approach if the system uses time-division multiple access (TDMA). In TDMA, each user is allotted a segment of time in which to transmit its information. The terminal transmitter must be synchronized with the system in order for its transmitted burst of data to arrive at the central node at the time when the node is prepared to receive the data. Synchronization of the terminal transmitter also makes sense with systems that combine signal processing at the central node with frequency-division multiple access (FDMA). If the terminals precorrect their transmission to be synchronized with the central node, the node can use a fixed set of channel filters and a single timing reference for the processing of all channels. Otherwise, the node would require a separate time and frequency acquisition and tracking capability for each incoming channel, and it would need to deal with the possibility of varying amounts of adjacent channel interference. It seems clear that terminal transmitter synchronization is often the cleaner, more reasonable system approach to synchronizing a network.

Transmitter synchronization procedures may be classified s being either open loop or closed loop. Open-loop techniques do not rely on any measurement of the arriving signal parameters at the central node. The terminal pre-corrects its transmission based on stored knowledge of link parameters that have been provided by some external authority but may possibly be modified by observations of a return signal from the central node. Open-loop techniques rely on link parameters being accurately known and predictable. They work best when link geometry is nominally fixed, and the links themselves operate continually for relatively long periods, once established. They tend to be difficult to use efficiently when the link geometry is not static or when the terminals access the system sporadically.

The main advantages of the open-loop methods are that acquisition is fast— the procedure can work without a return link, and the amount of real-time computation that is required is small. The disadvantages of open-loop methods are that they require the existence of the external authority that provides knowledge of the required link parameters, and that they are relatively inflexible. The lack of any direct real-time measure of system characteristics means that the system cannot adjust quickly to any unplanned change in conditions.

Closed-loop techniques, on the other hand, require little in the way of *a priori* knowledge of link parameters. Knowledge would be useful in reducing the time required for acquisition, but need not be precise as is required by open-loop methods. Closed-loop methods involve measurements of the synchronization accuracy of the incoming transmissions from the terminal upon their arrival at the central node, and the return of the results of these measurements to the terminal via a return path. Thus, closed-loop methods require a return path that provides a response to the terminal's transmission, the ability in the terminal to recognize the response for what it is, and the ability in the terminal to modify the transmitter characteristics appropriately, based on the response. This amounts to a requirement for a relatively large amount of real-time processing in the terminal, and two-way links between every terminal and the central node. The disadvantages of closed-loop methods are that they require a relatively large amount of real-time processing, require two-way links to every terminal, and that acquisition can take a relatively long time. The advantages are that no external source of knowledge is required for the system to work, and the responses on the return link allow the system to adapt easily and quickly to changing geometries and link conditions.

10.3.1 Open-Loop Transmitter Synchronization

Open-loop systems can be further subdivided into systems that employ information gained by observing a return link, and those that do not. Those that do not are the simplest of all, in terms of real-time processing requirements, but communication performance for these simple terminals is clearly very dependent on stable link characteristics.

All transmitter synchronization schemes attempt to precorrect the timing and transmission frequency of the signal in such a manner that the signal will arrive at a receiver with the expected frequency and at the expected time. Thus, to precorrect time, a transmitter would divide the distance between itself and the receiver by the speed of light to get the transmission transit time, and then shift the message transmission timing that much ahead. By transmitting the signal early, it will arrive at the receiver at the appropriate time. The time of arrival at the node is given by

$$T_A = T_t + \frac{d}{c} \tag{10.84}$$

where T_t is the actual transmission start time, d is the transmit distance, and c is the speed of light. Similarly, to precorrect the transmission frequency, the transmitter must allow for the Doppler shift caused by relative motion between the transmitter and the intended receiver. To be received correctly, the required transmission radian frequency is

$$\omega \approx \left(1 - \frac{V}{c}\right)\omega_0 \tag{10.85}$$

where c is the speed of light, V is the relative velocity (positive for decreasing transmission distance), and ω_0 is the nominal transmission radian frequency.

Unfortunately, in practice, neither the time nor the frequency precorrection can be done exactly. Even satellites in nominally geostationary orbits move slightly with respect to a point on the earth, and the behavior of the time and frequency references in the terminal and the central node are never entirely predictable. Thus, there will always be some time and frequency precorrection error. The time error may be expressed as

$$T_e = \frac{r_e}{c} + \Delta t \tag{10.86}$$

where r_e is the error in the range estimate and Δt is the difference between the time reference at the terminal and the reference at the receiver. The frequency error may be expressed as

$$\omega_e = \frac{V_e \omega_0}{c} + \Delta\omega \tag{10.87}$$

where V_e is the error in the measured or predicted relative velocity of the transmitter and receiver (the Doppler error) and $\Delta\omega$ is the frequency difference between the transmitter and the receiver frequency references. There are many other sources of time and frequency error in addition to those mentioned here, but they are typically much less important. Spilker [20] gives a reasonably complete accounting of sources of time and frequency error for satellite systems.

The error terms Δt and $\Delta\omega$ are typically due to random fluctuations in frequency references. The time reference for a transmitter or receiver is generally obtained by counting cycles of the frequency reference, so errors in the accuracy of the time and frequency references are related. The fluctuations in a frequency reference are very difficult to characterize statistically, although the power spectral density of he fluctuations is approximated by a sequence of power-law segments

[15]. Frequency references are often specified in terms of a maximum allowable fractional frequency change per day:

$$\delta = \frac{\Delta\omega}{\omega_0} \quad \text{hertz/hertz/day} \tag{10.88}$$

Typical values for δ range from 10^{-5} to 10^{-6}, for inexpensive crystal oscillators, to 10^{-9} to 10^{-11}, for high-quality crystal oscillators; to 10^{-12} for rubidium standards, to 10^{-13} for cesium standards. An effect of specifying system-frequency references by the maximum fractional frequency is that if there is no intervention, the offset form the nominal frequency ω_0 can grow linearly with time:

$$\Delta\omega(T) = \omega_0 \int_0^T \delta \, dt + \Delta t(0) = \omega_0 \, \delta T + \Delta\omega(0) \quad \text{hertz} \tag{10.89}$$

For a cycle-counting time reference, however, the cumulative time offset is related to the cumulative phase error of the reference:

$$\begin{aligned}
\Delta t(T) &= \int_0^T \frac{\Delta\omega(t)}{\omega_0} \, dt + \Delta t(0) \\
&= \int_0^T \delta t \, dt + \int_0^T \frac{\Delta\omega(0)}{\omega_0} \, dt + \Delta t(0) \\
&= \tfrac{1}{2} \delta T^2 + \frac{\Delta\omega(0) \, T}{\omega_0} + \Delta t(0)
\end{aligned} \tag{10.90}$$

Thus, without intervention, a time-reference error can grow quadratically with time. For open-loop transmitter synchronization systems, this quadratic growth in time error often sets limits on how often the external authority must intervene, either to update the terminal's knowledge of receiver timing, or to reset both the receiver's and the transmitter's time references to nominal. The quadratic error growth usually means that timing errors are more of an operational problem than are frequency errors, although this will depend on the system design.

If the transmitter does not have information from measurements on a return link, the time and frequency offsets as modeled by Equations (10.86) to (10.90) will allow a system designer the ability to determine the maximum interval between interventions on the basis of a probability-of-error criterion. Time- and frequency-reference recalibration is often a burdensome procedure; it should be done rarely as possible.

If a terminal has access to a return link from the central node and the ability to make comparative measurements between the local reference and incoming signal parameters, the interval between recalibrations can be made much longer. Large satellite control stations can measure and model the orbital parameters of nominally geostationary satellites to an accuracy of a few tens of feet in range and a few feet/second in velocity relative to the ground terminal. Thus, for the important special case of a synchronous satellite as the central node, the first terms on the right-hand side of Equations (10.86) and (10.87) are usually negligible. When this is true, the differences between the incoming signal parameters and those generated by the terminal's time and frequency references will approximate the error terms Δt and $\Delta \omega$. These error terms measured on the downlink can be used to compute appropriate corrections to the uplink transmissions. On the other hand, if the time and frequency references are known to be accurate but the link geometry is somewhat in question—perhaps because the terminal is mobile or the satellite is nongeostationary—the same sort of return link measurement could be used to resolve range or velocity uncertainties. These measures of range or relative velocity can then be used to precorrect uplink timing and frequency.

The case where a terminal is able to utilize measurements made on a return link signal is sometimes called quasi-closed-loop transmitter synchronization. The quasi-closed technique is clearly more adaptable to uncertainties in the communication system than is the purely open-loop system. The purely open-loop system requires complete *a priori* knowledge of all important link parameters in order to operate successfully. Unanticipated changes in the links cannot be tolerated. The quasi-closed-loop system, on the other hand, requires *a priori* knowledge of all but one of the important parameters in each of time and frequency, but the remaining term can be determined from observations of the return link. This adds complexity to the terminal, but it also adds the ability to adapt to certain types of unplanned link changes. This degree of adaptability can greatly reduce the frequency of required system calibration.

10.3.2 Closed-Loop Transmitter Synchronization

Closed-loop transmitter synchronization involves the transmission of special synchronization signals that are used to determine the signal's time or frequency error relative to the desired timing or frequency when the signal arrives at the receiver. The results of this determination are then fed back to the transmitter on a return link. The determination of synchronization errors can be either implicit or explicit. If the central node has sufficient processing capacity, the central node may make an actual error measurement. Such a measurement might be the amount and direction of offset, or perhaps simply the direction alone. This information would be formatted and returned to the transmitter on a return link. If the central node has little processing capability, the special synchronization signal may simply be turned around and returned to the transmitter on the return link. In this case, it becomes part of the transmitter's task to interpret the returned signal for itself. The design of a special synchronization signal that lends itself to easy unambiguous interpretation can be a challenge.

The relative advantages and disadvantages of the two types of closed-loop systems have to do with the location of the signal-processing capability and the efficiency of channel usage. A major advantage of having the processing at the central node is that results of the error measurements that are transmitted on the return link can be a short digital sequence. This efficient use of the return link can be important if a single return link is time-division multiplexed between a large number of terminals. A second potential advantage is that the error-measuring capability in the central node can be shared by all terminals communicating through the node. This can amount to a large saving in system processing capability. The principal potential advantage in having the processing at the terminal is that the central node may not be easily accessible, and reliability considerations may dictate a simple design. This has typically been the case when the central node is a space satellite. With continuing improvements in satellite technology, simplicity requirements can be expected to be less dominant in the future than in the past. Another potential advantage to having the processing in the terminal is that the response can be quicker because there is little processing delay in the central node. This may be important if link parameters are changing very rapidly. The primary disadvantages are the inefficient use of the return channel and that the return signals may be difficult to interpret. This difficulty would arise when the central node is not just a simple repeater but makes symbol decisions and transmits these decisions on the return link. This symbol decision capability can greatly improve the terminal-to-terminal error performance, but it complicates the synchronization procedure. This is because the effects of a time or frequency offset are resident in the return signal indirectly—that is, only as they have affected the symbol decisions. Consider the example of a BFSK transmission to a central node that makes noncoherent bit decisions. The decisions will be dependent on the detected signal energy in the mark and space detectors. If the transmitted signal is an alternating sequence of marks and spaces, the signal at the central node can be modeled as

$$r(t) = \begin{cases} \sin[(\omega_0 + \omega_s + \Delta\omega)t + \theta] & 0 \le t \le \Delta t \\ \sin[(\omega_0 + \Delta\omega)t + \theta] & \Delta t < t \le T \end{cases} \tag{10.91}$$

where T is the symbol interval, ω_0 is one symbol frequency, $(\omega_0 + \omega_s)$ is the other symbol frequency, $\Delta\omega$ is the frequency error at the central node, Δt is the signal arrival time error at the central node, and θ is an arbitrary phase angle. Now, if

$$x = \frac{1}{T} \int_0^T r(t) \cos \omega_0 t \, dt \tag{10.92}$$

and

$$y = \frac{1}{T} \int_0^T r(t) \sin \omega_0 t \, dt \tag{10.93}$$

represent the detector quadrature components, then the detected signal energy can be expressed as

$$
\begin{aligned}
z^2 &= x^2 + y^2 \\
&= \left(\frac{\sin[(\omega_s + \Delta\omega)\Delta t/2]}{(\omega_s + \Delta\omega)T} \right)^2 + \left(\frac{\sin[\Delta\omega(T - \Delta t)/2]}{\Delta\omega T} \right)^2 \\
&\quad + \frac{\cos(\Delta\omega \Delta t) + \cos[\Delta\omega T - (\omega_s + \Delta\omega)\Delta t] - \cos(\Delta\omega T) - \cos(\omega_s \Delta t)}{2\Delta\omega(\omega_s + \Delta\omega)T^2}
\end{aligned}
\tag{10.94}
$$

For the special case where the time error, Δt, is zero, Equation (10.94) simplifies to

$$z^2 = \left[\frac{\sin(\Delta\omega T/2)}{\Delta\omega T} \right]^2 \tag{10.95}$$

For the case where the frequency offset is zero,

$$z^2 = \left(\frac{T - \Delta t}{2T}\right)^2 + \left[\frac{\sin(\omega_s \Delta t/2)}{\omega_s T}\right]^2 \tag{10.96}$$

The important thing to notice in Equations (10.94) to (10.96) is that any time error or frequency offset or combination of both will decrease the detected signal energy in the correct symbol detector and introduce signal energy into the incorrect signal detector. This will reduce the effective distance between signals in signal space and degrade error performance. A measurement of error performance, however, which is all that is available on the return link, gives no insight into whether the problem is a frequency offset, a time error, or a combination of both. Thus the transmission of standard signals is not likely to provide a useful response for synchronization.

A useful technique for determining the correct frequency precorrection for our example of BFSK-signaling is to transmit a constant tone whose frequency is the average of the two symbol frequencies. Such a tone should produce a random binary sequence on the return link with equal numbers of marks and spaces. A frequency offset from the average would produce predominately marks or spaces. Finding the center frequency in this way allows accurate frequency precorrection of the signals. Once the correct frequency is found, the transmitter can transmit an alternating sequence of marks and spaces in order to discover correct timing. By varying the timing of the transmission through a range of half-a-symbol interval, the transmitter can look for the timing that provides the worst error performance. When the transmission arrival at the central node is displaced from correct timing by half-a-symbol interval, the two detectors will detect equal amounts of energy, and the binary sequence on the return link will be random. Determining the time when the transmitted and return signals are decorrelated will allow the transmitter to compute the correct transmission timing. Notice that this procedure works better than attempting to find the point at which error performance is the best. Any well-designed system will have sufficient transmission energy to allow for slight timing offsets, so an error-free return signal could be achieved with less than perfect timing. In fact, the larger the signal-to-noise ratio, the worse a best-finding procedure works. A worst-finding system, however, will work well for any well-designed system, and it will improve in potential accuracy with increasing signal-to-noise ratio. This can be seen intuitively, because increased signal-to-noise ratios will allow the system to tolerate larger timing errors, so the improvement in error performance as the timing error decreases from half-a-symbol time will be more rapid in the large signal-to-noise case than in the smaller signal-to-noise ratio case. This will allow a more precise determination of the half-symbol timing position.

10.4 CONCLUSION

This chapter has outlined the fundamental problems and issues associated with synchronization in digital communications. The trade-offs are generally between expense and complexity on the one hand, and error performance on the other. We have discussed receiver synchronization and phase-locked loops (PLL) in particular. Typically, it is the receiver that takes the most active role in the synchronization of a communications link. Even in cases where a terminal's transmitter assumes the more active role, as in some satellite links, the process is often aided by a return path that has been acquired by the terminal's receiver. Thus, receiver synchronization is more fundamental. Phase-locked loops and their variations are the primary control circuits used to track variations in phase of an incoming signal. The mathematics needed to describe the response of a PLL to a given input involves the solution to a nonlinear differential equation. It was shown, however, that under steady-state conditions, a linearized model provides a useful approximation to system performance. In circumstances where the linearized model cannot be accurately applied, results by Viterbi [8] for first-order loops were introduced. Although exact for first-order loops only, these results have been shown to be useful approximations to the performance of higher order loops as well [5].

The extremely important special case of suppressed-carrier loops was discussed. Suppressed-carrier loops are required to track the phase of an incoming signal that has no average energy at the carrier frequency. The common example of such a signal is one that has been modulated with standard antipodal BPSK. In this situation, a harmonic of the suppressed carrier is produced through the use of a nonlinearity, and the harmonic is tracked.

The next higher level of synchronization treated here was symbol synchronization. Two primary classes of symbol synchronization were discussed. Open-loop synchronizers operate directly on the modulated signal to produce a symbol transition indication. Closed-loop synchronizers use a closed-cycle control loop to acquire and track the symbol transitions.

The highest level of synchronization considered was frame synchronization. To receive the data in a useful form, the receiver must determine which symbols belong to which frames. This knowledge is equivalent to having frame sync, which is usually accomplished by including with the data symbols some recognizable pattern known to the receiver. The receiver scans the incoming data until it recognizes the pattern. Synchronization can be checked by looking for periodic repetitions of the pattern.

This chapter has necessarily been only an outline of the important problems, issues, and results relating to the synchronization of digital communication systems. The interested reader will find that the references listed are excellent works that will provide much greater depth of coverage than space has allowed here.

REFERENCES

1. Peterson, W. W., and Weldon, E. J., *Error-Correcting Codes,* The MIT Press, Cambridge, Mass., 1972.
2. Lee, E. A. and Messerschmitt, D. G., *Digital Communications,* Kluwer Academic Publications, Boston, 1988.
3. Mengali, U., and D'Andrea, A. N., *Synchronization Techniques for Digital Receivers,* Plenum Press, New York, 1997.

4. Meyr, H. Moeneclaey, M., and Fechtel, S. A., *Digital Communication Receivers,* John Wiley & Sons, Inc., New York, 1998.

5. Gardner, F. M., *Phaselock Techniques,* 2nd ed., John Wiley & Sons, Inc., New York, 1979.

6. Davenport, W. B., and Root, W. L., *Random Signals and Noise,* McGraw-Hill Book Company, New York, 1958.

7. Papoulis, A., *Probability, Random Variables, and Stochastic Processes,* McGraw-Hill Book Company, New York, 1965.

8. Viterbi, A. J., *Principles of Coherent Communications,* McGraw-Hill Book Company, New York, 1966.

9. Lindsey, W. C., *Synchronization Systems in Communication and Control,* Prentice-Hall, Inc., Englewood Cliffs, N.J., 1972.

10. Lindsey, W. C., and Simon, M. K., "Detection of Digital FSK and PSK Using a First-Order Phase-Locked Loop," *IEEE Trans. Commun.,* vol. COM25, no. 2, Feb. 1977, pp. 200–214.

11. Develet, J. A., Jr., "The Influence of Time Delay on Second-Order Phase Lock Loop Acquisition Range," *Int. Telem. Conf.,* London, 1963.

12. Johnson, W. A., "A General Analysis of the False-Lock Problem Associated with the Phase-Lock Loop," *The Aerospace Corp., Rep. TOR-269(4250-45)-1,* NASA Accession N64-13776, 1963.

13. Tausworthe, R. C., "Acquisition and False-Lock Behavior of Phase-Locked Loops with Noisy Inputs," *Jet Propulsion Laboratory, JPL SPS 37-46,* vol. 4, 1967.

14. Franks, L. E., "Synchronization Subsystems: Analysis and Design," in K. Feher, *Digital Communications, Satellite/Earth Station Engineering,* Prentice-Hall, Inc., Englewood Cliffs, N.J., 1981, Chap. 7.

15. Simon, M. K., and Yuen, J. H., "Receiver Design and Performance Characteristics," in J. H. Yuen, ed., *Deep Space Telecommunications Systems Engineering,* Plenum Press, New York, 1983.

16. Gardner, F. M., "Hangup in Phase-Lock Loops," *IEEE Trans. Commun.,* COM25, October 1977.

17. Blanchard, A., *Phase-Locked Loops,* John Wiley & Sons, Inc., New York, 1976.

18. Holmes, J. K., *Coherent Spread Spectrum Systems,* John Wiley & Sons, Inc., New York, 1982.

19. Lindsey, W. C., and Simon, M. K., eds., *Phase Locked Loops and Their Applications,* IEEE Press, New York, 1977.

20. Spilker, J. J., Jr., *Digital Communications by Satellite,* Prentice-Hall, Inc., Englewood Cliffs, N.J., 1977.

21. Wintz, P. A., and Luecke, E. J., "Performance of Optimum and Suboptimum Synchronizers," *IEEE Trans. Commun. Technol.,* June 1969, pp. 380–389.

22. Anderson, J. B., Aulin, T., and Sundberg, C. E., *Digital Phase Modulation,* Plenum Press, New York, 1986.

23. Laurent, P. A., "Exact and Approximate Construction of Digital Phase Modulations by Superposition of Amplitude Modulated Pulses", *IEEE Trans. Commun.,* COM–34, no. 2, pp. 150–160, Feb. 1986.

24. Lui, G. L., "Threshold Detection Performance of GMSK Signal with BT=0.5," *IEEE MILCOM 98 Proceedings,* vol. 2, Boston, October 18–21, 1998, pp. 515–519.

References

25. Kaleh, G., "Differentially Coherent Detection of Binary Partial Response Continuous Phase Modulation with Index 0.5," *IEEE Trans. Commun.,* COM–39, pp. 1335–40, Sept. 1991.

26. Barker, R. H., "Group Synchronization of Binary Digital Systems," in W. Jackson, ed., *Communicaiton Theory,* Academic Press, Inc., New York, 1953.

27. Willard, M. W., "Optimum Code Patterns for PCM Synchronization," *Proc. Natl. Telem. Conf.,* 1962, paper 5–5.

28. Newman, F., and Hofman, L., "New Pulse Sequences with Desirable Correlation Properties," *Proc. Natl. Telem. Conf.,* 1971, pp. 272–282.

29. Wu, W. W., *Elements of Digital Satellite Communications,* Vol. 1, Computer Science Press, Inc., Rockville, Md., 1984.

PROBLEMS

10.1. A transmitter is sending an unmodulated tone of constant energy (a beacon) to a distant receiver. The receiver and transmitter are in motion with respect to each other such that $d(t) = D[1 - \sin(mt)] + D_0$, where $d(t)$ is the distance between the transmitter and receiver (possibly this represents an aircraft doing "figure-eight" maneuvers over a ground station), and D, m, and D_0 are constants. This relative motion will cause a Doppler shift in the received transmitter frequency of

$$\Delta\omega_D(t) = \frac{\omega_0 V(t)}{c}$$

where $\Delta\omega_D$ is the Doppler shift, ω_0 is the nominal carrier frequency, $V(t) = d(t)$ is the relative velocity between the transmitter and receiver, and c is the speed of light. Assuming that the linearized loop equations hold and that the receiver's PLL is in lock (zero phase error) at $t = 0$, show that an appropriately designed first-order loop can maintain frequency lock.

10.2. Consider a transmitter and receiver that are in relative motion as in Problem 10.1. Once again assume that the linearized loop equations hold. Under this assumption determine the PLL phase error as a function of time for the all-pass and low-pass loop filters of Equations (10.13) and (10.14). Demonstrate that the validity of the assumption of the linearized loop equations depends on the value of the gain K_0.

10.3. A high-performance aircraft is transmitting an unmodulated carrier signal to a ground terminal. The ground terminal is initially in phase lock with the signal. The aircraft performs a maneuver whose dynamics are described by the equation for acceleration, $a(t) = At^2$, where A is a constant. Assuming that the linearized equations apply, determine the minimum order of the phase-locked loop required to track the signal from this aircraft.

10.4. Show that the loop bandwidth of a first-order phase-locked loop is given by $B_L = K_0/4$, where K_0 is the loop gain.

10.5. A second-order phase-locked loop has a low-pass loop filter given by

$$F(\omega) = \frac{\omega_1}{j\omega + \omega_1}$$

and a loop gain of K_0. Under the assumption that $K_0 \geq \omega_1/4$, show that the loop band-width of this phase-locked loop is given by $B_L = K_0/8$. [Hint:*

$$\int_{-\infty}^{\infty} \frac{dx}{R} = \frac{\pi \cos (h/2)}{2cq^3 \sin h} \quad \text{for } 4ac > b^2$$

where $R = a + bx^2 + cx^4$, $q = \sqrt[4]{a/c}$, and $\cos h = -b/2\sqrt{ac}$.]

10.6. A first-order phase-locked loop with loop gain K_0 is disturbed by additive white Gaussian noise of normalized (to unit signal energy) two-sided power spectral density of $N_0/2$ watts/hertz. Determine the necessary relationship between noise power spectral density and loop gain if the loop is designed to cycle slip no more often than once per day.

10.7. Viterbi [8] determined that the probability density function of the output phase of a first-order phase-locked loop disturbed by white Gaussian noise is given by

$$p(\phi) = \frac{\exp (\rho \cos \phi)}{2\pi I_0(\rho)}, \quad |\phi| \leq \pi, \quad \rho \geq 0$$

Demonstrate that $p(\phi)$ given above is in fact a probability density function, and compute the mean and variance of ϕ.

10.8. Computer simulations and laboratory measurements have indicated that the time between cycle slips is exponentially distributed; that is, the distribution function of the time between cycle slips, T, is given by

$$p(T) = 1 - \exp \left(-\frac{T}{T_m} \right)$$

Given this distribution function, find the mean time between cycle slips and the variance about this mean as functions of T_m. If the mean time between cycle slips is 1 day, what is the probability of cycle slips less than 1 hour apart? More than 3 days apart?

10.9. Consider a second-order phase-locked loop with a low-pass loop filter

$$F(\omega) = \frac{\omega_1}{j\omega + \omega_1}$$

During aided acquisition it is desired that this loop be scanned throughout a 1000-radian uncertainty region in 1 s. If the relationship between loop gain and filter constant is $K_0 = 2\omega_1$, what is the required relationship between loop gain and the single-sided additive white Gaussian noise power spectral density, N_0? Determine the largest value of N_0 that can be accommodated.

10.10. Consider the operation of an open-loop symbol synchronizer whose band pass filter (BPF) has a bandwidth of $0.1/T$ hertz, where T is the symbol period. For a bit energy-to-noise power spectral density ratio (E_b/N_0) of 10 dB, determine the magnitude of the approximate mean and variance of the fractional tracking error, and compute an upper bound on the probability that the tracking error exceeds three times its approximate fractional mean. (*Hint:* Consider the Chebyshev inequality [7].)

*I. S. Gradshteyn and I. M. Ryzhik, *Table of Integrals, Series and Products* (New York: Academic Press, 1965), 2.161.1.

10.11. A communication system is used to transmit commands to a payload at a data rate of 100 bits/s. Each command is preceded by an N-bit header that identifies it in the data stream. Assuming that except possibly for the header, the bits appear to be random $[P(1) = P(0) = \frac{1}{2}]$, what is the minimum-length header that would provide an expected frequency of false alarms of one per year? For a channel bit error probability of 10^{-5}, what is the probability of missing this header? What is the miss probability if the channel error probability is 2×10^{-2}? If the system is redesigned to accept the header with up to two errors, what is the minimum required length for an expected false alarm rate of one per year? What is the miss probability with this new system and a channel-error probability of 2×10^{-2}?

10.12. A deep-space probe is moving away from the earth at a nominal velocity of 15,000 m/s with a velocity uncertainty of ± 3 m/s. The probe frequency reference is specified to have a drift rate of no more than 10^{-9} Hz/Hz/day. The nominal downlink transmission frequency is 8 GHz. After a 1-month (30-day) silence the probe begins a scheduled transmission toward an earth terminal. The earth terminal contains a cesium standard. What center frequency and frequency search bandwidth should be used by the ground station? Assuming that the range to the probe was accurately known at the beginning of the month, and that the uncertainty in the probe's time and frequency references were zero $[\Delta t(0) = 0, \Delta\omega(0) = 0]$, what is the uncertainty in the time of arrival of the downlink transmission?

10.13. A communications link operates at a nominal center frequency of 10 GHz for a single brief period once a day. The receiver operates with a second-order PLL that has a pull-in range of ± 1 kHz. Assuming that the loop acquires using self-acquisition and that both the transmitter and receiver use the same type of frequency reference, what type of frequency reference must this be?

10.14. At some moment in time ($t = 0$), a clock's output signal has an error of -4×10^{-3} seconds with reference to a master clock. At this time, $t = 0$, the clock's driving oscillator is oscillating at the correct frequency f_r, but the oscillator is speeding up at the rate of 2-parts-in-10^{10} per day.
  a) How long will it take, in days, before the clock output signal has zero error?
  b) If the clock is allowed to operate for 30 days beyond the time of zero error, what will the error be then?

10.15. Under the usual assumptions of zero-mean AWGN and equal energy signals, verify that the right-hand side of Equation (10.67) is in the form of the likelihood function for the estimation of carrier phase and symbol timing.

10.16. Consider the case of full-response MSK signaling with a synchronization "training sequence" of alternating ones and zeros (i.e., $\alpha_k = 1$ for k even, and -1 for k odd).
  (a) Show that for this example there are only four distinct phase states $\{\Phi_k\}$.
  (b) Derive the form of the filters $h^{(\ell)}(t)$, given in Equation (10.64).
  (c) Using the results obtained in part b), derive expressions for Equations (10.68) and (10.69) for this simplified case.

10.17. Develop a rationale for why the iterative procedure suggested in Equations (10.70) and (10.71) will (or will not) work.

QUESTIONS

10.1. What is the definition of *synchronization* in the context of a digital communication system, and why is it important? (See Section 10.1.1.)

10.2. Why may a synchronization system that works well for a home radio receiver possibly be *inadequate* in a high performance aircraft? What modifications are likely to be required for adequate performance? (See Section 10.1.2.)

10.3. The *linearized loop equation* depends on an approximation. What is that approximation, why is it appropriate for loops that are in lock or near lock, and why is it not appropriate for acquisition analysis? (See Section 10.2.1.)

10.4. Second-order phase locked loops have several advantages in their performance, and as the basis for phase-tracking performance analysis. Name two such advantages. (See Section 10.2.1.1.)

10.5. Why are *continuous phase modulation* schemes of increasing importance in modern communications system, and what synchronization challenges do they pose? (See Section 10.2.3.1.)

10.6. What are the advantages and disadvantages of *data-aided* versus *non-data-aided* synchronizers? (See Section 10.2.3.2.)

10.7. Describe a situation in which it may be appropriate to require a transmitter to synchronize itself to the expectations of a receiver. (See Section 10.3.)

EXERCISES

Using the Companion CD, run the exercises associated with Chapter 10.

CHAPTER 11

Multiplexing and Multiple Access

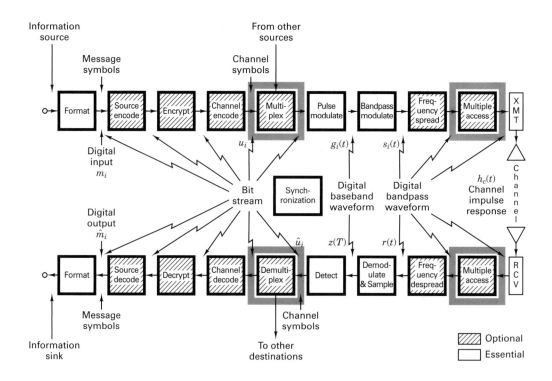

A communications resource (CR) represents the time and bandwidth that is available for communication signaling associated with a given system. It can be graphically envisioned with a plane, where the abscissa represents time, and the ordinate represents frequency. For the efficient development of a communication system, it is important to plan out the resource allocation among system users, so that no block of time/frequency is wasted, and so that the users can share the resource in an equitable manner.

The terms "multiplexing" and "multiple access" refers to the sharing of a CR. There is a subtle difference between multiplexing and multiple access. With *multiplexing,* users' requirements or plans for CR sharing are fixed, or at most, slowly changing. The resource allocation is assigned *a priori,* and the sharing is usually a process that takes place within the confines of a *local site* (e.g., a circuit board). *Multiple access,* however, usually involves the *remote sharing* of a resource, such as in the case of satellite communications. With a dynamically changing multiple access scheme, a system controller must become aware of each user's CR needs; the amount of time required for this information transfer constitutes an overhead and sets an upper limit on the efficiency of the utilization of the CR.

11.1 ALLOCATION OF THE COMMUNICATIONS RESOURCE

There are three basic ways to increase the throughput (total data rate) of a communications resource (CR). The first way is either to increase the transmitter's effective isotropic radiated power (EIRP) or to reduce system losses so that the received E_b/N_0 is increased. The second way is to provide more channel bandwidth. The third approach is to make the allocation of the CR more efficient. This third approach is the domain of communications multiple access. The problem, in the context of a satellite transponder, is to efficiently allocate portions of the transponder's fixed CR to a large number of users who seek to communicate digital information to each other at a variety of bit rates and duty cycles. The basic ways of distributing the communications resource, listed under the heading "multiplexing/multiple access" in Figure 11.1, are the following:

1. *Frequency division (FD).* Specified subbands of frequency are allocated.
2. *Time division (TD).* Periodically recurring time slots are identified. With some systems, users are provided a fixed assignment in time. With others, users may access the resource at random times.
3. *Code division (CD).* Specified members of a set of othogonal or nearly orthogonal spread spectrum codes (each using the full channel bandwidth) are allocated.
4. *Space division (SD) or multiple beam frequency reuse.* Spot beam antennas are used to separate radio signals by pointing in different directions. It allows for reuse of the same frequency band.
5. *Polarization division (PD) or dual polarization frequency reuse.* Orthogonal polarizations are used to separate signals, allowing for reuse of the same frequency band.

The key to *all* multiplexing and multiple access schemes is that various signals share a CR without creating unmanageable interference to each other in the detection process. The allowable limit of such interference is that signals on one CR channel should not significantly increase the probability of error in another channel. Orthogonal signals on separate channels will avoid interference between users. Signal waveforms $x_i(t)$, where $i = 1, 2, \ldots$, are defined to be orthogonal if they can be described in the time domain by

$$\int_{-\infty}^{\infty} x_i(t) x_j(t) \, dt = \begin{cases} K & \text{for } i = j \\ 0 & \text{otherwise} \end{cases} \tag{11.1}$$

where K is a nonzero constant. Similarly, the signals are orthogonal if they can be described in the frequency domain by

$$\int_{-\infty}^{\infty} X_i(f) X_j(f) \, df = \begin{cases} K & \text{for } i = j \\ 0 & \text{otherwise} \end{cases} \tag{11.2}$$

where the functions $X_i(f)$ are the Fourier transforms of the signal waveforms $x_i(t)$. Channelization, characterized by orthogonal waveforms, as shown in Equation (11.1), is called time-division multiplexing or time-division multiple access (TDM/TDMA), and that characterized by orthogonal spectra, as shown in Equation (11.2), is called frequency-division multiplexing or frequency-division multiple access (FDM/FDMA).

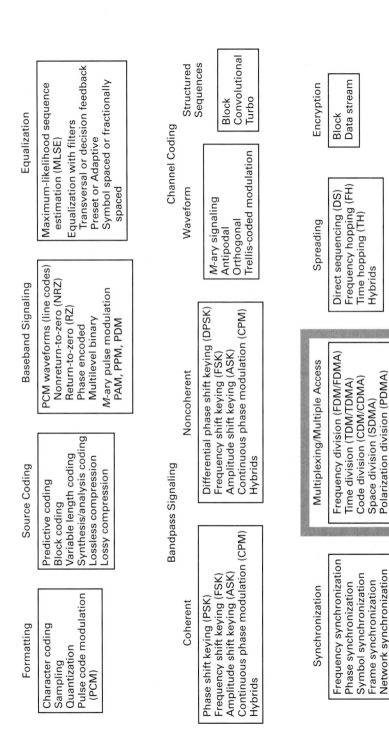

Figure 11.1 Basic digital communication transformations.

11.1.1 Frequency-Division Multiplexing/Multiple Access

11.1.1.1 Frequency-Division Mulitplex Telephony

In the early days of telephony, a separate pair of wires was needed for each telephone trunk circuit (trunk circuits interconnect intercity switching centers). As illustrated in Figure 11.2, the skies of all the major cities in the world grew dark with overhead wires as the demand for telephone service grew. A major develop-

Figure 11.2   In the early days of telephony a pair of wires was needed for each trunk circuit.

ment in the early 1900s, frequency-division multiplex (FDM) telephony, made it possible to transmit several telephone signals simultaneously on a single wire, and thereby transformed the methods of telephone transmission.

The communications resource (CR) is illustrated in Figure 11.3 as the frequency-time plane. The channelized spectrum shown here is an example of FDM or FDMA. The assignment of a signal or user to a frequency band is *long term* or *permanent;* the CR can simultaneously contain several spectrally separate signals. The first frequency band contains signals that operate between frequencies f_0 and f_1, the second between frequencies f_2 and f_3, and so on. The spectral regions between assignments, called *guard bands,* act as buffer zones to reduce interference between adjacent frequency channels. We might ask: How does one transform a baseband signal so that it occupies a higher frequency band? The answer: *heterodyning* or *mixing,* also called *modulating,* the signal with a fixed frequency from a sine-wave oscillator.

If two input signals to a mixer are sinusoids with frequencies f_A and f_B, the mixing or multiplication will yield new sum and difference frequencies at f_{A+B} and f_{A-B}. The trigonometric identity

$$\cos A \cos B = \tfrac{1}{2} \left[\cos (A + B) + \cos (A - B) \right] \qquad (11.3)$$

describes the effect of the mixer. Figure 11.4a illustrates the mixing of a typical voice-grade telephone signal $x(t)$ (baseband frequency range is 300 to 3400 Hz) with a sinusoid from a 20-kHz oscillator. The baseband two-sided magnitude spectrum, $|X(f)|$, is shown in Figure 11.4a. Can the mixer be a linear device? No. The output signal of a linear device will only consist of the *same* component frequencies as the input signal, differing only in amplitude and/or phase.

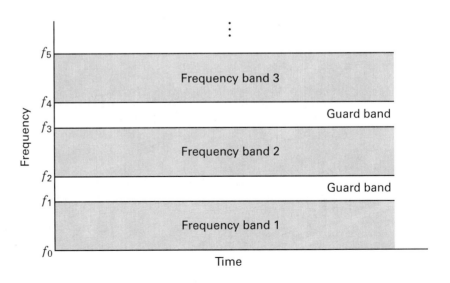

Figure 11.3   Frequency-division multiplexing.

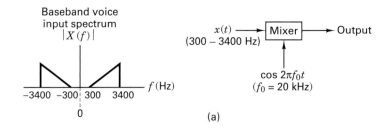

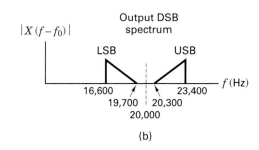

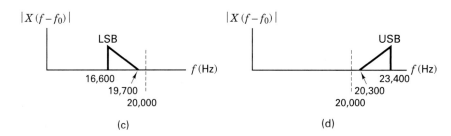

Figure 11.4 Heterodyning (mixing). (a) Mixing operation. (b) Mixer output spectrum. (c) Lower sideband. (d) Upper sideband.

Figure 11.4b illustrates the one-sided magnitude spectrum $|X(f - f_0)|$ at the mixer output. As a result of the mixing described by Equation 11.3), the output spectrum is a frequency-upshifted version of the baseband spectrum, centered at the oscillator frequency of 20 kHz. This spectrum is called a *double-sideband* (DSB) *spectrum* because the information appears in two different bands of the positive frequency domain. Figure 11.4c shows the lower sideband (LSB), whose frequency range is 16,600 to 19,700 Hz, the result of filtering the DSB spectrum. This sideband is sometimes referred to as the *inverted sideband* because the order of low-to-high frequency components is the reverse of that of the baseband components. Filtering can similarly be used to separate the upper sideband (USB), whose frequency range is 20,300 to 23,400 Hz, as shown in Figure 11.4d. This sideband is sometimes referred to as the *erect sideband* because the order of the low-to-high frequency components corresponds to that of the baseband components. Each sideband of the DSB spectrum contains the same information. Thus, only one sideband, either the USB or the LSB, is needed in order to retrieve the original baseband data.

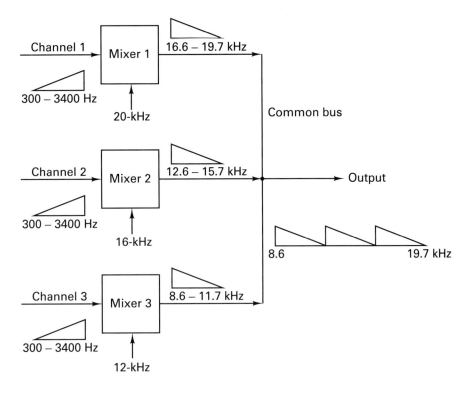

Figure 11.5 Simple FDM example. Three frequency-shifted voice channels.

A simple FDM example with three translated voice channels is seen in Figure 11.5. In channel 1, the 300- to 3400-Hz voice signal is mixed with a 20-kHz oscillator. In channels 2 and 3, a similar type of voice signal is mixed with a 16-kHz and 12-kHz oscillator, respectively. Only the lower sidebands are retained; the result of the mixing and filtering (to remove the upper sidebands) yields the frequency-shifted voice channels shown in Figure 11.5. The composite output waveform is just the sum of the three signals, having a total bandwidth in the range 8.6 to 19.7 kHz.

Figure 11.6 illustrates the two lowest levels of the FDM multiplex hierarchy for telephone channels. The first level consists of a *group* of 12 channels modulated onto subcarriers shown in the range 60 to 108 kHz. The second level is made up of five groups (60 channels) called a *supergroup* modulated onto the subcarriers shown in the range 312 go 552 kHz. The multiplexed channels are now treated as a composite signal that can be transmitted over cables or can be further modulated onto a carrier wave for radio transmission.

11.1.1.2 Frequency-Division Multiple Access of Satellite Systems

Most of the world's communication satellites are positioned in a *geostationary* or *geosynchronous* orbit. This means that the satellite is in a circular orbit, in the same plane as the earth's equatorial plane, and at such an altitude (approximately 19,330 nautical miles) that the orbital period is identical with the earth's rotational period.

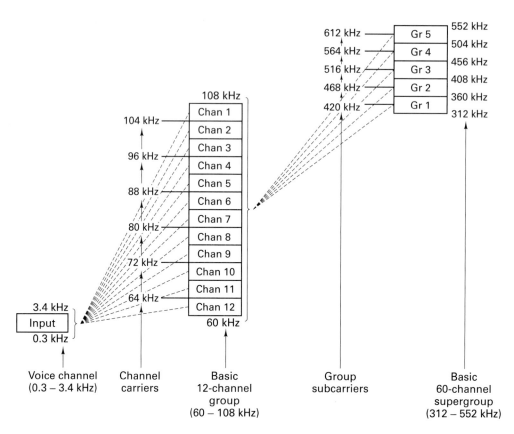

Figure 11.6 Modulation plan of a typical frequency-division multiplex system.

Since such satellites appear stationary when viewed from the earth, three of them spaced 120° apart can provide worldwide coverage (except for the polar regions). Most communication satellite systems are made up of nonregenerative repeaters or transponders. *Nonregenerative* means that the uplink (earth-to-satellite) transmissions are simply amplified, frequency shifted, and retransmitted on the downlink (satellite-to-earth) without any demodulation/remodulation or signal processing. The most popular frequency band for commercial satellite communications, called *C-band*, uses a 6-GHz carrier for the uplink and a 4-GHz carrier for the downlink. For C-band satellite systems, *each satellite* is permitted, by international agreement, to use a 500-MHz-wide spectral assignment. Typically, each satellite has 12 transponders with a bandwidth of 36 MHz each. The most common 36-MHz transponders operate in an FDM/FM/FDMA (frequency-division multiplex, frequency-modulated, frequency-division multiple access) multidestination mode. Let us consider each component of this name:

1. *FDM.* Signals such as telephone signals, each one having a single-sideband 4-kHz spectrum (including guard bands) are FDM'd to form a multichannel composite signal.

2. *FM.* The composite signal is frequency-modulated (FM) onto a carrier and transmitted to the satellite.
3. *FDMA.* Subdivisions of the 36-MHz transponder bandwidth may be assigned to different users. Each user receives a specific bandwidth allocation whereby he or she can access the transponder.

Thus, composite FDM channels are FM modulated and transmitted to the satellite within the bandwidth allocation of an FDMA plan. The major advantage of FDMA (compared with TDMA) is its simplicity. The FDMA channels require no synchronization or central timing; each channel is almost independent of all other channels. Later we discuss some advantages of TDMA compared to FDMA.

11.1.2 Time-Division Multiplexing/Multiple Access

In Figure 11.3, sharing of the communications resource (CR) is accomplished by allocating frequency bands. In Figure 11.7, the same CR is shared by assigning each of *M* signals or users the full spectral occupancy of the system for a short duration of time called a *time slot.* The unused time regions between slot assignments, called *guard times,* allow for some time uncertainty between signals in adjacent time slots, and thus act as buffer zones to reduce interference. Figure 11.8 is an illustration of a typical TDMA satellite application. Time is segmented into intervals called frames. Each frame is further partitioned into assignable user time slots. The frame structure repeats, so that a fixed TDMA assignment constitutes one or more slots that periodically appear during each frame time. Each earth station transmits its data in bursts, timed so as to arrive at the satellite coincident with its designated time slot(s). When the bursts are received by the satellite transponder, they are

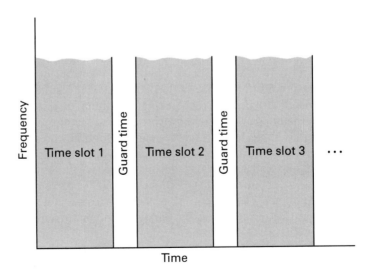

Figure 11.7 Time-division multiplexing.

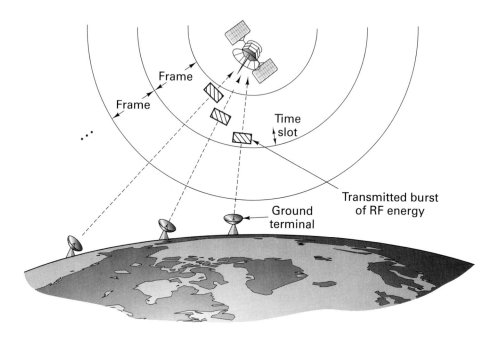

Figure 11.8 Typical TDMA configuration.

retransmitted on the downlink, together with the bursts from other stations. A receiving station detects and demultiplexes the appropriate bursts and feeds the information to the intended user.

11.1.2.1 Fixed-Assignment TDM/TDMA

The simplest TDM/TDMA scheme, called *fixed-assignment TDM/TDMA*, is so named because the M time slots that make up each frame are preassigned to signal sources, long term. Figure 11.9 illustrates, in block diagram form, the operation of such a system. The multiplexing operation consists of providing each source with an opportunity to occupy one or more slots. The demultiplexing operation consists of deslotting the information and delivering the data to the intended sink. The two commutating switches in Figure 11.9 have to be synchronized so that the message corresponding to source 1, for example, appears on the channel 1 output, and so on. The message itself generally comprises a preamble portion and a data portion. The preamble portion usually contains synchronization, addressing, and error-control sequences.

A fixed-assignment TDM/TDMA scheme is extremely efficient when the source requirements are predictable, and the traffic is heavy (the time slots are most always filled). However, for bursty or sporadic traffic, the fixed-assignment scheme is wasteful. Consider the simple example shown in Figure 11.10. In this example there are four time slots per frame; each slot is preassigned to users A, B, C, and D, respectively. In Figure 11.10a we see a typical activity profile of the four users. During the first frame time, user C has no data to transmit; during the second frame time, user B has none, and during the third frame time, user A has none. In a

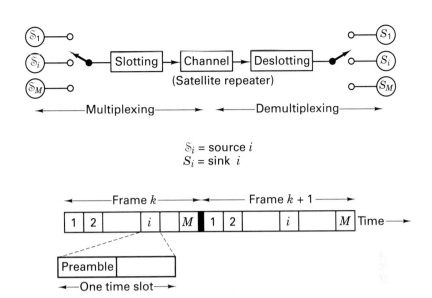

$$\mathcal{S}_i = \text{source } i$$
$$S_i = \text{sink } i$$

Figure 11.9   Fixed-assignment TDM.

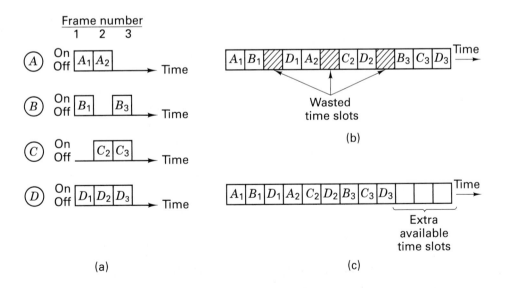

Figure 11.10   Fixed-assignment TDM versus packet switching. (a) Data source activity profiles. (b) Fixed-assignment time-division multiplexing. (c) Time-division packet switching (concentration).

fixed-assignment TDMA scheme, all of the slots within a frame are preassigned. If the "owner" of a slot has *no* data to send during a particular frame, that slot is wasted. The data stream, shown in Figure 11.10b, illustrates the wasted time slots in this example. When source requirements are unpredictable, as in this example, there can be more efficient schemes, involving the dynamic assignment of the slots rather than a fixed assignment. Such schemes are variously known as packet-switched systems, statistical multiplexers, or concentrators; the effect, shown in Figure 11.10c is to use all the slots in a frame in such a way that capacity is conserved.

11.1.3 Communications Resource Channelization

In Figure 11.3 we considered that the CR is partitioned into spectral bands, and in Figure 11.7 we viewed the same CR as being partitioned into time slots. Figure 11.11 represents a more general organization of the CR allowing for the assignment of a frequency band for a prescribed period of time. Such a multiple access scheme is referred to as *combined FDMA/TDMA*. For the assignments of frequency bands, let us assume an equal apportionment of the total bandwidth W, among M user groups or classes, so that M disjoint frequency bands of width W/M hertz are continuously available to their assigned group. Similarly, for the assignment of time slots, the time axis is partitioned into time frames, each of duration T, and the frames are partitioned into N slot times, each of duration T/N. We assume that the users are time synchronized and that the assigned slots are located periodically within the frames. Each user in each frequency band is permitted to transmit during each periodic appearance of the user's assigned slot, and is permitted to use the assigned channel bandwidth for the slot duration. A slot is uniquely determined as the mth slot within the nth frame. Referring to Figure 11.11, we can describe the time of a particular slot (n, m) with reference to time zero as follows:

$$\text{time of slot } (n, m) = nT + \frac{(m-1)T}{N} \le t \le nT + \frac{mT}{N}$$

$$n = 0, 1, \ldots; m = 1, 2, \ldots, N. \tag{11.4}$$

The nth frame time, T, is denoted by the time interval $[nT, (n+1)T]$. As can be seen in Figure 11.11, the domain of the unit signal is the intersection of the time slot (n, m) and the frequency band (j). Assume that a modulation/coding system is chosen so that the full bandwidth W of the CR can support R bits/s. In any frequency band having a bandwidth of W/M hertz, the associated bit rate will be R/M bits/s. FDMA alone would provide M bands each with a bandwidth of $1/M$ of the full bandwidth of the CR. TDMA alone would provide the full system bandwidth for each of the N slots, where the duration of each slot is $1/N$ of the frame time.

11.1.4 Performance Comparison of FDMA and TDMA

11.1.4.1 Bit Rate Equivalence of FDMA and TDMA

Figure 11.12 highlights the basic differences between an FDMA and TDMA system in a communications resource capable of supporting a total of R bits/s. In Figure 11.12a the system bandwidth is divided into M orthogonal frequency bands.

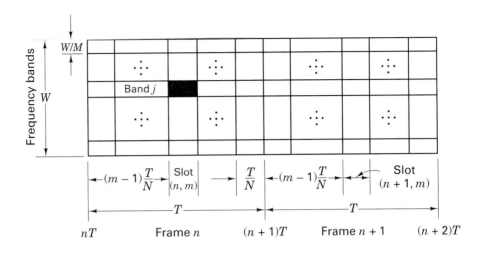

Figure 11.11 Communications resource: time/frequency channelization.

Hence each of the M sources, $\mathcal{S}_m (1 \le m \le M)$ can simultaneously transmit at a bit rate of R/M bits/s. In Figure 11.12b the frame is divided into M orthogonal time slots. Hence each of the M sources bursts its transmission at R bits/s, M times faster than the equivalent FDMA user for $(1/M)$th the time. In both cases, the source $\mathcal{S}_m$ transmits information at an average rate of R/M bits/s.

Let the information generated by each of the sources in Figure 11.12 be organized into b-bit groups, or *packets*. In the case of FDMA, the b-bit packets are transmitted in T seconds over each of the M disjoint channels. Therefore, the total bit rate required is

$$R_{\mathrm{FD}} = M \frac{b}{T} \quad \text{bits/s} \tag{11.5}$$

In the case of TDMA, the b bits are transmitted in T/M seconds from each source. Therefore, the bit rate required is

$$R_{\mathrm{TD}} = \frac{b}{T/M} \quad \text{bits/s} \tag{11.6}$$

Since Equations (11.5) and (11.6) yield identical results, we can conclude that

$$R_{\mathrm{FD}} = R_{\mathrm{TD}} = R = \frac{Mb}{T} \quad \text{bits/s} \tag{11.7}$$

Thus, both systems require the same full CR data rate, R bits/s.

11.1.4.2 Message Delays in FDMA and TDMA

From the previous sections it might appear that the duality between FDMA and TDMA will result in equivalent performance. This is not the case when the metric of performance is the average packet *delay*. It can be shown [1, 2] that

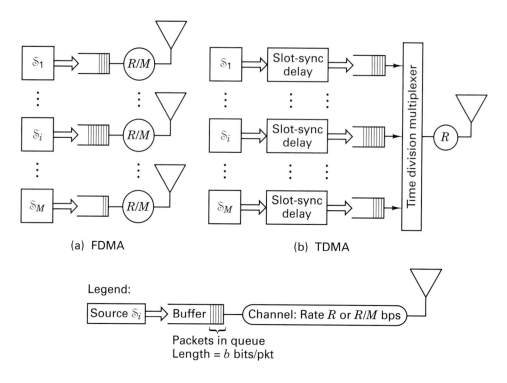

(a) FDMA   (b) TDMA

Legend:

Source $\mathcal{S}_i$ ⟹ Buffer ▥ ── Channel: Rate R or R/M bps

Packets in queue
Length = b bits/pkt

Figure 11.12 (a) FDMA: frequency divided into M orthogonal frequency bands. (b) TDMA: time divided into M orthogonal time slots (one packet per time slot).

TDMA is inherently superior to FDMA in the sense that the average packet delay using TDMA is less than the delay using FDMA.

As before, we assume that in the case of FDMA the system bandwidth is divided into M orthogonal frequency bands, and in the case of TDMA the frame is divided into M orthogonal time slots. For the analysis of message delay, the simplest case is that of deterministic data sources. It is assumed that the CR is 100% utilized, so that all frequency bands in the case of FDMA, and all time slots in the case of TDMA, are filled with data packets. For simplicity, it is also assumed that there are *no* overhead costs such as guard bands or guards times. The message delay can be defined as

$$D = w + \tau \tag{11.8}$$

where w is the average packet waiting time (prior to transmission) and τ is the packet transmission time. In the FDMA case, each packet is sent over a T-second interval, so the packet transmission time for FDMA is simply

$$\tau_{FD} = T \tag{11.9}$$

In the TDMA case, each packet is sent in slots of T/M seconds. We can thus write the TDMA packet transmission time with the use of Equation (11.7) as

$$\tau_{TD} = \frac{T}{M} = \frac{b}{R} \tag{11.10}$$

Since the FDMA channel is continuously available and packets are sent as soon as they are generated, the waiting time, w_{FD}, for FDMA is

$$w_{FD} = 0 \tag{11.11}$$

FDMA and TDMA bit streams are compared in Figure 11.13. For TDMA, Figure 11.13a illustrates that each user's slot begins at a different point in the T-second frame; that is, packet S_{mk} will start at $(m-1)T/M$ seconds ($1 \leq m \leq M$) after the packet generation instant. Therefore, the average waiting time that a TDMA packet sustains before transmission begins is

$$w_{TD} = \frac{1}{M} \sum_{m=1}^{M} (m-1) \frac{T}{M} = \frac{T}{M^2} \sum_{n=0}^{M-1} n = \frac{T}{M^2} \frac{(M-1)(M)}{2} \tag{11.12}$$

$$= \frac{T}{2} \left(1 - \frac{1}{M} \right)$$

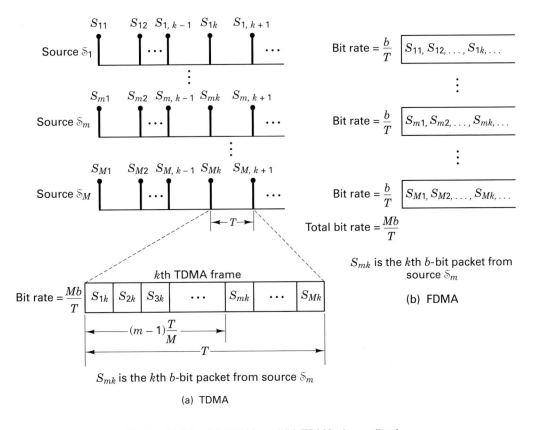

Figure 11.13 (a) TDMA and (b) FDMA channelization.

The maximum waiting time before transmission of a packet is $(M-1)T/M$ seconds, and on the average a packet will wait $\frac{1}{2}(M-1)(T/M) = (T/2)(1-1/M)$ seconds, as given by Equation (11.12).

To compare the average delay times, D_{FD} and D_{TD}, for FDMA and TDMA, respectively, we combine Equations (11.9) and (11.11) into Equation (11.8), and similarly combine Equations (11.10) and (11.12) into Equation (11.8), yielding

$$D_{FD} = T \tag{11.13}$$

$$D_{TD} = \frac{T}{2}\left(1 - \frac{1}{M}\right) + \frac{T}{M} = D_{FD} - \frac{T}{2}\left(1 - \frac{1}{M}\right) \tag{11.14}$$

Using Equation (11.7), Equation (11.14) can be written as

$$D_{TD} = D_{FD} - \frac{b}{2R}(M-1) \tag{11.15}$$

The result indicates that TDMA is inherently superior to FDMA, from a message-delay point of view. Although Equation (11.15) assumed that the data source is deterministic, the smaller average message delays for TDMA schemes hold up for any independent message arrival process [1, 2].

11.1.5 Code-Division Multiple Access

In Figure 11.3 the CR plane was illustrated as being shared by slicing it horizontally to form FDMA frequency bands, and in Figure 11.7 the same CR plane was illustrated as being shared by slicing it vertically to form TDMA time slots. These two techniques are the most common choices for multiple access applications. Figure 11.14 illustrates the CR being partitioned by the use of a hybrid combination of FDMA and TDMA known as *code-division multiple access* (CDMA). CDMA is an application of spread-spectrum (SS) techniques. Spread-spectrum techniques can be classified into two major categories: *direct-sequence* SS and *frequency hopping* SS. We introduce frequence hopping CDMA (FH-CDMA) in this chapter, and we treat direct-sequence CDMA together with the overall subject of spread-spectrum techniques in Chapter 12.

It is easiest to visualize *frequency hopping* CDMA, illustrated in Figure 11.14, as the short-term assignment of a frequency band to various signal sources. At each successive time slot, whose duration is usually brief, the frequency band assignments are reordered. In Figure 11.14, during time slot 1, signal 1 occupies band 1, signal 2 occupies band 2, and the signal 3 occupies band 3. During time slot 2, signal 1 hops to band 3, signal 2 hops to band 1, and signal 3 hops to band 2, and so on. The CR can thus be fully utilized, but the participants, having their frequency bands reassigned at each time slot, appear to be playing "musical chairs." Each user employs a pseudonoise (PN) code, orthogonal (or nearly orthogonal) to all the other user codes, that dictates the frequency hopping band assignments. Details of PN code sequences are treated in Section 12.2. Figure 11.14 is an oversimplified view of the way the CR is shared in frequency hopping CDMA, since the symmetry implies that each frequency hopping signal is in time synchronism with each of the

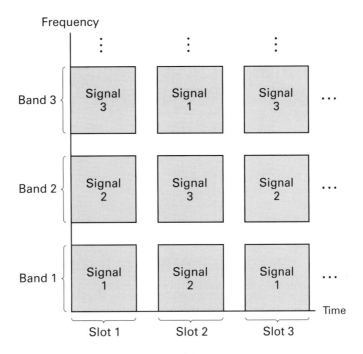

Frequency

Band 3

Band 2

Band 1

Slot 1   Slot 2   Slot 3

Time

Figure 11.14   Code-division multiplexing.

other signals. This is *not the case.* In fact, one of the attractions of CDMA compared to TDMA is that there is no need for synchronization among user groups (only between a transmitter and a receiver within a group).

The block diagram in Figure 11.15 illustrates the frequency hopping modulation process. At each frequency hop time the PN generator feeds a code sequence to a device called a *frequency hopper.* The frequency hopper synthesizes one of the allowable hop frequencies. Assume that the data modulation has an *M*-ary frequency shift keying (MFSK) format. The essential difference between a conventional MFSK system and a frequency hopping (FH) MFSK system is that in the conventional system, a data symbol modulates a carrier wave that is *fixed* in frequency, but in the hopping system, the data symbol modulates a carrier wave that *hops* across the total CR bandwidth. The FH modulation in Figure 11.15 can be thought of as a two-step process—data modulation and frequency hopping modulation—even though it can be implemented in a single step, where the modulator produces a transmission tone based on the simultaneous dictates of the PN code and the data. Frequency hopping systems are covered in detail in Section 12.4.

One might ask: Don't the FDMA and TDMA options provide sufficient multiple access flexibility? FDMA and TDMA methods can surely be relied on to apportion the communications resource equitably. Of what use is this hybrid technique? CDMA offers some unique advantages, as follows:

1. *Privacy.* When the code for a particular user group is only distributed among authorized users, the CDMA process provides communications privacy, since

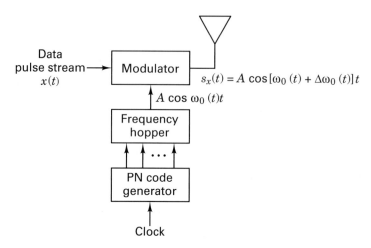

Data pulse stream → $x(t)$

Modulator

$s_x(t) = A \cos[\omega_0(t) + \Delta\omega_0(t)]t$

$A \cos \omega_0(t)t$

Frequency hopper

PN code generator

$\cdots$

Clock

Figure 11.15 CDMA frequency hopping modulation process.

the transmissions cannot easily be intercepted by unauthorized users without the code.

2. *Fading channels.* If a particular portion of the spectrum is characterized by fading, signals in that frequency range are attenuated. In an FDMA scheme, a user who was unfortunate enough to be assigned to the fading position of the spectrum might experience highly degraded communications for as long as the fading persists. However, in a FH-CDMA scheme, only during the time a user hops into the affected portion of the spectrum will the user experience degradation. Therefore, with CDMA, such degradation is shared among all the users.

3. *Jam resistance.* During a given CDMA hop, the signal bandwidth is identical to the bandwidth of conventional MFSK, which is typically equal to the minimum bandwidth necessary to transmit the MFSK symbol. However, over a duration of many time slots, the system will hop over a frequency band which is much wider than the data bandwidth. We refer to this utilization of bandwidth as spread spectrum. In Chapter 12 we develop, in detail, the resistance to jamming that spread spectrum affords a user.

4. *Flexibility.* The most important advantage of CDMA schemes, compared to TDMA, is that there need be no precise time coordination among the various simultaneous transmitters. The orthogonality between user transmissions on different codes is not affected by transmission-time variations. This will become clear upon closer examination of the autocorrelation and cross-correlation properties of the codes, considered in Chapter 12.

11.1.6 Space-Division and Polarization-Division Multiple Access

Figure 11.16a depicts the INTELSAT IVA application of space-division multiple access (SDMA), also called *multiple-beam frequency reuse.* INTELSAT IVA used a dual-beam receive antenna feeding two receivers to allow simultaneous access of

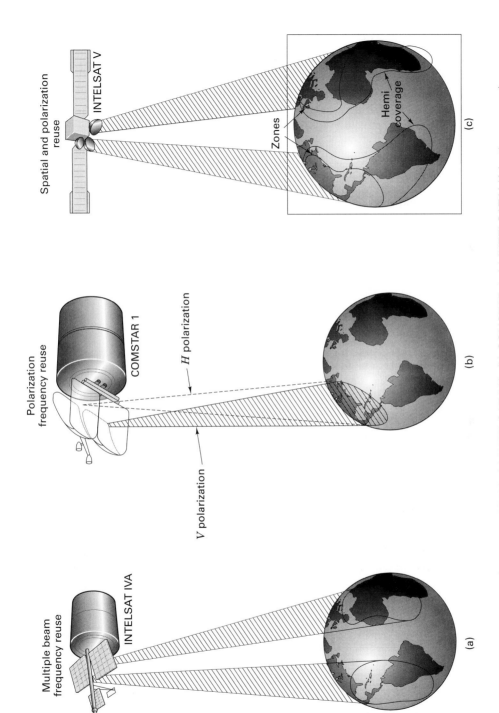

Figure 11.16 SDMA and PDMA. (a) INTELSAT IVA. (b) COMSTAR 1. (c) INTELSAT V (Atlantic coverage).

the satellite from two different regions of the earth. The frequency band allocated to each receive beam was identical because the uplink signals were spatially separated. In such cases, the frequency band is said to be *reused.*

Figure 11.16b depicts an application of polarization-division multiple access (PDMA), also called *dual-polarization frequency reuse,* from COMSTAR 1. Here separate antennas were used, each with different polarization and followed by separate receivers, allowing simultaneous access of the satellite from the same region of the earth. Each corresponding earth station antenna needs to be polarized in the same way as its counterpart in the satellite. (This is generally accomplished by providing each participating earth station with an antenna that has dual polarization.) The frequency band allocated to each antenna beam could be identical because the uplink signals were orthogonal in polarization. As with SDMA, the frequency band in PDMA is said to be reused. Figure 11.16c depicts an application of the simultaneous use of SDMA and PDMA in INTELSAT V. There are two separate hemispheric coverages, west and east. There are also two smaller zone beams; each zone beam overlaps a portion of one of the hemispheric beams and is separated from it by orthogonal polarization. Thus, there is a fourfold reuse of the spectrum.

11.2 MULTIPLE ACCESS COMMUNICATIONS SYSTEM AND ARCHITECTURE

A *multiple access protocol* or *multiple access algorithm* (MAA) is that rule by which a user knows how to use time, frequency, and code functions to communicate through a satellite to other users. A multiple access system is a combination of hardware and software that supports the MAA. The general goal of a multiple access system is to provide communications service in a timely, orderly, and efficient way.

Figure 11.17 illustrates some basic choices for the architecture of a satellite multiple access system. The legend indicates the symbols used for an earth station with and without an MAA controller, and a satellite with and without an MAA controller. Figure 11.17a illustrates the case where one earth station is designated as the master, or the controller. This earth station possesses an MAA computer and responds to the service requests of all other users. Notice that a user's request entails a transmission through the satellite and back down to the controller. The controller's response entails another transmission through the satellite; hence there are two up- and downlink transmissions required for each service assignment. Figure 11.17b illustrates the case where the MAA control is distributed among all the earth stations; there is no single controller. Each earth station uses the same algorithm and they each have identical knowledge regarding access requests and assignments; therefore, only one round trip is required for each service assignment. Figure 11.17c illustrates the case where the MAA controller is in the satellite. A service request goes from user to satellite, and the response from the satellite can follow immediately; therefore, only one round trip is required for each service assignment.

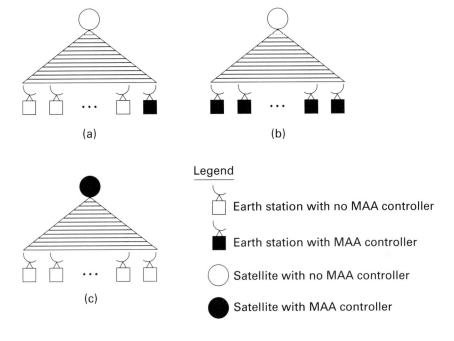

Figure 11.17 Satellite multiple access architecture. (a) Single earth station control. (b) Distributed earth station control. (c) Satellite control.

11.2.1 Multiple Access Information Flow

Figure 11.18 is a flow diagram describing the basic flow of information between the multiple access algorithm (MAA) or controller and an earth station; the numbers below correspond to those on the figure. Recall from the preceding section that the control may be lodged in the satellite, in a master station, or distributed among all the earth stations. The flow proceeds as follows:

1. *Channelization.* This term refers to the most general allocation information [e.g., channels 1 to N may be allocated for the Army and channels $(N+1)$ to M for the Navy]. This information seldom changes, and may be distributed to the earth stations by the use of a newsletter rather than via the communication system.

2. *Network state (NS).* This term refers to the state of the CR. A station is advised regarding the availability of the communications resource and where in the resource (e.g., time, frequency, code position) to transmit its service request(s).

3. *Service request.* Then the station makes its request(s) for service (e.g., allocation for m message slots).

4. Upon receipt of the service request(s), the controller sends the station a schedule regarding where and when to position its data in the CR.

5. The station transmits its data according to its assigned schedule.

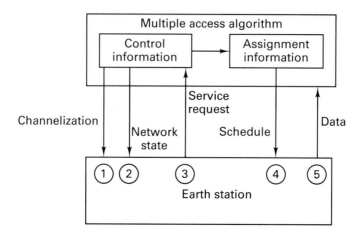

Figure 11.18 Multiple access information flow.

11.2.2 Demand-Assignment Multiple Access

Multiple access schemes are termed *fixed assignment* when a station has periodic access to the channel independent of its actual need. By comparison, dynamic assignment schemes, sometimes called *demand-assignment multiple access* (DAMA), give the station access to the channel only when it requests access. If the traffic from a station tends to be burst-like or intermittent, DAMA procedures can be much more efficient than fixed-assignment procedures. A DAMA scheme capitalizes on the fact that actual demand *rarely* equals the peak demand. If a system's capacity is equal to the total peak demand and if the traffic is bursty, the system will be underutilized most of the time. However, by using buffers and DAMA, a system with reduced average capacity can handle bursty traffic, at the cost of some queueing delay. Figure 11.19 summarizes the difference between a fixed system, whose capacity is equal to the sum of the user requirements, and a dynamic system, whose capacity is equal to the average of the user requirements.

11.3 ACCESS ALGORITHMS

11.3.1 Aloha

In 1971, the University of Hawaii began operation of its ALOHA system. A communication satellite was used to interconnect the several university computers by use of a random access protocol [3–7]. The system concept was extremely simple, consisting of the following modes:

1. *Transmission mode.* Users transmit at any time they desire, encoding their transmissions with an error detection code.

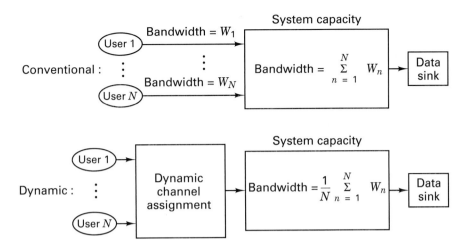

Figure 11.19 Bandwidth reduction for systems using dynamic channel assignment.

2. *Listening mode.* After a message transmission, a user listens for an acknowledgment (ACK) from the receiver. Transmissions from different users will sometimes overlap in time, causing reception errors in the data in each of the contending messages. We say that the messages have *collided*. In such cases, the errors are detected, and the users receive a negative acknowledgment (NAK).

3. *Retransmission mode.* When a NAK is received, the messages are simply retransmitted. Of course, if the colliding users were to retransmit immediately, they would collide again. Therefore, the users retransmit after a *random* delay.

4. *Timeout mode.* If, after a transmission, the user does not receive either an ACK or NAK within a specified time, the user retransmits the message.

11.3.1.1 Message Arrival Statistics

Assume that the total system demand requires an average message or packet arrival rate of λ successful or accepted messages per second. Because of the presence of collisions, some of the messages will be unsuccessful or rejected. Therefore, we define the total traffic arrival rate λ_t, as the acceptance rate λ, plus the rejection rate λ_r, as follows:

$$\lambda_t = \lambda + \lambda_r \tag{11.16}$$

Let us denote the length of each message or packet as b bits. Then we can define the average amount of successful traffic or *throughput*, ρ', on the channel in units of bits per second, as

$$\rho' = b\lambda \tag{11.17}$$

We can also define the *total traffic, G'*, on the channel, in units of bits per second, as

$$G' = b\lambda_t \tag{11.18}$$

With the channel capacity (maximum bit rate) designated as R bits per second, let us further define a *normalized throughput*

$$\rho = \frac{b\lambda}{R} \tag{11.19}$$

and a *normalized total traffic*

$$G = \frac{b\lambda_t}{R} \tag{11.20}$$

Normalized throughput, ρ, expresses throughput as a fraction ($0 \le \rho \le 1$) of channel capacity. Normalized total traffic, G, expresses total traffic as a fraction ($0 \le G \le \infty$) of the channel capacity. Notice that G can take on values greater than unity.

We can also define the transmission time of each packet as

$$\tau = \frac{b}{R} \quad \text{seconds/packet} \tag{11.21}$$

By substituting Equation (11.21) into Equations (11.19) and (11.20), we can write

$$\rho = \lambda\tau \tag{11.22}$$

and

$$G = \lambda_t \tau \tag{11.23}$$

A user can successfully transmit a message as long as no other user began one within the previous τ seconds or starts one within the next τ seconds. If another user began a message within the previous τ seconds, its tail end will collide with the current message. If another user begins a message within the next τ seconds, it will collide with the tail end of the current message. Thus a space of 2τ seconds is needed for each message.

The message arrival statistics for unrelated users of a communication system is often modeled as a Poisson process. The probability of having K new messages arrive during a time interval of τ seconds is given by the Poisson distribution [8] as

$$P(K) = \frac{(\lambda\tau)^K e^{-\lambda\tau}}{K!} \quad K \ge 0 \tag{11.24}$$

where λ is the average message arrival rate. Because the users transmit without regard for each other in the ALOHA system, this expression is useful for calculating the probability that exactly $K = 0$ other messages are transmitted during a time interval 2τ. This is the probability, P_s, that a user's message transmission was successful (experienced no collisions). To compute P_s, assuming that all traffic is Poisson, we use λ_t and 2τ in Equation 11.24. Thus,

$$P_s = P(K = 0) = \frac{(2\tau\lambda_t)^0 e^{-2\tau\lambda_t}}{0!} = e^{-2\tau\lambda_t} \qquad (11.25)$$

In Equation (11.16) we defined the total traffic arrival rate λ_t in terms of the successful portion λ and the repetition or unsuccessful portion λ_r; then, by definition, the probability of a successful packet can be expressed as

$$P_s = \frac{\lambda}{\lambda_t} \qquad (11.26)$$

By combining Equations (11.25) and (11.26), we have

$$\lambda = \lambda_t e^{-2\tau\lambda_t} \qquad (11.27)$$

By combining Equation (11.27) with Equations (11.22) and (11.23), we can write

$$\rho = G e^{-2G} \qquad (11.28)$$

Equation (11.28) relates the normalized throughput, ρ, to the normalized total traffic, G, on the channel for the ALOHA system. A plot of this relationship labeled "pure ALOHA" is shown in Figure 11.20; as G increases, ρ increases until a point is reached where further traffic increases create a large enough collision rate to cause

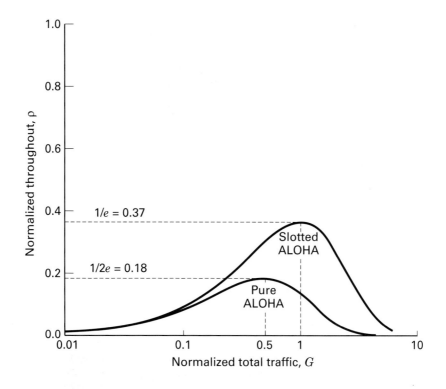

Figure 11.20 Throughput in ALOHA channels (successful transmissions versus total transmissions).

11.3   Access Algorithms

a reduction in the throughput. The maximum ρ, equal to $1/2e = 0.18$, occurs at a value of $G = 0.5$. Therefore, for a pure ALOHA channel, only 18% of the CR can be utilized. Simplicity of control is achieved at the expense of channel capacity [7, 9].

11.3.2 Slotted ALOHA

The pure ALOHA scheme can be improved by requiring a small amount of coordination among the stations. The slotted ALOHA (S-ALOHA) is such a system. A sequence of synchronization pulses is broadcast to all stations. As with pure ALOHA, packet lengths are constant. Messages are required to be sent in the slot time between synchronization pulses, and can be started only at the *beginning* of a time slot. This simple change reduces the rate of collisions by half, since only messages transmitted in the same slot can interfere with one another. It can be shown [9, 10] that for S-ALOHA, the reduction in the *collision window* from 2τ to τ results in the following relationship between normalized throughput ρ and normalized total traffic G:

$$\rho = Ge^{-G} \tag{11.29}$$

The plot of Equation (11.29) is shown in Figure 11.20 labeled "slotted ALOHA." Here the maximum value of ρ is $1/e = 0.37$, or an improvement of two times the pure ALOHA protocol.

The retransmission mode described for the pure ALOHA system was modified for S-ALOHA so that if a negative acknowledgment (NAK) occurs, the user retransmits after a *random* delay of an integer number of slot times. Figure 11.21 illustrates the S-ALOHA operation. A packet of data bits is shown transmitted by user k followed by the satellite acknowledgment (ACK). Also shown are users m and n simultaneously transmitting packets, which results in a collision; a NAK is returned. Each using station employs a random-number generator to select its retransmission time. The figure illustrates an example of the m and n retrans-

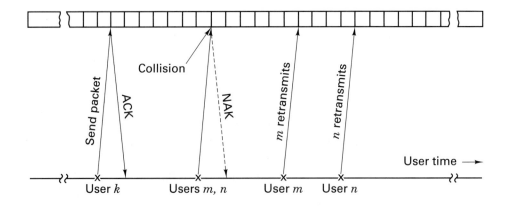

Figure 11.21 Random access scheme: slotted ALOHA operation.

mission at their respective randomly selected times. Of course, there is some probability that users m and n will recollide. However, in that case, they simply repeat the retransmission, using another random delay.

Example 11.1   Poisson Process

Assuming that packet transmissions and retransmission can both be described as a Poisson process, calculate the *probability* that a data packet transmission in an S-ALOHA system will experience a collision with *one other user*. Assume that the total traffic rate $\lambda_t = 10$ packets/s and the packet duration $\tau = 10$ ms.

Solution

$$P(K=1) = \left.\frac{(\tau\lambda_t)^K e^{-\tau\lambda_t}}{K!}\right|_{K=1}$$

$$= (10 \times 0.01)^1 e^{-0.1} = 0.1e^{-0.1}$$

$$= 0.09$$

11.3.3 Reservation-ALOHA

A significant improvement was made to the ALOHA system with the introduction of the reservation-ALOHA (R-ALOHA) [11] scheme. The R-ALOHA system has two basic modes: an unreserved mode and a reserved mode; each is described as follows:

Unreserved Mode (Quiescent State)

1. A time frame is established and divided into a number of small reservation subslots.
2. Users use these small subslots to reserve message slots.
3. After requesting a reservation, the user listens for an acknowledgment and a slot assignment.

Reserved Mode

1. The time frame is divided into $M + 1$ slots whenever a reservation is made.
2. The first M slots are used for message transmissions.
3. The last slot is subdivided into subslots to be used for reservation/requests.
4. Users send message packets only in their assigned portions of the M slots.

Consider the R-ALOHA example shown in Figure 11.22. In the quiescent state, with no reservations, time is partitioned into short subslots for making reservations. Once a reservation is made, the system is configured so that $M = 5$ message slots followed by $V = 6$ reservations subslots becomes the timing format. The figure illustrates a request and an acknowledgment in progress. In this example the station seeks to reserve three message slots. The reservation acknowledgment advises the using station where to locate its first data packet. Since the control is distributed so that all participants receive the downlink transmissions and are thus aware

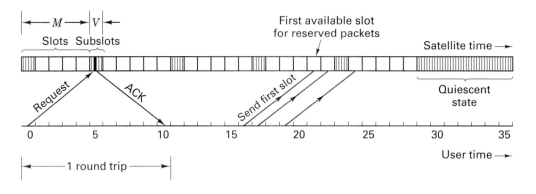

Figure 11.22 Example of reservation ALOHA. Station seeks to reserve three slots ($M = 5$ slots, $V = 6$ subslots).

of the reservations and time format, the acknowledgment need not disclose any more than the location of the first slot. As shown in Figure 11.22, the station sends its second packet in the slot following the first packet. The user further knows that the next slot is comprised of six subslots for reservations, so *no* packets are transmitted during this time. The third and final packet is sent in the following slot. When there are no reservations taking place, the system reverts back to its quiescent format of subslots only. Since the control is distributed, all the participants are made aware of the quiescent format by receiving appropriate synchronizing pulses on the downlink. Other interesting reservation schemes are discussed in Reference [12, 13].

11.3.4 Performance Comparison of S-ALOHA and R-ALOHA

From Chapters 3 and 4 the basic quality measure of a digital modulation scheme is its P_B versus E_b/N_0 curve. This measure is particularly useful because E_b/N_0 is a *normalized signal-to-noise ratio*; being normalized, the curves allow us to compare the performance of various modulation schemes. There is a similar performance measure for multiple access schemes. Here we are interested in the average delay versus normalized throughput. What would an *ideal delay–throughput curve* look like? Figure 11.23 illustrates such a curve. For normalized throughput values of, $0 \leq \rho < 1$, the delay equals zero until $\rho = 1$; then the delay increases without bound. Figure 11.23 also shows a *typical* delay–throughput curve and the direction in which the curve will move as delay performance improves.

Figure 11.24 compares the delay–throughput performance of S-ALOHA with that of R-ALOHA (formatted with two message slots and six reservation subslots). Knowing the location of the *ideal* curve it is easy to compare the delay performance of these two systems. For a throughput of less than approximately 0.20, the S-ALOHA manifests less average delay than does R-ALOHA. But for values of ρ between 0.20 and 0.67, it is apparent that R-ALOHA is superior, since the average delay is less. Why does the S-ALOHA perform better at low traffic intensity? The

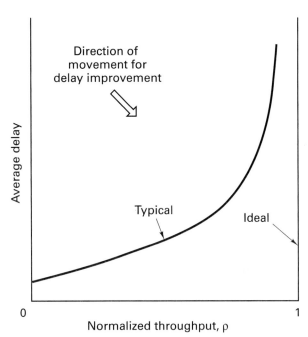

Direction of
movement for
delay improvement

Average delay

Typical

Ideal

Figure 11.23 Delay–throughput characteristic.

0

Normalized throughput, ρ

1

S-ALOHA algorithm does not require the overhead of the reservation subslots as does R-ALOHA. Therefore, at low values of ρ, R-ALOHA pays the price of greater delay due to the greater overhead. For $\rho > 0.2$, the collisions and retransmissions inherent in the S-ALOHA system cause it to incur greater delay (unbounded at $\rho = 0.37$), more quickly than the R-ALOHA system. At higher throughput ($0.2 < \rho < 0.67$), the overhead structure of R-ALOHA ensures that its delay degradation grows in a more orderly manner than S-ALOHA. For R-ALOHA, an unbounded delay is not reached until $\rho = 0.67$.

Example 11.2   Channel Utilization

(a) Normalized throughput, ρ, is a measure of channel utilization. It can be found by forming the ratio of the successfully transmitted message traffic, in bits per second to the total message traffic, including rejected messages, in bits per second. Calculate the normalized throughput of a channel that has a maximum data rate $R = 50$ kbits/s and operates with $M = 10$ ground stations, each station transmitting at the average rate of $\lambda = 2$ packets/second. The system format provides for $b = 1350$ bits/packet.

(b) Which of the three ALOHA schemes discussed—pure, slotted, and reservation— could be successfully used with this channel?

Solution

(a) Generalizing Equation (11.19) to allow for traffic from multiple stations, we have

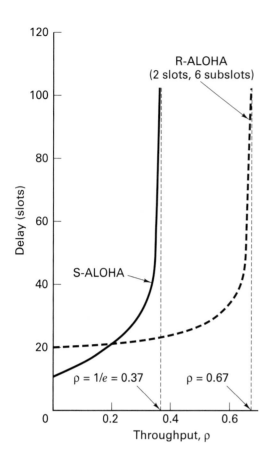

Figure 11.24 Delay–throughput comparison: S-ALOHA versus R-ALOHA on a satellite channel.

$$\rho = \frac{Mb\lambda}{R}$$

$$= \frac{10(1350)\,(2)}{50,000}$$

$$= 0.54$$

(b) Only the R-ALOHA scheme could be used for this system, since with each of the other schemes, 54% of the resource cannot be utilized.

11.3.5 Polling Techniques

One way to impose order on a system with multiple users having random access requirements is to institute a controller that periodically polls the user population to determine their service requests. If the user population is large (e.g., thousands of terminals) and the traffic is bursty, the time required to poll the population can be an excessive overhead burden. One technique for rapidly polling a user population [4, 14] is called a *binary tree search*. Figure 11.25 illustrates a satellite example of such a tree search to resolve contention among users. In this example, assume that

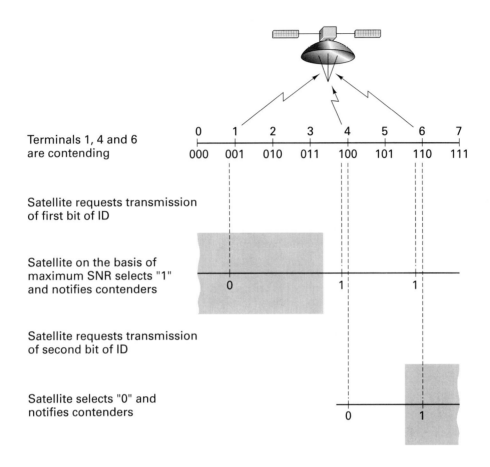

Terminals 1, 4 and 6 are contending

Satellite requests transmission of first bit of ID

Satellite on the basis of maximum SNR selects "1" and notifies contenders

Satellite requests transmission of second bit of ID

Satellite selects "0" and notifies contenders

Figure 11.25 Tree search to resolve contention (eight terminal example).

the total user population is eight terminals; let them be identified by the binary numbers 000 to 111 as shown in Figure 11.25. Assume that terminals 001, 100, and 110 are contending for the service of a single channel. The tree search operates by continually partitioning the population until there is just a single branch remaining. The terminal corresponding to that branch is the "winner" and hence the first terminal to access the channel. The operation is repeated and again yields a single terminal that may next use the channel. The algorithm proceeds according to the following steps (see Figure 11.25):

1. The satellite requests the transmission of the contending terminals' first (left-most) bit of their identification (ID) numbers.

2. Terminal 001 transmits a zero, and terminals 100 and 110 each transmit a one. The satellite, on the basis of received signal strength, selects one or zero as the bit it "heard." In this example the satellite chooses binary one and in-

forms the users accordingly. Half the user population now knows that it has not been selected. The terminals in the "losing" half "bow out" of contention during this pass through the tree. In this example terminal 001 bows out.

3. The satellite requests the transmission of the second identifying bit from the remaining contending terminals.

4. Terminal 100 transmits a zero, and terminal 110 transmits a one.

5. Assume that the satellite selects the zero and notifies the contenders accordingly. Terminal 110 bows out. The process continues until it is clear that terminal 100 is free to access the satellite.

6. When the channel becomes available, steps 1–5 are repeated.

Example 11.3 Comparison between Binary Tree Search and Straight Polling

(a) A binary tree search requires $n = \log_2 Q$ decisions for each pass through a population of Q terminals. A savings in time is possible with a tree search if the population is large and the average demand for service is small. Calculate the time needed for the straight polling of a population of 4096 terminals, to provide channel availability to 100 terminals requesting service. Compare the result with the time needed to perform a binary tree search 100 times, over the same population. Assume that the time required to poll one terminal and the time required for one decision of a binary tree search are each equal to 1 s.

(b) Develop an expression for Q', the largest number of terminals that results in the same (or less) time expended for binary tree searching as compared to straight polling.

(c) Compute Q' for part (a).

Solution

(a) Straight polling of 4096 terminals:

$$T = 4096 \times 1\ \text{s} = 4096\ \text{s}$$

Binary tree search for 100 terminals requires 100 passes through the binary tree:

$$T' = (100 \times \log_2 4096) \times 1\ \text{s} = 1200\ \text{s}$$

(b) Q' is the maximum number of terminals that will result in $T' \leq T$ in part (a). This will occur when

$$Q'' \log_2 Q \times 1\ \text{s/decision} = Q \times 1\ \text{s/poll}$$

$$Q' = \lfloor Q'' \rfloor = \left\lfloor \frac{Q}{\log_2 Q} \right\rfloor \qquad (11.30)$$

where $\lfloor x \rfloor$ is the largest integer no greater than x.

(c) Q' for part (a)

$$Q' = \left\lfloor \frac{4096}{\log_2 4096} \right\rfloor = 341\ \text{terminals}$$

A binary tree search for 341 terminals entails a search time of 4092 s.

11.4 MULTIPLE ACCESS TECHNIQUES EMPLOYED WITH INTELSAT

The first commercial, geostationary communication satellite (INTELSAT I, or Early Bird), launched in 1965, represented the start of a new telecommunications era. Its 240 voice circuits provided more capacity than the undersea cables laid between the United States and Europe during the previous 10 years [15].

Early Bird featured a hard-limiting nonlinear transponder using FDMA. When several signals having different carrier frequencies simultaneously occupy a nonlinear device, the result is the production of intermodulation products which are signals at all combinations of sum and difference frequencies [16–18]. The energy apportioned to these intermodulation or IM products represents a *loss* in the useful signal energy. In addition, if these IM products appear within the bandwidth of other signals, the effect is that of added *noise* for the other signals.

The nonlinear transponder in Early Bird allowed for only two earth stations (one in the United States and one in Europe) to simultaneously access the satellite. Figure 11.26 illustrates this satellite's operation between the United States and Europe. Three European earth stations were interconnected via a terrestrial network.

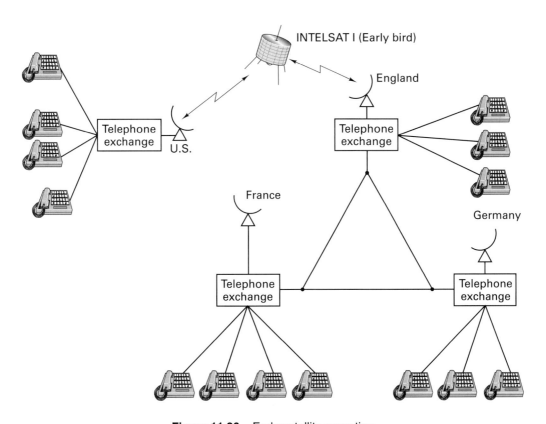

Figure 11.26 Early satellite operation.

Each month a different European station accessed the satellite and distributed the traffic to the other two stations.

11.4.1 Preassigned FDM/FM/FDMA or MCPC Operation

INTELSAT II and III improved multiple access capability by operating their travelling-wave tube amplifiers (TWTA) in the linear region instead of the hard-limiting region. This kept the IM products at an acceptable level, allowing more than two simultaneous accesses. (The price paid was a reduction in power amplifier efficiency.) Thus many FM carriers from various earth stations could simultaneously access these satellites. The operation is designated preassigned multidestination FDM/FM/FDMA or simply FDM/FM, or multichannel per carrier (MCPC), and is illustrated in Figure 11.27. Long-distance calls originating in country A enter the telephone exchange and are multiplexed into a supergroup (five groups of 12 voice circuits each). Country A transmits the supergroup on a single FM carrier at frequency f_A. Each group within the supergroup has been preassigned to an earth station in country A for telephone traffic destined to countries B through F. These countries each receive the signal on frequency f_A. The received signal is demodulated and demultiplexed at the destination country, selecting only those 12 channels preassigned to it.

11.4.2 MCPC Modes of Accessing an INTELSAT Satellite

INTELSAT has standardized the ways in which each 36-MHz transponder may be shared by specifying the occupied RF bandwidth and the number of 4-kHz channels per user. Some of these standard channels are shown in Table 11.1. Notice that the capacity of the transponder (last column in Table 11.1) drops as the number of carriers increases. The reasons are as follows:

1. Guard bands are needed between carrier bands; the more carriers there are, the more guard bands are needed. Hence capacity is reduced.
2. Multiple carriers in the nonlinear TWTA cause intermodulation (IM) products. If the TWTA is backed off into the linear region to reduce interference, the TWTA can provide less overall power. The channel becomes power limited and can service fewer carriers.

Table 11.1 indicates that a single carrier provides the most efficient use of the transponder. Why doesn't INTELSAT always operate its transponders in this mode? The answer is that not all earth stations have enough traffic to justify the assignment of an entire 36-MHz transponder. The other modes are needed so that various combinations of stations having less traffic will be able to share a transponder.

11.4.2.1 Bandwidth-Limited versus Power-Limited Conditions

In the preceding section it was stated that the backed-off transponder cannot support as many channels as the fully saturated transponder. It is useful to examine the two extreme transponder conditions, bandwidth limited and power limited, in

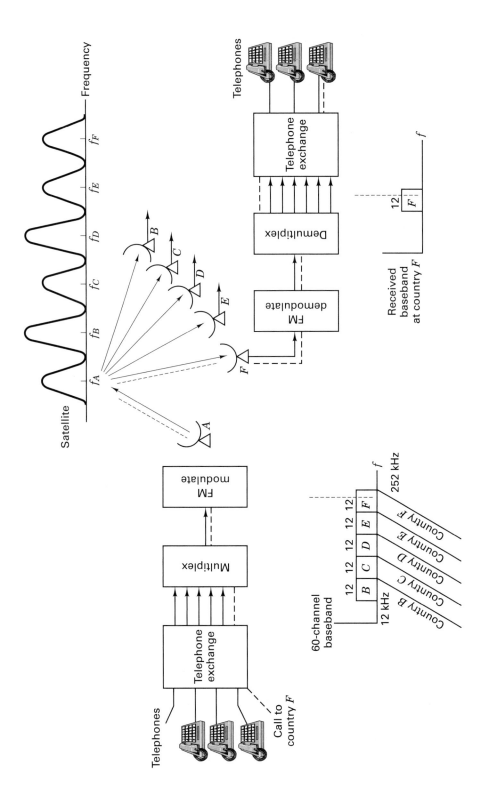

Figure 11.27 Preassigned multidestination FDM/FM carriers. (Reprinted with permission from J. G. Puente and A. M. Werth, "Demand-Assigned Service for the INTELSAT Global Network," *IEEE Spectrum*, Jan. 1971. © 1971 IEEE.)

TABLE 11.1   Standard INTELSAT MCPC Accessing Modes

Number of Carriers per Transponder	Carrier Bandwidth	Number of 4-kHz Channels per Carrier	Number of 4-kHz Channels per Transponder
1	36 MHz	900	900
4	3 at 10 MHz	132	456
	1 at 5 MHz	60	
7	5 MHz	60	420
14	2.5 MHz	24	336

the context of a satellite transponder. In Figure 11.28 we assume a 36-MHz transponder with a maximum power output of 20 W. Figure 11.28a illustrates an MCPC mode of operation whereby four carriers share the 36-MHz bandwidth. Assume that each carrier requires 4 W. The total output power is 16 W (less than the maximum capability of the amplifier); therefore, there is still power to spare. However, should another user want to access the transponder, the total 36-MHz bandwidth has already been allocated to the existing four carriers; there is no additional bandwidth to spare. Figure 11.28a illustrates this bandwidth-limited case.

Suppose that the previous example results in the production of serious IM products at the transponder. Assume that it is necessary to linearize the transponder by operating it at a reduced maximum power output of 12 W. With only a 12-W capability, the transponder can no longer support four users with 4 W each. One of the users must be "thrown off," as illustrated in Figure 11.28b. Therefore, we have bandwidth to support another user, but not sufficient power. Figure 11.28b illustrates this power-limited case.

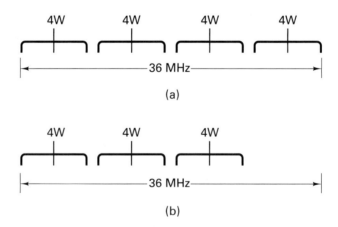

Figure 11.28   Bandwidth-limited versus power-limited configurations. (a) Bandwidth-limited example. (b) Power-limited example.

11.4.3 SPADE Operation

Thc preassigned MCPC multiple access scheme is very efficient when the traffic is heavy enough so that the channels are most always filled. However, if out of a 12-channel group, only one channel is active, the other 11 cannot be turned off. The FDM/FM transmission is made with or without actual telephone traffic on the channels. Therefore, the long-term preassignment of carriers to stations having light traffic is wasteful. Since there are many light traffic links, a flexible method to service them was needed. Also, an efficient way to handle overflow traffic from medium-capacity preassigned links was needed. Such was the motivation for a novel DAMA scheme known as SPADE, first used with INTELSAT IV. The acronym SPADE stands for "single-channel-per-carrier PCM multiple access demand assignment equipment." The principal features characterizing SPADE operation [15] are the following:

1. A single voice-grade channel is analog-to-digital (A/D) converted at a bit rate of 64 kbits/s.
2. This baseband digital signal modulates a carrier using quadrature phase shift keying (QPSK). Unlike the MCPC case, there is *only one* voice channel per carrier.
3. The channel spacing is 45 kHz. Within a transponder, there is bandwidth available for 800 channel carriers. Six carrier positions are vacant by design; thus there are 794 usable carriers.
4. The carrier is dynamically assigned, *upon demand.*
5. The dynamic assignment is accomplished over a 160-kHz common signaling channel (CSC) used as an "order-wire" or control circuit. The bit rate on the CSC is 128 kbits/s, and the modulation is binary phase shift keying (BPSK).

Figure 11.29 illustrates the frequency allocations for the CSC and the 800 carriers in the SPADE system. The SPADE operation can best be understood with the aid of Figure 11.30. The CSC operates in a fixed-assignment TDMA broadcast mode; that is, all earth stations monitor the CSC and are aware of the current state of channel assignments. Each earth station has a 1-ms time slot on the CSC (once every 50 ms) for requesting or releasing a channel. When an earth station needs a channel, it "seizes" a free one by requesting a frequency pair at random and transmitting its selection on the CSC. Random selection makes it unlikely that two stations will simultaneously request the same channel unless there are very few remaining. As soon as the channel is allocated, each of the other earth station processors deletes it from its list of available channels. The list is continually kept updated via the CSC. Thus, control of the SPADE access scheme is *distributed* among all the participating earth stations.

When the station finishes with the channel, the station indicates the channel's release by transmitting a signal in its time slot on the CSC. Each station receives this signal and designates the released channel as available. If two stations simultaneously seize the same channel, they each get a "busy" indication. They try again, selecting at random from the pool of available channels.

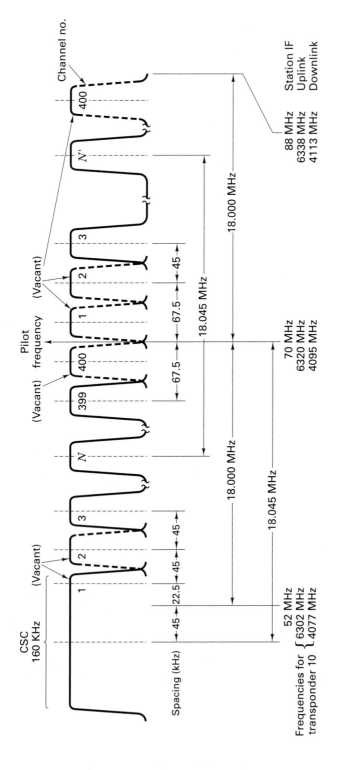

Figure 11.29 SPADE frequency allocations. (Reprinted with permission from J. G. Puente and A. M. Werth, "Demand-Assigned Services for the INTELSAT Global Network," *IEEE Spectrum*, Jan. 1971. © IEEE.)

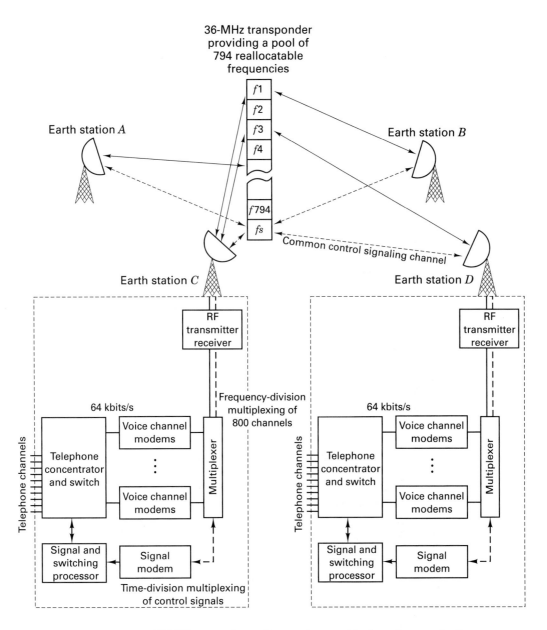

Figure 11.30 SPADE operation. (From James Martin, *Communications Satellite Systems,* © 1978, Fig. 15.2, p. 236. Reprinted by permission of Prentice-Hall, Englewood Cliffs, N.J.)

11.4.3.1 Transponder Capacity Utilization with SPADE

Table 11.2 is a continuation of Table 11.1. We see that the transponder bandwidth utilization with SPADE results in a total capacity of 800 voice channels per transponder. Compare Table 11.2 with Table 11.1. In Table 1.1, as the number of carriers increases from 1 to 14, the total number of channels decreases from 900 to 336. Why doesn't the SPADE system in Table 11.2 exhibit less capacity than the 336 channels associated with 14 carriers? The improved utilization comes about as follows. When there is only one voice channel per carrier, the carrier can be switched off when no speech is detected. Even with all channels operating, they can be switched off approximately 60% of the time. The transponders are power limited; power savings means that more channels can be transmitted. Also, SPADE uses digital voice transmission (QPSK); the bandwidth efficiency of the system is commensurate with the single-carrier FDM/FM case.

11.4.3.2 SPADE Efficiency

With MCPC, capacity is preassigned; a station's unused channels cannot be reallocated to other stations. SPADE is a DAMA system where all channels are shared. The channels are allocated to users as needed. An important telephone system quality measure, called the probability of blocking, is the probability that a requested circuit is not available. To achieve 1% probability of blocking requires four times as many MCPC channels as SPADE channels. A SPADE transponder with 800 channels is equivalent to 3200 MCPC channels [15].

11.4.3.3 Mixed-Size Earth Station Network Using SPADE

A standard-size INTELSAT earth station has a receiver sensitivity $G/T° = 40.7$ dB/K, whereas the smaller size stations have a $G/T° = 35$ dB/K. If 125 SPADE channels are destined for small stations, the total transponder capacity of 800 standard channels is reduced to 525 channels. This is the point at which half the available power is used to service the standard stations. The relationship between transponder capacity and channels allocated to small stations is shown in Figure 11.31. An explanation of this relationship can best be seen in Figure 11.32. When the total TWTA power provides service to large stations, Figure 11.32a illustrates that the 36-MHz bandwidth transponder is occupied by approximately 800 carriers each at a power level of x dBW (the bandwidth-limited case). When half the power is required to service small stations, Figure 11.32b illustrates that 400 carriers (half of the original 800) each at a power level of x dBW are reserved for the standard stations. Consider what happens to the remaining 400 carrier positions. From Chapter 5 we know that the error performance of a link is directly related to the

TABLE 11.2 SPADE Accessing

Number of Carriers per Transponder	Carrier Bandwidth	Number of 4-kHz Channels per Carrier	Numbers of 4-kHz Channels per Transponder
800	45 kHz	1	800

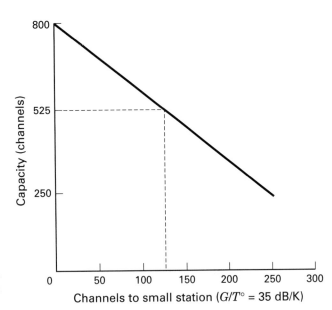

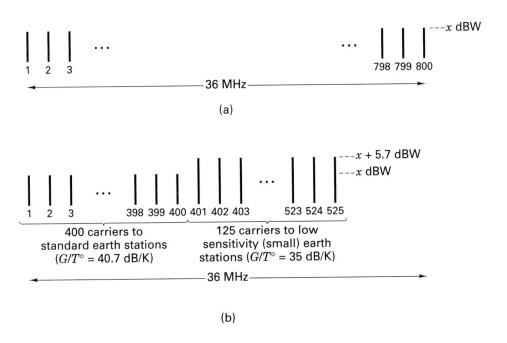

Figure 11.31 SPADE transponder capacity in a mixed-size earth station network.

Figure 11.32 Mixed-size earth station network. (a) When the total TWTA power services large stations: bandwidth limited (800 carriers). (b) When half the TWTA power services small stations: power limited (525 carriers).

product of EIRP and $G/T°$. For any link, one can trade off these two parameters, thereby maintaining a fixed level of performance. Since the small station has a $G/T°$ of 5.7 dB less than that of the standard station, it is necessary to supply the small station with 5.7 dB more EIRP for equivalent performance. The carrier power is *increased* by approximately 5.7 dB for each small station, thus the quantity of the remaining carriers serving these small stations is *decreased* by a similar amount. Therefore instead of 400 carriers, 125 (a reduction of 5.1 dB) are used to serve the small stations; the transponder is now power limited.

At the time a channel is assigned to a call, the transmitting station is apprised of the size of the destination station. Recall that these satellites are nonregenerative so that the apportionment of downlink EIRP is established by the transmitting station (see Section 5.7.1). The transmitting station sets its power level according to the needs of the receiving station.

11.4.4 TDMA in INTELSAT

The first generation of multiple access communication systems has been dominated by FDMA systems. The trend, however, is now in favor of TDMA systems, made possible by the availability of precise clocks and high-seed switching elements [19–24]. INTELSAT IV used a 128-kbits/s TDMA scheme for the common signaling channel that controls the SPADE network. Intelsat V introduced a 120-Mbits/s TDMA scheme for multiple-beam international digital service. One disadvantage or cost in implementing a TDMA scheme is the need for providing precise *synchronization* among the participating earth stations and the satellite. FDMA systems, not having such requirements, are less complex from a networking point of view. Comparisons of TDMA versus FDMA operation are summarized as follows:

1. FDMA can cause IM products. This can be avoided by operating the TWTA in its linear region, thereby reducing the available power output.
2. With TDMA, there is only one carrier present at a time in the TWTA. Thus, IM distortion cannot occur.
3. TDMA earth station equipment is more sophisticated and hence more costly than FDMA equipment. However, for earth stations providing multiple point-to-point channels, FDMA stations require separate radio-frequency (RF) up-conversion and down-conversion signal processing stages. Thus, with FDMA, the amount of equipment grows with the amount of simultaneous connectivity. With TDMA, such growth does not take place since channel selectivity is accomplished in time rather than frequency. Therefore, for a large multiply connected earth station, TDMA can be more cost-effective than FDMA.
4. In multiple-beam systems, each beam may need to communicate with every other beam. TDMA lends itself to conveniently forming connections sequentially as in satellite-switched TDMA (SS/TDMA). INTELSAT VI uses such satellite-switched TDMA (SS/TDMA), described in Section 11.4.5.

An example of the comparative performance of TDMA, FDM/FM, and SPADE is shown for an INTELSAT IV transponder as a plot of channel capacity versus earth station $G/T°$ in Figure 11.33. Figure 11.33a is for an earth coverage antenna, and Figure 11.33b is for a spot-beam antenna. From synchronous altitudes these antennas have half-power beamwidths of 17° and 4.5°, respectively. From these plots it is seen that single-carrier FDM/FM is as efficient as TDMA when the system is operated with standard earth stations ($G/T° = 40.7$ dB/K). For smaller earth stations ($G/T° \leq 31$ dB/K) working through earth-coverage transponders, SPADE is more efficient than TDMA and multicarrier FDM/FM (MCPC); only the four-carrier case is plotted. For earth stations having $G/T°$ in the range 19 to 40.7 dB/K working through a spot-beam transponder, TDMA is superior to SPADE and MCPC. For smaller earth stations having $G/T°$ in the range 6 to 19 dB/K working through a spot-beam transponder, SPADE is superior to TDMA and MCPC. In general, when working through *standard* earth stations it is seen [19] that TDMA was the most efficient multiple access scheme for INTELSAT IV.

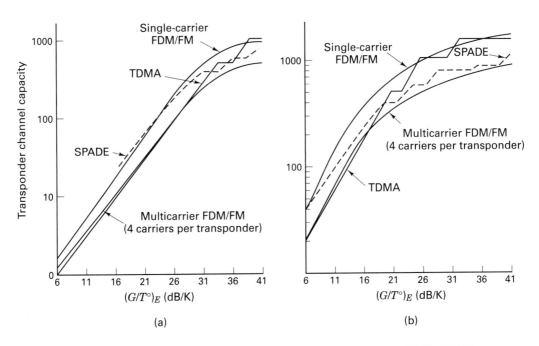

Figure 11.33 Channel capacity versus earth station $G/T°$ for FDMA, TDMA, and SPADE. (a) Global-beam transponder channel capacity as function of $(G/T°)_E$, where $(G/T°)_E$ means earth station $G/T°$. (b) Spot-beam transponder channel capacity as function of $(G/T°)_E$. [From D. Chakraborty, "INTELSAT IV Satellite System (Voice) Channel Capacity versus Earth Station Performance," *IEEE Trans. Commun. Tech.*, vol. COM19, no. 3, June 1971, pp. 355–362. © 1971 IEEE.]

11.4.4.1 PCM Multiplex Frame Structures

There are two digital telephony standards for PCM frame structures in operation. The North American standard is called *T-Carrier*; it is built around the 193-bit frame shown in Figure 11.34a. There are 24 channels; each channel contains an 8-bit voice sample. Also, there is one bit per frame with alternating value $1 \, 0 \, 1 \, 0 \ldots$ from frame to frame, used for frame alignment. Since a voice-grade telephone channel has a bandwidth of $W = 4$ kHz (including guard bands), the Nyquist sampling rate for recovering the analog information within 4 kHz is $f_s = 2W = 8000$ samples/s. Therefore, the basic PCM frame, called the *Nyquist frame*, which contains 24 voice samples from 24 different message sources, has a frame rate of 8000 frames/s (duration of 125 μs). Thus the basic T-Carrier bit rate is 193 bits/frame × 8000 frames/s = 1.544 Mbits/s.

The European standard is built around a 256-bit frame shown in Figure 11.34b. There are 30 message channels, each containing an 8-bit voice sample. Also, one 8-bit time slot is used for frame alignment and another 8-bit time slot is used for signaling (addressing) information. The European frame rate is the same

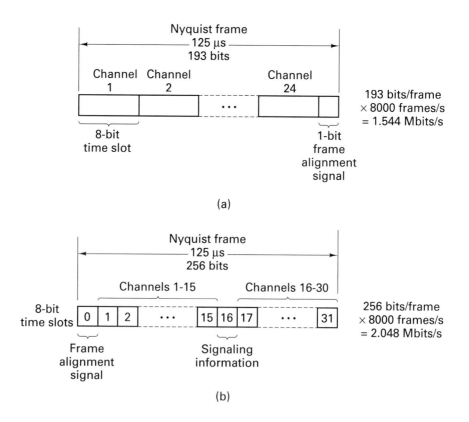

(a)

(b)

Figure 11.34 PCM multiplex frame structure. (a) Frame structure for T-Carrier (North American) PCM multiplex. (b) Frame structure for the European PCM multiplex.

as that of the T-Carrier. Therefore, the basic European bit rate is 256 bits/frame × 8000 frames/s = 2.048 Mbits/s.

11.4.4.2 The High-Rate TDMA Frame for Europe

Sixteen Nyquist frames of the European PCM Multiplex format are shown in Figure 11.35a. Each frame contains an 8-bit sample from each of 30 terrestrial channels, plus 8 bits of framing and 8 bits of signaling information. The TDMA frame duration is

$$16 \text{ Nyquist frames} \times 125 \text{ } \mu s/\text{Nyquist frame} = 2 \text{ ms}$$

Within this 2-ms frame are maintained

$$16 \text{ Nyquist frames} \times 256 \text{ bits/Nyquist frame} = 4096 \text{ bits}$$

The basic idea behind TDMA is that a user's low-rate data stream can share the CR with similar streams from other users by *bursting* the transmission at a

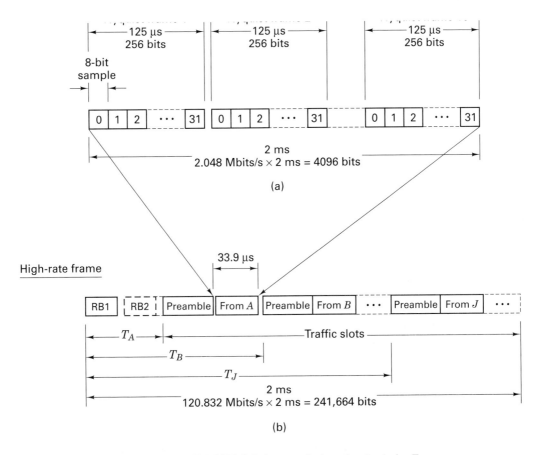

(a)

(b)

Figure 11.35 INTELSAT digital transmission standards for Europe. (a) Terrestrial PCM multiplex. (b) High-rate frame.

much faster rate than the rate at which it is generated. Figure 11.35b illustrates a 2-ms high-rate TDMA frame. The frame begins with a reference burst, RBI, emitted by a reference station. The burst contains information necessary to enable other stations to precisely position their message traffic bursts in the frame. There may be a second burst, RB2, for reliability, followed by a sequence of traffic slots. The traffic slots may be preassigned, or they may be assigned according to a DAMA protocol [20].

The PCM multiplex signal with a bit rate of $R_0 = 2.048$ Mbits/s and a frame duration of $T = 2$ ms is compressed (by a factor of 59) and transmitted using QPSK modulation at a burst rate of $R_T = 120.832$ Mbits/s (symbol rate of 60.416 megasymbols/s). The duration of the traffic data field T_{tr} in the high-rate TDMA frame is calculated as follows:

$$T_{tr} = \frac{R_0 T}{R_T} \tag{11.31}$$

$$= \frac{2.048 \times 10^6 \times 2 \times 10^{-3}}{120.832 \times 10^6}$$

$$= 33.9 \ \mu s$$

To obtain the total duration of a traffic burst, the time used for the preamble must be added. If the preamble contains S_P symbols, then, assuming QPSK modulation, the total length of the traffic burst measured in number of symbols is

$$S_T = \frac{R_0 T}{2} + S_P \tag{11.32}$$

and the burst-time duration is

$$T_T = \frac{2 S_T}{R_T} \tag{11.33}$$

If the preamble contains 300 symbols, then

$$S_T = \frac{2.048 \times 10^6 \times 2 \times 10^{-3}}{2} + 300$$

$$= 2348 \ \text{symbols}$$

Using this in Equation (11.33), we obtain

$$T_T = \frac{2 \times 2348}{120.832 \times 10^6} = 38.9 \ \mu s$$

11.4.4.3 The High-Rate TDMA Frame for North America

The INTELSAT TDMA burst (bit) rate of $R_T = 120.832$ Mbits/s was chosen to be compatible with both the European and North American standards. Figure 11.36 is similar to Figure 11.35 except that the PCM multiplex signal is the 24-channel T-Carrier instead of the 30-channel European standard. The essential

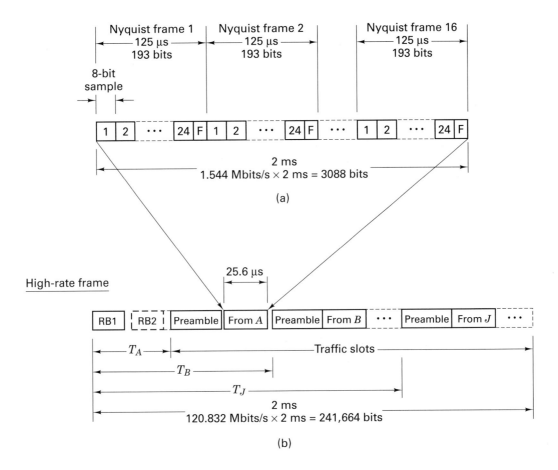

Figure 11.36 INTELSAT digital transmissions standards for T-Carrier.
(a) Terrestrial PCM multiplex. (b) High-rate frame.

T-Carrier features that are different from the European standard are listed as follows:

1. Each Nyquist frame comprises 24 channels or samples × 8 bits + 1 frame alignment bit = 193 bits.
2. The 16 Nyquist frames contain 16 × 193 = 3088 bits.
3. The T-Carrier data rate is 1.544 Mbits/s.
4. The duration of the traffic data field in the high-rate TDMA frame is calculated from Equation (11.31).

$$T_{\text{tr}} = \frac{1.544 \times 10^6 \times 2 \times 10^{-3}}{120.832 \times 10^6}$$

$$= 25.6 \ \mu\text{s}$$

11.4.4.4 INTELSAT TDMA Operation

At the transmitting earth station, the continuous low-rate data stream enters one of a pair of buffers illustrated in Figure 11.37a. When one buffer is filling at the low rate (1.544 Mbits/s or 2.048 Mbits/s), the other is emptying at the burst rate (120.832 Mbits/s). The buffers alternate functions at each TDMA frame. The time of application of the high-rate clock is controlled so that the traffic burst is transmitted in the proper interval to arrive at the satellite in its assigned position in the TDMA frame.

At the receiving station, the received traffic burst is routed to one of a pair of expansion buffers, shown in Figure 11.37b, that have the inverse function of the compression buffers in Figure 11.37a. When one buffer is filling at the high rate, the other is emptying at the desired output rate.

The most critical aspect of TDMA operation is the precise synchronization needed to assure orthogonality of the time slots [20]. Figure 11.38 illustrates the general idea behind most commercial satellite synchronization schemes. One station is designated as the master or control station. This station transmits periodic bursts of reference timing pulses. User stations also transmit their timing pulses, designated as slave pulses in Figure 11.38. On the downlink, the using station receives the master or reference pulses in addition to its own slave pulses. The time difference between the master and slave pulses corresponds to the timing error. The station adjusts its clock so as to reduce this timing error.

11.4.5 Satellite-Switched TDMA in INTELSAT

Modern communication satellites often employ several regional antenna beams. For a satellite based over the Atlantic Ocean, separate beams might be aimed at North America, Europe, South America, and Africa. Switches are used to allow

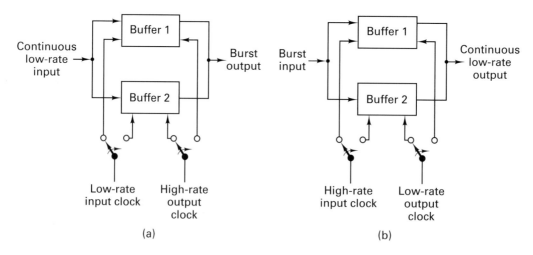

Figure 11.37 Burst compression and expansion buffers. (a) Compression buffers at transmitter. (b) Expansion buffers at receiver.

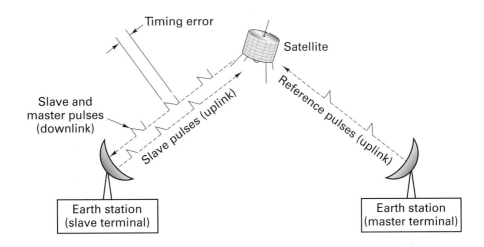

Figure 11.38 TDMA synchronization concept.

the interconnection of stations in one region to communicate with stations in another region. The basic goal of a satellite-switched TDMA (SS/TDMA) scheme is to provide an efficient way of cyclically providing interconnection of TDMA data among various coverage regions.

The heart of the system consists of a microwave switch matrix, located in the satellite, that is programmed via ground control to change states cyclically in rapid sequence, thus interconnecting distinct uplink beams to distinct downlink beams at each switching time. An earth station in the network communicates with those in other beams by transmitting TDMA bursts in the proper timing positions in the sequence. The pattern of switch states is selected so as to maximize the usable system capacity under the constraints of the traffic demands [21]. For complete interconnectivity between N beams, a total of $N!$ different satellite switch states or *modes* are required. Table 11.3 illustrates the six modes required for the full interconnectivity of a three-beam system.

In mode 1, the satellite receivers in beams A, B, and C are connected to the satellite transmitters for beams A, B, and C, respectively. An earth station in one of these beams can then communicate with other earth stations in the same beam. The beam is said to be *looped back* on itself.

TABLE 11.3 Three-Beam Satellite Switch Modes

Input	Output					
	Mode 1	Mode 2	Mode 3	Mode 4	Mode 5	Mode 6
A	A	A	B	B	C	C
B	B	C	A	C	A	B
C	C	B	C	A	B	A

Figure 11.39 illustrates a three-beam (beams *A*, *B*, and *C*) example of a SS/TDMA system. The satellite microwave switch matrix is configured in a *crossbar* design. This design can be thought of as being made up of row and column lines; when one row and one column are energized, contact is made at the intersection. A crossbar design only permits a single row to communicate with a single column at a time. If uplink A_U is connected to downlink B_D, *neither A_U nor B_D* can be simultaneously connected to any other beam.

In Figure 11.39; three different traffic patterns during time slot intervals T_1, interval T_2, and T_3, with three different switch states S_1, S_2, and S_3 are shown. During interval T_1, switch state S_1 interconnects the beams in a loop-back fashion which permits the uplink messages in slot T_1 to be delivered to their correct destinations. During time interval T_2, switch state S_2 interconnects uplink beam A_U to downlink beam B_D, uplink beam B_U to downlink beam C_D, and uplink beam C_U to downlink beam A_D. This connection pattern assures that the uplink messages in slot T_2 are delivered to their correct destinations. During time interval T_3, switch state S_3 similarly connects uplink transmissions to downlink beams to assure correct delivery of the data.

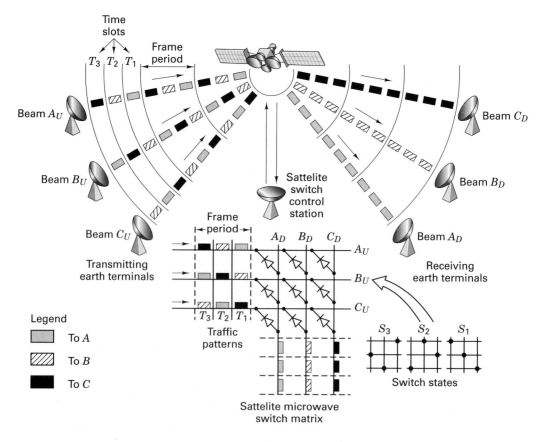

Figure 11.39 Satellite-switched TDMA (SS/TDMA).

The traffic patterns and their durations are programmed to optimize the resource capacity and to serve the users as efficiently as possible. The cyclic pattern can be reprogrammed by ground command to meet changing traffic requirements.

11.4.5.1 Traffic Matrix

Figure 11.40 is a matrix describing the communication traffic among N spot-beam coverages. In this figure, t_{ij} is the traffic volume from the ith beam to the jth beam. The subtotal

$$S_i = \sum_{j=1}^{N} t_{ij} \tag{11.34}$$

is the total traffic originating from the ith uplink beam, and

$$R_j = \sum_{i=1}^{N} t_{ij} \tag{11.35}$$

is the total traffic received in the jth downlink beam. When the traffic in a SS/TDMA system is controlled by a nonblocking switch (one that allows for the transmission of *all* messages, without any "busy" signals) a k-second time slot will be assigned to each channel in the TDMA frame. For efficient utilization of the CR, the total traffic in Figure 11.40 should be transmitted within a frame time T which should be made as short as possible. The minimum frame time for providing such nonblocking connectivity can be expressed [22] as

$$T_{\min} = k \max\left(\{S_i\}, \{R_j\}\right) \tag{11.36}$$

where $\max\left(\{S_i\}, \{R_j\}\right)$ is the maximum value taken over the set of all $\{S_i\}$ and $\{R_j\}$. Equation 11.36 describes the minimum time to communicate *all* of the traffic in the traffic matrix, for equal bandwidth per channel.

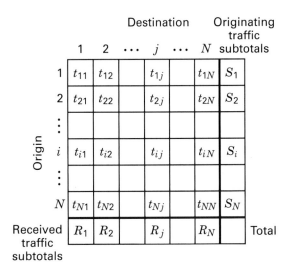

Figure 11.40 Traffic matrix.

11.5 MULTIPLE ACCESS TECHNIQUES FOR LOCAL AREA NETWORKS

A local area network (LAN) can be used to interconnect computers, terminals, printers, and so on, located within a building or a small set of buildings. While long-haul networks use the public telephone network for economic reasons, LAN designers usually lay their own high-bandwidth cables. Bandwidth is not as scarce as it is in the long-haul cases. Not being forced to optimize bandwidth, a LAN can use simple access algorithms [6, 25–27].

11.5.1 Carrier-Sense Multiple Access Networks

Ethernet is a LAN access scheme developed by the Xerox Corporation. The Ethernet scheme is based on the assumption that each local machine can sense the state of a common broadcast channel before attempting to use it. The technique is known as *carrier-sense multiple access with collision detection* (CSMA/CD). The word "carrier," here, means *any* electrical activity on the cable. Figure 11.41a illustrates the bit field format for the Ethernet specification; the details are listed as follows:

1. The maximum packet size is 1526 bytes, where a byte is 8 bits. The packet breakdown is 8-byte preamble + 14-byte header + 1500-byte data + 4-byte parity.

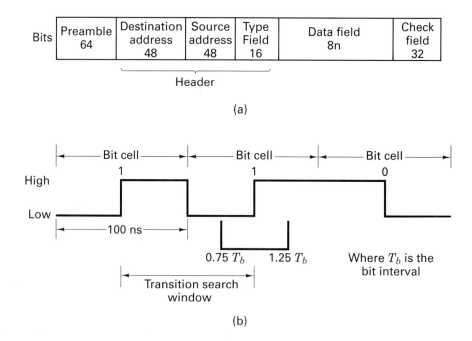

Figure 11.41 Ethernet bit field and PCM format. (a) Ethernet specification. (b) Manchester PCM format.

2. The minimum packet size is 72 bytes, consisting of an 8-byte preamble + 14-byte header + 46-byte data + 4-byte parity.

3. The minimum spacing between packets is 9.6 μs.

4. The preamble contains a 64-bit synchronization pattern of alternating ones and zeros, ending with two consecutive ones: (1 0 1 0 1 0 . . . 1 0 1 0 1 1).

5. The receiving station examines a destination address field in the header to see if it should accept a particular packet. The first bit indicates the type of address (0 = single address, 1 = group address); an entire field of ones means an all-station broadcast.

6. The source address is the unique address of the transmitting machine.

7. The type field determines how the data field is to be interpreted. For example, bits in the type field can be used to describe such things as data encoding, encryption, message priority, and so on.

8. The data field is an integer number of bytes from a minimum of 46 to a maximum of 1500.

9. The parity check field houses the parity bits which are generated by the following generating polynomial (see Section 6.7):

$$X^{32} + X^{26} + X^{23} + X^{22} + X^{16} + X^{12} + X^{11} + X^{10}$$

$$+ X^8 + X^7 + X^5 + X^4 + X^2 + X + 1$$

The Ethernet multiple access algorithm defines the following user action or response:

1. *Defer.* The user must not transmit when the carrier is present or within the minimum packet spacing time.

2. *Transmit.* The user may transmit if not deferring until the end of the packet or until a collision is detected.

3. *Abort.* If a collision is detected, the user must terminate packet transmission and transmit a short jamming signal to ensure that all collision participants are aware of the collision.

4. *Retransmit.* The user must wait a random delay time (similar to the ALOHA system) and then attempt retransmission.

5. *Backoff.* The delay before the nth attempt is a uniformly distributed random number from 0 to $2^n - 1$, for $(0 < n \leq 10)$. For $n > 10$, the interval remains 0 to 1023. The unit of time for the retransmission delay is 512 bits (51.2 μs).

Figure 11.41b illustrates a 10-Mbits/s data stream with Manchester PCM formatting from the Ethernet specification. Notice that with such formatting, each bit cell or bit position contains a transition. A binary one is characterized by transitioning from a low level to a high level, while a binary zero has the opposite transition. Therefore, the presence of data transitions denotes to all "listeners" that the carrier is present. If a transition is not seen between 0.75 and 1.25 bit times since the last transition, the carrier has been lost, indicating the end of a packet.

11.5.2 Token-Ring Networks

A carrier-sense network consists of a cable onto which all stations are passively connected. A *ring network,* by comparison, consists of a series of point-to-point cables between consecutive stations. The interfaces between the ring and the stations are active rather than passive. Figure 11.42a illustrates a typical unidirectional ring with interface connections to several stations. Figure 11.42b illustrates the state of the interface for the listen mode and the transmit mode. In the *listen mode* the input bits are copied to the output with a delay of one bit time. In the *transmit mode,* the connection is broken so that the station can enter its own data onto the ring. The token is defined as a special bit pattern (e.g., 1 1 1 1 1 1 1 1) which circulates on the ring whenever all stations are idle. How does the system ensure that message data do not contain a tokenlike sequence? *Bit stuffing* is used to prevent this pattern from occurring in the data. For the 8-bit token example shown, a bit-stuffing algorithm would insert a zero into the data stream after each sequence of seven consecutive ones. The data receiver would use a similar algorithm to dispose of the inserted bit following any sequence of seven consecutive ones. The token-ring access scheme works as follows:

1. A station wanting to send a message monitors the token appearing at the interface. When the last bit of the token appears, the station inverts it (e.g., 1 1 1 1 1 1 1 0). The station then breaks the interface connection and enters its own data onto the ring.

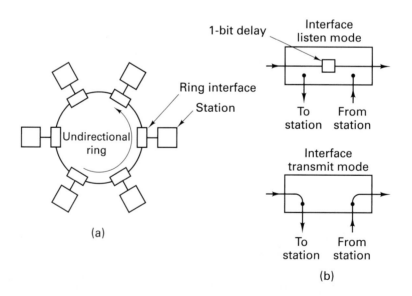

(a)

(b)

Figure 11.42 Token-ring network. (a) Network. (b) Listen and transmit modes.

2. As bits come back around the ring, they are removed by the sender. There is no limit on the size of the packets, because the entire packet never appears on the ring at one instant.

3. After transmitting the last bit of its message, the station must regenerate the token. After the last data bit has circled the ring and has been removed, the interface is switched back to the listen mode.

4. Contention is not possible with a token-ring system. During heavy traffic, as soon as a token is regenerated, the next downstream station requiring service will see and remove the token. Thereby, permission to transmit rotates smoothly around the ring. Since there is only one token, there is no contention.

The ring itself must have sufficient delay to enable a complete token to circulate when all stations are idle. A major issue in ring network design is the propagation distance or "length" of a bit. If the data rate is R Mbits/s, a bit is emitted every $(1/R)$ microseconds. Since the propagation rate along a typical coaxial cable is 200 m/μs, each bit occupies $200/R$ meters on the ring.

Example 11.4   Minimum Ring Size

If an 8-bit token is to be used on a 5-Mbits/s token-ring network, calculate the minimum *propagation distance, d_p,* needed for the ring circumference. Assume that the propagation velocity v_p is 200 m/μs.

Solution

$$R = 5 \text{ Mbits/s}$$

Time to emit one bit, t_b:

$$t_b = \frac{1}{5 \times 10^6} \text{ s}$$

Time to emit the 8-bit token, t_t:

$$t_t = \frac{8}{5 \times 10^6} \text{ s}$$

Propagation distance for the 8-bit token:

$$d_p = t_t \times v_p$$

$$= \tfrac{8}{5} \text{ μs} \times 200 \text{ m/μs}$$

$$= 320 \text{ m}$$

11.5.3 Performance Comparison of CSMA/CD and Token-Ring Networks

Figure 11.43 compares the delay-throughput characteristics of a CSMA/CD network with a token-ring network. In each case, the cable length is 2 km, there are 50 stations on the network, the average packet length is 1000 bits, and the header size

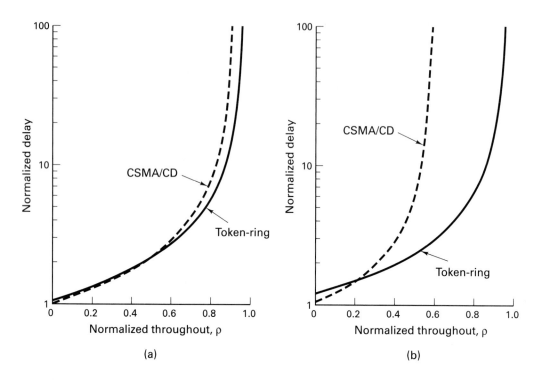

Figure 11.43 Delay versus throughput performance for CSMA/CD and token-ring networks. (a) Transmission rate = 1 Mbits/s. (b) Transmission rate = 10 Mbits/s. (Reprinted with permission from W. Bux, "Local-Area Subnetworks: A Performance Comparison," *IEEE Trans. Commun.*, vol. COM29, no. 10, Oct. 1981, pp. 1465–1473. © 1981 IEEE.)

is 24 bits. Figure 11.43a, the case where the transmission rate is 1 Mbits/s, illustrates that under these assumptions, CSMA/CD and token ring perform almost equally well. In Figure 11.43b, only one parameter has been changed as compared to Figure 11.43a; the transmission rate was increased to 10 Mbits/s. The difference for CSMA/CD is considerable; for normalized throughput, $\rho < 0.22$, CSMA/CD performs better than token ring. However, for $\rho > 0.22$, token ring clearly manifests better delay-throughput characteristics. To understand the reason for the poor CSMA/CD performance in Figure 11.43b, let us review the definition of ρ, described in Equations (11.17) and (11.19) and shown as

$$\rho = \frac{b\lambda}{R} = \frac{\rho'}{R}$$

where $\rho' = b\lambda$ is the channel throughput in bits per second and R is the channel capacity (maximum transmission bit rate). As R increases, channel throughput must increase accordingly for a given value of ρ. At higher channel throughput rates, a significant portion of the CSMA/CD transmission attempts ends in collision [26].

11.6 CONCLUSION

In this chapter we have outlined the concepts of resource sharing. The classical approaches of FDM/FDMA and TDM/TDMA were discussed in some detail. We also described a hybrid multiple access technique called CDMA, and introduced some of the satellite multiple access techniques that became popular in the 1970s and 1980s, known as multiple-beam frequency reuse and dual-polarization frequency reuse.

We described the demand-assignment (DAMA) techniques in the context of several versions of the ALOHA algorithm, and we considered several of the multiple-access techniques employed with INTELSAT, such as FDM/FM, SPADE, TDMA, and SS/TDMA. Finally, we examined two popular algorithms used for local area networks: carrier-sense multiple access with collision detection (CSMA/CD) and a token-ring network. The goals of the chapter were to introduce an assortment of multiple access techniques rather than attempting a rigorous treatment of any of them.

REFERENCES

1. Rubin, I., "Message Delays in FDMA and TDMA Communication Channels," *IEEE Trans. Commun.,* vol. COM27, no. 5, May 1979, pp. 769–777.

2. Nirenberg, L. M., and Rubin, I., "Multiple Access System Engineering—A Tutorial," *IEEE WESCON/78 Professional Program,* Modern Communication Techniques and Applications, session 21, Los Angeles, Sept. 13, 1978.

3. Abramson, N., "The ALOHA System—Another Alternative for Computer Communications," *Proc. Fall Joint Comput. Conf. AFIPS,* vol. 37, 1970, pp. 281–285.

4. Hayes, J. F., "Local Distribution in Computer Communications," *IEEE Commun. Mag.,* Mar. 1981, pp. 6–14.

5. Schwartz, M., *Computer-Communication Network Design and Analysis,* Prentice-Hall, Inc., Englewood Cliffs, N.J., 1977.

6. Tanenbaum, A. S., *Computer Networks,* Prentice-Hall, Inc., Englewood Cliffs, N.J., 1981.

7. Abramson, N., "The ALOHA System," in N. Abramson and F. F. Kuo, eds., *Computer Communication Networks,* Prentice-Hall, Inc., Englewood Cliffs, N.J., 1973.

8. Kleinrock, L., *Queueing Systems,* Vol. 1, *Theory,* John Wiley & Sons, Inc., New York, 1975.

9. Abramson, N., "Packet Switching with Satellites," *AFIPS Conf. Proc.,* vol. 42, June 1973, pp. 695–702.

10. Rosner, R. D., *Packet Switching,* Lifelong Learning Publications, Wadsworth Publishing Company, Inc., Belmont, Calif., 1982.

11. Crowther, W., Rettberg, R., Walden, D., Ornstein, S., and Heart, F., "A System for Broadcast Communication: Reservation ALOHA," *Proc. Sixth Hawaii Int. Conf. Syst. Sci.,* Jan. 1973, pp. 371–374.

12. Roberts, L., "Dynamic Allocation of Satellite Capacity through Packet Reservation," *AFIPS Conf. Proc.,* vol. 42, June 1973, p. 711.

13. Binder, R., "A Dynamic Packet-Switching System for Satellite Broadcast Channels," *Proc. Int. Conf. Commun.*, June 1975, pp. 41-1–41-5.

14. Capetanakis, J., "Tree Algorithms for Packet Broadcast Channels," *IEEE Trans. Inf. Theory*, vol. IT25, Sept. 1979, pp. 505–515.

15. Puente, J. G., and Werth, A. M., "Demand-Assigned Service for the INTELSAT Global Network," *IEEE Spectrum*, Jan. 1971, pp. 59–69.

16. Jones, J. J., "Hard Limiting of Two Signals in Random Noise," *IEEE Trans. Inf. Theory*, vol. IT9, Jan. 1963, pp. 34–42.

17. Bond, F. E., and Meyer, H. F., "Intermodulation Effects in Limiter Amplifier Repeaters," *IEEE Trans. Commun. Technol.*, vol. COM18, no. 2, Apr. 1970, pp. 127–135.

18. Shimbo, O., "Effects of Intermodulation, AM-PM Conversion, and Additive Noise in Multicarrier TWT Systems," *Proc. IEEE*, vol. 59, Feb. 1971, pp. 230–238.

19. Chakraborty, D. "INTELSAT IV Satellite System (Voice) Channel Capacity versus Earth-Station Performance," *IEEE Trans. Commun. Technol.*, vol. COM19, no. 3, June 1971, 355–362.

20. Campanella, S., and Schaefer, D., "Time Division Multiple Access Systems (TDMA)," in K. Feher, *Digital Communications, Satellite/Earth Station Engineering*, Prentice-Hall, Inc., Englewood Cliffs, N.J., 1983.

21. Scarcella, T., and Abbott, R. V., "Orbital Efficiency Through Satellite Digital Switching," *IEEE Commun. Mag.*, May 1983, pp. 38–46.

22. Muratani, T., Satellite-Switched Time-Domain Multiple Access," *Proc. IEEE Electron. and Aerosp. Conf. (EASCON)*, 1974, pp. 189–196.

23. Dill, G. D., "TDMA, The State-of the-Art," *Rec. IEEE Electron. Aerosp. Syst. Conv. (EASCON)*, Sept. 26–28, 1977, pp. 31-5A–31-5I.

24. Jarett, K., "Operational Aspects of Intelsat VI Satellite-Switched TDMA Communication System," *AIAA Tenth Commun. Satell. Syst. Conf.* Mar. 1984, pp.107–111.

25. Stallings, W., "Local Network Performance," *IEEE Commun. Mag.*, vol. 22, No. 2, Feb. 1984, pp. 27–36.

26. Bux, W., Local-Area Subnetworks: A Performance Comparison," *IEEE Trans. Commun.*, vol. COM29, no. 10, Oct. 1981, pp. 1465–1473.

27. Dixon, R. C., Strole, N. C., and Markov, J. D., "A Token-Ring Network for Local Data Communications," *IBM Syst. J.*, vol. 22, no. 1–2, 1983, pp. 47–62.

PROBLEMS

11.1. Design an FDM signal set consisting of five voice channels, each in the frequency range 300 to 3400 Hz. The multiplexed composite is to be made up of inverted sidebands and is to occupy the spectral region from 30 to 50 kHz.

 (a) Draw the composite spectrum, indicating individual spectrum and guard band frequency locations.

 (b) Draw a block diagram showing the heterodyning and filtering details and the required local oscillator values.

11.2. A receiver is tuned to receive the lower sideband (LSB) of a radio-frequency (RF) carrier wave with frequency, $f_c = 8$ MHz. The bandwidth of the LSB signal is 100 kHz. The receiver employs a local oscillator (LO) with frequency, f_{LO}, for heterodyning the received signal down to a lower intermediate frequency (IF). Assume that $f_{LO} > f_c$, and that the IF amplifier is centered at 2 MHz. Draw a block diagram of the heterodyning conversion, including the RF filter, the LO, and the IF filter. Indicate the center frequency of each filter and typical spectra of the signals at various points in the diagram.

11.3. Equations (11.13) to (11.15) demonstrate that the average message delay time for TDMA is less than that for FDMA. Discuss the practical benefits of such reduced delay in TDMA, as a function of frame time, for a satellite link with a one-way range of 36,000 km. For what values of frame time can there be a significant advantage of TDMA over FDMA?

11.4. A group of stations share a 56-kbits/s pure ALOHA channel. Each station outputs a packet on the average of once every 10 s, even if the previous one has not yet been sent (i.e., the stations buffer the packets). Each packet comprises 3000 bits. What is the maximum number of stations that can share this channel, assuming that the arrival process is Poisson?

11.5. A group of three stations share a 56-kbits/s pure ALOHA channel. The average bit rate transmitted from each of the three station is $R_1 = 7.5$ kbits/s, $R_2 = 10$ kbits/s, and $R_3 = 20$ kbits/s. The size of each packet is 100 bits/packet. Find the normalized total traffic on the channel, the normalized throughput, the probability of successful transmission, and the arrival rate of successful packets. Assume that the arrival process is Poisson.

11.6. Verify that for a pure ALOHA access scheme, the normalized throughput is bounded by $1/2e$ and that this maximum occurs when the normalized total traffic is equal to 0.5.

11.7. **(a)** Verify that Equation (11.24) is a valid probability density function (pdf) for a discrete random variable.
   (b) Calculate the mean of a discrete random variable having a pdf like the one given in Equation (11.24).
   (c) Show that your result in part (b) is consistent with the claim that λ is the average packet arrival rate.

11.8. Consider the pure ALOHA arrival scenario shown in Figure P11.1. The vertical arrows indicate packet arrival times. N_n is the number of arriving packets in the time interval $(T_{n-1}, T_n]$, where $(t_x, t_y]$ indicates the interval $t_x < t \leq t_y$. N_{n+1} is the number of arriving packets in $(T_n, T_{n+1}]$ and τ is the time duration per packet in seconds. The average arrival rate is λ_t. Assume the arrivals are independent of each other.
   (a) Write an expression for the joint pfd of N_n and N_{n+1}.

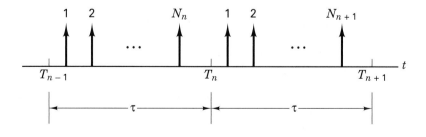

Figure P11.1

(b) Let T_n define the time at which user A's packet arrives. Express, in terms of the joint pdf of N_n and N_{n+1}, the probability that user A's transmission will be successful.

11.9. Let $N = N_n + N_{n+1}$, where N_n and N_{n+1} are as defined in Problem 11.8. Write an expression for the pdf of N, and give an interpretation for N.

11.10. Six thousand stations are competing for the use of a single slotted ALOHA channel. The average station makes 30 requests per hour, where each request is for one slot of 500-μs duration. Calculate the normalized total traffic on the channel.

11.11. Consider the arrival scenario of Figure P11.1; the location of the packet arrival times are permissible as shown under pure ALOHA, but not under slotted ALOHA, where arrivals are permitted only at the discrete times T_i, where $i = 0, 1, \ldots$. Assume that the average arrival rate is λ_t.
  (a) How would Figure P11.1 need to be modified if slotted ALOHA is used? How would the pdfs of N_n and N_{n+1} change?
  (b) If user A's packet arrives at time T_n, what is the probability of successful transmission?

11.12. A group of slotted-ALOHA stations generate a total of 120 requests per second, including both original and retransmissions. Each request is for a 12.5-ms duration slot.
  (a) What is the normalized total traffic on the channel?
  (b) What is the probability of a successful transmission on the first attempt?
  (c) What is the probability of exactly two collisions before a successful transmission?

11.13. Measurements of a slotted-ALOHA channel show that 20% of the slots are idle.
  (a) What is the normalized total traffic on the channel?
  (b) What is the normalized throughput?
  (c) Is the channel underloaded or overloaded?

11.14. Show that the sum of two Poisson processes, with rates λ_1 and λ_2, is also a Poisson process, with rate $\lambda_t = \lambda_1 + \lambda_2$. Generalize your result for the sum of n Poisson processes.

11.15. A 10-MHz transponder is occupied by 200 identical carriers, half servicing stations with $G/T = 40$ dB/K, the other half servicing stations with $G/T = 37$ dB/K. All stations have a requirement to operate with a bit error probability of 10^{-5}. The transponder is power limited under this configuration.
  (a) What is the maximum possible bandwidth for each carrier?
  (b) Suppose that each carrier has a bandwidth of 40 kHz, and the transponder is required to service a group of larger ($G/T = 40$ dB/K) stations only. How many stations can the transponder handle? Will the transponder be power or bandwidth limited?
  (c) Repeat part (b) for the case where the transponder is to service a group of small ($G/T = 37$ dB/K) stations only.

11.16. A TDMA system operates at 100 Mbits/s with a 2-ms frame time. Assume that all slots are of equal length and that a guard time of 1 μs is required between slots.
  (a) Compute the efficiency of the communications resource (CR) for the case of 1, 2, 5, 10, 20, 50, and 100 slots per frame.
  (b) Repeat part (a) assuming that a 100-bit preamble is required at the start of each slot. Compute the efficiency of the CR in terms of the desired information transmission.
  (c) Graph the results of parts (a) and (b).

11.17. With reference to Equation 11.36:

   (a) Discuss the efficiency of the CR use if all S_i and R_j are equal.

   (b) Discuss the effect of a few S_i or R_j being much larger than the majority. How can the efficiency of the CR be improved?

   (c) When are the distributions of S_i and R_j likely to be similar? Dissimilar?

11.18. (a) Consider a token-ring network operating at a transmission rate of 10 Mbits/s over a cable having a propagation velocity of 200 m/μs. How many meters of cable is equal to a delay of 1 bit at each ring interface?

   (b) If the token is 10 bits long, and all but three stations are switched off during evening hours, what is the minimum cable length needed for the ring?

QUESTIONS

11.1. What is typically meant by the terms *communications resource?* (See Chapter 11, introduction.)

11.2. What are the similarities and differences between the terms *multiplexing* and *multiple access?* (See Chapter 11, introduction.)

11.3. Why is it not possible to use a *linear device* as a mixer (See Section 11.1.1.1 and Appendix A.)

11.4. Is there any theoretical capacity advantage in providing users with FDMA service versus TDMA service? (See Section 11.1.4.1.)

11.5. What benefits might there be in using CDMA versus either FDMA or TDMA access schemes? (See Section 11.1.5.)

EXERCISES

Using the Companion CD, run the exercises associated with Chapter 11.

CHAPTER 12

Spread-Spectrum Techniques

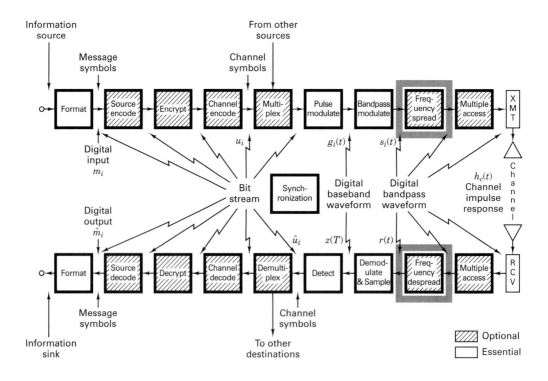

12.1 SPREAD-SPECTRUM OVERVIEW

The initial application of spread-spectrum (SS) techniques was in the development of military guidance and communication systems. By the end of World War II, spectrum spreading for jamming resistance was already a familiar concept to radar engineers [1], and during subsequent years, SS investigation was motivated primarily by the desire to achieve highly jam-resistant communication systems. As a result of this research, there emerged an assortment of other applications in such areas as energy density reduction, high-resolution ranging, and multiple access, which will be discussed in later sections. The techniques considered in this chapter are called *spread spectrum* because the transmission bandwidth employed is much greater than the minimum bandwidth required to transmit the information. A system is defined to be a spread-spectrum system if it fulfills the following requirements:

1. The signal occupies a bandwidth much in excess of the minimum bandwidth necessary to send the information.
2. Spreading is accomplished by means of a *spreading signal,* often called a *code signal,* which is independent of the data. The details of some spreading signals are described in later sections.
3. At the receiver, despreading (recovering the original data) is accomplished by the correlation of the received spread signal with a synchronized replica of the spreading signal used to spread the information.

Standard modulation schemes such as frequency modulation and pulse code modulation also spread the spectrum of an information signal, but they do not qualify as spread-spectrum systems since they do not satisfy all the conditions outlined above.

12.1.1 The Beneficial Attributes of Spread-Spectrum Systems

12.1.1.1 Interference Suppression Benefits

White Gaussian noise is a mathematical model that, by definition, has infinite power spread uniformly over all frequencies. Effective communication is possible with this interfering noise of infinite power because only the finite-power noise components that are present within the signal space (in other words, share the *same coordinates* as the signal components) can interfere with the signal. The balance of the noise power may be thought of as noise that is effectively tuned out by the detector (see Section 3.1.3). For a typical narrowband signal, this means that only the noise in the signal bandwidth can degrade performance. Since spread-spectrum (SS) techniques were initially developed as a military application to permit reliable communications in the face of an enemy interferer (jammer), we begin by focusing on the anti-jam (AJ) capabilities of SS. (Commercial applications are treated in Sections 12.7 and 12.8.)

The idea behind a spread-spectrum AJ system is as follows. Consider that many orthogonal signal coordinates or dimensions are available to a communication link and that only a small subset of these signal coordinates are used at any time. We assume that the jammer cannot determine the signal subset that is currently in use. For signals of bandwidth W and duration T, the number of signaling dimensions can be shown [2] to be approximately $2WT$. Given a specific design, the error performance of such a system is only a function of E_b/N_0. Against white Gaussian noise, with *infinite* power, the use of spreading (large $2WT$) offers no performance improvement. However, when the noise stems from a jammer with a *fixed finite* power and with uncertainty as to where in the signal space the signal coordinates are located, the jammer's choices are limited to the following:

1. Jam *all* the signal coordinates of the system, with an *equal* amount of power in each one, with the result that *little* power is available for each coordinate.
2. Jam a *few* signal coordinates with *increased* power in each of the jammed coordinates (or more generally, jam all the coordinates with various amounts of power in each).

Figure 12.1 compares the effect of spectrum spreading in the presence of white noise with spreading in the presence of an intentional jammer. The power spectral density of the signal is denoted $G(f)$ before spreading, and $G_{ss}(f)$ after spreading. For simplicity, the figure treats the frequency dimension only. In Figure 12.1a it can be seen that the single-sided power spectral density of white noise, N_0, is unchanged as a result of expanding the signal bandwidth from W to W_{ss}. The av-

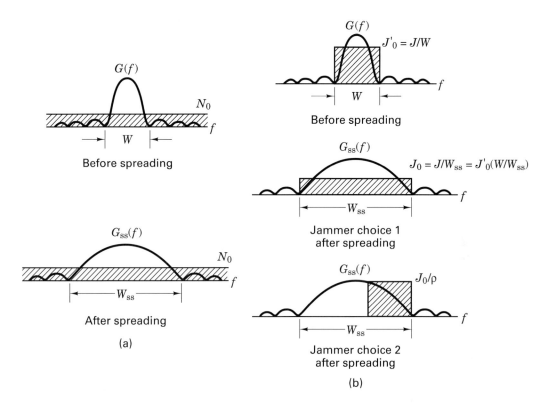

Figure 12.1 Effect of spectrum spreading. (a) Spectrum spreading in the presence of white noise. (b) Spectrum spreading in the presence of an intentional jammer.

erage power of white noise (area under the spectral density curve) is infinite. Hence, the use of spreading offers no performance improvement here. Figure 12.1b (upper diagram) illustrates the case of received (fixed finite) jammer power, J, and power spectral density, $J'_0 = J/W$, where W is the unspread bandwidth being jammed. Once the signal bandwidth is spread, the jammer can make one of the two choices listed earlier—choice 1 results in a reduction in jammer noise spectral density, J'_0, by a factor (W/W_{ss}) across the spread spectrum. The resulting noise spectral density, $J_0 = J/W_{ss}$, is referred to as the *broadband jammer noise spectral density*. Choice 2 results in a reduction in the number of signal coordinates that the jammer occupies. However, with choice 2 the jammer can increase its noise spectral density from J_0 to J_0/ρ $(0 < \rho \leq 1)$, where ρ is the portion of the spread-spectrum band the jammer elects to jam. If the jammer makes a poor choice in the coordinates to be jammed, the average effect of jamming will be less than if it makes a good choice. The larger the dimensionality of the signal set or the more signal coordinates the communicator can choose from, the greater is the jammer's uncertainty regarding the effectiveness of the jamming technique, and the better will be the protection against jamming. The comparison of unspread- versus spread-spectrum signaling should be done under the assumption that there is the same amount of total average power in both cases. Since the area under the power spectral density (psd) curves represent total average power, there should be equal area under each of the psd curves for the unspread and the spread examples. Hence, it should be clear that in Figure 12.1, the $G_{ss}(f)$ plots are not to scale in both parts a and b.

Jamming is not always the result of an intentional act. Sometimes, the jamming signal is caused by natural phenomena, and sometimes it is the result of self-interference caused by *multipath*, in which delayed versions of the signal, arriving via alternative paths, interfere with the direct path transmission.

12.1.1.2 Energy Density Reduction

One can imagine situations where it is desired that a communications link be operated without being detected by anyone other than the intended receiver. Systems designed for this special task are known as *low probability of detection* (LPD) or *low probability of intercept* (LPI) communication systems. These systems are designed to make the detection of their signals as difficult as possible by anyone but the intended receiver. The goal of such a system is to use the minimum signal power and the optimum signaling scheme that results in the minimum probability of being detected. Since, in spread-spectrum systems, the signal is spread over many more signaling coordinates than in conventional modulation schemes, the resulting signal power is, on average, spread thinly and uniformly in the spread domain. Therefore, not only can the spread-spectrum signal be made difficult to jam, but additionally, the signal's very existence may be rendered difficult to perceive. To anyone who does not possess a synchronized replica of the spreading signal, the spread-spectrum signal will seem "buried in the noise."

A *radiometer* is a simple power measuring instrument that can be used by an adversary to detect the presence of spread-spectrum signals within some bandwidth *W*. The radiometer, illustrated in Figure 12.2, consists of a bandpass filter (BPF) with bandwidth *W*, a squaring circuit to ensure a positive output value (since a measure of *signal energy* is being detected), and an integrating circuit. At time $t = T$, the output of the integrator is compared to a preset threshold. If the output of the integrator is larger than the threshold, a signal is declared present; otherwise, the signal is declared absent. References [3, 4] provide details on the detectability of spread-spectrum signals, using radiometers and other more complicated instruments that make use of the features of the SS signal itself.

Spread-spectrum systems that are designed to exhibit LPI may also exhibit a *low probability of position fix* (LPPF), which means that even if the presence of the signal is perceived, the direction of the transmitter is difficult to pinpoint. Some spread-spectrum systems also exhibit a *low probability of signal exploitation* (LPSE), which means that the identification of the source is difficult to ascertain.

Another, unrelated application of spread-spectrum signaling deals with the fact that in some cases energy density reduction may be required to meet national allocation regulations. Downlink transmissions from satellites must meet international regulations on the spectral density that impinges on the earth. By spreading the downlink energy over a wider bandwidth, the total transmitted power can be

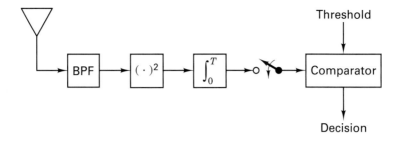

Figure 12.2 Radiometer.

increased and hence performance improved, while the energy density regulations are followed.

12.1.1.3 Fine Time Resolution

Spread-spectrum signals can be used for ranging or determination of position location. Distance can be determined by measuring the time delay of a pulse as it traverses the channel. Uncertainty in the delay measurement is inversely proportional to the bandwidth of the signal pulse. This can be seen by the illustration in Figure 12.3. The uncertainty of the measurement, Δt, is proportional to the rise time of the pulse, which is inversely proportional to the bandwidth of the pulse signal; that is,

$$\Delta t \approx \frac{1}{W} \tag{12.1}$$

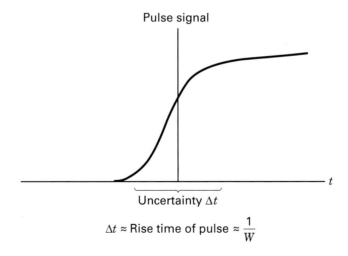

$$\Delta t \approx \text{Rise time of pulse} \approx \frac{1}{W}$$

Figure 12.3 Time-delay measurement.

The larger the bandwidth, the more precisely one can measure range. Over a Gaussian channel, a one-shot measurement on a single pulse is not very reliable. The spread-spectrum technique, however, uses a code signal consisting of a long sequence of polarity changes (e.g., a binary PSK-modulated signal) in place of the single pulse. Upon reception, the received sequence is correlated against a local replica and the results of the correlation are used to perform an accurate time-delay or range measurement.

12.1.1.4 Multiple Access

Spread-spectrum methods can be used as a multiple access technique, in order to share a communications resource among numerous users in a coordinated manner. The technique, termed *code-division multiple access* (CDMA), since each simultaneous user employs a unique spread-spectrum signaling code, was discussed briefly in Chapter 11. One of the by-products of this type of multiple access is the ability to provide communication privacy between users with different spreading signals. An unauthorized user (a user not having access to a spreading signal) cannot easily monitor the communications of the authorized users. (A more detailed treatment is presented in a later section.)

12.1.2 A Catalog of Spreading Techniques

Figure 12.4 highlights the popular techniques for spreading the information signal over a large number of signal coordinates or dimensions. For signals of bandwidth W and duration T, the dimensionality of the signaling space is approximately $2WT$. To increase the dimensionality, we can either increase W by spectrum spreading, or increase T by time spreading or time hopping (TH). With spectrum spreading the signal is spread in the frequency domain. With time hopping, a message with data rate R is allocated a longer transmission-time duration than would be used with a conventional modulation scheme. During this longer time the data are sent in bursts according to the dictates of a code. We can say that with time hopping the signal is spread in the time domain. For both cases, frequency spreading and time spreading, a jammer will be uncertain regarding the signaling subset that is currently in use.

In Figure 12.4, the first two items listed under the category of spreading, *direct sequencing* (DS) and *frequency hopping* (FH), are the most commonly used techniques for spectrum spreading. As a jamming-rejection technique, *time hopping* (TH), the third item in the list, is similar to spread spectrum in that the location of the signal coordinates is hidden from potential adversaries. Also, there are hybrid combinations of the spreading techniques, for example, DS/FH, FH/TH, and DS/FH/TH; however, these techniques can be viewed as simple extensions of the material presented here and we will not elaborate on them. In this chapter, we focus only on the two major spread-spectrum techniques: direct sequencing and frequency hopping.

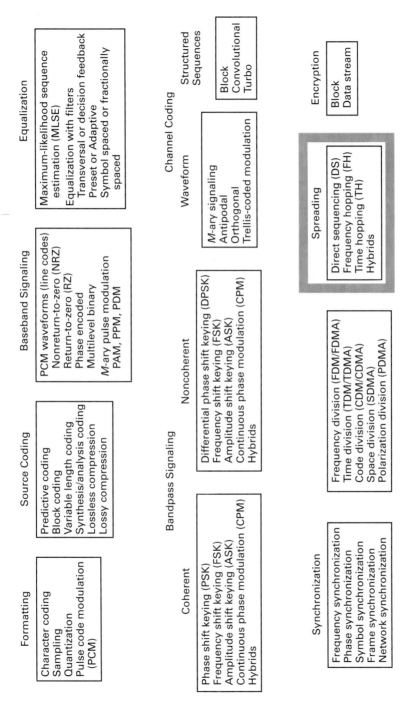

Figure 12.4 Basic digital communication transformations.

12.1.3 Model for Direct-Sequence Spread-Spectrum Interference Rejection

Figure 12.5 illustrates a model for direct-sequence spread-spectrum (DS/SS) interference rejection. At the modulator, the information signal $x(t)$, with a data rate of R bits/s, is multiplied by a spreading code signal $g(t)$, having a code symbol rate, usually called the code *chip rate*, R_{ch} chips/second. Assume that the transmission bandwidths for $x(t)$ and $g(t)$ are R hertz and R_{ch} hertz, respectively. Multiplication in the time domain transforms to convolution in the frequency domain:

$$x(t)g(t) \leftrightarrow X(\omega) * G(\omega) \tag{12.2}$$

Therefore, if the data signal is narrowband compared to the spreading signal, the resulting product signal $x(t)g(t)$ will have approximately the bandwidth of the spreading signal. (See Section A.5.)

At the demodulator, the received signal is ideally multiplied by a synchronized replica of the spreading code signal, $g(t)$, which results in the despreading of the signal. A filter with bandwidth R is used to remove any spurious higher-frequency components. If there is any undesired signal at the receiver, the multiplication by $g(t)$ will spread this undesired signal, in the same way that the multiplication by $g(t)$ at the transmitter spread the desired signal originally. Consider the effect on a jammer that attempts to position a narrowband jamming signal within the information bandwidth. The first operation at the receiver input is multiplication by the spreading signal. Hence, the jamming tone is spread to the bandwidth of the spreading signal.

The essence behind the interference rejection capability of a spread-spectrum system can be summarized as follows:

1. Multiplication by the spreading signal *once* spreads the signal bandwidth.
2. Multiplication by the spreading signal *twice*, followed by filtering, recovers the original signal.

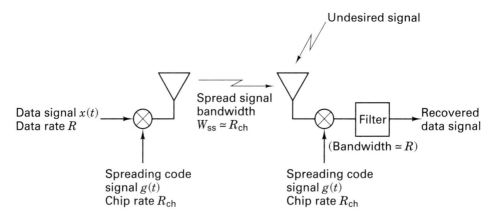

Figure 12.5   Basic spread-spectrum technique.

3. The desired signal gets multiplied *twice,* but the interference signal gets multiplied only *once.*

12.1.4 Historical Background

12.1.4.1 Transmitted Reference versus Stored Reference

During the early years of spread-spectrum investigation, one technique that was considered for operating a transmitter and receiver synchronously with a *truly random* spreading signal, such as wideband noise, was called a *transmitted reference* (TR) system. In a TR system, the transmitter would send two versions of an unpredictable wideband carrier—one modulated by data and the other unmodulated. These two signals were transmitted on separate channels. The receiver used the unmodulated carrier as the reference signal for despreading (correlating) the data-modulated carrier. The principal advantage of a TR system was that there were no significant synchronization problems at the receiver, since the data-modulated signal and the spreading signal used for despreading were transmitted simultaneously. The principal disadvantages of TR systems were that (1) the spreading code was sent in the clear and thus was available to any listener; (2) the system could be easily spoofed by a jammer sending a pair of waveforms acceptable to the receiver; (3) performance degraded at low signal levels since noise was present on both signals; and (4) twice the bandwidth and transmitted power were required because of the need to transmit the reference.

Modern spread-spectrum systems all use a technique called *stored reference* (SR), whereby the spreading code signal is independently generated at both the transmitter and the receiver. The main advantage of an SR system is that a well-designed code signal cannot be predicted by monitoring the transmission. Note that the noiselike code signal in an SR system cannot be truly random as it could in the case of a TR system. Since the same code must be generated independently at two or more sites, the code sequence must be deterministic, even though it should appear random to unauthorized listeners. Such random-appearing deterministic signals are called pseudonoise (PN), or pseudorandom signals; their generation is treated later in greater detail.

12.1.4.2 Noise Wheels

In the late 1940s and early 1950s, Mortimer Rogoff, working at ITT, demonstrated the fundamental operation of spectrum spreading systems with a novel experiment [5]. Using photographic techniques, Rogoff built a "noise wheel" for storing a noiselike signal. He randomly selected 1440 numbers not ending in 00 from the Manhattan telephone directory, and radially plotted the middle two of the last four digits so that the radius at every $\frac{1}{4}°$ represented a new random number. The drawing was transferred to the wheel-shaped film shown in Figure 12.6. When the wheel was rotated past a slit of light, the resulting intensity-modulated light beam provided a stored noiselike spreading signal to be sensed by a photocell.

Rogoff mounted two such identical wheels on a single axis driven by a 900-rpm synchronous motor. One wheel's noiselike spreading signal was modulated

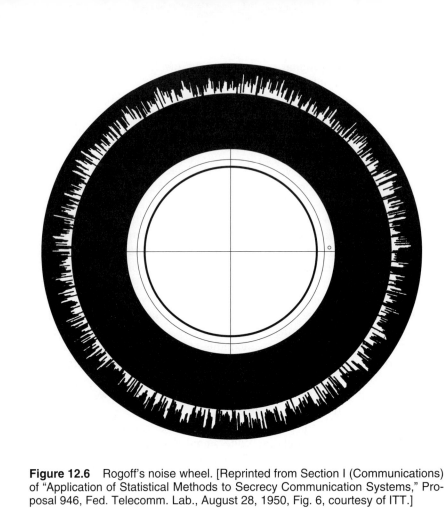

Figure 12.6   Rogoff's noise wheel. [Reprinted from Section I (Communications) of "Application of Statistical Methods to Secrecy Communication Systems," Proposal 946, Fed. Telecomm. Lab., August 28, 1950, Fig. 6, courtesy of ITT.]

with data (and interference) to provide one input to the receiving correlator, while the other wheel's unmodulated spreading signal provided the other input to the correlator. These baseband experiments, performed with the data rates of 1 bit/s, demonstrated the feasibility of conveying information hidden in noiselike signals [4].

12.2 PSEUDONOISE SEQUENCES

The spread-spectrum approach called *transmitted reference* (TR) can utilize a *truly* random code signal for spreading and despreading, since the code signal and the data-modulated code signal are simultaneously transmitted over different regions of the spectrum. The *stored reference* (SR) approach *cannot* use a truly random code signal since the code needs to be stored or generated at the receiver. For the SR system a *pseudonoise* or *pseudorandom* code signal must be used.

How does a pseudorandom signal differ from a random one? A random signal *cannot* be predicted; its future variations can only be described in a statistical sense. However, a pseudorandom signal is not random at all; it is a deterministic, periodic signal that is known to both the transmitter and receiver. Why the name "pseudonoise" or "pseudorandom"? Even though the signal is deterministic, it appears to have the statistical properties of sampled white noise. It appears, to an unauthorized listener, to be a truly random signal.

12.2.1 Randomness Properties

What are these randomness properties that make a pseudorandom signal appear truly random? There are three basic properties that can be applied to any periodic binary sequence as a test for the appearance of randomness. The properties, called *balance, run,* and *correlation,* are described for binary signals as follows:

1. *Balance property.* Good balance requires that in each period of the sequence, the number of binary ones differs from the number of binary zeros by at most one digit.

2. *Run property.* A *run* is defined as a sequence of a single type of binary digit(s). The appearance of the alternate digit in a sequence starts a new run. The length of the run is the number of digits in the run. Among the runs of ones and zeros in each period, it is desirable that about one-half the runs of each type are of length 1, about one-fourth are of length 2, one-eighth are of length 3, and so on.

3. *Correlation property.* If a period of the sequence is compared term by term with any cyclic shift of itself, it is best if the number of agreements differs from the number of disagreements by not more than one count.

In the next section, a PN sequence is generated to test these properties.

12.2.2 Shift Register Sequences

Consider the linear feedback shift register illustrated in Figure 12.7. It is made up of a four-stage register for storage and shifting, a modulo-2 adder, and a feedback path from the adder to the input of the register (modulo-2 addition has been defined in Section 2.9.3). The shift register operation is controlled by a sequence of clock pulses (not shown). At each clock pulse the contents of each stage in the register is shifted one stage to the right. Also, at each clock pulse the contents of stages X_3 and X_4 are modulo-2 added (a linear operation), and the result is fed back to stage X_1. The shift register sequence is defined to be the output of the last stage— stage X_4 in this example.

Assume that stage X_1 is initially filled with a one and the remaining stages are filled with zeros, that is, the initial state of the register is 1 0 0 0. From Figure 12.7 we can see that the succession of register states will be as follows:

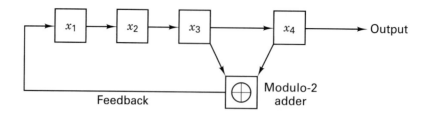

Figure 12.7 Linear feedback shift register example.

1000 0100 0010 1001 1100 0110 1011 0101

1010 1101 1110 1111 0111 0011 0001 1000

Since the last state, 1 0 0 0, corresponds to the initial state, we see that the register repeats the foregoing sequence after 15 clock pulses. The output sequence is obtained by noting the contents of stage X_4 at each clock pulse. The output sequence is seen to be

0 0 0 1 0 0 1 1 0 1 0 1 1 1 1

where the leftmost bit is the earliest bit. Let us test the sequence above for the randomness properties outlined in the preceding section. First, the balance property; there are seven zeros and eight ones in the sequence—therefore, the sequence meets the balance condition. Next, the run property; consider the zero runs—there are four of them. One-half are of length 1, and one-fourth are of length 2. The same is true for the one runs. The sequence is too short to go further, but we can see that the run condition is met. The correlation property is treated in Section 12.2.3.

The shift register generator produces sequences that depend on the number of stages, the feedback tap connections, and initial conditions. The output sequences can be classified as either *maximal length* or *nonmaximal length.* Maximal length sequences have the property that for an *n*-stage linear feedback shift register the sequence repetition period in clock pulses p is

$$p = 2^n - 1 \qquad (12.3)$$

Thus it can be seen that the sequence generated by the shift register generator of Figure 12.7 is an example of a maximal length sequence. If the sequence length is less than $(2^n - 1)$, the sequence is classified as a nonmaximal length sequence.

12.2.3 PN Autocorrelation Function

The autocorrelation function $R_x(\tau)$ of a periodic waveform $x(t)$, with period T_0, was given in Equation (1.23) and is shown below in normalized form.

$$R_x(\tau) = \frac{1}{K}\left(\frac{1}{T_0}\right) \int_{-T_0/2}^{T_0/2} x(t)x(t + \tau)dt \qquad \text{for } -\infty < \tau < \infty \qquad (12.4)$$

where

$$K = \frac{1}{T_0} \int_{-T_0/2}^{T_0/2} x^2(t)\, dt \tag{12.5}$$

When $x(t)$ is a periodic pulse waveform representing a PN code, we refer to each fundamental pulse as a *PN code symbol* or a *chip*. For such a PN waveform of unit chip duration and period p chips, the normalized autocorrelation function may be expressed as

$$R_x(\tau) = \frac{1}{p} \cdot \left(\begin{array}{l} \text{number of agreements less number of disagreements} \\ \text{in a comparison of one full period of the sequence} \\ \text{with a } \tau \text{ position cyclic shift of the sequence} \end{array} \right) \tag{12.6}$$

The normalized autocorrelation function for a maximal length sequence, $R_x(\tau)$, is shown plotted in Figure 12.8. It is clear that for $\tau = 0$, that is, when $x(t)$ and its replica are perfectly matched, $R(\tau) = 1$. However, for any cyclic shift between $x(t)$ and $x(t + \tau)$ with $(1 \leq \tau < p)$, the autocorrelation function is equal to $-1/p$ (for large p, the sequences are virtually decorrelated for a shift of a *single chip*).

It is now easy to test the output PN sequence of the shift register in Figure 12.7 for the third randomness property—correlation. The output sequence, as well as the same sequence with a single end-around shift, is as follows:

$$
\begin{array}{l}
0\ 0\ 0\ 1\ 0\ 0\ 1\ 1\ 0\ 1\ 0\ 1\ 1\ 1\ 1 \\
1\ 0\ 0\ 0\ 1\ 0\ 0\ 1\ 1\ 0\ 1\ 0\ 1\ 1\ 1 \\
\hline
d\ a\ a\ d\ d\ a\ d\ a\ d\ d\ d\ d\ a\ a\ a
\end{array}
$$

The digits that agree are labeled a and those that disagree are labeled d. Following Equation (12.6), the value of the autocorrelation function for this single one-chip shift is seen to be

$$R(\tau = 1) = \tfrac{1}{15}(7 - 8) = -\tfrac{1}{15}$$

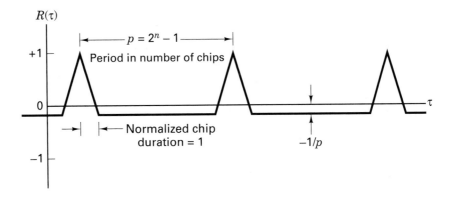

Figure 12.8 PN autocorrelation function.

Any cyclic shift yielding a mismatch from perfect synchronization results in the same autocorrelation value, $-1/p$. Hence the sequence meets the third randomness property.

12.3 DIRECT-SEQUENCE SPREAD-SPECTRUM SYSTEMS

The block diagram in Figure 12.9a depicts a *direct-sequence* (DS) modulator. "Direct sequence" is the name given to the spectrum spreading technique whereby a carrier wave is first modulated with a data signal $x(t)$, then the data-modulated signal is again modulated with a high-speed (wideband) spreading signal $g(t)$. Consider a constant-envelope data-modulated carrier having power P, radian frequency ω_0, and data phase modulation $\theta_x(t)$, given by

$$s_x(t) = \sqrt{2P} \cos \left[\omega_0 t + \theta_x(t) \right] \tag{12.7}$$

Upon further constant-envelope modulation by the spreading signal, $g(t)$, the transmitted waveform can be expressed as

$$s(t) = \sqrt{2P} \cos \left[\omega_0(t) + \theta_x(t) + \theta_g(t) \right] \tag{12.8}$$

where the phase of the carrier is now seen to have two components: $\theta_x(t)$ due to the data and $\theta_g(t)$ due to the spreading sequence.

In Chapter 4, it was shown that ideal suppressed carrier binary phase shift keying (BPSK) modulation results in instantaneous changes of π radians to the phase of the carrier, according to the dictates of the data. We can equivalently express Equation (12.7) as the multiplication of the carrier wave by $x(t)$, an antipodal pulse stream with pulse values of $+1$ or -1:

$$s_x(t) = \sqrt{2P} x(t) \cos \omega_0 t \tag{12.9}$$

If, like the data, the spreading sequence modulation is also BPSK, and $g(t)$ is an antipodal pulse stream with pulse values of $+1$ or -1, Equation (12.8) can be written as

$$s(t) = \sqrt{2P} x(t) g(t) \cos \omega_0 t \tag{12.10}$$

A modulator based on Equation (12.10) is illustrated in Figure 12.9b. The data pulse stream and the spreading pulse stream are first multiplied, and then the composite $x(t)$ modulates the carrier. If the assignment of pulse value to binary value is

Pulse value	Binary value
1	0
−1	1

then the initial step in the DS/BPSK modulation can be accomplished by the modulo-2 addition of the binary data sequence with the binary spreading sequence.

Demodulation of the DS/BPSK signal is accomplished by correlating or remodulating the received signal with a synchronized replica of the spreading signal

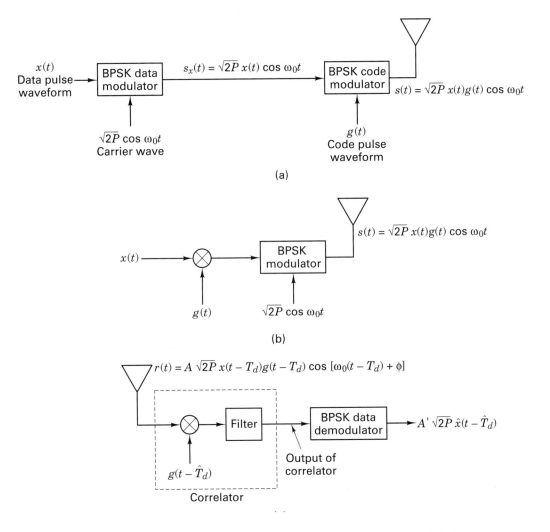

Figure 12.9 Direct-sequence spread-spectrum system. (a) BPSK direct-sequence transmitter. (b) Simplified BPSK direct-sequence transmitter. (c) BPSK direct-sequence receiver.

$g(t - \hat{T}_d)$ as seen in Figure 12.9c, where $\hat{T}_d$ is the receiver's estimate of the propagation delay T_d from the transmitter to the receiver. In the absence of noise and interference, the output signal from the correlator can be written as

$$A\sqrt{2P}x(t - T_d)g(t - T_d)g(t - \hat{T}_d) \cos[\omega_0(t - T_d) + \phi] \qquad (12.11)$$

where the constant A is a system gain parameter and ϕ is a random phase angle in the range $(0, 2\pi)$. Since $g(t) = \pm 1$, the product $g(t - T_d)g(t - \hat{T}_d)$ will be unity if $\hat{T}_d = T_d$, that is, if the code signal at the receiver is exactly synchronized with the code signal at the transmitter. When it is synchronized, the output of the receiver correlator is the despread data-modulated signal (except for a random phase ϕ and delay T_d). The despreading correlator is then followed by a conventional demodulator for recovering the data.

12.3.1 Example of Direct Sequencing

Figure 12.10 is an example of DS/BPSK modulation and demodulation following the block diagrams of Figure 12.9b and c. In Figure 12.10a are shown the binary data sequence (1, 0) and its bipolar pulse waveform equivalent $x(t)$, where the binary to pulse value assignments are the same as those described in the preceding section. Examples of a binary spreading sequence and its bipolar pulse waveform equivalent $g(t)$ are shown in Figure 12.10b. The modulo-2 addition of the data sequence and the code sequence, and the equivalent waveform of the product $x(t)g(t)$, is shown in Figure 12.10c.

For the BPSK modulation described by Equations (12.8) and (12.10), it is shown in Figure 12.10d that the phase of the carrier, $\theta_x(t) + \theta_g(t)$, equals π when the

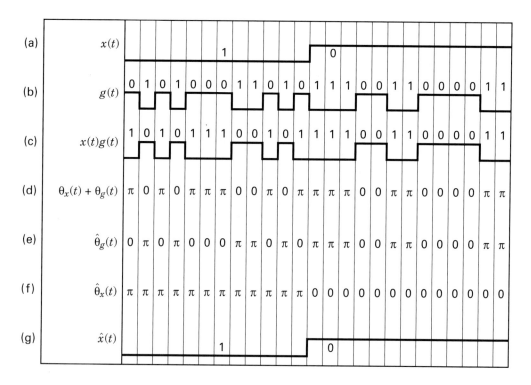

Figure 12.10 Spread-spectrum example using direct sequencing. (a) Binary data waveform to be transmitted. (b) Code sequence. (c) Transmitted sequence. (d) Phase of transmitted carrier. (e) Phase shift produced by receiver code. (f) Phase of received carrier after phase shifted (despread) by receiver code. (g) Demodulated data waveform.

value of the product waveform $x(t)g(t)$ equals -1 (or the modulo-2 sum of data and code is binary 1). Similarly, the phase of the carrier is zero when the value of $x(t)g(t)$ equals $+1$ (or the modulo-2 sum of data and code is binary 0). One can appreciate the *signal hiding* property of spread-spectrum signals by comparing the code waveform in Figure 12.10b with the composite waveform in Figure 12.10c. The latter has the signal $x(t)$ "hidden" within it. Just as your eyes have difficulty finding the slowly moving data signal in the rapidly moving code signal, it is similarly difficult for a receiver to recover a slowly moving signal from a rapidly moving code without having an exact replica of the code.

As shown in Figure 12.9c, DS/BPSK demodulation is a two-step process. The first step, despreading, is accomplished by correlating the received signal with a synchronized replica of the code. The second step, data demodulation, is accomplished with a conventional demodulator. In the example of Figure 12.10 we see the code replica $\hat{\theta}_g(t)$, in Figure 12.10e, as the phase shift (either 0 or π) that is produced at the receiver by the despreading code. Figure 12.10f illustrates the resulting estimate of the carrier phase $\hat{\theta}_x(t)$, after despreading or after $\hat{\theta}_g(t)$ has been added to $\theta_x(t) + \theta_g(t)$. At this point one can recognize the original data pattern in the phase terms of the carrier wave. The final step, shown in Figure 12.10g, is to recover an estimate of the data waveform, $\hat{x}(t)$, by the use of a BPSK demodulator.

12.3.2 Processing Gain and Performance

A fundamental issue in spread-spectrum systems is *how much* protection spreading can provide against interfering signals with finite power. Spread-spectrum techniques distribute a relatively low-dimensional signal in a large-dimensional signal space. The signal is "hidden" within the signal space, since we assume that a jammer does not know which signal coordinates are being transmitted at any time. The only recourse for the jammer, intent upon communication disruption, is to jam the entire space with its fixed total power, thus inducing a limited amount of interference in each signal coordinate, or to jam a portion of the signal space with its total power, thus leaving the remainder of the signal space free of interference.

Consider a set of D orthogonal signals, $s_i(t)$, $1 \leq i \leq D$, in an N-dimensional space, where in general, $D \ll N$. Following the development in Section 3.1.3, we can write.

$$s_i(t) = \sum_{j=1}^{N} a_{ij}\psi_j(t) \qquad \begin{matrix} i = 1, 2, \ldots, D; \\ D \ll N \end{matrix} \qquad 0 \leq t \leq T \qquad (12.12)$$

where

$$a_{ij} = \int_0^T s_i(t)\psi_j(t)\, dt \qquad\qquad (12.13)$$

and

$$\int_0^T \psi_j(t)\psi_k(t)\, dt = \begin{cases} 1 & \text{for } j = k \\ 0 & \text{otherwise} \end{cases} \qquad (12.14)$$

The $\{\psi_j(t)\}$ are linearly independent functions that *span* or characterize the N-dimensional orthonormal space and are called *basis* functions of the space. For every information symbol that is transmitted, a set of coefficients $\{a_{ij}\}$ is chosen independently, using a pseudorandom spreading code, in order to hide the D-dimensional signal set in the larger N-dimensional space. The set of random variables $\{a_{ij}\}$ assume the values $\pm a$, each with a probability of $\frac{1}{2}$. The receiver, of course, has access to each set of coefficients chosen in order to perform the necessary correlation despreading. Even if the same ith symbol is sent repeatedly, the set $\{a_{ij}\}$ used to transmit it is newly selected from symbol to symbol. The energy in each signal waveform of the D signal set will be assumed equal, so that we can write the average energy for each signal as

$$E_s = \int_0^T \overline{s_i^2(t)}\, dt = \sum_{j=1}^N \overline{a_{ij}^2} \quad i = 1, 2, \ldots, D \tag{12.15}$$

where the overbar means the expected value over the ensemble of many symbol transmissions. The independent coefficients have zero mean and correlation:

$$\overline{a_{ij}a_{ik}} = \begin{cases} \dfrac{E_s}{N} & \text{for } j = k \\ 0 & \text{otherwise} \end{cases} \tag{12.16}$$

The standard assumption is that the jammer has no a priori knowledge regarding the selection of the signaling coefficients $\{a_{ij}\}$. As far as the jammer is concerned, the coefficients are uniformly distributed over the N basis coordinates. If the jammer chooses to distribute its power uniformly over the total signal space, the jammer waveform $w(t)$ can be written

$$w(t) = \sum_{j=1}^N b_j \psi_j(t) \tag{12.17}$$

with total energy

$$E_w = \int_0^T w^2(t)\, dt = \sum_{j=1}^N b_j^2 \tag{12.18}$$

A reasonable goal for a jammer would be to devise a strategy for selecting the portions b_j^2, of its fixed total energy E_w so as to minimize the desired signal-to-noise ratio (SNR) at the receiver after demodulation.

At the receiver, the detector output (ignoring receiver noise),

$$r(t) = s_i(t) + w(t) \tag{12.19}$$

is correlated with the set of possible transmitted signals, so that the output of the ith correlator is

$$z_i = \int_0^T r(t)s_i(t)\, dt = \sum_{j=1}^N (a_{ij}^2 + b_j a_{ij}) \tag{12.20}$$

The second term on the right side of Equation (12.20) averages to zero over the ensemble of all possible pseudorandom code sequences, since the set of random variables $\{a_{ij}\}$ assume the values $\pm a$, each with probability $\frac{1}{2}$. Therefore, given that $s_m(t)$ was transmitted, the expected value of the output of the ith correlator can be written, following the development in References [6, 7],

$$\mathbf{E}(z_i|s_m) = \sum_{j=1}^{N} \overline{a_{ij}^2} = \begin{cases} E_s & \text{for } i = m \\ 0 & \text{otherwise} \end{cases} \tag{12.21}$$

In Equation (12.21), the term $\mathbf{E}(z_i|s_m)$ for $i = m$ is to be interpreted as follows. Given that $s_i(t)$ is to be transmitted, N coefficients a_{ij} $(1 \le j \le N)$ are chosen pseudo-randomly (the receiver is assumed to have access to each choice of the a_{ij} for correlation despreading). Hence, in computing $\mathbf{E}(z_i|s_i)$, even though the ith information symbol is specified at the transmitter, the pattern of coefficients used to send it appears random (to the unauthorized receiver) for each transmission. Equation (12.21) presumes that the jammer had not been successful in its attempt to employ some clever tactics (described in Section 12.6).

Let us assume that all D signals are equally likely. Then the expected value at the output of any of the D correlators is

$$\mathbf{E}(z_i) = \frac{E_s}{D} \tag{12.22}$$

Similarly, using Equations (12.15) to (12.21), we compute var $(z_i|s_i)$, the variance at the output of the ith correlator, given that the ith signal was transmitted:

$$\text{var}\,(z_i|s_i) = \sum_{j,k} b_j b_k \overline{a_{ij} a_{ik}} \tag{12.23}$$

$$= \sum_{j=1}^{N} b_j^2 \overline{a_{ij}^2}$$

$$= \sum_{j=1}^{N} b_j^2 \frac{E_s}{N} \tag{12.24}$$

$$= \frac{E_w E_s}{N}$$

For completeness, the variance at the output of the ith correlator, given that the mth signal was transmitted, where $i \ne m$, can similarly be computed to be

$$\text{var}\,(z_i|s_m) = \frac{E_w E_s}{N} + \frac{E_s^2}{N} \tag{12.25}$$

The signal-to-jammer ratio at the output of the ith correlator can be defined as

$$\text{SJR} = \sum_{m=1}^{D} \frac{\mathbf{E}^2(z_i|s_m)}{\text{var}\,(z_i|s_m)} P(s_m) = \frac{E_s^2/D}{E_w E_s/N} = \frac{E_s N}{E_w D} \tag{12.26}$$

where the probability of the mth signal $P(s_m) = 1/D$, since the signals are assumed to occur with equal probability, and where the signal energy and the jammer energy in the ith correlator are denoted by $\mathbf{E}^2(z_i)$ and var (z_i), respectively. Because of Equation (12.21), the only terms in the summation of Equation (12.26) not equal to zero are those for which $i = m$. The result is independent of the way in which the jammer chooses to distribute its energy. Therefore, regardless of how b_j is chosen, subject to $\Sigma_j\, b_j^2 = E_w$, the SJR in Equation (12.26) indicates that spreading gives the signal an advantage of a factor of N/D over the jammer. The ratio N/D is known as the *processing gain* G_p.

Since the approximate dimensionality of a signal with bandwidth W and duration T is $2WT$, we can express the processing gain as

$$ G_p = \frac{N}{D} \approx \frac{2W_{ss}T}{2W_{min}T} = \frac{W_{ss}}{R} \tag{12.27} $$

Where W_{ss} is the spread-spectrum bandwidth (the total bandwidth used by the spreading technique) and W_{min} is the minimum bandwidth of the data (taken to be the data rate, R). For direct sequence systems, W_{ss} is approximately the code chip rate R_{ch}, and W_{min} is similarly the data rate R, giving

$$ G_p = \frac{R_{ch}}{R} \tag{12.28} $$

where the *chip* is defined as the shortest uninterrupted waveform in the system. For direct-sequence spread-spectrum systems, the chip constitutes a PN-code pulse or signaling element.

Whatever the spread-spectrum application (e.g., interference rejection, fine-time resolution), processing gain is the parameter that expresses the performance advantage of the spread-spectrum system over a narrowband system. The modulation of choice for direct-sequence spread-spectrum systems is generally BPSK or QPSK. Thus, consider a single binary symbol comprising 1000 BPSK code chips. From Equation (12.28) it should be clear that the processing gain for such a signaling scheme is 1000. To appreciate that the processing gain of such a spread-spectrum system makes for more robust signaling than a similar narrowband scheme, consider the following. Imagine that the detection process makes a decision on each of the 1000 chips in our example. Of course it doesn't do that; the 1000 chips are accumulated and correlated to a code, yielding a single decision on the meaning of the bit. But, if you think of the process as 1000 individual decisions, you might say that 499 of those decisions can be incorrect, and you will still detect the bit correctly.

12.4 FREQUENCY HOPPING SYSTEMS

We now consider a spread-spectrum technique called frequency hopping (FH). The modulation most commonly used with this technique is M-ary frequency shift keying (MFSK), where $k = \log_2 M$ information bits are used to determine which one of M frequencies is to be transmitted. The position of the M-ary signal set is shifted pseudorandomly by the frequency synthesizer over a hopping bandwidth W_{ss}. A typical FH/MFSK system block diagram is shown in Figure 12.11. In a conventional MFSK system, the data symbol modulates a *fixed frequency* carrier; in an FH/MFSK system, the data symbol modulates a carrier whose frequency is *pseudorandomly* determined. In either case, a single tone is transmitted. The FH system in Figure 12.11 can be thought of as a two-step modulation process—data modulation and frequency-hopping modulation—even though it can be implemented as a single step whereby the frequency synthesizer produces a transmission tone based on the simultaneous dictates of the PN code and the data. At each frequency hop time, a PN generator feeds the frequency synthesizer a frequency word (a sequence of ℓ chips), which dictates one of 2^ℓ symbol-set positions. The frequency-hopping bandwidth W_{ss}, and the minimum frequency spacing between consecutive hop positions Δf, dictate the minimum number of chips necessary in the frequency word.

For a given hop, the occupied transmission bandwidth is identical to the bandwidth of conventional MFSK, which is typically much smaller than W_{ss}. However, averaged over many hops, the FH/MFSK spectrum occupies the entire spread-spectrum bandwidth. Spread-spectrum technology permits FH bandwidths of the order of several gigahertz, which is an order of magnitude larger than implementable DS bandwidths [8], thus allowing for larger processing gains in FH compared to DS systems. Since frequency hopping techniques operate over such wide bandwidths, it is difficult to maintain phase coherence from hop to hop. Therefore, such schemes are usually configured using noncoherent demodulation. Nevertheless, consideration has been given to coherent FH in Reference [9].

In Figure 12.11 we see that the receiver reverses the signal processing steps of the transmitter. The received signal is first FH demodulated (dehopped) by mixing it with the same sequence of pseudorandomly selected frequency tones that was used for hopping. Then the dehopped signal is applied to a conventional bank of M noncoherent energy detectors to select the most likely symbol.

Example 12.1 Frequency Word Size

A hopping bandwidth W_{ss} of 400 MHz and a frequency step size Δf of 100 Hz are specified. What is the minimum number of PN chips that are required for each frequency word?

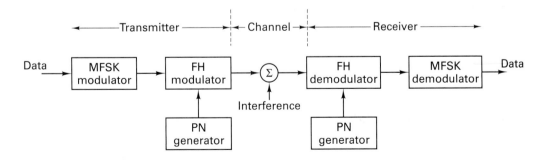

Figure 12.11 FH/MFSK system.

Solution

$$\text{Number of tones contained in } W_{ss} = \frac{W_{ss}}{\Delta f} = \frac{400 \text{ MHz}}{100 \text{ Hz}}$$

$$= 4 \times 10^6$$

$$\text{Minimum number of chips} = \lceil \log_2 (4 \times 10^6) \rceil$$

$$= 22 \text{ chips}$$

where $\lceil x \rceil$ indicates the smallest integer value not less than x.

12.4.1 Frequency Hopping Example

Consider the frequency hopping example illustrated in Figure 12.12. The input data consist of a binary sequence with a data rate of $R = 150$ bits/s. The modulation is 8-ary FSK. Therefore, the symbol rate is $R_s = R/(\log_2 8) = 50$ symbols/s (the symbol duration $T = 1/50 = 20$ ms). The frequency is hopped once per symbol, and the hopping is time synchronous with the symbol boundaries. Thus, the hopping rate is 50 hops/s. Figure 12.12 depicts the time–bandwidth plane of the communication resource; the abscissa represents time, and the ordinate represents the hopping bandwidth, W_{ss}. The legend on the right side of the figure illustrates a set of 8-ary FSK symbol-to-tone assignments. Notice that the tone separation specified is $1/T = 50$ Hz, which corresponds to the minimum required tone spacing for the orthogonal signaling of this noncoherent FSK example (see Section 4.5.4).

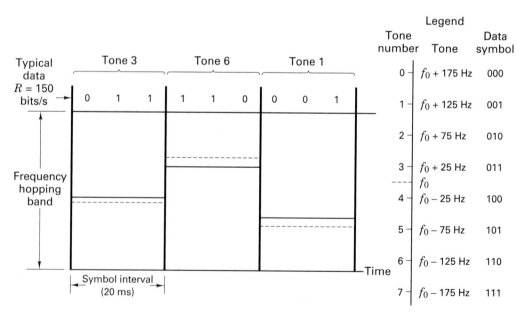

Figure 12.12   Frequency-hopping example using 8-ary FSK modulation.

A typical binary data sequence is shown at the top of Figure 12.12. Since the modulation is 8-ary FSK, the bits are grouped three at a time to form symbols. In a *conventional* 8-ary FSK scheme, a single-sideband tone (offset from f_0, the *fixed* center frequency of the data band), would be transmitted according to an assignment like the one shown in the legend. The only difference in this FH/MFSK example is that the center frequency of the data band f_0 is *not fixed*. For each new symbol, f_0 hops to a new position in the hop bandwidth, and the entire data-band structure moves with it. In the example of Figure 12.12, the first symbol in the data sequence, 0 1 1, yields a tone 25 Hz above f_0. The diagram depicts f_0 with a dashed line and the symbol tone with a solid line. During the second symbol interval, f_0 has hopped to a new spectral location, as indicated by the dashed line. The second symbol, 1 1 0, dictates that a tone indicated by the solid line, 125 Hz below f_0, shall be transmitted. Similarly, the final symbol in this example, 0 0 1, calls for a tone 125 Hz above f_0. Again, the center frequency has moved, but the relative positions of the symbol tones remain fixed.

12.4.2 Robustness

A common dictionary definition describes the term *robustness* as the state of being strong and healthy; full of vigor; hardy. In the context of communications, the usage is not too different. Robustness characterizes a signal's ability to withstand impairments from the channel, such as noise, jamming, fading, and so on. A signal configured with multiple replicate copies, each transmitted on a different frequency, has a greater likelihood of survival than does a single such signal with equal total power. The greater the diversity (multiple transmissions, at different frequencies, spread in time), the more robust the signal against random interference.

The following example should clarify the concept. Consider a message consisting of four symbols: s_1, s_2, s_3, s_4. The introduction of diversity starts by repeating the message N times. Let us choose $N = 8$. Then, the repeated symbols, called *chips,* can be written.

$$s_1 s_1 s_1 s_1 s_1 s_1 s_1 s_1 s_2 s_2 s_2 s_2 s_2 s_2 s_2 s_2 s_3 s_3 s_3 s_3 s_3 s_3 s_3 s_3 s_4 s_4 s_4 s_4 s_4 s_4 s_4 s_4$$

Each chip is transmitted at a different hopping frequency (the center of the data bandwidth is changed for each chip). The resulting transmissions at frequencies f_i, $f_j, f_k, \ldots$ yield a more robust signal than without such diversity. A target-shooting analogy is that a pellet from a barrage of shotgun pellets has a better chance of hitting a target, compared with the action of a single bullet.

12.4.3 Frequency Hopping with Diversity

In Figure 12.13 we extend the example illustrated in Figure 12.12, with the additional feature of a chip repeat factor of $N = 4$. During each 20-ms symbol interval, there are now four columns, corresponding to the four separate chips to be transmitted for each symbol. At the top of the figure we see the same data sequence, with $R = 150$ bps, as in the earlier example; and we see the same 3-bit partitioning to form the 8-ary symbols. Each symbol is transmitted four times, and for each transmission the center frequency of the data band is hopped to a new region of the hopping band, under the control of a PN code generator. Therefore, for this example, each chip interval, T_c, is equal to $T/N = 20$ ms/4 = 5 ms in duration, and the hopping rate is now

$$\frac{NR}{\log_2 8} = 200 \text{ hops/s}$$

Notice that the spacing between frequency tones must change to meet the changed requirement for orthogonality. Since the duration of each FSK tone is now equal to the chip duration, that is, $T_c = T/N$, the minimum separation between tones is $1/T_c = N/T = 200$ Hz. As in the earlier example, Figure 12.13 illustrates that the center of the data band (plus the modulation structure) is shifted at each new chip time. The position of the solid line (transmission frequency) has the same relationship to the dashed line (center of the data band) for each of the chips associated with a given symbol.

12.4.4 Fast Hopping versus Slow Hopping

In the case of direct-sequence spread-spectrum systems, the term "chip" refers to the PN code symbol (the symbol of shortest duration in a DS system). In a similar sense for frequency hopping systems, the term "chip" is used to characterize the

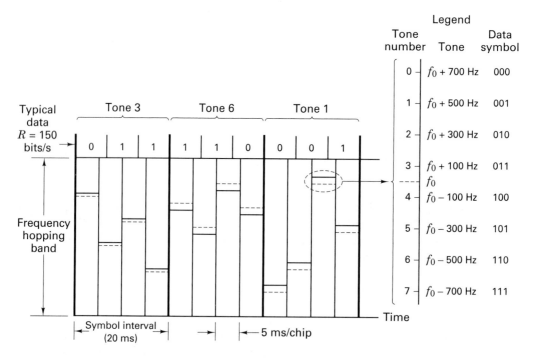

Figure 12.13 Frequency hopping example with diversity ($N = 4$).

Spread-Spectrum Techniques   Chap. 12

shortest uninterrupted waveform in the system. Frequency hopping systems are classified as *slow-frequency hopping* (SFH), which means there are several modulation symbols per hop, or as *fast-frequency hopping* (FFH), which means that there are several frequency hops per modulation symbol. For SFH, the shortest uninterrupted waveform in the system is that of the data symbol; however, for FFH, the shortest uninterrupted waveform is that of the hop. Figure 12.14a illustrates an example of FFH; the data symbol rate is 30 symbols/s and the frequency hopping rate is 60 hops/s. The figure illustrates the waveform $s(t)$ over one symbol duration ($\frac{1}{30}$ s). The waveform change in (the middle of) $s(t)$ is due to a new frequency hop. In this example, a chip corresponds to a hop since the hop duration is shorter than the symbol duration. Each chip corresponds to half a symbol. Figure 14.14b illustrates an example of SFH; the data symbol rate is still 30 symbols/s, but the frequency hopping rate has been reduced to 10 hops/s. The waveform $s(t)$ is shown over a duration of three symbols ($\frac{1}{10}$ s). In this example, the hopping boundaries appear only at the beginning and end of the three-symbol duration. Here, the changes in the waveform are due to the modulation state changes; therefore, in this

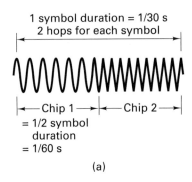

(a)

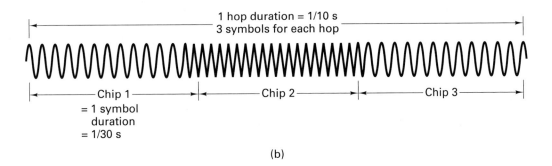

(b)

Figure 12.14 Chip—in the context of an FH/MFSK system. (a) Example 1: Frequency hopping MFSK system with symbol rate = 30 symbols/s and hopping rate = 60 hops/s. 1 chip = 1 hop. (b) Example 2: Same as part (a) except hopping rate = 10 hops/s. 1 chip = 1 symbol.

example a chip corresponds to a data symbol, since the data symbol is shorter than the hop duration.

Figure 12.15a illustrates an FFH example of a binary FSK system. The diversity is $N = 4$. There are 4 chips transmitted per bit. As in Figure 12.13, the dashed line in each column corresponds to the center of the data band and the solid line corresponds to the symbol frequency. Here, for FFH, the chip duration is the hop duration. Figure 12.15b illustrates an example of an SFH binary FSK system. In this case, there are 3 bits transmitted during the time duration of a single hop. Here, for SFH, the chip duration is the bit duration. If this SFH example were changed from a binary system to an 8-ary system, what would the chip duration then correspond to? If the system were implemented as an 8-ary scheme, each 3 bits would be transmitted as a single data symbol. The symbol boundaries and the hop boundaries would then be the same, and the chip duration, the hop duration, and the symbol duration would all be the same.

12.4.5 FFH/MFSK Demodulator

Figure 12.16 illustrates the schematic for a typical fast frequency hopping MFSK (FFH/MFSK) demodulator. First, the signal is dehopped using a PN generator identical to the one used for hopping. Then, after filtering with a low-pass filter that

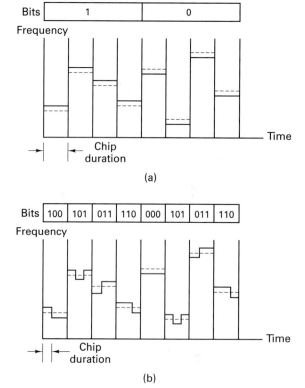

(a)

(b)

Figure 12.15 Fast hopping versus slow hopping in a binary system. (a) Fast-hopping example: 4 hops/bit. (b) Slow-hopping example: 3 bits/hop.

has a bandwidth equal to the data bandwidth, the signal is demodulated using a bank of M envelope or energy detectors. Each envelope detector is followed by a clipping circuit and an accumulator. The clipping circuit serves an important function in the presence of an intentional jammer or other strong unpredictable interference; it is treated in a later section. The demodulator does *not* make symbol decisions on a chip-by-chip basis. Instead, the energy from the N chips are accumulated, and after the energy from the Nth chip is added to the $N - 1$ earlier ones, the demodulator makes a symbol decision by choosing the symbol that corresponds to the accumulator, z_i $(i = 1, 2, \ldots, M)$, with maximum energy.

12.4.6 Processing Gain

Equation (12.27) shows the general expression for processing gain as $G_p = W_{ss}/R$. In the case of direct-sequence spread-spectrum, W_{ss} was set equal to the chip rate R_{ch}. In the case of frequency hopping, Equation (12.27) still expresses the processing gain, but here we set W_{ss} equal to the frequency band over which the system may hop. We designate this band as the *hopping band* $W_{hopping}$, and thus the processing gain for frequency hopping systems is written as

$$ G_p = \frac{W_{hopping}}{R} \tag{12.29} $$

12.5 SYNCHRONIZATION

For both DS and FH spread-spectrum systems, a receiver must employ a *synchronized* replica of the spreading or code signal to demodulate the received signal successfully. The process of synchronizing the locally generated spreading signal with

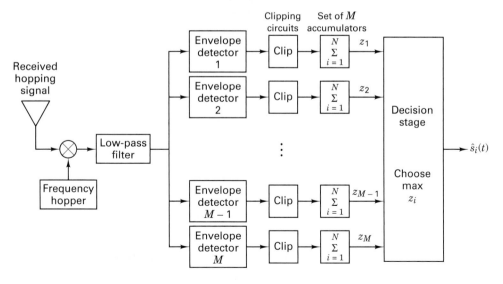

Figure 12.16 FFH/MFSK demodulator.

the received spread-spectrum signal is usually accomplished in two steps. The first step, called *acquisition,* consists of bringing the two spreading signals into *coarse* alignment with one another. Once the received spread-spectrum signal has been acquired, the second step, called *tracking,* takes over and continuously maintains the best possible waveform *fine* alignment by means of a feedback loop.

12.5.1 Acquisition

The acquisition problem is one of searching throughout a region of time and frequency uncertainty in order to synchronize the received spread-spectrum signal with the locally generated spreading signal. Acquisition schemes can be classified as coherent or noncoherent. Since the despreading process typically takes place before carrier synchronization, and therefore the carrier phase is unknown at this point, most acquisition schemes utilize noncoherent detection. When determining the limits of the uncertainty in time and frequency, the following items must be considered:

1. Uncertainty in the distance between the transmitter and the receiver translates into uncertainty in the amount of propagation delay.
2. Relative clock instabilities between the transmitter and the receiver result in phase differences between the transmitter and receiver spreading signals that will tend to grow as a function of elapsed time between synchronization.
3. Uncertainty of the receiver's relative velocity with respect to the transmitter translates into uncertainty in the value of Doppler frequency offset of the incoming signal.
4. Relative oscillator instabilities between the transmitter and the receiver result in frequency offsets between the two signals.

12.5.1.1 Correlator Structures

A common feature of all acquisition methods is that the received signal and the locally generated signal are first correlated to produce a measure of similarity between the two. This measure is then compared with a threshold to decide if the two signals are in synchronism. If they are, the tracking loop takes over.* If they are not, the acquisition procedure provides for a phase or frequency change in the locally generated code as a part of a systematic search through the receiver's phase and frequency uncertainty region, and another correlation is attempted.

Consider the direct-sequence *parallel-search* acquisition system shown in Figure 12.17. The locally generated code $g(t)$ is available with delays that are spaced one-half chip $(T_c/2)$ apart. If the time uncertainty between the local code and the received code is N_c chips, and a complete parallel search of the entire time uncertainty region is to be accomplished in a single search time, $2N_c$ correlators are used. Each correlator simultaneously examines a sequence of λ chips, after which the $2N_c$

---

*Quite often to maintain a small false alarm probability, the threshold crossing must be further verified by a suitable verification algorithm before the tracking loop takes over [4].

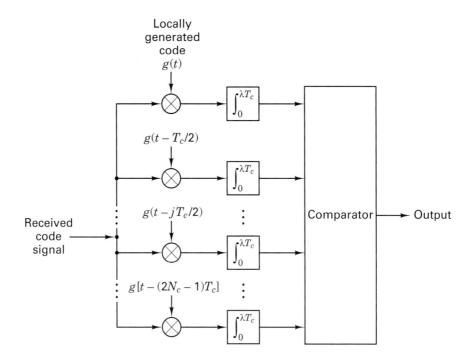

Figure 12.17 Direct-sequence parallel search acquisition.

correlator outputs are compared. The locally generated code, corresponding to the correlator with the largest output is chosen. Conceptually, this is the simplest of the search techniques; it considers all possible code positions (or fractional code positions) in parallel and uses a maximum likelihood algorithm for acquiring the code. Each detector output pertains to the identical observation of received signal plus noise. As λ increases, the synchronization error probability (i.e., the probability of choosing the incorrect code alignment) decreases. Thus, λ is chosen as a compromise between minimizing the probability of a synchronization error and minimizing the time to acquire.

Figure 12.18 illustrates a simple acquisition scheme for a frequency hopping system. Assume that a sequence of N consecutive frequencies from the hop sequence is chosen as a synchronization pattern (without data modulation). The N noncoherent matched filters each consists of a mixer followed by a bandpass filter (BPF) and a square-law envelope detector (an envelope detector followed by a square-law device). If the frequency hopping sequence is $f_1, f_2, \ldots, f_N$, delays are inserted into the matched filters so that when the correct frequency hopping sequence appears, the system produces a large output, indicating detection of the synchronization sequence. Acquisition can be accomplished rapidly because all possible code offsets are examined simultaneously. Note that the presence of bandpass filters (BPF) in Figure 12.18 indicates that the local oscillator frequencies $f_1, f_2, \ldots, f_N$ are chosen to have offsets by some intermediate frequency (IF) from

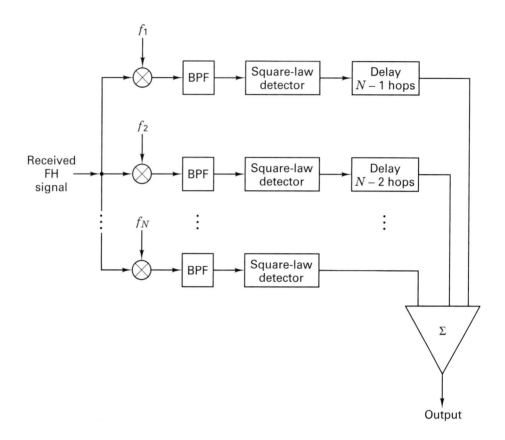

Figure 12.18 Frequency hopping acquisition scheme.

the expected received hop sequence. The same system can be implemented with local oscillator frequencies chosen (without offsets) so that the mixers yield baseband signals, and thus the filters would need to be low-pass filters (LPF). The mixers are typically complex, yielding in-phase and quadrature terms.

If, during each correlation, λ chips (each chip having a duration of T_c) are examined, the maximum time required for a fully parallel search is

$$(T_{\mathrm{acq}})_{\max} = \lambda T_c \tag{12.30}$$

The mean acquisition time of a parallel search system can be approximated by noting that after integrating over λ chips, a correct decision will be made with probability P_D, called the *probability of detection*. If an incorrect output is chosen, an additional λ chips are again examined to make a determination of the correct output. Therefore, on the average, the acquisition time is [4]

$$\overline{T}_{\mathrm{acq}} = \lambda T_c P_D + 2\lambda T_c P_D(1 - P_D) + 3\lambda T_c P_D(1 - P_D)^2 + \cdots \tag{12.31}$$

$$= \frac{\lambda T_c}{P_D}$$

Since the required number of correlators or matched filters can be prohibitively large, fully parallel acquisition techniques are not usually used. In place of Figures 12.17 and 12.18, a single correlator or matched filter can be implemented that will *serially search* until synchronization is achieved. Naturally, trade-offs between fully parallel, fully serial, and combinations of the two involve hardware complexity versus time to acquire for the same uncertainty and chip rate.

12.5.1.2 Serial Search

A popular strategy for the acquisition of spread-spectrum signals is to use a single correlator or matched filter to serially search for the correct phase of the DS code signal or the correct hopping pattern of the FH signal. A considerable reduction in complexity, size, and cost can be achieved by a serial implementation that repeats the correlation procedure for each possible sequence shift. Figures 12.19 and 12.20 illustrate the basic configuration for DS and FH spread-spectrum schemes, respectively. In a stepped serial acquisition scheme for a DS system, the timing epoch of the local PN code is set, and the locally generated PN signal is correlated with the incoming PN signal. At fixed examination intervals of λT_c (search dwell time), where $\lambda \gg 1$, the output signal is compared to a preset threshold. If the output is below the threshold, the phase of the locally generated code signal is incremented by a fraction (usually one-half) of a chip and the correlation is reexamined. When the threshold is exceeded, the PN code is assumed to have been acquired, the phase-incrementing process of the local code is inhibited, and the code tracking procedure will be initiated. In a similar scheme for FH systems, shown in Figure 12.20, the PN code generator controls the frequency hopper. Acquisition is accomplished when the local hopping is aligned with that of the received signal.

The maximum time required for a fully serial DS search, assuming that the search proceeds in half-chip increments, is

$$(T_{acq})_{max} = 2N_c \lambda T_c \qquad (12.32)$$

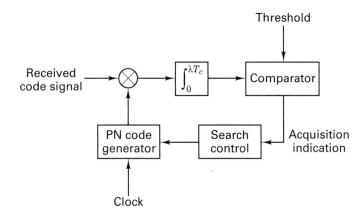

Figure 12.19 Direct-sequence serial search acquisition.

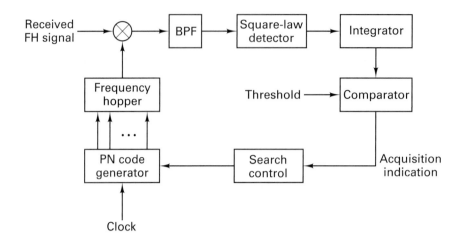

Figure 12.20 Frequency hopping serial search acquisition.

where the uncertainty region to be searched is N_c chips long. The mean acquisition time of a serial DS search system can be shown, for $N_c \gg \frac{1}{2}$ chip, to be [10]

$$\overline{T}_{\text{acq}} = \frac{(2 - P_D)(1 + KP_{\text{FA}})}{P_D}(N_c \lambda T_c) \qquad (12.33)$$

where λT_c is the search dwell time, P_D is the probability of correct detection, and P_{FA} is the probability of false alarm. We can regard the time interval $K\lambda T_c$, where $K \gg 1$, as the time needed to verify a detection. Therefore, in the event of a false alarm, $K\lambda T_c$ seconds is the time penalty incurred. For $N_c \gg \frac{1}{2}$ chip and $K \ll 2N_c$, the variance of the acquisition time is

$$(\text{var})_{\text{acq}} = (2N_c\lambda T_c)^2(1 + KP_{\text{FA}})\left(\frac{1}{12} + \frac{1}{P_D^2} - \frac{1}{P_D}\right) \qquad (12.34)$$

12.5.1.3 Sequential Estimation

Another search technique, called *rapid acquisition by sequential estimation* (RASE), proposed by Ward [11], is illustrated in Figure 12.21. The switch is initially in position 1. The RASE system enters its best estimate of the first n received code chips into the n stages of its local PN generator. The fully loaded register defines a starting state from which the generator begins its operation. A PN sequence has the property that the next combination of register states depends only on the present combination of states. Therefore, if the first n received chips are correctly estimated, all the following chips from the local PN generator will be correctly generated. The switch is next thrown to position 2. If the starting state had been correctly estimated, the local generator generates the same sequences as the incoming waveform, in the absence of noise. If the correlator output after λT_c exceeds a preset threshold level, we assume that synchronization has occurred. If the output is less than the threshold, the switch is returned to position 1, the register is reloaded

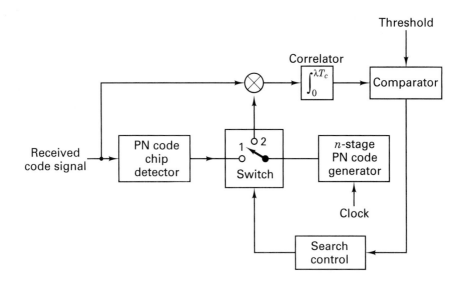

Figure 12.21 Rapid acquisition by sequential estimation.

with estimates of the next n received chips, and the procedure is repeated. Once synchronization has occurred, the system no longer needs estimates of the input code chips. We can calculate the *minimum* acquisition time for the case when no noise is present. The first n chips will be correctly loaded into the register, and therefore, the acquisition time is

$$T_{\text{acq}} = nT_c \qquad (12.35)$$

While the RASE system has a rapid acquisition capability it has the drawback of being highly vulnerable to noise and interference signals. The reason for this is that the estimation process consists of a simple chip-by-chip hard-decision demodulation, without using the interference rejection benefits of the PN code.

For an extensive treatment of sequential estimation, see Reference [4].

12.5.2 Tracking

Once acquisition or coarse synchronization is completed, tracking or fine synchronization takes place. Tracking code loops can be classified as coherent or noncoherent. A coherent loop is one in which the carrier frequency and phase are known exactly so that the loop can operate on a baseband signal. A noncoherent loop is one in which the carrier frequency is not known exactly (due to Doppler effects, for example), nor is the phase. In most instances, since the carrier frequency and phase are not known exactly, a priori, a noncoherent code loop is used to track the received PN code. Tracking loops are further classified as a *full-time* early-late tracking loop, often referred to as a *delay-locked loop* (DLL), or as a *time-shared* early-late tracking loop, frequently referred to as a *tau-dither loop* (TDL). A basic noncoherent DLL loop for a direct-sequence spread-spectrum system using binary phase shift keying (BPSK) is shown in Figure 12.22. The data $x(t)$ and the code $g(t)$ each modulate the carrier wave using BPSK, and as before in the absence of noise and interference, the received waveform can be expressed as

$$r(t) = A\sqrt{2P}\, x(t)g(t) \cos(\omega_0 t + \phi) \qquad (12.36)$$

where the constant A is a system gain parameter and ϕ is a random phase angle in the range $(0, 2\pi)$. The locally generated code of the tracking loop is offset in phase from the incoming $g(t)$ by a time τ, where $\tau < T_c/2$. The loop provides *fine* synchronization by first generating two PN sequences $g(t + T_c/2 + \tau)$ and $g(t - T_c/2 + \tau)$ delayed from each other by one chip. The two bandpass filters are designed to pass the data and to average the product of $g(t)$ and the two PN sequences $g(t \pm T_c/2 + \tau)$. (See Reference [4] for the optimum filter bandwidth for a given filter type.) The square-law envelope detector eliminates the data since $|x(t)| = 1$. The output of each envelope detector is given approximately by

$$E_D \approx \mathbf{E}\left\{ \left| g(t)g\left(t \pm \frac{T_c}{2} + \tau\right)\right|\right\} = \left|R_g\left(\tau \pm \frac{T_c}{2}\right)\right| \qquad (12.37)$$

where the operator $\mathbf{E}\{\cdot\}$ means *expected value* and $R_g(x)$ is the autocorrelation function of the PN waveform as shown in Figure 12.8. The feedback signal $Y(\tau)$ is shown in Figure 12.23. When τ is positive, the feedback signal $Y(\tau)$ instructs the voltage-controlled oscillator (VCO) to increase its frequency, thereby forcing τ to decrease, and when τ is negative, $Y(\tau)$ instructs the VCO to decrease, thereby forcing τ to increase. When τ is a suitably small number, $g(t)g(t + \tau) \approx 1$, yielding the

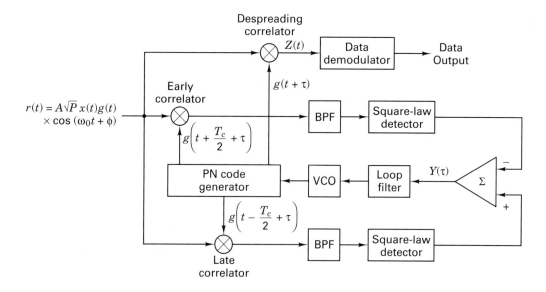

Figure 12.22 Delay-locked loop for tracking direct-sequence signals.

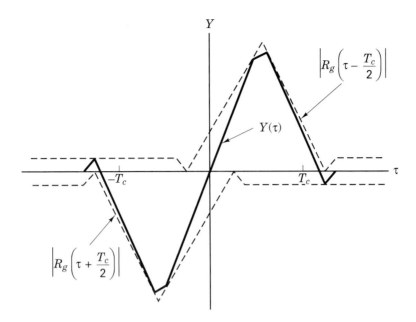

$$\left|R_g\left(\tau - \frac{T_c}{2}\right)\right|$$

$Y(\tau)$

$-T_c$

T_c

τ

$$\left|R_g\left(\tau + \frac{T_c}{2}\right)\right|$$

12.23   DLL feedback signal $Y(\tau)$.

despread signal $Z(t)$, which is then applied to the input of a conventional data demodulator. Detailed analysis of the DLL can be found in References [4, 12–14].

A problem with the DLL is that the early and late arms must be precisely gain balanced or else the feedback signal $Y(\tau)$ will be offset and will not produce a zero signal when the error is zero. This problem is solved by using a time-shared tracking loop in place of the full-time delay-locked loop. The time-shared loop time shares the use of the early-late correlators. The main advantages are that only one correlator need be used in the design of the loop, and further, that dc offset problems are reduced.

A problem with some control loops is that if things are going well and the loop is tracking accurately, the control signal is essentially zero. When the control signal is zero, the loop can get "confused" and do erratic things. This is especially the case in more sophisticated tracking loops that modify their own loop gain in response to the perceived environment. An offshoot of the time-shared tracking loop, called the *tau-dither loop* (TDL), shown in Figure 12.24, tends to deal with this potential problem by intentionally injecting a small error in the tracking correction, so that the loop kind of vibrates around the correct answer. This vibration is typically small, so that the loss in performance is minimal. This design has the advantage that only one correlator is needed to provide the code *tracking* function *and* the *despreading* function. Just as in the case of a DLL, the received signal is correlated with an early and a late version of the locally generated PN code. As shown in Figure 12.24, the PN code generator is driven by a clock signal whose phase is *dithered* back and forth with a square-wave switching function; this eliminates the necessity of ensuring identical transfer functions of the early and late

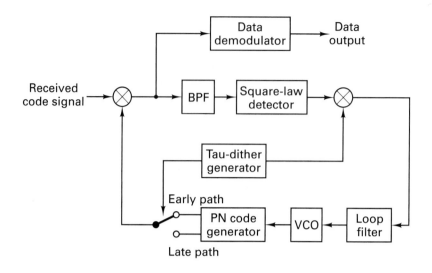

Figure 12.24 Tau-dither tracking loop.

paths. The signal-to-noise performance of the TDL is only about 1.1 dB worse than that of the DLL if the arm filters are designed properly [4]. (For a comprehensive treatment of synchronization of PN codes, see References [4, 15, 16].)

12.6 JAMMING CONSIDERATIONS

12.6.1 The Jamming Game

The goals of a jammer are to deny reliable communications to his adversary and to accomplish this at minimum cost. The goals of the communicator are to develop a jam-resistant communication system under the following assumptions: (1) complete invulnerability is not possible; (2) the jammer has a priori knowledge of most system parameters, such as frequency bands, timing, traffic, and so on; (3) the jammer has *no* a priori knowledge of the PN spreading or hopping codes. The signaling waveform should be designed so that the jammer cannot gain any appreciable jamming advantage by choosing a jammer waveform and strategy other than wideband Gaussian noise (i.e., being clever should gain nothing for the jammer). The fundamental design rule in specifying a jam-resistant system is to make it as costly as possible for the jammer to succeed in jamming the system.

12.6.1.1 Jammer Waveforms

There are many different waveforms that can be used for jamming communication systems. The most appropriate choice depends on the targeted system. Figure 12.25 shows power spectral density plots of examples of jammer waveforms versus a communicator's frequency hopped *M*-ary FSK (FH/MFSK) tone. The

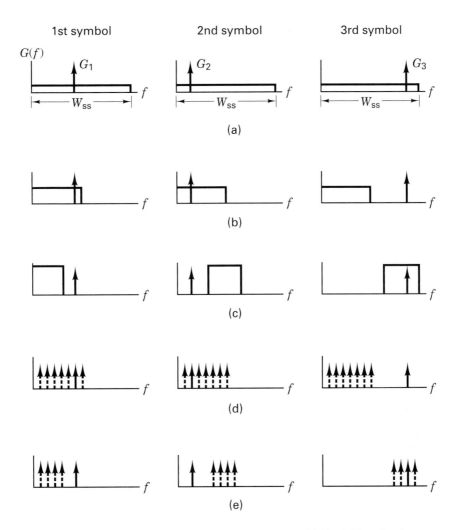

1st symbol   2nd symbol   3rd symbol

(a)

(b)

(c)

(d)

(e)

12.25 Jammer waveforms. (a) Full-band noise. (b) Partial-band noise. (c) Stepped noise. (d) Partial-band tones. (e) Stepped tones.

range of the abscissa represents the spread-spectrum bandwidth W_{ss}. The three columns in the figure represent three instances in time (three hop times) when symbols having spectra G_1, G_2, and G_3, respectively, are being transmitted. Figure 12.25a illustrates a relatively low-level noise jammer occupying the full spread-spectrum bandwidth. In Figure 12.25b the jammer strategy is to trade bandwidth occupancy for greater power spectral density (the total power, or area under the curve, remains the same). The figure indicates that in this case, the jammer noise does not always share the same bandwidth region as the signal, but when it does, the effect can be destructive. In Figure 12.25c the noise jammer strategy is again to jam only part of the band, so that the jammer power spectral density can be increased, but in this case the jammer steps through different regions of the band at random times, thus preventing the communicator from using adaptive techniques to avoid the jamming. In Figure 12.25d and e the jammer uses a group of tones, instead of a continuous frequency band, in partial-band (Figure 12.25d) and stepped fashion (Figure 12.25e). This is a technique most often used against FH systems. Another jamming technique, not shown in Figure 12.25, is a pulse jammer, consisting of pulse-modulated bandlimited noise. Unless otherwise stated, we shall assume that the jammer waveform is wideband noise and that the jammer strategy is to jam the entire bandwidth W_{ss} continuously. The effects of partial band jamming and pulse jamming are considered later.

12.6.1.2 Tools of the Communicator

The usual design goal for an anti-jam (AJ) communication system is to force a jammer to expend its resources over (1) a wide-frequency band, (2) for a maximum time, and (3) from a diversity of sites. The most prevalent design options are (1) frequency diversity, by the use of direct-sequence and frequency-hopping spread-spectrum techniques; (2) time diversity, by the use of time hopping; (3) spatial discrimination, by the use of a narrow-beam antenna, which forces a jammer to enter the receiver via an antenna sidelobe and hence suffer, typically, a 20- to 25-dB disadvantage, and (4) combinations of the previous three options.

12.6.1.3 *J/S* Ratio

In Chapter 5 we were concerned primarily with link error performance as a function of thermal noise interference. Emphasis was placed on the signal-to-noise ratio parameters—required E_b/N_0 and available E_b/N_0 for meeting a specified error performance. In this section we are similarly concerned with link error performance as a function of interference. However, here the source of interference is assumed to be wideband Gaussian noise power from a jammer in addition to thermal noise. Therefore, the SNR of interest is $E_b/(N_0 + J_0)$, where J_0 is the noise power spectral density due to the jammer. Unless otherwise specified, J_0 is assumed equal to J/W_{ss}, where J is the average received jammer power (jammer power referred to the receiver front end) and W_{ss} is the spread-spectrum bandwidth. Since the jammer power is generally much greater than the thermal noise power, the SNR of interest in a jammed environment is usually taken to be E_b/J_0. Therefore, similar to the thermal noise case, we define $(E_b/J_0)_{reqd}$ as the bit energy per jammer noise power spectral density *required* for maintaining the link at a specified error probability. The parameter E_b can be written as

$$E_b = ST_b = \frac{S}{R}$$

where S is the received signal power, T_b the bit duration, and R the data rate in bits/s. Then we can express $(E_b/J_0)_{reqd}$ as

$$\left(\frac{E_b}{J_0}\right)_{reqd} = \left(\frac{S/R}{J/W_{ss}}\right)_{reqd} = \frac{W_{ss}/R}{(J/S)_{reqd}} = \frac{G_p}{(J/S)_{reqd}} \qquad (12.38)$$

where $G_p = W_{ss}/R$ represents the *processing gain*, and $(J/S)_{reqd}$ can be written

$$\left(\frac{J}{S}\right)_{reqd} = \frac{G_p}{(E_b/J_0)_{reqd}} \qquad (12.39)$$

The ratio $(J/S)_{reqd}$ is a figure of merit that provides a measure of how *invulnerable* a system is to interference. Which system has better jammer-rejection capability: one with a larger $(J/S)_{reqd}$ or a smaller $(J/S)_{reqd}$? The *larger* the $(J/S)_{reqd}$, the *greater* is the system's noise rejection capability, since this figure of merit describes how much noise power relative to signal power is *required* in order to degrade the system's specified error performance. Of course, the communicator would like the communication system *not* to degrade at all.

Another way of describing the relationship in Equation (12.39) is as follows. An adversary would like to employ a jamming strategy that forces the effective $(E_b/J_0)_{reqd}$ to be as large as possible. The adversary may employ pulse, tone, or partial-band jamming rather than wideband noise jamming. A large $(E_b/J_0)_{reqd}$ implies a small $(J/S)_{reqd}$ ratio for a fixed processing again. This may force the communicator to employ a larger processing gain to increase the $(J/S)_{reqd}$. The system designer strives to choose a signaling waveform such that the jammer can gain no special advantage by using a jamming strategy other than wideband Gaussian noise.

12.6.1.4 Anti-Jam Margin

Sometimes the $(J/S)_{reqd}$ ratio is referred to as the *anti-jam* (AJ) *margin,* since it characterizes the system jammer-rejection capability. But this is not really a good use of the phrase since AJ margin usually means the safety margin against a *particular threat.* Using the same approach as in Chapter 5 (for calculating the margin against thermal noise), we can define the AJ margin as

$$M_{AJ}(dB) = \left(\frac{E_b}{J_0}\right)_r (dB) - \left(\frac{E_b}{J_0}\right)_{reqd} (dB) \qquad (12.40)$$

where $(E_b/J_0)_r$ is the E_b/J_0 *actually received.* Following the same format as Equation (12.38), we can express $(E_b/J_0)_r$ as

$$\left(\frac{E_b}{J_0}\right)_r = \frac{G_p}{(J/S)_r} \qquad (12.41)$$

where $(J/S)_r$, or simply J/S, is the ratio of the actually received jammer power to signal power. Later, we develop an expression for received E_b/I_0, similar to Equation (12.41), where I_0 is the interference power spectral density due to other users in a CDMA cellular system. The concept of computing such a bit-energy to interference ratio is the same, whether the interference stems from a jammer, an accidental interferer, or other users who are authorized to share the same spectral region.

We now combine Equation (12.41) with Equations (12.38) and (12.40), as follows:

$$M_{AJ}(dB) = \frac{G_p}{(J/S)_r}(dB) - \frac{G_p}{(J/S)_{reqd}}(dB) \qquad (12.42)$$

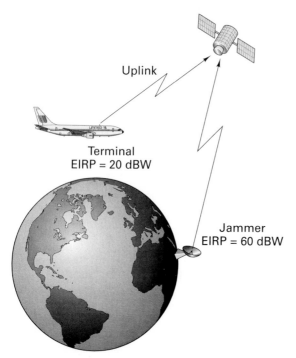

Uplink

Terminal
EIRP = 20 dBW

Jammer
EIRP = 60 dBW

Figure 12.26   Satellite jamming scenario.

$$= \left(\frac{J}{S}\right)_{\text{reqd}} (\text{dB}) - \left(\frac{J}{S}\right)_r (\text{dB}) \qquad (12.43)$$

Example 12.2   Satellite Jamming

Figure 12.26 illustrates a satellite jamming scenario. The airplane terminal is equipped with a frequency hopping (FH) spread-spectrum system transmitting with an $\text{EIRP}_T = 20$ dBW. The data rate is $R = 100$ bits/s. The jammer is transmitting wideband Gaussian noise, continually, with an $\text{EIRP}_J = 60$ dBW. Assume that $(E_b/J_0)_{\text{reqd}} = 10$ dB and that the path loss is identical for both the airplane terminal and the jammer.

(a) Should the communicators be concerned more with the jamming of the uplink or with that of the downlink?
(b) If it is desired to have an AJ margin of 20 dB, what should be the value of the hopping bandwidth W_{ss}?

Solution

(a) Jamming the uplink is of much greater concern, since such single-point interference could degrade the communications of a multitude of terminals that are simultaneously using the satellite transponder. To achieve an equivalent degradation by jamming the downlink, the jammer would have to jam each of the receiving terminals. Downlink jamming is of some concern for critical military missions, but of less concern than uplink jamming.

(b) With the assumption that the path loss is the same for both the communicator and the jammer, we can replace $(J/S)_r$ in Equation (12.43) with the ratio of *transmitted* jammer-to-signal power, $\text{EIRP}_J/\text{EIRP}_T$. Therefore, we can write

$$M_{\text{AJ}} (\text{dB}) = (J/S)_{\text{reqd}} (\text{dB}) + \text{EIRP}_T (\text{dBW}) - \text{EIRP}_J (\text{dBW})$$

$$= G_p (\text{dB}) - \left(\frac{E_b}{J_0}\right)_{\text{reqd}} (\text{dB}) + \text{EIRP}_T (\text{dBW}) - \text{EIRP}_J (\text{dBW})$$

$$G_p = 20 \text{ dB} + 10 \text{ dB} - 20 \text{ dBW} + 60 \text{ dBW} = 70 \text{ dB}$$

$$W_{\text{ss}} = G_p (\text{dB}) + R (\text{dB-Hz}) = 70 \text{ dB} + 20 \text{ dB-Hz}$$

$$= 90 \text{ dB-Hz} = 1 \text{ GHz}$$

Example 12.3 Satellite Downlink Jamming

In Example 12.2 the distance from the transmitting airplane to the receiving satellite and the distance from the jammer to the satellite were assumed identical. Certainly, the closer the jammer gets to the receiver, the greater will be the jamming interference. Consider a downlink jamming scenario where the satellite $\text{EIRP}_s = 35$ dBW, the jammer $\text{EIRP}_J = 60$ dBW, the space loss from the satellite to the receiving terminal is $L_s = 200$ dB, and the space loss from the jammer to the receiving terminal is $L'_s = 160$ dB. How much processing gain is needed to close the link with an AJ margin of 0 dB? Assume that $(E_b/J_0)_{\text{reqd}} = 10$ dB.

Solution

For the downlink jamming scenario the proximity of the jammer to the receiving airplane is much closer than that of the satellite to the airplane. These distances show up as the space losses in the $(J/S)_r$ term of Equation (12.43), as follows:

$$M_{\text{AJ}} (\text{dB}) = \left(\frac{J}{S}\right)_{\text{reqd}} (\text{dB}) - \left(\frac{J}{S}\right)_r (\text{dB})$$

where

$$\left(\frac{J}{S}\right)_r (\text{dB}) = \text{EIRP}_J (\text{dBW}) - L'_s (\text{dB}) - \text{EIRP}_s (\text{dBW}) + L_s (\text{dB})$$

and

$$\left(\frac{J}{S}\right)_{\text{reqd}} (\text{dB}) = \frac{W_{\text{ss}}}{R} (\text{dB}) - \left(\frac{E_b}{J_0}\right)_{\text{reqd}} (\text{dB})$$

Combining the above equations, and solving for processing gain, $G_p = W_{\text{ss}}/R$, yields

$$G_p (\text{dB}) = 75 \text{ dB}$$

12.6.2 Broadband Noise Jamming

If the jamming signal is modeled as a zero-mean wide-sense stationary Gaussian noise process with a flat power spectral density over the frequency range of interest, then for a fixed jammer received power, J, the jammer power spectral density J'_0 is equal to J/W, where W is the bandwidth that the jammer chooses to occupy. If the jammer strategy is to jam the entire spread-spectrum bandwidth, W_{ss}, with its fixed power, the jammer is referred to as a wideband or *broadband jammer,* and the jammer power spectral density is

$$J_0 = \frac{J}{W_{ss}} \qquad (12.44)$$

In Chapter 4 it was shown that the bit error probability P_B for a coherently demodulated BPSK system (without channel coding) is

$$P_B = Q\left(\sqrt{\frac{2E_b}{N_0}}\right) \qquad (12.45)$$

where $Q(x)$ is defined in Equations (3.43) and (3.44) and tabulated in Table B.1. The single-sided noise power spectral density N_0 represents thermal noise at the front end of the receiver. The presence of the jammer increases this noise power spectral density from N_0 to $(N_0 + J_0)$. Thus the average bit error probability for a coherent BPSK system in the presence of broadband jamming is

$$P_B = Q\left(\sqrt{\frac{2E_b}{N_0 + J_0}}\right) = Q\left[\sqrt{\frac{2E_b/N_0}{1 + (E_b/N_0)(J/S)/G_p}}\right] \qquad (12.46)$$

When P_B is plotted versus E_b/N_0 for a given J/S ratio, the resulting curves are such as those in Figure 12.27 [6, 21]. The curves in Figure 12.27, shown for two different

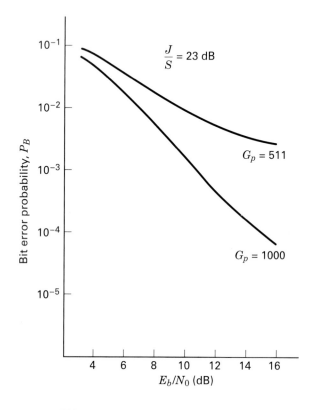

12.27 Bit-error probability versus E_b/N_0 for a given *J/S* ratio. (Reprinted with permission from R. L. Pickholtz, D. L. Schilling, and L. B. Milstein, "Theory of Spread-Spectrum Communications—A Tutorial," *IEEE Trans. Commun.,* vol. COM30, no. 5, May 1982, Fig. 11, p. 866 © 1982 IEEE.)

values of processing gain, *tend to flatten out* as E_b/N_0 increases, indicating that for a given ratio of jammer power to signal power, the jammer will cause some irreducible error probability. The only way to reduce this error probability is to increase the processing again.

12.6.3 Partial-Band Noise Jamming

A jammer can often increase the degradation to a FH system by employing *partial-band* jamming. Assuming that the frequency hopped modulation format is noncoherently detected binary FSK, the probability of a bit error, from Equation (4.96), is

$$P_B = \frac{1}{2} \exp\left(-\frac{E_b}{2N_0}\right) \tag{12.47}$$

Let us define a parameter, ρ, where $0 < \rho \le 1$, representing the fraction of the band being jammed. The jammer can trade bandwidth jammed for in-band jammer power, such that by jamming a band $W = \rho W_{ss}$, the jammer noise power spectral density can be concentrated to a level J_0/ρ, thus maintaining a constant average jamming received power J where $J = J_0 W_{ss}$.

In the case of partial-band jamming, a specific transmitted symbol will be received unjammed, with probability $(1 - \rho)$, and will be perturbed by jammer power with spectral density J_0/ρ, with probability ρ. Therefore, the average bit error probability can be written from Equation (12.47), as follows:

$$P_B = \frac{1 - \rho}{2} \exp\left(-\frac{E_b}{2N_0}\right) + \frac{\rho}{2} \exp\left[-\frac{E_b}{2(N_0 + J_0/\rho)}\right] \tag{12.48}$$

Since, in a jamming environment, it is often the case that $J_0 \gg N_0$, we can simplify Equation (12.48) to the form

$$P_B \approx \frac{\rho}{2} \exp\left(-\frac{\rho E_b}{2J_0}\right) \tag{12.49}$$

Figure 12.28 illustrates the probability of bit error versus E_b/J_0 for various values of the fraction, ρ. Clearly, the jammer would choose the fraction $\rho = \rho_0$ that maximizes P_B. Notice that ρ_0 decreases with increasing values of E_b/J_0 (see the ρ_0 locus in Figure 12.28). An expression for ρ_0 is easily found by differentiation (setting $dP_B/d\rho = 0$ and solving for ρ). This yields

$$\rho_0 = \begin{cases} \dfrac{2}{E_b/J_0} & \text{for } \dfrac{E_b}{J_0} > 2 \\[2ex] 1 & \text{for } \dfrac{E_b}{J_0} \le 2 \end{cases} \tag{12.50}$$

In this case, $(P_B)_{\max}$ is given by

$$(P_B)_{\max} = \begin{cases} \dfrac{e^{-1}}{E_b/J_0} & \text{for } \dfrac{E_b}{J_0} > 2 \\[3mm] \dfrac{1}{2}\exp\left(-\dfrac{E_b}{2J_0}\right) & \text{for } \dfrac{E_b}{J_0} \le 2 \end{cases} \qquad (12.51)$$

where e is the base of the natural logarithm ($e = 2.7183$). This result is dramatic; the effect of a worst-case partial-band jammer on a system with spread spectrum *but without coding* changes the exponential relationship of Equation (12.49) into the inverse linear one of Equation (12.51). The ρ_0 locus in Figure 12.28 illustrates the

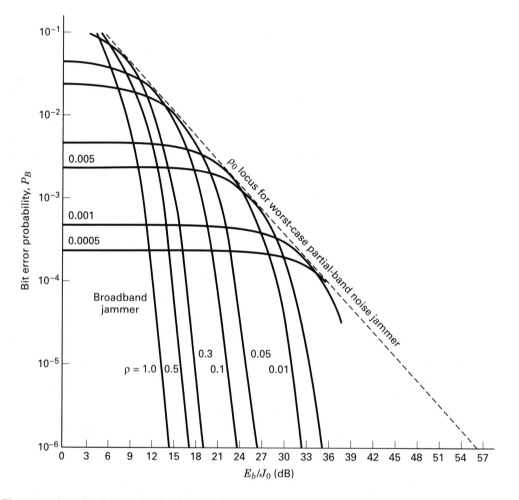

Figure 12.28 Partial-band noise jammer (FH/BFSK signaling). (Reprinted from M. K. Simon, J. K. Omura, R. A. Scholtz, and B. K. Levitt, *Spread Spectrum Communications,* Vol. 1, Fig. 3.24, p. 173. © 1985, with permission of the publisher, Computer Science Press, Inc., 1803 Research Blvd., Rockville, MD., 20850, USA.)

P_B versus E_b/J_0 performance for the worst-case partial-band jammer. Here at 10^{-6} bit-error probability there is over 40-dB difference between broadband noise jamming and the worst-case partial-band jamming for the same jamming power [4, 22]. Hence, an intelligent jammer, with fixed finite power, can produce significantly greater degradation with partial-band jamming than is possible with broadband jamming. Forward error correction (FEC) coding with appropriate interleaving can mitigate this degradation [9]. In fact, for codes with low enough rates, FEC can *force* a partial-band jammer to be a worst-case jammer only when operating as a broadband jammer [23, 24].

12.6.4 Multiple-Tone Jamming

In the case of *multiple-tone jamming,* the jammer divides its total received power, J, into distinct, equal-power, random-phase CW tones. These are distributed over the spread-spectrum bandwidth, W_{ss}, according to some strategy [9]. The analysis of the effects of tone jamming is more complicated than that of noise jamming, especially for DS systems. Therefore, the effect of a despread tone is often approximated as Gaussian noise. Reference [25] provides analysis of the performance of DS systems in the presence of multiple-tone interference. For a noncoherent FH/FSK system operating in the presence of partial-band tone jamming, the performance is often assumed the same as that of partial-band noise jamming [26]. However, multiple-CW-tone jamming can be more effective than partial-band noise against FH/MFSK signals because CW tones are the most efficient way for a jammer to inject energy into noncoherent detectors [8]. References, [8, 9, 26, 27] provide extensive treatment and analysis of the performance of various communications systems in the presence of various types of jammers.

In the FFH/MFSK demodulator of Figure 12.16, a chip-clipping circuit is shown between each envelope detector and accumulator. The function of such a circuit in a tone-jamming environment can best be understood with the aid of the example shown in Figure 12.29. An 8-ary FSK frequency-hopping system with no diversity, indicated in Figure 12.29a, is compared with a *fast* frequency-hopping system that combines chip repeating ($N = 4$ in this example) with the clipping of each chip, indicated in Figure 12.29b. Each row in the figures represents one of the $M = 8$ accumulators shown in Figure 12.16. The presence of a signal in the accumulator is indicated by a vector. In Figure 12.29a we see that, for a particular frequency hop, the data band is occupied by a received message symbol with received signal power S. If, by chance, a jamming tone with received power J, where $J \geq S$, falls on a different tone within this data band during the same hop, the detector would not be able to decide reliably on the correct symbol.

In Figure 12.29b, the communicator's four chips (the length of each vector is a measure of the clipped signal power, S') sum to the maximum capacity of the accumulator. If the jammer tones, by chance, fall in the same spectral region as that of the signal, they will not confuse the detector, since the jamming tones are also clipped to the same level, $J' = S'$, as the signal chips. In Figure 12.29b, two of the jamming tones fall in the data band, but because they are clipped, there is no confusion about the correct symbol decision.

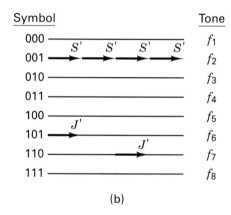

Symbol | Tone
000 —————— f_1
001 ——————→ S f_2
010 —————— f_3
011 —————— f_4
100 ——————→ J f_5
101 —————— f_6
110 —————— f_7
111 —————— f_8

(a)

Symbol | Tone
000 —————— f_1
001 S' S' S' S' ———→———→———→———→ f_2
010 —————— f_3
011 —————— f_4
100 —————— f_5
101 J' ———→ f_6
110 J' ———→ f_7
111 —————— f_8

(b)

Figure 12.29 Fast-hopping symbol repeat versus tone jamming. (a) One frequency hop. (b) Four frequency hops.

12.6.5 Pulse Jamming

Consider a spread-spectrum DS/BPSK communication system in the presence of a pulse-noise jammer. A pulse-noise jammer transmits pulses of bandlimited white Gaussian noise having a time-averaged received power J, although the actual power during a jamming pulse duration is larger. Assume that the jammer can choose the center frequency and bandwidth of the noise to be the same as the receiver's center frequency and bandwidth. Assume also that the jammer can trade duty cycle for increased (concentrated) jammer power, such that if the jamming is present for a fraction $0 < \rho < 1$ of the time, then during this time, the jammer power spectral density is increased to a level J_0/ρ, thus maintaining a constant time-averaged power J (where $J = J_0 W_{ss}$ and W_{ss} is the system spread-spectrum bandwidth).

The bit error probability P_B for a coherently demodulated BPSK system (without channel coding) was given in Equation (12.45):

$$P_B = Q\left(\sqrt{\frac{2E_b}{N_0}}\right)$$

The single-sided noise power spectral density N_0 represents thermal noise at the front end of the receiver. The presence of the jammer increases this noise power spectral density from N_0 to $(N_0 + J_0/\rho)$. Since the jammer transmits with duty cycle ρ, the average bit-error probability is

$$P_B = (1 - \rho)Q\left(\sqrt{\frac{2E_b}{N_0}}\right) + \rho Q\left(\sqrt{\frac{2E_b}{N_0 + J_0/\rho}}\right) \qquad (12.52)$$

We can generally assume that in a jamming environment, N_0 can be neglected. Therefore, we can write

$$P_B \approx \rho Q\left(\sqrt{\frac{2E_b\rho}{J_0}}\right) \qquad (12.53)$$

The jammer will, of course, attempt to choose the duty cycle ρ that maximizes P_B. Figure 12.30 illustrates P_B for various values of ρ. The value of $\rho = \rho_0$ that maximizes P_B decreases with increasing values of E_b/J_0, as was the case with partial-band jamming. This is seen by differentiating Equation (12.53) to obtain [4]

$$\rho_0 = \begin{cases} \dfrac{0.709}{E_b/J_0} & \text{for } \dfrac{E_b}{J_0} > 0.709 \\[3mm] 1 & \text{for } \dfrac{E_b}{J_0} \leq 0.709 \end{cases} \qquad (12.54)$$

which results in the maximum bit error probability

$$(P_B)_{\text{max}} = \begin{cases} \dfrac{0.083}{E_b/J_0} & \text{for } \dfrac{E_b}{J_0} > 0.709 \\[3mm] Q\left(\sqrt{\dfrac{2E_b}{J_0}}\right) & \text{for } \dfrac{E_b}{J_0} \leq 0.709 \end{cases} \qquad (12.55)$$

The effect of a worst-case pulse jammer upon a system with spread spectrum *but without coding* changes the complementary error function relationship of Equation (12.53) into the inverse linear one of Equation (12.55). As a result, at an error probability of 10^{-6}, there is about a 40-dB difference in E_b/J_0 between the broadband jammer and the worst-case pulse jammer (see Figure 12.30). For the same jammer power, the jammer can do considerably more harm to an uncoded DS/BPSK system with pulse jamming than with constant power jamming. The effect of a pulse-noise jammer on uncoded DS/BPSK is similar to the effect of a partial-band noise jammer on uncoded FH/BFSK, treated in Section 12.6.3. In both cases considerable degradation is brought about by concentrating more jammer power on a fraction of the transmitted uncoded symbols. Forward error correction coding with appropriate interleaving can almost fully restore this degraded performance [8, 23–25, 28].

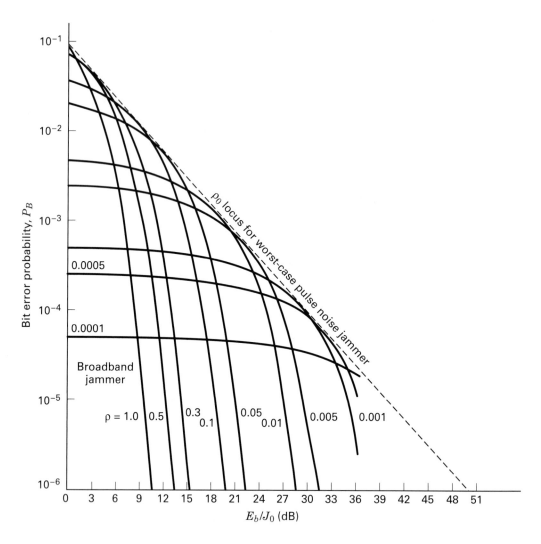

Figure 12.30 Pulse noise jammer (DS/BPSK signaling). (Reprinted from M. K. Simon, J. K. Omura, R. A. Scholtz, and B. K. Levitt, *Spread Spectrum Communications,* Vol. 1, Fig. 3.7, p. 150 © 1985, with permission of the publisher, Computer Science Press, Inc., 1803 Research Blvd., Rockville, Md. 20850 USA.)

12.6.6 Repeat-Back Jamming

In Examples 12.2 and 12.3 we considered an FH spread-spectrum system performance against a broadband Gaussian noise jammer. Notice that the frequency hopping rate did not enter into the margin computations. Isn't this disturbing? Intuitively, it would seem that the faster the frequency hops, the easier it is to "hide" the signal from the jammer. If the hopping rate truly does not enter into the computations, why not hop only once a day or once a week? The answer is that the measure of jammer-rejection capability, namely processing gain, G_p, is based on the assumption that the jammer is a "dumb" jammer; that is, the jammer knows the extent of the spread-spectrum bandwidth, W_{ss}, but does *not* know the exact spectral location of the signal at any moment in time. We assume that the hopping rate is *fast enough* to preclude the jammer from monitoring the transmitted signal so as to usefully change this jamming strategy. Under what condition is this assumption questionable? There are "smart" jammers that are known as *repeat-back jammers* or *frequency-follower (FF) jammers.* These jammers monitor a communicator's signal (usually via a sidelobe beam from the transmitting antenna). They possess wideband receivers and high-speed signal processing capability that enable them to rapidly concentrate their jamming signal power in the spectral vicinity of a communicator's FH/FSK signal. By so doing, the smart jammer can increase the jamming power in the communicator's instantaneous bandwidth, thereby gaining an advantage over a wideband jammer. Notice that this strategy is useful only against frequency-hopping spread-spectrum signals. In direct-sequence systems, there is no instantaneous narrowband signal for the jammer to detect.

What can be done to defeat the repeat-back jammer? One method is to simply hop so fast that by the time the jammer receives, detects, and transmits the jamming signal, the communicator is already transmitting at a *new* hop (which of course will be unaffected by jamming at the frequency of the prior hop). The following example should make this point clear.

Example 12.4 Fast Hopping to Evade the Repeat-Back Jammer

Assume that a repeat-back jammer is located $d = 30$ km away from the communicator. Assume further that the jammer can monitor any uplink transmission from the com-

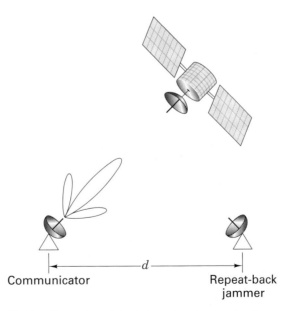

Figure 12.31 Example of fast hopping to evade the repeat-back jammer.

municator to a nearby satellite, as shown in Figure 12.31. How fast must the communicator hop his frequency to evade the repeat-back jammer? Assume that the jammer can change its jamming frequency in zero time, and that the only differential delay between the communicator's uplink signal and the jamming uplink signal is the propagation delay from the communicator to the jammer.

Solution

To ensure that the communicator's tone transmission and the jammer's attempt to disrupt that tone do not overlap in time, it is necessary that the duration of each hop have the value

$$T_{\text{hop}} \le \frac{d}{c} = \frac{3 \times 10^4 \text{m}}{3 \times 10^8 \text{ m/s}} = 10^{-4} \text{ s}$$

where c is the speed of light. Then $R_{\text{hop}} \ge 10{,}000$ hops/s.

12.6.7 BLADES System

Another technique capable of defeating the repeat-back jammer dates back to the mid-1950s, when Sylvania engineers developed a system named the *Buffalo Laboratories Application of Digitally Exact Spectra,* or BLADES. The system used its code generator to independently select two new frequencies for each bit; the *final choice* of the frequency tone actually transmitted was dictated by the data bit about to be transmitted. Figure 12.32 illustrates a typical data stream of binary ones and zeros, called *marks* and *spaces,* respectively, and a sequence of frequency pairs f_1 and f'_1, f_2 and $f'_2, \ldots$ The appearance of a mark dictates the choice of frequency f_i, while the appearance of a space dictates the choice of frequency f'_i. As shown in the figure, the data stream in this example gives rise to the sequence of transmitted tones, $f'_1, f_2, f'_3, f'_4, f_5$, and so on. How can such a system defeat a repeat-back jammer? The jammer monitors the transmissions and sends up energy in the neighborhood of the frequencies it perceives. The modulation of the BLADES system has no structure in the usual sense; either there *is* energy present or there is *no* energy

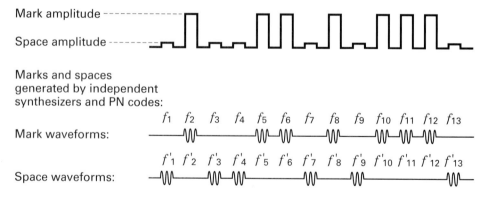

Figure 12.32   BLADES system.

present at a given frequency. The jammer sending narrowband energy in the same spectral neighborhood as the signal, is not destroying any modulation structure. For a noncoherent system, the jammer is only enhancing the communicator's signal. The only recourse for the repeat-back jammer is to change strategy by becoming a broadband jammer, and to jam the entire spread-spectrum bandwidth.

Notice that it is not really necessary to have a *pair* of frequencies for each bit. A *single* frequency will do. The communicator then transmits the pseudo-random frequency for a binary one and sends nothing for a binary zero. The receiver has the same code generator and therefore monitors the same pseudo-random frequencies. A binary one is detected by virtue of energy at the monitored frequency, and a binary zero is known by a lack of energy at the monitored frequency. Of course, the system is not as robust as when the marks and spaces are each transmitted on independently selected frequencies.

12.7 COMMERCIAL APPLICATIONS

12.7.1 Code-Division Multiple Access

Spread-spectrum multiple access techniques allow multiple signals occupying the same RF bandwidth to be transmitted simultaneously without interfering with one another. The application of spread-spectrum techniques to the problem of multiple access was discussed in Chapter 11 for a frequency hopped code-division multiple access (FH/CDMA) scheme. Here we consider CDMA using direct sequence (DS/CDMA). In these schemes, each of N user groups is given its own code, $g_i(t)$, where $i = 1, 2, \ldots, N$. The user codes are approximately orthogonal, so that the cross-correlation of two different codes is near zero. The main advantage of a CDMA system is that all the participants can share the full spectrum of the resource asynchronously; that is, the transition times of the different users' symbols do not have to coincide.

A typical DS/CDMA block diagram is shown in Figure 12.33. The first block illustrates the data modulation of a carrier, $A \cos \omega_0 t$. The output of the data modulator belonging to a user from group 1 is

$$s_1(t) = A_1(t) \cos \left[\omega_0 t + \phi_1(t) \right] \tag{12.56}$$

The waveform is very general in form; no restriction has been placed on the type of modulation that can be used.

Next, the data-modulated signal is multiplied by the spreading signal $g_1(t)$ belonging to user group 1, and the resulting signal $g_1(t)s_1(t)$ is transmitted over the channel. Simultaneously, users from group 2 through N multiply their signals by their own code functions. Frequently, each code function is kept secret, and its use is restricted to the community of authorized users. The signal present at the receiver is the linear combination of the emanations from each of the users. Neglecting signal delays, we show this linear combination as

$$g_1(t)s_1(t) + g_2(t)s_2(t) + \cdots + g_N(t)s_N(t) \tag{12.57}$$

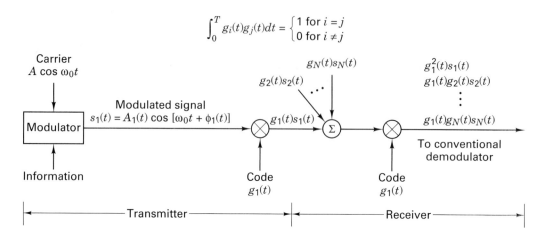

$$\int_0^T g_i(t)g_j(t)dt = \begin{cases} 1 \text{ for } i = j \\ 0 \text{ for } i \neq j \end{cases}$$

Figure 12.33   Code-division multiple access.

As mentioned earlier, multiplication of $s_1(t)$ by $g_1(t)$ produces a signal whose spectrum is the convolution of the spectrum of $s_1(t)$ with the spectrum of $g_1(t)$. Thus, assuming that the signal $s_1(t)$ is relatively narrowband compared with the code or spreading signal $g_1(t)$, the product signal $g_1(t)s_1(t)$ will have approximately the bandwidth of $g_1(t)$. Assume that the receiver is configured to receive messages from user group 1. Assume, too, that the $g_1(t)$ code, generated at the receiver, is perfectly synchronized with the received signal from a group 1 user. The first stage of the receiver multiplies the incoming signal of Equation (12.57) by $g_1(t)$. The output of the multiplier will yield the desired signal,

$$g_1^2(t)s_1(t)$$

plus a composite of undesired signals,

$$g_1(t)g_2(t)s_2(t) + g_1(t)g_3(t)s_3(t)$$

$$+ \cdots + g_1(t)g_N(t)s_N(t) \tag{12.58}$$

If the code functions $\{g_i(t)\}$ are chosen with orthogonal properties, similar to Equation (12.14), the desired signal can be extracted perfectly in the absence of noise since $\int_0^T g_i^2(t) = 1$, and the undesired signals are easily rejected, since $\int_0^T g_i(t)g_j(t)\,dt = 0$ for $i \neq j$. In practice, the codes are not perfectly orthogonal; hence, the cross-correlation between user codes introduces performance degradation, which limits the maximum number of simultaneous users.

Consider the frequency-domain view of the DS/CDMA receiver. Figure 12.34a illustrates the wideband input to the receiver; it consists of wanted and unwanted signals, each spread by its own code with code rate R_{ch}, and each having a power spectral density of the form $\text{sinc}^2(f/R_{ch})$. Receiver thermal noise is also shown as having a flat spectrum across the band. The combined waveform of Equa-

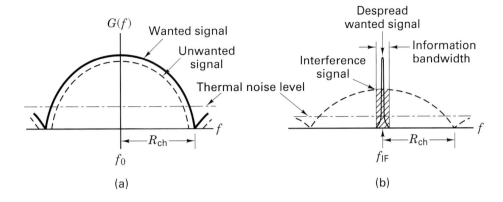

Figure 12.34 Spread-spectrum signal detection. (a) Spectrum at the input to receiver. (b) Spectrum after correlation with the correct and synchronized PN code.

tion (12.57) (desired plus undesired signals) is applied to the input of the receiver correlator driven by a synchronous replica of $g_1(t)$. Figure 12.34b illustrates the spectrum after correlation with the code $g_1(t)$ (despreading). The desired signal, occupying the information bandwidth centered at an intermediate frequency (IF), is then applied to a conventional demodulator, with bandwidth just wide enough to accommodate the despread signal. The undesired signals of Equation (12.58) remain effectively spread by $g_1(t)g_i(t)$. Only that portion of the spectrum of the unwanted signals falling in the information bandwidth of the receiver will cause interference with the desired signal.

Pursley [17] presents an excellent treatment on the performance of SSMA using DS, taking correlation properties of the code sequences into account. Also, Geraniotis [18] and Geraniotis and Pursley [19, 20] evaluate the performance of FH and DS multiple access systems subject to interference.

12.7.2 Multipath Channels

Consider a DS binary PSK communication system operating over a multipath channel that has more than one path from the transmitter to the receiver. Such multiple paths may be due to atmospheric reflection or refraction, or reflections from buildings or other objects, and may result in fluctuations in the received signal level. The different paths may consist of several discrete paths each with a different attenuation and time delay, or they might consist of a continuum of paths. Figure 12.35 illustrates a communication link with two discrete paths. The multipath wave is delayed by some time τ, compared with the direct wave. In television receivers, signals such as these cause "ghosts," or under extreme conditions, complete loss of picture synchronization.

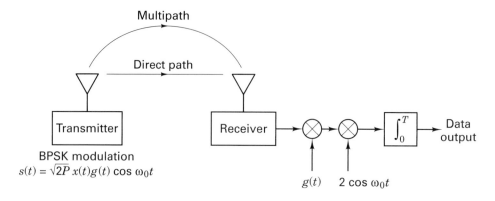

Multipath

Direct path

Transmitter

BPSK modulation
$s(t) = \sqrt{2P}\, x(t)g(t)\cos\omega_0 t$

Receiver

$\int_0^T$

Data
output

$g(t)$ $2\cos\omega_0 t$

Figure 12.35   Direct-sequence BPSK system operating over a multipath channel.

In a direct-sequence spread-spectrum system, if we assume that the receiver is synchronized to the time delay and RF phase of the direct path, the received signal can be expressed as

$$r(t) = Ax(t)\,g(t)\,\cos\omega_0 t + \alpha Ax(t-\tau)g(t-\tau)\cos(\omega_0 t + \theta) + n(t) \quad (12.59)$$

where $x(t)$ is the data signal, $g(t)$ is the code signal, $n(t)$ is a zero-mean Gaussian noise process, and τ is the differential time delay between the two paths, assumed to be in the interval $0 < \tau < T$. The angle θ is a random phase, assumed to be uniformly distributed in the range $(0, 2\pi)$, and α is the attenuation of the multipath signal relative to the direct path signal. For the receiver, synchronized to the direct path signal, the output of the correlator can be written as

$$z(t = T) = \int_0^T [Ax(t)g^2(t)\cos\omega_0 t \quad (12.60)$$

$$+ \alpha Ax(t-\tau)g(t)g(t-\tau)\cos(\omega_0 t + \theta) + n(t)g(t)]2\cos\omega_0 t\ dt$$

where $g^2(t) = 1$. Also, for $\tau > T_c$, $g(t)g(t-\tau) \approx 0$ (for codes with long periods), where T_c is the chip duration. Therefore, if T_c is less than the differential time delay between the multipath and direct path signals, we can write

$$z(t = T) = \int_0^T 2Ax(t)\cos^2\omega_0 t + 2n(t)g(t)\cos\omega_0 t\, dt = Ax(T) + n_0(T) \quad (12.61)$$

where $n_0(T)$ is a zero-mean Gaussian random variable. We see that the spread-spectrum system, similar to the case of CDMA, effectively eliminates the multipath interference by virtue of its code-correlation receiver.

If frequency hopping (FH) is used against the multipath problem, improvement in system performance is also possible but through a different mechanism. FH receivers avoid multipath losses by rapid changes in the transmitter frequency band, thus avoiding the interference by changing the receiver band position before the arrival of the multipath signal.

12.7.3 The FCC Part 15 Rules for Spread-Spectrum Systems

In the United States, the Federal Communications Commission (FCC) allows the general unlicensed operation of very lower power (less than 1 mW) radio equipment freely, except in certain restricted frequency bands. In 1985, Dr. Michael Marcus of the FCC was responsible for allowing higher power (up to 1 W) spread-spectrum radios in some of the bands, referred to as *Industrial, Scientific,* and *Medical* (ISM). The rules of allowable electromagnetic radiation for unlicensed devices appear in the *Code of Federal Regulations* (CFR) Title 47, Part 15; they are known simply as the Part-15 rules (wherein Section 15.247 covers spread spectrum).

The ISM frequency bands are used for instruments (e.g., medical diathermy equipment) as well as for critical government systems (e.g., radio location equipment) that radiate strong electromagnetic fields which can cause interference to other users. The ISM bands are particularly noisy bands. An unlicensed radio can be thought of as an "unwelcome guest" in a licensee's band. An unlicensed radio must be able to suffer interference but is not permitted to cause any interference to a licensed user.

For frequency-hopping systems, the Part-15 rules require that the average time of occupancy on any frequency shall not be greater than 0.4 second (or a minimum hopping rate of 2.5 hops/s). For direct sequence systems, the minimum required processing gain is 10 dB. For hybrid systems employing both direct sequence and frequency hopping, the minimum required processing gain is 17 dB. Three ISM spectral regions were designated for the operation of unlicensed spread-spectrum radios. Some of the details regarding the operation in these bands are shown in Table 12.1.

As a result of allowing higher power limits and no FCC licensing as described previously, commercial companies have introduced a wide range of innovative spread-spectrum radios capable of communications over greater distances than earlier low-power narrowband unlicensed radios. Some of these products include radios that link office equipment (e.g., shared printer, wireless local area net-

TABLE 12.1   Spread-Spectrum Operation Under Part-15 Rules

ISM Band	Total Bandwidth	Max. Bandwidth per Channel for FH*	Min. Number of Hopping Frequencies per Channel	Min. Bandwidth per Channel for DS*
902–928 MHz	26 MHz	500 kHz	25–50**	500 kHz
2.4000–2.4835 Ghz	83.5 MHz	1 MHz	75	500 kHz
5.7250–5.8500 GHz	125 MHz	1 MHz	75	500 kHz

*Maximum bandwidth per channel for frequency hopping is defined as the 20 dB bandwidth; minimum bandwidth per channel for direct sequence is defined as the 6 dB bandwidth.

**FH channels with bandwidth less than 250 kHz require at least 50 hopping frequencies per channel; FH channels with bandwidth greater than 250 kHz require at least 25 hopping frequencies per channel.

works), cordless telephones, wireless point-of-sales equipment (e.g., cash registers, bar-code readers).

12.7.4 Direct Sequence versus Frequency Hopping

Without interference from other radios and in free space, both direct-sequence (DS) and frequency-hopping (FH) spread-spectrum radios can, in theory, give the same performance. For mobile applications with large multipath delays, DS represents a reliable mitigation method, because such signaling renders all multipath signal copies that are delayed by more than one chip time from the direct signal as "invisible" to the receiver. (See Section 12.7.2.) FH systems can provide the same mitigation, only if the hopping rate is faster than the symbol rate, and if the hopping bandwidth is large. (See Chapter 15.)

Implementing a fast frequency-hopping (FFH) radio can be costly due to the need for high-speed frequency synthesizers. Consequently, hopping rates of commercial FH radios are generally slow compared with the data rate, and hence such systems behave like narrowband radios. Slow frequency hopping (SFH) and DS signaling each experience somewhat different interference. SFH radios typically suffer occasional strong bursty errors, while DS radios encounter more randomly distributed errors that are continuous and lower level. For high data rates, the impact of multipath tends to degrade such SFH radios more than DS radios. To mitigate the effects of bursty errors in FH radios, interleaving would have to be performed over long time durations. (See Chapter 15.) SFH is used for providing diversity in fixed wireless access applications or slowly moving systems, or merely to meet the Part-15 Rules. For commercial applications, implementation of DS radios with large processing gain can also be costly due to the need for high-speed circuits; thus, the processing gain for such radios is usually limited to less than 20 dB to avoid having to use high-speed circuits [29].

Example 12.5   Detection of Signals Buried in the Noise

In Section 12.1.1.1 it was shown that spread-spectrum techniques offer no error-performance advantage against thermal noise. In this example, we show that whatever value of received E_b/N_0 might be available to a narrowband system remains the same after spreading. Although there is no error-performance advantage against thermal noise, there is *no disadvantage* either, which makes it an attractive option for building systems to meet the FCC Part-15 Rules, as described in Section 12.7.3, and for multiple access systems (e.g., CDMA systems meeting Interim Standard IS-95).

Direct-sequence spread-spectrum techniques allow for the detection of signals that have power-spectral density (psd) levels well below the noise. Consider Figure 12.36a illustrating the received psd of a communication signal with an ideal rectangular shape, having an intensity of $S_0(f) = 10^{-5}$ W/Hz over a bandwidth of 1 MHz. Assume that the transmitted data rate is $R = 10^6$ bits/s. Also, assume AWGN noise and a received psd level of $N_0(f) = 10^{-6}$ W/Hz (not drawn to scale) at all frequencies. Find the received E_b/N_0 for this narrowband case. Next, consider that the communication signal is spread over a spread-spectrum bandwidth of $W_{ss} = 10^8$ Hz, as shown in Figure 12.36b, keeping the total average signal power the same as in the narrowband case

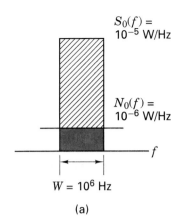

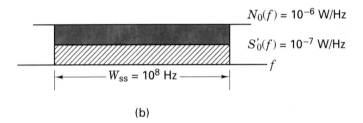

Figure 12.36 Signal and noise power spectral densities (a) Before spreading (b) After spreading.

(not drawn to scale). Show that with a spread-spectrum receiver, the received E_b/N_0 is the same as in the narrowband case, and hence the error performance is unchanged.

Solution

Before spreading, the total average power of the communication signal is $S = 10^{-5}$ W/Hz $\times 10^6$ Hz = 10 W, and the total average noise power is $N = 10^{-6}$ W/Hz $\times 10^6$ Hz = 1 Watt. The received E_b/N_0 is

$$\frac{E_b}{N_0} = \frac{S/R}{N_0} = \frac{10 \text{ W}/10^6 \text{ bits/s}}{10^{-6} \text{ W/Hz}} = 10 \text{ or } 10 \text{ dB}$$

After spreading, the psd, $S_0'(f)$, of the communication signal is reduced by the same factor (two orders of magnitude) as the increase in bandwidth; hence, the total average power of the communication signal is still 10 W. However, since the noise was assumed to be AWGN, its psd is not reduced; hence, the total average noise power is now equal to $N' = 10^{-6}$ W/Hz $\times 10^8$ Hz = 100 W. Therefore, after spreading, the received E_b/N_0 can be expressed as

$$\frac{E_b}{N_0} = \frac{S/R}{N'/W_{ss}} = \frac{S}{N'}\left(\frac{W_{ss}}{R}\right) = \frac{S}{N'}G_p = \frac{10 \text{ W}}{100 \text{ W}} \times 100 = 10 \text{ or } 10 \text{ dB}$$

where $G_p = W_{ss}/R = 100$ is the processing gain. The process of detecting direct-sequence spread-spectrum signals "buried in noise" does not lend itself to an intuitive

12.7   Commercial Applications

illustration, as can be verified in Figure 12.36b. Similarly, in the expression for received E_b/N_0 after spreading, the communication signal power is only 10 W and the noise power is 100 W, leading to the same lack of intuition regarding the detectability of the signal. It is the processing gain (which is not very visual) that allows us to receive the same value of E_b/N_0 as in the narrowband case.

12.8 CELLULAR SYSTEMS

Wireless personal communication systems, particularly cellular systems are relatively young applications of the communications technology. The following list includes some of the events that illustrate the evolution of this ever-growing business:

- 1921   Radio dispatch service initiated for police cars in Detroit, Michigan.
- 1934   Amplitude modulation (AM) mobile communication systems used by hundreds of state and municipal police forces in the U.S.
- 1946   Radiotelephone connections made to the public-switched telephone network (PSTN).
- 1968   Development of the cellular telephony concept at Bell Laboratories.
- 1981   Ericsson Corporation's Nordic Mobile Telephone (NMT) in Scandinavian countries becomes the first cellular system fielded.
- 1983   Cellular service in the United States—called the Advanced Mobile Phone System (AMPS) and using frequency modulation (FM)—placed in service in Chicago by Ameritech Corporation.
- 1990s   Second generation digital cellular deployed throughout the world. The Global system for Mobile (GSM) Communications becomes the pan-European standard. (Prior to GSM, many different cellular systems operating in Europe became operationally impractical.)
- 1990s   Second generation digital systems known as IS-54 and its successor IS-136 (TDMA), and IS-95 (CDMA) become operational in the United States.
- 2000s   Third-generation digital systems standardized at the network level to allow world-wide roaming start becoming operational. They offer enhanced services, such as connection to various PSTN systems with a single phone, and connecting to high data rate packet systems such as Internet Protocol (IP) networks.

12.8.1 Direct Sequence CDMA

Figures 11.3 and 11.7 depict sharing a communications resource via FDMA or TDMA. In the case of FDMA, different frequency bands are orthogonal to one another (assuming ideal filtering), and in the case of TDMA, different time slots are orthogonal to one another (assuming perfect timing). One can visualize a similar orthogonality among different channels in the case of frequency-hopping CDMA (as shown in Figure 11.14) if the codes that control the frequency hopping operate in

Power spectral density

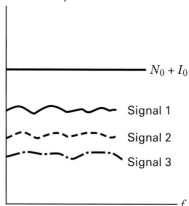

$N_0 + I_0$

Signal 1

Signal 2

Signal 3

f

Figure 12.37 Three DS/SS signals occupying the same spectral region.

such a way that no two users are collocated in time and frequency. It is easy to visualize the user transmissions hopping in frequency and time without any contention. However, in the case of direct-sequence spread-spectrum (DS/SS), visualization of the necessary orthogonality conditions (with multiple users simultaneously occupying the same spectrum) is not as easy. Figure 12.37 shows three different DS/SS signals that are spread over a broad range of frequencies below the level of noise-and-interference power spectral density, $N_0 + I_0$ (assumed to be wideband and Gaussian). An often asked question regarding Figure 12.37 is "How can one of these signals be detected when it is spectrally "buried" below the noise and interference, and it is collocated with other similar signals?" The answer is that the DS/SS receiver correlates the received signal to a particular user's PN code. If the PN codes are orthogonal to each other, then the other users' signals will average to zero during a long observation time. If the codes are not purely orthogonal, they will contribute some interference to the detection process.

In a mobile telephone system using CDMA, each of the users *do* interfere with one another for the following reasons: (1) Two different spreading codes from a family of perfectly orthogonal *long codes* may not yield zero correlation over a short interval of time, such as a symbol time. (2) Serving a large population of users, typically dictates the use of long codes. Such codes can be designed to have low cross-correlation properties but are not orthogonal. (3) Multipath propagation and imperfect synchronization cause interchip interference among the users.

Consider a reverse channel (mobile to base station) in a heavily loaded cell, where the interference caused by the many simultaneous CDMA signals typically outweighs the degradation caused by thermal noise. The assumption is generally made that the thermal noise can be neglected compared with the interference from other users. Thus, for $N_0 \ll I_0$, the following relationship for the received E_b/I_0, designated as $(E_b/I_0)_r$, can be obtained:

$$\left(\frac{E_b}{N_0 + I_0}\right)_r \approx \left(\frac{E_b}{I_0}\right)_r = \frac{S/R}{I/W_{ss}} = \frac{W_{ss}/R}{I/S} = \frac{G_p S}{I} \qquad (12.62)$$

where $G_p = W_{ss}/R$ is the processing gain, W_{ss} is the spread-spectrum bandwidth, S is one user's received power, and I is the interference power from all other users. Equation (12.62) shows that even when the received interference greatly exceeds a user's received power, it is the processing gain (via the mechanism of correlating to a code) that can yield an acceptable value of E_b/I_0. When the base station exercises power control so that each user's received power is balanced, then $I = S \times (M - 1)$, where M is the total number of users contributing to interference at the receiver. It is now possible to express $(E_b/I_0)_r$ in terms of the processing gain and the number of active users in the cell, as follows:

$$\left(\frac{E_b}{I_0}\right)_r \approx \frac{G_p S}{I} = \frac{G_p S}{S \times (M - 1)} = \frac{G_p}{M - 1} \tag{12.63}$$

Note that the received E_b/I_0 in Equation (12.63) is analogous to the E_b/J_0 for a jammed receiver in Equation (12.41), with J_0 and J replaced by I_0 and I, respectively. CDMA systems are affected by such interference (assumed wideband and Gaussian) in the same way, whether caused by jammers, accidental interferers, or authorized participants. In Equation (12.63), knowing G_p and the *required* E_b/I_0, designated $(E_b/I_0)_{\text{reqd}}$, for a given error performance, the *maximum* number of allowable users (interferers) per cell is

$$M_{\text{max}} \approx \frac{G_p}{(E_b/I_0)_{\text{reqd}}} \tag{12.64}$$

Note that Equation (12.63) indicates that for a heavily loaded cell, a CDMA system is interference limited. For example, if the number of active users occupying a cell were to suddenly double, then the *received* E_b/I_0 would essentially be halved. Also, by examining Equation (12.64), it can be seen that any reduction in $(E_b/I_0)_{\text{reqd}}$ has the effect of increasing the maximum allowable number of users in the cell. The following is a list of other factors that influence the final calculation for the maximum number of allowable users per cell:

- **Sectorizing or Antenna Gain (G_A).** Dividing the cell into three $120°$ sectors by using a separate directional antenna for each sector, provides a gain G_A of about 2.5 (or 4 dB) in the number of users that can be accommodated.
- **Voice Activity Factor (G_V).** The average speaker pauses about 60% of the time between words and sentences and for listening. Thus, for a CDMA voice circuit, transmission need take place only 40% of the time, whenever there is speech activity. For voice channels, this contributes an improvement factor G_V of about 2.5 (or 4 dB) in the number of users that can be accommodated.
- **Outer-Cell Interference Factor (H_0).** As described in Section 12.8.2, 100% frequency reuse can be employed for CDMA; all neighboring cells can use the same spectrum. Therefore, for a given level of interference I_x originating within a cell, there is additional interference originating outside of the cell. For signal-propagation loss that follows an $n = 4$th power exponent law (see Section 15.2.1), this additional interference is estimated at about 55% of the within-cell interference [30, 31]. The total interference is therefore approxi-

mately 1.55 I_x, resulting in a user capacity degradation factor H_0 of about 1.55 (or 1.9 dB).

- **Nonsynchronous Interference Factor (γ).** For estimating interference from other (within-cell and outer-cell) users, we assume an identical set of channels (e.g., all voice users requiring the same performance). We further assume that their despread interference can be approximated as a Gaussian random variable, that the users are spatially distributed in a uniform manner, and that power control within each cell is perfect. The worst-case interference comes about if all the interferers are chip and phase synchronized with the desired signal. For a nonsynchronous link, interference will not always be worst case. This lesser interference can be described by a factor γ that modifies Equation (12.64), thereby yielding more users per cell than that of the worst case. Assuming ideal rectangular-shaped chips, γ is equal to 1.5 [31–34]; this value will change for different chip shapes [31].

Using the factors G_A, G_V, H_0, and γ (and their typical values shown above) to determine the maximum possible number of simultaneous users per cell, M', yields

$$M' = \frac{\gamma \, G_A \, G_V}{H_0} \times M_{\max} = \frac{\gamma \, G_p \, G_A G_V}{(E_b/I_0)_{\text{reqd}} \, H_0} \approx 6 \times M_{\max} \qquad (12.65)$$

An accurate computation of capacity for a CDMA system is much more involved than Equation (12.65) suggests. The treatment leading to Equation (12.65) assumed perfect power control and a uniform distribution of users' location within cells. Thermal noise was neglected and no provision was made for traffic loading within cells. Terrain variations, which impact the accuracy of assuming an $n = 4$th power exponent law, were not considered. For lower values of n, there is potentially greater interference. The subject of CDMA capacity has been investigated in many publications, particularly in the context of systems designed to meet IS-95. The issue of capacity of CDMA systems is further addressed in [30–32, 35–38]. A very simplified analysis of three multiple access techniques that allow us to illustrate the capacity advantage of CDMA follows.

12.8.2 Analog FM versus TDMA versus CDMA

In 1976, prior to the implementation of cellular communication systems, New York City (with a population exceeding 10 million) could only support 543 simultaneous mobile users—3700 customers were on a waiting list. The cellular concept is illustrated in Figure 12.38 with a 7-cell configuration (one of several used). The idea of dividing a geographical region into cells and allowing the frequency allocation of one cell to be reused at other spatially separated cells represents one of the most important bandwidth-efficiency improvements in radio telephone systems.

In the United States, the frequency allocation for AMPS and other cellular systems is in the range of 869–894 MHz for base station transmit (mobile receive) channels, called *forward* or *downlink* channels, and 824–849 MHz for mobile transmit (base station receive) channels, called *reverse* or *uplink* channels. A single

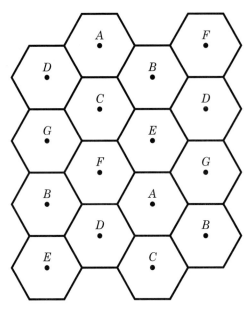

Figure 12.38 Seven-cell structure.

channel occupies a bandwidth of 30 kHz, sometimes called a *subband;* thus, a duplex pair (forward and reverse) occupy 60 kHz. The forward and reverse channels in each duplex pair are separated by 45 MHz. For mobile cellular service, the FCC has allocated each large metropolitan area (there about 750 such areas in the U.S.) 25 MHz for transmit and 25 MHz for receive. To foster competition, there are generally two service-provider companies allocated to each metropolitan area. Thus, each company has a total of 12.5 MHz for transmit and 12.5 MHz for receive.

When considering a wide geographical region made up of many cells, as seen in Figure 12.38, let's compare the capacity in units of channels per cell for three cellular systems: analog FM, TDMA, and CDMA. Computing capacity for the analog FM channels used in the AMPS system is quite straightforward. Consider the 12.5 MHz allocated to a service provider. In order to avoid interference between users operating in the same 12.5-MHz frequency band at comparable power levels, adjacent cells must operate at different frequencies. In the 7-cell configuration of Figure 12.38, communications within cell F may not operate in the same frequency band as communications in cells labeled A, B, C, D, E, and G. Although the service provider has been allocated 12.5 MHz, the frequency reuse pattern involved here dictates that only one-seventh of the allocation can be utilized within each cell. Thus, one-seventh of 12.5 MHz or equivalently 1.78 MHz can be used for transmit (and a similar amount for receive) within each cell. We say that such a 7-cell configuration has a *frequency-reuse factor* of $\frac{1}{7}$. Therefore, the number of 30-kHz subbands for analog FM channels is 1.78 MHz/30 kHz or approximately 57 channels per cell (not counting the control channels).

The U.S. standard describing the multiple access strategy for cellular TDMA is designated IS-54, which has been upgraded to IS-136. Systems designed to these standards must fit into the same frequency plan that was outlined for AMPS.

Therefore, each TDMA channel occupies 30 kHz. Fortunately, capacity improvements have come about only because the discipline of source coding has improved so dramatically since the 1950s. For *terrestrial* digital telephony, each voice signal is digitized to a bit rate of 64 kbits/s. Would a similar standard be used for *cellular* systems? Of course not, because cellular systems are so bandwidth limited. Source coding of speech can now produce telephone-quality fidelity at data rates of 8 kilobit/s, and it can even produce acceptable quality at lower data rates. For purposes of computation, if the often-chosen benchmark value of 10 kbit/s is used, then the capacity computation is again straightforward. Each of the 30 kHz channels can service 30 kHz/10 kbits/s = 3 users per 30 kHz subband. Thus, in TDMA, the number of simultaneous users per cell can be increased by a factor of 3 over the analog FM system. In other words, the number of TDMA channels is $57 \times 3 = 171$ channels per cell.

The main advantage of a CDMA cellular system over either analog FM or TDMA is that a frequency reuse factor of unity (100%) can be used. This means that the total FCC allocation of 12.5 MHz can be used for transmit (similarly for receive). In order to compare CDMA with the multiple access strategies in AMPS involving analog FM (which we can call FDMA) and IS-54-based TDMA, we start with Equation (12.65), but for a fair comparison, we eliminate the antenna gain factor G_A achieved through sectorizing the cell. The reason for this elimination is that G_A was not used in calculating the capacity for FDMA or TDMA, although both systems would also benefit from sectorization. Hence, the capacity of CDMA without sectorization becomes

$$M'' = \frac{\gamma \, G_p \, G_V}{(E_b/I_0)_{\text{reqd}} \, H_0} \tag{12.66}$$

Equation (12.28) then gives the processing gain

$$G_p = \frac{R_{\text{ch}}}{R} = \frac{12.5 \text{ Mchips/s}}{10 \text{ kbits/s}} = 1250 \tag{12.67}$$

Note that the chip rate of 12.5 Mchips/s is *not consistent* with IS-95 standards. It is used here to equitably compare CDMA across the entire allocation of the 12.5 MHz bandwidth, the same bandwidth used for analog FM and TDMA.

Selecting a nominal value of $(E_b/I_0)_{\text{reqd}}$ to be 7 dB (or the factor 5) [30], and for the factors G_V, γ, and H_0, using the values 2.5, 1.5 and 1.55, respectively, as described in Section 12.8.1, we then use Equation (12.66) to obtain

$$M'' = \frac{1.5 \times 1250 \times 2.5}{5 \times 1.55} \approx 605 \tag{12.68}$$

In summary, FDMA, using analog FM, TDMA, and CDMA, support 57, 171, and 605 channels per cell, respectively. Hence, it can be said that, in a given bandwidth, CDMA can exhibit about 10 times more user capacity than AMPS, and about 3.5 times the capacity of TDMA. It should be noted that the simple analysis leading to Equation (12.68) does not take into account other considerations, such as flat fading (see Chapter 15), which is sometimes encountered and may degrade the results of Equation (12.68). It should also be emphasized that the analysis was based on a CDMA reverse link, where unsynchronized users with long codes were assumed. In the forward direction (base station to mobile) orthogonal channelization can be used, which would improve the results of Equation (12.68).

It is difficult to compare CDMA with TDMA/FDMA in a fair way. On a single-cell basis, TDMA/FDMA capacity is dimension limited, while CDMA capacity is interference limited (discussed in the following section). From a multi-cell system view, all the systems are eventually interference limited. They attempt to optimize capacity with the following trade-offs. TDMA/FDMA systems trade-off larger reuse factors at the expense of greater interference. CDMA systems trade-off increased loading at the expense of greater interference.

12.8.3 Interference-Limited versus Dimension-Limited Systems

The interference in a properly designed and operating CDMA system is not severe; hence, all users can occupy the same spectrum. Nevertheless, from Equations (12.63) and (12.64), a CDMA system must be classified as interference-limited. Any reduction in $(E_b/I_0)_{reqd}$ can be translated almost directly into a larger number of simultaneous users. One can therefore see how important the incorporation of error-correction coding is to CDMA systems. An increase in coding gain by only 1 dB, which of course would reduce the $(E_b/I_0)_{reqd}$ by 1 dB, would yield a 25% increase in the number of allowable active users per cell.

In the context of single-cell operation, an FDMA and a TDMA system can be termed *frequency-dimension* and *time-dimension* limited, respectively. Consider a TDMA system. As time slots are assigned to an increasing number of users, there is no interference at the base station receiver caused by other mobile radios to the reception of a given user (assuming perfect synchronization). The user population can be increased until the number of time slots are exhausted. It is not possible to increase the number of users beyond the time-slot limit without intolerable interference. In a similar way, FDMA is frequency-dimension limited. It is not possible to increase the number of users beyond the frequency band limit without intolerable interference.

CDMA is interference limited because the introduction of each additional user raises the overall level of interference at the base station receivers. Each mobile radio introduces interference as a function of power level, synchronization, and code cross-correlation with other CDMA signals. The number of CDMA channels allowed depends on the level of total interference that can be tolerated. Figure 12.39 illustrates the basic difference between interference-limited systems, such as CDMA, and dimension-limited systems, such as TDMA. Assume that a fixed-size bandwidth is available for both. With TDMA in the context of a single cell, as time slots are filled by an increasing number of TDMA users, there is no interference at the base station receiver (caused by other mobile radios) to the reception of a given user. The number of TDMA users can increase until the number of available time slots is exhausted. It is then not possible to assign another time slot without causing an intolerable amount of interference. With CDMA, the introduction of each addi-

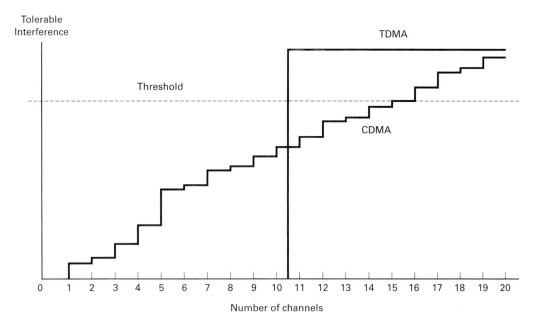

Figure 12.39   TDMA is time-dimension limited. CDMA is interference limited.

tional user raises the overall level of interference to the base station receivers. Each mobile radio might introduce a unique level of interference, owing to differences in power level, timing-synchronization, and cross-correlation with other code signals. Within a cell, channels are assigned to users until some predetermined interference threshold is reached [29]. Figure 12.39 shows that an interference-limited system is inherently more adaptive than a dimension-limited system. For example, on certain days of the year when it is well known that telephone traffic increases (such as Christmas Day and Mother's Day), a CDMA Operations Center can choose to tolerate a bit more interference in order to allow a larger number of users. With dimension-limited systems, no such dynamic trade-off can be made.

It is worth repeating that dimension-limited systems, such as FDMA and TDMA, are strictly dimension limited in the context of a single-cell operation. However, from a multi-cell perspective, one can trade-off frequency reuse factors versus the signal-to-interference (S/I) ratio to arrive at an interference-limited situation.

12.8.4 IS-95 CDMA Digital Cellular System

Interim Standard 95 (IS-95) specifies a wireless telephony system that uses direct-sequence spread-spectrum (DS/SS) as a multiple access technique. It was introduced by Qualcomm Corporation, and it was designed to operate in the same frequency band as the U.S. analog cellular system (AMPS), in which full duplex operation is achieved by using frequency division duplexing (FDD). The frequency allocation for AMPS provides 25 MHz in the range of 869–894 MHz for base station to mobile transmission (forward channels), and 25 MHz in the range of 824–849 MHz for mobile-to-base-station transmission (reverse channels). The IS-95 implementation strategy has been to introduce this code-division multiple-access (CDMA) system 1.25 MHz at a time, using dual-mode (AMPS and CDMA) mobile units. Being interference limited, systems designed to meet IS-95 specifications utilize various signal processing techniques to help reduce the $(E_b/N_0)_{reqd}$. The basic waveform, coding, and interference suppression features of such systems are outlined as follows:

- Each channel is spread across a bandwidth of about 1.25 MHz and filtered for spectral containment.
- The chip rate R_{ch} of the PN code is 1.2288 Mchips/s. The nominal data rate, known as Rate Set 1 (RS1), is 9.6 kbits/s, making the processing gain $G_p = R_{ch}/R = 128$. An extension to the original IS-95 introduced Rate Set 2 (RS2) at 14.4 kbits/s.
- The data modulation is binary phase-shift keying (BPSK), with quadrature phase-shift keying (QPSK) spreading. (Each quadrature component of the carrier wave is a BPSK signal modulated with the same data.)
- Convolutional coding with Viterbi decoding is used.
- Interleavers with a 20-ms time span are used for time diversity.
- Path diversity is exploited with a Rake receiver, and spatial diversity is implemented with two receive antennas per cell sector.
- Orthogonal code multiplexing is used for channelization.
- Power control is used to minimize transmitted power and thereby reduce interference.

The forward link comprises four types of channels: pilot, synchronization (SYNC), paging, and traffic. The reverse link comprises two types of channels: access and traffic. The history of IS-95 involves several standard committees and versions, with numbers such as IS-95A, JSTD-008, IS-95B, and IS-2000. IS-95B is a merging of IS-95-based methods for the cellular frequency band and the personal communication services (PCS) frequency band, for both voice and data. It provides data rates up to 115.2 kbits/s by aggregating up to eight RS2 channels. IS-2000 is a specification used to denote third-generation CDMA wireless systems, known as multi-carrier systems and having an assortment of new features. The treatment of CDMA in this section focuses on the original IS-95; the original structure remains valid for all IS-95-based variations, because they all share the basic architecture of the original system.

12.8.4.1 Forward Channel

The base station transmits a multiplex of 64 channels containing one pilot channel, one SYNC channel and at least one paging channel. The remaining 61 (or fewer) channels transmit user traffic. The IS-95 standard supports simultaneous transmission of voice, data, and signaling; variable rates for speech signals of 9600,

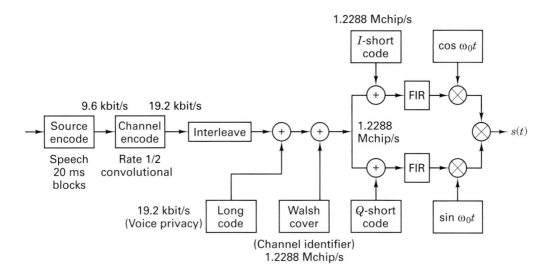

1.2288 Mchip/s

I-short code

$\cos \omega_0 t$

9.6 kbit/s   19.2 kbit/s

Source encode

Channel encode

Interleave

+

+

1.2288 Mchip/s

FIR

⊗

⊗ → s(t)

Speech 20 ms blocks

Rate 1/2 convolutional

FIR

⊗

19.2 kbit/s (Voice privacy)

Long code

Walsh cover

Q-short code

$\sin \omega_0 t$

(Channel identifier) 1.2288 Mchip/s

Figure 12.40   CDMA forward-traffic channel with full-rate speech.

4800, 2400, and 1200 bit/s are permitted. These rates are known as Rate Set 1. (Rate Set 2 supports up to 14.4 kbit/s.) Figure 12.40 is a simplified block diagram of the base station transmitter, implementing a typical 9.6 kilobits/s traffic channel. Using a linear predictive coding (LPC) algorithm (see Section 13.4.2), voice is first digitized to yield approximately 8 kilobits/s of raw digital speech. Error-detection bits are added, bringing the digital rate to 9.6 kilobits/s. The bit sequence is then processed in frame lengths of 20 ms. Hence, each 9.6 kilobits/s frame contains 192 bits. The next step shown in Figure 12.40 is convolutional coding (rate $\frac{1}{2}$, $K = 9$), where all information bits are equally protected. This brings the channel bit rate to 19.2 kilobits/s, which remains unchanged after interleaving by a block interleaver, with a span equal to one frame length of 20 ms. The next three steps involve the modulo-2 addition of binary digits representing different PN codes and orthogonal sequences for privacy, channelization, and base station identification. Each time a code is introduced, it can be thought of as a *barrier* or *door* that separates a specific message from others for a particular reason. Consider the privacy code. It is a *long* PN code implemented with a maximal-length, 42-stage shift register. at the system chip rate of 1.2288 Mchips/s, the code repeats approximately every 41 days. Systems designed to IS-95 specifications employ the same long-code hardware for all base stations and mobile units. To provide each mobile unit with its own unique code for privacy, each mobile is assigned a phase (time) offset of a privacy code. The parties carrying on a conversation do not need knowledge of each other's unique long-code offsets, since the base station demodulates and remodulates all traffic signals it processes. At the point that the privacy code is introduced in Figure 12.40, the channel-bit rate of 19.2 kilobits/s is not yet at the final chip rate. Hence, in the forward direction, the user's private code is applied in decimated fashion; that is, only every 64th bit of the sequence is used (which doesn't take away from the code's uniqueness).

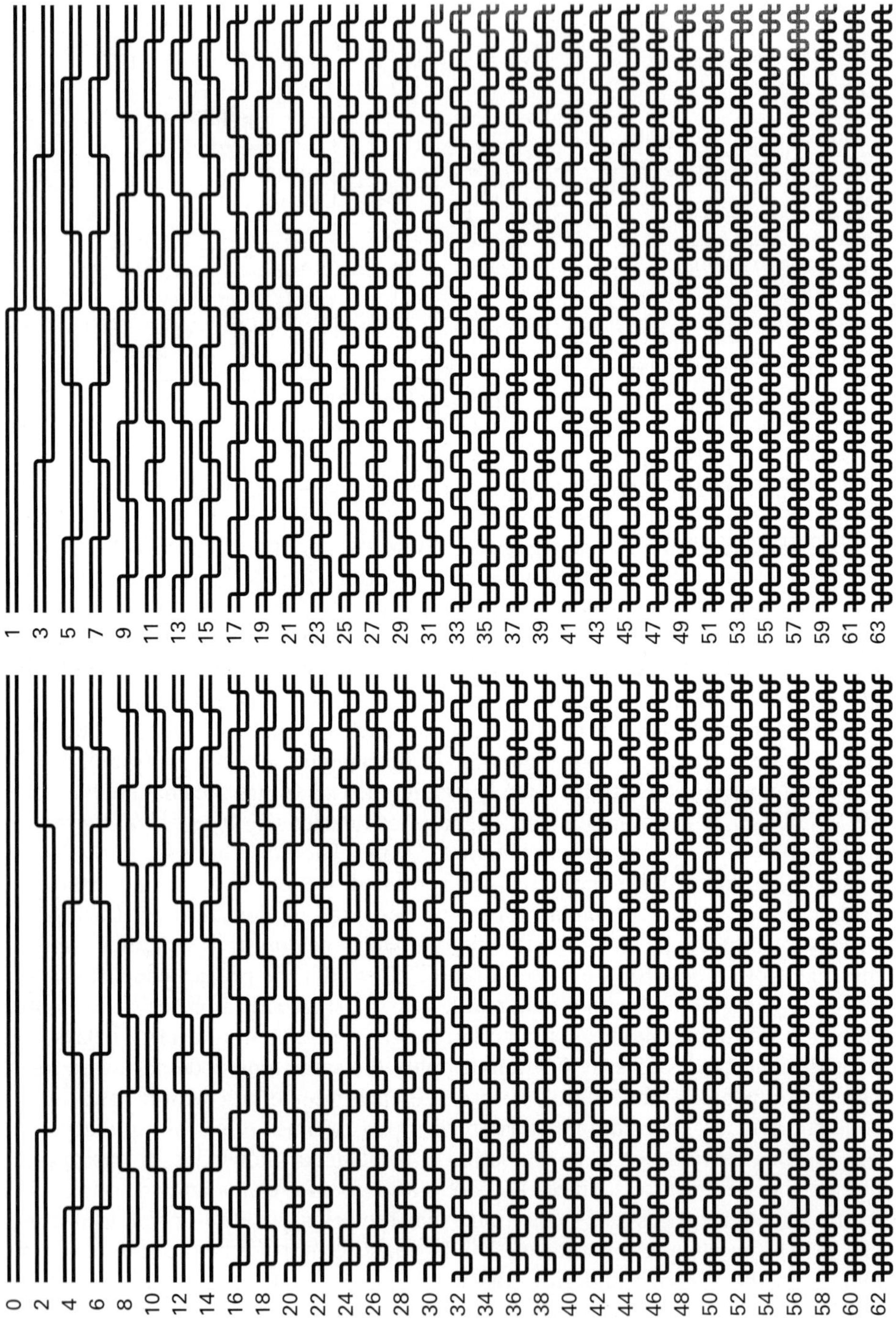

Figure 12.41 The set of 64 Walsh waveforms.

The next code, called a *Walsh cover,* is used for channelization plus spreading. It is an orthogonal code, which is mathematically constructed via the *Hadamard matrix.* (See Section 6.1.3.1 for the construction rules.) Using such a rule, one can form an orthogonal Walsh code of any desired dimension $2^k \times 2^k$, where k is a positive integer. The set of Walsh codes is described by a 64×64 array, where each row generates a different code. One of the 64 Walsh codes is modulo-2 added to the privacy-protected binary sequence, as shown in Figure 12.40. Because each of the 64 members of the Walsh code set are orthogonal to one another, their use in this manner channelizes the forward transmissions into 64 orthogonal signals. Channel number 0 is used as a pilot signal to assist coherent reception at the mobile unit, channel number 32 is used for synchronization, and at least one channel is reserved for paging. That leaves a maximum of 61 channels for traffic use. The Walsh cover is applied at the chip rate of 1.2288 Mchips/s. Thus, in the forward direction, each channel bit (at a rate of 19.2 kilobits/s) is transformed into 64 Walsh chips, producing a final chip rate of 1.2288 Mchips/s. Figure 12.41 illustrates the set of 64 Walsh waveforms. Figure 12.42 shows a simple channelization example using an orthogonal code such as a Walsh code. Unless the receiver applies the correct waveform for accessing a user's channel, the output is zero. Applying the correct waveform yields some nonzero value, A, that "unlocks the door" to that channel.

The next code in the forward direction (see Figure 12.40) is called the *short code* because it is configured with a 15-stage shift register, it repeats every $2^{15} - 1$ chips, and one period lasts 26.67 ms. This final "cloak" or "barrier," applied in quadrature at the chip rate of 1.2288 Mchips/s, provides scrambling of the signal. All base stations reuse the same Walsh channelization; without such scrambling,

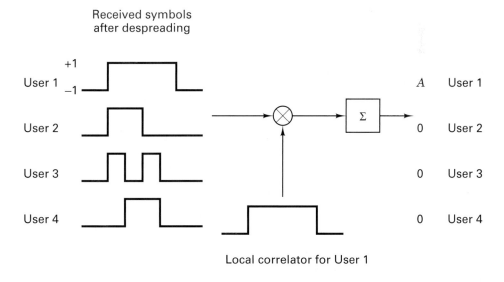

Figure 12.42 Example of channelizing transmissions with orthogonal functions.

the signals from different base stations would be somewhat correlated (which is certainly not desired). The short code can also be thought of as the address of the base station. Its implementation requires two different 15-stage shift registers: one for the inphase (I) channel, and one for the quadrature (Q) channel. Each base station uses a different 64-chip offset of the I and Q codes to identify its location; thus allowing for 512 unique addresses. This is deemed to be a sufficiently large number because addresses can be reused at base stations that are sufficiently separated from one another.

To summarize the functions of the three codes: the Walsh code provides orthogonality (for channelization) among all users located in the same cell; the short PN code maintains mutual randomness among users of different cells (for base station addressing); and the long PN code provides mutual randomness among different users of the system (for privacy). For the Walsh code to provide perfect orthogonality among channels, all the users must be synchronized in time with an accuracy corresponding to a small fraction of one chip. This is theoretically possible for the forward link because transmissions to all mobiles have a common origin at the base station. However, due to multipath effects, it is more accurate to say that the Walsh codes provide partial orthogonality. To obtain similar benefits on the reverse link would require closed-loop timing control, which is not implemented in IS-95. The reduced complexity is realized at the cost of greater within-cell interference. For third-generation wideband CDMA (WCDMA) systems, this option is present [38].

The last blocks of Figure 12.40 show wideband filtering (1.25 MHz) with finite impulse response (FIR) filters and the heterodyning of a carrier wave with BPSK modulation and QPSK spreading. The same coded bits are simultaneously present on the I and Q channels, but due to the short-code scrambling, the I and Q signals are different.

12.8.4.2 Reverse Channel

Each base station can transmit a multiplex of 64 channels, where 61 or fewer channels are used for traffic. But in the reverse direction (mobile to base station), there is just a single channel (signal) being transmitted (access request or traffic). Figure 12.43 depicts a simplified block diagram of a reverse-traffic channel transmission. The general structure is similar to the forward-traffic channel shown in Figure 12.40; however, there are several important differences. In IS-95, the reverse link does not support a pilot channel, since one would be required for each mobile unit. Thus, the reverse-channel signal is demodulated noncoherently at the base station. (In IS-2000, a pilot signal is provided for each reverse channel.) Since the reverse channel is less robust than the forward channel, a more powerful rate $\frac{1}{3}$ convolutional code is used to improve performance. Also, following the interleaver, notice that the channel bits modulate a 64-ary Walsh waveform. This is the same type of waveform that was used for channelization in the forward direction. However, in the reverse direction, Walsh waveforms are used for a totally different purpose: They become the modulating waveforms. Assuming a data rate of $R = 9.6$ kilobits/s, two information bits (which after coding are transformed into six channel bits, sometimes called code

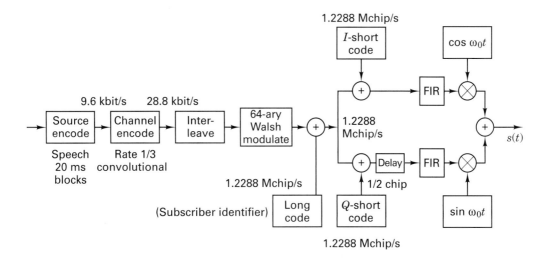

Figure 12.43 CDMA reverse-traffic channel with full-rate speech.

symbols), are mapped after interleaving into one of 64 orthogonal Walsh waveforms to be transmitted. Therefore, the Walsh waveform rate is

$$R_w = \frac{R_c}{\log_2 M} = \frac{R(n/k)}{\log_2 M} = \frac{9600 \times 3}{6} = 4{,}800 \text{ Walsh-symbols/s} \quad (12.69)$$

where the channel-bit rate R_c is equal to the data rate times the inverse of the code rate, namely, $R\,(n/k)$. Each of the 64-ary Walsh waveforms is made up of 64 elements, termed *Walsh chips*. Then, from Equation (12.69), we see that the Walsh chip rate is $64 \times 4800 = 307{,}200$ Walsh chips/s. Thus, the modulation has resulted in some spreading (not to the full bandwidth). The Walsh chips are then repeated 4 times to arrive at the final spread-spectrum rate of 1.2288 Mchips/s.

One might ask, "Why were 64-ary Walsh functions chosen as the modulation waveforms?" Consider the trade-offs described in Section 9.7.3 for power-limited channels. A natural choice for conserving power at the expense of bandwidth is *M*-ary orthogonal signaling such as MFSK. The larger the value of *M*, the greater will be the bandwidth expansion—yet, the greater will be the reduction in required E_b/N_0 for a specified level of performance. Choosing such a signaling scheme in a narrowband system is a true trade-off—for the price of expanded bandwidth, a reduction in required power is obtained. However, in spread-spectrum systems such as those that meet IS-95 specifications, the selection of 64-ary Walsh waveforms for modulation can be seen as "getting something for nothing," because the spread-spectrum system already occupies an expanded bandwith of 1.25 MHz. The choice of 64-ary orthogonal waveforms does not expand the bandwidth any further. When you look at the Walsh waveforms in Figure 12.41, imagine the pulse shapes to be somewhat rounded. Doesn't this waveform set remind you of MFSK? Well, they are in fact similar, and at the base station, a 64-ary Walsh waveform is (generally) detected noncoherently, much like the noncoherent detection of a 64-ary FSK tone. (Some base station receivers use coherent processing techniques, thereby providing 1-2 dB gain over noncoherent processing.)

One might ask, "Isn't channelization needed in the reverse direction?" Yes. It is always necessary to keep users separated; however, in the reverse direction, one user is distinguished from another via the long (privacy) code. In the forward direction, this code was used in a decimated fashion for privacy. In the reverse direction, as shown in Figure 12.43, the code is applied at the 1.2288 Mchips/s rate for channelization (addressing), and also for privacy, scrambling, and spreading. After spreading by this long code, the waveform is further spread by a pair of short PN codes to assure that the I and Q symbols are uncorrelated. The last blocks in Figure 12.43 show FIR filtering (1.25 MHz) and heterodyning a carrier with BPSK modulation in an offset QPSK (OQPSK) fashion. OQPSK is used in order to eliminate the possibility of the carrier wave changing phase by 180°. (See Section 9.8.1.) This feature reduces the peak-to-average power specification of the transmitter power amplifier, making its design easier. Notice that OQPSK is not used on the forward link since the transmitter sends a multiplex of 64 signals. Each forward transmission consists of a phasor representing the entire muliplex, whose resultant value can be one of a myriad of phase/amplitude possibilities. Hence, there would be no benefit from offsetting the I and Q channels, since carrier transitions through the origin could not be avoided. The final waveform is filtered to generate a spectrum with a 3-dB double-sided bandwidth of 1.25 MHz.

12.8.4.3 Receiver Structures

Mobile Receiver. The mobile receiver demodulates each of the forward-channel quadrature-BPSK waveforms coherently, using the pilot signal as a reference. The receiver structure implements a 3-finger Rake receiver to recover the three strongest multipath components (the minimum as defined in IS-95). The multipath components of the spread-spectrum signal are resolved and separated by the Rake receiver, provided that the differential path delays exceed one chip duration. FDMA waveforms cannot be so separated because they are inherently narrowband. Multipath components of TDMA waveforms can be better separated since each user transmits data in bursts. However, in a typical TDMA system, the bursts produce waveforms that are still too narrowband for multipath resolution at nominal delays. But for CDMA, if the spread-spectrum bandwidth exceeds 1 MHz, any multipath components that are separated by 1 μs delays or greater are separable. The Rake receiver tracks such paths rapidly and combines them constructively (even coherently at the mobile receiver). The operation of the Rake receiver is presented in Section 15.7.2. The soft-decision outputs of the demodulator are processed by a Viterbi decoder. The final step in recovering the information consists of determining which of the four possible data rates (9600, 4600, 2400, or 1200 bits/s) was actually utilized at the transmitter. This is accomplished without any overhead penalty by decoding the demodulated output four times, once for each of the four hypotheses. Several metrics are obtained from the decoding process and also from the pass/fail metrics of the error-detection bits. These are analyzed in order to select one final decoded sequence.

Base Station Receiver. The base station dedicates a separate channel in order to receive the transmissions of each active user in the cell. Each user's reverse-channel 64-ary Walsh-modulated signals are received noncoherently (much like the reception of noncoherent orthogonal MFSK). The receiver structure typically implements a 4-finger Rake receiver to demodulate the four strongest multipath components at the output of two antennas (see Section 15.7.2), which are spatially separated by several wavelengths for diversity reception. The soft-decision outputs of the demodulator are processed by a Viterbi decoder. The final step in recovering the information consists of decoding the demodulated data four times, using a procedure similar to that used in the forward direction, where metrics are compared in order to select one final decoded sequence.

12.8.4.4 Power Control

Power control is a necessity for a system in which many users simultaneously transmit to a base station using the same frequency. Without power control, users transmitting from locations near the base station would be received at power levels much higher than those of users transmitting from locations near the cell's edge. The main goal of power control is to adjust the users' transmitted power so as to provide at the base station an equal (and near constant) received power level from each mobile unit. In order to accomplish this, a key feature of the power-control algorithm is to command users to transmit at power levels that are inversely proportional to the received power level (from the base station). There are three power-control methods specified in IS-95: reverse-link open-loop control, forward/reverse-link closed-loop control, and forward link control.

Reverse-Link Open-Loop Control. The assumption is made that there are similar path losses on a forward and reverse channel, even though this is not completely true since they operate at frequencies that are separated by 45 MHz. The base station continually transmits a calibration constant (determined by its EIRP) on the SYNC channel. This information allows the mobile unit to use an estimated transmit power in order for the received power at the base station to be the same as that of other mobile units. Consider the following example of an open-loop control algorithm. The mobile transmission power is selected so that its transmission power plus the power received from the base station (reflecting path loss) should equal some value (for example, –73 dBm). This value is a function of the base station EIRP and appears on the SYNC channel. Before a mobile begins its transmission, it determines the power received on the forward link from the automatic gain control (AGC) circuit in its receiver. If the received power is, for example, –83 dBm, then the open-loop power-control algorithm dictates a transmit power of (–73 dBm) – (–83 dBm), or 10 dBm.

Forward/Reverse Link Closed-Loop Control. Power-control bits are sent on the forward link by "stealing" from the channel bits transmitting encoded traffic (resulting in a punctured code). Once every 6 Walsh waveforms, 2 channel bits are replaced with a power-control bit. Since the Walsh waveforms are transmitted at a rate of 4800 waveforms/s, the rate at which the power-control bits are transmitted is 800 bits/s. Thus, there are 16 such control bits in each 20 ms frame. The goal of this closed-loop power control is to correct the open-loop power-control estimate every 1.25 ms in steps of 1 dB. Later versions added the option for step sizes of 0.5 dB and 0.25 dB. The most significant benefit of such fast and accurate closed-loop power control is a significant reduction in the average transmitter power on the reverse channel. Analog mobile radios transmit enough power to maintain a link even during fading. Thus, most of the time such analog radios transmit excessive amounts of power. CDMA mobile radios operate at power levels no greater than what is needed to close the reverse link. A mobile unit using CDMA designed to meet IS-95 specifications requires approximately 20 dB to 30 dB less power than a mobile unit operating in an analog AMPS system [30].

Forward Link Control. The base station periodically reduces the power transmitted to the mobile unit. Whenever the mobile senses an increased frame-error rate, it requests additional power from the base station. Adjustments can be made periodically based on reported frame-error rates (FER).

Example 12.6   Signaling Elements (Numerology) Used in IS-95

There is a rich set of signaling elements used in CDMA systems that are designed to IS-95 specifications: data bits, channel bits, Walsh waveforms, Walsh chips, spread-spectrum chips, and BPSK waveforms. Consider a reverse traffic channel that is carrying full-rate digitized speech at 9.6 kbits/s, with a received $E_b/(N_0 + I_0) \approx E_b/I_0 = 7$ dB (assuming that $N_0 \ll I_0$). Find the values of the following received power-to-noise spectral-density and energy-to-noise spectral-density parameters: P_r/I_0, E_c/I_0, E_w/I_0, E_{wch}/I_0, and E_{ch}/I_0. Also, find the values of the following rates: R_c, R_w, R_{wch}, and R_{ch}, where c, w, wch, and ch represent channel bit, Walsh waveform, Walsh chip, and spread-spectrum chip, respectively. How many spread-spectrum (SS) chips correspond to one Walsh chip?

Solution

The key to this problem lies in the fundamental relationships between received power-to-noise spectral density and each of the signaling parameters, in a manner similar to that presented in Section 9.7.7. From that development, we can write

$$\frac{P_r}{I_0} = \frac{E_b}{I_0} R = \frac{E_c}{I_0} R_c = \frac{E_w}{I_0} R_w = \frac{E_{wch}}{I_0} R_{wch} = \frac{E_{ch}}{I_0} R_{ch} \qquad (12.69)$$

Since E_b/N_0 is 7 dB (or 5), and the data rate R is 9600 bits/s, from Equation (12.69) we obtain

$$\frac{P_r}{I_0} = \frac{E_b}{I_0} R = 48{,}000 \text{ Hz or } 46.8 \text{ dB-Hz}$$

For the reverse traffic channel, the code rate is $\frac{1}{3}$. Therefore,

$$\frac{E_c}{I_0} = \left(\frac{1}{3}\right) \frac{E_b}{I_0} = \frac{5}{3} \text{ or } 2.2 \text{ dB}$$

and

$$R_c = 3 \times R = 3 \times 9600 = 28{,}800 \text{ channel bits/s}$$

Each 64-ary Walsh waveform corresponds to 6 channel bits. Therefore,

$$\frac{E_w}{I_0} = 6 \times \frac{E_c}{I_0} = 6 \times \left(\frac{5}{3}\right) = 10 \text{ or } 10 \text{ dB}$$

and

$$R_w = \left(\frac{1}{6}\right) R_c = \left(\frac{1}{6}\right) 28{,}800 = 4800 \text{ Walsh waveforms/s}$$

A Walsh waveform is composed of 64 Walsh chips. Hence,

$$\frac{E_{wch}}{I_0} = \left(\frac{1}{64}\right) \frac{E_w}{I_0} = \left(\frac{1}{64}\right) \times 10 = \frac{10}{64} \text{ or } -8.1 \text{ dB}$$

and

$$R_{wch} = 64 \times R_w = 64 \times 4800 = 307{,}200 \text{ Walsh-chips/s}$$

In IS-95, the spread-spectrum chip rate is 1.2288 Mchips/s. Thus,

$$\frac{E_{ch}}{I_0} = \frac{P_r}{I_0} \times \left(\frac{1}{R_{ch}}\right) = \left(\frac{48{,}000}{1.2288 \times 10^6}\right) = 0.039 \text{ or } -14.1 \text{ dB}$$

SS-chips per Walsh chip:
$$\frac{R_{ch}}{R_{wch}} = \frac{1.2288 \times 10^6}{307{,}200} = 4$$

12.8.4.5 Typical Telephone Call Scenario

Turn on and Synchronization. Once power is applied to the mobile unit, the receiver scans continuously in search of available pilot signals. Such signals will be received from different base stations with different time-offsets of the short PN code (described in Section 12.8.4.1). The time-offset used by a base station differs by a multiple of 64 chips from all other base stations. Since the short code is maximal length, its 15-stage shift register produces $2^{15} - 1 = 32{,}767$ bits. After "bit stuffing" the sequence with one bit, 32,768 bits are produced before the whole process repeats itself. Thus, there are 32,768/64, or 512 available unique addresses. The 512 short PN codes can be generated by a simple time shift of a single PN sequence, because the base stations are time synchronized within a few microseconds of each other. At the chip rate of 1.2288 Mchips/s, there are 75 frames of the short code corresponding to a 2-second interval. The zero-offset address of the short code occurs on even second time marks. Consider the case of a base station whose address is represented by offset number 18. Then its transmission cycle begins at $(18 \times 64$ chips $\times (1/1.2288 \times 10^6)$ s/chip, or 937.5 μs after every even-second time mark.

Once the mobile unit completes its scan and is correlated to the strongest pilot signal, it is now synchronized with one of the 512 unique base station addresses. The mobile unit can now despread any of that base station's transmissions; however, it does not yet have system time, which is needed for access, paging, and traffic channels. Next, using the pilot signal as a reference, the mobile unit coherently demodulates the SYNC channel signal (Walsh 32), which the base station transmits continuously. The SYNC channel transmissions provide several system parameters, the key one being the state of a long code 320 ms in the future, giving the mobile unit time to decode, load its registers, and become system-time synchronized. This long code is one of a specific group of long codes used for access and paging. The mobile selects a predefined paging channel based on its serial number, and it monitors this paging channel for incoming calls. The mobile can now register with the base station, which allows for location-based paging rather than system-based paging when there is an incoming telephone call.

Idle-State Handoff. The mobile unit continually scans for alternative pilot signals. If it finds a stronger pilot signal from a different base station, the mobile locks on to the base station with the stronger pilot. Since there is no call in progress, the process simply serves to update the location of the mobile. The mobile has obtained system time from the SYNC channel. If there were only one base station, system time could be defined by whatever reference the base station chooses. With several operating base stations, the *handoff* process is facilitated if time is coordinated throughout the system. In IS-95, system time is specified to be *Universally Coordinated Time* (UTC) $\pm$ 3 µs. A practical way to implement this is with the use of a Global Positioning System (GPS) receiver at each base station.

Call Initiation. A call is initiated by the user keying in a telephone number and pushing the *send* button. This initiates an access probe. The mobile uses open-loop power control, choosing an initial transmission power level estimated from the pilot signal, as described in Section 12.8.4.4. All access channels use different long-code offsets. At the beginning of an access probe, the mobile pseudorandomly chooses one of the access channels associated with its paging channel. The transmission of an access probe is timed to begin at the start of an access channel slot, which is determined pseudorandomly. A key element of the access procedure involves the identification of the caller's serial number. Identification is needed because the base station cannot discriminate accesses from different users, since the access channel is a common channel.

The mobile-terminal time reference for transmission is determined by the earliest multipath component being used for demodulation. The mobile does not make transmission adjustments to account for propagation delay. Instead, the base station continually searches and tests for the presence of reverse channel signals. The mobile "listens" on the paging channel for a response from the base station. If there is none (collisions can occur during transmission on the access channels), the mobile attempts access again after waiting a pseudorandom time. When the mobile's access probe is successful, the base station response is a traffic channel assignment (Walsh code number).

Traffic channels use different long-code offsets than paging channels. Therefore, the mobile unit changes its long-code offset to one based on its serial number. After receiving the Walsh code assignment, the mobile begins an all-zeros data transmission on the traffic channel, and waits for a positive acknowledgment on the forward traffic channel. If the exchange is successful, the next step is ringing at the telephone that was called. Conversation can then commence.

Soft Handoff. During a call, the mobile may find an alternate strong pilot signal. It then transmits a control message to its base station, identifying the new base station with the stronger pilot signal and requesting a soft handoff. The original base station passes the request to a base station controller (BSC) that handles the radio resource control of the link; the BSC may or may not be collocated with a Mobile Switching Center (MSC) that handles the non-radio aspects of the link (e.g., switching). The BSC contacts the new base station and obtains a Walsh number assignment. This assignment is sent to the mobile via its original base station connection. During the transition, the mobile is supported by (connected to) both base stations, and a land link connection is maintained from the BSC to both base stations. The mobile combines the signals received from both base stations by using the two respective pilot signals as coherent phase references. Signal reception from two base stations simultaneously is facilitated by the Rake receiver, since the transmissions from both base stations appear as multipath components to the mobile receiver. At the BSC, where the signals are received noncoherently, the two received signals from the mobile are examined, and the better one is chosen in each 20-ms frame. The original base station drops the call when connection is firmly established in the new cell. Such dual connection, sometimes called "make before break," reduces the probability of a dropped call and of poor reception at a cell's edge.

12.9 CONCLUSION

Spread-spectrum (SS) technology has only emerged since the 1950s. Yet, this novel approach to applications, such as multiple access, ranging, and interference rejection, has rendered SS techniques extremely important to most current NASA and military communication systems. In this chapter we presented an overview enumerating the benefits and types of spread-spectrum techniques, as well as some historical background.

Since SS techniques were initially developed with military applications in mind, we started the treatment with discussions of anti-jam (AJ) systems. Pseudorandom sequences are at the heart of all present-day SS systems; we therefore treated PN generation and properties. Emphasis was placed on the two major spread-spectrum techniques; direct sequence and frequency hopping. Consideration was given to synchronization, a crucial aspect of spread-spectrum operation. Also, attention was devoted to the commercial use of spread-spectrum techniques for code-division multiple access (CDMA) systems, particularly direct-sequence CDMA, as it is specified in interim standard 95 (IS-95).

REFERENCES

1. Scholtz, R. A., "The Origins of Spread Spectrum Communications," *IEEE Trans. Commun.,* vol. COM30, no. 5, May 1982, pp. 822–854.

2. Shannon, C. E., "Communication in the Presence of Noise," *Proc. IRE,* Jan. 1949, pp. 10–21.

3. Dillard, R. A., "Detectability of Spread Spectrum Signals," *IEEE Trans. Aerosp. Electron. Syst.,* July 1979.

4. Simon, M. K., Omura, J. K., Scholtz, R. A., and Levitt, B. K., *Spread Spectrum Communications,* Computer Science Press, Inc., Rockville, Md., 1985.

5. de Rosa, L. A., and Rogoff, M., Sec. I (Communications) of *Application of Statistical Methods to Secrecy Communication Systems,* Proposal 946, Fed. Telecommun. Lab., Nutley, N.J., Aug. 28, 1950.

6. Pickholtz, R. L., Schilling, D. L., and Milstein, L. B., "Theory of Spread-Spectrum Communications—A Tutorial," *IEEE Trans. Commun.,* vol. COM30, no. 5, May 1982, pp. 855–884.

7. Pickholtz, R. L., Schilling, D. L., and Milstein, L. B., Revisions to "Theory of Spread-Spectrum Communications—A Tutorial," *IEEE Trans. Commun.,* vol. COM32, no. 2, Feb. 1984, pp. 211–212.

8. Simon, M. K., Omura, J. K., Scholtz, R. A., and Levitt, B. K., *Spread Spectrum Communications,* Vol. 2, Computer Science Press, Inc., Rockville, Md., 1985.

9. Simon, M. K., and Polydoros, A., "Coherent Detection of Frequency-Hopped Quadrature Modulations in the Presence of Jamming: Part I. QPSK and QASK; Part II. QPR class I Modulation," *IEEE Trans. Commun.,* vol. COM29, Nov. 1981, pp. 1644–1668.

10. Holmes, J. K., and Chen C. C., "Acquisition Time Performance of PN Spread-Spectrum Systems," *IEEE Trans. Commun.,* COM-25, August 1977, pp. 778–783.

11. Ward, R. B., "Acquisition of Pseudonoise Signals by Sequential Estimation," *IEEE Trans. Commun.,* COM13, Ded. 1965, pp. 475–483.

12. Spilker, J. J., and Magill, D. T., "The Delay-Lock Discriminator—An Optimum Tracking Device," *Proc. IRE,* Sept. 1961.

13. Spiler, J. J., "Delay-Lock Tracking of Binary Signals," *IEEE Trans. Space Electron. Telem.,* Mar. 1963.

14. Simon, M. K., "Noncoherent Pseudonoise Code Tracking Performance of Spread Spectrum Receivers," *Commun.,* vol. COM25, Mar. 1977.

15. Ziemer, R. E., and Peterson, R. L., *Digital Communications and Spread Spectrum Systems,* Macmillan Publishing Company, New York, 1985.

16. Holmes, J. K., *Coherent Spread Spectrum Systems,* John Wiley & Sons, Inc., New York, 1982.

17. Pursley, M. B., "Performance Evaluation for Phase-Coded Spread-Spectrum Multiple-Access Communication: Part I. System Analysis," *IEEE Trans. Commun.,* vol. COM25, no. 8, Aug. 1977, pp. 795–799.

18. Geraniotis, E., "Noncoherent Hybrid DS-SFH Spread-Spectrum Multiple-Access Communications," *IEEE Trans. Commun.,* vol. COM34, no. 9, Sept. 1986, pp. 862–872.

19. Greaniotis, E., and Pursley, M. B., "Error Probability for Direct-Sequence Spread-Spectrum Multiple-Access Communications: Part I. Upper and Lower Bounds," *IEEE Trans. Commun.,* vol. COM30, no. 5, May 1982, pp. 985–995.

20. Geraniotis, E., and Pursley, M. B., "Error Probabilities for Direct-Sequence Spread-Spectrum Multiple-Access Communications: Part II. Approximations," *IEEE Trans. Commun.,* vol. COM30, no. 5, May 1982, pp. 996–1009.

21. Schilling, D. L., Milstein, L. B., Pickholtz, R. L., and Brown, R. W., "Optimization of the Processing Gain of an *M*-ary Direct Sequence Spread Spectrum Communication System," *IEEE Trans. Commun.,* vol. COM28, no. 8, Aug. 1980, pp. 1389–1398.

22. Viterbi, A. J., and Jacobs, I. M., "Advances in Coding and Modulation for Noncoherent Channels Affected by Fading, Partial Band, and Multiple Access Interference," in A. S. Viterbi, ed., *Advances in Communication Systems,* Vol. 4, Academic Press, Inc., New York, 1975.

23. Stark, W. E., "Coding for Frequency-Hopped Spread-Spectrum Communication with Partial-Band Interference: Part I. Capacity and Cutoff Rate," *IEEE Trans. Commun.,* vol. COM33, no. 10, Oct. 1985, pp. 1036–1044.

24. Stark, W. E., "Coding for Frequency-Hopped Spread-Spectrum Communication with Partial-Band Interference: Part II. Coded Performance," *IEEE Trans. Commun.,* vol. COM33, no. 10, Oct. 1985, pp. 1045–1057.

25. Milstein, L. B., Davidovici, S., and Schilling, D. L., "The Effect of Multiple-Tone Interfering Signals on a Direct Sequence Spread Communication System," *IEEE Trans. Commun.,* vol. COM30, Mar. 1982, pp. 436–446.

26. Milstein, L. B., Pickholtz, R. L., and Schilling, D. L., "Optimization of the Processing Gain of an FSK-FH system," *IEEE Trans. Commun.,* vol. COM28, July 1980, pp. 1062–1079.

27. Huth, G. K., "Optimization of Coded Spread Spectrum Systems Performance," *IEEE Trans. Commun.,* vol. COM25, Aug. 1977, pp. 763–770.

28. Viterbi, A. J., "Spread Spectrum Communications—Myths and Realities," *IEEE Commun. Mag.,* May 1979, pp. 11–18.

29. Simon, M. K., Omura, J. K., Scholtz, R. A., and Levitt, B., *Spread Spectrum Communications Handbook,* Revised Edition, McGraw-Hill, Inc., New York, 1994.

30. Viterbi, A. J., "The Orthogonal-Random Waveform Dichotomy for Digital Mobile Personal Communication," *IEEE Personal Communications,* First Quarter 1994, pp. 18–24.

31. Kohno, R., Meidan, R., and Milstein, L. B., "Spread Spectrum Access Methods for Wireless Communications," *IEEE Communications Magazine,* January 1995, pp. 58–67.

32. Pickholtz, R. L., Milstein, L. B., and Schilling, D. L., "Spread Spectrum for Mobile Communications," *IEEE Trans. Vehicular Tech.,* vol. 40, no. 2, May 1991, pp. 313–321.

33. Morrow, R. K., Jr. and Lehnert, J. S., "Bit-to-Bit Error Dependence on Slotted DS/SSMA Packet Systems with Random Signature Sequences," *IEEE Trans. Commun.,* vol. 37, no. 10, October 1989, pp. 1052–1061.

34. Schilling, D. L., et al., "Spread Spectrum for Commercial Communications," *IEEE Communications Magazine,* April 1991, pp. 66–78.

35. Gilhousen, K. S., "On the Capacity of a Cellular CDMA System," *IEEE Trans. Vehicular Tech.,* vol. 40, no. 2, May 1991, pp. 303–312.

36. Viterbi, A. M., and Viterbi, A. J., "Erlang Capacity of a Power Controlled CDMA System," *IEEE JSAC, vol. 11, no. 6, pp. 892–899.*

37. Padovani, R., "Reverse Link Performance of IS-95 Based Cellular Systems, *IEEE Personal Communications,* Third Quarter 1994, pp. 28–34.

38. Wideband CDMA Special Issue, *IEEE Communications Magazine,* vol. 36, no. 9, September 1998.

PROBLEMS

12.1. Explain why a maximal-length n-stage linear feedback shift register can produce a sequence with a period no greater than $2^n - 1$.

12.2. Show that in a maximal-length n-stage linear feedback shift register the output stage must always be an input to the feedback network.

12.3. Consider the DS/BPSK spread-spectrum transmitter of Figure 12.9a or b. Let $x(t)$ be the sequence 1 0 0 1 1 0 0 0 1, arriving at a rate of 75 bits/s, where the leftmost bit is the earliest bit. Let $g(t)$ be generated by the shift register of Figure 12.7, with an initial state of 1 1 1 1 and a clock rate of 225 Hz.
(a) Sketch the final transmitted sequence $x(t)g(t)$.
(b) What is the bandwidth of the transmitted (spread) signal?
(c) What is the processing gain?
(d) Suppose that the estimated delay, $\hat{T}_d$, of Figure 12.9c is too large by one chip time. Sketch the despread chip sequence.
(e) Choose a decision rule for deciding on $\hat{x}(t)$ and identify the errors.

12.4. A total of 24 equal-power terminals are to share a frequency band through a code-division multiple access (CDMA) system. Each terminal transmits information at 9.6 kbits/s with a direct-sequence spread-spectrum BPSK modulated signal. Calculate the minimum chip rate of the PN code in order to maintain a bit error probability of 10^{-3}. Assume that the receiver noise is negligible with respect to the interference from the other users.

12.5. A feedback shift register PN generator produces a 31-bit PN sequence at a clock rate of 10 MHz. What are the equation and graphical form of the autocorrelation function and power spectral density of the sequence? Assume that the pulses have values of ± 1.

12.6. Consider an FH/MFSK system such as the one shown in Figure 12.11. Let the PN generator be defined by a 20-stage linear feedback shift register with a maximal length sequence. Each state of the register dictates a new center frequency within the hopping band. The minimum step size between center frequencies (hop to hop) is 200 Hz. The register clock rate is 2 kHz. Assume that 8-ary FSK modulation is used and that the data rate is 1.2 kbits/s.
(a) What is the hopping bandwidth?
(b) What is the chip rate?
(c) How many chips are there in each data symbol?
(d) What is the processing gain?

12.7. The block diagram of Figure 12.16 is described in Section 12.4.5 for a fast frequency hopping (FFH) demodulator. Draw a similar block diagram for a slow frequency hopping (SFH) demodulator, and explain how it would work.

12.8. Find the mean and the standard deviation of the time needed to acquire a 10-megachip/s BPSK modulated PN code sequence using a serial search where 100 chips are examined at a time. Assume that a correct detection results when all 100 received chips match the locally generated ones. The ratio of received chip energy to noise power spectral density is 9.6 dB, and the uncertainty time between the received and local code sequences is 1 ms. Assume that the probability of false lock (false alarm) is negligible.

12.9. There are 11 equal-power terminals in a CDMA communication system, transmitting signals toward a central node. Each terminal transmits information at 1 kbit/s on a 100-kchips/s direct-sequence spreading signal using BPSK modulation.
(a) If receiver noise is negligible with respect to the interference from other users, what is the ratio of bit energy to interference power spectral density (E_b/I_0) received by a receiving terminal?
(b) What is the effect on E_b/I_0 if all users double their output power?
(c) If the users wish to expand their service to 101 equal-power users, what must be done to the spreading codes to maintain the original E_b/I_0 ratio?

12.10. A CDMA system uses direct-sequence modulation with a data bandwidth of 10 kHz and a spread bandwidth of 10 MHz. With only one signal being transmitted, the received E_b/N_0 is 16 dB.

  (a) If the required $(E_b/N_0 + I_0)$ is 10 dB, how many equal-power users can share the band? Do not neglect receiver noise.

  (b) If each user's transmitted power is reduced by 3 dB, how many equal-power users can share the band?

  (c) If the received $E_b/N_0 \to \infty$ for each receiver, what is the maximum number of users that can share the band?

12.11. A DS/SS system is used to combat multipath. If the path length of the multipath wave is 100 m longer than that of the direct wave, what is the minimum chip rate necessary to reject the multipath interference?

12.12. A ground-to-synchronous satellite link must be closed in a jamming environment. The data rate is 1 kbit/s and the ground station has a 60-ft antenna. Antijam protection is provided by a 10-Mbits/s direct-sequence spread-spectrum code. The jammer has a 150-ft antenna and a transmitter with 400 kW of power. Assume equal space and propagation losses. How much power is required of the earth station transmitter to achieve an E_b/J_0 of 16 dB at the satellite receiver? Assume that the receiver noise is negligible.

12.13. Input data at 75 bits/s are channel encoded using a rate $\frac{1}{2}$ encoder. The coded bits are then modulated using 8-ary FSK. The FSK symbols are then spread by frequency hopping at a rate of 2000 hops/s.

  (a) What is the chip rate?

  (b) What is the order of diversity?

  (c) If there are two such signals, time-division multiplexed (TDM'd) on the channel at the same hopping rate, how would this affect the chip rate, symbol rate, and order of diversity?

  (d) If there are 80 such signals TDM's on the channel, how would this effect the chip rate, symbol rate, and order of diversity?

12.14. A frequency hopping noncoherent binary FSK system operates at an E_b/N_0 of 30 dB with a hopping bandwidth of 2 GHz. Assume that no channel coding is used. A jammer operating over the same broadband bandwidth yields a received $J_0 = 100N_0$.

  (a) What is the bit error probability, P_B?

  (b) If the jammer becomes a partial-band jammer, what bandwidth should it occupy to be most effective?

  (c) What is P_B as a result of such optimum partial-band jamming?

  (d) What is the unjammed P_B?

12.15. A noncoherent frequency hopping 8-ary FSK system hops at 12,000 hops/s over a bandwidth of 1 MHz. The symbol rate is 3000 symbols/s. Assume that channel coding is not used. The signal power at the input of the receiver is 10^{-12} W. A partial-band noise jammer occupies 50 kHz (assumed to be entirely within the hopping bandwidth of the signal). The received jammer power is 10^{-11} W. Assume that the system temperature is 290 K. What is the probability of bit error?

12.16. A coherent DS/BPSK system is transmitting at a data rate of 10 kbits/s in the presence of a broadband jammer. Assume that the system does not use channel coding. Also assume that the propagation losses are the same for the system and the jammer.

  (a) If the EIRP of the communicator is 20 kW and the EIRP of the jammer is 60 kW, calculate the required spread-spectrum bandwidth to achieve a bit error probability of $P_B = 10^{-5}$.

(b) If the jammer is a pulse jammer, calculate the pulse duty cycle that results in worst-case jamming. What is the value of P_B at this duty cycle?

12.17. A communicator intends to use frequency hopping at a hop rate of 10,000 hops/s to avoid a threat of repeat-back jamming.

(a) Ignoring the curvature of the earth, and assuming that the communicator is transmitting to a satellite of geosynchronous altitude (approximately 36,000 km) that is directly overhead, compute the *radius of vulnerability,* which is the radius outside of which the communicator is unconditionally safe from repeat-back jamming by a ground-based jammer.

(b) If the communicator knows that the jammer requires a minimum of 10 μs to identify the transmission frequency and tune the jammer output, compute the radius of vulnerability conditioned on this information.

12.18. Consider an airborne repeat-back jammer as shown in Figure P12.1. The communicator is using a FH/SS system. What is the minimum hop rate required in order that the repeat-back jamming does not degrade the message? What would be the minimum required hopping rate if the communicator and jammer switched positions (i.e., fixed land jammer and airborne communicator).

12.19. Spread-spectrum techniques can be used to meet government regulations regarding flux (power) density radiating the surface of the earth. If a satellite at synchronous altitude (36,000 km) transmit 4-kbits/s data using 100 W of EIRP, what spreading bandwidth is required to maintain a flux density on the earth's surface no greater than −151 dBW/m² in any 4-kHz band?

12.20. A communicator uses noncoherent BFSK modulation and frequency hopping to combat the effects of a jammer. The power of the communicator's signal at the re-

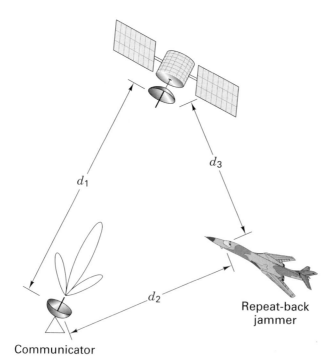

Communicator

Figure P12.1

Repeat-back
jammer

ceiver input is 10 μW. The SNR in the absence of jamming is assumed to be very large. The power of the jamming signal at the receiver input is 1 W.

(a) If the jammer jams the entire hopping bandwidth with equal amounts of Gaussian noise (the noise will be white within the band), what bandwidth expansion factor will allow the communicator to maintain a bit error probability of 10^{-4}?

(b) Assume that the jammer decides to "color" its jamming noise by reducing its energy by a fraction, α ($0 \le \alpha \le 1$), in half the hop bandwidth, and increasing it by a like amount in the other half (thereby keeping its transmitted energy constant). Assuming that the communicator does not modify his hopping pattern to avoid the jammer strategy, develop an expression for the bit error probability for this case of colored jamming.

(c) Determine the fraction, α, that is optimum from the jammer standpoint for each of the limiting cases (i) when the effective SNR is large and, (ii) when it is small.

12.21. Spread-spectrum (SS) techniques can provide impressive error-performance benefits against interfering signals. One might therefore think that such SS techniques might provide similar benefits against AWGN. Explain why this is not possible.

12.22. A hand-held direct-sequence spread-spectrum (DS/SS) radio is part of a cellular CDMA system. The system specifications are as follows: data and SS-code modulation is BPSK, data rate is 8,000 bits/s, carrier frequency is 1 GHz, chip rate is 25 Mchips/s, worst-case path loss is 138.6 dB, gain of transmitting antenna is 5 dBI, receiver figure-of-merit G/T is −18 dB/K, occasional deep small-scale fading loss is 30 dB, other losses are 4 dB, required E_b/N_0 is 4 dB. The factors G_A, G_V, H_0, and γ are 2.5, 2.5, 1.6, and 1, respectively. Hint: Refer to Chapter 5 for treatment of link parameters.

(a) Find the required transmitter power P_t during deep small-scale fading.

(b) To what level can P_t be powered down when there is no small-scale fading?

(c) What is the minimum required E_{ch}/N_0 to meet the specifications?

(d) What is the processing gain?

(e) What is the maximum number of users per cell?

12.23. A direct-sequence spread-spectrum system using BPSK modulation for both the data and the code is required to support a data rate of 9600 bits/s. The received pre-detection power-received versus N_0 (P_r/N_0) is 48-dB Hz, and the SS-processing gain is 1000. A BCH (63, 51) error-correcting code is used. Verify that these system specifications can provide a bit-error probability of 10^{-4}. Hint: Use Equation (6.46) in Chapter 6, for computing decoded bit-error probability.

12.24. (a) Consider a CDMA direct-sequence cellular telephone system, where each user requires an E_b/I_0 of 6 dB for acceptable voice quality. The chip rate is 3.68 Mchips/s, and the data rate is 14.4 kbits/s. Assume that the factors γ, G_V, and H_0, are 1.5, 2.5, and 1.5, respectively, and that transmission ceases during speech pauses. How many users per cell can be supported?

(b) If a powerful error-correcting code is used to lower the required E_b/I_0 by 1 dB, how many users per cell can be supported?

12.25. A direct-sequence spread-spectrum system uses QPSK modulation for transmitting data. It is required that the bit-error probability be 10^{-5} and that $E_{ch}/I_0 \le -30.4$ dB. Assuming perfect synchronization, what is the minimum number of chips/bit required?

12.26. A direct-sequence spread-spectrum system with a processing gain of 20 dB uses QPSK modulation for transmitting data. A rate $\frac{1}{2}$ error-correcting code is used, and the required bit-error probability is 10^{-5}. Assuming perfect synchronization, what is the minimum value of E_{ch}/I_0 and E_c/I_0 needed to support these requirements?

12.27. **(a)** A fast frequency-hopping spread-spectrum (FFH/SS) system uses 8-ary FSK modulation and a rate $\frac{1}{2}$ error-correcting code. Chips are transmitted with a repeat factor of $N = 4$. That is, each symbol is sent four times, each on a different hop. The required E_b/I_0 is 13 dB, the chip rate is 32 kchips/s, and the hopping bandwidth is 1.2 MHz. Find the data rate R, the processing gain G_p, the (P_r/I_0), the E_{ch}/I_0, the E_s/I_0, and the E_c/I_0.

(b) Will this system meet the FCC Part-15 requirements in the ISM band, for processing gain and bandwidth?

12.28. Consider a CDMA cellular telephone system designed to meet a standard similar to IS-95, with the following modifications. The spread-spectrum chip rate is 10.24 Mchips/s, and the data rate is 20 kbits/s. The reverse link uses a 256-ary Walsh waveform for modulating the rate $\frac{1}{2}$ coded data, requiring an E_b/I_0 of 6 dB. Find the values of the following received power-to-interference spectral density and energy-to-interference spectral density parameters: P_r/I_0, E_c/I_0, E_w/I_0, E_{wch}/I_0, and E_{ch}/I_0. Also, find the values of the following rates: R_c, R_w, and R_{wch}, where c, w, wch, and ch represent channel bit, Walsh waveform, Walsh chip, and spread-spectrum chip, respectively. What is the processing gain, and how many spread-spectrum chips correspond to one Walsh chip?

QUESTIONS

12.1. Frequency modulation (FM) and pulse code modulation (PCM) represent techniques that spread the spectrum of an information signal. Yet, such FM and PCM signals do not qualify as *spread-spectrum* signals. Why not? (See Section 12.1.)

12.2. List four *beneficial attributes* of spread-spectrum systems. (See Section 12.1.1.)

12.3. Describe three *randomness* properties that make pseudorandom signals appear to be random. (See Section 12.2.1.)

12.4. Define the term *chip,* in the context of a direct-sequence system, and in the context of a frequency-hopping system. (See Sections 12.3.2 and 12.4.4.)

12.5. What is meant by a *robust signal?* (See Section 12.4.2.)

12.6. What is the difference between *fast hopping* and *slow hopping?* (See Section 12.4.4.)

12.7. How does the *processing gain* parameter differ for direct-sequence systems compared with frequency-hopping systems? (See Sections 12.3.2 and 12.4.6.)

12.8. Explain how spread-spectrum systems can reliably receive signals that are "*buried in the noise.*" (See Example 12.5.)

12.9. For systems designed to meet IS-95, *Walsh codes* are used in the *forward* and *reverse* links for completely different functions. Explain the functions? (See Sections 12.8.4.1 and 12.8.4.2.)

EXERCISES

Using the Companion CD, run the exercises associated with Chapter 12.

CHAPTER 13

Source Coding

Fredric J. Harris
San Diego University
San Diego, California

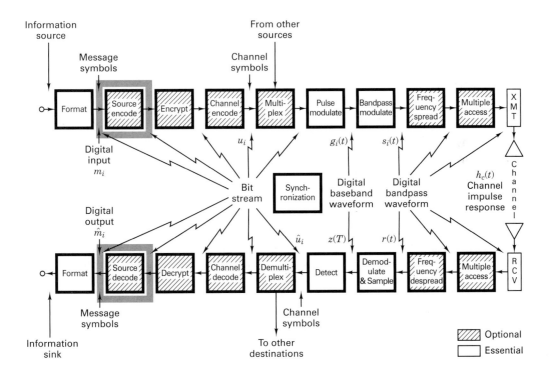

13.1 SOURCES

Source coding deals with the task of forming efficient descriptions of information sources. Efficient descriptions permit a reduction in the memory or bandwidth resources required to store or to transport sample realizations of the source data. For discrete sources, the ability to form reduced data-rate descriptions is related to the information content and the statistical correlation among the source symbols. For analog sources, the ability to form reduced data rate descriptions, subject to a stated fidelity criterion is related to the amplitude distribution and the temporal correlation of the source waveform. The goal of source coding is to form good fidelity description of the source for a given available bit rate, or to permit low bit rates to obtain a specified fidelity description of the source. To understand where the tools and techniques of source coding are effective, it is important to have common measures of source parameters. For this reason, in this section, we examine simple models of discrete and analog sources, and then we describe how source coding can be applied to these models.

13.1.1 Discrete Sources

A discrete source generates (or emits) a sequence of symbols $X(k)$, selected from a source alphabet at discrete time intervals kT, where $k = 1, 2, \ldots$ is a counting index. If the alphabet contains a finite number of symbols, say N symbols, the source is said to be a *finite discrete source*. An example of such a source is the output of a 12-bit digital-to-analog converter (which outputs one of 4096 discrete levels), or the output of a 10-bit analog-to-digital converter (which outputs one of 1024 binary 10-tuples). Another example of a discrete source is the succession of 8-bit ASCII characters emitted by a computer keyboard.

A finite discrete source is defined by the list of source symbols (sometimes called the alphabet) and the probability assigned to these symbols or (letters). We will assume that the source is short-term stationary—that is, the probability assignment is fixed over the observation interval. An example in which the alphabet is fixed but the probability assignment changes is found in the sequence of symbols emitted by a keyboard when someone is typing English text, followed by typing Spanish text, and finally by typing French text.

If we know that the probability of each symbol X_j is $P(X_j)$, we can determine the *self-information* $I(X_j)$ for each symbol in the alphabet set:

$$I(X_j) = -\log_2(p_j) \tag{13.1}$$

The average self-information for the symbols in an alphabet, also called the *source entropy*, is

$$H(X) = \mathbf{E}\{I(X_j)\} = -\sum_{j=1}^{N} p_j \log_2(p_j) \tag{13.2}$$

where $\mathbf{E}\{X\}$ is the expected value of X. The source entropy is defined as the average amount of information per source output. The source entropy is the average amount of uncertainty that is resolved by the use of the alphabet. It is thus the average amount of information that must be moved through the communication channel to resolve that uncertainty. It can be shown that this amount of information in bits per symbol is bounded below by zero if there is no uncertainty, and it is bounded above by $\log_2(N)$ if there is maximum uncertainty.

$$0 \leq H(X) \leq \log_2(N) \tag{13.3}$$

Example 13.1   Entropy of a Binary Source

Consider the binary source that generates independent symbols 0 and 1, with probabilities equal to p and $(1 - p)$, respectively. We described this source in Section 9.4.2 and presented its entropy function in Figure 9.5. If $p = 0.1$ and $(1 - p) = 0.9$, the source entropy is

$$H(X) = -[p \log_2(p) + (1 - p) \log_2(1 - p)] \tag{13.4}$$

$$= 0.47 \text{ bit/symbol}$$

Thus, this source can be described (with the use of appropriate coding) with less than half a bit per symbol, rather than with one-bit per symbol in its present form.

We note that the first reason that source coding works is because the information content of an N-symbol alphabet used in real communication systems is usually less than the upper bound of Equation (13.3). We know from experience, as we have noted in Example 9.2, that the symbols of English text are not equally likely. For instance, we use the high probability of certain letters in text as part of the strategy to initialize the game of Hangman. (In this game, a player must guess the letters, but not the positions, of a hidden word of known length. Penalties accrue to false guesses, and the letters of the entire word must be determined prior to the occurrence of six false guesses.)

A discrete source is said to be *memoryless* if the symbols emitted by the source are statistically independent. In particular, this means that for the symbols taken two at a time, the joint probability of the two elements is simply the product of their respective probabilities:

$$P(X_j, X_k) = P(X_j | X_k)P(X_k) = P(X_j)P(X_k) \qquad (13.5)$$

A result of statistical independence is that the information required to transmit a sequence of M symbols (called an M-tuple) from a given alphabet is precisely M times the average information required to transmit a single symbol. This happens because the probability of a statistically independent M-tuple is given by

$$P(X_1, X_2, \ldots, X_M) = \prod_{m=1}^{M} P(X_m) \qquad (13.6)$$

so that the average entropy per symbol of a statistically independent M-tuple is

$$H_M(X) = \frac{1}{M} \mathbf{E}\{-\log_2 P(X_1, X_2, \ldots, X_M)\}$$

$$= \frac{1}{M} \sum_{X_m} [-P(X_m) \log_2 P(X_m)] \qquad (13.7)$$

$$= H(X)$$

A discrete source is said to have memory if the source elements composing the sequence are not independent. The dependency between symbols means that in a sequence of M symbols, there is reduced uncertainty about the M-th symbol when we know the previous $(M-1)$ symbols. For instance, is there much uncertainty about the next symbol for the 10-tuple CALIFORNI_? The M-tuple with dependent symbols contains less information, or resolves less uncertainty, than does one with independent symbols. The entropy of a source with memory is the limit

$$H(X) = \lim_{M \to \infty} H_M(X) \qquad (13.8)$$

We observe that the entropy of an M-tuple from a source with memory is always less than the entropy of a source with the same alphabet and symbol probability, but without memory:

$$H_M(X)_{\text{memory}} < H_M(X)_{\text{no memory}} \qquad (13.9)$$

For example, given a symbol (or letter) "q" in English text, we know that the next symbol will probably be a "u". Hence, in a communication task, being told that the letter "u" follows a letter 'q' adds little information to our knowledge of the word being transmitted. As another example, given the letters 'th' the most likely symbol to follow is one of the following: a, e, i, o, u, r, and, space. Thus, adding the next symbol to the given set resolves some uncertainty, but not much. A formal statement of this awareness is that the average entropy per symbol of an *M*-tuple from a source with memory *decreases as the length M increases.* A consequence is that it is more efficient to encode symbols from a source with memory in groups of several symbols rather than to encode them one symbol at a time. For purposes of source encoding, encoder complexity, memory constraints, and delay considerations limit the size of symbol sequences treated as a group.

To help us understand the gains to be had in coding sources with memory, we form simple models of these sources. One such model is called a *first-order Markov source* [1]. This model identifies a number of states (or symbols in the context of information theory) and the conditional probabilities of transitioning to each next state. In the first-order model, the transition probabilities depend only on the present state. This is, $P(X_{i+1}|X_i, X_{i-1}, \ldots) = P(X_{i+1}|X_i)$. The model's memory does not extend beyond the present state. In the context of a binary sequence, this expression gives the probability of the next bit conditioned on the value of the current bit.

Example 13.2 Entropy of a Binary Source With Memory

Consider the binary (i.e., two symbol) first-order Markov source described by the state transition diagram shown in Figure 13.1. The source is defined by the state transition probabilities P(0|1) and P(1|0) of 0.45 and 0.05, respectively. The entropy of the source X is the weighted sum of the conditional entropies that correspond to the transition probabilities of the model. That is,

$$H(X) = P(0)H(X|0) + P(1)H(X|1) \tag{13.10}$$

where

$$H(X|0) = -[P(0|0) \log_2 P(0|0) + P(1|0) \log_2 P(1|0)]$$

and

$$H(X|1) = -[P(0|1) \log_2 P(0|1) + P(1|1) \log_2 P(1|1)]$$

The a priori probability of each state is found by the total probability equations

$$P(0) = P(0|0)P(0) + P(0|1)P(1)$$
$$P(1) = P(1|0)P(0) + P(1|1)P(1)$$
$$P(0) + P(1) = 1$$

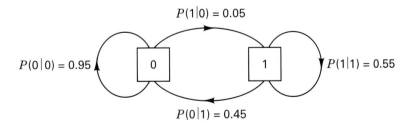

$P(1|0) = 0.05$

$P(0|0) = 0.95$ 0   1 $P(1|1) = 0.55$

$P(0|1) = 0.45$

Figure 13.1   State transition diagram for first-order Markov model.

Solving for the a priori probabilities using the transition probabilities, we have

$$P(0) = 0.9 \quad \text{and} \quad P(1) = 0.1$$

Solving for the source entropy using Equation (13.10), we have

$$H(X) = P(0)\,H(X|\,0) + P(1)\,H(X|\,1)$$
$$= (0.9)(0.286) + (0.1)(0.993)$$
$$= 0.357 \text{ bit/symbol}$$

Comparing this result with the result of Example 13.1, we see that the source with memory has lower *entropy* than the source without memory, even though the a priori symbol probabilities are the same.

Example 13.3 Extension Codes

The alphabet for the binary Markov source of Example 13.2 consists of 0 and 1, occurring with probabilities 0.9 and 0.1, respectively. Successive symbols are not independent, and we can define a new set of code symbols as binary 2-tuples (an extension code) to take advantage of this dependency:

Binary 2-tuple	Extension Symbol	Extension Symbol Probability
00	a	$P(a) = P(0|0)P(0) = (0.95)(0.9) = 0.855$
11	b	$P(b) = P(1|1)P(1) = (0.55)(0.1) = 0.055$
01	c	$P(c) = P(0|1)P(1) = (0.45)(0.1) = 0.045$
10	d	$P(d) = P(1|0)P(0) = (0.05)(0.9) = 0.045$

where the rightmost digit of the 2-tuple is the earliest digit. The entropy for this code extension alphabet is found by using an extension of Equation (13.10) as follows:

$$H(\mathbf{X}_2) = P(a)\,H(\mathbf{X}_2|\,a) + P(b)\,H(\mathbf{X}_2|\,b) + P(c)\,H(\mathbf{X}_2|\,c) + P(d)\,H(\mathbf{X}_2|\,d)$$

$$H(\mathbf{X}_2) = 0.825 \text{ bit/output symbol}$$

$$H(\mathbf{X}_2) = 0.412 \text{ bit/input symbol}$$

where $\mathbf{X}_k$ is the kth-order extension of the source $\mathbf{X}$. A longer extension code, which takes advantage of the adjacent symbol dependency, is of the following form:

Binary 3-tuple	Extension Symbol	Extension Symbol Probability
000	a	$P(a) = P(0|00)P(00) = (0.95)(0.855) = 0.8123$
100	b	$P(b) = P(1|00)P(00) = (0.05)(0.855) = 0.0428$
001	c	$P(c) = P(0|01)P(01) = (0.95)(0.045) = 0.0428$
111	d	$P(d) = P(1|11)P(11) = (0.55)(0.055) = 0.0303$
110	e	$P(e) = P(1|10)P(10) = (0.55)(0.045) = 0.0248$
011	f	$P(f) = P(0|11)P(11) = (0.45)(0.055) = 0.0248$
010	g	$P(g) = P(0|10)P(10) = (0.45)(0.045) = 0.0203$
101	h	$P(h) = P(1|01)P(01) = (0.05)(0.045) = 0.0023$

Again, using an extension of Equation (13.10), the entropy for this extension code is found to be

$$H(\mathbf{X}_3) = 1.223 \text{ bit/output symbol}$$

$$H(\mathbf{X}_3) = 0.408 \text{ bit/input symbol}$$

We note that the entropy of the one-symbol, two-symbol, and three-symbol descriptions of the source (0.470-, 0.412-, and 0.408-bit, respectively) are decreasing asymptotically toward the source entropy of 0.357 bit/input symbol. Remember that the source entropy is the lower bound in bits per input symbol for this (infinite memory) alphabet, and this bound cannot be achieved with finite length coding.

13.1.2 Waveform Sources

A waveform source is a random process of some random variable. We classically consider this random variable to be time, so that the waveform of interest is a time-varying waveform. Important examples of time-varying waveforms are the outputs of transducers used in process control, such as temperature, pressure, velocity, and flow rates. Examples of particularly high interest include speech and music. The waveform can also be a function of one or more spatial variables (e.g., displacement in x and y). Important examples of spatial waveforms include single images, such as a photograph, or moving images, such as successive images (at 24-frames/sec) of moving picture film. Spatial waveforms are often converted to time-varying waveforms by a scanning operation. This is done, for example, for facsimile and Joint Photographic Expert Group (JPEG) transmission, as well as for standard broadcast television transmission.

13.1.2.1 Amplitude Density Functions

Discrete sources were described by their list of possible elements (called letters of an alphabet) and their multidimensional probability density functions (pdf) of all orders. By analogy, waveform sources are similarly described in terms of their probability density functions as well as by parameters and functions derived from these functions. We model many waveforms as random processes with classical probability density functions and with simple correlation properties. In the modeling process, we distinguish between short-term or local (time) characteristics and long-term or global characteristics. This partition is necessary because many waveforms are nonstationary.

The probability density function of the actual process may not be available to the system designer. Sample density functions can of course be rapidly formed in real time during a short preliminary interval and used as reasonable estimates over the subsequent interval. A less ambitious task is simply to form short-term waveform-related averages. These include the sample mean (or time average value), the sample variance (or mean-square value of the zero mean process), and the sample correlation coefficients formed over the prior sample interval. In many applications of waveform analysis, the input waveform is converted into a zero-mean process by subtracting an estimate of its mean value. For instance, this happens, in comparators used in analog-to-digital converters for which auxiliary circuitry measures the internal dc-offset voltages and subtracts them in a process known as *autozero*. Further, the variance estimate is often used to scale the input waveform to match the dynamic amplitude range of subsequent waveform conditioning circuitry. This process, performed in a data collection process, is called *autoranging* or *automatic gain control* (AGC). The function of these signal-conditioning operations, mean removal, and variance control or gain adjustment (shown in Figure 13.2) is to normalize the probability density functions of the input waveforms. This normalization assures optimal utility of the limited dynamic range of subsequent recording, transmission, or processing subsystems.

Many waveform sources exhibit significant amplitude correlation in successive time intervals. This correlation means that signal levels in successive time intervals are not independent. If the time signal is independent over successive intervals, the autocorrelation function would be an impulse function. Many signals of engineering interest have finite width correlation functions. The effective width of the correlation function (in seconds) is called the correlation time of the process and is akin to the time constant of a low-pass filter. This time interval is an indication of how much shift along the time axis is required to decorrelate the data. If the correlation time is large we interpret this to mean that the waveform makes significant amplitude changes slowly. Conversely, if the correlation time is small, we infer that the waveform makes significant changes in amplitude very quickly.

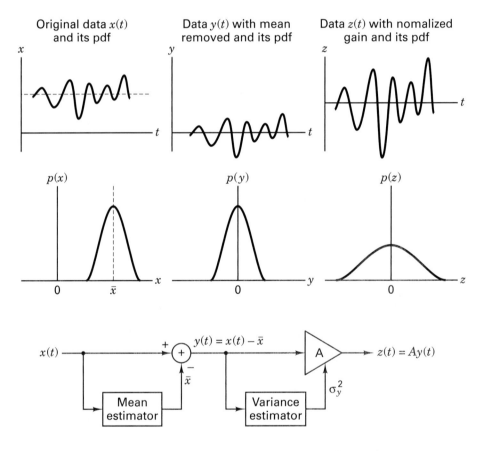

Figure 13.2 Mean removal and variance normalization (gain adjustment) for a data dependent signal conditioning system.

13.2 AMPLITUDE QUANTIZING

Amplitude quantizing is the task of mapping samples of a continuous amplitude waveform to a finite set of amplitudes. The hardware that performs the mapping is the analog-to-digital converter (ADC or A-to-D). The amplitude quantizing occurs after the sample-and-hold operation. The simplest quantizer to visualize performs an instantaneous mapping from each continuous input sample level to one of the preassigned equally spaced output levels. Quantizers that exhibit equally spaced increments between possible quantized output levels are called *uniform quantizers* or *linear quantizers*. Possible instantaneous input–output characteristics are easily visualized by a simple staircase graph consisting of risers and treads of the types shown in Figure 13.3. Figure 13.3a, b, and d show quantizers with uniform quantizing steps, while Figure 13.3 c is a quantizer with nonuniform quantizing steps. Figure 13.3a depicts a quantizer with midtread at the origin, while Figure 13.3b and d

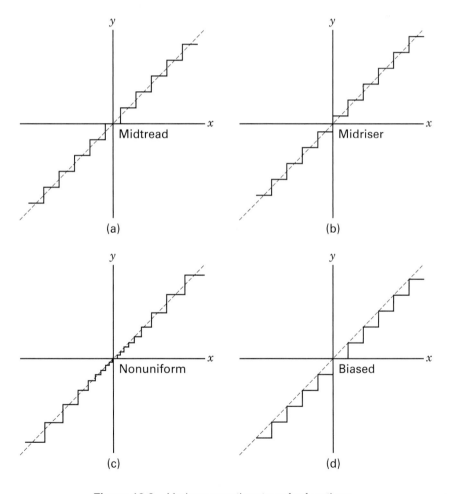

(a)   (b)

(c)   (d)

Figure 13.3   Various quantizer transfer functions.

present quantizers with midrisers at the origin. A distinguishing property of midriser and midtread converters is related to the presence or absence, respectively, of output level changes when the input to the converter is low-level idle noise. Further, Figure 13.3d presents a biased (i.e., truncation) quantizer, while the remaining quantizers in the figure are unbiased and are referred to as *rounding quantizers*. Such unbiased quantizers represent ideal models, but rounding is never implemented in A/D converters. Quantizers are typically implemented as truncation quantizers. The terms "midtread" and "midriser" are staircase terms used to describe whether the horizontal or vertical member of the staircase is at the origin. The unity-slope dashed line passing through the origin represents the ideal nonquantized input–output characteristic we are trying to approximate with the staircase. The difference between the staircase and the unity-slope line segment represents the approximation error made by the quantizer at each input level. Figure 13.4 illustrates the approximation error amplitude versus input amplitude function for each quantizer characteristic in Figure 13.3. Parts (a) through (d) of Figure 13.4 correspond to the same parts in Figure 13.3. This error is often modeled as quantizing noise because the error sequence obtained when quantizing a wideband random process is reminiscent of an additive noise sequence. Unlike true additive noise sources, however, the quantizing errors are signal dependent and are highly structured. It is desirable to break up this structure, which can be accomplished by introducing an independent noise perturbation, known as *dither,* prior to the quantization step. (This topic is discussed in Section 13.2.4.)

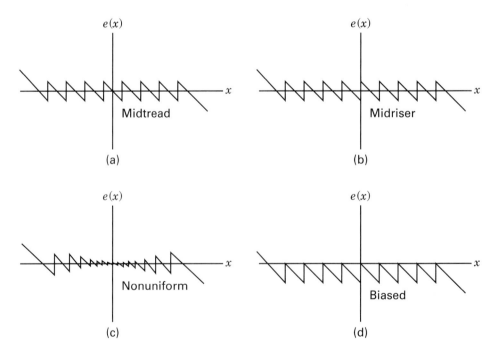

Figure 13.4 Instantaneous error for various quantizer transfer functions.

The linear quantizer is simple to implement and is particularly easy to understand. It is the universal form of the quantizer, in the sense that it makes no assumptions about the amplitude statistics and correlation properties of the input waveform, nor does it take advantage of user-related fidelity specifications. Quantizers that take advantage of these considerations are more efficient as source coders and are more task specific then the general linear quantizer; these quantizers are often more complex and more expensive, but they are justified in terms of improved system performance. There are applications for which the uniform quantizer is the most desirable amplitude quantizer. These include signal processing applications, graphics and display applications, and process control applications. There are other applications for which nonuniform adaptive quantizers are more desirable amplitude quantizers. These include waveform encoders for efficient storage and communication, contour encoders for images, vector encoders for speech, and analysis/synthesis encoders (such as the vocoder) for speech.

13.2.1 Quantizing Noise

The difference between the input and output of a quantizer is called the *quantizing error*. In Figure 13.5 we demonstrate the process of mapping the input sequence $x(t)$ to the quantized output sequence $\hat{x}(t)$. We can visualize forming $\hat{x}(t)$ by adding to each $x(t)$ an error sequence, $e(t)$:

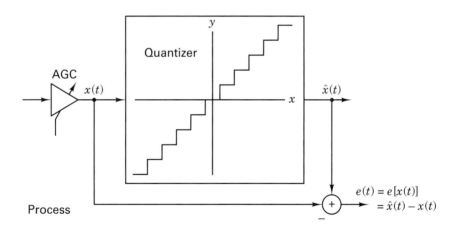

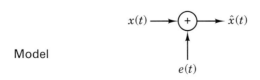

Figure 13.5   Process and model of quantizing noise corruption of input signal.

$$\hat{x}(t) = x(t) + e(t)$$

The error sequence $e(t)$ is deterministically defined by the input amplitude through the instantaneous error versus amplitude characteristic of the form in Figure 13.4. We note that the error sequence exhibits two distinct characteristics over different input operating regions.

The first operating interval is the granular error region corresponding to the input sawtooth-shaped error characteristic. Within this interval, the quantizer errors are confined by the size of the nearby staircase risers. The errors that occur in this region are called the *granular errors,* or sometimes the *quantizing errors.* The input interval for which the quantizing errors are granular defines the dynamic range of the quantizer. This interval is sometimes called the *region of linear operation.* Proper use of the quantizer requires that the input signal conditioning match the dynamic range of the input signal to the dynamic range of the quantizer. This is the function of the signal-dependent gain control system, called *automatic gain control* (AGC), indicated in the signal flow path of Figure 13.5.

The second operating interval is the nongranular error region corresponding to the linearly increasing (or decreasing) error characteristic. The errors that occur in this interval are called *saturation* or *overload* errors. When the quantizer operates in this region, we say that the quantizer is *saturated.* Saturation errors are larger than the granular errors and may have a more objectionable effect on reconstruction fidelity.

The quantization error corresponding to each value of input amplitude represents an error or noise term associated with that input amplitude. Under the assumptions that the quantization interval is small compared with the dynamic range of the input signal, and that the input signal has a smooth probability density function over the quantization interval, we can assume that the quantization errors are uniformly distributed over that interval, as illustrated in Figure 13.6. The pdf with zero mean corresponds to a rounding quantizer, while the pdf with a mean of $-q/2$ corresponds to a truncation quantizer.

A quantizer or analog-to-digital converter (ADC) is defined by the number, size, and location of its quantizing levels or step boundaries, and the corresponding step sizes. In a uniform quantizer, the step sizes are equal and are equally spaced. The number of levels N is typically a power of 2 of the form $N = 2^b$, where b is the number of bits used in the conversion process. This number of levels is equally distributed over the dynamic range of the possible input levels. Normally, this range is defined as $\pm E_{max}$, such as ± 1.0 V or ± 5.0 V. Thus, accounting for the full range of $2E_{max}$, the size of a quantization step is

$$q = \frac{2E_{max}}{2^b} \qquad (13.11)$$

As an example, using Equation (13.11), the quantizing step (hereafter called a *quantile.*) for a 10-bit converter operating over the ± 1.0 V range is 1.953 mV. Occasionally, the operating range of a converter is altered so that the quantile is a "whole" number. For example, changing the operating range of the converter to ± 1.024 V results in a quantizing step size of 2.0 mV. A useful figure of merit for the

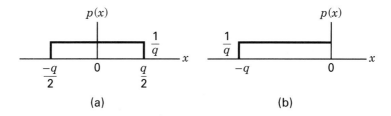

Figure 13.6 Probability density functions for quantizing error uniformly distributed over one quantile, q. (a) Probability density function for a rounding quantizer. (b) Probability density function for a truncating quantizer.

uniform quantizer is the quantizer output variance. If we assume that the quantization error is uniformly distributed over a single quantile interval q-wide, the quantizer variance (which represents the quantizer noise or error power) for the zero-mean error is found to be

$$\sigma^2 = \int_{-q/2}^{q/2} e^2 p(e)de = \int_{-q/2}^{q/2} e^2 \frac{1}{q} de = \frac{q^2}{12} \tag{13.12}$$

where $p(e) = 1/q$, over an interval of width q, is the probability density function (pdf) of the quantization error e. Thus, the rms quantizer noise in a quantile interval of width q is found to be $q \sqrt{12}$ or $0.29q$. Equation (13.12) determined the quantizing noise power over one quantile, assuming that the errors are equiprobable over the quantization interval. If we include operation in the saturation interval of a quantizer, or if we include nonuniform quantizers, we find that the quantization intervals are not of equal width over the range of the input variable, and that the amplitude density is not uniform over the quantization interval. We can account for this amplitude-dependent error power σ_q^2 by averaging the squared error over the amplitude variable weighting by the probability of that amplitude. This is expressed by

$$\sigma_q^2 = \mathbf{E}\{[x - q(x)]^2\} = \int_{-\infty}^{\infty} e^2(x)p(x)dx \tag{13.13}$$

where x is the input variable, $q(x)$ is its quantized version, $e(x) = x - q(x)$ is the error, and $p(x)$ is the pdf of the amplitude of x. We can partition the interval of integration in Equation (13.13) into two main intervals—one accounting for errors in the staircase or linear region of the quantizer, and the second accounting for errors in the saturation region. We define the saturation amplitude of the quantizer as E_{max}. Also, we assume an odd symmetric transfer function for the quantizer, and a symmetric pdf for the input signal. The error power σ_q^2, defined in Equation (13.13), is the total error power, which can be partitioned as

$$\sigma_q^2 = 2 \int_0^{\infty} e^2(x)p(x)dx \tag{13.14a}$$

$$= 2 \int_0^{E_{max}} e^2(x)p(x)dx + 2 \int_{E_{max}}^{\infty} e^2(x)p(x)dx \tag{13.14b}$$

$$= \sigma_{Lin}^2 + \sigma_{Sat}^2$$

where σ_{Lin}^2 is the error power in the linear region and σ_{Sat}^2 is the error power in the saturation region. The error power σ_{Lin}^2 can be further divided into subintervals corresponding to the successive discrete quantizer output levels (i.e., quantiles). If we assume that there are N such quantile levels, the integral becomes

$$\sigma_{Lin}^2 = 2 \sum_{n=0}^{N/2-1} \int_{x_n}^{x_{n+1}} e^2(x)p(x)dx \tag{13.15}$$

where x_n is a quantizer level and an interval or step size between two such levels is called a *quantile interval*. Recall that N is typically a power of 2. Thus, there are $N/2 - 1$ positive levels, $N/2 - 1$ negative levels, and a zero level, making a total of $N - 1$ levels and $N - 2$ intervals. If we now approximate the density function by a constant $q_n = (x_{n+1} - x_n)$, in each quantile interval, Equation (13.15) simplifies to

$$\sigma_{Lin}^2 = 2 \sum_{n=0}^{N/2-1} \frac{x^3}{3} \Big|_{x=-q_n/2}^{x=+q_n/2} p(x_n)$$

$$= 2 \sum_{n=0}^{N/2-1} \frac{q_n^2}{12} p(x_n)q_n \tag{13.16}$$

where $e(x)$ in Equation (13.15) has been replaced by x in Equation (13.16), since $e(x)$ is a linear function of x with unity slope and passes through zero at the midpoint of each interval. Also, the limits of integration in Equation (13.15) have been replaced by the change in x over a quantile interval. Since the change has been denoted as q_n, the lower and upper limits can be designated as $x = -q_n/2$ and $x = +q_n/2$, respectively. Equation (13.16) describes the error power in the linear region as a summation of error power $q_n^2/12$ in each quantile interval weighted by the probability $p(x_n)q_n$ of that error power.

13.2.2 Uniform Quantizing

If the quantizer has uniform quantiles equal to q and all intervals are equally likely, Equation (13.16) simplifies further to

$$\sigma_{Lin}^2 = \frac{2}{12} \sum_{n=0}^{N/2-1} q_n^2 \, p(x_n)q_n = \frac{2}{12} \sum_{n=0}^{N/2-1} q^2 \frac{1}{q(N-2)} q = \frac{q^2}{12} \tag{13.17}$$

If the quantizer does not operate in the saturation region (the quantization noise power), then $\sigma_q^2 = \sigma_{Lin}^2$; and these terms are often used interchangeably. Noise power alone will not fully describe the noise performance of the quantizer. A more meaningful measure of quality is the ratio of the second central moment (variance) of the quantizing noise and the input signal. Assuming that the input signal has zero mean, the signal variance is

$$\sigma_x^2 = \int_{-\infty}^{\infty} x^2 p(x)\, dx \tag{13.18}$$

Further insight into the average quantizer noise requires that we examine a specific density function and a specific quantizer.

Example 13.4   Uniform Quantizer

Determine the quantizer variance and the quantizer noise-to-signal power ratio (NSR) for a signal that is uniformly distributed over the full dynamic range of a uniform quantizer with 2^b equally spaced quantile levels. In this case there is no saturation noise and only the linear noise term must be computed. Each quantile interval is

$$q = (2E_{\max}) 2^{-b} \tag{13.19}$$

where $2E_{\max}$ is the input interval between the positive and negative boundaries of the linear quantizing range.

Solution

Substituting Equation (13.19) into Equation (13.12) or (13.17), we have the following quantizing noise power (in the linear region):

$$\sigma_q^2 = \frac{1}{12}(2E_{\max}2^{-b})^2 = \frac{1}{12}(2E_{\max})^2\, 2^{-2b} \tag{13.20}$$

The input signal power is found by performing the integration of Equation (13.18) for a uniform probability density function in the zero-mean interval spanning $2E_{\max}$, so that $p(x) = 1/(2E_{\max})$, and the signal variance is found to be

$$\sigma_x^2 = \int_{-E_{\max}}^{+E_{\max}} \frac{1}{2E_{\max}} x^2\, dx = \frac{1}{12}(2E_{\max})^2 \tag{13.21}$$

Taking the ratio of noise power to signal power (NSR), we have

$$\text{NSR} = \frac{\sigma_q^2}{\sigma_x^2} = 2^{-2b} \tag{13.22}$$

Now, converting the NSR to decibels, we have

$$\text{NSR}_{\text{dB}} = 10\log_{10}(\text{NSR}) = 10\log_{10}(2^{-2b}) \tag{13.23a}$$

$$= -20b\log_{10}(2) = -6.02\,b \text{ (dB)} \tag{13.23b}$$

Equation (13.23b) suggests that each bit used in the conversion process is worth –6.02 dB in noise-to-signal ratio. In fact, the NSR for any uniform quantizer not operating in saturation is of the form

$$\text{NSR}_{\text{dB}} = -6.02\,b + C \tag{13.24}$$

where the term C depends on the signal probability density function (pdf); it is positive for density functions that are narrow with respect to the saturation level of the quantizer.

13.2.2.1 Signal and Quantization Noise in the Frequency Domain

Up until now we have discussed quantization noise in terms of its effect on samples of the time series representing the sampled signal. The quantization noise can also be described in the frequency domain, which offers insight into effects of operating conditions that we will examine shortly. These include saturation (Section 13.2.3), dithering (Section 13.2.4), and noise feedback quantizers (Section 13.2.6).

Figure 13.7 presents the discrete Fourier transform of two sinusoids that have been sampled by a linear 10-bit ADC. The two sinusoids have relative amplitudes of 1.0 and 0.01 (i.e., one is 40 dB below the other). In Figure 13.7a the lower-frequency signal (labeled 0-dB) is scaled to 1 dB below full dynamic range of the 10-bit converter, which for convenience we set to be unity. Note that in Figure 13.7a, the full-scale 0-dB signal dwells an additional 6-dB below the input attenuation level of 1-dB. This is due to the factor of one-half in the spectral decomposition of a real signal over all nonzero frequencies. The average signal-to-quantization noise ratio (SNR) for a 10-bit quantizer is $60 + C$ dB. For a full-scale sinusoid, the constant C is 1.76 dB, resulting in a SNR of approximately 62 dB. The discrete Fourier transforms (DFTs) performed for Figure 13.7 were of length 256, and since the SNR of a transform increases in proportion to transform length (or integration time), there is an improvement in SNR of 24 dB due to the transform [2],

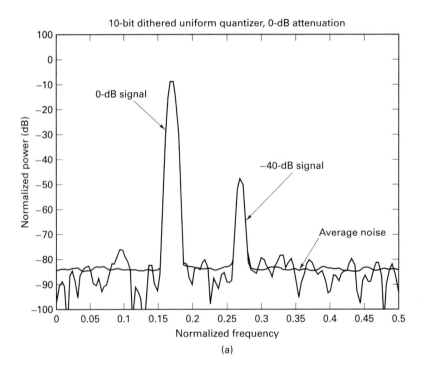

Figure 13.7 Power spectrum of signals quantized by uniform ADC.

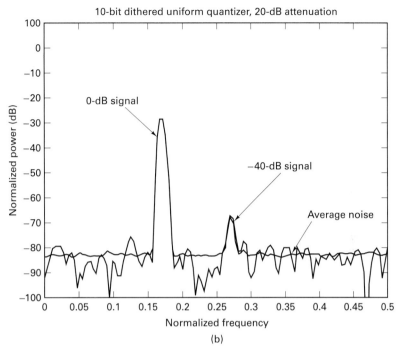

(b)

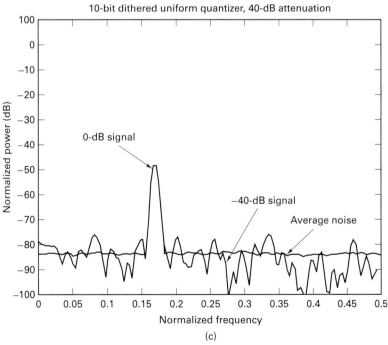

(c)

Figure 13.7 Continued

with a 3.0 dB loss due to windowing. Thus, at the output of the transform, the peak SNR due to quantizing is 62 + 24 − 3 = 83 dB. The noise signal, at each frequency of a DFT, can be represented as the square root of a sum of squared Gaussian random variables, which is characterized as a chi-square two-degrees-of-freedom process. The variance (noise power) is equal to the squared mean. Thus, there is significant variation about the expected noise power level. An ensemble average is required to obtain a stable estimate of the noise floor. The noise floor shown in figure 13.7, obtained from 400 averages, is seen to be −83 dB. A noise dither signal (described in Section 13.2.4), has been added to the signal prior to the quantization in order to randomize the quantization errors. In Figure 13.7b and c, the input signals are attenuated relative to the full-scale input, by 20 and 40 dB, respectively. This attenuation increases the constant C in Equation (13.24) by 20 and 40 dB, which we observe as reductions in the spectral levels of the input sinusoids by these same values. Note that the higher frequency input signal in Figure 13.7c, now attenuated 80 dB relative to full scale, is now 3-dB below the average noise level of the converter. The lower frequency sinusoid in Figure 13.7c is now attenuated 40 dB relative to full scale, and therefore exhibits an SNR that is 40 dB lower than the same signal of Figure 13.7a.

Given the task of minimizing the average quantizing noise-to-signal ratio in a quantizer, we are faced with a conflict of requirements. On one hand, we wish to keep the signals large with respect to the quantizing interval q, in order to achieve a large SNR. But we also find it necessary to keep the signal small to avoid saturating the quantizer. We resolve the opposing requirements by scaling the input signal so that its rms value is a specified fraction of the full-scale quantizer range. The specified fraction is chosen to balance the saturation errors (weighted by their probability of occurrence) against the quantizing errors (which are similarly weighted) and thus achieve a minimum noise-to-signal ratio. We discuss the location of this desired operating point for the converter in the next section.

13.2.3 Saturation

Figure 13.8 presents the average NSR of a uniform quantizer as a function of the ratio of quantizer saturation level to the rms signal value. The figure shows the NSR of signals with three different pdfs: arc-sine (sine wave density), uniform, and Gaussian.

In Figure 13.8, the abscissa is the ratio of the quantizer saturation level to the rms level of the input signal. For each of the three density functions shown, there is a value on the abscissa corresponding to a minimum NSR for a fixed number of bits; in other words, for a given input density function, we can determine the input signal level (relative to saturation) to achieve the minimum NSR. Reduced input signal levels correspond to larger NSR values on the abscissa, and represent movement to the right. Similarly, increased input signal levels also correspond to larger NSR values on the abscissa, and represent movement to the left. This increase is due to operating in the saturation region of the quantizer. We note that the rate of change of NSR as we move to the left of the optimum operating point is higher

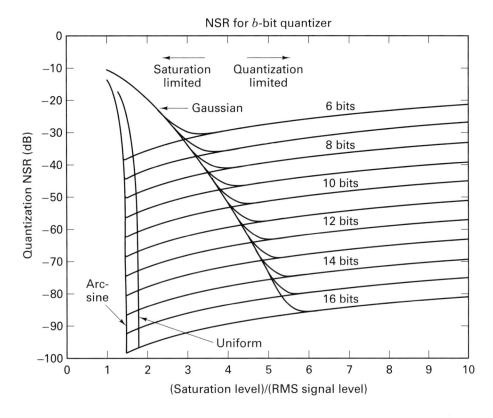

Figure 13.8 ADC NSR versus ratio of ADC saturation level to RMS signal level.

than that if we move to the right. This is particularly true, for instance, in the case of the arc-sine and uniform densities. This suggests that saturation noise is more objectionable that linear quantization noise. A consequence of this is that if we err in setting the operating point, called the *attack point* of a quantizer, it would be better to have the error on the side of excess attenuation rather than insufficient attenuation of the input signal. The onset of saturation occurs at different values of the abscissa. For sine waves (arc-sine density) it occurs at approximately $\sqrt{2}$. For triangle waves, (uniform density) it occurs at approximately $\sqrt{3}$. For noise-like signals, (Gaussian density) it occurs continuously, with decreasing probability as the signal level is reduced relative to saturation. As an example, a 10-bit converter has a minimum NSR of −60 dB for a uniform density when operating at the edge of saturation, and an NSR of − 62 dB for an arc-sine density, when similarly operating at the edge of saturation. On the other hand, the same 10-bit converter has a minimum NSR of approximately − 52 dB for all densities, when the rms level is set at $\frac{1}{4}$ of saturation (value of 4 on the abscissa). The figure dramatically demonstrates that saturation noise is more severe than is quantizing noise. This can be explained fairly simply by examining the instantaneous error characteristic, (as shown in Figure

13.4,) and noting that saturation errors are very large relative to the quantizing errors. Thus, a small amount of saturation, even if it occurs infrequently, will make a large contribution to the average noise level of the quantizer.

Saturation noise and quantization noise differ in another important way. Quantization noise tends to be white noise. Dither signals may be intentionally added to the analog signal prior to the quantizer to assure this property. Saturation noise, on the other hand, tends to be white only when the input signal has a broad bandwidth and tends to be harmonically related to the input signal if it has a narrow bandwidth, Thus, the effects of quantizing noise can be filtered or averaged because it has the characteristics of white noise. Saturation noise, on the other hand, is indistinguishable from signal content and generally cannot be reduced by subsequent averaging or filtering techniques.

Figure 13.9 presents the discrete Fourier transforms of the same signal set presented in Figure 13.7 quantized with a 10-bit ADC, except that in Figure 13.9 the peak signal amplitude was adjusted to be 10% (0.83 dB) above the ADC saturation level. Note the very many spectral artifacts caused by the saturation. These artifacts (the saturation noise) will grow larger as the signal excursions go deeper

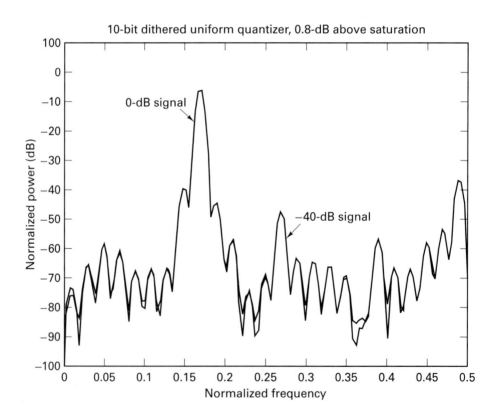

Figure 13.9 Power spectrum of uniformly quantized signals, with the quantizer saturating on signal peaks 0.8 dB beyond full-scale input level.

into saturation. Also note that some of these artifacts are larger than the −40 dB signal. Compare this figure with Figure 13.7 to see the dramatic difference that too little signal attenuation (hence, saturation) makes in the noise output of an ADC.

13.2.4 Dithering

Dithering is one of the cleverest applications of noise as a useful engineering tool. A dither signal is a small perturbation or disturbance added to a measurement process to reduce the effect of small local nonlinearities. The most familiar form of dither is the slight tapping we apply to the side of a d'Arsonval meter movement prior to taking the reading (before the days of digital meters). The tapping is a sequence of little impulses for displacing the needle movement beyond the local region, which exhibits a nonlinear coefficient of friction at low velocities. A more sophisticated example of this same effect is the mechanical dither applied to the counter-rotating laser beams of a laser beam gyro to break up low-level frequency entrapment, known as a *dead band* [3].

In the analog-to-digital converter application, the effect of the dither is to reduce or eliminate the local discontinuities (i.e., the risers and treads) of the instantaneous input–output transfer function. We can best visualize the effect of these discontinuities by listing the desired properties of the error sequence formed by the quantizer process and then examining the actual properties of the same sequence. The quantizer error sequence is modeled as additive noise. The desired properties of such a noise sequence $e(n)$ are as follows:

1. Zero mean: $\quad\quad\quad\quad\quad\quad\quad \mathbf{E}\{e(n)\} = 0$
2. White: $\quad\quad\quad\quad\quad\quad\quad\quad \mathbf{E}\{e(n)e(n + m)\} = \sigma^2\,\delta(m)$
3. Uncorrelated with data $x(n)$: $\quad \mathbf{E}\{e(n)\,x(n + m)\} = 0$

where n and m are sample indices, and $\delta(m)$ is a Dirac delta function. In Figure 13.10, we examine a sequence of samples formed by a truncating ADC and make the following observations:

1. The error sequence is all of the same polarity; therefore, it is not zero mean.
2. The error sequence is not independent, sample-to-sample; therefore, it is not white.
3. The error sequence is correlated with the input; therefore, it is not independent.

Repeated measurements of the same signal would result in the same noise, and thus no amount of averaging could reduce the deviation from the true input signal. Paradoxically, we would like this noise to be "noisier." If the noise were independent on successive measurements, averaging would reduce the deviation from the true values. Thus, faced with the problem that the noise we get is not the noise we want, we choose to alter that noise by adding our own. We add a perturbation to the measurement to override the undesired low-level structure of the quantizer noise. The added perturbation, in a sense, converts *bad noise to good noise* [4].

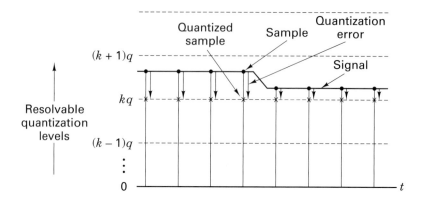

Figure 13.10 Sampled data sequence quantized to next-lowest quantile level by associated error sequence.

Example 13.5   Dither Linearization

We hypothesize a quantizer that can only measure integers and converts input data into the next lowest integer—a process called *truncation*. We make 10 measurements of a signal of, say, amplitude 3.7. In the absence of a dither we have readings all equal to 3.0. Now add a uniformly distributed (over 0 to 1) random number sequence to the input prior to performing the reading. The sequence of data has the following form:

Reading	Raw Signal	Quantized Raw Signal	Dither	Dithered Signal	Dith + Quant Signal
1	3.7	3.0	0.3485	4.0485	4.0
2	3.7	3.0	0.8685	4.5685	4.0
3	3.7	3.0	0.2789	3.9789	3.0
4	3.7	3.0	0.3615	4.0615	4.0
5	3.7	3.0	0.1074	3.8074	3.0
6	3.7	3.0	0.2629	3.9629	3.0
7	3.7	3.0	0.9252	4.6252	4.0
8	3.7	3.0	0.5599	4.2599	4.0
9	3.7	3.0	0.3408	4.0408	4.0
10	3.7	3.0	0.5228	4.2228	4.0

Means =		3.0	0.4576	4.1576	3.7
Dither mean =				0.4576	
Dithered Signal Mean − Dither mean =				3.7	

In this example, a biased dither was used to remove the quantizer bias. The average of the dithered and quantized readings if (for this example) a correct reading and, in general, will be closer to the true signal than will the nondithered and quantized measurements [5, 6].

To illustrate the effect that dithering has on the quantization process of a time-varying signal, consider the following experiment. Let us apply 60 dB of attenuation to a sinusoidal signal of amplitude 1.0. The attenuated signal, then, has full-scale amplitude of 0.001, which is approximately one half of the quantization interval of 0.001957 for a ten-bit uniform quantizer (obtained by dividing the peak-to-peak signal amplitude of 2 by $2^{10} - 2$). When the attenuated sinusoid is applied to a rounding quantizer, the output will be essentially all zeros except for an occasional count of ±1 quantile, which occurs when the input crosses the ±q/2 level of 0.000979 (corresponding to the least significant bit of the ADC). If the input signal were attenuated another 0.23 dB, the threshold levels of the least significant bit would never be crossed and the output sequence would be all zeros. Now, let us add a dither signal with rms amplitude of 0.001 to the attenuated sinusoid of 0.001, so that the signal plus dither regularly crosses the ±q/2 levels of the ADC. Figure 13.11 shows the power spectra obtained by transforming and averaging 400 realizations of this dithered signal. Lo and behold, the 60-dB attenuated signal, at the edge of the ADC's resolvability, is indeed present and has been accurately measured at –63 dB (–3 dB is due to the windowing). The dither signal has had the effect of extending the dynamic range of the ADC (typically by 9 to 12 dB or 1.5 to 2.0 bits) and has improved the effective linearity of the low-level ADC staircase approximation.

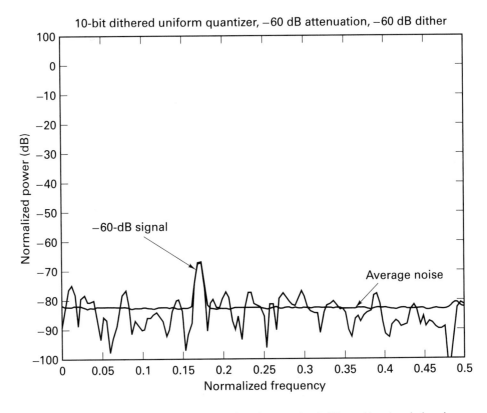

Figure13.11 Power spectrum of uniformly quantized dithered low level signal.

13.2.5 Nonuniform Quantizing

Uniform quantizers are the most common type of analog-to-digital converters because they are the most robust. By "robust" we mean that they are relatively insensitive to small changes in the input statistics. They achieve this robustness by not being finely tuned to one specific set of input parameters. This allows them to perform well even in the face of uncertain input parameters, and it means that small changes in input statistics will result in only small changes in output statistics.

When there is small uncertainty in the input signal statistics, it is possible to design a nonuniform quantizer that exhibits a smaller quantizer NSR than a uniform quantizer using the same number of bits. This is accomplished by partitioning the input dynamic range into nonuniform intervals such that the noise power, weighted by the probability of occurrence in each interval, is the same. Iterative solutions for the decision boundaries and step sizes for an optimal quantizer can be found for specific density functions and for a small number of bits. This task is simplified by modeling the nonuniform quantizer as a sequence of operators, as depicted in Figure 13.12. The input signal is first mapped, via a nonlinear function called a *compressor,* to an alternative range of levels. These levels are uniformly quantized and the quantized signal levels are then mapped, via a complementary nonlinear function called an *expander,* to the output range of levels. Borrowing part of the name from each of the operations COMpress and exPAND, we form the acronym by which this process is commonly identified: *companding.*

13.2.5.1 (Near) Optimal Nonuniform Quantizing

Examining the compressor characteristics $y = C(x)$ of Figure 13.13, we note that the quantizing step sizes for the output variable y are related to the quantizing step sizes for the input variable x through the slope $\dot{C}(x)$ [e.g., $\Delta y = \Delta x\, \dot{C}(x)$]. For an arbitrary pdf and arbitrary compressor characteristics, we can arrive at the output quantizing noise variance [7]

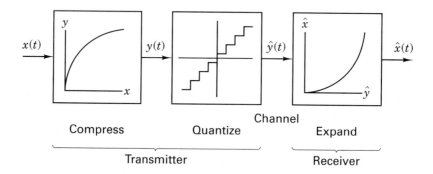

Figure 13.12 Nonuniform quantizer as a sequence of operators: compression, uniform quantization, and expansion.

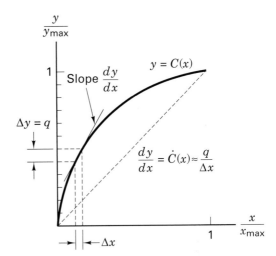

Figure 13.13 Compressor characteristics $C(x)$ and estimate to local slope $\dot{C}(x)$.

In the figure:
- $\dfrac{y}{y_{max}}$
- Slope $\dfrac{dy}{dx}$
- $y = C(x)$
- $\Delta y = q$
- $\dfrac{dy}{dx} = \dot{C}(x) \approx \dfrac{q}{\Delta x}$
- Δx
- $\dfrac{x}{x_{max}}$

$$\sigma_q^2 = \frac{q^2}{12} \int_{-x_{max}}^{x_{max}} \frac{p(x)}{|\dot{C}(x)|^2} \, dx \tag{13.25}$$

for a specific pdf, the compression characteristics $C(x)$ can be found which minimize σ_q^2. The optimal compressor law for a given pdf is [8]

$$C(x) = \int_0^x \sqrt[3]{Kp(z)} \, dz \tag{13.26}$$

We find that the optimal compressor characteristic is proportional to the integral of the cube root of the input probability density function. This is called *fine tuning*. If the compressor is designed to operate with one density function and it is used with some other density function (including scaled versions), the quantizer is said to be mismatched and there may be severe performance degradation due to the mismatch [6].

13.2.5.2 Logarithmic Compression

In the preceding section, we presented the compression law for the case in which the input signal's pdf is well defined. We now address the case for which little is known about the signal's pdf. This case occurs, for instance, when the average power of the input signal is a random variable. As an example, the voice level of a randomly chosen telephone user may vary from one extreme of a barely audible whisper to the other extreme of a bellowing shout.

For the case of an unknown pdf, the compressor characteristics of the nonuniform quantizer must be selected such that the resultant noise performance is independent of the specific density function. Although this is a worthy undertaking, it may not be possible to achieve this independence. We are willing to compromise, however, and we will settle for virtual independence over a large range of input variance and input density functions. An example of a quantizer that exhibits a

SNR independent of the input signal's pdf can be visualized with the aid of Figure 2.18. There we can see a very large difference in NSR ratio for different amplitude input signals when quantized with a uniform quantizer. By comparison, we can see that the nonuniform quantizer only permits large errors for large signals. This makes intuitive sense. If the SNR is to be independent of the amplitude distribution, the quantizing noise must be proportional to the input level. Equation (13.25) presented the quantizer noise variance for an arbitrary pdf and arbitrary compressor characteristics. The signal variance for any pdf is

$$\sigma_x^2 = \int_{-\infty}^{\infty} x^2 p(x) dx \tag{13.27}$$

In the absence of saturation, the quantizer SNR is of the form

$$\frac{\sigma_x^2}{\sigma_q^2} = \frac{\int_{-x_{max}}^{x_{max}} x^2 p(x) \, dx}{(q^2/12) \int_{-x_{max}}^{x_{max}} [p(x)/\dot{C}^2(x)] \, dx} \tag{13.28}$$

To have the SNR be independent of the specific density function, we require that the numerator be a scaled version of the denominator. This happens if the following is true:

$$[\dot{C}(x)]^2 = \left(\frac{K}{x}\right)^2 \tag{13.29}$$

or

$$\dot{C}(x) = \frac{K}{x} \tag{13.30}$$

from which we obtain by integration,

$$C(x) = \int_0^x \frac{K}{z} dz \tag{13.31}$$

or

$$C(x) = \log_e(x) + \text{constant} \tag{13.32}$$

This result is intuitively appealing. A *logarithmic compressor* allows for a *constant* SNR output, because with a logarithmic scale equal distances (or errors) are, in-fact, equal ratios, which is what we require in order for the SNR to remain fixed over the input signal range. In Equation (13.32), the constant is present to match the boundary conditions between x_{max} and y_{max}. Accounting for this boundary condition, we have the logarithmic converter of the form

$$\frac{y}{y_{max}} = \frac{C(x)}{y_{max}} = \log_e\left(\frac{x}{x_{max}}\right) \tag{13.33}$$

The form of the compression suggested by the logarithm function is shown in Figure 13.14a. The first difficulty with this function is that it does not map the negative input signals. We account for the negative signals by adding a reflected version of the log to the negative axis. This modification results in Figure 13.14 yielding

$$\frac{y}{y_{max}} = \log_e \left(\frac{|x|}{x_{max}} \right) \text{sgn}(x) \qquad (13.34)$$

where

$$\text{sgn } x = \begin{cases} +1 \text{ for } x \geq 0 \\ -1 \text{ for } x < 0 \end{cases}$$

The remaining difficulty we face is that the compression described by Equation (13.34) is not continuous through the origin; in fact, if completely misses the origin. We need to make a smooth transition between the logarithmic function and a linear segment passing through the origin. There are two standard compression functions that perform this transition—the μ-law and A-law companders.

μ-*Law Compander.* The μ-law compander, introduced by the Bell System for use in North America, is of the form

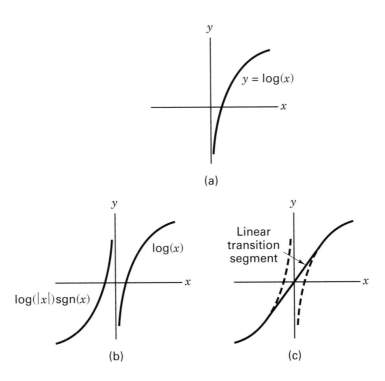

(a)

(b)   (c)

Figure 13.14 (a) Log function prototype for compression law. (b) Log | x | sgn x function prototype for compression law. (c) Log | x | sgn x function with a smooth transition between segments.

$$y = C(x) = y_{max} \frac{\log_e[1 + \mu(|x|/x_{max})]}{\log_e(1 + \mu)} \text{ sgn } x \qquad (13.35)$$

The approximate behavior of this compressor in the regions corresponding to small and large values of the argument are

$$y = C(x) = \begin{cases} y_{max} \dfrac{\mu(|x|/x_{max})}{\log_e(\mu)} & \text{for} \quad \mu\dfrac{|x|}{x_{max}} \ll 1 \\[3mm] y_{max} \dfrac{\log_e[\mu(|x|/x_{max})]}{\log_e(\mu)} & \text{for} \quad \mu\dfrac{|x|}{x_{max}} \gg 1 \end{cases} \qquad (13.36)$$

The parameter μ in the μ-law compander had originally been set to 100 for use with a 7-bit converter. It was later changed to 255 for use with an 8-bit converter. The 8-bit $\mu = 255$ μ-law converter has become the standard North American conversion law.

Example 13.6 Average SNR for μ-Law Compressor

The SNR for the μ-law compressor can be estimated by substituting the μ-law expression into Equation (13.28). For positive values of the input variable x, the compression law is

$$y = C(x) = y_{max} \frac{\log_e[1 + \mu(|x|/x_{max})]}{\log_e(1 + \mu)} \qquad (13.37)$$

Then the derivative is

$$\dot{y} = \dot{C}(x) = y_{max} \frac{1}{\log_e(1 + \mu)} \frac{\mu(1/x_{max})}{1 + \mu(|x|/x_{max})} \qquad (13.38)$$

For values of the input variable for which $\mu(x/x_{max})$ is large compared with unity, the derivative becomes

$$\dot{y} = \dot{C}(x) \approx \frac{1}{x} \frac{y_{max}}{\log_e(\mu)} \qquad (13.39)$$

Substituting for $1/\dot{C}(x)$ in Equation (13.28), we find that

$$\text{SNR} = \frac{\sigma_s^2}{\sigma_q^2} = \frac{1}{(q^2/12)[\log_e(\mu)/y_{max}]^2} \qquad (13.40)$$

$$= 3\left(\frac{2y_{max}}{q}\right)^2 \left(\frac{1}{\log_e(\mu)}\right)^2 \qquad (13.41)$$

The ratio $2y_{max}/q$ is approximately equal to the number of quantizing levels (2^b) of the b-bit compressed quantizer. For the 8-bit converter with $\mu = 255$,

$$\text{SNR} = 3\left[\frac{2^8}{\log_e(255)}\right]^2 = 3(46.166)^2 = 38.1 \text{ dB} \qquad (13.42)$$

For a comparison, the SNR of a μ-law quantizer is presented in Figure 13.15. Here, the SNR is plotted for input sinusoids of different amplitudes. Shown on the same fig-

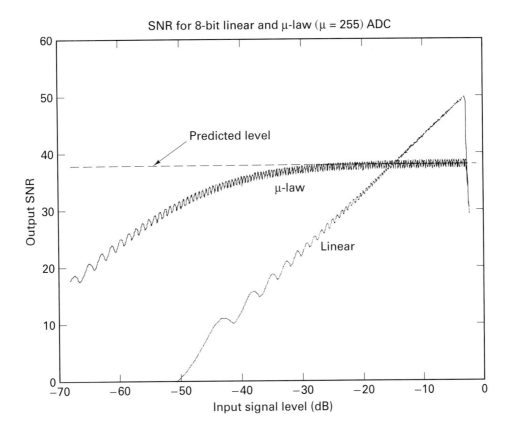

Figure 13.15   Predicted and measured SNR for a μ-law ADC.

ure is the 38.1 dB level computed in equation 13.42, and the SNR for a linear quantizer over the same range of input amplitudes. As predicted, the μ-law quantizer maintains a constant SNR over a significant range of input levels. The serration of the performance curve (granularity of the quantizer) is due to the log compression function. The actual converters exhibit additional serration due to piecewise linear approximation to the continuous μ-law curve. This is described shortly.

Figure 13.16 presents the discrete Fourier transform of the pair of input sinusoids of relative amplitude 1.0 (0-dB) and 0.01 (−40-dB). The input signal is quantized with a 10 bit μ-law (μ = 500) converter and in Figures 13.16a, b, and c, the signal levels are attenuated by 0, 20, and 40 dB, respectively, relative to full-scale input. Note that the quantizing noise levels for the full-scale signal in Figure 13.16a are higher than that of the uniform quantizer ADC, −72 dB as opposed to −83 dB seen in Figure 13.7. The improved SNR performance of the log-compressed ADC compared with the uniform ADC is to be seen for the attenuated signals. We see that as the input signal levels are attenuated, the quantizing noise is also reduced and that for 40 dB attenuation, the noise level drops to −108 dB. Thus, the log compressed ADC has no problem "seeing" the low-level input signal even with 40-dB attenuation in Figure 13.16c, while the same signal is lost in the noise of the uniform converter as seen in Figure 13.7c.

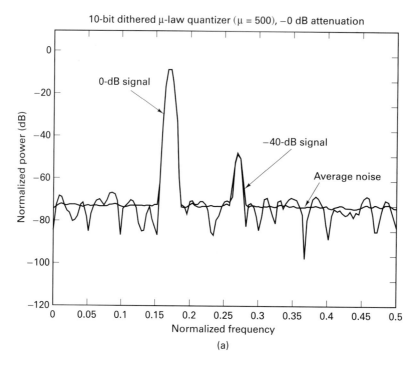

(a)

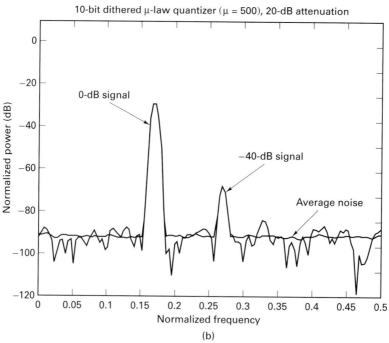

(b)

Figure 13.16 Power spectrum of signals quantized by μ-law ADC.

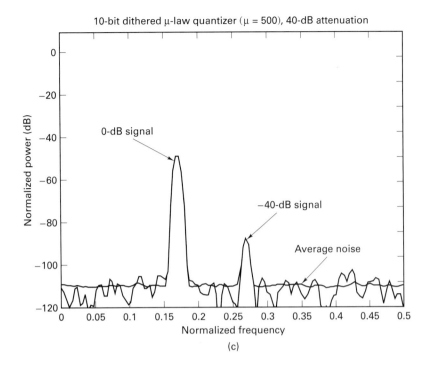

10-bit dithered μ-law quantizer (μ = 500), 40-dB attenuation

0-dB signal

−40-dB signal

Average noise

Normalized power (dB)

Normalized frequency

(c)

Figure 13.16   Continued

The μ-law compressor realization differs from the expression presented in Equation (13.35) in a minor way. As shown in Figure 13.17, 16 linear chord segments approximate the functional expression over the possible 256 output levels. Eight of these segments are in the first quadrant, eight are in the third quadrant, and the "0" segment has the same slope in both quadrants. Over each chord segment, the quantization is uniform in the four lower-order conversion bits. Thus the 8-bit compressed conversion format is of the form

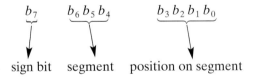

$$b_7 \qquad b_6\,b_5\,b_4 \qquad b_3\,b_2\,b_1\,b_0$$

sign bit   segment   position on segment

It is the piecewise chord approximation to the smooth function and a staircase approximation of each chord that accounts for additional serrations in the SNR curve presented in Figure 13.15.

A-Law Compander.   The *A*-law compander is the CCITT (hence the European) standard approximation to the logarithmic compression. The form of the compressor is

13.2   Amplitude Quantizing

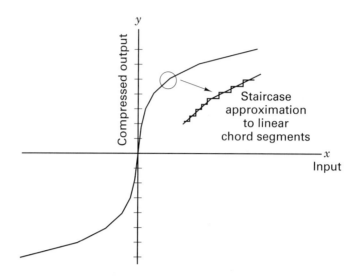

Figure 13.17 Seven-bit compressed quantization with 16-segment approximation to μ-law.

$$
y = C(x) = \begin{cases} y_{\max} \dfrac{A(|x|/x_{\max})}{1 + \log_e (A)} \, \mathrm{sgn}\, x & \text{for } \quad 0 < \dfrac{|x|}{x_{\max}} < \dfrac{1}{A} \\[3ex] y_{\max} \dfrac{1 + \log_e [A(|x|/x_{\max})]}{1 + \log_e A} \, \mathrm{sgn}\, x & \text{for } \dfrac{1}{A} < \dfrac{|x|}{x_{\max}} < 1 \end{cases}
$$

(13.43)

The standard value of the parameter A is 87.56, and for this value, using an 8-bit converter, the SNR is 38.0 dB. The A-law compression characteristic is approximated, in a manner similar to the μ-law compressor, by a sequence of 16 linear chords spanning the output range. The lower two chords in each quandrant are in fact a signal chord corresponding to the linear segment of the A-law compressor. One important difference between the A-law and the μ-law compression characteristics is that the A-law standard has a midriser at the origin, while the μ-law standard has a midtread at the origin. Thus, the A-law compressor has no zero value, and hence it exhibits no interval for which data are not being transmitted for zero input.

There are direct mappings from the A-law 8-bit compressed ADC format to a 12-bit linear binary code, and from the μ-law 8-bit compressed format to a 13-bit linear code [8]. This operation permits the A/D conversion to be performed with a uniform quantizer and then to be mapped to the smaller number of bits in a code converter. This also permits the inverse mapping at the receiver (i.e., the expansion) to be performed on the digital sample.

Pulse-Code Modulation. One of the tasks performed by a pulse-code modulation (PCM) process is the conversion of a waveform source to a binary sequence discrete source. This task is performed in a three-step process—sampling, quantizing, and encoding. We have addressed the sampling process in Chapter 2, and we have addressed the quantizing process in this chapter, as well as in Chapter 2. We note that the encoding process, following quantization (see Figure 2.2), is often embedded in the hardware that performs the quantization. It can be described as follows: successive approximation (SA) analog-to-digital (A/D) converters form the successive bits of the encoded data by a feedback, comparison, and decision process. In the feedback process, a binary search is conducted over the range of possible input levels by repeatedly asking, Is the input signal above or below the midpoint of the remaining uncertainty interval? By this technique, the uncertainty interval is reduced by one-half for each comparison and decision step until the uncertainty range matches the allowable quantizing interval.

In the SA conversion, the results of each previous decision reduce the uncertainty to be resolved during the next decision. In a similar manner, the results of the previous A/D conversions can be used to reduce the uncertainty to be resolved during the next conversion. This reduction in uncertainty is achieved by carrying forward to the next sample auxiliary information from earlier samples. This information is called the redundant part of the signal, and by carrying it forward we reduce the interval of uncertainty over which the quantizer and encoder must search for the next signal sample. Carrying data forward is one method of achieving *redundancy reduction.*

13.3 DIFFERENTIAL PULSE-CODE MODULATION

By the use of past data to assist in measuring (i.e., quantizing) new data, we leave ordinary PCM and enter the realm of differential PCM (DPCM). In DPCM, a prediction of the next sample value is formed from past values. This prediction can be thought of as instructions for the quantizer to conduct its search for the next sample value in a particular interval. By using the redundancy in the signal to form a prediction, the region of uncertainty is reduced and the quantization can be performed with a reduced number of decisions (or bits) for a given quantization level or with reduced quantization levels for a given number of decisions (or bits). The reduction in redundancy is realized by subtracting the prediction from the next sample value. This difference is called the *prediction error.*

The quantizing methods described in Section 13.2 are called *instantaneous* or *memoryless* quantizers because the digital conversion is based on the single (current) input sample. In Section 13.1 we identified the properties of sources that permitted source rate reductions. These properties were nonequiprobable source levels and nonindependent sample values. Instantaneous quantizers achieve source-coding gains by taking into account the probability density assignment for each sample. The quantizing methods that take account of sample-to-sample correlation are noninstantaneous quantizers. These quantizers reduce source redundancy by first converting the correlated input sequence into a related sequence with reduced correlation, reduced variance, or reduced bandwidth. This new sequence is then quantized with fewer bits.

The correlation characteristics of a source can be visualized in the time domain by samples of its autocorrelation function and in the frequency domain by its power spectrum. If we examine a power spectrum $G_x(f)$ of a short-term speech signal, as shown in Figure 13.18, we find that the spectrum has a global maxima in the neighborhood of 300 to 800 Hz and falls off at a rate of 6 to 12 dB/octave. By interpreting this power spectrum, we can infer certain properties of the time function from which it was derived. We observe that large changes in the signal occur slowly (low frequency) and that rapid changes in the signal (high frequency) must be of low amplitude. An equivalent interpretation can be found in the autocorrelation function $R_x(T)$ of the signal, as shown in Figure 13.19. Here a broad, slowly changing autocorrelation function suggests that there will be only slight change on a sample-to-sample basis, and that a time interval exceeding the correlation distance is required for a full amplitude change. The correlation distance seen in Figure 13.19 is the time difference between the peak correlation and the first zero correlation. In particular, correlation values for typical single-sample delay is on the order of 0.79 to 0.87, and the correlation distance is on the order of 4 to 6 sample intervals of T seconds per interval.

Since the difference between adjacent time samples for speech is small, coding techniques have evolved based on transmitting sample-to-sample differences rather than actual sample values. Successive differences are in fact a special case of a class of non-instantaneous converters called N-tap linear predictive coders. These coders, sometimes called predictor-corrector coders, predict the next input sample value based on the previous input sample values. This structure is shown in Figure 13.20. In this type of converter, the transmitter and the receiver have the same prediction model, which is derived from the signal's correlation characteristics. The encoder forms the prediction error (or the residue) as the difference between the next measured sample value and the predicted sample value. The equation for the prediction loop is

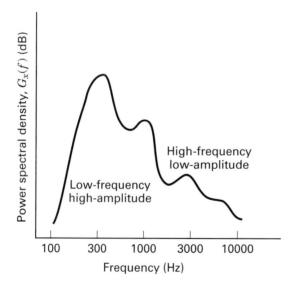

Figure 13.18 Typical power spectrum for speech signals.

Source Coding   Chap. 13

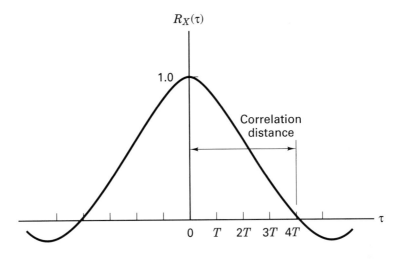

Figure 13.19 Autocorrelation function for typical speech signals.

$$d(n) = x(n) - \hat{x}(n)$$

where $x(n)$ is the nth input sample, $\hat{x}(n)$ is the predicted value of that sample, and $d(n)$ is the associated prediction error. This is performed in the predict-and-compare loop, the upper loop of the encoder shown in Figure 13.20. The encoder corrects its prediction by forming the sum of its prediction and the prediction error. The equations for the correction loop are

$$\tilde{d}(n) = \text{quant}\,[d(n)]$$

$$\tilde{x}(n) = \hat{x}(n) + \tilde{d}(n)$$

where quant $(\cdot)$ represents the quantization operation, $\tilde{d}(n)$ is the quantized version of the prediction error, and $\tilde{x}(n)$ is the corrected and quantized version of the input sample. This is performed in the predict-and-correct loop, the lower loop of the encoder, and the only loop of the decoder in Figure 13.20. The decoder must also be informed of the prediction error so that it can use its correction loop to correct its prediction. The decoder "mimics" the feedback loop of the encoder. The communication task is that of transmitting the difference (the error signal) between the predicted and the actual data sample. For this reason, this class of coder is often called a differential pulse code modulator (DPCM). If the prediction model forms predictions that are close to the actual sample values, the residues will exhibit reduced variance (relative to the original signal). From Section 13.2, we know that the number of bits required to move data through the channel with a given fidelity is related to the signal variance. Hence, the reduced variance sequence of residues can be moved through the channel with a reduced data rate.

The predictive converters must have a short-term memory that supports the real-time operations required for the prediction algorithm. In addition, they will

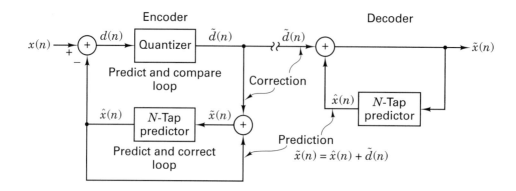

Encoder

Decoder

Figure 13.20 *N*-tap predictive differential pulse code modulator (DPCM).

often have a long-term memory that supports the slow time, often data-dependent operations, such as automatic gain control and filter coefficient adjustments. Predictors that incorporate the slower, data-dependent adjustment algorithms are called *adaptive predictors.*

13.3.1 One-Tap Prediction

The one-tap linear prediction coding (LPC) filter in the DPCM process predicts the next input sample value based on the previous input sample value. The prediction equation is of the form

$$x(n|n-1) = a\,x(n-1)|n-1) \tag{13.44}$$

where $x(n|m)$ is the estimate of x at time n given all the samples collected up through time m, and where "a" is a parameter used to minimize the prediction error. The prediction error available after the measurement is of the form

$$d(n) = [x(n) - x(n|n-1)] \tag{13.45a}$$

$$= [x(n) - a\,x(n-1|n-1)] \tag{13.45b}$$

The mean-squared error is of the form

$$\mathbf{E}\{d^2(n)\} = \mathbf{E}\{x(n)\,x(n) - 2\,a\,x(n)\,x(n-1\,|\,n-1) \tag{13.46}$$
$$+ a^2\,x(n-1\,|\,n-1)\,x(n-1\,|\,n-1)\}$$

If $x(n-1\,|\,n-1)$ is an unbiased estimate of $x(n-1)$, Equation (13.46) can be written as

$$R_d(0) = R_x(0) - 2\,a\,R_x(1) + a^2\,R_x(0) \tag{13.47a}$$

$$= R_x(0)\,[1 + a^2 - 2\,a\,C_x(1)] \tag{13.47b}$$

where $R_d(n)$ and $R_x(n)$ are the autocorrelation functions of the prediction error and the input signal, respectively, $R_d(0)$ is the power in the error, $R_x(0)$ is the power in the signal, and $C_x(n) = R_x(n)/R_x(0)$ is the normalized autocorrelation function. We can select the parameter a to minimize the prediction error power of Equation (13.47) by setting a zero the partial derivative of $R_d(0)$ with respect to a:

$$\frac{\partial R_d(0)}{\partial a} = R_x(0)\left[2a - 2C_x(1)\right] \tag{13.48}$$

Setting to zero and solving for a^{opt}, the optimal solution, we have

$$a^{\text{opt}} = C_x(1) \tag{13.49}$$

Substituting a^{opt} back into Equation (13.47), we obtain

$$R_d^{\text{opt}}(0) = R_x(0)\left[1 + a^{\text{opt}}C_x(1) - 2a^{\text{opt}}C_x(1)\right] \tag{13.50a}$$

$$= R_x(0)\left[1 - a^{\text{opt}}C_x(1)\right] \tag{13.50b}$$

$$= R_x(0)\left[1 - C_x^2(1)\right] \tag{13.50c}$$

We can define the *prediction gain* of the encoder as the ratio of input to output variances, $R_x(0)/R_d(0)$. For a fixed bit rate this gain represents an increase in output SNR, while for a fixed output SNR this gain represents a reduced bit-rate description. We note that the prediction gain for the optimal predictor is always greater than one for any value of signal correlation $R_x(0)$ as used in Equation (13.50b). On the other hand, the prediction gain is greater than one for the nonoptimum unity-gain, one-tap predictor, only if the signal correlation exceeds 0.5 as used in Equation (13.47b).

Example 13.7   Prediction Gain of a One-Tap LPC Filter

A signal with correlation coefficient $C_x(1)$ equal to 0.8 is to be quantized with a one-tap LPC filter. Determine the prediction gain when the prediction coefficient is (a) optimized with respect to the minimum prediction error, or (b) set to unity.

Solution

(a) From Equation (13.50c),

$$R_d^{\text{opt}}(0) = R_x(0)\,(1 - 0.64) = 0.36\,R_x(0) \tag{13.51a}$$

$$\text{Prediction gain} = 1/(0.36) = 2.78 \text{ or } 4.44 \text{ dB} \tag{13.51b}$$

(b) From Equation (13.47b),

$$R_d(0) = 2R_x(0)\,(1 - 0.8) = 0.40\,R_x(0) \tag{13.51c}$$

$$\text{Prediction gain} = 1/(0.40) = 2.50 \text{ or } 3.98 \text{ dB} \tag{13.51d}$$

13.3.2 *N*-Tap Prediction

The N-tap LPC filter predicts the next sample value based on a linear combination of the previous N sample values. We will assume that the quantized estimates used by the prediction filters are unbiased and error free. With this assumption, we can drop the double indices (used in Section 13.3.1) from the data in the filter but still use them for the predictions. Then the N-tap prediction equation takes the form

$$x(n \mid n - 1) = a_1\,x(n - 1) + a_2\,x(n - 2) + \cdots + a_N\,x(n - N) \tag{13.52}$$

The prediction error takes the form

$$d(n) = x(n) - x(n \mid n - 1) \tag{13.53a}$$

$$= x(n) - a_1 x(n-1) - a_2 x(n-2) - \cdots - a_N x(n-N) \tag{13.53b}$$

The mean-square prediction error is of the form

$$\mathbf{E}\{d(n)d(n)\} = \mathbf{E}\{[x(n) - x(n \mid n - 1)]^2\} \tag{13.54}$$

Clearly, the mean-square prediction error is quadratic in the filter coefficients a_j. As we did in Section 13.3.1, we can form the partial derivative of the mean-squared error with respect to each coefficient and solve for the coefficients that set the partials to zero. Formally, taking the partial derivative with respect to the jth coefficient prior to expanding $x(n|n-1)$, we have

$$\frac{\partial R_d(0)}{\partial a_j} = \mathbf{E}\left\{2[x(n) - x(n|n - 1)]\frac{\partial x(n|n - 1)}{\partial a_j} x(n|n - 1)\right\} \tag{13.55a}$$

$$= \mathbf{E}\{2[x(n) - x(n|n - 1)][-x(n - j)]\} \tag{13.55b}$$

$$= 2\mathbf{E}\{[x(n) - a_1 x(n-1) - a_2 x(n-2) - \cdots - a_N x(n-N)][-x(n-j)]\} \tag{13.55c}$$

$$= 2[R_x(j) - a_1 R_x(j-1) - a_2 R_x(j-2) - \cdots - a_N R_x(j-N)] \tag{13.55d}$$

This collection of equations (one for each j) can be arranged in matrix form known as the normal equations. This form is

$$
\begin{bmatrix} R_x(1) \\ R_x(2) \\ R_x(3) \\ \vdots \\ R_x(N) \end{bmatrix}
=
\begin{bmatrix}
R_x(0) & R_x(-1) & R_x(-2) & \cdots & R_x(-N+1) \\
R_x(1) & R_x(0) & R_x(-1) & \cdots & R_x(-N+2) \\
R_x(2) & R_x(1) & R_x(0) & \cdots & R_x(-N+3) \\
\vdots & \vdots & \vdots & \ddots & \vdots \\
R_x(N-1) & R_x(N-2) & R_x(N-3) & \cdots & R_x(0)
\end{bmatrix}
\begin{bmatrix} a_1 \\ a_2 \\ a_3 \\ \vdots \\ a_N \end{bmatrix}^{\text{opt}}
$$

$$\tag{13.56a}$$

The normal equations can be written more compactly as

$$\mathbf{r}_x(1, N) = \mathbf{R}_{xx}\, \mathbf{a}^{\text{opt}} \tag{13.56b}$$

where $\mathbf{r}_x(1, N)$ is the correlation vector of delays from 1 through N, $\mathbf{R}_{xx}$ is the correlation matrix (assuming a zero-mean process), and $\mathbf{a}^{\text{opt}}$ is the optimum filter weight vector.

To gain insight into the solution of the normal equations, we now recast the mean-square-error equation (13.54) in matrix form. We have

$$R_d(0) = \mathbf{E}\{[x(n) - \mathbf{a}^T\mathbf{x}(n - 1)][x(n) - \mathbf{x}^T(n - 1)\mathbf{a}]\} \tag{13.57a}$$

$$= R_x(0) - \mathbf{r}_x^T(1, N)\, \mathbf{a} - \mathbf{a}^T\mathbf{r}_x(-1,-N) + \mathbf{a}^T \mathbf{R}_{xx}\mathbf{a} \tag{13.57b}$$

where $\mathbf{r}^T$ is the transpose of $\mathbf{r}$. Substituting $\mathbf{a}^{\text{opt}}$ for $\mathbf{a}$ in Equation (13.57b), and then substituting $\mathbf{r}_x(1, N)$ for $\mathbf{R}_{xx}\mathbf{a}^{\text{opt}}$ in the resulting equation, yields

$$R_d(0) = R_x(0) - \mathbf{r}_x^T(1, N)\, \mathbf{a}^{\text{opt}} - \mathbf{a}^{\text{opt}\,T}\mathbf{r}_x(-1,-N) + \mathbf{a}^{\text{opt}\,T}\mathbf{r}_x(1, N) \tag{13.58a}$$

$$= R_x(0) - \mathbf{r}_x^T(-1,-N)\,\mathbf{a}^{\text{opt}} \qquad (13.58\text{b})$$

We can now bring the right-hand side of Equation (13.56) over to the left-hand side, and use Equation (13.58b) to augment the top row of the matrix to obtain the whitening form of the optimal predictor:

$$
\begin{bmatrix}
R_x(0) & R_x(-1) & R_x(-2) & R_x(-3) & \cdots & R_x(-N) \\
R_x(1) & R_x(0) & R_x(-1) & R_x(-2) & \cdots & R_x(-N+1) \\
R_x(2) & R_x(1) & R_x(0) & R_x(-1) & \cdots & R_x(-N+2) \\
R_x(3) & R_x(2) & R_x(1) & R_x(0) & \cdots & R_x(-N+3) \\
\vdots & \vdots & \vdots & \vdots & \ddots & \vdots \\
R_x(N) & R_x(N-1) & R_x(N-2) & R_x(N-3) & \cdots & R_x(0)
\end{bmatrix}
\begin{bmatrix}
1 \\ -a_1 \\ -a_2 \\ -a_3 \\ \vdots \\ -a_N
\end{bmatrix}^{\text{opt}}
=
\begin{bmatrix}
R_d(0) \\ 0 \\ 0 \\ 0 \\ \vdots \\ 0
\end{bmatrix}
$$

$$(13.59)$$

In this form, the only nonzero output of the matrix product occurs at time zero, which is akin to an output impulse.

The top row of Equation (13.59) states that the power in the prediction error is of the form

$$R_d(0) = R_x(0)[1 - a_1\,C_x(1) - a_2\,C_x(2) - \cdots - a_N\,C_x(N)] \qquad (13.60)$$

Compare this form with that of Equation (13.50b). An interesting property of the optimal N-tap predictor filter is that the coefficient set that obtains the minimum mean-square prediction error also predicts, with zero error, the next $N-1$ correlation samples from the previous $N-1$ correlation samples. For fixed filter coefficients, the DPCM coder can achieve a prediction gain, relative to linear quantizing, of 6 to 8 dB [9, 10]. This prediction gain is essentially independent of filter length once the length exceeds three or four taps. Additional gain is available if the coder has slow adaptive capabilities. Adaptive coders are introduced in Section 13.3.3 and discussed in some detail in Section 13.3.4.

13.3.3 Delta Modulation

The delta modulator, often denoted Δ modulator, is a process that embeds a low resolution A-to-D converter in a sampled data feedback loop that operates at rates far in excess of the signal's required Nyquist rate. The motivation for this technique is our awareness that in the conversion process, speed is less expensive than precision, and that by being clever one can use faster signal processing to obtain higher precision.

Equation (13.50c) demonstrated that the prediction gain for a one-tap predictor could be large if the normalized correlation coefficient, $C_x(1)$, is close to unity. Working toward the goal of high sample-to-sample correlation, the predictive filter is generally operated at a rate that far exceeds the Nyquist rate. For example, the sample rate might be chosen to be 64 times the Nyquist rate. Then for a 20 kHz bandwidth with a nominal sample rate of 48 kHz, the high correlation prediction filter would operate at a 3.072 MHz sample rate. The justification for the high sample rate is to insure that the sampled data is highly correlated so that a simple one-tap predictor will exhibit a small prediction error, which in turn permits the quantizer operating in the error loop to operate with a very small number of bits. The simplest form of the quantizer is a one-bit quantizer, which is, in fact, only a comparator that detects and reports the sign of the difference signal. As a result, the prediction error signal is a 1-bit word that has the interesting advantage of not requiring word framing in subsequent processing.

The block diagram of the one-tap linear predictor, illustrated in Figure 13.20, is shown in Figure 13.21, with slight modifications. Note that the one-tap predict and correct loop is now a simple integrator and that a low-pass reconstruction filter follows the predict and correct loop at the decoder. This filter removes the out-of-band quantizing noise that is generated by the two-level coding and that extends beyond the information bandwidth of this coding process. The coder is completely characterized by the sampling frequency, the quantizing step size to resolve the prediction error or *delta* of the loop, and the reconstruction filter. The equations for prediction and for the residual error of the modulator are of the form

$$x(n \mid n - 1) = x(n - 1 \mid n - 1) \tag{13.61a}$$

$$d(n) = x(n) - x(n \mid n - 1) \tag{13.61b}$$

where n is a sample index. This structure, sometimes called the *deltamodulator,* is a DPCM process for which the predict-and-correct loop consists of a digital accumulator.

13.3.4 Sigma-Delta Modulation

The structure of the Σ-Δ modulator can be examined from a number of perspectives, the most appealing being that of a modified one-tap DPCM converter, and that of an error-feedback converter. Let us start with the modified one-tap DPCM

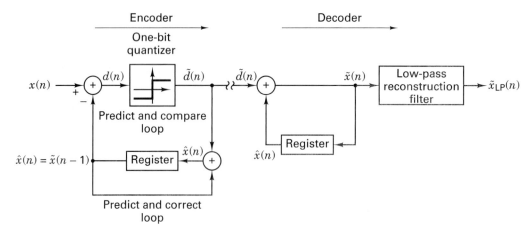

Figure 13.21   One-tap, one-bit DPCM coder (delta-modulator).

converter. As indicated earlier, the loop relies on high correlation of successive samples, a condition we assure by significant oversampling. We can enhance the correlation of the sampled data presented to the modulator by prefiltering the data with an integrator and then compensating for the prefilter with a postfiltering differentiator. This structure is shown in Figure 13.22, where the integrators, differentiator, and delay functions are expressed in terms of the z transform. (See Appendix E.) We can then rearrange the signal flow blocks to realize an economy of implementation. At the input to the encoder, there are the outputs of two digital integrators, which are summed and presented to the loop quantizer. Our first modification is that we can share a single digital integrator by sliding the two integrators through the summing junction in the encoder. Our second modification to the encoder is that the post filter differentiator can be moved to the decoder, which then cancels the digital integrator at the input to the decoder. All that remains of the decoder is the low-pass reconstruction filter. This simplified form of the modified DPCM system is shown in Figure 13.23. This form, called a *sigma-delta* modulator, contains an integrator (the *sigma*) and a DPCM modulator (the *delta*) [11].

The second perspective useful for understanding the Σ-Δ modulator is that of noise feedback loop. We understand that a quantizer adds an error to its input to form its output. When the signal is highly oversampled, not only are the samples highly correlated, the errors are as well. When errors are highly correlated, they are predictable, and thus they can be subtracted from the signal presented to the quantizer prior to the quantization process. When the signal and error are highly oversampled, the previous quantization error can be used as a good estimate of the current error. The previous error, formed as the difference between the input and output of the quantizer, is stored in a delay register for use as the estimate of the next quantization error. This structure is shown in Figure 13.24. The signal flow graph of Figure 13.24 can be redrawn to emphasize the two inputs, signal and quantization noise, as well as the two loops, one including the quantizer and one not including it. This form is shown in Figure 13.25 and except for explicitly showing the feedback leg of the digital integrator, this has the same structure as presented in Figure 13.23. From Figure 13.25, if follows that the output of the Σ-Δ modulator and its z-transform (see Appendix E) can be written as

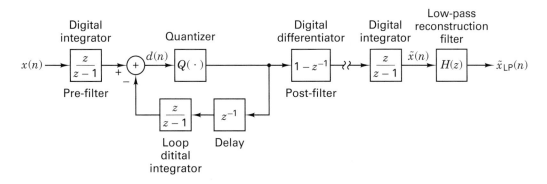

Figure 13.22   One-bit delta modulator, pre and post filtered form.

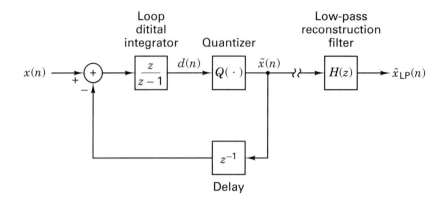

Figure 13.23 Σ-Δ modulator as a rearranged pre and post filtered Δ-modulator.

$$y(n) = \tilde{x}(n) = x(n) - q(n-1) + q(n) \qquad (13.62)$$
$$= x(n) + [q(n) - q(n-1)]$$

$$Y(Z) = X(Z) - Z^{-1}Q(Z) + Q(Z)$$
$$= X(Z) + Q(Z)[1 - Z^{-1}] \qquad (13.63)$$
$$= X(Z) + Q(Z)\frac{Z-1}{Z}$$

Intuitively, Equation (13.63) is pleasing, since it shows that the loop does not affect the input signal since only the noise is being circulated by the loop, and only the noise experiences the effect of the loop. The integrator in the feedback path of the

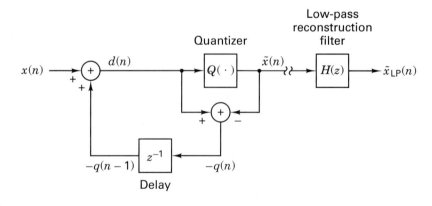

Figure 13.24 Σ-Δ modulator as a noise feedback process.

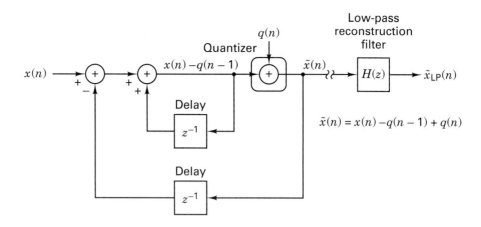

Figure 13.25 Noise feedback quantizer redrawn as Σ-Δ modulator.

noise signal is inverted by the unity-gain feedback loop to become a differentiator to the noise source.

The z-plane (like its analog counterpart, the s-plane) provides a convenient vehicle for mapping a frequency transfer function. (See Appendix E.) Such a transfer function is typically characterized in terms of a numerator and denominator polynomial whose roots are referred to as the *zeros* and *poles*, respectively, of the transfer function. These zeros and poles can be viewed as a surface emanating from the plane, representing the magnitude of the transfer function. The surface is easily visualized as being a stretched rubber sheet pulled off the ground by tent poles located at the pole positions and being held to the ground by tent stakes at the zero positions. The magnitude of the frequency response is the level of this surface observed as we traverse the unit circle in the z-plane (or $j\omega$ axis in the s-plane). Note that the *noise transfer function* (NTF), which is the frequency transfer function of the loop applied to the noise, has a pole at the origin and a transmission zero at dc ($z = e^{j\theta}$, $\theta = 0$, so $z = 1$). The pole zero plot of the NTF and the spectral response of the NTF along with a typical spectrum of the input signal are shown in Figure 13.26. Note that the zero of the noise transfer function is located at dc, and that in the neighborhood of dc, the quantization noise is suppressed by the NTF. Thus, there is no significant noise near dc due to the NTF, while concurrently, the signal spectrum is restricted by the significant oversampling to reside in a small neighborhood around dc with a typical width of approximately 1.5% of the sample rate. The function of the reconstruction filter is to suppress the quantization noise outside the signal bandwidth. The sample rate at the output of the filter is reduced to match the reduced bandwidth of the now nearly noise-free signal. Additional improvement in noise suppression can be obtained by raising the order of the zero in the NTF of the loop. Many Σ-Δ modulators are designed with NTFs displaying dual or triple zeros. Since the NTF zeros set the output noise power to zero, it hardly matters what level of noise power is presented to the feedback loop. Consequently, most Σ-Δ modulators are designed to operate with 1-bit converters with a number of very high performance modulators operating with 4-bit converters.

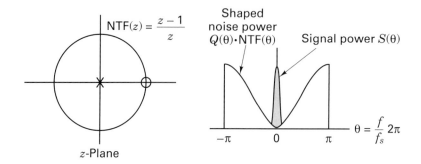

$$\left| \text{NTF}(z) = \frac{z-1}{z} \right|$$

Shaped noise power $Q(\theta) \cdot \text{NTF}(\theta)$

Signal power $S(\theta)$

$$\theta = \frac{f}{f_s} 2\pi$$

z-Plane

Figure 13.26 Noise transfer function in z-plane and power spectrum of signal and shaped noise of Σ-Δ modulator.

13.3.4.1 Noise Performance of Σ-Δ Modulator

In the previous section, we alluded to the ability of a Σ-Δ modulator to exchange excess sample rate for improvement in SNR of quantized data. We first consider how this trade occurs with oversampled and filtered data plus AWGN, and we will then examine the same process with shaped noise. If the quantization noise is white, and the signal is sampled at a rate that exceeds the Nyquist rate, the white noise is uniformly distributed over a spectral interval equal to the sample rate. This interval is denoted the *first Nyquist zone* or the *primary strip*. Since the quantizing noise power is fixed at $q^2/12$ (see Equation 13.12), the power spectral density of the quantizing noise for a signal sampled at the rate f_s must be $q^2/(12f_s)$ watts/Hz. Operating the quantizer at elevated sample rates reduces the power spectral density of the quantizing noise in the signal bandwidth. The oversampled data can be digitally filtered to reject the out-of-band quantizing noise and then down sampled to the signal's Nyquist rate. If the signal is oversampled by a factor of two, the filtering will reject half the noise power. Rejecting half the noise power reduces the rms value of the quantized noise amplitude by $\sqrt{2}$, or the power by 3 dB. To reduce the noise power by 6 dB, and thus improve quantization noise by 1-bit (see Equation 13.24), one must oversample by a factor of 4 and reject all but one-fourth of the quantizing noise with a filter. We conclude that we improve the SNR of an oversampled white noise quantizer, at the rate of 3-dB (or half a bit) per doubling of the sample rate relative to Nyquist rate.

We now address the rate at which we can improve SNR with an over sampled shaped noise quantizer. The NTF of the Σ-Δ shaping filter has a zero at dc which results in a double zero in the filter's power spectral response. If we expand the spectral response of the filter in a Taylor series, and truncate after the first nonzero term we obtain the following simple approximation to the filter response valid in the neighborhood of the signal spectrum:

$$H^2(\omega) = \left[2 \sin\left(\frac{\omega}{2\omega_s} \right) \right]^2$$

$$= 2\left[1 - \cos\left(\frac{\omega}{\omega_s}\right)\right] \tag{13.64}$$

$$= 2\left\{1 - \left[1 - \frac{1}{2!}\left(\frac{\omega}{\omega_s}\right)^2 + \cdots + \right]\right\} \approx \left(\frac{\omega}{\omega_s}\right)^2 = \left(\frac{f}{f_s}\right)^2$$

Here, f_s is the sample rate of the modulator.

The shaped noise power that survives the filtering action of the low-pass filter following the Σ-Δ modulator is of the form

$$N(\omega) = \frac{N_0}{2}\int_{-f_{BW}}^{f_{BW}}\left(\frac{f}{f_s}\right)^2 df = \frac{N_0}{2}\frac{1}{3}\left(\frac{f}{f_s}\right)^3\bigg|_{f=-f_{BW}}^{f=f_{BW}} = \frac{N_0}{3}\left(\frac{f_{BW}}{f_s}\right)^3 \tag{13.65}$$

The ratio of the noise term for a signal sampled at f_s, and one sampled at $2f_s$ and then filtered down to the same output signal bandwidth f_{BW}, is found to be 8 or 9 dB. Thus a Σ-Δ modulator with a single zero in its NTF improves its SNR by 9 dB or by 1.5 bits per doubling of the sample rate. Sigma-delta modulators built formed with multiple digital integrators and feedback loops have an increased number of transmission zeros in the NTF. By a similar derivation, we would find that the NTF of a Σ-Δ modulator with 2 and 3 zeros improves SNR by 15 and 21 dB (or 2.5 and 3.5 bits), respectively. Thus, a two-zero Σ-Δ operating at 64 times Nyquist or with sample rate doubled 6 times, exhibits a SNR improvement of 90 dB. The spectrum illustrated in Figure 13.27 was formed with a two-zero Σ-Δ modulator, and if we include the 6-dB loss due to the spectral decomposition of a real signal, 2-dB reduction due to amplitude back-off from full scale, and 3-dB loss due to the windowing of the DFT, the noise level is at the expected level of 79 dB below the spectral peak.

13.3.5 Sigma-Delta A-to-D Converter (ADC)

A Σ-Δ analog-to-digital (A-to-D) converter or ADC is usually implemented as an integrated circuit built around the Σ-Δ modulator. The circuit must contain auxiliary subsystems to form a complete system. These include an analog anti-alias filter, a sample-and-hold circuit, a switched-capacitor integrator for the modulator, a feedback digital-to-analog (D-to-A) converter or DAC, and a digital resampling filter. Due to the high oversampling rate of the process, the analog anti-alias filter can be simple RC or dual RC stage, with a large transition bandwidth extending over many octaves. The DAC converter is required to deliver an analog feedback signal from the output of the A-to-D quantizer. Since the DAC is in the feedback loop, it does not benefit from the loop gain, and consequently its linearity and precision must match the design performance levels of the entire system. The Σ-Δ modulator preserves the signal fidelity in a restricted segment of the sampled spectrum. The access this high-fidelity segment of the spectrum, the output of the modulator must be filtered and down sampled. The post-processing filter following the modulation process rejects the out-of-band noise located in the excess bandwidth that exists due to the oversampling. This us usually a linear-phase, finite-impulse response (FIR), resampling digital filter.

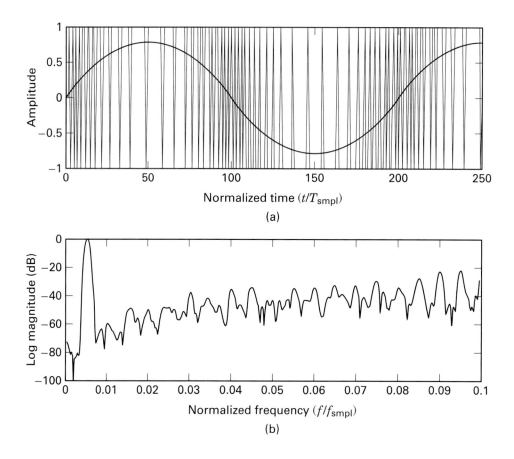

Figure 13.27 One bit Σ-Δ modulator: (a) input and output time series (b) spectral response.

Figure 13.27a presents an oversampled input sinusoidal signal and the corresponding output signal of a one-bit, dual-zero, Σ-Δ modulator. Figure 13.27b shows the spectral response of the output series. Note that the shaped noise spectrum in the neighborhood of the signal is approximately 80 dB below the peak spectrum of the input sinusoid. Note also that the output signal is restricted to ±1, and that the loop is essentially duty cycle modulating the square wave in proportion to the amplitude of the input signal. Figure 13.28 presents the time series and the spectrum obtained from the output of the down-sampling filter following the modulator.

13.3.6 Sigma-Delta D-to-A Converter (DAC)

The Σ-Δ modulator, originally designed to be embedded in an ADC, has become an essential building block of the reverse task—the digital-to-analog converter (DAC). Nearly all high-end audio equipment and most communication system

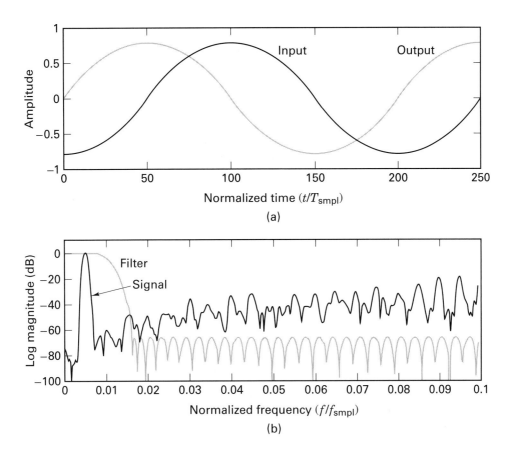

Figure 13.28 Post processing filter following Σ-Δ modulator: (a) input and output time series, (b) spectral response.

DACs are implemented with Σ-Δ converters. The process uses the Σ-Δ modulator as a digital-to-digital (D-to-D) transformation, which converts a high-precision (say, 16-bit) representation of oversampled digital data to a low-precision (say, 1-bit) representation of the same data. The oversampled one-bit data stream is then delivered to a 1-bit DAC with two analog output levels defined with the same precision as a 16-bit converter. Here, the advantage of a one-bit DAC operating at high speed but with only two levels is that speed is less expensive than precision. The 2-level, high-speed DAC replaces a low-speed DAC that would have to resolve 65,536 distinct levels. Very simple analog low-pass filtering following the 1-bit DAC suppresses the out-of-band noise spectrum and delivers the reduced bandwidth, high-fidelity version of the original digital data. The requantizing of the oversampled data is a straightforward signal processing task, using an all-digital Σ-Δ modulator. The only additional task to be performed when using a Σ-Δ DAC is the requirement to raise the sample rate of the data to 64 times its Nyquist rate.

This task is performed by a DSP-based interpolating filter, which is a standard signal-processing block found in most systems that use a DAC to transition between a digital signal source and an analog output [12].

As a standard illustration of the process, a CD player uses an interpolating filter to realize a 1-to-4 up-sampling, resulting in a separation of the periodic spectra associated with sampled data. This increased separation permits the smoothing filter following the DAC to have a wider transition bandwidth, and hence, a reduced component count and reduced implementation cost. The CD specification uses terms such as "four-to-one oversampled" to reflect the presence of the interpolating filters. Starting with the CD 1-to-4 interpolator, it is a simple task to again raise the sample rate by another factor of 1-to-16, with a second inexpensive interpolating filter. The data, now 64 times oversampled, is presented to the all-digital Σ-Δ modulator and one-bit DAC to complete the analog conversion process. This structure is shown in Figure 13.29.

There are many signals that are significantly oversampled with respect to the signal's bandwidth. These signals can easily be converted to analog representations by the use of a Σ-Δ modulator and a 1-bit DAC. Examples are control signals used in circuits such as AGC, carrier VCO, and timing VCO. Many systems employ the Σ-Δ modulator and 1-bit DAC to generate and deliver analog control signals throughout the system.

13.4 ADAPTIVE PREDICTION

The prediction gain to be had in classical predictive coders is proportional to the ratio of the signal variance to prediction-error variance. This is because for a fixed quantizing noise level, fewer bits are required to describe a signal with smaller energy. The utility of the predictive coder is limited by possible mismatches between the source signal and the predictor filter. The sources of mismatch are related to the time-varying behavior (i.e., nonstationarity) of the amplitude distribution and of the spectral or correlation properties of the signal. Adaptive encoders incorporate(slow-time) auxiliary loops to estimate the parameters required to obtain locally optimal performance. These auxiliary loops periodically schedule modifications to the prediction loop parameters and thus avoid predictor mismatch. The International Telegraph and Telephone Consultative Committee (CCITT) has selected a 32-kbits/s adaptive differential pulse-code modulation (ADPCM) coder

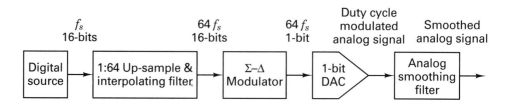

Figure 13.29 Signal flow graph of Σ-Δ digital-to-analog converter.

as a standard for toll-quality speech. This achieves a 2:1 savings in bit rate relative to 64-kbits/s logarithmic compressed PCM.

13.4.1 Forward Adaptation

In forward adaptation algorithms, the input data to be encoded are buffered and processed in order to estimate the local statistics, such as the first N samples of the autocorrelation function. The zero-delay correlation sample $R_x(0)$ is a short-term estimate of the variance. This estimate is used to adjust the automatic gain control (AGC) in order to obtain an optimal match of the scaled input signal to that of the quantizer dynamic range. This is denoted AQF, for *adaptive quantization forward control*. The remaining $N - 1$ correlation estimates are used to form new filter coefficients for the prediction filter. This adaptation is called adaptive prediction forward (APF) control. Figure 13.30 shows this form of the adaptive algorithm. This is an extension of the structure presented in Figure 13.20. Here the predictor coefficients are derived from the input data, now called *side information,* and must be transmitted along with the prediction errors from the encoder to the decoder. The update rate of these adaptive coefficients is related to the length of time the input signal can be considered locally stationary. For example, speech caused by mechanical displacement of the speech articulators (tongue, lips, teeth, etc.) cannot change characteristics more rapidly than 10 or 20 times per second. This suggests an update interval of 50 to 100 ms. Using arithmetically simple but suboptimal estimating algorithms to compute the local filter parameters makes a higher update rate necessary. An update every 20 ms to compute the parameters of a 10- to 12-tap filter has become a common rate. Prediction gains of 10 to 16 dB can be had with 10-tap filters, when feed-forward adaptation is used with predictive coders [13].

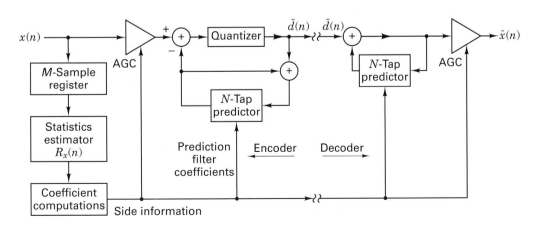

Figure 13.30 Forward adaptive prediction and quantization coding.

13.4.2 Synthesis/Analysis Coding

The encoding schemes we have examined until now can be classed as *waveform encoders*. They construct approximations to the input signals, which minimizes some distance measure between the signal and the approximation. These techniques are very general and can be applied to any signal source. Synthesis/analysis coders, on the other hand, are very signal specific; in particular, they are designed primarily for speech signals. These encoders take advantage of the fact that while the hearing mechanism responds to the amplitude content of a signal's short-term spectrum, it is fairly insensitive to its phase structure. Thus, this class of encoder forms a reconstructed signal that approximates the magnitudes and time-varying characteristic of a sequence of the signal's short-term spectra, but it makes no attempt to preserve its relative phase.

The spectral characteristic of speech appears to be stationary over periods between 20 and 50 ms. A number of techniques have evolved which analyze the spectral characteristics of voice every 20 ms and that use the results of that analysis to synthesize a waveform exhibiting the same short-term power spectrum. Some techniques employ a model of the speech-generation mechanism for which model parameters have to be estimated at the update rate. This type of encoder is best represented, in its various forms, as the linear predictive coder (LPC). Variations of LPC encoding manipulate the signal by combinations of spectral modifications and time partitions, which, with side information, reduce the number of time samples required to faithfully reconstruct the original spectrum. The common thread that runs through all synthesis/analysis encoders used for speech signals is that the voice signal is not required to "look" like the original signal, but rather, to "sound" like it.

13.4.2.1 Linear Predictive Coding

The adaptive predictors, described in Section 13.3.2, were designed to predict or form good estimates of an input signal. In the adaptive form, the prediction coefficients are recomputed as side information from periodic examination of the input data. Then, the difference between the input and the prediction is transmitted to the receiver to resolve the prediction error. *Linear predictive coders* (LPCs) are the natural extension of N-tap predictive coders. When the filter coefficients are periodically computed with a optimal algorithm, the prediction is so good that there is (essentially) no prediction error information worth transmitting to the receiver. Rather than transmit these low-level prediction errors, the LPC system transmits the filter coefficients and the voiced/unvoiced excitation decision for the model. Thus, the only data sent in LPC is the high-quality side information of the classic adaptive algorithm. An LPC model for voice synthesis is shown in Figure 13.31. The LPC encoders represent the core of hybrid coders that embed the coder and excitation generator in an analysis-by-synthesis loop that searches through excitation options to minimize the difference between the input signal and synthesized signal. The Regular-Pulse Excited (RPE), and the Codebook-Excited Linear Predictive (CELP) encoder are used in cellular phones to obtain toll-quality speech at

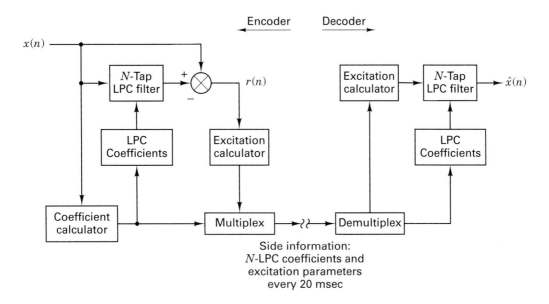

Encoder   Decoder

$x(n)$

N-Tap
LPC filter

+

$r(n)$

−

LPC
Coefficients

Excitation
calculator

Excitation
calculator

N-Tap
LPC filter

$\hat{x}(n)$

LPC
Coefficients

Coefficient
calculator

Multiplex

Demultiplex

Side information:
N-LPC coefficients and
excitation parameters
every 20 msec

Figure 13.31   Block diagram: Linear predictive coefficient (LPC) speech modeling.

data rates below 9.6 kbits/s. The Global System for Mobile (GSM) Communications uses RPE compression, while mobile phone systems, designed to meet code division multiple access (CDMA) standard IS-95, use a variant of CELP. Additional material on CELP is presented in section 13.8.1.3.

An application of this model employing a 12-tap LPC speech synthesizer is found in children's speaking games. Further treatment of LPC techniques used for speech are covered in Section 13.8.1.

13.5 BLOCK CODING

The quantizers we have examined up to now have been *scalar* in nature, since they form a *single output sample* based on the present input sample and (possibly) the N precious output samples. Block coders, on the other hand, form a *vector of output samples* based on the present and the N previous input samples. The *coding gain* of a waveform coder is the ratio of the input SNR to the output SNR. When the noise variances of the input and output are equal, this gain is simply the ratio of input-to-output signal variances. The ratio converts directly to 6 dB per bit for the difference between the number of input bits per sample and the average number of output bits per sample. Block coders can achieve impressive coding gains. On average, they can represent sequences quantized to 8 bits with only 1 or 2 bits per sample [8]. Block-coding techniques are varied, but a common thread that runs through block-coding techniques is the mapping of an input sequence to an alternative coordinate system. This mapping may be to a subspace of a larger space, so that the mapping may not be reversible [8]. Alternatively, a data-dependent editing scheme may be used to identify the subspace of the mapping from which the quantized data are extracted. Block-coding techniques are often classified by their mapping techniques, which include, for example, vector quantizers, various orthogonal transform coders, and channelized coders, such as subband coders. Block coders are further described by their algorithmic structures, such as codebook, tree, trellis, discrete Fourier transform, discrete cosine transform, discrete Walsh-Hadamard transform, discrete Karhunen-Loeve transform, and quadrature mirror filter-bank coders. We now examine examples of the various block-coding schemes.

13.5.1 Vector Quantizing

Vector quantizers represent an extension of conventional scalar quantization. In scalar quantization, a scalar value is selected from a finite list of possible values to represent an input sample. The value is selected to be close (in some sense) to the sample it is representing. The fidelity measures are various weighted mean-square measures that preserve our intuitive concept of distance in terms of ordinary vector lengths. By extension, in vector quantization, a vector is selected from a finite list of possible vectors to represent an input vector of samples. The selected vector is chosen to be close (in some sense) to the vector it is representing.

Each input vector can be visualized as a point in an N-dimensional space. The quantizer is defined by a partition of this space into a set of nonoverlapping volumes [14]. These volumes are called intervals, polygons, and polytopes, respectively, for one-, two-, and N-dimensional vector spaces. The task of the vector quantizer is to determine the volume in which an input vector is located; The output of the optimal quantizer is the vector identifying the centroid of that volume. As in the one-dimensional quantizer, the mean-square error is a function of the boundary locations for the partition and the multidimensional pdf of the input vector.

The description of a vector quantizer can be cast as two distinct tasks. The first is the code-design task, which deals with the problem of performing the multidimensional volume quantization (or partition) and selecting the allowable output sequences. The second task is that of using the code, and deals with searching for the particular volume with this partition that corresponds (according to some fidelity criterion) to the best description of the source. The form of the algorithm selected to control the complexity of encoding and decoding may couple the two tasks—the partition and the search. The standard vector coding methods are codebook-, tree-, and trellis-coding algorithms [15, 16].

13.5.1.1 Codebook, Tree, and Trellis Coders

The codebook coders are essentially table look-up algorithms; A list of candidate patterns (codewords) is stored in the codebook memory. Each pattern is identified by an address or pointer index. The coding routine searches through the list of patterns for the one that is closest to the input pattern and transmits to the receiver the address where that pattern can be found in its codebook. The tree and trellis coders are sequential coders. As such, the allowable codewords of the code cannot be selected independently but must exhibit a node-steering structure. This is similar to the structure of the sequential error-detection-and-correction algorithms, which traverse the branches of a graph while forming the branch weight approximation to the input sequence. (See Section 6.5.1.) A tree graph suffers from exponential memory growth as the dimension or depth of the tree increases. The trellis graph reduces the dimensionality problem by the simultaneous tracking of contender paths with an associated path-weight metric called *intensity,* with the use of a finite state trellis. (See Section 6.3.3.)

13.5.1.2 Code Population

The code vectors stored in the codebook, tree, or trellis are the likely or typical vectors. The first task, that of code design, in which the likely code vectors are identified, is called *populating* the code. The methods of determining the code population are classically *deterministic, stochastic,* and *iterative.* The deterministic population is a list of preassigned possible outputs based on a simple suboptimal or user-perception fidelity criterion or based on a simple decoding algorithm. An example of the former is the coding of the samples in 3-space of the red, green, and blue (RGB) components of a color TV signal. The eye does not have the same resolution to each color and it would appear that the coding could be applied independently to each color to reflect this different sensitivity. The resulting quantizing volumes would be rectangular parallelepipeds. The problem with independent quantizing is that we do not see images in this coordinate system; rather, we see images in the coordinates of luminance, hue, and saturation. A black-and-white photo, for example, uses only the luminance coordinate. Thus, quantizing RGB coordinates independently does not result in the smallest amount of user-perceived distortion for a given number of bits. To obtain improved distortion performance, the RGB quantizer should partition its space into regions that reflect the partitions in the alternate space. Alternatively, the quantization could be performed independently in the alternative space by the use of transform coding, treated in Section 13.6. Deterministic coding is the easiest to implement but leads to the smallest coding gain (smallest reduction in bit rate for a given SNR).

The stochastic population would be chosen based on an assumed underlying pdf of the input samples. Iterative solutions to the optimal partitions exist and can be determined for any assumed pdf. The overall samples are modeled by the assumed pdf. In the absence of an underlying pdf, iterative techniques based on a large population of training sequences can be used to form the partition and the output population. Training sequences may involve tens of thousands of representative input samples.

13.5.1.3 Searching

Given an input vector and a populated codebook, tree, or trellis, the coder algorithm must conduct a search to determine the best matching contender vector. An exhaustive search over all possible contenders will assure the best match. Coder performance improves for larger dimensional spaces, but so does complexity. An exhaustive search over a large dimension may be prohibitively time consuming. An alternative is to conduct a nonexhaustive, suboptimal search scheme, with acceptably small degradations form the optimal path. Memory requirements and computational complexity often are a driving consideration in the selection of search algorithms. Examples of search algorithms include single-path (best leaving branch) algorithms, multiple-path algorithms, and binary (successive approximation) codebook algorithms. Most of the search algorithms attempt to identify and discard unlikely patterns without having to test the entire pattern.

13.6 TRANSFORM CODING

In Section 13.5.1 we examined vector quantizers in terms of a set of likely patterns and techniques to determine the one pattern in the set closest to the input pattern. One measure of goodness of approximation is the weighted mean-square error of the form

$$d(\mathbf{X}, \hat{\mathbf{X}}) = (\mathbf{X} - \hat{\mathbf{X}})\mathbf{B}(\mathbf{X})(\mathbf{X} - \hat{\mathbf{X}})^T \qquad (13.76)$$

where $\mathbf{B}(\mathbf{X})$ is a weight matrix and $\mathbf{X}^T$ is the transpose of the vector $\mathbf{X}$. The minimization may be computationally simpler if the weighting matrix is a diagonal matrix. A diagonal weighting matrix implies a decoupled (or uncorrelated) coordinate set so that the error minimization due to quantization can be performed independently over each coordinate.

Thus, transform coding entails the following set of operations, which are shown in Figure 13.32:

1. An invertible transform is applied to the input vector.
2. The coefficients of the transform are quantized.
3. The quantized coefficients are transmitted and received.
4. The transform is inverted with quantized coefficients.

Note that the transform does not perform any source encoding, it merely allows for a more convenient description of the signal vector to permit ease of source encoding. The task of the transform is to map a correlated input sequence into a different coordinate system in which the coordinates have reduced correlation. Recall that this is precisely the task performed by predictive coders. The source encoding occurs with the bit assignment to the various coefficients of the transform. As part of this assignment, the coefficients may be partitioned into subsets that are quantized with a different number of bits but *not* with different quantizing step sizes. This assignment reflects the dynamic range (variance) of each coefficient and may be weighted by a measure that reflects the importance, relative to the human perception [17], of the basis element carried by each coefficient. A subset of the coefficients, for instance, may be set to zero amplitude, or may be quantized with 1 or 2 bits.

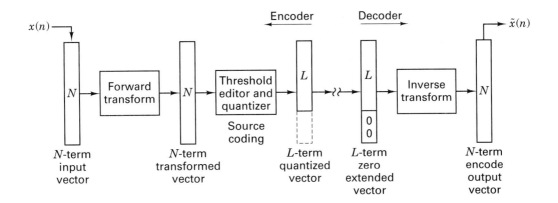

Figure 13.32 Block diagram: Transform coding.

The transformation can be chosen to be independent of the data vector. Examples of such transforms are the discrete Fourier transform (DFT), discrete Walsh-Hadamard transform (DWHT), discrete cosine transform (DCT), and the discrete slant transform (DST). The transformation can also be derived from the data vector, as is done in the discrete Karhunen-Loeve transform (DKLT), sometimes called the *principal component transform* (PCT) [18]. The data-independent transforms are easiest to implement but do not perform as well as the data-dependent transforms. Often, the attraction of computational simplicity is sufficient justification for using the data-independent transformations. The coding gain penalty for using a good suboptimal transformation is small (typically less than 2 dB), and the degradation is usually cited when demonstrating performance characteristics.

13.6.1 Quantization for Transform Coding

Transform coders are called spectral encoders because the signal is described in terms of a spectral decomposition (in a selected basis set). The spectral terms are computed for nonoverlapped successive blocks of input data. Thus, the output of a transform coder can be viewed as a set of time series, one series for each spectral term. The variance of each series can be determined and each can be quantized with a different number of bits. By permitting independent quantization of each transform coefficient, we have the option of allocating a fixed number of bits among the transform coefficients to obtain a minimum quantizing error.

13.6.2 Subband Coding

The transform coders of Section 13.6 were described as a partition of an input signal into a collection of slowly varying time series, each of which is associated with a particular basis vector of the transform. The spectral terms, the inner product of the data with the basis vectors, are computed by a set of inner products. The set of inner products can be computed by a set of *finite impulse response* (FIR) filters [19]. With this perspective, the transform coder can be considered to be performing a channelization of the input data. By extension, a *subband coder,* which performs a spectral channelization by a bank of contiguous narrowband filters, can be considered a special case of a transform coder. (A typical subband coder is shown in Figure 13.33.)

Casting the spectral decomposition of the data as a filtering task affords us the option of forming a class of custom basis sets (i.e., spectral filters)—in particular, basis sets that reflect out user-perception preferences and our source models. For example, the quantizing noise generated in a band with large variance will be confined to that band, not spilling into a nearby band with low variance and hence susceptible to low-level signals being masked by noise. We also have the option of forming filters with equal or unequal bandwidths. (See Figure 13.33.) Thus, we can independently assign to each subband the sample rate appropriate to its bandwidth and a number of quantizing bits appropriate to its variance. By comparison, in conventional transform coding, each basis vector amplitude is sampled at the same rate.

The subband coder can be designed as a transmultiplexer. Here the input signal is considered to be composed of a number of basis functions modeled as independent narrow-bandwidth subchannels. The encoder separates or channelizes the input signal into a set of low-data-rate, time-division multiplexed (TDM) channels.

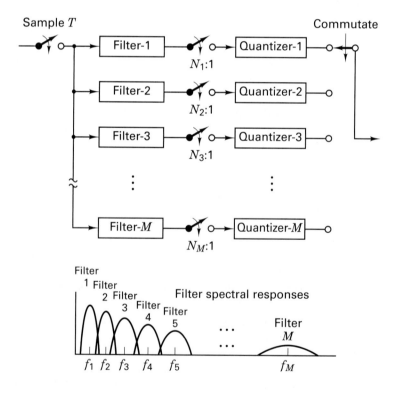

Figure 13.33 Subband coding performed by channelized spectral decomposition.

After quantization and transmission, the decoder reverses the filtering and resampling process, converting the TDM channels back to the original signal. In the classic approach to this process, one can implement a bank of narrow-band filters with the steps of heterodyning, low-pass filtering, and downsampling (often called *decimation*). This filtering operation reduces the input bandwidth to the selected channel bandwidth and resamples the signal to the lowest rate that avoids aliasing of the reduced bandwidth channelized data. At the receiver, the reverse process is performed. The channelized signals are passed through interpolating filters to increase their sample rate to the desired output sample rate are heterodyned back to their proper spectral position, and they are combined to form the original composite signal. For speech encoding—or, more generally, for signals that are related to mechanical resonance—filter banks with nonequal center frequencies and nonequal bandwidths are desirable. Such filters are called constant-Q (or proportional) filter banks. These filters have logarithmically spaced center frequencies with bandwidths proportional to the center frequencies. This proportional spacing appears as uniform spacing and bandwidth when viewed on a log scale, and it reflects the spectral properties of many physical acoustic sources.

13.7 SOURCE CODING FOR DIGITAL DATA

Coding to reduce the redundancy of a data source entails the selection of an (usually) efficient binary representation of that source. Often this requires the substitution of one binary representation of the source symbols with an alternative representation. The substitution is usually temporary and is performed to achieve an economy of storage or transmission of the discrete source symbols. The binary code assigned to each source symbol must satisfy certain constraints to permit reversal of the substitution. In addition, the code may be further constrained by system considerations, such as memory limits or implementation ease.

We are so used to assigning binary codes to represent source symbols that we may lose sight of the arbitrariness of this assignment. The most common example of this is the binary assignments to the cardinal numbers (let's not even consider the negative numbers). We can count in straight binary, binary-coded octal, binary-coded decimal, binary-coded hexadecimal, two-out-of-five decimal, excess-three decimal, and so forth. In this example, ease of computation, error detection, ease of display, or convenience of coding are the considerations for selecting the assignment. For the specific task of data compression, the primary goal is to *reduce the number of bits.*

Finite discrete sources are characterized by a set of distinct symbols. $X(n)$, where $n = 1, 2, \ldots, N$) is called the *source alphabet,* and where n is a data index. A complete characterization requires the probability of each symbol and the joint probabilities of the symbols taken two at a time, three at a time, and so on. The symbols may represent a two-level (binary) source, such as the black-and-white levels of a facsimile scan, or a many-symbol source, such as the 40 common characters of Sanskrit. Another common many-symbol alphabet is the keyboard of a computer terminal. These nonbinary symbols are mapped, via a dictionary called a character code, to a binary alphabet description. (See Figure 2.3 for the ASCII code and Figure 2.4 for the EBCDIC code.) The standard character codes are of fixed length, such as 5, 6, or 7 bits. The length is usually chosen so that there are enough binary characters to assign a unique binary sequence to each input alphabet character. These may include the upper and lowercase letters of the alphabet, numerals, punctuation, special characters, and control characters, such as backspace, return, and so forth. Fixed-length codes have the property that character boundaries are separated by a fixed bit count. This allows the conversion of a serial data stream to a parallel data stream by a simple bit counter.

Two-code standards may define the same symbol in different ways. For example, the ASCII (7-bit) code has enough bits to assign different binary sequences to the upper and lowercase versions of each letter. On the other hand, the *Baudot* (5-bit) code, with only 32 binary sequences, cannot do the same. To account for the full character set, the Baudot code defines two control characters, called *letter shift* (*LS*) and *figure shift* (*FS*), to be used as prefixes. When used, these control characters reassign the binary-to-symbol mapping. This works very much like the *shift key* on a typewriter—the shift key reassigns a completely new character set to the keyboard. In a similar fashion, the keyboards on some calculators have two prefix character keys, so that each keystroke can have three possible meanings. Also, some word processor instruction codes use double- and triple-stoke command functions. In a very real sense, these two- and three-word instructions represent a variable-length code assignment. These longer code words are assigned to characters (or instructions) that do not occur as often as those assigned single codewords. What we receive in exchange for using the occasional longer words in more efficient storage (smaller keyboard) or transmission of the source.

Data compression codes are often variable-length codes. Intuitively, we would expect the length of a binary sequence assigned to each alphabet symbol to be inversely related to the probability of that symbol. After all, if a symbol occurs with high probability, it contains little information and should not be assigned much of the system resources. In a similar manner, it would not seem unreasonable to find that when all symbols are equally likely, the code should be of fixed length. Perhaps the best known variable-length code is the Morse code. Samuel Morse counted the quantity of letters in a printer's font drawer to determine the relative frequency of letters in normal text. The variable-length code assignment reflects this relative frequency.

A significant amount of *data compression* can be realized when there is a wide difference in the probabilities of the symbols. To achieve this compression, there must also be a sufficiently large number of symbols. Sometimes, in order to have a large enough set of symbols, we form a new set of symbols derived from the original set, called an *extension code*. We have already seen this trick in Example 13.3, and we will examine the general technique in the next section.

13.7.1 Properties of Codes

Earlier we alluded to properties that a code must satisfy for it to be useful. Some of these properties are obvious, and some are not. It is worth listing and demonstrating the *desired properties*. Consider the following three-symbol alphabet with the probability assignments shown:

X_i	$P(X_i)$
a	0.73
b	0.25
c	0.02

Accompanying the input alphabet are the following six binary code assignments, where the rightmost bit is the earliest bit:

Symbol	Code 1	Code 2	Code 3	Code 4	Code 5	Code 6
a	00	00	0	1	1	1
b	00	01	1	10	00	01
c	11	10	11	100	01	11

Scan these for a moment and try to determine which codes are practical.

Uniquely Decodable Property. Uniquely decodable codes are those that allow us to invert the mapping to the original symbol alphabet. Obviously, code 1 in the preceding example is not uniquely decodable because the symbols a and b are assigned the same binary sequence. Thus, the first requirement of a useful code is that each symbol be assigned a unique binary sequence. By this condition, all the other codes appear satisfactory until we examine codes 3 and 6 carefully. These codes indeed have unique binary sequences assigned to each symbol. The problem occurs when these code sequences are strung together. For instance, try to decode the binary pattern 1 0 1 1 1 in code 3. Is it b, a, b, b, b or b, a, b, c or b, a, c, b? Trying to decode the same sequence in code 6 gives similar difficulties. These codes are not uniquely decodable, even though the individual characters have unique code assignments.

Prefix-Free Property. A sufficient (but not necessary) condition to assure that a code is uniquely decodeable is that no codeword be the prefix of any other code word. Codes that satisfy this condition are called prefix-free codes. Note that code 4 is not prefix-free, but it is uniquely decodable. Prefix-free codes also have the property that they are instantaneously decodable. Code 4 has a property that may be undesirable; it is not instantaneously decodable. An instantaneously decodable code is one for which the boundary of the present codeword can be identified by the end of the present codeword, rather than by the beginning of the next codeword. For instance, in transmitting the symbol b with the binary sequence 1 0 in code 4, the receiver cannot determine if this is the whole codeword for symbol b or the partial codeword for symbol c. By contrast, codes 2 and 5 are prefix free.

13.7.1.1 Code Length and Source Entropy

At the beginning of the chapter, we described the formal concept of information content and source entropy. We identified the self-information $I(X_n)$, in bits, about the symbol X_n denoted as $I(X_n) = \log[1/P(X_n)]$. From the perspective that information resolves uncertainty we recognize that the information content of a symbol goes to zero as the probability of that symbol goes to unity. We also defined the *entropy* of a finite discrete source as the average information of that source. From the perspective that information resolves uncertainty, the entropy is the average amount of uncertainty resolved per use of the alphabet. It also represents the average number of bits per symbol required to describe the source. In this sense, it is also the lower bound of what can be achieved with some variable length data compression codes. A number of considerations may prevent an actual code from achieving the entropy bound of the input alphabet. These include uncertainty in probability assignment and buffering constraints. The average bit length achieved by a given code is denoted by $\bar{n}$. This average length is computed as the sum of the binary code lengths n_i weighted by the probability of that code symbol $P(Xi)$.

$$\bar{n} = \sum_i n_i P(X_i)$$

A great deal in implied about the performance of a variable-length code when we say *average number of bits*. In a variable-length code assignment, some symbols will have code lengths which exceed the average length, while some will have code lengths which are smaller than the average. It may occur that a long pattern of symbols with long codewords is delivered to the coder. The short-term bit rate required to transmit these symbols will exceed the average bit rate of the code. If a channel is expecting data at the average rate, the local excess rate must be buffered in a memory. By the same token, a long pattern of symbols with short code words may be delivered to the coder. The short-term bit rate required to transmit these symbols will fall short of the average rate of the code. Here the channel will find itself waiting for bits that are not to be had. For this reason, data *buffering* is required to smooth the local statistical variations associated with the input alphabet.

The last caveat is that variable-length codes are designed to operate with a specified list of symbols and probabilities. If the data presented to the coder have a significantly different list of probabilities, the coder buffers may not be able to support the mismatch and buffer underflow or overflow will occur.

13.7.2 Huffman Code

The Huffman code [20] is a prefix-free, variable-length code that can achieve the shortest average code length $\bar{n}$ for a given input alphabet. The shortest average code length for a particular alphabet may be significantly greater than the entropy of the source alphabet. This inability to exploit the promised data compression is related to the alphabet, not to the coding technique. Often the alphabet can be modified to form an extension code, and the same coding technique is then reapplied to achieve better compression performance. Compression performance is measured by the *compression ratio*. This measure is equal to the ratio of the average number of bits per sample before compression to the average number of bits per sample after compression.

The Huffman coding procedure can be applied for transforming between any two alphabets. We will demonstrate the application of the procedure between an arbitrary input alphabet and a binary output alphabet. The Huffman code is generated as part of a tree-forming process. The process starts by listing the input alphabet symbols, along with their probabilities (or relative frequencies), in descending order of occurrence. These tabular entries correspond to the branch ends of a tree, as shown in Figure 13.34. Each branch is assigned a branch weight equal to the probability of that branch. The process now forms the tree that supports these branches. The two entries with the lowest relative frequency are merged (at a branch node) to form a new branch with their composite probability. After every merging, the new branch and the remaining branches are reordered (if necessary) to assure that the reduced table preserves the descending probability of occurrence. We call this reordering *bubbling* [21]. During the rearrangement after each merging, the new branch rises through the table until it can rise no further. Thus, if we form a branch with a weight of 0.2 and during the bubbling process find two other branches already with the 0.2 weight, the new branch is bubbled to the top of the 0.2 group, as opposed to simply joining it. The bubbling to the top of the group results in a code with reduced code length variance but otherwise a code with the same average length as that obtained by simply joining the group. This reduced code length variance lowers the chance of buffer overflow.

As an example of this part of the code process, we will apply the Huffman procedure to the input alphabet shown in Figure 13.34. The tabulated alphabet and the associated probabilities are shown on the figure. After forming the tree each branch node is labeled with a binary 1/0 decision to distinguish the two branches. The labeling is arbitrary, but for consistency, at each node we will label the branch going up with a "1" and the branch going down with a "0". After labeling the branch nodes, we trace the tree path from the base of the tree (far right) to each output branch (far left). The path contains the binary sequence to reach that branch. In the following table, we have listed at each end branch the path sequence corresponding to each path where $i = 1, \ldots, 6$:

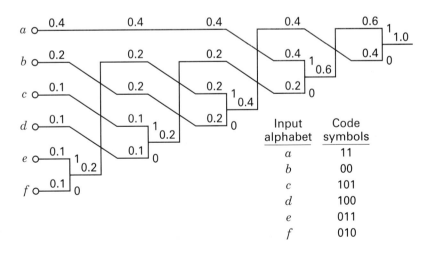

Figure 13.34 Huffman coding tree for a six-character set.

X_i	$P(X_i)$	Code	n_i	$n_i P(X_i)$
a	0.4	11	2	0.8
b	0.2	00	2	0.4
c	0.1	101	3	0.3
d	0.1	100	3	0.3
e	0.1	011	3	0.3
f	0.1	010	3	0.3
				$\bar{n} = 2.4$

We find that the average code length $\bar{n}$ for this alphabet *is* 2.4 bits per character. It does not mean that we have to find a way to transmit a noninteger number of bits. Rather, it means that, on average, 240 bits will have to be moved through the communication channel when transmitting 100 input symbols. For comparison, a fixed-length code required to span the six-character input alphabet would be of length 3 bits, and the entropy of the input alphabet, using Equation (13.2), is 2.32 bits. Thus, this code offers a compression ratio of 1.25 (3.0/2.4) and achieves 96.7% (2.32/2.40) of the possible compression ratio. As another example, one for which we can demonstrate the use of code extension, let us examine the three-character alphabet presented in Section 13.6.1:

X_j	$P(X_i)$
a	0.73
b	0.25
c	0.02

The Huffman code tree for this alphabet is shown in Figure 13.35, and the details are tabulated as

X_i	$P(X_i)$	Code	n_i	$n_j P(X_i)$
a	0.73	1	1	0.73
b	0.27	01	2	0.54
c	0.02	00	2	0.04
				$\bar{n} = 1.31$

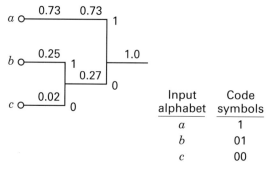

Input alphabet	Code symbols
a	1
b	01
c	00

Figure 13.35 Huffman coding tree for three-character set.

where $i = 1, \ldots, 3$. The average code length for this Huffman code is 1.31 bits; it would be 2 bits for a fixed-length code. The compression ratio for this code is 1.53. Again, using Equation (13.2), the entropy for the alphabet is 0.9443 bit, so that the efficiency $(0.944/1.31 = 72\%)$ of the code is significantly smaller than for the preceding example.

To improve coding efficiency or to achieve greater compression gain, we have to redefine the source alphabet. A larger source alphabet holds the promise of increased variability, which is one requirement to realize a reduction in average code length, and an increased number of tree branches for assigning the variable-length code. We do this by selecting characters two at a time from the source alphabet to be new characters in the extension alphabet. If we assume that the symbols are independent, the probability of each new element is the product of the individual probabilities. The extension alphabet is

X_i	$P(X_i)$	Code	n_i	$n_i P(X_i)$
aa	0.5329	1	1	0.5329
ab	0.1825	00	2	0.3650
ba	0.1825	011	3	0.5475
bb	0.0625	0101	4	0.2500
ac	0.0146	01000	5	0.0730
ca	0.0146	010011	6	0.0876
bc	0.0050	0100100	7	0.0350
cb	0.0050	01001011	8	0.0400
cc	0.0002	01001010	8	0.0016

$$\bar{n} = 1.9326 \text{ bits/two symbols}$$
$$= 0.9663 \text{ bit/symbol}$$

where $i = 1, \ldots, 9$, and the code sequence for each X_i has been found by the use of the Huffman procedure above. The compression ratio of this extension code is 2.07 and the coding efficiency is 97.7%.

Extension codes offer a very powerful technique to include the effects of non-independent symbol sets. For example, in English text, adjacent letters are highly correlated. Very common pairs include

th	re	in
sh	he	e_
de	ed	s_
ng	at	r_
te	es	d_

where the underscore represents a space. Similarly, common English three-tuples include

the	and	for
ing	ion	ess

Thus, rather than perform Huffman coding on the individual letters, it is more efficient to extend the alphabet to include all 1-tuples plus common 2-tuples and 3-tuples, and then to perform the coding on the extension code.

13.7   Source Coding for Digital Data

865

13.7.3 Run-Length Codes

In many applications, lengthy runs of specific symbols characterize a sequence of symbols to be transmitted or stored. Rather than code each symbol of a lengthy run, it makes sense to describe the run with an efficient substitution code. As an example, runs of spaces (the most common symbol in text) are encoded in many communication protocols by a control character followed by the character count. The IBM 3780 BISYNC protocol has an option to replace runs of spaces with an "IGS" character (if EBCDIC; or "GS" if ASCII), followed by a count of 2 to 63. Longer runs are partitioned into successive runs of 63 characters.

The run-length substitution coding can be applied to the original symbol alphabet or the binary representation of that alphabet. Run-length coding is particularly attractive for binary alphabets derived from specific sources. The most important commercial example is facsimile coding, used for transmitting documents by instant electronic mail [22].

13.7.3.1 Huffman Coding for Facsimile Transmission

Facsimile transmission is the process of transmitting a two-dimensional image as a sequence of successive line scans. The most common images are, in fact, documents containing text and figures. The position of the scan lines and the position along a scan line are quantized into spatial locations that define a two dimensional grid of picture elements, called *pixels.* The standard CCITT document is defined to be of width 8.27 in. (20.7 cm) and to have length 11.7 in. (29.2 cm), almost 8.5 in. by 11.0 in. The spatial quantization for normal resolution is 1728 pixels/line and 1188 lines/document. The standard also defines a high-resolution quantization with the same 1728 pixels/line but with 2376 lines/document. The total number of individual pixels for a normal-resolution facsimile transmission is 2,052,864, and it is doubled for high resolution. For comparison, the number of pixels in NTSC (National Television Standards Committee) standard commercial television is 480×640 or 307,200. Thus, facsimile has 6.7 or 13.4 times the resolution of a standard TV image.

The relative brightness or darkness of the scanned image at each position in the scan is quantized into two levels: B (for black) and W (for white). Thus, the signal observed during a scan line is a two-level pattern representing the B and W image intensity under the scan. It is easy to see that a horizontal scan line across this sheet of paper will exhibit a pattern consisting of long runs of B and W levels. The standard CCITT run-length coding scheme to compress the run of B and W levels is based on a modified variable-length Huffman code; it is listed in Table 13.1. Two types of patterns are identified, runs of W and runs of B. Each run length is described by a *partitioned codeword.* The first partition, called the *makeup codeword* or most significant bits (MSB), identifies runs with lengths that are multiples of 64. The second partition, called the *terminating codeword* of least significant bits (LSB), identifies the length of the remaining run. Each run of B (or W) of length from 0 through 63 is assigned a unique Huffman code word, as is each run of length $64 \times K$, where $K = 1, 2, \ldots, 27$. A unique END OF LINE (EOL) is also defined in

TABLE 13.1 Modified Huffman Code for CCITT Facsimile Standard

Run Length	White	Black	Run Length	White	Black
			Makeup Code Words		
64	11011	0000001111	960	011010100	0000001110011
128	10010	000011001000	1024	011010101	0000001110100
192	010111	000011001001	1088	011010110	0000001110101
256	0110111	000001011011	1152	011010111	0000001110110
320	00110110	000000110011	1216	011011000	0000001110111
384	00110111	000000110100	1280	011011001	0000001010010
448	01100100	000000110101	1344	011011010	0000001010011
512	01100101	0000001101100	1408	011011011	0000001010100
576	01101000	0000001101101	1472	010011000	0000001010101
640	01100111	0000001001010	1536	010011001	0000001011010
704	011001100	0000001001011	1600	010011010	0000001011011
768	011001101	0000001001100	1664	011000	0000001100100
832	011010010	0000001001101	1728	010011011	0000001100101
896	011010011	0000001110010	EOL	000000000001	000000000001

Run Length	White	Black	Run Length	White	Black
			Terminating Codewords		
0	00110101	000110111	32	00011011	000001101010
1	000111	010	33	00010010	000001101011
2	0111	11	34	00010011	000011010010
3	1000	10	35	00010100	000011010011
4	1011	011	36	00010101	000011010100
5	1100	0011	37	00010110	000011010101
6	1110	0010	38	00010111	000011010110
7	1111	00011	39	00101000	000011010111
8	10011	000101	40	00101001	000001101100
9	10100	000100	41	00101010	000001101101
10	00111	0000100	42	00101011	000011011010
11	01000	0000101	43	00101100	000011011011
12	001000	0000111	44	00101101	000001010100
13	000011	00000100	45	00000100	000001010101
14	110100	00000111	46	00000101	000001010110
15	110101	000011000	47	00001010	000001010111
16	101010	0000010111	48	00001011	000001100100
17	101011	0000011000	49	01010010	000001100101
18	0100111	0000001000	50	01010011	000001010010
19	0001100	00001100111	51	01010100	000001010011
20	0001000	00001101000	52	01010101	000001000100
21	0010111	00001101100	53	00100100	000000110111
22	0000011	00000110111	54	00100101	000000111000
23	0000100	00000101000	55	01011000	000000100111
24	0101000	00000010111	56	01011001	000000101000
25	0101011	00000011000	57	01011010	000001011000
26	0010011	000011001010	58	01011011	000001011001
27	0100100	000011001011	59	01001010	000000101011
28	0011000	000011001100	60	01001011	000000101100
29	00000010	000011001101	61	00110010	000001011010
30	00000011	000001101000	62	00110011	000001100110
31	00011010	000001101001	63	00110100	000001100111

the code, which indicates that no black pixels follow; hence, the next line should be started, which is akin to a carriage return on a typewriter [23].

Example 13.8   Run-Length Code

Use the modified Huffman code to compress the line

$$200 \text{ W, } 10 \text{ B, } 10 \text{ W, } 84 \text{ B, } 1424 \text{ W}$$

consisting of 1728 pixel elements.

Solution

Using Table 13.1, we determine the coding for this pattern to be (the spaces are for the ease of reading)

010111	10011	0000100	00111	0000001111	00001101000	000000000001
192W	8W	10B	10W	64B	20B	EOL

Only 56 bits are required to send this line containing a sequence of 1728 bits.

13.7.3.2 Lempel-Ziv (ZIP) Codes

A major difficulty in using the Huffman code is that the symbol probabilities must be known or estimated, and both the encoder and decoder must know the coding tree. If a tree is constructed from an unusual alphabet at the encoder, the channel connecting the coder and encoder must also deliver the coding tree as a header for the compressed file. This overhead would reduce the compression efficiency realized by building and applying the tree to the source alphabet. The Lempel-Ziv algorithm and its numerous variants use the text itself to iteratively construct a parsed sequence of variable length codewords that form a code dictionary.

The code assumes that a dictionary exists containing already-coded segments of a sequence of alphabet symbols. Data is encoded by looking through the existing dictionary for a match to the next short segment in the sequence being coded. If a match is found, the code operates under the following philosophy: since the receiver already has this code segment in its memory, there is no need to resend it, we only have to identify the address to retrieve the segment. The code references the location of the segment sequence and then appends the next symbol in the sequence to form a new entry in the code dictionary. The code starts with an empty dictionary, so the first elements are entries that do not refer to earlier entries. In one form of dictionary, we recursively form a running sequence of addresses and the segment of alphabet symbols contained therein. Encoded data is made up of a two-word packet of the form < dictionary address, next data character >, with each new data-character entry in the dictionary formed as a packet containing the address of that character, followed by the next character. An example of this coding technique follows:

Encode the sequence of symbols [a b a a b a b b b b b b b a b b b b b a]

Encoded packets:	<0,a>,	<0,b>,	<1,a>,	<2,a>,	<2,b>,	<5,b>,	<5,a>,	<6,b>,	<4,–>
Address:	1	2	3	4	5	6	7	8	
Content:	a	b	aa	ba	bb	bbb	bba	bbbb	

The initial packet <0,a> shows a zero address because there are no entries in the dictionary yet. In this packet, the character a is the first in the data sequence and it is assigned to address 1. The next packet <0,b> contains the second data character b which had not been in the dictionary yet (hence the address portion is 0); b is assigned to address 2. The next packet <1,a> represents the encoding of the next two data characters aa by calling out address 1 for the first a, and appending to that address the next character a. The pair of data characters aa is assigned to address 3. The next packet <2,a> represents the encoding of the next two data characters ba by calling out address 2 for the character b, and appending to that address the next character a. The pair of data characters ba is assigned to address 4, and so forth. Notice how the run-length coding is accomplished. The eighth packet is made up of address 6, containing three b characters followed by another b. For this example, the encoded data can be described with a three-bit address and a 0/1 to identify the appended character. There is a sequence of 9 symbols in the coded sequence, for a total of 36 bits to encode the data consisting of 20 characters. As in many compression schemes, coding gain is not realized for short sequences, as in this example, and is only seen for long sequences.

In another form of the Lempel-Ziv algorithm, the encoded data is represented as three word packets of the form <number of characters back, length, next character>. Here, the concept of an address is not used; instead, previous data sequences are referenced. Also, recursive references are permitted in the length parameter. This is demonstrated in the following example, presented as entry <1, 7, a>:

Encode the sequence of symbols [a b a a b a b b b b b b b a b b b b b a]

Encoded packets:	<0,0,a>,	<0,0,b>,	<2,1,a>,	<3,2,b>,	<1,7,a>,	<6,5,a>
Content:	a	b	aa	bab	bbbbbbba	bbbbba
Running text:	a	ab	abaa	abaabab	abaababbbbbbbba	whole seq.

Here too the coding gain of the process has not been demonstrated for the short run of data. Variations of the code limit the size of the back-pointer such as 12-bits for a maximum of a 4096 point back reference. This limitation constrains the size of memory required for the dictionary and reduces the likelihood of memory overflow. Variations may also constrain the length of the prefix or phrase defined by the first two arguments <back $n1$, forward $n2$, xxx > to be less than some value such as 16 to limit complexity of back search during encoding. The Lempel-Ziv algorithm exists in many commercial and share-ware forms, which include LZ77, Gzip, LZ78, LZW, and UNIX *compress*.

13.8 EXAMPLES OF SOURCE CODING

Source coding has become an essential subsystem in modern communication systems. High demand for bandwidth and storage capacity has been the motivator, while the integrated-circuit and signal-processing techniques have been the enablers. A secondary enabler for wide acceptance of a process in a communication system is the definition of industry-wide *standards* that permit multiple suppliers to field cost-effective and competitive realizations of the coding process. CCITT standards can be found for source coding or compression algorithms for speech, for audio, for still images, and for moving images. In this section, we will examine a number of standard-based source-coding algorithms to demonstrate the wide dissemination of source coding in communication systems as well as to illustrate typical levels of performance.

13.8.1 Audio Compression

Audio compression has become well entrenched in consumer and professional digital audio products such as the compact disc (CD), digital audio tape (DAT), the mini-disc (MD), the digital compact cassette (DCC), digital versatile disc (DVD), digital audio broadcasting (DAB), and motion picture experts group (MPEG) audio layer 3 (MP3) distribution on the internet. In addition, speech compression for telephony, and in particular, cellular telephony, required to preserve bandwidth and battery life has spawned a large number of speech compression standards. Different algorithms are applied to speech signals and wider bandwidth consumer entertainment signals. Audio and speech compression schemes can be conveniently partitioned into applications reflecting some measure of acceptable quality. The parameters describing this partition follow [24, 25]:

Typical parameters values for three classes of audio signals

	Frequency Range	Sampling Rate	PCM bits/ Sample	PCM bit-rate
Telephone Speech	300–3,400 Hz	8 kHz	8	64 kb/s
Wideband Speech	60–7,000 Hz	16 kHz	14	224 kb/s
Wideband Audio	10–20,000 Hz	48 kHz	16	768 kb/s

13.8.1.1 Adaptive Differential Pulse-Code Modulation

We start this discussion with telephone speech processing. One standard that addresses this area is the CCITT, G. 726 adaptive differential pulse-code modulation (ADPCM) codec. This standard encodes sample by sample, predicting the value of each sample from reconstructed speech of previous samples using an adaptive feedback predictor. It accepts toll-quality 8-bit linear, A-law, or μ-law sampled speech at 64-kbits/sec, and it outputs compressed speech at rates of 16, 24, 32, and 40-kbits/sec. The encoder uses a decoder in its feedback path to analyze and modify algorithm parameters in order to minimize the reconstruction errors. The predictor uses a sixth-order filter to model zeros and a second-order filter to model poles of the input signal source. A block diagram of the encoder is seen in Figure 13.36.

13.8.1.2 Subband-Partitioned, Adaptive Differential Pulse-Code Modulation

The CCITT G.722 is a wideband speech-coding standard. The wideband compression offers significant improvement in quality over that of telephone quality speech and is closer to broadcast quality speech and music signals. This coder uses complementary low-pass and high-pass filters to separate the 7-kHz input bandwidth sampled at 16-kHz into a higher and lower subband, each down-sampled to 8-kHz. Both low-pass and high-pass filters, as well as the resampling operation, are implemented in a digital filter known as a *quadrature mirror filter*. Independent ADPCM encoders process the reduced bandwidth time series from the two filters and form 48-kbit/s and 16-kbit/s data rates, respectively, as the low and high band outputs. These coders are a modified version of CCITT B.721 ADPCM speech coders that use backward prediction filters based on the coded difference signal. Discarding the least significant bit of the prediction filter coefficients permits this encoder to operate at 56 and 48 kbit/s, as well as the nominal 64 kbit/s. The reduced bit rate allows the communication system to assign the unused bits to an auxiliary data stream operating at 8 and 16-kbit/s when the channel supports a fixed output rate of 64 kbit/s. The predictor uses a pole-zero structure with 6-zeros and 2-poles. A block diagram of the wideband 64-kbit/s audio coder is shown in Figure 13.37.

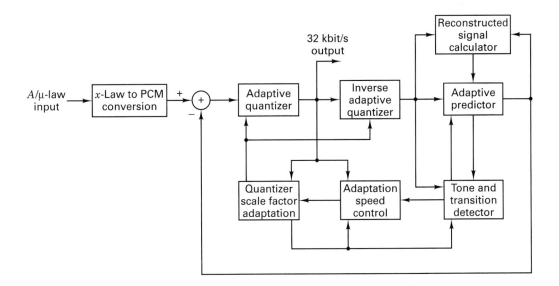

Figure 13.36 Adaptive differential pulse code modulator speech codec (G.726).

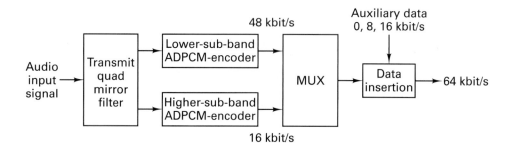

Figure 13.37 Wideband QMF-ADPCM (64-kbit/s) codec (G.722).

13.8.1.3 Codebook-Excited Linear-Predictive Coding (CELP)

Speech coders using linear predictive filters can provide high quality encoded speech at rates above 16 kbits/sec, but they degrade quickly at lower rates. The LPC encoders can be modified to obtain high-quality speech compression at rates on the order of 4.8 to 9.6 kbits/s by recasting the synthesis problem as a two-step procedure called *synthesis by analysis*. In the first step, we form a 10-th order LPC model of the signal that is valid over a short interval—say, every 20 ms. In the second step, we find the waveform that when applied to the LPC model forms an output signal as close as possible to the original signal being synthesized. We accomplish this task by sequentially applying candidate excitation waveforms to the model and comparing each synthesized wave-shape with the original signal and then selecting the one that minimizes the error between the original signal and the output of the driven model.

From the physiology of the speech-generation process we know that the excitation for the speech process often consists of periodic pulses (formed by the vibrating vocal chords). The period P of the periodic pulses is related to the speakers pitch. A one-tap recursive filter is defined by two parameters; P for the number of delay intervals in the feedback path and g, the feedback gain. The impulse response of this filter is a decaying sequence with P zero-value output samples between successive non-zero output samples. The output of this filter generates the periodic excitation signal applied to the input of the LPC model (see Section 13.3.2). The synthesis algorithm has to test possible values of P from a list of candidate values. The two pitch parameters are estimated every 5-milliseconds. The input to the pitch filter is extracted from a table of candidate excitation sequences. The output of the pitch filter in turn drives the LPC model. The table, typically containing 1024 entries, is called a codebook. The codebook is accessed every 2.5 ms. When the best combination of codebook entries and pitch period is determined by exhaustive search, a frame is assembled containing the sequence of pitch parameters, the sequence of codebook addresses, and information about the LPC coefficients.

The encoder must deliver parameters describing the LPC model to the decoder. The LPC filter spectral response is very sensitive to coefficient quantization and as such would have to be represented with an unacceptably large number of bits. The LPC coefficients are transformed to a different set of parameters called *Line Spectral Pairs* [10] that are insensitive to quantization

Systems designed to the IS-95 standard use the following LPC frame format. The frame required to describe 20 ms of data contains 192 bits assigned to represent encoded parameters in the following manner:

10 LPC Coefficients	40 bits
4 Lag and Gain Parameters	40 bits
8 Codebook Addresses	80 bits
Parity, Check bits, and Overhead	32 bits

The overall bit rate for this system is 192 bits per 20 ms, or 9600 bits/s. The bit rate can be reduced if the encoder detects speech pauses.

13.8.1.4 MPEG Layers I, II and III

The International Standards Organizations (ISO) and Motion Picture Experts Group (MPEG) audio coding standard describes audio compression for synchronized audio to accompany the compressed video known as MPEG. It combines features of MUSICAM (Masking pattern adapted Universal Subband Integrated Coding And Multiplexing) and ASPEC (Adaptive Spectral Perceptual Entropy Coding). It consists of three layers (codes) of increasing complexity and improving subjective performance, and it operates with input sampling rates of 32, 44.1 and 48 kHz, and it outputs bit rates per monophonic channel between 32 and 192 kbit/s, or per stereophonic channel between 64 and 384 kbit/s. The standard supports single channel mode, stereo mode, dual channel mode (for bilingual audio programs), and an optional joint stereo mode. In this later mode, the two coders for the left and right channel can support each other by exploiting common statistics between these channels in order to compress the audio bit rate to an even higher degree than is possible in monophonic transmission [26].

The encoder operates in conjunction with a real-time model of the *human spectral perception threshold*. This threshold is a frequency-dependent boundary or threshold that marks sound pressure levels (SPL) below which the human ear cannot detect sounds. This curve, called the *acuity threshold* is generated during a hearing test. The acuity threshold is normally presented as amplitude levels as a function of spectral position, in much the same way as a power-spectrum curve. This threshold is a time-varying function of the short-time power spectral density and is locally peaked in response to high-level tones and tone-like signals (called tonals). The lifting of the threshold due to the presence of strong tonals results in a local masking of spectral components below the new threshold level. Signal spectral components below the threshold level that cannot be heard are declared irrelevant, and they are not encoded in the compression process. Signals above the frequency-dependent threshold are encoded with sufficient fidelity to keep the approximation error below the acuity threshold. This is accomplished by partitioning the spectrum with a set of narrow-band filters and assigning a sufficient number of bits to describe each filter output in proportion to its amplitude above the threshold. Thus, a signal that is 30 dB above threshold in a particular band will be assigned 5-bits for its quantization, for which case the quantization noise falls below the threshold, since the quantization noise-to-signal ratio is reduced by 6 dB per bit. A typical threshold acuity curve is presented in Figure 13.38.

The encoder operates in this manner. The standard 16-bit PCM audio signal is windowed and converted into spectral subband components by a polyphase filter bank consisting of 32 equally spaced bandpass filters. The filter bank is designed with adjacent channel attenuation exceeding 96 dB, a level required to supress perceptual distortion caused by quantization noise. The filtered output signals are downsampled to the Nyquist rate for each bandpass bandwidth. In the decoder, this process is reversed. The sampling rate of each subband filter is increased to that of the original source signal by interpolating the subband signals formed at the bandpass outputs of the synthesis filter bank. Figure 13.39 presents a block diagram of the Layer I and II MPEG Audio Encoder and Decoder.

In Layer III of the MPEG/ISO (MP3) standard, a higher frequency resolution is achieved that closely matches critical frequency-resolution bandwidths of the human auditory process. This enhanced partition is achieved by post-processing the 32 subband signals with an overlapped and windowed 6-point or 18-point modified Discrete Cosine Transform (MDCT). (A short description of the DCT is presented in the next section on image compression.) The resulting number of frequency bands that can be resolved in Layer 3 is 32×18 or 576, which each filter representing a bandwidth of 24,000/576 or 41.67 Hz. Layer III differs from layer I and II by the addition of a modified DCT to the analysis bank, a Huffman coder to the output of the quantizer, and a side information data path.

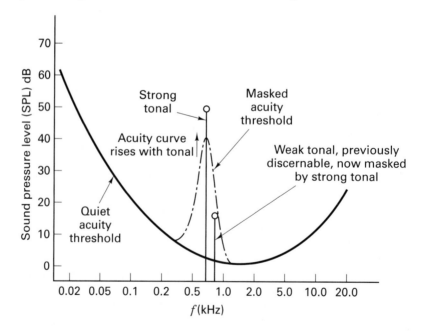

Figure 13.38 Acuity threshold and auditory masking.

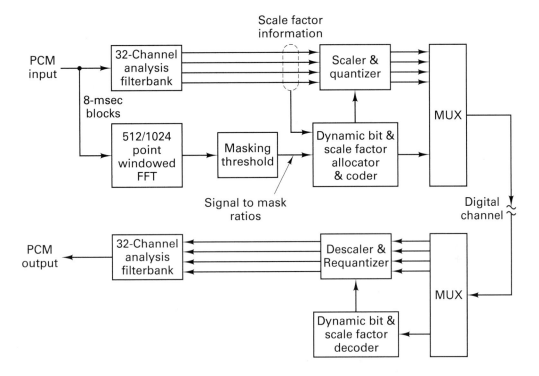

Figure 13.39 Block diagram of ISO/MPEG audio encoder and decoder, layers I and II.

13.8.2 Image Compression

An old saying we have heard many times is that, *A picture is worth a thousand words.* Is this saying true? A 1000 words contain 6000 characters, which, when coded as 7-bit ASCII symbols, require a total of 42,000 bits. What size image (or picture) can we describe with 42,000 bits? If we use a monochrome (i.e. black and white) image with a standard 8-bit grey scale the image would be restricted to 5250 pixels (or picture elements). This image would be of dimension 70 by 75 pixels, and if we assume a medium quality image is formed with 300 pixels per inch, we find that our image is approximately $\frac{1}{4}$ inch by $\frac{1}{4}$ inch. We certainly have need for some image coding.

Let us approach the problem from the opposite direction. How big is an image? Selecting an 8.5 by 11.0 inch sheet of paper containing an image resolved at 300 pixals per inch, we find the image contains $8.5 \times 300 \times 11.0 \times 300$ or 8.4×10^6 picture elements. If the picture is full color with three colors per element, each described by 8-bit words, we find the image contains 2×10^8 bits which is equivalent to 4.8×10^6 6-character ASCII words. Perhaps the old saying should be updated to reflect that *A picture is worth about five million words.* For comparison with other image formats, we note that a single frame of a high definition televison image contains approximately 1.8×10^6 pixels, a standard television image contains approximately 0.33×10^6 pixels, and high-end computer monitors contain 1.2-to-3.1 $\times 10^6$ picture elements.

Technology has presented us with low-cost, high-resolution color printers, scanners, cameras, and monitors, enabling us to capture and present images for commerce and entertainment. The storage and transport of these images rely quite heavily on source coding to reduce the demands on bandwidth and memory. There are numerous standards that have been developed to compress images. In the next section, we examine elements of two primary compression schemes [26, 27].

13.8.2.1 Joint Photographic Experts Group (JPEG)

JPEG is the common name given to the ISO/JPEG international standard 10918-1 or ITU-T Recommendation T.81 standard for "Digital Compression of Continuous-Tone Still Images." JPEG is primarily known as a transform-based lossy compression scheme, and this is the mode that is examined here. Lossy compression permits errors in the signal construction. The error levels are restricted to be below the perception threshold of a human observer. JPEG supports three modes of operation related to the Discrete Cosine Transform (DCT): Sequential DCT, Progressive DCT, and Hierarchical, as well as a lossless mode using differential prediction and entropy coding of the prediction error. The DCT is a numerical transform related to the Discrete Fourier transform (DFT) for obtaining the spectral decomposition of even symmetric sequences. When the input sequence is even symmetric, there is no need for the sine components of the transform; hence, a DCT can replace the FFT.

Let us start with an introduction to the 2-D 8-by-8 DCT. We first comment on the use of the DCT to form a spectral description of an 8-by-8 pixel block. The 2-D DCT is a separable transform that can be written as a double sum over the two dimensions. The separable DCT performs eight 8-point DCTs in each direction; hence, the basic building block is a single 8-point DCT. The first question to address is, Why use a DCT rather than some other transform, such as the DFT? The answer is related to the interaction between the sampling theorem and Fourier transforms. Sampling in one domain induces periodicity in the other. When we sample a time series, its spectrum becomes periodic. On the other hand, when we sample the spectrum of a time series, the time series is periodically extended. We acknowledge this effect by labeling this process the *periodic extension* and by labeling the resulting processing a *periodogram*. The periodic extension of the original data, (shown in Figure 13.40) exhibits discontinuities at the boundaries that limit the rate of spectral decay in the spectrum to $1/f$. We can form the even-extension of the data by reflecting the data about one of its boundaries. When this data is periodically extended, as indicated in Figure 13.40, discontinuities no longer reside in the amplitude of the data but rather in its first derivative, so that the rate of spectral decay increases to $1/f^2$. The faster rate of spectral decay means that there will be a reduced number of significant spectral terms. A second advantage to the DCT is that since the data is even symmetric, its transform is also real and symmetric and hence has no need for the odd symmetric basis terms, the sine functions.

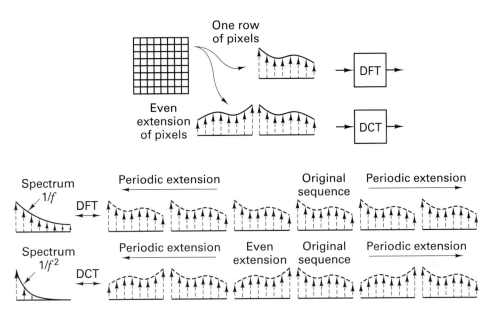

Figure 13.40 Spectral decay & periodic extension of time series by DFT & DCT.

Since the amplitude of an image is highly correlated over short spatial intervals, the DCT of an 8-by-8 block of pixels is dominated by the dc term and further contains relatively few significant terms. A typical set of amplitudes and their DCT is suggested in Figure 13.41. Note that the spectral terms fall at least as fast as $1/f^2$, and that most of the high frequency terms are essentially zero. The spectrum is passed to a quantizer that uses standard quantization tables to assign bits to the spectral terms in proportion to their amplitude, as well as their psychovisual contribution. Different quantization tables are used for the luminance and chrominance components.

To take advantage of the large number of zero-valued entries in the quantized DCT, the spectral addresses of the DCT are scanned in a zigzag pattern, as shown in Figure 13.42. The zigzag pattern assures a long run of zeros. This improves the coding efficiency of the run-length Huffman code describing the spectral samples. Figure 13.43 presents the block diagram of a JPEG encoder. The signal delivered to the encoder is normally presented as raster-scanned and sampled primary additive colors, red, green and blue, (R-G-B). The color planes are transformed to luminance (Y) and chrominance components $0.564 \times (B\text{-}Y)$ (denoted C_B) and $0.713 \times (R\text{-}Y)$ (denoted C_R), using a variation of the color difference transformation developed for color TV. This mapping is

$$\begin{bmatrix} Y \\ C_B \\ C_R \end{bmatrix} = \begin{bmatrix} 0.299 & 0.587 & 0.114 \\ -0.169 & -0.331 & 0.500 \\ 0.500 & -0.419 & -0.081 \end{bmatrix} \begin{bmatrix} R \\ G \\ B \end{bmatrix}$$

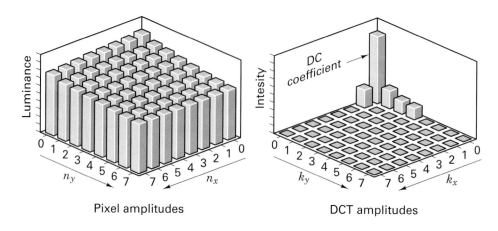

Pixel amplitudes   DCT amplitudes

Figure 13.41   Pixel and DCT amplitudes describing same 8-by-8 block of pixels.

where the Y component is formed to reflect the human-eye sensitivity to the primary colors.

The human eye exhibits a different acuity for the color components of an image than it does for the luminance (black and white) component. This difference in resolution capability is the result of the distribution of the color receptors (rods) and luminance receptors (cones) on the retina. The human eye can resolve 1-inch alternating black and white stripes at 180 ft (1/40-th degree). By comparison, 1-inch color stripes of blue-red or blue-green are not resolvable at distances greater than 40 feet (1/8-th degree). Hence, three-color images require about 1/25-th (1/5-th in each direction) more data than is required to form a black-and-white image. Early photographers knew the eye required very little color detail to color an image, and a lively industry existed in which artists hand-colored black and white photographs and postal-cards. Both analog and digital color TV takes advantage of this acuity difference to deliver the additional color components in a significantly reduced bandwidth. NTSC delivers all three colors in a 0.5 MHz bandwidth, rather than the

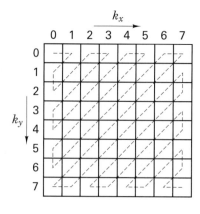

Figure 13.42   Zigzag scan pattern for DCT spectral terms.

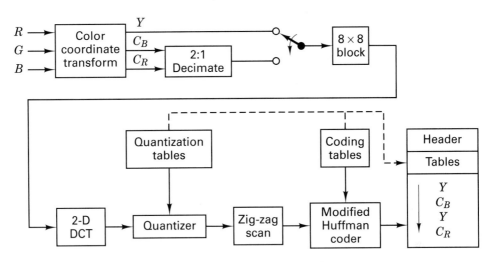

Figure 13.43 Block diagram of JPEG encoder.

4.2-MHz bandwidth required for the luminance component. JPEG similarly takes advantage of the color acuity difference and down samples the color difference components 2-to-1 in the scan direction (x), but not in the successive line (y) direction.

The luminance and down-sampled color difference signals are sequentially presented as 8-by-8 blocks to the 2-D DCT. The outputs of the DCT are quantized by the appropriate table and then zigzag scanned for presentation to the Huffman coder. JPEG uses the Huffman coder to encode the ac-coefficients of the 2D-DCT, but since the DC components are highly correlated over adjacent blocks, uses differential encoding for the DC component. The decoder, of course, reverses these operations to form an image.

13.8.2.1.1 Variations of JPEG Decoding.
During reconstruction of the image, the decoder can operate sequentially, starting in the upper left corner of the image and forming 8-by-8 pixel blocks as they arrive. This is the sequential mode of JPEG. In the progressive coding mode, the image is first assembled in 8-by-8 blocks formed by only the dc component in each block. This is a very quick process that presents a coarse but recognizable preview image, a process often seen on the internet when downloading GIF (Graphic Interchange Format) files which has delivered the only dc components at the beginning of the data transfer. The image is then updated with each 8-by-8 block formed from the dc component and the first two dc components which are the next set of data delivered to the decoder. Finally, the image is updated at full resolution formed by the full set of coefficients associated with each 8-by-8 block.

In Hierarchal Coding, the image is encoded and decoded as overlapped frames. A low resolution, down sampled 4-to-1 in each direction image is encoded using the DCT and quantized coefficient processing to form the first frame. The image formed by this frame is up sampled and compared with a higher resolution,

13.8   Examples of Source Coding   **879**

down-sampled (2-to-1 in each direction) version of the original image and the difference, representing an error in the formation of the image, is again coded as an MPEG image. The two frames formed by the two layers of coding are used to form a composite image that is up-sampled and compared with the original image. The difference between the original image and two lower level resolution reconstructions is formed at the highest resolution and JPEG-encoded once again. This process is useful for delivering images with successively higher quality reconstruction in a way similar to the progressive coding. The difference here is that the additional resolution is available but may not be sent unless requested. An example might be a user scanning an image library and requesting the final quality image after reviewing a set of images. Another application might be the delivery of one level of quality to a display on a personal computer but a higher level of quality to a display on a high-resolution workstation.

As a final comment on JPEG, JPEG-2000 is a JPEG initiative to define a *New Image Coding System,* addressing internet and mobile applications. This system offers low bandwidth, multiple resolution, error resilience, image security, and low complexity. It is based on wavelet compression algorithms, and relative to JPEG, it offers improved compression efficiency with multiple resolution capabilities [28].

13.8.2.2 Motion Picture Experts Group (MPEG)

MPEG is a set of standards designed to support *Coding of Moving Pictures and Associated Audio—for digital storage media at up to 1.5 Mbits/s.* MPEG-1, ISO standard 11172 approved in November 1992, was designed to permit full motion video recordings on CD players originally designed for stereo audio playback. MPEG-2, ISO standard 13818 or ITU T-recommendation H.262, *Generic Coding of Moving Pictures and Associated Audio,* approved in November 1994, addressed greater input-output format flexibility, data rates, and system considerations such as transport and synchronization, topics neglected in MPEG-1. MPEG-2 supports variations of Digital TV covering digitized video that matches existing analog formats with selectable quality through DVD (digital video disc) and HDTV (high definition television) with different aspect ratio, line rate, pixel span, scan conversion options, and various resampling options for the color difference components. The following section describes the basic theory of operation of the simplest version of MPEG-2.

MPEG-2. MPEG compresses a sequence of moving images by taking advantage of the high correlation that exists between successive pictures of a moving picture. MPEG constructs three types of pictures: Intra pictures (*I*-pictures), Predicted pictures (*P*-pictures), and Bi-directional Prediction pictures (*B*-pictures). In MPEG, every *M*-th picture in a sequence can be fully compressed using a standard JPEG algorithm; these are the *I*-pictures. The process then compares successive *I*-pictures and identifies portions of the image that have moved. The image sections that didn't move are carried forward in time to intermediate pictures by the decoder memory. The process then selects a subset of intermediate pictures, then predicts (via linear interpolation between *I*-pictures) and corrects the location of the image sections that were found to have moved. These predicted and corrected images are the *P*-pictures. The pictures between the *I*- and *P*-pictures are the *B*-pictures that incorporate the stationary image sections uncovered by moving sections. The relative position of these picture is shown in Figure 13.44. The relative position of these pictures is shown in Figure 13.44. Note that *P* and *B*-pictures are permitted but not required, and their quantity is each variable. A sequence can be formed without any *P* or *B* pictures, but a sequence containing only *P* or *B*-pictures cannot exist.

The *I*-pictures are compressed as if they were JPEG images. This compression is applied to four contiguous 8-by-8 blocks called a macro block. The macro blocks can be down sampled for subsequent compression of the chrominance components. The macro blocks and their down sampling options are shown in Figure 13.45. Compression of the *I*-frame is performed without reference to any earlier or later pictures in the frame sequence. The distance in sequence count between *I*-pictures is adjustable, and it can be as small as 1 so that the *I*-pictures are adjacent, or as large as reconstruction memory would allow. Editing cuts in a sequence of pictures and local program insertion can only occur at *I*-pictures. Since one-half second is acceptable time accuracy for executing such editing, then the distance between *I*-pictures is usually limited to approximately 15 image pictures for the NTSC standard of 30 pictures per second, or 12 image pictures for the British standard PAL (phase alternating lines) of 25 pictures per second.

The first processing step performed by MPEG is the task of determining which macro blocks have moved between *I*-pictures. This is determined by carrying each macro block from one *I*-frame forward to the next and performing a two-dimensional cross correlation in the neighborhood of its original location. Motion vectors are determined for each macro block that identifies the direction and the amount of movement of each shifted macro block. Macro blocks that have not shifted are stationary in the image pictures between the *I*-picturs and can be carried forward in the intermediate image pictures.

The next processing step in MPEG is to form the *P*-frame between the *I*-pictures. We first assume that the shifted macro blocks have moved linearly in time between the two positions identified in the first processing step. Each macro block is placed at its predicted location on the *P*-frame and is cross-correlated in its neighborhood to determine the true location of the macro block in the *P*-frame. The difference between the predicted and true position of the macro block is an error in the prediction and this error is compressed using the DCT and is used to correct the *P*-frame. The same information is forwarded to the decoder so that it

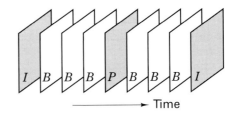

Figure 13.44   Sequence of pictures in MPEG compression.

Time

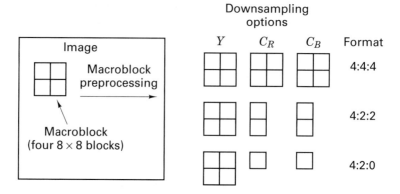

Figure 13.45   Preprocessing of macro block for chrominance down sampling.

can correct its predictions. Figure 13.46 presents the macro block shift between *I*-pictures and an intermediate *P*-picture.

B-pictures are image pictures located between *I*-pictures and *P*-pictures. In these pictures the motion vectors move the shifted macro blocks linearly through time to their bi-directional-interpolated positions at each successive *B*-frame in the sequence. *I*-pictures require the maximum amount of data to describe their DCT compressed contents. *P*-pictures require less data, having only to describe the error pixels in the motion predicted macro blocks in the frame. The rest of the pixels in the frame are carried forward in memory from the previous *I*-frame. *B*-pictures are the most efficient pictures in the set. They only have to *linearly shift and correct* the pixels newly covered and uncovered due to the motion of macro blocks on that image frame.

Reconstruction of the images at the decoder requires that the sequence of images be delivered in the order necessary for appropriate processing. For instance, since the computation of *B*-pictures require information from the *I* and *P* or between the *P*- and *P*-pictures on either side, the *I* and *P* pictures must be delivered first. The following is an example of the required ordering of images at the input and output of the encoder and decoder.

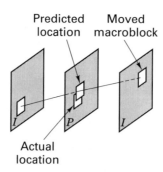

Predicted location   Moved macroblock

Actual location

Figure 13.46   Macro block movement between *I*-pictures and *P*-picture.

Picture Order at Encoder Input

1	2	3	4	5	6	7	8	9	10	11	12	13
I_n	B_1	B_2	P_1	B_3	B_4	P_2	B_5	B_6	I_{n+1}	B_1	B_2	P_1

Encoded Picture Order at Encoder Output and Decoder Input

1	2	3	4	5	6	7	8	9	10	11	12	13
I_n	P_1	B_1	B_2	P_2	B_3	B_4	I_{n+1}	B_5	B_6	P_1	B_1	B_2

Picture Order at Decoder Output

1	2	3	4	5	6	7	8	9	10	11	12	13
I_n	B_1	B_2	P_1	B_3	B_4	P_2	B_5	B_6	I_{n+1}	B_1	B_2	P_1

Figure 13.47 presents a block diagram of an MPEG encoder. Notice its structure is the standard predictor-corrector model. An interesting relationship exists between the eye-brain measure of image quality and image activity. On one hand, when the image contains considerable movement, the eye will accept lower quality images; and on the other hand, when the image contains little movement, the eye is very sensitive to artifacts. In the coder, absence of movement affects coding activity and results in lower data rate delivered to the output buffer. The buffer recognizes this as an indicator of stationary images and controls the image by permitting higher quality quantization of the DCT. The output rate of the output buffer is fixed by communication link considerations. Flow control is used to match the average input rate to the fixed output rate. The flow control detects low encoder activity by noting that its buffer is emptying faster than it is filling. A simple indicator of the difference between input and output rates is the location of the output-address pointer. If this pointer is moving towards the beginning of buffer memory, a precursor of memory underflow, the system increases the input rate by selecting a quantization table that delivers a larger

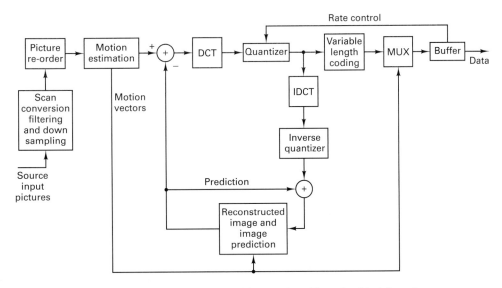

Figure 13.47 Block diagram of MPEG encoder with embedded decoder.

number of bits per DCT. Similarly, if the pointer is moving towards the end of buffer memory, a precursor of over-flow, the system decreases the input rate by selecting a quantization table that delivers a smaller number of bits per DCT. This process matches the image quality to the eye-brain quality threshold while sustaining the average output rate to the channel.

13.9 CONCLUSION

In this chapter, we have presented some of the highlights of source coding. We saw that source coding can be applied to digital data and to waveform signals. Digital data can be reconstructed exactly from a reduced data description of a source, when the source exhibits correlation between alphabet elements, or if the elements are not equally likely. Waveform signals, in general, experience distortion when represented by a digital description. This distortion can be made arbitrarily small by an appropriate increase in bit rate required to describe the source. Source coding can also be applied to waveform sources to obtain reduced-data-rate descriptions, when the source exhibits a long correlation interval of if the possible amplitudes are not equally probable.

The system advantage of source coding is the reduced need for the system resources of bandwidth and/or energy per bit required to deliver a description of the source. This advantage is available in exchange for a third system resource computation and memory. With the cost of these latter resources continuing to fall as they have over the past decades, source coding promises to fill an ever-increasing role in future communication and storage systems. The interested reader is encouraged to examine References [8, 17, 24–26] dealing with source coding.

REFERENCES

1. Papoulis, A., *Probability, Random Variables, and Stochastic Processes,* McGraw-Hill Book Company, New York, *1965.*
2. Harris, F. J., "Windows, Harmonic Analysis, and the Discrete Fourier Transform," *Proc. IEEE,* vol. 67, no., Jan. 1979.
3. Martin, G., "Gyroscopes May Cease Spinning," *IEEE Spectrum,* vol. 23, no. 2, Feb. 1986, pp. 48–53.
4. Vanderkooy, J., and Lipshitz, S. T., "Resolution beyond the Least Significant Bit with Dither," *J. Audio Eng. Soc.,* no. 3, Mar. 1984, pp. 106–112.
5. Blesser, B. A., "Digitization of Audio: A Comprehensive Examination of Theory, Implementation, and Current Practice," *J. Audio Eng. Soc.,* vol. 26, no. 10, Oct. 1978, pp. 739–771.
6. Sluyter, R. J., "Digitization of Speech," *Phillips Tech. Rev.,* vol. 41, no. 7–8, 1983–84, pp. 201–221.
7. Bell Telephone Laboratories Staff, *Transmission Systems for Communications,* Western Electric Co. Technical Publications, Winston-Salem, N.C., 1971.
8. Jayant, N. S., and Noll, P., *Digital Coding of Waveforms,* Prentice-Hall, Inc., Englewood Cliffs, N.J., 1984.

9. Markel, J. D., and Gray, A. H., Jr., *Linear Prediction of Speech,* Springer-Verlag, New York, 1976.

10. Deller, J., Proakis, J., and Hansen, J., *Discrete-Time Processing of Speech Signals,* Macmillan, New York, 1993.

11. Candy, J., and Temes, G., *Oversampling Delta-Sigma Data Converters,* IEEE Press, 1991.

12. Dick, C., and Harris, F., *FPGA Signal Processing Sigma-Delta Modulation,* IEEE Signal Proc. Mag., Vol. 17, No. 1, Jan 2000, pp 20–35.

13. Cummisky, P., Jayant, N., and Flanagan, J., *Adaptive Quantization in Differential PCM Coding of Speech,* Bell Syst. Tech J., Vol. 52, 1973, pp. 115–119.

14. Gersho, A., "Asymptotically Optimal Block Quantization," *IEEE Trans. Inf. Theory,* vol. IT25, no. 4, July 1979, pp. 373–380.

15. Gersho, A., "On the Structure of Vector Quantizers," *IEEE Trans. Inf. Theory,* vol. IT28, no. 2, Mar. 1982, pp. 157–166.

16. Abut, H, *Vector Quantization,* IEEE Press, 1990.

17. Jefffress, L., "Masking," in J. Tobias, ed., *Foundations of Modern Auditory Theory,* Academic Press, Inc., New York, 1970.

18. Lynch, T. J., *Data Compression Techniques and Applications,* Lifetime Learning Publications, New York, 1985.

19. Schafer, R. W., and Rabiner, L. R., "Design of Digital Filter Banks for Speech Analysis," *Bell Syst. Tech. J.,* vol. 50, no. 10, Dec. 1971, pp. 3097–3115.

20. Huffman, D. A., "A Method for the Construction of Minimum Redundancy Codes," *Proc. IRE.,* vol. 40, Sept. 1952, pp. 1098–1101.

21. Hamming, R. W., *Coding and Information Theory,* Prentice-Hall, Inc., Englewood Cliffs, N.J., 1980.

22. Hunter, R., and Robinson, A., *International Digital Facsimile Coding Standard,* Proc. IEEE, Vol. 68, No. 7, July 1980, pp. 854–867.

23. McConnel, K., Bodson, D., and Urban, S., *FAX: Facsimile Technology and Systems,* Artech House, 1999.

24. Cox, R., *Three New Speech Coders From the ITU Cover a Range of Applications,* IEEE Comm. Mag., Vol. 35, NO. 9, Sept 1997, pp. 40–47.

25. Noll, P., *Wideband Speech and Audio Coding,* IEEE Comm. Mag., Vol. 31, No. 11, Nov. 1993, pp. 34–44.

26. Solari, S., *Digital Video and Audio Compression,* McGraw-Hill, New York, 1997

27. Rzeszewski, T., *Digital Video: Concepts and Applications Across Industries,* IEEE Press, 1995

28. Ebrahimi, T., Santa Cruz, D., Christopoulos, C., Askelöf, J., Larsson, M. "JPEG 2000 Still Image Coding Versus Other Standards", *SPIE International Symposium,* 30 July–4 August 2000, Special Session on JPEG2000, San Diego, CA.

PROBLEMS

13.1. A discrete source generates three independent symbols *A, B,* and C with probabilities 0.9, 0.08, and 0.02, respectively. Determine the entropy of the source.

13.2. A discrete source generates two dependent symbols *A* and *B* with conditional probabilities

$$P(A|A) = 0.8 \quad P(B|A) = 0.2$$
$$P(A|B) = 0.6 \quad P(B|B) = 0.4$$

(a) Determine the probabilities of symbols A and B.

(b) Determine the entropy of the source.

(c) Determine the entropy of the source if the symbols were independent with the same probabilities.

13.3. A 16-bit linear analog-to-digital converter operates over an input range of ±5.0 V.

(a) Determine the size of a quantile.

(b) Determine the rms quantizing noise voltage.

(c) Determine the average SNR (due to quantizing) for a full-scale sinusoidal input signal.

(d) Consider that the distance traveled on a 100-mile automobile trip is measured to the same accuracy as that of the 16-bit converter. What is the rms error in feet?

13.4. A 10-bit A-to-D converter (ADC) is designed to operate over a full-scale range of ±5.0 V.

(a) Determine the size of a single quantile step.

(b) For a 5.0-V (full-scale) sinusoid, determine the output signal-to-quantizing noise ratio.

(c) For a 0.050-V ($\frac{1}{100}$ of full-scale) sinusoid, determine the output signal-to-quantizing noise ratio.

(d) For an input signal with a Gaussian-distributed amplitude, the probability of saturation is controlled by adjusting the input attenuator so that the saturation level corresponds to four standard deviations. Determine the output signal-to-quantizing noise ratio for this case.

(e) Determine the probability of signal saturation for the signal described in part (d).

13.5. Determine the optimal compression characteristic for the input density function shown in Figure P13.1 (an approximation to a continuous density function).

13.6. A 10-bit μ-law converter is designed to operate over a full-scale range of ±5.0 V.

(a) If $\mu = 100$, determine the output signal-to-quantizing noise ratio for a 5.0-V (full-scale) sinusoid.

(b) If $\mu = 100$, determine the output signal-to-quantizing noise ratio for a 0.050-V (1/100-th of full scale) sinusoid.

(c) Repeat parts (a) and (b) for $\mu = 250$.

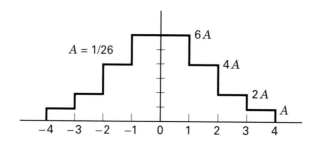

Figure P13.1

13.7. A compact disc (CD) recording system samples each of two stereo signals with a 16-bit A-to-D converter at 44.1 kilosamples/s.

    (a) Determine the output signal-to-noise ratio for a full-scale sinusoid.

    (b) If the recorded music is designed to have a crest factor (peak-to-rms ratio) of 20, determine the average output signal-to-quantizing noise ratio.

    (c) The bit stream of digitized data is augmented by the addition of error-correcting bits, substitution bits to aid the clock extraction by a phase-locked loop (PLL), and display and control bit fields. These additional bits represent 100% overhead; that is, 2 bits are stored for each bit generated by the ADC. Determine the output bit rate of the CD recorder system.

    (d) The CD can record an hour's worth of music. Determine the number of bits recorded on a CD.

    (e) For a comparison, a good collegiate dictionary may contain 1500 pages, 2 columns/page, 100 lines/column, 7 words/line, 6 letters/word, and 6 bits/letter. Determine the number of bits required to describe the dictionary and estimate the number of comparable books that can be stored on a CD.

13.8. A 1-bit quantizer is being designed to sample an input sinusoid of amplitude A with uniformly distributed phase. Determine the amplitude x_0, the output level of the 1-bit quantizer, which minimizes the mean-square quantization error.

13.9. A one-step linear predictive filter is to be used to sample a constant-amplitude sinusoid. The ratio of sample frequency to sinusoid frequency is 10.0. Determine the prediction coefficient of the filter. Determine the ratio of output power to input power for the one-tap predictor.

13.10. A two-tap linear predictor filter is being designed to operate in a DPCM system. The predictor is of the form $\hat{x}(n) = a_1 x(n-1) + a_2 x(n-2)$

    (a) Determine the values a_1^{opt} and a_2^{opt} which minimize the mean-square prediction error.

    (b) Determine the expression for the mean-square prediction error.

    (c) Determine the prediction error power if the correlation coefficient of the input signal is of the form

$$c(n) = \begin{cases} 1 - \dfrac{|n|}{4} & \text{for } n = -4, -3, -2, -1, 0, 1, 2, 3, 4 \\ 0 & \text{otherwise} \end{cases}$$

    (d) Determine the prediction error power if the correlation coefficient of the input signal is of the form $C(n) = \cos \theta_0 n$

13.11. A single loop sigma-delta modulator is designed to operate at 20 times the Nyquist rate for a signal with a 10-kHz bandwidth. The converter is a 1-bit ADC.

    (a) Determine the maximum SNR for a 8.0-kHz input signal.

    (b) Determine the maximum SNR for the same signal if the modulator is operated at 50 times the Nyquist rate.

    (c) Determine the maximum SNR for the same signal if the modulator is replaced with a 2-zero modulator operating at 20 times the Nyquist rate.

13.12. Design a binary Huffman code for a discrete source of three independent symbols A, B, and C with probabilities 0.9, 0.08, and 0.02, respectively. Determine the average code length for the code.

13.13. Design a binary first-order extension code (two symbols at a time) for the discrete source described in Problem 13.12. Determine the average code length per symbol for this code.

13.14. An input alphabet (a keyboard on a word processor) consists of 100 characters.
  (a) If the keystrokes are encoded by a fixed-length code, determine the required number of bits for the encoding.
  (b) We make the simplifying assumption that 10 of the keystrokes are equally likely and that each occurs with probability 0.05. We also assume that the remaining 90 keystrokes are equally likely. Determine the average number of bits required to encode this alphabet using a variable-length Huffman code.

13.15. Use the CCITT-modified Huffman facsimile code to encode the following single-line sequence of 2047 black-and-white pixels. Determine the ratio of coded bits to input bits.

1W 1B 2W 2B 4W 4B 8W 8B 16W 16B 32W 32B

64W 64B 128W 128B 256W 256B 512W 512B 1W

13.16. JPEG quantizes the spectral terms formed by the DCT of the even extension of its processed data. To demonstrate the relative losses in a DCT and an FFT, form the even extension and the replicate extension of the series {10 12 14 16 18 20 22 24} to obtain {10 12 14 16 18 20 22 24 10 12 14 16 18 20 22 24} and (10 12 14 16 18 20 22 24 24 22 20 18 16 14 12 10} respectively. FFT the two time series and compare the relative sizes of the spectral components (other than the dc term). Now edit the spectra, setting to zero all but 5 spectral bins: In the even extension we keep bins [1 2 3 15 16] while in the periodic extension we keep bins [1 3 5 13 15]. Take the inverse FFT of each and compare the relative size of the reconstruction error for the two transforms.

13.17. JPEG uses a zigzag scan pattern to address the DCT spectral terms delivered by its quantizer. An alternate scan pattern would be a raster scan, scanning successive rows commonly done in an image scan. Compare the coding efficiency of the zigzag scan to the raster scan when the non-zero spectral terms are $S(0, 0) = 11001100$, $S(1, 0) = 10101$, and $S(0, 1) = 110001$. Use the modified Huffman code of Table 13.2 to identify run lengths of zeros and assume the following table defines the bit assignment per spectral bin

8	6	5	4	3	2	2	2
6	5	4	3	2	2	1	1
5	4	3	2	2	1	1	1
4	3	2	2	1	1	1	1
3	2	2	1	1	1	1	1
2	2	1	1	1	1	1	1
2	1	1	1	1	1	1	1
2	1	1	1	1	1	1	1

13.18. The DCT transforms a block of 8×8 pixels containing 8-bit words into a block of 8×8 spectral samples containing the number of bits indicated in the quantization table of problem 13.17. Assuming there is no run length of zeros, so that every bit is present in the output of the DCT, compute the compression ratio (input/output bits) attributed to the DCT. Then compute the compression ratio assuming that the number of significant DCT coefficients is limited to the upper triangle of the quantization table consisting of one 8-bit word, two 6-bit words, and three 5-bit words with the remaining bits carried by the run length code for 95 zeros.

QUESTIONS

13.1. Prior to transmission or storage, why are signals subjected to *source coding* operations? (See Sections 13.1, and 13.7.)

13.2. What properties of a *continuous signal* permit the signal to be represented with a reduced number of bits per sample? (See Sections 13.1, 13.3, and 13.7.)

13.3. What properties of a *discrete signal* permit the signal to be represented with a reduced number of bits per symbol? (See Sections 13.1, and 13.7.)

13.4. Most quantizers are *uniform* in step size. Applications exist for which the quantizers have *non-uniform* step sizes. These are sometimes called *companded* quantizers. Why would we want a quantizer with unequal step sizes? (See Section 13.2.5.)

13.5. An analog-to-digital converter (ADC) represents the levels of a sampled data signal with a number of bits per sample selected to satisfy a required fidelity requirement. Most ADCs are *memoryless,* which means each quantization (conversion) is performed independently of other conversions. How can memory be used to reduce the number of bits per sample? (See Section 13.3.)

13.6. Source coding reduces *redundancy* and discards *irrelevant* content? What is the difference between redundancy and irrelevancy? (See Section 13.7.)

13.7. An often-heard saying is that "A picture is worth a thousand words." Is a picture really worth a thousand words? (See Section 13.8.2.)

EXERCISES

Using the Companion CD, run the exercises associated with Chapter 13.

CHAPTER 14

Encryption
and
Decryption

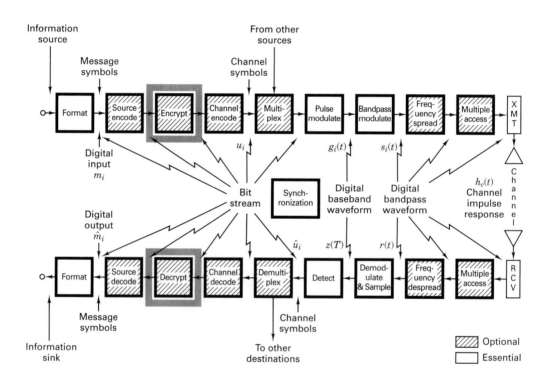

14.1 MODELS, GOALS, AND EARLY CIPHER SYSTEMS

14.1.1 A Model of the Encryption and Decryption Process

The desire to communicate privately is a human trait that dates back to earliest times. Hence the history of secret communications is rich with unique inventions and colorful anecdotes [1]. The study of ways to disguise messages so as to avert unauthorized interception is called *cryptography*. The terms *encipher* and *encrypt* refer to the message transformation performed at the transmitter, and the terms *decipher* and *decrypt* refer to the inverse transformation performed at the receiver. The two primary reasons for using cryptosystems in communications are (1) *privacy,* to prevent unauthorized persons from extracting information from the channel (eavesdropping); and (2) *authentication,* to prevent unauthorized persons from injecting information into the channel (spoofing). Sometimes, as in the case of electronic funds transfer or contract negotiations, it is important to provide the electronic equivalent of a *written signature* in order to avoid or settle any dispute between the sender and receiver as to what message, if any, was sent.

Figure 14.1 illustrates a model of a cryptographic channel. A message, or plaintext, M, is encrypted by the use of an invertible transformation, E_K, that produces a ciphertext, $C = E_K(M)$. The ciphertext is transmitted over an insecure or *public channel.* When an authorized receiver obtains C, he decrypts it with the inverse transformation, $D_K = E_K^{-1}$, to obtain the original plaintext message, as follows:

$$D_K(C) = E_K^{-1}[E_K(M)] = M \qquad (14.1)$$

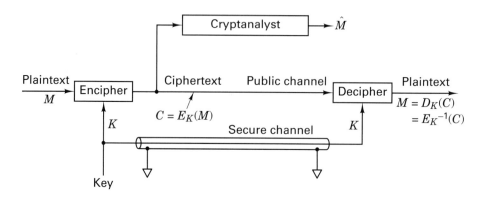

Figure 14.1   Model of a cryptographic channel.

The parameter K refers to a set of symbols or characters called a *key*, which dictates a specific encryption transformation, E_K, from a family of cryptographic transformations. Originally, the security of cryptosystems depended on the secrecy of the entire encryption process, but eventually systems were developed for which the general nature of the encryption transformation or algorithm could be publicly revealed, since the security of the system depended on the specific key. The key is supplied along with the plaintext message for encryption, and along with the ciphertext message for decryption. There is a close analogy here with a general-purpose computer and a computer program. The computer, like the cryptosystem, is capable of a large variety of transformations, from which the computer program, like the specific key, selects one. In most cryptosystems, anyone with access to the key can both encrypt and decrypt messages. The key is transmitted to the community of authorized users over a secure channel (as an example, a courier may be used to hand-carry the sensitive key information); the key usually remains unchanged for a considerable number of transmissions. The goal of the *cryptanalyst* (eavesdropper or adversary) is to produce an estimate of the plaintext, $\hat{M}$, by analyzing the ciphertext obtained from the public channel, without benefit of the key.

Encryption schemes fall into two generic categories: *block encryption,* and *data-stream* or simply *stream encryption.* With block encryption, the plaintext is segmented into blocks of fixed size; each block is encrypted independently from the others. For a given key, a particular plaintext block will therefore be carried into the same ciphertext block each time it appears (similar to block encoding). With data-stream encryption, similar to convolutional coding, there is no fixed block size. Each plaintext bit, m_i, is encrypted with the ith element, k_i, of a sequence of symbols (key stream) generated with the key. The encryption is *periodic* if the key stream repeats itself after p characters for some fixed p; otherwise, it is nonperiodic.

In general, the properties desired in an encryption scheme are quite different from those desired in a channel coding scheme. For example, with encryption, plaintext data should never appear directly in the ciphertext, but with channel coding, codes are often in *systematic form* comprising unaltered message bits plus parity bits (see Section 6.4.5). Consider another example of the differences between encryption and channel coding. With block encryption, a single bit error at the input of the decryptor might change the value of many of the output bits in the block. This effect, known as *error propagation,* is often a desirable cryptographic property since it makes it difficult for unauthorized users to succeed in spoofing a system. However, in the case of channel coding, we would like the system to correct as many errors as possible, so that the output is relatively unaffected by input errors.

14.1.2 System Goals

The major requirements for a cryptosystem can be stated as follows:

1. To provide an *easy* and *inexpensive* means of encryption and decryption to all authorized users in possession of the appropriate key
2. To ensure that the cryptanalyst's task of producing an estimate of the plaintext without benefit of the key is made *difficult* and *expensive*

Successful cryptosystems are classified as being either *unconditionally secure* or *computationally secure.* A system is said to be *unconditionally secure* when the amount of information available to the cryptanalyst is insufficient to determine the encryption and decryption transformations, no matter how much computing power the cryptanalyst has available. One such system, called a *one-time pad,* involves encrypting a message with a random key that is used one time only. The key is never reused; hence the cryptanalyst is denied information that might be useful against subsequent transmissions with the same key. Although such a system is unconditionally secure (see Section 14.2.1), it has limited use in a conventional communication system, since a new key would have to be distributed for each new message—a great logistical burden. The distribution of keys to the authorized users is a major problem in the operation of any cryptosystem, even when a key is used for an extended period of time. Although some systems can be proven to be unconditionally secure, currently there is no known way to demonstrate security for an arbitrary cryptosystem. Hence the specifications for most cryptosystems rely on the less formal designation of *computational security* for x number of years, which means that under circumstances favorable to the cryptanalyst (i.e., using state-of-the-art computers) the system security could be broken in a period of x years, but could not be broken in less than x years.

14.1.3 Classic Threats

The weakest classification of cryptanalytic threat on a system is called a *ciphertext-only attack.* In this attack the cryptanalyst might have *some* knowledge of the general system and the language used in the message, but the only significant data available to him is the encrypted transmission intercepted from the public channel.

A more serious threat to a system is called a *known plaintext attack;* it involves knowledge of the plaintext *and* knowledge of its ciphertext counterpart. The rigid structure of most business forms and programming languages often provides an opponent with much a priori knowledge of the details of the plaintext message. Armed with such knowledge and with a ciphertext message, the cryptanalyst can mount a known plaintext attack. In the diplomatic arena, if an encrypted message directs a foreign minister to make a particular public statement, and if he does so without paraphrasing the message, the cryptanalyst may be privy to both the ciphertext *and* its exact plaintext translation. While a known plaintext attack is not always possible, its occurrence is frequent enough that a system is not considered secure unless it is designed to be secure against the plaintext attack [2].

When the cryptanalyst is in the position of *selecting* the plaintext, the threat is termed a *chosen plaintext attack*. Such an attack was used by the United States to learn more about the Japanese cryptosystem during World War II. On May 20, 1942, Admiral Yamamoto, Commander-in-Chief of the Imperial Japanese Navy, issued an order spelling out the detailed tactics to be used in the assault of Midway island. This order was intercepted by the Allied listening posts. By this time, the Americans had learned enough of the Japanese code to decrypt most of the message. Still in doubt, however, were some important parts, such as the *place* of the assault. They suspected that the characters "AF" meant Midway island, but to be sure, Joseph Rochefort, head of the Combat Intelligence Unit, decided to use a chosen plaintext attack to trick the Japanese into providing concrete proof. He had the Midway garrison broadcast a distinctive plaintext message in which Midway reported that its fresh-water distillation plant had broken down. The American cryptanalysts needed to wait only two days before they intercepted a Japanese ciphertext message stating that AF was short of fresh water [1].

14.1.4 Classic Ciphers

One of the earliest examples of a monoalphabetic cipher was the *Caesar Cipher,* used by Julius Caesar during the Gallic wars. Each plaintext letter is replaced with a new letter obtained by an *alphabetic shift.* Figure 14.2a illustrates such an encryption transformation, consisting of three end-around shifts of the alphabet. When using this Caesar's alphabet, the message, "now is the time" is encrypted as follows:

Plaintext:	N	O	W	I	S	T	H	E	T	I	M	E
Ciphertext:	Q	R	Z	L	V	W	K	H	W	L	P	H

The decryption key is simply the number of alphabetic shifts; the code is changed by choosing a new key. Another classic cipher system, illustrated in Figure 14.2b, is called the *Polybius square.* Letters I and J are first combined and treated as a single character since the final choice can easily be decided from the context of the message. The resulting 25 character alphabet is arranged in a 5×5 array. Encryption of any character is accomplished by choosing the appropriate row-column (or column-row) number pair. An example of encryption with the use of the Polybius square follows:

Plaintext:	A B C D E F G H I J K L M N O P Q R S T U V W X Y Z
Chiphertext:	D E F G H I J K L M N O P Q R S T U V W X Y Z A B C

(a)

	1	2	3	4	5
1	A	B	C	D	E
2	F	G	H	IJ	K
3	L	M	N	O	P
4	Q	R	S	T	U
5	V	W	X	Y	Z

(b)

Figure 14.2   (a) Caesar's alphabet with a shift of 3. (b) Polybius square.

Plaintext:	N	O	W	I	S	T	H	E	T	I	M	E
Ciphertext:	33	43	25	42	34	44	32	51	44	42	23	51

The code is changed by a rearrangement of the letters in the 5×5 array.

The *Trithemius progressive key,* shown in Figure 14.3, is an example of a *polyalphabetic cipher.* The row labeled shift 0 is identical to the usual arrangement of the alphabet. The letters in the next row are shifted one character to the left with an end-around shift for the leftmost position. Each successive row follows the same pattern of shifting the alphabet one character to the left as compared to the prior row. This continues until the alphabet has been depicted in all possible arrangements of end-around shifts. One method of using such an alphabet is to select the first cipher character from the shift 1 row, the second cipher character from the shift 2 row, and so on. An example of such encryption is

Plaintext:	N	O	W	I	S	T	H	E	T	I	M	E
Ciphertext:	O	Q	Z	M	X	Z	O	M	C	S	X	Q

There are several interesting ways that the Trithemius progressive key can be used. One way, called the *Vigenere key method,* employs a keyword. The key dictates the row choices for encryption and decryption of each successive character in the message. For example, suppose that the word "TYPE" is selected as the key; then an example of the Vigenere encryption method is

Key:	T	Y	P	E	T	Y	P	E	T	Y	P	E
Plaintext:	N	O	W	I	S	T	H	E	T	I	M	E
Ciphertext:	G	M	L	M	L	R	W	I	M	G	B	I

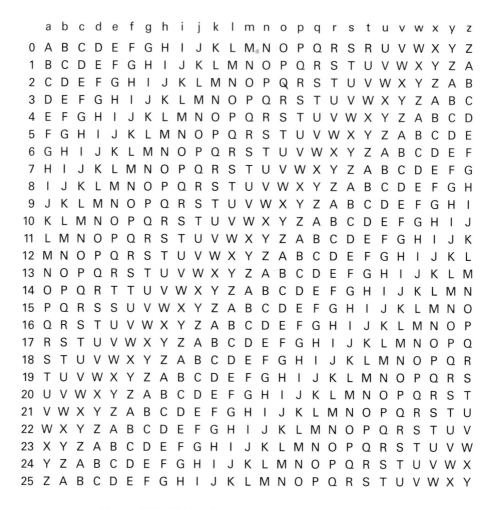

Figure 14.3 Trithemius progressive key.

where the first letter, T, of the key indicates that the row choice for encrypting the first plaintext character is the row starting with T (shift 19). The next row choice starts with Y (shift 24), and so on. A variation of this key method, called the *Vigenere auto (plain) key method,* starts with a single letter or word used as a *priming key.* The priming key dictates the starting row or rows for encrypting the first or first few plaintext characters, as in the preceding example. Next, the *plaintext characters* themselves are used as the key for choosing the rows for encryption. An example using the letter "F" as the priming key is

Key:	F	N	O	W I	S	T	H	E	T	I	M
Plaintext:	N	O	W	I S	T	H	E	T	I	M	E
Ciphertext:	S	B	K	E A	L	A	L	X	B	U	Q

With the auto key method, it should be clear that feedback has been introduced to the encryption process. With this feedback, the choice of the ciphertext is dictated by the contents of the message.

A final variation of the Vigenere method, called the *Vigenere auto (cipher) key method,* is similar to the plain key method in that a priming key and feedback are used. The difference is that after encryption with the priming key, each successive key character is the sequence is obtained from the prior *ciphertext character* instead of from the plaintext character. An example should make this clear; as before, the letter "F" is used as the priming key:

Key:	F	S	G	C	K	C	V	C	G	Z	H	T
Plaintext:	N	O	W	I	S	T	H	E	T	I	M	E
Ciphertext:	S	G	C	K	C	V	C	G	Z	H	T	X

Although each key character can be found from its preceding ciphertext character, it is functionally dependent on *all* the preceding characters in the message plus the priming key. This has the effect of diffusing the statistical properties of the plaintext across the ciphertext, making statistical analysis very difficult for a cryptanalyst. One weakness of the cipher key example depicted here is that the ciphertext contains key characters which will be exposed on the public channel "for all to see." Variations of this method can be employed to prevent such overt exposure [3]. By today's standards Vigenere's encryption schemes are not very secure; his basic contribution was the discovery that nonrepeating key sequences could be generated by using the messages themselves or functions of the messages.

14.2 THE SECRECY OF A CIPHER SYSTEM

14.2.1 Perfect Secrecy

Consider a cipher system with a finite message space $\{M\} = M_0, M_1, \ldots, M_{N-1}$ and a finite ciphertext space $\{C\} = C_0, C_1, \ldots, C_{U-1}$. For any M_i, the a priori probability that M_i is transmitted is $P(M_i)$. Given that C_j is received, the a posteriori probability that M_i was transmitted is $P(M_i|C_j)$. A cipher system is said to have *perfect secrecy* if for every message M_i and every ciphertext C_j, the a posteriori probability is equal to the a priori probability:

$$P(M_i|C_j) = P(M_i) \tag{14.2}$$

Thus, for a system with perfect secrecy, a cryptanalyst who intercepts C_j obtains no further information to enable him or her to determine which message was transmitted. A necessary and sufficient condition for perfect secrecy is that for every M_i and C_j,

$$P(C_j|M_i) = P(C_j) \tag{14.3}$$

The schematic in Figure 14.4 illustrates an example of perfect secrecy. In this example, $\{M\} = M_0, M_1, M_2, M_3, \{C\} = C_0, C_1, C_2, C_3, \{K\} = K_0, K_1, K_2, K_3, N = U = 4,$

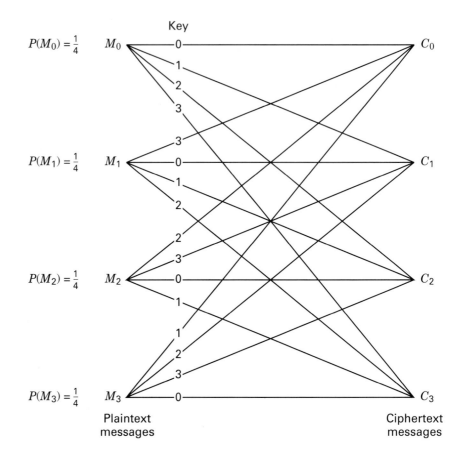

Figure 14.4 Example of perfect secrecy.

and $P(M_i) = P(C_j) = \frac{1}{4}$. The transformation from message to ciphertext is obtained by

$$C_s = T_{K_j}(M_i) \tag{14.4}$$

$$s = (i + j) \text{ modulo-}N$$

where T_{K_j} indicates a transformation under the key, K_j, and x modulo-y is defined as the remainder of dividing x by y. Thus $s = 0, 1, 2, 3$. A cryptanalyst intercepting one of the ciphertext messages $C_s = C_0, C_1, C_2,$ or C_3 would have no way of determining which of the four keys was used, and therefore whether the correct message is $M_0, M_1, M_2,$ or M_3. A cipher system in which the number of messages, the number of keys, and the number of ciphertext transformations are all equal is said to have perfect secrecy if and only if the following two conditions are met:

1. There is only one key transforming each message to each ciphertext.
2. All keys are equally likely.

If these conditions are not met, there would be some message M_i such that for a given C_j, there is no key that can decipher C_j into M_i, implying that $P(M_i|C_j) = 0$ for some i and j. The cryptanalyst could then eliminate certain plaintext messages from consideration, thereby simplifying the task. Perfect secrecy is a very desirable objective since it means that the cipher system is unconditionally secure. It should be apparent, however, that for systems which transmit a large number of messages, the amount of key that must be distributed for perfect secrecy can result in formidable management problems, making such systems impractical. Since in a system with perfect secrecy, the number of different keys is at least as great as the number of possible messages, if we allow messages of unlimited length, perfect secrecy requires an infinite amount of key.

**Example 14.1   Breaking a Cipher System When the Key Space is Smaller
Than the Message Space**

Consider that the 29-character ciphertext

G R O B O K B O D R O R O B Y O C Y P I O C D O B I O K B

was produced by a Caesar cipher (see Section 14.1.4) such that each letter has been shifted by K positions, where $1 \le K \le 25$. Show how a cryptanalyst can break this code.

Solution

Because the number of possible keys (there are 25) is smaller than the number of possible 29-character meaningful messages (there are a myriad), perfect secrecy cannot be achieved. In the original polyalphabetic cipher of Figure 14.3, a plaintext character is replaced by a letter of increasingly higher rank as the row number (K) increases. Hence, in analyzing the ciphertext, we reverse the process by creating rows such that each ciphertext letter is replaced by letters of decreasing rank. The cipher is easily broken by trying all the keys, from 1 to 25, as shown in Figure 14.5, yielding only one key ($K = 10$) that produces the meaningful message: WHERE ARE THE HEROES OF YESTERYEAR (The spaces have been added.)

Example 14.2   Perfect Secrecy

We can modify the key space of Example 14.1 to create a cipher having perfect secrecy. In this new cipher system each character in the message is encrypted using a *randomly selected* key value. The key, K, is now given by the sequence $k_1, k_2, \ldots, k_{29}$, where each k_i is a random integer in the range (1, 25) dictating the shift used for the ith character; thus there are a total of $(25)^{29}$ different key sequences. Then the 29-character ciphertext in Example 14.1 could correspond to *any* meaningful 29-character message. For example, the ciphertext could correspond to the plaintext (the spaces have been added)

ENGLISH AND FRENCH ARE SPOKEN HERE

derived by the key 2, 4, 8, 16, 6, 18, 20, ... Most of the 29-character possibilities can be ruled out because they are not meaningful messages (this much is known without the ciphertext). Perfect secrecy is achieved because interception of the ciphertext in this system reveals no additional information about the plaintext message.

Key   Text

Key																													
0	G	R	O	B	O	K	B	O	D	R	O	R	O	B	Y	O	C	Y	P	I	O	C	D	O	B	I	O	K	B
1	F	Q	N	A	N	J	A	N	C	Q	N	Q	N	A	X	N	B	X	O	H	N	B	C	N	A	H	N	J	A
2	E	P	M	Z	M	I	Z	M	B	P	M	P	M	Z	W	M	A	W	N	G	M	A	B	M	Z	G	M	I	Z
3	D	O	L	Y	L	H	Y	L	A	O	L	O	L	Y	V	L	Z	V	M	F	L	Z	A	L	Y	F	L	H	Y
4	C	N	K	X	K	G	X	K	Z	N	K	N	K	X	U	K	Y	U	L	E	K	Y	Z	K	X	E	K	G	X
5	B	M	J	W	J	F	W	J	Y	M	J	M	J	W	T	J	X	T	K	D	J	X	Y	J	W	D	J	F	W
6	A	L	I	V	I	E	V	I	X	L	I	L	I	V	S	I	W	S	J	C	I	W	X	I	V	C	I	E	V
7	Z	K	H	U	H	D	U	H	W	K	H	K	H	U	R	H	V	R	I	B	H	V	W	H	U	B	H	D	U
8	Y	J	G	T	G	C	T	G	V	J	G	J	G	T	Q	G	U	Q	H	A	G	U	V	G	T	A	G	C	T
9	X	I	F	S	F	B	S	F	U	I	F	I	F	S	P	F	T	P	G	Z	F	T	U	F	S	Z	F	B	S
10	W	H	E	R	E	A	R	E	T	H	E	H	E	R	O	E	S	O	F	Y	E	S	T	E	R	Y	E	A	R
11	V	G	D	Q	D	Z	Q	D	S	G	D	G	D	Q	N	D	R	N	E	X	D	R	S	D	Q	X	D	Z	Q
12	U	F	C	P	C	Y	P	C	R	F	C	F	C	P	M	C	Q	M	D	W	C	Q	R	C	P	W	C	Y	P
13	T	E	B	O	B	X	O	B	Q	E	B	E	B	O	L	B	P	L	C	V	B	P	Q	B	O	V	B	X	O
14	S	D	A	N	A	W	N	A	P	D	A	D	A	N	K	A	O	K	B	U	A	O	P	A	N	U	A	W	N
15	R	C	Z	M	Z	V	M	Z	O	C	Z	C	Z	M	J	Z	N	J	A	T	Z	N	O	Z	M	T	Z	V	M
16	Q	B	Y	L	Y	U	L	Y	N	B	Y	B	Y	L	I	Y	M	I	Z	S	Y	M	N	Y	L	S	Y	U	L
17	P	A	X	K	X	T	K	X	M	A	X	A	X	K	H	X	L	H	Y	R	X	L	M	X	K	R	X	T	K
18	O	Z	W	J	W	S	J	W	L	Z	W	Z	W	J	G	W	K	G	X	Q	W	K	L	W	J	Q	W	S	J
19	N	Y	V	I	V	R	I	V	K	Y	V	Y	V	I	F	V	J	F	W	P	V	J	K	V	I	P	V	R	I
20	M	X	U	H	U	Q	H	U	J	X	U	X	U	H	E	U	I	E	V	O	U	I	J	U	H	O	U	Q	H
21	L	W	T	G	T	P	G	T	I	W	T	W	T	G	D	T	H	D	U	N	T	H	I	T	G	N	T	P	G
22	K	V	S	F	S	O	F	S	H	V	S	V	S	F	C	S	G	C	T	M	S	G	H	S	F	M	S	O	F
23	J	U	R	E	R	N	E	R	G	U	R	U	R	E	B	R	F	B	S	L	R	F	G	R	E	L	R	N	E
24	I	T	Q	D	Q	M	D	Q	F	T	Q	T	Q	D	A	Q	E	A	R	K	Q	E	F	Q	D	K	Q	M	D
25	H	S	P	C	P	L	C	P	E	S	P	S	P	C	Z	P	D	Z	Q	J	P	D	E	P	C	J	P	L	C

Figure 14.5 Example of breaking a cipher system when the key space is smaller than the message space.

14.2.2 Entropy and Equivocation

As discussed in Chapter 9, the amount of information in a message is related to the probability of occurrence of the message. Messages with probability of either 0 or 1 contain no information, since we can be very confident concerning our prediction of their occurrence. The more uncertainty there is in predicting the occurrence of a message, the greater is the information content. Hence when each of the messages in a set is equally likely, we can have *no* confidence in our ability to predict the occurrence of a particular message, and the uncertainty or information content of the message is maximum.

Entropy, $H(X)$, is defined as the average amount of information per message. It can be considered a measure of how much *choice* is involved in the selection of a message X. It is expressed by the following summation over all possible messages:

$$H(X) = -\sum_X P(X) \log_2 P(X) = \sum_X P(X) \log_2 \frac{1}{P(X)} \qquad (14.5)$$

When the logarithm is taken to the base 2, as shown, $H(X)$ is the *expected number of bits* in an *optimally encoded* message X. This is not quite the measure that a cryptanalyst desires. He will have intercepted some ciphertext and will want to know how confidently he can predict a message (or key) given that this particular ciphertext was sent. *Equivocation,* defined as the conditional entropy of X given Y, is a more useful measure for the cryptanalyst in attempting to break the cipher and is given by

$$H(X|Y) = -\sum_{X,Y} P(X,Y) \log_2 P(X|Y) \qquad (14.6)$$

$$= \sum_Y P(Y) \sum_X P(X|Y) \log_2 \frac{1}{P(X|Y)}$$

Equivocation can be thought of as the uncertainty that message X was sent, having received Y. The cryptanalyst would like $H(X|Y)$ to approach zero as the amount of intercepted ciphertext, Y, increases.

Example 14.3 Entropy and Equivocation

Consider a sample message set consisting of eight equally likely messages $\{X\} = X_1, X_2, \dots, X8$.

(a) Find the entropy associated with a message from the set $\{X\}$.
(b) Given another equally likely message set $\{Y\} = Y_1, Y_2$. Consider that the occurrence of each message Y narrows the possible choices of X in the following way:

$$\text{If } Y_1 \text{ is present: only } X_1, X_2, X_3, \text{ or } X_4 \text{ is possible}$$

$$\text{If } Y_2 \text{ is present: only } X_5, X_6, X_7, \text{ or } X_8 \text{ is possible}$$

Find the equivocation of message X conditioned on message Y.

Solution

(a) $P(X) = \frac{1}{8}$
$H(X) = 8[(\frac{1}{8}) \log_2 8] = 3$ bits/message
(b) $P(Y) = \frac{1}{2}$. For each Y, $P(X|Y) = \frac{1}{4}$ for four of the X's and $P(X|Y) = 0$ for the remaining four X's. Using Equation (14.6), we obtain

$$H(X|Y) = 2[(\tfrac{1}{2})4(\tfrac{1}{4} \log_2 4)] = 2 \text{ bits/message}$$

We see that knowledge of Y has reduced the uncertainty of X from 3 bits/message to 2 bits/message.

14.2.3 Rate of a Language and Redundancy

The *true rate* of a language is defined as the average number of *information bits* contained in each character and is expressed for messages of length N by

$$r = \frac{H(X)}{N} \tag{14.7}$$

where $H(X)$ is the message entropy, or the number of bits in the *optimally encoded* message. For large N, estimates of r for written English range between 1.0 and 1.5 bits/character [4]. The *absolute rate* or maximum entropy of a language is defined as the maximum number of information bits contained in each character assuming that all possible sequences of characters are equally likely. The absolute rate is given by

$$r' = \log_2 L \tag{14.8}$$

where L is the number of characters in the language. For the English alphabet $r' = \log_2 26 = 4.7$ bits/character. The true rate of English is, or course, much less than its absolute rate since, like most languages, English is highly redundant and structured.

The *redundancy* of a language is defined in terms of its true rate and absolute rate as

$$D = r' - r \tag{14.9}$$

For the English language with $r' = 4.7$ bits/character and $r = 1.5$ bits/character, $D = 3.2$, and the ratio $D/r' = 0.68$ is a measure of the redundancy in the language.

14.2.4 Unicity Distance and Ideal Secrecy

We stated earlier that perfect secrecy requires an infinite amount of key if we allow messages of unlimited length. With a finite key size, the equivocation of the key $H(K|C)$ generally approaches zero, implying that the key can be uniquely determined and the cipher system can be broken. The *unicity distance* is defined as the smallest amount of ciphertext, N, such that they key equivocation $H(K|C)$ is close to zero. Therefore, the unicity distance is the amount of ciphertext needed to uniquely determine the key and thus break the cipher system. Shannon [5] described an *ideal secrecy* system as one in which $H(K|C)$ does not approach zero as the amount of ciphertext approaches infinity; that is, no matter how much ciphertext is intercepted, the key cannot be determined. The term "ideal secrecy" describes a system that does not achieve perfect secrecy but is nonetheless unbreakable (unconditionally secure) because it does not reveal enough information to determine the key.

Most cipher systems are too complex to determine the probabilities required to derive the unicity distance. However, it is sometimes possible to approximate unicity distance, as shown by Shannon [5] and Hellman [6]. Following Hellman, assume that each plaintext and ciphertext message comes from a finite alphabet of L symbols.

Thus there are $2^{r'N}$ possible messages of length, N, where r' is the absolute rate of the language. We can consider the total message space partitioned into two classes, meaningful messages, M_1, and meaningless messages M_2. We then have

$$\text{number of meaningful messages} = 2^{rN} \tag{14.10}$$

$$\text{number of meaningless messages} = 2^{r'N} - 2^{rN} \tag{14.11}$$

where r is the true rate of the language, and where the a priori probabilities of the message classes are

$$P(M_1) = \frac{1}{2^{rN}} = 2^{-rN} \quad M_1 \text{ meaningful} \tag{14.12}$$

$$P(M_2) = 0 \quad\quad\quad M_2 \text{ meaningless} \tag{14.13}$$

Let us assume that there are $2^{H(K)}$ possible keys (size of the key alphabet), where $H(K)$ is the entropy of the key (number of bits in the key). Assume that all keys are equally likely; that is,

$$P(K) = \frac{1}{2^{H(K)}} = 2^{-H(K)} \tag{14.14}$$

The derivation of the unicity distance is based on a *random cipher* model, which states that for each key K and ciphertext C, the decryption operation $D_K(C)$ yields an independent random variable distributed over all the possible $2^{r'N}$ messages (both meaningful and meaningless). Therefore, for a given K and C, the $D_K(C)$ operation can produce any one of the plaintext messages with equal probability.

Given an encryption described by $C_i = E_{K_i}(M_i)$, a *false solution* F arises whenever encryption under another key K_j could also produce C_i either from the message M_i or from some other message M_j; that is,

$$C_i = E_{K_i}(M_i) = E_{K_j}(M_i) = E_{K_j}(M_j) \tag{14.15}$$

A cryptanalyst intercepting C_i would not be able to pick the correct key and hence could not break the cipher system. We are not concerned with the decryption operations that produce *meaningless* messages because these are easily rejected.

For every correct solution to a particular ciphertext there are $2^{H(K)} - 1$ incorrect keys, each of which has the same probability $P(F)$ of yielding a false solution. Because each meaningful plaintext message is assumed equally likely, the probability of a false solution, is the same as the probability of getting a meaningful message, namely,

$$P(F) = \frac{2^{rN}}{2^{r'N}} = 2^{(r-r')N} = 2^{-DN} \tag{14.16}$$

where $D = r' - r$ is the redundancy of the language. The expected number of false solutions $\bar{F}$ is then

$$\bar{F} = [2^{H(K)} - 1]P(F) = [2^{H(K)} - 1]2^{-DN} \tag{14.17}$$

$$\approx 2^{H(K)-DN}$$

Because of the rapid decrease of $\bar{F}$ with increasing N,

$$\log_2 \bar{F} = H(K) - DN = 0 \qquad (14.18)$$

is defined as the point where the number of false solutions is sufficiently small so that the cipher can be broken. The resulting unicity distance is therefore

$$N = \frac{H(K)}{D} \qquad (14.19)$$

We can see from Equation (14.17) that if $H(K)$ is much larger than DN, there will be a large number of meaningful decryptions, and thus a small likelihood of a cryptanalyst distinguishing which meaningful message is the correct message. In a loose sense, DN represents the number of equations available for solving for the key, and $H(K)$ the number of unknowns. When the number of equations is smaller than the number of unknown key bits, a unique solution is not possible and the system is said to be unbreakable. When the number of equations is larger than the number of unknowns, a unique solution is possible and the system can no longer be characterized as unbreakable (although it may still be computationally secure).

It is the predominance of meaningless decryptions that enables cryptograms to be broken. Equation (14.19) indicates the value of using *data compression* techniques prior to encryption. Data compression removes redundancy, thereby increasing the unicity distance. Perfect data compression would result in $D = 0$ and $N = \infty$ for any key size.

Example 14.4   Unicity Distance

Calculate the unicity distance for a written English encryption system, where the key is given by the sequence $k_1, k_2, \ldots, k_{29}$, where each k_i is a random integer in the range $(1, 25)$ dictating the shift number (Figure 14.3) for the ith character. Assume that each of the possible key sequences is equally likely.

Solution

There are $(25)^{29}$ possible key sequences, each of which is equally likely. Therefore, using Equations (14.5), (14.8), and (14.19) we have

$$\text{Key entropy:} \quad H(K) = \log_2 (25)^{29} = 135 \text{ bits}$$

$$\text{Absolute rate for English:} \quad r' = \log_2 26 = 4.7 \text{ bits/character}$$

$$\text{Assumed true rate for English:} \quad r = 1.5 \text{ bits/character}$$

$$\text{Redundancy:} \quad D = r' - r = 3.2 \text{ bits/character}$$

$$N = \frac{H(K)}{D} = \frac{135}{3.2} \approx 43 \text{ characters}$$

In Example 14.2, perfect secrecy was illustrated using the same type of key sequence described here, with a 29-character message. In this example we see that if the available ciphertext is 43 characters long (which implies that some portion of the key sequence must be used twice), a unique solution may be possible. However, there is no indication as to the computational difficulty in finding the solution. Even though we have estimated the theoretical amount of ciphertext required to break the cipher, it might be computationally infeasible to accomplish this.

14.3 PRACTICAL SECURITY

For ciphertext sequences greater than the unicity distance any system can be solved, in principle, merely by trying each possible key until the unique solution is obtained. This is completely impractical, however, except when the key is extremely small. For example, for a key configured as a permutation of the alphabet, there are $26! \approx 4 \times 10^{26}$ possibilities (considered small in the cryptographic context). In an exhaustive search, one might expect to reach the right key at about halfway through the search. If we assume that each trial requires a computation time of 1 μs, the total search time exceeds 10^{12} years. Hence techniques other than a brute-force search (e.g., statistical analysis) must be employed if a cryptanalyst is to have any hope of success.

14.3.1 Confusion and Diffusion

A statistical analysis using the frequency of occurrence of individual characters and character combinations can be used to solve many cipher systems. Shannon [5] suggested two encryption concepts for frustrating the statistical endeavors of the cryptanalyst. He termed these encryption transformations confusion and diffusion. *Confusion* involves substitutions that render the final relationship between the key and ciphertext as complex as possible. This makes it difficult to utilize a statistical analysis to narrow the search to a particular subset of the key variable space. Confusion ensures that the majority of the key is needed to decrypt even very short sequences of ciphertext. *Diffusion* involves transformations that smooth out the statistical differences between characters and between character combinations. An example of diffusion with a 26-letter alphabet is to transform a message sequence $M = M_0, M_1, \ldots$ into a new message sequence $Y = Y_0, Y_1, \ldots$ according to the relationship

$$Y_n = \sum_{i=0}^{s-1} M_{n+i} \quad \text{modulo-26} \tag{14.20}$$

where each character in the sequence is regarded as an integer modulo-26, s is some chosen integer, and $n = 0, 1, 2, \ldots$. The new message, Y, will have the same redundancy as the original message, M, but the letter frequencies of Y will be more uniform than in M. The effect is that the cryptanalyst needs to intercept a longer sequence of ciphertext before any statistical analysis can be useful.

14.3.2 Substitution

Substitution encryption techniques, such as the Caesar cipher and the Trithemius progressive key cipher, are widely used in puzzles. Such simple substitution ciphers offer little encryption protection. For a substitution technique to fulfill Shannon's concept of *confusion,* a more complex relationship is required. Figure 14.6 shows one example of providing greater substitution complexity through the use of a nonlinear transformation. In general, n input bits are first represented as one of 2^n different characters (binary-to-octal transformation in the example of Figure 14.6. The set of 2^n characters are then permuted so that each character is transposed to one of the others in the set. The character is then converted back to an n-bit output.

It can be easily shown that there are $(2^n)!$ different substitution or connection patterns possible. The cryptanalyst's task becomes computationally unfeasible as n gets large, say $n = 128$; then $2^n = 10^{38}$, and $(2^n)!$ is an astronomical number. We recognize that for $n = 128$, this substitution box (S-box) transformation is complex (confusion). However, although we can identify the S-box with $n = 128$ as ideal, its implementation is not feasible because it would require a unit with $2^n = 10^{38}$ wiring connections.

To verify that the S-box example in Figure 14.6 performs a *nonlinear transformation,* we need only use the superposition theorem stated below as a test. Let

$$C = Ta + Tb$$
$$C' = T(a + b)$$

$$(14.21)$$

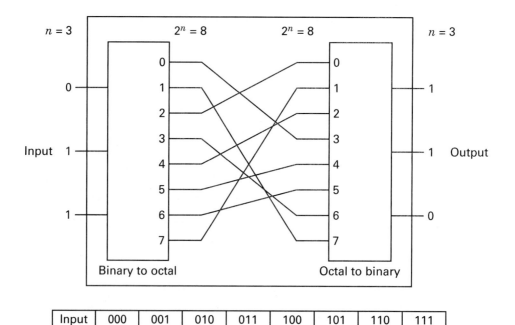

Input	000	001	010	011	100	101	110	111
Output	011	111	000	110	010	100	101	001

Figure 14.6   Substitution box.

where a and b are input terms, C and C' are output terms, and T is the transformation. Then

$$\text{If } T \text{ is linear: } C = C' \text{ for all inputs}$$

$$\text{If } T \text{ is nonlinear: } C \neq C'$$

Suppose that $a = 001$ and $b = 010$; then, using T as described in Figure 14.6, we obtain

$$C = T(001) \oplus T(010) = 111 \oplus 000 = 111$$

$$C' = T(001 \oplus 010) = T(011) = 110$$

where the symbol $\oplus$ represents modulo-2 addition. Since $C \neq C'$, the S-box is nonlinear.

14.3.3 Permutation

In permutation (transposition), the positions of the plaintext letters in the message are simply rearranged, rather than being substituted with other letters of the alphabet as in the classic ciphers. For example, the word THINK might appear, after permutation, as the ciphertext HKTNI. Figure 14.7 represents an example of binary data permutation (a linear operation). Here we see that the input data are simply rearranged or permuted (P-box). The technique has one major disadvantage when used alone; it is vulnerable to trick messages. A trick message is

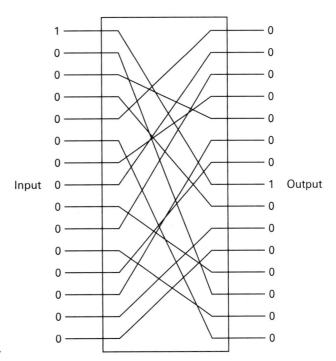

Figure 14.7 Permutation box.

illustrated in Figure 14.7. A single 1 at the input and all the rest 0 quickly reveals one of the internal connections. If the cryptanalyst can subject the system to a plaintext attack, he will transmit a sequence of such trick messages, moving the single 1 one position for each transmission. In this way, each of the connections from input to output is revealed. This is an example of why a system's security should not depend on its architecture.

14.3.4 Product Cipher System

For transformation involving reasonable numbers of n-message symbols, both of the foregoing cipher systems (the S-box and the P-box) are by themselves wanting. Shannon [5] suggested using a *product cipher* or a combination of S-box and P-box transformations, which together could yield a cipher system more powerful than either one alone. This approach of alternately applying substitution and permutation transformations has been used by IBM in the LUCIFER system [7, 8], and has become the basis for the national Data Encryption Standard (DES) [9]. Figure 14.8 illustrates such a combination of P-boxes and S-boxes. Decryption is accomplished by running the data backward, using the inverse of each S-box. The system as pictured in Figure 14.8 is difficult to implement since each S-box is different, a randomly generated key is not usable, and the system does not lend itself to repeated use of the same circuitry. To avoid these difficulties, the LUCIFER system [8] used two different types of S-boxes, S_1 and S_0, which could be publicly revealed. Figure 14.9 illustrates such a system. The input data are transformed by the sequence of S-boxes and P-boxes under the dictates of a key. The 25-bit key in this example designates, with a binary one or zero, the choice (S_1 or S_0) of each of the 25 S-boxes

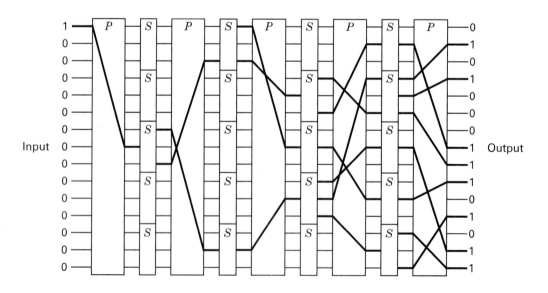

Figure 14.8   Product cipher system.

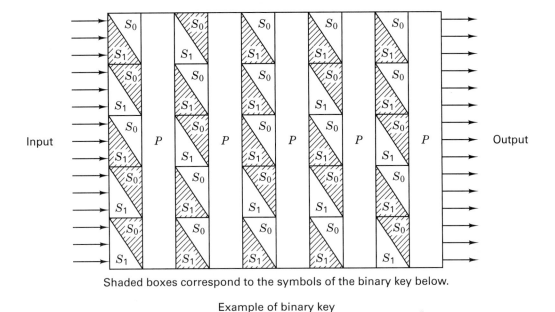

Shaded boxes correspond to the symbols of the binary key below.

Example of binary key

1 0 1 0 0 0 1 0 1 1 1 1 1 0 1 1 0 1 0 1 1 1 0 1 0

Figure 14.9   Individual keying capability.

in the block. The details of the encryption devices can be revealed since security of the system is provided by the key.

The iterated structure of the product cipher system in Figure 14.9 is typical of most present-day block ciphers. The messages are partitioned into successive blocks of n bits, each of which is encrypted with the same key. The n-bit block represents one of 2^n different characters, allowing for $(2^n)!$ different substitution patterns. Consequently, for a reasonable implementation, the substitution part of the encryption scheme is performed in parallel on small segments of the block. An example of this is seen in the next section.

14.3.5 The Data Encryption Standard

In 1977, the National Bureau of Standards adopted a modified Lucifer system as the national Data Encryption Standard (DES) [9]. From a system input-output point of view, DES can be regarded as a block encryption system with an alphabet size of 2^{64} symbols, as shown in Figure 14.10. An input block of 64 bits, regarded as a plaintext symbol in this alphabet, is replaced with a new ciphertext symbol. Figure 14.11 illustrates the system functions in block diagram form. The encryption algorithm starts with an initial permutation (IP) of the 64 plaintext bits, described in the IP-table (Table 14.1). The IP-table is read from left to right and from top to bottom, so that bits $x_1, x_2, \ldots, x_{64}$ are permuted to $x_{58}, x_{50}, \ldots, x_7$. After this initial permutation, the heart of the encryption algorithm consists of 16 iterations using

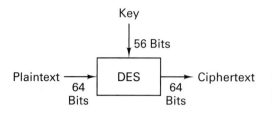

Key

56 Bits

Plaintext → DES → Ciphertext
64 Bits   64 Bits

Figure 14.10 Data encryption standard (DES) viewed as a block encryption system.

the standard building block (SBB) shown in Figure 14.12. The standard building block uses 48 bits of key to transform the 64 input data bits into 64 output data bits, designated as 32 left-half bits and 32 right-half bits. The output of each building block becomes the input to the next building block. The input right-half 32 bits (R_{i-1}) are copied unchanged to become the output left-half 32 bits (L_i). The R_{i-1} bits are also *extended* and transformed into 48 bits with the E-table (Table 14.2), and then modulo-2 summed with the 48 bits of the key. As in the case of the IP-table, the E-table is read from left to right and from top to bottom. The table expands bits

$$R_{i-1} = x_1, x_2, \ldots, x_{32}$$

into

$$(R_{i-1})_E = x_{32}, x_1, x_2, \ldots, x_{32}, x_1 \qquad (14.22)$$

Notice that the bits listed in the first and last columns of the E-table are those bit positions that are used twice to provide the 32 bit-to-48 bit expansion.

Next, $(R_{i-1})_E$ is modulo-2 summed with the ith key selection, explained later, and the result is segmented into eight 6-bit blocks

$$B_1, B_2, \ldots, B_8$$

That is,

$$(R_{i-1})_E \oplus K_i = B_1, B_2, \cdots, B_8 \qquad (14.23)$$

Each of the eight 6-bit blocks, B_j, is then used as an input to an S-box function which returns a 4-bit block, $S_j(B_j)$. Thus the input 48 bits are transformed by the S-box to 32 bits. The S-box mapping function, S_j, is defined in Table 14.3. The transformation of $B_j = b_1, b_2, b_3, b_4, b_5, b_6$ is accomplished as follows. The integer corresponding to bits, b_1, b_6 selects a row in the table, and the integer corresponding to bits $b_2 b_3 b_4 b_5$ selects a column in the table. For example, if $b_1 = 110001$, then S_1 returns the value in row 3, column 8, which is the integer 5 and is represented by the bit sequence 0101. The resulting 32-bit block out of the S-box is then permuted using the P-table (Table 14.4). As in the case of the other tables, the P-table is read from left to right and from top to bottom, so that bits $x_1, x_2, \ldots, x_{32}$ are permuted to $x_{16}, x_7, \ldots, x_{25}$. The 32-bit output of the P-table is modulo-2 summed with the input left-half 32 bits (L_{i-1}), forming the output right-half 32 bits (R_i).

The algorithm of the standard building block can be represented by

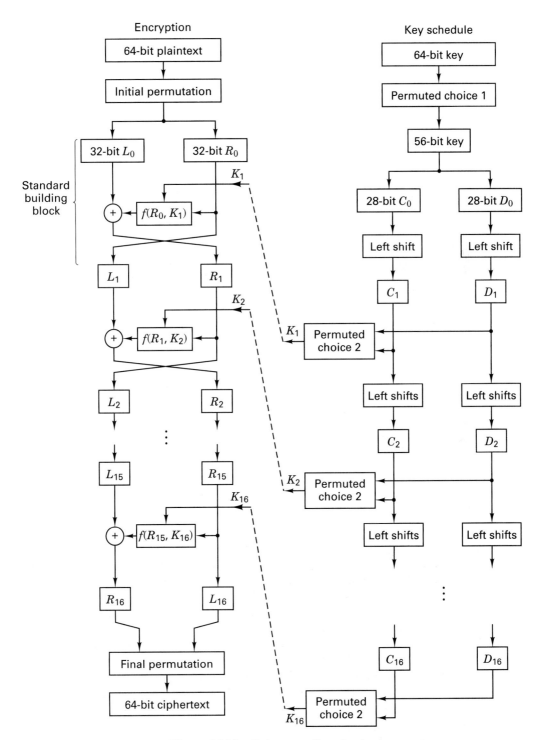

Figure 14.11 Data encryption standard.

TABLE 14.1   Initial Permutation (IP)

58	50	42	34	26	18	10	2
60	52	44	36	28	20	12	4
62	54	46	38	30	22	14	6
64	56	48	40	32	24	16	8
57	49	41	33	25	17	9	1
59	51	43	35	27	19	11	3
61	53	45	37	29	21	13	5
63	55	47	39	31	23	15	7

$$L_i = R_{i-1} \tag{14.24}$$

$$R_i = L_{i-1} \oplus f(R_{i-1}, K_i) \tag{14.25}$$

where $f(R_{i-1}, K_i)$ denotes the functional relationship comprising the E-table, S-box, and P-table we have described. After 16 iterations of the SBB, the data are transposed according to the final inverse permutation (IP^{-1}) described in the IP^{-1}-table (Table 14.5), where the output bits are read from left to right and from top to bottom, as before.

To decrypt, the same algorithm is used but the key sequence that is used in the standard building block is taken in the reverse order. Note that the value of $f(R_{i-1}, K_i)$ which can also be expressed in terms of the output of the ith block as $f(L_i, K_i)$, makes the decryption process possible.

14.3.5.1 Key Selection

Key selection also proceeds in 16 iterations, as seen in the key schedule portion of Figure 14.11. The input key consists of a 64-bit block with 8 parity bits in positions 8, 16, ... , 64. The permuted choice 1 (PC-1) discards the parity bits and permutes the remaining 56 bits as shown in Table 14.6. The output of PC-1 is split into two halves, C and D, of 28 bits each. Key selection proceeds in 16 iterations in

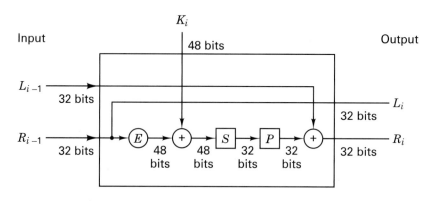

Figure 14.12   Standard building block (SBB).

TABLE 14.2 *E*-Table Bit Selection

32	1	2	3	4	5
4	5	6	7	8	9
8	9	10	11	12	13
12	13	14	15	16	17
16	17	18	19	20	21
20	21	22	23	24	25
24	25	26	27	28	29
28	29	30	31	32	1

TABLE 14.3 *S*-Box Selection Functions

Row	Column																
	0	1	2	3	4	5	6	7	8	9	10	11	12	13	14	15	
0	14	4	13	1	2	15	11	8	3	10	6	12	5	9	0	7	
1	0	15	7	4	14	2	13	1	10	6	12	11	9	5	3	8	S_1
2	4	1	14	8	13	6	2	11	15	12	9	7	3	10	5	0	
3	15	12	8	2	4	9	1	7	5	11	3	14	10	0	6	13	
0	15	1	8	14	6	11	3	4	9	7	2	13	12	0	5	10	
1	3	13	4	7	15	2	8	14	12	0	1	10	6	9	11	5	S_2
2	0	14	7	11	10	4	13	1	5	8	12	6	9	3	2	15	
3	13	8	10	1	3	15	4	2	11	6	7	12	0	5	14	9	
0	10	0	9	14	6	3	15	5	1	13	12	7	11	4	2	8	
1	13	7	0	9	3	4	6	10	2	8	5	14	12	11	15	1	S_3
2	13	6	4	9	8	15	3	0	11	1	2	12	5	10	14	7	
3	1	10	13	0	6	9	8	7	4	15	14	3	11	5	2	12	
0	7	13	14	3	0	6	9	10	1	2	8	5	11	12	4	15	
1	13	8	11	5	6	15	0	3	4	7	2	12	1	10	14	9	S_4
2	10	6	9	0	12	11	7	13	15	1	3	14	5	2	8	4	
3	3	15	0	6	10	1	13	8	9	4	5	11	12	7	2	14	
0	2	12	4	1	7	10	11	6	8	5	3	15	13	0	14	9	
1	14	11	2	12	4	7	13	1	5	0	15	10	3	9	8	6	S_5
2	4	2	1	11	10	13	7	8	15	9	12	5	6	3	0	14	
3	11	8	12	7	1	14	2	13	6	15	0	9	10	4	5	3	
0	12	1	10	15	9	2	6	8	0	13	3	4	14	7	5	11	
1	10	15	4	2	7	12	9	5	6	1	13	14	0	11	3	8	S_6
2	9	14	15	5	2	8	12	3	7	0	4	10	1	13	11	6	
3	4	3	2	12	9	5	15	0	11	14	1	7	6	0	8	13	
0	4	11	2	14	15	0	8	13	3	12	9	7	5	10	6	1	
1	13	0	11	7	4	9	1	10	14	3	5	12	2	15	8	6	S_7
2	1	4	11	13	12	3	7	14	10	15	6	8	0	5	9	2	
3	6	11	13	8	1	4	10	7	9	5	0	15	14	2	3	12	
0	13	2	8	4	6	15	11	1	10	9	3	14	5	0	12	7	
1	1	15	13	8	10	3	7	4	12	5	6	11	0	14	9	2	S_8
2	7	11	4	1	9	12	14	2	0	6	10	13	15	3	5	8	
3	2	1	14	7	4	10	8	13	15	12	9	0	3	5	6	11	

TABLE 14.4 *P*-Table Permutation

16	7	20	21
29	12	28	17
1	15	23	26
5	18	31	10
2	8	24	14
32	27	3	9
19	13	30	6
22	11	4	25

order to provide a different set of 48 key bits to each SBB encryption iteration. The C and D blocks are successively shifted according to

$$C_i = \mathrm{LS}_i(C_{i-1}) \quad \text{and} \quad D_i = \mathrm{LS}_i(D_{i-1}) \qquad (14.26)$$

where LS_i is a left circular shift by the number of positions shown in Table 14.7. The sequence C_i, D_i is then transposed according to the permuted choice 2 (PC-2) shown in Table 14.8. The result is the key sequence K_i, which is used in the ith iteration of the encryption algorithm.

The DES can be implemented as a block encryption system (see Figure 14.11), which is sometimes referred to as a *codebook* method. A major disadvantage of this method is that a given block of input plaintext will always result in the same output ciphertext (under the same key). Another encryption mode, called the *cipher feedback* mode, encrypts single bits rather than characters, resulting in a stream encryption system [3]. With the cipher feedback scheme (described later), the encryption of a segment of plaintext not only depends on the key and the current data, but also on some of the earlier data.

Since the late 1970s, two points of contention have been widely publicized about the DES [10]. The first concerns the key variable length. Some researchers felt that 56 bits are not adequate to preclude an exhaustive search. The second concerns the details of the internal structure of the *S*-boxes, which were never released by IBM. The National Security Agency (NSA), which had been involved in the testing of the DES algorithm, had requested that the information not be publicly discussed, because it was sensitive. The critics feared that NSA had been involved in design selections that would allow NSA to "tap into" any DES-encrypted messages [10]. DES is no longer a viable choice for strong encryption. The 56-bit key

TABLE 14.5 Final Permutation (IP^{-1})

40	8	48	16	56	24	64	32
39	7	47	15	55	23	63	31
38	6	46	14	54	22	62	30
37	5	45	13	53	21	61	29
36	4	44	12	52	20	60	28
35	3	43	11	51	19	59	27
34	2	42	10	50	18	58	26
33	1	41	9	49	17	57	25

TABLE 14.6 Key Permutation PC-1

57	49	41	33	25	17	9
1	58	50	42	34	26	18
10	2	59	51	43	35	27
19	11	3	60	52	44	36
63	55	47	39	31	23	15
7	62	54	46	38	30	22
14	6	61	53	45	37	29
21	13	5	28	20	12	4

can be found in a matter of days with relatively inexpensive computer tools [11]. (Some alternative algorithms are discussed in Section 14.6.)

14.4 STREAM ENCRYPTION

Earlier, we defined a *one-time pad* as an encryption system with a random key, used one time only, that exhibits unconditional security. One can conceptualize a stream encryption implementation of a one-time pad using a truly random key stream (the key sequence never repeats). Thus, perfect secrecy can be achieved for an infinite number of messages, since each message would be encrypted with a different portion of the random key stream. The development of stream encryption schemes represents an attempt to emulate the one-time pad. Great emphasis was placed on generating key streams that appeared to be random, yet could easily be implemented for decryption, because they could be generated by algorithms. Such stream encryption techniques use pseudorandom (PN) sequences, which derive their name from the fact that they appear random to the casual observer; binary

TABLE 14.7 Key Schedule of Left Shifts

Iteration, i	Number of left shifts
1	1
2	1
3	2
4	2
5	2
6	2
7	2
8	2
9	1
10	2
11	2
12	2
13	2
14	2
15	2
16	1

TABLE 14.8 Key Permutation PC-2

14	17	11	24	1	5
3	28	15	6	21	10
23	19	12	4	26	8
16	7	27	20	13	2
41	52	31	37	47	55
30	40	51	45	33	48
44	49	39	56	34	53
46	42	50	36	29	32

pseudorandom sequences have statistical properties similar to the random flipping of a fair coin. However, the sequences, of course, are deterministic (see Section 12.2). These techniques are popular because the encryption and decryption algorithms are readily implemented with feedback shift registers. At first glance it may appear that a PN key stream can provide the same security as the one-time pad, since the period of the sequence generated by a maximum-length linear shift register is $2^n - 1$ bits, where n is the number of stages in the register. If the PN sequence were implemented with a 50-stage register and a 1-MHz clock rate, the sequence would repeat every $2^{50} - 1$ microseconds, or every 35 years. In this era of large-scale integrated (LSI) circuits, it is just as easy to provide an implementation with 100 stages, in which case the sequence would repeat every 4×10^{16} years. Therefore, one might suppose that since the PN sequence does not repeat itself for such a long time, it would appear truly random and yield perfect secrecy. There is one important difference between the PN sequence and a truly random sequence used by a one-time pad. The PN sequence is generated by an algorithm; thus, knowing the algorithm, one knows the entire sequence. In Section 14.4.2 we will see that an encryption scheme that uses a linear feedback shift register in this way is very vulnerable to a *known plaintext attack*.

14.4.1 Example of Key Generation Using a Linear Feedback Shift Register

Stream encryption techniques generally employ shift registers for generating their PN key sequence. A shift register can be converted into a pseudorandom sequence generator by including a feedback loop that computes a new term for the first stage based on the previous n terms. The register is said to be linear if the numerical operation in the feedback path is linear. The PN generator example from Section 12.2 is repeated in Figure 14.13. For this example, it is convenient to number the stages as shown in Figure 14.13, where $n = 4$ and the outputs from stages 1 and 2 are modulo-2 added (linear operation) and fed back to stage 4. If the initial state of stages (x_4, x_3, x_2, x_1) is 1 0 0 0, the succession of states triggered by clock pulses would be 1 0 0 0, 0 1 0 0, 0 0 1 0, 1 0 0 1, 1 1 0 0, and so on. The output sequence is made up of the bits shifted out from the rightmost stage of the register, that is, 1 1 1 1 0 1 0 1 1 0 0 1 0 0 0, where the rightmost bit in this sequence is the earliest output and the leftmost bit is the most recent output. Given any linear feedback shift register of degree n, the output sequence is ultimately periodic.

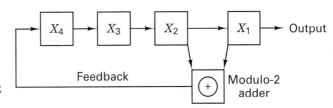

Figure 14.13 Linear feedback shift register example.

14.4.2 Vulnerabilities of Linear Feedback Shift Registers

An encryption scheme that uses a linear feedback shift register (LFSR) to generate the key stream is very vulnerable to attack. A cryptanalyst needs only $2n$ bits of plaintext and its corresponding ciphertext to determine the feedback taps, the initial state of the register, and the entire sequence of the code. In general, $2n$ is very small compared with the period $2^n - 1$. Let us illustrate this vulnerability with the LFSR example illustrated in Figure 14.13. Imagine that a cryptanalyst who knows nothing about the internal connections of the LFSR manages to obtain $2n = 8$ bits of ciphertext and its plaintext equivalent:

$$\text{Plaintext:} \quad 0\ 1\ 0\ 1\ 0\ 1\ 0\ 1$$
$$\text{Ciphertext:} \quad 0\ 0\ 0\ 0\ 1\ 1\ 0\ 0$$

Where, the rightmost bit is the earliest received and the leftmost bit is the most recent that was received.

The cryptanalyst adds the two sequences together, modulo-2, to obtain the segment of the key stream, 0 1 0 1 1 0 0 1, illustrated in Figure 14.14. The key stream sequence shows the contents of the LFSR stages at various times. The rightmost border surrounding four of the key bits shows the contents of the shift register at time t_1. As we successively slide the "moving" border one digit to the left, we see the shift register contents at times $t_2, t_3, t_4, \ldots$. From the linear structure of the four-stage shift register, we can write

$$g_4 x_4 + g_3 x_3 + g_2 x_2 + g_1 x_1 = x_5 \tag{14.27}$$

where x_5 is the digit fed back to the input and g_i (= 1 or 0) defines the ith feedback connection. For this example, we can thus write the following four equations with four unknowns, by examining the contents of the shift register at the four times shown in Figure 14.14:

$$g_4(1) + g_3(0) + g_2(0) + g_1(1) = 1$$
$$g_4(1) + g_3(1) + g_2(0) + g_1(0) = 0$$
$$g_4(0) + g_3(1) + g_2(1) + g_1(0) = 1 \tag{14.28}$$
$$g_4(1) + g_3(0) + g_2(1) + g_1(1) = 0$$

The solution of Equations (14.28) is $g_1 = 1$, $g_2 = 1$, $g_3 = 0$, $g_4 = 0$, corresponding to the LFSR shown in Figure 14.13. The cryptanalyst has thus learned the connections of

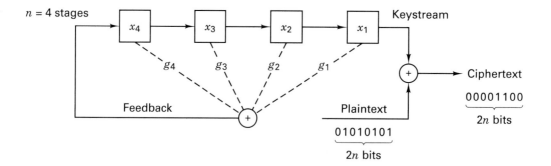

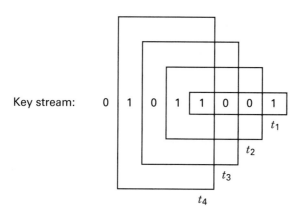

Figure 14.14 Example of vulnerability of a linear feedback shift register.

the LFSR, together with the starting state of the register at time t_1. He can therefore know the sequence for all time [3]. To generalize this example for any n-stage LFSR, we rewrite Equation (14.27) as follows:

$$x_{n+1} = \sum_{i=1}^{n} g_i x_i \qquad (14.29)$$

We can write Equation (14.29) as the matrix equation

$$\mathbf{x} = \mathbf{X}\mathbf{g} \qquad (14.30)$$

where

$$\mathbf{x} = \begin{bmatrix} x_{n+1} \\ x_{n+2} \\ \vdots \\ x_{2n} \end{bmatrix} \qquad \mathbf{g} = \begin{bmatrix} g_1 \\ g_2 \\ \vdots \\ g_n \end{bmatrix}$$

and

$$\mathbf{X} = \begin{bmatrix} x_1 & x_2 & \cdots & x_n \\ x_2 & x_3 & \cdots & x_{n+1} \\ \vdots & \vdots & & \vdots \\ x_n & x_{n+1} & \cdots & x_{2n-1} \end{bmatrix}$$

It can be shown [3] that the columns of $\mathbf{X}$ are linearly independent; thus $\mathbf{X}$ is non-singular (its determinant is nonzero) and has an inverse. Hence,

$$\mathbf{g} = \mathbf{X}^{-1} \mathbf{x} \qquad (14.31)$$

The matrix inversion requires at most on the order of n^3 operations and is thus easily accomplished by computer for any reasonable value of n. For example, if $n = 100$, $n^3 = 10^6$, and a computer with a 1-μs operation cycle would require 1 s for the inversion. The weakness of a LFSR is caused by the linearity of Equation (14.31). The use of *nonlinear feedback* in the shift register makes the cryptanalyst's task much more difficult, if not computationally intractable.

14.4.3 Synchronous and Self-Synchronous Stream Encryption Systems

We can categorize stream encryption systems as either *synchronous* of *self-synchronous*. In the former, the key stream is generated independently of the message, so that a lost character during transmission necessitates a resynchronization of the transmission and receiver key generators. A synchronous stream cipher is shown in Figure 14.15. The starting state of the key generator is initialized with a known input, I_0. The ciphertext is obtained by the modulo addition of the ith key character, k_i, with the ith message character, m_i. Such synchronous ciphers are generally designed to utilize *confusion* (see Section 14.3.1) but not *diffusion*. That is, the encryption of a character is not diffused over some block length of message. For this reason, synchronous stream ciphers do not exhibit *error propagation*.

In a *self-synchronous* stream cipher, each key character is derived from a fixed number, n, of the preceding ciphertext characters, giving rise to the name *cipher feedback*. In such a system, if a ciphertext character is lost during

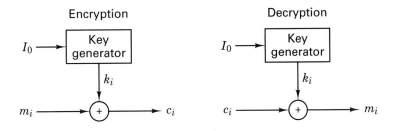

Figure 14.15   Synchronous stream cipher.

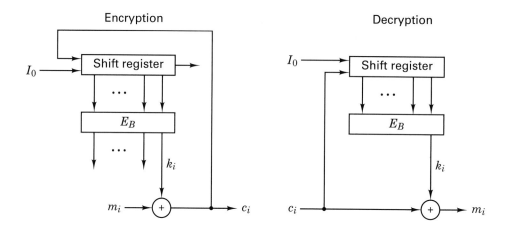

Figure 14.16 Cipher feedback mode.

transmission, the error propagates forward for n characters, but the system resynchronizes itself after n correct ciphertext characters are received.

In Section 14.1.4 we looked at an example of cipher feedback in the Vigenere auto key cipher. We saw that the advantages of such a system are that (1) a nonrepeating key is generated, and (2) the statistics of the plaintext message are diffused throughout the ciphertext. However, the fact that they key was exposed in the ciphertext was a basic weakness. This problem can be eliminated by passing the ciphertext characters through a nonlinear block cipher to obtain the key characters. Figure 14.16 illustrates a shift register key generator operating in the cipher feedback mode. Each output ciphertext character, c_i (formed by the modulo addition of the message character, m_i, and the key character, k_i), is fed back to the input of the shift register. As before, initialization is provided by a known input, I_0. At each iteration, the output of the shift register is used as input to a (nonlinear) block encryption algorithm E_B. The low-order output character from E_B becomes the next key character, k_{i+1}, to be used with the next message character, m_{i+1}. Since, after the first few iterations, the input to the algorithm depends only on the ciphertext, the system is self-synchronizing.

14.5 PUBLIC KEY CRYPTOSYSTEMS

The concept of public key cryptosystems was introduced in 1976 by Diffie and Hellman [12]. In conventional cryptosystems the encryption algorithm can be revealed since the security of the system depends on a safeguarded key. The same key is used for both encryption and decryption. Public key cryptosystems utilize *two different* keys, one for encryption and the other for decryption. In public key cryptosystems, not only the encryption algorithm but also the encryption key can be publicly revealed without compromising the security of the system. In fact, a public directory, much like a telephone directory, is envisioned, which contains the

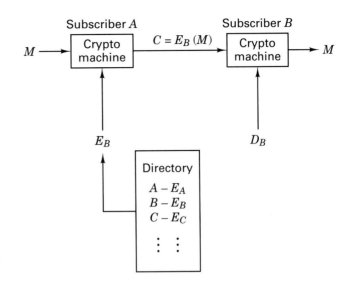

Figure 14.17 Public key cryptosystem.

encryption keys of all the subscribers. Only the decryption keys are kept secret. Figure 14.17 illustrates such a system. The important features of a public key cryptosystem are as follows:

1. The encryption algorithm E_K and the decryption algorithm D_K are invertible transformations on the plaintext M, or the ciphertext C, defined by the key K. That is, for each K and M, if $C = E_K(M)$, then $M = D_K(C) = D_K[E_K(M)]$.

2. For each K, E_K and D_K are easy to compute.

3. For each K, the computation of D_K from E_K is computationally intractable.

Such a system would enable secure communication between subscribers who have never met or communicated before. For example, as seen in Figure 14.17, subscriber A can send a message, M, to subscriber B by looking up B's encryption key in the directory and applying the encryption algorithm, E_B, to obtain the ciphertext $C = E_B(M)$, which he transmits on the public channel. Subscriber B is the only party who can decrypt C by applying his decryption algorithm, D_B, to obtain $M = D_B(C)$.

14.5.1 Signature Authentication Using a Public Key Cryptosystem

Figure 14.18 illustrates the use of a public key cryptosystem for signature authentication. Subscriber A "signs" his message by first applying his decryption algorithm, D_A, to the message, yielding $S = D_A(M) = E_A^{-1}(M)$. Next, he uses the encryption algorithm, E_B, of subscriber B to encrypt S, yielding $C = E_B(S) = E_B[E_A^{-1}(M)]$, which he transmits on a public channel. When subscriber B receives C, he first decrypts it using his private decryption algorithm, D_B, yielding $D_B(C) = E_A^{-1}(M)$. Then he applies the encryption algorithm of subscriber A to produce $E_A[E_A^{-1}(M)] = M$.

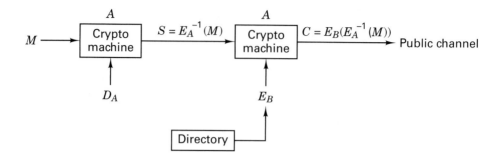

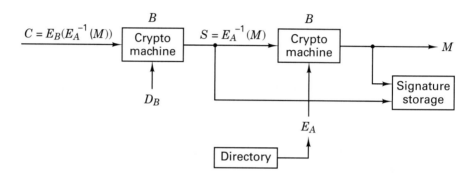

Figure 14.18 Signature authentication using a public key cryptosystem.

If the result is an intelligible message, it must have been initiated by subscriber A, since no one else could have known A's secret decryption key to form $S = D_A(M)$. Notice that S is both message dependent and signer dependent, which means that while B can be sure that the received message indeed came from A, at the same time A can be sure that no one can attribute any false messages to him.

14.5.2 A Trapdoor One-Way Function

Public key cryptosystems are based on the concept of trapdoor one-way functions. Let us first define, a *one-way function* as an easily computed function whose inverse is computationally infeasible to find. For example, consider the function $y = x^5 + 12x^3 + 107x + 123$. It should be apparent that given x, y is easy to compute, but given y, x is relatively difficult to compute. A *trapdoor one-way function* is a one-way function whose inverse is easily computed if certain features, used to design the function, are known. Like a trapdoor, such functions are easy to go through in one direction. Without special information the reverse process takes an impossibly long time. We will apply the concept of a trapdoor in Section 14.5.5, when we discuss the Merkle–Hellman scheme.

14.5.3 The Rivest–Shamir–Adelman Scheme

In the Rivest–Shamir–Adelman (RSA) scheme messages are first represented as integers in the range $(0, n - 1)$. Each user chooses his own value of n and another pair of positive integers e and d, in a manner to be described below. The user places his encryption key, the number pair (n, e), in the public directory. The decryption key consists of the number pair (n, d), of which d is kept secret. Encryption of a message M and decryption of a ciphertext C are defined as follows:

$$\text{Encryption:} \quad C = E(M) = (M)^e \text{ modulo-}n$$
$$\text{Decryption:} \quad M = D(C) = (C)^d \text{ modulo-}n \tag{14.32}$$

They are each easy to compute and the results of each operation are integers in the range $(0, n - 1)$. In the RSA scheme, n is obtained by selecting *two large prime numbers* p and q and multiplying them together:

$$n = pq \tag{14.33}$$

Although n is made public, p and q are kept hidden, due to the great difficulty in factoring n. Then

$$\phi(n) = (p - 1)(q - 1) \tag{14.34}$$

called *Euler's totient function*, is formed. The parameter $\phi(n)$ has the interesting property [12] that for any integer X in the range $(0, n - 1)$ and any integer k,

$$X = X^{k\phi(n)+1} \text{ modulo-}n \tag{14.35}$$

Therefore, while all other arithmetic is done modulo-n, arithmetic in the exponent is done modulo-$\phi(n)$. A large integer, d, is randomly chosen so that it is relatively prime to $\phi(n)$, which means that $\phi(n)$ and d must have no common divisors other than 1, expressed as

$$\gcd[\phi(n), d] = 1 \tag{14.36}$$

where gcd means "greatest common divisor." Any prime number greater than the larger of (p, q) will suffice. Then the integer e, where $0 < e < \phi(n)$, is found from the relationship

$$ed \text{ modulo-}\phi(n) = 1 \tag{14.37}$$

which, from Equation (14.35), is tantamount to choosing e and d to satisfy

$$X = X^{ed} \text{ modulo-}n \tag{14.38}$$

Therefore,

$$E[D(X)] = D[E(X)] = X \tag{14.39}$$

and decryption works correctly. Given an encryption key (n, e), one way that a cryptanalyst might attempt to break the cipher is to factor n into p and q, compute $\phi(n) = (p - 1)(q - 1)$, and compute d from Equation (14.37). This is all straightforward except for the factoring of n.

The RSA scheme is based on the fact that it is easy to generate two large prime numbers, p and q, and multiply them together, but it is very much more difficult to factor the result. The product can therefore be made public as part of the encryption key, without compromising the factors that would reveal the decryption key corresponding to the encryption key. By making each of the factors roughly 100 digits long, the multiplication can be done in a fraction of a second, but the exhaustive factoring of the result should take billions of years [2].

14.5.3.1 Use of the RSA Scheme

Using the example in Reference [13], let $p = 47$, $q = 59$. Therefore, $n = pq = 2773$ and $\phi(n) = (p - 1)(q - 1) = 2668$. The parameter d is chosen to be relatively prime to $\phi(n)$. For example, choose $d = 157$. Next, the value of e is computed as follows (the details are shown in the next section):

$$ed \text{ modulo } \phi(n) = 1$$

$$157e \text{ modulo } 2688 = 1$$

Therefore, $e = 17$. Consider the plaintext example

ITS ALL GREEK TO ME

By replacing each letter with a two-digit number in the range (01, 26) corresponding to its position in the alphabet, and encoding a blank as 00, the plaintext message can be written as

0920 1900 0112 1200 0718 0505 1100 2015 0013 0500

Each message needs to be expressed as an integer in the range $(0, n - 1)$; therefore, for this example, encryption can be performed on blocks of four digits at a time since this is the maximum number of digits that will always yield a number less than $n - 1 = 2772$. The first four digits (0920) of the plaintext are encrypted as follows:

$$C = (M)^e \text{ modulo-}n = (920)^{17} \text{ modulo-2773} = 948$$

Continuing this process for the remaining plaintext digits, we get

$$C = 0948 \ 2342 \ 1084 \ 1444 \ 2663 \ 2390 \ 0778 \ 0774 \ 0219 \ 1655$$

The plaintext is returned by applying the decryption key, as follows:

$$M = (C)^{157} \text{ modulo-2773}$$

14.5.3.2 How to Compute e

A variation of Euclid's algorithm [14] for computing the gcd of $\phi(n)$ and d is used to compute e. First, compute a series $x_0, x_1, x_2, \ldots$, where $x_0 = \phi(n)$, $x_1 = d$, and $x_{i+1} = x_{i-1} \text{ modulo-}x_i$, until an $x_k = 0$ is found. Then the gcd $(x_0, x_1) = x_{k-1}$. For each x_i compute numbers a_i and b_i such that $x_i = a_i x_0 + b_i x_1$. If $x_{k-1} = 1$, then b_{k-1} is the multiplicative inverse of x_1 modulo-x_0. If b_{k-1} is a negative number, the solution is $b_{k-1} + \phi(n)$.

Example 14.5 Computation of e from d and $\phi(n)$

For the previous example, with $p = 47$, $q = 59$, $n = 2773$, $\phi(n) = 2688$, and d chosen to be 157, use the Euclid algorithm to verify that $e = 17$.

Solution

i	x_i	a_i	b_i	y_i
0	2668	1	0	
1	157	0	1	16
2	156	1	−16	1
3	1	−1	17	

where

$$y_i = \left\lfloor \frac{x_{i-1}}{x_i} \right\rfloor$$

$$x_{i+1} = x_{i-1} - y_i x_i$$

$$a_{i+1} = a_{i-1} - y_i a_i$$

$$b_{i+1} = b_{i-1} - y_i b_i$$

Hence

$$e = b_3 = 17$$

14.5.4 The Knapsack Problem

The classic knapsack problem is illustrated in Figure 14.19. The knapsack is filled with a subset of the items shown with weights indicated in grams. Given the weight of the filled knapsack (the scale is calibrated to deduct the weight of the empty knapsack), determine which items are contained in the knapsack. For this simple example, the solution can easily be found by trial and error. However, if there are 100 possible items in the set instead of 10, the problem may become computationally infeasible.

Let us express the knapsack problem in terms of a knapsack vector and a data vector. The knapsack vector is an n-tuple of distinct integers (analogous to the set of possible knapsack items)

$$\mathbf{a} = a_1, a_2, \ldots, a_n$$

The data vector is an n-tuple of binary symbols

$$\mathbf{x} = x_1, x_2, \ldots, x_n$$

The knapsack, S, is the sum of a subset of the components of the knapsack vector:

$$S = \sum_{i=1}^{n} a_i x_i \quad \text{where } x_i = 0, 1 \tag{14.40}$$

$$= \mathbf{a}\mathbf{x}$$

The knapsack problem can be stated as follows: Given S and knowing $\mathbf{a}$, determine $\mathbf{x}$.

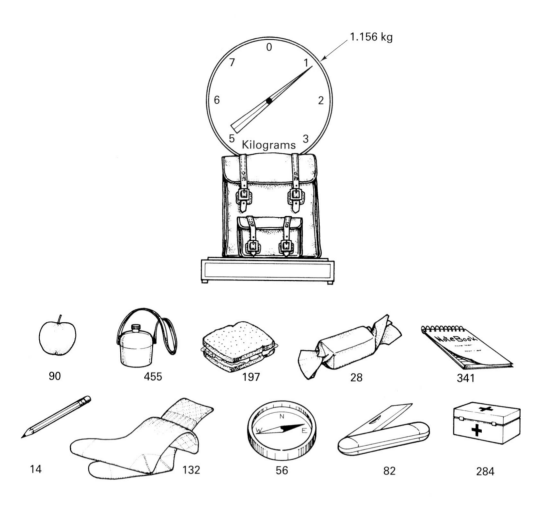

Figure 14.19 Knapsack problem.

Example 14.6 Knapsack Example

Given $\mathbf{a} = 1, 2, 4, 8, 16, 32$ and $S = \mathbf{ax} = 26$, find $\mathbf{x}$.

Solution

In this example $\mathbf{x}$ is seen to be the *binary* representation of S. The decimal-to-binary conversion should appear more familiar with $\mathbf{a}$ expressed as $2^0, 2^1, 2^2, 2^3, 2^4, 2^5$. The data vector $\mathbf{x}$ is easily found since $\mathbf{a}$ in this example is *super-increasing,* which means that each component of the n-tuple $\mathbf{a}$ is larger than the sum of the preceding components. That is,

$$a_i > \sum_{j=1}^{i-1} a_j \quad i = 2, 3, \dots, n \tag{14.41}$$

When $\mathbf{a}$ is super-increasing, the solution of $\mathbf{x}$ is found by starting with $x_n = 1$ if $S \geq a_n$ (otherwise $x_n = 0$) and continuing according to the relationship

$$x_i = \begin{cases} 1 & \text{if } S - \sum_{j=i+1}^{n} x_j a_j \geq a_i \\ 0 & \text{otherwise} \end{cases} \qquad (14.42)$$

where $i = n - 1, n - 2, \ldots, 1$. From Equation (14.42) it is easy to compute $\mathbf{x} = 0\,1\,0\,1\,1\,0$.

Example 14.7 Knapsack Example

Given $\mathbf{a} = 171, 197, 459, 1191, 2410, 4517$ and $S = \mathbf{ax} = 3798$, find $\mathbf{x}$.

Solution

As in Example 14.6, $\mathbf{a}$ is super-increasing; therefore, we can compute $\mathbf{x}$ using Equation (14.42), which again yields

$$\mathbf{x} = 0\,1\,0\,1\,1\,0$$

14.5.5 A Public Key Cryptosystem Based on a Trapdoor Knapsack

This scheme, also known as the Merkle–Hellman scheme [15], is based on the formation of a knapsack vector that is not super-increasing and is therefore not easy to solve. However, an essential part of this knapsack is a *trapdoor* that enables the authorized user to solve it.

First, we form a super-increasing n-tuple $\mathbf{a}'$. Then we select a prime number M such that

$$M > \sum_{i=1}^{n} a_i' \qquad (14.43)$$

We also select a random number W, where $1 < W < M$, and we form W^{-1} to satisfy the following relationship:

$$WW^{-1} \text{ modulo-}M = 1 \qquad (14.44)$$

the vector $\mathbf{a}'$ and the numbers M, W, and W^{-1} are all kept hidden. Next, we form $\mathbf{a}$ with the elements from $\mathbf{a}'$, as follows:

$$a_i = Wa_j' \text{ modulo-}M \qquad (14.45)$$

The formation of $\mathbf{a}$ using Equation (14.45) constitutes forming a knapsack vector with a *trapdoor*. When a data vector $\mathbf{x}$ is to be transmitted, we multiply $\mathbf{x}$ by $\mathbf{a}$, yielding the number S, which is sent on the public channel. Using Equation (14.45), S can be written as follows:

$$S = \mathbf{ax} = \sum_{i=1}^{n} a_i x_i = \sum_{i=1}^{n} (Wa_i' \text{ modulo-}M)x_i \qquad (14.46)$$

The authorized user receives S and, using Equation (14.44), converts it to S':

$$S' = W^{-1}S \text{ modulo-}M = W^{-1} \sum_{i=1}^{n} (Wa_i' \text{ modulo-}M)x_i \text{ modulo-}M$$

$$= \sum_{i=1}^{n} (W^{-1} W a_i' \text{modulo-}M) x_i \text{ modulo-}M \qquad (14.47)$$

$$= \sum_{i=1}^{n} a_i' x_i \text{ modulo-}M$$

$$= \sum_{i=1}^{n} a_i' x_i$$

Since the authorized user knows the secretly held super-increasing vector $\mathbf{a}'$, he or she can use S' to find $\mathbf{x}$.

14.5.5.1 Use of the Merkle–Hellman Scheme

Suppose that user A wants to construct public and private encryption functions. He first considers the super-increasing vector $\mathbf{a}' = (171, 197, 459, 1191, 2410, 4517)$

$$\sum_{i=1}^{6} a_i' = 8945$$

He then chooses a prime number M larger than 8945, a random number W, where $1 \leq W < M$, and calculates W^{-1} to satisfy $WW^{-1} = 1$ modulo-M.

$$\left. \begin{array}{l} \text{Choose } M = 9109 \\ \text{choose } W = 2251 \\ \text{then } W^{-1} = 1388 \end{array} \right\} \text{kept hidden}$$

He then forms the trapdoor knapsack vector as follows:

$$a_i = a_i' 2251 \text{ modulo-}9109$$

$$\mathbf{a} = 2343,\ 6215,\ 3892,\ 2895,\ 5055,\ 2123$$

User A makes public the vector $\mathbf{a}$, which is clearly not super-increasing. Suppose that user B wants to send a message to user A.

If $\mathbf{x} = 0\,1\,0\,1\,1\,0$ is the message to be transmitted, user B forms

$$S = \mathbf{ax} = 14{,}165 \text{ and transmits it to user } A$$

User A, who receives S, converts it to S':

$$S' = \mathbf{a}'\mathbf{x} = W^{-1}S \text{ modulo-}M$$

$$= 1388 \cdot 14{,}165 \text{ modulo-}9109$$

$$= 3798$$

Using $S' = 3798$ and the super-increasing vector $\mathbf{a}'$, user A easily solves for $\mathbf{x}$.

The Merkle–Hellman scheme is now considered broken [16], leaving the RSA scheme (as well as others discussed later) as the algorithms that are useful for implementing public key cryptosystems.

14.6 PRETTY GOOD PRIVACY

Pretty Good Privacy (PGP) is a security program that was created by Phil Zimmerman [17] and published in 1991 as free-of-charge shareware. It has since become the "de facto" standard for electronic mail (e-mail) and file encryption. PGP, widely used as version 2.6, remained essentially unchanged until PGP version 5.0 (which is compatible with version 2.6) became available. Table 14.9 illustrates the algorithms used in versions 2.6, 5.0, and later.

As listed in Table 14.9, PGP uses a variety of encryption algorithms, including both private-key- and public-key-based systems. A private-key algorithm (with a new session key generated at each session) is used for encryption of the message. The private-key algorithms offered by PGP are International Data Encryption Algorithm (IDEA), Triple-DES (Data Encryption Standard), and CAST (named after the inventors Carlisle Adams and Stafford Tavares [19]). A public-key algorithm is used for the encryption of each session key. The public-key algorithms offered by PGP are the RSA algorithm, described in Section 14.5.3, and the Diffie-Hellman algorithm.

Public-key algorithms are also used for the creation of digital signatures. PGP version 5.0 uses the Digital Signature Algorithm (DSA) specified in the NIST Digital Signature Standard (DSS). PGP version 2.6 uses the RSA algorithm for its digital signatures. If the available channel is insecure for key exchange, it is safest to use a public-key algorithm. If a secure channel is available, then private-key encryption is preferred, since it typically offers improved speed over public-key systems.

The technique for message encryption employed by PGP version 2.6 is illustrated in Figure 14.20. The plaintext is compressed with the ZIP algorithm prior to encryption. PGP uses the ZIP routine written by Jean-Loup Gailly, Mark Alder, and Richard B. Wales [18]. If the compressed text is shorter than the uncompressed text, the compressed text will be encrypted, otherwise the uncompressed text is encrypted.

TABLE 14.9   PGP 2.6 versus PGP 5.0 and Later

Function	PGP Version 2.6 Algorithm Used [17]	PGP Version 5.0 and Later Algorithm Used [18]
Encryption of message using private-key algorithm with private-session key	IDEA	Triple-DES, CAST, or IDEA
Encryption of private-session key with public-key algorithm	RSA	RSA or Diffie-Hellman (the Elgamal variation)
Digital Signature	RSA	RSA and NIST[1] Digital Signature Standard (DSS)[2]
Hash Function used for creating message digest for Digital Signatures	MD5	SHA-1

[1] National Institute of Standards and Technology, a division of the U.S. Department of Commerce.

[2] Digital Signature Standard selected by NIST.

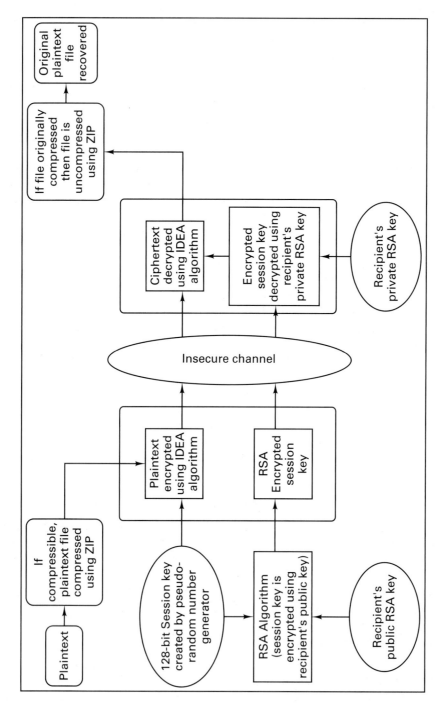

Figure 14.20 The PGP technique.

Small files (approximately 30 characters for ASCII files) will not benefit from compression. Additionally, PGP recognizes files previously compressed by popular compression routines, such as PKZIP, and will not attempt to compress them. Data compression removes redundant character strings in a file and produces a more uniform distribution of characters. Compression provides a shorter file to encrypt and decrypt (which reduces the time needed to encrypt, decrypt, and transmit a file), but compression is also advantageous because it can hinder some cryptanalytic attacks that exploit redundancy. If compression is performed on a file, it should occur *prior to* encryption (never afterwards). Why is that a good rule to follow? Because a *good* encryption algorithm yields ciphertext with a nearly statistically uniform distribution of characters; therefore, if a data compression algorithm came after such encryption, it should result in no compression at all. If any ciphertext can be compressed, then the encryption algorithm that formed that ciphertext was a poor algorithm. A compression algorithm should be *unable* to find redundant patterns in text that was encrypted by a good encryption algorithm.

As shown in Figure 14.20, PGP Version 2.6 begins file encryption by creating a 128-bit session key using a pseudo-random number generator. The compressed plaintext file is then encrypted with the IDEA private-key algorithm using this random session key. The random session key is then encrypted by the RSA public-key algorithm using the *recipient's public key*. The RSA-encrypted session key and the IDEA-encrypted file are sent to the recipient. When the recipient needs to read the file, the encrypted session key is first decrypted with RSA using the *recipient's private key*. The ciphertext file is then decrypted with IDEA using the decrypted session key. After uncompression, the recipient can read the plaintext file.

14.6.1 Triple-DES, CAST, and IDEA

As listed in Table 14.9, PGP offers three block ciphers for message encryption, Triple-DES, CAST, and IDEA. All three ciphers operate on 64-bit blocks of plaintext and ciphertext. Triple-DES has a key size of 168-bits while CAST and IDEA use key lengths of 128 bits.

14.6.1.1 Description of Triple-DES

The Data Encryption Standard (DES) described in Section 14.3.5 has been used since the late 1970s, but some have worried about its security because of its relatively small key size (56 bits). With Triple-DES, the message to be encrypted is run through the DES algorithm 3 times (the second DES operation is run in decrypt mode); each operation is performed with a different 56-bit key. As illustrated in Figure 14.21, this gives the effect of a 168-bit key length.

14.6.1.2 Description of CAST

CAST is a family of block ciphers developed by Adams and Tavares [19]. PGP version 5.0 uses a version of CAST known as CAST5, or CAST-128. This version has a block size of 64-bits and a key length of 128-bits. The CAST algorithm uses six *S*-boxes with an 8-bit input and a 32-bit output. By comparison, DES uses

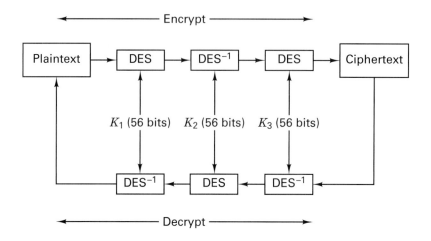

Figure 14.21 Encryption/decryption with triple-DES.

eight *S*-boxes with a 6-bit input and a 4-bit output. The *S*-boxes in Cast-128 were designed to provide highly nonlinear transformations, making this algorithm particularly resistant to cryptanalysis [11].

14.6.1.3 Description of IDEA

The International Data Encryption Algorithm (IDEA) is a block cipher designed by Xuejia Lai and James Massey [19]. It is a 64-bit iterative block cipher (involving eight iterations or rounds) with a 128-bit key. The security of IDEA relies on the use of three types of arithmetic operations on 16-bit words. The operations are addition modulo 2^{16}, multiplication modulo $2^{16} + 1$, and bit-wise exclusive-OR (XOR). The 128-bit key is used for the iterated encryption and decryption in a re-ordered fashion. As shown in Table 14.10, the original key K_0 is divided into eight 16-bit subkeys $Z_x^{(R)}$, where x is the subkey number of the round R. Six of these subkeys are used in round 1, and the remaining two are used in round 2. K_0 is then rotated 25 bits to the left yielding K_1, which is in turn divided into eight subkeys; the first 4 of these subkeys are used in round 2, and the last four in round 3. The process continues, as shown in Table 14.10, yielding a total of 52 subkeys.

The subkey schedule for each round is listed in Table 14.11 for both encryption and decryption rounds. Decryption is carried out in the same manner as encryption. The decryption subkeys are calculated from the encryption subkeys, as shown in Table 14.11, where it is seen that the decryption subkeys are either the additive or multiplicative inverses of the encryption subkeys.

The message is divided into 64-bit data blocks. These blocks are then divided into four 16-bit subblocks: M_1, M_2, M_3, and M_4. A sequence of such four subblocks becomes the input to the first round of IDEA algorithm. This data is manipulated for a total of eight rounds. Each round uses a different set of six subkeys as specified in Table 14.11. After a round, the second and third 16-bit data subblocks are

TABLE 14.10 IDEA formation of Subkeys

128-bit key (divided into eight 16-bit subkeys)	Bit string from which keys are derived
$Z_1^1 Z_2^1 Z_3^1 Z_4^1 Z_5^1 Z_6^1 Z_1^2 Z_2^2$	K_0 = Original 128-bit key
$Z_3^2 Z_4^2 Z_5^2 Z_6^2 Z_1^3 Z_2^3 Z_3^3 Z_4^3$	K_1 = 25-bit rotation of K_0
$Z_5^3 Z_6^3 Z_1^4 Z_2^4 Z_3^4 Z_4^4 Z_5^4 Z_6^4$	K_2 = 25-bit rotation of K_1
$Z_1^5 Z_2^5 Z_3^5 Z_4^5 Z_5^5 Z_6^5 Z_1^6 Z_2^6$	K_3 = 25-bit rotation of K_2
$Z_3^6 Z_4^6 Z_5^6 Z_6^6 Z_1^7 Z_2^7 Z_3^7 Z_4^7$	K_4 = 25-bit rotation of K_3
$Z_5^7 Z_6^7 Z_1^8 Z_2^8 Z_3^8 Z_4^8 Z_5^8 Z_6^8$	K_5 = 25-bit rotation of K_4
$Z_1^{out} Z_2^{out} Z_3^{out} Z_4^{out}$	First 64 bits of K_6 where K_6 = 25-bit rotation of K_5

swapped. After the completion of the eighth round, the four subblocks are manipulated in a final output transformation. For the representation of $Z_x^{(R)}$ shown in Tables 14.10 and 14.11, the round number is shown without parentheses for ease of notation.

Each round consists of the steps shown in Table 14.12. The final values from steps 11–14 form the output of the round. The two inner 16-bit data subblocks (except for the last round) are swapped, and then these four subblocks are the input to the next round. This technique continues for a total of 8 rounds. After round 8, the final output transformation is as follows:

1. $M_1 \times Z_1^{out}$ (first subkey of output transformation)
2. $M_2 + Z_2^{out}$
3. $M_3 + Z_3^{out}$
4. $M_4 \times Z_4^{out}$

TABLE 14.11 IDEA Subkey Schedule

Round	Set of Encryption Subkeys	Set of Decryption Subkeys
1	$Z_1^1 Z_2^1 Z_3^1 Z_4^1 Z_5^1 Z_6^1$	$(Z_1^{out})^{-1} -Z_2^{out} -Z_3^{out} (Z_4^{out})^{-1} Z_5^8 Z_6^8$
2	$Z_1^2 Z_2^2 Z_3^2 Z_4^2 Z_5^2 Z_6^2$	$(Z_1^8)^{-1} -Z_2^8 -Z_3^8 (Z_4^8)^{-1} Z_5^7 Z_6^7$
3	$Z_1^3 Z_2^3 Z_3^3 Z_4^3 Z_5^3 Z_6^3$	$(Z_1^7)^{-1} -Z_2^7 -Z_3^7 (Z_4^7)^{-1} Z_5^6 Z_6^6$
4	$Z_1^4 Z_2^4 Z_3^4 Z_4^4 Z_5^4 Z_6^4$	$(Z_1^6)^{-1} -Z_2^6 -Z_3^6 (Z_4^6)^{-1} Z_5^5 Z_6^5$
5	$Z_1^5 Z_2^5 Z_3^5 Z_4^5 Z_5^5 Z_6^5$	$(Z_1^5)^{-1} -Z_2^5 -Z_3^5 (Z_4^5)^{-1} Z_5^4 Z_6^4$
6	$Z_1^6 Z_2^6 Z_3^6 Z_4^6 Z_5^6 Z_6^6$	$(Z_1^4)^{-1} -Z_2^4 -Z_3^4 (Z_4^4)^{-1} Z_5^3 Z_6^3$
7	$Z_1^7 Z_2^7 Z_3^7 Z_4^7 Z_5^7 Z_6^7$	$(Z_1^3)^{-1} Z_2^3 -Z_3^3 (Z_4^3)^{-1} Z_5^2 Z_6^2$
8	$Z_1^8 Z_2^8 Z_3^8 Z_4^8 Z_5^8 Z_6^8$	$(Z_1^2)^{-1} -Z_2^2 -Z_3^2 (Z_4^2)^{-1} Z_5^1 Z_6^1$
Output Transformation	$Z_1^{out} Z_2^{out} Z_3^{out} Z_4^{out}$	$(Z_1^1)^{-1} -Z_2^1 -Z_3^1 (Z_4^1)^{-1}$

Example 14.8 The First Round of the IDEA Cipher

Consider that the message is the word "HI", which we first transform to hexadecimal (hex) notation. We start with the ASCII code table in Figure 2.3, where bit 1 is the least significant bit (LSB). We then add an eighth zero-value most significant bit (MSB), which might ordinarily be used for parity, and we transform four bits at a time reading from MSB to LSB. Thus, the letter H in the message transforms to 0048 and

TABLE 14.12   IDEA Operational Steps in Each Round

1. $M_1 \times Z_1^{(R)}$
2. $M_2 + Z_2^{(R)}$
3. $M_3 + Z_3^{(R)}$
4. $M_4 \times Z_4^{(R)}$
5. XOR[3] the results from steps 1 and 3.
6. XOR the results from steps 2 and 4.
7. Result from step 5 and $Z_5^{(R)}$ are multiplied.
8. Result from step 6 and 7 are added.
9. Result from step 8 and $Z_6^{(R)}$ are multiplied.
10. Results from steps 7 and 9 are added.
11. XOR the results from steps 1 and 9.
12. XOR the results from steps 3 and 9.
13. XOR the results from steps 2 and 10.
14. XOR the results from steps 4 and 10.

---

[3] The exclusive-OR (XOR) operation is defined as: 0 XOR 0 = 0, 0 XOR 1 = 1, 1 XOR 0 = 1, and 1 XOR 1 = 0

the letter I transforms to 0049. For this example, we choose a 128-bit key, K_0, expressed with eight groups or *subkeys* of 4-hex digits each, as follows: K_0 = 0008 0007 0006 0005 0004 0003 0002 0001, where the rightmost subkey is the least significant. Using this key and the IDEA cipher, find the output of round 1.

Solution

The message is first divided into 64-bit data blocks. Each of these blocks is then divided into subblocks, M_i, where $i = 1, \ldots 4$, each subblock containing 16-bits or 4-hex digits. In this example the message "HI" is only 16-bits in length, hence (using hex notation) $M_1 = 4849$ and $M_2 = M_3 = M_4 = 0000$. Addition is performed modulo 2^{16}, and multiplication is performed modulo $2^{16} + 1$. For the first round, the specified 128-bit key is divided into eight 16-bit subkeys starting with the least significant group of hex digits, as follows: $Z_1^{(1)} = 0001$, $Z_2^{(1)} = 0002$, $Z_3^{(1)} = 0003$, $Z_4^{(1)} = 0004$, $Z_5^{(1)} = 0005$, $Z_6^{(1)} = 0006$, $Z_1^{(2)} = 0007$, and $Z_2^{(2)} = 0008$.

The steps outlined in Table 14.11 yield:

1. $M_1 \times Z_1 = 4849 \times 0001 = 4849$.
2. $M_2 + Z_2 = 0000 + 0002 = 0002$.
3. $M_3 + Z_3 = 0000 + 0003 = 0003$.
4. $M_4 \times Z_4 = 0000 \times 0004 = 0000$.
5. The result from step (1) is XOR'ed with the result from step (3) yielding 4849 XOR 0003 = 484A, as follows:

$$\begin{array}{rllll}
 & 0100 & 1000 & 0100 & 1001 \quad \text{(4849 hex converted to binary)} \\
\text{XOR} & 0000 & 0000 & 0000 & 0011 \quad \text{(0003 hex converted to binary)} \\
\hline
 & 0100 & 1000 & 0100 & 1010
\end{array}$$

Converting back to hex yields: 484A (where A is the hex notation for 1010 binary)

6. Results from steps (2) and (4) are XOR'ed: 0002 XOR 0000 = 0002.
7. Result from step (5) and Z_5 are multiplied: 484A × 0005 = 6971.
8. Results from steps (6) and (7) are added: 0002 + 6971 = 6973.

9. Result from step (8) and Z_6 are multiplied: $6973 \times 0006 = 78B0$.
10. Results from steps (7) and (9) are added: $6971 + 78B0 = E221$.
11. Results from steps (1) and (9) are XOR'ed: $4849 \text{ XOR } 78B0 = 30F9$.
12. Results from steps (3) and (9) are XOR'ed: $0003 \text{ XOR } 78B0 = 78B3$.
13. Results from steps (2) and (10) are XOR'ed: $0002 \text{ XOR } E221 = E223$.
14. Results from steps (4) and (10) are XOR'ed: $0000 \text{ XOR } E221 = E221$.

The output of round 1 (the result from steps 11–14) is: 30F9 78B3 E223 E221. Prior to the start of round 2, the two inner words of the round 1 output are swapped. Then, seven additional rounds and a final output transformation are performed.

14.6.2 Diffie-Hellman (Elgamal Variation) and RSA

For encryption of the session key, PGP offers a choice of two public-key encryption algorithms, RSA and the Diffie-Hellman (Elgamal variation) protocol. PGP allows for key sizes of 1024 to 4096 bits for RSA or Diffie-Hellman algorithms. The key size of 1024 bits is considered safe for exchanging most information. The security of the RSA algorithm (see Section 14.5.3) is based on the difficulty of factoring large integers.

The Diffie-Hellman protocol was developed by Whitfield Diffie, Martin E. Hellman, and Ralph C. Merkle in 1976 [19, 20] for public-key exchange over an insecure channel. It is based on the difficulty of the discrete logarithm problem for finite fields [21]. It assumes that it is computationally infeasible to compute g^{ab} knowing only g^a and g^b. U.S. Patent 4,200,770, which expired in 1997, covers the Diffie-Hellman protocol and variations such as Elgamal. The Elgamal variation, which was developed by Taher Elgamal, extends the Diffie-Hellman protocol for message encryption. PGP employs the Elgamal variation of Diffie-Hellman for the encryption of the session-key.

14.6.2.1 Description of Diffie-Hellman, Elgamal Variant:

The protocol has two-system parameter n and g that are both public. Parameter n is a large prime number, and parameter g is an integer less than n that has the following property: for every number p between 1 and $n - 1$ inclusive, there is a power k of g such that $g^k = p \bmod n$. The Elgamal encryption scheme [19, 21] that allows user B to send a message to user A is described below:

- User A randomly chooses a large integer, a (this is user A's private key).
- User A's public key is computed as: $y = g^a \bmod n$.
- User B wishes to send a message M to user A. User B first generates a random number k that is less than n.
- User B computes the following:

$$y_1 = g^k \bmod n$$

$$y_2 = M \times (y^k \bmod n) \text{ (recall that } y \text{ is users A's public key).}$$

- User B sends the ciphertext (y_1, y_2) to user A.

- Upon receiving ciphertext (y_1, y_2), user A computes the plaintext message M as follows:

$$M = \frac{y_2}{y_1^a \bmod n}$$

Example 14.9   Diffie-Hellman (Elgamal variation) for Message Encryption

Consider that the public-system parameters are $n = 11$ and $g = 7$. Suppose that user A chooses the private key to be $a = 2$. Show how user A's public key is computed. Also, show how user B would encrypt a message $M = 13$ to be sent to user A, and how user A subsequently decrypts the ciphertext to yield the message.

Solution

User A's public key $(y = g^a \bmod n)$ is computed as: $y = 7^2 \bmod 11 = 5$. User B wishes to send message $M = 13$ to user A. For this example, let user B randomly choose a value of k (less than $n = 11$) to be $k = 1$. User B computes the ciphertext pair

$$y_1 = g^k \bmod n = 7^1 \bmod 11 = 7$$

$$y_2 = M \times (y^k \bmod n) = 13 \times (5^1 \bmod 11) = 13 \times 5 = 65$$

User A receives the ciphertext $(7, 65)$, and computes message M as follows:

$$M = \frac{y_2}{y_1^a \bmod n} = \frac{65}{7^2 \bmod 11} = \frac{65}{5} = 13$$

14.6.3 PGP Message Encryption

The private-key algorithms that PGP uses for message encryption were presented in Section 14.6.1. The public-key algorithms that PGP uses to encrypt the private-session key were presented in Section 14.6.2. The next example combines the two types of algorithms to illustrate the PGP encryption technique shown in Figure 14.20.

Example 14.10   PGP Use of RSA and IDEA for Encryption

For the encryption of the session key, use the RSA public-key algorithm with the parameters taken from Section 14.5.3.1, where $n = pq = 2773$, the encryption key is $e = 17$, and the decryption key is $d = 157$. The encryption key is the recipient's public key, and the decryption key is the recipient's private key. From Example 14.8, use the session key $K_0 = 0008\ 0007\ 0006\ 0005\ 0004\ 0003\ 0002\ 0001$, and the ciphertext of 30F9 78B3 E223 E221 representing the message "HI", where all the digits are shown in hexadecimal notation. (Note that the ciphertext was created by using only one round of the IDEA algorithm. In the actual implementation, 8 rounds plus an output transformation are performed.) Encrypt the session key, and show the PGP transmission that would be made.

Solution

Following the description in Section 14.5.3.1, the session key will be encrypted using the RSA algorithm with the recipient's public key of 17. For ease of calculation with a simple calculator, let us first transform the session key into groups made up of base-10 digits. In keeping with the requirements of the RSA algorithm, the value ascribed to any group may not exceed $n - 1 = 2772$. Therefore, let us express the 128-bit key in terms of 4-digit groups, where we choose the most significant group (leftmost) to represent 7 bits, and the balance of the 11 groups to represent 11 bits each. The transformation from base-16 to base-10 digits can best be viewed as a two-step process, (1) conversion to binary and, (2) conversion to base 10. The result is $K_0 = 0000\ 0032\ 0000\ 1792\ 0048\ 0001\ 0512\ 0064\ 0001\ 1024\ 0064\ 0001$. Recall from Equation 14.32, that $C = (M)^e$ modulo-n where M will be one of the 4-digit groups of K_0. The leftmost four groups are encrypted as:

$$C_{12} = (0000)^{17} \bmod 2773 = 0.$$
$$C_{11} = (0032)^{17} \bmod 2773 = 2227.$$
$$C_{10} = (0000)^{17} \bmod 2773 = 0.$$
$$C_9 = (1792)^{17} \bmod 2773 = 2704.$$

An efficient way to compute modular exponentiation is to use the Square-and-Multiply algorithm. This algorithm [21] reduces the number of modular multiplications needed to be performed from $e - 1$ to at most 2ℓ, where ℓ is the number of bits in the binary representation. Let us demonstrate the use of the Square-and-Multiply algorithm by encrypting one of the session-key decimal groups (the eleventh group from the right, $M_{11} = 0032$), where $n = 2773$ and $e = 17$. In using this algorithm, we first convert e to its binary representation (17 decimal = 10001 binary).

The calculations are illustrated in Table 14.13. Modulo-n math is used, where $n = 2773$ in this example. The second column contains the binary code, with the most significant bit (MSB) in row 1. Each bit value in this column acts to control a result in column 3. The starting value, placed in column 3 row 0, is always 1. Then, the result for any row in column 3 depends on the value of the bit in the corresponding row in column 2; if that entry contains a "1," then the previous row-result is squared and multiplied by the plaintext (32 for this example). If a row in the second column contains a "0", then the result of that row in column 3 equals only the square of the previous row's result. The final value is the encrypted ciphertext ($C = 2227$). Repeating this method for each of the twelve decimal groups that comprise K_0 results in the ciphertext of the session key to be: $C = 0000\ 2227\ 0000\ 2704\ 0753\ 0001\ 1278\ 0272\ 0001\ 1405\ 0272\ 0001$. This RSA-encrypted session key (represented here in decimal) together with the IDEA-encrypted message of 30F9 78B3 E223 E221 (represented here in hex) can now be transmitted over an insecure channel.

TABLE 14.13 The Square-and-Multiply Algorithm with Plaintext = 32

Row Number	Binary representation of e (MSB first)	Modulo multiplication (modulo 2773)
0		1
1	1	$1^2 \times 32 = 32$
2	0	$32^2 = 1024$
3	0	$1024^2 = 382$
4	0	$382^2 = 1728$
5	1	$1728^2 \times 32 = 2227$

14.6.4 PGP Authentication and Signature

The public key algorithms can be used to authenticate or "sign" a message. As illustrated in Figure 14.18, a sender can encrypt a document with his private key (which no one else has access to) prior to encrypting it with the recipient's public

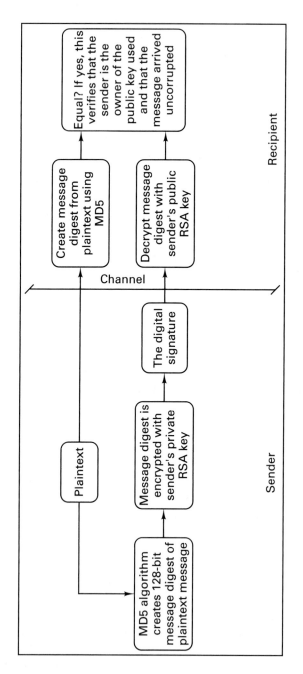

Figure 14.22 PGP signature technique.

key. The recipient must first use his private key to decrypt the message, followed by a second decryption using the sender's public key. This technique encrypts the message for secrecy and also provides authentication of the sender.

Because of the slowness of public-key algorithms, PGP allows for a different method of authenticating a sender. Instead of the time-consuming process of encrypting the entire plaintext message, the PGP approach encrypts a fixed-length message digest created with a one-way hash function. The encryption of the message digest is performed using a public-key algorithm. This method is known as a *digital signature* and is shown in Figure 14.22. A digital signature is used to provide authentication of both *the sender* and *the message*. Authentication of the message provides a verification that the message was not altered in some way. Using this technique, if a message has been altered in any way (i.e. by a forger), its message digest will be different.

PGP version 2.6 uses the MD5 (Message Digest 5) algorithm to create a 128-bit message digest (or hash value) of the plaintext. This hash value is then encrypted with the sender's private key and sent with the plaintext. When the recipient receives the message, he will first decrypt the message digest with the sender's public key. The recipient will then apply the hash function to the plaintext and compare the two message digests. If they match, the signature is valid. In Figure 14.22, the message is sent without encryption (as plaintext), but it may be encrypted by the method illustrated in Figure 14.20.

14.6.4.1 MD5 and SHA-1

MD5 and SHA-1 are hash functions. A hash function $H(x)$ takes an input and returns a fixed-size string h, called the hash value (also known as a message digest). A cryptographic hash function has the following properties:

1. The output length is fixed.
2. The hash value is relatively simple to compute.
3. The function is one way—in other words, it is hard to invert. If given a hash value h, it is computationally infeasible to find the function's input x.
4. The function is *collision free*. A collision-free hash function is a function for which it is infeasible that two different messages will create the same hash value.

The MD5 algorithm used in PGP version 2.6 creates a 128-bit message digest. The MD5 algorithm processes the text in 512-bit blocks through four rounds of data manipulation. Each round uses a different nonlinear function that consists of the logical operators AND, OR, NOT or XOR. Each function is performed 16 times in a round. Bit shifts and scalar additions are also performed in each round [19]. Hans Dobbertin [18] has determined that collisions may exist in MD5. Because of this potential weakness, the PGP specification recommends using the Digital Signature Standard (DSS). DSS uses the SHA-1 (Secure Hash Algorithm-1) algorithm. The SHA-1 algorithm takes a message of less than 2^{64} bits in length and produces a 160-bit message digest. SHA-1 is similar to MD5 in that it uses a different nonlinear function in each of its 4 rounds. In SHA-1, each function is performed 20 times per round. SHA-1 also uses various scalar additions and bit shifting. The algorithm is slightly slower than MD5 but the larger message digest (160-bit versus 128 bit) makes it more secure against brute-force attacks [19]. A brute-force attack consists of trying many input combinations in an attempt to match the message digest under attack.

14.6.4.2 Digital Signature Standard and RSA

For digital signatures, PGP version 2.6 uses the RSA algorithm for encryption of the hash value produced by the MD5 function; however, versions 5.0 and later adhere to the NIST Digital Signature Standard (DSS) [22]. The NIST DSS requires the use of the SHA-1 hash function. The hash value is then encrypted using the Digital Standard Algorithm (DSA). Like the Diffie-Hellman protocol, DSA is based on the discrete logarithm problem. (Reference [22] contains a detailed description of DSA.)

14.7 CONCLUSION

In this chapter we have presented the basic model and goals of the cryptographic process. We looked at some early cipher systems and reviewed the mathematical theory of secret communications established by Shannon. We defined a system that can exhibit perfect secrecy and established that such systems can be implemented but that they are not practical for use where high-volume communications are required. We also considered practical security systems that employ Shannon's techniques (known as confusion and diffusion) to frustrate the statistical endeavors of a cryptanalyst.

The outgrowth of Shannon's work was utilized by IBM in the LUCIFER system, which later grew into the National Bureau of Standards' Data Encryption Standard (DES). We outlined the DES algorithm in detail. We also considered the use of linear feedback shift registers (LFSR) for stream encryption systems, and demonstrated the intrinsic vulnerability of an LFSR used as a key generator.

We also looked at the area of public-key cryptosystems and examined two schemes, the Rivest–Shamir–Adelman (RSA) scheme, based on the product of two large prime numbers, and the Merkle-Hellman scheme, based on the classical knapsack problem. Finally, we looked at the novel scheme of Pretty Good Privacy (PGP), developed by Phil Zimmerman and published in 1991. PGP utilizes the benefits of both private and public-key systems and has proven to be an important file-encryption method for sending data via electronic mail.

REFERENCES

1. Kahn, D., *The Codebreakers*, Macmillan Publishing Company, New York, 1967.
2. Diffie, W., and Hellman, M.E., "Privacy and Authentication: An Introduction to Cryptography," *Proc. IEEE,* vol. 67, no. 3, Mar. 1979, pp. 397–427.

3. Beker,H., and Piper, F., *Cipher Systems,* John Wiley & Sons, Inc., New York, 1982.

4. Denning, D.E.R., *Cryptography and Data Security,* Addison-Wesley Publishing Company, Reading, Mass., 1982.

5. Shannon, C.E., "Communication Theory of Secrecy Systems," *Bell Syst. Tech. J.,* vol. 28, Oct. 1949, pp. 656–715.

6. Hellman, M. E., "An Extension of the Shannon Theory Approach to Cryptography," *IEEE Trans. Inf. Theory,* vol. IT23, May 1978, pp. 289–294.

7. Smith, J. L., "The Design of Lucifer, a Cryptographic Device for Data Communications," *IBM Research Rep. RC-3326,* 1971.

8. Feistel, H. "Cryptography and Computer Privacy," *Sci. Am.,* vol. 228, no. 5, May 1973, pp. 15–23.

9. National Bureau of Standards, "Data Encryption Standard," *Federal Information Processing Standard (FIPS),* Publication no. 46, Jan. 1977.

10. United States Senate Select Committee on Intelligence, "Unclassified Summary: Involvement of NSA in the Development of the Data Encryption Standard," *IEEE Commun. Soc. Mag.,* vol. 16, no. 6, Nov. 1978, pp. 53–55.

11. Stallings, W., *Cryptography and Network Security, Second Edition,* Prentice Hall, Upper Saddle River, NJ. 1998.

12. Diffie, W., and Hellman, M. E., "New Directions in Cryptography," *IEEE Trans. Inf. Theory,* vol. IT22, Nov. 1976, pp. 644–654.

13. Rivest, R.L., Shamir, A., and Adelman, L., "On Digital Signatures and Public Key Cryptosystems," *Commun. ACM,* vol. 21, Feb. 1978, pp. 120–126.

14. Knuth, D. E., *The Art of Computer Programming,* Vol. 2, *Seminumerical Algorithms,* 2nd ed., Addison-Wesley Publishing Company, Reading, Mass., 1981.

15. Merkel, R. C., and Hellman, M. E., "Hiding Information and Signatures in Trap-Door Knapsacks," *IEEE Trans. Inf. Theory,* vol. IT24, Sept. 1978, pp. 525–530.

16. Shamir, A., "A Polynomial Time Algorithm for Breaking the Basic Merkle-Hellman Cryptosystem," *IEEE 23rd Ann. Symp. Found. Comput. Sci.,* 1982, pp. 145–153.

17. Zimmerman, P. *The Official PGP User's Guide,* MIT Press, Cambridge, 1995.

18. *PGP Freeware User's Guide, Version 6.5,* Network Associates, Inc., 1999.

19. Schneier, B., *Applied Cryptography,* John Wiley & Sons, New York, 1996.

20. Hellman, M. E., Martin, Bailey, Diffie, W., and Merkle, R. C., *United States Patent 4,200,700: Cryptographic Apparatus and Method,* United States Patent and Trademark Office, Washington, DC, 1980.

21. Stinson, Douglas, *Cryptography Theory and Practice.* CRC Press, Boca Raton, FL, 1995.

22. *Digital Signature Standard* (Federal Information Processing Standards Publication 186–1), Government Printing Office, Springfield, VA, Dec. 15, 1998.

PROBLEMS

14.1. Let X be an integer variable represented with 64 bits. The probability is $\frac{1}{2}$ that X is in the range $(0, 2^{16} - 1)$, the probability is $\frac{1}{4}$ that X is in the range $(2^{16}, 2^{32} - 1)$, and the probability is $\frac{1}{4}$ that X is in the range $(2^{32}, 2^{64} - 1)$. Within each range the values are equally likely. Compute the entropy of X.

14.2. A set of equally likely weather messages are: sunny (S), cloudy (C), light rain (L), and heavy rain (H). Given the added information concerning the time of day (morning or afternoon), the probabilities change as follows:

$$\text{Morning:} \quad P(S) = \tfrac{1}{8}, \quad P(C) = \tfrac{1}{8}, \quad P(L) = \tfrac{3}{8}, \quad P(H) = \tfrac{3}{8}$$

$$\text{Afternoon:} \quad P(S) = \tfrac{3}{8}, \quad P(C) = \tfrac{3}{8}, \quad P(L) = \tfrac{1}{8}, \quad P(H) = \tfrac{1}{8}$$

    (a) Find the entropy of the weather message.
    (b) Find the entropy of the message conditioned on the time of day.

14.3. The Hawaiian alphabet has only 12 letters—the vowels, a, e, i, o, u, and the consonants, h, k, l, m, n, p, w. Assume that each vowel occurs with probability 0.116, and that each consonant occurs with probability 0.06. Also assume that the average number of *information bits* per letter is the same as that for the English language. Calculate the unicity distance for an encrypted Hawaiian message if the key sequence consists of a random permutation of the 12-letter alphabet.

14.4. Estimate the unicity distance for an English language encryption system that uses a key sequence made up of 10 random alphabetic characters:
    (a) Where each key character can be any one of the 26 letters of the alphabet (duplicates are allowed).
    (b) Where the key characters may not have any duplicates.

14.5. Repeat Problem 14.4 for the case where the key sequence is made up of ten integers randomly chosen from the set of numbers 0 to 999.

14.6. **(a)** Find the unicity distance for a DES system which encrypts 64-bit blocks (eight alphabetic characters) using a 56-bit key.
    (b) What is the effect on the unicity distance in part (a) if the key is increased to 128 bits?

14.7. In Figures 14.8 and 14.9, *P*-boxes and *S*-boxes alternate. Is this arrangement any more secure than if all the *P*-boxes were first grouped together, followed by all the *S*-boxes similarly grouped together? Justify your answer.

14.8. What is the output of the first iteration of the DES algorithm when the plaintext and the key are each made up of zero sequences?

14.9. Consider the 10-bit plaintext sequence 0 1 0 1 1 0 1 0 0 1 and its corresponding ciphertext sequence 0 1 1 1 0 1 1 0 1 0, where the rightmost bit is the earliest bit. Describe the five-stage linear feedback shift register (LFSR) that produced the key sequence and show the initial state of the register. Is the output sequence of maximal length?

14.10. Following the RSA algorithm and parameters in Example 14.5, compute the encryption key, e, when the decryption key is chosen to be 151.

14.11. Given e and d that satisfy ed modulo-$\phi(n) = 1$, and a message that is encoded as an integer number, M, in the range $(0, n-1)$ such that the gcd $(M, n) = 1$. Prove that $(M^e$ modulo-$n)^d$ modulo-$n = M$.

14.12. Use the RSA scheme to encrypt the message $M = 3$. Use the prime numbers $p = 5$ and $q = 7$. Choose the decryption key, d, to be 11, and calculate the value of the encryption key, e.

14.13. Consider the following for the RSA scheme.
    (a) If the prime numbers are $p = 7$ and $q = 11$, list five allowable values for the decryption key, d.

(b) If the prime numbers are $p = 13$, $q = 31$, and the decryption key is $d = 37$, find the encryption key, e, and describe how you would use it to encrypt the word "DIGITAL."

14.14. Use the Merkle–Hellman public key scheme with the super-increasing vector, $\mathbf{a}' = 1$, 3, 5, 10, 20. Use the following additional parameters: a large prime number $M = 51$ and a random number $W = 37$.

    (a) Find the nonsuper-increasing vector, $\mathbf{a}$, to be made public, and encrypt the data vector 1 1 0 1 1.

    (b) Show the steps by which an authorized receiver decrypts the ciphertext.

14.15. Using the Diffie-Hellman (Elgamal variation) protocol, encrypt the message $M = 7$. The system parameters are $n = 17$ and $g = 3$. The recipients private key is $a = 4$. Determine the recipient's public key. For message encryption with the randomly selected k, use $k = 2$. Verify the accuracy of the ciphertext by performing decryption using the recipient's private key.

14.16. Find the hexadecimal (hex) value of the message "no" after one round of the IDEA algorithm. The session key in hex notation is = 0002 0003 0002 0003 0002 0003 0002 0003, where the rightmost 4-digit group represents the subkey Z_1. For the message "no," let each ASCII character be represented by a 16-bit data subblock, where "n" = 006E and "o" = 006F.

14.17. In the PGP Example 14.10, the IDEA session key is encrypted using the RSA algorithm. The resulting encrypted session key (in base-10 notation) was: 0000 2227 0000 2704 0753 0001 1278 0272 0001 1405 0272 0001, where the least significant (rightmost) group is group 1. Using the decryption key, decrypt group 11 of this session key using the Square-and-Multiply technique.

QUESTIONS

14.1. What are the two major requirements for a useful *cryptosystem*? (See Section 14.1.2.)

14.2. Shannon suggested two encryption concepts that he termed *confusion* and *diffusion*. Explain what these terms mean. (See Section 14.3.1.)

14.3. If *high-level security* is desired, explain why a linear feedback shift register (LFSR) would not be used. (See Section 14.4.2.)

14.4. Explain the major difference between conventional cryptosystems and *public key cryptosystems*. (See Section 14.5.)

14.5. Describe the steps used for message encryption employed by the *Data Encryption Standard* (DES). How different is the operation when using Triple-DES? (See Sections 14.3.5 and 14.6.1.1)

14.6. Describe the steps used for message encryption employed by version 2.6 of the *Pretty Good Privacy* (PGP) technique. (See Section 14.6.1.3.)

EXERCISES

Using the Companion CD, run the exercises associated with Chapter 14.

CHAPTER 15

Fading Channels

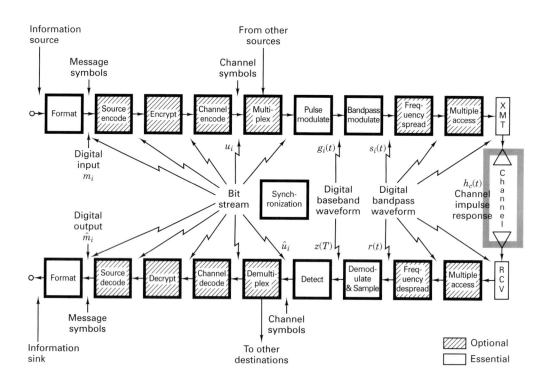

When the mechanisms that cause fading in communication channels were first modeled in the 1950s and 1960s, the principles developed were primarily applied to over-the-horizon communications covering a wide range of frequency bands. The 3-30 MHz high-frequency (HF) band used for ionospheric propagation, and the 300 MHz-3 GHz ultra-high-frequency (UHF), as well as the 3-30 GHz super-high-frequency (SHF) bands used for tropospheric scatter are examples of channels affected by fading phenomena. Although the fading effects in mobile radio channels are somewhat different than those encountered in ionospheric and tropospheric channels, the early models are still quite useful in helping to characterize the fading effects in mobile digital communication systems. This chapter emphasizes so-called *Rayleigh fading*, primarily in the UHF band, that affects mobile systems, such as cellular and personal communication systems (PCS). Emphasis is also placed on the fundamental fading manifestations, the types of degradation, and the methods for mitigating the degradation. Two examples of specific mitigation techniques are examined—the Viterbi equalizer, implemented in the Global System for Mobile (GSM) Communications, and the Rake receiver, used in CDMA systems built to meet Interim Standard-95 (IS-95).

15.1 THE CHALLENGE OF COMMUNICATING OVER FADING CHANNELS

In the analysis of communication system performance, the classical (ideal) additive-white-Gaussian-noise (AWGN) channel, with statistically independent Gaussian noise samples corrupting data samples free of intersymbol interference

(ISI), is the usual starting point for developing basic performance results. An important source of performance degradation is thermal noise generated in the receiver. Another source of degradation stems from both natural and man-made sources of noise and interference that enter the receiving antenna, and can be quantified by a parameter called *antenna temperature.* (See Section 5.5.5.) Thermal noise typically has a flat power spectral density over the signal band and a zero-mean Gaussian voltage probability density function (pdf). In mobile communication systems, the external noise and interference are often more significant than the receiver thermal noise. When modeling practical systems, the next step is the introduction of bandlimiting filters. Filtering in the transmitter usually serves to satisfy some regulatory requirement on spectral containment. Filtering in the receiver is often the result of implementing a matched filter, as treated in Section 3.2.2. Due to the bandlimiting and phase-distortion properties of filters, special signal design and equalization techniques may be required to mitigate the filter-induced ISI.

If a radio channel's propagating characteristics are not specified, one usually infers that the signal attenuation versus distance behaves as if propagation takes place over ideal free space. The model of free space treats the region between the transmit and receive antennas as being free of all objects that might absorb or reflect radio frequency (RF) energy. It also assumes that, within this region, the atmosphere behaves as a perfectly uniform and nonabsorbing medium. Furthermore, the earth is treated as being infinitely far away from the propagating signal (or, equivalently, as having a reflection coefficient that is negligible). Basically, in this idealized free-space model, the attenuation of RF energy between the transmitter and receiver behaves according to an inverse-square law. The received power expressed in terms of transmitted power is attenuated by a factor $L_s(d)$. This factor, shown below, is called *path loss* or *free space loss*, and is predicated on the receiving antenna being isotropic (see Section 5.3.1.1).

$$L_s(d) = \left(\frac{4 \pi d}{\lambda} \right)^2 \tag{15.1}$$

In Equation (15.1), d is the distance between the transmitter and the receiver, and λ is the wavelength of the propagating signal. For this case of idealized propagation, received signal power is very predictable. For most practical channels, where signal propagation takes place in the atmosphere and near the ground, the free-space propagation model is inadequate to describe the channel behavior and predict system performance. In a wireless mobile communication system, a signal can travel from transmitter to receiver over multiple reflective paths. This phenomenon, referred to as *multipath propagation*, can cause fluctuations in the received signal's amplitude, phase, and angle of arrival, giving rise to the terminology *multipath fading.* Another name, *scintillation*, which originated in radio astronomy, is used to describe the fading caused by physical changes in the propagating medium, such as variations in the electron density of the ionosopheric layers that reflect high frequency (HF) radio signals. Both fading and scintillation refer to a signal's random fluctuations; the main difference is that scintillation involves mechanisms (e.g., electrons) that are much smaller than a wavelength. The end-to-end modeling and design of systems that incorporate techniques to mitigate the effects of fading are usually more challenging than those whose sole source of performance degradation is AWGN.

15.2 CHARACTERIZING MOBILE-RADIO PROPAGATION

Figure 15.1 represents an overview of fading-channel manifestations. It starts with two types of fading effects that characterize mobile communications: large-scale fading and small-scale fading. Large-scale fading represents the average signal power attenuation or the path loss due to motion over large areas. In Figure 15.1, the large-scale fading manifestation is shown in blocks 1, 2, and 3. This phenomenon is affected by prominent terrain contours (e.g., hills, forests, billboards, clumps of buildings, etc.) between the transmitter and receiver. The receiver is often said to be "shadowed" by such prominences. The statistics of large-scale fading provide a way of computing an estimate of path loss as a function of distance. This is often described in terms of a mean-path loss (nth-power law) and a log-normally distributed variation about the mean. Small-scale fading refers to the dramatic changes in signal amplitude and phase that can be experienced as a result of small changes (as small as a half-wavelength) in the spatial positioning between a receiver and transmitter. As indicated in Figure 15.1, blocks 4, 5, and 6, small-scale fading manifests itself in two mechanisms—time-spreading of the signal (or signal dispersion) and time-variant behavior of the channel. For mobile-radio applications, the channel is time-variant because motion between the transmitter and receiver results in propagation path changes. The rate of change of these propagation conditions accounts for the fading rapidity (rate of change of the fading impairments). Small-scale fading is called *Rayleigh fading* if there are multiple reflective paths that are large in number, and if there is no line-of-sight signal component; the envelope of such a received signal is statistically described by a Rayleigh pdf. When there is a dominant nonfading signal component present, such as a line-of-sight propagation path, the small-scale fading envelope is described by a Rician pdf [1]. In other words, the small-scale fading statistics are said to be Rayleigh whenever the line of sight path is blocked, and Rician otherwise. A mobile radio roaming over a large area must process signals that experience both types of fading: small-scale fading superimposed on large-scale fading.

Large-scale fading (attenuation or path loss) can be considered as a spatial average over the small-scale fluctuations of the signal. It is generally evaluated by averaging the received signal over 10 to 30 wavelengths, in order to decouple the small-scale (mostly Rayleigh) fluctuations from the large-scale shadowing effects (typically log-normal). There are three basic mechanisms that impact signal propagation in a mobile communication system [1]:

- *Reflection* occurs when a propagating electromagnetic wave impinges upon a smooth surface with very large dimensions relative to the RF signal wavelength (λ).
- *Diffraction* occurs when the propagation path between the transmitter and receiver is obstructed by a dense body with dimensions that are large relative to λ, causing secondary waves to be formed behind the obstructing body.

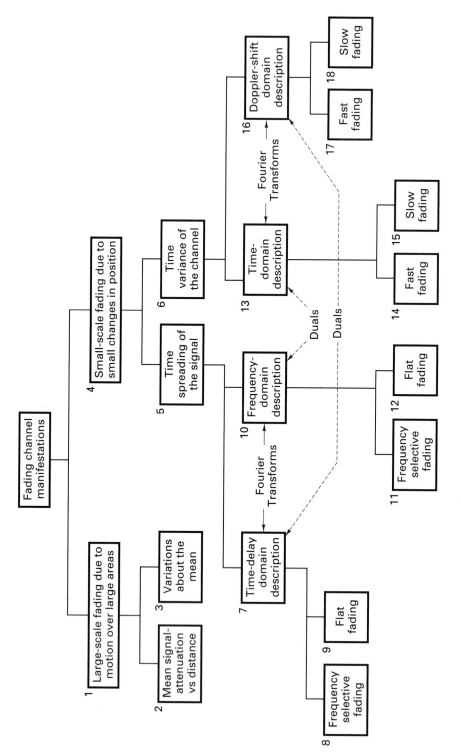

Figure 15.1   Fading channel manifestations.

Diffraction is a phenomenon that accounts for RF energy traveling from transmitter to receiver without a line-of-sight path between the two. If is often termed *shadowing* because the diffracted field can reach the receiver even when shadowed by an impenetrable obstruction.

- *Scattering* occurs when a radio wave impinges on either a large, rough surface or any surface whose dimensions are on the order of λ or less, causing the energy to be spread out (scattered) or reflected in all directions. In an urban environment, typical signal obstructions yielding scattering include lampposts, street signs, and foliage. The name *scatterer* applies to any obstruction in the propagation path that causes a signal to be reflected or scattered.

Figure 15.1 may serve as a table of contents for the sections that follow. The two manifestations of small-scale fading, signal time-spreading (signal dispersion) and the time-variant nature of the channel, will be examined in two domains: time and frequency, as indicated in Figure 15.1, blocks 7, 10, 13, and 16. For signal dispersion, the fading degradation types are categorized as being frequency-selective or frequency-nonselective (flat), as listed in blocks 8, 9, 11, and 12. For the time-variant manifestation, the fading degradation types are categorized as fast fading or slow fading, as listed in blocks 14, 15, 17, and 18. The labels indicating Fourier transforms and duals will be explained later.

Figure 15.2 is a convenient pictorial (not a precise graphical representation) showing the various contributions that must be considered when estimating path loss for link budget analysis in a mobile radio application [2]: (1) mean path loss as a function of distance, due to large-scale fading, (2) near-worst-case variations about the mean path loss or large-scale fading margin (typically 6–10 dB), and (3) near-worst-case Rayleigh or small-scale fading margin (typically 20–30 dB). In Figure 15.2, the annotations "≈ 1–2%" indicate a suggested area (probability) under the tail of each pdf as a design goal. Hence, the amount of margin indicated is intended to provide adequate received signal power for approximately 98 to 99 percent of each type of fading variation (large- and small-scale).

Using complex notation, a transmitted signal is written as:

$$s(t) = \mathrm{Re}\{g(t)e^{j2\pi f_c t}\} \tag{15.2}$$

where $\mathrm{Re}\{\cdot\}$ denotes the real part of $\{\cdot\}$, and f_c is the carrier frequency. The baseband waveform $g(t)$ is called the complex envelope of $s(t)$ (see Section 4.6), and can be expressed as

$$g(t) = |g(t)|e^{j\phi(t)} = R(t)e^{j\phi(t)} \tag{15.3}$$

where $R(t) = |g(t)|$ is the envelope magnitude, and $\phi(t)$ is its phase. For a purely phase- or frequency-modulated signal, $R(t)$ will be constant, and in general, will vary slowly compared to $t = 1/f_c$.

In a fading environment, $g(t)$ will be modified by a complex dimensionless multiplicative factor $\alpha(t)e^{-j\theta(t)}$. (Later, we show this derivation.) The *modified* baseband waveform can be written as $\alpha(t)e^{-j\theta(t)}g(t)$; but for now, let us examine the magnitude, $\alpha(t)R(t)$ of this envelope, which can be expressed in terms of three positive terms, as follows [3]:

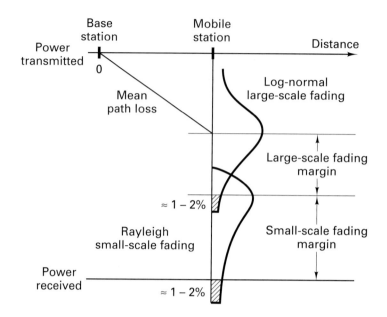

Figure 15.2 Link-budget considerations for a fading channel.
Source: Greenwood, D. and Hanzo, L., "Characterization of Mobile Radio Channels," *Mobile Radio Communications,* edited by R. Steele, Chapter 2, Pentech Press, London, 1994.

$$\alpha(t)\, R(t) = m(t) \times r_0(t) \times R(t) \qquad (15.4)$$

where $m(t)$ is called the *large-scale-fading component* of the envelope, and $r_0(t)$ is called the *small-scale-fading component.* Sometimes, $m(t)$ is referred to as the *local mean* or *log-normal fading*, because generally its measured values can be statistically described by a log-normal pdf; or equivalently, when measured in decibels, $m(t)$ has a Gaussian pdf. Furthermore, $r_0(t)$ is sometimes referred to as *multipath* or *Rayleigh fading.* For the case of a mobile radio, Figure 15.3 illustrates the relationship between $\alpha(t)$ and $m(t)$. In this figure, we consider that an *unmodulated* carrier wave is being transmitted, which in the context of Equation (15.4) means that for all time, $R(t) = 1$. Figure 15.3a is a representative plot of signal power received versus antenna displacement (typically in units of wavelength). The signal power received is of course a function of the multiplicative factor $\alpha(t)$. Small-scale fading superimposed on large-scale fading can be readily identified. The typical antenna displacement between adjacent signal-strength nulls due to small-scale fading is approximately a half wavelength. In Figure 15.3b, the large-scale fading or local mean $m(t)$ has been removed in order to view the small-scale fading $r_0(t)$, referred to some average constant power. Recall that $m(t)$ can generally be evaluated by averaging the received envelope over 10 to 30 wavelengths. The log-normal fading is a relatively slow varying function of position, while the Rayleigh fading is a relatively fast varying function of position. Note that for an application involving motion, such as the case of a radio in a moving vehicle, a function of position is

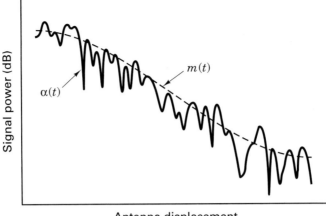

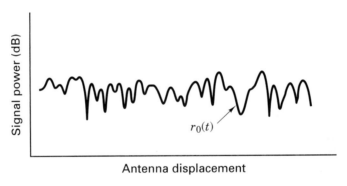

Figure 15.3 Large-scale fading and small-scale fading.

tantamount to a function of time. In the selections that follow, some of the details regarding the statistics and mechanisms of large-scale and small-scale fading are enumerated.

15.2.1 Large-Scale Fading

For mobile radio applications, Okumura [4] made some of the earlier comprehensive path-loss measurements for a wide range of antenna heights and coverage distances. Hata [5] transformed Okumura's data into parametric formulas. In general, propagation models for both indoor and outdoor radio channels indicate that the mean path loss $\overline{L_p}(d)$, as a function of distance d between a transmitter and receiver is proportional to an nth-power of d relative to a reference distance d_0 [1], or, mathematically,

$$\overline{L}_p(d) \propto \left(\frac{d}{d_0}\right)^n \tag{15.5}$$

$\overline{L}_p(d)$ is often stated in decibels

$$\overline{L}_p(d)\ (\text{dB}) = L_s(d_0)\ (\text{dB}) + 10\, n \log\left(\frac{d}{d_0}\right) \tag{15.6}$$

The reference distance d_0 corresponds to a point located in the far field of the transmit antenna. Typically, the value of d_0 is taken to be 1 km for large cells, 100 m for microcells, and 1 m for indoor channels. Moreover, $L_s(d_0)$ is evaluated using Equation (15.1) or by conducting measurements. $\overline{L}_p(d)$ is the average path loss (over a multitude of different sites) for a given value of d. When plotted on a log-log scale, $\overline{L}_p(d)$ versus d (for distances greater than d_0) yields a straight line with a slope equal to $10\, n$ dB/decade. The value of the path-loss exponent n depends on the frequency, antenna heights, and propagation environment. In free space where signal propagation follows an inverse-square law (as described in Section 5.3.1), n is equal to 2, as seen in Equation (15.1). In the presence of a very strong guided wave phenomenon (like urban streets), n can be lower than 2. When obstructions are present, n is larger. Figure 15.4 shows a scatter plot of path loss versus distance for measurements made at several sites in Germany [6]. Here, the path loss has been measured relative to a reference distance $d_0 = 100$ m. Also shown are straight-line fits to various exponent values.

The path loss versus distance expressed in Equation (15.6) is an average, and therefore not adequate to describe any particular setting or signal path. It is necessary to provide for variations about the mean because the environment of different sites may be quite different for similar transmitter-receiver (T-R) separations. Figure 15.4 illustrates that path-loss variations can be quite large. Measurements have shown that for any value of d, the path loss L_p is a random variable having a log-normal distribution about the mean distant-dependent value $\overline{L}_p(d)$ [7]. Thus, path loss $L_p(d)$ can be expressed in terms of $\overline{L}_p(d)$, in Equation (15.6), plus a random variable X_σ, as [1]

$$L_p(d)\ (\text{dB}) = L_s(d_0)\ (\text{dB}) + 10\, n \log_{10}(d/d_0) + X_\sigma\ (\text{dB}) \tag{15.7}$$

where X_σ denotes a zero-mean, Gaussian random variable (in decibels) with standard deviation σ (also in decibels). X_σ is site and distance dependent. Since X_σ and $L_p(d)$ are random variables, if Equation (15.7) is used as the basis for computing an estimate of path loss or link margin, some value for X_σ must first be chosen. The choice of the value is often based on measurements (made over a wide range of locations and T-R separations). It is not unusual for X_σ to take on values as high as 6 to 10 dB or greater. Thus, the parameters needed to statistically describe path loss due to large-scale fading, for an arbitrary location with a specific transmitter-receiver separation are (1) the reference distance, (2) the path-loss exponent, and (3) the standard deviation of X_σ. (There are several good references dealing with

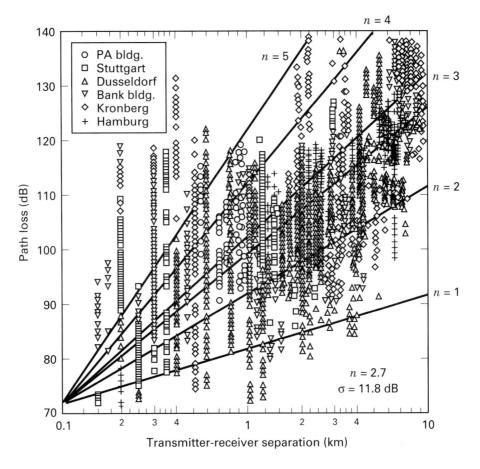

Figure 15.4 Path loss versus distance measured in several German cities.

Source: Seidel, S.Y., et. al., "Path Loss, Scattering and Multipath Delay Statistics in Four European Cities for Digital Cellular and Microcellular Radiotelephone," *IEEE Transactions on Vehicular Technology,* Vol. 40, No. 4, pp. 721–730, November 1991.

the measurement and estimation of propagation path loss for many different applications and configurations [1, 5–9].)

15.2.2 Small-Scale Fading

Here we develop the small-scale fading component, $r_0(t)$. Analysis proceeds on the assumption that the antenna remains within a limited trajectory, so that the effect of large-scale fading $m(t)$ is a constant (assumed unity). Assume that the antenna is traveling, and that there are multiple scatterer paths, each associated with a time-variant propagation delay $\tau_n(t)$, and a time-variant multiplicative factor $\alpha_n(t)$. Neglecting noise, the received bandpass signal, can be written as

$$r(t) = \sum_n \alpha_n(t) s[t - \tau_n(t)] \tag{15.8}$$

Substituting Equation (15.2) into Equation (15.8), we write the received bandpass signal as follows:

$$r(t) = \mathrm{Re}\left(\left\{ \sum_n \alpha_n(t) g[t - \tau_n(t)] \right\} e^{j2\pi f_c [t - \tau_n(t)]} \right)$$

$$= \mathrm{Re}\left(\left\{ \sum_n \alpha_n(t) e^{-j2\pi f_c \tau_n(t)} g[t - \tau_n(t)] \right\} e^{j2\pi f_c t} \right) \tag{15.9}$$

From Equation (15.9), it follows that the equivalent received baseband signal is

$$z(t) = \sum_n \alpha_n(t) e^{-j2\pi f_c \tau_n(t)} g[t - \tau_n(t)] \tag{15.10}$$

Consider the transmission of an *unmodulated* carrier at frequency f_c. In other words, for all time, $g(t) = 1$. Then, the received baseband signal, for this case of an unmodulated carrier and discrete multipath components given by Equation (15.10), reduces to

$$z(t) = \sum_n \alpha_n(t) e^{-j2\pi f_c \tau_n(t)} = \sum_n \alpha_n(t) e^{-j\theta_n(t)} \tag{15.11}$$

where $\theta_n(t) = 2\pi f_c \tau_n(t)$. The baseband signal $z(t)$ consists of a sum of time-variant phasors having amplitudes $\alpha_n(t)$ and phases $\theta_n(t)$. Notice that $\theta_n(t)$ will change by 2π radians whenever τ_n changes by $1/f_c$ (typically, a very small delay). For a cellular radio operating at $f_c = 900$ Mhz, the delay $1/f_c = 1.1$ nanoseconds. In free space, this corresponds to a change in propagation distance of 33 cm. Thus, in Equation (15.11), $\theta_n(t)$ can change significantly with relatively small propagation delay changes. In this case, when two multipath components of a signal differ in path length by 16.5 cm, one signal will arrive 180 degrees out of phase with respect to the other signal. Sometimes the phasors add constructively and sometimes they add destructively, resulting in amplitude variations or fading of $z(t)$. Equation (15.11) can be expressed more compactly as the net received envelope, which is the summation over all the scatterers, as

$$z(t) = \alpha(t) e^{-j\theta(t)} \tag{15.12}$$

where $\alpha(t)$ is the resultant amplitude and $\theta(t)$ is the resultant phase. The right side of Equation (15.12) represents the same complex multiplicative factor that was described earlier in Section 15.2. Equation (15.12) is an important result because it tells us that even though a *bandpass* signal $s(t)$, as expressed in Equation (15.2), is the signal that experienced the fading effects and gave rise to the received signal $r(t)$, these effects can be described by analyzing $r(t)$ at the *baseband* level.

Figure 15.5 illustrates the primary mechanism that causes fading in multipath channels, as described by Equations (15.11) and (15.12). In the figure, a reflected signal has a phase delay (a function of additional path length) with respect to a

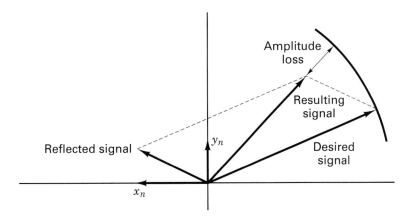

Figure 15.5 Effect of a multipath reflected signal on a desired signal.

desired signal. The reflected signal also has reduced amplitude (a function of the reflection coefficient of the obstruction). Reflected signals can be described in terms of orthogonal components $x_n(t)$ and $y_n(t)$, where $x_n(t) + jy_n(t) = \alpha_n(t)e^{-j\theta_n(t)}$. If the number of such stochastic components is large and none are dominant, then *at a fixed time*, the variables $x_r(t)$ and $y_r(t)$, resulting from their addition, will have a Gaussian pdf. These orthogonal components yield the small-scale (fading) magnitude $r_0(t)$, that was defined in Equation (15.4). For the case of an unmodulated carrier wave, as shown in Equation (15.12), $r_0(t)$ is the magnitude of $z(t)$, as follows:

$$r_0(t) = \sqrt{x_r^2(t) + y_r^2(t)} \tag{15.13}$$

When the received signal is made up of multiple reflective rays plus a significant line-of-sight (nonfaded) component, the received envelope amplitude has a Rician pdf as shown below, and the fading is referred to as *Rician Fading* [2]:

$$p(r_0) = \begin{cases} \dfrac{r_0}{\sigma^2} \exp\left[-\dfrac{(r_0^2 + A^2)}{2\sigma^2} \right] I_0\left(\dfrac{r_0 A}{\sigma^2} \right) & \text{for } r_0 \geq 0, A \geq 0 \\ 0 & \text{otherwise} \end{cases} \tag{15.14}$$

Although $r_0(t)$ varies dynamically with motion (time), at any *fixed time* it is a random variable, whose value stems from the ensemble of real positive numbers. Hence, in describing probability density functions, it is appropriate to drop the functional dependence on time. The parameter σ^2 is the predetection mean power of the multipath signal, A denotes the peak magnitude of the *non-faded* signal component (called the *specular component*) and $I_0 (\cdot)$ is the modified Bessel function of the first kind and zero order [11]. The Rician distribution is often described in terms of a parameter K, which is defined as the ratio of the power in the specular component to the power in the multipath signal. It is given by $K = A^2/(2\sigma^2)$. As the magnitude of the specular component approaches zero, the Rician pdf approaches a Rayleigh pdf, expressed as

$$p(r_0) = \begin{cases} \dfrac{r_0}{\sigma^2} \exp \left[-\dfrac{r_0^2}{2\sigma^2} \right] & \text{for } r_0 \geq 0 \\ 0 & \text{otherwise} \end{cases} \qquad (15.15)$$

The Rayleigh faded component is sometimes called the *random* or *scatter* or *diffuse* component. The Rayleigh pdf results from having no specular signal component; thus, for a single link (no diversity), it represents the pdf associated with the worst case of fading per mean received signal power. For the remainder of this chapter, unless stated otherwise, it will be assumed that loss of signal-to-noise ratio (SNR) due to fading follows the Rayleigh model. It will also be assumed that the propagating signal is in the UHF band, encompassing cellular and personal communications services (PCS), with nominal frequency allocations of 1 GHz and 2 GHz, respectively.

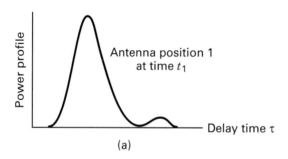

(a)

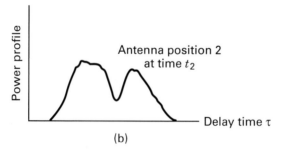

(b)

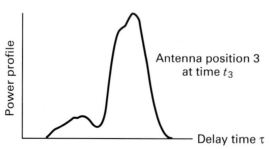

Figure 15.6 Response of a multipath channel to a narrow pulse versus delay, as a function of antenna position.

As indicated in Figure 15.1, blocks 4, 5, and 6, small-scale fading manifests itself in two mechanisms: (1) time-spreading of the underlying digital pulses within the signal, and (2) a time-variant behavior of the channel due to motion (e.g., a receive antenna on a moving platform). Figure 15.6 illustrates the consequences of both manifestations by showing the response of a multipath channel to a narrow pulse versus delay, as a function of antenna position (or time, assuming a mobile travelling at a constant velocity). In Figure 15.6, it is important to distinguish between two different time references—delay time τ and transmission or observation time t. Delay time refers to the time-spreading effect resulting from the fading channel's nonoptimum impulse response. The transmission time, however, is related to the antenna's motion or spatial changes, accounting for propagation path changes that are perceived as the channel's time-variant behavior. Note that for constant velocity, as is assumed in Figure 15.6, either antenna position or transmission time can be used to illustrate this time-variant behavior. Figures 15.6a–15.6c show the sequence of received pulse-power profiles as the antenna moves through a succession of equally spaced positions. Here, the interval between antenna positions is 0.4λ [12], where λ is the wavelength of the carrier frequency. For each of these three cases shown, the response pattern differs significantly in the delay time of the largest signal component, the number of signal copies, their magnitudes, and the total received power (area in each received power profile). Figure 15.7

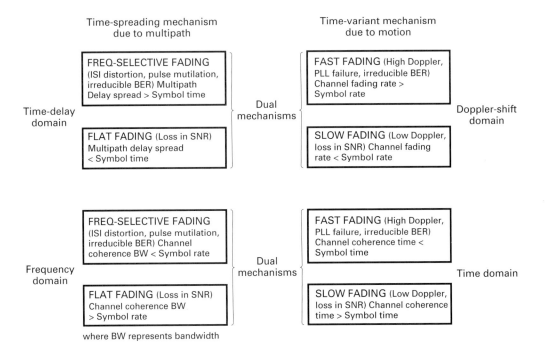

Figure 15.7 Small-scale fading: Mechanisms, degradation categories, and effects.

summarizes these two small-scale fading mechanisms, the two domains (time or time-delay and frequency or Doppler shift) for viewing each mechanism, and the degradation categories each mechanism can exhibit. Note that any mechanism characterized in the time domain can be characterized equally well in the frequency domain. Hence, as outlined in Figure 15.7, the time-spreading mechanism will be characterized in the time-delay domain as a multipath delay spread, and in the frequency domain it will be characterized as a channel coherence bandwidth. Similarly, the time-variant mechanism will be characterized in the time domain as a channel coherence time and in the Doppler-shift (frequency) domain as a channel fading rate or Doppler spread. These mechanisms and their associated degradation categories are examined in the sections that follow.

15.3 SIGNAL TIME-SPREADING

15.3.1 Signal Time-Spreading Viewed in the Time-Delay Domain

A simple way to model the fading phenomenon was introduced by Bello [13] in 1963; he proposed the notion of wide-sense stationary uncorrelated scattering (WSSUS). The model treats signals arriving at a receive antenna with different delays as uncorrelated. It can be shown [2, 13] that such a channel is effectively WSS in both the time and frequency domains. With such a model of a fading channel, Bello was able to define functions that apply for all time and all frequencies. For the mobile channel, Figure 15.8 contains four functions that make up this model [2, 10, 13–15]. These functions are examined here and in later sections, starting with Figure 15.8a and proceeding counter-clockwise toward Figure 15.8d.

In Figure 15.8a, a *multipath-intensity profile* is plotted ($S(\tau)$ versus time delay τ). Knowledge of $S(\tau)$ helps answer the question, "For a transmitted impulse, how does the average received power vary as a function of time delay τ?" The term "time delay" is used to refer to the excess delay. It represents the signal's propagation delay that exceeds the delay of the first signal arrival at the receiver. For a typical wireless channel, the received signal usually consists of several discrete multipath components, causing $S(\tau)$ to exhibit multiple isolated peaks, sometimes referred to as *fingers* or *returns*. For some channels, such as the tropospheric scatter channel, received signals are often seen as a continuum of multipath components [10, 15]. In such cases, $S(\tau)$ is a relatively smooth (continuous) function of τ. For making measurements of the multipath intensity profile, wideband signals (impulses or spread spectrum) need to be used [15]. For a single transmitted impulse, the time T_m between the first and last received component represents the *maximum excess delay,* after which the multipath signal power falls below some threshold level relative to the strongest component. The threshold level might be chosen at 10 dB or 20 dB below the level of the strongest component. Note that for an ideal system (zero excess delay) the function $S(\tau)$ would consist of an ideal impulse with weight equal to the total average received signal power.

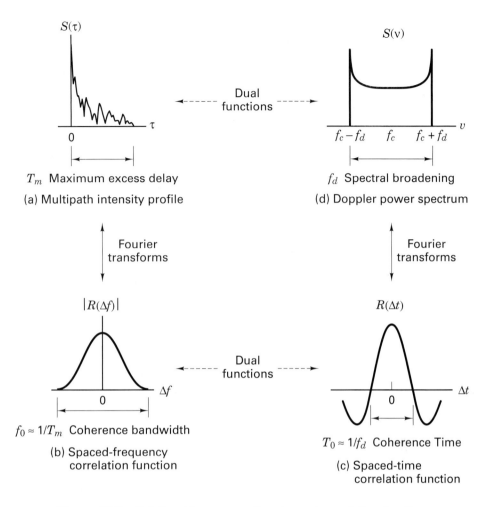

$S(\tau)$

$S(v)$

$\longleftarrow$ ------ Dual functions ------ $\longrightarrow$

0   τ

$f_c - f_d$ f_c $f_c + f_d$ v

T_m Maximum excess delay

f_d Spectral broadening

(a) Multipath intensity profile

(d) Doppler power spectrum

Fourier transforms

Fourier transforms

$|R(\Delta f)|$

$R(\Delta t)$

$\longleftarrow$ ------ Dual functions ------ $\longrightarrow$

0   Δf

0   Δt

$f_0 \approx 1/T_m$ Coherence bandwidth

$T_0 \approx 1/f_d$ Coherence Time

(b) Spaced-frequency correlation function

(c) Spaced-time correlation function

Figure 15.8 Relationships among the channel correlation functions and power density functions.

15.3.1.1 Degradation Categories due to Signal Time-Spreading Viewed in the Time-Delay Domain

In a fading channel, the relationship between maximum excess delay time T_m and symbol time T_s can be viewed in terms of two different degradation categories: *frequency-selective fading* and *frequency nonselective* or *flat fading*. (See Figure 15.1, blocks 8 and 9, and Figure 15.7.) A channel is said to exhibit frequency-selective fading if $T_m > T_s$. This condition occurs whenever the received multipath components of a symbol extend beyond the symbol's time duration. Such multipath dispersion of the signal yields the same kind of ISI distortion that is caused by an electronic filter. In fact, another name for this category of fading degradation is *channel-induced ISI*. In the case of frequency-selective fading, mitigating the distortion is possible because many of the multipath components are resolvable by the receiver. (Several such mitigation techniques are described in later sections.)

A channel is said to exhibit *frequency nonselective* or *flat fading* if $T_m < T_s$. In this case, all of the received multipath components of a symbol arrive within the symbol time duration; hence, the components are not resolvable. Here, there is no channel-induced ISI distortion, since the signal time spreading does not result in significant overlap among neighboring received symbols. There is still performance degradation, since the unresolvable phasor components can add up destructively to yield a substantial reduction in SNR. Also, signals that are classified as exhibiting flat fading can sometimes experience the distortion effects of frequency-selective fading. This will be explained later when viewing degradation in the frequency domain, where the phenomenon is more easily described. For loss in SNR due to flat fading, the mitigation technique called for is to improve the received SNR (or to reduce the required SNR). For digital systems, introducing some form of signal diversity and using error-correction coding is the most efficient way to accomplish this.

15.3.2 Signal Time-Spreading Viewed in the Frequency Domain

A completely analogous characterization of signal dispersion can be specified in the frequency domain. In Figure 15.8b, the function $|R(\Delta f)|$, designated a *spaced-frequency* correlation function, can be seen; it is the Fourier transform of $S(\tau)$. The function $R(\Delta f)$ represents the correlation between the channel's response to two signals as a function of the frequency difference between the two signals. It can be thought of as the channel's frequency transfer function. Therefore, the time-spreading manifestation can be viewed as if it were the result of a filtering process. Knowledge of $R(\Delta f)$ helps answer the question, "What is the correlation between received signals that are spaced in frequency $\Delta f = f_1 - f_2$?" The function $R(\Delta f)$ can be measured by transmitting a pair of sinusoids separated in frequency by Δf, cross-correlating the complex spectra of the two separately received signals, and repeating the process many times with ever-larger separation Δf. Therefore, the measurement of $R(\Delta f)$ can be made with a sinusoid that is swept in frequency across the band of interest (a wideband signal). The *coherence bandwidth* f_0 is a statistical measure of the range of frequencies over which the channel passes all spectral components with approximately equal gain and linear phase. Thus, the coherence bandwidth represents a frequency range over which a signal's frequency components have a strong potential for amplitude correlation. That is, spectral components in that range are affected by the channel in a similar manner—as, for example, exhibiting fading or no fading. Note that f_0 and T_m are reciprocally related (within a multiplicative constant). As an approximation, we can say that

$$f_0 \approx 1/T_m \qquad (15.16)$$

The maximum excess delay T_m is not necessarily the best indicator of how any given system will perform when signals propagate on a channel, because different channels with the same value of T_m can exhibit very different signal-intensity profiles over the delay span. A more useful parameter is the delay spread, most often characterized in terms of its root-mean-squared (rms) value, called the rms delay spread, as

$$\sigma_\tau = \sqrt{\overline{\tau^2} - (\overline{\tau})^2} \qquad (15.17)$$

where $\overline{\tau}$ is the mean excess delay, $(\overline{\tau})^2$ is the mean squared, $\overline{\tau^2}$ is the second moment, and σ_τ is the square root of the second central moment of $S(\tau)$ [1].

A universal relationship between coherence bandwidth and delay spread that would be useful for all applications does not exist. However, using Fourier transform techniques an approximation can be derived from actual signal dispersion measurements in various channels. Several approximate relationships have been developed. If coherence bandwidth is defined as the frequency interval over which the channel's complex frequency transfer function has a correlation of at least 0.9, the coherence bandwidth is approximately [16]

$$f_0 \approx \frac{1}{50\sigma_\tau} \qquad (15.18)$$

For the case of a mobile radio, an array of radially uniformly spaced scatterers, all with equal-magnitude reflection coefficients but independent, randomly occurring reflection phase angles [17, 18], is generally accepted as a useful model for an urban propagation environment. This model is referred to as the *dense-scatterer* channel model. With the use of such a model, coherence bandwidth has similarly been defined [17], for a bandwidth interval over which the channel's complex frequency transfer function has a correlation of at least 0.5, to be

$$f_0 = \frac{0.276}{\sigma_\tau} \qquad (15.19)$$

Studies involving ionospheric effects often employ the following definition [19]:

$$f_0 = \frac{1}{2\pi\sigma_\tau} \qquad (15.20)$$

A more popular approximation of f_0, corresponding to a bandwidth interval having a correlation of at least 0.5, is [1]

$$f_0 \approx \frac{1}{5\sigma_\tau} \qquad (15.21)$$

The delay spread and coherence bandwidth are related to a channel's multipath characteristics, differing for different propagation paths (such as metropolitan areas, suburbs, hilly terrain, indoors, etc.). It is important to note that neither of the parameters in Equation (15.21) depend on signaling speed. A system's signaling speed only influences its transmission bandwidth W.

15.3.2.1 Degradation Categories due to Signal Time-Spreading Viewed in the Frequency Domain

A channel is referred to as frequency-selective if $f_0 < 1/T_s \approx W$, where the symbol rate $1/T_s$ is nominally taken to be equal to the signaling rate or signal bandwidth W. In practice, W may differ from $1/T_s$, due to system filtering or data modulation type (e.g., QPSK, MSK, spread spectrum, etc.) [20]. Frequency-selective fading distortion occurs whenever a signal's spectral components are not all affected equally by the channel. Some of the signal's spectral components falling outside the coherence bandwidth will be affected differently (independently), compared with those components contained within the coherence bandwidth. Figure 15.9 contains three examples. Each one illustrates the spectral density versus frequency of a transmitted signal having a bandwidth of W Hz. Superimposed on the plot in Figure 15.9a is the frequency transfer function of a frequency-selective channel ($f_0 < W$). Figure 15.9a shows that various spectral components of the transmitted signal will be affected differently.

Frequency-nonselective of flat-fading degradation occurs whenever $f_0 > W$. Hence, all of the signal's spectral components will be affected by the channel in a similar manner (e.g., fading or no fading). This is illustrated in Figure 15.9b, which features the spectral density of the same transmitted signal having a bandwidth of W Hz. However, superimposed on this plot is the frequency transfer function of a flat-fading channel ($f_0 > W$). Figure 15.9b illustrates that all of the spectral components of the transmitted signal will be affected in approximately the same way. Flat fading does not introduce channel-induced ISI distortion, but performance degradation can still be expected due to the loss in SNR whenever the signal is fading. In order to avoid channel-induced ISI distortion, the channel is required to exhibit flat fading. This occurs, provided that

$$f_0 > W \approx \frac{1}{T_s} \tag{15.22}$$

Hence, the channel coherence bandwidth f_0 sets an upper limit on the transmission rate that can be used without incorporating an equalizer in the receiver.

For the flat-fading case, where $f_0 > W$ (or $T_m < T_s$), Figure 15.9b shows the usual flat-fading pictorial representation. However, as a mobile radio changes its position, there will be times when the received signal experiences frequency-selective distortion even though $f_0 > W$. This is seen in Figure 15.9c, where the null of the channel's frequency transfer function occurs near the band center of the transmitted signal's spectral density. When this occurs, the baseband pulse can be especially mutilated by deprivation of its low-frequency components. One consequence of such loss is the absence of a reliable pulse peak on which to establish the timing synchronization or from which to sample the carrier phase carried by the pulse [17]. Thus, even through a channel is categorized as flat fading (based on rms relationships), it can still manifest frequency-selective fading on occasions. It is fair to say that a mobile radio channel, classified as exhibiting flat-fading degradation, cannot exhibit flat fading all of the time. As f_0 becomes much larger than W (or T_m becomes much smaller than T_s), less time will be spent exhibiting the type of

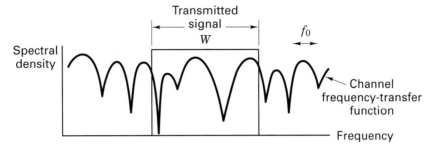

(a) Typical frequency-selective fading case $(f_0 < W)$

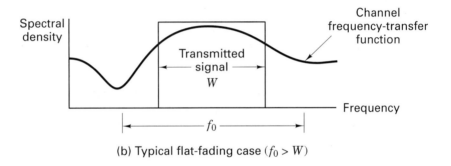

(b) Typical flat-fading case $(f_0 > W)$

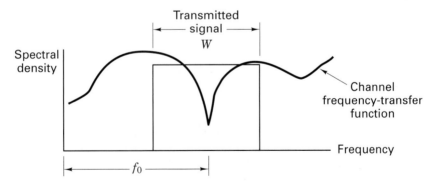

(c) Null of channel frequency-transfer function
occurs at signal band center $(f_0 > W)$

Figure 15.9 Relationships between the channel frequency-transfer function and a transmitted signal with bandwidth *W*.

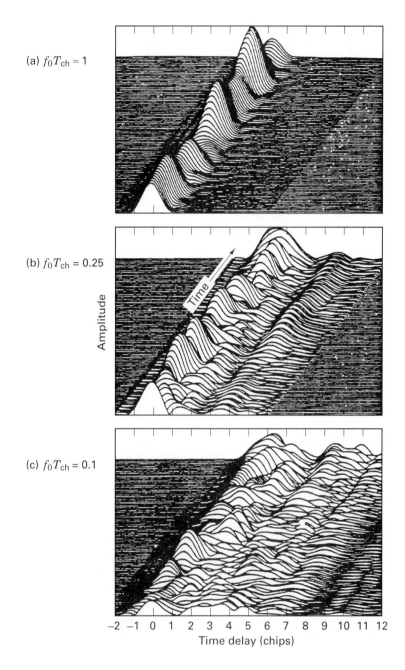

(a) $f_0 T_{\text{ch}} \approx 1$

(b) $f_0 T_{\text{ch}} = 0.25$

Amplitude

Time

(c) $f_0 T_{\text{ch}} = 0.1$

-2 -1 0 1 2 3 4 5 6 7 8 9 10 11 12
Time delay (chips)

Figure 15.10 DS/SS matched-filter output time-history examples for 3 levels of channel conditions, where T_{ch} is the time duration of a chip.
Source: Bogusch, R.L., *Digital Communications in Fading Channels: Modulation and Coding,* Mission Research Corp., Santa Barbara, California, Report no. MRC-R-1043, March 11, 1987.

condition shown in Figure 15.9c. By comparison, it should be clear that in Figure 15.9a the fading is independent of the position of the signal band, and frequency-selective fading occurs all the time and not just occasionally.

15.3.3 Examples of Flat Fading and Frequency-Selective Fading

Figure 15.10 shows some examples of flat fading and frequency-selective fading for a direct-sequence spread-spectrum (DS/SS) system [19, 20]. In Figure 15.10, there are three plots of the output of a pseudonoise (PN) code correlator versus delay as a function of time (transmission or observation time). Each amplitude-versus-delay plot is akin to $S(\tau)$ versus τ, as shown in Figure 15.8a. The key difference is that the amplitudes shown in Figure 15.10 represent the output of a correlator; hence, the waveshapes are a function not only of the impulse response of the channel but also of the impulse response of the correlator. The delay time is expressed in units of chip durations (chips), where the chip is defined as the spread-spectrum, minimal-duration keying element. For each plot, the observation time is shown on an axis perpendicular to the amplitude-versus-time-delay plane. Figure 15.10 is drawn from a satellite-to-ground communications link exhibiting scintillation because of atmospheric disturbances. However, Figure 15.10 is still a useful illustration of three different channel conditions that might apply to a mobile radio situation. A mobile radio that moves along the observation-time axis is affected by changing multipath profiles along the route, as seen in the figure. The scale along the observation-time axis is also in units of chips. In Figure 15.10a, the signal dispersion (one "finger" of return) is on the order of a chip time duration, T_{ch}. In a typical DS/SS system, the spread-spectrum signal bandwidth is approximately equal to $1/T_{ch}$; hence, the normalized coherence bandwidth $f_0 T_{ch}$ of approximately unity in Figure 15.10a, implies that the coherence bandwidth is about equal to the spread-spectrum bandwidth. This describes a channel that can be called frequency-nonselective or slightly frequency-selective. In Figure 15.10b, where $f_0 T_{ch} = 0.25$, the signal dispersion is more pronounced. There is definite interchip interference, due to the coherence bandwidth being approximately 25 percent of the spread-spectrum bandwidth. In Figure 15.10c, where $f_0 T_{ch} = 0.1$, the signal dispersion is even more pronounced, with greater interchip-interference effects, due to the coherence bandwidth being approximately 10 percent of the spread-spectrum bandwidth. The coherence bandwidths (relative to the spread-spectrum signaling speed) shown in Figures 15.10b and 15.10c depict channels that can be categorized, respectively, as moderately and highly frequency selective. Later, it is shown that a DS/SS system operating over a frequency-selective channel at the chip level does not necessarily experience frequency-selective distortion at the symbol level.

The signal dispersion manifestation of a fading channel is analogous to the signal spreading that characterizes an electronic filter. Figure 15.11a depicts a wideband filter (narrow impulse response) and its effect on a signal in both the time domain and the frequency domain. This filter resembles a flat-fading channel yielding an output that is relatively free of distortion. Figure 15.11b shows a narrowband filter (wide impulse response). The output signal suffers much distortion, as shown in both time and frequency. Here, the process resembles a frequency-selective channel.

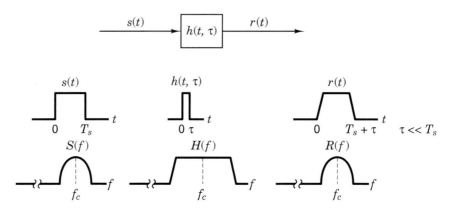

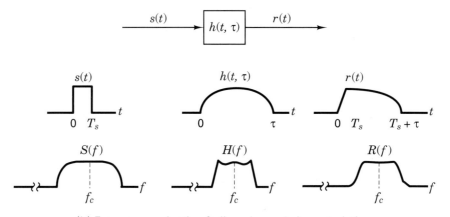

(a) Flat fading channel characteristics

(b) Frequency selective fading channel characteristics

Figure 15.11 Flat-fading and frequency-selective fading characteristics.
Source: Rappaport, T.S., *Wireless Communications,* Prentice-Hall, Upper Saddle
River, New Jersey, 1996.

15.4 TIME VARIANCE OF THE CHANNEL CAUSED BY MOTION

15.4.1 Time Variance Viewed in the Time Domain

Signal dispersion and coherence bandwidth characterize the channel's time-spreading properties in a local area. However, they do not offer information about the time-varying nature of the channel caused by relative motion between a transmitter and receiver or by movement of objects within the channel. For mobile-radio applications, the channel is time variant because motion between the transmitter and receiver results in propagation-path changes. For a transmitted continuous wave (CW) signal, such changes cause variations in the signal's amplitude and phase at the receiver. If all scatterers making up the channel are stationary, whenever motion ceases the amplitude and phase of the received signal remains constant—that is, the channel appears to be time invariant. Whenever motion begins again, the channel appears time variant. Since the channel characteristics are dependent on the positions of the transmitter and receiver, time variance in this case is equivalent to spatial variance.

Figure 15.8c shows the function $R(\Delta t)$, designated the *spaced-time* correlation function; it is the autocorrelation function of the channel's response to a sinusoid. This function specifies the extent to which there is correlation between the channel's response to a sinusoid sent at time t_1 and the response to a similar sinusoid sent at time t_2, where $\Delta t = t_2 - t_1$. The *coherence time* T_0 is a measure of the expected time duration over which the channel's response is essentially invariant. Earlier, measurements of signal dispersion and coherence bandwidth were made by using wideband signals. Now, to measure the time-variant nature of the channel, a narrowband signal is used [15]. To measure $R(\Delta t)$, a single sinusoid ($\Delta f = 0$) can be transmitted at times t_1 and t_2, and the cross-correlation function of the received signals is determined. The function $R(\Delta t)$ and the parameter T_0 provide knowledge about the fading rapidity of the channel. Note that for an ideal *time-invariant* channel (e.g., transmitter and receiver exhibiting no motion at all), the channel's response would be highly correlated for all values of Δt; thus, $R(\Delta t)$ as a function of Δt would be a constant. For example, if a stationary user's location is characterized by a multipath null, then that null remains unchanged until there is some movement (either by the transmitter or receiver or by objects within the propagation path). When using the dense-scatterer channel model described earlier, with constant mobile velocity V and an unmodulated CW signal having wavelength λ, the normalized $R(\Delta t)$ is described as [18]

$$R(\Delta t) = J_0(kV\Delta t) \tag{15.23}$$

where $J_0(\cdot)$ is the zero-order Bessel function of the first kind [11], $V\Delta t$ is distance traversed, and $k = 2\pi/\lambda$ is the free-space phase constant (transforming distance to radians of phase). Coherence time can be measured in terms of either time or distance traversed (assuming some fixed velocity). Amoroso described such a measurement using a CW signal and a dense-scatterer channel model [17]. He measured the statistical correlation between the combination of received magnitude and phase sampled at a particular antenna location x_0 and the corresponding combination sampled at some displaced location $x_0 + \zeta$, with displacement measured in units of wavelength λ. For a displacement ζ of 0.4λ between two antenna locations, the combined magnitudes and phases of the received CW are statistically uncorrelated. In other words, the signal observation at x_0 provides no information about the signal at $x_0 + \zeta$. For a given velocity, this displacement is readily transformed into units of time (coherence time).

15.4.1.1 The Basic Fading Manifestations are Independent of One Another

For a moving antenna, the fading of a received carrier wave is usually regarded as a random process, even though the fading record may be completely predetermined from the disposition of scatterers and the propagation geometry from the transmitter to the receiving antenna. This is because the same waveform received by two antennas that are displaced by at least 0.4λ are statistically uncorrelated [17, 18]. Since such a small distance (about 13 cm for a carrier wave at 900 MHz) corresponds to statistical decorrelation in received signals, the basic fading manifestations of signal dispersion and fading rapidity can be considered to be independent of each other. Any of the cases in Figure 15.10 can provide some insight here. At each instant of time (corresponding to a spatial location) we see a multipath intensity profile $S(\tau)$ as a function of delay τ. The multipath profiles are primarily determined by the surrounding terrain (buildings, vegetation, etc.). Consider Figure 15.10b, where the direction of motion through regions of differing multipath profiles is indicated by an arrow labeled *time* (it might also be labeled *antenna displacement*). As the mobile moves to a new spatial location characterized by a different profile, there will be changes in the fading state of the channel as characterized by the profile at the new location. However, because one profile is decorrelated with another profile at a distance as short as 13 cm (for a carrier at 900 MHz), the rapidity of such changes only depends on the speed of movement, not on the underlying geometry of the terrain.

15.4.1.2 The Concept of Duality

The mathematical concept of duality can be defined as follows: Two processes (functions, elements, or systems) are *dual* to each other if their mathematical relationships are the same even though they are described in terms of different parameters. In this chapter, it is interesting to note duality when examining time-domain versus frequency-domain relationships.

In Figure 15.8, we can identify functions that exhibit similar behavior across domains. For the purpose of understanding the fading-channel model, it is useful to refer to such functions as *duals*. For example, the phenomenon of signal dispersion can be characterized in the frequency domain by $R(\Delta f)$, as shown in Figure 15.8b. It yields knowledge about the range of frequencies over which two spectral components of a received signal have a strong potential for amplitude and phase correlation. Fading rapidity is characterized in the time domain by $R(\Delta t)$, as shown in Figure 15.8c. It yields knowledge about the span of time over which two received signals have a strong potential for amplitude and phase correlation. These two correlation functions, $R(\Delta f)$ and $R(\Delta t)$, have been labeled as duals. This is also noted in Figure 15.1 as the duality between blocks 10 and 13, and in Figure 15.7 as the duality between the time-spreading mechanism in the frequency domain and the time-variant mechanism in the time domain.

15.4.1.3 Degradation Categories due to Time Variance, Viewed in the Time Domain

The time-variant nature or fading rapidity mechanism of the channel can be viewed in terms of two degradation categories, as listed in Figure 15.7: *fast fading* and *slow fading*. The term "fast fading" is used for describing channels in which $T_0 < T_s$, where T_0 is the channel coherence time, and T_s is the time duration of a transmission symbol. Fast fading describes a condition where the time duration in which the channel behaves in a correlated manner is short compared with the time duration of a symbol. Therefore, it can be expected that the fading character of the channel will change several times during the time span of a symbol, leading to distortion of the baseband pulse shape. Analogous to the distortion previously described as channel-induced ISI, here distortion takes place because the received signal's components are not all highly correlated throughout time. Hence, fast fading can cause the baseband pulse to be distorted, often resulting in an irreducible error rate. Such distorted pulses cause synchronization problems (failure of phase-locked-loop receivers), in addition to difficulties in adequately designing a matched filter.

A channel is generally referred to as introducing slow fading if $T_0 > T_s$. Here, the time duration in which the channel behaves in a correlated manner is long compared with the time duration of a transmission symbol. Thus, one can expect the channel state to virtually remain unchanged during the time in which a symbol is transmitted. The propagating symbols will likely not suffer from the pulse distortion described earlier. The primary degradation in a slow-fading channel, as with flat fading, is loss in SNR.

15.4.2 Time Variance Viewed in the Doppler-Shift Domain

A completely analogous characterization of the time-variant nature of the channel can be presented in the Doppler-shift (frequency) domain. Figure 15.8d shows a *Doppler power spectral density* (or Doppler spectrum) $S(v)$, plotted as a function of Doppler-frequency shift, v. For the case of the dense-scatterer model, a vertical receive antenna with constant azimuthal gain, a uniform distribution of signals arriving at all arrival angles throughout the range $(0, 2\pi)$, and an unmodulated CW signal, the signal spectrum at the antenna terminals is [18]

$$S(v) = \frac{1}{\pi f_d \sqrt{1 - \left(\dfrac{v - f_c}{f_d}\right)^2}} \tag{15.24}$$

The equality holds for frequency shifts of v that are in the range $\pm f_d$ about the carrier frequency f_c, and would be zero outside that range. The shape of the RF Doppler spectrum described by Equation (15.24) is classically bowl-shaped, as seen in Figure 15.8d. Note that the spectral shape is a result of the dense-scatterer channel model. Equation (15.24) has been shown to match experimental data gathered for mobile radio channels [22]; however, different applications yield different spectral shapes. For example, the dense-scatterer model does not hold for the indoor radio channel; the channel model for an indoor area assumes $S(v)$ to be a flat spectrum [23].

In Figure 15.8d, the sharpness and steepness of the boundaries of the Doppler spectrum are due to the sharp upper limit on the Doppler shift produced by a vehicular antenna travelling among the stationary scatterers of the dense scatterer model. The largest magnitude (infinite) of $S(v)$ occurs when the scatterer is directly ahead of the moving antenna platform or directly behind it. In that case, the magnitude of the frequency shift is given by

$$f_d = \frac{V}{\lambda} \tag{15.25}$$

where V is relative velocity and λ is the signal wavelength. When the transmitter and receiver move toward each other, f_d is positive, and when they move away from each other, f_d is negative. For scatterers directly broadside of the moving platform, the magnitude of the frequency shift is zero. The fact that Doppler components arriving at exactly $0°$ and $180°$ have an infinite power spectral density is not a problem, since the angle of arrival is continuously distributed and the probability of components arriving at exactly these angles is zero [1, 18].

$S(v)$ is the Fourier transform of $R(\Delta t)$. It is known that the Fourier transform of the autocorrelation function of a time series equals the magnitude squared of the Fourier transform of the original time series. Therefore, measurements can be made by simply transmitting a sinusoid (narrowband signal) and using Fourier analysis to generate the power spectrum of the received amplitude [15]. This Doppler power spectrum of the channel yields knowledge about the spectral spreading of a transmitted sinusoid (impulse in frequency) in the Doppler-shift domain. As indicated in Figure 15.8, $S(v)$ can be regarded as the dual of the multipath intensity profile $S(\tau)$, since the latter yields knowledge about the time spreading of a transmitted impulse in the time-delay domain. This is also noted in Figure 15.1 as the duality between blocks 7 and 16, and in Figure 15.7 as the duality between the time-spreading mechanism in the time-delay domain and the time-variant mechanism in the Doppler-shift domain.

Knowledge of $S(v)$ makes it possible to estimate how much spectral broadening is imposed on the signal as a function of the rate of change in the channel state. The width of the Doppler power spectrum (denoted f_d) is referred to in the literature by several different names: *Doppler spread, fading rate, fading bandwidth,* or *spectral broadening.* Equation (15.24) describes the Doppler frequency shift. In a typical multipath environment, the received signal travels over several reflected paths, each with a different distance and a different angle of arrival. The Doppler shift of each arriving path is generally different from that of other paths. The effect on the received signal manifests itself as a Doppler spreading of the transmitted signal frequency, rather than as a shift. Note that the Doppler spread f_d and the coherence time T_0 are reciprocally related (within a multiplicative constant), resulting in an approximate relationship between the two parameters given by

$$T_0 \approx \frac{1}{f_d} \tag{15.26}$$

Hence, the Doppler spread f_d (or $1/T_0$) is regarded as the typical *fading rate* of the channel. Earlier, T_0 was described as the expected time duration over which the channel's response to a sinusoid is essentially invariant. When T_0 is defined more precisely as the time duration over which the channel's response to sinusoids yields a correlation between them of at least 0.5, the relationship between T_0 and f_d is approximately [2]

$$T_0 \approx \frac{9}{16\pi f_d} \tag{15.27}$$

A popular rule of thumb is to define T_0 as the geometric mean of Equations (15.26) and (15.27). This yields

$$T_0 = \sqrt{\frac{9}{16\pi f_d^2}} = \frac{0.423}{f_d} \tag{15.28}$$

For the case of a 900 MHz mobile radio, Figure 15.12 illustrates the typical effect of Rayleigh fading on a signal's envelope amplitude versus time [1]. The figure shows that the distance traveled by the mobile in a time interval corresponding to two adjacent nulls (small-scale fades) is on the order of a half-wavelength ($\lambda/2$). Thus, from Figure 15.12 and Equation (15.25), the time required to traverse a distance $\lambda/2$ (approximately the coherence time), when traveling at a constant velocity V, is

$$T_0 \approx \frac{\lambda/2}{V} = \frac{0.5}{f_d} \tag{15.29}$$

Thus, when the interval between fades is approximately $\lambda/2$, as shown in Figure 15.12, the resulting expression for T_0 in Equation (15.29) is quite close to the geometric mean shown in Equation (15.28). From Equation (15.29), and using the parameters shown in Figure 15.12 (velocity = 120 km/hr, and carrier frequency = 900 MHz), it is straightforward to determine that the channel coherence time is approximately 5 ms and the Doppler spread (channel fading rate) is approximately 100 Hz. Therefore, if this example represents a channel over which digitized speech signals are transmitted with a typical rate of 10^4 symbols/s, the fading rate is considerably less than the symbol rate. Under such conditions, the channel would manifest slow-fading effects. Note that if the abscissa of Figure 15.12 were labeled in units of wavelength instead of time, the plotted fading characteristics would look the same for any radio frequency and any antenna speed.

15.4.2.1 Analogy for Spectral Broadening in Fading Channels

Let us discuss the reason why a signal experiences spectral broadening as it propagates from or is received by a moving platform, and why this spectral broadening (also called the fading rate of the channel) is a function of the speed of motion. An analogy can be used to explain this phenomenon. Figure 15.13 shows the keying of a digital signal (such as amplitude-shift-keying or frequency-shift-keying)

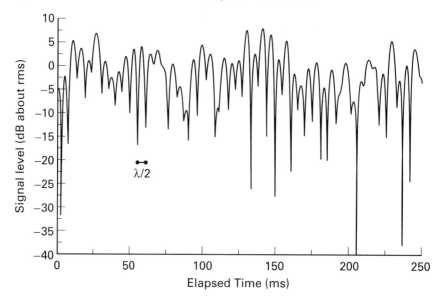

Figure 15.12 A typical Rayleigh fading envelope at 900 MHz.
Rappaport, T.S., *Wireless Communications,* Chapter 4, Prentice Hall, Upper Saddle River, New Jersey, 1996.

where a single tone $\cos 2\pi f_c t$ defined for $-\infty < t < \infty$, is characterized in the frequency domain in terms of impulses (at $\pm f_c$). This frequency domain representation is ideal (i.e., zero bandwidth), since the tone is a single frequency with infinite time duration. In practical applications, digital signaling involves switching (keying) signals on and off at a required rate. The keying operation can be viewed as multiplying the infinite-duration tone in Figure 15.13a by an ideal rectangular on-off (switching) function in Figure 15.13b. The frequency-domain description of this switching function is of the form sinc fT. (See Appendix A, Table A.1.)

In Figure 15.13c, the result of the multiplication yields a tone, $\cos 2\pi f_c t$, that is time-duration limited. The resulting spectrum is obtained by convolving the spectral impulses shown in part (a) of Figure 15.13 with the sinc fT function of part (b), yielding the broadened spectrum depicted in part (c). It is further seen that, if the signaling occurs at a faster rate characterized by the rectangle of shorter duration in part (d), the resulting signal spectrum in part (e) exhibits greater spectral broadening. The changing state of a fading channel is somewhat analogous to the on-off keying of digital signals. The channel behaves like a switch, turning the signal "on and off." The greater the rapidity of the change in the channel state, the greater spectral broadening experienced by signals propagating over such a channel. The analogy is not exact because the on and off switching of signals may result in phase discontinuities, while the typical multipath-scatterer environment induces phase-continuous effects.

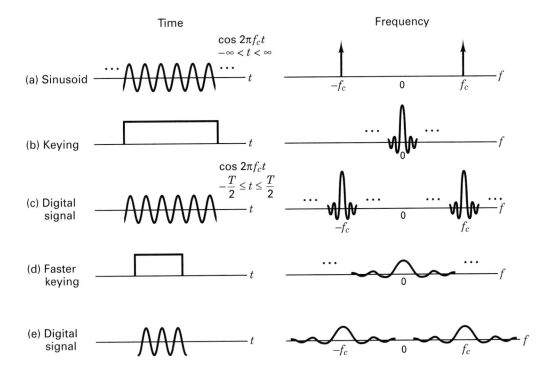

Figure 15.13 Analogy between spectral broadening in fading and spectral broadening in keying a digital signal.

15.4.2.2 Degradation Categories due to Time Variance, Viewed in the Doppler-Shift Domain

A channel is said to be fast fading if the symbol rate $1/T_s$ (approximately equal to the signaling rate or bandwidth W) is less than the fading rate $1/T_0$ (approximately equal to f_d); that is, fast fading is characterized by

$$W < f_d \tag{15.30}$$

or

$$T_s > T_0 \tag{15.31}$$

Conversely, the channel is referred to as slow fading if the signaling rate is greater than the fading rate. Thus, in order to avoid signal distortion caused by fast fading, the channel must be made to exhibit slow fading characteristics by insuring that the signaling rate exceeds the channel fading rate. That is,

$$W > f_d \tag{15.32}$$

or

$$T_s < T_0 \tag{15.33}$$

15.4   Time Variance of the Channel Caused by Motion   **973**

In Equation (15.22), it was shown that due to signal dispersion, the coherence bandwidth f_0 sets an *upper limit* on the signaling rate that can be used without suffering frequency-selective distortion. Similarly, Equation (15.32) shows that due to Doppler spreading, the channel fading rate f_d sets a *lower limit* on the signaling rate that can be used without suffering fast-fading distortion. For HF communication systems, when teletype or Morse-coded messages were transmitted at low data rates, the channels often exhibited fast-fading characteristics. However, most present-day terrestrial mobile-radio channels can generally be characterized as slow fading.

Equations (15.32) and (15.33) do not go far enough in describing the desirable behavior of the channel. A better way to state the requirement for mitigating the effects of fast fading would be that we desire $W \gg f_d$ (or $T_s \ll T_0$). If this condition is not satisfied, the random frequency modulation (FM) due to varying Doppler shifts will degrade system performance significantly. The Doppler effect yields an irreducible error rate that cannot be overcome by simply increasing E_b/N_0 [24]. This irreducible error rate is most pronounced for any transmission scheme that involves modulating the carrier phase. A single specular Doppler path, without scatterers, registers an instantaneous frequency shift, classically calculated as $f_d = V/\lambda$. However, a combination of specular and multipath components yields a rather complex time dependence of instantaneous frequency, which can cause frequency swings much larger than $\pm V/\lambda$ when the information is recovered by an instantaneous frequency detector (a nonlinear device) [25]. Figure 15.14 illustrates how this can happen. Owing to vehicle motion, at time t_1 the specular phasor has rotated through an angle θ, while the net phasor has rotated through an angle ϕ, which is about four times greater than θ. The rate of change of phase at a time near this particular fade is about 4 times that of the specular Doppler alone. Therefore, the instantaneous frequency shift $d\phi/dt$ would be about 4 times that of the specular doppler shift. The peaking of instantaneous frequency shifts at a time near deep fades is akin to the phenomenon of FM "clicks" or "spikes." Figure 15.15 illustrates the seriousness of this problem. The figure shows bit-error rate versus E_b/N_0 performance plots for $\pi/4$ DQPSK signaling at $f_c = 850$ MHz for various simulated mobile speeds [26]. It should be clear that at high speeds, the performance curve bottoms out at an error-rate level that may be unacceptably high. Ideally, coherent demodulators that lock onto and track the information signal should suppress the effect of this FM noise and thus cancel the impact of Doppler shift. However, for large values of f_d, carrier recovery is difficult to implement because very wideband (relative to the data rate) phase-lock loops (PLLs) need to be designed. For voice-grade applications with bit-errors rates of 10^{-3} to 10^{-4}, a large value of Doppler shift is considered to be on the order of $0.01 \times W$. Therefore, to avoid fast-fading distortion and the Doppler-induced irreducible error rate, the signaling rate should exceed the fading rate by a factor of 100 to 200 [27]. The exact factor depends on the signal modulation, receiver design, and required error rate [1, 25–29]. Davarian [29] showed that a frequency-tracking loop can help lower but not completely remove the irreducible error rate in a mobile system by using differential minimum-shift keying (DMSK) modulation.

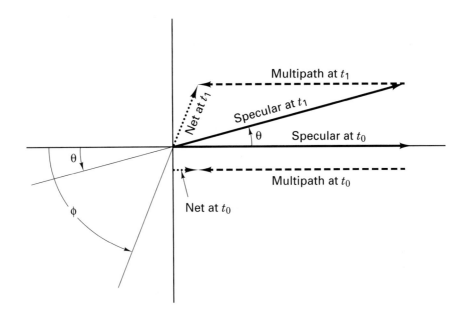

Figure 15.14 A combination of specular and multipath components can register much larger frequency swings than $\pm v/\lambda$.
Source: Amoroso, F. "Instantaneous Frequency Effects in a Doppler Scattering Environment", *IEEE International Conference on Communications*, June 7–10, 1987, pages 1458–1466

15.4.3 Performance over a Slow- and Flat-Fading Rayleigh Channel

For the case of a discrete multipath channel, with a complex envelope $g(t)$ described by Equation (15.3), a demodulated signal (neglecting noise) is described by Equation (15.10), which is rewritten as follows:

$$z(t) = \sum_n \alpha_n(t) e^{-j2\pi f_c \tau_n(t)} R[t - \tau_n(t)] e^{j\phi(t-\tau_n)} \qquad (15.34a)$$

where $R(t) = |g(t)|$ is the envelope magnitude, and $\phi(t)$ is its phase. Assume that the channel exhibits flat fading, so that the multipath components are not resolvable. Then the $\{\alpha_n(t)\}$ terms in Equation 15.34a, in one signaling interval T, need to be expressed as a resultant amplitude $\alpha(T)$ of all the n phasors received in that interval. Similarly, the phase terms in Equation 15.34a, in one signaling interval, need to be expressed as the resultant phase $\theta(T)$ of all the n fading phasors plus the information phase received in that interval. Assume also that the channel exhibits slow fading, so that the phase can be estimated from the received signal without significant error using phase-lock loop (PLL) circuitry or some other appropriate techniques. Therefore, for a slow- and flat-fading channel, we can express a received

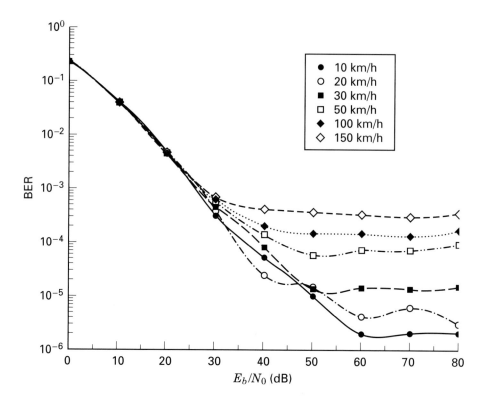

Figure 15.15 Error performance versus E_b/N_0 for $\pi/4$ DQPSK for various mobile speeds: $f_c = 850$ MHz, $R_s = 24$ ksymbol/s.
Source: Fung, V., Rappaport T.S., and Thoma, B., "Bit-Error Simulation for $\pi/4$ DQPSK Mobile Radio Communication Using Two-Ray and Measurement-Based Impulse Response Models," *IEEE Journal on Selected Areas in Communication,* Vol. 11, No. 3, April 1993, pp. 393–405.

test statistic $z(T)$ out of the demodulator in each signaling interval, including the noise $n_0(T)$, as

$$z(T) = \alpha(T) \, R(T) e^{-j[\theta(T) - \phi(T)]} + n_0(T) \qquad (15.34b)$$

For simplicity, we now replace $\alpha(T)$ with α. For binary signaling over an AWGN channel with a fixed attenuation of $\alpha = 1$, the bit-error probabilities for the basic coherent and noncoherent PSK and orthogonal FSK are given in Chapter 4, Table 4.1. The plots of bit-error probability versus E_b/N_0 for these signaling schemes each manifest the classical exponential relationship (a waterfall-shape associated with AWGN performance). However, for multipath conditions, if there is no specular signal component, α is a Rayleigh distributed random variable; or equivalently, α^2 is described by a chi-square pdf. Under these Rayleigh fading conditions, Figure 15.16 depicts the performance curves. When $(E_b/N_0) \, \mathbf{E}(\alpha^2) \gg 1$, where $\mathbf{E}(\cdot)$ represents

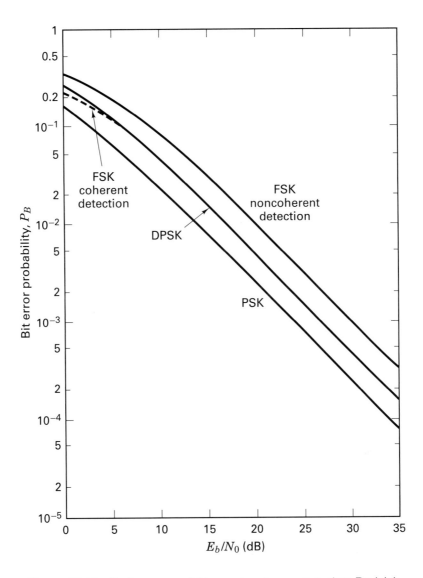

Figure 15.16 Performance of binary signaling over a slow Rayleigh fading channel.
Source: Proakis, J.G., *Digital Communications*, McGraw-Hill Book Company, New York, 1983.

statistical expectation, then the bit-error probability expressions for the basic binary signaling schemes shown in Figure 15.16 are given in Table 15.1. Each of the signaling schemes that had manifested a waterfall-shaped performance plot under AWGN interference in Figure 4.25 of Chapter 4 now exhibits performance that takes the form of an inverse linear function, as a result of the Rayleigh fading.

15.4   Time Variance of the Channel Caused by Motion   **977**

TABLE 15.1   Rayleigh-Limit Bit-Error Performance (where $(E_b/N_0)\,\mathbf{E}(\alpha^2) \gg 1$)

Modulation	P_B
PSK (Coherent)	$\dfrac{1}{4(E_b/N_0)\,\mathbf{E}(\alpha^2)}$
DPSK (Differentially Coherent)	$\dfrac{1}{2(E_b/N_0)\,\mathbf{E}(\alpha^2)}$
Orthogonal FSK (Coherent)	$\dfrac{1}{2(E_b/N_0)\,\mathbf{E}(\alpha^2)}$
Orthogonal FSK (Noncoherent)	$\dfrac{1}{(E_b/N_0)\,\mathbf{E}(\alpha^2)}$

Proakis, J.G., *Digital Communications,* McGraw-Hill, New York, 1983.

15.5 MITIGATING THE DEGRADATION EFFECTS OF FADING

Figure 15.17 subtitled "The Good, The Bad, and The Awful," highlights three major performance categories in terms of bit-error probability P_B versus E_b/N_0. The leftmost exponentially shaped curve highlights the performance that can be expected when using any nominal modulation scheme in AWGN interference. Observe that at a reasonable E_b/N_0 level, good performance can be expected. The middle curve, referred to as the *Rayleigh limit,* shows the performance degradation resulting from a loss in E_b/N_0 that is characteristic of flat fading or slow fading when there is no line-of-sight signal component present. The curve is a function of the reciprocal of E_b/N_0 (an inverse-linear function), so for practical values of E_b/N_0, performance will generally be "bad." In the case of Rayleigh fading, parameters with overbars are often introduced to indicate that an average is being taken over the "ups" and "downs" of the fading experience. Therefore, one often sees such bit-error probability plots with averaged parameters denoted by $\overline{P_B}$ and $\overline{E_b/N_0}$. This notation emphasizes the fact that the fading channel has memory; thus, received samples of the signal are correlated to one another in time. Therefore, when producing such error-probability plots for a fading channel, one needs to examine the process over a window of time that is much larger than the channel coherence time. The curve that reaches an irreducible error-rate level, sometimes called an *error floor,* represents "awful" performance, where the bit-error probability can level off at values nearly equal to 0.5. This shows the severe performance degrading effects that are possible with frequency-selective fading or fast fading.

   If the channel introduces signal distortion as a result of fading, the system performance can exhibit an irreducible error rate at a level higher than the desired error rate. In such cases, no amount of E_b/N_0 will help achieve the desired level of performance, and the only approach available for improving performance is to use some other form of mitigation to remove or reduce the signal distortion. The mitigation method depends on whether the distortion is caused by frequency-selective fading or fast fading. Once the signal distortion has been mitigated, the

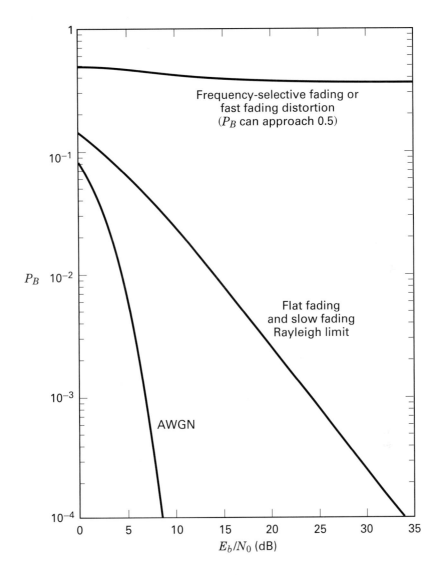

Figure 15.17 Error performance: The good, the bad, and the awful.

P_B-versus-E_b/N_0 performance can transition from the "awful" category to the merely "bad" Rayleigh-limit curve. Next, it is possible to further ameliorate the effects of fading and strive to approach AWGN system performance by using some form of diversity to provide the receiver with a collection of uncorrelated replicas of the signal, and by using a powerful error-correction code.

In Figure 15.18, several mitigation techniques for combating the effects of both signal distortion and loss in SNR are listed. Just as Figures 15.1 and 15.7 serve as a guide for characterizing fading phenomena and their effects, Figure 15.18 can

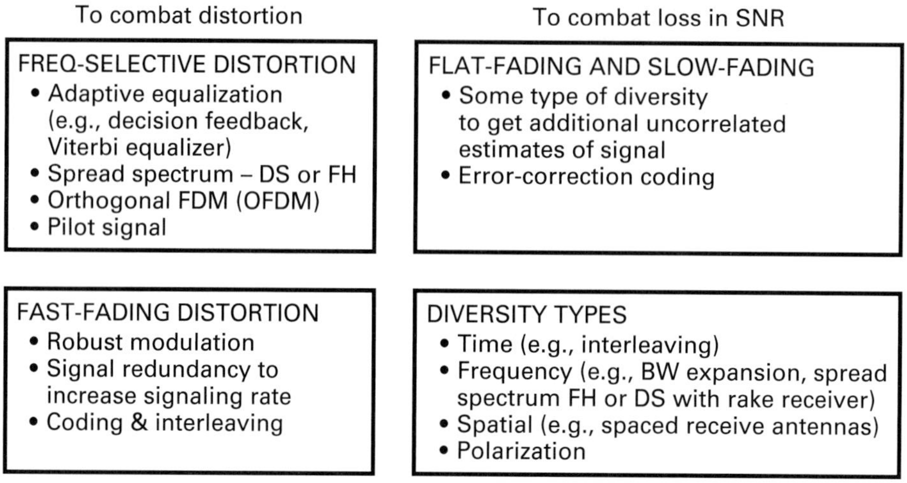

To combat distortion	To combat loss in SNR

FREQ-SELECTIVE DISTORTION
• Adaptive equalization
  (e.g., decision feedback,
  Viterbi equalizer)
• Spread spectrum – DS or FH
• Orthogonal FDM (OFDM)
• Pilot signal
	FLAT-FADING AND SLOW-FADING
	• Some type of diversity
	  to get additional uncorrelated
	  estimates of signal
	• Error-correction coding

FAST-FADING DISTORTION
• Robust modulation
• Signal redundancy to
  increase signaling rate
• Coding & interleaving
	DIVERSITY TYPES
	• Time (e.g., interleaving)
	• Frequency (e.g., BW expansion, spread
	  spectrum FH or DS with rake receiver)
	• Spatial (e.g., spaced receive antennas)
	• Polarization

Figure 15.18   Basic mitigation types.

similarly serve to describe mitigation methods that can be used to ameliorate the effects of fading. The mitigation approaches to be used when designing a system should be considered in two basic steps: first, choose the type of mitigation to reduce or remove any distortion degradation; next, choose a diversity type that can best approach AWGN system performance.

15.5.1 Mitigation to Combat Frequency-Selective Distortion

Equalization can mitigate the effects of channel-induced ISI that is brought on by frequency-selective fading. That is, it can help modify system performance described by the curve that is "awful" in Figure 15.17 to the one that is merely "bad." The process of equalizing for mitigating ISI effects involves using methods to gather the dispersed symbol energy back into its original time interval. In effect, an equalizer is an inverse filter of the channel. If the channel is frequency selective, the equalizer enhances the frequency components with small amplitudes and attenuates those with large amplitudes. The goal is for the combination of channel and equalizer filter to provide a flat composite received frequency response and linear phase [30]. Because in a mobile system the channel response varies with time, the equalizer filter must also change or adapt to the time-varying channel characteristics. Such equalizer filters are therefore adaptive devices that accomplish more than distortion mitigation; they also provide diversity. Since distortion mitigation is achieved by gathering the dispersed symbol's energy back into the symbol's original time interval (so that it doesn't hamper the detection of other symbols), the equalizer is simultaneously providing the receiver with symbol energy that would otherwise be lost.

The decision feedback equalizer (DFE) has a feedforward section that is a linear transversal filter [30] whose stage length and tap weights are selected to coherently combine virtually all of the current symbol's energy. The DFE also has a feedback section, which removes energy remaining from previously detected symbols [10, 30–32]. The basic idea behind the DFE is that once an information symbol has been detected, the ISI that it induces on future symbols can be estimated and subtracted before the detection of subsequent symbols. (See Section 3.4.3.2.)

A maximum-likelihood sequence estimation (MLSE) equalizer tests all possible data sequences (rather than detecting each received symbol by itself) and chooses the data sequence that is the most probable of all the candidates. The MLSE equalizer was first proposed by Forney [33] and implemented by using the Viterbi decoding algorithm [34]. The MLSE is optimal in the sense that it minimizes the probability of a sequence error. Because the Viterbi decoding algorithm is typically used in the implementation of the MLSE equalizer, this device is often referred to as the *Viterbi equalizer.* Later in this chapter, we illustrate the adaptive equalization performed in the Global System for Mobile (GSM) Communications using the Viterbi equalizer.

Direct-sequence spread-spectrum (DS/SS) techniques can be used to mitigate frequency-selective ISI distortion because the hallmark of spread-spectrum systems is their capability to reject interference, and ISI is a type of interference. Consider a DS/SS binary phase-shift keying (PSK) communication channel comprising one direct path and one reflected path. Assume that the propagation from transmitter to receiver results in a multipath wave that is delayed by τ compared with the direct wave. Neglecting noise, the received signal can be expressed as

$$r(t) = Ax(t)p(t)\cos(2\pi f_c t) + \alpha Ax(t - \tau)p(t - \tau)\cos(2\pi f_c t + \theta) \quad (15.35)$$

where $x(t)$ is the data signal, $p(t)$ is the pseudonoise (PN) spread code, and τ is the differential time delay between the two paths. The angle θ is a random phase, assumed to be uniformly distributed in the range $(0, 2\pi)$, and α is the attenuation of the multipath signal relative to the direct path signal. The receiver multiples the incoming $r(t)$ by the code $p(t)$. If the receiver is synchronized to the direct path signal, multiplication by the code signal yields

$$r(t)p(t) = Ax(t)p^2(t)\cos(2\pi f_c t) + \alpha Ax(t - \tau)p(t)p(t - \tau)\cos(2\pi f_c t + \theta) \quad (15.36)$$

where $p^2(t) = 1$. If τ is greater than the PN chip duration, then

$$\left| \int p(t)p(t - \tau)dt \right| \ll \left| \int p^2(t)dt \right| \quad (15.37)$$

over some appropriate interval of integration (correlation). Thus, the spread spectrum system effectively eliminates the multipath interference by virtue of its code-correlation receiver. Even though channel-induced ISI is typically transparent to DS/SS systems, such systems suffer from the loss in energy contained in the multipath components rejected by the receiver. The need to gather up this lost energy belonging to such multipath chips was the motivation for developing the Rake receiver [35–37]. The Rake receiver dedicates a separate correlator to each multipath component (finger) and coherently adds the energy from each finger by selectively delaying each (the earliest component gets the longest delay), so that they can all the coherently combined.

Earlier, we described a channel that could be classified as flat fading but occasionally exhibits frequency-selective distortion when the null of the channel's frequency transfer function occurs at the center of the signal band. The use of DS/SS is a practical way of mitigating such distortion because the wideband SS signal can span many lobes of the selectively faded channel frequency response. Hence, a great deal of pulse energy is passed by the scatterer medium, in contrast to the channel nulling effect on a relatively narrowband signal [17]. (See Figure 15.9c.) The ability of the signal spectrum to span over many lobes of the frequency-selective channel transfer function is the key to how DS/SS signaling can overcome the degrading effects of a multipath environment. This requires that the spread-spectrum bandwidth W_{ss} (or the chip rate, R_{ch}) be greater than the coherence bandwidth f_0. The larger the ratio of W_{ss} to f_0, the more effective will be the mitigation. The time-domain view of such mitigation was expressed in Equations (15.36) and (15.37). That is, to resolve multipath components (and either reject them or, with a Rake receiver, exploit them) requires that the spread-spectrum signal dispersion be greater than a chip time.

Frequency hopping spread spectrum (FH/SS) can be used as a technique to mitigate the distortion caused by frequency-selective fading, provided the hopping rate is at least equal to the symbol rate. Compared with DS/SS, mitigation takes place through a different mechanism. FH systems avoid the degradation effects due to multipath by rapidly changing the carrier-frequency band in the transmitter. Interference is avoided by similarly changing the band position in the receiver before the arrival of the multipath signal.

Orthogonal frequency-division multiplexing (OFDM) can be used for signal transmission in frequency-selective fading channels to avoid the use of an equalizer by lengthening the symbol duration. The approach is to partition (demultiplex) a high symbol-rate sequence into N symbol groups, so that each group contains a sequence of a lower symbol rate (by the factor $1/N$) than the original sequence. The signal band is made up of N orthogonal carrier waves, and each one is modulated by a different symbol group. The goal is to reduce the symbol rate (signaling rate) $W \approx 1/T_s$ on each carrier such that it is less than the channel's coherence bandwidth f_0. OFDM, originally referred to as *Kineplex*, is a technique that has been implemented in the United States in mobile radio systems [38] and has been chosen by the European community, under the name Coded OFDM (COFDM), for high-definition television (HDTV) broadcasting [39].

Pilot Signal is the name given to a signal intended to facilitate the coherent detection of waveforms. Pilot signals can be implemented in the frequency domain as in-band tones [40], or they can be implemented in the time domain as digital sequences, which can also provide information about the channel state and thus improve performance in fading conditions [41].

15.5.2 Mitigation to Combat Fast-Fading Distortion

Fast-fading distortion calls for the use of a robust modulation (noncoherent or differentially coherent) scheme that does not require phase tracking, and that reduces the detector integration time [19]. Another technique is to increase the symbol rate

$W \approx 1/T_s$, such that it is greater than the fading rate $f_d \approx 1/T_0$, by adding signal redundancy. Error-correction coding can also provide mitigation; instead of providing more signal energy, a code reduces the required E_b/N_0 for a desired error performance. For a given E_b/N_0, with coding present, the error floor out of the demodulator will not be lowered, but a lower error rate out of the decoder can be achieved [19]. Thus, with coding, one can get acceptable error performance and in effect withstand a large error floor from the demodulator that might have otherwise been unacceptable. To realize these coding benefits, errors out of the demodulator should be uncorrelated (which will generally be the case in a fast-fading environment), or an interleaver must be incorporated into the system design.

An interesting filtering technique can provide mitigation when fast-fading distortion and frequency-selective distortion occur simultaneously. The frequency-selective distortion can be mitigated by the use of an OFDM signal set. Fast fading, however, will typically degrade conventional OFDM because the Doppler spreading corrupts the orthogonality of the OFDM subcarriers. A polyphase filtering technique [42] is used to provide time-domain shaping and partial-response coding (see Section 2.9) to reduce the spectral sidelobes of the signal set and thus help preserve its orthogonality. The process introduces known ISI and adjacent channel interference (ACI), which are then removed by a post-processing equalizer and canceling filter [43].

15.5.3 Mitigation to Combat Loss in SNR

After implementing some mitigation technique to combat signal distortion due to frequency-selective or fast fading, the next step is to use diversity methods to move the system operating point from the error-performance curve labeled as "bad" in Figure 15.17 to a curve that approaches AWGN performance. The term "diversity" is used to denote the various methods available for providing the receiver with uncorrelated renditions of the signal of interest. Uncorrelated is the important feature here, since it would not help the receiver to have additional copies of a signal if the copies were all equally poor. Following are some of the ways in which diversity methods can be implemented:

- *Time diversity* can be provided by transmitting the signal on L different time slots with time separation of at least T_0. Interleaving, when used along with error-correction coding, is a form of time diversity.
- Frequency diversity can be provided by transmitting the signal on L different carriers with frequency separation of at least f_0. Bandwidth expansion is a form of frequency diversity. The signal bandwidth W is expanded so that it is greater than f_0, thus providing the receiver with several independently fading signal replicas. This achieves frequency diversity of the order $L = W/f_0$. Whenever W is made larger than f_0, there is the potential for frequency-selective distortion unless mitigation in the form of equalization is provided. Thus, an expanded bandwidth can improve system performance (via diversity) only if the frequency-selective distortion that the diversity may have introduced is mitigated.

- *Spread Spectrum Systems* are systems that excel at rejecting interfering signals by the use of bandwidth expansion techniques. In the case of direct-sequence spread-spectrum (DS/SS), it was demonstrated earlier that multipath components are rejected if they are time delayed by more than the duration of one chip. However, in order to approach AWGN performance, it is necessary to compensate for the loss in energy contained in those rejected components. The Rake receiver (described later) makes it possible to coherently combine the energy from several of the multipath components arriving along different paths (with sufficient differential delay). Thus, used with a Rake receiver, DS/SS modulation can be said to achieve path diversity. The Rake receiver is needed in phase-coherent reception; but in differentially coherent bit detection, a simple delay—equivalent to the duration of one bit with complex conjugation—can be implemented [44].

- *Frequency-hopping spread-spectrum* (FH/SS) is sometimes used as a diversity mechanism. The GSM system uses slow FH (217 hops/s) to compensate for those cases where the mobile unit is moving very slowly (or not at all) and experiencing deep fading due to a spectral null.

- *Spatial diversity* is usually accomplished through the use of multiple receive antennas, separated by a distance of at least 10 wavelengths when located at a base station (and less when located at a mobile unit). Signal processing techniques must be employed to choose the best antenna output or to coherently combine all the outputs. Systems have also been implemented with multiple transmitters, each at a different location, as in the Global Positioning System (GPS).

- *Polarization diversity* [45] is yet another way to achieve additional uncorrelated samples of the signal.

- Any diversity scheme may be viewed as a trivial form of repetition coding in space or time. However, techniques exist for improving the loss in SNR in a fading channel that are more efficient and more powerful than repetition coding. Error-correction coding represents a unique mitigation technique, because, instead of providing more signal energy, it reduces the required E_b/N_0 needed to achieve a desired performance level. Error-correction coding, coupled with interleaving [19, 46–51], is probably the most prevalent of the mitigation schemes used to provide improved system performance in a fading environment. Note that the time diversity mechanism obtained through interleaving relies on the vehicle motion to spread the errors during the fading. The faster the speed of the mobile unit, the more effective is the interleaver. The interleaver is less effective at slow speeds. (This vehicle speed-versus-interleaver performance is demonstrated in Section 15.5.6.)

15.5.4 Diversity Techniques

The goal in implementing diversity techniques is to utilize additional independent (or at least uncorrelated) signal paths to improve the received SNR. Diversity can provide improved system performance at relatively low cost; unlike equalization, diversity requires no training overhead. In this section, we show the error-performance improvements that can be obtained with the use of diversity techniques. The bit-error-probability, $\overline{P}_B$, averaged through all the "ups and downs" of the fading experience in a slow-fading channel can be computed as follows:

$$\overline{P}_B = \int_0^\infty P_B(x)\, p(x)\, dx \qquad (15.38)$$

where $P_B(x)$ is the bit-error probability for a given modulation scheme at a specific value of SNR $= x$, where $x = \alpha^2\, E_b/N_0$, and $p(x)$ is the pdf of x due to the fading conditions. With E_b and N_0 constant, α is used to represent the amplitude variations due to fading (see Section 15.2.2).

For Rayleigh fading, α has a Rayleigh distribution so α^2, and consequently x, have a chi-squared distribution. Thus, following the form of Equation (15.15),

$$p(x) = \frac{1}{\Gamma} \exp\left(-\frac{x}{\Gamma}\right) \qquad x \geq 0 \qquad (15.39)$$

where $\Gamma = \overline{\alpha^2}\, E_b/N_0$ is the SNR averaged through the "ups and downs" of fading. If each diversity (signal) branch, $i = 1, \ldots, M$, has an instantaneous SNR $= \gamma_i$, and we assume that each branch has the same average SNR given by Γ, then

$$p(\gamma_i) = \frac{1}{\Gamma} \exp\left(-\frac{\gamma_i}{\Gamma}\right) \qquad \gamma_i \geq 0 \qquad (15.40)$$

The probability that a single branch has SNR less than some threshold γ, is

$$P(\gamma_i \leq \gamma) = \int_0^\gamma p(\gamma_i)d\gamma_i = \int_0^\gamma \frac{1}{\Gamma} \exp\left(-\frac{\gamma_i}{\Gamma}\right) d\gamma_i \qquad (15.41)$$

$$= 1 - \exp\left(-\frac{\gamma}{\Gamma}\right)$$

The probability that all M independent signal diversity branches are received simultaneously with an SNR less than some threshold value γ, is

$$P(\gamma_1, \ldots, \gamma_M \leq \gamma) = \left[1 - \exp\left(-\frac{\gamma}{\Gamma}\right)\right]^M \qquad (15.42)$$

The probability that any single branch achieves SNR $> \gamma$ is

$$P(\gamma_i > \gamma) = 1 - \left[1 - \exp\left(-\frac{\gamma}{\Gamma}\right)\right]^M \qquad (15.43)$$

This is the probability of exceeding a threshold when selection diversity is used.

Example 15.1   Benefits of Diversity

Assume that 4-branch diversity is used and that each branch receives an independently Rayleigh-fading signal. If the average SNR is $\Gamma = 20$ dB, determine the probability that all 4 branches are received simultaneously with an SNR less than 10 dB (and also, the probability that this threshold will be exceeded). Compare the results with the case in which no diversity is used.

Solution

Using Equation (15.42) with $\gamma = 10$ dB, and $\gamma/\Gamma = 10$ dB $- 20$ dB $= -10$ dB $= 0.1$, we solve for the probability that the SNR will drop below 10 dB, as follows:

$$P(\gamma_1, \gamma_2, \gamma_3, \gamma_4 \leq 10 \text{ dB}) = [1 - \exp(-0.1)]^4 = 8.2 \times 10^{-5}$$

Or, using selection diversity, we can say that

$$P(\gamma_i > 10 \text{ dB}) = 1 - 8.2 \times 10^{-5} = 0.9999$$

Without diversity,

$$P(\gamma_1 \leq 10 \text{ dB}) = [1 - \exp(-0.1)]^1 = 0.095$$

$$P(\gamma_1 > 10 \text{ dB}) = 1 - 0.095 = 0.905$$

15.5.5.1 Diversity Combining Techniques

The most common techniques for combining diversity signals are *selection, feedback, maximal ratio, and equal gain.* For systems using spatial diversity, selection involves the sampling of M antenna signals and sending the largest one to the demodulator. Selection diversity combining is relatively easy to implement, however, it is not optimal because it does not make use of all the received signals simultaneously.

With feedback or scanning diversity, instead of using the largest of M signals, the M signals are scanned in a fixed sequence until one that exceeds a given threshold is found. This one becomes the chosen signal until it falls below the established threshold and the scanning process starts again. The error-performance of this technique is somewhat inferior to the other methods, but feedback diversity is quite simple to implement.

In the case of maximal-ratio combining, the signals from all of the M branches are weighted according to their individual SNRs and then summed. The individual signals must be cophased before being summed. The control algorithms for setting gains and delays are similar to those used in equalizers and in Rake receivers. Maximal-ratio combining produces an average SNR $\overline{\gamma_M}$ equal to the sum of the individual average SNRs, as shown below [30].

$$\overline{\gamma_M} = \sum_{i=1}^{M} \overline{\gamma_i} = \sum_{i=1}^{M} \Gamma = M\Gamma \tag{15.44}$$

where we assume that each branch has the same average SNR given by $\overline{\gamma_i} = \Gamma$. Thus, maximal ratio combining can produce an acceptable average SNR, even when none of the individual $\overline{\gamma_i}$ is acceptable. It uses each of the M branches in a cophased and weighted manner such that the largest possible SNR is available at the receiver. Equal-gain combining is similar to maximal-ratio combining, except that the weights are all set to unity. The possibility of achieving an acceptable output SNR from a number of unacceptable inputs is still retained. The performance is marginally inferior to maximal ratio combining. (See reference [52] for a detailed treatment of diversity combining.)

15.5.5 Modulation Types for Fading Channels

It should be apparent that an amplitude-based signaling scheme, such as amplitude shift keying (ASK) or quadrature amplitude modulation (QAM), is inherently vulnerable to performance degradation in a fading environment. Thus, for fading channels, the preferred choice for a signaling scheme is a frequency or phase-based modulation type.

In considering orthogonal FSK modulation for fading channels, the use of MFSK (with $M = 8$ or larger) is useful because its error performance is better than binary signaling. In slow Rayleigh fading channels, binary DPSK and 8-FSK perform within 0.1 dB of each other [19]. At first glance, one might argue that a higher order orthogonal alphabet expands the transmission bandwidth, which at some point may cause the coherence bandwidth of the channel to be exceeded, leading to frequency-selective fading. However, for MFSK, the transmission bandwidth that must be available is much larger than the bandwidth of the propagating signal. For example, consider the case of 8-FSK and a symbol rate of 10,000 symbols/s. The transmission bandwidth is $MR_s = 80,000$ Hertz. This is the bandwidth that must be available for the system's use. However, each time that a symbol is transmitted, only one single-sideband tone (having a spectral occupancy of 10,000 hertz) is sent, not the whole alphabet. In considering PSK modulation for fading channels, higher order modulation alphabets perform poorly. MPSK with $M = 8$ or larger should be avoided [19]. Example 15.2 examines a mobile communication system to substantiate such avoidance.

Example 15.2   Variations in a Mobile Communication System

The Doppler spread $f_d = V/\lambda$ shows that the fading rate is a direct function of velocity. Table 15.2 shows the Doppler spread versus vehicle speed at carrier frequencies of 900 MHz and 1800 MHz. Calculate the phase variation per symbol for the case of signaling with QPSK modulation at the rate of 24.3 kilosymbols/s. Assume that the carrier frequency is 1800 MHz and that the velocity of the vehicle is 50 miles/hr (80 km/hr). Repeat for a vehicle speed of 100 miles/hr.

Solution

$$\Delta\theta/\text{symbol} = \frac{f_d \text{ Hz}}{R_s \text{ symbols/s}} \times 360°$$

$$= \frac{132 \text{ Hz}}{24.3 \times 10^3 \text{ symbols/s}} \times 360°$$

$$= 2°/\text{symbol}$$

At a velocity of 100 miles/hr: $\Delta\theta/\text{symbol} = 4°/\text{symbol}$

Thus, it should be clear why MPSK with a value of $M > 4$ is not generally used to transmit information in a multipath environment.

TABLE 15.2 Doppler Spread versus Vehicle Speed

Velocity		Doppler (Hz)	Doppler (Hz)
miles/hr	km/hr	900 MHz ($\lambda = 33$ cm)	1800 MHz ($\lambda = 16.6$ cm)
3	5	4	8
20	32	27	54
50	60	66	132
80	108	106	212
120	192	160	320

15.5.6 The Role of an Interleaver

In Section 8.2, the various attributes of an interleaver are described. For transmission in a multipath environment, the primary benefit of an interleaver is to provide time diversity (when used along with error-correction coding). The larger the time span over which the channel symbols are separated, the greater chance that contiguous bits (after deinterleaving) will have been subjected to uncorrelated fading manifestations, and thus the greater chance of achieving effective diversity. Figure 15.19 illustrates the benefits of providing an interleaver time span T_{IL}, which is large compared with the channel coherence time T_0, for the case of DBPSK modulation with soft-decision decoding of a rate $\frac{1}{2}$, $K = 7$ convolutional code, over a slow Rayleigh-fading channel. It should be apparent that an interleaver having the largest ratio of T_{IL}/T_0 is the best performing (large demodulated BER leading to small decoded BER). This leads to the conclusion that T_{IL}/T_0 should be some large number—say, 1000 or 10,000. However, in a real-time communication system, this is not possible because the inherent time delay associated with an interleaver would be excessive. As described in Section 8.2.1 for the case of a block interleaver, before the first row of an array can be transmitted, virtually the entire array must be loaded. Similarly at the receiver, before the array can be deinterleaved, virtually the entire array must be stored. This leads to a delay of one block of data each at the transmitter and receiver. In Example 15.3, it is shown that for a cellular telephone system with a carrier frequency of 900 MHz, a T_{IL}/T_0 ratio of 10 is about as large as one can implement without suffering excessive delay.

It is interesting to note that the interleaver provides no benefit against multipath unless there is motion between the transmitter and receiver (or motion of objects within the signal-propagating paths). As the motion increases in velocity, so does the benefit of a given interleaver to the error performance of the system. (Don't use this as an excuse for exceeding a highway speed limit.) This is shown in Figure 15.20, where part (a) of the figure shows a terrain that is mapped out with attenuation factors $\{\alpha_i\}$, for a particular mobile communications link over a particular terrain. In the region between the points d_0 and d_1, the attenuation factor is α_1. Between the points d_1 and d_2, the attenuation factor is α_2, and so forth. Assume that the points d_i are equally separated by a distance Δd. Part (b) of the figure shows an automobile that is traveling at a slow speed; as the vehicle traverses a distance Δd, nine symbols are emitted from its transmitter. Assume that the inter-

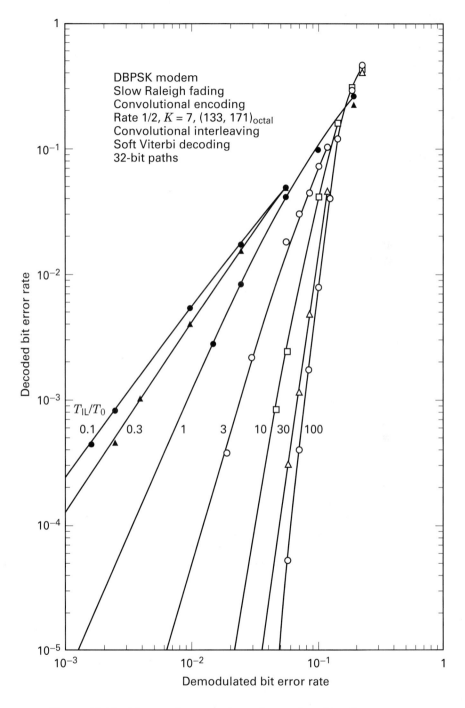

Figure 15.19 Error performance for various ratios of interleaver span to coherence time.

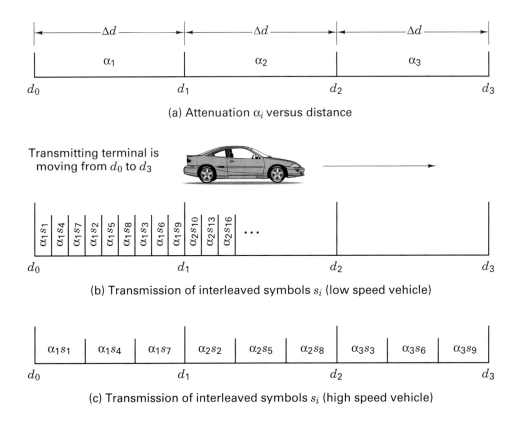

(a) Attenuation α_i versus distance

Transmitting terminal is moving from d_0 to d_3

(b) Transmission of interleaved symbols s_i (low speed vehicle)

(c) Transmission of interleaved symbols s_i (high speed vehicle)

Figure 15.20   The benefits of interleaving improve with increased vehicle speed.

leaver has a span of three symbol intervals, so that symbols s_1 through s_9 appear in the permuted order shown in part (b) of the figure. Notice that all nine of the symbols experience the same attenuation α_1, so that after deinterleaving there is no benefit obtained by using an interleaver with this small a span. Now consider part (c) of the figure, where the vehicle is moving 3 times faster than it is in part (b); thus, as the vehicle traverses a distance Δd, only three symbols are emitted from its transmitter. As before, the symbols are affected by the regional attenuation yielding the nine symbol sequence shown in part (c) of the figure. After deinterleaving of the sequence shown in part (c), we have the following attenuation factor-symbol pairs result: $\alpha_1 s_1$, $\alpha_2 s_2$, $\alpha_3 s_3$, $\alpha_1 s_4$, $\alpha_2 s_5$, $\alpha_3 s_6$, $\alpha_1 s_7$, $\alpha_2 s_8$, $\alpha_3 s_9$. It can be seen that adjacent symbols are affected by different attenuation factors. Thus, the interleaver with too small of a span to yield any benefit at low speeds does provide benefit at faster speeds.

Figure 15.21 also provides evidence that, although communications degrade with increased speed of the mobile unit (the fading rate increases), the benefit of an interleaver is enhanced with increased speed. Figure 15.21 shows the results of field testing performed on a CDMA system meeting the Interim Specification 95

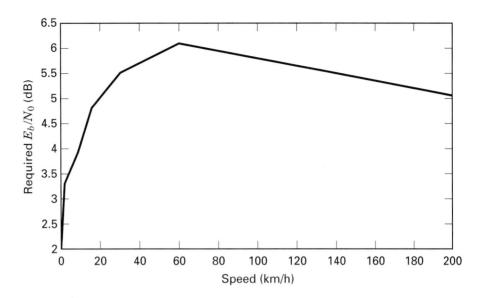

Figure 15.21 Typical E_b/N_0 performance versus vehicle speed for 850 MHz links to achieve a frame-error rate of 1 percent over a Rayleigh channel with two independent paths.

(IS-95) over a link comprising a moving vehicle and a base station [53]. The figure shows a plot of required E_b/N_0 versus vehicle speed to maintain a frame (20 ms of data) error rate of 1%. The best performance (smallest E_b/N_0 requirement) is achieved at low speeds from 0 to 20 km/hr. This slow-speed region is where the system power-control methods can most effectively compensate for the effects of slow fading; at these slow speeds the interleaver cannot provide any benefit, and the plot shows a steep degradation as a function of speed. For velocity in the range of 20–60 km/hr, the steepness of this degradation is reduced. This is the range in which the dynamics of the system power control cannot quite keep up with the increase in fading rapidity; and at the same time, the interleaver does not yet provide sufficient benefit. The speed of 60 km/hr represents the worst error-performance case for this system. As the vehicle goes faster than 60 km/hr, the power control no longer provides any benefits against fading, but the interleaver provides a steadily increasing (with speed) performance improvement. The interleaver's task of transforming the effects of a deep fade (time-correlated degradation events) into random events becomes easier with increased speed. In summary, the system error performance over a fading channel typically degrades with increased speed because of the increase in Doppler spread or fading rapidity. However, the action of an interleaver in the system provides mitigation, which becomes more effective at higher speeds. This trend toward improved error performance doesn't continue indefinitely. Eventually, the irreducible error floor seen in Figure 15.15 dominates. Therefore, if the type of measurements shown in Figure 15.21 were made at speeds beyond 200 km/hr, there would be a point at which the curve would turn around and steadily show the degradation effects due to the increased Doppler.

15.6 SUMMARY OF THE KEY PARAMETERS CHARACTERIZING FADING CHANNELS

We summarize the conditions that must be met so the channel does not introduce frequency-selective distortion and fast-fading distortion. Combining the expressions given by Equations (15.22), (15.32) and (15.33), we obtain

$$f_0 > W > f_d \tag{15.45}$$

or

$$T_m < T_s < T_0 \tag{15.46}$$

In other words, it is desired that the channel coherence bandwidth exceed the signaling rate, which in turn should exceed the fading rate of the channel. Recall that without distortion mitigation, f_0 sets an upper limit and f_d sets a lower limit on the signaling rate.

15.6.1 Fast-Fading Distortion: Case 1

If the conditions of Equations (15.45) and (15.46) are not met, distortion will result unless appropriate mitigation is provided. Consider the fast-fading case in which the signaling rate is less than the channel fading rate. That is,

$$f_0 > W < f_d \tag{15.47}$$

Mitigation consists of using one or more of the following methods (see Figure 15.18):

- Choose a modulation/demodulation technique that is most robust under fast-fading conditions. That means, for example, avoiding schemes that require PLLs for carrier recovery, since the fast fading could keep a PLL from achieving lock conditions.
- Incorporate sufficient redundancy so that the transmission symbol rate exceeds the channel fading rate but at the same time does not exceed the coherence bandwidth. The channel can then be classified as flat fading. However, as pointed out in Section 15.3.3, even flat-fading channels will experience frequency-selective distortion whenever a channel transfer function exhibits a spectral null near the signal band center. Since this happens only occasionally, mitigation can be accomplished by adequate error-correction coding and interleaving.
- The previously mentioned two mitigation approaches should result in the demodulator operating at the Rayleigh limit [19]. (See Figure 15.17.) However, the probability of error versus E_b/N_0 curve may exhibit flattening (as seen in

Figure 15.15) due to the FM noise that results from the random Doppler spreading. The use of an in-band pilot tone and a frequency-control loop can decrease the level at which the performance curve exhibits the flattening effect.

- To avoid the error floor due to random Doppler spreading, the signaling rate should be increased to about 100–200 times the fading rate [27]. This is one motivation for designing mobile communication systems to operate in a time-division multiple access (TDMA) mode.
- Incorporate error-correction coding and interleaving to further improve system performance.

15.6.2 Frequency-Selective Fading Distortion: Case 2

Consider the frequency-selective case in which the coherence bandwidth is less than the symbol rate, while the symbol rate is greater than the Doppler spread. That is,

$$f_0 < W > f_d \tag{15.48}$$

Since the transmission symbol rate exceeds the channel fading rate, there is no fast-fading distortion. However, mitigation of frequency-selective effects is necessary. One or more of the following techniques may be considered (see Figure 15.18):

- Adaptive equalization, spread spectrum (DS or FH), OFDM, pilot signal. The European GMS system uses a midamble training sequence in each transmission time slot, so that the receiver can estimate the impulse response of the channel. A Viterbi equalizer (explained later) is implemented for mitigating the frequency-selective distortion.
- Once the distortion effects have been reduced, diversity techniques (as well as error-correction coding and interleaving) should be introduced in order to approach AWGN performance. For direct-sequence spread-spectrum (DS/SS) signaling, the use of a Rake receiver (explained later) can be used for providing diversity by coherently combining multipath components that would otherwise be lost.

14.6.3 Fast-Fading and Frequency-Selective Fading Distortion: Case 3

Consider the case in which the channel coherence bandwidth is less than the signaling rate, which in turn is less than the fading rate. This condition is mathematically described by

$$f_0 < W < f_d \tag{15.49}$$

or

$$f_0 < f_d \tag{15.50}$$

Clearly, the channel exhibits both fast fading and frequency-selective fading. Recall from Equations (15.45) and (15.46) that f_0 sets an upper limit and f_d sets a lower limit on the signaling rate. Thus, the condition described by Equation (15.50) presents a difficult design problem, because, unless distortion mitigation is provided, the *maximum* allowable signaling rate is, strictly speaking, *less* than the *minimum* allowable signaling rate. Mitigation in this case is similar to the initial approach outlined in case 1:

- Choose a modulation/demodulation technique that is most robust under fast fading conditions.
- Use transmission redundancy in order to increase the transmitted symbol rate.
- Provide some form of frequency-selective fading mitigation in a manner similar to that outlined in case 2.
- Once the distortion effects have been reduced, introduce some form of diversity (as well as error-correction coding and interleaving), in order to approach AWGN performance.

Example 15.3 Equalizers and Interleavers for Mobile Communications

Consider a cellular telephone that is located in a vehicle travelling at 60 miles per hour (96 km/hr). The carrier frequency is 900 MHz. Use the GSM equalizer test profile shown in Figure 15.22 to determine the following: (a) the rms delay spread, σ_τ, (b) the maximum allowable signal bandwidth $W \approx 1/T_s$ that does not require the use of an equalizer, (c) when operating over a channel with the delay spread found in part (a), which of the following systems requires an equalizer—the United States Digital Cellular Standard (USDC) known as IS-54 (updated to IS-136), the Global System for Mobile (GSM) Communications, CDMA systems designed to meet IS-95? The bandwidths and symbol rates of these systems are USDC: $W = 30$ kHz, $1/T_s = 24.3$ kilosymbols/s; GSM: $W = 200$ kHz, $1/T_s = 271$ kilosymbols/s; and IS-95: $W = 1.25$ MHz, $1/T_s = 9.6$ kilosymbols/s. (d) The total (transmitter plus receiver) time delay caused by the interleaver, when the ratio of interleaver span to coherence time T_{IL}/T_0 is equal to 10. If the total tolerable time delay (transmitter plus receiver) for speech is 100 ms, can such an interleaver be implemented for speech? (e) Repeat parts (a) though (d) for a carrier frequency of 1900 MHz.

Solution

(a) In Figure 15.22, the GSM test profile shows an idealized multipath component (finger) located at each of six delay times $\{\tau_k\}$ from 0 to 16 μs. Each finger can be designated by $S(\tau_k)$, its average relative power, which in this profile is unity (0 dB). The profile represents a *fictitious* multipath environment used for equalization testing [15]. With the finger locations shown on the figure, the mean delay spread is computed as follows:

$$\bar{\tau} = \frac{\sum\limits_k S(\tau_k)\tau_k}{\sum\limits_k S(\tau_k)} = \frac{0 + 3.2 + 6.4 + 9.6 + 12.8 + 16.0}{6} = 8 \text{ μs}$$

The second moment of delay spread, $\overline{\tau^2}$, and the rms delay spread, σ_τ, are respectively computed as

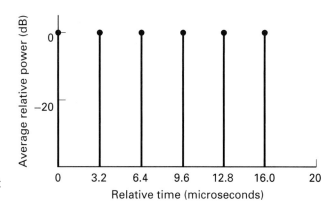

Figure 15.22 GSM equalizer test profile.

$$\overline{\tau^2} = \frac{\sum_k S(\tau_k)\,\tau_k^2}{\sum_k S(\tau_k)} = \frac{0 + 3.2^2 + 6.4^2 + 9.6^2 + 12.8^2 + 16.0^2}{6} = 93.87\ \mu s^2$$

and using Equation (15.17), we compute

$$\sigma_\tau = \sqrt{\overline{\tau^2} - (\overline{\tau})^2} = \sqrt{93.87 - 8^2} = 5.5\ \mu s$$

(b) Using Equation (15.21), the channel coherence bandwidth is determined to be

$$f_0 \approx \frac{1}{5\sigma_\tau} = \frac{1}{5 \times 5.5\ \mu s} = 36.4\ \text{kHz}$$

Thus, the maximum allowable signal bandwidth that will not require the implementation of an equalizer is $W = 36.4$ kHz.

(c) For the various system bandwidths given in this example, it is apparent that the need for an equalizer in USDC is marginal, while in GSM it is definitely required. With regard to systems that are designed to meet IS-95, since the signaling rate or transmission bandwidth W of 1.25 MHz is much larger than the coherence bandwidth f_0 of 36.4 kHz, the system exhibits frequency-selective fading. However, in such direct-sequence spread-spectrum (DS/SS) systems, W is purposely spread with the intent of exceeding f_0 and thus mitigating the effects of frequency-selective fading. An equalizer is only required if ISI poses a problem, and ISI is not a problem if the symbol rate is smaller than the coherence bandwidth (or the symbol duration is larger than the multipath spread). Hence, in the IS-95 example, since the symbol rate of 9.6 ksymbols/s is considerably smaller than the coherence bandwidth, an equalizer is not needed. A Rake receiver, which is described in Section 15.7.2, is used for exploiting path diversity; at the chip level its implementation resembles that of an equalizer.

(d) To determine the interleaver delay, we compute the Doppler spread and coherence time using Equations (15.25) and (15.29), as follows:

$$f_d = \frac{V}{\lambda} = \frac{\dfrac{96{,}000\ \text{m/hr}}{3600\ \text{s/hr}}}{\dfrac{3 \times 10^8\ \text{m/s}}{9 \times 10^8\ \text{Hz}}} = 80\ \text{Hz}, \qquad \text{Thus, } T_0 \approx \frac{0.5}{f_d} = 6.3\ \text{ms}$$

Based on the requirement that $T_{IL}/T_0 = 10$, the interleaver span is $T_{IL} = 63$ ms, making the total transmitter plus receiver delay time equal to 126 ms. For speech, this may be in the marginally acceptable range. Mobile systems often use interleavers with shorter spans that produce one-way delays in the range of 20–40 ms.

(e) Repeating for a carrier frequency of 1900 MHz, the coherence bandwidth calculations are unaffected by the change in carrier frequency, but the Doppler spread, coherence time, and interleaver delay must be computed again. The results are

$$f_d = \frac{V}{\lambda} = 169 \text{ Hz}, \qquad \text{Thus, } T_0 \approx \frac{0.5}{f_d} = 3 \text{ ms}$$

Thus, the interleaver span is $T_{IL} = 30$ ms, making the total transmitter plus receiver delay equal to 60 ms, which is acceptable for speech signals.

15.7 APPLICATIONS: MITIGATING THE EFFECTS OF FREQUENCY-SELECTIVE FADING

15.7.1 The Viterbi Equalizer as Applied to GSM

Figure 15.23 shows the GSM time-division multiple access (TDMA) frame, having a duration of 4.615 ms and comprising 8 slots, one assigned to each active mobile user. A normal transmission burst occupying one time slot contains 57 message bits on each side of a 26-bit midamble, called a *training* or *sounding sequence*. The slot-time duration is 0.577 ms (or the slot rate is 1733 slots/s). The purpose of the midamble is to assist the receiver in estimating the impulse response of the channel adaptively (during the time duration of each 0.577 ms slot). In order for the technique to be effective, the fading characteristics of the channel must not change appreciably during the time interval of one slot. In other words, there cannot be any fast-fading degradation during a time slot when the receiver analyzes the midamble distortion; otherwise, efforts to compensate for the channel's fading characteristics will not be effective. Consider for example a GSM receiver used aboard a high-speed train, traveling at a constant velocity of 200 km/hr (55.56 m/s). Assume the carrier frequency to be 900 MHz, (the wavelength is $\lambda = 0.33$ m). From Equation (15.29), the distance corresponding to a half-wavelength is traversed in

$$T_0 \approx \frac{\lambda/2}{V} \approx 3 \text{ ms} \tag{15.51}$$

which, as indicated in Equation (15.51), corresponds approximately to the coherence time. Therefore, the channel coherence time is more than 5 times greater than the slot time of 0.577 ms. The time needed for a significant change in channel fading characteristics is relatively long compared with the time duration of one slot. Note that the choices made for GSM in the design of its TDMA slot time and midamble were undoubtedly influenced by the need to preclude fast-fading effects, which could cause the equalizer to be ineffective. The GSM symbol rate (or bit rate, since the modulation is binary) is 271 kilosymbols/s and the bandwidth W is 200 kHz. Since the typical rms delay spread σ_τ in an urban environment is on the order of

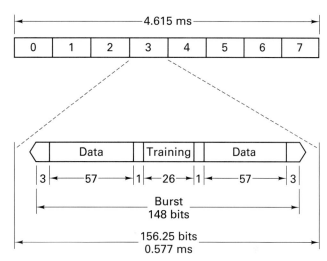

Figure 15.23 The GSM TDMA frame and time slot containing a normal burst.

2 μs, then using Equation (15.21), it can be seen that the resulting coherence bandwidth f_0 is approximately 100 kHz. It should therefore be apparent that since $f_0 < W$, the GSM receiver must utilize some form of mitigation to combat frequency-selective distortion. To accomplish this goal, the Viterbi equalizer is typically implemented.

Figure 15.24 illustrates the basic functional blocks used in a GSM receiver for estimating the channel impulse response. This estimate is used to provide the detector with channel-corrected reference waveforms [54], as explained below. In the final step, the Viterbi algorithm is used to compute the MLSE of the message bits. A received signal can be described in terms of the transmitted signal convolved with the impulse response of the channel. Let $s_{tr}(t)$ denote the transmitted midamble training sequence, and $r_{tr}(t)$ denote the corresponding received midamble training sequence. Thus,

$$r_{tr}(t) = s_{tr}(t) * h_c(t) \qquad (15.52)$$

where $*$ denotes convolution and noise has been neglected. At the receiver, since $r_{tr}(t)$ is part of the received normal burst, it is extracted and sent to a filter having impulse response $h_{mf}(t)$ that is matched to $s_{tr}(t)$. This matched filter yields at its output an estimate of $h_c(t)$, denoted $h_e(t)$, and developed from Equation (15.52) as

$$h_e(t) = r_{tr}(t) * h_{mf}(t)$$
$$= s_{tr}(t) * h_c(t) * h_{mf}(t) \qquad (15.53)$$
$$= R_s(t) * h_c(t)$$

where $R_s(t) = s_{tr}(t) * h_{mf}(t)$ is the autocorrelation function of $s_{tr}(t)$. If $s_{tr}(t)$ is designed to have a highly peaked (impulse-like) autocorrelation function $R_s(t)$, then $h_e(t) \approx h_c(t)$. Next, using a windowing function $w(t)$, we truncate $h_e(t)$ to form a computationally affordable function $h_w(t)$. The time duration of $w(t)$, denoted L_o,

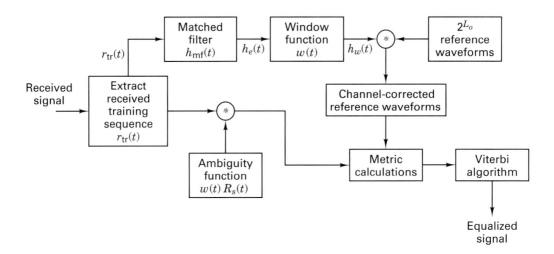

Figure 15.24 The Viterbi equalizer as applied to GSM.

must be large enough to compensate for the effect of typical channel-induced ISI. The term L_o consists of the sum of two contributions—namely, L_{CISI}, corresponding to the controlled ISI caused by Gaussian filtering of the baseband waveform (which then modulates the carrier using MSK), and L_C, corresponding to the channel-induced ISI caused by multipath propagation. Therefore, L_o can be written as

$$L_o = L_{CISI} + L_C \qquad (15.54)$$

The GSM system is required to provide distortion mitigation caused by signal dispersion having delay spreads of approximately 15–20 µs. Since, in GSM, the bit duration is 3.69 µs, we can express L_o in units of bit intervals. Thus, the Viterbi equalizer used in GSM has a memory of 4 to 6 bit intervals. For each L_o-bit interval in the message, the function of the Viterbi equalizer is to find the most likely L_o-bit sequence out of the 2^{L_o} possible sequences that might have been transmitted. Determining the most likely transmitted L_o-bit sequence requires that 2^{L_o} meaningful reference waveforms be created by modifying (or distorting) the 2^{L_o} ideal waveforms (generated at the receiver) in the same way that the channel has distorted the transmitted slot. Therefore, the 2^{L_o} reference waveforms are convolved with the windowed estimate of the channel impulse response $h_w(t)$, in order to generate the distorted or so-called channel-corrected reference waveforms. Next, the channel-corrected reference waveforms are compared with the received data waveforms to yield metric calculations. However, before the comparison takes place, the received data waveforms are convolved with the known windowed autocorrelation function $w(t)R_s(t)$, transforming them in a manner comparable to the transformation applied to the reference waveforms. This filtered message signal is compared with all possible 2^{L_o} channel-corrected reference signals, and metrics are computed in a manner similar to that used in the Viterbi decoding algorithm (VDA).

The VDA yields the maximum likelihood estimate of the transmitted data sequence [34].

Note that most equalizing techniques use filters in order to compensate for the non-optimum properties of $h_c(t)$; that is, equalizing filters attempt to modify the distorted pulse shapes. However, the operation of a Viterbi equalizer is quite different. It entails making measurements of $h_c(t)$, and then providing a means of adjusting the receiver to the channel environment. The goal of such adjustments is to enable the detector to make good estimates from the distorted pulse sequence. With a Viterbi equalizer, the distorted samples are not reshaped or directly compensated in any way; instead the mitigating technique is for the receiver to adjust itself in such a way that it can better deal with the distorted samples.

15.7.2 The Rake Receiver Applied to Direct-Sequence Spread-Spectrum (DS/SS) Systems

Interim Specification 95 (IS-95) describes a DS/SS cellular system that uses a Rake receiver [35–37] to provide path diversity. The Rake receiver searches through the different multipath delays for code correlation and thus recovers delayed signals, which are then optimally combined with the output of other independent correlators. In Figure 15.25, the power profiles associated with the five chip transmissions of the code sequence 1 0 1 1 1 are shown, where the observation times are labeled t_{-4} for the earliest transmission and t_0 for the latest. Each abscissa shows three "fingers" or components arriving with delays τ_1, τ_2, and τ_3. Assume that the intervals between the transmission times t_i and the intervals between the delay times τ_i are each one chip in duration. From this, one can conclude that the finger arriving at

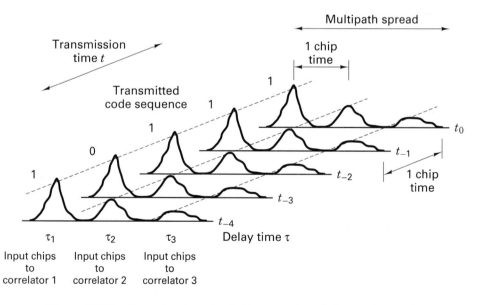

Figure 15.25   Example of received chips seen by a 3-finger rake receiver.

the receiver at time t_{-4}, with delay τ_3, is time coincident with two other fingers, namely the fingers arriving at times t_{-3} and t_{-2}, with delays τ_2 and τ_1, respectively. Since, in this example, the delayed components are separated by at least one chip time, they can be resolved. At the receiver, there must be a sounding device that is dedicated to estimating the τ_i delay times. Note that, for a terrestrial mobile radio system, the fading rate is relatively slow (in the order of milliseconds) or the channel coherence time is relatively large, compared with the chip time duration $(T_0 > T_{ch})$. Hence, the changes in τ_i occur slowly enough that the receiver can readily adapt to them.

Once the τ_i delays are estimated, a separate correlator is dedicated to recovering each resolvable multipath finger. In this example, there would be three such dedicated correlators, each one processing a delayed version of the same chip sequence 1 0 1 1 1. In Figure 15.25, each correlator receives chips with power profiles represented by the sequence of fingers shown along a diagonal line. For simplicity, the chips are all shown as positive signaling elements. In reality, these chips form a PN sequence, which of course, contains both positive and negative pulses. Each correlator attempts to correlate these arriving chips with the same appropriately synchronized PN code. At the end of a symbol interval (typically, there are hundreds or even thousands of chips per symbol), the outputs of the correlators are coherently combined, and a symbol detection is made. Figure 15.26 illustrates the

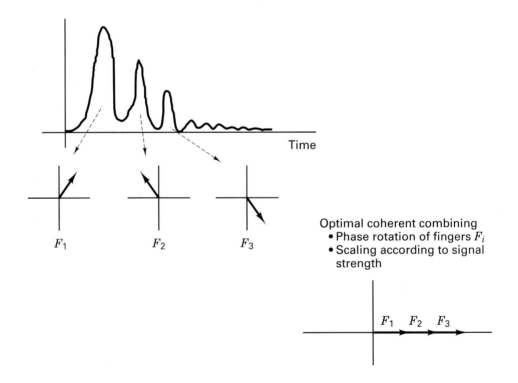

Figure 15.26   Coherent combining of multipath returns in the rake receiver.

phase rotation of fingers (F_i) provided by the Rake receiver, in order to facilitate coherent combining. At the chip level, the Rake receiver resembles an equalizer, but its real function is to exploit the path diversity.

The interference-suppression capability of DS/SS systems stems from the fact that a code sequence arriving at the receiver time-shifted by merely one chip will have very low correlation to the particular PN code with which the sequence is correlated. Therefore, any code chips that are delayed by one or more chip times will be suppressed by the correlator. The delayed chips only contribute to raising the interference level (correlation sidelobes). The mitigation provided by the Rake receiver can be termed *path diversity*, since it allows the energy of chips that arrive via multiple paths to be combined coherently. Without the Rake receiver, this energy would be transparent and therefore lost to the DS/SS receiver. In Figure 15.25, looking vertically above point τ_3 it is clear that there is interchip interference due to different fingers arriving simultaneously. The spread-spectrum processing gain allows the system to endure such interference at the chip level. No other equalization is deemed necessary in IS-95.

15.8 CONCLUSION

In this chapter, the major elements that contribute to fading in certain communication channels have been characterized. Figure 15.1 was presented as a guide for characterizing the fading phenomena. Two types of fading, large scale and small scale, were described. Two manifestations of small-scale fading (signal dispersion and fading rapidity) were examined. Each examination involved two views, one in time and the other in frequency. Two degradation categories were defined for dispersion: frequency-selective fading and flat fading. Two degradation categories were defined for fading rapidity: fast and slow fading. The small-scale fading degradation categories were summarized in Figure 15.7. A mathematical model using correlation and power density functions was presented in Figure 15.8. This model yields a useful symmetry to help us visualize the Fourier transform and duality relationships that describe the fading phenomena. Further, mitigation techniques for ameliorating the effects of each degradation category were presented; these were summarized in Figure 15.18. Finally, mitigation methods that have been applied in two different systems, GSM and CDMA systems meeting IS-95, were described.

REFERENCES

1. Rappaport, T. S., *Wireless Communications,* Chapters 3 and 4, Prentice Hall, Upper Saddle River, New Jersey, 1996.

2. Greenwood, D. and Hanzo, L., "Characterization of Mobile Radio Channels," *Mobile Radio Communications,* edited by R. Steele, Chapter 2, Pentech Press, London, 1994.

3. Lee, W. C. Y., "Elements of Cellular Mobile Radio Systems," *IEEE Trans. on Vehicular Technology,* vol. V–35, no. 2, May 1986, pp. 48–56.

4. Okumura, Y., et. al., "Field Strength and its Variability in VHF and UHF Land Mobile Radio Service," *Review of the Elec. Comm. Lab,* vol. 16, nos. 9 & 10, 1968, pp. 825–73.

5. Hata, M., "Empirical Formulae for Propagation Loss in Land Mobile Radio Services," *IEEE Trans. on Vehicular Technology,* vol. VT–29, no. 3, 1980, pp. 317–25.

6. Seidel, S. Y., et. al., "Path Loss, Scattering and Multipath Delay Statistics in Four European Cities for Digital Cellular and Microcellular Radiotelephone," *IEEE Transactions on Vehicular Technology,* Vol. 40, No. 4, November 1991, pp. 721–30.

7. Cox, D. C., Murray, R., and Norris, A., "800 MHz Attenuation Measured in and around Suburban Houses," *AT&T Bell Laboratory Technical Journal,* Vol. 673, No. 6, July–August 1984, pp. 921–54.

8. Schilling, D. L., et al., "Broadband CDMA for Personal Communications Systems," *IEEE Communications Magazine,* vol. 29, no. 11, November 1991, pp. 86–93.

9. Andersen, J. B., Rappaport, T. S., Yoshida, S., "Propagation Measurements and Models for Wireless Communications Channels, *IEEE Communications Magazine,* vol. 33, no. 1, Jan. 1995, pp. 42–49.

10. Proakis, J. G., *Digital Communications,* Chapter 7, McGraw–Hill Book Company, New York, 1983.

11. Schwartz, M., *Information, Transmission, Modulation, and Noise,* Second Edition, McGraw Hill, New York, 1970.

12. Amoroso, F., "Investigation of Signal Variance, Bit Error Rates and Pulse Dispersion for DSPN Signalling in a Mobile Dense Scatterer Ray Tracing Model," *Int'l Journal of Satellite Communications,* vol. 12, 1994, pp. 579–88.

13. Bello, P. A., "Characterization of Randomly Time-Variant Linear Channels," *IEEE Trans. on Commun. Syst.,* Dec. 1963, pp. 360–93.

14. Green, P. E., Jr., "Radar Astronomy Measurement Techniques," *MIT Lincoln Laboratory,* Lexington, Mass., Tech. Report No. 282, December 1962.

15. Pahlavan, K., and Levesque, A. H., *Wireless Information Networks,* Chapters 3 and 4, John Wiley and Sons, New York, 1995.

16. Lee, W. Y. C., *Mobile Cellular Communications,* McGraw-Hill Book Co., New York, 1989.

17. Amoroso, F. "Use of DS/SS Signaling to Mitigate Rayleigh Fading in a Dense Scatterer Environment," *IEEE Personal Communications,* vol. 3, no. 2, April 1996, pp. 52–61.

18. Clarke, R. H., "A Statistical Theory of Mobile Radio Reception," *Bell System Technical J.,* vol. 47, no. 6, July–August 1968, pp. 957–1000.

19. Bogusch, R. L., *Digital Communications in Fading Channels: Modulation and Coding,* Mission Research Corp., Santa Barbara, California, Report no. MRC-R-1043, March 11, 1987.

20. Amoroso, F., "The Bandwidth of Digital Data Signals," *IEEE Communications Magazine,* vol. 18, no. 6, November 1980, pp. 13–24.

21. Bogusch, R. L., et al., "Frequency Selective Propagation Effects on Spread-Spectrum Receiver Tracking," *Proceedings of the IEEE,* vol. 69, no. 7, July 1981, pp. 787–96.

22. Jakes, W. C. (Ed.), *Microwave Mobile Communications,* John Wiley & Sons, New York, 1974.

23. *Joint Technical Committee of Committee T1 R1P1.4 and TIA TR46.33/TR45.4.4 on Wireless Access,* "Draft Final Report on RF Channel Characterization," Paper No. JTC(AIR)/94.01.17–238R4, January 17, 1994.

24. Bello, P. A. and Nelin, B. D., "The Influence of Fading Spectrum on the Binary Error Probabilities of Incoherent and Differentially Coherent Matched Filter Receivers," *IRE Transactions on Commun. Syst.,* vol. CS-10, June 1962, pp. 160–68.

25. Amoroso, F., "Instantaneous Frequently Effects in a Doppler Scattering Environment," *IEEE International Conference on Communications,* June 7–10, 1987, pp. 1458–66.

26. Fung, V., Rappaport, T. S., and Thoma, B., "Bit-Error Simulation for π/4 DQPSK Mobile Radio Communication Using Two-Ray and Measurement-Based Impulse Response Models," *IEEE J. Sel. Areas Commun.,* vol. 11, no. 3, April 1993, pp. 393–94.

27. Bateman, A. J. and McGeehan, J. P., "Data Transmission over UHF Fading Mobile Radio Channels," *IEEE Proceedings,* vol. 131, Pt. F, No. 4, July 1984, pp. 364–74.

28. Feher, K. *Wireless Digital Communications,* Prentice Hall, Upper Saddle River, New Jersey, 1995.

29. Davarian, F., Simon, M., And Sumida, J., "DMSK: A Practical 2400–bps Receiver for the Mobile Satellite Service," *Jet Propulsion Laboratory* Publication 85–51 (MSAT–X Report No. 111), June 15, 1985.

30. Rappaport, T. S., *Wireless Communications,* Chapter 6, Prentice Hall, Upper Saddle River, New Jersey, 1996.

31. Bogusch, R. L., Guigliano, F. W., and Knepp, D. L., "Frequency-Selective Scintillation Effects and Decision Feedback Equalization in High Data-Rate Satellite Links," *Proceedings of the IEEE,* vol. 71, no. 6, June 1983, pp. 754–67.

32. Qureshi, S. U. H., "Adaptive Equalization," *Proceedings of the IEEE,* vol. 73, no. 9, September 1985, pp. 1340–87.

33. Forney, G. D., "The Viterbi Algorithm," *Proceedings of the IEEE,* vol. 61, no. 3, March 1978, pp. 268–78.

34. Viterbi, A. J. and Omura, J. K., *Principles of Digital Communication and Coding,* McGraw-Hill, New York, 1979.

35. Price, R. and Green P. E. Jr., "A Communication Technique for Multipath Channels," *Proceedings of the IRE,* March 1958, pp. 555–70.

36. Turin, G. L., "Introduction to Spread-Spectrum Antimultipath Techniques and their Application to Urban Digital Radio," *Proceedings of the IEEE,* vol. 68, no. 3, March 1980, pp. 328–53.

37. Simon, M. K., Omura, J. K., Scholtz, R. A., and Levitt, B. K., *Spread Spectrum Communications Handbook,* McGraw Hill Book Co., 1994.

38. Birchler, M. A. and Jasper, S. C., "A 64 kbps Digital Land Mobile Radio System Employing M-16QAM," *Proceedings of the 1992 IEEE Int'l. Conference on Selected Topics in Wireless Communications,* Vancouver, British Columbia, June 25–26, 1992, pp. 158–62.

39. Sari, H., Karam, G., and Jeanclaude, I., "Transmission Techniques for Digital Terrestrial TV Broadcasting," *IEEE Communications Magazine,* vol. 33, no. 2, February 1995, pp. 100–109.

40. Cavers, J. K., The Performance of Phase Locked Transparent Tone-in-Band with Symmetric Phase Detection, *IEEE Trans. on Commun.,* vol. 39, no. 9, Sept. 1991, pp. 1389–99.

41. Moher, M. L. and Lodge, J. H., "TCMP—A Modulation and Coding Strategy for Rician Fading Channel," *IEEE Journal on Selected Areas in Communications,* vol. 7, no. 9, December 1989, pp. 1347–55.

42. Harris, F., "On the Relationship Between Multirate Polyphase FIR Filters and Windowed, Overlapped FFT Processing," *Proceedings of the Twenty Third Annual Asilomar Conference on Signals, Systems, and Computers,* Pacific Grove, California, Oct. 30 to Nov. 1, 1989, pp. 485–88.

43. Lowdermilk, R. W., and Harris, F., "Design and Performance of Fading Insensitive Orthogonal Frequency Division Multiplexing (OFDM) using Polyphase Filtering

Techniques," *Proceedings of the Thirtieth Annual Asilomar Conference on Signals, Systems, and Computers,* Pacific Grove, California, November 3–6, 1996.

44. Kavehrad, M. and Bodeep, G. E., "Design and Experimental Results for a Direct-Sequence Spread-Spectrum Radio Using Differential Phase-Shift Keying Modulation for Indoor Wireless Communications," *IEEE JSAC* vol. SAC-5, no. 5, June 1987, pp. 815–23.

45. Hess, G. C., *Land-Mobile Radio System Engineering,* Artech House, Boston, 1993.

46. Hagenauer, J., and Lutz, E., "Forward Error Correction Coding for Fading Compensation in Mobile Satellite Channels," *IEEE JSAC,* vol. SAC–5, no. 2, February 1987, pp. 215–25.

47. McLane, P. I., et al., "PSK and DPSK Trellis Codes for Fast Fading, Shadowed Mobile Satellite Communication Channels," *IEEE Trans. on Comm.,* vol. 36, no. 11, November 1988, pp. 1242–46.

48. Schlegel, C., and Costello, D. J., Jr., "Bandwidth Efficient Coding for Fading Channels: Code Construction and Performance Analysis," *IEEE JSAC,* vol. 7, no. 9, December 1989, pp. 1356–68.

49. Edbauer, F., "Performance of Interleaved Trellis-Coded Differential 8-PSK Modulation over Fading Channels," *IEEE J. on Selected Areas in Comm.,* vol. 7, no. 9, December 1989, pp. 1340–46.

50. Soliman, S., and Mokrani, K., "Performance of Coded Systems over Fading Dispersive Channels," *IEEE Trans. on Communications,* vol. 40, no. 1, January 1992, pp. 51–59.

51. Divsalar, D. and Pollara, F. "Turbo Codes for PCS Applications," *Proc. ICC '95,* Seattle, Washington, June 18–22, 1995, pp. 54–59.

52. Simon, M and Alouini, M-S., *Digital Communications over Fading Channels: A Unified Approach to Performance Analysis,* John Wiley, New York, 2000.

53. Padovani, R., "Reverse Link Performance of IS-95 Based Cellular Systems," *IEEE Personal Communications,* Third Quarter 1994, pp. 28–34.

54. Hanzo, L. and Stefanov, J., "The Pan-European Digital Cellular Mobile Radio System—Known as GSM," *Mobile Radio Communications,* edited by R. Steele, Chapter 8, Pentech Press, London, 1992.

PROBLEMS

15.1. The probability density function for a Rayleigh continuous random variable, is given in Equation (15.15).

(a) Find an expression for the *distribution function,* as described in Section 1.5.1.

(b) For a signal transmitted over a mobile wireless channel, experiencing Rayleigh fading, use the distribution function to estimate the percentage of time that the signal level is 15 dB below the rms value.

(c) Repeat part b) for a signal level that is 5 dB below the rms value.

15.2. A signal within a mobile wireless system undergoes time spreading. The symbol rate $R_s = 20$ ksymbols/sec. Channel measurements indicate that the mean excess delay is 10 μs, while the second moment of the excess delay is 1.8×10^{-10} s^2.

(a) Calculate the coherence bandwidth, f_0, if it is defined as the frequency interval over which the channel's complex frequency transfer function has a correlation of at least 0.9.

(b) Repeat part a) if f_0 is defined for a bandwidth interval having a correlation of at least 0.5.

(c) Determine whether the signaling will undergo frequency-selective fading.

15.3. Consider a channel whose power-density profile is made up of three impulse functions with relative power and time-delay locations as follows: −20-dB at 0 μs, 0-dB at 2 μs, and −10-dB at 3 μs.

(a) Calculate the mean excess delay.

(b) Calculate the second moment of the excess delay.

(c) Calculate the rms delay spread.

(d) Estimate the coherence bandwidth (corresponding to a correlation ≥ 0.9).

(e) If the receiver is located in an airplane traveling at 800 km/hr, and the time required to traverse a half wavelength is 100 μs, calculate an approximate value for the transmission frequency.

15.4. Given a mobile wireless system with a carrier frequency of $f_c = 900$ MHz, and a Doppler frequency of $f_d = 50$ Hz. Assuming that the dense-scatterer model applies,

(a) Plot the *Doppler power spectral density S(v)* in the range of $f_c \pm f_d$ (use about 10 points).

(b) Explain why $S(v)$ has the response it does at the boundaries.

(c) Calculate the coherence time T_0, assuming that the channel's response to sinusoids yields a correlation of at least 0.5.

15.5. For each of the fading-effect categories below, name an application that generally fits that category. Provide numerical justification.

(a) frequency-selective, fast-fading

(b) frequency-selective, slow-fading

(c) flat-fading, fast-fading

(d) flat-fading, slow-fading

15.6. **(a)** What is the relationship between the signal power-density profile, as measured by the rms delay σ_τ, and the Doppler power-spectral density, as measured by the fading bandwidth f_d?

(b) What is the relationship between the spaced-frequency correlation function, as measured by the coherence bandwidth f_0, and the spaced-time correlation function, as measured by the coherence time T_0?

15.7. Consider an indoor narrowband mobile communication system that is characterized by a power density profile made up of four impulse functions with relative power and time-delay locations as follows: 0-dB at 0 ns, −3-dB at 100 ns, −3-dB at 200 ns, and −6-dB at 300 ns. What is the maximum symbol rate that such a system can support without using an equalizer? To find coherence bandwidth, use the criterion that spaced frequency tones are correlated to at least 0.5.

15.8. Consider a mobile wireless system, using QPSK modulation at 24.3 ksymbols/s, and a carrier frequency of 1900 MHz. What is the fastest speed in km/hr that is permissible for a vehicle using such a system, if it is required that the change in phase, $\Delta\theta$, due to spectral broadening (Doppler spread) does not exceed 5°/symbol?

15.9. For an interleaver to provide meaningful time diversity, a rule of thumb requires that the span of the interleaver, T_{IL}, should be at least ten times greater than the channel coherence time, T_0. Show a graph of T_{IL} versus frequency. (Plot three frequency points: 300 MHz, 3 GHz, and 30 GHz) for the following mobile telephone users.

(a) A pedestrian walking at a speed of 1 m/s.

(b) A high-speed train with a speed of 50 m/s.

(c) If the telephone handles real-time speech, which of the six plotted points represent cases where meaningful time diversity can be achieved by using an interleaver span of ten times T_0?

(d) What general conclusions can you draw?

15.10. A transmitted signal having a bandwidth of 5 kHz, and propagating over a channel having a coherence bandwidth of 50 kHz, is clearly an example of a flat-fading channel. Explain how such a channel can occasionally experience frequency-selective fading.

15.11. Consider a TDMA mobile wireless system, having a carrier frequency of 1900 MHz, that operates on a high-speed train at a speed of 180 km/hr. For learning the channel's impulse response in order to provide equalization, each user's transmission consists of training bits in addition to data bits. It is required that the training sequence should consist of 20 bits, should not occupy more than 20% of the total bits, and that these training bits should be embedded in the data, at least every $T_0/4$ s. Assuming binary modulation, what is the slowest transmission rate that would support these requirements without fast fading?

15.12. (a) In the late 1990's in Japan, a digital portable telephone system, called the Portable Handyphone System (PHS) was specified. The specification for PHS includes a carrier spacing of 300 kHz. Is the standard susceptible to frequency-selective fading in environments for which the channel has an rms delay spread on the order of 300 ns?

(b) A telephone standard called Digital Enhanced Cordless Telecommunications (DECT) was designed for high traffic density and short range (indoor) communications. The specification for DECT includes a carrier spacing of 1.728 MHz. Consider that the rms delay spread equals 150 ns. Calculate whether an equalizer needs to be included within the DECT receiver.

15.13. An interleaver span should be at least 10 times greater than the channel coherence time in order to provide significant time diversity for a mobile wireless system. Consider using such an interleaver in the design of a 1 GHz mobile system for pedestrians walking at a speed of 0.5 m/s. How large would the interleaver span have to be? Is this feasible for a real-time voice system?

15.14. If we desire to keep the transmitter-plus-receiver interleaver delay below 100 ms, what is the largest ratio of interleaver span to channel coherence time, T_{IL}/T_0, that can be used, for the following cases?

(a) The channel has a fading rate of 100 Hz.

(b) The channel has a fading rate of 1000 Hz.

15.15. A mobile wireless system is configured to support a data rate of 200 kbits/s using QPSK modulation and a carrier frequency of 1900 MHz. It is intended for use in a vehicle that typically travels at a speed of 96 km/hr.

(a) What change in phase angle value, $\Delta\theta$ per symbol, can be expected?

(b) What is the value of $\Delta\theta$ per symbol, if the data rate is decreased to 100 kbits/s?

(c) Repeat part (b) for a speed of 48 km/hr.

(d) Draw some general conclusions.

15.16. A channel exhibiting multipath fading has an rms delay spread of $\sigma_\tau = 10$ μs, and a Doppler spread of $f_d = 1$ Hz. The baseband pulse duration is chosen to be $T_s = 1$ μs.

(a) What is the value of the channel coherence bandwidth?

(b) What is the value of the channel coherence time?

(c) How would you classify the channel with regard to frequency selectivity and fading rapidity.

(d) How would you change the pulse duration (or data rate) to help mitigate the effects of fading?

15.17. For a mobile wireless application, a phase-based modulation scheme is particularly susceptible to phase distortion. Such distortion can be avoided if the rate of signaling is at least 100 times greater than the fading rate [27]. Consider a radio system operating at a carrier frequency of 1900 MHz, and moving at a velocity of 96 km/hr. In order to avoid degradation due to fast fading, what is the slowest symbol rate that should be sent over such a system?

15.18. Consider a mobile communication system which has a framing and time-slot arrangement as shown below in Figure P15.1.

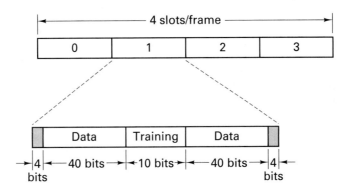

Figure P15.1 TDMA frame format.

There are a total of 4 time slots per frame; each user is assigned a single time slot per frame. Each time slot consists of 98 bits as shown in Figure P15.1. The signaling consists of QPSK modulation with a carrier frequency of 700 MHz. The symbol rate is 33.6 ksymbols/s, and the transmission bandwidth is 47 kHz. The system must perform normally at velocities up to 100 km/hr. Measurements performed on the physical channel conclude that the typical rms delay spread is in the order of 4 μs.

(a) Assuming that the training sequence yields an estimate of the channel impulse response during each slot time, will this design suffer the degradation effects of fast fading?

(b) Will this design suffer the degradation effects of frequency-selective fading?

15.19. The total permissible delay for data transmitted over a particular mobile wireless channel is limited to 340 ms. The data rate is equal to 19.2 ksymbols/s, and the data is interleaved to provide time diversity. The system exhibits delays as shown in Table P15.1.

TABLE P15.1 Delay values in ms

Delay T	Value (in ms)
Encoder	2
Modulator	10
Channel	0.3
Demodulator	25
Decoder	$2 \times 10^8 / f_{clk}$

The delay in milliseconds for the decoder, is given by $2 \times 10^8/f_{clk}$, where f_{clk} is the decoder clock speed. Calculate the minimum decoder clock speed that would be required for the following interleaver sizes.

(a) 100 bits

(b) 1000 bits

(c) 2850 bits

(d) What can you conclude about the decoder speed as the interleaver size increases?

15.20. Consider a mobile wireless orthogonal FDM (OFDM) system that is intended to operate in a vehicle with speeds up to 80 km/hr in an urban environment having a coherence bandwidth of 100 kHz. The carrier frequency is 3 GHz, and it is required that data be transmitted at a symbol rate of 1024 ksymbols/s. Choose an appropriate subcarrier plan, with the following goals: 1) avoid the use of an equalizer, and 2) minimize any effects due to fast fading. The plan should address how many subcarriers should be used, how far apart in frequency they should be separated, and what value of symbol rate per subcarrier should be used.

15.21. A mobile wireless system uses direct-sequence spread-spectrum (DS/SS) signaling to mitigate the effects of a received signal having two components—a direct path and a reflected path. The reflected path is 120 m longer than the direct path. What should the chip rate be, in order for such a system to provide mitigation against the effects of multipath?

15.22. It is well known that direct-sequence spread-spectrum (DS/SS) signaling can be used as a mitigation technique to protect against the channel-induced ISI of a frequency-selective channel. Yet, when you examine Figure 15.25 at some instant of time, say time τ_3, there exists interchip interference. Doesn't there need to be further equalization techniques used, in order to overcome this chip-level interference? Explain.

15.23. CDMA and TDMA are unique in the sense that each of these multiple-access schemes is associated with a property for mitigating one of the fading degradation types. Against which type of degradation does each of these schemes provide a "natural" protection.

15.24. Consider a diversity scheme consisting of four branches as shown in Figure P15.2. Each branch is responsible for processing an independent Rayleigh faded signal, $r(t)$. At a particular instant in time, the received signal can be expressed as a four branch vector $\mathbf{r} = [r_1 \, r_2 \, r_3 \, r_4]$, where r_i represents the voltage signal at branch i. Also, the gain

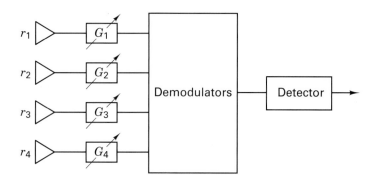

Figure P15.2 Four branch diversity receiver.

in the branches can be expressed with a four-branch vector $\mathbf{G} = [G_1 \ G_2 \ G_3 \ G_4]$, where G_i represents the voltage gain at branch i. Consider a moment in time when the measured values of $\mathbf{r}$ equal $[0.87, 1.21, 0.66, 1.90]$, and the associated gains of $\mathbf{G}$ equal $[0.5, 0.8, 1.0, 0.8]$. The average noise power of each branch N is equal to 0.25.

(a) Calculate the SNR applied to the detector?

(b) It can be shown [1] that the SNR is maximized when the value of each G_i is equal to r_i^2/N. Based on this condition, determine the maximum achievable SNR.

15.25. A system makes use of branch diversity in order to improve the receiver SNR. It is assumed that each branch receives an independently Rayleigh-faded signal. The receiver must meet a requirement that the probability of all branches being received with an SNR less than some threshold is equal to 10^{-4}, where the threshold is chosen to be 5 dB, and the average SNR is equal to 15 dB.

(a) Calculate the number of diversity branches M required in the receiver to meet this requirement?

(b) Based on the result from part a), calculate the probability that any single branch achieves an SNR > 5 dB?

15.26. A diversity scheme used in a receiver consists of two single branches. The following is received from each branch:

$$\begin{bmatrix} \text{Branch 1} \\ \text{Branch 2} \end{bmatrix} = \begin{bmatrix} 1.85 & 1.91 & -1.31 & -1.58 & 1.21 & 1.93 & 1.11 & -1.67 & 2.13 & -2.25 \\ 1.67 & 1.69 & -2.13 & -1.26 & 1.74 & 1.76 & 1.29 & -1.93 & 2.31 & -1.08 \end{bmatrix}$$

The first row shows the voltage samples received from the first branch and the second row shows the voltage samples received from the second branch. Each column corresponds to a particular instant in time. Consider that the average noise power in each branch is equal to 0.25 W and assume that the above values have been co-phased for use in both maximum ratio and equal gain combining. The instantaneous voltage gains provided by an attenuator for branch 1 and branch 2 are $G_1 = 1.2$ and $G_2 = 1.4$ respectively. Also with feedback diversity assume that the threshold has been set to an SNR of 5 dB. Calculate which branch is output to the detector if the following diversity-combining techniques are used:

(a) Selection.

(b) Feedback.

Calculate the value of SNR that is outputted to the detector when the following diversity combining schemes are used.

(c) Maximum Ratio.

(d) Equal Gain.

15.27. The response of a channel to ideal positive and negative pulses are spread over three pulse-time durations, as shown in Figure P15.3. Thus, for a sequence of transmitted pulses, the received waveform consists of the superposition of $L = 3$ events (segments from three pulses)—the current pulse, plus the memory of two past pulses. Use an *encoding trellis diagram* to represent this channel-induced ISI, and annotate each trellis branch with a voltage value resulting from that transition. Imagine that the system is initially flushed into the 00 state by transmitting two negative-polarity pulses. Then, consider the transmission of the sequence 11011 using the ideal pulses in Figure P15.3. Determine the amplitude values of the received distorted signal, and show its path along the trellis diagram. Hint: This binary finite-state machine has 2^{L-1} states. Use quadrille paper to perform the superposition needed to represent the distorted waveforms characterizing the channel. The trellis diagram is formed as

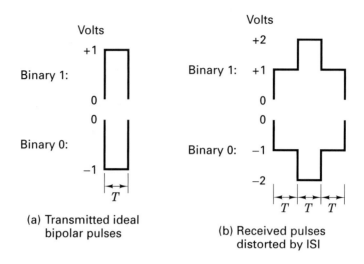

Volts	Volts
+1	+2
Binary 1:	Binary 1: +1
0	0
0	0
Binary 0:	Binary 0: −1
−1	−2
T	T T T
(a) Transmitted ideal bipolar pulses	(b) Received pulses distorted by ISI

Figure P15.3

in Section 7.2.3, with the exception that voltage levels are used here instead of code bits.

15.28. Use the channel characteristics and training sequence outlined in Problem 15.27, and add noise voltages of {+1 −1 +1 −1 +1} to the received distorted waveform. Use a *decoding trellis diagram* to illustrate how the Viterbi decoding algorithm is used in this equalization process, and show the computations that yield the first message bit. Hint: The process resembles the decoding of convolutionally encoded bits, with the exception that voltage levels are used here instead of code bits.

15.29. A mobile communication system uses a Viterbi equalizer in order to combat the effects of fading. The symbol rate is equal to 160 ksymbols/s, and BPSK modulation is used. The signal dispersion resulting from channel-induced ISI is equal to 25 μs.
  (a) Calculate the approximate amount of memory, L_o, in units of bit intervals required to be included in the Viterbi equalizer.
  (b) How is the memory size affected if the symbol rate is doubled?

QUESTIONS

15.1. What are the two fading mechanisms that characterize small-scale fading? Explain the *Fourier transform* and *duality* relationships among the time and frequency descriptions of these mechanisms. (See Sections 15.2 through 15.4.)

15.2. What are the differences between *Rician* fading and *Rayleigh* fading? (See Section 15.2.2.)

15.3. Define the following parameters: rms delay spread, coherence bandwidth, coherence time, Doppler spread. How are they related? (See Sections 15.3 and 15.4.)

15.4. What are the two *degradation categories* that characterize signal time-spreading, and what are the two degradation categories that characterize the time-variant nature of the channel? (See Sections 15.3 and 15.4.)

15.5. Why are the two basic fading mechanisms that characterize small-scale fading considered to be *independent* of one another? (See Section 15.4.1.1.)

15.6. Why does *signal distortion* caused by fading cause more serious degradation effects compared with a *loss* in SNR? (See Section 15.5.)

15.7. What techniques can be used for mitigating the effects of frequency-selective fading? What techniques can be used for mitigating the effects of fast fading? (See Section 15.5.)

15.8. What are the various ways of introducing signal *diversity*? (See Section 15.5.3.)

15.9. If there is *no motion* between a transmitter and receiver, how large would the interleaver span need to be in order to provide any time-diversity benefits? (See Section 15.5.6.)

EXERCISES

Using the Companion CD, run the exercises associated with Chapter 15.

APPENDIX A

A Review
of Fourier Techniques

A.1 SIGNALS, SPECTRA, AND LINEAR SYSTEMS

Electrical communication signals consist of time-varying voltage or current waveforms, typically described in the time domain. It is also convenient to describe such signals in the frequency domain. A signal's frequency-domain description is called its *spectrum*. Spectral concepts are important in communication analysis and design; they can describe a signal by its average power or energy content at various frequencies, and they illustrate how much of the electromagnetic spectrum (bandwidth) the signal occupies. Broadcast stations are required by the Federal Communications Commission (FCC) to operate at their assigned frequency with very tight tolerances on the occupied bandwidth; for example, amplitude-modulated (AM) radio channels are spaced 10 kHz apart, and television channels are spaced 6 MHz apart. Our interest in spectra and Fourier techniques has to do with the real-world constraints of ensuring that our communication signals are confined to specified spectral boundaries.

Frequency spectral characteristics can be ascribed to both signal waveforms and to circuits. When we say that a particular spectrum describes a *signal,* we mean that one way of characterizing the signal waveform is to specify its amplitude and phase as a function of frequency. However, when we talk about the spectral attributes of a *circuit* we are referring to the output versus input frequency-domain transfer function of the circuit; in other words, we are characterizing the circuit by how much of a specific input signal spectrum is allowed to pass through it.

A.2 FOURIER TECHNIQUES FOR LINEAR SYSTEM ANALYSIS

Fourier techniques are often used for analyzing linear circuits or systems in the following ways: (1) by predicting the system response, (2) by determining the system dynamic specification (transfer function), and (3) by evaluating or interpreting test results. Item 1, predicting system response, is illustrated schematically in Figure A.1. Let the input be an arbitrary periodic waveform with period equal to T_0 seconds. Fourier techniques allow us to describe such an input as a sum of sinusoidal waveforms, as shown in the figure. The lowest-frequency sinusoid, or the *fundamental* frequency of the input periodic, has frequency $1/T_0$ hertz; the balance of the sinusoids have frequencies that are integral *harmonics* $(2/T_0, 3/T_0, \ldots)$ of this fundamental frequency. An important attribute of a linear system is that *superposition* applies, which means that the response to the sum of excitations is the sum of the responses to the individually applied excitations. In fact, this is used as a definition of linearity. Specifically, if

$$y_1(t) = \text{system response to } x_1(t)$$

$$y_2(t) = \text{system response to } x_2(t)$$

and

$$ay_1(t) + by_2(t) = \text{system response to } ax_1(t) + bx_2(t)$$

for all a, b, $x_1(t)$, and $x_2(t)$, then the system is linear. A consequence of this definition is that the output response of a *linear* system with sinusoidal input waveforms must be made up of sinusoidal waveforms having the *same frequencies* as the input waveforms; such a system is typically specified by an output versus input *frequency transfer function* (magnitude and phase versus frequency) as shown in Figure A.2. Figure A.2a illustrates a typical example of signal magnitude versus

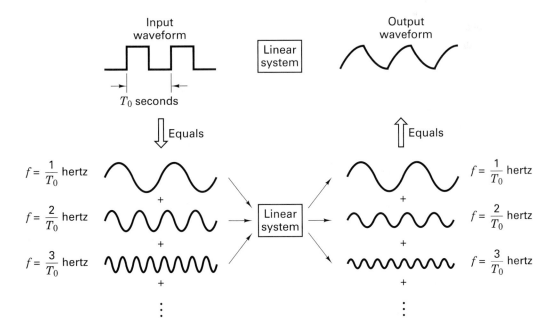

Figure A.1 Predicting system response.

A.2   Fourier Techniques for Linear System Analysis

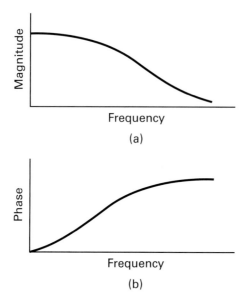

Frequency

(a)

Phase

Frequency

(b)

Figure A.2 System transfer function. (a) Magnitude response. (b) Phase response.

frequency; similarly, Figure A.2b illustrates a typical example of signal phase versus frequency.

The system transfer function serves as a performance specification; it describes the system response to each of the component sinusoids. Therefore, with the system transfer function in hand, one can predict each of the resulting output components. Using the principle of superposition, the final step of the analysis is to sum the individual output responses, thus forming the resulting overall response to the input periodic. (See Figure A.1.) In a similar manner, one can determine a system's transfer function, or evaluate a system's test results from knowledge of the input and output waveforms.

The development of Fourier methods had a major impact on the analysis of linear systems; it provided the translation between transient phenomena and sinusoidal techniques, and it simplified the analysis of linear systems under the excitation of any arbitrary input waveform. Just as logarithms allow the operation of multiplication to be treated as addition, so Fourier techniques allow the replacement of complex waveforms with sinusoidal components and sinusoidal methods.

A.2.1 Fourier Series Transform

Signals that are periodic with finite energy within each period can be represented by the *Fourier series*. The following equation describes such an arbitrary periodic waveform $x(\lambda)$ in terms of an infinite number of increasing harmonic sine and cosine components:

$$x(\lambda) = \tfrac{1}{2}a_0 + a_1 \cos \lambda + a_2 \cos 2\lambda + a_3 \cos 3\lambda \qquad (A.1)$$

$$+ \cdots + b_1 \sin \lambda + b_2 \sin 2\lambda + b_3 \sin 3\lambda + \cdots$$

The terms $\cos\lambda$ and $\sin\lambda$ are called the *fundamental terms*; the terms $\cos n\lambda$ and $\sin n\lambda$, for $n > 1$, are called *harmonic terms*, where n is an integer. The terms a_n and b_n represent the coefficients of the fundamental and harmonics, and $\frac{1}{2}a_0$ is the constant or dc term.

The function $x(\lambda)$ must have a period of 2π, or a submultiple thereof, and it must be single valued. The Fourier series can be thought of as a "recipe" for synthesizing any *arbitrary periodic* waveform using sinusoidal components. To be useful, the series must converge; that is, the sum of the series, as more and more of the higher harmonics are added, must approach a limit.

The process of synthesizing an arbitrary periodic waveform, from the coefficient values describing the mix of harmonics, is termed *synthesis*. The inverse process of calculating the coefficient values is termed *analysis*. Calculation of the coefficients is facilitated by the fact that the average of the sine and cosine cross-products is zero, as well as the average of any sinusoid. The following equations illustrate the basic averaging properties of the sine, cosine, and their products and cross-products:

$$\left.\begin{aligned}
\int_{-\pi}^{\pi} \sin m\lambda \, d\lambda &= 0 \\[1em]
\int_{-\pi}^{\pi} \cos m\lambda \, d\lambda &= 0 \\[1em]
\int_{-\pi}^{\pi} \sin m\lambda \cos n\lambda \, d\lambda &= 0
\end{aligned}\right\} \text{ where } m \text{ and } n \text{ are any integers} \qquad (A.2)$$

$$\left.\begin{aligned}
\int_{-\pi}^{\pi} \sin m\lambda \sin n\lambda \, d\lambda &= 0 \\[1em]
\int_{-\pi}^{\pi} \cos m\lambda \cos n\lambda \, d\lambda &= 0
\end{aligned}\right\} \text{ for } m \neq n \qquad (A.3)$$

$$\left.\begin{aligned}
\int_{-\pi}^{\pi} (\sin m\lambda)^2 \, d\lambda &= \pi \\[1em]
\int_{-\pi}^{\pi} (\cos m\lambda)^2 \, d\lambda &= \pi
\end{aligned}\right\} \text{ for } m = n \qquad (A.4)$$

Consider how one could go about finding the value of the coefficient, a_n or b_n, in Equation (A.1). To find the coefficient a_3, for example, we can multiply both sides of Equation (A.1) by $\cos 3\lambda \, d\lambda$ and integrate, as follows:

$$\int_{-\pi}^{\pi} x(\lambda) \cos 3\lambda \, d\lambda = \int_{-\pi}^{\pi} \tfrac{1}{2} a_0 \cos 3\lambda \, d\lambda + \int_{-\pi}^{\pi} a_1 \cos \lambda \cos 3\lambda \, d\lambda$$

$$+ \int_{-\pi}^{\pi} a_2 \cos 2\lambda \cos 3\lambda \, d\lambda + \int_{-\pi}^{\pi} a_3 (\cos 3\lambda)^2 \, d\lambda + \cdots$$

$$+ \int_{-\pi}^{\pi} b_1 \sin \lambda \cos 3\lambda \, d\lambda + \int_{-\pi}^{\pi} b_2 \sin 2\lambda \cos 3\lambda \, d\lambda$$

$$+ \int_{-\pi}^{\pi} b_3 \sin 3\lambda \cos 3\lambda \, d\lambda + \cdots$$

$$\int_{-\pi}^{\pi} x(\lambda) \cos 3\lambda \, d\lambda = \int_{-\pi}^{\pi} a_3 (\cos 3\lambda)^2 \, d\lambda = a_3 \pi$$

$$a_3 = \frac{1}{\pi} \int_{-\pi}^{\pi} x(\lambda) \cos 3\lambda \, d\lambda$$

We can generalize the preceding analysis to get

$$a_n = \frac{1}{\pi} \int_{-\pi}^{\pi} x(\lambda) \cos n\lambda \, d\lambda \tag{A.5}$$

$$b_n = \frac{1}{\pi} \int_{-\pi}^{\pi} x(\lambda) \sin n\lambda \, d\lambda \tag{A.6}$$

a_0 is found by solving Equation (A.5) with $n = 0$. This results in

$$\tfrac{1}{2} a_0 = \frac{1}{2\pi} \int_{-\pi}^{\pi} x(\lambda) \, d\lambda \tag{A.7}$$

which represents the zero-frequency term, or the average value of the periodic waveform. The synthesis process of Equation (A.1) can be expressed in more compact form as follows:

$$x(\lambda) = \tfrac{1}{2} a_0 + \sum_{n=1}^{\infty} (a_n \cos n\lambda + b_n \sin n\lambda) \tag{A.8}$$

There are several ways to express the *transform pair* (analysis and synthesis) of the Fourier series. The most common form makes use of the following identities to express the sine and cosine in exponential form:

$$\cos \lambda = \frac{e^{j\lambda} + e^{-j\lambda}}{2} \tag{A.9}$$

$$\sin \lambda = \frac{e^{j\lambda} - e^{-j\lambda}}{2j} \tag{A.10}$$

A periodic function with period T_0 seconds has frequency components of f_0, $2f_0$, $3f_0$, $\ldots$, where $f_0 = 1/T_0$ is called the *fundamental frequency*. We also refer to the frequency components as ω_0, $2\omega_0$, $3\omega_0$, $\ldots$, where $\omega_0 = 2\pi/T_0$ is called the fundamental *radian* frequency. The terms f and ω are each used to denote frequency. When f is used, frequency in hertz is intended; when ω is used, frequency is radians/second is intended. Let us replace the $n\lambda$ terms of Equations (A.5) to (A.8) with $2\pi nf_0 t = 2\pi nt/T_0$ as the general argument of the sinusoidal components, where n is an integer. For $n = 1$, nf_0 represents the fundamental frequency; for $n > 1$, nf_0 represents harmonics of the fundamental frequency. Using Equations (A.8) to (A.10), we can express $x(t)$ in exponential form as follows:

$$x(t) = \frac{a_0}{2} + \frac{1}{2}\sum_{n=1}^{\infty} [(a_n - jb_n)e^{j2\pi nf_0 t} + (a_n + jb_n)e^{-j2\pi nf_0 t}] \qquad (A.11)$$

Let c_n denote the complex coefficients, or spectral components of $x(t)$, related to a_n and b_n by

$$c_n = \begin{cases} \frac{1}{2}(a_n - jb_n) & \text{for } n > 0 \\ \dfrac{a_0}{2} & \text{for } n = 0 \\ \frac{1}{2}(a_n + jb_n) & \text{for } n < 0 \end{cases} \qquad (A.12)$$

Then we can simplify Equation (A.11), writing

$$x(t) = \sum_{n=-\infty}^{\infty} c_n e^{j2\pi nf_0 t} \qquad (A.13)$$

where the coefficients of the exponential harmonics are

$$c_n = \frac{1}{T_0}\int_{-T_0/2}^{T_0/2} x(t)e^{-j2\pi nf_0 t}\,dt \qquad (A.14)$$

To verify Equation (A.14), we multiply both sides of Equation (A.13) by $e^{-j2\pi mf_0 t}\,dt/T_0$, integrate over the interval $(-T_0/2, T_0/2)$, and use the relationship

$$\frac{1}{T_0}\int_{-T_0/2}^{T_0/2} e^{j(n-m)2\pi f_0 t}\,dt = \delta_{nm} = \begin{cases} 1 & \text{for } n = m \\ 0 & \text{for } n \neq m \end{cases} \qquad (A.15)$$

where δ_{nm} is known as the *Kronecker delta*. By multiplying and integrating in this way we obtain, for all integers m,

$$\frac{1}{T_0}\int_{-T_0/2}^{T_0/2} x(t)e^{-j2\pi mf_0 t}\,dt = \sum_{n=-\infty}^{\infty} c_n \delta_{nm} = c_m \qquad (A.16)$$

In general, the coefficient c_n is a complex number; it can be expressed in the form

$$c_n = |c_n|e^{j\theta_n} \qquad (A.17)$$

$$c_{-n} = |c_n|e^{-j\theta_n} \qquad (A.18)$$

where

$$|c_n| = \tfrac{1}{2} \sqrt{a_n^2 + b_n^2} \qquad\qquad (A.19)$$

$$\theta_n = \tan^{-1} - \frac{b_n}{a_n} \qquad\qquad (A.20)$$

$$b_0 = 0 \quad \text{and} \quad c_0 = \frac{a_0}{2}$$

The value of $|c_n|$ defines the magnitude of the nth harmonic component of the periodic waveform, so that a plot of $|c_n|$ versus frequency, called the *magnitude spectrum,* yields the magnitude of each of the n discrete harmonics in the signal. Similarly, a plot of θ_n versus frequency, called the *phase spectrum,* yields the phase of each harmonic component in the signal.

The Fourier coefficients of a real-valued periodic time function exhibit the relationship

$$c_{-n} = c_n^* \qquad\qquad (A.21)$$

where c_n^* is the complex conjugate of c_n. We therefore have

$$|c_{-n}| = |c_n| \qquad\qquad (A.22)$$

and the magnitude spectrum is an even function of frequency. Similarly, the phase spectrum θ_n is an odd function of frequency, because from Equation (A.20),

$$\theta_{-n} = -\theta_n \qquad\qquad (A.23)$$

The Fourier series is particularly useful in characterizing arbitrary periodic waveforms, with finite energy in each period, as presented above. The Fourier series can also be used to characterize nonperiodic signals having finite energy over a finite interval. However, a more convenient frequency-domain representation for such signals uses the Fourier integral transform. (See Section A.2.3.)

A.2.2 Spectrum of a Pulse Train

A signal of great interest in digital communications is an ideal periodic sequence of rectangular pulses, called a *pulse train,* illustrated in Figure A.3. For the pulse train, $x_p(t)$, with pulse amplitude A, pulse width T, and period T_0, the reader can verify, using Equations (A.14) and (A.10), the following expression for the Fourier series coefficients:

$$c_n = \frac{AT}{T_0} \frac{\sin(\pi n T/T_0)}{\pi n T/T_0} = \frac{AT}{T_0} \operatorname{sinc} \frac{nT}{T_0} \qquad\qquad (A.24)$$

In this expression,

$$\operatorname{sinc} y = \frac{\sin \pi y}{\pi y}$$

The sinc function, as shown in Figure A.4, has a maximum value of unity at $y = 0$ and approaches zero as y approaches infinity, oscillating through positive and

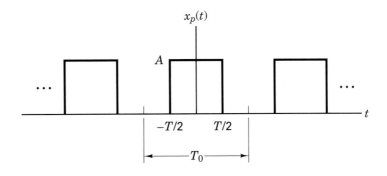

Figure A.3 Pulse train.

negative values. It goes through zero at $y = \pm 1, \pm 2, \ldots$ The pulse train magnitude spectrum, $|c_n|$ as a function of n/T_0, is plotted in Figure A.5a, and the phase spectrum, θ_n, is plotted in Figure A.5b. The positive and negative frequencies of the two-sided spectrum represent a useful way of expressing the spectrum mathematically; of course, only the positive frequencies can be reproduced in a laboratory.

Synthesis is performed by substituting the coefficients of Equation (A.24) into Equation (A.13). The resulting series yields the original ideal pulse train, $x_p(t)$, synthesized from its component parts:

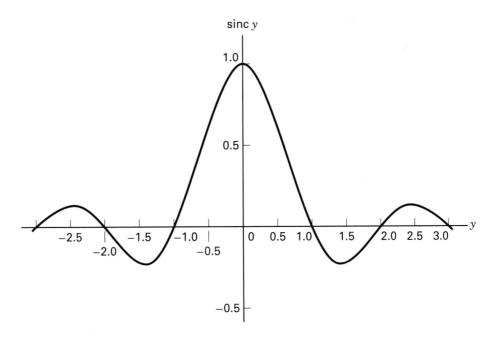

Figure A.4 Sinc function.

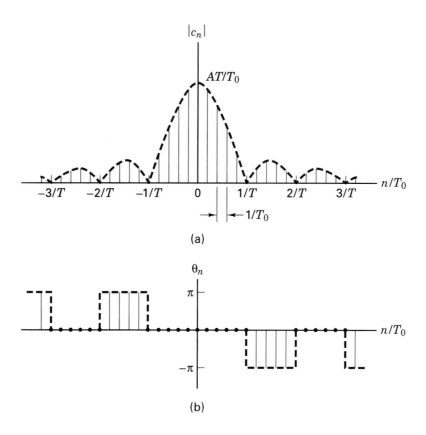

Figure A.5 Spectrum of a pulse train. (a) Magnitude spectrum. (b) Phase spectrum.

$$x_p(t) = \frac{AT}{T_0} \sum_{n=-\infty}^{\infty} \text{sinc} \, \frac{nT}{T_0} \, e^{j2\pi n f_0 t} \qquad (A.25)$$

The ideal periodic pulse train contains frequency components at all integer multiples of the fundamental. In communication systems the significant portion of a baseband signal's power or energy is often assumed to be contained within the frequencies from zero to the first null of the magnitude spectrum (see Figure A.5a). Therefore, $1/T$ is often used as a measure of signal *bandwidth*, in hertz, for a pulse train with pulse width T. Note that bandwidth is inversely proportional to pulse width; the narrower are the pulses, the wider is the bandwidth associated with these pulses. Also, notice that the spacing between spectral lines $\Delta f = 1/T_0$ is inversely proportional to the pulse period; as the period increases, the lines move closer together.

A.2.3 Fourier Integral Transform

In communication systems we often encounter nonperiodic signals having finite energy in a finite interval, and having zero energy outside this interval. Such signals can be conveniently characterized using the Fourier integral transform, or simply the *Fourier transform*. We can describe the nonperiodic signal as a periodic one, in the limiting sense. For example, consider the pulse train shown in Figure A.3. As $T_0 \to \infty$ and the pulse train approaches a single pulse, $x(t)$, the number of spectral lines approaches infinity and the spectral plot approaches a smooth frequency spectrum $X(f)$. For this limiting case, we can define the Fourier integral transform pair.

$$X(f) = \int_{-\infty}^{\infty} x(t)e^{-j2\pi ft}\, dt \qquad (A.26)$$

and

$$x(t) = \int_{-\infty}^{\infty} X(f)e^{j2\pi ft}\, dt \qquad (A.27)$$

where f is frequency measured in hertz. This pair can be used to describe the time–frequency relationship for nonperiodic signals.

Henceforth, the Fourier integral transform operation will be designated by the notation $\mathcal{F}\{\cdot\}$, and the inverse Fourier integral transform will be designated by $\mathcal{F}^{-1}\{\cdot\}$. The relationship between the time and frequency domains will be indicated by using the double arrow as follows:

$$x(t) \leftrightarrow X(f)$$

This notation indicates that $X(f)$ is the Fourier transform of $x(t)$ and that $x(t)$ is the inverse Fourier transform of $X(f)$. In the typical communications context, $x(t)$ is a real-valued function and $X(f)$ is a complex function, having real and imaginary components; in polar form, the spectrum, $X(f)$, can be specified by a magnitude characteristic and a phase characteristic:

$$X(f) = |X(f)|\, e^{j\theta(f)} \qquad (A.28)$$

The properties of $X(f)$, the spectrum of a nonperiodic waveform, are similar to those of the spectrum for a periodic waveform, presented in Equations (A.17) to (A.23); that is, when $x(t)$ is real valued,

$$X(-f) = X^*(f) \qquad (A.29)$$

$$= |X(f)|\, e^{-j\theta(f)} \qquad (A.30)$$

where X^* is the complex conjugate of X. The magnitude spectrum $|X(f)|$ is an even function of f and the phase spectrum is an odd function of f. In many cases $X(f)$ is either purely real or purely imaginary, and only one plot suffices to describe it.

A.3 FOURIER TRANSFORM PROPERTIES

There are many excellent references dealing with the details of Fourier transforms and their properties [1–4]. In this appendix we will emphasize the properties that are fundamental to communication systems. Some of the key features affecting signal transmission in communication systems are time delay, phase shift, multiplication by other signals, frequency translation, waveform convolution, and spectral convolution. We shall focus on the Fourier properties (shifting and convolution) needed to describe these key communication features.

A.3.1 Time Shifting Property

If $x(t) \leftrightarrow X(f)$, then

$$\mathcal{F}\{x(t - t_0)\} = \int_{-\infty}^{\infty} x(t - t_0)e^{-j\,2\pi ft}\,dt \qquad (A.31)$$

Let $\mu = t - t_0$; then

$$\mathcal{F}\{x(t - t_0)\} = \int_{-\infty}^{\infty} x(\mu)e^{-j\,2\pi f\,(\mu + t_0)}\,d\mu$$

$$= X(f)e^{-j\,2\pi ft_0}$$

As a signal is delayed in time, the magnitude of its frequency spectrum remains unchanged, but its phase spectrum experiences a phase shift. A time shift of t_0 in the time domain is equivalent to multiplication by $e^{-j2\pi ft_0}$ (a phase shift of $-2\pi ft_0$) in the frequency domain.

A.3.2 Frequency Shifting Property

If $x(t) \leftrightarrow X(f)$, then

$$\mathcal{F}\{x(t)e^{j\,2\pi f_0 t}\} = \int_{-\infty}^{\infty} x(t)e^{j\,2\pi f_0 t}\,e^{-j\,2\pi ft}\,dt$$

$$= \int_{-\infty}^{\infty} x(t)e^{-j\,2\pi (f - f_0)t}\,dt \qquad (A.32)$$

$$= X(f - f_0)$$

This is the basic *frequency translating* property that describes the shifted spectrum resulting from multiplying a signal by $e^{j2\pi f_0 t}$. Equation (A.32) can be used in conjunction with Equation (A.9) to yield the Fourier transform of a waveform multiplied by a cosine wave, as follows:

$$x(t) \cos 2\pi f_0 t = \tfrac{1}{2}[x(t)e^{j\,2\pi f_0 t} + x(t)e^{-j\,2\pi f_0 t}] \qquad (A.33)$$

$$x(t) \cos 2\pi f_0 t \leftrightarrow \tfrac{1}{2}[X(f - f_0) + X(f + f_0)]$$

This property is also called the *mixing* or *modulation* theorem. Multiplication of an arbitrary signal by a sinusoid of frequency f_0 translates the original signal spectrum by f_0, and also by $-f_0$.

A.4 USEFUL FUNCTIONS

A.4.1 Unit Impulse Function

A useful function in communication theory is the unit impulse or *Dirac delta* function, $\delta(t)$. The impulse function can be developed from any of several fundamental functions (e.g., a rectangular pulse or a triangular pulse). In each development, the impulse function is defined in the limiting sense (the pulse amplitude approaches infinity, the pulse width approaches zero, but the area under the pulse is constrained to be unity) [5]. The unit impulse function has the following important properties:

$$\int_{-\infty}^{\infty} \delta(t)\, dt = 1 \tag{A.34}$$

$$\delta(t) = 0 \quad \text{for } t \neq 0 \tag{A.35}$$

$$\delta(t) \text{ is unbounded at } t = 0 \tag{A.36}$$

$$\mathscr{F}\{\delta(t)\} = \mathscr{F}^{-1}\{\delta(f)\} = 1 \tag{A.37}$$

$$\int_{-\infty}^{\infty} x(t)\delta(t - t_0)\, dt = x(t_0) \tag{A.38}$$

Equation (A.38) is known as the *sifting* or *sampling property;* the unit impulse multiplier selects a sample of the function $x(t)$ evaluated at $t = t_0$.

In some problems, it is useful to use the following equivalent integrals for an impulse function, defined in the time domain or the frequency domain [3].

$$\delta(t) = \int_{-\infty}^{\infty} e^{j2\pi ft}\, df \tag{A.39}$$

$$\delta(f) = \int_{-\infty}^{\infty} e^{-j2\pi ft}\, dt \tag{A.40}$$

A.4.2 Spectrum of a Sinusoid

For the purpose of representing a sinusoidal waveform by a Fourier transform, the waveform may be assumed to exist only in the interval $(-T_0/2 < t < T_0/2)$. Under these conditions the function has a Fourier transform as long as T_0 is finite. In the limit, T_0 is made very large, but finite. The spectrum of the waveform $x(t) = A \cos 2\pi f_0 t$ can be found by using Equations (A.9) and (A.26):

$$X(f) = \int_{-\infty}^{\infty} \frac{A}{2} (e^{j2\pi f_0 t} + e^{-j2\pi f_0 t})e^{-j2\pi ft}\, dt$$

$$= \frac{A}{2} \int_{-\infty}^{\infty} e^{-j2\pi(f-f_0)t} + e^{-j2\pi(f+f_0)t}\, dt$$

As described in Equation (A.40), the foregoing integral expression can be equated to unit impulse functions located at frequencies $\pm f_0$ as follows:

$$X(f) = \frac{A}{2}[\delta(f - f_0) + \delta(f + f_0)] \tag{A.41}$$

Similarly, the spectrum of a sine waveform $y(t) = A \sin 2\pi f_0 t$ can be shown to be equal to

$$Y(f) = \frac{A}{2j}[\delta(f - f_0) - \delta(f + f_0)] \tag{A.42}$$

The cosine waveform spectrum is shown in Figure A.6, and the sine waveform spectrum is shown in Figure A.7. Each of the impulse functions shown on these spectral plots is depicted as a spike with a weight of $A/2$ or $-A/2$.

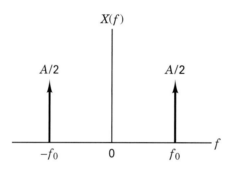

Figure A.6 Spectrum for $x(t) = A \cos 2\pi f_0 t$.

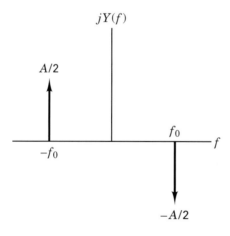

Figure A.7 Spectrum for $y(t) = A \sin 2\pi f_0 t$.

A.5 CONVOLUTION

Convolution was used by Oliver Heaviside in the late nineteenth century to calculate electrical circuit output current when the input voltage waveform was more complicated than a simple battery source. The use of the methods of Heaviside predates the use of the analytical methods developed by Fourier and Laplace (even though publications by Fourier and Laplace came earlier).

The response of a circuit to an impulse voltage $v(t) = \delta(t)$ is called the *impulse response* and is denoted by $h(t)$, as shown in Figure A.8; it is simply the output voltage that would result if the input were a delta function. Heaviside approximated an arbitrary voltage waveform, like the one shown in Figure A.9a, by a set of equally spaced pulses. Such pulses of finite height and nonzero duration are shown in Figure A.9b. In the limit as the pulse width $\Delta\tau$ approaches zero, each pulse approaches an impulse function with weight equal to the area under that pulse. In the following discussion we shall refer to these equally spaced pulses as *impulses* even though they are impulses *only in the limit.*

Care needs to be taken with the notation of time, since we are interested in the times at which impulses are applied and also the times at which their output responses are observed. We need to identify these two different time sequences; we shall use the following notation:

1. Time of the input application will be termed τ, so that the input voltage impulses are designated $v(\tau_1), v(\tau_2), \ldots, v(\tau_N)$.
2. Time of the output response will be termed t, so that the output currents are designated $i(t_1), i(t_2), \ldots, i(t_N)$.

Heaviside found the response or current produced by each input impulse independently; then he added the individual responses to get the total current. The weight of the impulse produced by the rectangular voltage at time τ_1 is the product $v(\tau_1)\,\Delta\tau$. The series of impulses can approximate the arbitrary input voltage as closely as desired by allowing $\Delta\tau$ to approach zero. Note again that the instant at which an impulse is applied is called τ_i, and the instance at which the system response is determined is called t_i, where τ is the input time variable, t is the output time variable, and $i = 1, \ldots, N$.

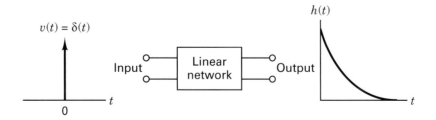

Figure A.8 Impulse response of a linear system.

A.5   Convolution

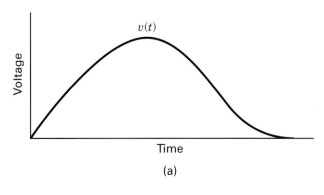

(a)

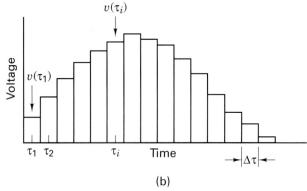

(b)

Figure A.9 (a) Input voltage wave-form. (b) Approximate input voltage waveform.

Figure A.10 illustrates the output response $i(t) = A_1h(t - \tau_1)$ to an impulse with height $v(\tau_1)$. Since the input impulse at τ_1 is *not* a *unit* impulse, we weight it with its strength or area, $A_1 = v(\tau_1)\,\Delta\tau$. At some time t_1, where $t_1 > \tau_1$, the output response to the impulse $v(\tau_1)$ is expressed as

$$i(t_1) = A_1h(t_1 - \tau_1) \quad \text{for } t_1 > \tau_1$$

as shown in Figure A.10. When there are several input impulses, the total output response for a linear system is simply the sum of the individual responses. Figure A.11 illustrates the response of the network to two input impulses. For N impulses, the output current measured at time t_1 can be expressed as

$$i(t_1) = A_1h(t_1 - \tau_1) + A_2h(t_1 - \tau_2) + \cdots + A_N(t_1 - \tau_N)$$

where the impulses are applied at $\tau_1, \tau_2, \ldots, \tau_N$ and where $t_1 > \tau_N$.

Any impulses applied at times greater than t_1 are disregarded, for they contribute nothing to $i(t_1)$. This corresponds to the *causality* requirement for physically realizable systems, which states that the system response must be zero prior to the application of the excitation. By generalizing, we get the output current at any time t, namely,

A Review of Fourier Techniques   App. A

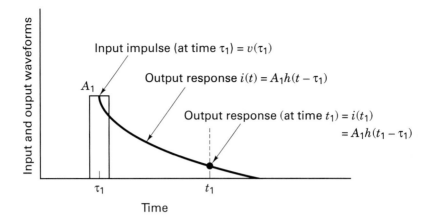

Figure A.10   Output response to an impulse at time τ_1.

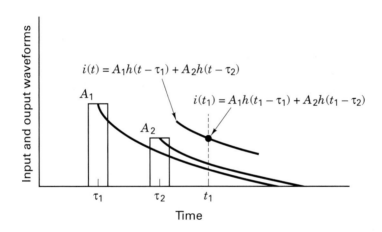

Figure A.11   Output response to two impulses.

$$i(t) = A_1 h(t - \tau_1) + A_2 h(t - \tau_2) + \cdots + A_N h(t - \tau_N)$$

or

$$i(t) = \sum_{j=1}^{N} v(\tau_j)\,\Delta\tau h(t - \tau_j) \tag{A.43}$$

since the height of the impulse at τ_j is equal to $v(\tau_j)$. As $\Delta\tau$ approaches zero, the sum of the input impulses approaches the actual applied voltage $v(\tau)$; we can re-place $\Delta\tau$ with $d\tau$, and the summation becomes the *convolution integral:*

$$i(t) = \int_{-\infty}^{\infty} v(\tau)h(t - \tau)\,d\tau \tag{A.44a}$$

A.5   Convolution

or

$$i(t) = \int_{-\infty}^{\infty} v(t - \tau)\, h(\tau)\, d\tau \qquad\qquad \text{(A.44b)}$$

In shorthand notation, this is expressed as

$$i(t) = v(t) * h(t) \qquad\qquad \text{(A.45)}$$

In summary, $i(t)$ is the sum of the individual impulse responses as a function of output time t. Each impulse response is due to an impulse applied at some input time τ and is weighted by the strength of that impulse.

A.5.1 Graphical Illustration of Convolution

Consider that an input square pulse $v(t)$ is applied to a linear network whose impulse response is labeled $h(t)$ as shown in Figure A.12a. The output response is characterized by the convolution integral expressed in Equation (A.44).

The independent variable in the convolution integral is τ. The functions $v(\tau)$ and $h(-\tau)$ are shown in Figure A.12b. Note that $h(-\tau)$ is obtained by folding $h(\tau)$ about $\tau = 0$. The term $h(t - \tau)$ represents the function $h(-\tau)$ shifted by t seconds along the positive τ axis. Figure A.12c shows the function $h(t_1 - \tau)$. The value of the convolution integral at $t = t_1$ is given by Equation (A.44) evaluated at $t = t_1$. This is simply the area under the product curve of $v(\tau)$ and $h(t_1 - \tau)$, shown shaded in Figure A.12d. Similarly, the convolution integral evaluated at $t = t_2$ is equal to the shaded area in Figure A.12e. Figure A.12f is a plot of the output response as a result of the square pulse input to the circuit with impulse response shown in Figure A.12a. Each evaluation of the convolution integral, at some time t_i, yields one point, $i(t_i)$, on the plot of Figure A.12f.

A.5.2 Time Convolution Property

If $x_1(t) \leftrightarrow X_1(f)$, and $x_2(t) \leftrightarrow X_2(f)$, then

$$x_1(t) * x_2(t) = \int_{-\infty}^{\infty} x_1(\tau)\, x_2(t - \tau)\, d\tau$$

$$\mathscr{F}\{x_1(t) * x_2(t)\} = \int_{-\infty}^{\infty} \int_{-\infty}^{\infty} x_1(\tau)\, x_2(t - \tau)\, d\tau\, e^{-j2\pi ft}\, dt$$

For linear systems, we may exchange the order of integration as follows:

$$\mathscr{F}\{x_1(t) * x_2(t)\} = \int_{-\infty}^{\infty} x_1(\tau)\, d\tau \int_{-\infty}^{\infty} x_2(t - \tau) e^{-j2\pi ft}\, dt \qquad \text{(A.46)}$$

By the Fourier *time shifting property,* the second integral expression of the right-hand side is equal to $X_2(f)e^{-j2\pi f\tau}$:

$$\mathscr{F}\{x_1(t) * x_2(t)\} = X_2(f) \int_{-\infty}^{\infty} x_1(\tau) e^{-j2\pi f\tau}\, d\tau \qquad \text{(A.47)}$$

$$= X_1(f)X_2(f)$$

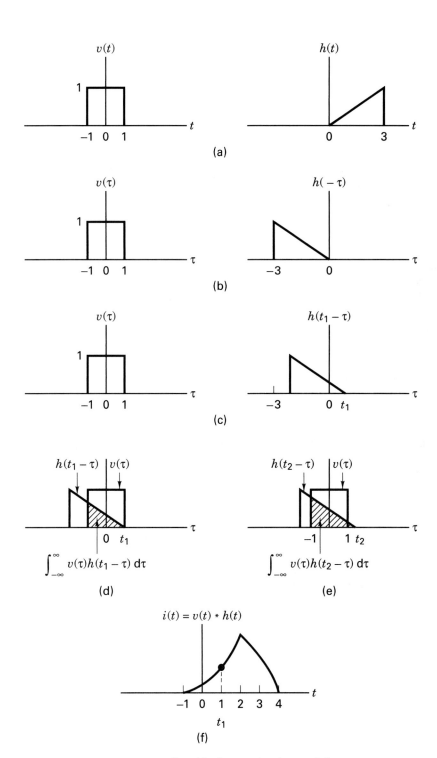

Figure A.12 Graphical example of convolution.

Therefore, the operation of *convolution* in the time domain can be replaced by *multiplication* in the frequency domain.

A.5.3 Frequency Convolution Property

Because of the symmetry of the Fourier transform pair in Equations (A.26) and (A.27), it can be shown that multiplication in the time domain transforms to convolution in the frequency domain:

$$x_1(t)x_2(t) \leftrightarrow X_1(f) * X_2(f) \tag{A.48}$$

The properties that transform multiplication in one domain to convolution in the other domain are particularly useful, since one operation is often easier to perform than the other. For example, we discussed earlier that Heaviside used convolution to solve for the output current of a linear system when the input was excited by an arbitrary voltage waveform. Such methods involve the (sometimes tedious) convolution of an input waveform with the impulse response of a system. Since convolution in the time domain is transformed into multiplication in the frequency domain, as shown in Equation (A.47), for a linear system we can simply multiply the input waveform spectrum by the system transfer function. The output waveform is then found by taking the inverse Fourier transform of the product:

$$i(t) = \mathscr{F}^{-1}\{V(f)H(f)\} \tag{A.49}$$

Solutions of the form shown in Equation (A.49) are often much easier to perform than those described by Equation (A.45). However, under certain circumstances, the operation of convolution is so simple that it can be performed graphically, by inspection. For example, suppose that we wished to multiply an arbitrary waveform by some fixed frequency cosine wave, such as a carrier wave, in the case of modulation. By applying Equation (A.48), we can convolve the spectrum of the arbitrary waveform with the spectrum of the cosine wave. This is easily accomplished, as is shown in the next section.

A.5.4 Convolution of a Function with a Unit Impulse

By the property shown in Equation (A.47), it should be clear that if

$$x(t) \leftrightarrow X(f)$$

and since

$$\delta(t) \leftrightarrow 1$$

then

$$x(t) * \delta(t) \leftrightarrow X(f) \tag{A.50}$$

It should also be evident that

$$x(t) * \delta(t) = x(t) \tag{A.51}$$

and

$$X(f) * \delta(f) = X(f) \tag{A.52}$$

We therefore conclude that convolution of a function with a unit impulse function reproduces the original function. A simple extension of Equation (A.52) yields

$$X(f) * \delta(f - f_0) = X(f - f_0) \tag{A.53}$$

Figure A.13 illustrates the ease of convolving the spectrum of an arbitrary waveform with the spectrum of a cosine wave. Figure A.13a shows an arbitrary baseband spectrum $X(f)$. Figure A.13b shows a spectrum, $Y(f) = \delta(f - f_0) + \delta(f + f_0) = \mathcal{F}\{2 \cos 2\pi f_0 t\}$. The output, $Z(f) = X(f) * Y(f)$, in Figure A.13c is obtained by convolving the waveform spectrum with the impulse functions of $Y(f)$ according to Equation (A.53), where the impulses act as sampling functions. Hence, for this simple example, convolution can be performed graphically by sweeping the sampling impulses past the waveform spectrum. Multiplication by the impulse functions at each step in the sweep yields replications of the waveform spectrum. The result, shown in Figure A.13c, is a shifted version of the original spectrum $X(f)$ to the locations of the impulse functions in Figure A.13b.

A.5.5 Demodulation Application of Convolution

In Section A.5.4 we examined a waveform multiplied by $2 \cos 2\pi f_0 t$. We illustrated the frequency-domain view of convolving the waveform spectrum with a cosine-wave spectrum. In this section we look at the reverse process. A waveform that has been multiplied by $2 \cos 2\pi f_0 t$ is to be demodulated (the waveform is to be restored to its baseband frequency range).

Figure A.14a represents the spectrum, $Z(f)$, of the waveform that has been upshifted in frequency. We can demodulate this upshifted waveform and recover the baseband waveform, by multiplying it by $2 \cos 2\pi f_0 t$. Instead, we shall illustrate the detection process in the frequency domain by convolving $Z(f)$ with the spectrum of the carrier, $Y(f) = \delta(f - f_0) + \delta(f + f_0)$, shown in Figure a.14b.

A simple extension of Equations (A.52) and (A.53) yields

$$X(f - f_0) * \delta(f - f_1) = X(f - f_0 - f_1) \tag{A.54}$$

Therefore, the result of demodulation, $X(f) = Z(f) * Y(f)$ is obtained by applying Equation (A.54). The resulting signal spectrum appears at baseband (detected) and also at frequencies $\pm 2f_0$, as shown in Figure A.14c. As in the previous section, the convolution can be performed graphically. The resulting Figure A.14c contains the following terms:

$$[Z(f - f_0) + Z(f + f_0)] * [\delta(f - f_0) + \delta(f + f_0)]$$
$$= Z(f - f_0) * \delta(f - f_0) + Z(f - f_0) * \delta(f + f_0) \tag{A.55}$$
$$+ Z(f + f_0) * \delta(f - f_0) + Z(f + f_0) * \delta(f + f_0)$$
$$= 2Z(f) + Z(f - 2f_0) + Z(f + 2f_0)$$

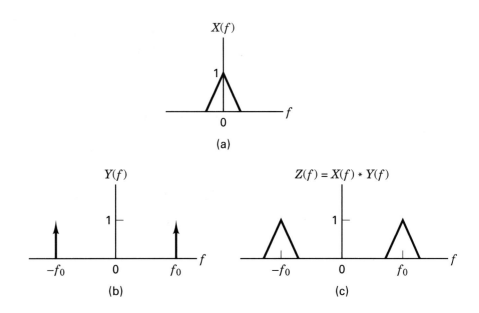

Figure A.13 Convolving a signal spectrum with a cosine-wave spectrum.

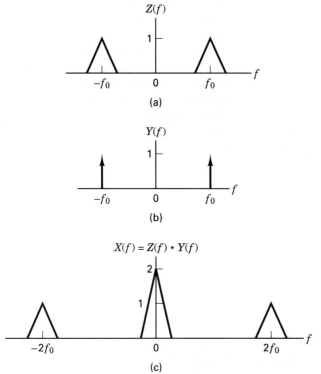

Figure A.14 Demodulation application.

Notice that the resulting terms consist of the baseband spectrum plus terms associated with higher-frequency components. The result is typical of the detection process; the higher-frequency terms are filtered and discarded, leaving the demodulated baseband spectrum.

A.6 TABLES OF FOURIER TRANSFORMS AND OPERATIONS

Commonly used Fourier transforms are listed in Table A.1 and Fourier operations in Table A.2.

TABLE A.1 Fourier Transforms

$x(t)$	$X(f)$
1. $\delta(t)$	1
2. 1	$\delta(f)$
3. $\cos 2\pi f_0 t$	$\dfrac{1}{2}[\delta(f - f_0) + \delta(f + f_0)]$
4. $\sin 2\pi f_0 t$	$\dfrac{1}{2j}[\delta(f - f_0) - \delta(f + f_0)]$
5. $\delta(t - t_0)$	$\exp(-j2\pi f t_0)$
6. $\exp(j2\pi f_0 t)$	$\delta(f - f_0)$
7. $\exp(-a|t|), \quad a > 0$	$\dfrac{2a}{a^2 + (2\pi f)^2}$
8. $\exp\left[-\pi\left(\dfrac{t}{T}\right)^2\right]$	$T \exp[-\pi(fT)^2]$
9. $u(t) = \begin{cases} 1 & \text{for } t > 0 \\ 0 & \text{for } t < 0 \end{cases}$	$\dfrac{1}{2}\delta(f) + \dfrac{1}{j2\pi f}$
10. $\exp(-at)\,u(t), \quad a > 0$	$\dfrac{1}{a + j2\pi f}$
11. $t \exp(-at)\,u(t), \quad a > 0$	$\dfrac{1}{(a + j2\pi f)^2}$
12. $\mathrm{rect}\left(\dfrac{t}{T}\right)$	$T \,\mathrm{sinc}\, fT$
13. $\cos 2\pi f_0 t\left[\mathrm{rect}\left(\dfrac{t}{T}\right)\right]$	$\dfrac{T}{2}[\mathrm{sinc}\,(f - f_0)T + \mathrm{sinc}\,(f + f_0)T]$
14. $W \,\mathrm{sinc}\, Wt$	$\mathrm{rect}\left(\dfrac{f}{W}\right)$
15. $\begin{cases} 1 - \dfrac{|t|}{T} & \text{for } |t| \le T \\ 0 & \text{for } |t| > T \end{cases}$	$T \,\mathrm{sinc}^2 fT$
16. $\displaystyle\sum_{m=-\infty}^{\infty} \delta(t - mT_0)$	$\dfrac{1}{T_0}\displaystyle\sum_{n=-\infty}^{\infty} \delta\left(f - \dfrac{n}{T_0}\right)$

Note: rect $(f/2W) = 1$ for $-W < f < W$, 0 for $|f| > W$, and sinc $x = (\sin \pi x)/\pi x$.

TABLE A.2 Fourier Operations

Operation	$x(t)$	$X(f)$
1. Scaling	$x(at)$	$\dfrac{1}{|a|} X\left(\dfrac{f}{a}\right)$
2. Time shifting	$x(t - t_0)$	$X(f) \exp(-j2\pi f t_0)$
3. Frequency shifting	$x(t) \exp(j2\pi f_0 t)$	$X(f - f_0)$
4. Time differentiation	$\dfrac{d^n x}{dt^n}$	$(j2\pi f)^n X(f)$
5. Frequency differentiation	$(-j2\pi t)^n (x(t))$	$\dfrac{d^n X}{df^n}$
6. Time integration	$\displaystyle\int_{-\infty}^{t} x(\tau)\,d\tau$	$\dfrac{1}{j2\pi f} X(f) + \dfrac{1}{2} X(0)\delta(f)$
7. Time convolution	$x_1(t) * x_2(t)$	$X_1(f)X_2(f)$
8. Frequency convolution	$x_1(t)x_2(t)$	$X_1(f) * X_2(f)$

REFERENCES

1. Papoulis, A., *Signal Analysis,* McGraw-Hill Book Company, New York, 1977.
2. Panter, P. F., *Modulation, Noise, and Spectral Analysis,* McGraw-Hill Book Company, New York, 1965.
3. Bracewell, R., *The Fourier Transform and Its Applications,* McGraw-Hill Book Company, New York, 1978.
4. Haykin, S., *Communication Systems,* John Wiley & Sons, Inc., New York, 1983.
5. Schwartz, M., *Information, Transmission, Modulation, and Noise,* McGraw-Hill Book Company, New York, 1980.

APPENDIX B

Fundamentals
of Statistical
Decision Theory

The basic elements of a statistical decision problem are (1) a set of hypotheses that characterize the possible true states of nature, (2) a test in which data are obtained from which we wish to infer the truth, (3) a decision rule that operates on the data to decide in an optimal fashion which hypothesis best describes the true state of nature, and (4) a criterion of optimality. These fundamental steps are treated in the material that follows. The *optimality criterion* we will choose for the decision rule is to minimize the probability of making an erroneous decision, although other criteria are possible [1].

The subject of statistical decision theory and hypothesis testing builds on the mathematical discipline of probability theory and random variables. It is assumed that the reader has a familiarity with these subjects; if not, Reference [2] is a suggested resource.

B.1 BAYES' THEOREM

The mathematical foundations of hypothesis testing rest on Bayes' theorem, which is derived from the definition of the relationship between the conditional and joint probability of the random variables A and B:

$$P(A|B)P(B) = P(B|A)P(A) = P(A, B) \tag{B.1}$$

A statement of the theorem is

$$P(A|B) = \frac{P(B|A)P(A)}{P(B)} \tag{B.2}$$

Bayes' theorem allows us to infer the conditional probability $P(A|B)$, from the conditional probability $P(B|A)$.

B.1.1 Discrete Form of Bayes' Theorem

Bayes' theorem can be expressed in discrete form as

$$P(s_i|z_j) = \frac{P(z_j|s_i)P(s_i)}{P(z_i)} \quad \begin{matrix} i = 1, \dots, M \\ j = 1, \dots \end{matrix} \tag{B.3}$$

where

$$P(z_j) = \sum_{i=1}^{M} P(z_j|s_i)P(s_i)$$

In a communications application, s_i is the ith signal class, from a set of M classes, and z_j is the jth sample of a received signal. Equation (B.3) can be thought of as the description of an experiment involving a received sample and some statistical knowledge of the signal classes to which the received sample may belong. The probability of occurrence of the ith signal class, $P(s_i)$, before the experiment, is called the *a priori probability*. As a result of examining a particular received sample, z_j, we can find a statistical measure of the *likelihood* that z_j belongs to class s_i from the conditional probability density function (pdf) $P(z_j|s_i)$. *After* the experiment, we can compute the *a posteriori probability, $P(s_i|z_j)$,* which can be thought of as a "refinement" of our prior knowledge. Thus we enter into the experiment with some a priori knowledge concerning the probability of the state of nature, and after examining a sample signal, we are provided with an "after-the-fact" a posteriori probability. The parameter $P(z_j)$ is the probability of the received sample, z_j, over the entire space of signal classes. This term, $P(z_j)$, can be thought of as a scaling factor, since its value is the same for *each* signal class.

Example B.1 Use of Bayes' Theorem (Discrete Form)

Given two boxes of parts. Box 1 contains 1000 parts, of which 10% are defective, and box 2 contains 2000 parts, of which 5% are defective. If a box is randomly chosen and then a part is randomly chosen from it, tested, and found to be good, what is the probability that the part came from box 1?

Solution

$$P(\text{box } 1|\text{GP}) = \frac{P(\text{GP}|\text{box } 1)P(\text{box } 1)}{P(\text{GP})}$$

where GP means "good part."

$$P(\text{GP}) = P(\text{GP}|\text{box } 1)P(\text{box } 1) + P(\text{GP}|\text{box } 2)P(\text{box } 2)$$

$$= (0.90)(0.5) + (0.95)(0.5)$$

$$= 0.450 + 0.475 = 0.925$$

$$P(\text{box } 1|\text{GP}) = \frac{0.450}{0.925} = 0.486$$

Before the experiment, the a priori probability of having chosen either box 1 or box 2 was equally likely. After obtaining a good part, the Bayesian computation can be regarded as a way of "fine tuning" our thinking that $P(\text{box } 1) = 0.5$ to yield the a posteriori probability of 0.486. The Bayes' theorem is simply a formalization of common sense. Having selected a good part from one of the two boxes, isn't it intuitively reasonable that there is a higher probability that the part came from the box with the larger concentration of good parts, and a lower probability that it came from the box with the smaller concentration of good parts? The Bayes' theorem has refined the a priori statistic into an a posteriori statistic for the probability of box selection.

Example B.2   Decision Theory Applied to a Betting Game

A box has three coins: a fair coin, a two-headed coin, and a two-tailed coin. You are asked to pick one coin at random, look at one side only, and guess head or tail for the other side. What is the optimum decision strategy for this game?

Solution

We can view this problem as a signal detection problem. A signal is transmitted, but because of the channel noise, the received signal is somewhat obscured. Not being able to look at the other side of the coin is tantamount to receiving a noise perturbed signal. Let H_i represent the hypotheses ($i = F, H, T$), where F, H, and T, stand for fair, head, and tail, respectively:

$$H_F: H, T \text{ (fair coin)}$$

$$H_H: H, H \text{ (two-headed coin)}$$

$$H_T: T, T \text{ (two-tailed coin)}$$

Let z_j represent the received sample ($j = H, T$), where z_H is a head and z_T is a tail. Let the priori probabilities of the hypotheses be equally likely, so that $P(H_F) = P(H_H) = P(H_T) = \frac{1}{3}$. Using Bayes' theorem,

$$P(H_i | z_j) = \frac{P(z_j | H_i) P(H_i)}{\sum_i P(z_j | H_i) P(H_i)}$$

we need to compute the probability for each hypothesis, given each signal class. Thus, we need to examine the results of *six* computations before we can establish an optimum decision strategy. In each case, the value of $P(z_j|H_i)$ can be obtained from the conditional probabilities drawn in Figure B.1. Consider that we choose a coin and view a head (z_H), we compute the following three a posteriori probabilities:

$$P(H_F | z_H) = \frac{(\frac{1}{2})(\frac{1}{3})}{(\frac{1}{2})(\frac{1}{3}) + (1)(\frac{1}{3}) + 0} = \frac{1}{3}$$

$$P(H_H | z_H) = \frac{(1)(\frac{1}{3})}{(\frac{1}{2})(\frac{1}{3}) + (1)(\frac{1}{3}) + 0} = \frac{2}{3}$$

$$P(H_T | z_H) = 0$$

If the received sample is a tail (z_T), we similarly compute

$$P(H_F | z_T) = \frac{1}{3}$$

$$P(H_H | z_T) = 0$$

$$P(H_T | z_T) = \frac{2}{3}$$

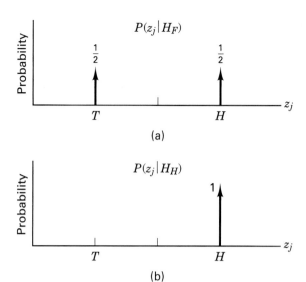

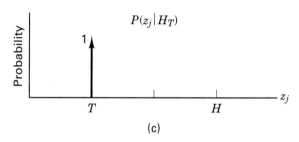

Figure B.1 Conditional probability $P(z_j | H_i)$.
(a) Conditioned on the fair-coin hypothesis.
(b) Conditioned on the two-headed-coin hypothesis. (c) Conditioned on the two-tailed-coin hypothesis.

The optimum decision strategy then is as follows: If a head, z_H, is received, choose hypothesis H_H (that the other side is also a head). If a tail, z_T, is received, choose hypothesis H_T (that the other side is also a tail).

B.1.2 Mixed Form of Bayes' Theorem

For most communication engineering applications of interest, the possible values of the received samples are *continuous* in range, because of the additive Gaussian noise in the channel. Therefore, the most useful form of Bayes' theorem contains a continuous- instead of a discrete-valued pdf. We shall rewrite Equation (B.3) to emphasize this change:

$$P(s_i | z) = \frac{p(z | s_i) P(s_i)}{p(z)} \quad i = 1, \dots, M \tag{B.4}$$

$$p(z) = \sum_{i=1}^{M} p(z | s_i) P(s_i)$$

Here, $p(z | s_i)$ is the conditional pdf of the received continuous-valued sample, z, conditioned on the signal class, s_i.

Example B.3 A Pictorial View of Bayes' Theorem

Consider two signal classes, s_1 and s_2, characterized by the triangular-shaped conditional pdfs, $p(z|s_1)$ and $p(z|s_2)$, illustrated in Figure B.2. A signal is received; it might have any value on the z-axis. If the pdfs did not overlap, we could classify the signal with certainty. For the example shown in Figure B.2, we need a rule to help us classify received signals, since some signals will fall in the region where the two pdfs overlap. Consider a received signal, z_a. Assume that the two signal classes, s_1 and s_2, are equally likely, and calculate the two alternative a posteriori probabilities. Suggest a decision rule that the receiver should use for deciding to which signal class z_a belongs. Repeat this for signal z_b.

Solution

From Figure B.2 we can see that $p(z_a|s_1) = 0.5$ and $p(z_a|s_2) = 0.3$. Thus,

$$P(s_1|z_a) = \frac{p(z_a|s_1)P(s_1)}{p(z_a|s_1)P(s_1) + p(z_a|s_2)P(s_2)}$$

$$= \frac{(0.5)(0.5)}{(0.5)(0.5) + (0.3)(0.5)} = \frac{5}{8}$$

and

$$P(s_2|z_a) = \frac{(0.3)(0.5)}{(0.5)(0.5) + (0.3)(0.5)} = \frac{3}{8}$$

One rule is to decide that the received signal belongs to the class with the maximum a posteriori probability (class s_1). An equivalent rule, for the case of equal a priori probabilities, is to examine the value of the pdf conditioned on each signal class (referred to as the likelihood of the signal class) and choose the class with the maximum. Examine Figure B.2 and notice that this *maximum likelihood rule* parallels our intuition. The likelihood that signal z_a belongs to each class corresponds to an encircled point on each pdf. The maximum likelihood rule is to choose the signal class that yields the largest conditional probability of all the alternatives. We repeat the computations for the received signal z_b, as follows:

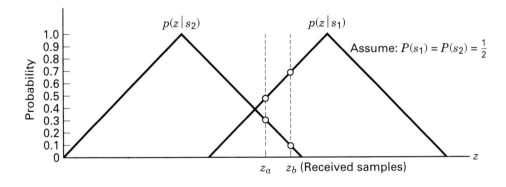

Figure B.2 Pictorial view of Bayes' theorem.

$$P(s_1|z_b) = \frac{(0.7)(0.5)}{(0.7)(0.5) + (0.1)(0.5)} = \frac{7}{8}$$

$$P(s_2|z_b) = \frac{(0.1)(0.5)}{(0.7)(0.5) + (0.1)(0.5)} = \frac{1}{8}$$

As before, the maximum likelihood rule dictates that we choose signal class s_1. Notice that in the case of received sample z_b, we can have greater confidence in the correctness of our choice compared to the case of signal z_a. This is because the ratio of $p(z_b|s_1)$ to $p(z_b|s_2)$ is considerably larger than the ratio of $p(z_a|s_1)$ to $p(z_a|s_2)$.

B.2 DECISION THEORY

B.2.1 Components of the Decision Theory Problem

Having reviewed hypothesis testing based on Bayesian statistics, let us examine more carefully the components of the decision theory problem in the context of a communication system, as shown in Figure B.3. The signal source at the transmitter consists of a set $\{s_i(t)\}$, $i = 1, \ldots, M$, of waveforms (or hypotheses). A signal waveform $r(t) = s_i(t) + n(t)$ is received, where $n(t)$ is an additive white Gaussian noise (AWGN) process introduced in the channel. At the receiver, the waveform is reduced to a single number $z(t = T)$, which may appear anywhere on the z-axis.

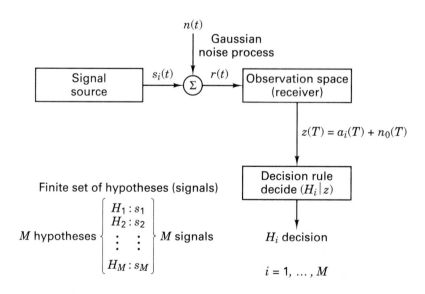

Figure B.3 Components of the decision theory problem in the context of a communication system.

Because the noise is a Gaussian process and the receiver is assumed linear, the output, $z(t)$, is also a Gaussian process [1], and the number, $z(T)$, is a *continuous-valued random variable.*

$$z(T) = a_i(T) + n_0(T) \tag{B.5}$$

The sample $z(T)$ is made up of a signal component, $a_i(T)$, and a noise component, $n_0(T)$. The time T is the symbol duration. At each kT, where k is an integer, the receiver uses a decision rule for deciding which signal class has been received. For ease of notation, Equation (B.5) is sometimes written simply as $z = a_i + n_0$, where the functional dependence on T is implicit.

B.2.2 The Likelihood Ratio Test and the Maximum A Posteriori Criterion

A reasonable starting point for establishing the receiver decision rule for the case of *two* signal classes is

$$P(s_1|z) \overset{H_1}{\underset{H_2}{\gtrless}} P(s_2|z) \tag{B.6}$$

Equation (B.6) states that we should choose hypothesis H_1 if the a posteriori probability $P(s_1|z)$ is greater than the a posteriori probability $P(s_2|z)$. Otherwise, we should choose hypothesis H_2.

We can replace the posteriori probabilities of Equation (B.6) with their equivalent expressions from Bayes' theorem [Equation (B.4)], yielding

$$p(z|s_1) P(s_1) \overset{H_1}{\underset{H_2}{\gtrless}} P(z|s_2) P(s_2) \tag{B.7}$$

We now have a decision rule in terms of pdfs (likelihoods). If we rearrange Equation (B.7) and put it in the form

$$\frac{p(z|s_1)}{p(z|s_2)} \overset{H_1}{\underset{H_2}{\gtrless}} \frac{P(s_2)}{P(s_1)} \tag{B.8}$$

then the left-hand ratio is known as the *likelihood ratio* and the entire equation is often referred to as the *likelihood ratio test.* Equation (B.8) corresponds to making a decision based on a comparison of a measurement of a received signal to a threshold. Since the test is based on choosing the signal class with maximum a posteriori probability, the decision criterion is called the *maximum a posteriori* (MAP) criterion. It is also called the *minimum error criterion,* since on the average, this criterion yields the minimum number of incorrect decisions. It should be emphasized that this criterion is optimum only when each of the error types are equally harmful or costly. When some of the error types are more costly than others, a criterion that incorporates relative cost of the errors should best be employed [1].

B.2.3 The Maximum Likelihood Criterion

Very often there is no knowledge available about the a priori probabilities of the hypotheses or signal classes. Even when such information is available, its accuracy is sometimes mistrusted. In those instances, decisions are usually made by assuming the most conservative a priori probabilities possible; that is, the values of the a priori probabilities are selected so that the classes are *equally likely*. When this is done, the MAP criterion is known as the *maximum likelihood criterion,* and Equation (B.8) can be written as

$$\frac{p(z|s_1)}{p(z|s_2)} \underset{H_2}{\overset{H_1}{\gtrless}} 1 \tag{B.9}$$

Notice that the maximum likelihood criterion of Equation (B.9) is the same as the maximum likelihood rule that was described in Example B.3.

B.3 SIGNAL DETECTION EXAMPLE

B.3.1 The Maximum Likelihood Binary Decision

The pictorial view of the decision process in Example B.3 dealt with triangular-shaped probability density functions as a convenient example. Figure B.4 illustrates the conditional pdfs for the binary noise-perturbed output signals, $z(T) = a_1 + n_0$ and $z(T) = a_2 + n_0$ from a typical receiver. The signals a_1 and a_2 are mutually independent and are equally likely. The noise n_0 is assumed to be an independent Gaussian random variable with zero mean, variance σ_0^2, and pdf given by

$$p(n_0) = \frac{1}{\sigma_0 \sqrt{2\pi}} \exp\left[-\frac{1}{2}\left(\frac{n_0^2}{\sigma_0^2}\right)\right] \tag{B.10}$$

We can therefore write the likelihood ratio, described in Equation (B.8), as

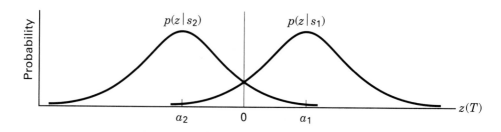

Figure B.4 Conditional pdfs for a typical binary receiver.

$$\Lambda(z) = \frac{p(z|s_1)}{p(z|s_2)}$$

$$= \frac{\dfrac{1}{\sigma_0\sqrt{2\pi}}\exp\left[-\dfrac{1}{2}\left(\dfrac{z-a_1}{\sigma_0}\right)^2\right]}{\dfrac{1}{\sigma_0\sqrt{2\pi}}\exp\left[-\dfrac{1}{2}\left(\dfrac{z-a_2}{\sigma_0}\right)\right]} \overset{H_1}{\underset{H_2}{\gtrless}} \frac{P(s_2)}{P(s_1)}$$

(B.11)

$$= \frac{\exp\left(-\dfrac{z^2}{2\sigma_0^2}\right)\exp\left(-\dfrac{a_1^2}{2\sigma_0^2}\right)\exp\left(\dfrac{2za_1}{2\sigma_0^2}\right)}{\exp\left(-\dfrac{z^2}{2\sigma_0^2}\right)\exp\left(-\dfrac{a_2^2}{2\sigma_0^2}\right)\exp\left(\dfrac{2za_2}{2\sigma_0^2}\right)} \overset{H_1}{\underset{H_2}{\gtrless}} \frac{P(s_2)}{P(s_1)}$$

$$= \exp\left[\frac{z(a_1-a_2)}{\sigma_0^2} - \frac{a_1^2-a_2^2}{2\sigma_0^2}\right] \overset{H_1}{\underset{H_2}{\gtrless}} \frac{P(s_2)}{P(s_1)}$$

where a_1 is the receiver output signal component when $s_1(t)$ is sent and a_2 is the output signal component when $s_2(t)$ is sent. The inequality relationship described by Equation (B.11) is preserved for any *monotonically* increasing (or decreasing) transformation. Therefore, to simplify Equation (B.11), we take the natural logarithm of both sides, resulting in the log-likelihood ratio.

$$L(z) = \frac{z(a_1-a_2)}{\sigma_0^2} - \frac{a_1^2-a_2^2}{2\sigma_0^2} \overset{H_1}{\underset{H_2}{\gtrless}} \ln\frac{P(s_2)}{P(s_1)}$$

(B.12)

When the classes are equally likely,

$$\ln\frac{P(s_2)}{P(s_1)} = 0$$

so that

$$z \overset{H_1}{\underset{H_2}{\gtrless}} \frac{a_1^2-a_2^2}{2(a_1-a_2)}$$

$$z \overset{H_1}{\underset{H_2}{\gtrless}} \frac{a_1+a_2}{2} = \gamma_0$$

(B.13)

For *antipodal signals*, $s_1(t) = -s_2(t)$ and $a_1 = -a_2$; thus, we can write

$$z \overset{H_1}{\underset{H_2}{\gtrless}} 0$$

(B.14)

Therefore, the maximum likelihood rule for the case of equally likely antipodal signals compares the received sample to a zero threshold, which is tantamount to deciding $s_1(t)$ if the sample is positive, and $s_2(t)$ if the signal is negative.

B.3   Signal Detection Example

B.3.2 Probability of Bit Error

For the binary example in Section B.3.1, we want to compute the bit error probability, P_B, using the decision rule in Equation (B.13). The probability of an error is calculated by summing the probabilities of the various ways that an error can be made:

$$P_B = P(H_2|s_1)P(s_1) + P(H_1|s_2)P(s_2) \tag{B.15}$$

That is, given that class $s_1(t)$ was transmitted, an error results if hypothesis H_2 is chosen; or, given that class $s_2(t)$ was transmitted, an error results if hypothesis H_1 is chosen. For the special case of symmetric probability density functions, and for $P(s_1) = P(s_2) = 0.5$, we can write

$$P_B = P(H_2|S_1) = P(H_1|S_2) \tag{B.16}$$

The probability of an error, P_B, is equal to the probability that an incorrect hypothesis, H_1, will be decided when $s_2(t)$ is sent, or that H_2 will be decided when $s_1(t)$ is sent. Thus P_B is numerically equal to the area under the "tail" of either pdf, $p(z|s_1)$ or $p(z|s_2)$, falling on the *incorrect side* of the threshold. We can therefore compute P_B by integrating $p(z|s_1)$ between the limits $-\infty$ and γ_0 or by integrating $p(z|s_2)$ between the limits γ_0 and ∞, as follows:

$$P_B = \int_{\gamma_0=(a_1+a_2)/2}^{\infty} p(z|s_2)\, dz \tag{B.17}$$

$$= \int_{(a_1+a_2)/2}^{\infty} \frac{1}{\sigma_0\sqrt{2\pi}} \exp\left[-\frac{1}{2}\left(\frac{z-a_2}{\sigma_0}\right)^2\right] dz$$

Let

$$u = \frac{z-a_2}{\sigma_0}$$

Then $\sigma_0\, du = dz$, and

$$P_B = \int_{u=(a_1-a_2)/2\sigma_0}^{u=\infty} \frac{1}{\sqrt{2\pi}} \exp\left(-\frac{u^2}{2}\right) du = Q\left(\frac{a_1-a_2}{2\sigma_0}\right) \tag{B.18}$$

where $Q(x)$, called the *complementary error function* or *co-error function,*[*] is tabulated in Table B.1.

Another form of the co-error function which is frequently used is

$$\text{erfc}\,(x) = \frac{2}{\sqrt{\pi}} \int_{x}^{\infty} \exp\,(-u^2)\, du \tag{B.19}$$

The two co-error functions, $Q(x)$ and erfc (x), are related as follows:

---

[*]Note that the co-error function is defined in several ways; however, all definitions are essentially equivalent.

TABLE B.1 Complementary Error Function $Q(x) = \int_x^\infty (1/\sqrt{2\pi}) \exp(-u^2/2)\,du$

					$Q(x)$					
x	0.00	0.01	0.02	0.03	0.04	0.05	0.06	0.07	0.08	0.09
0.0	0.5000	0.4960	0.4920	0.4880	0.4840	0.4801	0.4761	0.4721	0.4681	0.4641
0.1	0.4602	0.4562	0.4522	0.4483	0.4443	0.4404	0.4364	0.4325	0.4286	0.4247
0.2	0.4207	0.4168	0.4129	0.4090	0.4052	0.4013	0.3974	0.3936	0.3897	0.3859
0.3	0.3821	0.3783	0.3745	0.3707	0.3669	0.3632	0.3594	0.3557	0.3520	0.3483
0.4	0.3446	0.3409	0.3372	0.3336	0.3300	0.3264	0.3228	0.3192	0.3156	0.3121
0.5	0.3085	0.3050	0.3015	0.2981	0.2946	0.2912	0.2877	0.2843	0.2810	0.2776
0.6	0.2743	0.2709	0.2676	0.2643	0.2611	0.2578	0.2546	0.2514	0.2483	0.2451
0.7	0.2420	0.2389	0.2358	0.2327	0.2296	0.2266	0.2236	0.2206	0.2168	0.2148
0.8	0.2169	0.2090	0.2061	0.2033	0.2005	0.1977	0.1949	0.1922	0.1894	0.1867
0.9	0.1841	0.1814	0.1788	0.1762	0.1736	0.1711	0.1685	0.1660	0.1635	0.1611
1.0	0.1587	0.1562	0.1539	0.1515	0.1492	0.1469	0.1446	0.1423	0.1401	0.1379
1.1	0.1357	0.1335	0.1314	0.1292	0.1271	0.1251	0.1230	0.1210	0.1190	0.1170
1.2	0.1151	0.1131	0.1112	0.1093	0.1075	0.1056	0.1038	0.1020	0.1003	0.0985
1.3	0.0968	0.0951	0.0934	0.0918	0.0901	0.0885	0.0869	0.0853	0.0838	0.0823
1.4	0.0808	0.0793	0.0778	0.0764	0.0749	0.0735	0.0721	0.0708	0.0694	0.0681
1.5	0.0668	0.0655	0.0643	0.0630	0.0618	0.0606	0.0594	0.0582	0.0571	0.0559
1.6	0.0548	0.0537	0.0526	0.0516	0.0505	0.0495	0.0485	0.0475	0.0465	0.0455
1.7	0.0446	0.0436	0.0427	0.0418	0.0409	0.0401	0.0392	0.0384	0.0375	0.0367
1.8	0.0359	0.0351	0.0344	0.0336	0.0329	0.0322	0.0314	0.0307	0.0301	0.0294
1.9	0.0287	0.0281	0.0274	0.0268	0.0262	0.0256	0.0250	0.0244	0.0239	0.0233
2.0	0.0228	0.0222	0.0217	0.0212	0.0207	0.0202	0.0197	0.0192	0.0188	0.0183
2.1	0.0179	0.0174	0.0170	0.0166	0.0162	0.0158	0.0154	0.0150	0.0146	0.0143
2.2	0.0139	0.0136	0.0132	0.0129	0.0125	0.0122	0.0119	0.0116	0.0113	0.0110
2.3	0.0107	0.0104	0.0102	0.0099	0.0096	0.0094	0.0091	0.0089	0.0087	0.0084
2.4	0.0082	0.0080	0.0078	0.0075	0.0073	0.0071	0.0069	0.0068	0.0066	0.0064
2.5	0.0062	0.0060	0.0059	0.0057	0.0055	0.0054	0.0052	0.0051	0.0049	0.0048
2.6	0.0047	0.0045	0.0044	0.0043	0.0041	0.0040	0.0039	0.0038	0.0037	0.0036
2.7	0.0035	0.0034	0.0033	0.0032	0.0031	0.0030	0.0029	0.0028	0.0027	0.0026
2.8	0.0026	0.0025	0.0024	0.0023	0.0023	0.0022	0.0021	0.0021	0.0020	0.0019
2.9	0.0019	0.0018	0.0018	0.0017	0.0016	0.0016	0.0015	0.0015	0.0014	0.0014
3.0	0.0013	0.0013	0.0013	0.0012	0.0012	0.0011	0.0011	0.0011	0.0010	0.0010
3.1	0.0010	0.0009	0.0009	0.0009	0.0008	0.0008	0.0008	0.0008	0.0007	0.0007
3.2	0.0007	0.0007	0.0006	0.0006	0.0006	0.0006	0.0006	0.0005	0.0005	0.0005
3.3	0.0005	0.0005	0.0005	0.0004	0.0004	0.0004	0.0004	0.0004	0.0094	0.0003
3.4	0.0003	0.0003	0.0003	0.0003	0.0003	0.0003	0.0003	0.0003	0.0003	0.0002

$$\text{erfc}(x) = 2Q(x\sqrt{2}) \qquad \text{(B.20)}$$

$$Q(x) = \frac{1}{2}\,\text{erfc}\left(\frac{x}{\sqrt{2}}\right) \qquad \text{(B.21)}$$

REFERENCES

1. Van Trees, H. L., *Detection, Estimation, and Modulation Theory,* Part 1, John Wiley & Sons, Inc., New York, 1968.

2. Papoulis, A., *Probability, Random Variables, and Stochastic Processes,* McGraw-Hill Book Company, New York, 1965.

APPENDIX C

Response
of Correlators
to White Noise

The inputs to a bank of N correlators represent a white Gaussian noise process, $n(t)$, with zero mean and two-sided power spectral density, $N_0/2$. The output of each correlator at time $t = T$ is a *Gaussian random variable* defined by

$$n_j = \int_0^T n(t)\psi_j(t)\,dt \quad j = 1, \ldots, N \tag{C.1}$$

where $\{\psi_j(t)\}$ forms an orthonormal set. Since n_j is Gaussian, it is characterized completely by its mean and variance. The mean is

$$\bar{n}_j = \mathbf{E}\{n_j\} = \mathbf{E}\left\{\int_0^T n(t)\psi_j(t)\,dt\right\} \tag{C.2}$$

where $\mathbf{E}\{\cdot\}$ is the expected value operator. The variance of n_j is

$$\sigma_j^2 = \mathbf{E}\{n_j^2\} - \bar{n}_j^2 \tag{C.3}$$

$$= \mathbf{E}\left\{\int_0^T n(t)\psi_j(t)\,dt \int_0^T n(s)\psi_j(s)\,ds\right\} - \bar{n}_j^2 \tag{C.4}$$

$$= \int_0^T\int_0^T \mathbf{E}\{n(t)n(s)\psi_j(t)\psi_j(s)\}\,dt\,ds - \bar{n}_j^2 \tag{C.5}$$

Since $n(t)$ is a zero-mean process,

$$\mathbf{E}\{n(t)\} = 0 \tag{C.6}$$

which implies that

$$\bar{n}_j = \mathbf{E}\{n_j\} = 0 \tag{C.7}$$

The autocorrelation function of the process $n(t)$ is

$$R_n(t, s) = \mathbf{E}\{n(t)n(s)\} \tag{C.8}$$

If the noise $n(t)$ is assumed stationary, then $R_n(t, s)$ is a function of only the time difference $\tau = t - s$. From Equation (C.5), we then have

$$\sigma_j^2 = \text{var}\{n_j\} = \int_0^T \int_0^T R_n(\tau)\psi_j(t)\psi_j(s)\, dt\, ds \tag{C.9}$$

For a stationary random process, the power spectral density $G_n(f)$ and the autocorrelation function $R_n(\tau)$ form a Fourier transform pair. Thus, we can write

$$R_n(\tau) = \int_{-\infty}^{\infty} G_n(f)e^{j2\pi f\tau}df \tag{C.10}$$

Since $n(t)$ is white noise, its power spectral density $G_n(f)$ is $N_0/2$ for all f, and we can write Equation (C.10) as

$$R_n(\tau) = \int_{-\infty}^{\infty} \frac{N_0}{2} e^{j2\pi f\tau}df = \frac{N_0}{2}\delta(\tau) \tag{C.11}$$

where $\delta(\tau)$ is the unit impulse function defined in Section A.4.1. Substituting Equation (C.11) into Equation (C.9), we get

$$\sigma_j^2 = \frac{N_0}{2} \int_0^T \int_0^T \delta(t - s)\psi_j(t)\psi_j(s)\, dt\, ds \tag{C.12}$$

$$= \frac{N_0}{2} \int_0^T \psi_j^2(t)\, dt = \frac{N_0}{2} \quad j = 1, \ldots, N \tag{C.13}$$

where we have utilized the *sifting property* of the unit impulse function (see Section A.4.1) and the fact that $\{\psi_j(t)\}$, $j = 1, \ldots, N$, constitutes an orthonormal set. Thus, for white Gaussian noise with two-sided power spectral density $N_0/2$ watts/hertz, the output noise power from each of the N correlators is equal to $N_0/2$ watts.

Often-Used Identities

$$\cos x \cos y = \tfrac{1}{2}\cos(x+y) + \tfrac{1}{2}\cos(x-y) \qquad \text{(D.1)}$$

$$\sin x \sin y = -\tfrac{1}{2}\cos(x+y) + \tfrac{1}{2}\cos(x-y) \qquad \text{(D.2)}$$

$$\sin x \cos y = \tfrac{1}{2}\sin(x+y) + \tfrac{1}{2}\sin(x-y) \qquad \text{(D.3)}$$

$$\cos x \sin y = \tfrac{1}{2}\sin(x+y) - \tfrac{1}{2}\sin(x-y) \qquad \text{(D.4)}$$

$$\sin(x \pm y) = \sin x \cos y \pm \cos x \sin y \qquad \text{(D.5)}$$

$$\cos(x \pm y) = \cos x \cos y \mp \sin x \sin y \qquad \text{(D.6)}$$

$$\cos^2 x = \tfrac{1}{2}(1 + \cos 2x) \qquad \text{(D.7)}$$

$$\sin^2 x = \tfrac{1}{2}(1 - \cos 2x) \qquad \text{(D.8)}$$

$$\sin x \cos x = \tfrac{1}{2}\sin 2x \qquad \text{(D.9)}$$

$$\sin x + \sin y = 2 \sin \tfrac{1}{2}(x+y) \cos \tfrac{1}{2}(x-y) \qquad \text{(D.10)}$$

$$\sin x - \sin y = 2 \cos \tfrac{1}{2}(x+y) \sin \tfrac{1}{2}(x-y) \qquad \text{(D.11)}$$

$$\cos x + \cos y = 2 \cos \tfrac{1}{2}(x+y) \cos \tfrac{1}{2}(x-y) \qquad \text{(D.12)}$$

$$\cos x - \cos y = -2 \sin \tfrac{1}{2}(x+y) \sin \tfrac{1}{2}(x-y) \qquad \text{(D.13)}$$

$$\sin x = \frac{e^{jx} - e^{-jx}}{2j} \qquad \text{(D.14)}$$

$$\cos x = \frac{e^{jx} + e^{-jx}}{2} \qquad \text{(D.15)}$$

$$P_B = \frac{1}{n} \sum_{j=2}^{n} j \binom{n}{j} p^j (1-p)^{n-j} = p - p(1-p)^{n-1} \qquad \text{(D.16)}$$

Proof:

$$j \binom{n}{j} = j \frac{n!}{j! \, (n-j)!} = \frac{n!}{(j-1)! \, (n-j)!} = n \frac{(n-1)!}{(j-1)! \, [(n-1) - (j-1)]!}$$

$$= n \binom{n-1}{j-1}$$

$$P_B = \sum_{j=2}^{n} \binom{n-1}{j-1} p^j (1-p)^{n-j} = p \sum_{j=2}^{n} \binom{n-1}{j-1} p^{j-1} (1-p)^{(n-1)-(j-1)}$$

Change of parameter: $i = (j-1)$

Therefore, $(j = 2)$ becomes $(i = 1)$, and $(j = n)$ becomes $(i = n-1)$.

$$P_B = p \sum_{i=1}^{n-1} \binom{n-1}{i} p^i (1-p)^{(n-1)-i}$$

$$= p \sum_{i=0}^{n-1} \left[\binom{n-1}{i} p^i (1-p)^{(n-1)-i} \right] - p \binom{n-1}{0} p^0 (1-p)^{(n-1)-0}$$

$$= p[1 - (1-p)^{n-1}]$$

$$= p - p(1-p)^{n-1}$$

s-Domain, z-Domain and Digital Filtering

Robert W. Stewart
Department of Electronic and Electrical Engineering
University of Strathclyde, Glasgow, Scotland, UK

In Appendix A, Equation (A.26) and (A.27) defined the Fourier and inverse Fourier transform, respectively. Although the Fourier transform is useful for a steady-state frequency analysis of a system, it is not particularly useful for transient analysis; and for some functions, the Fourier integral does not in fact exist, whereas the Laplace transform, discussed in this appendix, does. Hence, to allow a more in-depth analysis of linear systems, we often choose to use the Laplace transform. In terms of Laplace and Fourier defining equations, it is straightforward to show that the Laplace transform is an extension of the Fourier transform. If the system under analysis is a discrete time system rather than continuous time system, then we can use the notationally simpler z-transform, which can be directly derived from the Laplace transform. Another reason for using the Laplace transform (for continuous time analysis) and z-transform (for discrete time analysis) is that operations that are "awkward" in the time domain, such as convolution, can be performed more easily in the s-domain or z-domain.

Hence, in this appendix the s-domain, z-domain, and discrete frequency transform are reviewed and thereafter the ubiquitous digital filter is introduced with varied reference to the Laplace- and z-transforms.

E.1 THE LAPLACE TRANSFORM

Recall the Fourier transform in Equation (A.26) of Appendix A, namely,

$$X(f) = \int_{-\infty}^{\infty} x(t)e^{-j2\pi ft}\,dt \quad \text{or} \quad X(\omega) = \int_{-\infty}^{\infty} x(t)e^{-j\omega t}\,dt \qquad \text{(E.1)}$$

where $\omega = 2\pi f$.

If we define a new function $v(t)$ that is $x(t)$ multiplied by the real exponential time function $e^{-\sigma t}$, where σ is a real number, then $v(t) = x(t)e^{-\sigma t}$. Hence, the Fourier transform of the function $v(t)$ is

$$V(\omega) = \int_{-\infty}^{\infty} v(t)e^{-j\omega t}\, dt = \int_{-\infty}^{\infty} x(t)e^{-\sigma t}e^{-j\omega t}\, dt = \int_{-\infty}^{\infty} x(t)e^{-(\sigma + j\omega)t}\, dt \quad \text{(E.2)}$$

Therefore, we can rewrite Equation (E.1) as

$$X(\sigma + j\omega) = \int_{-\infty}^{\infty} x(t)e^{-(\sigma + j\omega)t}\, dt \quad \text{(E.3)}$$

Letting s be the complex frequency $s = \sigma + j\omega$, we can now define the *Laplace transform* of a time domain signal $x(t)$ as

$$X(s) = \int_{-\infty}^{\infty} x(t)e^{-st}\, dt \quad \text{(E.4)}$$

where "s" is the Laplace variable. If we rewrite the inverse Fourier transform in Equation (A.27) in terms of angular frequency $\omega = 2\pi f$, then $d\omega/df = 2\pi$ and

$$x(t) = \int_{-\infty}^{\infty} X(\omega)\, e^{j\omega t}\, \frac{d\omega}{2\pi} \quad \text{(E.5)}$$

Since $s = \sigma + j\omega$, it follows that $ds/d\omega = j$, and we can therefore specify the *inverse Laplace transform* as

$$x(t) = \frac{1}{j2\pi} \int_{\sigma - j\infty}^{\sigma + j\infty} X(s)e^{st}\, ds \quad \text{(E.6)}$$

Equations (E.4) and (E.6) are referred to as a *Laplace transform pair* $[x(t) \leftrightarrow X(s)]$, or more precisely, the *two-sided or bilateral Laplace transform pair*. If we (reasonably) assume that the signal $x(t)$ does not exist before $t = 0$ (i.e., it is causal), then the transform can thus be termed *one sided* or *unilateral* and written as

$$X(s) = \int_{0}^{\infty} x(t)e^{-st}\, dt \quad \text{(E.7)}$$

The *inverse unilateral Laplace transform* is the same as Equation (E.6). Hence, it is also appropriate to refer to Equation (E.6) and (E.7) as the *one-sided Laplace transform pair,* or *unilateral Laplace transform pair.*

E.1.1 Standard Laplace Transforms

Table E.1 shows some standard one-sided Laplace transforms. Note that the (two-sided) Laplace transform of Equation (E.4) is identical to the Fourier transform of Equation (A.26) if we set $s = j\omega$, where $\omega = 2\pi f$. In order to develop the Laplace transform, $x(t)$ was multiplied by a "convergence factor" $e^{-\sigma t}$, where σ is any real number. Hence, when actually evaluating the integration, the Laplace transform may exist for many functions that do not have a corresponding Fourier transform.

TABLE E.1 Laplace Transforms

Waveform Type	Time Function	Laplace Transform
Impulse	$\delta(t)$	1
Unit Step function	$u(t)$	$\dfrac{1}{s}$
Ramp function	$tu(t)$	$\dfrac{1}{s^2}$
Exponential	$e^{at}u(t)$	$\dfrac{1}{s-a}$
	$te^{at}u(t)$	$\dfrac{1}{(s-a)^2}$
Sine wave	$\sin(\omega t)u(t)$	$\dfrac{\omega}{(s^2+\omega^2)}$
Cosine wave	$\cos(\omega t)u(t)$	$\dfrac{s}{s^2+\omega^2}$
Damped sine wave	$e^{at}\sin(\omega t)u(t)$	$\dfrac{\omega}{(s-a)^2+\omega^2}$
Damped cosine wave	$e^{at}\cos(\omega t)u(t)$	$\dfrac{(s-a)}{(s-a)^2+\omega^2}$

One of the key advantages of the Laplace transform is the capability of transforming functions that are not absolutely integrable.

E.1.2 Laplace Transform Properties

If we know the Laplace transform pair $y(t) \leftrightarrow Y(s)$, then it can be shown that for a delayed version of the signal denoted as $y(t - t_0)$,

$$y(t - t_0) \leftrightarrow e^{-st_0} Y(s) \qquad (E.8)$$

This is the simple time-shift property of Laplace transforms. Table (E.2) shows other properties of Laplace transforms. All of these properties can be verified by direct evaluation of the Laplace transform property of interest. Note that the relationship of $s = j\omega$ between the Laplace and Fourier transform means that there is a simple equivalence between like transforms of Table E.1 with those in Table A.1, as well as like operations in Table E.2 with those in Table A.2.

TABLE E.2 Laplace Operations

Property	Time Function	Laplace Transform
General or arbitrary function	$x(t)$	$X(s)$
General or arbitrary function	$y(t)$	$Y(s)$
Linearity	$ax(t) + by(t)$	$aX(s) + bY(s)$
Time shift $(\tau > 0)$	$x(t - \tau)$	$e^{-s\tau}X(s)$
Time scaling	$x(at)$	$\dfrac{1}{a}X\left(\dfrac{s}{a}\right)$
Modulation	$e^{-at}x(t)$	$X(s - a)$
Differentiation	$\dfrac{dx(t)}{dt}$	$sX(s) - x(0)$
Integration	$\int_{-\infty}^{t} x(\tau)d\tau$	$\dfrac{X(s)}{s}$
Convolution	$x(t)*y(t)$	$X(s)Y(s)$

E.1.3 Using the Laplace Transform

Laplace transforms are useful where (time) differential equations have to be solved or convolution operations have to be performed. For example, to find the current $i(t)$ in the simple RC circuit in Figure E.1, note that the voltage across the capacitor (integrating component) and resistor is the input voltage

$$v_{in}(t) = i(t)R + \frac{q}{C} = i(t)R + \frac{1}{C}\int_{0}^{t} i(t)\,dt \qquad (E.9)$$

If the input voltage is a unit-step power source, $v_{in}(t) = u(t)$, and q is the charge across the capacitor (in Coulombs); taking Laplace transforms of Equation (E.9), using Tables E.1 and E.2, we obtain

$$V_{in}(s) = RI(s) + \frac{I(s)}{sC} \text{ and } \Rightarrow I(s) = \frac{V_{in}(s)}{R + 1/(sC)} = \frac{1/R}{s + 1/(RC)} \qquad (E.10)$$

(Note that for the unit step, $V_{in}(s) = 1/s$.) Next, transforming back to the time domain (based again on the Laplace transform tables) gives

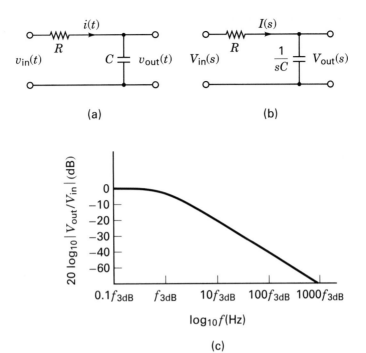

Figure E.1 (a) Resistor capacitor circuit, (b) Laplace transform representation, (c) Magnitude response.

$$i\left(t\right) = \frac{1}{R}e^{-t/(RC)} \tag{E.11}$$

E.1.4 Transfer Function

Using the Laplace transform, we can specify the (s-variable) transfer function of a linear system. From Equation (E.10), if the resistance $R = 0$, then the impedance of the capacitor can be calculated as

$$Z_c = \frac{V_{in}\left(s\right)}{I(s)} = \frac{1}{sC} \tag{E.12}$$

The input and output Laplace voltages can be specified as

$$V_{in}(s) = I(s)\,R + \frac{I(s)}{sC} \qquad V_{out}(s) = \frac{I(s)}{sC} \tag{E.13}$$

and, therefore, the (Laplace) transfer function can be specified as

E.1   The Laplace Transform

$$H(s) = \frac{V_{out}(s)}{V_{in}(s)} = \frac{\dfrac{I(s)}{sC}}{I(s)R + \dfrac{I(s)}{sC}} = \frac{1}{sRC + 1} \tag{E.14}$$

E.1.5 RC Circuit Low Pass Filtering

Let the input to the RC circuit be a complex sinusoid $v_{in}(t) = e^{j\omega t}$. From Appendix A, we note that we can equate this to the Fourier transform by setting $s = j\omega$, where $\omega = 2\pi f$. Thus, the frequency response can be realized from the transfer function

$$\frac{V_{out}(f)}{V_{in}(f)} = \frac{1}{j\omega RC + 1} = \frac{1}{j2\pi fRC + 1} = \frac{1}{\sqrt{(2\pi fRC)^2 + 1}} e^{-j[\arctan(2\pi fRC)]} \tag{E.15}$$

For small values of f, then $|H(f)| \approx 1$; and for large values of f, then $|H(f)| \approx 0$. When $f = f_0 = 1/(2\pi RC)$, then $|H(f)| \approx 1/\sqrt{2}$. Noting that $20 \log_{10}(1/\sqrt{2}) = -3$ dB, then f_0 is the "–3-dB" frequency when the output power is half of the input power. Therefore, Equation (E.15) specifies a low-pass filter that is the same as was presented in Equation (1.63). Low frequencies are passed and high frequencies are attenuated; this is shown in the magnitude frequency response of Figure E.1c.

E.1.6 Poles and Zeroes

Linear systems and hence (linear) analog filters can be represented by time-domain differential equations. For example, consider the second-order example

$$y(t) = A\frac{d^2x(t)}{dt^2} + B\frac{dx(t)}{dt} + Cx(t) + D\frac{d^2y(t)}{dt^2} + E\frac{dy(t)}{dt} \tag{E.16}$$

The various orders of differentiation and/or integration are implemented in the real world by using capacitive and inductive components in combination with feedback amplifiers of given orders [2]. Taking the Laplace transform of Equation (E.16) yields the more mathematically (and notationally) convenient Laplace equation

$$Y(s) = As^2X(s) + BsX(s) + CX(s) + Ds^2Y(s) + EsY(s) \tag{E.17}$$

and the transfer function

$$H(s) = \frac{Y(s)}{X(s)} = \frac{As^2 + Bs + C}{-Ds^2 - Es + 1} = \frac{A(s - a_0)(s - a_1)}{-D(s - b_0)(s - b_1)} \tag{E.18}$$

The roots $\{a_0, a_1\}$ of the numerator s-polynomial are termed the *zeroes*, and the roots $\{b_0, b_1\}$ of the denominator s-polynomial are termed the *poles*. Note that if A, B, and C are real values, then the zeroes $\{a_0, a_1\}$ will be complex conjugates.

E.1.7 Linear System Stability

Consider briefly the single-pole equation corresponding to an arbitrary linear system:

$$H(s) = \frac{1}{s - \sigma} \tag{E.19}$$

The impulse response of this circuit can be found (by using Table E.1) to be the inverse of the transfer function in Equation (E.19); and if $\sigma = \rho + j\zeta$, then the impulse response is

$$h(t) = e^{\sigma t} = e^{\rho t} e^{j\zeta t} \tag{E.20}$$

Noting that $\text{Re}[\sigma] = \rho$, then the impulse response is diverging for increasing t (time) if $\rho > 0$. However, for $\rho < 0$, the impulse response is converging for increasing t. The term $e^{j\zeta t}$ is a complex (oscillating) sinusoid (as discussed in Section A.2.1). In slightly different language, we could state that *a system is stable if all s-domain poles have a real component less than zero.*

Thus, if the poles were plotted on the complex s-plane, then all poles must be on the left-hand side of the complex plane. Figure E.2 illustrates the complex s-plane regions of stability and an example of a *stable* 3rd-order transfer function, where the poles are in the left-hand side of the plane, i.e. the poles have a negative real component. Note that the zeroes may be in either the left- or right-hand side of the s-plane and have no bearing on stability.

For a circuit with more than one pole, one can think of the transfer function as a cascade of single-pole circuits, as follows:

$$H(s) = \frac{(s-a_0)(s-a_1)(s-a_2)}{(s-b_0)(s-b_1)(s-b_2)} = (s-a_0)(s-a_1)(s-a_2) \left[\frac{1}{s-b_0} \right] \left[\frac{1}{s-b_1} \right] \left[\frac{1}{s-b_2} \right]$$

$$\tag{E.21}$$

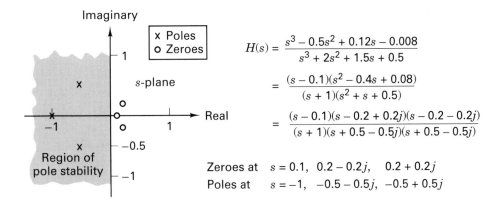

Figure E.2 Plotting the poles and zeroes of an s-domain transfer function.

For stability, all poles must lie on the left-hand side of the complex plane. Note that for real circuits with real Laplace coefficients (i.e. in Equation (E.16) A, B, C, D, E would be all real valued), the poles and zeroes will either be real or occur as complex conjugate pairs as illustrated in Figure (E.2).

For our previous example of an RC circuit, the transfer function in Equation (E.14) is unconditionally stable since $2\pi RC$ is always a positive value—which is of course the expected result. Instability will only be seen in linear systems that have feedback (recursion), such as filters implemented with inverting or noninverting operational amplifier circuits.

E.2 THE z-TRANSFORM

The z-transform is essentially the discrete equivalent of the Laplace transform. It allows for conventional mathematical analysis (transient and steady state) and manipulation of signals and systems. Probably the most frequent modern use of the z-transform is for specifying discrete systems and analyzing their stability.

The z-transform allows the convolution of an input signal and a discrete linear system to be calculated in a mathematically tractable form. Also, the "poles" and "zeroes" of a system can be observed, leading to information regarding the dynamic behavior and stability of a discrete system. It is important to note that the poles and zeroes of the z-transform are different from the poles and zeroes of the Laplace transform.

E.2.1 Calculating the z-Transform

We can derive the z-transform from the Laplace transform of Equation (E.4) by considering a signal $x(t)$ that is sampled every T seconds, thus yielding a discrete time signal of samples $x(0), x(T), x(2T), \ldots = \{x(kT)\}$. The sampled data represent a set of weighted and translated delta functions whose Laplace transform (by the time shifting property) can be written as

$$X(s) = \sum_{k=0}^{\infty} x(kT)e^{-skT} \qquad (E.22)$$

If we employ the parameter $z = e^{sT}$ and replace discrete time kT with a sample number k, we get

$$X(z) = \sum_{k=0}^{\infty} x(k)z^{-k} \qquad (E.23)$$

As an example, the z-transform of a simple unit step function $u(k)$ is

$$U(z) = \sum_{k=0}^{\infty} u(k)z^{-k} = 1 + z^{-1} + z^{-2} + z^{-3} + \ldots = \frac{1}{1 - z^{-1}} \qquad (E.24)$$

Note the geometric series form and the assumption that $|z| < 1$ (the region of convergence). As in the case of the Laplace transform, the z-transform can be

TABLE E.3 z-Transforms of Some Simple Functions

Waveform Type	Time Function	z-Transform
Impulse	$\delta(k)$	1
Delayed impulse	$\delta(k - m)$	z^{-m}
Unit step function	$u(k)$	$\dfrac{z}{z - 1}$
Ramp function	$k u(k)$	$\dfrac{z}{(z - 1)^2}$
Exponential	$e^{ak} u(k)$	$\dfrac{z}{z - e^a}$
Sine wave	$\sin(\omega k) u(k)$	$\dfrac{z\sin(\omega)}{z^2 - 2z\cos(\omega) + 1}$
Cosine wave	$\cos(\omega k) u(k)$	$\dfrac{z[z - \cos(\omega)]}{z^2 - 2z\cos(\omega) + 1}$

tabulated in tables, such as Table E.3, and operations of the z-transform can be tabulated in tables, such as Table E.4.

E.2.2 The Inverse z-Transform

We can transform back from the z-domain to the time domain using the inverse z-transform [2], as follows:

$$x(k) = z^{-1}\{X(z)\} = \frac{1}{j2\pi} \oint_C X(z)\, z^{k-1} dz \qquad \text{(E.25)}$$

TABLE E.4 z-Transform Operations

Property	Time Function	Z Transform
General or arbitrary function	$x(t)$	$X(z)$
General or arbitrary function	$y(t)$	$Y(z)$
Linearity	$ax(t) + by(t)$	$aX(z) + bY(z)$
Time shift	$x(k - m)$	$z^{-m}X(z)$
Modulation	$e^{-j\omega k}x(k)$	$X(e^{j\omega}z)$
Exponential scaling	$a^k x(k)$	$X(z/a)$
Ramp scaling	$kx(k)$	$-z\dfrac{d}{dz}X(z)$
Convolution	$x(k)*h(k)$	$X(z)H(z)$

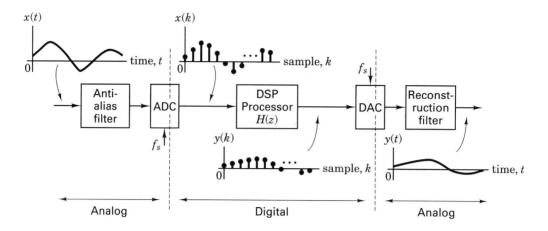

Figure E.3 Digital filter equations are implemented on the DSP Processor which processes the time sampled data signal to produce a time sampled output data signal.

Here, the complex integration $\oint$ is over any simple contour in the region of convergence of $X(z)$ that circles the point $z = 0$. The evaluation of the inverse z-transform would appear to be somewhat more complicated than for the z-transform. Typical operations required include partial fraction evaluation, polynomial division, the residue theorem, and difference equation synthesis. Hence, most z-transformations, and inverse z-transformations are calculated using tables of standard integrals and properties, and explicit evaluation of Equation (E.25) can usually be avoided. In modern-day DSP analysis, software packages such as SystemView [1] are used to manipulate discrete signals and systems, and the z-transform is largely used simply as a convenient analytical notation for the specification and stability of discrete signals and systems.

E.3 DIGITAL FILTERING

Using appropriate analog and digital components, a digital filter can be set up to perform a desired frequency discrimination or phase-modifying function. Figure E.3 shows the components required for a digital filter that produces the filtered sequence $y(k)$ for the input sequence $x(k)$ [2]. A general digital filter output $y(k)$ is produced from a weighted sum of past inputs $x(k)$ and past outputs $y(k-n)$, with $n > 0$. A signal flow graph (composed of only adders, multipliers and sample delays), for a four-feedforward weight and three-weight feedback digital filter, is shown in Figure E.4. (Note that a single sample time delay is indicated by the symbol Δ. Often, one sees such time-domain signal flow graphs drawn with a mixture of time-domain and z-domain notation, using z^{-1} to represent a delay; although it is a common practice, it is not precise.

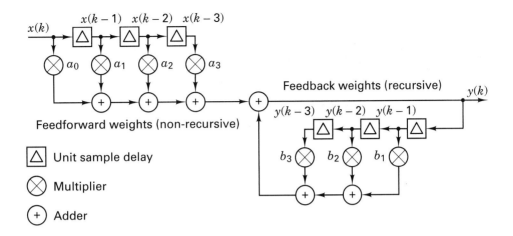

Figure E.4   A generic digital filter.

The output of this filter is given as

$$y(k) = a_0 x(k) + a_1 x(k-1) + a_2 x(k-2) + a_3 x(k-3) +$$
$$+ b_1 y(k-1) + b_2 y(k-2) + b_3 y(k-3) \tag{E.26}$$

$$= \sum_{n=0}^{3} a_n x(k-n) + \sum_{m=1}^{3} b_m y(k-m)$$

Taking the z-transform of Equation (E.26) gives

$$Y(z) = a_0 X(z) + a_1 X(z)z^{-1} + a_2 X(z)z^{-2} + a_3 X(z)z^{-3} \tag{E.27}$$
$$+ b_1 Y(z)z^{-1} + b_2 Y(z)z^{-2} + b_3 Y(z)z^{-3}$$

E.3.1 Digital Filter Transfer Function

The transfer function of the digital filter in Figure E.4 is produced by rearranging Equation (E.27) to yield

$$H(z) = \frac{Y(z)}{X(z)} = \frac{a_0 + a_1 z^{-1} + a_2 z^{-2} + a_3 z^{-3}}{1 - b_1 z^{-1} - b_2 z^{-2} - b_3 z^{-3}}$$

$$= \frac{a_0(1 - \alpha_1 z^{-1})(1 - \alpha_2 z^{-1})(1 - \alpha_3 z^{-1})}{(1 - \beta_1 z^{-1})(1 - \beta_2 z^{-1})(1 - \beta_3 z^{-1})} \tag{E.28}$$

$$= \frac{a_0(z - \alpha_1)(z - \alpha_2)(z - \alpha_3)}{(z - \beta_1)(z - \beta_2)(z - \beta_3)} = \frac{A(z)}{B(z)}$$

where the α values are the z-domain *zeroes* of the filter and the β values are the z-domain *poles* of the filter, found by finding the roots of the numerator polynomial $A(z) = 0$ and the denominator polynomial $B(z) = 0$, respectively. For a digital filter similar to the signal flow graph in Figure E.4 but with N feedforward weights and $M - 1$ feedback weights, the numerator and denominator polynomials will be of order N and M, respectively, in the transfer function analogous to Equation (E.28).

E.3.2 Single Pole Filter Stability

Due to the presence of numerical feedback in the signal flow graph, the digital filter may be (numerically) unstable. For example, consider the single-feedback weight filter of Figure E.5:

$$y(k) = x(k) + by(k - 1) \tag{E.29}$$

The impulse response of this filter (i.e., applying a discrete unit impulse $\delta(k)$, following the principles of convolution described in Section A.5) is

$$h(k) = b^k \tag{E.30}$$

If $|b| < 1$, then the filter impulse response is converging (stable); and if $|b| > 1$, the filter impulse response is diverging (unstable). Figure E.5 illustrates a converging impulse response where $|b| < 1$, and, more precisely, $-1 < b < 1$. Taking the z-transform of Equation (E.29) yields

$$H(z) = \frac{Y(z)}{X(z)} = \frac{1}{1 - bz^{-1}} = \frac{z}{z - b} \tag{E.31}$$

From Equation (E.31), we show the z-domain signal flow graph in Figure E.5b which corresponds to the time domain signal flow graph of Figure E.5a. The delay elements in Figure E.5b (represented by Δ in Figure E.5a) are now represented by z^{-1}, and the inputs and outputs are specified as the z-transforms $X(z)$ and $Y(z)$. Note, however, that the general topology of the two signal flow graphs is the same. (This perhaps indicates why digital filter signal flow graphs are often drawn (imprecisely) with a mixed time-domain and z-domain notation). We can state the

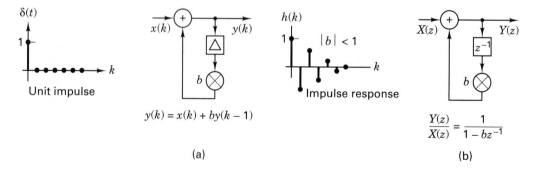

Figure E.5 (a) A single feedback weight filter time domain signal flow graph, and (b) the z-domain signal flow graph equivalent.

stability criteria of $|b| < 1$ as follows: *for stability, the magnitude of the poles (or roots of the denominator polynomial) of the digital filter transfer function must be less than one.*

E.3.3 General Digital Filter Stability

From the factorized transfer function in Equation (E.28), we can redraw the (time domain) signal flow graph of Figure E.4 and the z-domain signal flow graph of Figure E.6. This signal flow graph essentially corresponds to rewriting Equation (E.28) in the form

$$H(z) = a_0(1 - \alpha_1 z^{-1}) \cdot (1 - \alpha_2 z^{-1}) \cdot (1 - \alpha_3 z^{-1}) \tag{E.32}$$

$$\cdot \left[\frac{1}{1 - \beta_1 z^{-1}} \right] \cdot \left[\frac{1}{1 - \beta_2 z^{-1}} \right] \cdot \left[\frac{1}{1 - \beta_3 z^{-1}} \right]$$

This explicitly shows the first-order sections for each zero and pole of the filter. In order for the filter to be stable, then all of the pole values $\{\beta_1, \beta_2, \beta_3\}$ in the cascade must have a magnitude of less than 1. If any one first-order section is unstable (or diverging), then so is the entire cascade. As was noted for the Laplace transform, the z-domain poles (and zeroes) may be complex, which is the reason that magnitude rather than amplitude is the stability criterion. (Note that the signal flow graph implementation of Figure E.6 is only for analysis purposes, and a digital filter would never be practically implemented in this factored form because some of the multiplier values may be complex and thus computational requirements would be unnecessarily increased for the implementation of a real coefficient filter.)

E.3.4 *z*-Plane Pole-Zero Diagram and the Unit Circle

If the complex poles and zeroes of a filter or linear system are plotted on the real and imaginary axes, the resulting plane can be referred to as the z-plane. We can verify the stability of a system by observing that all poles lie within the *unit circle*. Figure E.7 shows the z-plane for a filter with transfer function

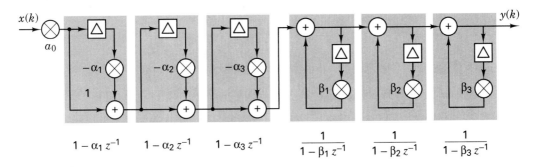

Figure E.6   Digital filter as a cascade of feedforward and feedback first order sections.

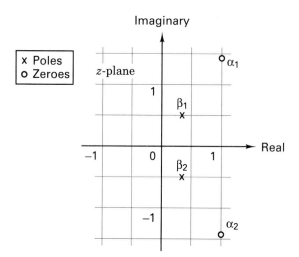

Imaginary

x Poles
o Zeroes

z-plane

α_1

β_1
×

1

β_2
×

-1 0   1   Real

α_2
o

-1

Figure E.7 Poles and zeroes plotted on the z-plane.

$$H(z) = \frac{1 - 2z^{-1} + 3z^{-2}}{1 - \frac{2}{3}z^{-1} + \frac{1}{3}z^{-2}}$$

$$= \frac{(1 - (1 + j\sqrt{2})z^{-1})(1 - (1 - j\sqrt{2})z^{-1})}{(1 - (1/3 + j\sqrt{2}/3)z^{-1})(1 - (1/3 - j\sqrt{2}/3)z^{-1})} \qquad \text{(E.33)}$$

$$= \frac{(1 - \alpha_1 z^{-1})(1 - \alpha_2 z^{-1})}{(1 - \beta_1 z^{-1})(1 - \beta_2 z^{-1})}$$

with zeroes at $z = 1 + j\sqrt{2}$ and $z = 1 - j\sqrt{2}$ and poles at $z = 1/3 + j\sqrt{2}/3$ and $z = 1/3 - j\sqrt{2}/3$. This filter is clearly stable, as all poles lie within the unit circle.

E.3.5 Discrete Fourier Transform of Digital Filter Impulse Response

The frequency response of a digital filter is calculated from the discrete Fourier transform (DFT) of the filter impulse response. Recall, from the Fourier transform Equation (A.26), that

$$X(f) = \int_{-\infty}^{\infty} x(t)e^{-j2\pi ft}\, dt \qquad \text{(E.34)}$$

This equation can be used to evaluate the Fourier transform of the impulse response of a filter. It can be simplified by realizing that we are now using a sampled version of a signal $x(t)$ with samples every $T_s = 1/f_s$ seconds:

$$X(f) = \int_{-\infty}^{\infty} x(kT_s)e^{-j2\pi fkT_s}d(kT_s) = \sum_{k=-\infty}^{\infty} x(kT_s)e^{-j2\pi fkT_s} = \sum_{k=-\infty}^{\infty} x(kT_s)e^{-(j2\pi fk)/f_s}$$

$$\text{(E.35)}$$

Of course, the digital filter impulse response is causal and the first sample of the impulse response is at $k = 0$, and the last sample is at $k = N - 1$, giving a total of N samples in the transform. Hence, for this finite number of samples, rewriting Equation (E.35) with respect to the sample number k rather than explicit time kT_s gives

$$X(f) = \sum_{k=0}^{N-1} x(k)e^{-(j 2\pi f k)/f_s} \tag{E.36}$$

Note that Equation (E.36) is actually calculated for a continuous frequency variable f. In reality we need only evaluate this equation at certain discrete frequencies, which are the zero frequency (dc) and harmonics of the "fundamental" frequency, for a total of N discrete frequencies $0, f_0, 2f_0$, up to f_s, where $f_0 = 1/NT_s$:

$$X\left(\frac{nf_s}{N}\right) = \sum_{k=0}^{N-1} x(k)e^{-(j\, 2\pi k f_s n)/N f_s} \quad \text{for } n = 0 \text{ to } N - 1 \tag{E.37}$$

Simplifying to use only the time index k and the frequency index n gives the *discrete Fourier transform (DFT)*

$$X(n) = \sum_{k=0}^{N-1} x(k)e^{-(j\, 2\pi kn)/N} \quad \text{for } n = 0 \text{ to } N - 1 \tag{E.38}$$

Given that the discrete signal $x(k)$ was sampled at the rate of f_s samples/s, then the signal has image (or alias) components above $f_s/2$. Hence, when evaluating Equation (E.38), it is only necessary to evaluate up to $f_s/2$. Note that Equation (E.38) is the same as the z-transform of Equation (E.23) if we set $z = e^{(j2\pi n)/N}$ for a sequence of length N samples.

E.4 FINITE IMPULSE RESPONSE FILTER DESIGN

By far the most common type of digital filter is the finite impulse response (FIR) filter, which as the name suggests has an impulse response of finite duration. This filter has no feedback weights (recall Figure E.4) and therefore we can conclude that it is unconditionally stable. The output of the FIR filter in Figure E.8 is

$$y(k) = a_0 x(k) + a_1 x(k-1) + a_2 x(k-2) + a_3 x(k-3) + \cdots + a_{N-1} x(k-N+1)$$

$$= \sum_{n=0}^{N-1} a_n x(k-n) \tag{E.39}$$

and the transfer function therefore contains only zeros and no poles:

$$H(z) = a_0 + a_1 z^{-1} + a_2 z^{-2} + a_3 z^{-3} + \cdots + a_{N-1} z^{-N+1} \tag{E.40}$$

$$= a_0(1 - \alpha_1 z^{-1})(1 - \alpha_2 z^{-1})(1 - \alpha_3 z^{-1})\ldots(1 - \alpha_N z^{-1})$$

The FIR filter is essentially a moving average calculation, whereby the output is a weighted average of the last N input samples. Hence, this type of filter is often

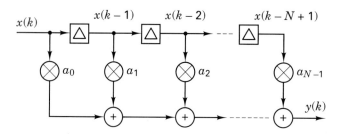

$x(k)$ $x(k-1)$ $x(k-2)$ $x(k-N+1)$

a_0 a_1 a_2 a_{N-1}

$y(k)$

Figure E.8 Finite impulse response digital filter.

termed a *moving average filter.* Other names include *tapped delay line,* and *transversal filter.*

E.4.1 FIR Filter Design

Using modern DSP analysis software, such as SystemView [1], FIR digital filters are designed based on a magnitude frequency plot with specified tolerances and user requirements, as illustrated in Figure E.9 for low-pass filter. Classical filter design techniques, such as Parks-McLellan, Remez Exchange, Kaiser window and so on [4], are then used to realize a suitable frequency response that satisfies the user requirements with a minimum number of weights. Unless otherwise specified in the design process, most FIR filters are designed to have linear phase or a constant group delay (corresponding to a symmetrical impulse response).

Figure E.10 shows the impulse response and frequency response of a digital filter design with parameters of cut-off frequency 1000 Hz, stopband attenuation of 20 dB, passband ripple of 3 dB and transition band of 500 Hz; the sampling frequency, f_s is 10,000 Hz. If a filter with a more stringent frequency response is required (such as more stopband attenuation) then it is likely that the filter design procedure will produce an FIR with more filter weights [4].

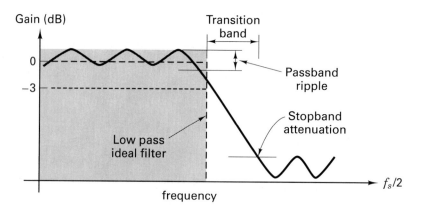

Figure E.9 Generic low pass filter magnitude response. The more stringent the filter requirements of stopband attenuation, transition bandwidth and to a lesser extent passband ripple, the more weights that are required.

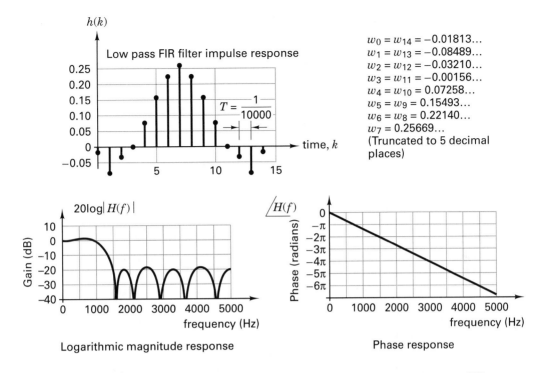

$$w_0 = w_{14} = -0.01813...$$
$$w_1 = w_{13} = -0.08489...$$
$$w_2 = w_{12} = -0.03210...$$
$$w_3 = w_{11} = -0.00156...$$
$$w_4 = w_{10} = 0.07258...$$
$$w_5 = w_9 = 0.15493...$$
$$w_6 = w_8 = 0.22140...$$
$$w_7 = 0.25669...$$
(Truncated to 5 decimal places)

Figure E.10 The impulse response $h(n) = w_n$ and the frequency response $H(f)$ of a low-pass filter with 15 weights, a sampling rate of 10,000 Hz, designed from specified cut-off frequency designed at around 1000Hz.

E.4.2 The FIR Differentiator

Consider the simple digital filtering *differentiator* shown in Figure E.11. We can intuitively reason that this filter is a high-pass filter by observing the output for both low- and high-frequency input sinusoids. The output sequence of this filter is given by

$$y(k) = [x(k) - x(k-1)] \qquad (E.41)$$

Taking the z-transform of Equation (E.41) gives

$$Y(z) = [X(z) - X(z)z^{-1}] \qquad (E.42)$$

Therefore, the transfer function is given by

$$\frac{Y(z)}{X(z)} = (1 - z^{-1}) \qquad (E.43)$$

Figure E.12 illustrates why this filter acts to produce a high-pass function. Essentially the output of the filter is the difference between the last two samples. If the

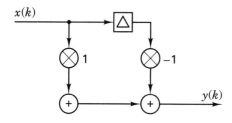

x(k)

△

⊗ 1   ⊗ −1

+   +   y(k)

Figure E.11   Differentiator/high pass filter.

difference between successive samples was small (as for a low frequency), then the output is small. If the difference is large (as for a high frequency), then the output is large. If a dc signal was input, then the output amplitude would be zero, i.e. infinite attenuation. The frequency response can also be calculated from the Fourier transform of the impulse response.

If the filter weights are changed to $\{1/T, -1/T\}$ from $\{1, -1\}$ where the sampling frequency is $f_s = 1/T$, then, for low-frequency inputs, $y(k)$ is approximately the differential of the input:

$$y(k) \approx \frac{x(k) - x(k-1)}{T} \approx \frac{dx(t)}{dt} \quad \text{and} \quad \frac{Y(z)}{X(z)} = \frac{1}{T}(1 - z^{-1}) \quad (E.44)$$

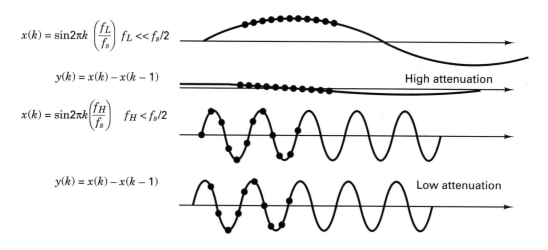

$x(k) = \sin 2\pi k \left(\dfrac{f_L}{f_s}\right) \ f_L \ll f_s/2$

$y(k) = x(k) - x(k-1)$ High attenuation

$x(k) = \sin 2\pi k \left(\dfrac{f_H}{f_s}\right) \ f_H < f_s/2$

$y(k) = x(k) - x(k-1)$ Low attenuation

Figure E.12   Digital filter differentiator acting as a high pass filter.

E.5 INFINITE IMPULSE RESPONSE FILTER DESIGN

Infinite impulse response (IIR) filters are usually designed from analog prototypes using a mapping from the s-plane to the z-domain. As their name would suggest, the impulse response (assuming infinite precision arithmetic) can be infinite in duration. IIR filters have both feedforward and feedback weights, as was shown in Figure E.4. IIR filters can have a very long impulse response for just a few weights, due to the recursive nature of the signal flow graph. Hence, it may be possible to design an IIR filter with fewer weights than an FIR filter, for the same functional magnitude response. In general, an IIR digital filter will not have linear phase.

E.5.1 Backward Difference Operator

Equation (E.44) represents a means of relating the (continuous time) Laplace transform variable s to the (discrete time) z-transform variable z. Given that the Laplace domain representation of time differentiation (d/dt) is the variable s, for example,

$$y(t) = \frac{dx(t)}{dt} \quad \Rightarrow Y(s) = sX(s) \tag{E.45}$$

then, given, for instance, a general low-pass Butterworth characteristic

$$H(s) = \frac{1}{s^2 + \sqrt{2}\,s + 1} \tag{E.46}$$

we could produce a discrete approximation to this analog circuit by substituting the approximation

$$s \approx \frac{1}{T}(1 - z^{-1}) \tag{E.47}$$

into Equation (E.46) to produce a z-domain equation

$$\begin{aligned} H(z) = H(s)\Big|_{s=\frac{1}{T}(1-z^{-1})} &= \frac{1}{\frac{1}{T^2}(1 - z^{-1})^2 + \sqrt{2}\frac{1}{T}(1 - z^{-1}) + 1} \\[2mm] &= \frac{T^2}{(1 - 2z^{-1} + z^{-2}) + \sqrt{2}\,T(1 - z^{-1}) + T^2} \\[2mm] &= \frac{T^2}{z^{-2} - (\sqrt{2}T + 2)z^{-1} + (1 + \sqrt{2}T + T^2)} \end{aligned} \tag{E.48}$$

At low frequencies, where the approximation of Equation (E.47) is "good," then it may be that this transform produces a "reasonable" digital filter equivalent of the analog low-pass Butterworth. (Equation (E.47) is sometimes called the "backward

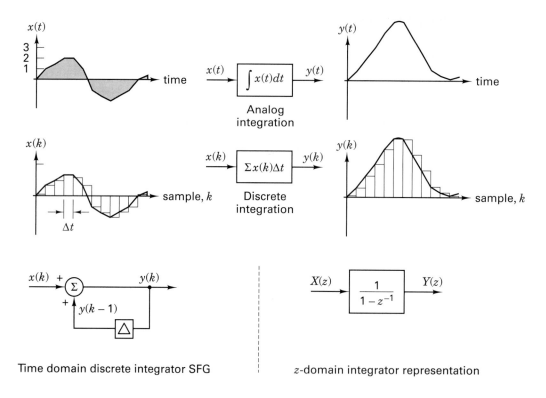

Time domain discrete integrator SFG | z-domain integrator representation

Figure E.13 Single pole filter acting as an integrator. Note that a feedback weight of just less than 1 is often included in the feedback loop to introduce some "forgetting" into the integrator.

difference operator"). Unfortunately, this mapping is very poor at high frequencies, and consequently, it cannot be used to produce high-pass filters. Hence, it is rarely used.

E.5.2 IIR Filter Design using the Bilinear Transform

The bilinear transform is obtained by replacing s with the approximation

$$s \approx \frac{2}{T} \frac{(1 - z^{-1})}{(1 + z^{-1})} \qquad (E.49)$$

This substitution provides a mapping that preserves the stability of the analog prototype and will produce filters that are greatly improved over the backward difference operator of Equation (E.47) [2]. SystemView [1] uses the bilinear transform to produce digital filters from standard analog prototypes, such as Butterworth, Elliptic, and Chebychev. Note that the bilinear transform will always yield a filter that has both poles and zeroes, and it therefore represents an infinite impulse response (IIR) design.

E.5.3 The IIR Integrator

A digital *integrator* is essentially a one-weight IIR filter:

$$y(k) = x(k) + y(k - 1) = \sum_{i=0}^{k} x(i) \qquad \text{(E.50)}$$

In the z-domain, the transfer function of a discrete integrator is obtained from

$$Y(z) = X(z) + z^{-1} Y(z) \qquad \text{(E.51)}$$

yielding

$$\frac{Y(z)}{X(z)} = \frac{1}{1 - z^{-1}} \qquad \text{(E.52)}$$

Figure E.13 shows the implementation of the simple digital integrator and a graphical representation of the relationship with continuous time integration (i.e., area under the graph).

If a weight value of less than one is included in the feedback loop (say 0.99), then the integrator is often called a *leaky integrator*. When viewed in the frequency domain, a (leaky) integrator has the characteristics of a simple low-pass filter.

REFERENCES

1. SystemView DSP Communications Software. Elanix, Westlake Village, CA, 2000.
2. B. Porat. A Course in Digital Signal Processing. John Wiley and Sons, 1997.
3. T. K. Moon, W. C. Stirling. Mathematical Methods and Algorithms for Digital Signal Processing. Prentice Hall, 2000.
4. R. W. Stewart. The DSPedia: A Multimedia Resource for DSP, BlueBox Multimedia, UK, 2000.

APPENDIX F

List of Symbols

a_{ij} Coefficient of jth basis function

a_j Signal component output of jth correlator

A Peak amplitude of a waveform

A_e Effective area (of an antenna)

B_L Single-sided loop bandwidth

c Speed of light $\approx 3 \times 10^8$ m/s

C Channel capacity

C Electrical capacitance

$C/_k T°$ Ratio of average carrier power to noise power spectral density

d Distance

d_0 Reference distance

d_f Free distance

d_{min} Minimum distance

D Delay time (of message)

D Redundancy of a language

D Decryption transformation

e The natural number 2.7183

$\mathbf{e}$ Error pattern vector

$e(t)$ Error signal

$\mathbf{e}(X)$ Error pattern polynomial

E Encryption transformation

E_x Energy of waveform $x(t)$

$\mathbf{E}\{X\}$ Expected value of the random variable X

EIRP Effective radiated power with reference to an isotropic source

E_b/J_0 Ratio of bit energy to jammer power spectral density

E_b/N_0 Ratio of bit energy to noise power spectral density

E_c/N_0 Ratio of channel symbol energy to noise power spectral density

f Frequency (hertz)

f_c Carrier-wave frequency

f_m Maximum frequency

f_0 Coherence bandwidth

f_d Doppler frequency spread

f_s Sampling frequency

f_l Lower cutoff filter frequency

f_u Upper cutoff filter frequency

F Noise figure

$\mathscr{F}\{x\}$ Fourier transform of the function $x(t)$

$\mathscr{F}^{-1}\{X\}$ Inverse Fourier transform of the function $X(f)$

F Field

F^* Finite field

$g(t)$ Pseudorandom code function

$g(t)$ Signal waveform (baseband)

$\mathbf{g}(X)$ Generator polynomial (for a cyclic code)

G Antenna gain

G Coding gain

$\mathbf{G}$ Generator matrix (for a linear block code)

G Normalized total message traffic

G_p Processing gain

$G_x(f)$ Power spectral density of waveform $x(t)$

$h(t)$ Impulse response of a network

$h_c(t)$ Channel impulse response

$\mathbf{H}$ Parity-check matrix for a code

H_i The ith hypothesis

$\mathbf{H}_k$ Hadamard matrix

$H(f)$ Frequency transfer function of a network

$H_0(f)$ Optimum frequency transfer function

$H(X)$ Entropy of information source X

$H(X|Y)$ conditional entropy (entropy of X, given Y)

$i(t)$ Electrical current waveform

I Electrical current

$I_0(x)$ Zero-order modified Bessel function of the first kind

$I(X)$ Self-information of information source X

J Received average jammer power

J_0 Jammer power spectral density

J/S Ratio of received average jammer power to average signal power

k Number of bits per M-ary signal set

k/n Code rate (ratio of number of data bits to total bits in codeword)

K Constraint length of a convolutional encoder

K Key, dictating a specific encryption or decryption transformation

ℓ Number of quantization bits

$\ell(d_k)$ Likelihood of data bit d_k

L Look-ahead length for convolutional feedback decoding

L Number of branch words in sequence

L Number of quantization levels

$L(d_k)$ Log-likelihood ratio of data bit d_k

L_e Extrinsic log-likelihood ratio

L_s Free space path loss

L_o Other losses

L_p Path loss

L_o	Observation time
L_c	Observation time for channel-induced ISI
L_{CISI}	Observation time for controlled ISI
L_c	Channel log-likelihood ratio
$\mathbf{m}$	Message vector
$\mathbf{m}(X)$	Message polynomial
m_i	Data bit
M	Margin
M	Waveform- or signal-set size
(n, k)	Code designation by number of total bits (n) and data bits (k) in codeword
$\bar{n}$	Average number of bits per character
n_0	Noise random variable output of correlator at symbol time $t = T$
$n(t)$	Gaussian noise process
N	Noise power
N	Unicity distance
N_0	Level of single-sided power spectral density of white noise
NSR	Ratio of average noise power to average signal power
p_c	Probability of channel symbol error
p_i	parity bit
$p(t)$	Instantaneous power
$p(x)$	Probability density function of a continuous random variable
$p(x\|y)$	Probability density function of x conditioned on y
$\mathbf{P}$	Parity array
P_B	Probability of bit error
P_E	Probability of symbol error
P_{FA}	Probability of false alarm
P_m	Probability of miss
P_M	Probability of message or block error
P_{nd}	Probability of undetected error
P_r/N_0	Ratio of received average signal power to noise power spectral density
$\mathbf{p}(X)$	Remainder polynomial
$P(X)$	Probability of a discrete random variable
P_x	Average power in waveform $x(t)$
q	Quantization step size (quantile interval)
$\mathbf{q}(X)$	Quotient polynomial
$Q(x)$	Complementary error function (integral of the tail beyond x of the Gaussian density function)
r	Filter roll-off factor
r	True rate of a language
r'	Absolute rate of a language
$r(t)$	Received signal waveform
R	Data rate (bits/second)
$R(\Delta f)$	Spaced-frequency correlation function
$R(\Delta t)$	Spaced-time correlation function
R_c	Code-bit rate or channel-bit rate (code bits/second)
R_{ch}	Chip rate (chips/second)
R_s	Symbol rate (symbols/second)
$R_x(\tau)$	Autocorrelation function of waveform $x(t)$
$\mathscr{R}$	Electrical resistance
$s(t)$	Signal waveform
$\hat{s}(t)$	Estimate of signal waveform
$S(v)$	Doppler power spectral density
$S(\tau)$	Multipath intensity profile
$\mathbf{s}$	Signal vector
sgn x	Sign function of x
S_k	State at time k
S	Signal power
$\mathbf{S}$	Syndrome vector
SJR	Ratio of average signal power to average jammer power
SNR	Ratio of average signal power to average noise power
S/N	Ratio of signal power to noise power
$S(f)$	Fourier transform of the waveform $s(t)$
$\mathbf{S}(X)$	Syndrome polynomial
t	Number of errors correctable in an error-correcting code
t	Independent time variable
t_0	Time delay
t_{ij}	Amount of message traffic from i to j
T	Pulse width
T	Symbol interval
$T(D)$	Transfer function or generating function of convolutional code
T_{ch}	Duration of a chip
T_{hop}	Duration of a hop
T_s	Sampling interval
T°	Temperature
T_A°	Antenna temperature
T_L°	Effective line temperature
T_m	Multipath delay time (maximum)
T_0	Coherence time
T_{IL}	Interleaver span
T_R°	Effective receiver temperature
T_S°	System temperature
T_{acq}	Time to acquire
u_i	Code symbol
$u(t)$	Unit step function
$\mathbf{U}$	Codeword vector
$\mathbf{U}(X)$	Codeword polynomial
v	Relative velocity
$v(t)$	Electrical voltage waveform
var (X)	Variance of random variable X
V	Velocity
$w(t)$	Jammer waveform
W	Bandwidth
W_f	Filter bandwidth
W_{DSB}	Double-sideband bandwidth
W_N	Noise equivalent bandwidth
W_{ss}	Spread-spectrum bandwidth
$z(t)$	Output of matched filter or correlator
α_k	Forward state metric at time k
β_k	Reverse state metric at time k
Γ	SNR averaged through the "ups and downs" of fading
Γa	State metric for state a
γ	threshold (decision) level
γ_0	Optimum threshold level
$\gamma_{\mathbf{U}^{(m)}}$	Likelihood of codeword $\mathbf{U}^{(m)}$
δ	Fractional frequency drift per day
δ_k	Branch metric at time k
δ_{mn}	Kronecker delta function
$\delta(t)$	Impulse (Dirac delta) function
ϵ	Error
ζ	Loop damping characteristic (second-order loop)
η	Antenna efficiency
$\theta(t)$	Time-varying phase
$\Theta(\omega)$	Fourier transform of $\theta(t)$
κ	Boltzmann's constant, 1.38×10^{-23}J/K
$\Lambda(d_k)$	Likelihood ratio of data bit d_k
λ	Joint probability
λ	Wavelength
λ	Packet arrival rate
π	Pi, 3.14159
ρ	Fraction of the frequency band being jammed
ρ	Fraction of the time the jammer is "on"
ρ	Normalized loop signal-to-noise ratio
ρ	Normalized message throughput
ρ	Number of erasures correctable in an error-correcting code
ρ	Time-correlation coefficient
ρ_0	Value of ρ that maximizes bit error probability (worst-case jamming)
σ_τ	rms delay spread
σ_X	Standard deviation of random variable X
σ_X^2	Variance of random variable X
τ	Pulse width
τ	Time shift (independent variable of the autocorrelation function)
$\psi_j(t)$	Basis function
$\Psi_x(f)$	Energy spectral density of waveform $x(t)$
ω	Radian frequency (radians per second)

Index

Index   **1079**